PRACTICAL FUNCTIONAL GROUP SYNTHESIS

T0310169

PRACTICAL FUNCTIONAL GROUP SYNTHESIS

ROBERT A. STOCKLAND, Jr.

Department of Chemistry
Bucknell University
Lewisburg, PA

Copyright © 2016 by John Wiley & Sons, Inc. All rights reserved

Published by John Wiley & Sons, Inc., Hoboken, New Jersey
Published simultaneously in Canada

No part of this publication may be reproduced, stored in a retrieval system, or transmitted in any form or by any means, electronic, mechanical, photocopying, recording, scanning, or otherwise, except as permitted under Section 107 or 108 of the 1976 United States Copyright Act, without either the prior written permission of the Publisher, or authorization through payment of the appropriate per-copy fee to the Copyright Clearance Center, Inc., 222 Rosewood Drive, Danvers, MA 01923, (978) 750-8400, fax (978) 750-4470, or on the web at www.copyright.com. Requests to the Publisher for permission should be addressed to the Permissions Department, John Wiley & Sons, Inc., 111 River Street, Hoboken, NJ 07030, (201) 748-6011, fax (201) 748-6008, or online at http://www.wiley.com/go/permissions.

Limit of Liability/Disclaimer of Warranty: While the publisher and author have used their best efforts in preparing this book, they make no representations or warranties with respect to the accuracy or completeness of the contents of this book and specifically disclaim any implied warranties of merchantability or fitness for a particular purpose. No warranty may be created or extended by sales representatives or written sales materials. The advice and strategies contained herein may not be suitable for your situation. You should consult with a professional where appropriate. Neither the publisher nor author shall be liable for any loss of profit or any other commercial damages, including but not limited to special, incidental, consequential, or other damages.

For general information on our other products and services or for technical support, please contact our Customer Care Department within the United States at (800) 762-2974, outside the United States at (317) 572-3993 or fax (317) 572-4002.

Wiley also publishes its books in a variety of electronic formats. Some content that appears in print may not be available in electronic formats. For more information about Wiley products, visit our web site at www.wiley.com.

Library of Congress Cataloging-in-Publication Data:

Stockland, Robert.
Practical functional group synthesis / Robert A. Stockland, Jr.
    pages   cm
  Includes bibliographical references and index.
  ISBN 978-1-118-61280-4 (pbk.)
1. Chemical bonds.   2. Functional groups.   3. Chemistry, Organic.   I. Title.
  QD461.S85 2016
  547′.2–dc23

                                                            2015024767

Set in 10/12pt Times by SPi Global, Pondicherry, India

Printed in the United States of America

10 9 8 7 6 5 4 3 2 1

1   2016

CONTENTS

PREFACE

When faced with a new reaction, chemists often stand in front of the bench asking themselves, "...OK, how am I going to do this?" During their education, scientists are taught how to predict the products of countless reactions "on paper." While these discussions are critically important, little time is devoted to making the connection between paper reactions and actually synthesizing a new compound. This is due, in part, to the large number of variations for each reaction.

In many cases, synthetic chemists simply need a reasonable way to make a linkage or new compound. Often, the incorporation of functional groups can be accomplished by a range of catalysts and conditions. Given the myriad of reactions that are added to the literature every week, selecting a suitable approach for the synthesis of the desired compound can be a daunting process.

This text provides the synthetic community with a collection of reactions for the synthesis of a wide range of functional groups through the formation of carbon–heteroelement bonds. Special attention has been devoted to the inclusion of operationally straightforward reactions. For example, reactions that do not require the use of a glove box and can be carried out under an atmosphere of air are particularly attractive. In addition to conventionally heated reactions, a variety of microwave-assisted processes have been included.

Chapter 1 provides an introduction to modern synthetic chemistry with emphasis on topics such as solvent selection, catalyst choice, and a discussion of common challenges associated with functional group synthesis. This chapter also includes a detailed discussion of how to adapt conventionally heated reactions to microwave-assisted versions.

Each of the remaining chapters is focused on a single element such as oxygen or nitrogen and is organized by the type of bond generated in the carbon–heteroelement bond-forming step. In addition to general descriptions of the reactions, troubleshooting guides are included for each section and over 100 detailed preparations are included.

ROBERT A. STOCKLAND, Jr.

ACKNOWLEDGMENTS

I thank Jack McKenna for giving me the opportunity to work in his research lab as an undergraduate student. Working with him was extremely valuable and had instilled a sense of excitement for chemical discovery. I thank Gordon Anderson and Richard Jordan for giving me an incredible amount of freedom as a graduate student and postdoctoral fellow. While some of my ideas were a little crazy, they graciously allowed me to try a wide range of different synthetic approaches and encouraged me to be creative when solving problems. I give a tremendous amount of credit to my wife for her patience and help with this project. In addition to helping with the organization and flow of the text, she created the index and served as the editor/producer of the tutorial videos.

1

INTRODUCTION TO PRACTICAL FUNCTIONAL GROUP SYNTHESIS

1.1 GENERAL APPROACHES FOR DESIGNING SYNTHESES

The construction of functionalized organic compounds remains one of the most challenging areas of synthetic chemistry, and scientists continue to redefine the limits of chemical reactions through the development of processes with increased chemoselectivity, enantioselectivity, and operational simplicity. In many cases, classic reactions are still the most effective means to generate the functional group of interest. Many of these reactions have been modernized to increase the tolerance to preexisting functional groups, decrease the required catalyst loading, increase the selectivity of the process, or minimize the waste generated from the process.

Despite years of education, many novice researchers flounder when trying to design a successful synthesis. They often get stuck on a specific step in a synthesis or are unable to purify an intermediate or product due to contamination by secondary products or solvent. While there are "chemistry" pitfalls that plague new synthetic methods, there are also a range of practical considerations that could render a clever synthesis unachievable. Other syntheses never get started because the starting materials are inaccessible. To help address these issues, a list of general questions for the design of a successful synthesis is provided below:

- What is the goal of the synthesis?
- How pure do the intermediates and products need to be?
- Are all of the reagents/catalysts/additives commercially available?
- If the starting materials are not commercially available, how long will it take to make them?
- Are there any reaction specific issues that need to be considered?

Practical Functional Group Synthesis, First Edition. Robert A. Stockland, Jr.
© 2016 John Wiley & Sons, Inc. Published 2016 by John Wiley & Sons, Inc.

- How long will it take to complete the proposed synthesis?
- Is the shortest route really the most practical?
- Will the chemistry benefit from microwave irradiation?
- Is a glovebox required?
- Does the chemistry require the use of a vacuum/inert gas manifold?
- What are the safety concerns for this approach?
- Are the reagents/catalysts sensitive to light?
- Is a solvent needed?
- Do the solvents need to be rigorously dried?
- Will the solvents be easy to remove?
- Do the solvents need to be degassed/deoxygenated?
- Are the products stable to air?
- How will the products be purified?
- How will the products be stored?
- How will the products be characterized?

What is the goal of the synthesis?   This might seem like an obvious question, but many researchers get bogged down with parts of the synthesis that are not relevant to why the compound will be prepared. If a few milligrams of a target compound are all that is needed in order to screen for a specific activity/property, it is not an appropriate use of time and resources to spend weeks searching the literature or running dozens of screening reactions to optimize the conditions. Simply find a decent synthesis, make the amount that is needed, and submit it for screening. Alternatively, if the project is focused on method development, and the product yields are critical for establishing the scope of the reaction, time needs to be spent optimizing the conditions so an accurate comparison between the new method and established protocols can be made.

How pure do the intermediates and products need to be?   This might seem like another question with an obvious answer, but there are several aspects that need to be considered. It is rarely a good use of time and effort to prepare analytically pure samples of an intermediate in a multistep synthesis if that intermediate will simply be transformed into something else. If the next step in the synthesis will not be inhibited by the impurities in the crude reaction mixture, do not spend time purifying the intermediate. Instead, wait until the end of the synthesis and rigorously purify the final compound.

Are all of the reagents/catalysts/additives commercially available?   If all of the starting materials are commercially available, the chemist will be able to begin work on the proposed synthesis quickly. Given the vast array of reagents and catalysts that are commercially available, the likelihood that the specific materials needed for the proposed synthesis is high. Arguably, this is the most important contributing factor when adopting a new synthesis.

If the materials are not commercially available, how long will it take to make the starting materials?   If the starting materials are not available, the literature preparations must be carefully analyzed to determine how long it will take to generate usable quantities of the starting materials and catalysts.

Are there any reaction specific issues that need to be considered?   In some cases, the unintended reactivity of substrates, catalysts, and additives can complicate a reaction that looks reasonable on paper. Each component of the reaction needs to be evaluated against the remainder to anticipate unintended reaction pathways.

How long will it take to complete the proposed synthesis?   Naturally, this is a bit of a tricky question. The level of difficulty of each step in the synthesis needs to be evaluated as well as how long it will take to make/purify the starting materials and any intermediates. After analyzing the individual steps and calculating a time frame, add 30% to the total because something will not work as planned. Once the overall calculation is complete, an accurate assessment of the approach can be made.

Is the shortest route really the most practical?   In many cases, adding one or two operationally trivial steps to a synthesis is much easier than fighting with a single challenging reaction.

Will the chemistry benefit from microwave irradiation?   Fundamentally, if a reaction needs to be heated, it is likely to be more efficient, cleaner, and faster in a microwave reactor. Since time is one of the most precious commodities in the modern synthetic lab, getting to the target compound quickly is critical.

Is a glovebox required?   For most reactions, needing to use a glovebox will be a guaranteed hassle. All of the glassware needs to be flame dried before taking it into the glovebox, all solvents need to be rigorously dry and degassed, and everything needs to be pumped into the box. Some of the issues with solvents are mitigated by connecting a Grubbs style solvent drying system to the glovebox and pumping dry deoxygenated solvent into the box under pressure. Additionally, unwanted volatile organics are often found in gloveboxes. Unless your research team rigorously maintains the glovebox and routinely checks the quality of the atmosphere, it is often easier to keep a vacuum manifold free of contaminants.

Does the chemistry require the use of a vacuum/inert gas manifold?   Most modern synthetic laboratories have several manifolds dedicated to synthetic chemistry. As a result, most modern preparations assume that one will be available. As a result, if a vacuum/inert atmosphere manifold is not available for the proposed synthesis, each step must be carefully screened to ensure that one will not be needed.

What are the safety concerns for this approach?   While most chemists associate safety with flammability or risk of explosion, the toxicity of the reagents and products needs to be evaluated. For example, if a published preparation using phosgene as a reagent would cut the total synthesis time by 50%, it should still never be adopted by researchers who are not specifically trained on how to handle such a dangerous reagent.

Are the reagents/catalysts sensitive to light?   While it is relatively rare that an organic product will be sensitive to light, it is quite common for metal catalysts to be light sensitive. Many gold(I) compounds are quite sensitive and will degrade upon exposure to light.

Naturally, this is quite substrate dependent, and while some catalysts will degrade within a few seconds upon exposure to light, others are quite stable. As a result, the individual compounds need to be evaluated for stability. Many researchers have spent far too much time attempting to determine why a catalyst was not as active as it should be when it was simply partially decomposed due to exposure to light.

Is a solvent needed?   This is a critical question that is worth investigating. Most historical preparations of organic compounds employ a solvent. However, that does not mean that those are the highest-yielding procedures or that the solvent is really required. If the reaction is successful without the addition of solvent, removing it will significantly simplify the operational procedures since issues surrounding the use of a solvent will be eliminated. Carrying out the reaction under solvent-free conditions has its own challenges; however, they are often offset by the advantages of the approach.

Do the solvents need to be rigorously dried?   With the widespread adoption of solvent purification systems, drying solvents is significantly easier than it was a few decades ago. It should be noted that some solvents are unable to be effectively dried using these systems and must still be dried using alternative methods. If anhydrous solvents are needed and a solvent purification system is not available, activated molecular sieves could be the most practical solution. Studies have shown that the water content of the solvents dried using molecular sieves is similar or lower than solvents dried with traditional methods [1]. Remember to take into account the time that it will take to activate the sieves prior to adding them to the solvent.

Will the solvents be easy to remove?   If your reaction requires the use of a solvent, the ability to remove it after the reaction is over is a critical concern. Low-boiling solvents such as diethyl ether are often easily evaporated under vacuum; however, high-boiling solvents can be more challenging to remove.

Do the solvents need to be degassed/deoxygenated?   If the answer is yes, there are two common approaches that are used. If oxygen is the only problematic gas, simply sparging the solvent with nitrogen or argon for 30 min will likely be sufficient to deoxygenate the solvent. Care must be taken not to introduce water into the vessel when sparging since the vessel could become quite cold due to solvent evaporation and moisture could condense. This is normally avoided by performing the sparging in septa-sealed vessels using a long needle or a specially designed glass frit to bubble the nitrogen or argon through the reaction mixture. A smaller needle is inserted into the septa at top of the vessel that will serve as the vent. If the solvent needs to be completely degassed, this is commonly achieved through several freeze–pump–thaw cycles. These steps consist freezing the dry solvent in a specially designed flask. Once frozen, the flask is placed under a static vacuum and warmed until the solvent melts. This process is repeated several times until no gas is evolved. This endpoint can be challenging to determine when low-boiling solvents are used. In general, three to four freeze–pump–thaw cycles are needed to degas the solvent. The prerequisite for this work is an outstanding vacuum line. If the vacuum is poor, this approach will not work as effectively. Additionally, the vacuum line needs to be pristine since contaminants from inside the vacuum line could contaminate the solvent. Special glassware has been designed specifically for this application. While deoxygenating solvents using a nitrogen purge is quite successful using needles, *degassing* solvents using several freeze–pump–thaw cycles should not be attempted using needles plunged through septa.

Are the products stable to air?   This is commonly encountered with organophosphines since many of these compounds often oxidize in air to generate organophosphine oxides. If the target phosphine is not stable to air, the purification will be quite challenging, and purification by column chromatography will be nearly impossible. For compounds that are especially sensitive to air, specific details will be provided in the sections dealing with those compounds. It is important to identify when an air-sensitive product will be generated. If the equipment is not in place to handle and store the air-sensitive materials, it might be advantageous to devise an alternative approach.

How will the products be purified?   The most common methods for the purification of organic compounds are column chromatography, distillation, sublimation, and crystallization. Column chromatography can be the most time-consuming of the purification methods, but it can be the most effective. Distillation is often an excellent method for the separation of volatile compounds from a crude reaction mixture. In many cases, compounds with relatively high boiling points can still be separated using a vacuum distillation. Sublimation is also a relatively straightforward approach to the purification of many organic compounds, and a host of sublimators are commercially available. Crystallization is a popular method for the purification of solid organic compounds. However, devising a successful approach to the crystallization of a new compound can be challenging.

How will the products be stored?   This is a critical concern when working with compounds that readily oxidize in air, absorb moisture, or decompose at room temperature or upon exposure to light. When reading a literature synthesis for a catalyst/ starting material/product that is air sensitive and needs to be stored at low temperature, it would be logical to store the compound in the glovebox freezer. However, if your glovebox does not have a freezer or you do not have a glovebox, the synthesis will be much more complex and challenging. It might be more practical to search for a different approach.

How will the products be characterized?   The characterization methods need to be addressed when planning a synthesis. If the piece of equipment is not available for a critical characterization of the starting material, intermediate, or final product, it will not be possible to complete the synthesis. The type and complexity of the characterization methods will vary according to the synthesis. The author instructions for journals such as *Organic Letters* and the *Journal of Organic Chemistry* are often excellent places to find a list of the characterizations typically required for reporting the synthesis of new compounds. Additionally, it is important to use as many techniques as possible when characterizing synthetic targets. Trying to publish IR data for stable new organic compounds as the only method of characterization is typically unacceptable. The connectivity of new compounds must be rigorously investigated and questioned until all methods of characterization have been exhausted. This will aid in ensuring that the connectivity and chirality have been correctly assigned.

1.2   NEW VERSIONS OF "CLASSIC" ORGANIC REACTIONS

Synthetic chemistry is quite fluid and researchers are constantly devising new versions of classic reactions and generating catalysts with higher activities and increased selectivities. While many classic reactions remain the best methods for the preparation of a specific group of compounds, modifications of these classic reactions are constantly improving various aspects of the chemistry.

As a representative example of how the modernization of classic reactions can alter the design of a synthesis, consider the preparation of ethers through a copper-catalyzed coupling reaction. The Ullmann-type synthesis of ethers is one of the most well-known versions of this reaction. Historically, an Ullmann synthesis used a stoichiometric (or higher) amount of copper bronze to promote the reaction [2–4]. This resulted in a significant amount of copper waste that needed to be separated from the desired reaction products and disposed of. Furthermore, the temperatures required for the copper bronze-promoted Ullmann coupling were typically above 200 °C. Thus, it would be advantageous to develop a catalytic method for the synthesis of these compounds that functioned at lower temperatures. Reducing the copper to catalytic levels would make the removal of the copper as well as the isolation/purification of the desired diaryl ether significantly easier, while the lower temperature would facilitate the use of temperature-sensitive substrates. A significant step toward this goal was recently achieved by Buchwald and Venkataraman [5,6]. Buchwald's approach used common copper salts along with a solubilizing ligand to generate a catalytically active copper complex that promoted the coupling reaction (Scheme 1.1). Venkataraman used a discrete copper(I) species containing neocuproine as the chelating ligand (Scheme 1.2) to catalyze the reaction. The conditions for both approaches were quite mild (110 °C) relative to the classic Ullmann synthesis, and both approaches generated the desired ethers in excellent yield. The advantages of these modifications to the classic reaction included significantly reduced reaction temperatures and a significant reduction in the amount of copper needed to promote the reaction. The use of organotrifluoroborates as the coupling partners has also been reported (Scheme 1.3) [7] along with a microwave-assisted version of this reaction [8].

One of the most active areas of research is the conversion of reactions that are classically catalyzed by transition metals into metal-free versions. The economic issues that are driving this area of research revolve around the high cost of many catalysts as well as the challenges encountered when trying to recycle and reuse these catalysts in subsequent reactions. There are also toxicity issues to consider with some metals. However, significant reactivity issues can be encountered when trying to reproduce the unique reactivity that transition metal catalysts exhibit using nonmetallic systems. Several of the following chapters and sections will highlight recent developments in this area.

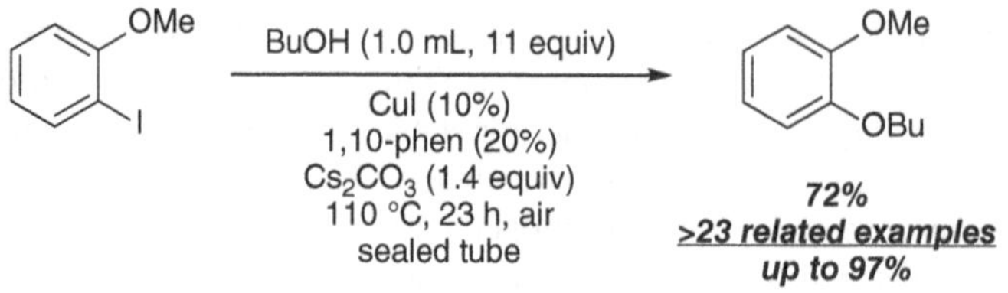

SCHEME 1.1   Synthesis of alkyl aryl ethers using the cuprous iodide/phen catalyst system [5].

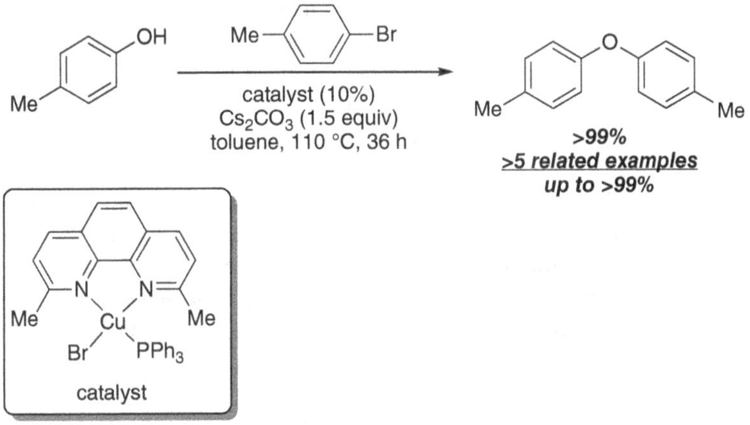

SCHEME 1.2   Synthesis of diaryl ethers using a discrete copper complex [6].

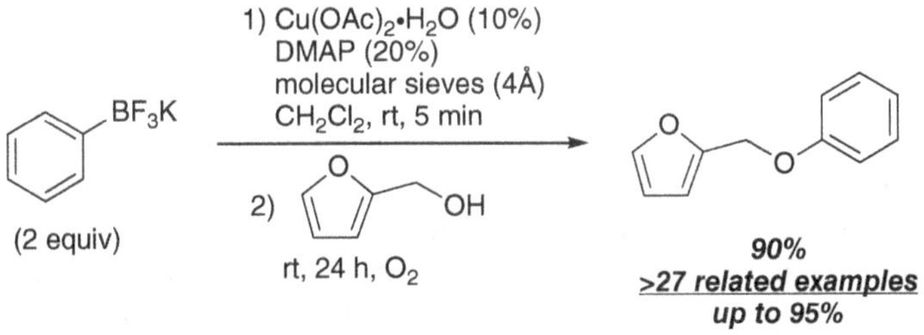

SCHEME 1.3   Synthesis of alkyl aryl ethers using aryltrifluoroborate salts [7].

1.3   SOLVENT SELECTION AND SOLVENT-FREE REACTIONS

One of the biggest issues encountered when devising a synthetic strategy is the solvent selection. From a practical perspective, a good solvent will dissolve the components of a reaction that the chemist wishes to dissolve, not react with either the reagents or products, be easily removed, not be flammable, and not be toxic or harmful to the environment. Some solvents will satisfy all of those requirements, while others will only match a few.

How easily the solvent can be removed from the crude reaction mixture remains one of the most important issues when considering a solvent. It will not be successful to design an elegant synthesis for a target molecule only to discover that small amounts of solvent cannot be removed. Low-boiling solvents such as dichloromethane and pentane are often easily removed under vacuum, although small amounts can be encapsulated in solid products. Trituration of the solid while under high vacuum can be an effective way to solve this issue. This is quite common when the products are foams. High-boiling solvents such as DMSO and NMP can be very challenging to remove from crude reaction mixtures. While it is common practice to remove these solvents under vacuum, this is often not practical. For very polar solvents such as these, it might be possible to remove the solvents using an aqueous extraction. DMSO and NMP are quite soluble in water, while many of the organic products are not.

In some cases, it will be possible to recycle a solvent for use in additional reactions. Low-boiling solvents often work well since they can simply be distilled from the reaction mixture. For most synthetic laboratories, recycling the comparatively small amounts of solvents used for in synthesis of small molecules is often not economically feasible. However, recycling the liters of solvents used for column chromatography is quite common. Since the most common solvents used for this application are hexane/pentane and ethyl acetate, it is common to find a dedicated still for these solvents in many synthetic laboratories. It is also noteworthy to point out that many research groups do not take the time to try and separate hexane from ethyl acetate. The composition of the solvent mixture is determined after distillation and adjusted to provide the eluent for the next separation.

Before starting a synthesis, the chemistry should be evaluated to determine whether or not a solvent is actually needed. In many cases, reactions are faster, cleaner, and more efficient in the absence of a solvent. In the extreme case, the desired reaction will *only* occur in the absence of a solvent. The iron-catalyzed addition of secondary phosphines to alkynes is an example of such a reaction (Scheme 1.4) [9]. In the absence of solvent, this double addition reaction cleanly affords the 1,2-bisphosphine in excellent yield. When solvents were added to the reaction, the addition reaction was completely suppressed.

There are several operational issues to consider when performing solvent-free reactions. Most laboratory-scale reactions are carried out on 0.1–0.5 g scales. Under solvent-free conditions, the small amounts of reagents might be problematic for common round-bottom flasks since much of the material could be spun onto the sides of the flask and not actually be in contact with the other reagents. This is not typically encountered when using a solvent since the solvent will typically rinse down the sides of the flask. Instead of using round-bottom flasks, small vials are ideally suited to carry out solvent-free reactions. There are also a number of issues to consider when using vials. One of the most important issues is

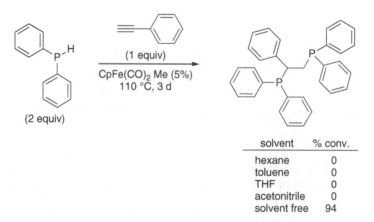

solvent	% conv.
hexane	0
toluene	0
THF	0
acetonitrile	0
solvent free	94

SCHEME 1.4   Effect of solvent on the double hydrophosphination of alkynes [9].

the physical state of the reagents and products. If all of the reagents and products are liquids, using small flat-bottom vials and small stirring bars will likely be effective at mixing the reagents and the reaction will be successful. If one of the reagents or products is a solid and does not entirely dissolve in the other reagent, or tends to precipitate out of the reaction, flat-bottom vials with small stirring bars will not be as effective since the material will collect in the edges of the vial.

A reaction that contains a number of solid reagents, catalysts, additives, or products can be one of the most challenging to convert into a solvent-free version. For these reactions, achieving effective mixing could be difficult. In these cases, gently heating the reaction mixture could be successful or adding a very small amount of a solvent might also aid in the reaction. While this approach is a bit substrate dependent, a good general place to start is to add one mass equivalent of solvent per reagent.

A general approach to solvent-free reactions carried out on scales between 0.1 and 0.5 g is to use small vials with rounded bottoms and large stirring bars. Using small stirring bars will not effectively stir the material in the bottom of the vial. There are many inexpensive vials that are readily available from a range of vendors. One of the more attractive options is to use the vials sold by microwave manufacturers for running microwave-assisted reactions. As a representative example, the CEM reactor vials are shown in Figure 1.1. These are useful for solvent-free reactions since they have rounded bottoms, the tops can be crimp sealed with septa, and they will fit into standard centrifuges. It is worth noting that these vials can be used for reactions carried out in and out of the microwave. As mentioned earlier, one of the most important components to a solvent-free reaction is making sure all the reagents are thoroughly mixed and the desired stoichiometry is achieved. Spinning the vial in a centrifuge will concentrate the reagents at the bottom of the vial. This can be repeated many times if needed. Naturally, this is challenging to achieve with a round-bottom flask.

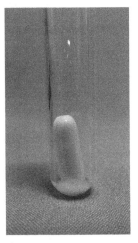

FIGURE 1.1   Solvent-free reaction using both a solid and a liquid reagent. *Source*: Image of the CEM microwave reactor vials was produced with permission from CEM Corporation.

Troubleshooting solvent-free reactions.   *While it is not possible to predict every challenge that will arise during the course of a synthesis, the following observations and reaction requirements are typically encountered during a solvent-free synthesis.*

Observation/Issue/Goal	Possible Solution/Suggestion
The stirring bar is flinging the reagents onto the sides of the round-bottom flask under solvent-free conditions	Change from a round-bottom flask to a narrow vial with a rounded bottom and use a large stirring bar to triturate the reagents
All of the reagents and products are liquids	This is the ideal case for a solvent-free reaction—small vials with small stirring bars would be a good place to start
The reagents keep getting stuck on the bottom edges of the flat-bottom vial and do not appear to be properly mixing	Change to a small vial with a rounded bottom
After injecting the liquid reagents into the vial, they are sticking to the sides of the vial and not interacting with the other reagents at the bottom of the flask	Centrifuge the reaction tube to concentrate all of the reagents in the bottom of the tube
Despite being very careful, some of the solid is sticking to the sides of the vial	Use a rolled piece of weighing paper or a sleeve to aid in delivery of the solid catalyst to the bottom of the reactor vial. It might help to centrifuge the vial for a few minutes although this tends to be less effective with small amounts of solid catalysts
The reaction is not going to completion and is no longer stirring due to the formation of a gel or solid	If you have already changed to a larger stirring bar, it might be necessary to add a small amount of a solvent to the reaction mixture to inhibit the formation of the gel. While the amount of solvent needed to promote the reaction will vary, a good place to start is 1 mass equivalent per reagent
Improper mixing of solid reagents	Use as large a stirring bar as possible. If this is a microwave-assisted reaction, the length of the stirring bar should not exceed ¼ of the standing wave
At least one of the reagents or the product is a solid	There could be incomplete mixing of the solids. Vials with rounded bottoms and large stirring bars are a good place to start. If this is not effective, try adding a small amount of solvent to the reaction mixture
Sublimation of one or more of the reagents into the neck of the flask is occurring during the course of a reaction using conventional heating	Try to submerge as much of the vial as possible into the oil bath or block heater. Alternatively, use an external heating source to prevent the sublimation. In some cases, adding a small amount of solvent might be successful
Sublimation of one or more of the reagents is occurring during the course of the reaction using microwave heating	Since the reaction vessel is fully contained by the reactor, applying an external heating source to prevent the sublimation is not possible. Adding a small amount of solvent could circumvent the problem

(Continued)

(Continued)

Observation/Issue/Goal	Possible Solution/Suggestion
The mineral base is not fully mixing with the reagents	Change to a liquid base
A gas is generated during the reaction and pressure builds inside the vessel	This could generate an extremely dangerous situation. One solution would be to add an active purge to the reaction. A slow purge using an inert gas is often sufficient to inhibit a pressure buildup. If septa capped vessels are used for the reaction, simply adding a nitrogen/argon feed (needle) and vent needle could solve the problem

1.4   OPERATIONAL SIMPLICITY

The operational simplicity of a synthetic method remains one of the most critical parameters that should be considered when designing a synthesis. If the protocol is so complex that only an expert can get it to work 50% of the time, the approach is not very practical for a typical synthetic laboratory. A synthetic approach that is straightforward to carry out is more practical and likely to be adopted by the general community.

In many cases, the operational difficulties of a specific approach can be mitigated through the careful selection of solvents, catalysts, and reagents. However, in some cases, the complexity of the synthesis is a result of the chemistry itself. In these cases, cleverly devising completely different strategies might be the only way to circumvent a problematic step. As listed previously, increasing the overall length of a synthesis by the addition of a few easy steps can often be much more practical than struggling with one challenging reaction.

An example of how a chemical discovery has greatly reduced the operational difficulty of a synthetic approach is provided by the synthesis of dialkylbiarylphosphine ligands (Buchwald ligands) [10]. Dialkylbiarylphosphines are a valuable class of monodentate phosphine ligands that stabilize/solubilize a number of metal centers and promote transition metal-catalyzed reactions (Figure 1.2). These supporting ligands have been successfully used in a range of carbon–heteroatom bond-forming reactions [11–14]. Although these

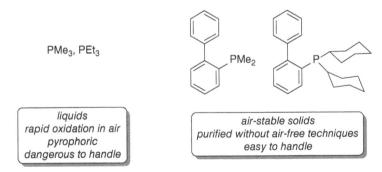

FIGURE 1.2   Stability of common trialkylphosphines and dialkylbiarylphosphines.

compounds are useful, the discovery that they are remarkably stable to air [15] has catapulted this class of ligands to be one of the most practical choices for synthetic chemists. This stability stands in stark contrast to the extreme air sensitivity of many alkylphosphines. For example, tricyclohexylphosphine is a very strong ligand for a host of metal centers, but it must be stored and handled under an inert atmosphere to prevent rapid oxidation. Many trialkylphosphines are so sensitive to air that they will spontaneously ignite. The vast majority of the dialkylbiarylphosphines can be handled, loaded into reaction flasks in air, and stored at ambient temperature for years. Virtually every synthetic lab working on metal-catalyzed coupling chemistry has several of these ligands on the shelf.

This discovery reaches far beyond the use of dialkylbiarylphosphine ligands in catalytic reactions. It has changed the way in which chemists think about the synthesis of new alkylphosphines. As long as the biaryl fragment is present, there is a reasonable expectation of some air stability [15]. The other groups on the phosphorus center could be highly electron donating, and the resulting phosphine would still have some stability to air. The dialkylbiarylphosphine that might be predicted to be highly reactive toward oxygen would contain small donating alkyl groups such as methyl. Indeed, high-yielding syntheses of methyl-containing phosphines remain rare, and trimethylphosphine is known to spontaneously ignite in air. However, when one of the methyl groups on PMe_3 was exchanged for the biaryl fragment, the resulting compound, dimethylbiphenylphosphine, was quite stable to air [16]. Naturally, the level of stability will vary from group to group; however, dialkylbiarylphosphines are unlikely to be pyrophoric. Why this chemistry has had such an impact can be found in the classic approaches to the synthesis of alkylphosphine ligands. The historical approaches to the preparation of these compounds are complicated by the sensitivity of the reagents, intermediates, and products to air. This requires all of the various steps and purifications required to form the desired phosphine to be carried out with rigorous exclusion of air. Naturally, many of the approaches to the synthesis of the dialkylbiarylphosphine ligands still require an oxygen-free environment; however, their isolation/purification can often be carried out without rigorous exclusion of oxygen. Setting up the reactions in a deoxygenated environment without a glovebox can be difficult, but it is still possible for many synthetic laboratories. However, purification of an air-sensitive phosphine such as PCy_3 or PMe_3 by column chromatography without a glovebox is remarkably difficult. Historically, most chemists might have looked upon the lack of a glovebox as a reason to not even consider alkylphosphine syntheses. Currently, if a synthetic laboratory does not have a glovebox and is interested in making new phosphine ligands, a very practical approach would be to simply try and add a biaryl group to the phosphorus center and modify the remainder of the fragments on the phosphorus center to suit the needs of the desired application.

1.5 METAL-CATALYZED TRANSFORMATIONS

The use of transition metal catalysts to promote carbon–carbon and carbon–heteroelement bond-forming reactions has significantly increased over the past few decades. These metal complexes often promote unique reactions that are challenging or impossible to achieve using classic organic methodology. Representative examples of metal-catalyzed carbon–heteroelement bond-forming reactions are shown in Schemes 1.5–1.9. One of the current challenges in this field includes the design of metal complexes that exhibit high catalytic activities and selectivities while being stable to moisture and oxygen.

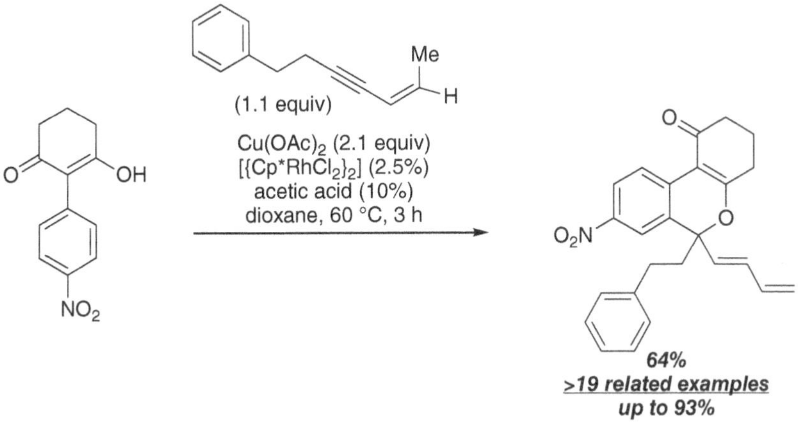

SCHEME 1.5   Example of a rhodium-catalyzed carbon–oxygen bond-forming reaction [17].

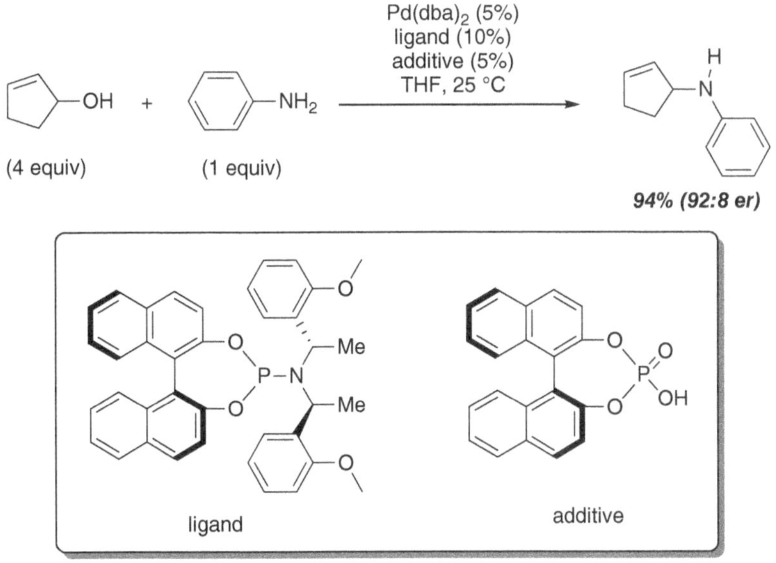

SCHEME 1.6   Example of an asymmetric palladium-catalyzed carbon–nitrogen bond-forming reaction [18].

The importance of designing operationally simple catalyst syntheses cannot be overstated. Ideally, the metal complexes should be robust species that will remain intact until converted into the active species during the course of the reaction. It is also advantageous for the purity of the metal complexes to be easily determined. In some cases, researchers add a "catalyst" of unknown purity and composition to reactions and are confused when they obtain strange results. This is often the case when using highly insoluble metal precursors such as cuprous iodide or ferric chloride. The former tends to readily generate the oxide upon exposure to air, and if the contaminated copper iodide is used in a Sonogashira coupling reaction, it could promote the formation of

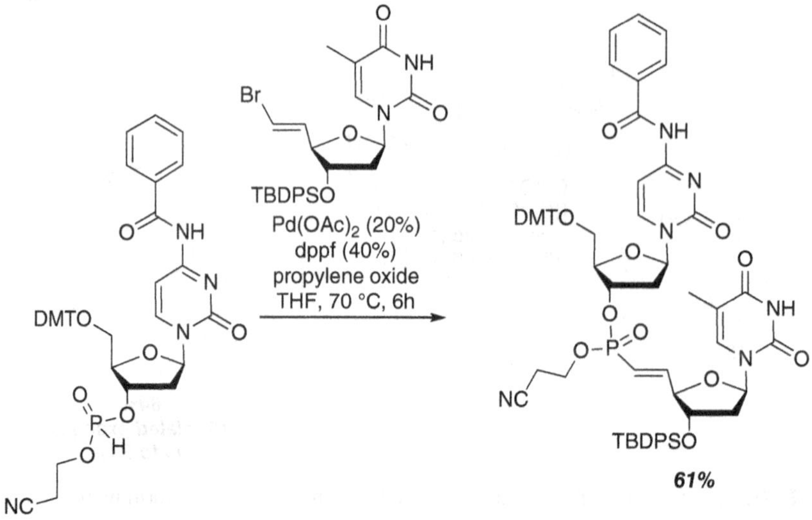

SCHEME 1.7 Example of a palladium-catalyzed carbon–phosphorus bond-forming reaction [19].

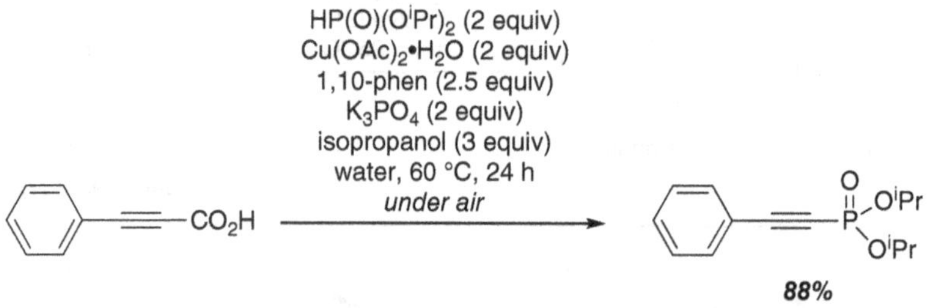

SCHEME 1.8 Example of a decarboxylative copper-promoted carbon–phosphorus bond-forming reaction [20].

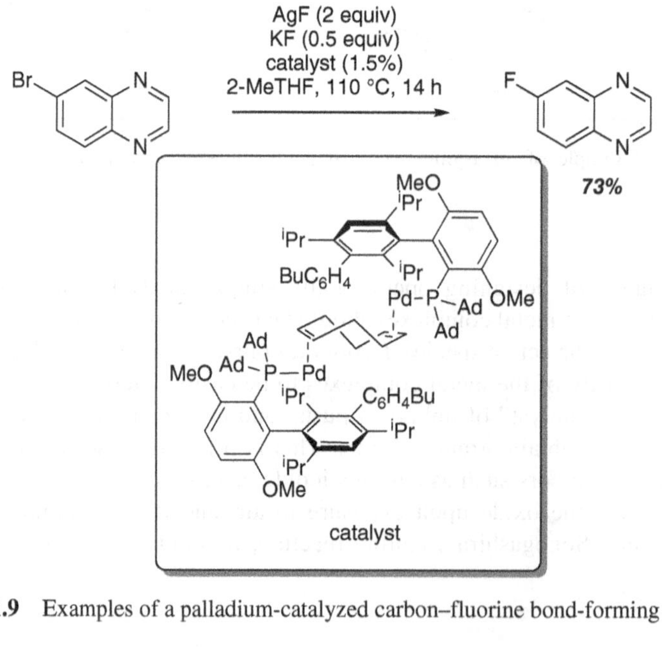

SCHEME 1.9 Examples of a palladium-catalyzed carbon–fluorine bond-forming reaction [21].

diynes instead of the desired cross-coupling. If the metal complexes are thermally unstable or rapidly decompose in the light, the evaluation of the reaction results will always be suspect. The worst-case scenario would be for the results to be unrepeatable in other laboratories.

Chemists often devote considerable amounts of time and resources on the design and synthesis of new and more exotic ligands that will increase the activity or selectivity of a specific transition metal-catalyzed reaction. A researcher may spend years optimizing the synthesis of the new ligand. However, if the ligand dissociates from the metal center during the course of the reaction, interpreting the results of the catalysis will be challenging since the dissociated ligand may not behave as an innocent species in the reaction mixture. This is also true for additives, cocatalysts, and promoters. As a general rule, it is critically important to make sure that all materials are clean and free of contaminants. It will be well worth the time, and the effects of the foreign materials can always be determined by adding controlled amounts of materials to the reactions mixtures after the baseline reactivity with clean materials has been established.

Several examples highlight the importance trying to rigorously characterize catalysts. The first example involves the use of $Pd_2(dba)_3$ (dba = dibenzylideneacetone) solvates. These compounds are commercially available and remain popular sources of palladium(0). Thousands of reactions have been reported using this palladium source. In a typical reaction, this compound is treated with a supporting ligand such as a phosphine in order to displace the dba and generate the desired metal–ligand complex. In addition to the possibility that the dba could reversibly coordinate to the metal center and alter the activity of the catalyst, the possibility that commercially available samples as well as "in-house" prepared material could be contaminated by nanoparticle palladium is a significant concern [22]. While some of the contamination levels have varied according to the sample, levels as high as 40% have been found. Depending upon the reaction medium, the solubility of this material can be quite different from the $Pd_2(dba)_3$ solvate. If the nanoparticle palladium is never solubilized, this results in a lower level of available palladium being added to the reaction and could significantly alter the outcome of the catalysis. Naturally, this altered stoichiometry could result in unrepeatable results since many palladium-catalyzed reactions are quite sensitive to the Pd/L ratio. This contamination could also lead to TON and TOF that are not a true reflection of the activity of the ligated palladium species. In addition to these issues, if the nanoparticle palladium is truly insoluble and a heterogeneous solution is formed, this material could conceivably catalyze an entirely different reaction pathway.

While it might seem reasonable to attempt to characterize the catalyst using ^1H NMR spectroscopy, the ^1H NMR spectrum of Pd_2dba_3 solvates can contain a vast number of resonances due to coordinated and free dba. An elegant solution to this NMR problem that combined detailed one- and two-dimensional NMR investigations including diffusion studies was reported by Ananikov [22]. Elemental analysis is also an option; however, unless the analyzer is located in-house, there will always be uncertainty associated with the results due to potential decomposition while in transit. In the ideal case, both elemental analysis and the spectroscopic data should be used to provide a more accurate picture of the composition of these metal complexes. As with many metal catalysts, it is advantageous to prepare them immediately prior to use. While some reactions are not as sensitive as others, this issue could cause problems for enantioselective synthesis due to the presence of resolved Lewis base (chiral ligand) in solution.

A second example that illustrates the importance of rigorously characterizing catalysts and is focused on the use of iron salts as catalysts for organic transformations. One of the most popular ways of generating a catalytically active species was to solubilize a common iron salt such as $FeCl_3$ with a chelating ligand. A variety of diamines have proven successful in this role, and a number of organic reactions have been catalyzed using this approach. The issue with this chemistry is the potential for trace contaminants in the ferric chloride to serve as the true catalyst in the reaction. Buchwald and Bolm demonstrated that even 10 ppm copper(I) oxide could dramatically increase the yields in a number of reactions [23]. Key to this investigation was the observation that 99.99% pure $FeCl_3$ was much less active, and only upon the addition of the copper salt was significant conversion observed. Thus, even small amounts of trace metal contamination could lead to incorrect conclusions regarding the outcome of the reactions.

Ideally, when metal salts are used to generate active catalysts in solution, they should be of the highest grade when establishing baselines for reactivity, selectivity, and substrate scope. A representative sample of each lot should also be analyzed for the presence of trace impurities. For contamination due to metal salts, this is most commonly accomplished by ICP-AES. For many metal salts, the level of detection for this technique is in the ppm range. Although this will not completely eliminate the possibility of contamination, it will provide other researchers with an idea of what is present. Ideally, the results of contamination screen should be included in the supplementary materials of published reports. Once the baselines have been established, it would be valuable to the synthetic community to screen the metal salts of lower purity in order to determine if there are any differences between the grades.

It remains critically important to determine the precise identities of the active species and elucidate as many of the mechanistic details as possible. Such investigations enable the design of second- and third-generation catalysts with enhanced reactivity. The undeniable conclusion resulting from even a cursory survey of the literature is that it is often remarkably difficult to determine the precise identity of the active species that catalyzes a specific reaction. The metal complex that is added to the flask is rarely the active species. However, following common practice, the compound that is added to the flask will be referred to as the "catalyst" for the purposes of this text. Naturally, it would be more accurate to refer to these compounds as precatalysts or precursors.

1.6   ORGANOCATALYSIS

The ability of a transition metal catalyst to promote unique reactions that are not accessible with classic organic methodology as well as the potential to direct the selectivity of a reaction remains one of the major driving forces for the development of new metal catalysts. However, there are considerable challenges that must be overcome when using a few of the popular catalyst systems. Many transition metal catalysts must be handled and stored over an inert atmosphere such as nitrogen or argon. Some of these catalysts can be rendered inactive upon exposure to moisture. For a number of catalysts, the toxicity of the metal can be a concern. Over the past few years, a considerable amount of effort has been devoted to the conversion of metal-promoted synthetic strategies into approaches that use entirely organic compounds as catalysts [24–34].

Practical advantages of an organocatalyst can include an increased tolerance of the catalyst and catalytic reactions to moisture and air, greater selectivity, and the minimization

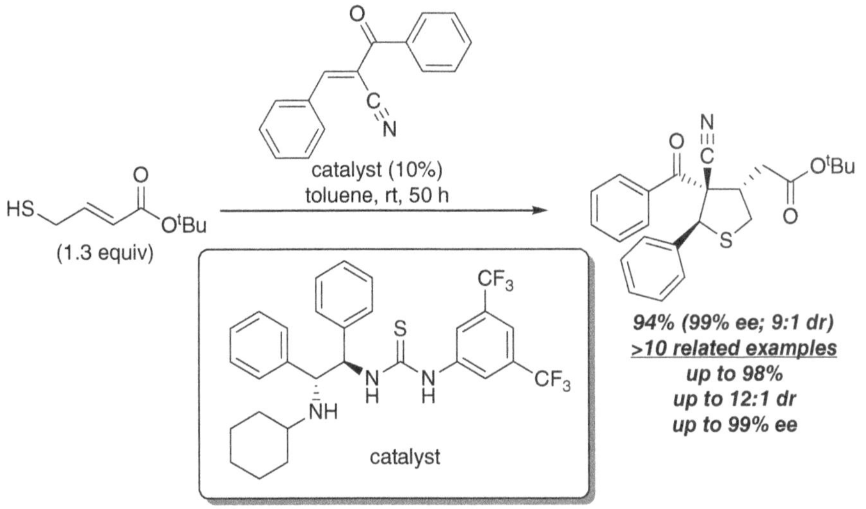

SCHEME 1.10 Organocatalytic asymmetric synthesis of functionalized tetrahydrothiophenes [35].

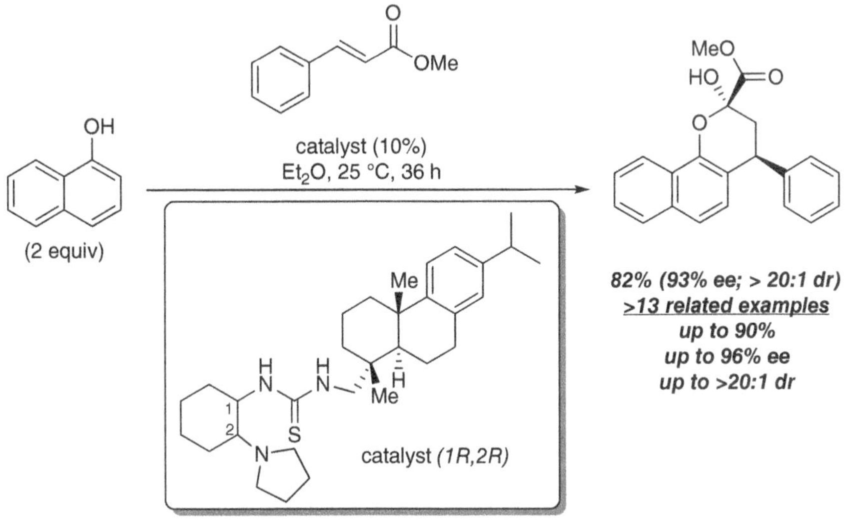

SCHEME 1.11 Organocatalytic asymmetric synthesis of chromane derivatives [36].

of secondary reaction pathways. A few representative examples of organocatalyzed carbon–heteroelement bond-forming reactions are shown in Schemes 1.10 and 1.11.

1.7 MICROWAVE- AND ULTRASOUND-ASSISTED CHEMISTRY

Over the past decade, focused microwave reactors have revolutionized how synthetic chemists approach the preparation of chemical compounds. Due to their ability to rapidly heat reactions on small to moderate scales, microwave-assisted chemistry has become

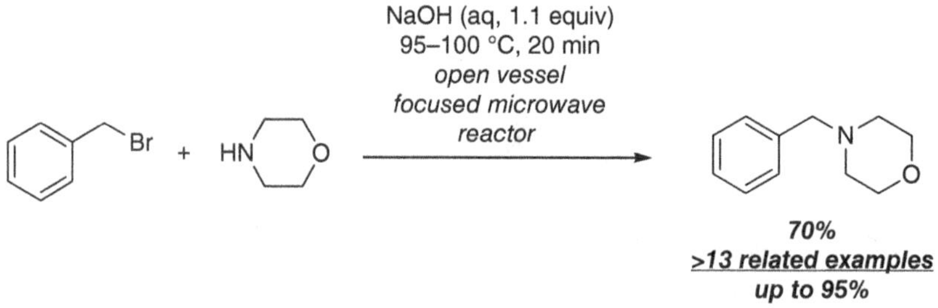

SCHEME 1.12 Microwave-assisted aqueous alkylation of amines using an open vessel [47].

"first choice" for synthetic chemists. A number of excellent reviews have been published on microwave-assisted synthesis [37–46]. The following sections provide a brief outline of the practical considerations when designing a microwave-assisted approach.

Many of the early examples of microwave-assisted chemistry used domestic microwaves with holes cut through the casing in order to accommodate condensers. This approach suffered from poor reproducibility due to the nonhomogeneity of the microwave field. Rather than cutting holes in the microwave oven, reactions were performed under solvent-free conditions on silica or other solid supports. A typical reaction was carried out by suspending the silica in a solution of the reagents followed by removal of the solvent. Irradiation of the resulting mixture generated products. This silica-supported approach also suffered from poor reproducibility, and some of the reactions carried out on dry silica that were reported to be microwave-assisted reactions were actually complete prior to irradiation. The heating needed to remove the solvent promoted the reaction.

Once dedicated microwave reactors were designed and made commercially available, one of the early design considerations was the ability of the reactor to accommodate a condenser. Indeed, several commercial microwave reactors are able to accommodate standard glassware for refluxing solutions. A number of reactions have been reported using this approach including the N-alkylation of primary and secondary amines by alkyl halides (Scheme 1.12) [47]. The reaction was operationally simple to set up and simply consisted of adding all the reagents/solvent to a flask, capping with a condenser, and irradiating for 20 min.

1.7.1 General Checklist for Performing a Microwave-Assisted Reaction

Is microwave chemistry an option?   For reactions that need to be heated, the answer to this question is almost always yes. However, care must be taken when determining the conditions for a specific reaction. The following sections will aid in making an educated first attempt rather than simply blindly irradiating the sample to see what would happen.

Solvent selection.   One of the most important considerations when designing a microwave-assisted reaction is whether or not a solvent is actually needed for the reaction. Some reactions will not be successful under solvent-free conditions; however, since the solvents are typically disposed of at the end of reactions, the elimination of solvents from chemical reactions is a step forward when designing sustainable chemical reactions. The vast majority of molecules containing functional groups will have a dipole moment and absorb microwave irradiation without the addition of a solvent. If a solvent is required for the success of the reaction, a minimal amount of solvent should be used. These "near-solvent-free" reactions still significantly reduce the amount of solvent used by the synthesis.

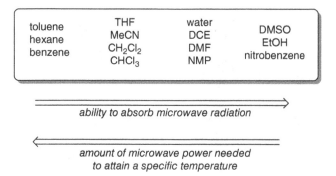

FIGURE 1.3   General relationship between the polarity of the solvent and the ability of the solvent to absorb microwave energy.

For microwave-assisted reactions, polar solvents such as alcohols and DMSO absorb microwave radiation well, while nonpolar solvents such as hexane and toluene are poor microwave absorbers and cannot be heated effectively (Figure 1.3). A number of different additives can be used to help heat reaction mixtures where both the reagents and solvents are nonpolar. Detailed information regarding the behavior of solvents in a microwave field as well as examples of reactions in various solvents has been summarized in several excellent reviews and texts [37,38,42,48,49].

Although commercial microwave reactors are able to accommodate condensers and other pieces of glassware, the vast majority of reported reactions are carried out using sealed reactor vessels. Using these sealed vessels, many organic solvents can be heated well above their boiling points. Some solvents can exhibit significantly different properties when heated above the boiling point of the solvent. For example, while water has a high dielectric constant at 25 °C, the dielectric constant is significantly lower when water is heated to 250 °C [50].

Many synthetic chemists are accustomed to designing reactions where a solvent is used to dissolve all of the reagents. If reagents or catalysts are not fully dissolved, this can lead to confusing conclusions when trying to interpret the results of subsequent runs. In some cases, heating the reactions will solubilize all of the reagents and create a homogeneous reaction. Given the precise control over the reaction time temperature and microwave power that commercial microwave reactors provide, reactions can routinely be carried out under neat conditions, near-solvent-free conditions, or diluted with solvent (Figure 1.4).

In order to gain insight into the physical state of the reaction mixture, it would be advantageous to be able to visually inspect the reaction mixture while it is being irradiated. While this might seem like a challenging task, CEM has been able to incorporate a camera into the microwave cavity that allows visual inspection of the reaction mixture before, during, and following irradiation. This camera provides insight into the physical state of the reaction mixture and facilitates the determination of whether all the materials are dissolved, the effectiveness of the stirring, and whether decomposition is occurring. Figure 1.5 shows an example of this approach. The reaction pictured is a rhodium-catalyzed addition of secondary phosphine oxides to a testosterone derivative [51]. The reaction solvent was THF and $(Ph_3P)_3RhCl$ was the catalyst used for the reaction. Without the camera, a heterogeneous reaction mixture was placed into the reactor, and a heterogeneous reaction mixture was removed from the reactor at the end. The chemist would be

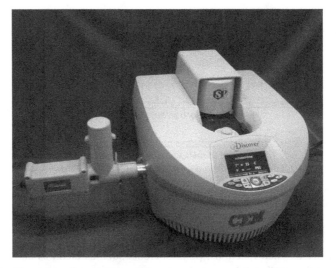

FIGURE 1.4 Example of a commercial microwave reactor with an integrated camera. *Source*: Acknowledgment is given to CEM Corporation for the generous loan of the microwave reactor for the construction of this text.

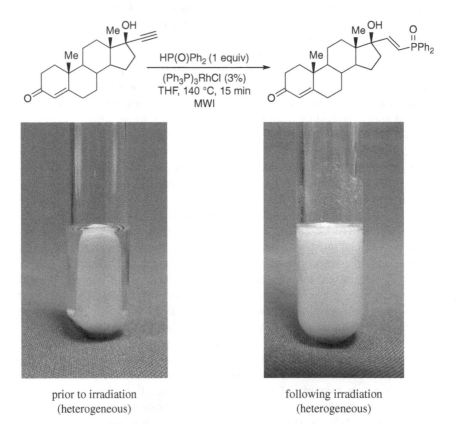

prior to irradiation
(heterogeneous)

following irradiation
(heterogeneous)

FIGURE 1.5 Before and after pictures of the reaction mixture. It should be noted that the picture of the final product has a slight yellow tint to the solution that is not shown in the B&W photo. *Source*: Images of the CEM microwave reactor vials were produced with permission from CEM Corporation.

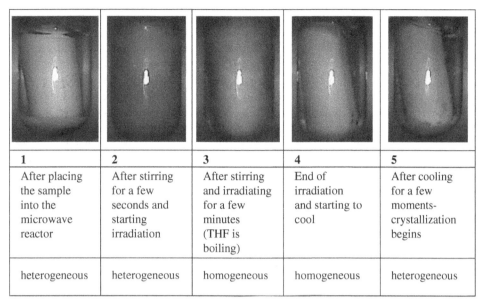

1	2	3	4	5
After placing the sample into the microwave reactor	After stirring for a few seconds and starting irradiation	After stirring and irradiating for a few minutes (THF is boiling)	End of irradiation and starting to cool	After cooling for a few moments- crystallization begins
heterogeneous	heterogeneous	homogeneous	homogeneous	heterogeneous

FIGURE 1.6 Pictures from inside the microwave cavity before, during, and following irradiation. It should be noted that the color of the reaction mixture is slightly yellow after a few minutes of irradiation. The color is not illustrated in this B&W photo. The bright spot in the middle of the frame is due to the LED light used to provide light in the otherwise dark cavity of the microwave reactor. *Source*: Acknowledgment is given to CEM Corporation for the generous loan of a microwave reactor with integrated camera for the construction of this text.

unable to determine whether or not the reaction was ever homogeneous during irradiation. Figure 1.6 shows the progression of the reaction using the integrated camera. Before irradiation, the reaction is clearly heterogeneous, and the red (dark) catalyst can be seen in the bottom right side of the reactor vial (picture 1). As the reaction begins, it is still clearly heterogeneous, and the stirring bar is barely visible (picture 2). After reaching 140 °C and boiling for a few minutes, the solution has become homogeneous (picture 3). At the end of the irradiation period, all of the materials are still dissolved (picture 4). After standing for a few moments and cooling to 50 °C, the product begins to precipitate from the reaction mixture (picture 5). It should be noted that the microwave reactor will be unlikely to homogenize truly heterogeneous reactions. Mineral bases, molecular sieves, supported catalysts, or silicon carbide disks would be cases when the microwave reactor would not be effective at bringing all of the reaction components into the same phase. The most valuable aspect of camera is the ability to evaluate the reaction in real time. Additionally, the reaction can be remotely monitored and evaluated by multiple people simultaneously. This can be particularly attractive when collaborators or research facilities are in different locations.

Time, temperature, and power.   Modern microwave reactors are able to provide precise control over the amount of microwave energy that is used to heat the reaction mixture. Through careful selection of the reagents, catalysts, and solvents, many reactions can be designed to rapidly generate the target compounds in high yields. With a little experimentation, conventionally heated reactions can be readily converted into microwave-assisted versions.

Irradiating the sample with the highest available power for the shortest period of time in order to rapidly heat the reaction mixture to the highest possible temperature is a very popular method and has been used to generate countless compounds. While this might seem like a natural approach to microwave-assisted reactions, it could present problems for some systems. Unfortunately, this is often how chemists who are new to microwave chemistry approach their first reactions. They are often excited about the published results using microwave reactors and have read about how microwave reactors decrease reaction times, increase yields, and decrease secondary products. They take a reaction they are having trouble with and irradiate it at the maximum power level thinking that more power is better. This can result in the formation of an intractable mixture that is not salvageable. When working with a new reaction or catalyst, it often takes some experimentation to find the optimal conditions. For example, if the reaction is being carried out under solvent-free conditions, one of the reagents could preferentially absorb more of the applied microwaves and be preferentially heated. In the extreme case, this heated reagent could decompose before it ever has a chance to react with the other substrates. In some cases, the preferential heating leads to the formation of side reactions and secondary products. Additionally, metal-containing catalysts and additives could absorb more than the other reagents and become preferentially heated/destroyed. This leads to a significant amount of frustration on the part of the scientist trying to synthesize a desired compound. When a transition metal-catalyzed microwave-assisted reaction fails, the scientist is left to wonder if the catalyst was the problem, whether or not oxygen was introduced, or could the reagents be contaminated. While all of these reasons are entirely valid, it is also entirely possible that the failure of the reaction was simply due to using too much microwave power.

In some cases, only minimal amounts of microwave power are needed to promote a specific reaction. This is often the case when irradiating reactions carried out under solvent-free conditions with polar reagents or reactions carried out with polar solvents. Determining how much power is needed can often be the critical aspect of the synthetic design. The good news is that due to the operational simplicity of modern microwave reactors, finding a good set of conditions often takes less than a day. For high-absorbing materials such as DMF or ionic liquids, this power level could be quite low. Once the minimum power level has been determined, increasing/decreasing the power setting for subsequent screening reactions in 10–20% increments and analyzing the results will rapidly provide an experimental determination of how much microwave power is needed to achieve a desired result.

The microwave-assisted addition of secondary phosphine oxides to activated alkenes is an excellent example of how the power setting can significantly affect the outcome of a reaction (Table 1.1) [52]. This reaction was operationally trivial and consisted of adding the solid reagents to a reactor vial (in air), exchanging of the atmosphere for nitrogen, injecting the liquid reagents, and irradiating. For most of the examples, the neat reaction mixture absorbed microwaves well and heated to the desired temperature rapidly. When a high amount of microwave power (250 W) was applied to rapidly heat the sample to the desired temperature of 125 °C, the yield of the desired alkylphosphine oxide was low and a significant amount of diphenylphosphinic acid was observed. Using the optimization approach described in the preceding paragraph, the authors discovered that reducing the initial microwave power led to reduced amounts of the undesired phosphinic acid. Using this low power level, outstanding yields of the target alkylphosphine oxides were obtained after only 30 min of irradiation. While it might be tempting to irradiate reaction mixtures at

TABLE 1.1   Effect of Microwave Power on the Product Distribution of a *Phospha*-Michael Addition [52].

Run	Power (W)	Time (Min)	Temperature (°C)[a]	%A[b]	%B[b]
1	250	5	125	60	30
2	200	10	125	65	29
3	100	10	125	73	22
4	25	30	125	95	0

[a] Temperatures measured using an IR sensor at the bottom of the reaction vessel.
[b] The remaining compounds were not identified.

the highest available power for the shortest amount of time, it might be worth decreasing to lower power levels to obtain a cleaner product.

A solvent-free reaction using all solid reagents is another reaction that might tempt a researcher to increase the amount of microwave power. The researcher may believe that irradiating at a high power level will melt the sample rapidly. Typically, solid reagents need to "melt" or gel in order to absorb significant amounts of the microwave irradiation. While this does happen in some cases, it is also likely that the melted reagents will be exposed to high levels of microwave radiation for a few moments while the reactor lowers the power to prevent overheating. Even a few moments at high power could be enough to generate secondary products and decompose the sample. When working with a mixture of solids under solvent-free conditions, the obvious initial thought is to check the melting point of the solids to give an approximate temperature where melting will occur. However, when two or more solids are mixed, the melting point of the mixture is typically lowered, sometimes significantly. Researchers can take advantage of this observation and design a heating profile that might take advantage of this observation.

Another issue that could complicate the interpretation of results from microwave-assisted reactions involves syntheses that generate an ionic material or products that have a higher dipole than the starting materials/solvent. These products will absorb well and heat the sample rapidly. Common examples of this include the synthesis of ionic liquids and phosphonium salts. Consider the reaction of triphenylphosphine with benzyl chloride. It could be argued that these reagents might absorb some of the applied microwave radiation; however, neither will absorb as well as the ionic phosphonium salt generated in the reaction. Once even a small amount of this product was formed, it will be rapidly heated by even a small amount of applied microwave radiation. Even if the microwave reactor is programmed to decrease the microwave power in response to an increase in temperature, it may not be able to compensate if this ionic material is generated in the reaction. In such cases, a vessel failure is possible. These issues could be very important when trying to convert a conventionally heated reaction into a microwave-assisted reaction.

1 inch stirring bar	**micro stirring bar**
Advantages	Advantages
• good with heterogeneous mixtures	• good with homogeneous reactions
• good with homogeneous mixtures	• occupies a very small volume
• excellent for triturating solids	
• excellent for solvent free and near	Disadvantages
solvent free reactions	• ineffective with heterogeneous reactions
	• poor for triturating solids
Disadvantages	• poor for solvent free reactions unless all
• occupies a considerable volume	reactants, additives, and products are fluid

FIGURE 1.7   Stirring options when using small reactor vials. *Source*: Images of the CEM microwave reactor vials were produced with permission from CEM Corporation.

Stirred or not stirred.   One of the considerations to address when designing a microwave-assisted synthesis is whether or not the reaction needs to be stirred. If all of the reagents and products are liquids or everything is fully dissolved in a solvent (homogeneous reaction), stirring may not be needed provided that the solution is well mixed prior to irradiation. If one or more of the reagents, catalysts, or additives are solids and not dissolved (heterogeneous reaction), best results will be obtained if the reaction is stirred. There are several options for effective stirring. In some cases, small stirring bars will provide effective stirring; however, the formation of gels or precipitation of products or reagents can inhibit the stirring process. In those cases, it is best to change to a larger stirring bar for effective stirring. Examples of the two scenarios are shown in Figure 1.7. For microwave-assisted reactions, make sure that the stirring bar is less than ¼ of the standing wave for the microwave reactor.

Safety considerations.   This is dependent upon the type of synthesis that will be carried out in the microwave reactor. Arguably, one of the biggest concerns is the possibility for a pressure buildup that results in a vessel failure. This can happen when volatile compounds are used as reagents or solvents or when volatile compounds are generated from the reaction. Most modern microwave reactors have safety protocols built into the system to vent the cavity in case of a vessel failure. In general, it is advisable to review all of the boiling points of the compounds that are added and generated from the reaction to ensure that a low-boiling compound will be minimized.

In addition to the use of volatile organic compounds, another possibility for a vessel failure occurs when small amounts of a transition metal catalyst become stuck to the side of the reactor vial. If the metal is reduced to its elemental state, arcing is possible. This is more commonly encountered in solvent-free reactions. The solution to this issue is to use a sleeve to make sure that all of the reagents are transferred to the bottom of the reaction vial.

Additives.   As mentioned previously, some reaction mixtures may be poor absorbers and heat very slowly in the microwave. In those cases, it is possible to include an additive to promote heating. The two main types of additives are ionic materials and silicon carbide disks. When considering ionic materials, a wide variety of versions to choose from are commercially available. The best choice for the specific reaction under investigation will vary from reaction to reaction and can range from mineral salts to ionic liquids. Ionic liquids are attractive since their polarity can be manipulated so they are immiscible with the main reaction solvent and serve only to transfer heat to the nonpolar reaction medium. This is a very effective method since ionic liquids absorb microwave energy very efficiently. They also have the advantage that they do not interfere with the stirring. Detailed investigations by Leadbeater have shown this to be an effective means of heating a poorly absorbing mixture [53]. Once the reaction is complete, simply spinning the reactor in a centrifuge will concentrate the ionic liquid and the reaction mixture can be decanted or removed with a pipette. The second type of additive is a silicon carbide disk [54]. These solid materials have an extremely high absorption of microwaves and will heat very rapidly. Since the disks are insoluble in the reaction medium, they are readily removed from once the reaction is over. It should be noted that in many cases, the disks will interfere with the stirring of the reaction mixture. Both approaches have advantages and disadvantages, and the approach that will work best will depend upon your specific reaction.

The use of ultrasound to promote organic reactions has also proven to be effective [55–59]. While several theories were initially proposed to explain the observations, the generally accepted explanation for the accelerating effect of ultrasound on chemical reactions is due to the formation of "hot spots" resulting from collapse of microbubbles in the reaction mixture. The local temperatures and pressures of these hot spots have been calculated to be extremely high (>4000 K and >400 atm) [56]. An excellent review by Thompson and Doraiswamy summarizes the critical parameters that should be consulted when designing an ultrasound-assisted reaction [56]. It is also noteworthy to mention that several groups have been working to combine ultrasound and microwave chemistry [57,58]. While only a few reports have been published, combining the two techniques is an interesting approach. Despite the potential advantages to ultrasound-assisted reactions, it has not been as widely adopted as microwave chemistry.

1.8   SUSTAINABILITY

There are a range of areas that need to be addressed when designing sustainable organic syntheses. Essentially, sustainable processes will be easier to adopt if they are operationally simple and convenient to implement. It is easy to think about trying to take the philosophical high road and try to force scientists to adapt. Practically, this will be nearly impossible. If the new techniques/approaches are overly time intensive, cumbersome, or low yielding, widespread adoption will be less likely. The following sections will highlight a number of areas for further development.

Designing catalysts and reactions that are compatible with water. Organic solvents often need to be anhydrous and deoxygenated; thus, one of the most time-intensive components of a synthesis involves the drying and deoxygenating of reaction solvents. While some of this is mitigated through the use of solvent purification systems, the system must be constantly maintained and the solvent checked for trace amounts of moisture. If a solvent purification system is not available, solvents can be distilled from reactive metals. In addition to the safety hazard this introduces to the laboratory, the stills must be maintained and checked daily. As an alternative, organic solvents can also be dried using activated molecular sieves. A recent report compared the effectiveness of molecular sieves and reactive metals and concluded that molecular sieves were just as successful at removing water from common organic solvents as the reactive metals [1]. While using molecular sieves is easier, it creates additional waste since the molecular sieves are rarely recycled and reused. A significant reduction in the preparation time and amount of waste generated would be achieved through the development of reactions that will tolerate small to large amounts of water. While a number of reactions have been reported that are successful in the presence of small amounts of water, a vast number of syntheses are not. Thus, the challenge for the synthetic community is to develop a substantial number of water-tolerant reactions that would enable researchers to prepare a vast array of functional groups. The need to dry and deoxygenate organic solvents would then become the exception rather than the rule.

A reduction in the amount of solvents needed to isolate target compounds is one of the top priorities when designing sustainable practices. While the conversion to solvent-free reactions or near-solvent-free reactions will eliminate the solvent from the carbon–heteroelement bond-forming step, a significant amount of solvent will often be used to isolate the target compound by column chromatography. There are a couple of general strategies that are often used to address the use of hazardous solvents in column chromatography. The first approach entails simply exchanging commonly used solvents such as hexane for less hazardous versions such as heptane. Blends of ethyl acetate and ethanol have been used as replacements for dichloromethane. While this is not ideal, it allows researchers to continue to use the equipment already in place in their laboratories. A number of excellent reviews on green solvents have appeared [49,60–67]. The use of automated chromatography systems is attractive since they provide more precise control of the separation. The second option entails changing to supercritical CO_2 as the mobile phase. This requires a substantial economic commitment since new equipment will be needed. A significant amount of time will be required to understand the strengths and limitations of the new system.

Strategies to overcome functional group intolerance. Historically, the most common approach to dealing with functional group intolerance entailed the conversion of sensitive groups into more stable derivatives for the course of the reaction (Figure 1.8). Later in the synthesis, removal of the protecting group generated the target compound. While this appears straightforward, the removal of protecting groups can be quite cumbersome and result in significant reductions of the isolated yields. Thus, the development of high-yielding protection/deprotection strategies continues to be a challenge for the synthetic community.

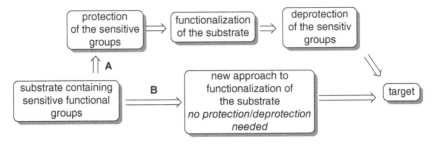

FIGURE 1.8   Development of new approaches versus protection/deprotection strategies.

From another perspective, the design of new reactions that would enable the substrate to be functionalized without protection of the sensitive groups would remove both the protection and deprotection steps from the synthesis entirely.

An example of this entails the synthesis of air-sensitive phosphines. Exposing many phosphines to oxygen results in the formation of phosphine oxides. These compounds can be quite difficult to separate from the desired phosphines. While researchers are often quite adept at removing the oxygen from reaction mixtures, it is often much more challenging to remove all traces of oxygen from every step in the isolation and purification process. As a result, several strategies have been developed to circumvent the formation of unwanted phosphine oxides. One of the most promising ways to accomplish this is through the formation of borane-protected phosphines [68]. The formation of these masked phosphines is trivial and typically consists of simply injecting a solution of $Me_2S\text{-}BH_3$ to the reaction mixture. Once isolated, the free phosphine can be generated from the borane-protected species through several approaches. Arguably, the simplest of which entails simply refluxing the phosphine-borane in ethanol [69]. Thus, the protection of the sensitive phosphine is straightforward, and removal of the protecting group can be trivial.

Developing one-pot reactions or multicomponent reactions.   Syntheses that are able to combine several sequential reactions into a single vessel without purification of the intermediates are quite valuable [70–74]. These reactions often save time as well as eliminate the solvents that would have normally been used for chromatographic separation of the intermediates. These processes are often referred to as "telescoping" or "one-pot" reactions and have been steadily growing in popularity. The trick with these reactions is to design the syntheses so that the steps are chemoselective for the desired transformation and are not affected by any of the catalysts and additives from the previous reactions.

An elegant one-pot approach to the synthesis of functionalized pyridines has recently been provided by Menendez (Scheme 1.13) [75]. His approach started with a Lewis acid-promoted coupling reaction to generate an intermediate enamine. Microwave heating followed by air oxidation generated the heterocycle. A key reagent in the mechanism was 2-furylmethylamine. Once it was incorporated in the initial step, it underwent an elimination reaction upon heating in the absence of solvent to generate a cyclic imine followed by air oxidation to generate the functionalized pyridine. It was also noteworthy that the chemistry was successful using air as the sole oxidant. No additional oxidizing agents were required. A range of fused ring systems including quinolones, isoquinolines, and phenanthridines were successfully prepared in moderate to excellent yields (up to 93%) following this procedure.

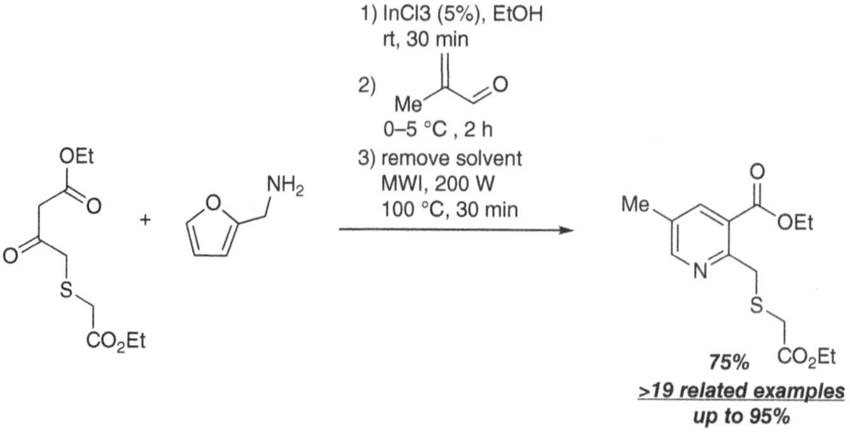

SCHEME 1.13   Lewis acid-catalyzed sequential synthesis of a pyridine [75].

While a number of organic reactions can be linked together in a sequential fashion, it is often challenging to link several metal-catalyzed reactions into a one-pot reaction. A successful example of this was recently reported by Nolan [76]. He was able to combine a gold-catalyzed hydrophenoxylation reaction with a palladium-catalyzed C–H activation (Scheme 1.14) or Mizoroki–Heck coupling in a tandem one-pot reaction. Furthermore, he was able to accomplish the initial hydrophenoxylation reaction under solvent-free conditions. The chemistry was quite tolerant to a range of electron-donating and electron-withdrawing groups on the phenol, and moderate to excellent yields of the benzo[c]chromenes were obtained. This remarkable tolerance of the palladium-catalyzed reaction to the presence of gold provides the synthetic community with a proof of concept for the development of a myriad of sequential reactions given the wide range of gold-catalyzed reactions combined with the staggering number of palladium-catalyzed processes.

Flow chemistry.   There has been a growing interest in the development of continuous flow reactors for the preparation of organic compounds [77–84]. Performing the reaction under flow conditions typically uses considerably less solvent than a typical batch reaction and can afford higher yields of target compounds. The ability to generate moderate to large amounts of material as needed is a significant benefit to using continuous flow reactions. The conversion of amines into amides using esters as substrates serves as a representative example of this chemistry (Scheme 1.15) [85]. The reaction was carried out under mild conditions (25 °C) using THF or DMF as the solvent, and the base used in this chemistry was a bulky amide $(LiN(SiMe_3)_2)$ in order to mitigate potential secondary reactions resulting from nucleophilic behavior of the base.

In addition to flow systems that use conventional heating, microwave-assisted flow chemistry has also been reported [78,86–88]. Organ reported the synthesis of benzimidazole through microwave-assisted flow chemistry (Scheme 1.16) [86]. One of the critical components of the flow system was a silicon carbide reactor tube. As mentioned previously, these materials have a very high absorption of microwave radiation and heat very rapidly upon irradiation. The ability of the silicon carbide to transfer heat is also very high; thus, this material is ideally suited for the construction of a reactor tube that will be used

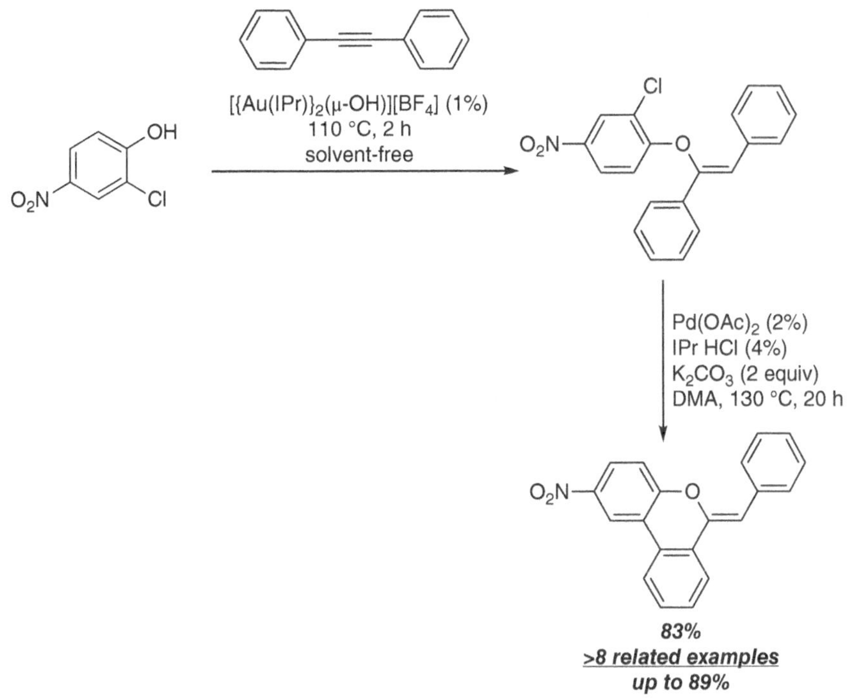

SCHEME 1.14   One-pot sequential gold- and palladium-catalyzed reactions [76].

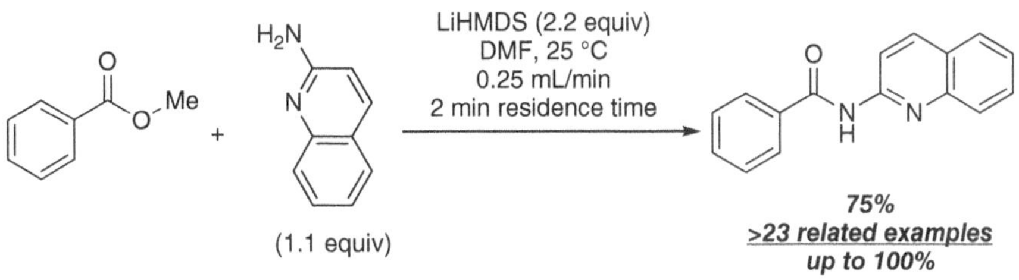

SCHEME 1.15   Amide synthesis at 25 °C using a continuous flow reactor [85].

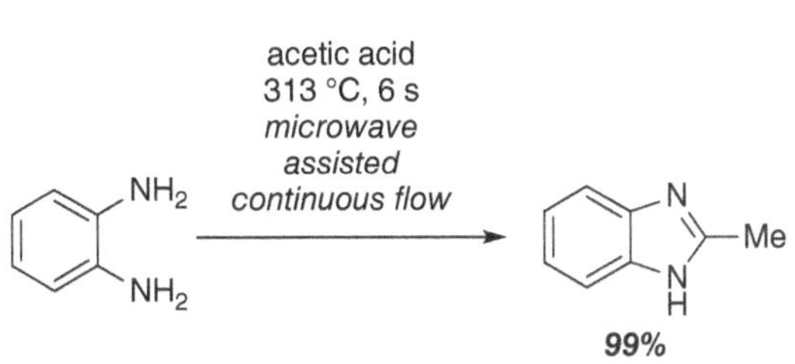

SCHEME 1.16   Benzimidazole synthesis through microwave-assisted flow chemistry [86].

in a microwave-assisted experiment. Historically, one of the limitations with focused microwave reactors was the requirement that all the reactions needed to be run in batches and the scales were relatively small. While this is appropriate for most small synthetic laboratories, the small batches are a limitation when larger amounts are needed. The microwave-assisted flow approach directly addresses this issue.

Given the potential advantages to performing reactions under continuous flow, a variety of chemistries have been modified for flow conditions. A recent example of this was reported by Ley and combined flow and electrochemistry for the methoxylation of cyclic amines (Scheme 1.17) [89]. The chemistry was carried out in a microfluidic electrolytic cell using a substoichiometric amount of $[Et_4N][BF_4]$ as an electrolyte. The authors used a steel cathode and screened a range of materials for the anode and determined that carbon was the most effective for the methoxylation reaction. The chemistry was selective for the α-position, and a range of cyclic amines were successfully functionalized using this approach.

The generation of solids that can clog a flow reactor remains one of the most significant issues that must be circumvented when designing a reaction that will be performed in a flow reactor. As a result of this, not every reaction will be amenable to this approach. It should be noted that this is a very active area of research and new developments that address the current difficulties are being published with increasing frequency. For example, Organ has recently reported the use of a pressure control device that is able to provide very high back pressures (1100 psi) [86]. Such pressures will aid in preventing tube clogging.

Transition metal-catalyzed reactions that use ultralow catalyst loadings have also been adapted for continuous flow. Leadbeater reported an elegant example of this chemistry and was able to decrease the palladium loading to 10 ppm and still achieve a quantitative conversion in the desired biaryl (Scheme 1.18) [90]. This was a microwave-assisted

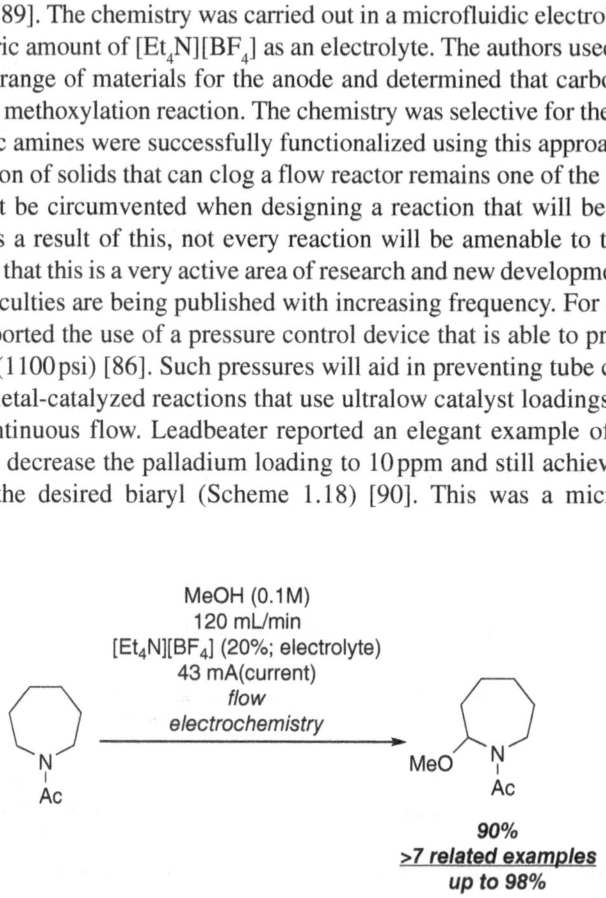

SCHEME 1.17   Methoxylation reactions using flow electrochemistry [89].

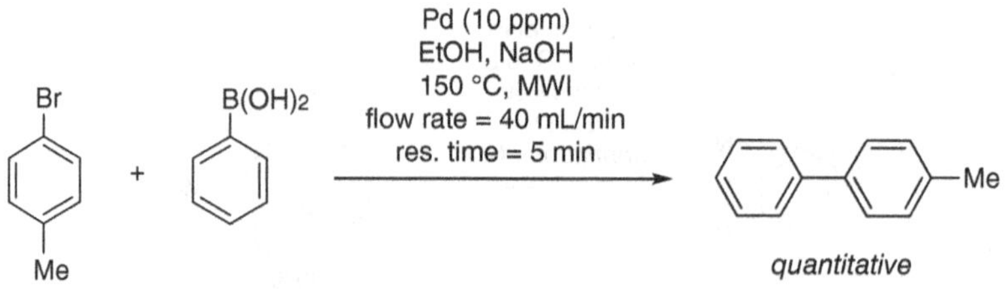

SCHEME 1.18   Microwave-assisted flow chemistry using ultralow catalyst loading [90].

reaction and used a residence time of 5 min in the reactor. As with many of the continuous flow approaches described previously, the authors noted that the buildup of solids in the tubing could inhibit the processing of large amount of material. Ideally, the synthesis would be designed such that the precipitation of the products or reagents does not occur. However, this is not always possible, and alternative strategies such as the use of systems that provide high back pressures as described in the previous paragraph might be successful.

For those reactions that are well suited to this style of chemistry, performing the reaction under flow is a very practical approach to the synthesis of new compounds. There are start-up costs to consider when setting up a flow system, as well as time lost in order to understand how the system operates. Given the potential advantages, this is likely to be time well spent.

Transition metal-free addition and coupling reactions.   Due to the high toxicity and high cost of many transition metal catalysts, the conversion of what are typically thought of as metal-catalyzed reactions into metal-free processes is of current interest to the synthetic community. The example shown in Scheme 1.19 shows the formation of an interesting tetrazine through metal-free S_NAr substitutions [91]. While most aryl halides are unreactive in classic nucleophilic substitution reactions and need to be generated through metal-catalyzed cross-coupling reactions, a number of highly activated systems are susceptible to S_NAr chemistry.

There is considerable interest in the development of metal-free C–H activation reactions that lead to the formation of carbon–heteroelement bonds [31,92]. The potential to transform unfunctionalized hydrocarbons into useful precursors and products remains the driving force in this area. This area of research is quite challenging, and significant hurdles have slowed the progress in this area. Many of the reactions exhibit a high degree of substrate specificity, and general procedures are rare. For the formation of nitrogen–carbon bonds, oxidative amination has shown some promise for the construction of azides, amides, and related compounds [92]. The conversion of unfunctionalized aryl ethers into azides through C–H activation (Scheme 1.20) serves as a representative example of this transformation [93]. The synthesis was achieved through the use of a hypervalent iodine reagent (PIFA) and $TMSN_3$ as the azide precursor. While many similar reactions proceed through the formation of diaryliodonium salts, the authors proposed that this chemistry

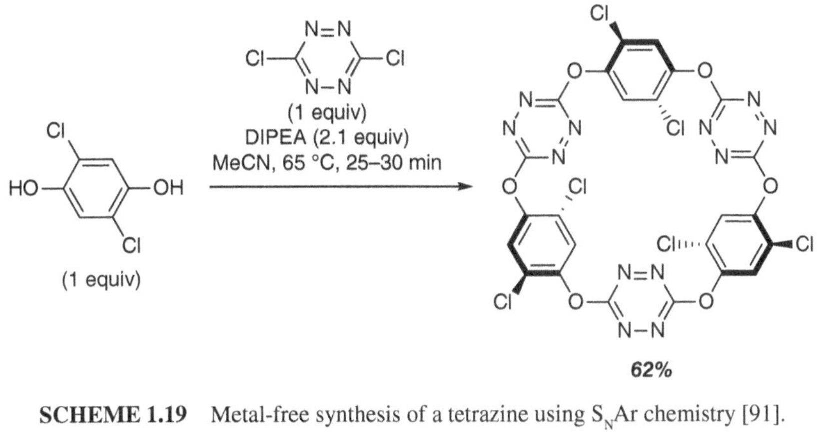

SCHEME 1.19   Metal-free synthesis of a tetrazine using S_NAr chemistry [91].

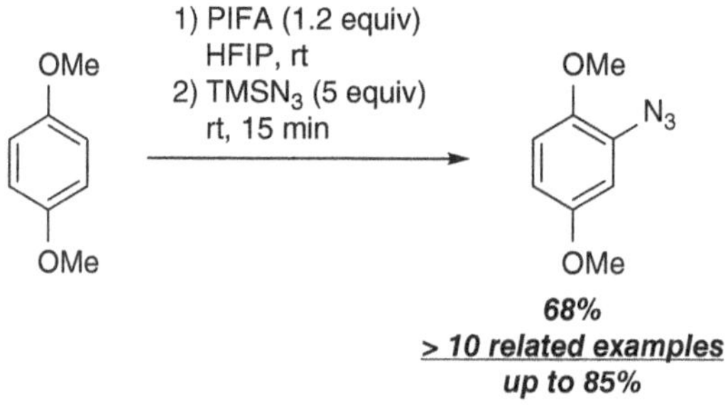

SCHEME 1.20   Metal-free C–H azidation of electron-rich arenes [93].

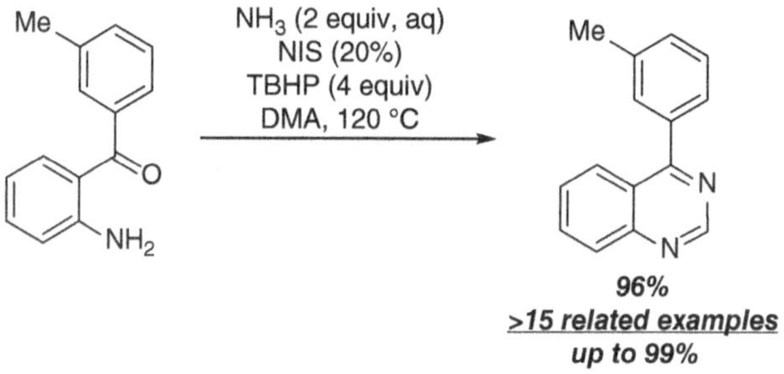

SCHEME 1.21   C–H oxidative amination reaction [94].

generated the functionalized materials through the formation of radical cations. The resulting aryl azides are valuable precursors for a range of nitrogen-rich compounds such as triazoles.

The development of effective methods for the functionalization of primary amines and ammonia remains a standing goal for the synthetic community. Although ammonia is readily available and inexpensive, it has been very challenging to involve in nitrogen–carbon bond-forming reactions. Combing such a transformation with a metal-free C–H activation reaction would be a valuable methodology. Recently, Wang has reported the synthesis of heterocycles through such a reaction (Scheme 1.21) [94]. He used aqueous ammonia as the nitrogen source and a substoichiometric amount of NIS along with an excess of TBHP to promote the reaction. While it was tempting to propose the involvement of a hypervalent iodine species in this chemistry, the authors discounted that possibility through carefully designed control reactions. This chemistry is specific for ortho-substituted anilines, tolerant of electron-donating and electron-withdrawing groups, and generated outstanding yields of the quinazolines.

1.9   ASYMMETRIC SYNTHESIS

The manipulation of the stereochemical outcome of a specific reaction is commonly achieved through the use of transition metal catalysts bearing a chiral ligand or through the use of a resolved organocatalyst. This requirement adds considerable challenges to the synthesis, and in some cases, the selective preparation of a target substrate with the desired stereochemistry is not possible. Both transition metal and organocatalyzed approaches have their merits, and the most successful system will depend upon the reaction as well as the nature of the substrates. One of the consistent issues with asymmetric synthesis remains the substrate specificity inherent to many catalysts. It is quite common for the enantioselectivity of a new reaction to be extremely high for only three to four substrates. Furthermore, reports that list the selectivity for only one to two substrates are of limited value unless those specific substrates are known to be troublesome.

To this end, the burden is on the synthetic community to continue to develop active catalysts with greater selectivity and increased substrate scope. The following chapters will highlight asymmetric reactions for the formation of functional groups through the formation of carbon–heteroelement bonds with special attention paid to reactions that are highly selective but also to those that show the largest substrate scope.

REFERENCES

[1] Williams, D. B. G.; Lawton, M. *J. Org. Chem.* **2010**, *75*, 8351–8354.

[2] Ullmann, F. *Ber. Dtsch. Chem. Ges.* **1904**, *37*, 853–854.

[3] Hassan, J.; Sevignon, M.; Gozzi, C.; Schulz, E.; Lemaire, M. *Chem. Rev.* **2002**, *102*, 1359–1470.

[4] Ley, S. V.; Thomas, A. W. *Angew. Chem. Int. Ed.* **2003**, *42*, 5400–5449.

[5] Wolter, M.; Nordmann, G.; Job, G. E.; Buchwald, S. L. *Org. Lett.* **2002**, *4*, 973–976.

[6] Gujadhur, R. K.; Bates, C. G.; Venkataraman, D. *Org. Lett.* **2001**, *3*, 4315–4317.

[7] Quach, T. D.; Batey, R. A. *Org. Lett.* **2003**, *5*, 1381–1384.

[8] Manbeck, G. F.; Lipman, A. J.; Stockland, R. A., Jr.; Freidl, A. L.; Hasler, A. F.; Stone, J. J.; Guzei, I. A. *J. Org. Chem.* **2005**, *70*, 244–250.

[9] Kamitani, M.; Itazaki, M.; Tamiya, C.; Nakazawa, H. *J. Am. Chem. Soc.* **2012**, *134*, 11932–11935.

[10] Kaye, S.; Fox, J. M.; Hicks, F. A.; Buchwald, S. L. *Adv. Synth. Catal.* **2001**, *343*, 789–794.

[11] Surry, D. S.; Buchwald, S. L. *Chem. Sci.* **2011**, *2*, 27–50.

[12] Cheung, C. W.; Buchwald, S. L. *J. Org. Chem.* **2014**, *79*, 5351–5358.

[13] Friis, S. D.; Skrydstrup, T.; Buchwald, S. L. *Org. Lett.* **2014**, *16*, 4296–4299.

[14] Ikawa, T.; Barder, T. E.; Biscoe, M. R.; Buchwald, S. L. *J. Am. Chem. Soc.* **2007**, *129*, 13001–13007.

[15] Barder, T. E.; Buchwald, S. L. *J. Am. Chem. Soc.* **2007**, *129*, 5096–5101.

[16] Kendall, A. J.; Salazar, C. A.; Martino, P. F.; Tyler, D. R. *Organometallics* **2014**, *33*, 6171–6178.

[17] Burns, D. J.; Lam, H. W. *Angew. Chem. Int. Ed.* **2014**, *53*, 9931–9935.

[18] Banerjee, D.; Junge, K.; Beller, M. *Angew. Chem. Int. Ed.* **2014**, *53*, 13049–13053.

[19] Abbas, S.; Bertram, R. D.; Hayes, C. J. *Org. Lett.* **2001**, *3*, 3365–3367.

[20] Li, X.; Yang, F.; Wu, Y.; Wu, Y. *Org. Lett.* **2014**, *16*, 992–995.

[21] Lee, H. G.; Milner, P. J.; Buchwald, S. L. *J. Am. Chem. Soc.* **2014**, *136*, 3792–3795.

[22] Zaleesskiy, S. S.; Ananikov, V. P. *Organometallics* **2012**, *31*, 2302–2309.

[23] Buchwald, S. L.; Bolm, C. *Angew. Chem. Int. Ed.* **2009**, *48*, 5586–5587.

[24] Aleman, J.; Cabrera, S. *Chem. Soc. Rev.* **2013**, *42*, 774–793.

[25] Gaunt, M. J.; Johansson, C. C. C.; McNally, A.; Vo, N. T. *Drug Discov. Today* **2007**, *12*, 8–27.

[26] Akiyama, T.; Itoh, J.; Fuchibe, K. *Adv. Synth. Catal.* **2006**, *348*, 999–1010.

[27] Allemann, C.; Gordillo, R.; Clemente, F. R.; Cheong, P. H.-Y.; Houk, K.N. *Acc. Chem. Res.* **2004**, *37*, 558–569.

[28] Dalko, P. I.; Moisan, L. *Angew. Chem. Int. Ed.* **2004**, *43*, 5138–5175.

[29] Dalko, P. I.; Moisan, L. *Angew. Chem. Int. Ed.* **2001**, *40*, 3726–3748.

[30] Allais, C.; Grassot, J.-M.; Rodriguez, J.; Constantieux, T. *Chem. Rev.* **2014**, *114*, 10829–10868.

[31] Sun, C.-L.; Shi, Z.-J. *Chem. Rev.* **2014**, *114*, 9219–9280.

[32] Chauhan, P.; Mahajan, S.; Enders, D. *Chem. Rev.* **2014**, *114*, 8807–8864.

[33] Zhu, Y.; Wang, Q.; Cornwall, R. G.; Shi, Y. *Chem. Rev.* **2014**, *114*, 8199–8256.

[34] Volla, C. M. R.; Atodiresei, I.; Rueping, M. *Chem. Rev.* **2014**, *114*, 2390–2431.

[35] Meninno, S.; Croce, G.; Lattanzi, A. *Org. Lett.* **2013**, *15*, 3436–3439.

[36] Jiang, X.; Wu, L.; Xing, Y.; Wang, L.; Wang, S.; Chen, Z.; Wang, R. *Chem. Commun.* **2012**, *48*, 446–448.

[37] Dallinger, D.; Kappe, C. O. *Chem. Rev.* **2007**, *107*, 2563–2591.

[38] Kappe, C. O. *Chem. Soc. Rev.* **2008**, *37*, 1127–1139.

[39] Caddick, S.; Fitzmaurice, R. *Tetrahedron* **2009**, *65*, 3325–3355.

[40] Antonio, C.; Deam, R. T. *Phys. Chem. Chem. Phys.* **2007**, *9*, 2976–2982.

[41] Perreux, L.; Loupy, A. *Tetrahedron* **2001**, *57*, 9199–9223.

[42] Gabriel, C.; Gabriel, S.; Grant, E. H.; Halstead, B. S. J.; Mingos, M. P. *Chem. Soc. Rev.* **1998**, *27*, 213–224.

[43] Strauss, C. R.; Rooney, D. W. *Green Chem.* **2010**, *12*, 1340–1344.

[44] Kappe, C. O. *Angew. Chem. Int. Ed.* **2004**, *43*, 6250–6284.

[45] Nilsson, P.; Olofsson, K.; Larhed, M. *Top. Curr. Chem.* **2006**, *266*, 103–144.

[46] Pillai, U. R.; Sahle-Demessie, E.; Varma, R. S. *J. Mater. Chem.* **2002**, *12*, 3199–3207.

[47] Ju, Y.; Varma, R. S. *Green Chem.* **2004**, *6*, 219–221.

[48] Hayes, B. L. *In Microwave synthesis: chemistry at the speed of light;* CEM Publishing: Matthews, NC, **2002**.

[49] Polshettiwar, V.; Varma, R. S. *Acc. Chem. Res.* **2008**, *41*, 629–639.

[50] Patrick, H. R.; Griffith, K.; Liotta, C. L.; Eckert, C. A.; Glaser, R. *Ind. Eng. Chem. Res.* **2001**, *40*, 6063–6067.

[51] Stockland, R. A., Jr.; Lipman, A. J.; Bawiec, J. A., III; Morrison, P. E.; Guzei, I. A.; Findeis, P. M.; Tamblin, J. F. *J. Organomet. Chem.* **2006**, *691*, 4042–4053.

[52] Stockland, R. A. Jr.; Taylor, R. I.; Thompson, L. E.; Patel, P. B. *Org. Lett.* **2005**, *7*, 851–853.

[53] Leadbeater, N. E.; Torenius, H. M. *J. Org. Chem.* **2002**, *67*, 3145–3148.

[54] Kremsner, J. M.; Kappe, C. O. *J. Org. Chem.* **2006**, *71*, 4651–4658.

[55] Cravotto, G.; Gaudino, E. C.; Cintas, P. *Chem. Soc. Rev.* **2013**, *42*, 7521–7534.

[56] Thompson, L. H.; Doraiswamy, L. K. *Ind. Eng. Chem. Res.* **1999**, *38*, 1215–1249.

[57] Barge, A.; Tagliapietra, S.; Tei, L.; Cintas, P.; Cravotto, G. *Curr. Org. Chem.* **2008**, *12*, 1588–1612.

[58] Cravotto, G.; Cintas, P. *Chem. Eur. J.* **2007**, *13*, 1902–1909.

[59] Bruckmann, A.; Krebs, A.; Bolm, C. *Green Chem.* **2008**, *10*, 1131–1141.

[60] Sheldon, R. A. *Green Chem.* **2005**, *7*, 267–278.

[61] Li, C.-J. *Chem. Rev.*. **2005**, *105*, 3095–3166.

[62] Polshettiwar, V.; Varma, R. S. *Chem. Soc. Rev.* **2008**, *37*, 1546–1557.

[63] Liu, S.; Xiao, J. *J. Mol. Catal. A Chem.* **2007**, *270*, 1–43.

[64] Dunn, J. B.; Savage, P. E. *Green Chem.* **2003**, *5*, 649–655.

[65] Hoffmann, J.; Nuchter, M.; Ondruschka, B.; Wasserscheid, P. *Green Chem.* **2003**, *5*, 296–299.

[66] Nolen, S. A.; Liotta, C. L.; Eckert, C. A.; Glaser, R. *Green Chem.* **2003**, *5*, 663–669.

[67] Miao, W.; Chan, T. H. *Acc. Chem. Res.* **2006**, *39*, 897–908.

[68] Staubitz, A.; Robertson, A. P. M.; Sloan, M. E.; Manners, I. *Chem. Rev.* **2010**, *110*, 4023–4078.

[69] Moulin, D.; Bago, S.; Bauduin, C.; Darcel, C.; Jugé, S. *Tetrahedron: Asymmetry* **2000**, *11*, 3939–3956.

[70] Zeng, X. *Chem. Rev.* **2013**, *113*, 6864–6900.

[71] Yang, Z.-P.; Zhang, W.; You, S.-L. *J. Org. Chem.* **2014**, *79*, 7785–7798.

[72] Cantillo, D.; Damm, M.; Dallinger, D.; Bauser, M.; Berger, M.; Kappe, C. O. *Org. Process Res. Dev.* **2014**, *18*, 1360–1366.

[73] Kuethe, J. T.; Tellers, D. M.; Weissman, S. A.; Yasuda, N. *Org. Process Res. Dev.* **2009**, *13*, 471–477.

[74] Tejedor, D.; Garcia-Tellado, F. *Chem. Soc. Rev.* **2007**, *36*, 484–491.

[75] Raja, V. P. A.; Tenti, G.; Perumal, S.; Menendez, J. C. *Chem. Commun.* **2014**, *50*, 12270–12272.

[76] Oonishi, Y.; Gómez-Suárez, A.; Martin, A. R.; Makida, Y.; Slawin, A. M. Z.; Nolan, S. P. *Chem. Eur. J.* **2014**, *20*, 13507–13510.

[77] Noel, T.; Buchwald, S. L. *Chem. Soc. Rev.* **2011**, *40*, 5010–5029.

[78] Baxendale, I. R.; Hayward, J. J.; Ley, S. V. *Comb. Chem. High Throughput Screen.* **2007**, *10*, 802–836.

[79] Ley, S. V. *Chem. Rec.* **2012**, *12*, 378–390.

[80] Mason, B. P.; Price, K. E.; Steinbacher, J. L.; Bogdan, A. R.; McQuade, D. T. *Chem. Rev.* **2007**, *107*, 2300–2318.

[81] Webb, D.; Jamison, T. F. *Chem. Sci.* **2010**, *1*, 675–680.

[82] Frost, C. G.; Mutton, L. *Green Chem.* **2010**, *12*, 1687–1703.

[83] Ley, S.; Baxendale, I. R. *Nat. Rev. Drug Discov.* **2002**, *1*, 573–586.

[84] Hartman, R. L.; McMullen, J. P.; Jensen, K. F. *Angew. Chem. Int. Ed.* **2011**, *50*, 7502–7519.

[85] Vrijdag, J. L.; Delgado, F.; Alonso, N.; De Borggraeve, W. M.; Perez-Macias, N.; Alcazar, J. *Chem. Commun.* **2014**, *50*, 15094–15097.

[86] Sauks, J. M.; Mallik, D.; Lawryshyn, Y.; Bender, T.; Organ, M. *Org. Process Res. Dev.* **2014**, *18*, 1310–1314.

[87] Comer, E.; Organ, M. G. *J. Am. Chem. Soc.* **2005**, *127*, 8160–8167.

[88] Zang, Q.; Javed, S.; Porubsky, P.; Ullah, F.; Neuenswander, B.; Lushington, G. H.; Basha, F. Z.; Organ, M. G.; Hanson, P. R. *ACS Comb. Sci.* **2012**, *14*, 211–217.

[89] Kabeshov, M. A.; Musio, B.; Murray, P. R. D.; Browne, D. L.; Ley, S. V. *Org. Lett.* **2014**, *16*, 4618–4621.

[90] Bowman, M. D.; Holcomb, J. L.; Kormos, C. M.; Leadbeater, N. E.; Williams, V. A. *Org. Process Res. Dev.* **2008**, *12*, 41–57.

[91] Guo, Q.-H.; Fu, Z.-D.; Zhao, L.; Wang, M.-X. *Angew. Chem. Int. Ed.* **2014**, *53*, 13548–13552.

[92] Samanta, R.; Matcha, K.; Antonchick, A. P. *Eur. J. Org. Chem.* **2013**, *2013*, 5769–5804.

[93] Kita, Y.; Tohma, H.; Hatanaka, K.; Takada, T.; Fujita, S.; Mitoh, S.; Sakurai, H.; Oka, S. *J. Am. Chem. Soc.* **1994**, *116*, 3684–3691.

[94] Yan, Y.; Zhang, Y.; Feng, C.; Zha, Z.; Wang, Z. *Angew. Chem. Int. Ed.* **2012**, *51*, 8077–8081.

2

PREPARATION OF ALCOHOLS, ETHERS, AND RELATED COMPOUNDS

2.1 PREPARATION OF ALCOHOLS, ETHERS, AND RELATED COMPOUNDS THROUGH THE FORMATION OF OXYGEN–CARBON(SP³) BONDS

Alcohols are arguably the simplest and most common oxygen-containing functional groups. For secondary and tertiary alcohols, one of the most popular synthetic approaches entails the addition of Grignard and organolithium reagents to carbonyl-containing compounds. For these reactions, the oxygen is already attached to the framework. In contrast, the hydration of alkenes generates alcohols through the formation of oxygen–carbon(sp³) bonds. Fundamentally, the addition of the oxygen can occur on either side of the alkene, and considerable effort has been devoted to the development of processes that are selective for one mode of addition. For Markovnikov selective reactions, the acid-catalyzed reactions of alkenes has been used to generate a vast assortment of alcohols [1].

The hydration of alkenes with anti-Markovnikov selectivity is a much more challenging reaction [1]. Although a few scattered reports have appeared, a practical process with wide substrate scope remains quite rare. A clever approach to the synthesis of these primary alcohols was recently reported by Herzon (Scheme 2.1) [2]. Instead of focusing on alkene hydration, his method started with alkynes and entailed a tandem hydration/reduction process. This one-pot synthesis used a ruthenium complex to catalyze the reaction. Although the need to generate this catalyst added to the operational complexity of the overall process, this ruthenium complex was air stable and easily prepared and handled. The reductive hydration reaction was carried out under very mild conditions (room temperature) and afforded good to excellent yields of the primary alcohols (up to 94%). The presence of a vast array of functional groups was well tolerated by this process. Crowded alkynes such as 2-ethynyl-1,3,5-trimethylbenzene and propargyl alcohols such as ethynyl estradiol were also successfully hydrated (Example 2.1). Due to the wide substrate scope

Practical Functional Group Synthesis, First Edition. Robert A. Stockland, Jr.
© 2016 John Wiley & Sons, Inc. Published 2016 by John Wiley & Sons, Inc.

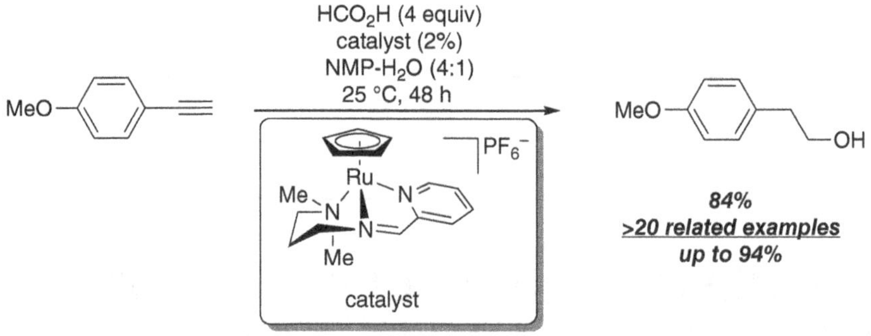

SCHEME 2.1   Ruthenium-catalyzed preparation of primary alcohols from alkynes [2].

Example 2.1   Formal Reductive Hydration of Crowded Alkynes Using Ruthenium Catalysts [2].

Chemical Safety Instructions

<u>Before starting this synthesis, all safety, health, and environmental concerns must be evaluated using the most recent information, and the appropriate safety protocols followed. All appropriate personal safety equipment must be used. All waste must be disposed of in accordance with all current local and government regulations.</u>

The individual performing these procedures and techniques assumes all risks and is responsible for ensuring the safety of themselves and those around them. This chemistry is not intended for the novice and should only be attempted by professionals who have been well trained in synthetic organic chemistry. The authors and Wiley assume no risk and disclaim any liabilities for any damages or injuries associated in any way with the chemicals, procedures, and techniques described herein.

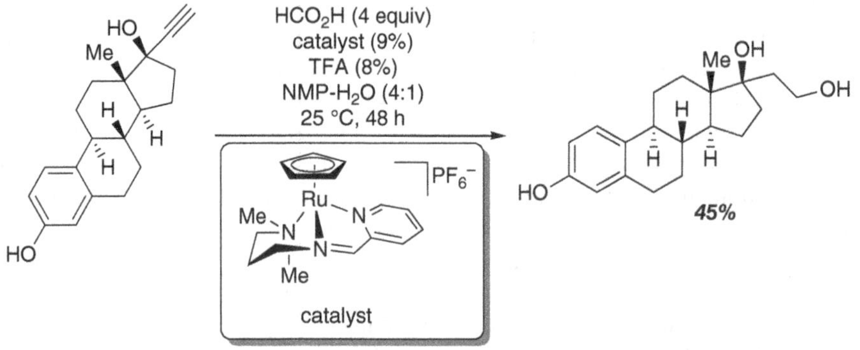

In a nitrogen-filled drybox, a 4-mL vial was charged sequentially with a Teflon-coated stirbar, the ruthenium complex (22.6 mg, 45.0 μmol, 0.0900 equiv), N-methyl-2-pyrrolidinone (2.1 mL), and ethynyl estradiol (148 mg, 500 μmol). The vial was sealed and removed from the drybox. Water (350 μL), formic acid (75.5 μL, 2.00 mmol, 4.00 equiv), and trifluoroacetic acid (3.0 μL, 39 μmol, 0.078 equiv) were then added in sequence with exclusion of oxygen (nitrogen-filled bag). The vial was sealed with a Teflon-lined cap, and the sealed vial was removed from the nitrogen-filled bag. The reaction mixture was stirred for 48 h at 25°C. The product mixture was diluted with ethyl acetate (2 mL), and the diluted solution was transferred to a separatory funnel. Saturated aqueous sodium chloride solution (10 mL) was added, and the layers that formed were separated. The aqueous layer was extracted with ethyl acetate (3 × 10 mL). The organic layer was dried over sodium sulfate, and the dried solution was filtered. The filtrate was concentrated, and the residue obtained was purified by flash column chromatography (eluting with 50% ethyl acetate–hexanes initially, grading to 75% ethyl acetate–hexanes) to afford the alcohol as a white solid (70.5 mg, 45%).

Adapted with permission from M. Zeng, L. Li, and S. B. Herzon, *J. Am. Chem. Soc.*, **2014**, *136*, 7058–7067. *Copyright (2014) American Chemical Society.*

SCHEME 2.2   Conversion of alkyltrifluoroborates into alcohols [3].

and functional group tolerance, this is one of the most attractive one-pot approaches for the synthesis of primary alcohols.

The conversion of alkyltrifluoroborates into alcohols has been achieved using Oxone as the oxidizing agent (Scheme 2.2) [3]. The reactions were quite practical since they were carried out in a mixture of water and acetone at room temperature in reaction vessels that were open to the atmosphere. The reactions were also remarkably fast, and most substrates were converted into products within a few minutes. Using this approach, a range of primary and secondary borates was converted into alcohols in moderate to excellent yield. When resolved alkyltrifluoroborates were used, they were converted into alcohols with retention of configuration. One of the more valuable aspects of this chemistry was the ability to generate the alcohols while retaining alkyl bromides that could be functionalized in subsequent reactions. To widen the scope of the chemistry, the authors carried out a tandem process that resulted in the formal conversion of an alkene into an alcohol. This is an attractive approach to the laboratory scale synthesis of alcohols given the stability and availability of the alkyltrifluoroborate salts coupled with the selectivity and operational simplicity of the method.

The nickel-catalyzed synthesis of Z-allylic alcohols in a two-step, one-pot reaction from conjugated dienes has been reported (Scheme 2.3) [4]. The key with this chemistry was the initial hydroboration of the 1,3-diene promoted by common nickel catalysts. Once the intermediate allylboronate was formed, treatment with $NaOH/H_2O_2$ provided the alcohols. This chemistry was an attractive route to allylic alcohols as a range of functional groups including preexisting primary alcohols, esters, ethers, and phthalimides were well tolerated.

In principle, alkyl halides should be able to be converted into alcohols through substitution chemistry. Although this might seem conceptually straightforward, the generation of significant amounts of elimination products remains one of the challenges

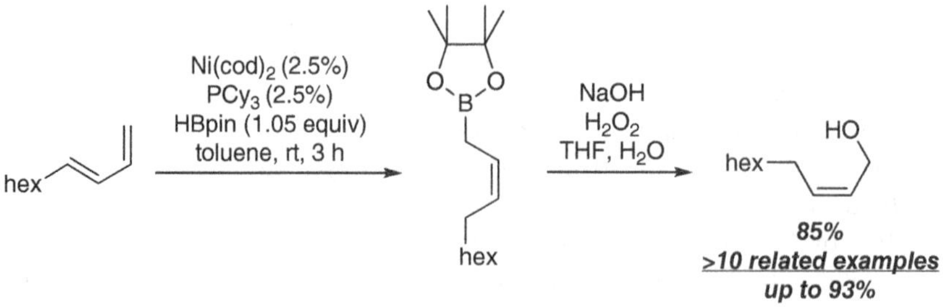

SCHEME 2.3 Synthesis of Z-allylic alcohols from conjugated dienes [4].

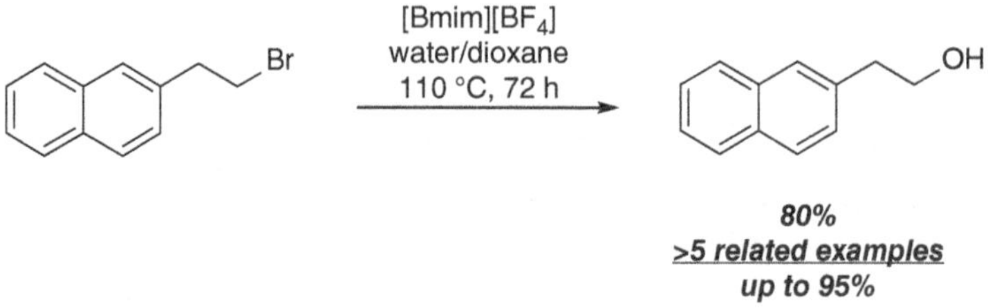

SCHEME 2.4 Preparation of alcohols from alkyl halides in ionic liquids [5].

encountered in the hydroxylation of alkyl halides. A successful version of this reaction with minimal elimination products was reported using an ionic liquid-assisted approach (Scheme 2.4) [5]. The reactions needed to be heated, and moderate to excellent conversions were obtained after stirring for 3–72 h. Of the systems screened, benzyl bromides were the most reactive, and 91% conversion into the alcohol was observed after 3 h. While alkyl bromides and iodides were well tolerated by the substitution process, alkyl chlorides afforded lower yields of the alcohols.

The C—H hydroxylation of alkanes is an efficient approach to the synthesis of alcohols since no prefunctionalization of the substrate is needed. This is a challenging reaction, and the development of general approaches with wide substrate scope continues to be the subject of intense research. In addition to the inherent low reactivity of C—H bonds, achieving high selectivity can be a considerable problem. Most organic compounds have many C—H bonds, and developing catalyst systems that will selectively functionalize a selected C—H bond and ignore the remainder of the substrate is challenging. For the selective hydroxylation of tertiary C—H bonds, a practical approach was reported using simple ruthenium catalysts (Scheme 2.5) [6]. In addition to ruthenium trichloride, an oxidant was required, and the authors found that $KBrO_3$ was the most effective under the reaction conditions. The hydroxylation reaction was insensitive to the presence of moisture, and indeed the optimized conditions used a mixture of acetonitrile and water. A vast array of functional groups was well tolerated by the hydroxylation reaction, and moderate to good yields were obtained for most substrates. It should be noted that the process was also selective for specific tertiary C—H bonds (Scheme 2.5). In related work, a discrete ruthenium catalyst has been used for the selective hydroxylation of tertiary and benzylic C—H bonds (Scheme 2.6) [7].

Secondary alcohols have been prepared through a three-component coupling reaction catalyzed by a gold compound (Scheme 2.7) [8]. In addition to the boronic acid and water, the third component of the coupling reaction was an unactivated terminal alkene.

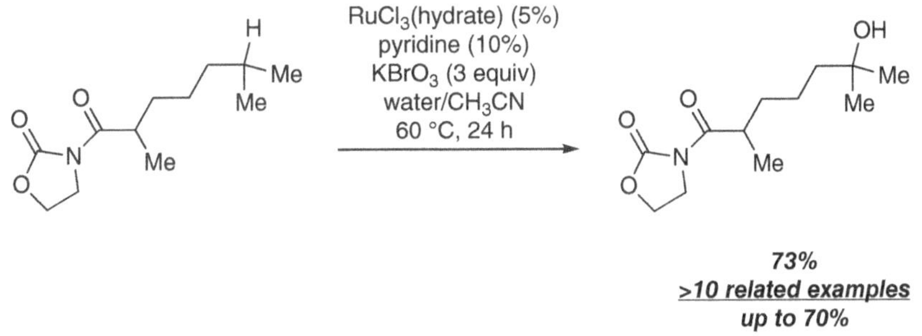

SCHEME 2.5 Ruthenium-catalyzed C—H hydroxylation of tertiary C—H bonds [6].

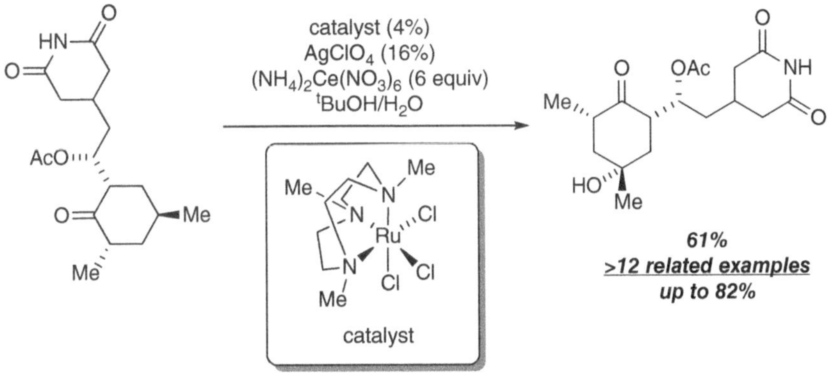

SCHEME 2.6 Ruthenium-catalyzed C—H hydroxylation using a discrete ruthenium catalyst [7].

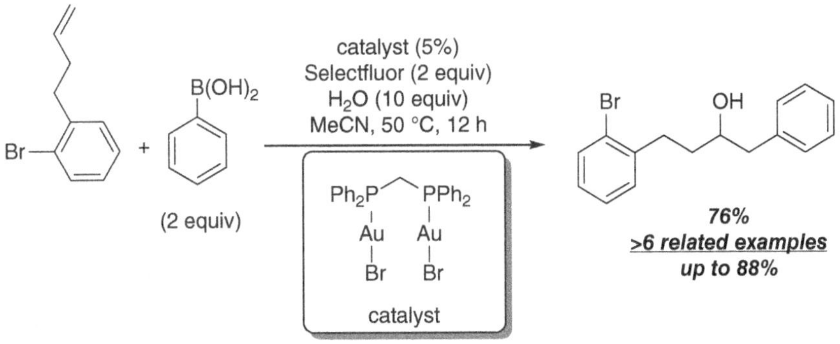

SCHEME 2.7 Gold-catalyzed synthesis of alcohols through a three-component coupling reaction [8].

The gold complex that served as the catalyst contained two gold atoms held in close proximity to each other due to the constraints of a ligand framework. The authors also noted that the reaction benefited from addition of the boronic acid and catalyst in two portions that were separated by 1 h. After some experimentation, the reaction was found to be tolerant to an assortment of preexisting functional groups.

In addition to transition metal-catalyzed C—H hydroxylation reactions, considerable effort has been devoted to the development of metal-free processes. One such approach entailed the use of dioxiranes for the selective functionalization of tertiary C—H bonds (Scheme 2.8) [9]. The active species in the hydroxylation reaction was proposed to be a dioxirane generated from the reaction of the hydrogen peroxide with a ketone. Following the screening of a range of ketones, 2,2,2-trifluoroacetophenone with a fluorine in the 4-position of the aromatic ring was found to be effective catalyst precursor. Using this system, a range of tertiary C—H bonds was hydroxylated in moderate to good yield. Of the substrates screened, adamantane was the most responsive substrate for the chemistry. In related work, the selective hydroxylation of tertiary C—H bonds has been achieved using an Oxone-promoted process in water/hexafluoroisopropanol [10].

The asymmetric hydroxylation of enals has been achieved using a resolved N-heterocyclic carbene catalyst (Scheme 2.9) [11]. The chemistry was carried out under very mild conditions using 4-nitropyridine N-oxide as the oxidant and was highly selective (up to 92% ee). An array of enals bearing a range of alkyl and aryl groups was successfully functionalized in this work. The authors proposed a radical pathway for the chemistry based upon a series of mechanistic investigations.

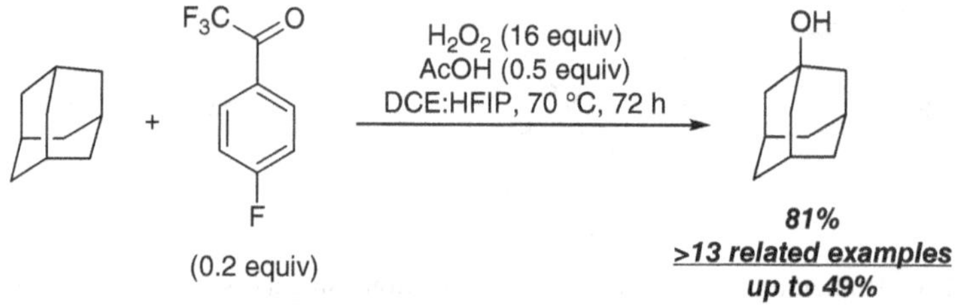

SCHEME 2.8 Metal-free C—H hydroxylation of tertiary alkanes [9].

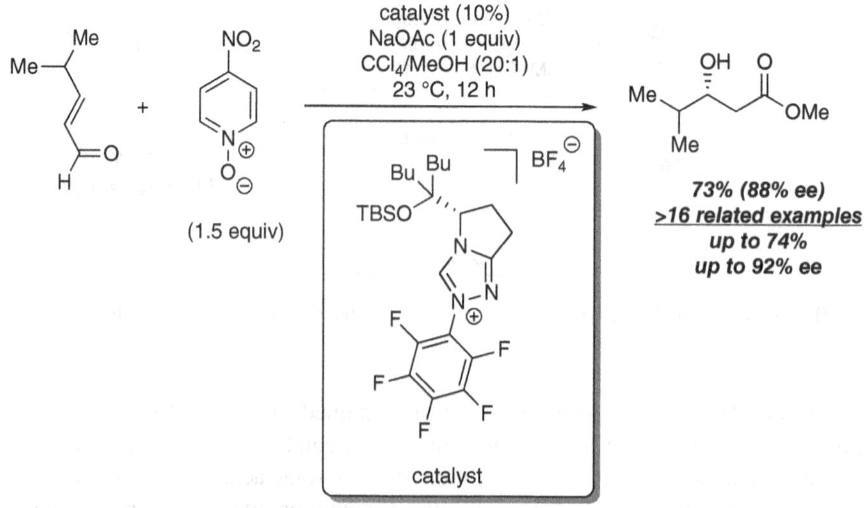

SCHEME 2.9 Asymmetric hydroxylation of enals catalyzed by an organocatalyst [11].

Work by Hartwig demonstrated that cyclometalated iridium complexes bearing phosphoramidite ligands promoted the enantioselective conversion of allylic carbonates into allylic ethers in excellent yields with outstanding selectivities [12]. In related work, chiral allylic alcohols were prepared using an iridium-catalyzed reaction between allylic carbonates and silanolates (Scheme 2.10) [13]. The catalyst system was comprised of a common iridium complex along with a chiral phosphoramidite. While the initial studies were carried out using a cinnamyl carbonate derivative, the optimized system was extended to a host of allylic carbonates. The chemistry was regioselective for the formation of the branched isomer and was highly enantioselective (up to 99% ee). An assortment of electron-rich and electron-poor arenes were tolerated along with several heteroaromatic containing substrates.

One of the most popular approaches to the laboratory scale synthesis of ethers is the addition of alkoxides and phenoxides to a suitable substrate such as an alkyl bromide. This reaction is known as the Williamson ether synthesis. For primary substrates, this approach tended to work quite well, and a host of ethers have been prepared using this method. The chemistry is less straightforward when secondary or tertiary alkyl halides were used due to competing elimination processes. As a representative example, the successful synthesis of an alkyl aryl ether is shown in Example 2.2 [14]. The reaction was carried out in acetone using allyl bromide and a functionalized phenol as the substrates and potassium carbonate as the base. While many bases have been used in Williamson ether syntheses, a mild base was critical for this work since it was needed to deprotonate the phenol without deprotonating the alkyne. This was critical for the success of the chemistry as the alkyne was needed for later steps in the reaction. In related work, potassium carbonate promoted the synthesis of photoactivatable fluorescein derivatives through a Williamson ether synthesis (Scheme 2.11) [15]. It also promoted the synthesis of a morphine precursor as well (Example 2.3) [16].

The ribofuranosylation of alcohols has been achieved using a base/triphenylphosphine promoted addition of ribofuranosyl iodide to alkyl iodides (Scheme 2.12) [17]. The chemistry was carried out under mild conditions (0 °C) and was highly alpha-selective. Triphenylphosphine was a critical component of the chemistry, and although the precise role of this additive was not fully elucidated, its presence nearly completely suppressed the formation of products resulting from beta-addition. A range of alcohols were successfully functionalized using this approach with excellent selectivity.

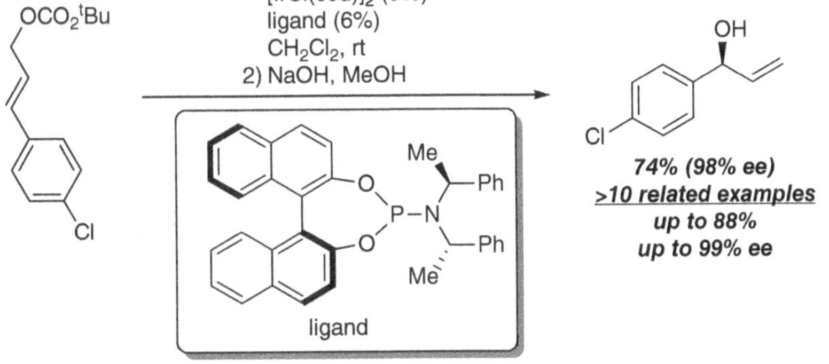

SCHEME 2.10 Asymmetric synthesis of allylic alcohols using an iridium-catalyzed process [13].

Example 2.2   Synthesis of Alkyl Aryl Ethers through Substitution Chemistry [14].

Before starting this synthesis, carefully follow the safety instructions listed in Example 2.1.

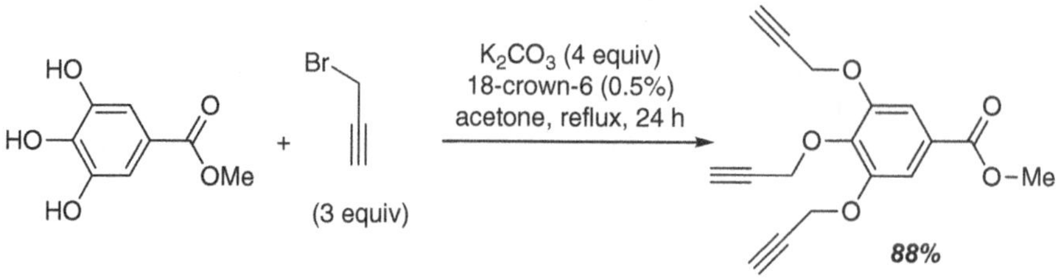

A mixture of methyl 3,4,5-trihydroxybenzoate (1.84 g, 10 mmol), propargyl bromide (80% wt in toluene, 3.4 mL, 31 mmol), K_2CO_3 (5.53 g, 40 mmol), and 18-crown-6 (13.2 mg, 0.05 mmol) in acetone (50 mL) was heated to reflux for 24 h. The solid was removed by filtration, and the acetone was removed by rotary evaporation. Water (50 mL) was added, and the mixture was extracted with CH_2Cl_2. The organic phase was dried over Na_2SO_4, and the solvent was removed in vacuo to give the desired product as a white powder (2.63 g, 88%).

Adapted with permission from S. Zhang and Y. Zhao, Bioconjug. Chem., 2011, 22, 523–528. Copyright (2011) American Chemical Society.

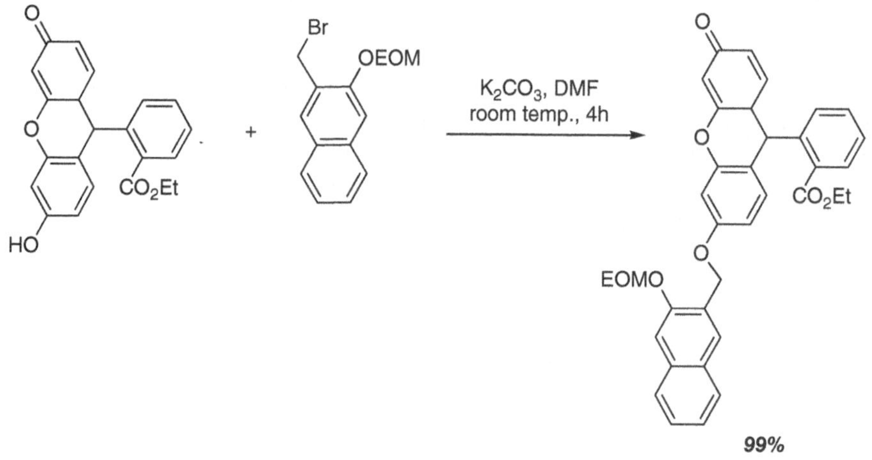

SCHEME 2.11   Preparation of photoactivatable fluorescein derivatives [15].

An interesting approach to the synthesis of unsymmetrical ethers has been reported using a ruthenium complex to catalyze a dehydrative etherification (Scheme 2.13) [18]. A ruthenium hydride compound served as the catalyst for the reaction, and a vast array of alcohols were screened in this study. Primary and secondary as well as benzyl alcohols were successfully coupled with phenols. Furthermore, a number of functional groups were tolerated by this chemistry due to the lack of strong acids or bases needed to promote the

Example 2.3   Alkyl Aryl Ethers through Nucleophilic Substitution Chemistry [16].

Before starting this synthesis, carefully follow the safety instructions listed in Example 2.1.

Caution! Diethyl ether is highly flammable.

To a mixture of methyl 3-(3-hydroxy-4-methoxyphenyl)propanoate (11.0 g, 52 mmol), 2-bromo-1-(4-((tetrahydro-2H-pyran-2-yl)oxy)phenyl)ethanone (18.8 g, 63 mmol), tetra-n-butylammonium bromide (840 mg, 2.6 mmol), and potassium carbonate (10.9 g, 79 mmol) was added dichloromethane and water (500 mL, 1 : 1), and the reaction mixture was heated at 55 °C for 15 h. The reaction mixture was cooled to room temperature, and the aqueous separated and further extracted with dichloromethane (2 × 50 mL). The combined organic layer was washed with brine (150 mL), dried (MgSO$_4$), and filtered, and the solvent was removed under reduced pressure. The residue was triturated with ice-cold methanol. The resulting precipitate was recrystallized from room temperature dichloromethane and diethyl ether, and the solid washed with cold ether. This yielded the title compound (20.6 g, 48.1 mmol, 92%) as a colorless solid.

Adapted with permission from M. Tissot, R. J. Phipps, C. Lucas, R. M. Leon, R. D. M. Pace, T. Ngouansavanh, M. J. Gaunt, *Angew. Chem. Int. Ed.*, **2014**, *53*, 13498–13501. *Copyright (2014) Wiley-VCH Verlag GmbH & Co. KGaA, Weinheim.*

oxygen–carbon(sp^3) bond-forming reaction. This is perhaps one of the most efficient methods for the synthesis of ethers when the substrates were highly functionalized.

The addition of alcohols to alkenes is a valuable approach to the synthesis of ethers. One of the most popular alkenes for this addition reaction is dihydropyran. The addition reaction is commonly catalyzed by p-toluenesulfonic acid, and a vast array of alcohols including tertiary systems are converted into the THP ethers under mild conditions. When activated alkenes are used in the O—H addition reactions, the process is commonly referred to as an oxa-Michael addition and is a valuable approach to the synthesis of ethers due to the atom-efficient nature of the chemistry. While Michael additions using carbon-based nucleophiles have been well studied, less attention has been given to these oxa-Michael processes due to issues related to the reversibility of the addition reaction as well as the poor nucleophilicity of the alcohols and related neutral oxygen nucleophiles. The reaction can be promoted by a wide range of catalysts and has been reviewed [19].

An example of a successful approach to the oxa-Michael addition reaction used copper salts to catalyze the addition of primary alcohols to acrylamide derivatives (Scheme 2.14) [20]. The reaction was carried out under mild conditions using a

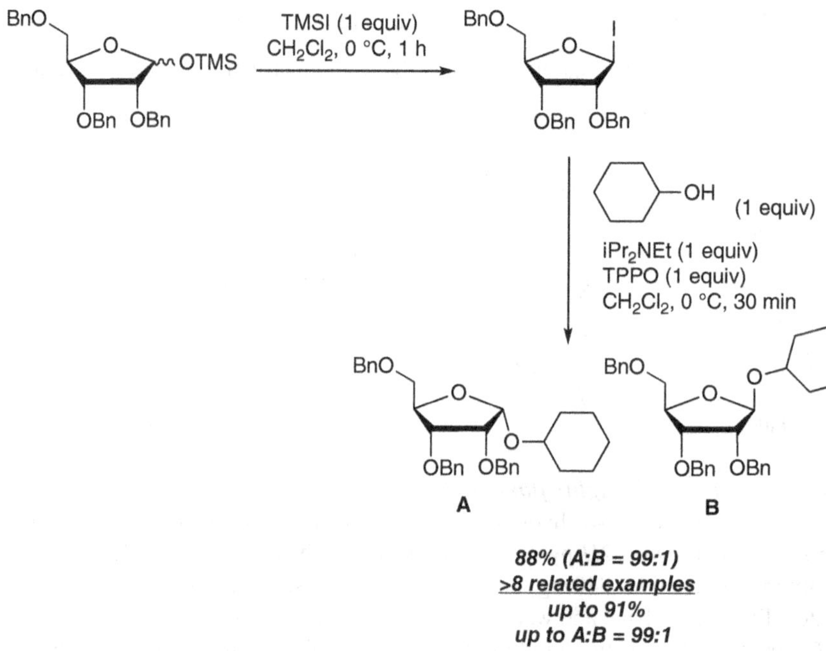

SCHEME 2.12   Ribofuranosylation of alcohols using ribofuranosyl iodides as substrates [17].

catalytic amount of copper chloride and afforded moderate to good yields of the ethers from a range of functionalized substrates. The ability to carry out the reaction under an atmosphere of air makes this one of the most practical approaches to the synthesis of this group of ethers.

While activated alkenes can be converted into ethers through oxa-Michael additions, analogous reactions using unactivated alkenes are significantly more challenging [21], and a considerable amount of research has been devoted to developing general and practical approaches. One such process entailed the addition of phenols to alkenes using a gold compound as the catalyst (Scheme 2.15) [22]. The catalyst system used for this reaction was comprised of triphenylphosphine gold(I) chloride along with silver(I) triflate. The addition chemistry was insensitive to the electronic composition of the phenol as a range of electron-donating and electron-withdrawing groups were well tolerated by the approach. It should be noted that a method outlining the use of triflic acid alone for the Markovnikov addition of phenols to unactivated alkenes has been reported (Scheme 2.16) [23].

The iodine-catalyzed addition of carboxylic acids to styrene derivatives generated high yields of esters and related compounds (Scheme 2.17) [24]. This transition metal-free reaction used TBHP as the oxidant and was carried out under very mild conditions using an assortment of alkenes and carboxylic acids as substrates. The chemistry was quite tolerant of preexisting functional groups, and even aryl bromides were retained through the oxidation process. In addition to the styrene derivatives, aliphatic alkenes such as 1-octene were also successfully converted into esters using this approach.

While the synthesis of ethers through the addition of alkoxides to secondary alkyl halides can be problematic due to the formation of elimination side products, a

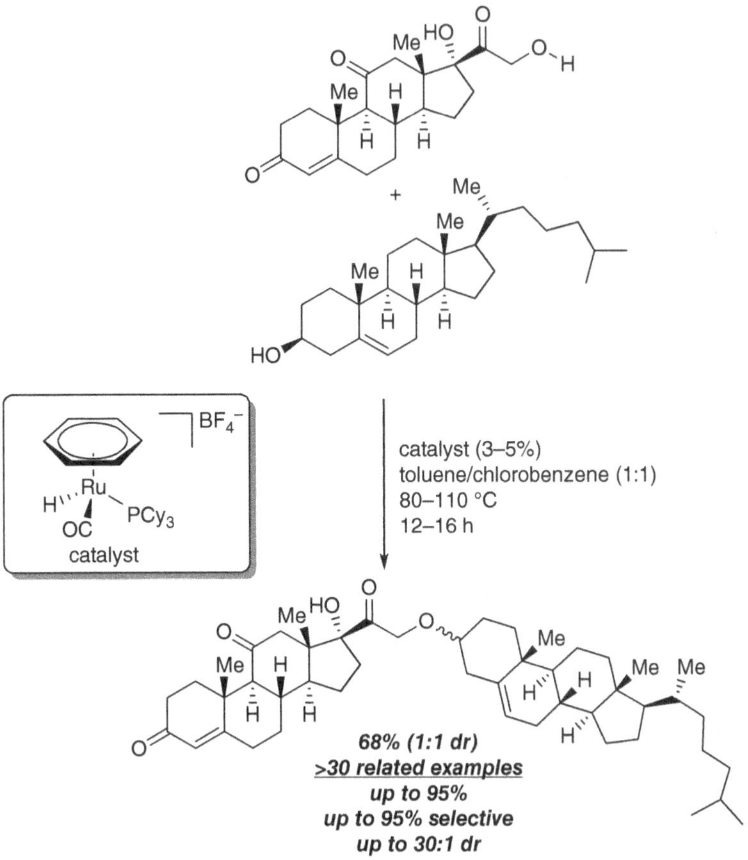

SCHEME 2.13 Ether synthesis through ruthenium catalyzed dehydrative etherification [18].

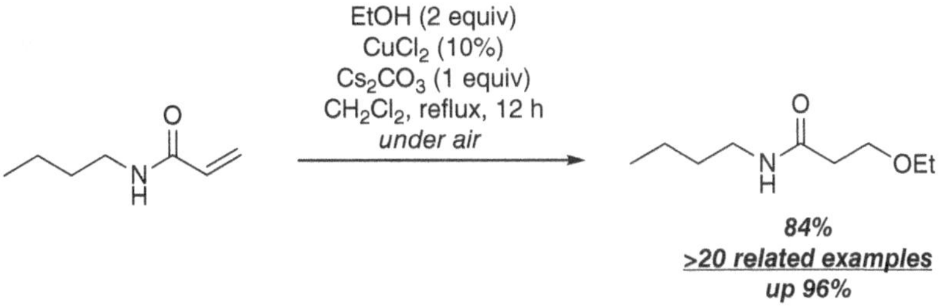

SCHEME 2.14 Copper-catalyzed oxa-Michael additions to acrylamide derivatives [20].

three-component coupling reaction between an unactivated terminal alkene, boronic acid, and alcohol generated moderate to high yields of ethers with minimal side products (Scheme 2.18) [8]. The reaction was catalyzed by a gold dimer and used Selectfluor as an activating agent. In addition to the generation of ethers using this approach, esters could also be prepared in moderate yield (up to 69%) if the alcohol was replaced by a carboxylic acid.

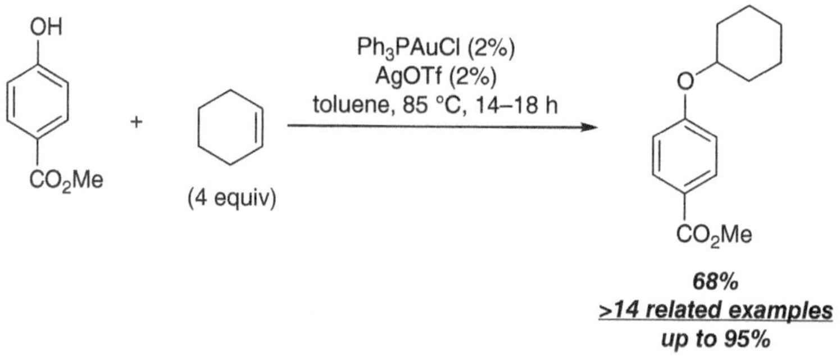

SCHEME 2.15 Gold-catalyzed addition of phenols to unactivated alkenes [22].

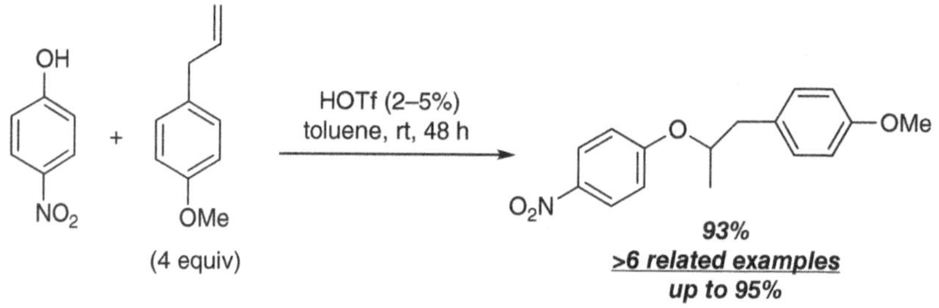

SCHEME 2.16 Triflic acid-promoted addition of phenols to unactivated alkenes [23].

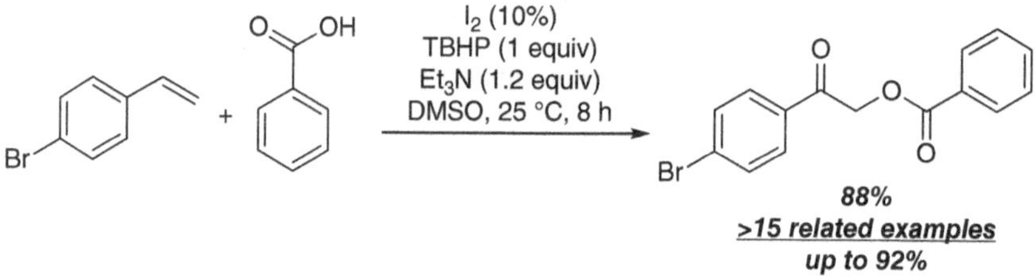

SCHEME 2.17 Iodine-catalyzed synthesis of esters [24].

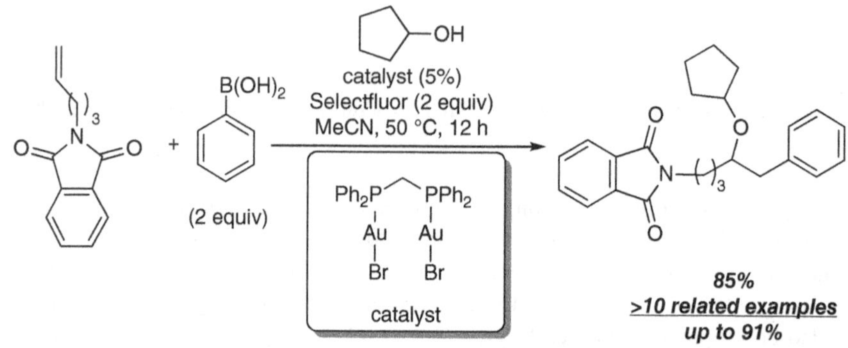

SCHEME 2.18 Gold-catalyzed synthesis of ethers through a three-component coupling reaction [8].

The atom-efficient addition of carboxylic acids to olefins is one of the more attractive routes to the generation of esters. A challenging version of this reaction involves the addition to unactivated alkenes. A convenient and practical solution to this problem was developed based upon a gold-catalyzed process (Scheme 2.19) [22]. The combination of a triphenylphosphine gold(I) complex and silver(I) triflate generated the catalytically active species in solution, and a variety of alkenes were successfully converted into esters at 85 °C in toluene. An assortment of phenols bearing electron-donating and electron-withdrawing groups were used as the oxygen nucleophiles.

In addition to the gold-catalyzed chemistry described in the preceding section, ruthenium complexes also catalyzed the addition of carboxylic acids to alkenes (Scheme 2.20) [25]. A common air-stable ruthenium compound served as the catalyst for the reaction along with silver(I) triflate as an additive. While a range of monosubstituted terminal alkenes were successfully converted into esters in moderate to excellent yields (up to 95%), analogous reactions with internal or disubstituted alkenes were less successful. Both the gold- and ruthenium-catalyzed additions were attractive due to the availability of the catalysts used for the reactions.

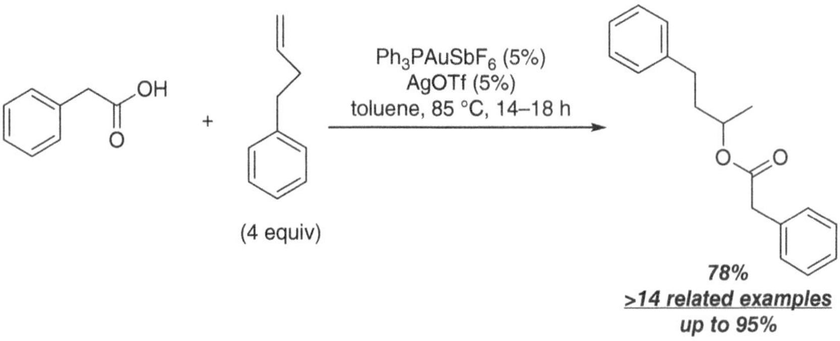

SCHEME 2.19 Preparation of esters through the gold-catalyzed addition of carboxylic acids to alkenes [22].

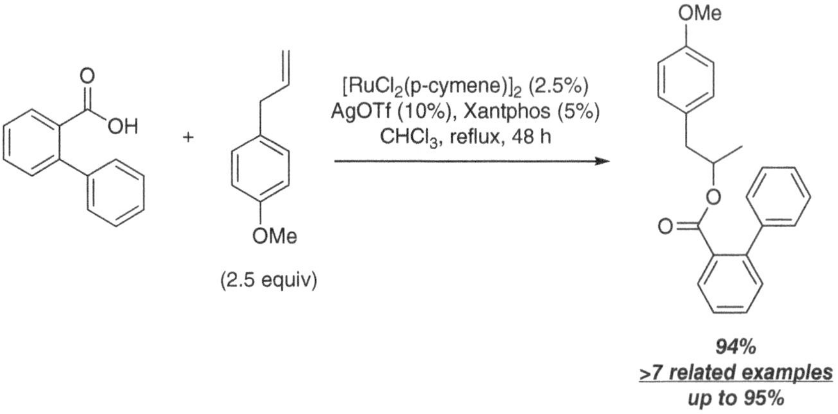

SCHEME 2.20 Ruthenium-catalyzed addition of carboxylic acids to alkenes [25].

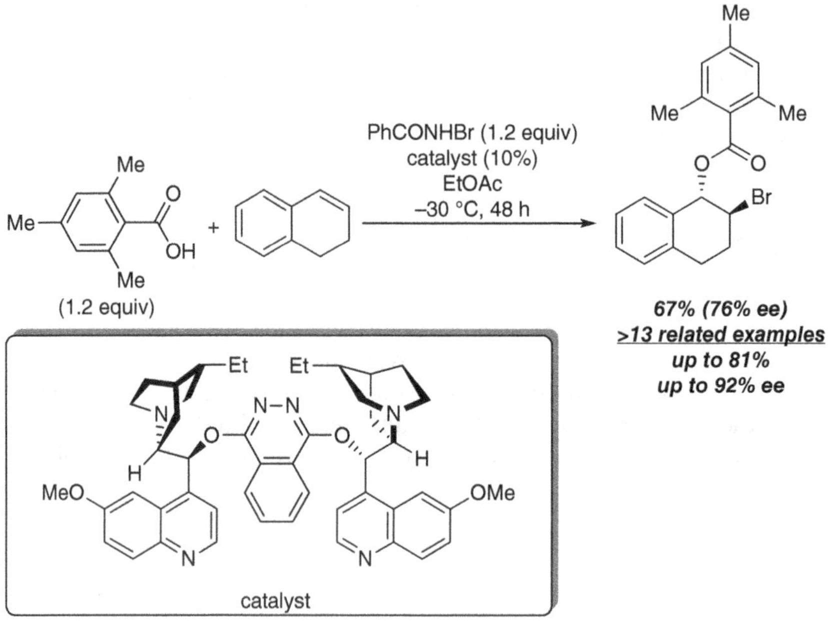

SCHEME 2.21   Asymmetric bromoesterification of alkenes [26].

The bromoesterification of unactivated alkenes has been achieved using an organocatalytic approach (Scheme 2.21) [26]. After screening a range of organocatalysts for their efficacy toward the selective haloesterification reaction, the authors determined that cinchona alkaloids generated brominated esters in high yields with high selectivities (up to 92% ee). While the use of NBS did generate some of the product, the selectivity was low. When they switched to *N*-bromobenzamide as the bromine source, high enantioselectivity was obtained [26]. Using this optimized system, a variety of unactivated alkenes were functionalized with the highest enantioselectivities observed for substrates bearing electron-withdrawing groups.

A series of functionalized chromanes were generated from a C—H alkylation reaction between 1-naphthol and a series of α,β-unsaturated-α-keto esters (Scheme 2.22) [27]. The chemistry was carried out under very mild conditions using an organocatalyst and was remarkably insensitive to the electronic composition of the α-keto ester. Heteroaromatic examples were also successfully functionalized using this approach. The chemistry was successfully extended to 2-naphthol and excellent conversions (up to 90%), and high enantioselectivities were obtained (up to 96% ee). In related work, a simple proline derivative was also able to promote the asymmetric synthesis of chromane derivatives (Scheme 2.23) [28].

The interest in developing new synthetic approaches for the preparation of oxetanes has grown significantly due to the impact of these compounds in organic synthesis. A two-step method for their synthesis used a rhodium-catalyzed process for the formation of an intermediate with a C-O bond (Scheme 2.24 and Example 2.4) [29]. Once the new oxygen–carbon bond was generated, the cyclization was promoted by the addition of sodium hydride in DMF. The chemistry was tolerant to a variety of functional groups, and a range of oxetanes were generated following this approach.

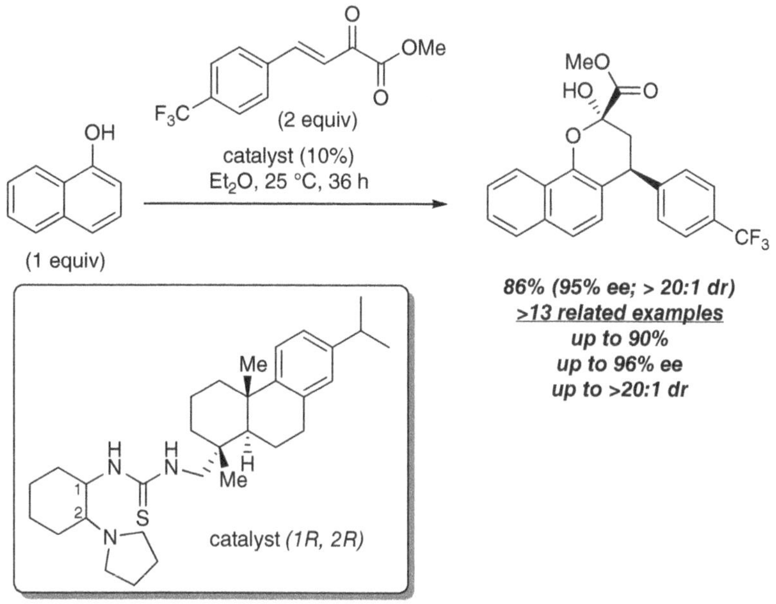

SCHEME 2.22 Enantioselective synthesis of functionalized chromanes [27].

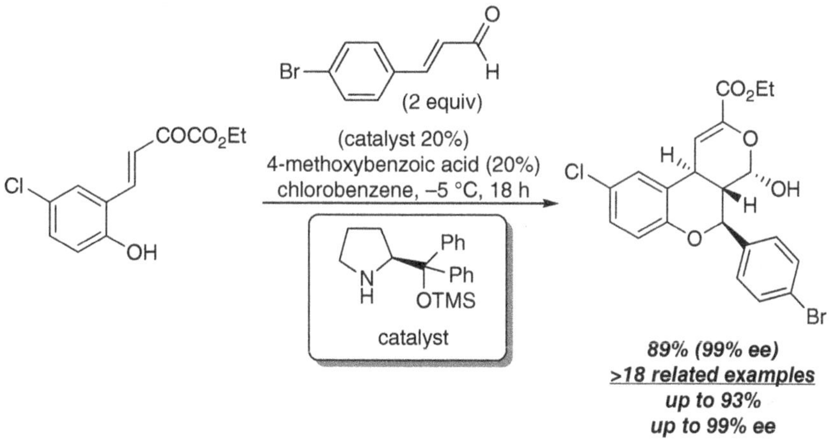

SCHEME 2.23 Asymmetric synthesis of chromane derivatives using a proline derivative as an organocatalyst [28].

The conversion of carbon dioxide into useable substrates for further functionalization remains a current and challenging goal in synthetic chemistry. To this end, the direct conversion of diols into carbonates was achieved using cerium oxide/cyanopyridine as the catalyst system (Scheme 2.25) [30]. Both five- and six-membered carbonates can be generated in excellent yields from the corresponding diols. It should be noted that the preparation of six-membered carbonates through this type of reaction was a rare conversion. In related work, the conversion of epoxides into cyclic carbonates has been accomplished using a porphyrin-based catalyst (Scheme 2.26) [31].

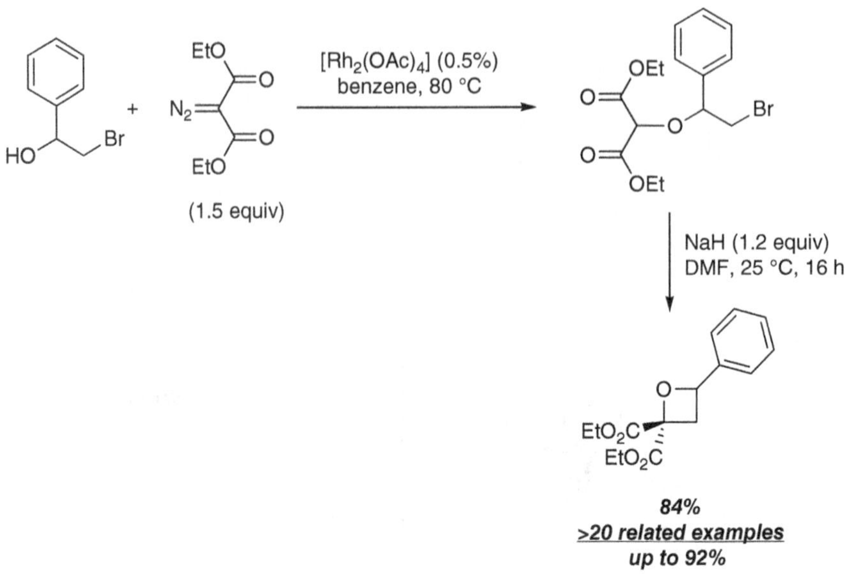

SCHEME 2.24   Preparation of oxetanes through a rhodium-catalyzed process [29].

Example 2.4   Synthesis of Functionalized Oxetanes [29].

Part 1

Before starting this synthesis, carefully follow the safety instructions listed in Example 2.1.

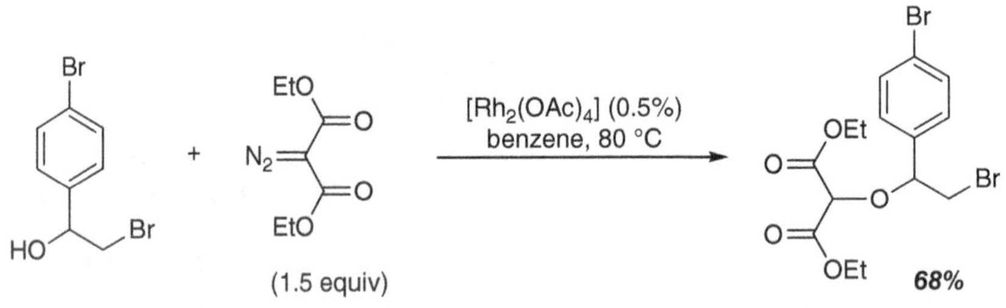

Caution! Ethyl acetate and hexanes are highly flammable.

A mixture of diazomalonate (838 mg, 4.5 mmol), alcohol (839 mg, 3.0 mmol), and dirhodium(II), tetraacetate (6.2 mg, 0.014 mmol) in benzene (30 mL) was heated at 80°C for 2 h. The reaction mixture was allowed to cool to rt. Water (30 mL) was added, and the layers were separated. The aqueous layer was extracted with $CHCl_3$ (3 × 30 mL). The organic extracts were combined, dried (Na_2SO_4), and concentrated under vacuum. Purification by flash chromatography (10% EtOAc in hexanes) afforded the acyclic ether as a colorless oil (890 mg, 68%).

Part II

Before starting this synthesis, carefully review the chemical safety guidelines and follow the instructions listed in Example 2.1.

83%

Caution! Sodium hydride reacts violently with moisture.

Caution! Ethyl acetate and hexanes are highly flammable.

DMF (32.5 mL) was added to a flask containing sodium hydride (60% w/v, 49 mg, 1.2 mmol), which had been cooled to 0°C. The ether from part 1 (440 mg, 1.0 mmol) in DMF (7.5 mL) was added dropwise to the stirred suspension of sodium hydride in DMF at 0°C over the course of 10 min. The reaction mixture was stirred at 25°C for 16 h. Saturated aq. NH_4Cl (40 mL) was added followed by EtOAc (40 mL). The layers were separated, and the aqueous layer was extracted with EtOAc (3×40 mL). The organic extracts were combined, dried (Na_2SO_4), and concentrated in vacuo. Purification by flash chromatography (20% EtOAc in hexanes) afforded oxetane as a colorless oil (299 mg, 83%).

Adapted with permission from O. A. Davis, J. A. Bull, Angew. Chem. Int. Ed., **2014**, *53,* 14230–14234. *Copyright (2014) Wiley-VCH Verlag GmbH & Co. KGaA, Weinheim.*

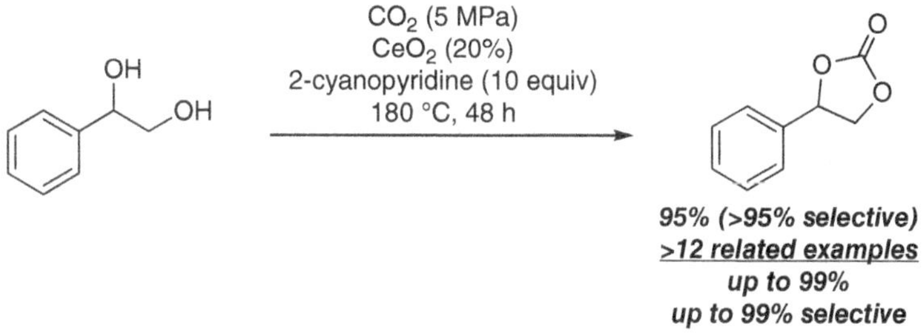

95% (>95% selective)
>12 related examples
up to 99%
up to 99% selective

SCHEME 2.25   Conversion of a diol into a carbonate using carbon dioxide [30].

Considerable effort has been devoted to developing asymmetric approaches to the preparation of epoxides. Elegant work by Sharpless [32] and others has facilitated the preparation of a host of resolved examples [33, 34]. The asymmetric epoxidation of chalcones has been achieved using a Yb-based catalyst and TBHP as the oxidant (Scheme 2.27) [35]. The ligand system for the Yb was based upon a chiral diphenylprolinolate ligand, and moderate to excellent yields of the chiral epoxides were obtained along with outstanding selectivity.

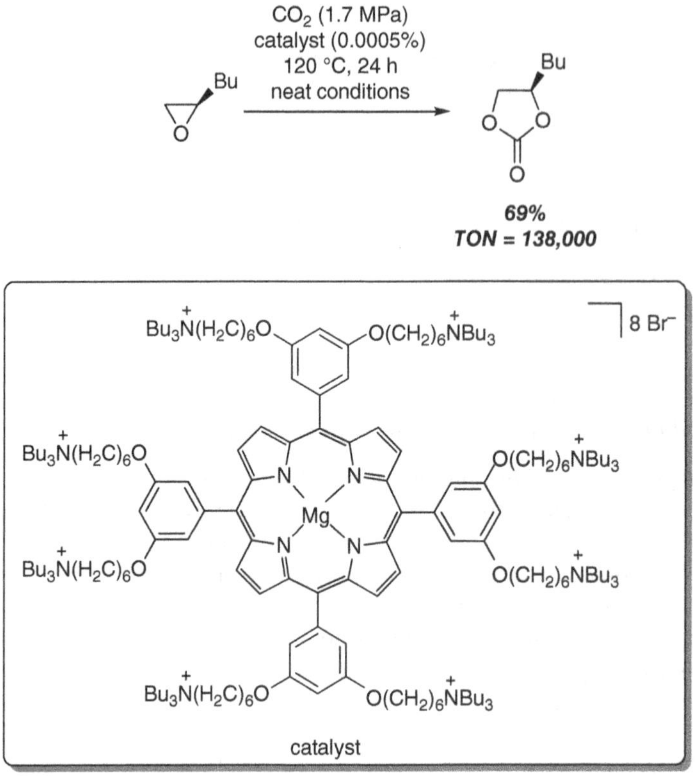

SCHEME 2.26   Synthesis of cyclic carbonates using a porphyrin based catalyst [31].

The stereoselectivity of the reaction varied with the electronic composition of the aryl rings on the chalcone. The use of a *p*-methyl-substituted chalcone afforded the highest enantiose-lectivity (99% ee) while chalcones bearing a *p*-trifluoromethyl group resulted in the lowest selectivities (80% ee). For the right system, this is a very efficient and selective approach to the synthesis of chiral epoxides.

The Darzens reaction is typically carried out in an organic solvent using an alpha-halogenated carbonyl compound and an aldehyde as substrates, and although this process tends to be high yielding, the reaction times can be long (days to weeks). Increasing the temperature to reduce the reaction time tends to result in a loss of selectivity. An interesting approach to this reaction was reported using water as the solvent (Scheme 2.28) [36]. The reaction time was decreased significantly (1 h), and the yields of the epoxides were still excellent, and the stereoselectivity was outstanding.

In addition to alcohols and related compounds, a variety of heteroatom-containing substrates also undergo oxygen–carbon(sp^3) bond-forming reactions. As a representative example, an asymmetric intramolecular iodinative cyclization of phosphorami-dates has been achieved using an organocatalyst (Scheme 2.29) [37]. The reaction was carried out at −20 °C using NIS as the iodine source and a chiral Bronsted acid

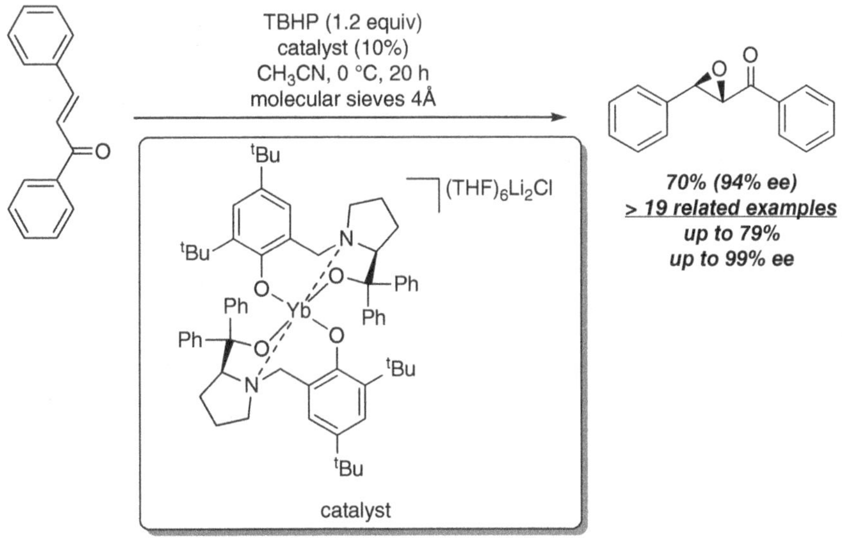

SCHEME 2.27   Asymmetric epoxidation of chalcones using a Yb-based catalyst [35].

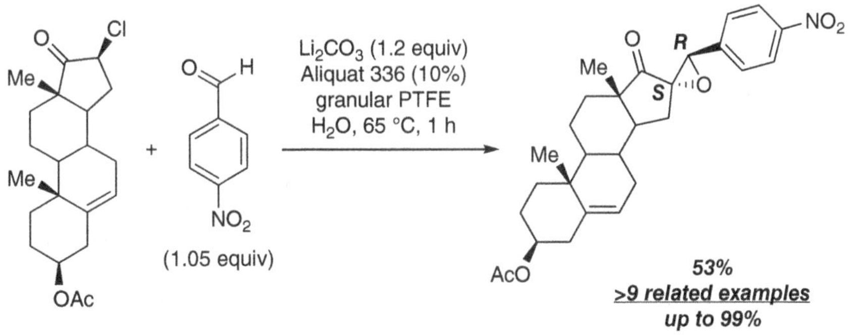

SCHEME 2.28   Synthesis of resolved steroidal epoxides through an aqueous Darzens-type reaction [36].

as the catalyst. An assortment of phosphoramidates bearing tethered alkenes was successfully cyclized using this catalyst system with excellent regio- and stereoselectivities. The chemistry tolerated a range of alkyl as well as electron-rich and electron-deficient aryl substituents on the alkene fragment.

2.1.1   Troubleshooting the Preparation of Alcohols, Ethers, and Related Compounds

While it is not possible to predict all of the challenges and difficulties that will arise during the preparation of these compounds, the following suggestions will hopefully provide some direction for common issues and reaction requirements.

Problem/Issue/Goal	Possible Solution/Suggestion
Need a good way to make a primary alcohol from an alkyne	One of the most attractive approaches to the synthesis of primary alcohols is the ruthenium-catalyzed hydration/reduction of terminal alkynes [2]
Can an alkylboronic acid be converted into an alcohol?	While versions of this reaction have appeared in the literature, a practical metal-free approach for this transformation uses Oxone in acetone [3]
The starting material has a Bpin group and not a boronic acid. Does it need to be converted into the boronic acid before converting it into an alcohol?	The direct conversion of the alkylboronates has been achieved, and a convenient metal-free approach uses a combination of sodium hydroxide and hydrogen peroxide [4]
The starting material has an unactivated alkene. Can it be converted into an ether?	A reasonable approach would be to try a gold-catalyzed addition reaction. This works quite well under fairly mild conditions when phenols were used as coupling partners [22]
Is it possible to convert alkyl halides into alcohols using simple substitution chemistry with hydroxide?	One of the difficulties typically encountered with this approach is the formation of side products due to competing elimination processes. Naturally, the extent of the elimination reaction will be substrate specific. It should be noted that the direct conversion of alkyl halides into alcohols has been achieved in an ionic liquid/water mixture [5]
Can a hydrocarbon fragment be converted into an alcohol through C—H activation?	This challenging reaction has been accomplished with a number of different systems; however, general approaches are rare. As a result, each catalyst system should be evaluated against the substrate under investigation. For example, if the target substrate is a tertiary C—H bond, there are several catalysts that will promote the chemoselective C—H hydroxylation of tertiary C—H bonds in the presence of other C—H bonds [6, 7]
Are there any practical metal-free approaches for selective C—H hydroxylation reactions?	An approach using a combination of hydrogen peroxide and an electron-deficient ketone was successfully applied to C—H hydroxylation reactions [9]. This approach has been reported to be highly substrate dependent, but it would be a reasonable starting point
Need to generate a chiral allylic alcohol	This can be a very challenging reaction. One of the more successful syntheses used iridium-based catalysts and allylic carbonates as substrates [13]
What is a practical approach to the synthesis of alkyl aryl ethers?	The classic approach to the preparation of these compounds is through a Williamson ether synthesis. Due to the nature of the chemistry, this method is most successful with primary alkyl systems with good leaving groups such as bromide/iodide [14, 15]
The substrate has an activated alkene. Can it be converted into an ether?	This can be accomplished through an oxa-Michael addition reaction [19, 20]
Need to convert an unactivated alkene into an ether	There are several approaches for this transformation. The direct addition of phenols and alcohols to unactivated alkenes has been accomplished using a range of methods. Practical approaches include a gold-catalyzed process [22] as well as an acid-promoted reaction [23]

(Continued)

(Continued)

Problem/Issue/Goal	Possible Solution/Suggestion
Need to convert an unactivated alkene into an ester	This can be accomplished through the gold- or ruthenium-catalyzed addition of carboxylic acids to alkenes [22, 25]
Are the methods for ether/ester synthesis limited to alcohols and carboxylic acids?	A number of heteroatom-containing compounds have been involved in oxygen–carbon(sp³) bond-forming reactions. The intramolecular iodinative cyclization of phosphoramidates serves as a representative example [37]

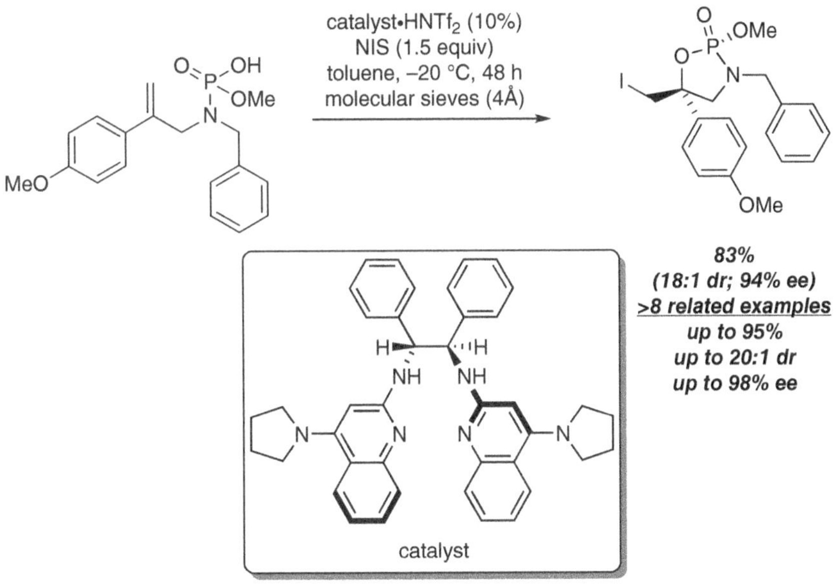

SCHEME 2.29 Iodinative cyclization of phosphoramidates [37].

2.2 PREPARATION OF PHENOLS, ARYL ETHERS, AND RELATED COMPOUNDS THROUGH THE FORMATION OF OXYGEN–CARBON(sp²) BONDS

Phenols are high-value compounds with a myriad of applications. Modern approaches to the synthesis of these compounds have largely focused on oxidations and metal-mediated hydroxylation reactions due to their high yields, selectivity, and functional group tolerance [38–40]. One of the most attractive routes to the generation of phenols entailed the treatment of aryltrifluoroborate salts with Oxone (Scheme 2.30) [3]. These reactions were operationally straightforward and could be carried out under an atmosphere of air. The conversion was also remarkably fast, and most reactions were complete within a few minutes at room temperature. The oxidation chemistry was remarkably tolerant to a wide variety of preexisting electron-donating and electron-withdrawing groups as well as heteroaryl substrates. This chemistry is also attractive due to the wide assortment of

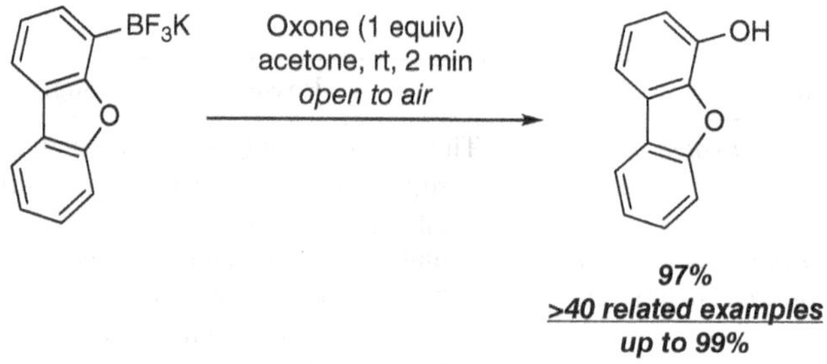

SCHEME 2.30   Oxidation of aryltrifluoroborate salts using Oxone [3].

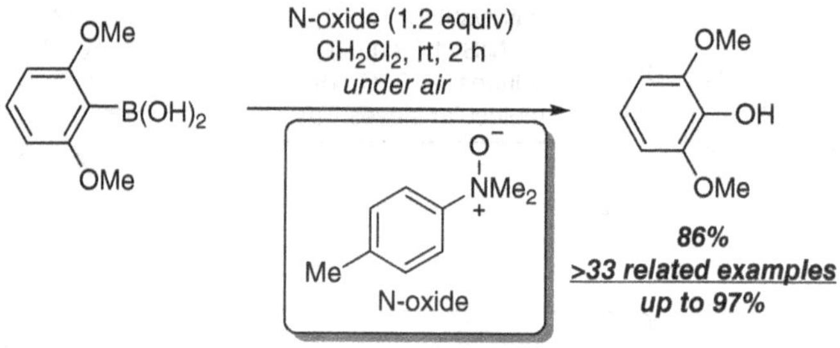

SCHEME 2.31   *N*-oxide promoted conversion of arylboronic acids into phenols [41].

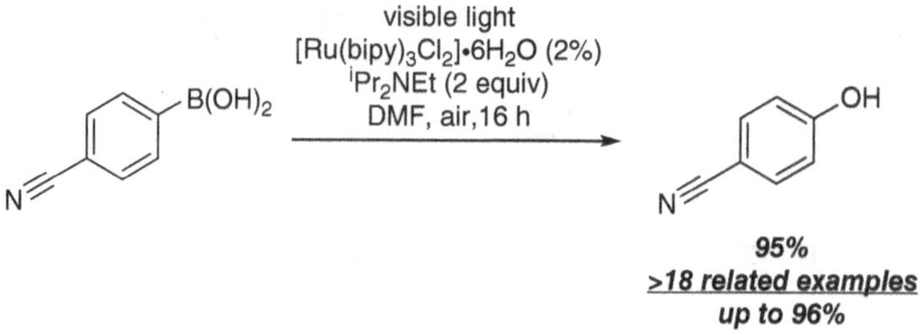

SCHEME 2.32   Conversion of arylboronic acids into phenols using visible light catalysis [42].

aryltrifluoroborate salts that are readily available. *N*-oxides can also be used to convert arylboronic acids into phenols (Scheme 2.31) [41].

The conversion of boronic acids into phenols has been achieved using visible light catalysis (Scheme 2.32) [42]. In addition to light, the catalyst components consisted of a ruthenium bipyridine species, an amine, and DMF. All the components were needed for a successful reaction since low conversions were observed if any of the components were absent. The ability to use air as the oxidant was one of the practical aspects of this chemistry. While a wide range of arylboronic acids were successfully transformed into phenols, a pinacol-derived arylboronate also served as a substrate in these reactions (up to 94% conversion). In addition to the ruthenium-catalyzed reaction, a metal-free version of the chemistry was also developed using an organic dye.

An interesting approach to the synthesis of phenols proceeds through the oxidation of arylhydrosilanes (Scheme 2.33) [43]. The conditions were quite mild, and moderate to good yields of the phenols were obtained using several modifications to a standard set of conditions. While part of the goal of the chemistry was to develop reactions that did not require the use of a fluoride source, the authors did screen the oxidations in the presence of fluoride (TBAF) and found that the reactions were faster. The yields varied between the two methods, and some substrate-specific reactivity was noted. Substrates bearing electron-withdrawing groups were especially active. For arylsilanes bearing electron-donating groups, their reactivity under the standard conditions was sluggish. To circumvent this reactivity problem, these compounds were converted into the arylmethoxysilane prior to the oxidation step. Using these two methods, a host of arylhydrosilanes were converted into phenols in high yield.

Transition metal-catalyzed routes are also known for the conversion of arylboronic acids into phenols. Of the copper-catalyzed approaches, the methods that are tolerant to air and water are particularly practical due to the operational simplicity. One example of this chemistry entailed the use of copper sulfate as the catalyst (Scheme 2.34 and Example 2.5) [44]. Using 1,10-phenanthroline as a supporting ligand for the copper center, a wide array of electron-deficient and electron-rich arylboronic acids were converted into phenols in excellent yields. Even bulky arylboronic acids such as 2,6-dimethylphenylboronic acid were successfully converted (87%). In related work, copper oxide has also been successfully used to prepare phenols in water and under air from arylboronic acids bearing an assortment of electron-donating and electron-withdrawing groups (Scheme 2.35) [45].

While the copper coupling of arylboronic acids and hydroxide sources successfully generated a host of phenols, an arylboronic acid or derivative was a prerequisite for this

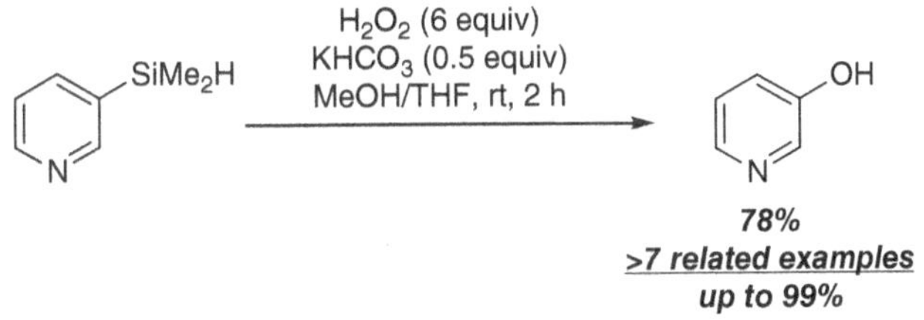

SCHEME 2.33   Conversion of arylhydrosilanes into phenols through a fluorine-free procedure [43].

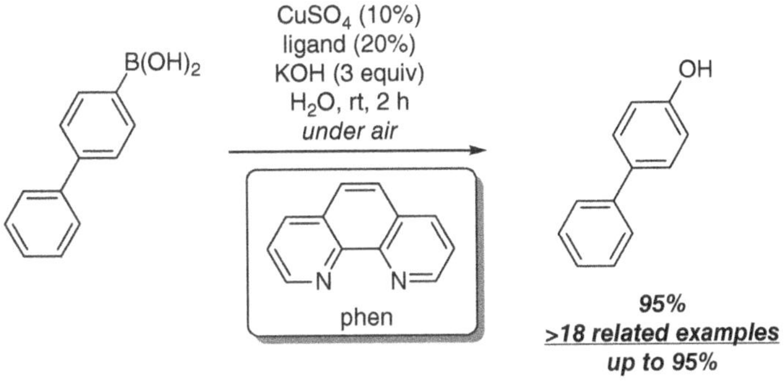

SCHEME 2.34   Synthesis of phenols in water and air with copper(II) sulfate [44].

Example 2.5   Copper-Catalyzed Conversion of ArylBoronic Acids into Phenols [44].

Before starting this synthesis, carefully follow the safety instructions listed in Example 2.1.

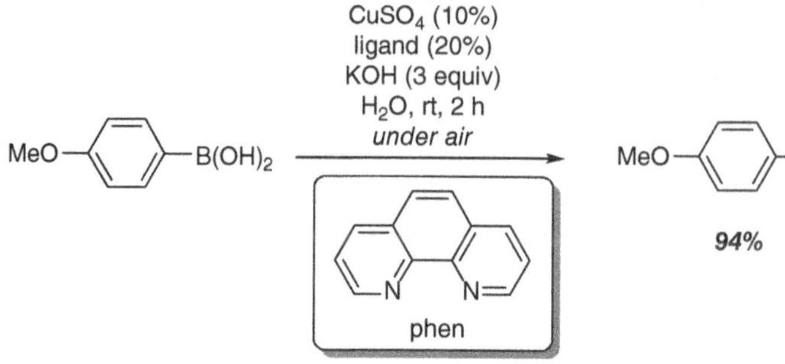

A mixture of $CuSO_4$ (16 mg, 0.1 mmol), 1,10-phenanthroline (36 mg, 0.2 mmol), 4-methoxyphenylboronic acid (152 mg, 1 mmol), and KOH (168 mg, 3.0 mmol) in H_2O (5 mL) was stirred for 2 h at room temperature open to air (without bubbling air). After the boronic acid was exhausted completely (monitored by TLC), the reaction was cooled to 0°C and quenched carefully by aqueous solution of HCl (2 M, 15 mL). The resultant mixture was extracted with EtOAc (3 × 20 mL). The combined organic layers were washed with brine (20 mL) and dried over Na_2SO_4. After removal of the solvent in vacuum, the residue was purified by chromatography (silica gel, EtOAc : PE = 1 : 5) to give desired phenol (117 mg, 94%) as a colorless solid.

Adapted with permission from J. Xu, X. Wang, C. Shao, D. Su, G. Cheng, and Y. Hu, Org. Lett., **2010,** *12, 1964–1967. Copyright (2010) American Chemical Society.*

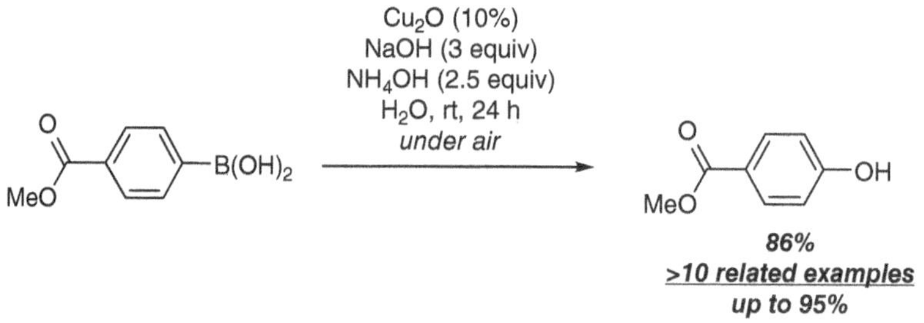

SCHEME 2.35   Use of copper oxide in the synthesis of phenols [45].

chemistry. The ability to effect the transformation starting from aryl halides would be an attractive route to the preparation of phenols. To this end, a number of metal-catalyzed approaches for this reaction have been developed [38–40]. Predictably, the challenges encountered in this chemistry mirrored classic cross-coupling chemistry. An attractive catalyst would be general for a wide range of electron-rich and electron-deficient arenes; highly active at ultralow catalyst levels; tolerant of substitution patterns; able to functionalize aryl chlorides, bromides, and iodides; environmentally friendly; inexpensive; and insensitive to air and water. Naturally, it will be difficult for a single catalyst system to satisfy all of these goals.

For copper-catalyzed conversions, a number of catalysts, conditions, and approaches have been designed to functionalize a range of electron-rich and electron-poor aryl halides as well as bulky aryl halides. One approach to the synthesis of phenols used copper iodide and a β-diketone as the supporting ligand for the catalyst system (Scheme 2.36) [46]. Cesium hydroxide served as the hydroxide source for this chemistry, and moderate to excellent yields of the phenols were obtained (up to 97%). A range of preexisting functional groups including alkyl, nitro, nitrile, phenyl, ketone, and ether were all well tolerated by this chemistry. Aryl bromides bearing electron-withdrawing groups as well as aryl iodides were successfully used in the coupling reaction under the author's standard conditions; however, the general system was not extended to aryl chlorides, and electron-rich aryl bromides were sluggish. Fortunately, the authors were able circumvent the low reactivity of the latter by adding sodium iodide to the reaction mixture prior the addition of the hydroxide source. Presumably, the sodium iodide converts the aryl bromide into an aryl iodide through an aromatic Finkelstein reaction. Using this approach, a number of aryl bromides were successfully converted into phenols. In related work, a range of aryl iodides and electron-deficient aryl bromides as well as extremely crowded aryl iodides were converted into phenols using a copper-catalyzed reaction (Scheme 2.37) [47]. In addition to these methods, copper-catalyzed approaches that function in pure water were also achieved through the use of TBAB and Bu_4NOH as a phase-transfer catalysts [48, 49]. A heterogeneous copper-catalyzed process was also developed that could be reused for multiple reactions [50].

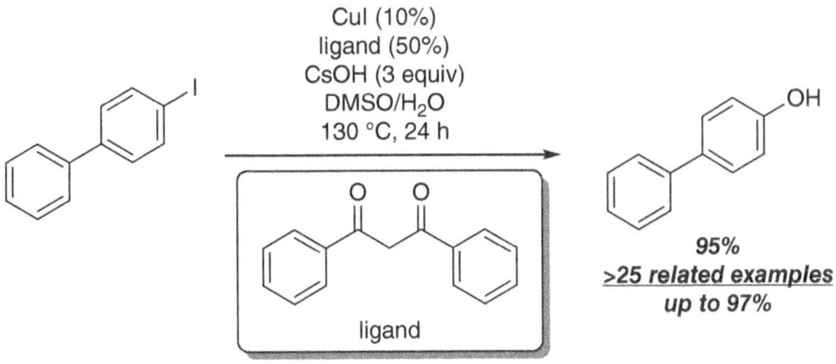

SCHEME 2.36   Copper-catalyzed coupling of aryl bromides/iodides with CsOH [46].

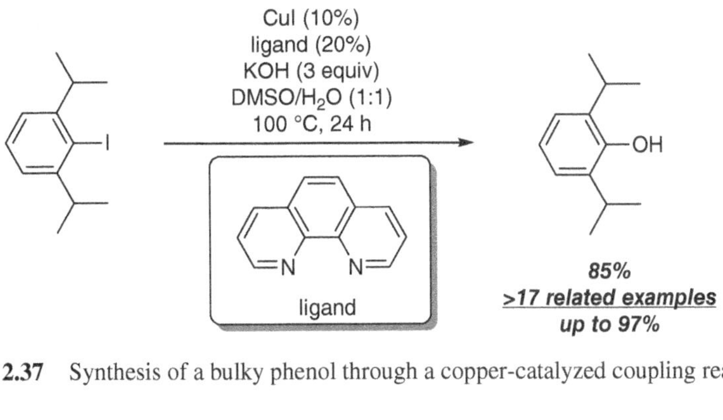

SCHEME 2.37   Synthesis of a bulky phenol through a copper-catalyzed coupling reaction [47].

The palladium-catalyzed coupling of hydroxide sources with aryl halides has been achieved [51]. Pd_2dba_3 was used as the palladium source, and XPhos or Me_4XPhos was used as the supporting ligand. One of the impressive aspects of this chemistry was the ability to hydroxylate aryl chlorides as well as aryl bromides. A biphasic solvent system was used to solubilize both the KOH and the organic components, and heating to 100 °C afforded outstanding yields of the phenols. The authors built upon this system and reported a modified catalyst for the coupling of aryl chlorides and bromides with hydroxide sources (Scheme 2.38) [52]. While their initial report on this chemistry used Pd_2dba_3 as the palladium source and a bulky dialkylbiarylphosphine as the supporting ligand, the modified catalyst system was comprised of a palladacycle along with 1 equivalent of a dialkylbiarylphosphine ligand. Using this revised system, a range of aryl chlorides and bromides were hydroxylated with potassium hydroxide in good to outstanding yield upon heating at 80 °C (up to 100%). In addition to these reactions, the authors determined that changing the hydroxide source to cesium hydroxide enabled the development of a room temperature coupling of aryl bromides (Example 2.6) [52]. This chemistry is one of the most practical approaches to the synthesis of phenols due to its ability to hydroxylate aryl bromides at room temperature as well as a range of aryl chlorides at elevated temperatures. In related work, several microwave-assisted palladium-catalyzed conversions of aryl chlorides and bromides into phenols have been reported (Scheme 2.39 and Example 2.7) [53, 54].

Imidazole-functionalized phosphines have also been used as supporting ligands in hydroxylation reactions (Scheme 2.40) [55]. Similar to the dialkylbiarylphosphine ligands, the imidazole-based ligands were quite stable to air and could be readily handled. The hydroxylation reactions were carried out in a mixture of dioxane and water and afforded good to outstanding yields of a range of functionalized phenols (up to 99%). The chemistry was also insensitive to crowded substrates, and 2,6-dimethyl-substituted aryl halides were converted into the phenols in high yield (up to 90%).

In addition to the synthesis of phenols through metal-catalyzed cross-coupling chemistry between aryl halides and hydroxide sources, the C—H hydroxylation of arenes has also been used to prepare phenols. This is an attractive approach to the synthesis of these valuable compounds since it does not require prior functionalization of the arene.

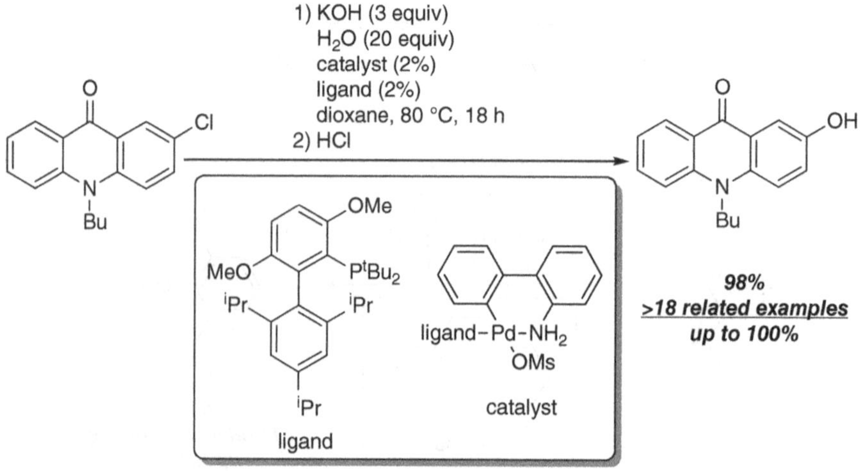

SCHEME 2.38   Palladium-catalyzed synthesis of phenols from aryl chlorides [52].

Example 2.6   Room Temperature Palladium-Catalyzed Hydroxylation of Aryl Chlorides [52].

Before starting this synthesis, carefully follow the safety instructions listed in Example 2.1.

Caution! Hexane and ethyl acetate are highly flammable.

An oven-dried 20 mL resealable screw-cap test tube (A) equipped with a Teflon-coated magnetic stir bar was charged with the ligand (9.7 mg, 0.020 mmol, 2 mol %) and (hetero) aryl halide (if solid) (1.0 mmol, 1 equiv). Tube A was evacuated and backfilled with argon (this process was repeated a total of three times), and standard aqueous CsOH solution (mole ratio of CsOH to $H_2O = 3:10$; CsOH (450 mg, 3.0 mmol, 3 equiv) dissolved in deionized water (180 mg, 10.0 mmol, 10 equiv) as prepared in the preceding procedures) was then added into tube A via syringe, followed by the addition of (hetero)aryl halide (if liquid) (1.0 mmol, 1 equiv). A second oven-dried 10 mL resealable screw-cap test tube (B) equipped with a Teflon-coated magnetic stir bar was charged with the catalyst (17.1 mg, 0.020 mmol, 2 mol %). Tube B was then evacuated and backfilled with argon (this process was repeated a total of three times), and 1,4-dioxane (2.0 mL) was added into tube B via syringe. The reaction mixture in tube B was stirred at rt for approximately 1 min to form a homogeneous solution. The precatalyst solution from tube B was transferred into tube A via syringe. The resulting reaction mixture in tube A was stirred at rt for 18 h. The crude product was diluted with EtOAc (5 mL) and then acidified with aqueous HCl solution (1 M, 5 mL). The resulting reaction mixture in the capped test tube was agitated until all of the solid was dissolved. The reaction mixture was transferred into a separatory funnel and then neutralized with saturated NaHCO$_3$ solution (5 mL). The organic fraction was isolated, and the aqueous fraction was rinsed with EtOAc (2 × 10 mL). The combined organic fractions were concentrated in vacuo. The crude product residue was purified by flash chromatography using a solvent mixture (EtOAc/hexanes) as an eluent to afford the isolated product. The reported yields are of the isolated product and averages of two runs. The title compound was prepared using 1-bromo-2,4-dimethylbenzene (185 mg, 1.0 mmol); EtOAc/hexanes (1:6); 87 mg, 71%.

Adapted with permission from C. W. Cheung and S. L. Buchwald, *J. Org. Chem.*, **2014**, *79*, 5351–5358. *Copyright (2014) American Chemical Society.*

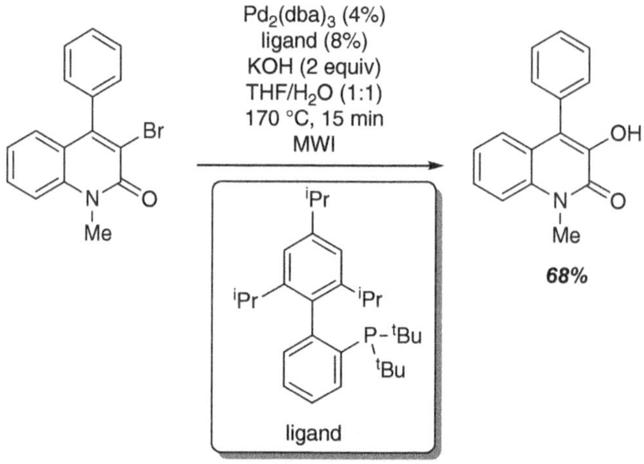

SCHEME 2.39   Microwave-assisted conversion of vinyl bromides into alcohols [53].

Example 2.7   Microwave-Assisted Palladium-Catalyzed Synthesis of Phenols [54].

Before starting this synthesis, carefully follow the safety instructions listed in Example 2.1.

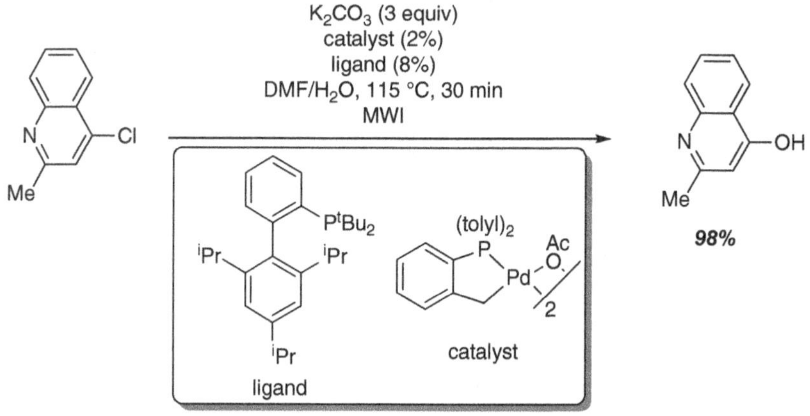

Glass tubes were each charged with Herrmann's palladacycle (19 mg, 0.02 mmol), ligand (35 mg, 0.08 mmol), and a base (3 mmol). If the aryl chloride (1 mmol) was a solid, it was added at the same time. Then, DMF (4.5 mL) and H_2O (0.5 mL) were added into the tube. If the aryl chloride was a liquid, it was added prior to adding DMF and H_2O. Each tube was equipped with a condenser, then evacuated, and backfilled with argon (evacuation and backfilling were each repeated three times). The reaction mixtures were each microwave irradiated (200 W) for 30 min at 115 °C. The mixtures were then cooled to room temperature and evaporated to dryness under vacuum. Water (15 mL) was added to each residual. The solutions were carefully acidified with 1 N HCl (4–5 mL) and extracted with ethyl acetate (5 × 10 mL). The organic layers were combined and evaporated. The crude products were each chromatographed through a column of silica gel with ethyl acetate/hexane as the eluent. The title compound was isolated as a white solid (98%).

Adapted with permission from C.-W. Yu, G. S. Chen, C.-W. Huang, and J.-W. Chern, *Org. Lett.*, **2012**, *14*, 3688–3691. *Copyright (2012) American Chemical Society.*

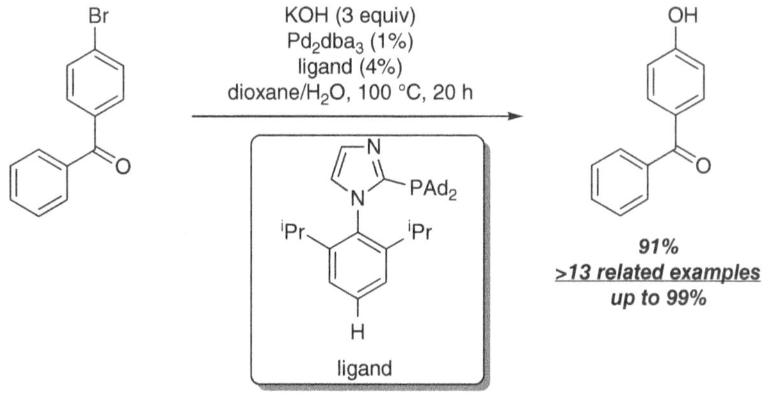

SCHEME 2.40 Imidazole-functionalized phosphines as supporting ligands in phenol synthesis [55].

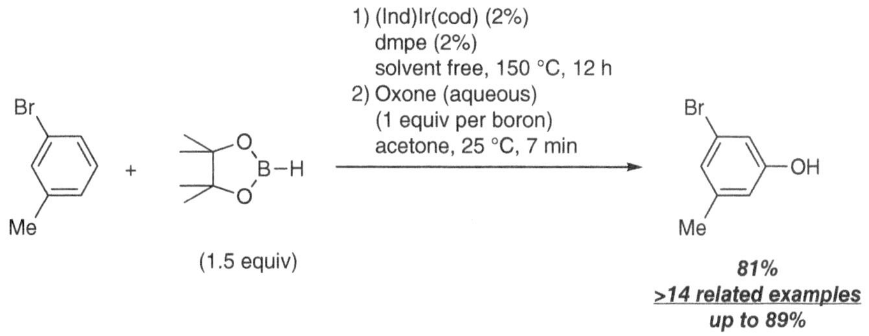

SCHEME 2.41 Tandem iridium-catalyzed C—H borylation and oxidation for the synthesis of phenols [56].

Naturally, this is a challenging reaction. One of the first versions of this reaction used a two-step one-pot approach (Scheme 2.41) [56]. The overall process was comprised of an initial iridium-catalyzed C—H borylation to generate an arylboronate followed by the addition of Oxone to convert the arylboronate into a phenol. It should be noted that this was quite a practical approach since both steps were completed in a single reaction vessel. The reaction was tolerant of a range of electron-donating and electron-withdrawing groups on the arene and afforded moderate to excellent yields of the phenols [56].

A metal-catalyzed C—H hydroxylation reaction has been achieved using a common ruthenium catalyst (Schemes 2.42 and 2.43) [57–59]. The reaction was carried under mild conditions and afforded moderate yields of the phenols. The use of a carbamate directing group enabled the regioselective addition of the hydroxyl group to the ortho-position. It should also be noted that the para-selective hydroxylation of anisole derivatives was accomplished without a directing group. The authors also carried out competition experiments between carbamates and esters and discovered that the carbamate was a more potent directing group. In related work, the C—H hydroxylation of arenes was accomplished using a Ru/Selectfluor system (Scheme 2.44) [60].

While the majority of ruthenium-catalyzed C—H hydroxylations use catalyst loadings between 5 and 10% Ru, a system has been developed that promoted the reaction at 1% catalyst loading (Scheme 2.45) [61]. The authors approach involved the use of a discrete

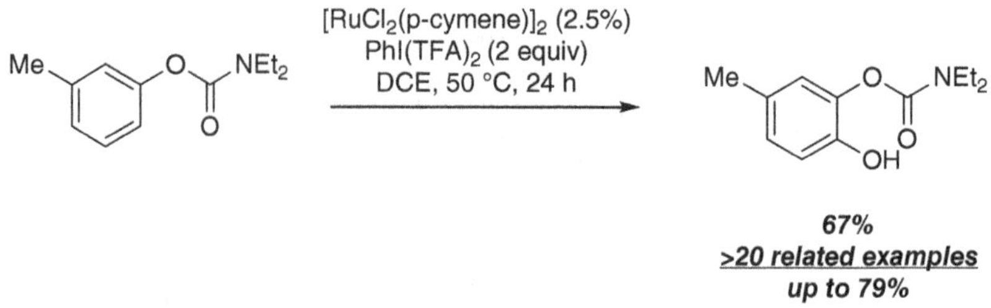

SCHEME 2.42 Ruthenium-catalyzed ortho selective C—H hydroxylation of arenes [57].

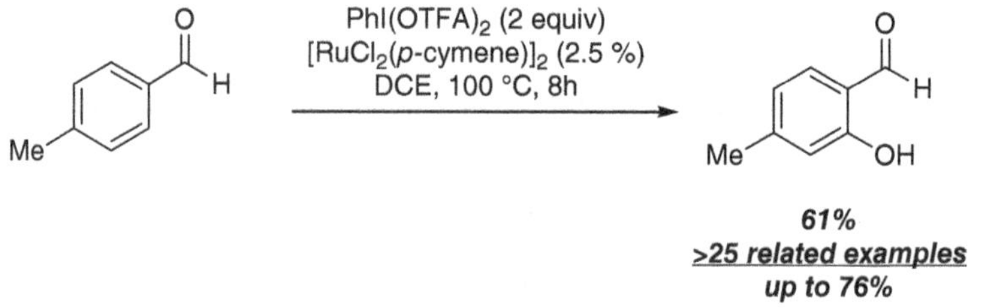

SCHEME 2.43 C—H hydroxylation of arenes using ruthenium catalysts [58].

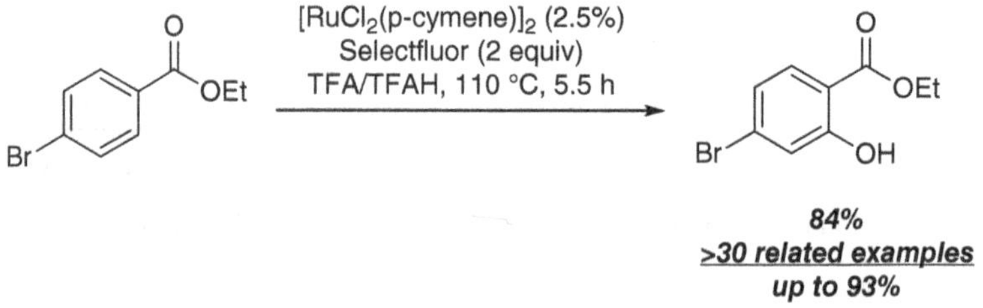

SCHEME 2.44 Synthesis of phenols using ruthenium/Selectfluor combinations [60].

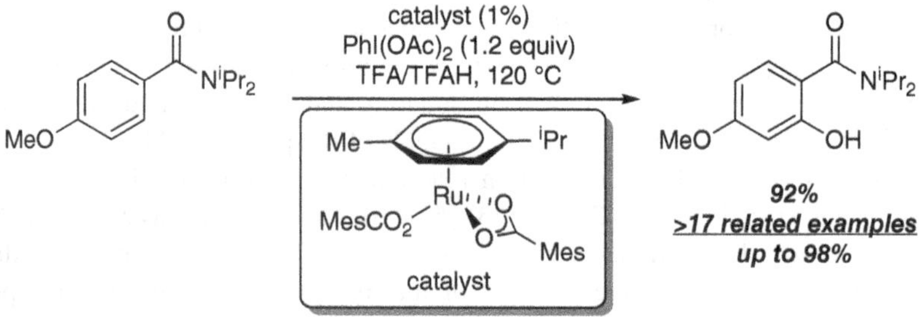

SCHEME 2.45 Ruthenium-catalyzed hydroxylation of benzamides [61].

ruthenium carboxylate catalyst instead of the halide-bridged dimer as the precatalyst. Using PhI(OAc)$_2$ as the oxidant in the system, a range of benzamides were hydroxylated in moderate to excellent yield (up to 98%).

The palladium-catalyzed ortho-hydroxylation of arenes generated acylphenols through a C—H hydroxylation reaction (Scheme 2.46 and Example 2.8) [62]. While a range of oxidants were screened for activity, BTI ([bis(trifluoroacetoxy)iodo]

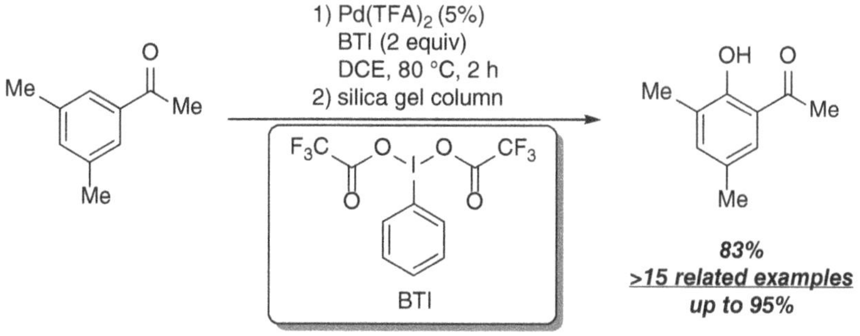

SCHEME 2.46   Synthesis of *ortho*-acylphenols [62].

Example 2.8   Palladium-Catalyzed C—H Hydroxylation of an Arene [62].

Before starting this synthesis, carefully follow the safety instructions listed in Example 2.1.

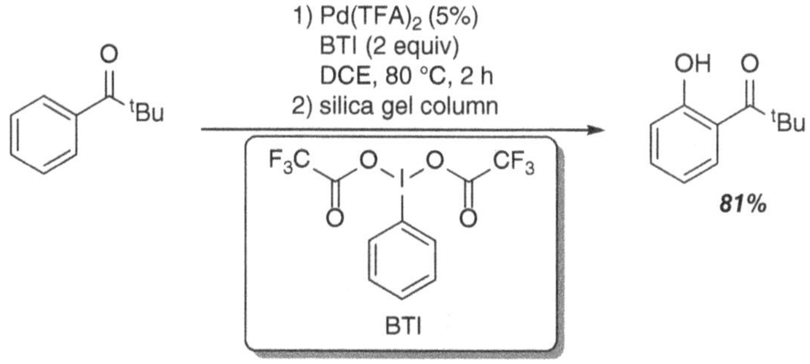

This reaction was carried out in air. Pd(TFA)$_2$ (6.6 mg, 0.02 mmol) and [bis(trifluoroace-toxy)iodo]benzene (344 mg, 0.8 mmol) were charged in a 4-mL scintillation vial, followed by 2 mL DCE and ketone substrate (0.4 mmol). The reaction was sealed with a Teflon-lined cap and heated at 80 °C on a pie block. When the reaction was done as monitored by TLC, solvent was removed in vacuo. The residue was subject to flash chromatography (silica gel) using Hex/DCM to give desired product. The title compound was isolated as a colorless oil (81%).

Adapted with permission from F. Mo, L. J. Trzepkowski, *and* G. Dong, *Angew. Chem. Int. Ed.*, **2012**, *51*, 13075–13079. *Copyright (2012) Wiley-VCH Verlag GmbH & Co. KGaA, Weinheim.*

benzene) was the most effective for the synthesis of the phenols. Once the trifluoroac-etate derivatives were generated, they were converted into phenols during the isolation and purification steps. This chemistry was a rare example of the successful use of ketones as directing groups. One of the most attractive aspects of this chemistry was the ability to carry out the reactions in air. In related work, the palladium-catalyzed hydroxylation of ketone derivatives has been achieved and was an attractive approach to the synthesis of phenols due to its wide substrate scope and the observation that many of the hydroxylations could be carried out at room temperature [63]. Elemental oxygen has also been used in the palladium-catalyzed C—H hydroxylation of arenes (Scheme 2.47) [64].

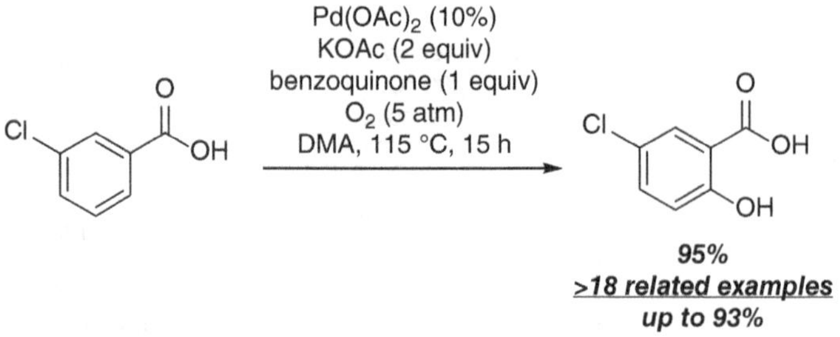

SCHEME 2.47   Palladium-catalyzed C—H hydroxylation using molecular oxygen [64].

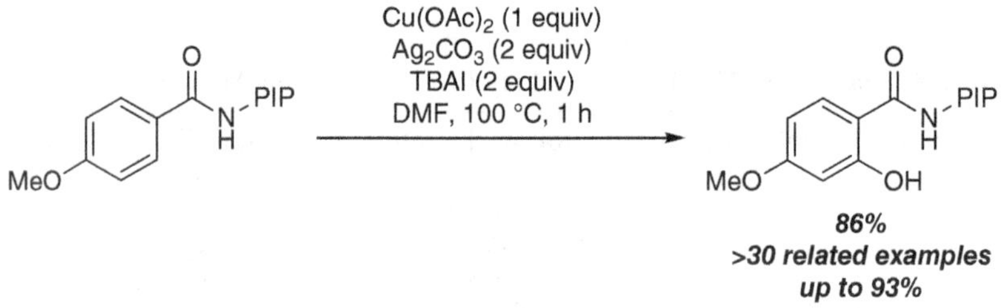

SCHEME 2.48   Copper-promoted C—H hydroxylation [65].

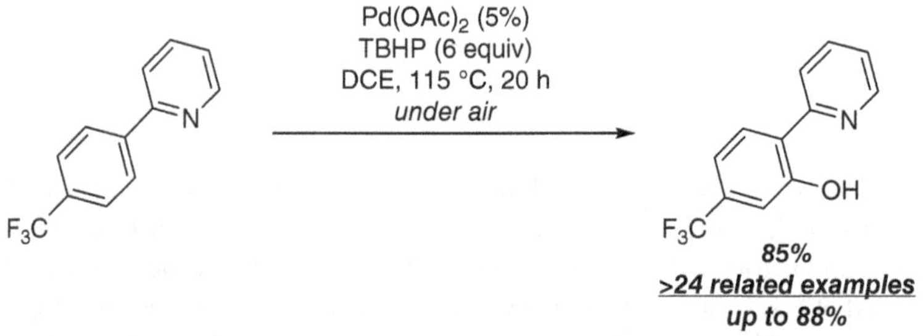

SCHEME 2.49   Palladium-catalyzed C—H hydroxylation of 2-arylpyridines [66].

The copper-promoted C—H hydroxylation of arenes has also been achieved (Scheme 2.48) [65]. Using copper(II) acetate and silver(I) carbonate as the catalyst system, a range of arenes were hydroxylated in moderate to excellent yield (up to 93%). In general, both electron-rich and electron-poor arenes were well tolerated. An assortment of arenes was functionalized using this approach.

In addition to the use of ketones, amides, carbamates, and other directing groups, pyridine has recently been used to direct an ortho C—H hydroxylation (Scheme 2.49) [66]. The hydroxylation chemistry was highly selective and catalyzed by a common palladium compound. No supporting ligand was needed for a successful reaction, although both the starting material and the product could serve as a supporting ligand for the metal center. One of the most attractive aspects of this approach was the ability to carry out the reaction under an atmosphere of air.

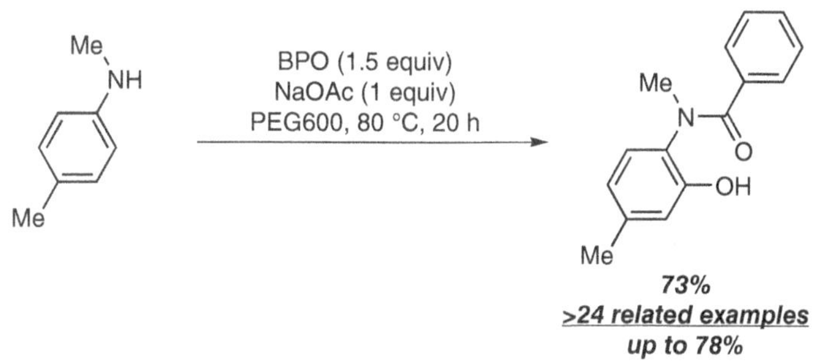

SCHEME 2.50   Benzoyl peroxide-promoted C—H hydroxylation [67].

A metal-free approach to the C—H hydroxylation of arenes has been developed using benzoyl peroxide as the promoter (Scheme 2.50) [67]. The authors initially began their investigation using palladium compounds to promote the reaction, but discovered that the palladium was not needed through a series of control reactions. Using benzoyl peroxide along with sodium acetate as an additive, the C—H hydroxylation of an assortment of *N*-alkyl anilines was achieved in polyethylene glycol. The reaction was also chemoselective for the arene bearing the nitrogen, and little to no hydroxylation was observed on the other arene group. For meta-substituted systems, a mixture of isomers was formed with the least hindered site preferentially functionalized for most substrates.

The generation of ethers through the formation of oxygen–carbon(sp^2) bonds (Williamson approach) is rare due to the sluggish reactivity of vinyl and aryl halides in nucleophilic substitution. Furthermore, while nucleophilic substitution chemistry tends to work well for primary systems, significant amounts of elimination products are often observed with secondary alkyl halides, and tertiary substrates are typically unresponsive. As a result, a number of transition metal-catalyzed routes to the formation of ethers through the formation of oxygen–carbon(sp^2) bonds have been developed.

Copper complexes have received considerable interest in the synthetic community for their ability to promote the formation of oxygen–carbon(sp^2) bonds [68–74]. As a result, a host of copper species and conditions have been developed for the formation of alkyl aryl ethers. A practical approach to these compounds entailed the generation of the catalytically active species in solution from the combination of common copper salts with chelating ligands (Example 2.9) [75]. Using this catalyst system, an assortment of primary and secondary alcohols were successfully coupled with aryl iodides to generate the alkyl aryl ethers. The chemistry was tolerant of a wide range of electronically diverse aryl iodides, and even an unprotected iodoaniline was converted into the ether, albeit in reduced yield. The operational simplicity was very high as all the reagents, additives, and catalysts could be loaded into the reaction vial in air. Indeed the oxygen–carbon bond-forming reaction could be carried out under air in sealed vials. No glovebox or nitrogen line was needed for this chemistry.

Copper compounds also promoted the coupling between aryltrifluoroborate salts and alcohols (Scheme 2.51) [76]. A vast array of primary, secondary, benzyl, and allylic alcohols were arylated using a catalyst system comprised of copper acetate and DMAP. Heterocyclic substrates were also well tolerated by this approach. The reactions were

Example 2.9   Copper-Catalyzed Synthesis of Alkyl Aryl Ethers in Air [75].

Before starting this synthesis, carefully follow the safety instructions listed in Example 2.1.

Caution! Ethyl acetate and hexane are highly flammable.
This reaction was carried out under an atmosphere of air in a sealed tube. A test tube was charged with CuI (20 mg, 0.10 mmol, 0.10 equiv), 1,10-phenanthroline (36 mg, 0.20 mmol, 0.20 equiv), Cs$_2$CO$_3$ (652 mg, 2.0 mmol, 2.0 equiv), 3,5-dimethyliodobenzene (144 µL, 1.0 mmol, 1.0 equiv), and *n*-heptanol (1.0 mL, 7.1 mmol, 7.1 equiv). The test tube was sealed, and the reaction mixture was stirred at 110 °C for 18 h. The resulting suspension was cooled to room temperature and filtered through a 0.5 × 1 cm pad of silica gel, eluting with ethyl acetate. The filtrate was concentrated in vacuo. Purification of the residue by flash chromatography on silica gel (2 × 20 cm; hexane) provided 178 mg (81% yield) of the title compound as a colorless oil.

Adapted with permission from M. Wolter, G. Nordmann, G. E. Job, and S. L. Buchwald, *Org. Lett.*, **2002**, *4*, 973–976. *Copyright (2002) American Chemical Society.*

carried out under mild conditions under an atmosphere of oxygen. The authors noted that air could be used as the oxidant; however, the yields were higher with pure oxygen. Work by Molander outlined the O-arylation of several amino acids bearing unprotected alcohols (Scheme 2.52 and Example 2.10) [77]. Serine contains a primary alcohol in the pendant group, while threonine bears a secondary alcohol. After suitable protection of the other functional groups, both of these alcohols were successfully converted into alkyl aryl ethers through a copper-catalyzed cross-coupling reaction with aryltrifluoroborate salts. After screening a range of catalysts, bases, and additives, the authors determined that the catalyst system comprised of copper(II) acetate, and DMAP was the most efficient for the chemistry. Curiously, copper sulfate was completely inactive, and using DBU as the base resulted in significant homocoupling. Using the optimized system, a range of a functionalized aryltrifluoroborate salts were coupled with protected serine and threonine. There was a bit of substrate-specific reactivity, and electron-donating groups tended to afford higher yields of the coupling products. Aryltrifluoroborates bearing electron-deficient aryl groups afforded the coupling products, albeit in reduced yields. In addition to the excellent yields, several of the attractive aspects of this chemistry included the ability to carry out the reactions under an atmosphere of air in an open flask at room temperature.

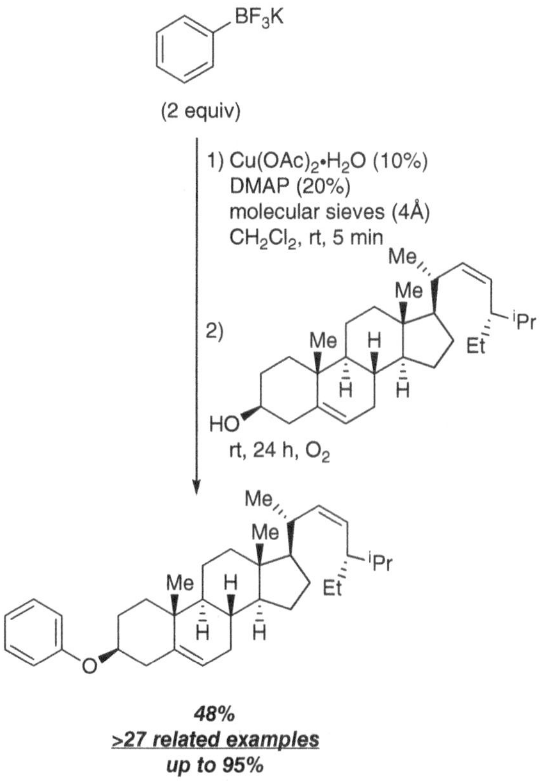

SCHEME 2.51   A-ring functionalization using copper catalysis [76].

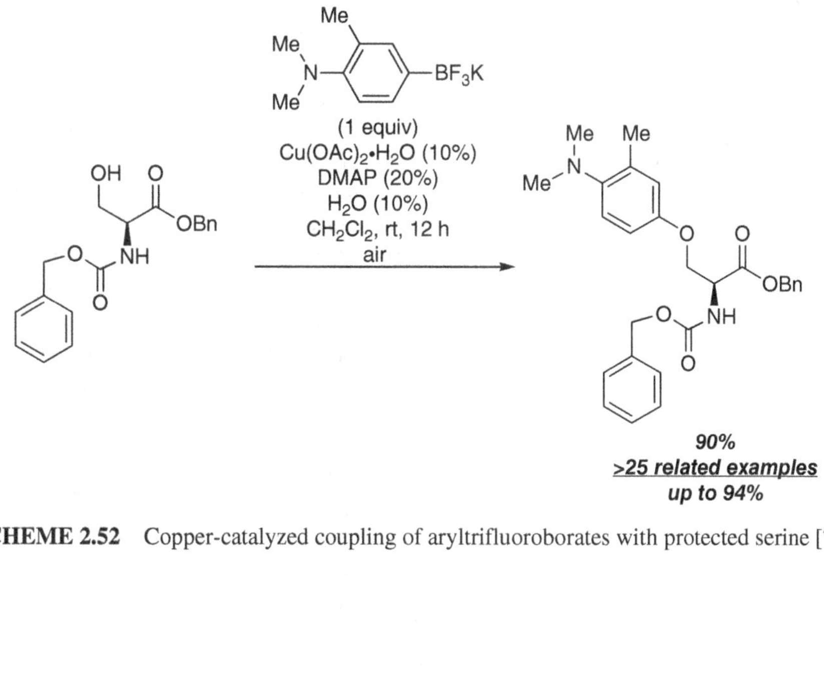

SCHEME 2.52   Copper-catalyzed coupling of aryltrifluoroborates with protected serine [77].

Example 2.10   Preparation of Alkyl Aryl Ethers from Protected Amino Acids [77].

Before starting this synthesis, carefully follow the safety instructions listed in Example 2.1.

(1 equiv)

83%

Caution! Ethyl acetate and hexane are highly flammable.

To the aryl/heteroarylboronic acid (1 mmol), Pg–Ser–OR' (or Pg–Thr–OR', 1 mmol) and DMAP (24 mg, 20 mol %) was added $Cu(OAc)_2 \cdot H_2O$ (20 mg, 10 mol %) in a glass Biotage vial equipped with a magnetic stirring bar. CH_2Cl_2 (3.5 mL, 0.275 M) and water (0.1 mL) were added, and the reaction was stirred overnight at rt (monitored by TLC). The suspension was then concentrated under reduced pressure and purified by Combiflash column chromatography using EtOAc and hexanes to obtain the desired product. The product was characterized by spectroscopic techniques. The target compound was isolated as a yellow oil (83%).

Adapted with permission from M. El Khatib and G. A. Molander, *Org. Lett.*, **2014**, *16*, 4944–4947. *Copyright (2014) American Chemical Society.*

One of the alcohols that has been challenging to involve in copper-catalyzed cross-coupling reactions is tetrahydrofurfuryl alcohol. Under classic Ulmann conditions, this alcohol tended to undergo secondary reactions leading to an intractable mixture of products. In order to develop a more general approach to the synthesis of alkyl aryl ethers containing this fragment, a coupling reaction starting from aryl halides was devised (Scheme 2.53 and Example 2.11) [78]. The catalyst used for this reaction was an octanuclear copper complex consisting of eight copper atoms in a cube with a chloride in the middle of the cube. This copper complex was unusually stable to oxygen and moisture and could be stored in the light and in the presence of air indefinitely. The catalyst was also quite soluble in a wide range of alcohols, and while the specific benefit of having all of the copper atoms sequestered in the organosoluble cube was not elucidated, it was an active catalyst for the coupling of aryl halides with tetrahydrofurfuryl alcohol. The need to generate a noncommercially available catalyst for the chemistry was an added step, but the synthesis of this catalyst was simple and high yielding. Aryl iodides were significantly more reactive than aryl bromides and chlorides, and a wide range of electronically varied substrates were successfully used in this chemistry. The authors also carried out an investigation of conventional versus microwave heating and found that both methods generated excellent yields of the alkyl aryl ethers. An attractive aspect to this chemistry was the observation that no glove box or nitrogen line was needed for this chemistry as the reactions could be carried out under an atmosphere of air. This chemistry is an attractive approach to the synthesis of these compounds due to the use of an air and light stable copper catalyst as well as the high level of operational simplicity.

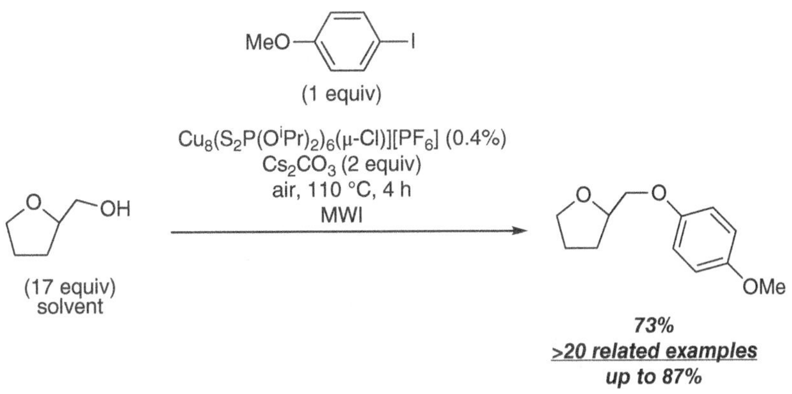

SCHEME 2.53   Microwave-assisted synthesis of alkyl aryl ethers [78].

Example 2.11   Microwave-Assisted Synthesis of Tetrahydrofurfuryl Aryl Ethers [78].

Before starting this synthesis, carefully follow the safety instructions listed in Example 2.1.

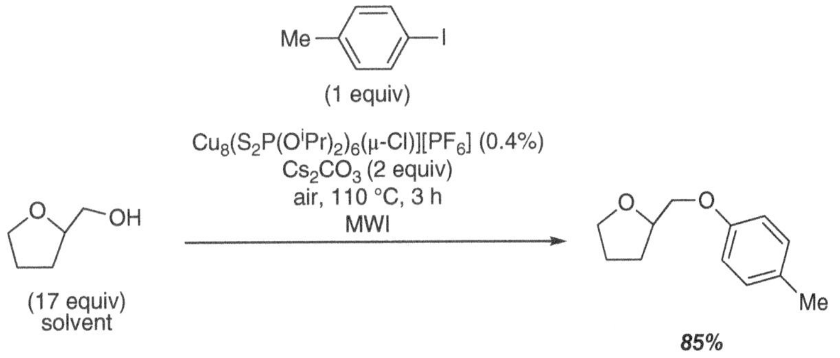

This reaction was carried out under an atmosphere of air. A 10-mL reactor tube was charged (in air) with $[Cu_8(S_2P(O^iPr)_2)_6(\mu\text{-Cl})][PF_6]$ (0.008 g, 4.1 µmol), 4-iodotoluene (0.20 g, 0.92 mmol), Cs_2CO_3 (0.60 g, 1.84 mmol), tetrahydrofurfuryl alcohol (1.50 mL, 15.5 mmol), and a magnetic stirring bar. The reactor tube was crimped closed (under air), placed in the microwave reactor, and connected to the monitoring system. The reaction was heated to 110 °C for 3 h. The initial microwave power level setting was 50 W. This power level was maintained until the desired temperature was reached. For the remainder of the experiment, a power level of 9 W (pressure ~ 19 psi) was used to maintain the temperature. The vessel was cooled to room temperature, filtered through a small plug of silica gel, and dried under vacuum. The excess tetrahydrofurfuryl alcohol was removed under vacuum with gentle heating. Purification of the residue by column chromatography (silica, hexane/ethyl acetate 3 : 1) afforded 0.15 g (85%) of the title compound as a dull yellow oil.

The chemistry was repeated under an atmosphere of nitrogen along with the addition of crushed molecular sieves (0.5 g) to afford 0.155 g (88%) of the ether after irradiating for 2 h purification by column chromatography.

Caution: for reactions involving low boiling alcohols, a significant amount of pressure builds up in the microwave reactor vial. Cooling the vial to room temperature before opening the flask is highly recommended.

Adapted with permission from G. F. Manbeck, A. J. Lipman, R. A. Stockland, Jr., A. L. Freidl, A. F. Hasler, J. J. Stone, I. A. Guzei, *J. Org. Chem.*, **2005**, *70*, 244–250. *Copyright (2005) American Chemical Society.*

The copper-catalyzed coupling of glycolic acids with aryl iodides has been accomplished using common copper catalysts (Scheme 2.54) [79]. This was a chemoselective transformation that resulted in the preferential formation of the alkyl aryl ether instead of the ester. The carboxylic acid was retained in the product and was able to be further transformed in subsequent reactions. The electronic composition of the aryl iodide did not significantly influence the outcome of the reaction, and a range of electron-rich and electron-poor arenes were well tolerated by the chemistry.

The copper-catalyzed decarboxylative etherification of carboxylate salts is an interesting approach to the synthesis of ethers (Scheme 2.55) [80]. As mentioned in previous sections, carboxylic acids are attractive entry points for syntheses due to the vast number of examples that can be readily obtained or synthesized. Additionally, they are often quite stable and easily handled. The catalyst system developed by the authors consisted of the copper(II) acetate along with silver(I) carbonate with elemental oxygen as an oxidant. After screening a range of oxygen nucleophiles, the authors determined that tetraalkyl orthosilicates were the most effective. It was noteworthy that changing to triphenylboronate enabled the preparation of a diaryl ether. The chemistry was quite tolerant to the presence of electron-donating and electron-withdrawing groups in the ortho-position of the carboxylate salts.

The copper-catalyzed conversion of arenes into ethers through the direct C—H activation of arenes is an attractive approach to the synthesis of these compounds since prefunctionalization of the arenes would not be required. However, developing a general procedure for this reaction has been quite challenging. One successful version of this reaction entailed the use of directing groups to promote the oxygen–carbon(sp^2) bond formation at a specific position. In the vast majority of cases, this position was adjacent to the directing group [81]. Using this approach, a copper-catalyzed dehydrogenative alkoxylation of arenes has been developed (Scheme 2.56) [82]. The catalyst system was comprised of a catalytic amount of

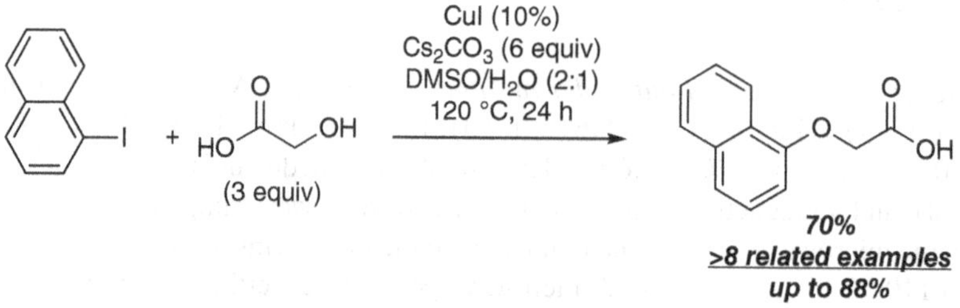

SCHEME 2.54   Chemoselective copper-catalyzed formation of alkyl aryl ethers [79].

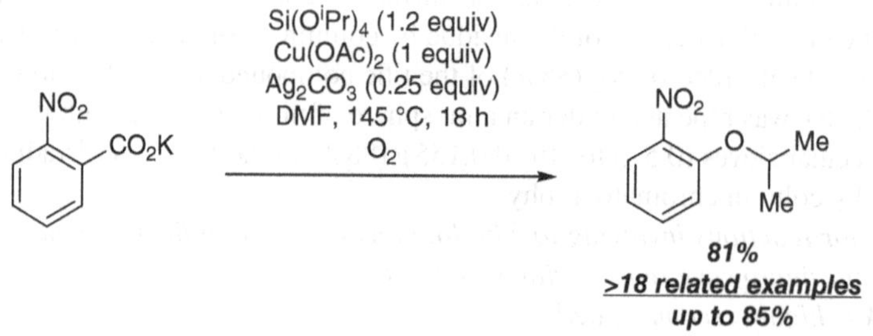

SCHEME 2.55   Decarboxylative etherification of carboxylate salts [80].

copper(II) acetate and an excess of a silver salt under an atmosphere of oxygen. After screening a range of silver salts, the authors found that silver triflate was the most effective for the synthesis of the alkyl aryl ethers. It was noteworthy that an atmosphere of oxygen was needed for a successful synthesis, even if an elevated amount of the silver salt was used. The chemistry was tolerant to a wide range of electronically and sterically diverse substrates and even a heteroaromatic example was described.

The palladium-catalyzed coupling of aryl halides with alkoxides is a convenient approach to the synthesis of ethers and has been carried out with a host of catalysts, aryl halides, and alkoxides [83–86]. For most methods, palladium(II) acetate or Pd$_2$dba$_3$ served as the palladium source, and a vast array of monodentate and bidentate phosphines have been used as supporting ligands. It should be noted that the same phosphine ligand was rarely the most effective for all oxygen–carbon(sp^2) bond-forming reactions. Often, each set of substrates needed to be evaluated for activity using a range of palladium sources and phosphine ligands to determine an effective system. For the synthesis of alkyl aryl ethers, one example of this chemistry used bulky dialkylbiarylphosphine ligands along with palladium(II) acetate to promote the reaction (Scheme 2.57) [86]. The reactions were carried out under solvent-free conditions with heating to generate moderate to excellent yields of the ethers. The functional group tolerance was moderate with the highest yields obtained with electron-neutral or electron-poor aryl halides. Electron-rich system such as anisoles tended to generate secondary products and afforded lower yields of the

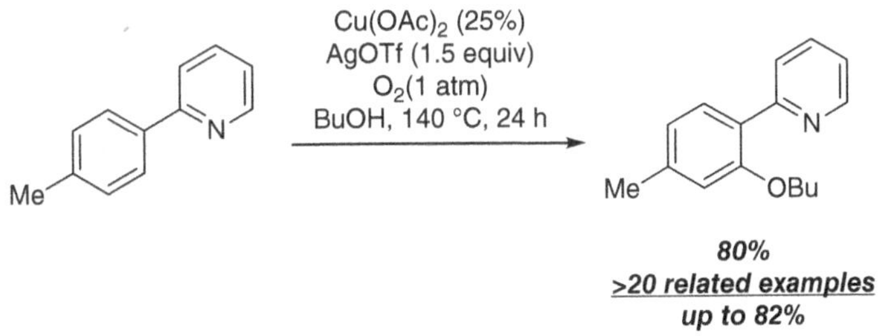

SCHEME 2.56   Copper-catalyzed dehydrogenative alkoxylation of 2-arylpyridines [82].

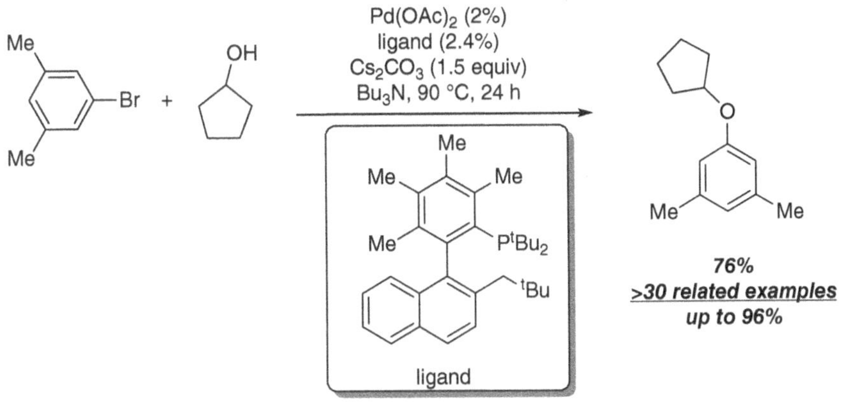

SCHEME 2.57   Palladium-catalyzed preparation of alkyl aryl ethers [86].

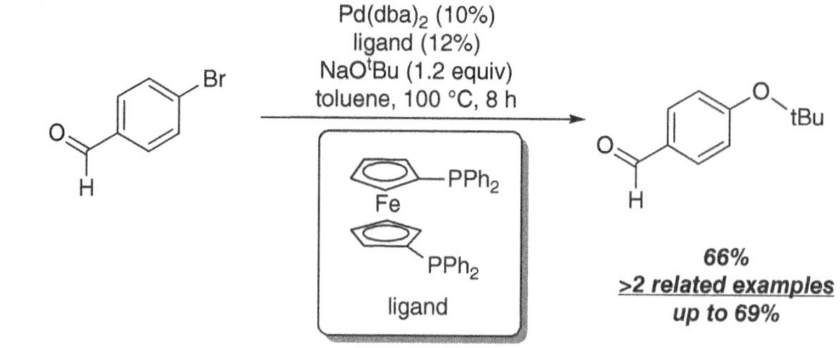

SCHEME 2.58   Synthesis of alkyl aryl ethers using a dppf-supported palladium catalyst [85].

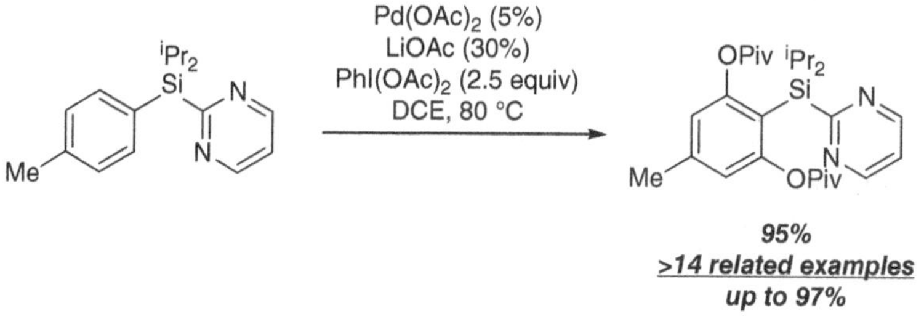

SCHEME 2.59   Palladium-catalyzed double pivaloxylation of arenes [88].

unsymmetrical ethers. In related work, Hartwig outlined the coupling of alkoxides with aryl halides using a catalyst system comprised of Pd(dba)$_2$ and dppf (Scheme 2.58) [85]. The palladium-catalyzed ortho-alkoxylation of arylnitrile has also been described [87]. Following these reports, a vast array of modifications to the core system has been described, and a few representative examples will be outlined in the following sections.

As mentioned earlier, the direct formation of alkyl aryl ethers through C—H activation is an attractive process for the formation of alkyl aryl ethers since the prefunctionalization of the substrates was not required. In addition to the reactions using copper catalysts, palladium complexes are also quite active in this reaction, and a range of catalysts and conditions have been developed to promote the formation of oxygen–carbon(sp^2) bonds [81]. While a wide range of directing groups have been used to promote the addition at a specific site, these groups are often tightly bound to the substrate and are challenging to remove. As a result, there is considerable interest in the development of successful C—H activation processes that employ labile directing groups. One example of this chemistry used 2-pyrimidyldiisopropylsilyl as the directing group (Scheme 2.59) [88]. Using palladium(II) acetate as the catalyst for the reaction, a range of oxidants and additives were screened for activity toward the C—H activation reaction. Using the optimized system, an array of arenes bearing the 2-pyrimidyldiisopropylsilyl directing group as well as a host of electron-donating and electron-withdrawing groups were converted into alkyl aryl ethers in moderate to excellent yields.

Many of the practical routes to the synthesis of diaryl ethers are focused on the use of transition metal complexes as catalysts for cross-coupling reactions [68–70]. Copper has played a central role in this chemistry and has been used extensively in the synthesis of diaryl ethers.

The Ulmann coupling is perhaps the most well known of the copper-catalyzed reactions [70, 89]. The classic Ulmann coupling reaction uses copper bronze to promote the coupling of phenols to generate the diaryl ethers. Although this reaction has been used with great success for the formation of oxygen–carbon(sp^2) bonds, it suffers from harsh conditions and limited substrate scope and requires a stoichiometric or greater amount of copper to be added. While a number of versions were periodically developed through the 1900s, significant advancements were made in the late 1990s [68, 71–74].

Elegant work by Evans demonstrated that copper compounds were effective promoters for the coupling of phenols with arylboronic acids to generate diaryl ethers (Scheme 2.60) [71]. A number of copper catalysts were screened for activity, and copper(II) acetate was found to be the most effective for the coupling reaction. One of the challenges encountered with this and related chemistry was the propensity of the arylboronic acid to react with itself and generate a triphenylboroxine accompanied by the release of water. As water could interfere with the desired coupling reaction, the authors determined that the addition of molecular sieves sequestered trace moisture and promoted clean coupling chemistry. One of the attractive aspects of this chemistry was the ability to retain aryl halides and selectively react at the phenolic alcohol without significant side reactions due to coupling between the boronic acid and the aryl iodide. In related work, the authors developed a similar copper-promoted intramolecular oxygen–carbon(sp^2) bond-forming reaction for the synthesis of macrocycles (Scheme 2.61) [90].

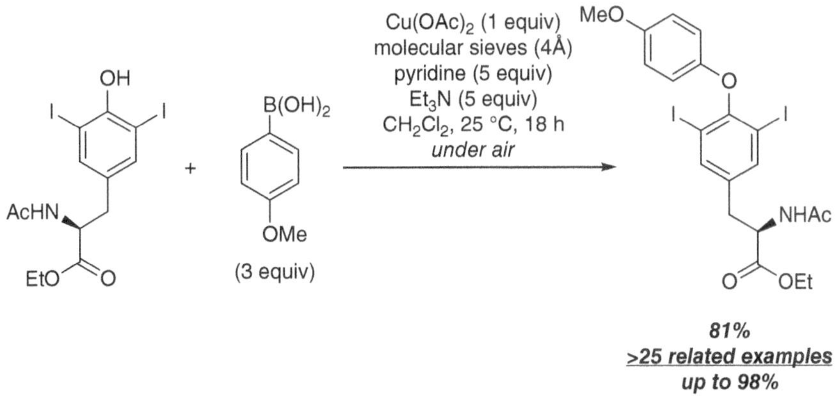

SCHEME 2.60 Copper-catalyzed coupling between phenols and arylboronic acids [71].

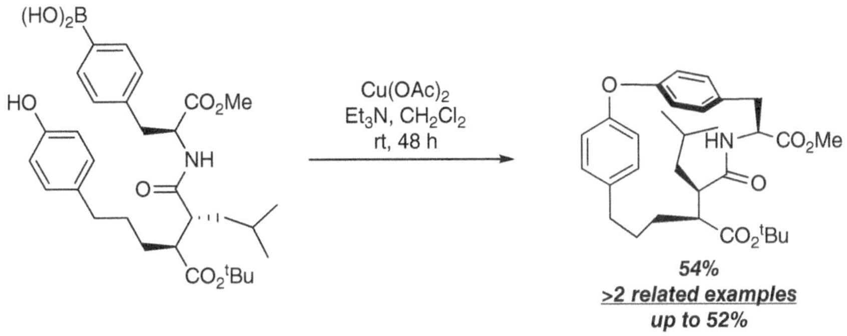

SCHEME 2.61 Intramolecular oxygen–carbon(sp^2) bond formation through cross-coupling [90].

Buchwald reported the use of a copper triflate complex in the cross-coupling of phenols with aryl halides (Scheme 2.62) [74]. This chemistry generated moderate to excellent yields of the diaryl ethers bearing a range of functional groups. Aryl iodides were more reactive than the aryl bromides, and substrates bearing a single ortho-substituent were tolerated using this approach. Although considerable advances in the synthesis of ethers have been made using copper salts, it should be noted that the use of copper powder has not completely disappeared since it was used to generate a crowded diaryl ether (Scheme 2.63) [91].

Venkataraman used a well-defined coordination complex as the catalyst for the cross-coupling of aryl halides with phenols. Cesium carbonate was the base for the reaction (Scheme 2.64) [92]. The synthesis of the catalyst was trivial and consisted of simply

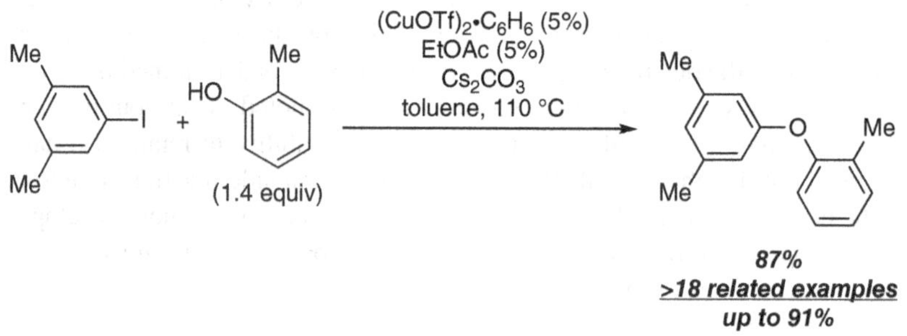

SCHEME 2.62 Copper-catalyzed synthesis of diaryl ethers through cross-coupling [74].

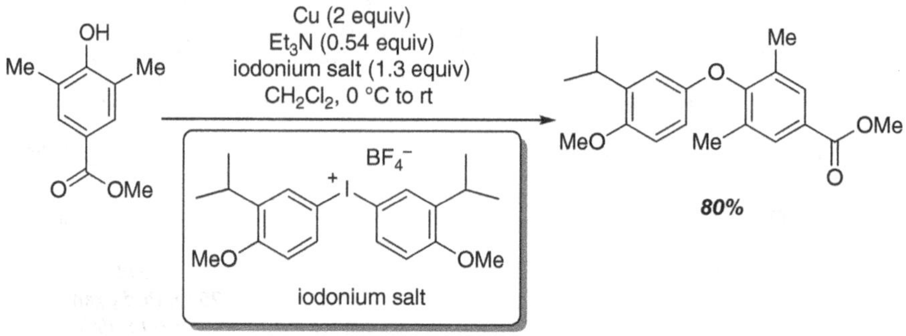

SCHEME 2.63 Copper-catalyzed coupling of iodonium salts with phenols [91].

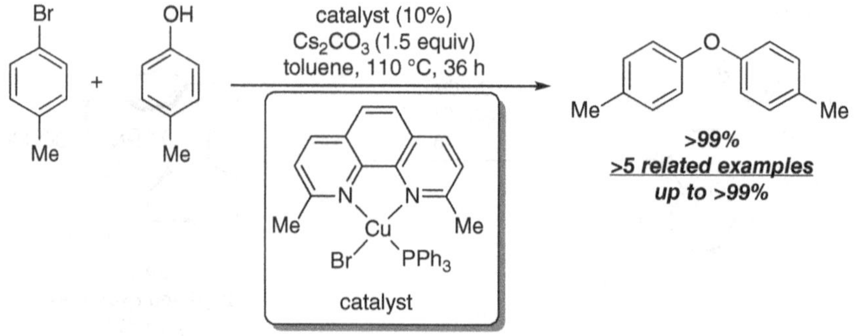

SCHEME 2.64 Synthesis of diaryl ethers using a well-defined copper complex [92].

dissolving CuBr(PPh$_3$)$_3$ and adding neocuproine. Using this discrete species as the catalyst, a variety of phenols were coupled with aryl bromides. The lowest yields were obtained with ortho-substituted aryl halides. An attractive aspect of this work was the observation that both electron-deficient and electron-rich arenes were successfully used to generate the diaryl ethers.

An organocatalytic approach to the synthesis of benzoxazoles using 4-iodonitrobenzene as the catalyst has been reported (Scheme 2.65) [93]. The chemistry was carried out under very mild conditions and afforded excellent yields of the heterocycles. Using Oxone as the oxidant, a vast array of benzamides bearing an assortment of preexisting functional groups including electron-donating and electron-withdrawing substituents were converted into benzoxazoles.

The synthesis of a group of benzofuranones has been accomplished through an intramolecular C—H activation/cyclization sequence catalyzed by palladium(II) acetate (Scheme 2.66) [94]. An iodonium salt was used as the oxidant, and a protected amino acid derivative was used as an additive in the reaction. The authors proposed that the mechanism of the reaction included a Pd(II)/Pd(IV) cycle and that the oxidant was needed to oxidize the palladium during the catalysis.

An interesting approach to the preparation of macrocycles has been accomplished through a series of oxygen–carbon(sp^2) bond-forming reactions (Scheme 2.67) [95]. While many of the reactions described previously needed a metal catalyst to promote the formation of the ethers, this chemistry was successful without the addition of such a catalyst. Treatment of 1,4-dihydroxybenzenes with 3,6-dichlorotetrazine in acetonitrile generated the macrocycles in moderate to good yield. A range of functionalized 1,4-dihydroxybenzenes

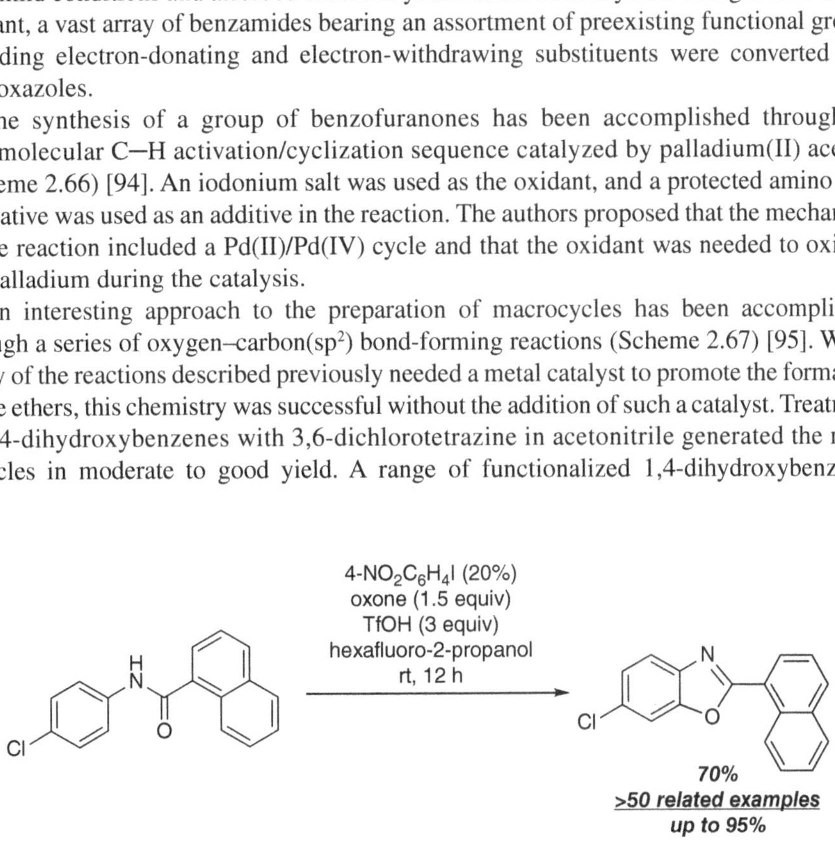

SCHEME 2.65   Synthesis of benzoxazoles through an organocatalytic approach [93].

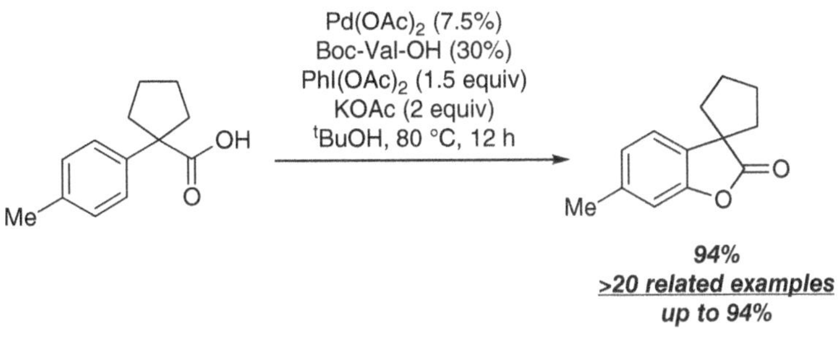

SCHEME 2.66   Palladium-catalyzed synthesis of benzofuranones [94].

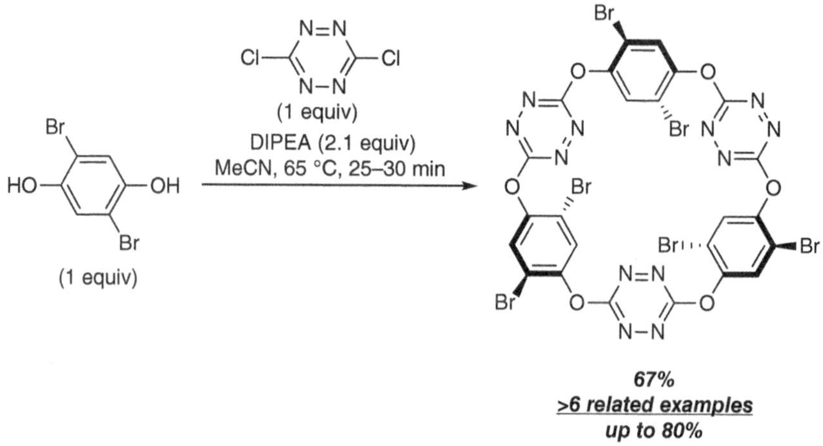

67%
>6 related examples
up to 80%

SCHEME 2.67   Synthesis of macrocycles through a series of S$_N$Ar reactions [95].

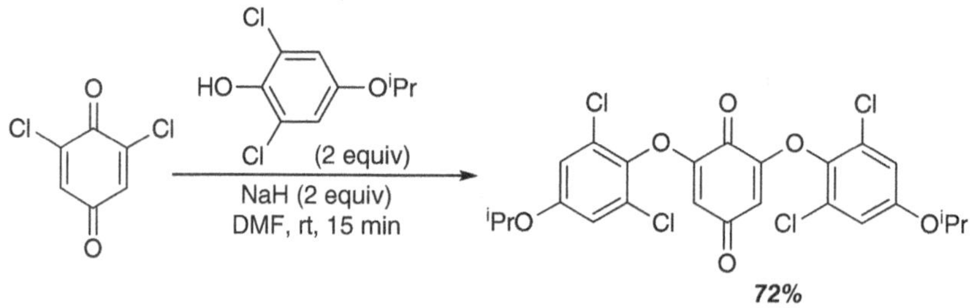

72%

SCHEME 2.68   Preparation of a crowded ether [96].

including halogenated derivatives was well tolerated by the chemistry. In related work, the synthesis of very bulky ethers was accomplished (Schemes 2.68 and 2.69) [96].

The synthesis of heterocycles through a series of direct annulation reactions has been achieved using a palladium-catalyzed and copper-assisted process (Scheme 2.70) [97]. While palladium(II) acetate generated moderate amounts of a related heterocycle (65%), Pd$_2$dba$_3$ was a much more effective catalyst. The chemistry required elevated temperatures, and a range of preexisting functional groups including alkyl, alkoxy, and halogens were well tolerated.

The synthesis of functionalized furan derivatives was achieved through a two-step, one-pot reaction (Scheme 2.71) [98]. The first step in the process consisted of preparing 1,3-dienyl ethers through a ruthenium-catalyzed process. The second step generated the furan derivatives through a copper-promoted cyclization. A range of alkynes bearing electron-donating groups and electron-withdrawing groups were screened in this study, and both groups were transformed into the furan derivatives. The second step was carried out while being aerated.

While a considerable amount of effort has been devoted to developing copper-mediated reactions that mimic palladium-catalyzed processes due to the reduced cost of copper versus palladium, their reactivity can be markedly different in certain reactions and give rise to a different product distribution. Such a difference has been observed in the reaction of carboxylic acids with arylboronic acids [99]. When palladium catalysts were

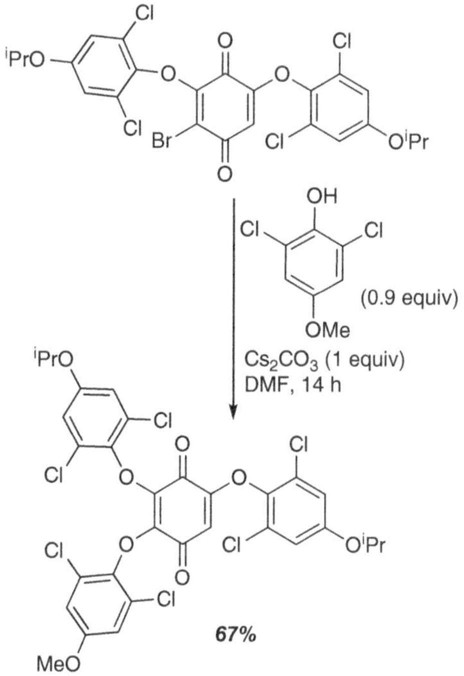

SCHEME 2.69 Synthesis of a crowded ether [96].

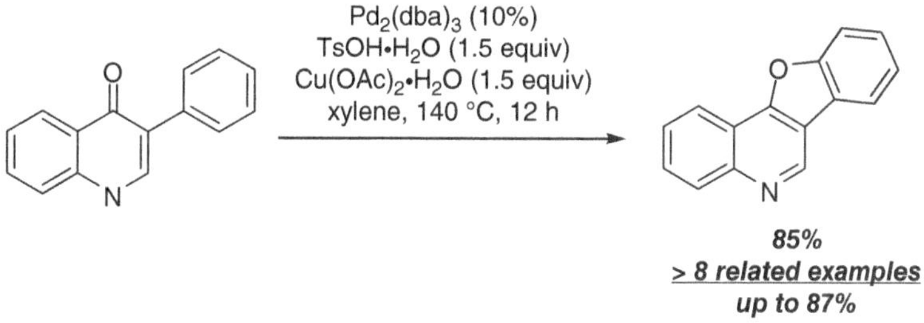

SCHEME 2.70 Palladium-catalyzed synthesis of heterocycles through C—H activation [97].

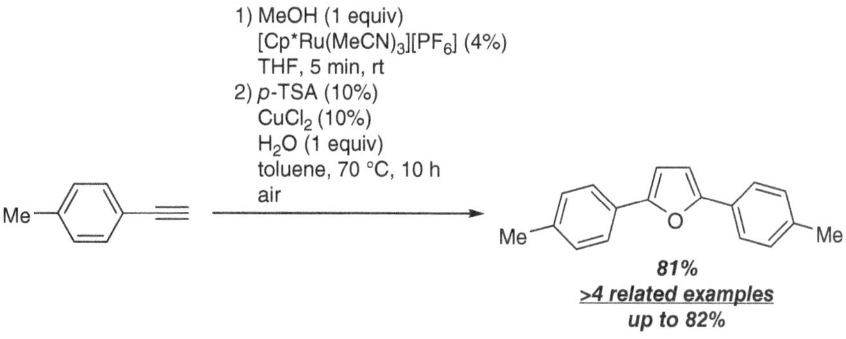

SCHEME 2.71 Preparation of 2,5-disubstituted furans [98].

used, a decarboxylative process occurred to generate carbon–carbon bonds along with the concomitant formation of CO_2. When analogous reactions were carried out using copper catalysts, a nondecarboxylative coupling reaction occurred to afford a series of aryl esters (Scheme 2.72). The chemistry was remarkably tolerant to bulky groups on the carboxylic acid, and a range of aryl, vinyl, and heteroaromatic carboxylic acids were successfully converted into esters using this approach. The authors did observe an electronic effect, and electron-rich aryl carboxylic acids afforded significantly higher yields of the coupling products. In related work, the copper-catalyzed synthesis of aryl esters was accomplished under an atmosphere of air using urea as an additive (Example 2.12) [100].

The palladium-catalyzed and copper-promoted synthesis of polycyclic aromatic hydrocarbons has been achieved using an oxidative carbonylation reaction (Scheme 2.73) [97]. While palladium was proposed to be the catalyst in these reactions, the copper(II)

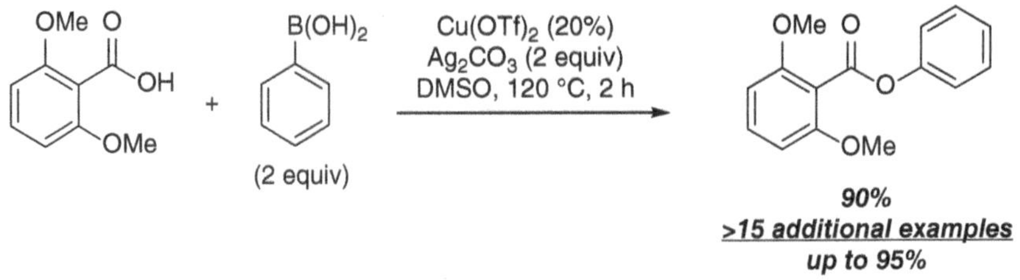

SCHEME 2.72 Copper-catalyzed cross-coupling between arylboronic acids and carboxylic acids [99].

Example 2.12 Synthesis of Aryl Esters Using Copper Catalysts in Air [100].

Before starting this synthesis, carefully follow the safety instructions listed in Example 2.1.

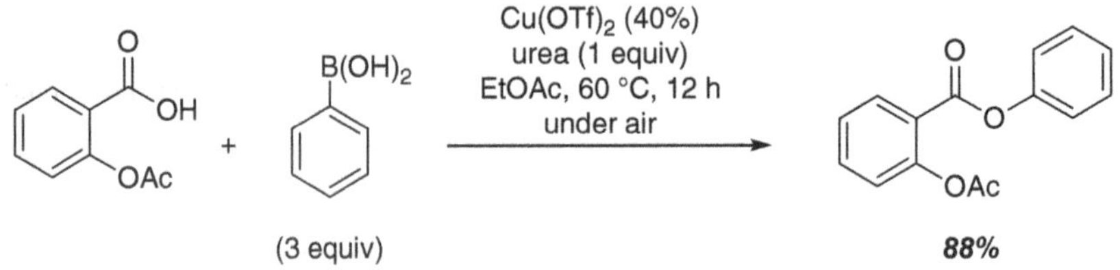

Caution! Ethyl acetate is highly flammable.

Under air, a reaction tube was charged with carboxylic acid (0.2 mmol), arylboronic acid (0.6 mmol), Cu(OTf)$_2$ (29 mg, 40 mol%), urea (0.2 mmol), and ethyl acetate (2 mL). The mixture was stirred at 60 °C. After the mixture was kept stirring for 12 or 24 h, the solvent was evaporated under reduced pressure, and the residue was purified by flash column chromatography on silica gel to give the product. The title compound was isolated as a white solid (88%).

Adapted with permission from L. Zhang, G. Zhang, M. Zhang, and J. Cheng, J. Org. Chem., **2012,** *75,* 7472–7474. *Copyright (2012) American Chemical Society.*

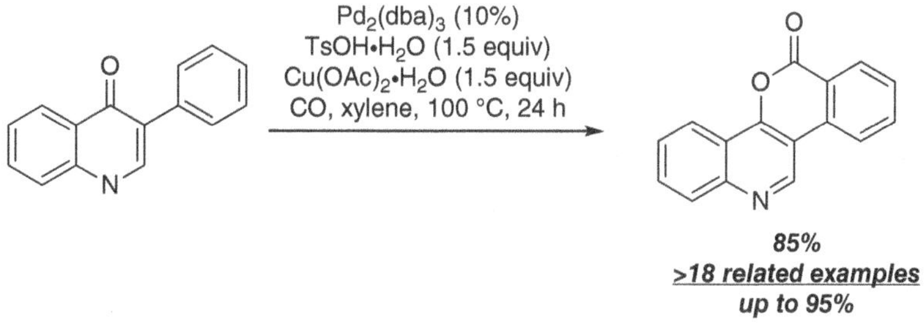

SCHEME 2.73   Palladium-catalyzed copper-promoted oxidative carbonylation [97].

species was suggested to function as an oxidant. Although a range of alternative oxidizing agents including oxygen, benzoquinone, and MnO_2 were screened, none of them were as effective as copper(II) acetate. It should be noted that benzoquinone did generate a moderate yield of the desired ester (65%). The functional group tolerance of this chemistry was moderate, and a range of substrates bearing halogens, ethers, alkyl, trifluoromethyl, as well as fused heterocycles were all successfully converted into the desired cyclic esters. In the absence of carbon monoxide, these starting materials undergo annulation reactions to generate benzofuro[3,2-c]quinolines.

2.2.1   Troubleshooting the Preparation of Phenols, Aryl Ethers, and Related Compounds

While it is not possible to predict all of the challenges and difficulties that will arise during the preparation of these compounds, the following suggestions will hopefully provide some direction for common issues and reaction requirements.

Problem/Issue/Goal	Possible Solution/Suggestion
Do not have access to a glove box or vacuum line and need to make a phenol	There are several approaches that do not require the use of a glove box or nitrogen line. If an arylboronic acid can be obtained, several metal-catalyzed [44, 45] and metal-free [3, 41] approaches are known to effect this transformation in air and water
The starting material can easily be converted into an aryltrifluoroborate. Can ArBF₃K species be converted into phenols?	This reaction works quite well. Treatment of ArBF₃K with Oxone in acetone under air affords excellent yields of the phenol [3]
The substrate has a very crowded boronic acid. What would be a reasonable method for converting it into a phenol?	The metal-free *N*-oxide-promoted conversion of arylboronic acids into phenols has demonstrated considerable tolerance to bulky groups [41]
The substrate has a very crowded aryl iodide. Can it be directly converted into a phenol without going through the boronic acid?	The copper-catalyzed coupling of aryl iodides with hydroxide sources [47, 79] would be a reasonable entry point to this chemistry
Need to use pure water as the solvent in the synthesis of a phenol	A copper-catalyzed process converts aryl iodides and bromides into phenols in pure water using TBAB as a phase-transfer catalyst [48]

(Continued)

(Continued)

Problem/Issue/Goal	Possible Solution/Suggestion
Despite best efforts, water is apparently contaminating oxygen–carbon bond reactions using arylboronic acids	This could be due to the reaction of the boronic acid with itself to generate a triarylboroxine along with the release of water. If this is problematic for the desired chemistry, the addition of molecular sieves has been shown to be successful at removing the water from related reactions [71]
Need a practical approach to the synthesis of alkyl aryl ethers through cross-coupling	A copper-catalyzed process is perhaps the most operationally simple since several versions can be carried out under air [75]
Can an alkyl aryl ether be generated in a microwave-assisted reaction?	This has been achieved using a copper catalyst [78]
Can an alkyl aryl ether be prepared through a decarboxylative etherification process?	A copper catalyst has been shown to be active in this reaction [80]. The authors used carboxylate salts as the starting materials in this work
The starting material can be converted into an arylboron species fairly easily. Will arylboron species react with alcohols in a cross-coupling reaction to generate ethers?	This has been established for several organoboron compounds. Perhaps the most attractive are the aryltrifluoroborate salts due to their stability [77]
What is a good directing group that can be easily removed after a C—H alkoxylation reaction?	The 2-pyrimidyldiisopropylsilyl group has been shown to be an effective directing group for the C—H functionalization reaction and is readily removed after the formation of the ether [88]
Need a good method for the preparation of a diaryl ether	Both palladium and copper catalysts are known to promote this reaction [68–70]
Need a metal-free preparation of a diaryl ether	This is $S_N Ar$ chemistry and will be successful with substrates that are amenable to this reaction. An interesting version of this chemistry has been reported for the preparation of macrocycles [95]
The starting phenol has groups at the 2 and 6 positions, and it needs to be converted into a diaryl ether. What is a good approach for preparing crowded diaryl ethers?	There are several approaches that have successful generated very crowded diaryl ethers. A copper-catalyzed method would be a reasonable approach [91]
Can aryl esters be generated from a cross-coupling approach, or will a decarboxylation process occur?	That depends upon the metal and supporting ligands used in the catalysis. For most palladium complexes, decarboxylation tends to occur; however, a copper complex has been shown to promote the coupling instead of decarboxylation [99, 100]

2.3 PREPARATION OF VINYL ETHERS AND RELATED COMPOUNDS THROUGH THE FORMATION OF OXYGEN–CARBON(sp²) BONDS

Industrially, large quantities of vinyl ethers can be prepared following Reppe's ethynylation reaction that involved reacting acetylene gas with alcohols [101]. Due to the challenges and hazards associated with the experiment as well as the required handling of acetylene gas (under pressure) alternative approaches to the synthesis of vinyl ethers are typically used in small laboratory settings. One such approach used common iridium complexes to catalyze the reaction between vinyl acetate and alcohols (Scheme 2.74) [102]. During the course of the investigation, the authors screened a variety of iridium, platinum, rhodium, and palladium complexes for activity and determined that [IrCl(cod)]$_2$ was the most active catalyst under the reaction conditions. Using that catalyst, they successfully converted a range of aliphatic alcohols as well as phenols into vinyl ethers. Yields were moderate to excellent for a range of alcohols, and the chemistry was also able to functionalize bulks alcohols in a few hours at 100 °C. One of the challenges with using transition metal catalysts for this type of reaction was the generation of acetals due to multiple additions of alcohols to alkynes. This undesired reaction was minimized using this Ir-catalyzed process. It was also noteworthy that this process generated an ethenyl ether. Most of the additional examples described in the sections later afforded internal vinyl ethers.

Palladium complexes have also been used to generate vinyl ethers through transetherification reactions [103]. In addition to simple substrates, this chemistry has been extended to steroidal substrates (Scheme 2.75) [104]. The catalyst was a phenanthroline complex of palladium acetate, and ethyl vinyl ether served as the source of the vinyl fragment. The reaction conditions were mild, but the duration of the reaction was long (4 d).

One of the most attractive approaches to the laboratory scale synthesis of vinyl ethers remains the copper-catalyzed cross-coupling between alcohols and vinyl halides (Scheme 2.76) [105, 106]. For the copper(I) iodide approach, the most effective stabilizing/solubilizing ligand was tetramethyl-1,10-phenanthroline, and cesium carbonate was an effective base for the chemistry. Moderate to good yields of a range of vinyl ethers were obtained after heating the reaction for 12–48 h. One of the most attractive aspects about this chemistry was the ability to perform the reactions under an atmosphere of air. This is a significant practical advantage of this work.

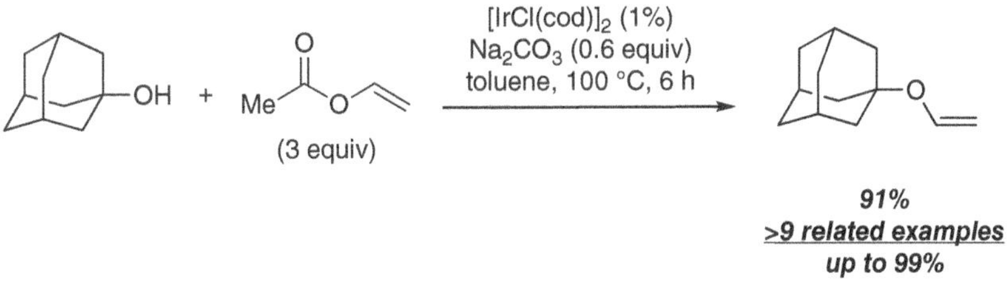

SCHEME 2.74   Iridium-catalyzed synthesis of vinyl ethers from vinyl acetate [102].

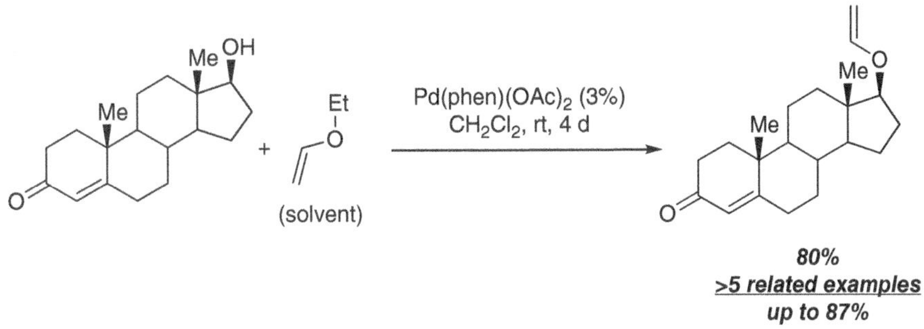

SCHEME 2.75   Synthesis of vinyl ethers through palladium-mediated transetherification [104].

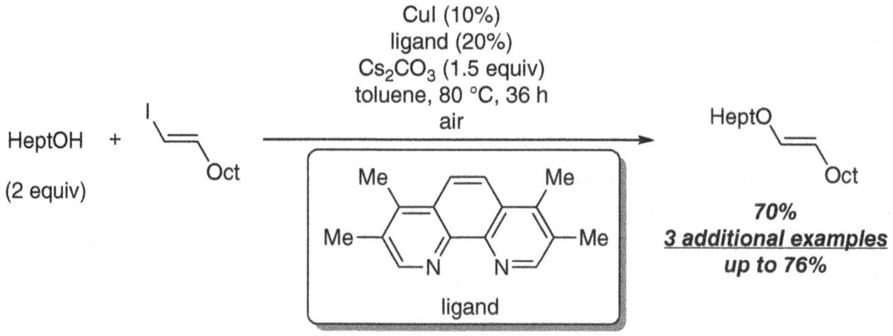

SCHEME 2.76   Copper-catalyzed coupling of vinyl iodides with alcohols in air [105].

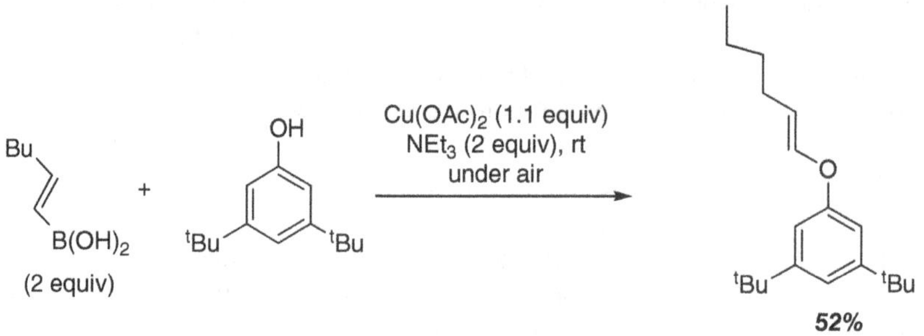

SCHEME 2.77   Copper-catalyzed coupling of vinylboronic acids with a phenol [107].

Vinylboronic acids can be used in place of the vinyl halides in copper-promoted syntheses of vinyl ethers (Scheme 2.77) [107]. In contrast to the catalytic amount of copper that was needed for the cross-coupling reactions described previously, this chemistry needed a stoichiometric amount of the copper. A significant advantage to the system was the observation that it could be carried out at room temperature and was successful under air. While only a single substrate was presented, this demonstrated that the approach had potential for the selective synthesis of vinyl ethers.

The extension of this chemistry to ethenylboronic acid would be attractive since the process would generate a terminal alkene. However, achieving this goal has been challenging due to reported stability problems with ethenylboronic acid. A clever solution to this problem has been devised by O'Shea [108]. Instead of struggling with ethenylboronic acid, they converted it into 2,4,6-trivinylcyclotriboroxane through a cyclodehydration reaction and isolated it as it is a pyridine adduct. Gratifyingly, it still maintained high activity in metal-mediated cross-coupling reactions. Using this bench-stable source of the vinylboron fragment, the authors devised a copper-mediated oxygen–carbon bond-forming reaction and generated a range of aryl vinyl ethers (Scheme 2.78). This chemistry was carried out in dichloromethane and still needed a significant amount of copper to be successful; however, the reactions were carried out at room temperature under air. Given the challenges associated with the use of vinylboronic acid, this boroxine ring equivalent is an effective alternative.

In addition to the copper-catalyzed cross-coupling of vinylboronic acids or their equivalents with alcohols, vinylboronates can also be used to generate vinyl ethers (Scheme 2.79) [109].

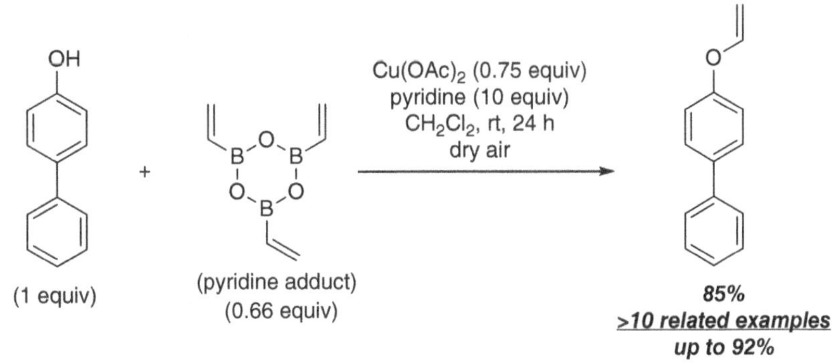

SCHEME 2.78 Copper-catalyzed synthesis of alkyl aryl ethers [108].

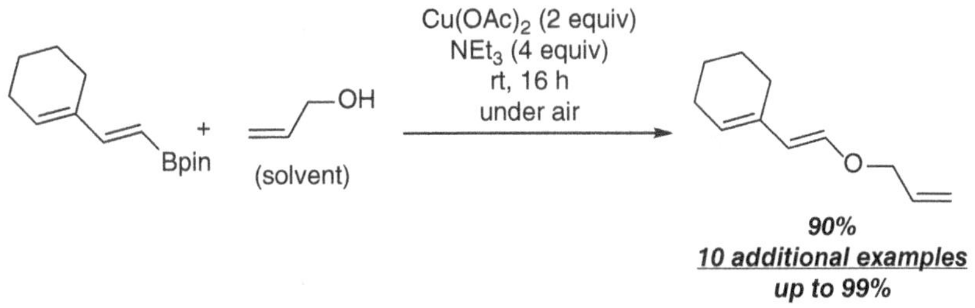

SCHEME 2.79 Copper-catalyzed coupling of vinylboronates with allylic alcohols [109].

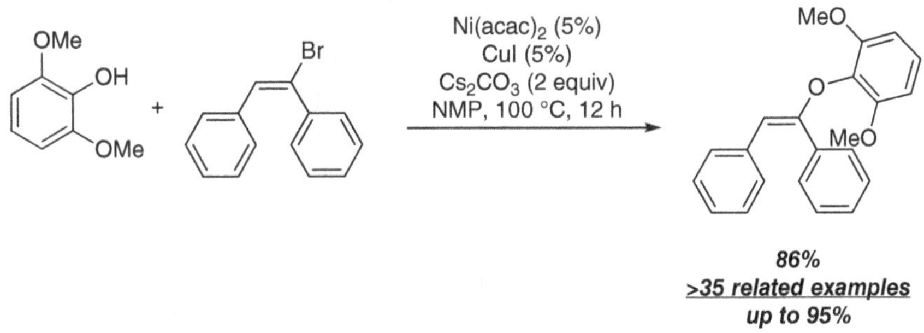

SCHEME 2.80 Synthesis of crowded vinyl ethers using copper/nickel compounds [110].

Similar to the results described earlier, the chemistry afforded remarkably high yields of the vinyl ethers when an organic base was added to the reaction mixture, and the reactions could be carried under an atmosphere of air.

An additional approach to the cross-coupling reaction has been developed using a dual metal catalyst system (Scheme 2.80 and Example 2.13) [110]. The overall reaction coupled a phenol with a vinyl halide to generate vinyl ethers in good to excellent yield. The process was catalyzed by the combination of copper and nickel, and while the precise nature of the active species was not determined, both metals were critical to a successful coupling reaction as low yields of the vinyl ethers were obtained if either metal was absent from the

Example 2.13   Preparation of a Styrenyl Ether through Cross-Coupling Chemistry [110].

Before starting this synthesis, carefully follow the safety instructions listed in Example 2.1.

Caution! Ethyl acetate and hexane are highly flammable.
In a 10-mL round bottom flask, a mixture of (*E*)-1-(2-bromovinyl)-4-methylbenzene (197 mg, 1 mmol), 4-methoxy phenol (124 mg, 1 mmol), Cs_2CO_3 (650 mg, 2 mmol), Ni(acac)$_2$ (13 mg, 0.05 mmol),CuI (10 mg, 0.05 mmol), and NMP (3 mL) was heated at 100 °C under argon for 8 h (TLC). The reaction mixture was then allowed to cool and was extracted with ethyl acetate (3×20 mL). The extract was washed with water (10 mL) and brine (10 mL). Then, the organic phase was dried over Na_2SO_4 and evaporated to leave the crude product, which was purified by column chromatography over silica gel (hexane/ethyl acetate 95:5) to provide the pure (*E*)-1-methoxy-4-((4-methylstyryl)oxy)benzene as a viscous liquid (221 mg, 92%).

Adapted with permission from D. Kundu, P. Maity, and B. C. Ranu, *Org. Lett.*, **2014**, *16*, 1040–1043. *Copyright (2014) American Chemical Society.*

reaction mixture. The reaction was remarkably tolerant to bulky groups on both the phenol and the vinyl halide. The authors did observe a geometry limitation for the vinyl halides. While *E*-halogenated alkenes afforded high yields of the cross-coupling product, analogous reactions using *Z*-isomer generated none of the vinyl ether and converted the starting compounds into 1,3-diynes.

In addition to cross-coupling chemistry, one of the most direct routes to the formation of vinyl ethers entails the addition of alcohols and phenols to alkynes. This atom-efficient transformation is referred to as hydroalkoxylation and can be tuned to generate vinyl ethers in excellent yields under very mild conditions. Both intermolecular and intramolecular versions of this reaction are well known, and a vast array of catalysts and conditions has been used for the successful synthesis of vinyl ethers [111, 112]. The following sections will highlight a number of advances in this area with special attention paid to modifications such as the development of fast reactions, "green" processes, as well as the ability of the reactions to tolerate moisture or air.

Although examples of mercury-catalyzed hydroalkoxylation reactions are well known, there are considerable issues to consider prior to adopting such a reaction including health concerns, waste disposal, and catalyst activity. To circumvent these challenges, other metal catalysts were investigated for their activity toward the addition reaction. Among them, gold compounds are attractive alternatives. One of the first reports of a practical gold-catalyzed hydroalkoxylation of alkynes was provided by Teles [113]. Using an air-stable gold complex (Ph$_3$PAuMe) and a strong acid, he was able to promote the addition of alcohols to alkynes. For most substrates, when excess alcohol was used, a

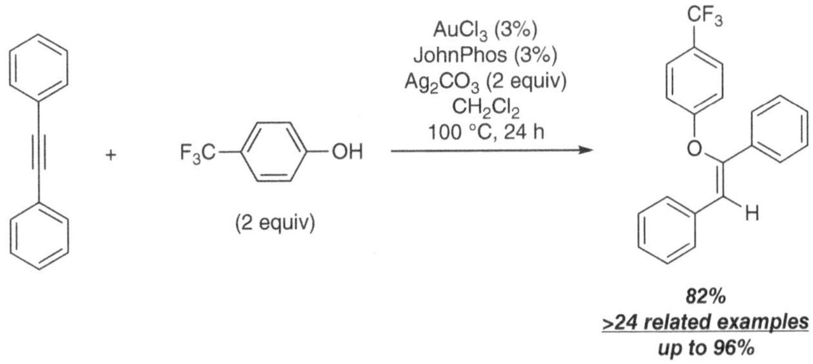

SCHEME 2.81   Gold-catalyzed synthesis of vinyl ethers [114].

double addition to the alkyne occurred to afford acetals. When a deficiency of the alcohol was used, the vinyl ethers were observed. Following this work, a tremendous amount of effort has been devoted to developing more active and selective gold catalysts for this transformation.

The hydrophenoxylation of internal alkynes has been accomplished using gold catalysts bearing bulky dialkylbiarylphosphine ligands (Scheme 2.81) [114]. One of the most attractive aspects of the chemistry was the ability to generate the gold catalyst in solution from commercially available air-stable precursors. The authors screened a range of additives and discovered that silver(I) carbonate was superior for some substrate combinations, while potassium carbonate was successful for others. Using the optimized system, heating a phenol and an internal alkyne to 100 °C for 12 h to several days generated moderate to excellent yields of the vinyl ethers. *Caution! The chemistry involved heating dichloromethane mixtures to well above the boiling point of dichloromethane. Considerable pressure could buildup in the reaction vessel resulting in hazardous situations. All appropriate safety measures must be taken when performing this addition reaction.* A range of phenols bearing a variety of electron-donating and electron-withdrawing groups were well tolerated by this approach. While the chemistry was fairly insensitive to the electronic composition of the phenol, it was sensitive to the presence of bulky groups. No addition product was obtained when the addition reaction was performed with 2,6-dimethylphenol (100 °C, 146 h).

Gold catalysts have been used to generate a host of different oxygen-containing compounds. As mentioned in other sections, the active species in the gold reactions was proposed to be a coordinatively unsaturated gold cation. This species can be generated through the treatment of organogold compounds with strong acids or through the addition of silver salts to chlorogold precursors. For the latter method, one of the persistent questions that has yet to be fully addressed is the role of the silver salt after it removed the halogen from the gold center. The discussion typically revolves around the proposed innocence of the silver salt generated in the reaction. The influence the silver salt could have over the outcome of the reaction has been called the "silver effect" [115–117]. As a result, the development of a process for the generation of an active catalyst without the addition of a silver salt would be advantageous. This has been achieved through the treatment of organogold compounds with acids. However, this approach is generally incompatible with substrates bearing acid sensitive groups. As a result, it would be useful to develop a

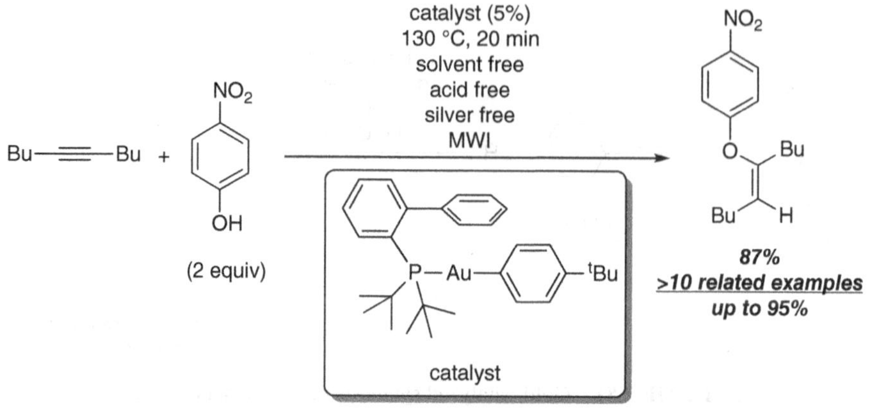

SCHEME 2.82   Gold-catalyzed synthesis of vinyl ethers using microwave heating [119].

process for the generation of the active species without the need for a silver salt or acid promoters. Elegant work by Nolan has shown that hydroxide bridged digold compounds are able to promote the hydrophenoxylation of alkynes without the addition of added acids or silver salts [118].

The hydrophenoxylation of internal alkynes has been achieved using a single-component gold catalyst without the addition of solvent, silver additives, or acid promoters (Scheme 2.82 and Example 2.14) [119]. The idea with this chemistry was to incorporate a group on the gold center that would be sensitive to one of the reagents and generate the coordinatively unsaturated gold(I) catalyst through reaction with the reagents instead of with a silver salt or acid. This was accomplished by generating an arylgold complex with an electron-donating group on the aryl ring. Using a compound containing this fragment as well as a bulky dialkylbiarylphosphine ligand, phenols were added to internal alkynes in moderate to excellent yield. A focused microwave reactor was used to heat the reactions, and both diarylalkynes and dialkylalkynes were functionalized using this approach. It should be noted that the chemistry was also successful using conventional heating. The chemistry was operationally straightforward as all of the reagents could be weighed out in air and only purged with nitrogen immediately prior to irradiation. Although the gold catalyst needed to be prepared in a separate step, this extra step is mitigated by the ability to perform the addition reaction without the need for acids, solvents, or silver additives. In related work, Nolan was able to adapt his hydrophenoxylation and hydroalkoxylation chemistry catalyzed by hydroxide bridged digold compounds to solvent-free approaches [120, 121]. The hydroalkoxylation chemistry was particularly attractive since it could be carried out in air.

An extremely rapid synthesis of 1,3-dienyl ethers from terminal alkynes and alcohols has been reported (Scheme 2.83) [98]. The reaction was catalyzed by a ruthenium complex bearing a bulky cyclopentadienyl derivative and acetonitrile as supporting ligands. The reaction was highly selective for the formation of the E,E-1,3 dienyl ethers and tolerated a wide range of electron-donating and electron-withdrawing groups. One of the most attractive aspects of this chemistry was the speed of the reactions. Many of the reactions were complete within 1 min, and all were finished within 5 min. In related work, a gold-catalyzed intermolecular hydroalkoxylation reaction between allylic alcohols and alkynes generated allyl vinyl ethers that underwent a Claisen rearrangement [123].

Example 2.14 Microwave-Assisted Gold-Catalyzed Hydrophenoxylation of Internal Alkynes [119].

Before starting this synthesis, carefully follow the safety instructions listed in Example 2.1.

Part 1: Preparation of the Arylgold Catalyst [122]

Caution! Hexane and tetrahydrofuran are highly flammable!
A reactor vial (10 mL) was charged with the LAuCl precursor (L = JohnPhos), 2 equiv of the arylboronic acid, 2 equiv of Cs_2CO_3, and a magnetic stirring bar. After exchanging the air for nitrogen, 1.5 mL of THF was added by syringe. The samples were triturated for 5 min on a stirring plate and irradiated. The initial power setting listed for each reaction was maintained until the desired temperature was reached. The power was then reduced for the remainder of the reaction to maintain the temperature. No ramping periods were used in these reactions; thus, the reaction time listed is the total irradiation time (not the time at the desired temperature). Care must be taken not to use initial high levels of microwave power (to rapidly heat the sample) as significant decomposition was observed under those conditions. Following irradiation, the residue was cooled to room temperature and purified by column chromatography. After drying ($CH_2Cl_2/MgSO_4$), the arylgold compounds were isolated as powders. For the title compound, the general procedure was followed with (JohnPhos)AuCl (0.10 g, 0.19 mmol), 4-*tert*-butylphenylboronic acid (0.067 g, 0.38 mmol), and cesium carbonate (0.13 g, 0.40 mmol). Temperature = 70 °C, time = 30 min, initial power level = 30 W. Chromatography: basic alumina (37.5 g), hexane/THF 75 : 25. R_f = 0.68 (hexane/THF 75 : 25), yield = 0.095 g of a white powder (80%).

*Adapted from H. K. Lenker, T. G. Gray, R. A. Stockland Jr. Dalton Trans. **2012**, 41, 13274–13276 with permission of The Royal Society of Chemistry.*

Part 2: Microwave-Assisted Hydrophenoxylation of Alkynes [119]

Before starting this synthesis, carefully follow the safety instructions listed in Example 2.1.

Caution! Hexane and ethyl acetate are highly flammable!
A reactor vial (10 mL) was charged with diphenylacetylene (0.050 g, 0.28 mmol), 4-nitrophenol (0.078 g, 0.56 mmol), the gold catalyst (0.0088 g, 0.014 mmol; shown in part 1), and a magnetic stirring bar. After exchanging the air for nitrogen, the samples were irradiated in a focused microwave reactor or heated in an oil bath. Microwave settings: 130 °C, 20 min, initial power level = 50 W. The initial power setting listed for each reaction was maintained until the desired temperature was reached. No ramping periods were used in these reactions; thus, the 20-min reaction time is the total irradiation period (not the time at the desired temperature). Once cooled to room temperature, the vinyl ether was purified by column chromatography (silica gel) using hexane/EtOAc (90 : 10) as the eluent. After drying using molecular sieves (hexane/EtOAc solution), the vinyl ether was isolated as a white powder following removal of the volatiles.

Adapted from M. E. Richard, D. V. Fraccica, K. J. Garcia, E. J. Miller, R. M. Ciccarelli, E. C. Holahan, V. L. Resh, A. Shah, P. M. Findeis and R. A. Stockland Jr. *Beilstein J. Org. Chem.* **2013**, *9*, 2002–2008 *with permission.*

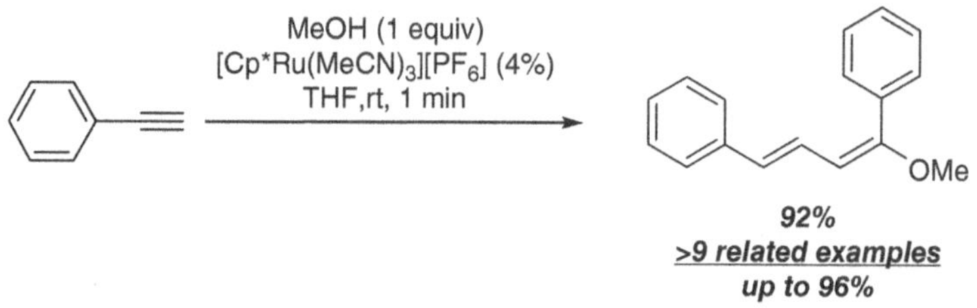

SCHEME 2.83   Ruthenium-catalyzed synthesis of 1,3-dienyl ethers [98].

The synthesis of cyclic ethers through an intramolecular hydroalkoxylation reaction is a valuable transformation. A host of metal catalysts have been developed to promote this reaction [111, 112]. While a number of these catalysts are highly active for the reaction, many are intolerant of the presence of moisture or atmosphere. To address this issue, a range of late transition metal complexes have been used to promote the cyclization reaction. As an example, a range of NHC-ligated copper complexes were used as catalysts for the intramolecular cyclization of alkynyl alcohols resulting in the generation of cyclic vinyl ethers (Scheme 2.84) [124]. The chemistry tended to be selective for the formation of the *exo*-cyclization product, and excellent yields were generally obtained. Perhaps the most practical advantage to this chemistry was the observation that the reactions could be exposed to air prior to the start of the reactions.

The intramolecular hydroalkoxylation of internal alkynes has been used to generate a group of oxazocenones (Scheme 2.85) [125]. The reactions were carried out at room temperature using a catalyst system comprised of an electron-deficient phosphine ligated gold complex along with a silver salt. The reaction was highly selective for the formation of the oxazocenone, and only minor amounts of the regioisomer were observed. The selectivity of the reaction was sensitive to the nature of the supporting phosphine ligand, and while the use of $(C_6F_5)_3PAuCl$ as the catalyst afforded moderate to high selectivities, analogous reactions carried out with more electron-donating

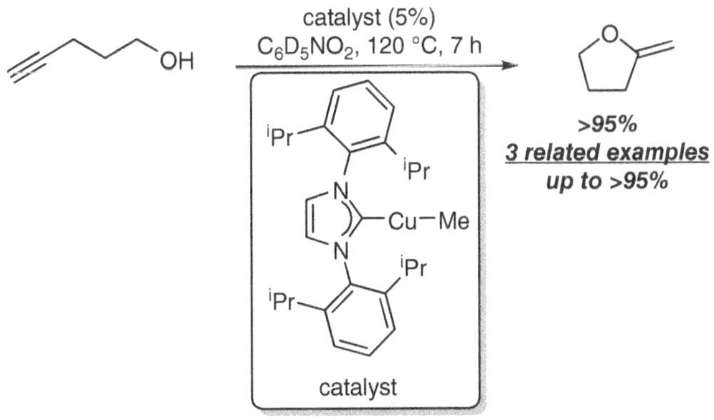

SCHEME 2.84 Copper-catalyzed intramolecular cyclization of alkynyl alcohols [124].

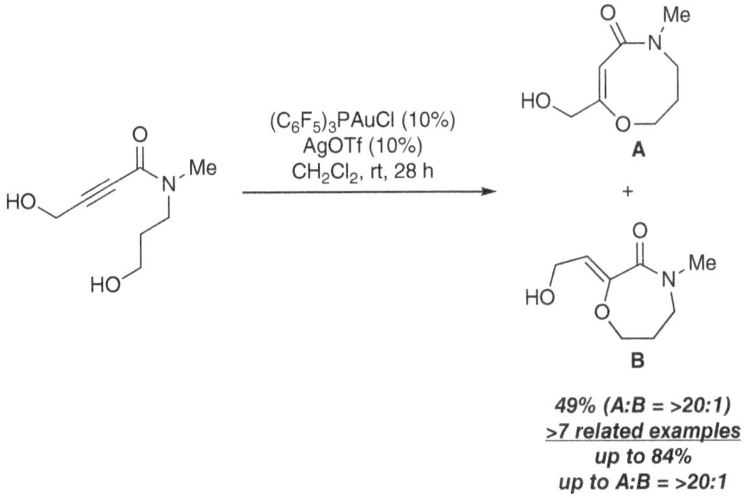

SCHEME 2.85 Gold-catalyzed intramolecular hydroalkoxylation of internal alkynes [125].

phosphine ligands were high yielding, but less selective. It should be noted that the extension of the chemistry to terminal alkynes and secondary amides resulted in low conversions.

While a number of oxygen–carbon(sp^2) bond-forming reactions utilize terminal and internal alkynes, addition reactions using cyclohexyne, benzyne, and related compounds have received less attention. To provide a facile pathway for the synthesis of a variety of heterocycles, cyclohexyne was generated and trapped by a variety of nitrogen- and oxygen-containing reagents (Scheme 2.86) [126]. For the synthesis of isoxazolines, cyclohexyne was trapped with a nitrone to generate the heterocycle. While only a few substrates were screened in this work, it provides the foundation for further development.

A series of functionalized 1-arylisochromenes have been prepared from 2-alkynylbenzaldehydes through a hydroarylation/cycloisomerization reaction (Scheme 2.87) [127]. A variety of silver salts and solvent combinations were screened for activity, and

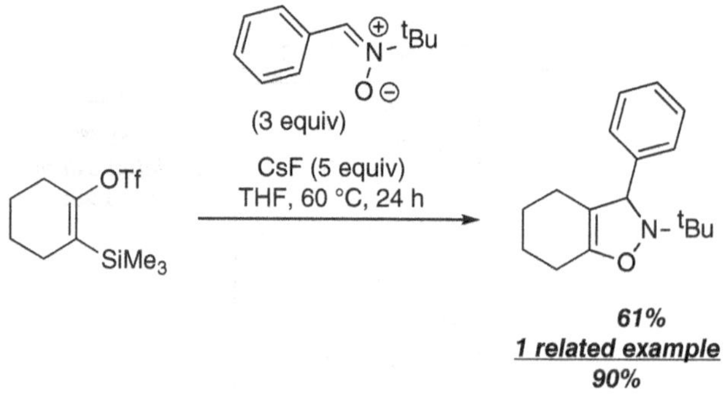

SCHEME 2.86 Synthesis of five-membered heterocycles by formally trapping cyclohexyne with a nitrone [126].

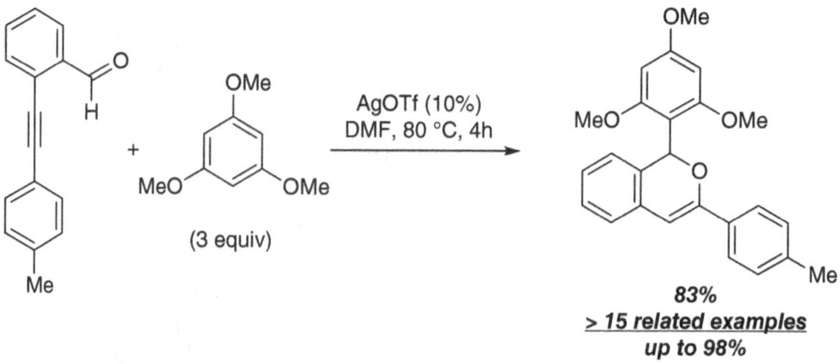

SCHEME 2.87 Synthesis of arylisochromenes through a hydroarylation/cycloisomerization reaction [127].

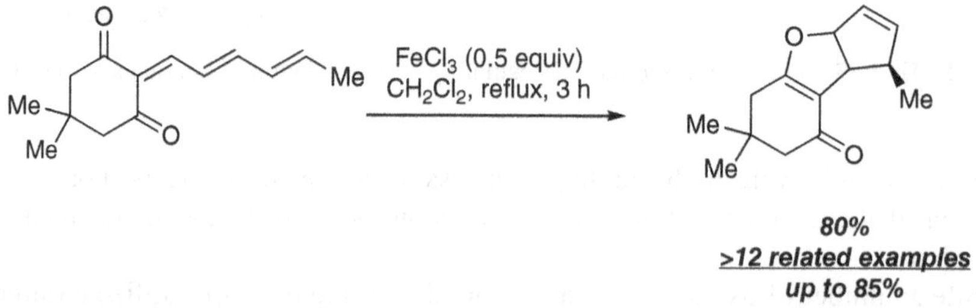

SCHEME 2.88 Synthesis of furan derivatives through an iron-catalyzed vinylogous iso-Nazarov reaction [128].

silver(I) triflate was determined to be the most active when used with DMF as the solvent. Reaction times were short, and most were complete within a few hours. One of the more attractive aspects of the chemistry was the ability to retain an aryl bromide through the hydroarylation/cycloisomerization process. This provides an entry point for further functionalization of the substrates. In related work, the iron-catalyzed vinylogous iso-Nazarov reaction generated a host of furan derivatives in moderate to good yield (Scheme 2.88) [128].

The preparation of functionalized oxazoles was achieved in a two-step process starting from *N*-sulfonyl propargylamides (Scheme 2.89) [129]. An iodocyclization of the *N*-sulfonyl propargylamides using NIS in DMF comprised the first step in the process. Yields of this reaction were generally good to excellent, and the reaction was selective for the 5-*exo*-dig product. Treatment of this iodoalkylidenedihydrooxazole with elemental oxygen generated the functionalized oxazole in moderate yield. After some experimentation, the authors devised a one-pot approach to the synthesis of the oxazoles. This transition metal-free heterocycle synthesis was attractive due to simplicity of the approach.

As mentioned earlier, carboxylic acids are attractive substrates for syntheses. While the majority of metal-catalyzed cross-coupling reactions have involved decarboxylative processes resulting in the formal loss of CO_2 along with the formation of carbon–carbon or carbon–heteroelement bonds, nondecarboxylative processes can generate a host of new compounds through coupling of the carboxylic acid. To this end, the formation of vinyl esters was developed using the copper-catalyzed coupling of carboxylic acids with vinylboronic acids (Scheme 2.90) [99]. The catalyst system was comprised of a common copper salt along with 2 equivalents of a silver salt to mediate the process. No added ligands for the copper were added. The cross-coupling chemistry was particularly effective with electron-rich aromatic rings, while lower yields of the esters were obtained electron-deficient substrates. The chemistry was also quite tolerant to bulky groups (Scheme 2.90), and has been extended to a vast array of carboxylic acids and vinyltrifluoroborate salts [130].

The addition of carboxylic acids to alkynes is an atom-efficient transformation that generates vinyl esters. This reaction is typically catalyzed by transition metals, and one

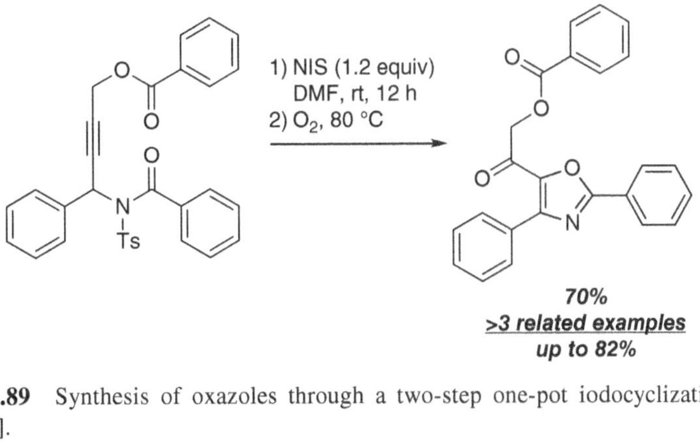

70%
>3 related examples
up to 82%

SCHEME 2.89   Synthesis of oxazoles through a two-step one-pot iodocyclization/oxidation process [129].

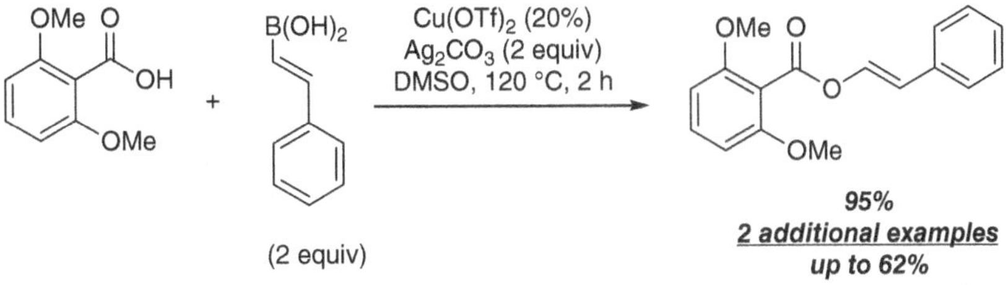

95%
2 additional examples
up to 62%

SCHEME 2.90   Copper-catalyzed cross-coupling of vinylboronic acids with carboxylic acids [99].

of the first examples was provided by Shvo [131]. Using $Ru_3(CO)_{12}$ as the catalyst, a series of alkyl and aryl carboxylic acids were added to several internal and terminal alkynes with heating (145 °C) to afford vinyl esters. Following this report, a number of metal catalysts and conditions have been developed for the addition reaction [132–135]. One of the most practical advances involved switching from the ruthenium carbonyl catalyst to a much more soluble *p*-cymene ligated ruthenium dimer (Schemes 2.91 and 2.92) [136]. This ruthenium complex is quite stable to oxygen and moisture and is readily handled and stored. One of the valuable observations the authors made during their investigation was that the addition of small amounts of base significantly increased the amount of vinyl ester generated from the addition reaction. Furthermore, they discovered that the identity of the base was tied to the selectivity of the reaction. When common inorganic bases were used, the reaction was highly regioselective for the formation of the anti-Markovnikov products and stereoselective for the Z-isomer. When organic amines were used as the base, the regioselectivity was reversed, and the reaction was highly selective for the Markovnikov product. Furthermore, the supporting ligand for the ruthenium also played a significant role in the reaction, and after a series of screening experiments, the authors determined that $P(Fur)_3$ was the most effective ligand for the anti-Markovnikov selective addition reactions, while $P(C_6H_4Cl)_3$ generated the highest yields of the Markovnikov products. In related work, a detailed study using both conventional and microwave heating was provided by Demonceau [137]. Using $(RuCl_2(p\text{-cymene})(PPh_3))$ as the catalyst, a range of electron-rich and electron-deficient carboxylic acids were successfully converted into vinyl ethers with predominantly Markovnikov selectivity without the addition of a base.

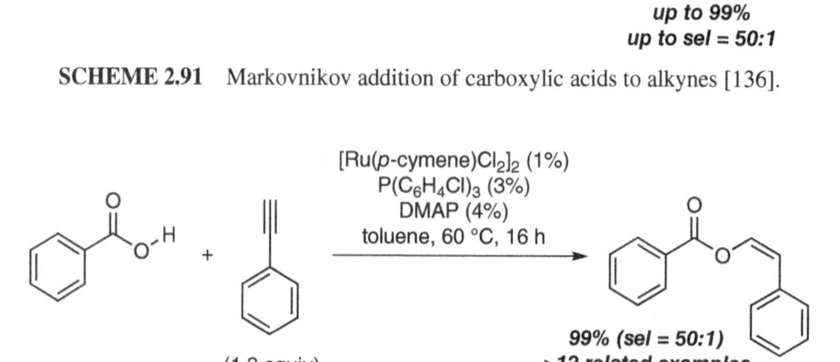

SCHEME 2.91   Markovnikov addition of carboxylic acids to alkynes [136].

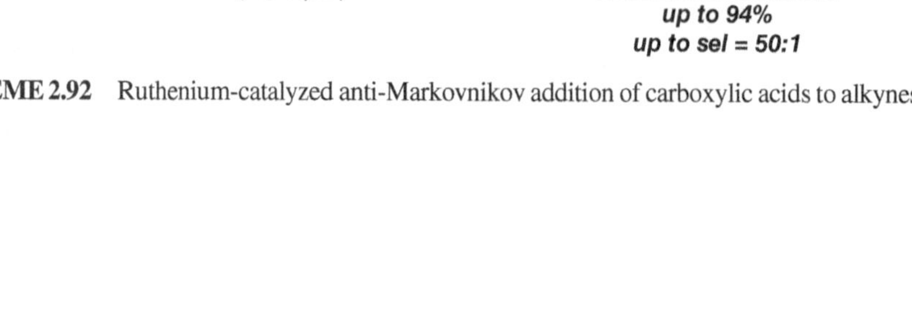

SCHEME 2.92   Ruthenium-catalyzed anti-Markovnikov addition of carboxylic acids to alkynes [136].

In addition to ruthenium-catalyzed reactions, a range of other transition metal catalysts have shown activity toward the addition reaction. A series of air-stable gold compounds promoted the addition of carboxylic acids to alkynes (Scheme 2.93) [138]. A variety of gold and silver compounds were screened as catalysts for the reaction, and the most effective pair under the mildest conditions was (Ph$_3$P)AuCl and AgPF$_6$. Under the reactions conditions, the reaction was highly selective for the formation of the Markovnikov addition product, and minimal or none of the anti-Markovnikov products were observed.

In addition to reactions involving alkynes such as hexyne and phenyl acetylene, a range of functionalized alkynes have been used in this chemistry. Ynamides were converted into vinyl ethers using palladium catalysts (Scheme 2.94) [139]. Palladium acetate was used as the catalyst in these reactions, and no added solubilizing/stabilizing ligand was needed. The yields were moderate to good with the lowest yields observed for an ynamide bearing an electron-rich arene (43%). The chemistry was highly selective for the addition of the carboxylic acid to the end of the alkyne adjacent to the nitrogen atom. Ynol ethers can also be functionalized using metal-catalyzed hydroacyloxylation reactions (Scheme 2.95) [140]. A variety of gold-, ruthenium-, silver-, and copper-containing compounds were screened for activity toward the addition reaction, and the silver salts were clearly the most active. Similar to the ynamide addition reactions, the ester was formed on the carbon adjacent to the oxygen. Yields were good to excellent, and a broad range of preexisting functional groups were well tolerated by this chemistry.

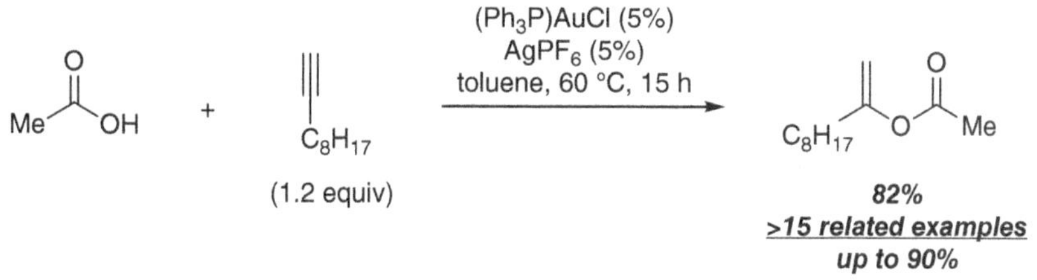

SCHEME 2.93   Gold-catalyzed addition of carboxylic acids to alkynes [138].

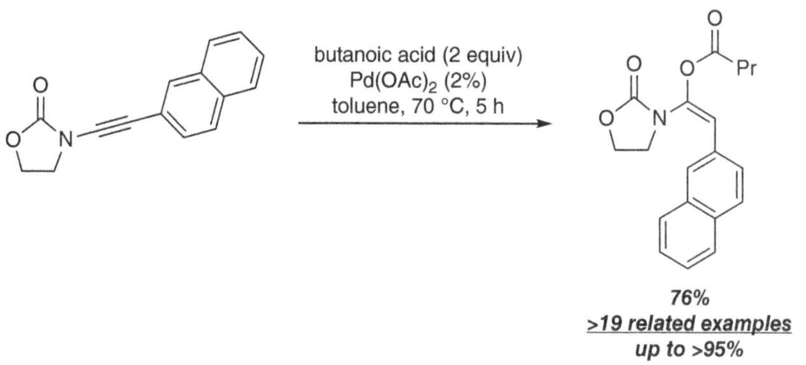

SCHEME 2.94   Palladium-catalyzed addition of carboxylic acids to ynamides [139].

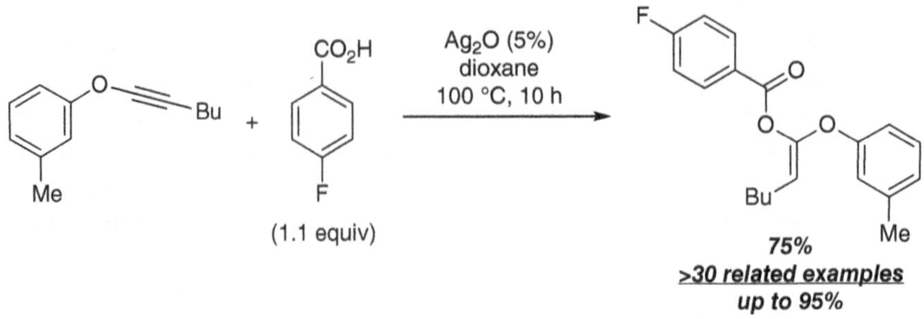

SCHEME 2.95 Silver-catalyzed addition of carboxylic acids to ynol ethers [140].

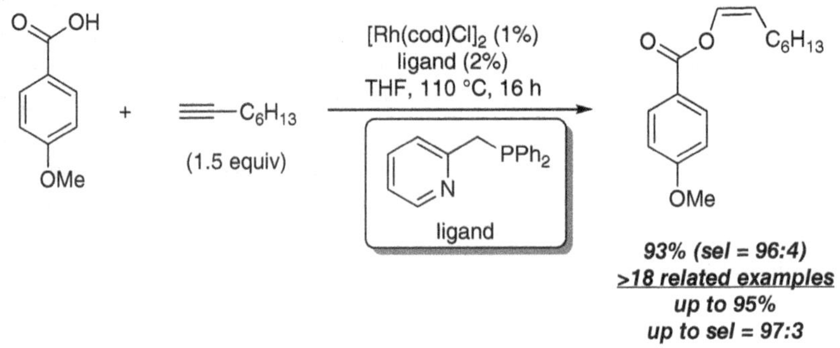

SCHEME 2.96 Rhodium-catalyzed anti-Markovnikov addition of carboxylic acids to alkynes [141].

While the majority of metal catalysts generated the Markovnikov addition product, a few catalyst systems were able to generate the anti-Markovnikov vinyl esters with Z-stereochemistry [136]. In related work, rhodium catalysts were also able to promote the addition reaction to regioselectively generate the anti-Markovnikov vinyl esters with Z-stereochemistry (Scheme 2.96) [141]. A wide range of carboxylic acids and a number of alkylalkynes were used in this chemistry. Curiously, phenylacetylene was unable to be converted into the vinyl ester following this approach. Additionally internal alkynes were also unresponsive under the reaction conditions.

Most of the metal-catalyzed systems described earlier required the use of organic solvents for a successful reaction. THF, toluene, dichloromethane, and dioxane are examples of solvents commonly used in this chemistry. As mentioned in Chapter 1 of this text, the development of reactions that are successful in the presence of small amounts of water or in pure water is highly desired from an operational simplicity perspective. To this end, the metal-catalyzed synthesis of vinyl esters in pure water using an intriguing ruthenium catalyst has been developed (Scheme 2.97) [142]. The chemistry is highly selective for the Markovnikov product, and good to excellent yields of the vinyl esters were obtained with gentle heating for a few hours in most cases. It should be noted that the reactions were carried out under an atmosphere of nitrogen. Although this work required the preparation of a new ruthenium catalyst, the ability to perform the reaction in pure water is a significant advancement for the chemistry.

In some cases, the treatment of terminal acetylenes with carboxylic acids in the presence of ruthenium catalysts did not generate a simple hydroalkoxylation product.

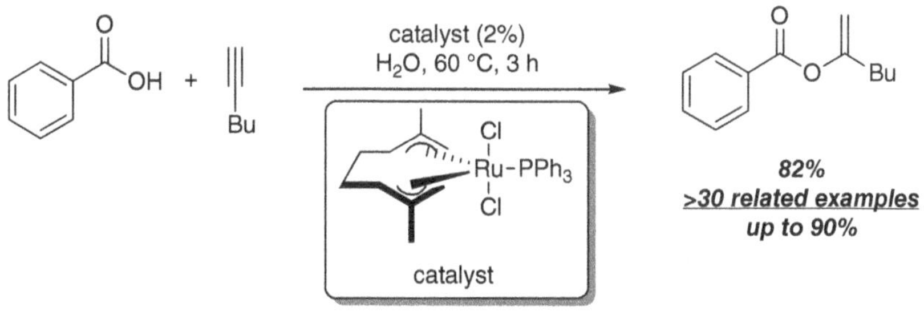

SCHEME 2.97 Ruthenium-catalyzed synthesis of vinyl esters in water [142].

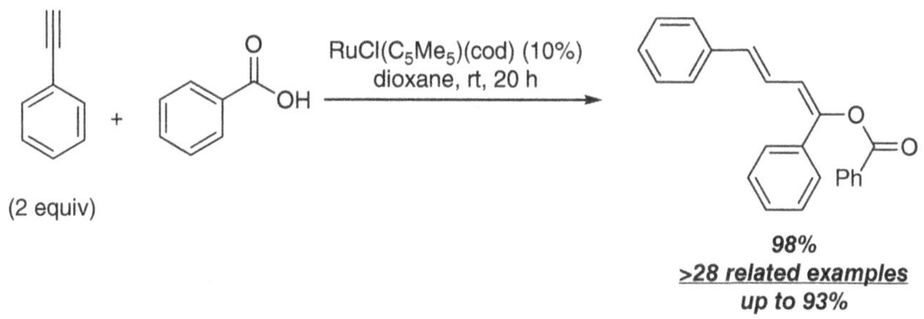

SCHEME 2.98 Ruthenium-catalyzed synthesis of dienyl acetates [143].

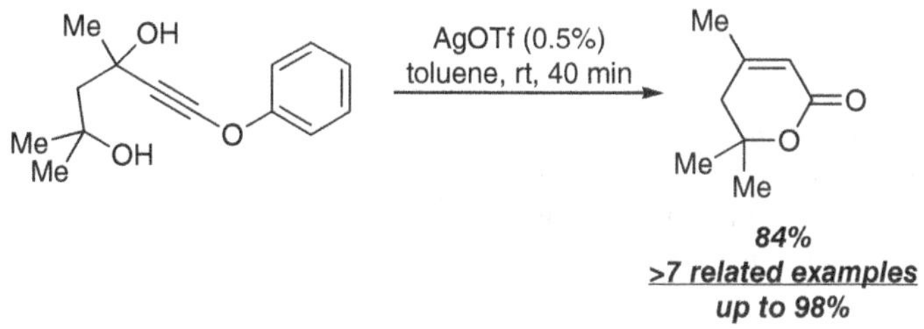

SCHEME 2.99 Silver-promoted cyclization to generate vinyl esters [144].

When some ruthenium catalysts were used, the reaction was much more complex. An example of this variability involved the use of RuCl(C_5Me_5)(COD) as the catalyst. This compound catalyzed the formation of a 1,3-dienyl ester instead of a simple vinyl ester (Scheme 2.98) [143]. The overall reaction can be thought of as a three-component reaction where two of the components were the same. One of the most attractive aspects of the chemistry was the observation that the reaction was highly selective for the formation of the *E,E*-1,4-disubstituted-1,3-diene.

The intramolecular cyclization of phenoxyethynyl diols into unsaturated lactones has been achieved using a silver salt as the catalyst for the reaction (Scheme 2.99) [144]. The authors screened a variety of transition metal catalysts for activity toward the cyclization reaction including gold/silver combinations and determined that silver triflate alone was a successful catalyst. Furthermore, the catalyst loading was quite low (0.5%). The reaction conditions were quite mild, and moderate to excellent yields of the target compounds

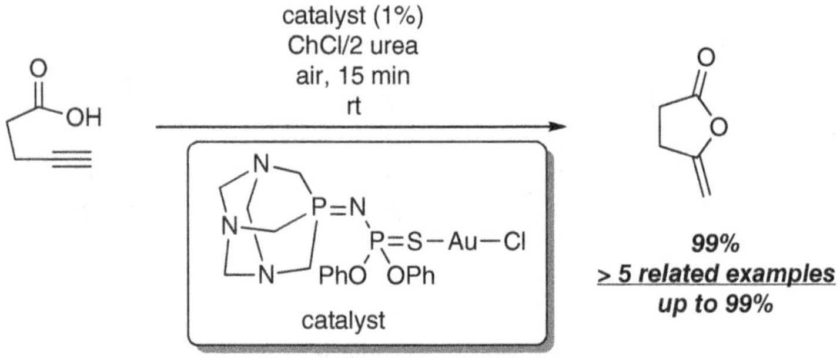

SCHEME 2.100   Gold-catalyzed synthesis of enol lactones using eutectic mixtures [145].

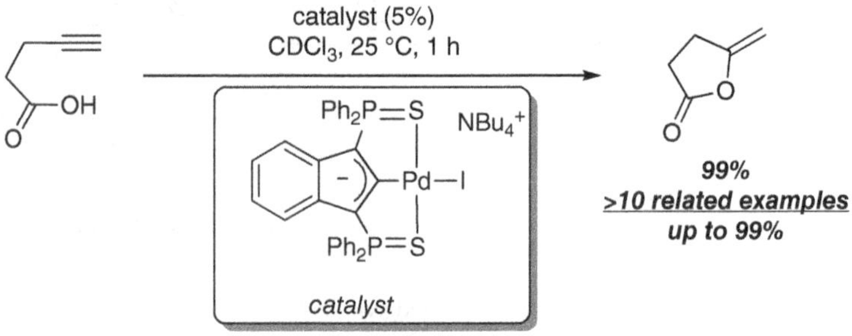

SCHEME 2.101   Cyclization of alkynoic acids using palladium pincer compounds to generate lactones [146, 147].

were obtained after stirring for less than 1 h. In related work, an unusual phosphine sulfide ligated gold catalyst was used to promote the cyclization of alkynoic acids to generate vinyl cyclic esters (enol lactones) in eutectic mixtures at room temperature (Scheme 2.100) [145]. One of the attractive aspects of this chemistry was that the reactions could be carried out under air. Chelated indenediide pincer complexes of palladium have also shown high activity toward the conversion of alkynoic acids into lactones (Scheme 2.101) [146, 147].

The synthesis of seven- or eight-membered lactones can be challenging. A gold/copper-catalyzed reaction for the formation of these valuable compounds has been reported starting from chiral alkynyl esters (Scheme 2.102) [148]. The authors screened a variety of gold and silver combinations for activity toward the cyclization, and while some generated moderate yields of the heterocycle, none of them were as effective as a gold/copper pair. This was an unusual pairing, and the precise role each of the compounds played in the catalysis was not elucidated. A variety of oxazepanones were generated using the catalyst system with the highest yields observed with substrates bearing electron-deficient arene rings. The catalyst system was also highly selective for the 7-*exo*-dig cyclization, and any initial chirality was retained through the cyclization.

Isocoumarins were generated through a copper-catalyzed three-component coupling reaction (Scheme 2.103) [149]. The chemistry combined a terminal alkyne with a benzyne (generated in solution) along with carbon dioxide to form the isocoumarins. It was

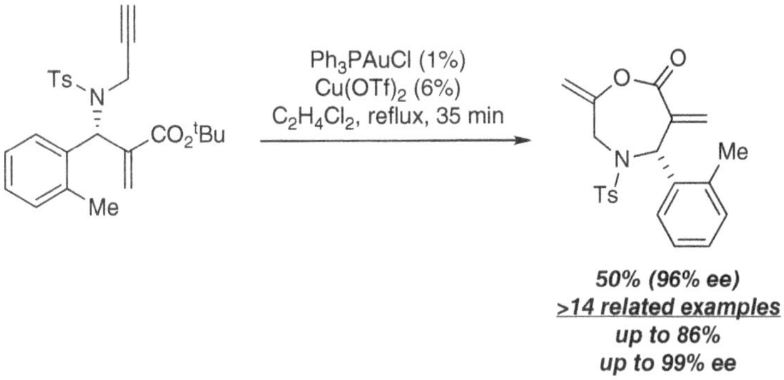

SCHEME 2.102 Synthesis of oxazepanones through a gold-catalyzed cyclization reaction [148].

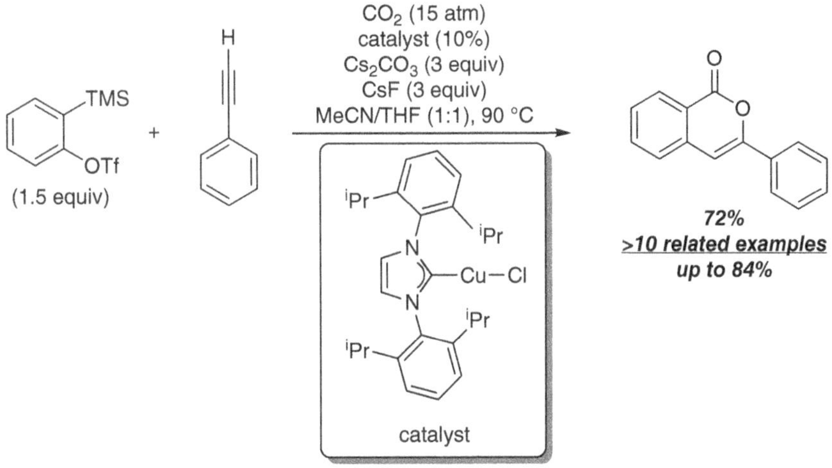

SCHEME 2.103 Preparation of isocoumarins through a copper-catalyzed coupling reaction [149].

noteworthy that carbon dioxide was successfully used in this synthesis. Devising strategies for the utilization of this greenhouse gas remains an important goal for the synthetic community. The authors screened a range of supporting ligands for the copper center and determined that an N-heterocyclic carbene was the most effective. Although phosphine ligands such as dppe had been effective ligands in similar reactions [150], they were unsuccessful with this chemistry. A wide range of terminal arylalkynes were used in this chemistry with the highest yields obtained with electron-rich arenes and the lowest obtained with electron-poor arylalkynes.

The synthesis of bicyclic heterocycles was achieved through an organocatalytic annulation between α-keto esters and 2-butynoate (Scheme 2.104) [151]. The reaction was carried out under very mild conditions, and the product distribution was highly dependent upon the selection of organocatalyst and additives. In the presence of methanol, tricyclohexylphosphine promoted the formation of cyclopentene rings. In the absence of methanol and with molecular sieves added to the reaction, the annulation reaction afforded a bicyclic ring system. The authors proposed that the molecular sieves acted as a methanol scavenging agent. It should

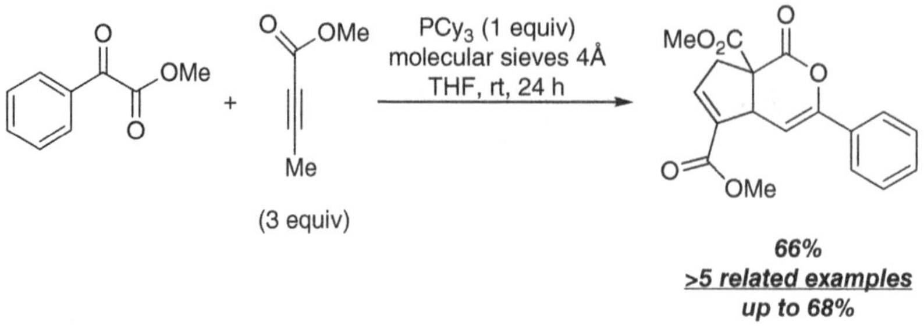

SCHEME 2.104   Synthesis of bicyclic heterocycles through a phosphine-catalyzed annulation reaction [151].

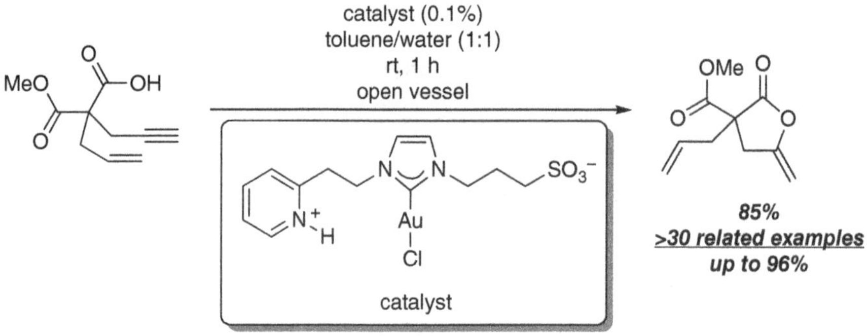

SCHEME 2.105   Gold-catalyzed synthesis of enol lactones through cycloisomerization [152].

be noted that the authors screened a range of phosphines for activity toward the annulation reaction, and tricyclohexylphosphine was the most effective. Overall, this is a clever approach to the transition metal-free synthesis of bicyclic heterocycles.

Gold complexes catalyzed the preparation of enol lactones through cycloisomerization reactions (Scheme 2.105 and Example 2.15) [152]. The reactions were carried out at room temperature using low concentrations of the gold catalyst. It should be noted that only the chlorogold complex was needed to promote the reaction. No silver salts or acids were needed to generate the enol lactone products. One of the more impressive aspects of the catalyst system was its tolerance to a significant amount of water. Indeed, the conditions were optimized in a 1:1 toluene/water biphasic mixture. Furthermore, this reaction was successful under air in an open vessel. This is one of the most attractive approaches to the synthesis of enol esters given the ability to perform the reaction on the benchtop without the need for a vacuum manifold or glove box.

The asymmetric copper-catalyzed synthesis of cyclic vinyl ethers has been achieved (Scheme 2.106) [153]. The overall reaction starts with β-keto esters and propargyl acetates and generates the 2,3-dihydropyrans in moderate to good yield with excellent selectivity. The process was catalyzed by the combination of a copper salt with a resolved ketimine-type supporting ligand. In general, the functional group tolerance of this system was good, and an array of propargyl acetates bearing electron-rich and electron-deficient aryl groups were successfully functionalized.

Example 2.15 Cycloisomerization of Alkynoic Acids Using Water-Soluble Gold Catalysts [152].

Part 1: Synthesis of the Water Soluble Gold Catalyst

Before starting this synthesis, carefully follow the safety instructions listed in Example 2.1.

In the absence of light and under N_2 atmosphere, the corresponding zwitterionic imidazolium derivative (1.5 mmol), tetrabutylammonium chloride (0.417 g, 1.5 mmol), and Ag_2O (0.394 g, 1.7 mmol) were refluxed for 24 h in 30 mL of dichloromethane. The reaction mixture was then filtered through Celite, and $[AuCl(SMe_2)]$ (0.442 g, 1.5 mmol) was added to the filtrate, leading to the extensive precipitation of AgCl. The mixture was stirred at room temperature for an additional hour and filtered again through Celite. The colorless filtrate was treated overnight at room temperature with p-TsOH·H_2O (0.285 g, 1.63 mmol), thus generating a white solid precipitate, which was filtered, washed with dichloromethane (3×10 mL), and dried in vacuo to afford the gold complex (0.633 g, 80%).

Part 2: Gold-Catalyzed Synthesis of Enol Lactones in a Biphasic Mixture Under Air

Before starting this synthesis, carefully follow the safety instructions listed in Example 2.1.

To a biphasic system composed of 1 mL of toluene and 1 mL of distilled water, 0.3 mmol of the corresponding alkynoic acid (0.047 g, 0.3 mmol) and the appropriate gold complex

(0.1 mol%; shown in Scheme 2.105) were added. The resulting mixture was stirred under air, at room temperature, until complete conversion of the alkynoic acid was observed (TLC). The organic phase was then separated, the aqueous one extracted with diethyl ether (2×2 mL), and the combined organic extracts dried over anhydrous $MgSO_4$ and filtered over a short pad of silica gel using CH_2Cl_2 as eluent. The volatiles were removed under vacuum to yield the corresponding lactone in high yield and purity. When required, lactones were further purified by column chromatography over silica gel using hexane: EtOAc (9 : 1) as eluent.

Adapted with permission from E. Tomás-Mendivil, P. Y. Toullec, J. Borge, S. Conejero, V. Michelet, V. Cadierno, *ACS Catal.*, **2013**, *3*, 3086–3098. *Copyright (2013) American Chemical Society.*

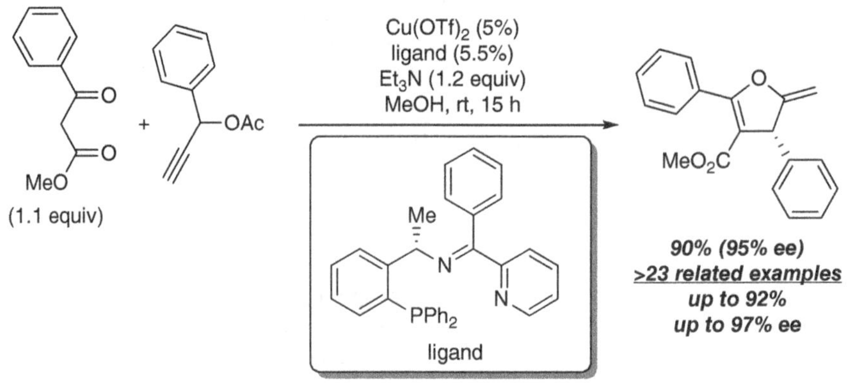

SCHEME 2.106   Copper-catalyzed asymmetric synthesis of functionalized 2,3-dyhydropyrans [153].

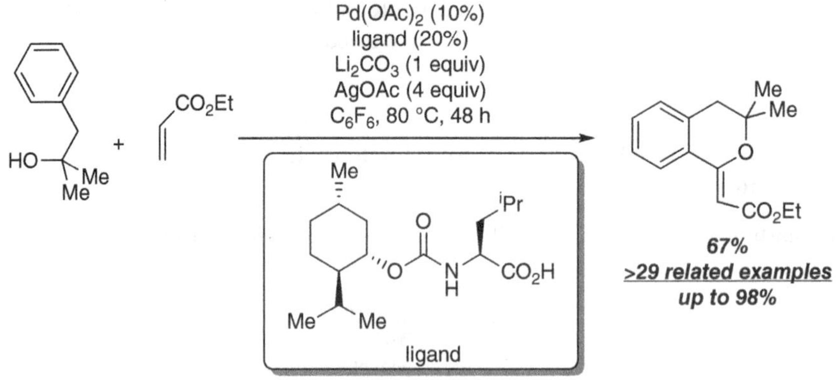

SCHEME 2.107   Palladium-catalyzed intramolecular C—H olefination [154].

The palladium-catalyzed preparation of vinyl ethers through a tandem reaction consisting of an initial C—H olefination of a tertiary alcohol by an activated alkene followed by an intramolecular oxidative cyclization has been achieved using a palladium catalyst and an unusual supporting ligand (Scheme 2.107) [154]. This ligand was a leucine derivative bearing a menthyl substituent. Using this catalyst system, a host of tertiary alcohols were converted into the vinyl ethers in moderate to excellent yields with the most successful substrates bearing electron-donating substituents on the aryl fragments. While there was some substrate specificity, many of the electron-deficient examples afforded lower yields of the vinyl ethers. A range of activated alkenes were screened during this investigation, and while acrylates and vinylphosphonates generated the desired vinyl ethers, vinyl sulfones underwent the C—H olefination reaction with retention of the tertiary alcohols to afford acyclic products.

The rhodium-catalyzed acetoxylation of enamides generated vinyl acetates (Scheme 2.108) [155]. In addition to the rhodium complex and silver additive, the catalyst system used an excess of $Cu(OAc)_2$ to promote the reaction. The authors proposed that the copper(II) acetate served as the acetate source as well as the oxidizing agent. The reaction generated moderate to excellent yields of the vinyl acetates and was highly selective for the formation of Z-isomer. The chemistry was also tolerant of a wide range of preexisting functional groups including ethers, halogens, trifluoromethyl, and esters. In addition to the high yields, selectivity, and substrate scope, one of the most impressive aspects of the chemistry was its tolerance to air. In related work, the copper(II) acetate catalyzed synthesis of vinyl alkoxyamines through the addition of TEMPO to propargyl alcohols under air (Scheme 2.109) [156].

A recent report described the synthesis of phosphaisocoumarins based upon a C—H activation/hydroelementation sequence (Scheme 2.110) [157]. Using readily available internal alkynes and $R_2P(O)(OH)$ substrates, moderate to excellent yields of the

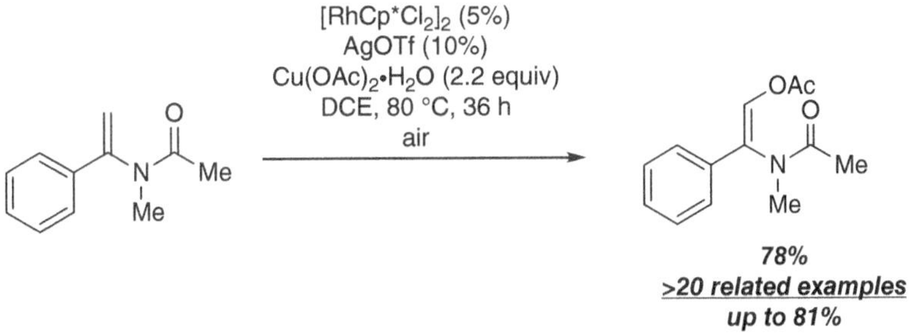

SCHEME 2.108   Rhodium-catalyzed synthesis of Z-vinyl acetates from enamides [155].

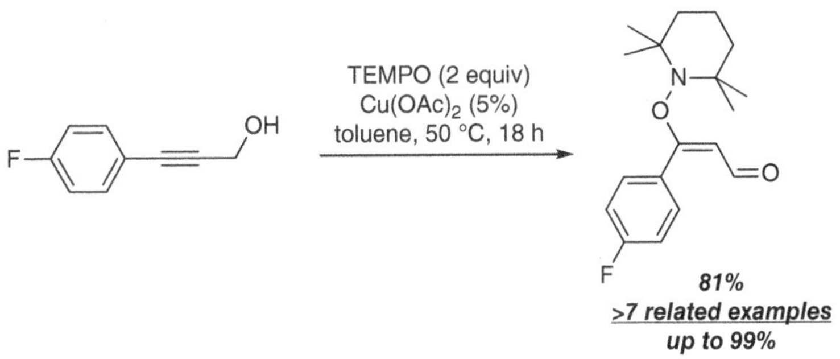

SCHEME 2.109   Copper-catalyzed synthesis of alkoxyamines [156].

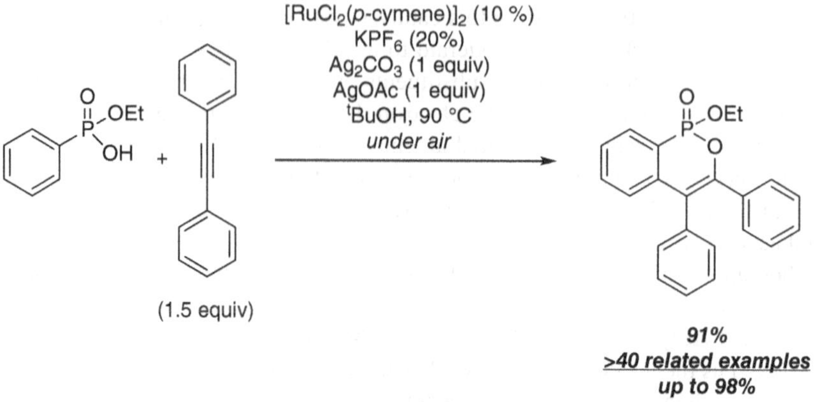

SCHEME 2.110   Ruthenium-catalyzed synthesis of phosphaisocoumarins [157].

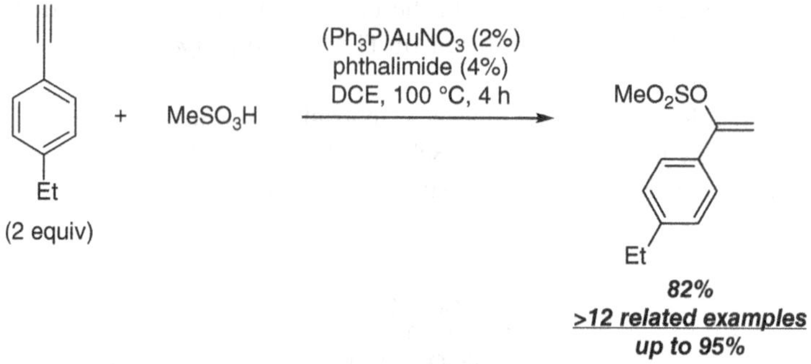

SCHEME 2.111   Gold-catalyzed addition of sulfonic acids to alkynes [160].

heterocycles were obtained. While a range of metal catalysts are known to promote this reaction, the successful application of easily prepared ruthenium catalysts was attractive due to the significant cost savings (Ru vs. Au/Pd/Pt). The reaction was proposed to proceed through the formation of an arylruthenium intermediate followed by insertion of the alkyne into the newly formed Ru–Ar bond. Reductive elimination generated the phosphaisocoumarins. The chemistry was operationally trivial and could be carried out under an atmosphere of air. In terms of the reaction scope, this chemistry was successful with both electronic and steric manipulation of the alkyne and O–H donor. Rhodium-catalyzed versions of this reaction has been reported and generated good to excellent yields of the phosphaisocoumarins [158, 159].

The gold-catalyzed addition of sulfonic acids to alkynes generates vinyl sulfonates (Scheme 2.111) [160]. The most effective gold catalyst screened in this investigation was Ph_3PAuNO_3. Other gold complexes and combinations of gold and silver compounds were significantly less effective toward the addition reaction. The other component of the catalytic system was phthalimide (2 equiv per gold). The precise role of this additive was not elucidated, but the authors proposed that it could serve to stabilize the gold complex through the generation of $Ph_3PAu(NPhth)$. The reaction was largely regioselective for the formation of the Markovnikov product, and selectivities were >90:10. Curiously, a meta-fluorinated arylalkynes predominately generated the anti-Markovnikov product (93 : 7). Internal alkynes could also be functionalized using this approach. This single-component gold-catalyzed process is an attractive approach to the synthesis of vinyl sulfonates.

Hydroformylation is one of the most popular approaches for the formation of aldehydes from alkenes and is typically used for the large-scale synthesis of these compounds [161, 162]. The Wacker oxidation of alkenes affords ketones and aldehydes through the formation of oxygen–carbon(sp^2) bonds and remains one of the most storied reactions in organometallic chemistry [163–166]. The classic version of the Wacker oxidation converts ethylene into acetaldehyde using a palladium catalysts and is typically encountered in large-scale industrial settings. Selectivity is not an issue when ethylene is the substrate; however, when a choice exists, the process is typically selective for the formation of the Markovnikov product [165]. With the goal of making this valuable reaction practical enough for small synthetic laboratories to adopt, a host of modifications have been made to the classic system. The following sections will highlight a number of these alterations.

While considerable progress has been made in Wacker and related oxidation processes, reducing the catalyst loading and eliminating the need for additives such as copper salts remains a continuing goal in the field. To this end, Sigman developed a convenient approach to the Markovnikov selective process (Scheme 2.112) [167]. The process used a low loading of a palladium catalyst along with O$_2$ as the oxidant to effect the oxidation of the alkene. The reaction conditions were mild, and moderate to excellent yields of the methyl ketones were obtained. The authors also screened several chiral substrates and found that any stereochemical enrichment was retained in the product. Due to several reasons including the ability to carry out the reaction with a relatively low pressure of oxygen, low catalyst loading, and the circumvention of the need for copper additives, this is an attractive approach for the laboratory scale Wacker reaction. In related work, the use of iron(III) sulfate as the oxidant in alkene oxidation was presented as a greener approach to the Markovnikov selective reaction [168].

While the vast majority of catalyst systems are selective for the formation of the Markovnikov product, considerable effort has been devoted to the development of catalysts that will reverse the regioselectivity and generate the anti-Markovnikov products. A number of approaches have been developed; however, most afforded lower yields or exhibited moderate selectivities. Grubbs reported a highly selective and efficient process for the

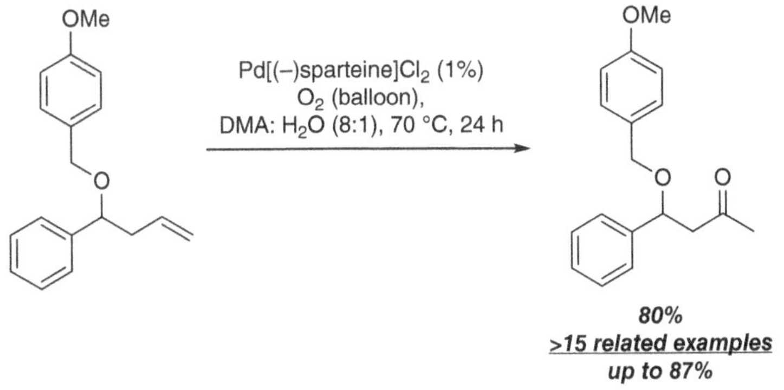

80%
>15 related examples
up to 87%

SCHEME 2.112   Wacker oxidation of alkenes using oxygen as the oxidant [167].

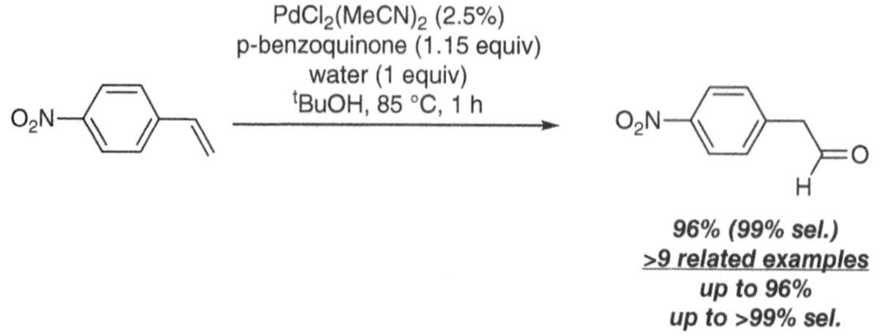

PdCl$_2$(MeCN)$_2$ (2.5%)
p-benzoquinone (1.15 equiv)
water (1 equiv)
^{t}BuOH, 85 °C, 1 h

96% (99% sel.)
>9 related examples
up to 96%
up to >99% sel.

SCHEME 2.113   An anti-Markovnikov selective Wacker-type oxidation [169].

formation of the anti-Markovnikov product using styrene derivatives as substrates (Scheme 2.113) [169]. His catalyst system was comprised of a common air-stable palladium source, benzoquinone, and water (1 equiv). A range of styrene derivatives bearing both electron-donating groups and electron-withdrawing groups were converted into aldehydes using this approach with some of the highest yields afforded by substrates bearing electron-withdrawing groups. The chemistry was highly selective for the formation of the anti-Markovnikov product, and many substrates were transformed with 99% selectivity. Although the chemistry could not directly make use of elemental oxygen as the oxidant, oxygen could be indirectly involved through the treatment of the hydroquinone by-product with O$_2$ in order to regenerate the benzoquinone oxidant used in the reaction. In related work, the oxidation of highly functionalized alkenes with anti-Markovnikov selectivity was achieved using sodium nitrite as an additive [170]. Mechanistic studies revealed that the oxygen found in the aldehyde originated from the sodium nitrite [171].

In addition to Wacker-type oxidation chemistry, the hydration of alkynes is another popular approach to the synthesis of carbonyl-containing compounds [172]. Classic versions of this reaction are catalyzed by mercury salts and are less attractive due to issues surrounding the use of mercury compounds. As a result, a host of new catalysts have been developed for the hydration of alkynes. Building on Teles's observation that an active catalyst for the hydroalkoxylation of alkynes could be generated by the treatment of organogold complexes with strong acids, a modified version of the catalyst system was developed and screened for activity toward the addition of water to alkynes [173]. A single-component gold catalyst was also developed for acid-sensitive substrates [174]. Work by Nolan demonstrated that an air-stable *N*-heterocyclic carbene ligated gold compound promoted the addition of water to alkynes (Scheme 2.114) [175]. The chemistry was highly selective for the formation of the Markovnikov product, and both internal and terminal alkynes were functionalized using this approach. Even the typically sluggish diphenylacetylene was hydrated in 77% yield. In related work, the authors were able to devise a catalyst system for the conversion of propargyl acetates into enones [176].

Ruthenium complexes are often inexpensive, readily available, and active catalysts for the hydration of alkynes [177–179]. In contrast to most mercury and gold systems, several of the Ru-based catalysts displayed a propensity to generate the anti-Markovnikov hydration products. One of the more practical approaches to this anti-Markovnikov hydration chemistry was developed by Herzon (Scheme 2.115) [180]. His approach entailed the use of a ruthenium compound bearing a 5,5′-(bistrifluoromethyl)-2,2′-(bipyridine) as the supporting ligand. A variety of ruthenium complexes could be isolated, and at least one

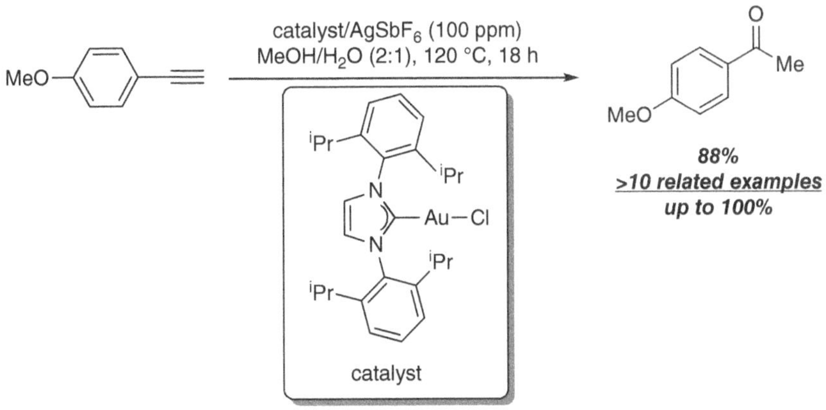

SCHEME 2.114 Gold-catalyzed Markovnikov hydration of alkynes using low catalyst loading [175].

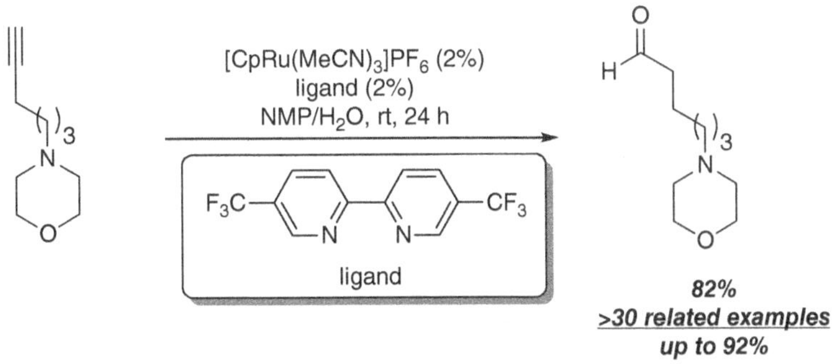

SCHEME 2.115 Ruthenium-catalyzed anti-Markovnikov hydration of alkynes [180].

was found to be quite stable to air for extended periods of time. The authors also found it convenient to generate this catalyst in solution from readily available precursors (Scheme 2.115). Using this catalyst, a number of alkynes were converted into aldehydes in moderate to excellent yields with outstanding selectivity. The ability to carry out the reactions at room temperature was another practical advantage to this chemistry. Overall, this is an efficient approach to the anti-Markovnikov hydration of alkynes.

An intriguing approach to the synthesis of 1,2-diaryl-1,2-diketones starting from alkynones has been reported (Scheme 2.116) [181]. While metal-catalyzed oxidation of alkynes into diketones is a well-known reaction, transition metal-free versions are rare. Alkynones are attractive entry points for the synthesis of diketones since a large number of them are readily available bearing a host of functional groups. This chemistry used molecular oxygen and a base to promote the conversion and was successful for a wide assortment of alkyones bearing electron-rich and electron-poor arenes. The highest yields were obtained when electron-donating groups were present on both arenes. In related work, ozone was able to convert a range of alkenes into diketones (Scheme 2.117) [182]. Molecular oxygen has also been successfully used to oxidize glycine derivatives (Scheme 2.118) [183].

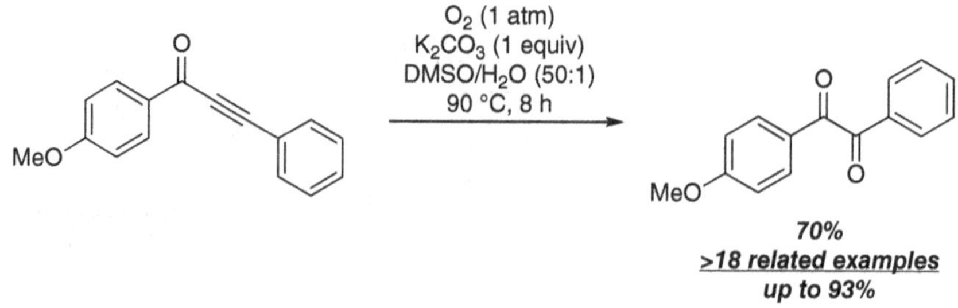

SCHEME 2.116 Conversion of alkynones into diketones [181].

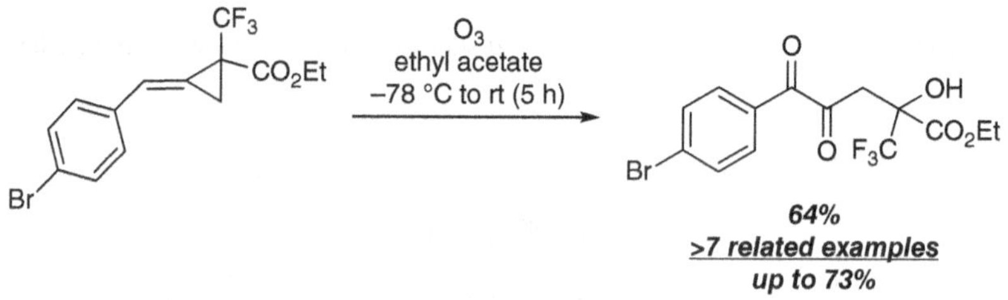

SCHEME 2.117 Ozonation of methylenecyclopropanes [182].

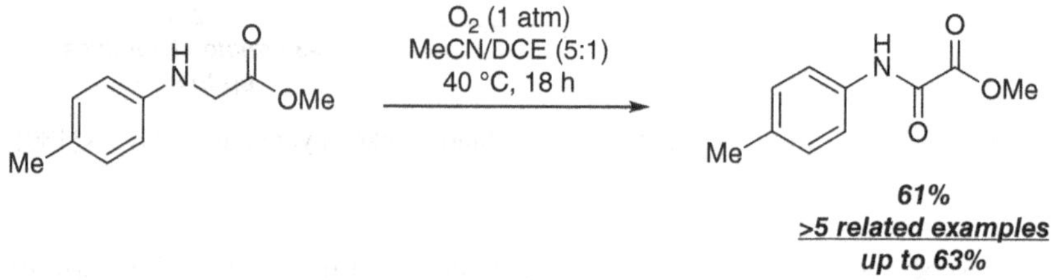

SCHEME 2.118 Oxidation of glycine derivatives by molecular oxygen [183].

An interesting approach to the conversion of alkynes into maleic anhydrides has been reported using a nickel-catalyzed double-carboxylation reaction (Scheme 2.119) [184]. While a number of the CO_2-utilization methods require elevated pressures of carbon dioxide, this methodology only needs 1 atm of carbon dioxide. The catalyst system has a number of components, but all of them are readily available. While terminal alkynes would be attractive substrates for these reactions, both these and internal arylalkynes did not generate the target compounds under the standard reaction conditions developed by the authors. It should be noted that a cyclic alkyne was successfully functionalized with retention of the ring system.

The direct oxidation of C—H bonds is an attractive approach for the preparation of oxygenated compounds. This is a challenging reaction, and general approaches with moderate substrate scope are rare. Most systems are only able to functionalize a few specific substrates. An interesting approach to this chemistry entailed the use of iron

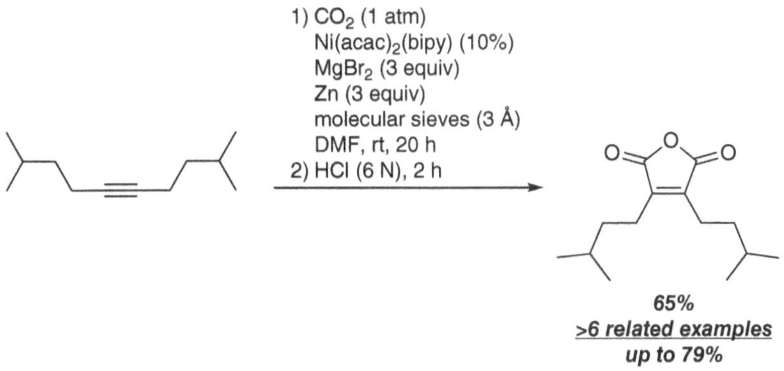

1) CO$_2$ (1 atm)
   Ni(acac)$_2$(bipy) (10%)
   MgBr$_2$ (3 equiv)
   Zn (3 equiv)
   molecular sieves (3 Å)
   DMF, rt, 20 h
2) HCl (6 N), 2 h

65%
>6 related examples
up to 79%

SCHEME 2.119   Conversion of internal alkynes into maleic anhydrides [184].

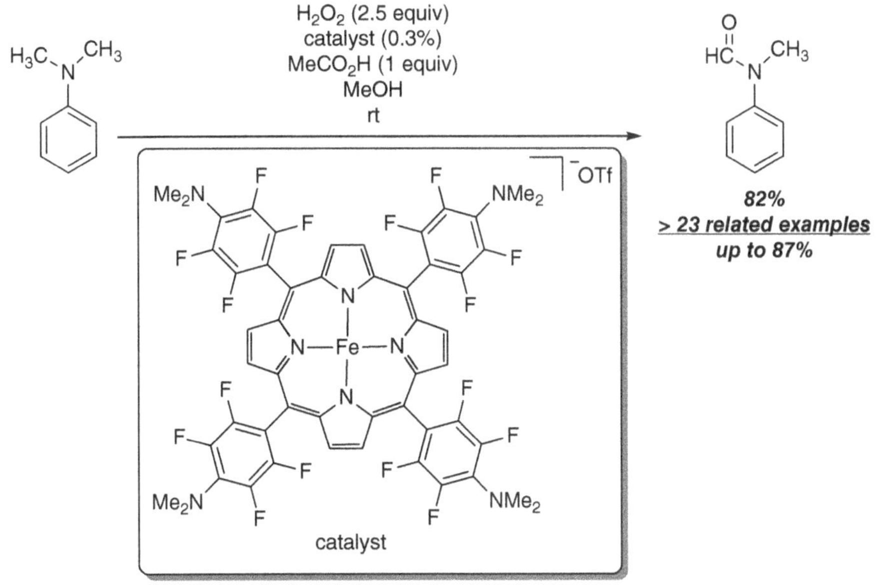

H$_2$O$_2$ (2.5 equiv)
catalyst (0.3%)
MeCO$_2$H (1 equiv)
MeOH
rt

82%
> 23 related examples
up to 87%

catalyst

SCHEME 2.120   Direct oxidation of C—H bonds using an iron catalyst [185].

catalysts containing a functionalized porphyrin core (Scheme 2.120) [185]. Hydrogen peroxide served as the oxidant in this chemistry, and reaction conditions were quite mild (room temperature). The substrates in this reaction were dialkylaniline derivatives, and the products were formamides. The process had excellent functional group tolerance, and an array of electron-donating and electron-withdrawing groups such as alkyl, aldehyde, halogens, ethers, and even an unprotected alkyne were retained. The chemistry was also fairly insensitive to bulky groups, and N,N-dimethyl-2,4,6-trimethylaniline was converted into the formamide in good yield (70%). When longer alkyl chains were attached to the nitrogen, the chemistry was selective for oxidation of the carbon attached to the nitrogen.

2.3.1 Troubleshooting the Preparation of Vinyl Ethers and Related Compounds

While it is not possible to predict all of the challenges and difficulties that will arise during the preparation of these compounds, the following suggestions will hopefully provide some direction for common issues and reaction requirements.

Problem/Issue/Goal	Possible Solution/Suggestion
Need to make a vinyl ether and do not have a glovebox or nitrogen line	Vinyl ethers can be generated by a copper-catalyzed cross-coupling reaction between alcohols and vinyl halides under air [105]
Need to use a vinylboronate as the starting material for the synthesis of a vinyl ether	The copper-promoted coupling of alcohols with vinyl boronates and boronic acids has been reported with yields as high as 99% [109]
Can a vinyl acetate be converted into a vinyl ether?	This reaction has been developed and uses a common iridium catalyst for the conversion [102]. This process is also attractive as it could be used to generate ethenyl ethers
Need to generate an ethenyl ether	This is a challenging reaction and has been accomplished through copper-catalyzed cross-coupling chemistry using a trivinylcyclotriboroxane as a vinylboronic acid equivalent [108]. It has also been achieved using iridium catalysis [102]
During the purification of vinyl ethers generated through cross-coupling chemistry, the stabilizing/solubilizing ligand such as a phosphine was difficult to remove and contaminated the products	A reasonable solution would be not to use a supporting ligand for the coupling reaction. Several successful approaches to this have been developed including an interesting copper/nickel coupling of vinyl halides with phenols [110]
Can a ketone or aldehyde be converted into a vinyl ether?	This chemistry has been achieved through the olefination of alkoxy sulfones [186]
Need to generate a vinyl ether with anti-Markovnikov selectivity and E-stereochemistry	A reasonable approach to this chemistry would be to use a cross-coupling reaction between a vinyl halide and phenol. Copper- and nickel-catalyzed reactions have been developed [110]
Can cyclic vinyl ethers be generated through hydroalkoxylation chemistry	There are several approaches to this chemistry using a host of different transition metals and lanthanides as catalysts [111, 112]. A ruthenium-catalyzed process is attractive since it generated cyclic ethers in air [157]. A copper catalyst has also shown the ability to generate high yields of cyclic ethers [124]
Need to make a vinyl ester with anti-Markovnikov selectivity	Both ruthenium- and rhodium-catalyzed routes are known for to generate the anti-Markovnikov addition products [136, 141]
Can any of the hydroalkoxylation reactions be carried out under air?	Several methods are tolerant to air. Nolan has developed a gold-catalyzed process for the addition of alcohols to internal alkynes that tolerates air [121]. A ruthenium-catalyzed hydroalkoxylation/annulation was carried out under air to generate an interesting class of phosphaisocoumarins [157]

(Continued)

(Continued)

Problem/Issue/Goal	Possible Solution/Suggestion
Can functionalized alkynes such as ynol ethers and ynamides be used in a hydroacyloxylation reaction?	Yes, many functionalized alkynes such as ynamides and ynol ethers are well tolerated in several metal-catalyzed vinyl ester syntheses [139, 140]
Can carboxylic acids be coupled directly with boronic acids or related species to generate vinyl esters?	Decarboxylation is a common side reaction with this chemistry; however, a copper-catalyzed approach has minimized this secondary process and generated good to excellent yields of the vinyl esters with electron-rich aryl carboxylic acids [130]
Need to use water as the solvent for the synthesis of a vinyl ester	If a carboxylic acid can be generated or obtained to serve as the substrate, a ruthenium-catalyzed addition of carboxylic acids to alkynes has been reported [142]
Prepare a vinyl ester through hydroalkoxylation chemistry with anti-Markovnikov selectivity and Z-stereochemistry	The ruthenium-catalyzed approach to this chemistry is particularly attractive due to the mild conditions, high selectivity, and the use of an air-stable catalyst [136]. A rhodium-catalyzed process using an easily handled catalyst precursor has also been reported [141]
Can cyclic esters (vinyl lactones) be generated through metal-catalyzed intramolecular cyclizations?	Several catalysts have shown promise for this reaction including one that function in the presence of air [145]
Need a practical version of the Wacker oxidation of alkenes with Markovnikov selectivity	While several versions of this reaction are known, a modification that uses low palladium loadings and functions at low O_2 pressures has been reported [167]
Does a "greener" version of the Wacker oxidation chemistry exist?	While a number of approaches have eliminated several of the more environmentally harmful additives, the use of iron(III) sulfate as the terminal oxidant in the Markovnikov-selective process is particularly attractive due to its operational simplicity and functional group tolerance [168]
Need a practical version of the Wacker oxidation of alkenes with anti-Markovnikov selectivity	This is a challenging reaction and general approaches with wide substrate scope are rare. A catalyst system for a highly anti-Markovnikov process has been developed for styrene derivatives as well as a range of functionalized alkenes [169, 170]
Need to hydrate an alkyne with Markovnikov selectivity	Gold catalysts are highly active for this transformation. The use of an air-stable chlorogold complex as the catalyst precursor is particularly attractive as it is effective at ultralow catalyst loadings (as low as 10 ppm) [175]
Need to hydrate an alkyne with anti-Markovnikov selectivity	This transformation is much less common than the Markovnikov selective version. A reasonable approach would be to use a ruthenium catalyst to promote the anti-Markovnikov hydration reaction [180]
Need to make a 1,2-diketone from an alkyne	While a number of transition metal-catalyzed routes are known for this reaction, the metal-free conversion of alkynones into 1,2-diketones is an attractive approach [181]

2.4 PREPARATION OF ALKYNYL ETHERS AND RELATED COMPOUNDS THROUGH THE FORMATION OF OXYGEN–CARBON(SP) BONDS

In contrast with the vast array of known alkyl or aryl alcohols/ethers, relatively scant attention has been paid to ynols and related compounds. This is due, in part, to the rapid conversion of formal ynols into ketene tautomers. It should be noted that metal ynolates can be readily generated and used for organic transformations [187]. For ynol ethers, a relatively small number of approaches are known for the preparation of these compounds. The following sections will highlight modifications made to this system as well as new methods for generating oxygen–carbon (sp³) bonds.

If it is convenient to start from an alcohol, these compounds can be transformed into ynol ethers following a procedure outlined by Greene [188]. As a representative example, conversion of cyclohexanol into ynol ethers proceeded in good yield following this approach (Scheme 2.121) [189]. Initial deprotonation of the alcohol followed by addition of the trichloroethylene and subsequent injection of BuLi generated the lithium acetylide that was converted into the ynol ether upon aqueous workup. While there is a bit of operational complexity in this approach, it is mitigated by the observation that all of the steps occurred in the same flask. It is simply a matter of injecting reagents along with heating and cooling the flask. Naturally, there will be some functional group incompatibility due to the use of KH and BuLi, but if the alcohol is well protected, this is an attractive approach to the conversion into an ynol ether.

While there are a number of routes to metal ynolates and siloxy alkynes, terminal alkynes are convenient entry points to this chemistry [187]. Elegant work by Stang demonstrated that metal ynolates can be generated from terminal alkynes using alkynyliodonium reagents [190]. The direct products of this reaction were siloxy alkynes, which were converted into the lithium ynolates using MeLi. Alternative approaches to the synthesis of siloxy alkynes are known and also start from terminal alkynes (Scheme 2.122) [191, 192]. Deprotonation of the terminal acetylide with a strong base followed by the addition of *tert*-butyl hydroperoxide and subsequent trapping with triisopropyltrifluoromethanesulfonate afforded the siloxy alkynes in good yield. The typical reactivity that was normally associated with smaller amides was mitigated by using the bulky MHMDS bases. While only a few examples were reported, this work outlines a proof-of-concept for the chemistry and provides the groundwork for expansion to other substrates.

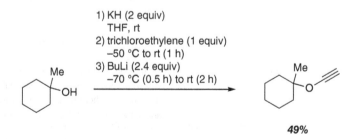

SCHEME 2.121 Synthesis of ynol ethers using an elimination approach [188].

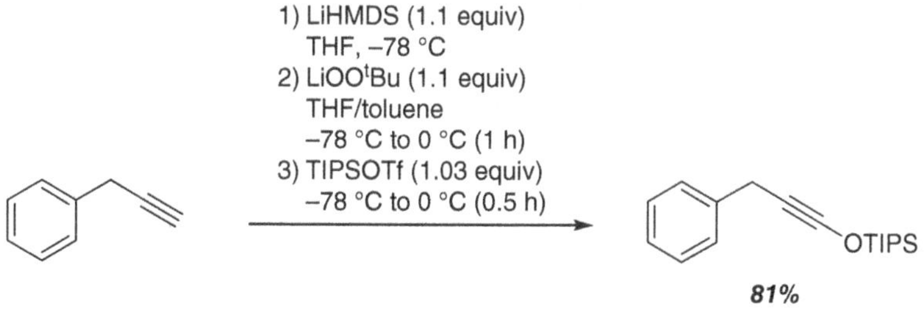

1) LiHMDS (1.1 equiv)
   THF, −78 °C
2) LiOO^{t}Bu (1.1 equiv)
   THF/toluene
   −78 °C to 0 °C (1 h)
3) TIPSOTf (1.03 equiv)
   −78 °C to 0 °C (0.5 h)

81%

SCHEME 2.122   Synthesis of a siloxy alkyne starting from a terminal acetylene [191].

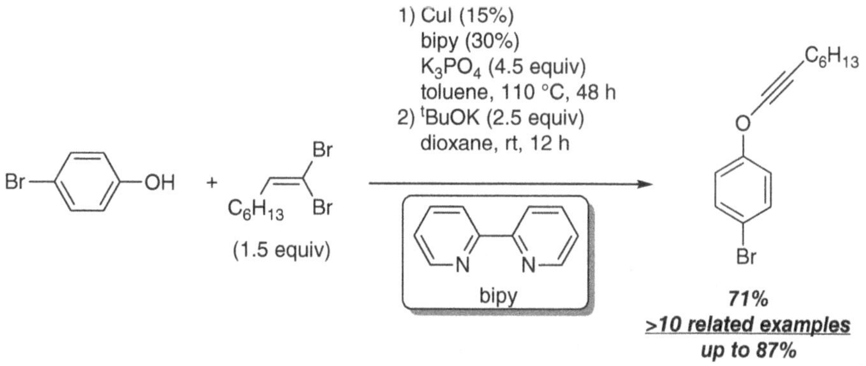

1) CuI (15%)
   bipy (30%)
   K$_3$PO$_4$ (4.5 equiv)
   toluene, 110 °C, 48 h
2) ^{t}BuOK (2.5 equiv)
   dioxane, rt, 12 h

Br─⟨ ⟩─OH   +

C$_6$H$_{13}$

(1.5 equiv)

bipy

C$_6$H$_{13}$

Br

71%
≥10 related examples
up to 87%

SCHEME 2.123   Copper-promoted synthesis of ynol ethers [193].

The synthesis of ynol ethers from *gem*-dibromoalkenes has been achieved through a two-step process (Scheme 2.123) [193, 194]. The first step involved the use of copper catalysts to promote the conversion of the dibromoalkene starting material into a brominated vinyl ether. The authors screened a range of supporting ligands and discovered that an inexpensive bipyridine was among the most effective for the chemistry. The second step used a strong base to promote the elimination reaction and generate the ynol ether. A range of phenols bearing electron-donating and electron-withdrawing groups were successfully transformed in this work including a protected estrogen derivative (Example 2.16) [193].

The synthesis of ynol ethers from sulfonamides using potassium alkoxides proceeded in moderate yields under mild conditions (Scheme 2.124) [195]. The chemistry entailed initial conversion of alcohols into potassium alkoxides through the addition of potassium to solutions of the alcohol, followed by the sulfonamide. An amine was needed for a successful conversion and dimethyl amine was a convenient choice. There was a level of operational complexity to this approach in that elemental potassium needed to be handled. Provided that the commercially available solutions were active in the chemistry, it would be more practical to obtain solutions of the alkoxides.

In addition to these methods, ynol ethers and metal ynolates can be generated from dibromocarboxylic acids using strong bases [196–200]. While this methodology generated the desired ynol species in good to excellent yield, the process needs to start from a dibrominated carboxylic acid.

Example 2.16 Synthesis of Ynol Ethers from *gem*-Dibromoalkenes [193].

Before starting this synthesis, carefully follow the safety instructions listed in Example 2.1.

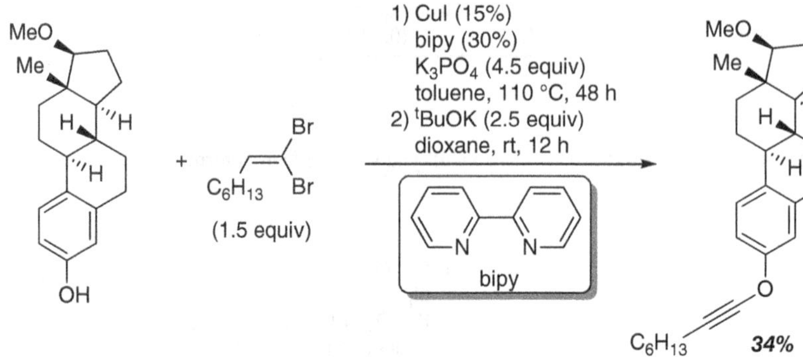

A 15-mL pressure tube was charged with the phenol (0.67 mmol), 2,2'-bipyridine (31 mg, 0.2 mmol), K_3PO_4 (650 mg, 3.1 mmol), and copper(I) iodide (19 mg, 0.1 mmol); if solid, the 1,1-dibromo-1-alkene (1.0 mmol) was also introduced at this stage. The tube was fitted with a rubber septum, evacuated under high vacuum, and backfilled with argon. Dry and degassed toluene (2 mL) was next added as well as the 1,1-dibromo-1-alkene (1.0 mmol), which was added at this stage if liquid. The rubber septum was replaced by Teflon-coated screw cap and the heterogeneous suspension heated at 110 °C for 2 days, cooled to room temperature, diluted with dry dioxane (2 mL), and treated with potassium *tert*-butoxide (188 mg, 1.67 mmol). The resulting mixture was stirred overnight at room temperature, filtered through a plug of silica gel (washed with EtOAc), and concentrated in vacuo. The crude residue was purified by flash column chromatography.

Adapted with permission from K. Jouvin, A. Bayle, F. Legrand, and G. Evano, *Org. Lett.*, **2012**, *14*, 1652–1655. *Copyright (2012) American Chemical Society.*

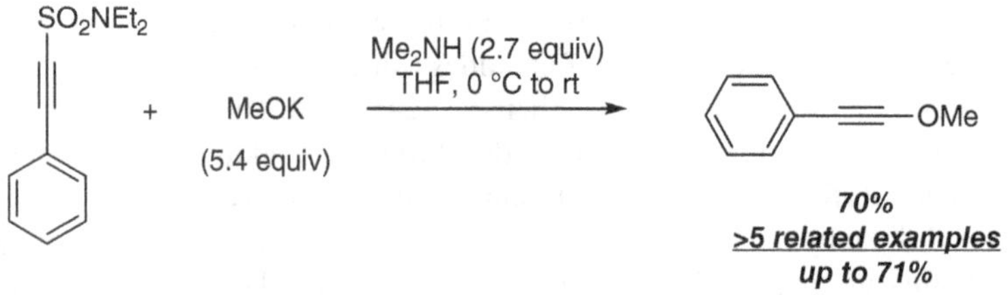

SCHEME 2.124 Synthesis of ynol ethers from sulfonamides [195].

2.4.1 Troubleshooting the Preparation of Alkynyl Ethers and Related Compounds

While it is not possible to predict all of the challenges and difficulties that will arise during the preparation of alkynyl ethers and related compounds, the following suggestions will hopefully provide some direction for common issues and reaction requirements.

Problem/Issue/Goal	Possible Solution/Suggestion
Need to generate an ynol ether and have to start from an alcohol	The conversion of alcohols into ynol ethers has been achieved [188, 189]. This methodology requires the use of powerful bases; thus, the potential for functional group incompatibility needs to be addressed prior to implementation of this approach
Need to generate an alkynyl ether and it would be convenient to start from the gem-dibromo alkene	Ynol ethers have been generated from *gem*-dibromoalkenes through initial formation of a brominated vinyl ether followed by elimination [193]
Need to generate an ynol ether	Terminal alkynes are convenient starting points for this chemistry and several excellent approaches are known [187, 191, 192]

REFERENCES

[1] Beller, M.; Seayad, J.; Tillack, A.; Jiao, H. *Angew. Chem. Int. Ed.* **2004**, *43*, 3368–3398.

[2] Zeng, M.; Li, L.; Herzon, S. B. *J. Am. Chem. Soc.* **2014**, *136*, 7058–7067.

[3] Molander, G. A.; Cavalcanti, L. N. *J. Org. Chem.* **2011**, *76*, 623–630.

[4] Ely, R. J.; Morken, J. P. *J. Am. Chem. Soc.* **2010**, *132*, 2534–2535.

[5] Kim, D. W.; Hong, D. J.; Seo, J. W.; Kim, H. S.; Kim, H. K.; Song, C. E.; Chi, D. Y. *J. Org. Chem.* **2004**, *69*, 3186–3189.

[6] McNeill, E.; Du Bois, J. *J. Am. Chem. Soc.* **2010**, *132*, 10202–10204.

[7] McNeill, E.; Du Bois, J. *Chem. Sci.* **2012**, *3*, 1810–1813.

[8] Melhado, A. D.; Brenzovich, W. E. Jr.; Lackner, A. D.; Toste, F. D. *J. Am. Chem. Soc.* **2010**, *132*, 8885–8887.

[9] Pierce, C. J.; Hilinski, M. K. *Org. Lett.* **2014**, *16*, 6504–6507.

[10] Adams, A. M.; Du Bois, J. *Chem. Sci.* **2014**, *5*, 656–659.

[11] White, N. A.; Rovis, T. *J. Am. Chem. Soc.* **2014**, *136*, 14674–14677.

[12] Leitner, A.; Shu, C.; Hartwig, J. F. *Org. Lett.* **2005**, *7*, 1093–1096.

[13] Lyothier, I.; Defieber, C.; Carreira, E. M. *Angew. Chem. Int. Ed.* **2006**, *45*, 6204–6207.

[14] Zhang, S.; Zhao, Y. *Bioconjug. Chem.* **2011**, *22*, 523–528.

[15] Nekongo, E. E.; Popik, V. V. *J. Org. Chem.* **2014**, *79*, 7665–7671.

[16] Tissot, M.; Phipps, R. J.; Lucas, C.; Leon, R. M.; Pace, R. D. M.; Ngouansavanh, T.; Gaunt, M. J. *Angew. Chem. Int. Ed.* **2014**, *53*, 13498–13501.

[17] Oka, N.; Kajino, R.; Takeuchi, K.; Nagakawa, H.; Ando, K. *J. Org. Chem.* **2014**, *79*, 7656–7664.

[18] Kim, J.; Lee, D.-H.; Kalutharage, N.; Yi, C. S. *ACS Catal.* **2014**, *4*, 3881–3885.

[19] Nising, C. F.; Brase, S. *Chem. Soc. Rev.* **2012**, *41*, 988–999.

[20] Wang, F.; Yang, H.; Fu, H.; Pei, Z. *Chem. Commun.* **2013**, *49*, 517–519.

[21] Barreiro, E. M.; Adrio, L. A.; Hii, K. K.; Brazier, J. B. *Eur. J. Org. Chem.* **2013**, *2013*, 1027–1039.

[22] Yang, C.-G.; He, C. *J. Am. Chem. Soc.* **2005**, *127*, 6966–6967.

[23] Li, Z.; Zhang, J.; Brouwer, C.; Yang, C.-G.; Reich, N. W.; He, C. *Org. Lett.* **2006**, *8*, 4175–4178.

[24] Reddi, R. N.; Prasad, P. K.; Sudalai, A. *Org. Lett.* **2014**, *16*, 5674–5677.

[25] Oe, Y.; Ohta, T.; Ito, Y. *Tetrahedron Lett.* **2010**, *51*, 2806–2809.

[26] Li, L.; Su, C.; Liu, X.; Tian, H.; Shi, Y. *Org. Lett.* **2014**, *16*, 3728–3731.

[27] Jiang, X.; Wu, L.; Xing, Y.; Wang, L.; Wang, S.; Chen, Z.; Wang, R. *Chem. Commun.* **2012**, *48*, 446–448.

[28] Geng, Z.-C.; Zhang, S.-Y.; Li, N.-K.; Li, N.; Chen, J.; Li, H.-Y.; Wang, X.-W. *J. Org. Chem.* **2014**, *79*, 10772–10785.

[29] Davis, O. A.; Bull, J. A. *Angew. Chem. Int. Ed.* **2014**, *53*, 14230–14234.

[30] Honda, M.; Tamura, M.; Nakao, K.; Suzuki, K.; Nakagawa, Y.; Tomishige, K. *ACS Catal.* **2014**, *4*, 1893–1896.

[31] Ema, T.; Miyazaki, Y.; Shimonishi, J.; Maeda, C.; Hasegawa, J. *J. Am. Chem. Soc.* **2014**, *136*, 15270–15279.

[32] Katsuki, T.; Sharpless, K. B. *J. Am. Chem. Soc.* **1980**, *102*, 5974–5976.

[33] Wong, O. A.; Shi, Y. *Chem. Rev.* **2008**, *108*, 3958–3987.

[34] Zhu, Y.; Wang, Q.; Cornwall, R. G.; Shi, Y. *Chem. Rev.* **2014**, *114*, 8199–8256.

[35] Qian, Q.; Tan, Y.; Zhao, B.; Feng, T.; Shen, Q.; Yao, Y. *Org. Lett.* **2014**, *16*, 4516–4519.

[36] Li, B.; Li, C. *J. Org. Chem.* **2014**, *79*, 8271–8277.

[37] Toda, Y.; Pink, M.; Johnston, J. N. *J. Am. Chem. Soc.* **2014**, *136*, 14734–14737.

[38] Enthaler, S.; Company, A. *Chem. Soc. Rev.* **2011**, *40*, 4912–4924.

[39] Willis, M. C. *Angew. Chem. Int. Ed.* **2007**, *46*, 3402–3404.

[40] Alonso, D. A.; Nájera, C.; Pastor, I. M.; Yus, M. *Chem. Eur. J.* **2010**, *16*, 5274–5284.

[41] Zhu, C.; Wang, R.; Falck, J. R. *Org. Lett.* **2012**, *14*, 3494–3497.

[42] Zou, Y.; Chen, J.; Liu, X.; Lu, L.; Davis, R. L.; Jørgensen, K. A.; Xiao, W. *Angew. Chem. Int. Ed.* **2012**, *51*, 784–788.

[43] Rayment, E. J.; Summerhill, N.; Anderson, E. A. *J. Org. Chem.* **2012**, *77*, 7052–7060.

[44] Xu, J.; Wang, X.; Shao, C.; Su, D.; Cheng, G.; Hu, Y. *Org. Lett.* **2010**, *12*, 1964–1967.

[45] Yang, H.; Li, Y.; Jiang, M.; Wang, J.; Fu, H. *Chem. Eur. J.* **2011**, *17*, 5652–5660.

[46] Tlili, A.; Xia, N.; Monnier, F.; Taillefer, M. *Angew. Chem. Int. Ed.* **2009**, *48*, 8725–8728.

[47] Zhao, D.; Wu, N.; Zhang, S.; Xi, P.; Su, X.; Lan, J.; You, J. *Angew. Chem. Int. Ed.* **2009**, *48*, 8729–8732.

[48] Yang, D.; Fu, H. *Chem. Eur. J.* **2010**, *16*, 2366–2370.

[49] Xu, H.-J.; Liang, Y.-F.; Cai, Z.-Y.; Qi, H.-X.; Yang, C.-Y.; Feng, Y.-S. *J. Org. Chem.* **2011**, *76*, 2296–2300.

[50] Ding, G.; Han, H.; Jiang, T.; Wu, T.; Han, B. *Chem. Commun.* **2014**, *50*, 9072–9075.

[51] Anderson, K. W.; Ikawa, T.; Tundel, R. E.; Buchwald, S. L. *J. Am. Chem. Soc.* **2006**, *128*, 10694–10695.

[52] Cheung, C. W.; Buchwald, S. L. *J. Org. Chem.* **2014**, *79*, 5351–5358.

[53] Arshad, N.; Hashim, J.; Kappe, C. O. *J. Org. Chem.* **2008**, *73*, 4755–4758.

[54] Yu, C.-W.; Chen, G. S.; Huang, C.-W.; Chern, J.-W. *Org. Lett.* **2012**, *14*, 3688–3691.

[55] Schulz, T.; Torborg, C.; Schäffner, B.; Huang, J.; Zapf, A.; Kadyrov, R.; Börner, A.; Beller, M. *Angew. Chem. Int. Ed.* **2009**, *48*, 918–921.

[56] Maleczka, R. E. Jr.; Shi, F.; Holmes, D.; Smith, M. R., III. *J. Am. Chem. Soc.* **2003**, *125*, 7792–7793.

[57] Liu, W.; Ackermann, L. *Org. Lett.* **2013**, *15*, 3484–3486.

[58] Yang, F.; Rauch, K.; Kettelhoit, K.; Ackermann, L. *Angew. Chem. Int. Ed.* **2014**, *53*, 11285–11288.

[59] Thirunavukkarasu, V. S.; Kozhushkov, S. I.; Ackermann, L. *Chem. Commun.* **2014**, *50*, 29–39.

[60] Yang, Y.; Lin, Y.; Rao, Y. *Org. Lett.* **2012**, *14*, 2874–2877.

[61] Thirunavukkarasu, V. S.; Hubrich, J.; Ackermann, L. *Org. Lett.* **2012**, *14*, 4210–4213.

[62] Mo, F.; Trzepkowski, L. J.; Dong, G. *Angew. Chem. Int. Ed.* **2012**, *51*, 13075–13079.

[63] Shan, G.; Yang, X.; Ma, L.; Rao, Y. *Angew. Chem. Int. Ed.* **2012**, *51*, 13070–13074.

[64] Zhang, Y.-H.; Yu, J.-Q. *J. Am. Chem. Soc.* **2009**, *131*, 14654–14655.

[65] Li, X.; Liu, Y.-H.; Gu, W.-J.; Li, B.; Chen, F.-J.; Shi, B.-F. *Org. Lett.* **2014**, *16*, 3904–3907.

[66] Dong, J.; Liu, P.; Sun, P. *J. Org. Chem.* **2015**, *80*, 2925–2929.

[67] Zhang, Z.-J.; Quan, X.-J.; Ren, Z.-H.; Wang, Y.-Y.; Guan, Z.-H. *Org. Lett.* **2014**, *16*, 3292–3295.

[68] Qiao, J. X.; Lam, P. Y. S. *Synthesis* **2011**, *6*, 829–856.

[69] Prim, D.; Campagne, J.-M.; Joseph, D.; Andrioletti, B. *Tetrahedron*, **2002**, *58*, 2041–2075.

[70] Ley, S. V.; Thomas, A. W. *Angew. Chem. Int. Ed.* **2003**, *42*, 5400–5449.

[71] Evans, D. A.; Katz, J. L.; West, T. R. *Tetrahedron Lett.* **1998**, *39*, 2937–2940.

[72] Lam, P. Y. S.; Vincent, G.; Clark, C. G.; Deudon, S.; Jadhav, P. K. *Tetrahedron Lett.* **2001**, *42*, 3415–3418.

[73] Chan, D. M. T.; Monaco, K. L.; Wang, R.-P.; Winters, M. P. *Tetrahedron Lett.* **1998**, *39*, 2933–2936.

[74] Marcoux, J.-F.; Doye, S.; Buchwald, S. L. *J. Am. Chem. Soc.* **1997**, *119*, 10539–10540.

[75] Wolter, M.; Nordmann, G.; Job, G. E.; Buchwald, S. L. *Org. Lett.* **2002**, *4*, 973–976.

[76] Quach, T. D.; Batey, R. A. *Org. Lett.* **2003**, *5*, 1381–1384.

[77] El Khatib, M.; Molander, G. A. *Org. Lett.* **2014**, *16*, 4944–4947.

[78] Manbeck, G. F.; Lipman, A. J.; Stockland, R. A. Jr.; Freidl, A. L.; Hasler, A. F.; Stone, J. J.; Guzei, I. A. *J. Org. Chem.* **2005**, *70*, 244–250.

[79] Xiao, Y.; Xu, Y.; Cheon, H.-S.; Chae, J. *J. Org. Chem.* **2013**, *78*, 5804–5809.

[80] Bhadra, S.; Dzik, W. I.; Goossen, L. J. *J. Am. Chem. Soc.* **2012**, *134*, 9938–9941.

[81] Lyons, T. W.; Sanford, M. S. *Chem. Rev.* **2010**, *110*, 1147–1169.

[82] Bhadra, S.; Matheis, C.; Katayev, D.; Gooßen, L. J. *Angew. Chem. Int. Ed.* **2013**, *52*, 9279–9283.

[83] Palucki, M.; Wolfe, J. P.; Buchwald, S. L. *J. Am. Chem. Soc.* **1996**, *118*, 10333–10334.

[84] Palucki, M.; Wolfe, J. P.; Buchwald, S. L. *J. Am. Chem. Soc.* **1997**, *119*, 3395–3396.

[85] Mann, G.; Hartwig, J. F. *J. Am. Chem. Soc.* **1996**, *118*, 13109–13110.

[86] Vorogushin, A. V.; Huang, X.; Buchwald, S. L. *J. Am. Chem. Soc.* **2005**, *127*, 8146–8149.

[87] Li, W.; Sun, P. *J. Org. Chem.* **2012**, *77*, 8362–8366.

[88] Gulevich, A. V.; Melkonyan, F. S.; Sarkar, D.; Gevorgyan, V. *J. Am. Chem. Soc.* **2012**, *134*, 5528–5531.

[89] Monnier, F.; Taillefer, M. *Angew. Chem. Int. Ed.* **2009**, *48*, 6954–6971.

[90] Decicco, C. P.; Song, Y.; Evans, D. A. *Org. Lett.* **2001**, *3*, 1029–1032.

[91] Boyer, S. H.; Jiang, H.; Jacintho, J. D.; Reddy, M. V.; Li, H.; Li, W.; Godwin, J. L.; Schulz, W. G.; Cable, E. E.; Hou, J.; Wu, R.; Fujitaki, J. M.; Hecker, S. J.; Erion, M. D. *J. Med. Chem.* **2008**, *51*, 7075–7093.

[92] Gujadhur, R. K.; Bates, C. G.; Venkataraman, D. *Org. Lett.* **2001**, *3*, 4315–4317.

[93] Alla, S. K.; Sadhu, P.; Punniyamurthy, T. *J. Org. Chem.* **2014**, *79*, 7502–7511.

[94] Cheng, X.-F.; Li, Y.; Su, Y.-M.; Yin, F.; Wang, J.-Y.; Sheng, J.; Vora, H. U.; Wang, X.-S.; Yu, J.-Q. *J. Am. Chem. Soc.* **2013**, *135*, 1236–1239.

[95] Guo, Q.-H.; Fu, Z.-D.; Zhao, L.; Wang, M.-X. *Angew. Chem. Int. Ed.* **2014**, *53*, 13548–13552.

[96]   Salih, M. Q.; Beaudry, C. M. *Org. Lett.* **2014**, *16*, 4964–4966.

[97]   Ji, F.; Li, X.; Wu, W.; Jiang, H. *J. Org. Chem.* **2014**, *79*, 11246–11253.

[98]   Zhang, M.; Jiang, H.-F.; Neumann, H.; Beller, M.; Dixneuf, P. H. *Angew. Chem. Int. Ed.* **2009**, *48*, 1681–1684.

[99]   Dai, J.-J.; Liu, J.-H.; Luo, D.-F.; Liu, L. *Chem. Commun.* **2011**, *47*, 677–679.

[100] Zhang, L.; Zhang, G.; Zhang, M.; Cheng, J. *J. Org. Chem.* **2010**, *75*, 7472–7474.

[101] Reppe, W. *Justus Liebigs Ann Chem* **1956**, *601*, 81–138.

[102] Okimoto, Y.; Sakaguchi, S.; Ishii, Y. *J. Am. Chem. Soc.* **2002**, *124*, 1590–1591.

[103] Mckeon, J. E.; Fitton, P.; Griswold, A. A. *Tetrahedron* **1972**, *28*, 227–232.

[104] Weintraub, P. M.; King, C.-H. R. *J. Org. Chem.* **1997**, *62*, 1560–1562.

[105] Nordmann, G.; Buchwald, S. L. *Am. Chem. Soc.* **2003**, *125*, 4978–4979.

[106] Keegstra, M. A. *Tetrahedron* **1992**, *48*, 2681–2690.

[107] Lam, P. Y. S.; Vincent, G.; Bonne, D.; Clark, C. G. *Tetrahedron Lett.* **2003**, *44*, 4927–4931.

[108] McKinley, N. F.; O'Shea, D. F. *J. Org. Chem.* **2004**, *69*, 5087–5092.

[109] Shade, R. E.; Hyde, A. M.; Olsen, J.-C.; Merlic, C. A. *J. Am. Chem. Soc.* **2010**, *132*, 1202–1203.

[110] Kundu, D.; Maity, P.; Ranu, B. C. *Org. Lett.* **2014**, *16*, 1040–1043.

[111] Alonso, F.; Beletskaya, I. P.; Yus, M. *Chem. Rev.* **2004**, *104*, 3079–3160.

[112] McDonald, F. E. *Chem. Eur. J.* **1999**, *5*, 3103–3106.

[113] Teles, J. H.; Brode, S.; Chabanas, M. *Angew. Chem. Int. Ed.* **1998**, *37*, 1415–1418.

[114] Kuram, M. R.; Bhanuchandra, M.; Sahoo, A. K. *J. Org. Chem.* **2010**, *75*, 2247–2258.

[115] Al-Amin, M.; Johnson, J. S.; Blum, S. A. *Organometallics* **2014**, *33*, 5448–5456.

[116] Shi, Y.; Peterson, S. M.; Haberaecker, W. W. III.; Blum, S. A. *J. Am. Chem. Soc.* **2008**, *130*, 2168–2169.

[117] Wang, D.; Cai, R.; Sharma, S.; Jirak, J.; Thummanapelli, S. K.; Akhmedov, N. G.; Zhang, H.; Liu, X.; Petersen, J. L.; Shi, X. *J. Am. Chem. Soc.* **2012**, *134*, 9012–9019.

[118] Oonishi, Y.; Gómez-Suárez, A.; Martin, A. R.; Nolan, S. P. *Angew. Chem. Int. Ed.* **2013**, *52*, 9767–9771.

[119] Richard, M. E.; Fraccica, D. V.; Garcia, K. J.; Miller, E. J.; Ciccarelli, R. M.; Holahan, E. C.; Resh, V. L.; Shah, A.; Findeis, P. M.; Stockland, R. A., Jr. *Beilstein J. Org. Chem.* **2013**, *9*, 2002–2008.

[120] Oonishi, Y.; Gómez-Suárez, A.; Martin, A. R.; Makida, Y.; Slawin, A. M. Z.; Nolan, S. P. *Chem. Eur. J.* **2014**, *20*, 13507–13510.

[121] Veenboer, R. M. P.; Dupuy, S.; Nolan, S. P. *ACS Catal.* **2015**, *5*, 1330–1334.

[122] Lenker, H. K.; Gray, T. G.; Stockland, R. A. Jr. *Dalton Trans.* **2012**, *41*, 13274–13276.

[123] Ketcham, J. M.; Biannic, B.; Aponick, A. *Chem. Commun.* **2013**, *49*, 4157–4159.

[124] Pouy, M. J.; Delp, S. A.; Uddin, J.; Ramdeen, V. M.; Cochrane, N. A.; Fortman, G. C.; Gunnoe, T. B.; Cundari, T. R.; Sabat, M.; Myers, W. H. *ACS Catal.* **2012**, *2*, 2182–2193.

[125] Scully, S. S.; Zheng, S.-L.; Wagner, B. K.; Schreiber, S. L. *Org. Lett.* **2015**, *17*, 418–421.

[126] Medina, J. M.; McMahon, T. C.; Jimenez-Oses, G.; Houk, K. N.; Garg, N. K. *J. Am. Chem. Soc.* **2014**, *136*, 14706–14709.

[127] Mariaule, G.; Newsome, G.; Toullec, P. Y.; Belmont, P.; Michelet, V. *Org. Lett.* **2014**, *16*, 4570–4573.

[128] Riveira, M. J.; Mischne, M. P. *J. Org. Chem.* **2014**, *79*, 8244–8254.

[129] Hu, Y.; Yi, R.; Wang, C.; Xin, X.; Wu, F.; Wan, B. *J. Org. Chem.* **2014**, *79*, 3052–3059.

[130] Huang, F.; Quach, T. D.; Batey, R. A. *Org. Lett.* **2013**, *15*, 3150–3153.

[131] Rotem, M.; Shvo, Y. *Organometallics* **1983**, *2*, 1689–1691.

[132] Doucet, H.; Hofer, J.; Bruneau, C.; Dixneuf, P. H. *J. Chem. Soc., Chem. Comm.* **1993**, 850–851.

[133] Douchet, H.; Martin-Vaca, B.; Bruneau, C.; Dixneuf, P. H. *J. Org. Chem.* **1995**, *60*, 7247–7255.

[134] Mitsudo, T.; Hori, Y.; Wantabe, Y. *J. Org. Chem.* **1985**, *50*, 1566–1568.

[135] Mitsudo, T.; Hori, Y.; Yamakawa, Y.; Watanabe, Y. *J. Org. Chem.* **1987**, *52*, 2230–2239.

[136] Goossen, L. J.; Paetzold, J.; Koley, D. *Chem. Commun.* **2003**, 706–707.

[137] Nicks, F.; Libert, L.; Delaude, L.; Demonceau, A. *Aust. J. Chem.* **2009**, *62*, 227–231.

[138] Chary, B. C.; Kim, S. *J. Org. Chem.* **2010**, *75*, 7928–7931.

[139] Smith, D. L.; Goundry, W. R. F.; Lam, H. W. *Chem. Commun.* **2012**, *48*, 1505–1507.

[140] Yin, J.; Bai, Y.; Mao, M.; Zhu, G. *J. Org. Chem.* **2014**, *79*, 9179–9185.

[141] Lumbroso, A.; Vautravers, N. R.; Breit, B. *Org. Lett.* **2010**, *12*, 5498–5501.

[142] Cadierno, V.; Francos, J.; Gimeno, J. *Organometallics* **2011**, *30*, 852–862.

[143] Le Paih, J.; Monnier, F.; Derien, S.; Dixneuf, P. H.; Clot, E.; Eisenstein, O. *J. Am. Chem. Soc.* **2003**, *125*, 11964–11975.

[144] Egi, M.; Ota, Y.; Nishimura, Y.; Shimizu, K.; Azechi, K.; Akai, S. *Org. Lett.* **2013**, *15*, 4150–4153.

[145] Rodriguez-Alvarez, M. J.; Vidal, C.; Diez, J.; Garcia-Alvarez, J. *Chem. Commun.* **2014**, *50*, 12927–12929.

[146] Espinosa-Jalapa, N. A.; Ke, D.; Nebra, N.; Le Goanvic, L.; Mallet-Ladeira, S.; Monot, J.; Martin-Vaca, B.; Bourissou, D. *ACS Catal.* **2014**, *4*, 3605–3611.

[147] Nebra, N.; Monot, J.; Shaw, R.; Martin-Vaca, B.; Bourissou, D. *ACS Catal.* **2013**, *3*, 2930–2934.

[148] Kamimura, A.; Yamane, Y.; Yo, R.; Tanaka, T.; Uno, H. *J. Org. Chem.* **2014**, *79*, 7696–7702.

[149] Yoo, W.-J.; Nguyen, T. V. Q.; Kobayashi, S. *Angew. Chem. Int. Ed.* **2014**, *53*, 10213–10217.

[150] Xie, C.; Liu, L.; Zhang, Y.; Xu, P. *Org. Lett.* **2008**, *10*, 2393–2396.

[151] Mbofana, C. T.; Miller, S. J. *ACS Catal.* **2014**, *4*, 3671–3674.

[152] Tomas-Mendivil, E.; Toullec, P. Y.; Borge, J.; Conejero, S.; Michelet, V.; Cadierno, V. *ACS Catal.* **2013**, *3*, 3086–3098.

[153] Zhu, F.-L.; Wang, Y.-H.; Zhang, D.-Y.; Xu, J.; Hu, X.-P. *Angew. Chem. Int. Ed.* **2014**, *53*, 10223–10227.

[154] Lu, Y.; Wang, D.-H.; Engle, K. M.; Yu, J.-Q. *J. Am. Chem. Soc.* **2010**, *132*, 5916–5921.

[155] Yu, W.; Chen, J.; Gao, K.; Liu, Z.; Zhang, Y. *Org. Lett.* **2014**, *16*, 4870–4873.

[156] Kang, Y.-W.; Choi, Y.-J.; Jang, H.-Y. *Org. Lett.* **2014**, *16*, 4842–4845.

[157] Park, Y.; Jeon, I.; Shin, S.; Min, J.; Lee, P. H. *J. Org. Chem.* **2013**, *78*, 10209–10220.

[158] Seo, J.; Park, Y.; Jeon, I.; Ryu, T.; Park, S.; Lee, P. H. *Org. Lett.* **2013**, *15*, 3358–3361.

[159] Unoh, Y.; Hashimoto, Y.; Takeda, D.; Hirano, K.; Satoh, T.; Miura, M. *Org. Lett.* **2013**, *15*, 3258–3261.

[160] Cui, D.-M.; Meng, Q.; Zheng, J.-Z.; Zhang, C. *Chem. Commun.* **2009**, *12*, 1577–1579.

[161] Chen, C.; Li, P.; Hu, Z.; Wang, H.; Zhu, H.; Hu, X.; Wang, Y.; Lv, H.; Zhang, X. *Org. Chem. Front.* **2014**, *1*, 947–951.

[162] Chen, C.; Qiao, Y.; Geng, H.; Zhang, X. *Org. Lett.* **2013**, *15*, 1048–1051.

[163] Smidt, J.; Sedlmeier, J.; Hafner, W.; Sieber, R.; Sabela, A.; Jira, R. *Angew. Chem.* **1962**, *74*, 93–102.

[164] Jira, R. *Angew. Chem. Int. Ed.* **2009**, *48*, 9034–9037.

[165] Stirling, A.; Nair, N. N.; Lledos, A.; Ujaque, G. *Chem. Soc. Rev.* **2014**, *43*, 4940–4952.

[166] Keith, J. A.; Henry, P. M. *Angew. Chem. Int. Ed.* **2009**, *48*, 9038–9049.

[167] Cornell, C. N.; Sigman, M. S. *Org. Lett.* **2006**, *8*, 4117–4120.

[168] Fernandes, R. A.; Chaudhari, D. A. *J. Org. Chem.* **2014**, *79*, 5787–5793.

[169] Teo, P.; Wickens, Z. K.; Dong, G.; Grubbs, R. H. *Org. Lett.* **2012**, *14*, 3237–3239.

[170] Wickens, Z. K.; Skakuj, K.; Morandi, B.; Grubbs, R. H. *J. Am. Chem. Soc.* **2014**, *136*, 890–893.

[171] Wickens, Z. K.; Morandi, B.; Grubbs, R. H. *Angew. Chem. Int. Ed.* **2013**, *52*, 11257–11260.

[172] Hintermann, L.; Labonne, A. *Synthesis* **2007**, *8*, 1121–1150.

[173] Mizushima, E.; Sato, K.; Hayashi, T.; Tanaka, M. *Angew. Chem. Int. Ed.* **2002**, *41*, 4563–4565.

[174] Leyva, A.; Corma, A. *J. Org. Chem.* **2009**, *74*, 2067–2074.

[175] Marion, N.; Ramon, R. S.; Nolan, S. P. *J. Am. Chem. Soc.* **2009**, *131*, 448–449.

[176] Marion, N.; Carlqvist, P.; Gealageas, R.; de Frémont, P.; Maseras, F.; Nolan, S. P. *Chem. Eur. J.* **2007**, *13*, 6437–6451.

[177] Tokunaga, M.; Wakatsuki, Y. *Angew. Chem. Int. Ed.* **1998**, *37*, 2867–2869.

[178] Suzuki, T.; Tokunaga, M.; Wakatsuki, Y. *Org. Lett.* **2001**, *3*, 735–737.

[179] Grotjahn, D. B.; Lev, D. A. *J. Am. Chem. Soc.* **2004**, *126*, 12232–12233.

[180] Li, L.; Zeng, M.; Herzon, S. B. *Angew. Chem. Int. Ed.* **2014**, *53*, 7892–7895.

[181] Wang, X.; Cheng, G.; Shen, J.; Yang, X.; Wei, M.; Feng, Y.; Cui, X. *Org. Chem. Front.* **2014**, *1*, 1001–1004.

[182] Sang, R.; Tang, X.-Y.; Shi, M. *Org. Chem. Front.* **2014**, *1*, 770–773.

[183] Huo, C.; Yuan, Y.; Wu, M.; Jia, X.; Wang, X.; Chen, F.; Tang, J. *Angew. Chem. Int. Ed.* **2014**, *53*, 13544–13547.

[184] Fujihara, T.; Horimoto, Y.; Mizoe, T.; Sayyed, F. B.; Tani, Y.; Terao, J.; Sakaki, S.; Tsuji, Y. *Org. Lett.* **2014**, *16*, 4960–4963.

[185] Du, Y.-D.; Tse, C.-W.; Xu, Z.-J.; Liu, Y.; Che, C.-M. *Chem. Commun.* **2014**, *50*, 12669–12672.

[186] Surprenant, S.; Chan, W. Y.; Berthelette, C. *Org. Lett.* **2003**, *5*, 4851–4854.

[187] Shindo, M. *Tetrahedron* **2007**, *63*, 10–36.

[188] Moyano, A.; Charbonnier, F.; Greene, A. E. *J. Org. Chem.* **1987**, *52*, 2919–2922.

[189] Yamasaki, R.; Terashima, N.; Sotome, I.; Komagawa, S.; Saito, S. *J. Org. Chem.* **2010**, *75*, 480–483.

[190] Stang, P. J.; Surber, B. W. *J. Am. Chem. Soc.* **1985**, *107*, 1452–1453.

[191] Sweis, R. F.; Schramm, M. P.; Kozmin, S. A. *J. Am. Chem. Soc.* **2004**, *126*, 7442–7443.

[192] Julia, M.; Saint-jalmes, V. P.; Verpeaux, J.-N. *Synlett* **1993**, 233–234.

[193] Jouvin, K.; Bayle, A.; Legrand, F.; Evano, G. *Org. Lett.* **2012**, *14*, 1652–1655.

[194] Jouvin, K.; Coste, A.; Bayle, A.; Legrand, F.; Karthikeyan, G.; Tadiparthi, K.; Evano, G. *Organometallics* **2012**, *31*, 7933–7947.

[195] Gray, V. J.; Cuthbertson, J.; Wilden, J. D. *J. Org. Chem.* **2014**, *79*, 5869–5874.

[196] Yoshikawa, T.; Shindo, M. *Org. Lett.* **2009**, *11*, 5378–5381.

[197] Shindo, M.; Matsumoto, K.; Shishido, K. *Org. Synth.* **2007**, *84*, 11–21.

[198] Yoshikawa, T.; Mori, S.; Shindo, M. *J. Am. Chem. Soc.* **2009**, *131*, 2092–2093.

[199] Shindo, M. *Tetrahedron Lett.* **1997**, *38*, 4433–4436.

[200] Shindo, M.; Sato, Y.; Shishido, K. *Tetrahedron* **1998**, *54*, 2411–2422.

3

SYNTHESIS OF AMINES, AMIDES, AND RELATED COMPOUNDS

ALKYLAMINES

3.1 SYNTHESIS OF ALKYLAMINES AND RELATED COMPOUNDS THROUGH NITROGEN–CARBON(SP³) BOND-FORMING REACTIONS

The preparative routes leading to the formation of alkylamines and related compounds are quite diverse. The following section will highlight a number of these approaches with special attention paid to methods that are operationally simple, high yielding, or exhibit broad functional group tolerance.

Substitution chemistry remains one of the most popular and simplest approaches to the formation of nitrogen–carbon(sp³) bonds. An example of this method involved the addition of pyrimidines to brominated esters (Scheme 3.1) [1]. The first step in this sequence was the protection of the primary amine in order to direct the reaction to the $N(1)$-H of the pyrimidine. Once the protection was complete, addition of a strong base deprotonated the $N(1)$-H and generated the amide that attacked the brominated ester to generate the new nitrogen–carbon(sp³) bond.

Another example of this chemistry generated seven-membered heterocycles through a double substitution reaction (Scheme 3.2 and Example 3.1) [2]. Sodium hydroxide served as the base in this chemistry, and moderate to excellent yields of the N-alkylation products were obtained. In addition to the double substitution reaction, a range of alkylamines were generated as part of this investigation. The reaction was carried out in a focused microwave reactor and was quite practical since a glovebox or nitrogen manifold was not needed as the reactions were carried out under air.

Practical Functional Group Synthesis, First Edition. Robert A. Stockland, Jr.
© 2016 John Wiley & Sons, Inc. Published 2016 by John Wiley & Sons, Inc.

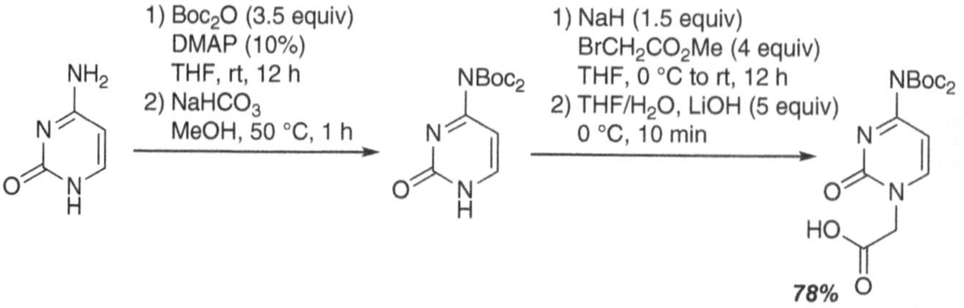

SCHEME 3.1 Classic synthesis of nitrogen–carbon(sp³) bonds through substitution chemistry [1].

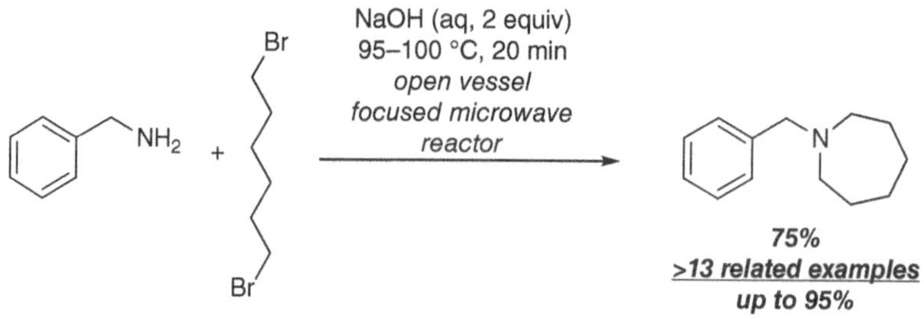

SCHEME 3.2 Microwave-assisted alkylation of benzylamine in an open flask [2].

Example 3.1 N-Alkylation through Substitution Chemistry [2].

Chemical Safety Instructions

Before starting this synthesis, all safety, health, and environmental concerns must be evaluated using the most recent information, and the appropriate safety protocols followed. All appropriate personal safety equipment must be used. All waste must be disposed of in accordance with all current local and government regulations.

The individual performing these procedures and techniques assumes all risks and is responsible for ensuring the safety of themselves and those around them. This chemistry is not intended for the novice and should only be attempted by professionals who have been well trained in synthetic organic chemistry. The authors and Wiley assume no risk and disclaim any liabilities for any damages or injuries associated in any way with the chemicals, procedures, and techniques described herein.

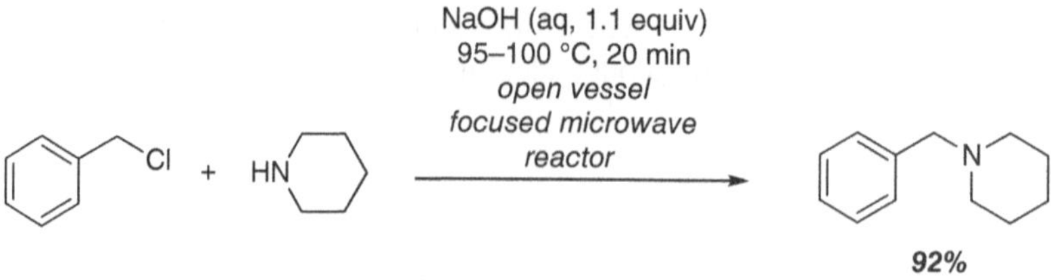

Caution! Hexane and ethyl acetate are highly flammable.

1 mmol benzyl chloride (0.127 g), 1 mmol piperidine (0.085 g), and 1.1 mmol NaOH in water (2.20 mL 0.5 M solution) were placed in a round-bottom glass flask equipped with a condenser and a magnetic stirrer. The flask was placed in a CEM Discover Focused Microwave Synthesis System and subjected to MW irradiation at 80–100 °C (power 250 W) for 25 min. After completion of the reaction (monitored by TLC), the product was extracted into ethyl acetate. Removal of the solvent under reduced pressure followed by flash column chromatography using hexane/ethyl acetate (4/1) as eluent afforded 1-benzylpiperidine (0.161 g, 92%).

Adapted from Y. Ju *and* R. S. Varma, *Green Chem.*, **2004**, *6*, 219–221 *with permission of The Royal Society of Chemistry.*

In many cases, monoalkylation of a primary amine can be a challenging reaction; however, this has been achieved using a ruthenium-catalyzed process (Scheme 3.3) [3]. Using an air-stable ruthenium species and an aminoamide as the solubilizing/stabilizing ligand for the ruthenium center, several anilines were converted into secondary amines in good to excellent yield. The other notable feature of this system was the ability to use alcohols as the alkylating agents instead of alkyl halides.

An iridium-catalyzed process has also been developed for the monoalkylation of aniline derivatives using alcohols as the alkylating agents (Scheme 3.4) [4]. The iridium catalyst functions at relatively low loadings using AgNTf$_2$ as an activating agent and cesium carbonate as the base. Benzyl alcohols, aliphatic alcohols, and heteroaromatic substrates were successfully used in these reactions. The reaction was also relatively insensitive to the electronic composition of the aniline derivative. In addition to their work with alcohols, the authors also screened a range of simple alkylamines for activity and discovered that triethylamine successfully alkylated a range of aniline derivatives (Scheme 3.5).

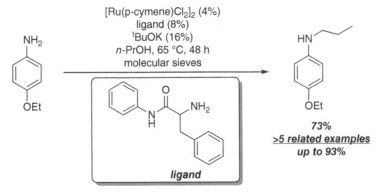

SCHEME 3.3   Selective monoalkylation of aniline derivatives using alcohols and a ruthenium catalyst [3].

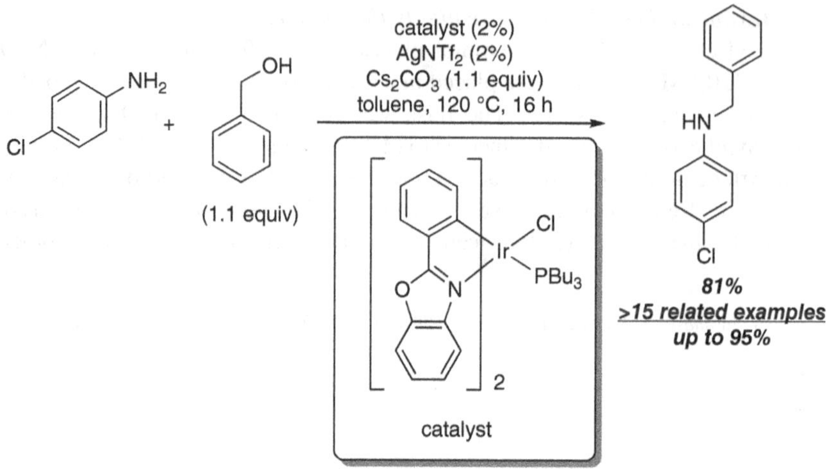

SCHEME 3.4   Iridium-catalyzed N-alkylation using alcohols [4].

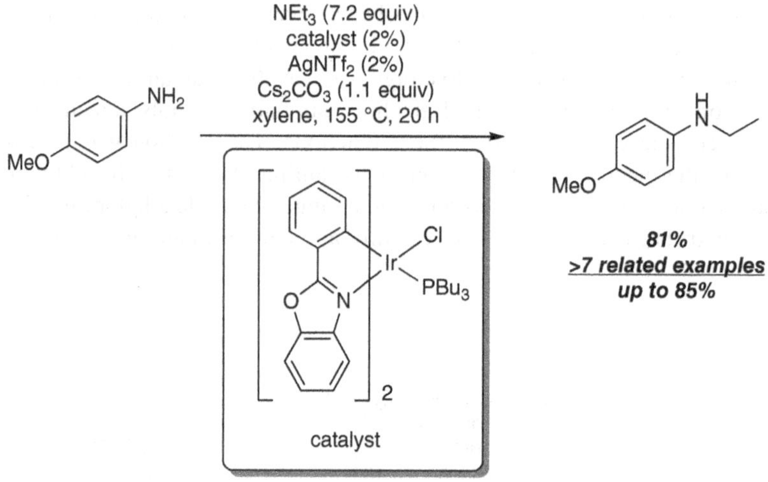

SCHEME 3.5   Iridium-catalyzed N-alkylation using trialkylamines [4].

In related work, rhenium(VII) oxide catalyzed the chemoselective N-alkylation of hydroxycarbamates (Scheme 3.6 and Example 3.2) [5]. Although alkylation of the oxygen was possible, the chemistry was selective for attachment to the nitrogen. This reaction is attractive since it could be carried out at room temperature in an open vessel under an atmosphere of air. It should also be noted that the authors discovered that a trace of moisture was needed to promote the reaction; thus, rigorous drying of the solvents and reagents was not needed for this reaction.

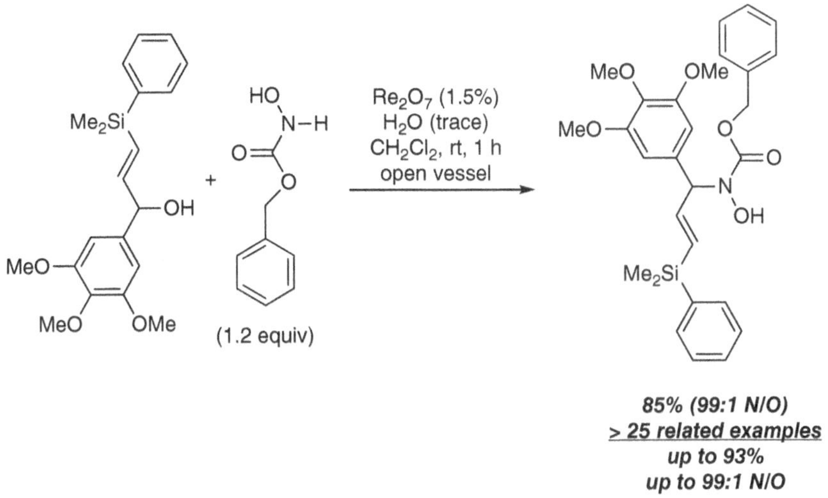

SCHEME 3.6   Rhenium-catalyzed alkylation of hydroxycarbamates [5].

Example 3.2   Chemoselective N-Alkylation Reactions [5].

Before starting this synthesis, carefully follow the safety instructions listed in Example 3.1.

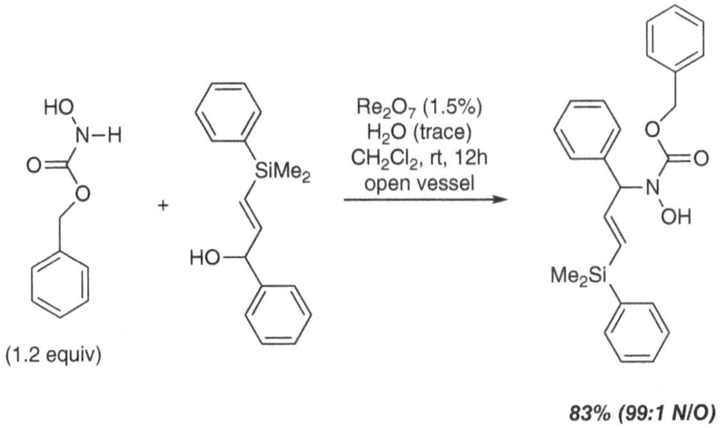

Caution! Hexane and ethyl acetate are highly flammable.

This reaction was carried out in an open vessel. To a stirred solution of alcohol (1 equiv) and amine (1.2 equiv) in dichloromethane (2 mL) was added Re_2O_7 (1.5 mol%), and the solution was stirred at room temperature or at 20 °C, in a flask open to the atmosphere. After the completion of the reaction, as judged by TLC, saturated ammonium chloride (1 mL) was added and the solution was extracted with dichloromethane (3 × 10 mL). The combined organic extract was dried over anhydrous sodium sulfate, and then the crude product was purified by column chromatography on silica gel using hexane/EtOAc (9 : 1) as the eluent to afford the (E)-allylic hydroxylamine (83%) as a colorless oil.

Adapted with permission from S. W. Chavhan, C. A. McAdam, M. J. Cook, *J. Org. Chem.*, **2014**, *79*, 11234–11240. *Copyright (2014) American Chemical Society.*

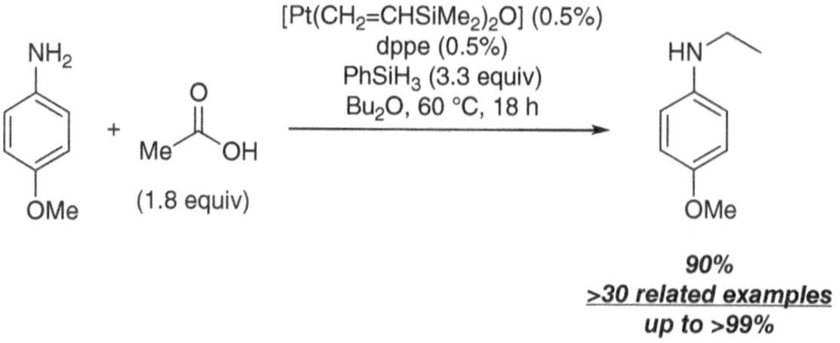

SCHEME 3.7   Alkylation of aniline derivatives using carboxylic acids [6].

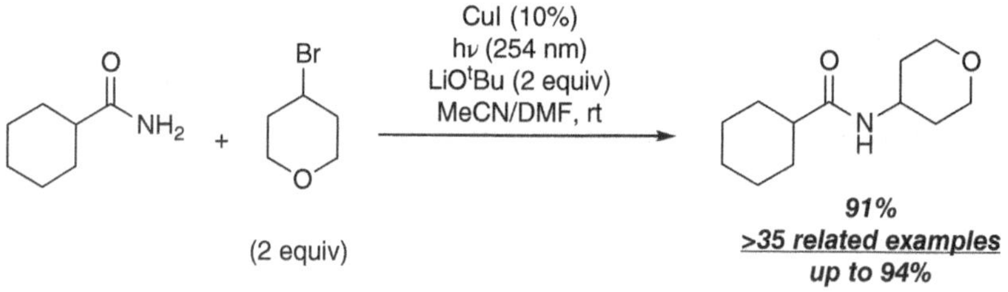

SCHEME 3.8   Preparation of amides through a copper-catalyzed photoinduced reaction [7].

In addition to the use of alcohols and amines as alkylating agents, carboxylic acids have been used (Scheme 3.7) [6]. Carboxylic acids are typically quite stable and can be stored for long periods of time with minimal decomposition. Using a platinum precursor along with a common bisphosphine ligand and phenylsilane as a reducing agent, an assortment of aniline derivatives as well as alkylamines were alkylated using carboxylic acids. The reactions were carried out with gentle heating and afforded good to excellent yields of the N-alkylated compounds.

The N-alkylation of a variety of amides using secondary alkyl halides as substrates has been achieved through a copper-catalyzed photoinduced process (Scheme 3.8) [7]. The reactions were carried out under mild conditions using UV light to promote the transformation. The chemistry had broad tolerance to preexisting functional groups and afforded good to excellent yields of the amides. It should be noted that secondary alkyl halides often are challenging to successfully functionalize due to competing side reactions.

Alkylboronates can be converted into protected amines using methoxyamine as the nitrogen source (Scheme 3.9) [8]. Amines were initially generated in this work but were protected in order to facilitate purification by column chromatography. While the protection does add another step to the process, the overall process can be carried out in a single reaction vessel.

A rhodium-catalyzed approach to the intramolecular addition of N–H bonds to activated alkenes generated cyclic sulfonamides through a tandem olefination/aza-Michael-type addition reaction (Scheme 3.10) [9]. Using a variety of acrylates and sulfonamides, a series of sulfur- and nitrogen-containing heterocycles were generated in outstanding yield. The reactions proceeded through an initial rhodium-catalyzed double olefination reaction followed by the N–H addition. There was a considerable amount of substrate-specific

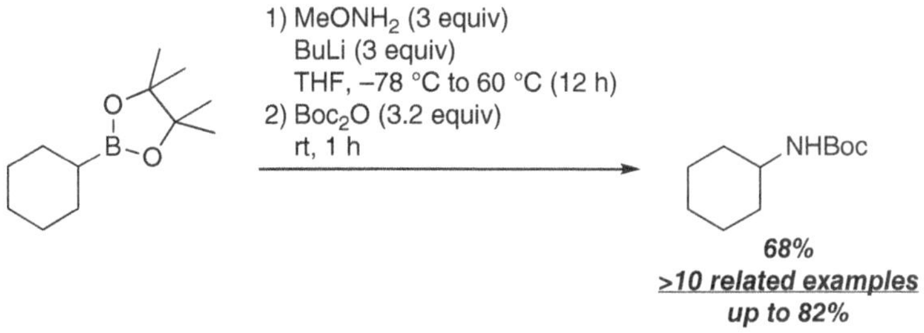

SCHEME 3.9   Amine synthesis from alkylboronates [8].

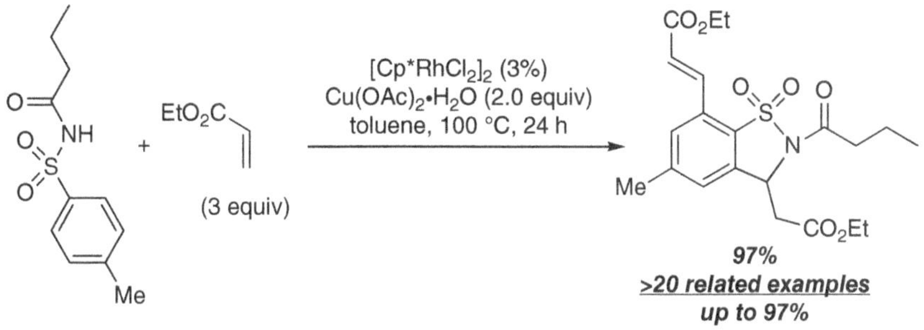

SCHEME 3.10   Rhodium-catalyzed synthesis of cyclic sulfonamides [9].

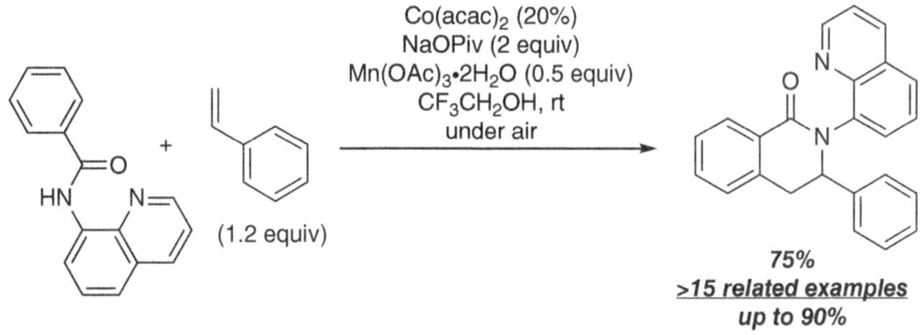

SCHEME 3.11   Cobalt-catalyzed synthesis of nitrogen heterocycles under air [10].

reactivity noted by the authors. For example, while ethyl acrylate and a number of other acrylates were successfully used in this chemistry, styrene and a range of activated alkenes including vinylphosphonate diesters and ethyl vinyl ketone afforded lower yields of the addition products. For the right system, this is an efficient approach to the synthesis of these heterocycles. In related work, the cobalt-catalyzed synthesis of nitrogen heterocycles under air was achieved (Scheme 3.11) [10, 11].

Oxidative amination chemistry has been used to generate a series of nitrogen heterocycles (Scheme 3.12) [12]. The reaction was catalyzed by palladium and used elemental oxygen

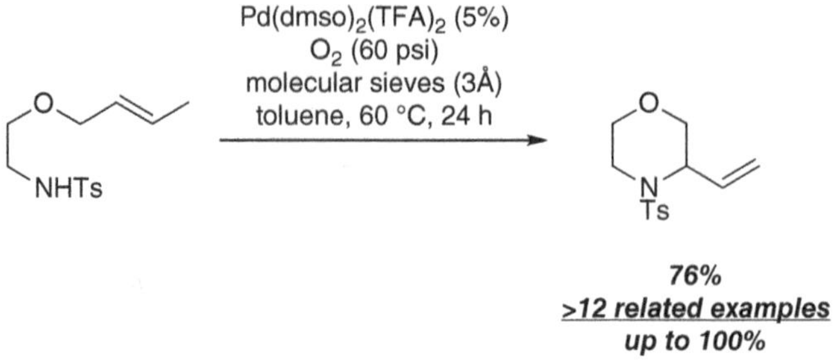

SCHEME 3.12   Synthesis of heterocycles through oxidative amination chemistry [12].

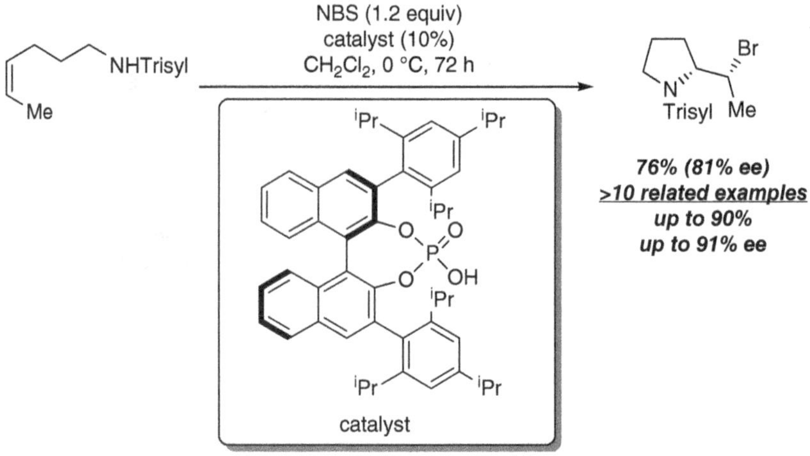

SCHEME 3.13   Preparation of tetrahydropyrroles through bromocyclization [13].

as the terminal oxidant. The conditions were mild, and an assortment of heterocycles including piperazine and piperidines were generated in good to outstanding yields.

Bromocyclization reactions are well known for their ability to facilitate the construction of chiral compounds. An example this chemistry outlined the preparation of tetrahydropyrroles (Scheme 3.13) [13]. A chiral phosphoric acid was used to provide the stereocontrol, and NBS served as the bromine source. Using this system, a range of gamma-amino alkenes were converted into nitrogen heterocycles in high yield with moderate to good selectivity.

An effective synthesis of 1,2-dihydroquinolines has been devised using a Lewis acid-mediated bicyclization reaction between alkynylanilines and aldehydes (Scheme 3.14 and Example 3.3) [14]. The authors screened a range of Lewis acids for activity toward the reaction, and scandium(III) triflate was the most effective under the reaction conditions. The reaction proceeded under mild conditions and afforded moderate to good yields of the polycyclic compounds when electron-neutral or electron-rich aldehydes were used. Electron-deficient aldehydes were sluggish under the standard conditions; however, the authors discovered that the use of benzoic acid as a cocatalyst enabled the use of these substrates in the bicyclization reaction.

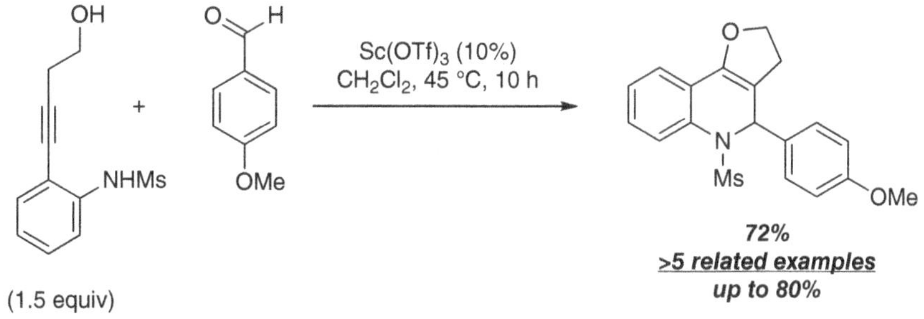

SCHEME 3.14   Preparation of 1,2-dihydroquinolines through a bicyclization process [14].

Example 3.3   Coupling of Aldehydes with Alkynylanilines to Generate 1,2-Dihydroquinolines [14].

Before starting this synthesis, carefully follow the safety instructions listed in Example 3.1.

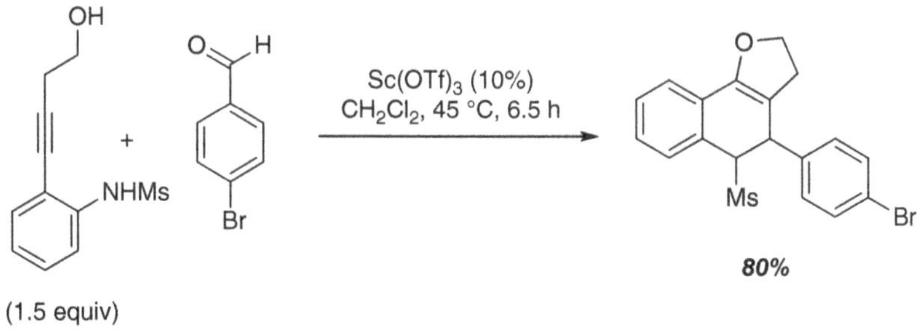

Caution! Ethyl acetate and petroleum ether are highly flammable.
To a Schlenk tube were added Sc(OTf)$_3$ (49.6 mg, 0.10 mmol), aldehyde (186.0 mg, 1.0 mmol)/dichloromethane (2.5 mL), and alkyne (359.8 mg, 1.5 mmol)/dichloromethane (2.5 mL) sequentially. The mixture was stirred at 45 °C. After 6.5 h, the reaction was complete as monitored by TLC. After evaporation, the mixture was purified by column chromatography on silica gel (eluent: petroleum ether/ethyl acetate/triethylamine = 600/100/1) to afford the heterocycle (327.0 mg, 80%).

Adapted with permission from C. Zhu, S. Ma, Angew. Chem. Int. Ed., 2014, 53, 13532–13535. Copyright (2014) Wiley-VCH Verlag GmbH & Co. KGaA, Weinheim.

A multicomponent reaction has been used to generate β-lactams and pyrrolidine-2,5-dione derivatives (Scheme 3.15) [15]. The initial step in the process entailed stirring the four components in methanol at room temperature for 12–24 h to generate the Ugi adducts. Addition of potassium carbonate to this intermediate and heating for a few hours afforded pyrrolidine-2,5-dione derivatives in good to excellent yields. This process was relatively insensitive to the electronic composition of aldehydes, and a range of electron-donating and electron-withdrawing groups were well tolerated. If the second step of the reaction was carried out in acetonitrile instead of methanol, β-lactams are formed instead

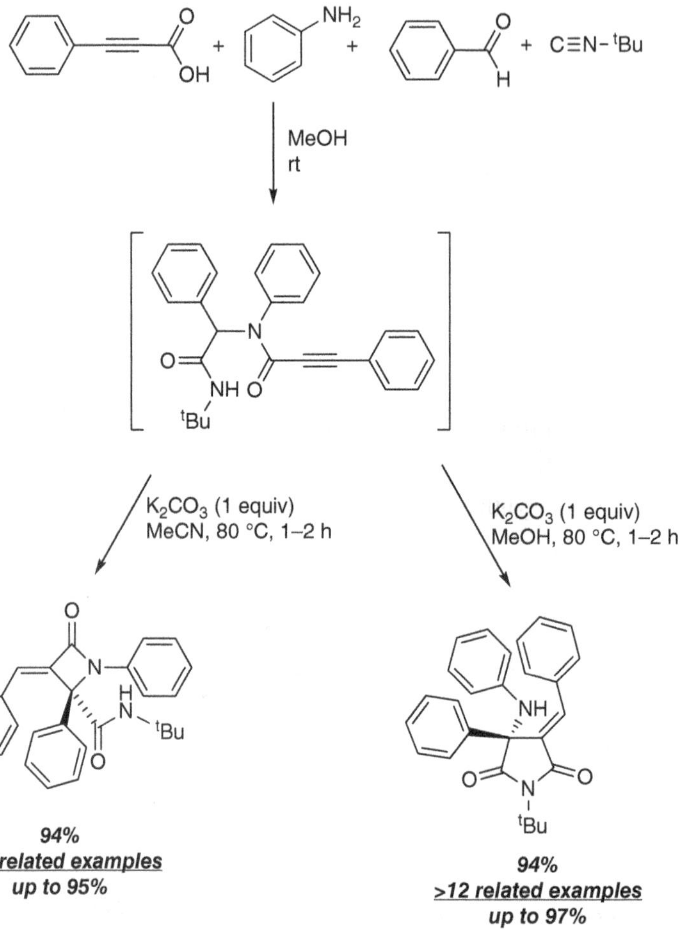

SCHEME 3.15 Preparation of β-lactams and pyrrolidine-2,5-dione derivatives through multicomponent reactions [15].

of the pyrrolidine-2,5-dione derivatives. While the precise mechanism behind this switch in reactivity was not fully elucidated by the authors, they proposed that it could originate with the differences between the basicity of the potassium carbonate in methanol versus acetonitrile.

The rhodium-catalyzed synthesis of 2-picolylamines and imidazo[1,5-a]pyridines from pyridotriazoles has been achieved (Scheme 3.16) [16]. For the preparation of 2-picolylamines (Example 3.4), the reaction was operationally simple and consisted of stirring a dinuclear rhodium catalyst with the two reagents at 120 °C. This approach tolerated a broad spectrum of preexisting functional groups, and even 3(2-H)-pyridazinone was successfully used. For the preparation of the imidazo[1,5-a]pyridines, a two-stage approach was used for the preparation of the heterocycles that included an initial N–H insertion followed by cyclization.

Propargylamines have been generated through a multicomponent reaction between a terminal alkyne, (R)-glyceraldehyde acetonide, and dibenzylamine (Scheme 3.17) [17].

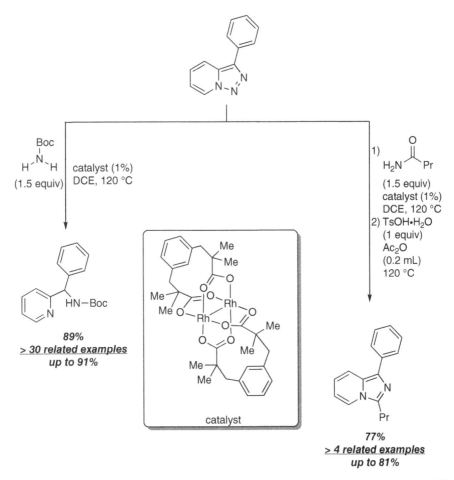

SCHEME 3.16   Rhodium-catalyzed reactions of pyridotriazoles with amines and amides [16].

The process was catalyzed by copper(I) bromide and proceeded under mild conditions with high diastereoselectivity. The authors noted that no trace of the other diastereomer was observed. It is also worth noting that no chiral ligands (for the copper) were added to the reaction mixture, and yet high selectivity was still obtained. In related work, the propargylamines were generated through a silver-catalyzed multicomponent reaction (Scheme 3.18) [18].

The addition of N–H bonds to alkenes is an atom-efficient method for the preparation of amines. When activated alkenes such as acrylates were used as substrates, this process is referred to as an aza-Michael addition. Using this approach, a great variety of amines and related compounds have been generated through systematic variation of the nitrogen-bearing substrate and alkene. In many cases, a catalyst was not needed for the addition reaction [19].

When unactivated alkenes were used as substrates, the addition process is referred to as hydroamination [19]. This process is extremely valuable for the preparation of new amines and related compounds and is often catalyzed by transition metal complexes or bases [19].

Example 3.4 Synthesis of 2-Picolylamines from Pyridotriazoles Using Rhodium Catalysts [16].

Before starting this synthesis, carefully follow the safety instructions listed in Example 3.1.

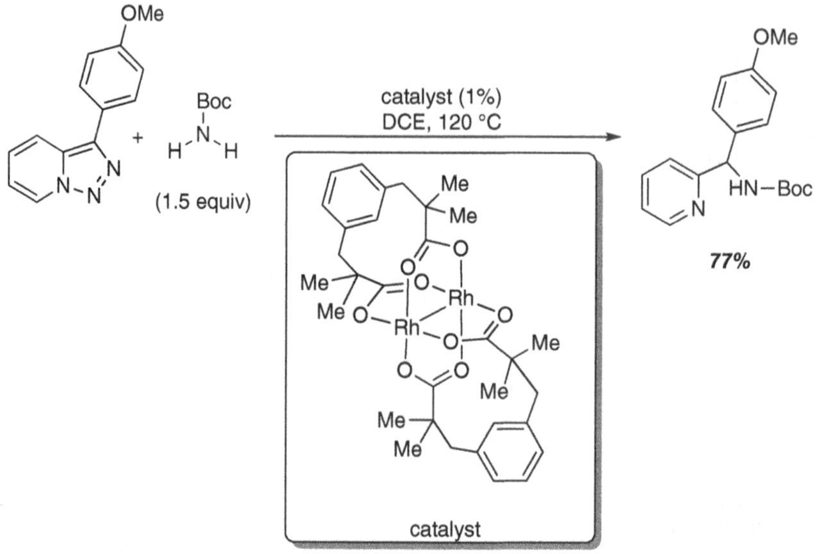

An oven-dried 3.0 mL V-vial equipped with a stirring bar was charged with the catalyst (Rh$_2$(esp)$_2$; 1 mol%), pyridotriazole (0.2 mmol), amine (1.5 equiv), and DCE (2 mL) under N$_2$ atmosphere. The reaction vessel was capped with a Mininert syringe valve, and the reaction mixture was stirred at 120 °C. Upon completion, the reaction mixture was cooled to room temperature, concentrated under reduced pressure, and the crude product was purified by column chromatography to afford the corresponding N–H insertion product.

*Adapted with permission from Y. Shi, A. V. Gulevich, V. Gevorgyan, Angew. Chem. Int. Ed., **2014**, 53, 14191–14195. Copyright (2014) Wiley-VCH Verlag GmbH & Co. KGaA, Weinheim.*

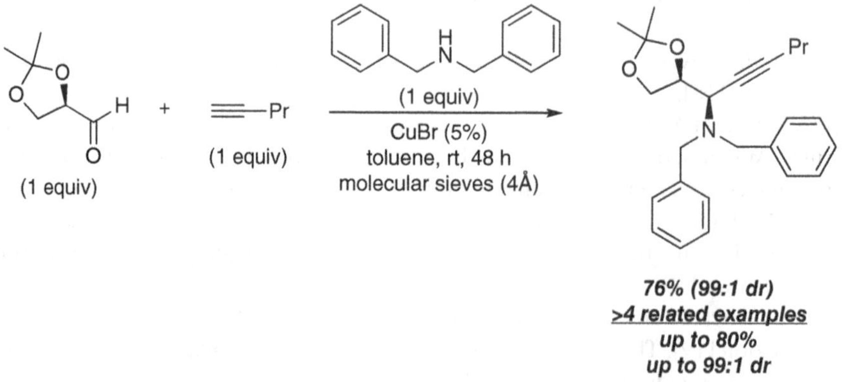

SCHEME 3.17 Stereoselective synthesis of propargylamines [17].

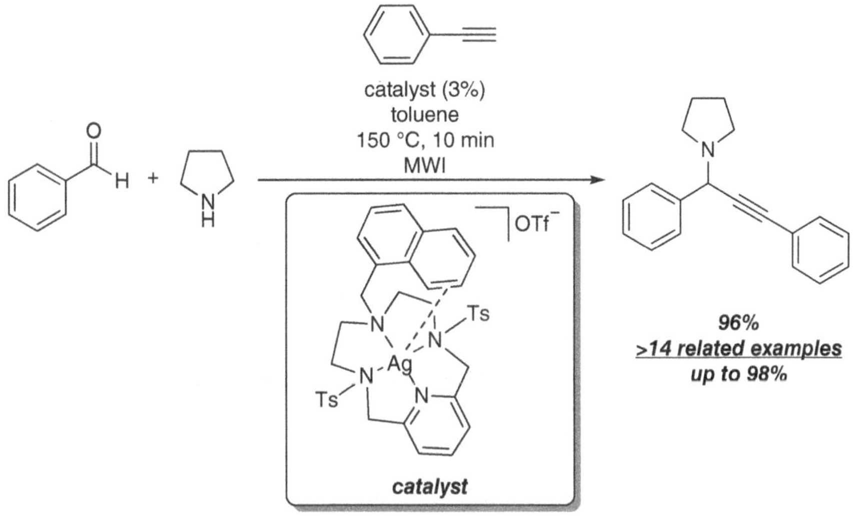

SCHEME 3.18   Silver-catalyzed synthesis of propargylamines through a multicomponent reaction [18].

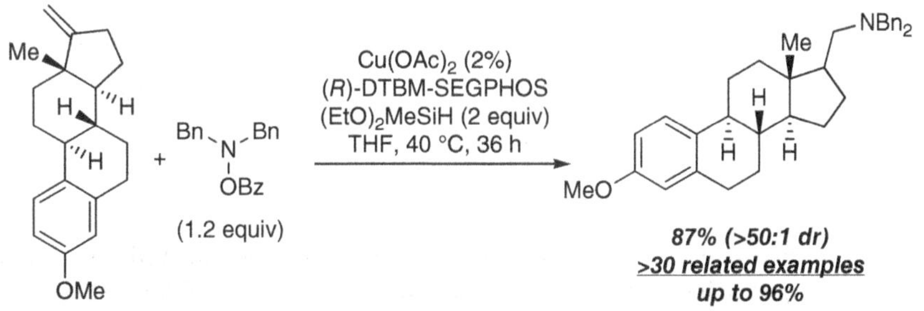

SCHEME 3.19   Copper-catalyzed anti-Markovnikov hydroamination of alkenes [20].

Hydroamination is also a valuable process since it incorporates a "synthetic handle" into a hydrocarbon fragment that can be used to template or promote further functionalization. Although this approach is remarkably popular, there are considerable issues with regioselectivity and stereoselectivity, and each set of substrates should be evaluated against the array of available catalysts when designing a synthesis.

One of the most general processes for the hydroamination of 1,1-disubstituted alkenes uses copper(II) acetate as the catalyst and hydroxylamine esters as the nitrogen precursors (Scheme 3.19) [20]. The reactions were carried out under mild conditions and were highly selective for the formation of the anti-Markovnikov amines. For many of the appropriately designed substrates, the chemistry was also highly stereoselective, and diastereomeric ratios as high as >50:1 and enantiomeric excesses as high as 99% were obtained. The functional group tolerance was outstanding, and an assortment of alkenes bearing a range of electronically and sterically diverse substituents were hydroaminated using this approach.

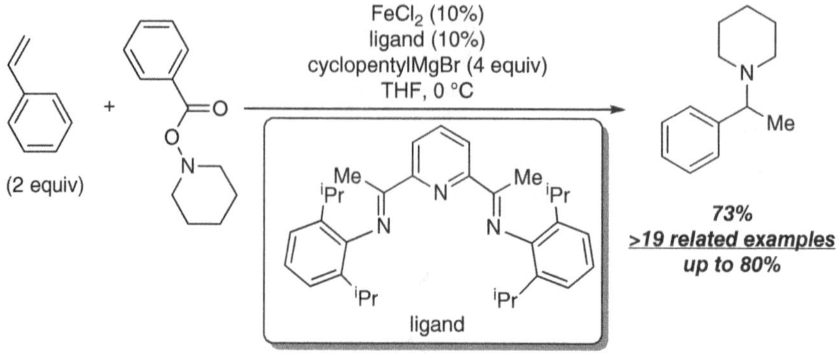

SCHEME 3.20   Iron-catalyzed Markovnikov hydroamination of styrenes [21].

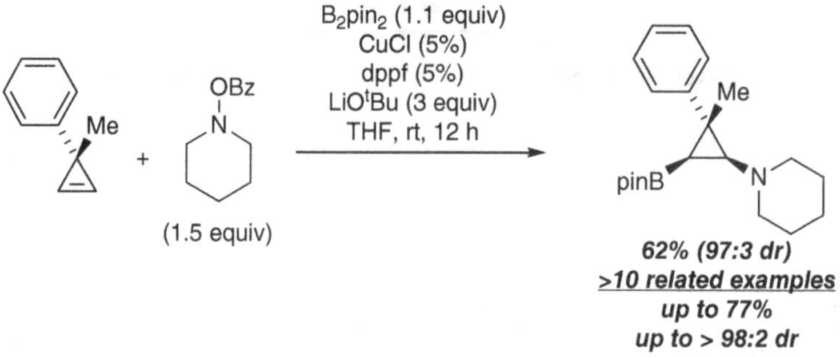

SCHEME 3.21   Copper-catalyzed preparation of cyclopropylaminoboronic esters [22].

The Markovnikov selective hydroamination of alkenes can be achieved using a range of catalysts. One approach used inexpensive iron salts to promote the addition (Scheme 3.20) [21]. In addition to the iron source, a bulky pyridine bis(imine) was used as the solubilizing/stabilizing ligand for the iron center, and cyclopentylmagnesium bromide acted as a reducing agent. A variety of *O*-benzoyl-*N*,*N*-dialkylhydroxylamines served as the nitrogen sources, and the reactions were carried out under mild conditions affording moderate yields of the Markovnikov addition products. In related work, the copper-catalyzed synthesis of cyclopropylaminoboronic esters was achieved under mild conditions (Scheme 3.21) [22].

Despite the attention this area has received, a general approach to the hydroamination of unfunctionalized simple aliphatic alkenes has remained a challenging goal for the synthetic community [19]. To address this issue, an iridium-catalyzed process has been developed for the Markovnikov selective hydroamination of alkenes such as octene (Scheme 3.22) [23]. The catalyst system was comprised of a common iridium complex and a Segphos-type bisphosphine along with 1 equivalent of ethyl acetate. While the precise role of the ethyl acetate was not fully elucidated, its presence resulted in a significant increase in the reaction rate. Using this approach, a variety of indoles were successfully added to octene and related alkenes in moderate to good yields. The addition was selective for the formation of the Markovnikov adducts and was moderately stereoselective (up to 74% ee).

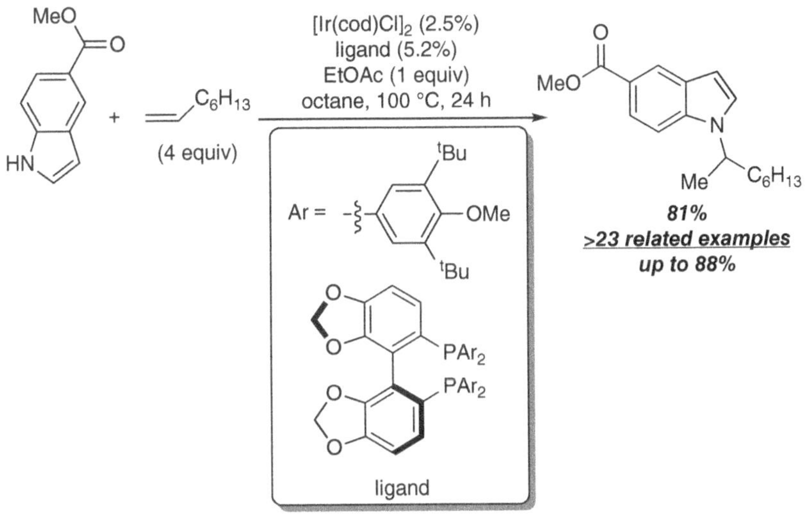

SCHEME 3.22   Markovnikov addition of indoles to unactivated aliphatic alkenes [23].

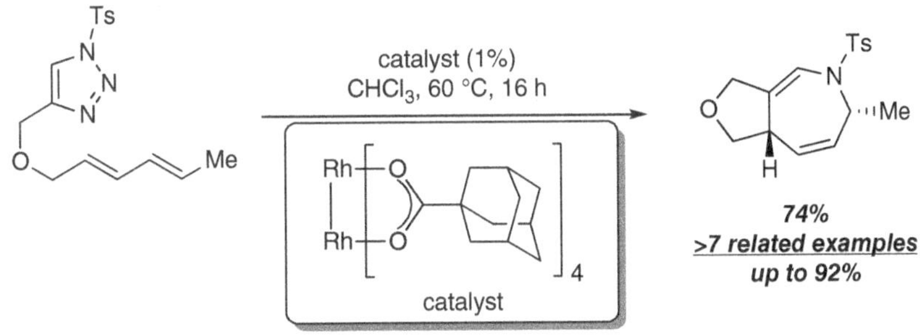

SCHEME 3.23   Rhodium-catalyzed intramolecular cyclization to generate dihydroazepine derivatives [24].

Intramolecular cyclization reactions involving nitrogen nucleophiles and alkenes have been used to generate a vast number of heterocycles. For unactivated alkenes, this process was typically catalyzed by a transition metal catalyst. As an example of this approach, the synthesis of dihydroazepine derivatives from dienyltriazoles has been described (Scheme 3.23) [24]. A dinuclear rhodium complex served as the catalyst for the reaction, and moderate to good yields of the heterocycles were obtained. Additionally, a microwave-assisted version of the reaction was reported using a slightly modified catalyst (Scheme 3.24) as well as a version using visible light catalysis (Scheme 3.25) [25].

The synthesis of nitrogen heterocycles has been achieved through variations of intramolecular imino-Nazarov cyclizations (Schemes 3.26 and 3.27) [26]. A silver-catalyzed version of the reaction generated indolines in moderate to good yields from a range of functionalized aldehydes and secondary amines, while a gadolinium-promoted reaction afforded fused tetrahydroquinoline derivatives in moderate yields.

The preparation of a range of functionalized indenes and dihydronaphthalenes has been achieved through a one-pot multistep reaction (Scheme 3.28) [27]. One of the key steps in

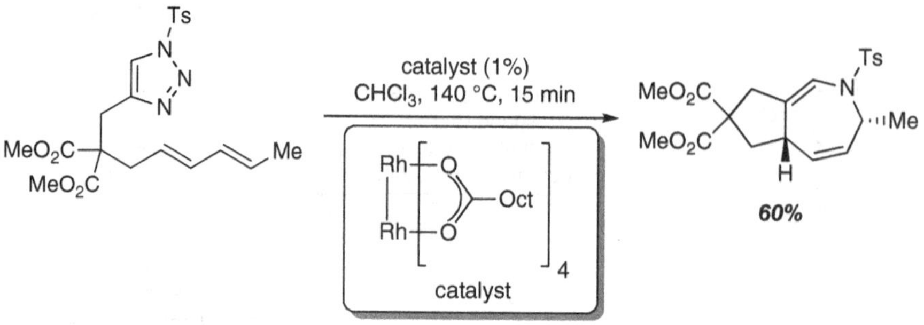

SCHEME 3.24 Microwave-assisted synthesis of dihydroazepine derivatives [24].

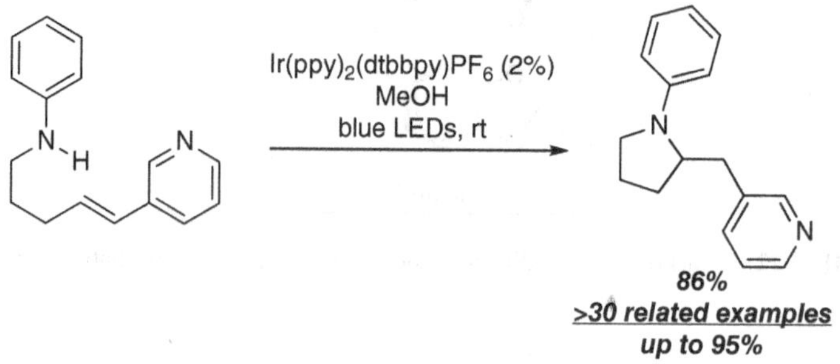

SCHEME 3.25 Intramolecular alkene hydroamination using photoredox catalysis [25].

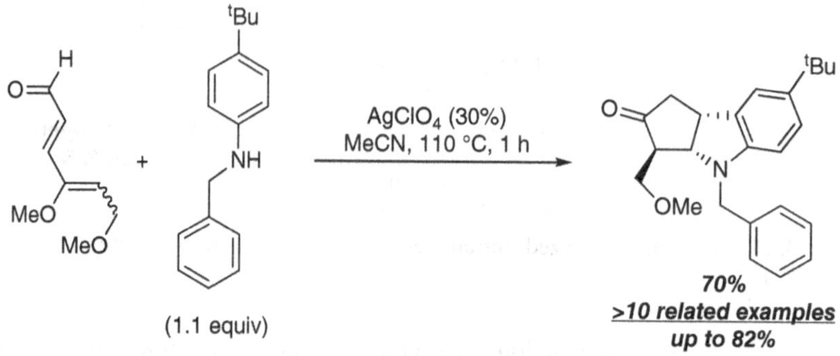

SCHEME 3.26 Silver-promoted synthesis of fused indolines [26].

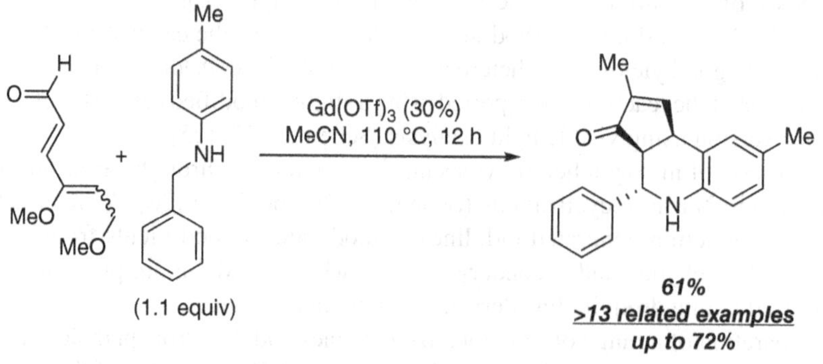

SCHEME 3.27 Gadolinium-promoted synthesis of fused tetrahydroquinoline derivatives [26].

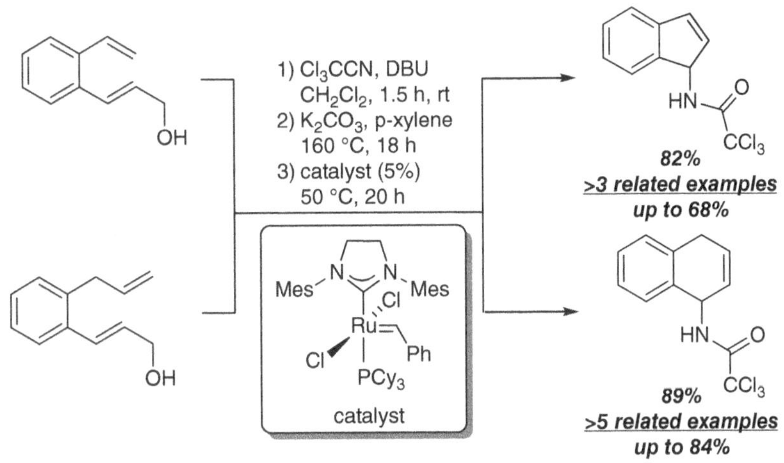

SCHEME 3.28 Ruthenium-catalyzed synthesis of functionalized dihydronaphthalenes and indenes [27].

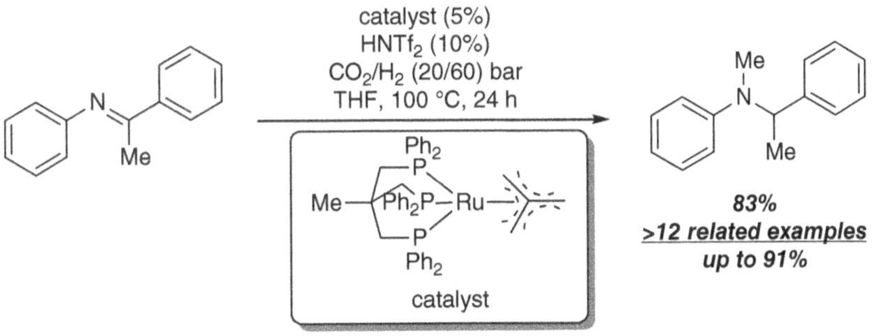

SCHEME 3.29 Ruthenium-catalyzed formal N-methylation of imines [28].

this synthesis was a ruthenium-catalyzed ring closing metathesis step to generate five- or six-membered rings. A carbene-ligated ruthenium catalyst was effective for this step. Overall, moderate to good yields of the amides were generated from allylic alcohols using this three-step one-pot approach.

The N-methylation of imines has been achieved using carbon dioxide and elemental hydrogen as the formal source of a methyl group (Scheme 3.29) [28]. The reductive process was catalyzed by a ruthenium catalyst bearing a tridentate supporting ligand that was activated by an acid bearing a weakly coordinating anion. One of the most attractive aspects of this chemistry was ability to utilize carbon dioxide as a formal source of a single carbon atom. An assortment of electronically and sterically diverse imines were functionalized using this approach with yields as high as 91%. This reductive methylation chemistry was successfully incorporated into a multicomponent reaction, and a range of unsymmetrical methylamines were generated from a mixture of an aldehyde, aniline, and carbon dioxide (Scheme 3.30).

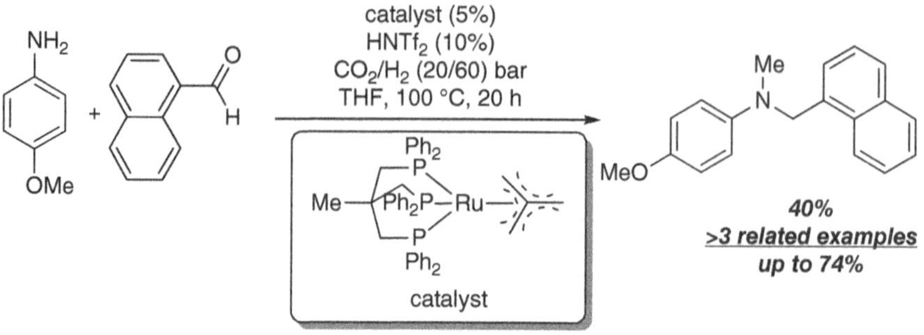

SCHEME 3.30   Amine synthesis through a multicomponent coupling reaction [28].

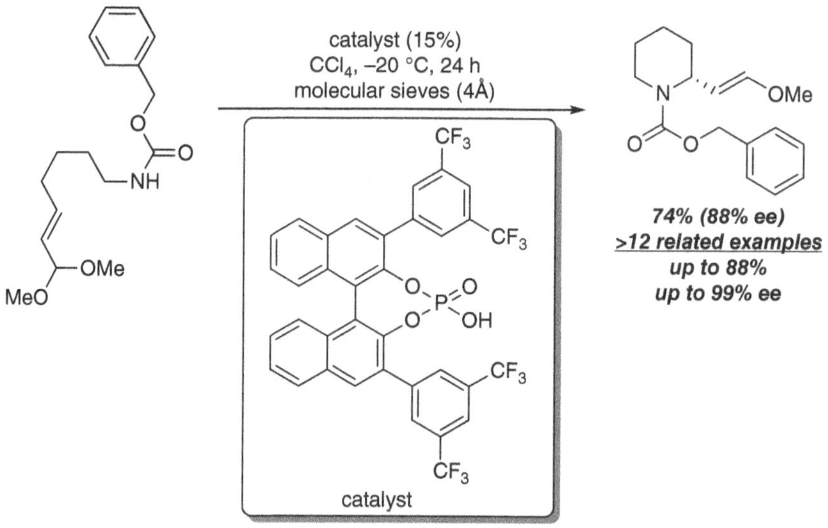

SCHEME 3.31   Asymmetric synthesis of piperidines using phosphoric acid catalysts [29].

The synthesis of chiral piperidines has been accomplished through an intramolecular cyclization reaction using chiral phosphoric acid catalysts (Scheme 3.31) [29]. Treatment of protected amines bearing tethered unsaturated acetals with catalytic amounts of a chiral phosphoric acid catalyzed the formation of chiral piperidines with high selectivity.

An isomerization reaction was used to convert functionalized oxindoles into lactams (Scheme 3.32 and Example 3.5) [30]. This intramolecular process was promoted by the presence of electron-withdrawing groups on the oxindole and successfully generated small to large lactams. Indeed, a 21-membered macrocycle was successfully generated following this approach. This was an impressive accomplishment since classic approaches to the preparation of these large lactams were often complicated by oligomerization instead of cyclization.

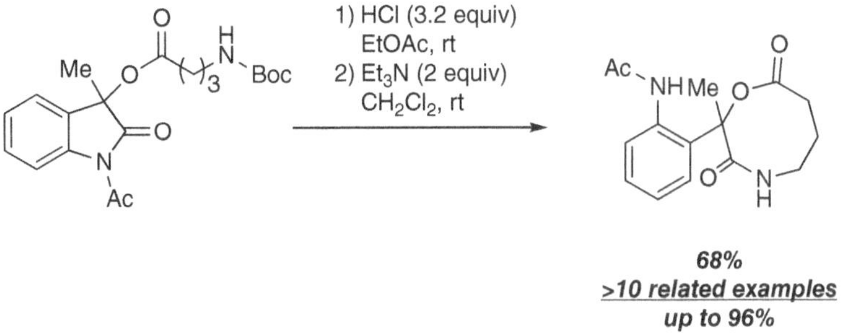

SCHEME 3.32 Synthesis of macrocyclic lactams through intramolecular cyclizations [30].

Example 3.5 Preparation of Macrocycles through N–C Bond-Forming Reactions [30].

Before starting this synthesis, carefully follow the safety instructions listed in Example 3.1.

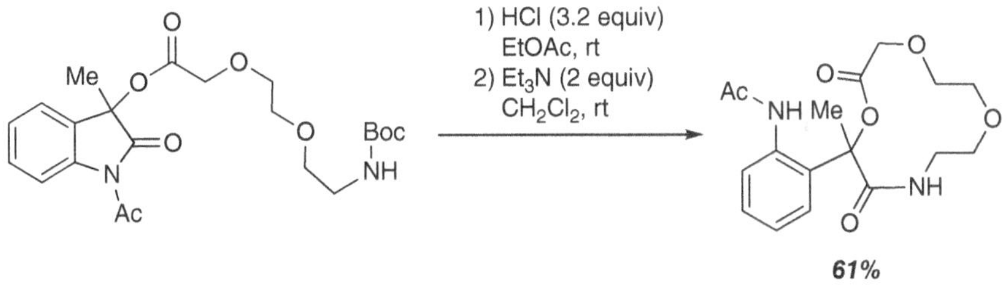

To a solution of Boc-protected oxindole (82.8 mg, 0.184 mmol) in EtOAc (0.2 mL) was added a 3 M solution of anhydrous HCl in EtOAc (0.59 mmol). The mixture was stirred at room temperature overnight. The solvent was removed in vacuo. To the resulting solid were added CH_2Cl_2 (0.6 mL) and triethylamine (0.37 mmol). The mixture was stirred for 8 h at room temperature, diluted with CH_2Cl_2 (10 mL), and washed with aqueous 10% citric acid. The organic phase was dried over sodium sulfate and concentrated in vacuo. Purification by chromatography over silica gel (eluent 1% MeOH in AcOEt) afforded the lactam as an amorphous solid (39.3 mg, 61%).

Adapted with permission from D. Sarraf, N. Richy, J. Vidal, *J. Org. Chem.*, **2014**, *79*, 10945–10955. *Copyright (2014) American Chemical Society.*

A series of chiral pyrrolidines have been generated through a two-step one-pot reaction (Scheme 3.33) [31]. The first step in the synthesis was comprised of an organocatalyzed nitro-Mannich reaction, while the second step consisted of a gold-catalyzed hydroamination. The two-step process generated the heterocycles in good to excellent yields. The substrate scope was fairly broad, and a range of electron-donating and electron-withdrawing groups were tolerated with the lowest yields obtained using electron-rich arenes.

A multicomponent copper-catalyzed reaction has been used to generate a group of chiral isoindolinones (Scheme 3.34) [32]. Using a common copper precursors and a

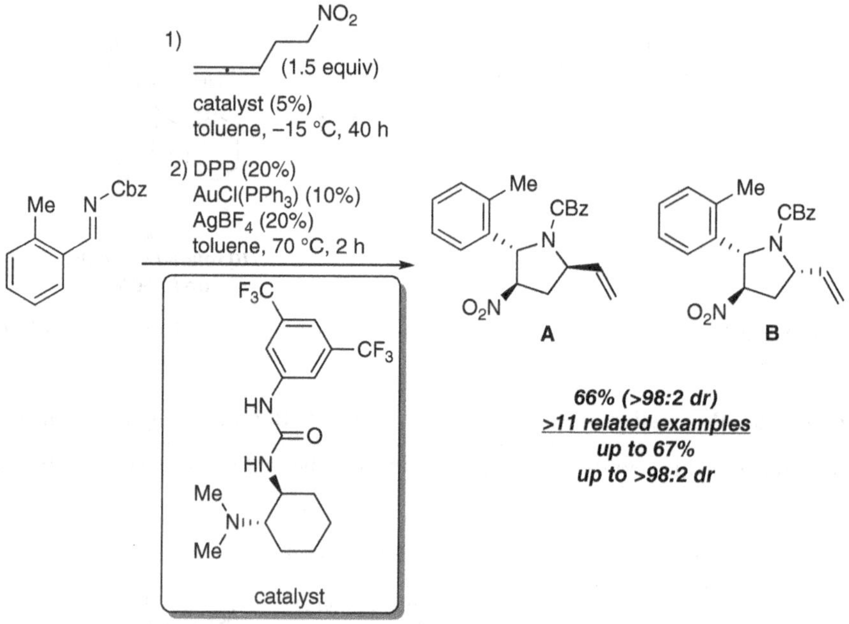

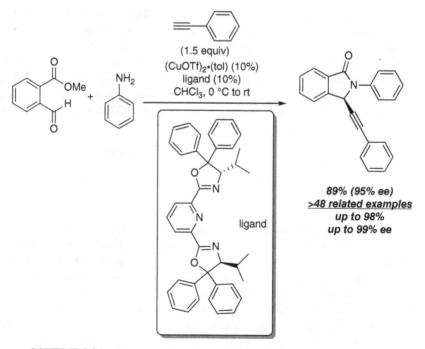

SCHEME 3.33   Enantioselective synthesis of pyrrolidines [31].

SCHEME 3.34   Copper-catalyzed preparation of chiral isoindolinones [32].

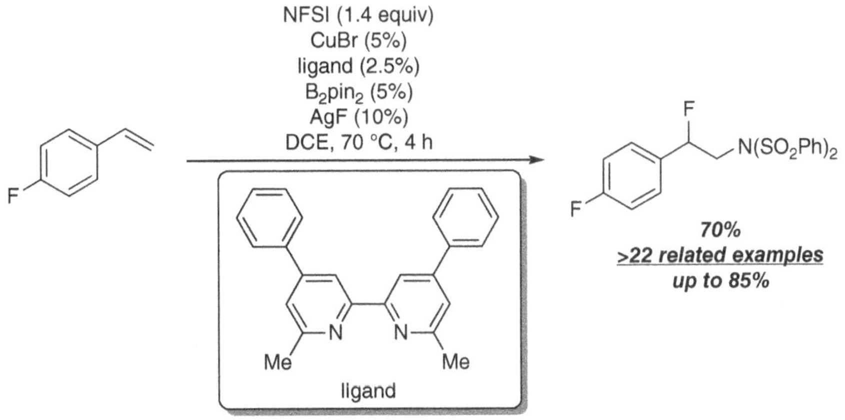

SCHEME 3.35   Highly regioselective aminofluorination of styrene [33].

resolved supporting ligand, a vast assortment of esters, anilines, and alkynes were used to generate over 48 heterocycles in good to excellent yields (up to 98%) with outstanding selectivities (up to 99% ee).

Aminofluorination is a valuable approach to the simultaneous incorporation of multiple functional groups. Using this approach, a variety of styrene derivatives were functionalized (Scheme 3.35) [33]. While a number of copper compounds promoted the reaction, the combination of a simple copper salt along with a bathocuproine ligand was the most effective. The reaction was highly regioselective and added to the fluorine to the α-position.

The metal-free asymmetric amination of benzofuranones has been achieved using binol-derived P-spiro quaternary phosphonium salts as organocatalysts (Scheme 3.36) [34]. The amination reactions were carried out under mild conditions using (E)-dibenzyl diazene-1,2-dicarboxylate as the nitrogen source. After stirring for several days, outstanding yields of the aminated compounds were obtained.

Cinnamic acids and related compounds can be converted into pyrrolidinone or dihydropyridinones through cyclocondensation reactions (Scheme 3.37) [35]. While both pathways were catalyzed by an *N*-heterocyclic carbene, pyrrolidinones were generated through the addition of α-amino ketones to cinnamic acid, while the dihydropyridinones were formed upon the addition of cyclic imines. Both pathways afforded good to excellent yields of the heterocycles with high selectivities.

The synthesis of carbazoles through an intramolecular oxidative C–H amination was achieved through either a transition metal-catalyzed pathway or through the use of an iodonium reagent (Scheme 3.38) [36]. For both routes, N-substituted 2-amidobiphenyls served as the substrates and generated moderate yields of the heterocycles. For optimal results, the authors determined that the combination of the copper salt and the phenyliodonium diacetate reagent was superior to the use of either (individual) reagent. After performing several screening reactions, the authors suggested that the synergistic effect observed when the two reagents were present could be due to copper enhancing the activity of the iodonium reagent.

The palladium-catalyzed intermolecular annulation reaction between secondary amines and alkenes afforded *N*-arylindoles (Scheme 3.39) [37]. The process was catalyzed by palladium salts using phenanthroline as the supporting/stabilizing ligand. There was some

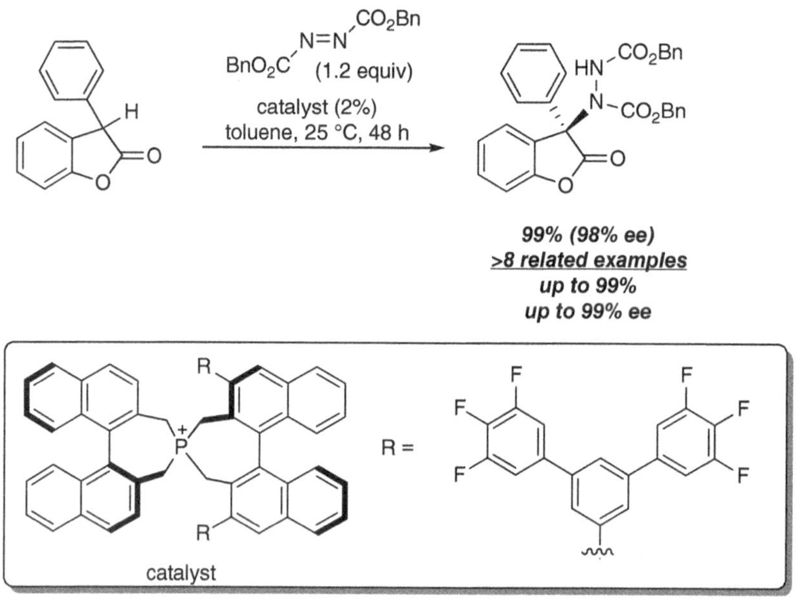

SCHEME 3.36 Metal-free amination reaction using chiral phosphonium salts [34].

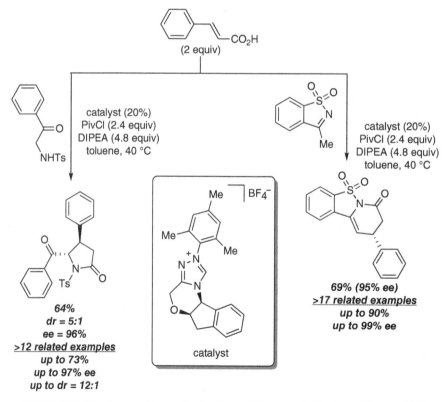

SCHEME 3.37 Asymmetric synthesis of pyrrolidinone and dihydropyridinones [35].

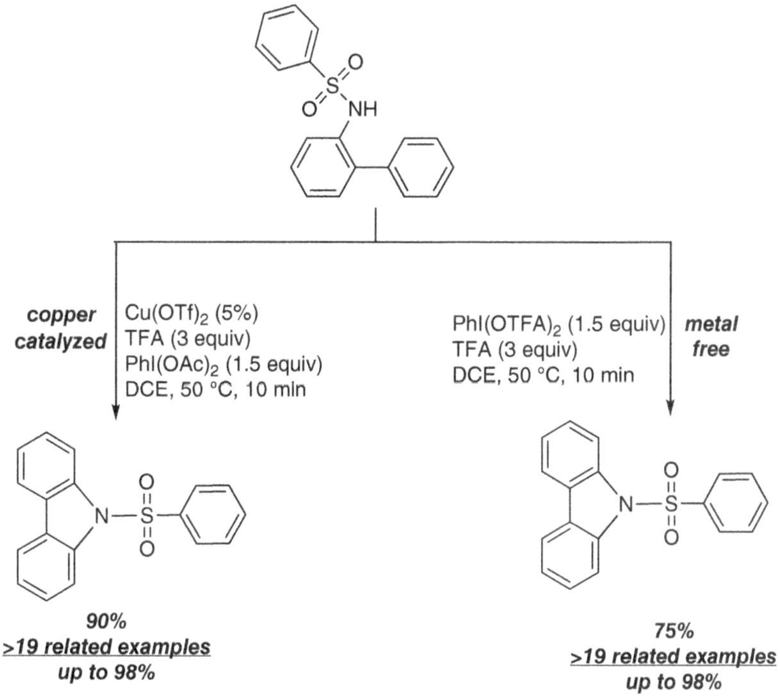

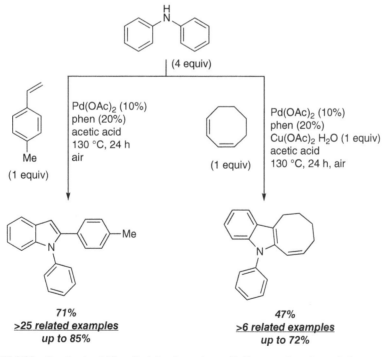

SCHEME 3.38   Synthesis of carbazoles through intramolecular oxidative C–H amination [36].

SCHEME 3.39   Synthesis of *N*-arylindoles through a palladium-catalyzed annulation process [37].

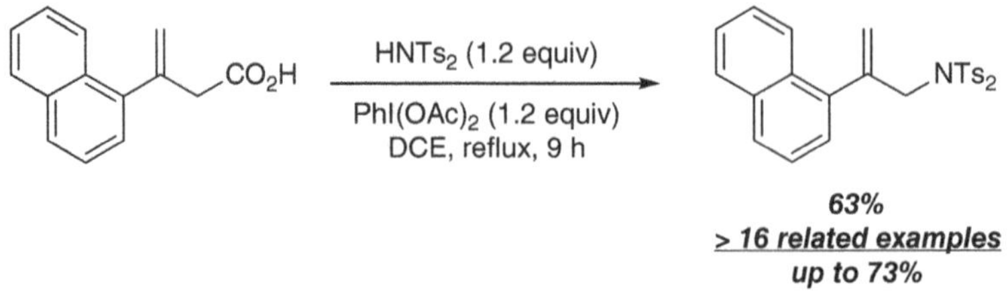

SCHEME 3.40   Synthesis of amides from carboxylic acids through decarboxylative amidation [38].

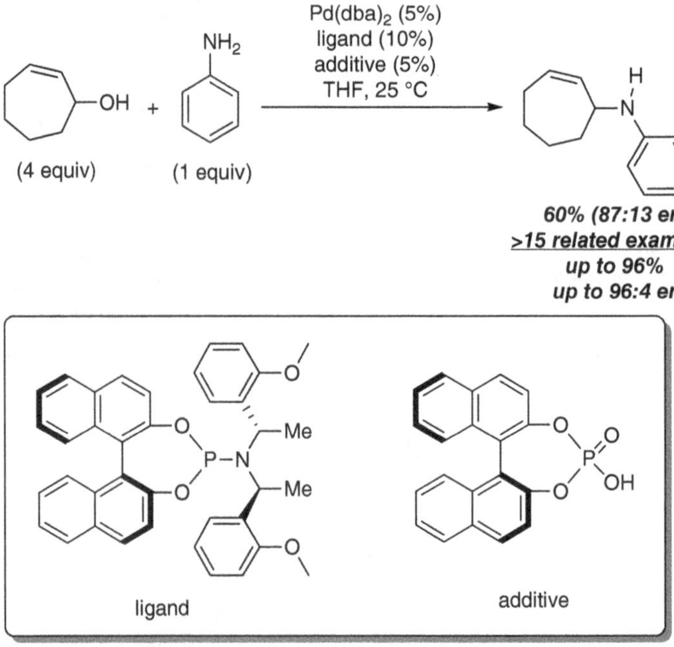

SCHEME 3.41   Enantioselective synthesis of allylic amines [39].

substrate-specific reactivity noted by the authors, and several of the substrates were able to be successfully functionalized/cyclized simply using air as the terminal oxidant, while other substrates required the use of copper acetate as the oxidant.

Carboxylic acids can be converted into amides through a transition metal-free decarboxylative amidation process (Scheme 3.40) [38]. The reaction was promoted by hypervalent iodine reagents, and moderate to good yields of the amides were obtained for a group of structurally and electronically diverse carboxylic acids.

The synthesis of resolved allylic amines has been achieved using the combination of a chiral palladium catalyst and a resolved phosphoric acid catalyst (Scheme 3.41 and Example 3.6) [39]. The overall reaction generated highly resolved allylic amines after simply stirring at 25 °C. While the precise interaction between the catalyst components

Example 3.6   Asymmetric Amination of Racemic Allylic Alcohols [39].

Before starting this synthesis, carefully follow the safety instructions listed in Example 3.1.

Caution! THF, hexane, and ethyl acetate are highly flammable.
Under an argon atmosphere, an oven-dried Schlenk tube (10 mL) was charged with 2-cyclopenten-1-ol (1 mmol), aniline (0.25 mmol), followed by freshly prepared Pd(dba)$_2$ (5 mol%), chiral phosphoric acid (5 mol%), and then the phosphoramidite ligand (10 mol%). Freshly distilled THF (1 mL) and a magnetic stir bar were added, and the reaction mixture was stirred at 25 °C. After completion, the reaction mixture was diluted with ethyl acetate (10 mL) and dried over anhydrous Na$_2$SO$_4$. The filtrate was concentrated under reduced pressure and the residue was purified by silica gel column chromatography using ethyl acetate/hexane as an eluent to afford the corresponding allylic amine (94%).

Adapted with permission from D. Banerjee, K. Junge, M. Beller, *Angew. Chem. Int. Ed.*, **2014**, *53*, 13049–13053. *Copyright (2014) Wiley-VCH Verlag GmbH & Co. KGaA, Weinheim.*

was not fully elucidated, it was highly selective and removal of any of the components reduced the selectivity. A vast array of cyclic and acyclic allylic alcohols bearing a range of electron-donating and electron-withdrawing groups were successfully functionalized using this approach.

The synthesis of *N,N*-bicyclic pyrazolidin-3-one derivatives has been accomplished through a 1,3-dipolar cycloaddition between *N,N'*-cyclic azomethine imines and hydroxyl styrenes (Scheme 3.42) [40]. The chemistry was carried out under mild conditions and afforded moderate yields of the heterocycles with good to excellent selectivity.

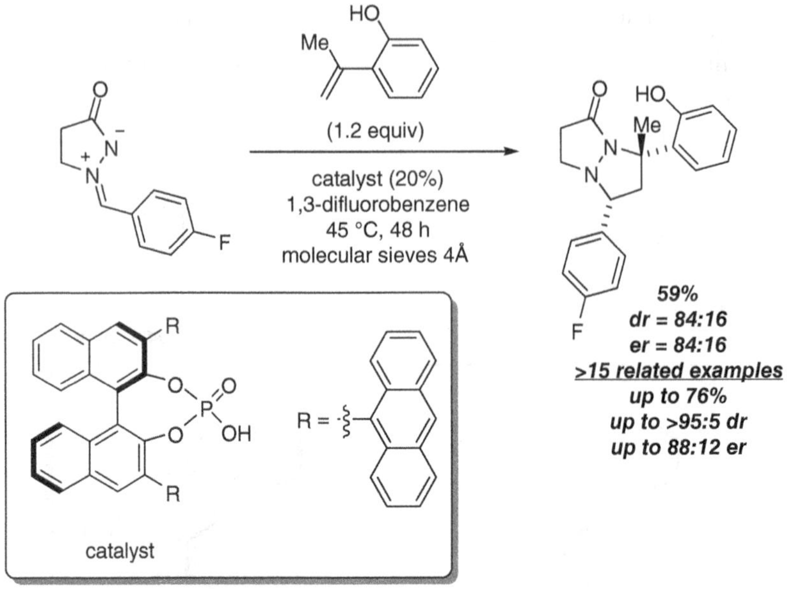

SCHEME 3.42   Synthesis of *N,N*-bicyclic pyrazolidin-3-one derivatives [40].

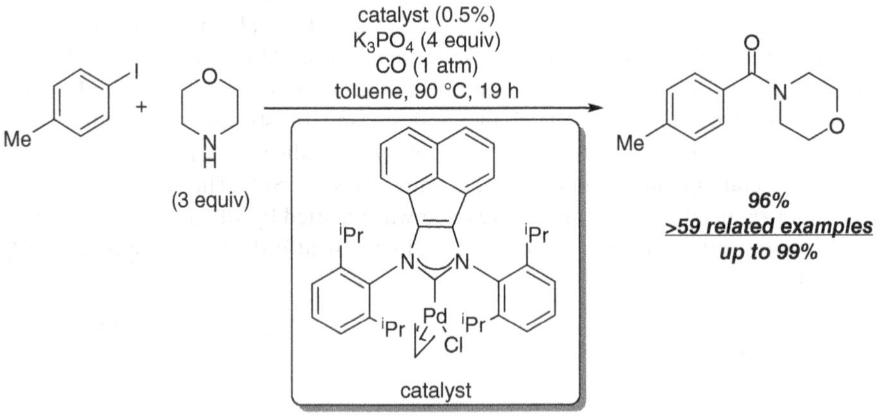

SCHEME 3.43   Palladium-catalyzed aminocarbonylation using low-pressure CO (1 atm) [41].

One of the most direct ways to construct amides is through metal-catalyzed aminocarbonylation. Classically, aminocarbonylation has been impractical for most synthetic laboratories due to the need for a high pressure of carbon monoxide in order to drive reactions to completion. To address this issue, a carbene-ligated palladium complex was developed that successfully promoted the aminocarbonylation of aryl iodides using a low CO pressure (1 atm) (Scheme 3.43) [41]. The chemistry successfully used both primary and secondary amines and a range of iodoarenes to generate an assortment of amides in moderate to excellent yields. The authors demonstrated the practical aspects of the chemistry through the synthesis of anticancer agent tamibarotene from a readily available aniline derivative.

3.1.1 Troubleshooting the Synthesis of Alkylamines and Related Compounds

While it is not possible to predict all of the challenges and difficulties that will arise during the preparation of these compounds, the following suggestions will hopefully provide some direction for common issues and reaction requirements.

Problem/Issue/Goal	Possible Solution/Suggestion
Can alkylamines be generated through substitution chemistry?	This is arguably the most popular route to alkylamines. The chemistry is largely governed by the rules of nucleophilic substitution and generally works quite well [1, 2]
Need an approach to the monoalkylation of an aniline derivative	This can be a challenging reaction. Using ruthenium catalysts to promote the monoalkylation of aniline derivatives with primary alcohols as the alkylating agents would be a reasonable starting point [3]. An iridium-catalyzed monoalkylation reaction has also been reported [4]
Will tertiary amines alkylate arylamines?	This reaction has been reported using iridium catalysts [4]
The substrate has both nucleophilic nitrogen and oxygen groups. Is it possible to chemoselectively alkylate the nitrogen?	Naturally, there will be a considerable amount of substrate-specific reactivity to consider when designing such a reaction. It should be noted that chemoselective N-alkylation (over O-alkylation) has been achieved using rhenium oxide as the catalyst [5]
Can carboxylic acids be used as N-alkylating agents?	Using platinum-based catalysts, carboxylic acids served as efficient alkylating agents for a group of anilines and alkylamines [6]
Can secondary alkyl halides be used in cross-coupling reactions with nitrogen nucleophiles?	Secondary alkyl halides are often challenging to successfully use in coupling reactions; however, a light-assisted copper-catalyzed process has successfully coupled primary amides with secondary alkyl halides [7]
Can alkylboronates be converted into alkylamines?	This has been achieved using methoxyamine as the nitrogen source [8]
Need a good approach for the preparation of propargylamines	Both copper- and silver-catalyzed three-component reactions between an aldehyde, terminal alkyne, and secondary amine have been used to generate propargylamines [17, 18]
Need a good system for the anti-Markovnikov hydroamination of an unactivated alkene	The copper-catalyzed hydroamination between hydroxylamine esters and unactivated alkenes generates amines with the desired regiochemistry for a wide range of diverse substrates [20]

(Continued)

(Continued)

Problem/Issue/Goal	Possible Solution/Suggestion
Need a good system for the Markovnikov hydroamination of unactivated and 1,1-disubstituted alkenes	An iron-catalyzed process generates amines with the desired regiochemistry [21]
What is a good approach for the Markovnikov hydroamination of simple unactivated alkenes such as 1-hexene?	An iridium-catalyzed reaction process has successfully promoted the addition of indoles to unactivated alkenes with Markovnikov selectivity [23]
The substrate as an imine core. Can the imine be alkylated in a reductive alkylation reaction to generate a tertiary amine?	The reductive N-methylation of imines has been accomplished using a mixture of carbon dioxide and elemental hydrogen along with a ruthenium catalyst [28]
Is it possible to generate large macrocycles incorporating amide linkages?	The cyclization of oxindoles generated cyclic lactams with large ring sizes [30]. There was considerable substrate-specific reactivity noted in this work, and each substrate should be evaluated against the reported work for best results
The starting material is a styrene derivative. Can an aminofluorination reaction be carried out on these substrates?	A copper-catalyzed aminofluorination works quite well for these substrates and incorporates the fluorine in the α-position [33]
Can carboxylic acids be converted into amides?	A transition metal-free process for this reaction utilized hypervalent iodine reagents to promote the transformation [38]
Most aminocarbonylation reactions need high pressures of carbon monoxide. Are approaches known that are successful with low pressures of CO?	A palladium-catalyzed aminocarbonylation reaction was devised that only used 1 atm of CO pressure [41]

3.2 SYNTHESIS OF ARYLAMINES AND RELATED COMPOUNDS THROUGH NITROGEN–CARBON(sp²) BOND-FORMING REACTIONS

Aniline derivatives are perhaps the simplest examples of arylamines, and a wide variety of approaches have been used to generate these compounds. While electrophilic aromatic substitution is a well-known route, there are considerable issues with this approach including harsh conditions, low reactivity of common arenes, and a significant number of substrate-specific reactions. To circumvent many of the challenges associated with these reactions, a number of alternative approaches have been developed. The following sections will highlight representative reactions that have been used to generate arylamines through the formation of nitrogen–carbon(sp²) bonds.

The arylation of ammonia remains one of the most attractive entry points for the synthesis of arylamines. Ammonia is inexpensive and readily available; however, it is a gas at room temperature under atmospheric pressure. This significantly increases the operational complexity of its use and requires high-pressure reaction vessels. To address these issues, a range of ammonia alternatives have been used in the synthesis of arylamines. While some were more successful than others, many tended to suffer from an additional complication due to a competing diarylation reaction that led to the formation of diarylamines. A practical solution to the selective synthesis of aniline derivatives was reported by Hartwig

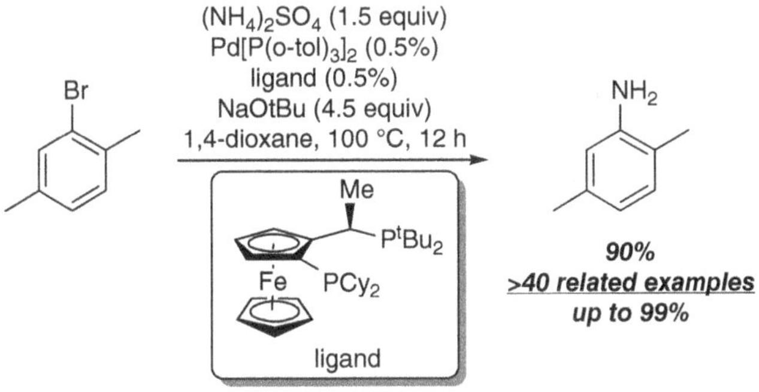

SCHEME 3.44   Palladium-catalyzed synthesis of aniline derivatives [42].

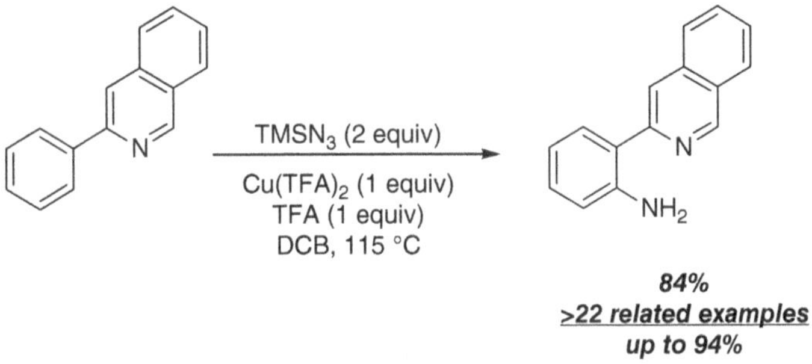

SCHEME 3.45   Aniline synthesis through copper-promoted C–H amination [43].

(Scheme 3.44) [42]. His approach used ammonium sulfate in place of ammonia with great success. The catalyst system for the aniline synthesis was comprised of a palladium precursor as well as a JosiPhos ligand. While several phosphine ligands promoted the N-arylation process, the combination of Pd[P(o-tolyl)$_3$]$_2$ and the JosiPhos ligand was highly active and selective at low catalyst loadings. Indeed, the chemistry was successful with less than 1% catalyst loading. Using this catalyst system, a vast array of aryl halides were converted into the aniline derivatives in good to excellent yields (up to 99%). Aryl chlorides tended to be more selective than aryl bromides for the synthesis of primary arylamines, and heteroaromatic substrates were well tolerated. In addition to the preparation of primary arylamines, the chemistry could be modified through the replacement of the ammonium sulfate with alkylamine hydrochloride salts in order to generate secondary alkyl arylamines. Overall, this is a practical approach to the preparation of primary arylamines.

A variety of other reagents have been used as substitutes for gaseous ammonia including trimethylsilyl azide. An example outlining the use of this reagent in the synthesis of functionalized aniline derivatives through C–H amination has been reported (Scheme 3.45) [43]. The reaction was promoted by a copper salt, and after screening a range of suitable species, Cu(TFA)$_2$ was found to be the most effective. The authors chose to screen a group of arenes bearing a pyridine-type directing group and were able to selectively incorporate the amino group in the position adjacent to the directing group. The chemistry was relatively insensitive to the presence of electron-donating or electron-withdrawing groups on the

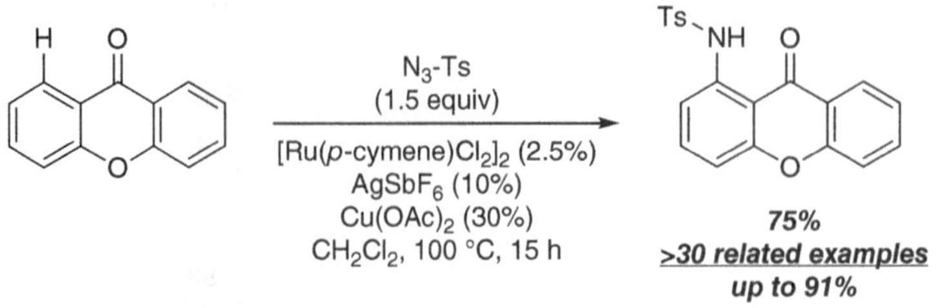

SCHEME 3.46   C–H amidation of xanthones using sulfonyl azides [44].

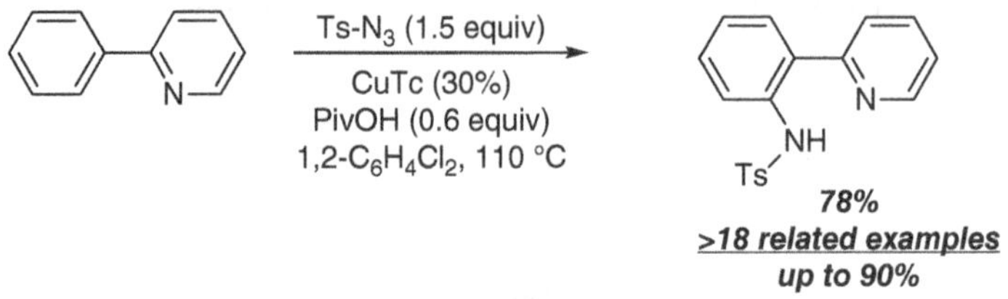

SCHEME 3.47   C–H amidation of 2-phenylpyridine [45].

arene. The overall reaction was redox neutral, and no added oxidant was needed to generate the aniline. In related work, the ruthenium C–H amidation of xanthones with sulfonyl azides has been achieved (Scheme 3.46) [44], as well as the copper-catalyzed C–H amidation of phenylpyridines (Scheme 3.47) [45].

Due to the challenges associated with classic syntheses, alternative routes to the synthesis of arylamines were investigated. One such alternate route entailed the use of palladium complexes as catalysts for the cross-coupling of aryl halides with primary and secondary amines. Following the pioneering work of Buchwald and Hartwig [46, 47], a vast array of metal complexes and supporting ligands have been used to prepare arylamines. As a result, more active and robust catalyst systems were developed, and a staggering number of arylamines were generated. Despite the attention this area has received, there are still a number of substrate-specific reactivity profiles that are challenging to overcome. Furthermore, the selection of solvent, palladium precursor, and supporting ligand plays a critical role in the design of a successful coupling reaction. In an effort to alleviate some of the confusion, an outstanding user guide has been generated by Buchwald that provides some direction when planning a synthesis [48].

One of the most effective combinations for coupling aryl halides with primary amines is a BrettPhos-ligated palladium precursor along with added BrettPhos as a supporting ligand (Scheme 3.48) [49]. This combination is remarkably effective for the cross-coupling using aryl chlorides at very low catalyst loadings (0.01%). The most effective base for this specific reaction was NaO^{t}Bu, and good to excellent yields of the secondary arylamines were isolated after heating for an hour. Primary alkylamines and arylamines were both well tolerated by the coupling chemistry. When aryl chlorides were used as coupling partners, electron-deficient arenes were sluggish; however, excellent yields were obtained with electron-poor mesylates. Overall, this is an effective system for the coupling of a variety of aryl halides, mesylates, and related compounds with primary amines.

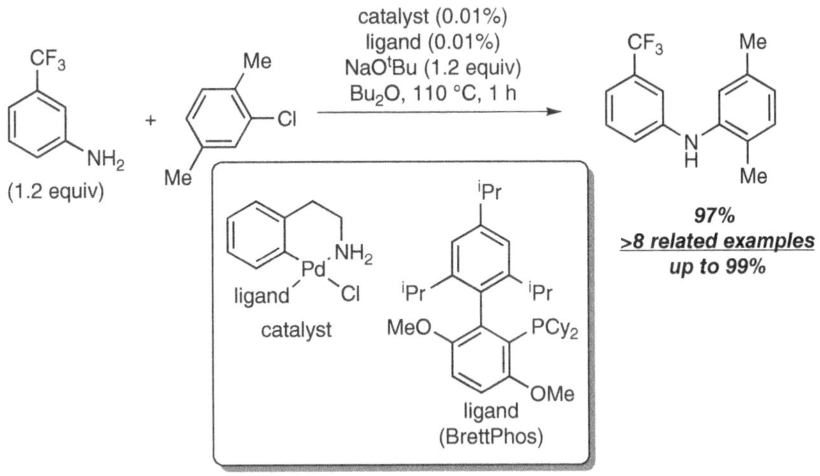

SCHEME 3.48   Using BrettPhos as the supporting ligand in the coupling of aryl chlorides with primary amines [49].

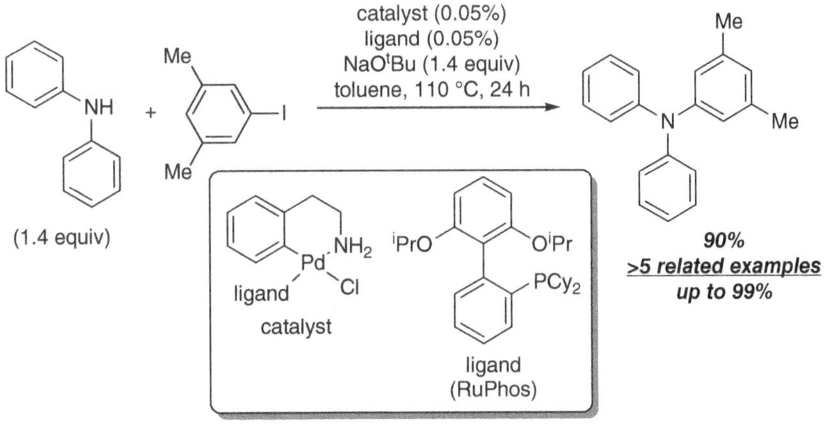

SCHEME 3.49   Using RuPhos as the supporting ligand in the palladium-catalyzed coupling of aryl iodides with secondary amines [50].

When secondary amines were used as coupling partners, a catalyst system comprised of a RuPhos-ligated palladium precursor and an extra equivalent of RuPhos was found to be an effective approach (Scheme 3.49) [50]. One of the most attractive aspects of this chemistry was the observation that the catalyst system functioned at very low catalyst loadings. A range of aryl iodides were coupling with secondary amines in excellent yield. Secondary arylamines and alkylamines were successfully converted into tertiary amines using sodium butoxide or cesium carbonate as the base.

In addition to the work using palladium catalysts, a wide array of copper catalysts have been developed for the synthesis of arylamines. Some of the earliest examples of the copper-catalyzed coupling chemistry used simple copper salts such as copper acetate

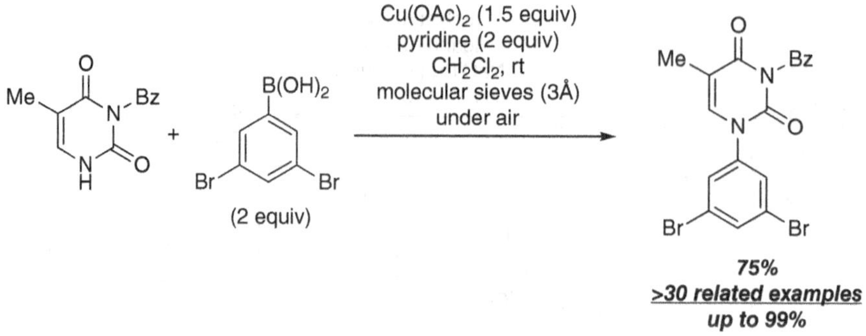

SCHEME 3.50   Copper-catalyzed coupling of pyrimidines with arylboronic acids [51].

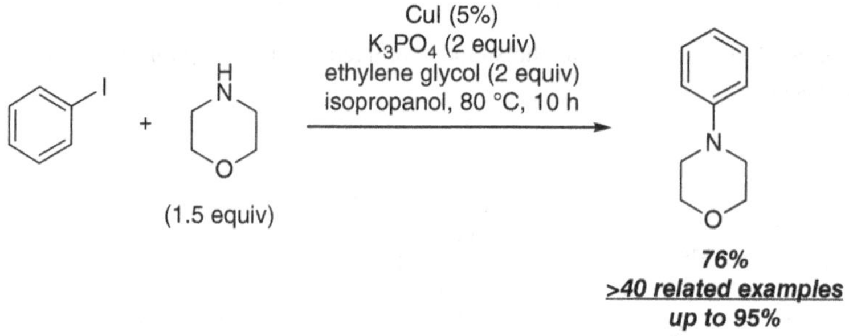

SCHEME 3.51   Using copper catalysts to generate arylamines [57].

to promote the cross-coupling reactions (Scheme 3.50) [51–53]. Most of the efforts in this area have been devoted to devising different solubilizing/stabilizing ligands for the copper center [54–56]. One of the practical versions of these reactions entailed the use of copper iodide as the catalyst system and ethylene glycol as the supporting ligand (Scheme 3.51) [57].

Discrete copper compounds can also be used as catalysts for the synthesis of aryl-amines (Scheme 3.52) [58]. Venkataraman used a neocuproine-ligated copper(I) species to promote the coupling of aryl halides with secondary amines. A practical advantage to this chemistry was that only air-stable materials were needed to construct the catalyst needed for the cross-coupling. A base was needed to promote the reaction, and potassium *tert*-butoxide was found to be more effective in the cross-coupling than other common bases including potassium phosphate, sodium methoxide, or cesium carbonate. Curiously, cesium carbonate was not as active in this chemistry, but it was quite effective in the preparation of diaryl ethers. Several aryl halides were screened for activity, and aryl bromides and iodides afforded moderate to good yields of the arylamines. It should be noted that an electron-neutral aryl chloride was converted into the triarylamine, albeit in lower yield (49%).

The ability to retain groups that enable the functionalization of a substrate after the initial transformation is a critical aspect of modern synthetic chemistry. These synthetic "handles" can be a variety of groups; however, halogenated fragments are attractive due to the vast array of synthetic approaches that have been developed for their transformation.

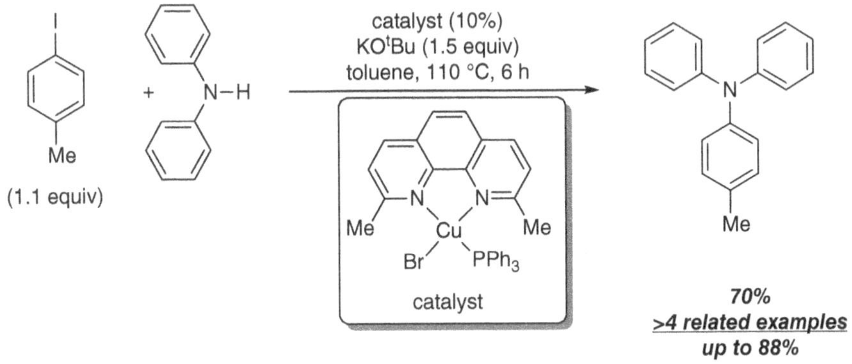

SCHEME 3.52   Synthesis of arylamines using a discrete copper complex [58].

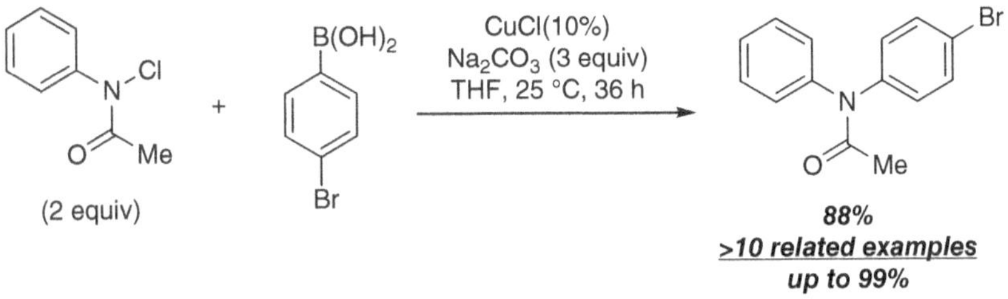

SCHEME 3.53   Copper-catalyzed coupling of arylboronic acids with chloroamides [59].

To this end, the copper-catalyzed coupling of chloroamides with halogenated arylboronic acids was developed (Scheme 3.53) [59]. The catalyst system was comprised of copper(I) chloride and sodium carbonate. No supporting ligand for the copper center was added in this work, although it should be noted that the starting material and product could solubilize/stabilize the copper center. The conditions were remarkably mild (25 °C), and outstanding yields of the arylamides were obtained after stirring for 36 h (up to 99%). Aryl bromides and iodides are typically the most desired halogens for subsequent functionalizations. To this end, brominated and iodinated arylboronic acids and chloroamides were screened for activity in the coupling reaction with special attention to the ability to retain the aryl halide. The cross-coupling reaction retained the halogenated arene regardless of which side (amide vs. arylboronic acid) was halogenated.

From the discussion earlier, the modern synthesis of nitrogen–carbon(sp^2) bonds is typically achieved by transition metal-catalyzed reactions. Using this approach, a vast array of nitrogen-containing compounds and heterocycles were generated from an array of nitrogen nucleophiles and carbon-based electrophiles. The following section will highlight a representative sampling of these reactions as well as a range of metal-free transformations that have been developed.

An intriguing approach to the synthesis of pyrroles has been reported using an iron-promoted domino reaction (Scheme 3.54 and Example 3.7) [60]. Starting from cyclopropanes and using a common iron salt as the catalyst for the reaction, a series of cyclopropanes were converted into pyrroles in good to excellent yields. One of the more attractive aspects of this chemistry was the ability to carry out the functionalization under air.

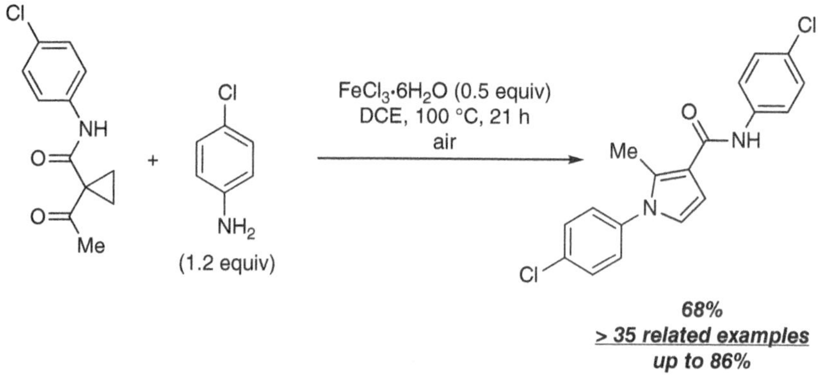

SCHEME 3.54 Synthesis of pyrroles through an iron-mediated domino reaction [60].

Example 3.7 Preparation of Functionalized Pyrroles from Cyclopropanes [60].

Before starting this synthesis, carefully follow the safety instructions listed in Example 3.1.

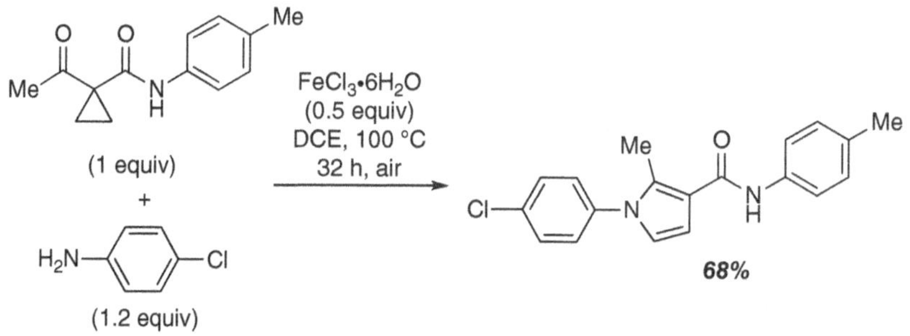

This reaction was carried out under an atmosphere of air. To a round-bottom flask (25 mL) equipped with a spherical condenser (20 cm length) were added 1-acetyl-*N*-(*p*-tolyl) cyclopropane-1-carboxamide (217 mg, 1.0 mmol), 4-chloroaniline (154 mg, 1.2 mmol), FeCl$_3$6H$_2$O (135 mg, 0.5 mmol), and DCE (4.0 mL). Then the mixture was well stirred at 100 °C in air until the carboxamide was completely consumed (TLC monitor). After cooling, the mixture was filtered through a pad of Celite, eluting with CH$_2$Cl$_2$ (6 mL × 3). The volatiles were removed under reduced pressure, and the residue was purified by short flash silica gel column chromatography to give compound the pyrrole (221 mg, 68%) (eluent: petroleum ether/ethyl acetate = 10/1).

Adapted with permission from Z. Zhang, W. Zhang, J. Li, Q. Liu, T. Liu, G. Zhang, J. Org. Chem., 2014, 79, 11226–11233. Copyright (2014) American Chemical Society.

The palladium-catalyzed synthesis of functionalized indoles through intramolecular nitrogen–carbon(sp^2) bond formation has been reported (Scheme 3.55) [61]. This was an oxidative cycloisomerization process, and the authors found that molecular oxygen (1 atm) was an effective oxidant for the process. The substrate scope for this reaction was quite

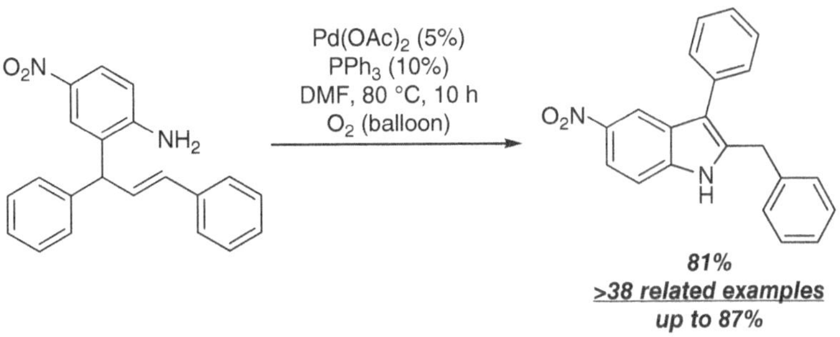

SCHEME 3.55   Palladium-catalyzed synthesis of pyrroles using molecular oxygen [61].

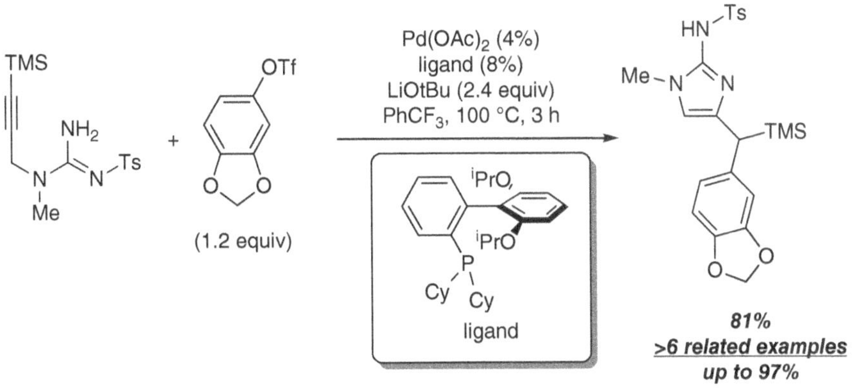

SCHEME 3.56   Synthesis of functionalized 2-aminoimidazoles through carboamination [62].

broad, and an assortment of electron-deficient and electron-rich aniline derivatives were functionalized in moderate to good yields.

The synthesis of functionalized 2-aminoimidazoles has been reported using a palladium-catalyzed approach (Scheme 3.56) [62]. The catalyst system was comprised of palladium acetate and a bulky dialkylbiarylphosphine ligand. A range of these ligands were screened, and RuPhos was the most effective for the carboamination chemistry. Along with a base, this system promoted the formation of the 2-aminoimidazole derivatives through the carboamination of *N*-propargyl guanidines with aryl triflates.

The preparation of a wide assortment of pyrrole derivatives used a phosphine-ligated gold complex as the catalyst precursor along with terminal alkynes and *N*-(2,2-dialkoxyethyl)benzamide as substrates (Scheme 3.57) [63]. A variety of supporting ligands were screened for activity, and a bulky dialkylbiarylphosphine (RuPhos) was found to be the most effective ligand for the system. The authors generated the catalytically active gold catalyst in solution through the addition of silver salts, and moderate to good yields of the pyrrole derivatives were obtained.

The synthesis of 1-amino-isoquinoline-*N*-oxides from 2-alkynylbenzaldoxime and related compounds has been achieved using a silver/copper-cocatalyzed reaction in air (Scheme 3.58) [64]. While silver triflate was used as the silver source, a range of copper salts were screened for activity, and copper iodide was found to be an effective cocatalyst

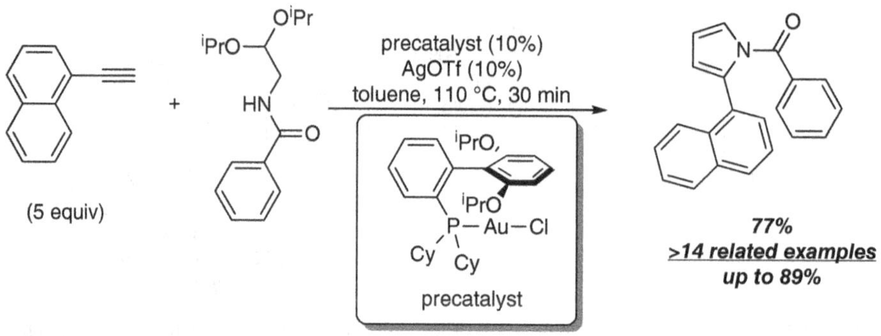

SCHEME 3.57 Gold-catalyzed synthesis of pyrroles from terminal alkynes [63].

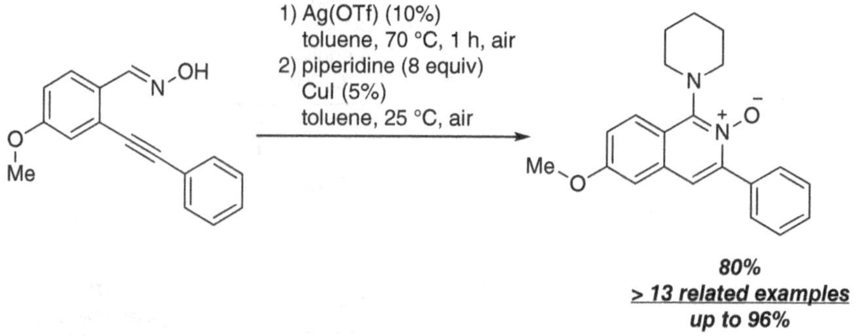

SCHEME 3.58 Preparation of 1-amino-isoquinoline-*N*-oxides from 2-alkynylbenzaldoximes [64].

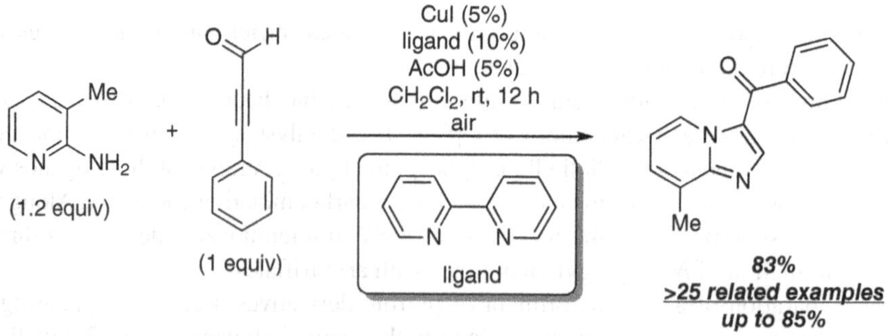

SCHEME 3.59 Synthesis of imidazo[1,2-*a*]pyridines through copper-catalyzed cyclization/oxidation process [65].

for the reaction. Using this catalyst system, a range of 1-amino-isoquinoline-*N*-oxides were prepared in good to excellent yields.

The preparation of imidazo[1,2-*a*]pyridines has been achieved using a copper-catalyzed cyclization/oxidation process (Scheme 3.59 and Example 3.8) [65]. Using a common copper salt and bipyridine as the stabilizing/solubilizing ligand, outstanding yields of the

Example 3.8   Copper-Catalyzed Synthesis of Imidazo[1,2-*a*]Pyridines [65].

Before starting this synthesis, carefully follow the safety instructions listed in Example 3.1.

This reaction was carried out under an atmosphere of air. 3-Phenylpropiolaldehyde (0.5 mmol), pyridin-2-amine(0.6 mmol), AcOH (5 mol%), CuI (5 mol%), and bipy (10 mol%) were stirred for 12 h in CH$_2$Cl$_2$ (3 mL) at room temperature. After reaction was complete, as monitored by TLC and GC/MS analysis, the solvent was then removed, and the product was isolated by column chromatography (eluted with petroleum ether/ethyl acetate = 2 : 1) to give a pure sample (80%, 88.8 mg).

Adapted with permission from H. Cao, X. Liu, J. Liao, J. Huang, H. Qiu, Q. Chen, Y. Chen, *J. Org. Chem.*, **2014**, *79*, 11209–11214. *Copyright (2014) American Chemical Society.*

heterocycles were obtained after simply stirring at room temperature for 12 h. One of the practical aspects of this chemistry was the use of elemental oxygen as the sole oxidant in the reaction. Additionally, pure oxygen was not needed for this reaction, and an air atmosphere was effective.

A series of nitrogen-containing heterocycles were generated through a four-component bicyclization process (Scheme 3.60 and Example 3.9) [66]. Two approaches to the skeletally diverse heterocycles were developed, with good to excellent yields obtained for a vast array of substrates. One of the strengths of this approach was the remarkable tolerance of the process to a range of functional groups.

The synthesis of α-benzofuranylacetamides and indolylacetamides has been achieved through a palladium-catalyzed process (Scheme 3.61 and Example 3.10) [67]. The overall reaction consisted of an intramolecular oxy/amino-palladation/isocyanide insertion process starting from functionalized propargyl alcohols. Although a vast array of compounds were successfully generated using this approach, the conversion of nitrated anilines was unsuccessful under the reaction conditions.

The transition metal-free synthesis of *N*-acyl diarylamines has been achieved through a three-step process (Scheme 3.62) [68]. Although many steps were involved in this approach, moderate to good yields of the desired compounds were obtained for most substrates.

Functionalized indoles were prepared in high yield through a ruthenium-catalyzed reaction between aryl azides and terminal alkynes (Scheme 3.63) [69]. The results of this reaction were in stark contrast to the normal reactivity of organoazides with alkynes that generated triazoles. The change in the direction of the reaction was proposed to be due to the presence of powerful electron-withdrawing groups on the aryl azide.

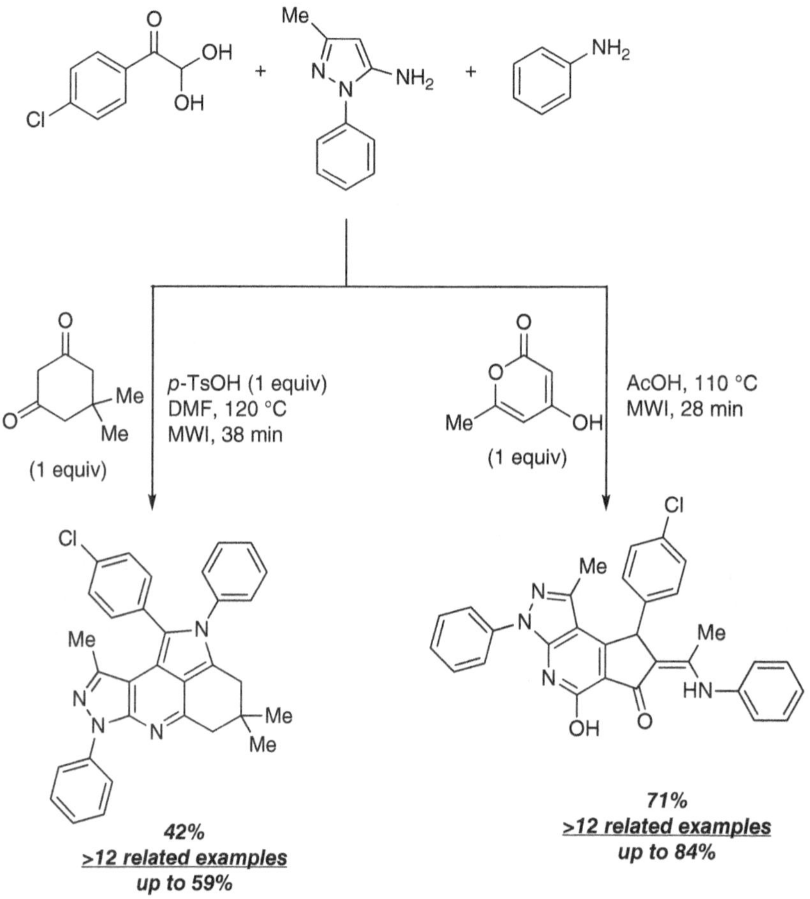

SCHEME 3.60 Generation of nitrogen-containing heterocycles through a 4-component coupling reaction [66].

Example 3.9 Preparation of Functionalized Pyridine Derivatives through Bicyclization [66].

Before starting this synthesis, carefully follow the safety instructions listed in Example 3.1.

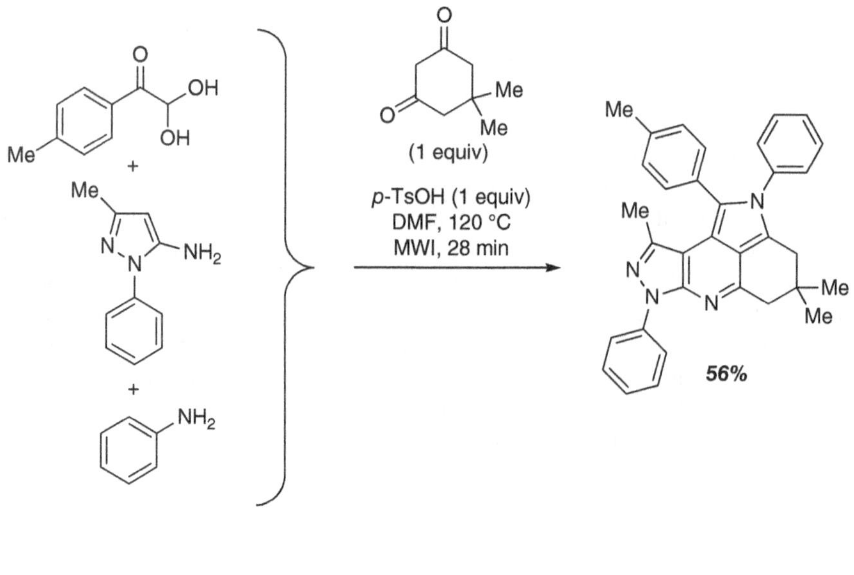

Caution! Petroleum ether, acetone, and ethyl acetate are highly flammable.

5,5-Dimethylcyclohexane-1,3-dione (140 mg, 1.0 mmol) and aniline (93 mg, 1.0 mmol) were added to an Initiator (10 mL) reaction vial, and 2,2-dihydroxy-1-(*p*-tolyl) ethanone (166 mg, 1.0 mmol), 3-methyl-1-phenyl-1*H*-pyrazol-5-amine (173 mg, 1.0 mmol), and *p*-TsOH (172 mg, 1.0 mmol) as well as *N,N*-dimethylformamide (DMF, 1.5 mL) were then successively added. Subsequently, the reaction vial was capped, and then the contents were stirred for 20 s. The mixture was irradiated (time, 28 min; temperature, 120 °C; absorption level, high; fixed hold time) until TLC [4/1 (v/v) petroleum ether/acetone] revealed that conversion of the 3-methyl-1-phenyl-1*H*-pyrazol-5-amine was complete. The system was neutralized with diluted hydrochloric acid and then diluted with cold water (40 mL). The solid product was collected by filtration and purified by flash column chromatography (silica gel, petroleum ether/ethyl acetate) to afford the title compound as a yellow solid (270 mg, 56%).

Adapted with permission from X.-J. Tu, W.-J. Hao, Q. Ye, S.-S. Wang, B. Jiang, G. Li, S.-J. Tu, *J. Org. Chem.*, **2014**, *79*, 11110–11118. *Copyright (2014) American Chemical Society.*

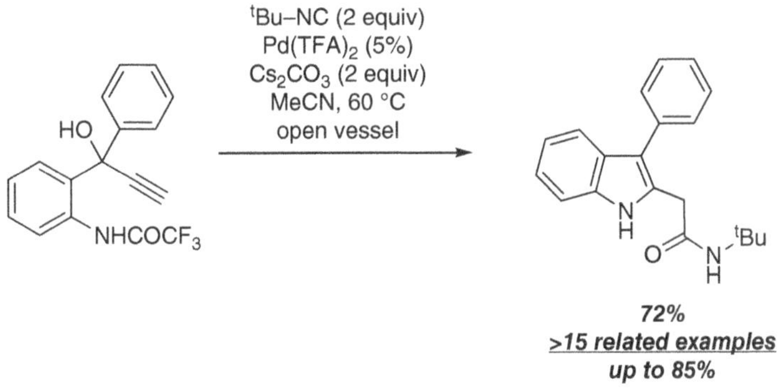

SCHEME 3.61 Generation of indolylacetamides in an open vessel [67].

Example 3.10 Preparation of an Indolylacetamides in an Open Vessel [67].

Before starting this synthesis, carefully follow the safety instructions listed in Example 3.1.

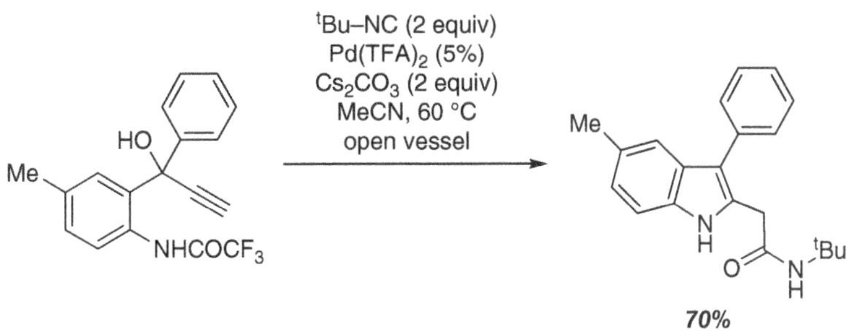

Caution! Ethyl acetate, acetonitrile, and hexane are highly flammable.

This reaction was carried out in an open vessel. To a stirred solution of the protected 2-aminophenylpropargyl alcohol (0.333 g, 1 mmol, 1 equiv) in acetonitrile (5 mL) were added Pd(TFA)$_2$ (0.0132 g, 0.04 mmol, 0.04 equiv), Cs$_2$CO$_3$ (0.648 g, 2 mmol, 2 equiv), and tert-butyl isocyanide (0.166 g, 2 mmol, 2 equiv) at room temperature, and the reaction mixture was stirred at 60°C for 24 h. After cooling to room temperature, the reaction mixture was diluted with water (20 mL) and extracted with EtOAc (2×15 mL). The combined EtOAc extracts were washed with brine (15 mL) and dried over Na_2SO_4. After removal of the volatiles under vacuum, the crude material was purified on silica (using 15% EtOAc/hexane) to obtain the title compound (0.224 g, 70%) as a yellow gum.

Adapted with permission from N. Thirupathi, M. H. Babu, V. Dwivedi, R. Kant, M. S. Reddy, *Org. Lett.,* **2014,** *16,* 2908–2911. *Copyright (2014) American Chemical Society.*

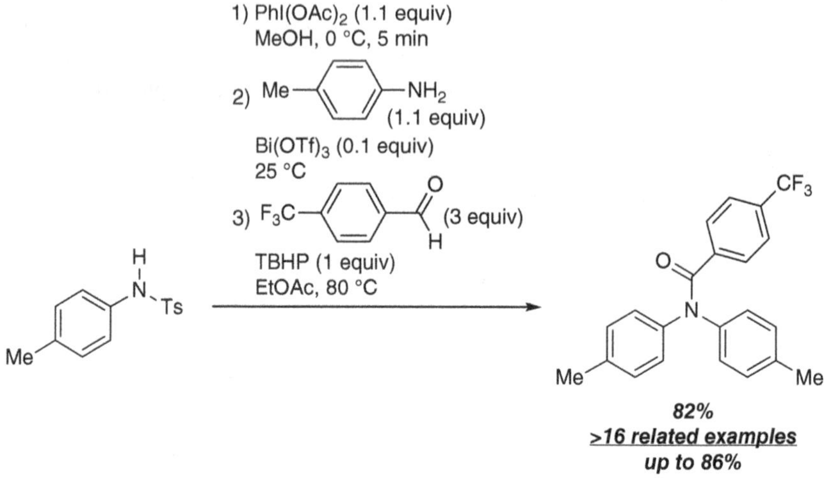

SCHEME 3.62   Preparation of amides from anilines and aldehydes [68].

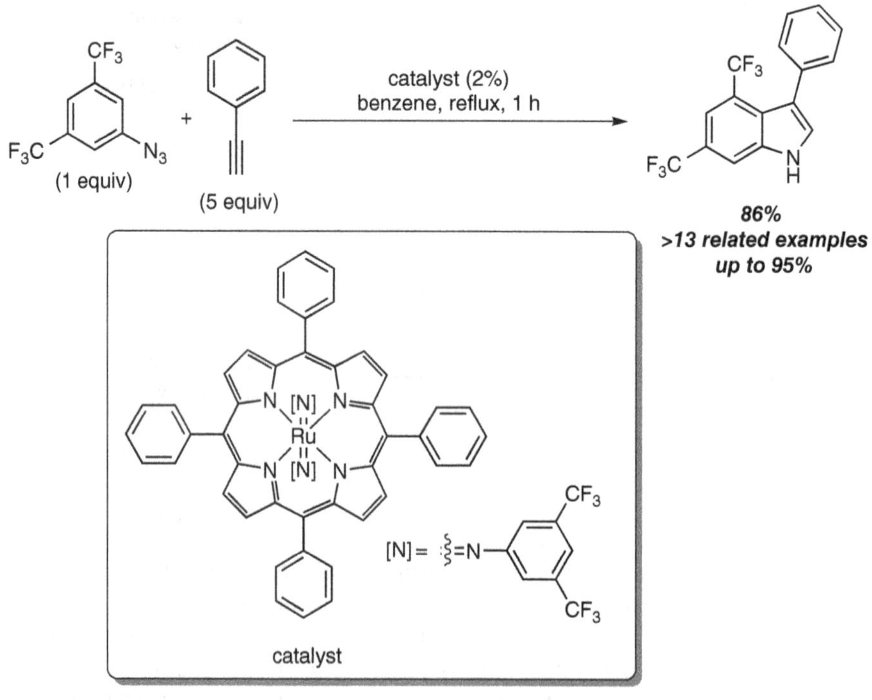

SCHEME 3.63   Ruthenium-catalyzed synthesis of functionalized indoles [69].

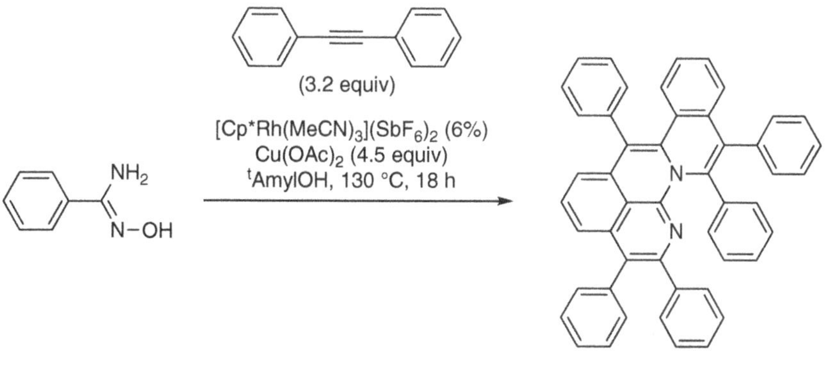

(3.2 equiv)

[Cp*Rh(MeCN)$_3$](SbF$_6$)$_2$ (6%)
Cu(OAc)$_2$ (4.5 equiv)
^{t}AmylOH, 130 °C, 18 h

83%
>15 related examples
up to 88%

SCHEME 3.64   Rhodium-catalyzed preparation of polyheteroaromatic compounds [70].

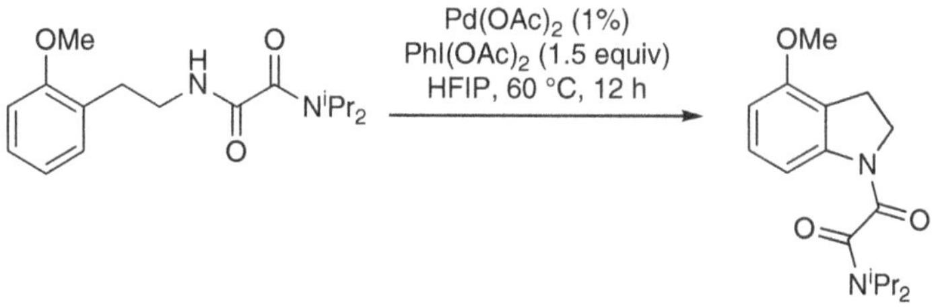

Pd(OAc)$_2$ (1%)
PhI(OAc)$_2$ (1.5 equiv)
HFIP, 60 °C, 12 h

76%
>30 related examples
up to 92%

SCHEME 3.65   Palladium-catalyzed synthesis of indolines [71].

The preparation of polyheteroaromatic compounds was achieved through a double C–H activation reaction using a crowded rhodium catalyst and an excess of copper acetate to promote the reaction (Scheme 3.64) [70]. The copper was proposed to act as an oxidant in the reaction. Using this catalyst system, a vast array of polyheteroaromatic compounds were generated in good to excellent yield. The process was also quite insensitive to the electronic composition of the substrates as electron-donating and electron-withdrawing groups were well tolerated by this approach.

The synthesis of indolines through an intramolecular C–H amination reaction has been reported (Scheme 3.65) [71]. The reaction was catalyzed by a common palladium compound and promoted by an iodine(III) reagent. This catalyst system generated good to excellent yields of the heterocycles at 60 °C.

Quinoxaline *N*-oxides were generated in a metal-free reaction through a dehydrogenative cyclization between imines and *tert*-butyl nitrite (Scheme 3.66) [72]. The reaction was significantly accelerated in the presence of TBAB, and an assortment of imines were successfully converted into the heterocycles. One of the attractive aspects of this chemistry was the ability to carry out the reactions in air. It should also be noted that the reactions were extremely fast and high yields of the desired *N*-oxides were obtained in 15 min at room temperature.

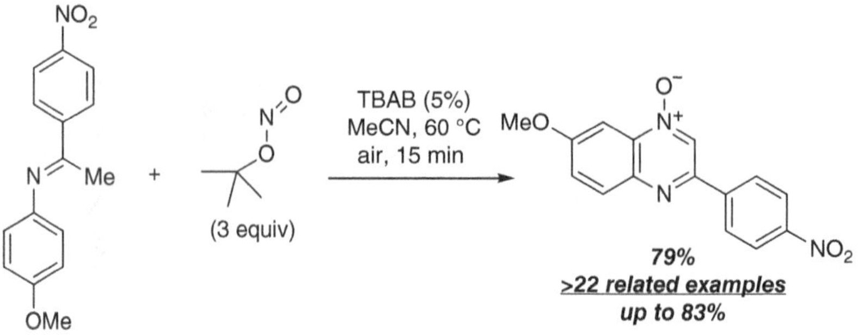

SCHEME 3.66 Metal-free synthesis of quinoxaline *N*-oxides [72].

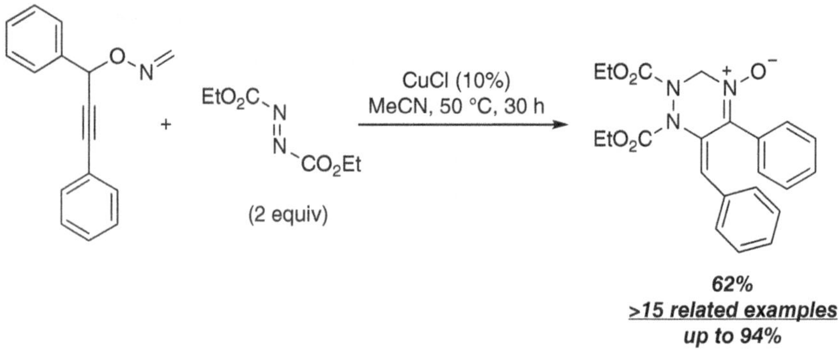

SCHEME 3.67 Preparation of triazine oxides from propargylic oximes [73].

The copper-catalyzed reaction between azodicarboxylates and formaldoximes generated triazine oxides (Scheme 3.67) [73]. While a number of copper salts were effective at promoting the reaction, copper chloride was among the most efficient and was selected for the remainder of the studies. Using this catalyst, an assortment of azodicarboxylates and formaldoximes were converted into the triazine oxides with the highest yields obtained when the azodicarboxylates contained alkyl substituents. When dibenzyl azodicarboxylates were screened, lower yields of the desired triazine oxide was obtained.

A multicomponent reaction has also been used to generate indoles, pyrazoles, and pyridazinones from arenediazonium salts (Scheme 3.68) [74]. An inexpensive silver salt served as the catalyst for this transformation. The reaction was generally regioselective and insensitive to the presence of electron-donating or electron-withdrawing groups. The synthesis of linked indoles has also been accomplished using this approach (Scheme 3.69).

A palladium-catalyzed approach to the synthesis of polycyclic indolines has been developed using Pd(PPh$_3$)$_4$ as the catalyst (Scheme 3.70) [75]. The authors screened a range of bases for activity toward the cyclization and found that cesium carbonate was the most effective. Using the optimized system, a great host of aryl halides were successfully converted into the target compounds in good to excellent yield.

A series of phosphonylpyrazoles have been generated through 1,3-dipolar cycloaddition reactions starting from diethyl 1-diazo-2-oxopropylphosphonate (Scheme 3.71) [76].

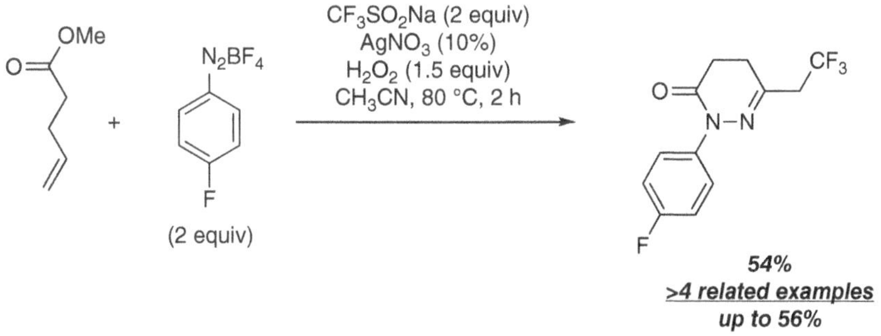

SCHEME 3.68   Dihydropyridazinone synthesis using a multicomponent reaction [74].

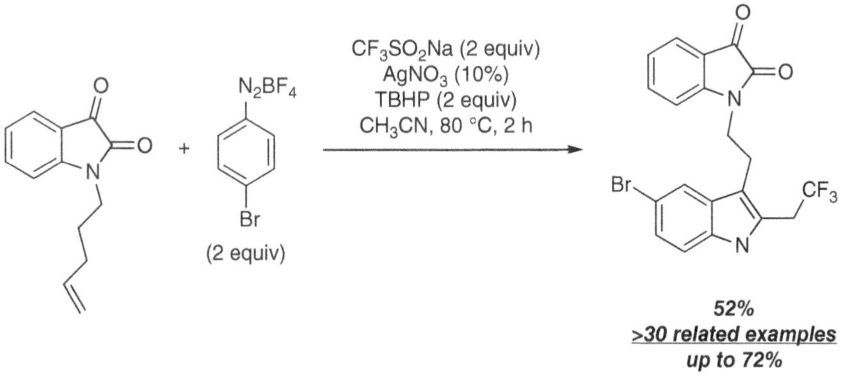

SCHEME 3.69   Synthesis of linked indoles [74].

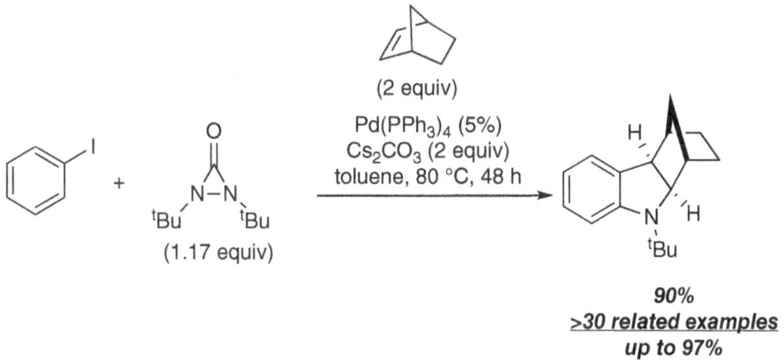

SCHEME 3.70   Palladium-catalyzed synthesis of polycyclic indolines [75].

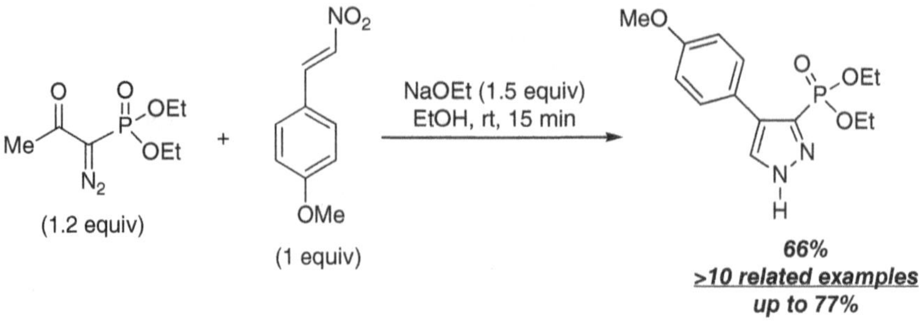

SCHEME 3.71 Synthesis of functionalized phosphonylpyrazoles through 1,3-dipolar cycloaddition reactions [76].

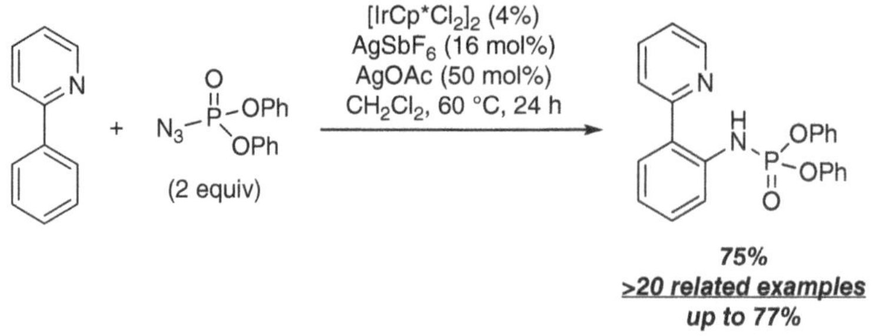

SCHEME 3.72 Synthesis of *N*-aryl phosphoramidates through the phosphoramidation of C–H bonds [77].

Nitroalkenes were the addition partners for these reactions, and a number of nitrostyrenes bearing a range of electron-donating and electron-withdrawing groups as well as several heteroaromatic substrates were successfully transformed into phosphonylpyrazoles.

A directed C–H activation has been used to generate *N*-aryl phosphoramidates (Scheme 3.72) [77]. The process was catalyzed by an iridium complex, and a pyridine group was used to direct the C–H phosphoramidation reaction. Using this approach, a diverse set of phosphoramidites bearing a range of preexisting functional groups were generated.

The synthesis of a range of functionalized pyrroles was accomplished through an oxidative cyclization reaction starting from β-enamino ketones or esters and alkynoates (Scheme 3.73) [78]. The chemistry was catalyzed by copper complexes, and after screening a range of copper precursors, copper iodide was found to be the most effective. Although significant amounts of products could be generated under an atmosphere of air, higher yields were obtained using pure oxygen.

The preparation of functionalized pyrroles has been accomplished through titanium-mediated reactions (Scheme 3.74) [79]. The first step in the process involves the generation of a metallocycle from the reaction of a diyne with the titanium reagent that was generated in solution. Once formed, treatment of the titanacycle with an activated imine followed by hydrolysis generated the pyrrole. The overall reaction can be described as an imino aza-Nazarov cyclization, and the process generated good to excellent yields of the 3-aminopyrroles.

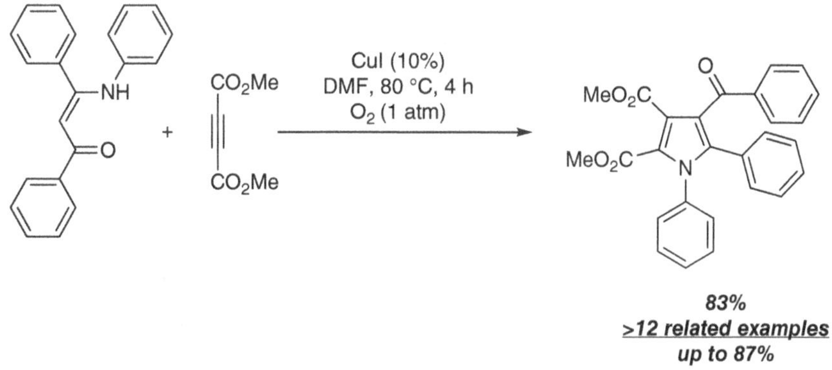

SCHEME 3.73 Synthesis of functionalized pyrroles through an oxidative cyclization process [78].

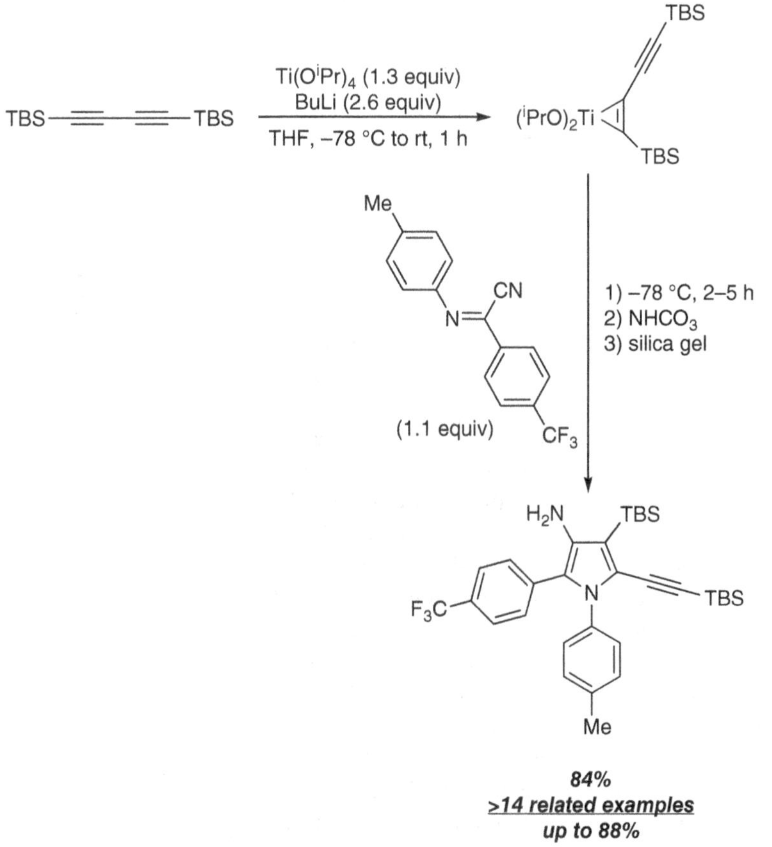

SCHEME 3.74 Synthesis of functionalized pyrroles through titanium-mediated reactions [79].

The selective amidation of unactivated C–H bonds continues to be a challenging reaction. One approach that has been successful involved the addition of directing groups such that the incoming nitrogen-containing fragment was added to a position adjacent to this directing group. Using this approach, 2-phenylpyridine was amidated through a C–H

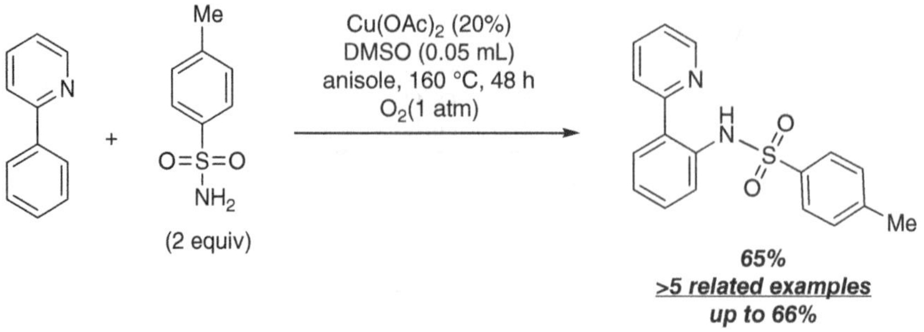

SCHEME 3.75 Copper-catalyzed C–H amidation of 2-phenylpyridine [80].

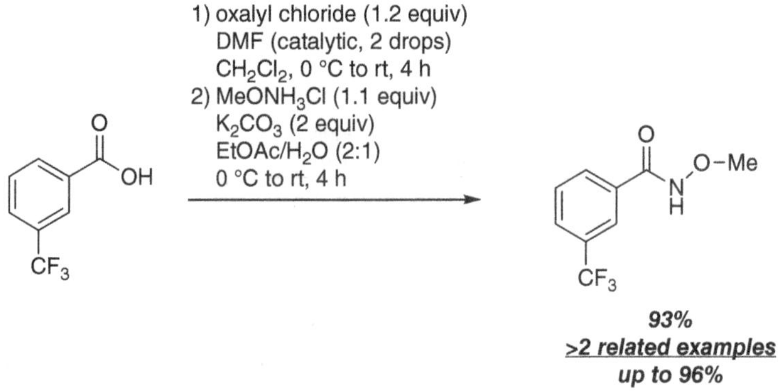

SCHEME 3.76 Synthesis of *O*-methyl hydroxamates from benzoic acid derivatives [81].

activation process using anilines, sulfonamides, and carboxamides (Scheme 3.75) [80]. The process was catalyzed by copper salts, and after screening a range of compounds, copper acetate was found to be the most effective. The authors also found that air could promote the reaction but the yield was moderate. Pure oxygen was a more potent oxidant.

The synthesis of *O*-methyl hydroxamates has been accomplished starting from carboxylic acids (Scheme 3.76 and Example 3.11) [81]. The multistep process was carried out under mild conditions and afforded outstanding yields of the desired hydroxamate (up to 96%). A related report outlined the conversion of unactivated anilines into amides under continuous flow conditions (Scheme 3.77) [82]. The reactions were quite fast and only required a 2 min residence time.

Annulation reactions between unactivated alkynes and a range of heteroatom-containing substrates have facilitated the preparation of a host of heterocycles. For example, the preparation of isoquinolones was achieved through a rhodium-catalyzed annulation of internal acetylenes with *O*-methyl hydroxamates (Scheme 3.78 and Example 3.12) [81]. The reaction conditions were quite mild, and most reactions were complete within 16 h. The overall process was redox neutral and used a catalytic amount of cesium acetate to promote the reaction. Additionally, the process was not appreciably sensitive to the electronic composition of the *O*-methyl hydroxamates. One of the more attractive aspects of this chemistry was the observation that the rhodium-catalyzed reaction

Example 3.11   Amidation Reaction: Carboxylic Acid into an Amide [81].

Before starting this synthesis, carefully follow the safety instructions listed in Example 3.1.

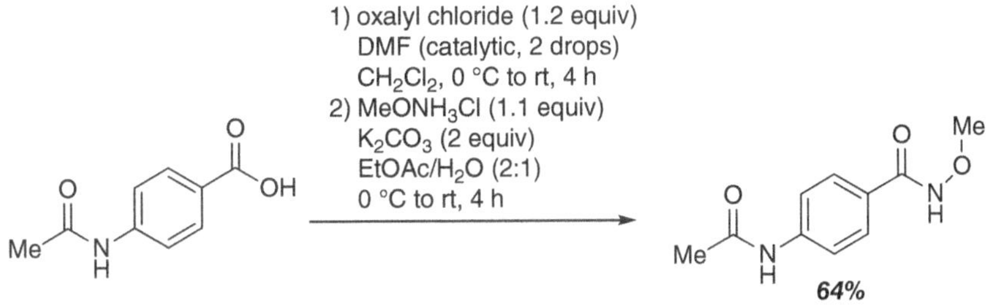

1) oxalyl chloride (1.2 equiv)
   DMF (catalytic, 2 drops)
   CH₂Cl₂, 0 °C to rt, 4 h
2) MeONH₃Cl (1.1 equiv)
   K₂CO₃ (2 equiv)
   EtOAc/H₂O (2:1)
   0 °C to rt, 4 h

64%

Caution! Ethyl acetate and petroleum ether are highly flammable.
Part 1: To a solution of the carboxylic acid (269 mg, 1.5 mmol, 1 equiv) in CH_2Cl_2 (0.3 M) at 0 °C under Ar was added dropwise oxalyl chloride (1.2 equiv) followed by a catalytic amount of DMF (two drops). The reaction was allowed to stir at room temperature until completion (typically 4 h). The solvent was then removed under reduced pressure to afford the corresponding crude acid chloride.

Part 2: Methoxyamine hydrochloride (1.1 equiv) was added to a biphasic mixture of K_2CO_3 (2 equiv) in a 2:1 mixture of $EtOAc:H_2O$ (0.2 M). The resulting solution was cooled to 0 °C followed by dropwise addition of the unpurified acid chloride dissolved in a minimum amount of EtOAc. The flask containing the acid chloride was then rinsed with additional EtOAc. The reaction was allowed to stir for 4 h while reaching room temperature. The phases were separated and the aqueous phase was extracted twice with ethyl acetate. The combined organic layers were dried over $MgSO_4$, filtered, and evaporated under reduced pressure. The residue was purified by flash column chromatography on silica gel using ethyl acetate/petroleum ether (1:1) as the eluent to afford the amide as a pale orange/yellow solid (201 mg, 64%).

Adapted with permission from N. Guimond, C. Gouliaras, K. Fagnou, *J. Am. Chem. Soc.* **2010**, *132*, 6908–6909. *Copyright (2010) American Chemical Society.*

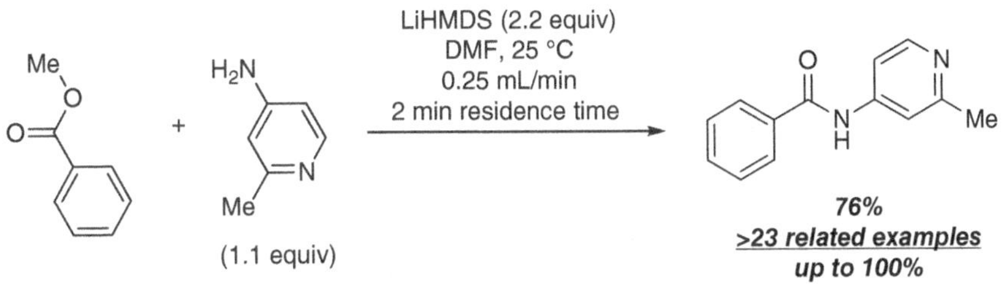

LiHMDS (2.2 equiv)
DMF, 25 °C
0.25 mL/min
2 min residence time

(1.1 equiv)

76%
>23 related examples
up to 100%

SCHEME 3.77   Conversion of poorly nucleophilic amines into amides under continuous flow conditions [82].

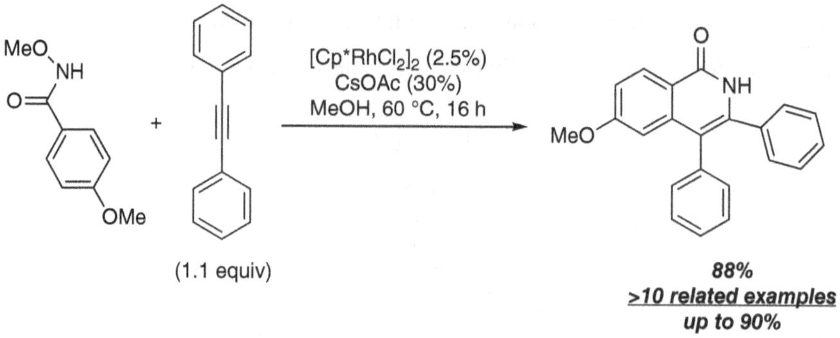

SCHEME 3.78   Synthesis of isoquinolones through rhodium-catalyzed annulation [81].

Example 3.12   Rhodium-Catalyzed Annulation Reaction [81].

Before starting this synthesis, carefully follow the safety instructions listed in Example 3.1.

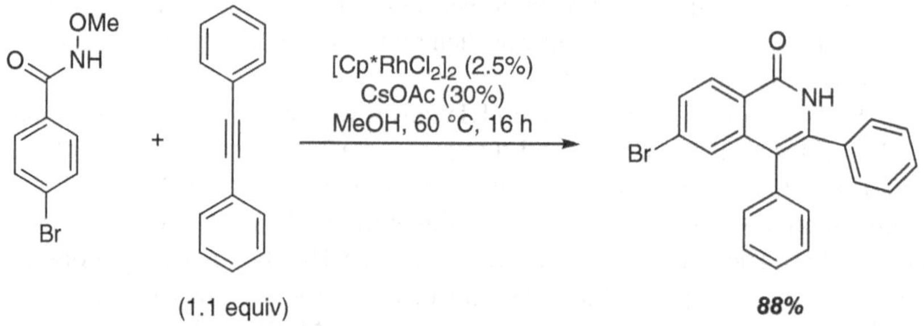

Caution! Diethyl ether and petroleum ether are highly flammable.

Without any particular precautions to extrude oxygen or moisture, the hydroxamic acid derivative (46 mg, 0.20 mmol, 1 equiv), diphenylacetylene (1.1 equiv), [Cp*RhCl$_2$]$_2$ (2.5 mol%), and CsOAc (30 mol%) were added to a 13×100 mm test tube along with a magnetic stirring bar. MeOH was added and the mixture was stirred at 60 °C. After stirring for 16 h, the reaction mixture was diluted with CH$_2$Cl$_2$ and transferred in a round-bottom flask. Silica was added to the flask and volatiles were evaporated under reduced pressure. The residue was purified by flash column chromatography using silica gel (gradient 50–75% Et$_2$O in petroleum ether) and afforded the title compound as a tan-colored solid (66.3 mg, 88%).

Adapted with permission from N. Guimond, C. Gouliaras, K. Fagnou, *J. Am. Chem. Soc.* **2010**, *132*, 6908–6909. *Copyright (2010) American Chemical Society.*

could be carried out without rigorous removal of oxygen. In related work, the metal-free synthesis of isoquinolones has been achieved through the annulation of *O*-methyl hydroxamates with internal alkynes (Scheme 3.79 and Example 3.13) [83].

Primary amides have been generated through the hydration of nitriles using a simple rhodium catalyst to promote the reaction (Scheme 3.80) [84]. The rhodium catalyst was ligated by cyclooctadiene and tris-(dimethylamino)phosphine. The latter was selected to

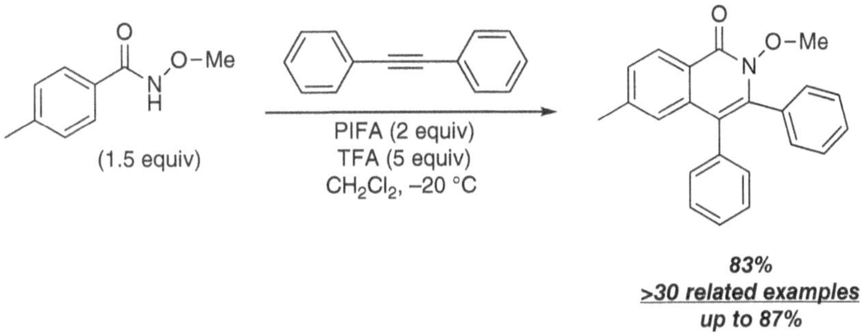

SCHEME 3.79 Metal-free annulation reactions between *O*-methyl hydroxamates and internal alkynes [83].

Example 3.13 A Metal-Free Annulation Reaction [83].

Before starting this synthesis, carefully follow the safety instructions listed in Example 3.1.

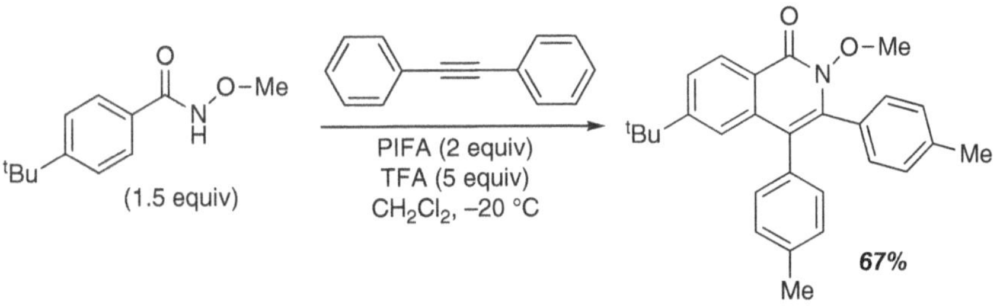

Caution! Ethyl acetate and petroleum ether are highly flammable.

To an oven-dried reaction tube (10 mL) were added amide (0.25 mmol, 1.5 equiv) and alkyne (0.17 mmol, 1 equiv) under an argon atmosphere. Dry CH_2Cl_2 (2 mL) was added by syringe, and the resultant mixture was cooled to -20 °C. A CH_2Cl_2 solution of PIFA (0.34 mmol, 2 equiv) and TFA (0.85 mmol, 5 equiv) in 1.4 mL of dry DCM was then dropped into the reaction. After that, the mixture was stirred at designated temperature (-20 °C or room temperature) and monitored by TLC. Upon completion, 3 mL of saturated $NaHCO_3$ was added, and the mixture was stirred for 10 min followed by extracted with CH_2Cl_2 (3×5 mL). The combined organic solution was washed with brine (20 mL) and dried over Na_2SO_4. The solvent was removed under vacuum, and the residue was purified by silica gel chromatography (eluent: EtOAc/petroleum ether) to afford the annulation product as a white amorphous solid (46.9 mg, 67%).

Adapted with permission from Z.-W. Chen, Y.-Z. Zhu, J.-W. Ou, Y.-P. Wang, J.-Y. Zheng, J. Org. Chem., **2014**, *79*, 10988–10998. *Copyright (2014) American Chemical Society.*

provide an increased solubility of the rhodium complex in water. Using this catalyst, a vast array of alkyl- and arylnitriles were hydrated. The electronic composition of the arylnitriles affected the rate of the reaction, and substrates bearing electron-donating groups were sluggish. Additionally, heptanenitrile and acetonitrile gave low yields of the amides. For arylnitriles, this is an attractive approach to the synthesis of primary amides.

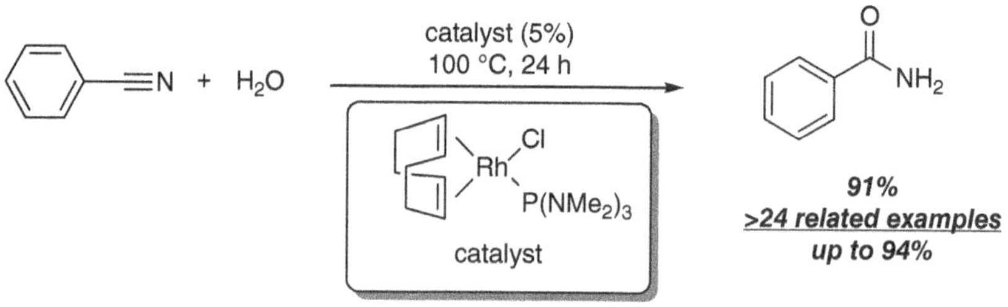

SCHEME 3.80   Rhodium-catalyzed hydration of nitriles [84].

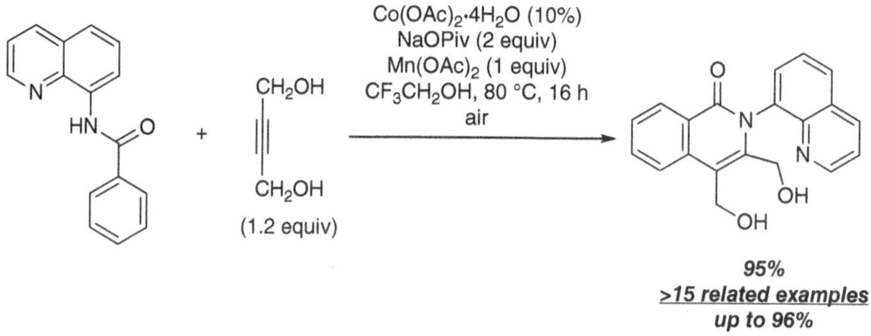

SCHEME 3.81   Synthesis of tertiary amides through annulation [85].

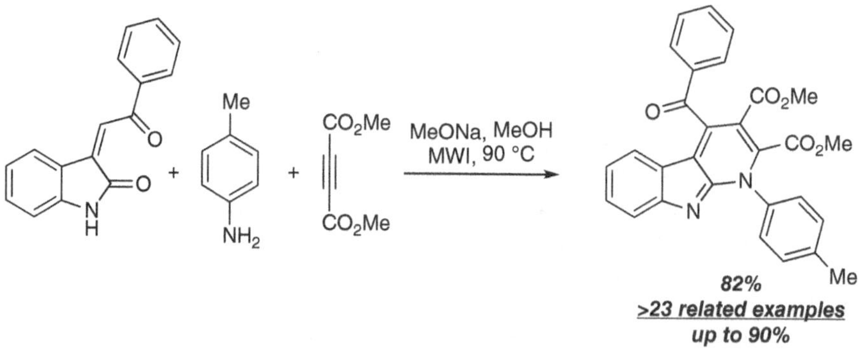

SCHEME 3.82   Synthesis of functionalized indoles through a multicomponent reaction [86].

The preparation of tertiary amides has been achieved through a cobalt-catalyzed annulation reaction (Scheme 3.81) [85]. Cobalt catalysts are attractive since they are inexpensive and readily available. The reactions were carried out in air (as the terminal oxidant) and afforded an assortment of the functionalized heterocycles.

As shown earlier, multicomponent reactions are powerful approaches to the synthesis of heteroatom-containing compounds. Using this approach, functionalized indoles were generated through the reaction of an internal alkyne with a 3-aroylmethylidene-2-oxindole and an aniline derivative (Scheme 3.82) [86]. The process was carried out using microwave heating, and a vast array of fused heterocycles were generated. In related work, a

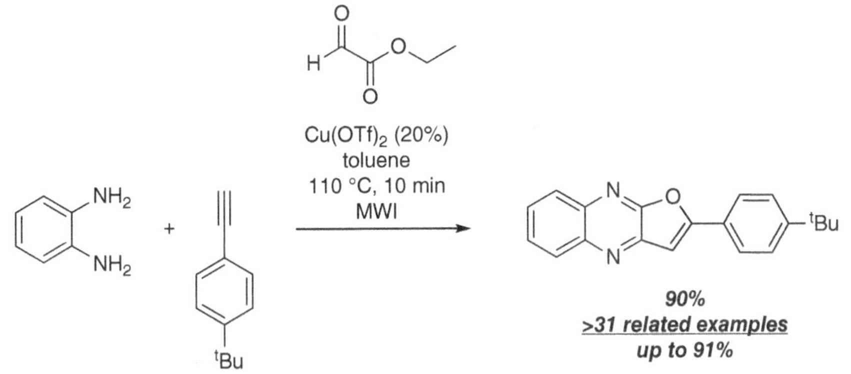

SCHEME 3.83   Synthesis of furoquinoxalines through a three-component coupling [87].

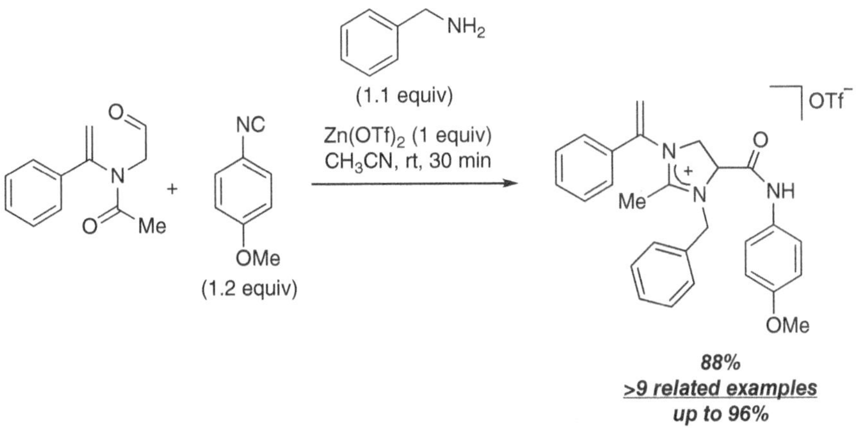

SCHEME 3.84   Synthesis of *N*-heterocyclic carbenes through a three-component coupling reaction [88].

copper-catalyzed multicomponent reaction has been used to prepare an array of functionalized furoquinoxalines (Scheme 3.83) [87].

Using zinc triflate as a promoter, a series of functionalized *N*-heterocyclic carbenes were generated through a three-component coupling reaction (Scheme 3.84) [88]. In addition to serving as the promoter for this reaction, the zinc salt also provided the triflate fragment. Treatment of amines with *N*-formylmethylcarboxamides and isocyanides along with zinc(II) triflate under mild conditions generated the target compounds in good to outstanding yield. It was also noteworthy that the reactions were remarkably fast under these mild conditions.

An interesting approach to the synthesis of phthalimides has been reported that starts from aminoquinoline benzamides (Scheme 3.85) [89]. The reaction entailed the carbonylation of the substrate using a cobalt(II) acetylacetonate as the catalyst. The reaction conditions were remarkably mild and could be carried out without exclusion of air. Indeed, air was critical to the success of the reaction, and the oxygen in air was proposed to act as a terminal oxidant for the process.

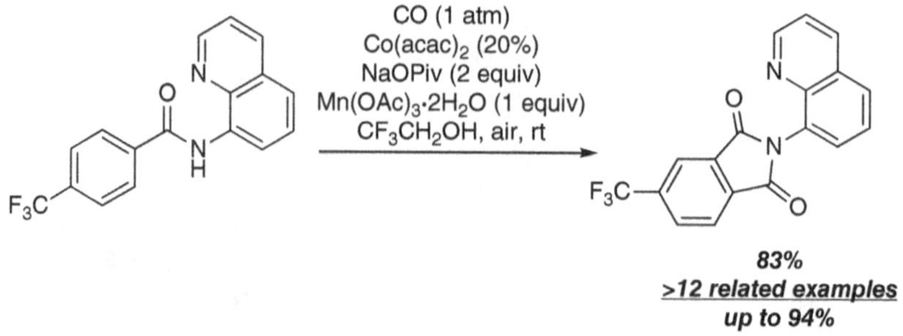

SCHEME 3.85 Preparation of N-arylated phthalimides through carbonylation [89].

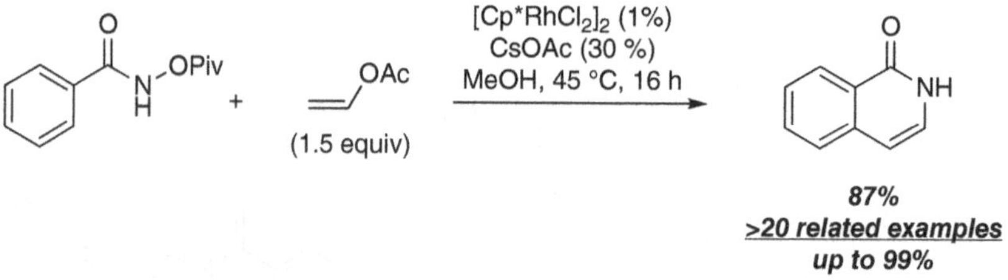

SCHEME 3.86 Preparation of isoquinolone using vinyl acetate as an acetylene equivalent [90].

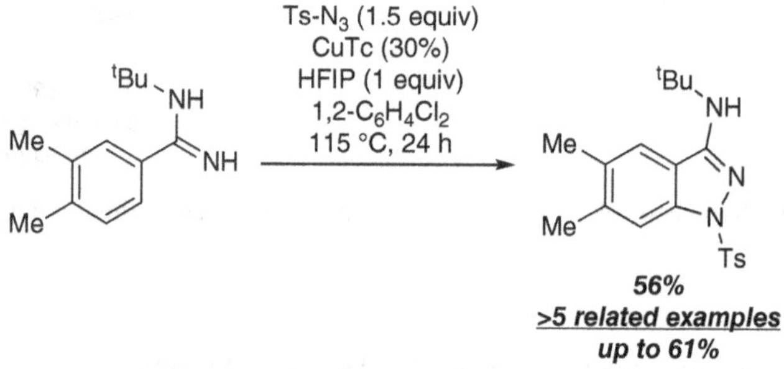

SCHEME 3.87 Synthesis of indazole derivatives using copper catalysts [45].

Vinyl acetate has been used as an acetylene equivalent in the synthesis of isoquinolones (Scheme 3.86) [90]. The reaction was catalyzed by a common rhodium compound and proceeded under very mild conditions to generate a diverse group of isoquinolones bearing a range of electron-donating and electron-withdrawing groups.

Indazoles can be generated through a copper-catalyzed process using TsN_3 as the amino source (Scheme 3.87) [45]. Using CuTc as the catalyst for the reaction, moderate yields of the indazoles were obtained. The chemistry was tolerant to several functional groups including halogens and ethers.

The gold-catalyzed synthesis of indole derivatives bearing a range of functional groups has been achieved using a tandem cycloisomerization/functionalization process using ynamides and alkynylanilines as substrates (Scheme 3.88) [91]. A carbene-ligated gold

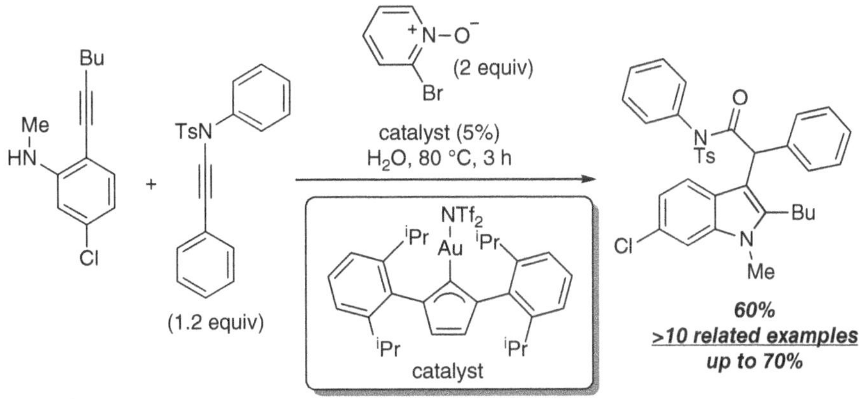

SCHEME 3.88   Gold-catalyzed synthesis of indoles from ynamides [91].

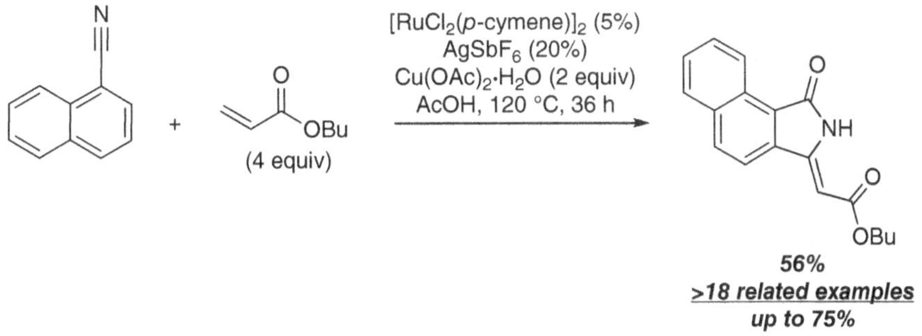

SCHEME 3.89   Preparation of (Z)-3-methyleneisoindolin-1-ones [92].

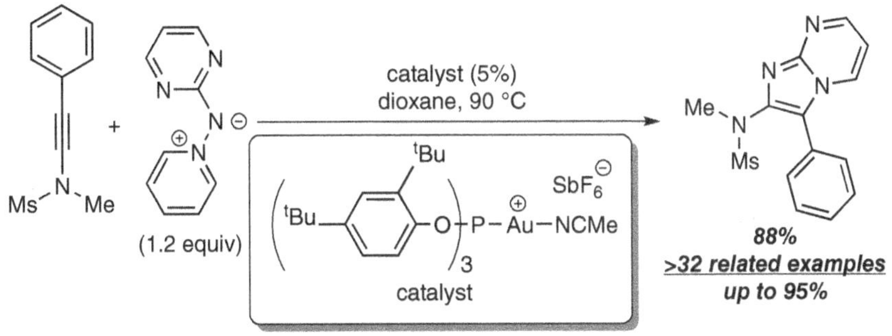

SCHEME 3.90   Gold-catalyzed synthesis of imidazopyrimidines from ynamides [93].

complex was used as the catalyst for the reaction, and a range of electronically and steri-cally diverse indole derivatives were generated using this chemistry.

A series of (Z)-3-methyleneisoindolin-1-ones were generated using a ruthenium-catalyzed cyclization (Scheme 3.89) [92]. Benzonitriles and activated alkenes such as acrylates served as the substrates in these reactions, and moderate to good yields were obtained.

Ynamides were converted into imidazopyrimidines through a gold-catalyzed reaction (Scheme 3.90) [93]. Pyridinium *N*-(2-pyrimidinyl)aminide served as the second

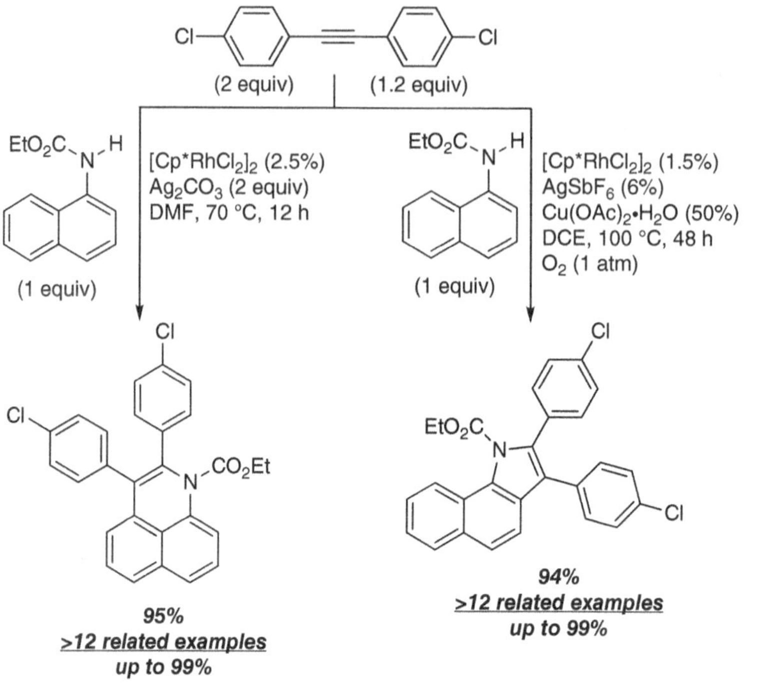

SCHEME 3.91   Annulation of internal alkynes with naphthylcarbamates [94].

component in the reaction, and a vast array of preexisting functional groups were well tolerated by this chemistry.

The rhodium-catalyzed annulation of internal alkynes with naphthylcarbamates was found to afford either benzoindole or benzoquinoline derivatives depending upon the additives used in the reaction (Scheme 3.91) [94]. The key component in this chemistry was copper acetate. In the presence of this copper salt (0.5 equiv), benzoindoles were preferentially formed in high yield. If the copper salt was never added to the reaction mixture, benzoquinolines were formed in high yield. This chemoselective reaction is an attractive approach to the synthesis of these compounds due to the operational simplicity of the process and the ease in which either substrate could be selected.

The preparation of 3-aminopyrazoles from aryl hydrazides has been accomplished through a three-component coupling reaction (Scheme 3.92) [95]. While a number of copper salts were active for the cyclization reaction, the combination of copper(I) chloride and Et$_3$N was the most efficient. The reaction conditions were quite mild, and simply stirring the reagents for a few hours at room temperature afforded good to excellent yields of the pyrazoles. While there was some substrate-specific reactivity observed in these reactions, a number of diynes and aryl hydrazides were successfully used to generate a host of 3-aminopyrazoles. The authors also found that 4-iminopyrimines could be generated if the aryl hydrazide was replaced with *N*-phenylbenzimidamides (Scheme 3.93) [95].

The palladium-catalyzed coupling of primary amides with halogenated heterocycles affords secondary amides in moderate to good yields (Scheme 3.94 and Example 3.14) [96].

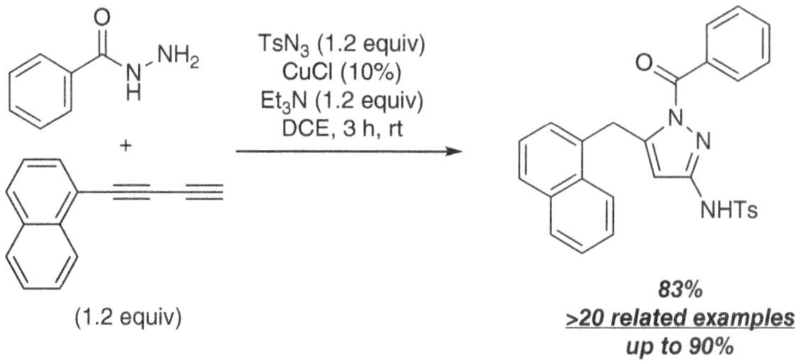

SCHEME 3.92 Copper-catalyzed synthesis of 3-aminopyrazoles from aryl hydrazides [95].

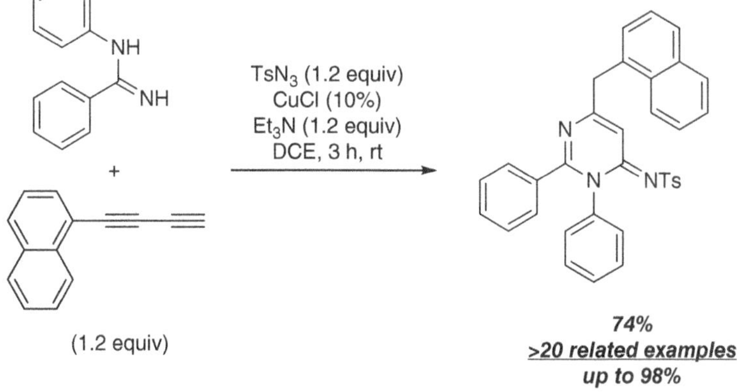

SCHEME 3.93 Copper-catalyzed synthesis of 4-iminopyrimines from *N*-phenylbenzimidamides [95].

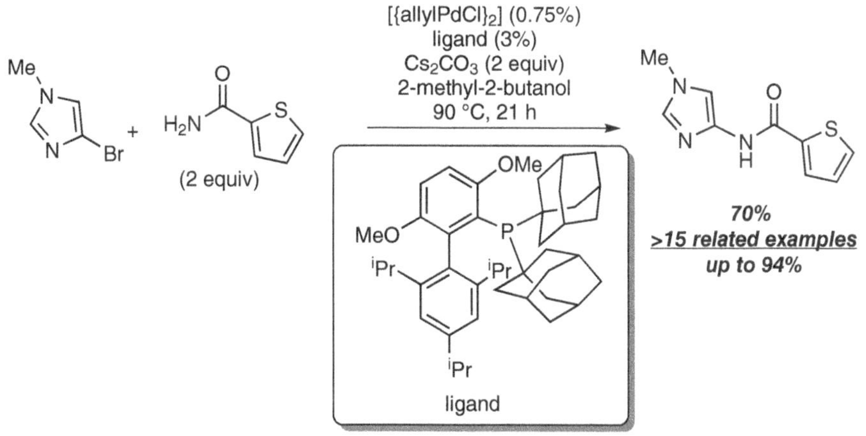

SCHEME 3.94 Coupling primary amides with aryl halides using palladium catalysts [96].

Example 3.14   Palladium-Catalyzed Synthesis of Heterocyclic Amides [96].

Before starting this synthesis, carefully follow the safety instructions listed in Example 3.1.

An oven-dried test tube was equipped with a magnetic stir bar and charged with [{(allyl) PdCl}$_2$] (0.75%), phosphine ligand (3.0%), Cs$_2$CO$_3$ (650 mg, 2 mmol), nicotinamide (244 mg, 2 mmol), and 4-bromo-1-tritylimidazole (389 mg, 1 mmol). The test tube was sealed with a screw-cap septum and then evacuated and backfilled with argon (this process was repeated a total of three times). 2-Methyl-2-butanol (2 mL) was then added by syringe. The reaction mixture was heated at 90 °C for 21 h. The reaction mixture was cooled to room temperature, diluted with EtOAc, washed with a saturated solution of sodium bicarbonate, dried over Na$_2$SO$_4$, concentrated in vacuo, and purified by flash chromatography (silica gel) using MeOH/CH$_2$Cl$_2$ (1 : 24) as the eluent to afford the amide as a white powder (387 mg, 90%).

Adapted with permission from M. Su, S. L. Buchwald, Angew. Chem. Int. Ed., 2012, 51, 4710–4713. Copyright (2012) WILEY-VCH Verlag GmbH & Co. KGaA, Weinheim.

An effective catalyst system was comprised of a simple palladium(II) precursor and a bulky dialkylbiarylphosphine ligand. An attractive aspect of this chemistry was the observation that the catalyst system was active at very low catalyst loadings (0.75%). A variety of brominated heterocycles were successfully coupled with a diverse group of primary amides using this catalyst.

Highly substituted indoles were generated through a one-pot sequential process from acrylonitrile derivatives (Scheme 3.95) [97]. The first step consisted of treating the precursor with an aniline derivative in the presence of a strong base to generate the allylic amine. The second step was comprised of a copper-catalyzed cross-coupling reaction between the tethered aryl bromide and the allylic amine generated in the first step of the process. The overall yields were moderate to good, and a variety of indoles were generated.

An interesting approach to the preparation of highly functionalized pyrroles was accomplished in a three-step one-pot reaction starting from benzonitrile derivatives (Scheme 3.96). The initial step consisted of treating a Reformatsky reagent with a terminal

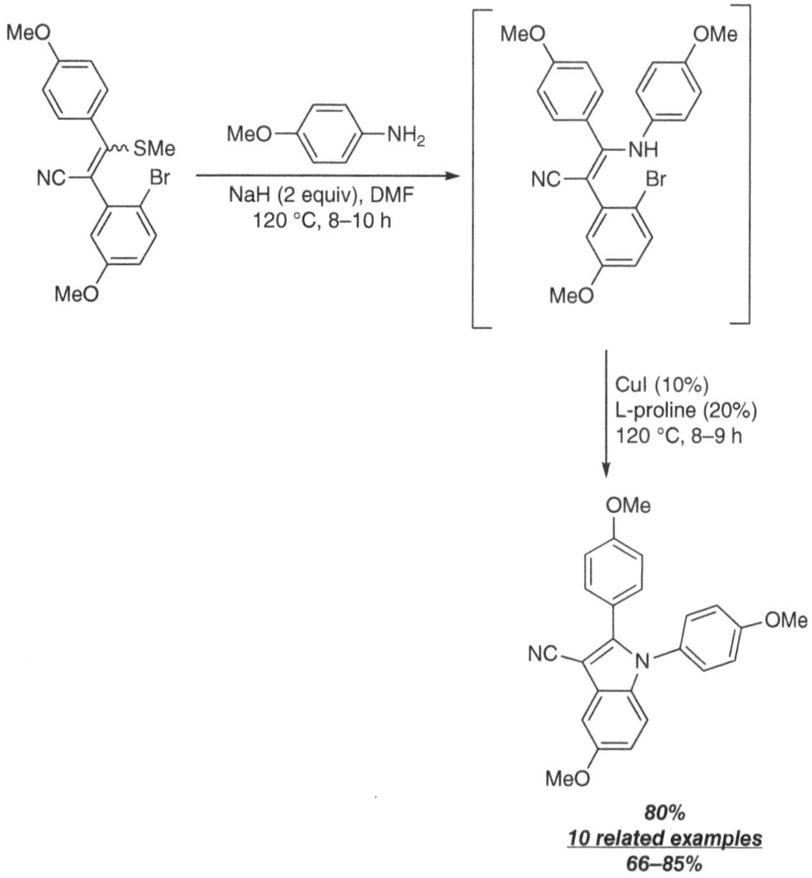

SCHEME 3.95   One-pot synthesis of functionalized indoles [97].

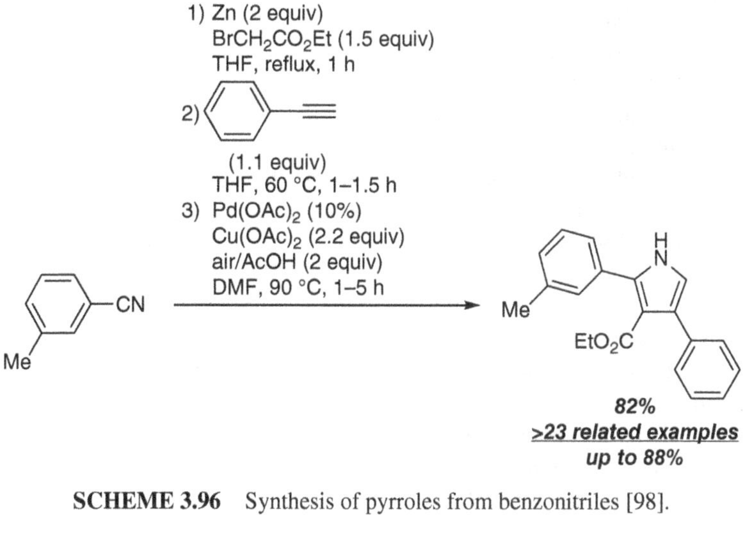

SCHEME 3.96   Synthesis of pyrroles from benzonitriles [98].

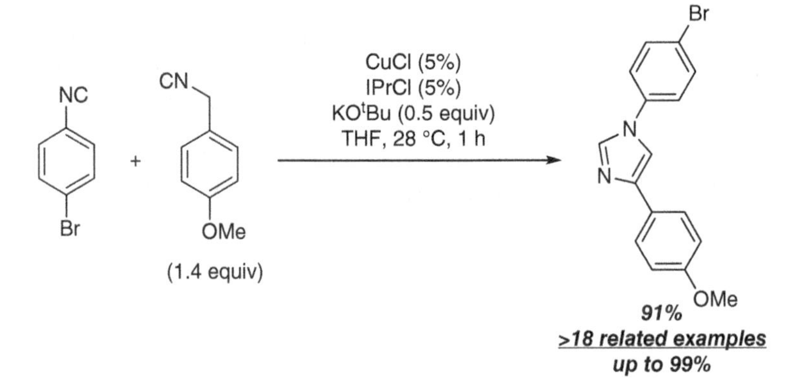

SCHEME 3.97   Synthesis of 1,4-diaryl-1*H*-imidazoles from isocyanides [99].

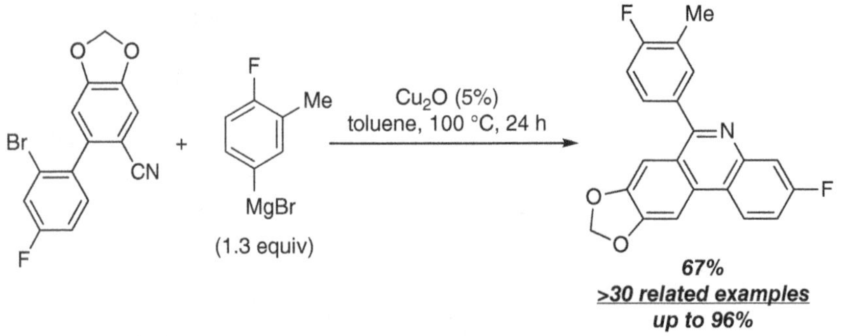

SCHEME 3.98   Preparation of polysubstituted phenanthridines [100].

alkyne to generate α-vinylated-β-enamino esters as intermediates. A palladium-catalyzed cyclization afforded the functionalized pyrroles in moderate to good yield.

1,4-Diaryl-1*H*-imidazoles remain a challenging class of compounds to prepare without the use of harsh reaction conditions. To address this issue, a low-temperature synthesis of these compounds was developed starting from common aryl isocyanides (Scheme 3.97) [99]. Treatment of the aryl isocyanides with a benzyl isocyanides with a copper catalyst afforded good to excellent yields of a wide assortment of 1,4-diaryl-1*H*-imidazoles.

Polysubstituted phenanthridines were generated through a copper-catalyzed process using Grignard reagents and 2′-bromo-[1,1′-biphenyl]-2-carbonitrile derivatives as substrates (Scheme 3.98) [100]. While a range of copper salts generated various amounts of the target compounds, copper(I) oxide was the most effective. This was interesting since copper(II) oxide is very insoluble and the standard conditions did not employ a solubilizing/stabilizing ligand for the copper center. It should be noted that the cyclization reaction proceeded under an atmosphere of air, albeit in lower yield.

An oxidative amination reaction has been used to generate 2-nitroimidazopyridines (Scheme 3.99) [101]. Treatment of nitrostyrenes with various 2-aminopyridines using iron nitrate as the catalyst generated the heterocycles in moderate to good yields after stirring for a few hours. One of the interesting aspects of this chemistry was the observation that

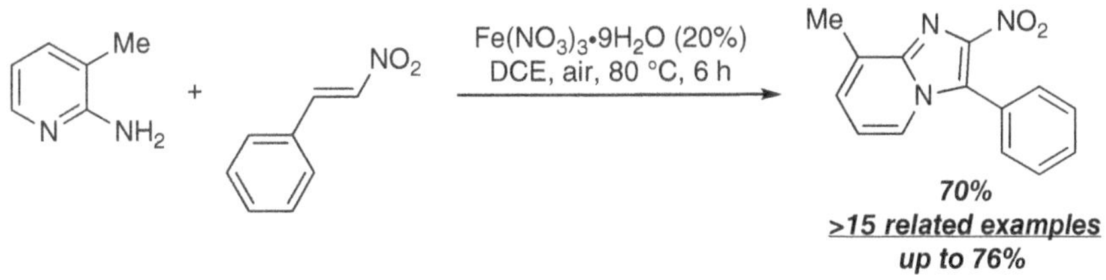

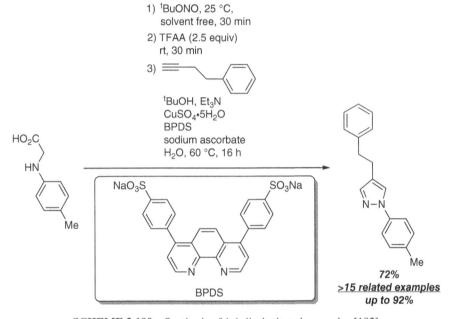

SCHEME 3.99   Regioselective synthesis of 2-nitroimidazopyridines [101].

SCHEME 3.100   Synthesis of 1,4-disubstituted pyrazoles [102].

regioselectivity of the process was very high and reversed from what was normally observed in similar reactions. The ability to carry out the reactions under an atmosphere of air was a practical advantage to this chemistry.

A series of pyrazoles have been generated through a one-pot multistep process (Scheme 3.100) [102]. Despite the number of steps needed for the process, the overall yields of the heterocycles were quite good. One of the key aspects of the chemistry was the ability to carry out the copper-catalyzed reaction in water. To aid the coupling reaction, a water-soluble sulfonated phosphine ligand was used to solubilize/stabilize the copper catalyst.

The copper-catalyzed coupling of 1*H*-pyrazolo[3,4-*b*]pyridin-3-amines with arylboronic acids affords N-arylated products in good to excellent yields (Scheme 3.101) [103]. The chemistry was insensitive to the electronic composition of the arylboronic acid, and an array of electron-donating and electron-withdrawing groups were well tolerated. One of the most attractive aspects of this chemistry was the ability to carry out the arylation reaction under air at room temperature.

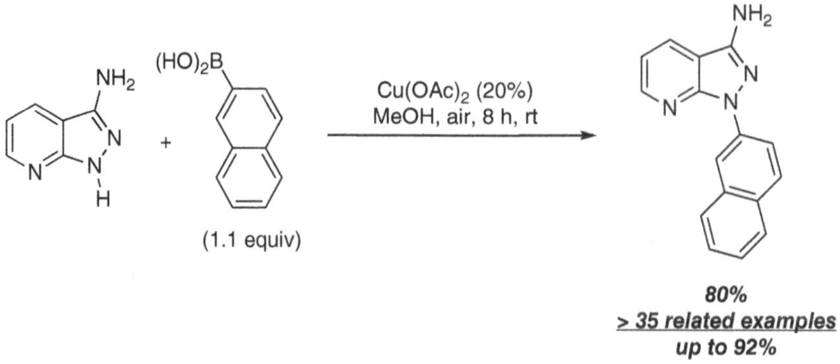

SCHEME 3.101   Copper-catalyzed N-arylation of 1*H*-pyrazolo[3,4-*b*]pyridin-3-amine [103].

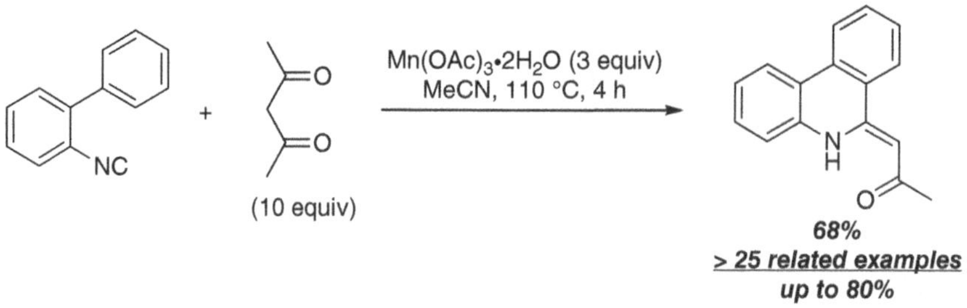

SCHEME 3.102   Synthesis of functionalized phenanthridines from 2-isocyanobiaryls [104].

A series of functionalized phenanthridines have been generated from 2-isocyanobiaryls using manganese(III) acetate to assist the process by promoting a radical cascade cyclization reaction between β-dicarbonyl derivatives and the starting isocyanides (Scheme 3.102) [104]. In general, the process tolerated the changes in the electronics of the isocyanides and β-dicarbonyl compounds fairly well. The authors noted a steric effect in the chemistry as the incorporation of a methyl group in the ortho-position on the 2-isocyanobiaryl fragment only afforded 20% of the phenanthridine.

Enamines have been used to convert quinoxalin-2(1*H*)-ones into a series of structurally and electronically diverse *N*-pyrrolylbenzimidazol-2-ones by simply refluxing the reagents in acetic acid for 6h (Scheme 3.103) [105]. The substrate scope was broad, and the isolated yields were outstanding (up to 99%). While both regioisomers were possible, there was a preference for the formation of *N*-(pyrrol-3-yl)benzimidazol-2-ones.

The conversion of benzyl azides into isoquinolines through a two-step process ending with an annulation reaction has been accomplished (Scheme 3.104) [106]. A key aspect of this chemistry was the ability to generate an N–H ketimine in solution through the reaction of the benzyl azide with a ruthenium catalyst. Trapping this ketimine with an internal alkyne generated isoquinolones. The authors noted that a wide range of electron-donating and electron-withdrawing groups were generally well tolerated by this tandem reaction.

The synthesis of polycyclic indoles has been accomplished through a two-step one-pot reaction (Scheme 3.105) [107]. A copper-catalyzed coupling reaction between an arylboronic acid and the brominated 2-alkynylanilines comprised the first step in the process. A practical aspect of this chemistry was the ability to carry out the coupling reaction under air.

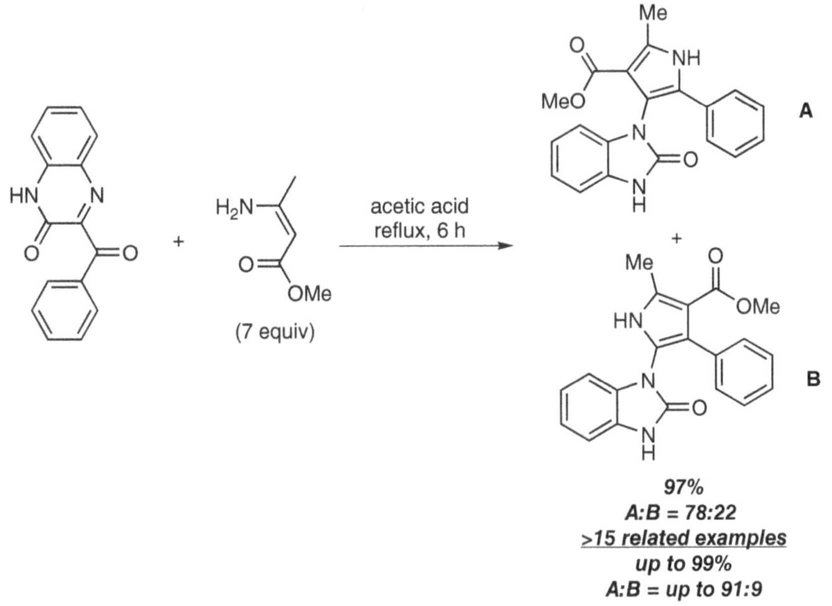

SCHEME 3.103 Preparation of *N*-pyrrolylbenzimidazol-2-ones [105].

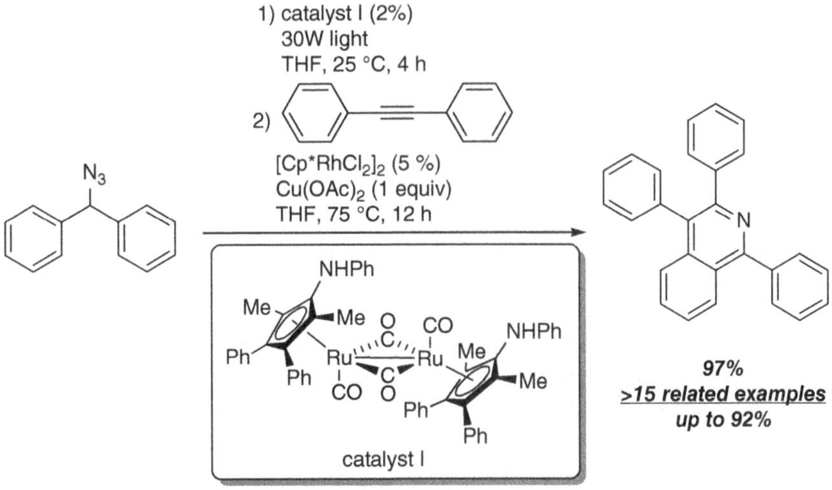

SCHEME 3.104 Synthesis of isoquinolones using visible light catalysis [106].

The second step in the process consisted of a palladium-catalyzed intramolecular cycliza-tion followed by an intramolecular C–H arylation reaction to afford the polycyclic indoles. While only a few examples were reported, the tolerance of the first step in the process to air coupled with the availability of all the reagents makes this an attractive approach to the syn-thesis of this interesting class of compounds.

Compounds bearing a variety of heteroatoms are often attractive entry points for the construction of complex organic architectures due to the ability to chemoselectively

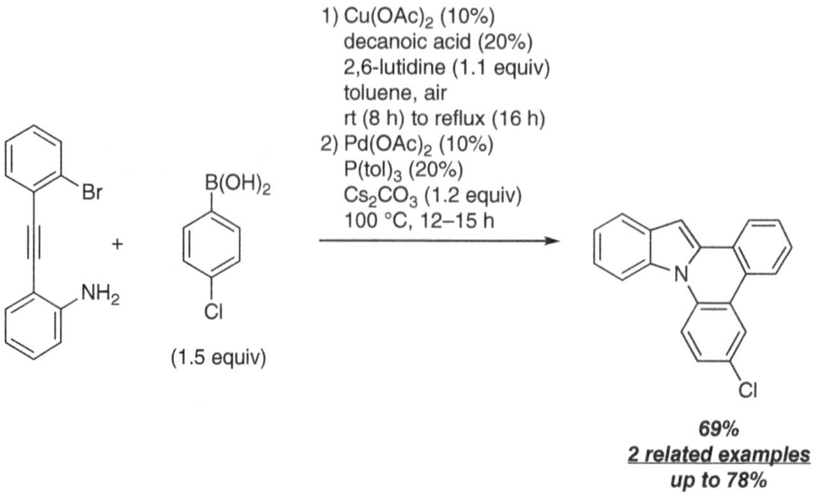

SCHEME 3.105   Preparation of polycyclic indoles through a two-step one-pot reaction [107].

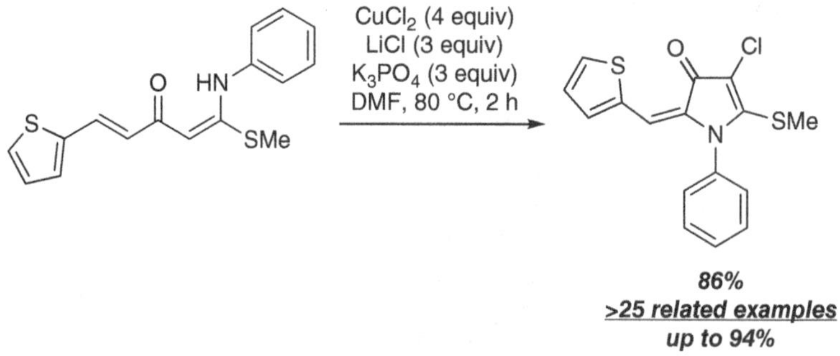

SCHEME 3.106   Copper-catalyzed synthesis of heteroatom-rich pyrrolones [108].

transform specific fragments of the substrate while retaining others for subsequent functionalization. To this end, a heteroatom-rich series of pyrrolones were generated through a copper-promoted process (Scheme 3.106) [108]. α-Alkenoyl ketene N,S-acetals served as the substrates for this chemistry. Through systematic variation of the substrates, the authors were able to determine that the chemistry was largely insensitive to the electronic composition of the arene attached to the nitrogen in the substrates, although the addition of substituents in the ortho-positions reduced the yield of the pyrrolones.

The ability to prepare triazoles through the copper-catalyzed coupling of organoazides with terminal alkynes has made a considerable impact on synthetic chemistry [109–111]. The influence of this chemistry can be seen in a number of fields. This reaction is quite practical due to its high yields and operational simplicity. Additionally, the tolerance of the approach to preexisting functional groups is outstanding. As a representative example of this chemistry, the synthesis of tethered glucose units through click reactions is shown in Scheme 3.107 [112]. The reaction was carried out in two steps using D_2O/acetonitrile as the solvent mixture at 0 °C. The classic catalyst for this reaction is copper sulfate, and

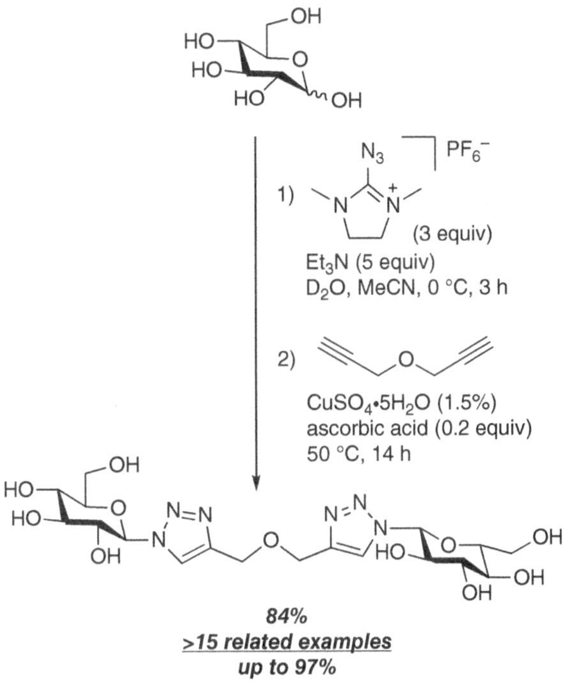

SCHEME 3.107   Synthesis of linked triazoles [112].

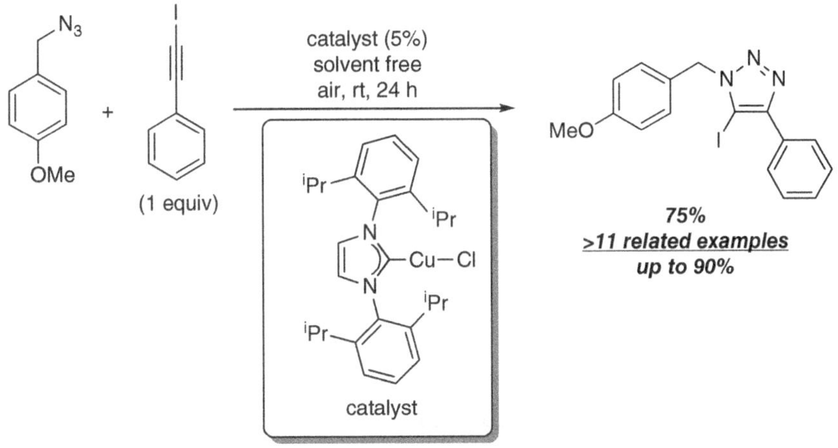

SCHEME 3.108   Synthesis of triazoles under solvent-free conditions in air [113].

ascorbic acid is commonly used as an additive. Using this catalyst system, good to excellent yields of the linked triazoles were isolated.

In addition to the use of copper salts as catalysts, a variety of discrete organosoluble copper(I) compounds have been generated and used to promote the synthesis of triazoles. An example of this chemistry entailed the use of an *N*-heterocyclic carbene as the stabilizing/solubilizing ligand for the copper (Scheme 3.108) [113]. Iodoalkynes were used as the substrates in this chemistry, and good yields of the triazoles were obtained with a diverse assortment of organoazides. This chemistry was noteworthy since the iodide was retained in the product and could be further functionalized in subsequent reactions. Additionally, this represented a rare example of a successful click reaction using a nonterminal alkyne. One of the attractive aspects of this chemistry was that it was carried out under solvent-free conditions under an atmosphere of air.

While the vast majority of reports describing the synthesis of triazoles used copper salts or discrete complexes to promote the reactions, a number of metal-free approaches have been developed. One example of this group used DBU in chloroform to effect the cycloaddition reaction (Scheme 3.109 and Example 3.15) [114,115]. A host of alkyl and

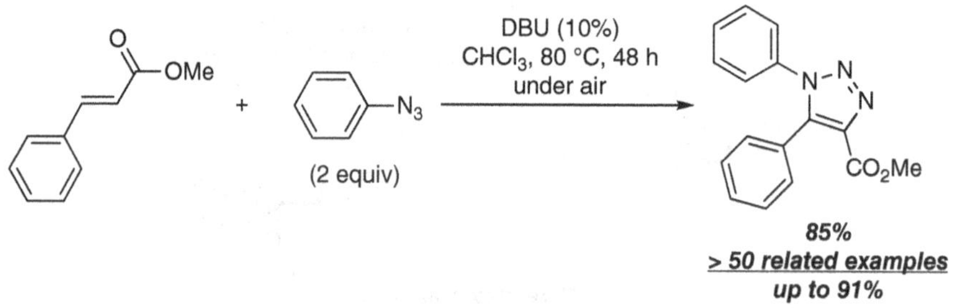

SCHEME 3.109   Synthesis of triazoles from methyl cinnamate under copper-free conditions [114].

Example 3.15   Metal-Free Triazole Synthesis Using Activated Alkenes [114].

Before starting this synthesis, carefully follow the safety instructions listed in Example 3.1.

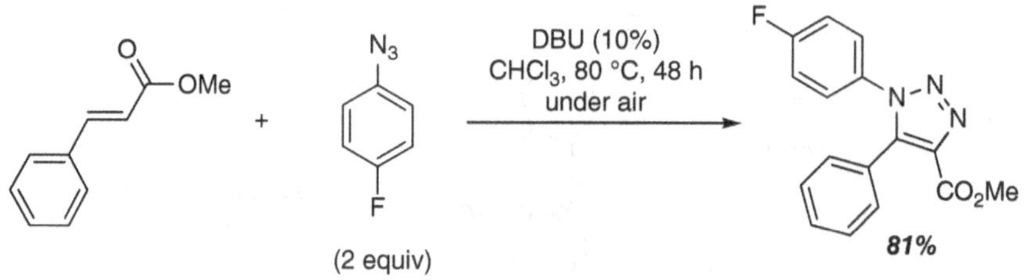

Caution! Organic azides can be explosive.

Caution! Hexane and ethyl acetate are highly flammable.

To a solution of $CHCl_3$ (0.3 mL) were added α,β-unsaturated ester (0.10 mmol), azide (0.20 mmol), and DBU (0.01 mmol). The reaction mixture was stirred under air at 80 °C until the α,β-unsaturated ester was completely consumed (monitored by TLC). The mixture was concentrated under reduced pressure and purified by flash chromatography eluted with hexane/EtOAc = 20 : 1 to afford the corresponding desired product as a yellow solid (81%).

Adapted with permission from W. Li, J. Wang, *Angew. Chem. Int. Ed.*, **2014**, *53*, 14186–14190. *Copyright (2014) Wiley-VCH Verlag GmbH & Co. KGaA, Weinheim.*

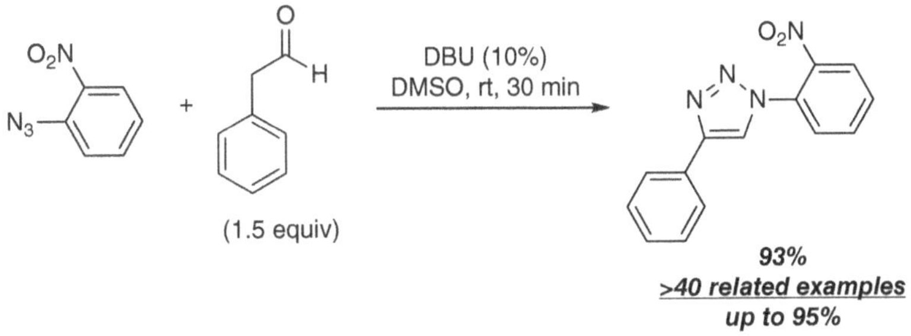

SCHEME 3.110   Metal-free synthesis of a triazole in DMSO [116].

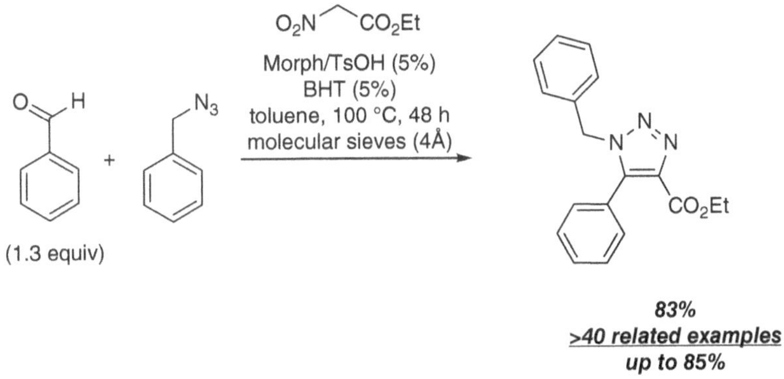

SCHEME 3.111   Metal-free synthesis of triazoles using the morpholinium/TsOH/BHT system [117].

aryl azides were active in the reaction, and over 50 triazoles were generated with isolated yields generally above 80%. A practical benefit to this approach was the observation that these reactions could be carried out under an atmosphere of air. In related work, a version of this reaction was carried out in DMSO (Scheme 3.110) [116], and a metal-free synthesis of triazoles was achieved using the morpholinium/TsOH/BHT system (Scheme 3.111) [117].

Nitroarenes are some of the most widely used compounds in synthetic chemistry; however, the majority of synthetic routes still rely on electrophilic aromatic substitution procedures using harsh conditions (H_2SO_4/HNO_3). These harsh conditions automatically eliminate a host of potential substrates due to functional group incompatibility. With the goal of developing a much more mild approach to the synthesis of nitroarenes, a transition metal-free approach to the synthesis of these valuable compounds has been developed using arylboronic acids as substrates and bismuth nitrate as the nitrating agent (Scheme 3.112) [118]. Even crowded arylboronic acids were successfully nitrated in good yields using this approach (Example 3.16) [118]. The chemistry was relatively insensitive to the electronic composition of the arylboronic acid, and an assortment of arylboronic acids bearing electron-donating and electron-withdrawing substituents were successfully converted into nitroarenes.

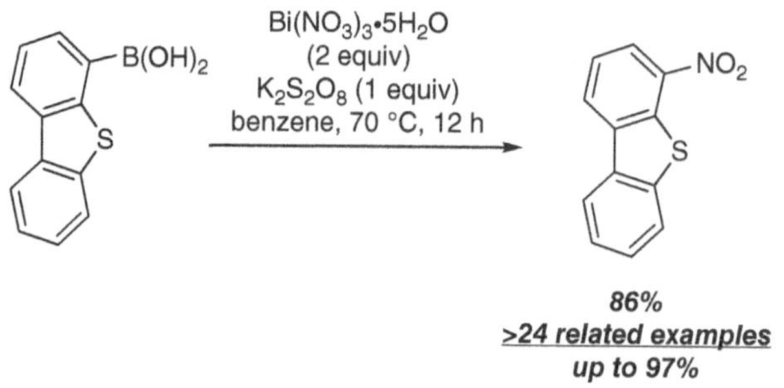

SCHEME 3.112 Preparation of nitroarenes through nitration of arylboronic acids [118].

Example 3.16 Preparation of a Crowded Nitroarene [118].

Before starting this synthesis, carefully follow the safety instructions listed in Example 3.1.

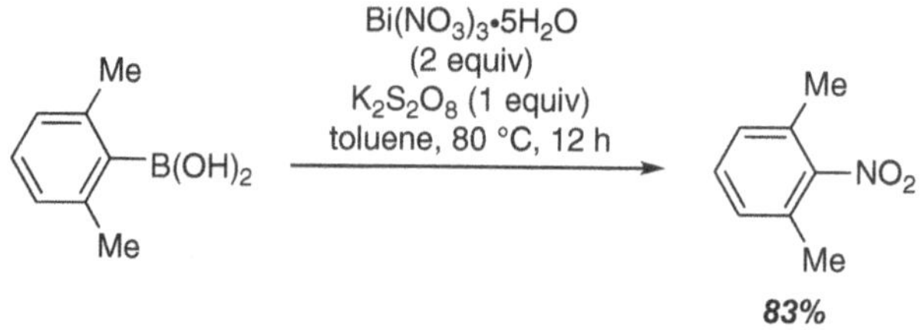

Caution! Ethyl acetate and petroleum ether are highly flammable.
To an oven-dried screw-cap reaction tube charged with a magnetic stir bar were added bismuth(III) nitrate (485 mg, 1.0 mmol), potassium perdisulfate (135 mg, 0.5 mmol), and arylboronic acid (0.5 mmol). The tube was then evacuated and backfilled with nitrogen with the help of a syringe. This evacuation/backfill sequence was repeated two additional times. Under a counterflow of nitrogen, solvent (toluene, 2 mL) was added. The tube was placed in a preheated oil bath at 70 °C, and the reaction mixture was stirred vigorously for 12 h. Then the reaction mixture was cooled to room temperature. Dichloromethane (4 mL) and ethyl acetate (2 mL) were added, and the mixture was stirred. Finally, organic extract obtained upon filtration through a Celite bed was concentrated and was purified by column chromatography through a silica gel column (mesh 60–120) (eluent: ethyl acetate/petroleum ether (1 : 99 v/v); yellow liquid; yield: 83% (63 mg)).

Adapted with permission from S. Manna, S. Maity, S. Rana, S. Agasti, *and* D. Maiti, *Org. Lett.*, **2012**, *14*, 1736–1739. *Copyright (2012) American Chemical Society.*

The synthesis of *N*-aryl sulfoximines has been achieved through a copper-catalyzed arylation reaction (Scheme 3.113) [119]. While a range of copper salts were active under the reaction conditions, copper iodide was determined to be the most effective. Using this copper source, an assortment of sulfoximines were arylated using trimethoxyphenylsilane in the presence of TBAF. It should be noted that an atmosphere of oxygen was needed for a successful arylation reaction. No conversion was observed in air.

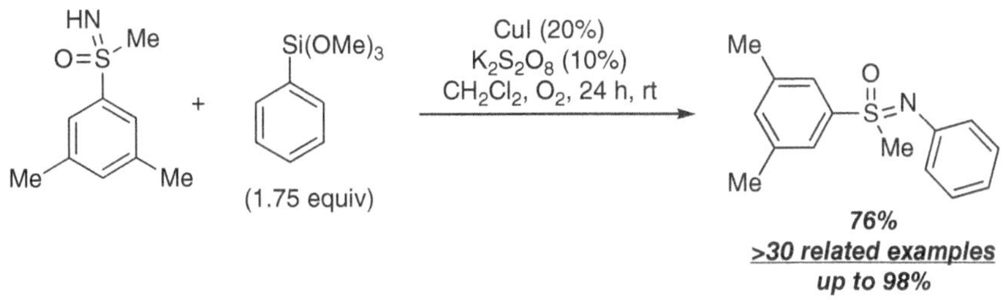

SCHEME 3.113   Copper-catalyzed synthesis of *N*-aryl sulfoximines [119].

3.2.1   Troubleshooting the Synthesis of Arylamines and Related Compounds

While it is not possible to predict all of the challenges and difficulties that will arise during the preparation of these compounds, the following suggestions will hopefully provide some direction for common issues and reaction requirements.

Problem/Issue/Goal	Possible Solution/Suggestion
Ammonia is impractical for us to use. Are there surrogates for ammonia?	One of the most practical replacements for gaseous ammonia is ammonium sulfate. It has been used with great success in the palladium-catalyzed synthesis of aniline derivatives [42]
Need a reliable way to generate arylamines from aryl halides and primary amines	One of the most effective catalysts for this reaction is a palladium complex with BrettPhos as the supporting ligand [49]
Need a reliable way to generate arylamines from aryl halides and secondary amines	This coupling reaction has been successful with a palladium catalyst containing RuPhos as the supporting ligand [50]
Can copper catalysts be used for the synthesis of arylamines through cross-coupling?	Several copper catalyst systems are known to promote this reaction [51, 57, 58]
It would be convenient to start from a boronic acid. Are methods known for the conversion of arylboronic acids into arylamines?	Several metal-catalyzed routes are known to promote this reaction including copper-catalyzed processes [51]. Some of the reported syntheses can be carried out under air [103]
The starting material is a terminal alkyne. Can it be converted into a pyrrole?	This transformation has been achieved through a gold-catalyzed reaction [63]
Need a good approach to the preparation of indolylacetamides	A palladium-catalyzed intramolecular cyclization reaction would be a reasonable place to start [67]. This reaction has been carried out in reactor vessel that was open to air
Do not have a glovebox and need to synthesize a pyrrole	There are several approaches to this synthesis that are tolerant to oxygen. Several palladium- and iron-catalyzed reactions actually use oxygen as the terminal oxidant [60, 61]
Can aldehydes be used as a source of the RC(O)- fragment in the synthesis of amides?	This has been achieved through a transition metal-free three-component coupling reaction [68]

(Continued)

(Continued)

Problem/Issue/Goal	Possible Solution/Suggestion
Can organoazides be converted into indoles?	There are several successful approaches to this reaction including a ruthenium-catalyzed approach [69]
Need to generate a nitrogen-containing polyheteroaromatic compound	This is a challenging reaction that is very substrate dependent. A rhodium-catalyzed approach that is particularly attractive started from internal alkynes [70]
The starting arene has been challenging to halogenate (in preparation for a cross-coupling reaction) while retaining a secondary amide. Can indoles be generated through C–H activation reactions?	A palladium-catalyzed approach using an arene bearing a tethered secondary amide would be a reasonable place to start [71]
Need a good way to make a pyrrole bearing synthetic "handles" that can be further functionalized	The preparation of a variety of pyrroles bearing esters and other functional groups has been accomplished using a copper-mediated oxidative cyclization [78]
Need a good way to make an O-methyl hydroxamate	Benzoic acids are convenient entry points for this chemistry as they can be converted in a one-pot reaction into an O-methyl hydroxamates [81]
The starting material has an internal alkyne. Can it be converted into an isoquinolone?	This reaction has been developed using rhodium complexes as catalysts and O-methyl hydroxamates as the substrates along with an internal alkyne [81]
Are metal-free methods known for annulation reactions between alkynes and hydroxamates?	These reactions have been developed using PIFA along with TFA in dichloromethane [83]
The starting material has a benzonitrile. Can it be converted into a primary amide?	A reasonable place to start would be the rhodium-catalyzed hydration of nitriles [84]
Need to do an annulation with acetylene, but the gas is impractical for us to use	Vinyl acetate has been used as an acetylene equivalent [90]
Can primary amides be used in cross-coupling reactions to selectively generate secondary amides?	This can be a challenging reaction. A palladium-catalyzed process has been developed that is highly selective for the formation of the secondary amide [96]
It would be convenient to be able to convert a benzonitrile into a pyrrole	This conversion has been achieved through a multicomponent reaction between a terminal alkyne, benzonitrile, and a halogenated ester [98]
Need to make a 1,4-diaryl imidazole	A practical copper-catalyzed method for the synthesis of imidazoles with this challenging substitution pattern has been developed [99]
Need a good method for the synthesis of a triazole	The copper-catalyzed version of this reaction is perhaps the most robust and high yielding [112]. Additionally, a number of methods have been reported to be stable to air [113]
Need a metal-free route to the preparation of triazoles	Several reports have outlined metal-free approaches to the synthesis of triazoles. Stirring an α,β-unsaturated ester with an organoazide and DBU is a reasonable place to start due to its operational simplicity [114]
Can an arylboronic acid be converted into a nitroarene?	Arylboronic acids can be directly nitrated using bismuth(III) nitrate [118]

3.3 SYNTHESIS OF VINYLAMINES AND RELATED COMPOUNDS THROUGH NITROGEN–CARBON(sp²) BOND-FORMING REACTIONS

Enamines and enamides are valuable compounds since they are convenient precursors for a myriad of nitrogen-containing compounds. As a result, a significant number of synthetic routes are known for the preparation of these compounds. The following sections will high-light practical routes to these compounds with special attention devoted to reactions that are atom efficient, operationally simple, and have broad substrate scope.

The palladium-catalyzed cross-coupling of vinyl halides with secondary amines leads to the formation of tertiary enamines (Scheme 3.114) [120]. The catalyst system was com-prised of a soluble source of Pd(0) along with a bisphosphine ligand as well as a base to consume the acid generated in the reaction. Several secondary amines were successfully used along with a host of vinyl bromides to afford the enamines in moderate to good yields. Even the use of tetrasubstituted alkenes such as 2-bromo-3-methyl-2-butene participated in the N-vinylation reaction.

In addition to palladium-catalyzed processes, an assortment of copper-promoted N-vinylation reactions have been developed. As an example, an approach using copper(I) thiophenecarboxylate (CuTC) as the catalyst for the preparation of enamides has been developed (Scheme 3.115) [121]. The overall reaction consisted of coupling vinyl iodides with primary or secondary amides using the copper catalyst and cesium carbonate as the base. Similar to the reactions earlier, the *E*-stereochemistry of the vinyl halide was retained through the coupling reaction. In related work, the coupling of vinyl iodides with primary and secondary amides was achieved using $Cu(NCMe)_4PF_6$ as the copper source and tetramethyl-1,10-phenanthroline as the solubilizing/stabilizing ligand (Scheme 3.116) [122]. Curiously, the authors found that the rubidium carbonate was superior to cesium carbonate for the generation of the amides.

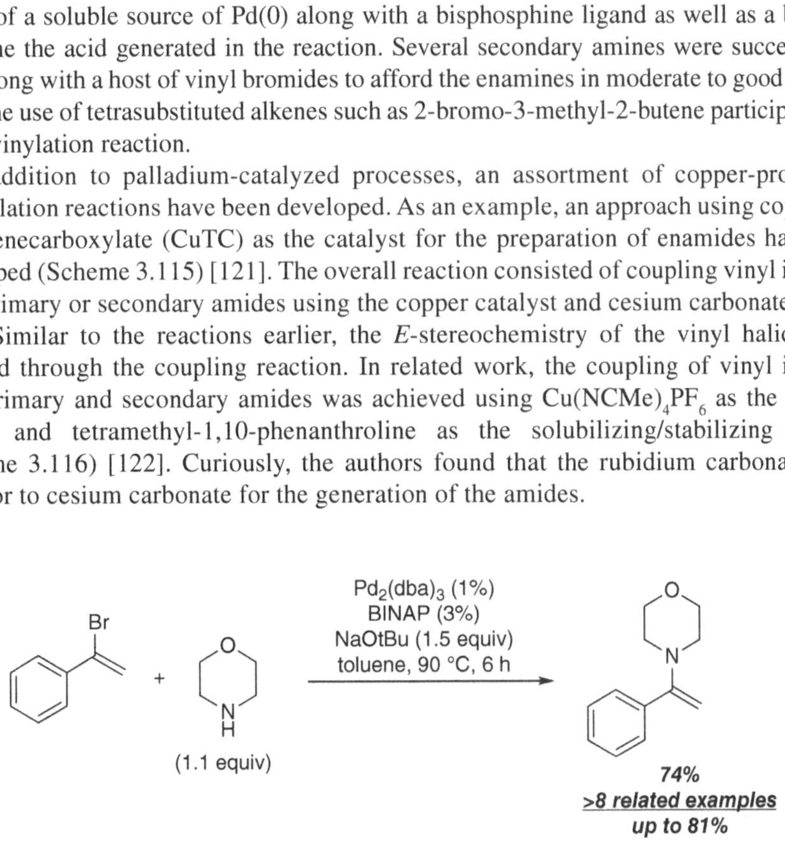

SCHEME 3.114 Preparation of enamines through palladium-catalyzed cross-coupling [120].

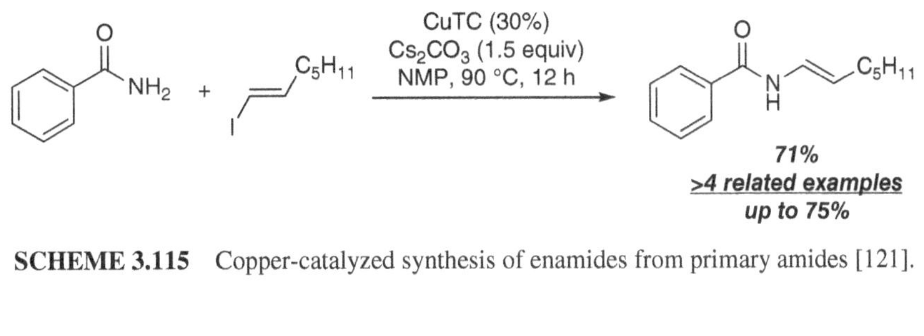

SCHEME 3.115 Copper-catalyzed synthesis of enamides from primary amides [121].

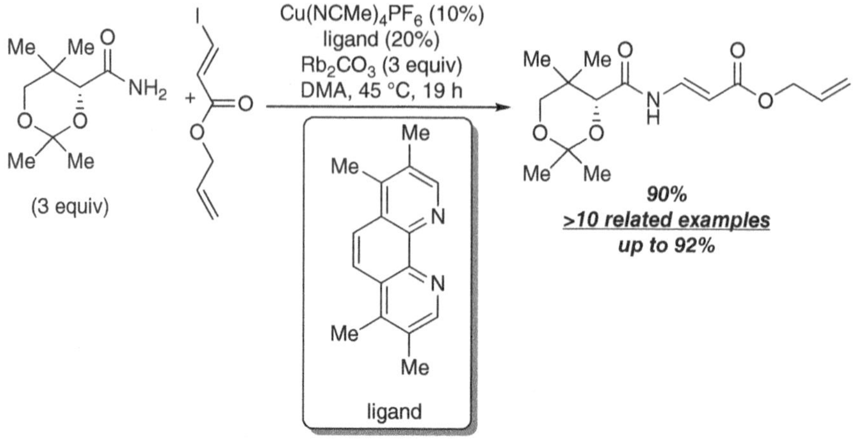

SCHEME 3.116   Copper-catalyzed synthesis of enamides using a rubidium carbonate as the base [122].

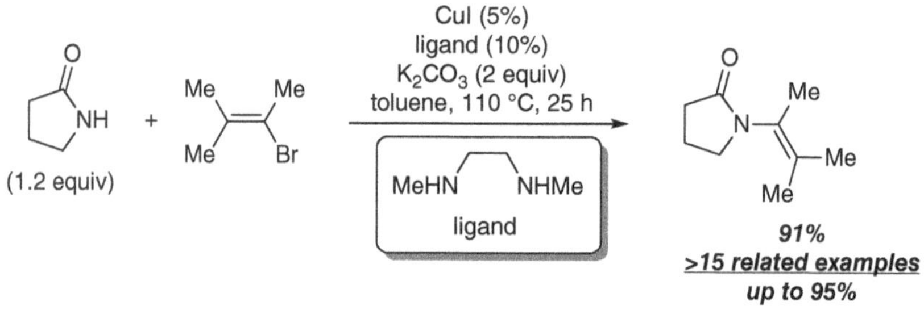

SCHEME 3.117   Preparation of enamides using copper salts and DMEDA [123].

One of the most general approaches to the preparation of enamides and related compounds through cross-coupling chemistry used copper iodide as the metal source and DMEDA as the solubilizing ligand (Scheme 3.117 and Example 3.17) [123]. A variety of substituted vinyl bromides and iodides were coupled with an array of primary and secondary amides to generate the enamides in moderate to excellent yields. Even tetrasubstituted vinyl bromides were successfully used in this chemistry. In related work, amino acid derivatives were used as the solubilizing/stabilizing ligands in the copper iodide-catalyzed synthesis of enamides [124].

The copper-promoted coupling of vinylboronic acids with amines and amides bearing reactive N–H groups has been achieved (Scheme 3.118) [125]. Using the combination of copper acetate and pyridine along with air as an oxidant, a number of enamines and enamides were prepared in moderate to excellent yield. One practical advantage to this chemistry was the observation that it could be carried out in air.

The development of new approaches for the functionalization of nucleobases remains a current and challenging goal for the synthetic community. To this end, an efficient process for the N-vinylation of a variety of purine and pyrimidines has been devised using a copper-promoted coupling reaction (Scheme 3.119 and Example 3.18) [51]. A variety of

Example 3.17   Cross-Coupling of Bromobutenes with 2-Azetidinone [123].

Before starting this synthesis, carefully follow the safety instructions listed in Example 3.1.

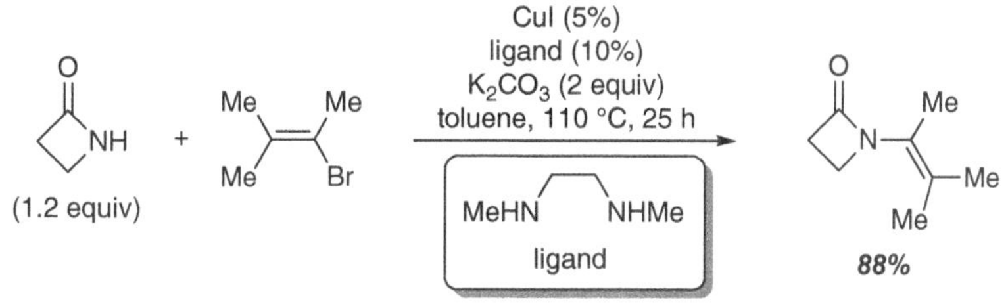

Caution! Ethyl acetate is highly flammable.
A resealable Schlenk tube was charged with CuI (9.6 mg, 0.050 mmol, 5.0 mol%), K_2CO_3 (280 mg, 2.0 mmol), and amide (1.2 mmol), evacuated, and backfilled with argon. N,N'-Dimethylethylenediamine (11 μL, 0.10 mmol, 10 mol%), vinyl bromide (1.00 mmol), and toluene (1.0 mL) were added under argon. The Schlenk tube was sealed with a Teflon valve and immersed in a preheated oil bath; the reaction mixture was stirred at the indicated temperature until the complete consumption of starting material was observed as indicated by GC analysis. The reaction vessel was removed from the oil bath, and the resulting pale tan suspension was allowed to reach room temperature; then, it was filtered through a plug silica gel (1 × 0.5 cm) eluting with ethyl acetate (50 mL). The filtrate was concentrated and the residue was purified by column chromatography on silica gel (2 × 10 cm, ethyl acetate) to provide the desired product as a clear oil (88%).

Adapted with permission from L. Jiang, G. E. Job, A. Klapars, *and* S. L. Buchwald, *Org. Lett.*, **2003**, *5*, 3667–3669. *Copyright (2003) American Chemical Society.*

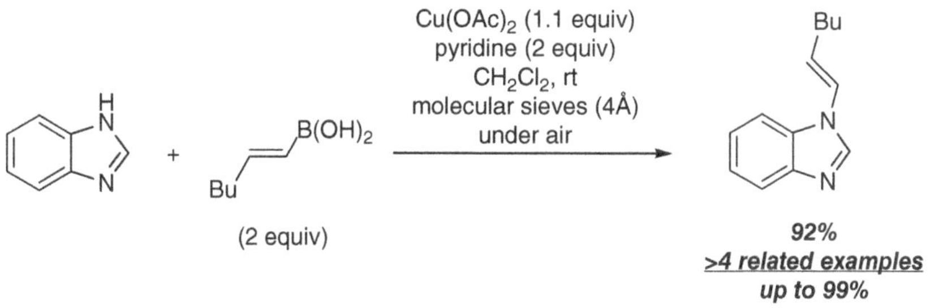

SCHEME 3.118   Coupling of a vinylboronic acid with benzimidazole [125].

alkenylboronic acids served as the electrophiles, and a number of new derivatives were generated after stirring at room temperature under air.

The preparation of *N*-alkenyl nitrones has been achieved through the treatment of fluorenone oxime with vinylboronic acids (Scheme 3.120) [126]. The nitrogen–carbon(sp^2) bond-forming reaction was carried out under mild conditions using copper acetate as the

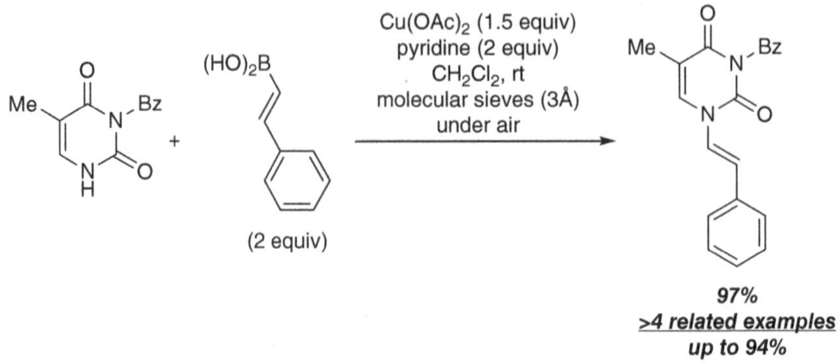

SCHEME 3.119   N-vinylation of pyrimidines through copper-catalyzed cross-coupling [51].

Example 3.18   Preparation of Enamides through Cross-Coupling in Air [51].

Before starting this synthesis, carefully follow the safety instructions listed in Example 3.1.

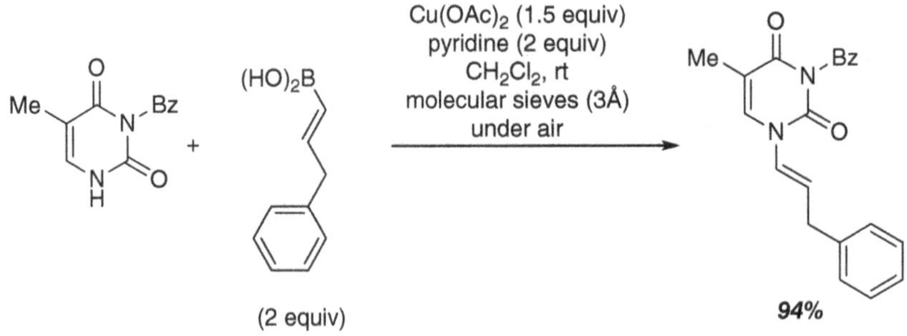

To a stirred suspension of dry Cu(OAc)$_2$ (0.60 mmol), N^3-benzoylthymine (0.40 mmol), E-3-phenyl-1-propen-1-ylboronic acid (0.80 mmol), and activated 3Å molecular sieves (200 mg) in dry CH$_2$Cl$_2$ (3.0 mL) was added pyridine (0.80 mmol) at rt. The mixture was stirred vigorously for 93 h at room temperature in the presence of air. The reaction mixture was diluted with CH$_2$Cl$_2$ (15 mL), filtered through a pad of Celite, and washed with water (15 mL) in the presence of EDTA (200 mg). The colorless organic phase was dried over MgSO$_4$ and was evaporated to dryness under vacuum. The residue was subjected to flash chromatography to afford the title compound (94%) as a white solid.

Adapted with permission from M. F. Jacobsen, M. M. Knudsen, *and* K. V. Gothelf, *J. Org. Chem.*, **2006**, *71*, 9183–9190. *Copyright (2006) American Chemical Society.*

catalyst. An assortment of functionalized boronic acids were successfully used in this work including a number of electronically diverse styrene derivatives. Similar to the other copper-promoted reactions described in this section, the reactions were able to be carried out under air. In related work, vinyltrifluoroborate salts have also been used in N-vinylation

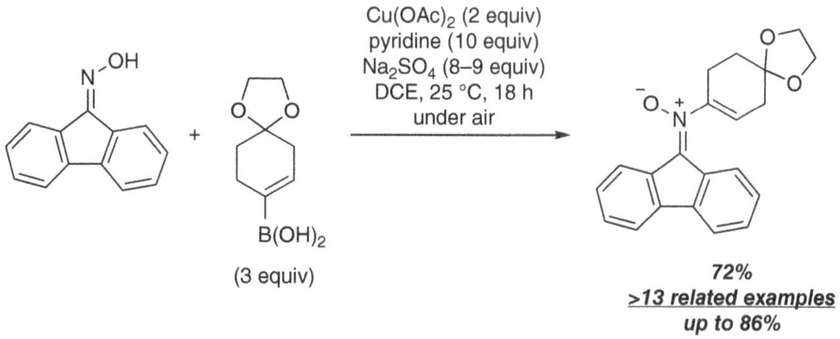

SCHEME 3.120   Synthesis of vinyl nitrones using alkenylboronic acids [126].

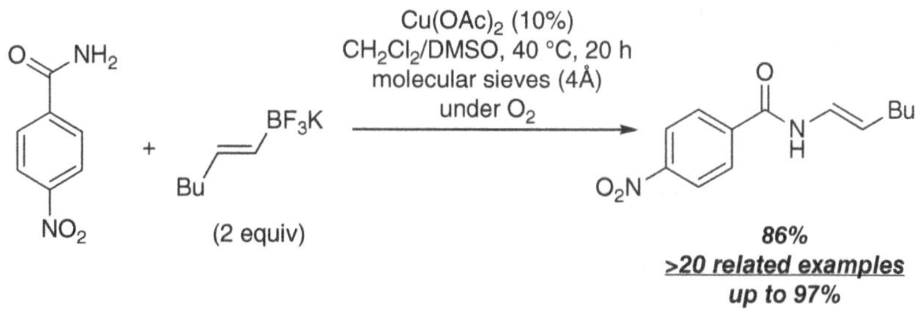

SCHEME 3.121   Using alkenyltrifluoroborate salts in N-vinylation reactions [127].

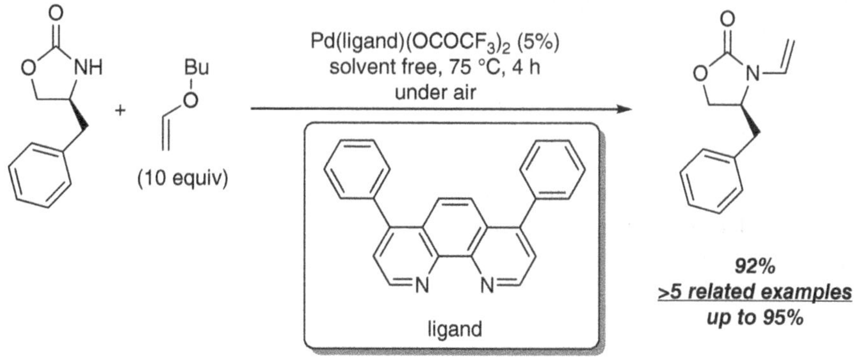

SCHEME 3.122   Synthesis of enamides through vinyl transfer reactions [130].

reactions with great success (Scheme 3.121) [127–129]. These borylating agents are remarkably stable; however, they exhibit a high level of reactivity in metal-catalyzed cross-coupling reactions.

An interesting approach to the preparation of enamides entailed a formal vinyl transfer from vinyl ethers (Scheme 3.122) [130]. The reaction was catalyzed by palladium complexes bearing a phenanthroline derivative as the solubilizing/stabilizing ligand. Several secondary amides including cyclic and acyclic substrates were screened, and moderate to good yields of the enamides were obtained. This was an attractive approach to these compounds as it enabled the N-ethenylation of secondary amides. Many of the coupling strategies outlined earlier facilitated the incorporation of larger vinyl fragments; however, the addition of a simple –C_2H_3 has often been more challenging. Furthermore, this process was quite practical as it was carried out under an atmosphere of air and did not require the use of a glovebox or nitrogen manifold.

In addition to cross-coupling approaches, enamines and enamides can be generated through hydroamination reactions. These reactions typically involve an alkyne and a nucleophilic nitrogen source as well as a catalyst and are valuable synthetic methods since they are typically atom efficient and high yielding. Many of the catalysts needed for this reaction are transition metal based, although considerable effort has been devoted to developing transition metal-free processes. As with many of the hydroelementation reactions discussed in other chapters, developing efficient catalysts that provide a high level of control over the regioselectivity and stereoselectivity of the process remains a current and challenging goal for the synthetic community. It should be noted that general processes with wide functional group tolerance are quite rare. Many catalyst systems are effective for a specific group of alkynes or nitrogen nucleophiles. As a result, the choice of catalysts, solvents, substrates, and additives needs to be carefully considered when designing a selective hydroamination reaction. In addition to these concerns, the conversion from a secondary enamine into an imine further complicates the product distribution. If a saturated amine is the desired product, reduction of the enamine/imine mixture can mitigate some of the issues. If all of these concerns are addressed, the possibility of functional group incompatibility still exists due to the propensity of the nitrogen nucleophile to react with a wide range of electrophiles. Despite these challenges, this chemistry remains a popular approach to the synthesis of enamines and enamides [19, 131–134]. The following sections will highlight a number of these valuable reactions.

The rhodium-catalyzed addition of aniline derivatives to terminal alkynes has been achieved using a crowded rhodium complex (Scheme 3.123) [135]. One of the practical advantages of this chemistry was the fact that the catalyst successfully promoted the hydroamination reactions at low catalyst loadings (0.5%). Only arylalkynes were successfully functionalized using this approach, and alkylalkynes were unresponsive under the reaction conditions. It was noteworthy that no polymerization of phenylacetylene was observed. In related work, the gold-catalyzed Markovnikov selective addition of aniline to phenylacetylene was reported using a very low catalyst loading and a bulky monodentate phosphine as the supporting ligand (Scheme 3.124) [136].

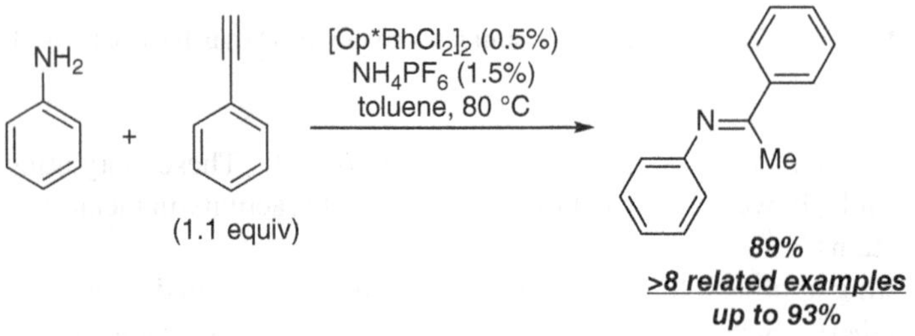

SCHEME 3.123   Rhodium-catalyzed Markovnikov addition of aniline to phenylacetylene [135].

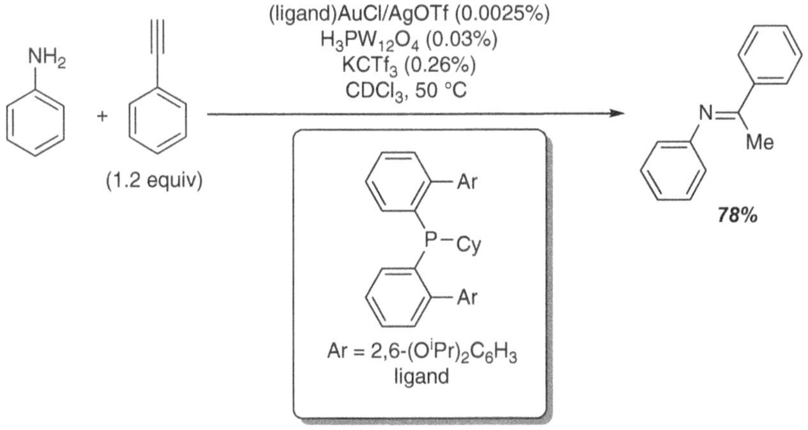

SCHEME 3.124   Gold-catalyzed Markovnikov hydroamination [136].

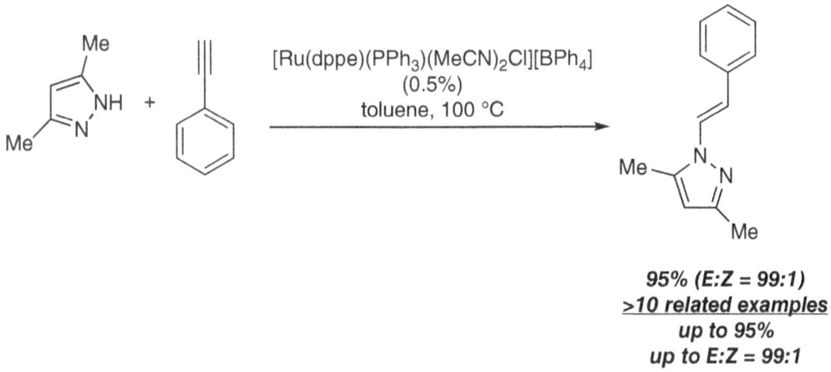

SCHEME 3.125   Ruthenium-catalyzed anti-Markovnikov hydroamination [137].

The anti-Markovnikov addition of nitrogen nucleophiles to alkynes has been accomplished using ruthenium catalysts (Scheme 3.125) [137]. During the screening process, the authors discovered that when the reactions were carried out at 80 °C, moderate yields of the hydroamination products were obtained (50%); however, the stereocontrol was poor and a $4:1$ ($E:Z$) ratio of the enamines was obtained. When the temperature was increased to 100°C, the E-isomer was obtained exclusively. At the higher temperatures, most substrates exclusively generated the E-isomer, although some substrate-specific reactivity was observed.

Common nickel compounds have been used to catalyze the hydroamination of internal activated alkenes (Scheme 3.126) [138]. The reactions were carried out under a high pressure of argon and afforded moderate yields of the enamines. While activated internal alkenes were successfully functionalized using this approach, unactivated substrates such as diphenylacetylene were unreactive and afforded low conversions (~30%).

Extremely crowded imines have been prepared using a rhodium-catalyzed process (Scheme 3.127) [139]. The key to the catalyst system was the use of an iminopyridine

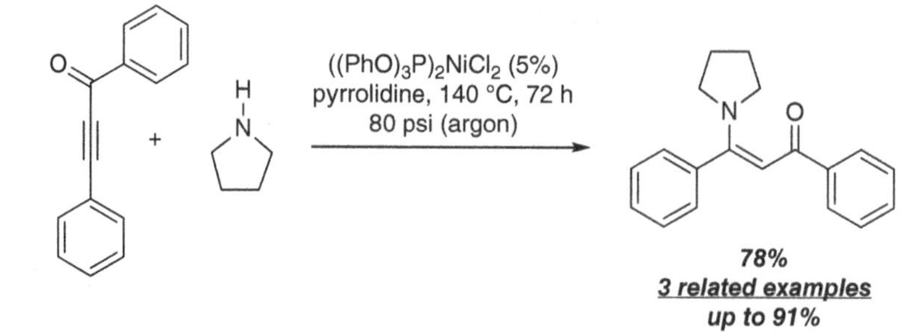

SCHEME 3.126 Nickel-catalyzed hydroamination of activated internal alkynes [138].

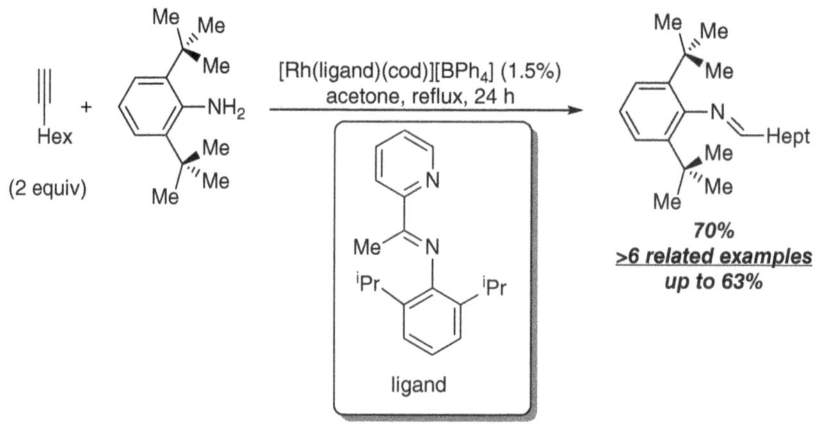

SCHEME 3.127 Rhodium-catalyzed anti-Markovnikov hydroamination of bulky anilines [139].

ligand on the metal center as the solubilizing/stabilizing ligand. Using this catalyst system, a range of aniline derivatives including the extremely bulky 2,6-ditert-butylaniline were successfully converted into imines with the bulkier aniline derivatives affording the highest yields of the imines. The chemistry was highly regioselective and exclusively generated the anti-Markovnikov products. The authors also noted that similar yields of the products were obtained without rigorous exclusion of oxygen and moisture.

As mentioned earlier, control over the stereoselectivity as well as the regioselectivity of the addition process remains a current challenge. Despite significant efforts, most approaches only provide access to one regio- or stereoisomer. During an investigation of the ruthenium-catalyzed hydroamination of alkynes, the ability to switch the stereochemistry of the addition through a minor modification of the catalyst system was discovered (Schemes 3.128 and 3.129) [140]. When chelating bisphosphines were used as the supporting ligands, the chemistry regioselectively afforded the anti-Markovnikov products with predominately Z-stereochemistry. Changing the ligand to a monodentate trialkylphosphine retained the regioselectivity; however, the stereochemistry was reversed and the E-enamides were formed with $E:Z$ ratios of up to 30:1. The authors also noted that the presence of air had only a minimal effect on the yields and product distributions.

An interesting gold-catalyzed approach to the preparation of enamides has been devised starting from propargyl alcohols (Scheme 3.130) [141]. The overall process was

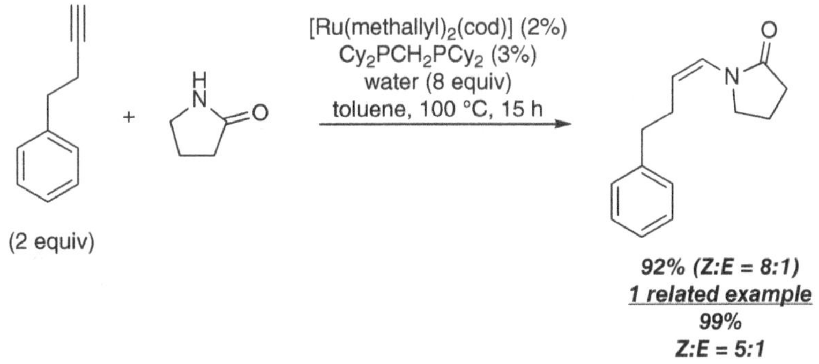

SCHEME 3.128   Z-selective anti-Markovnikov hydroamination using ruthenium catalysts [140].

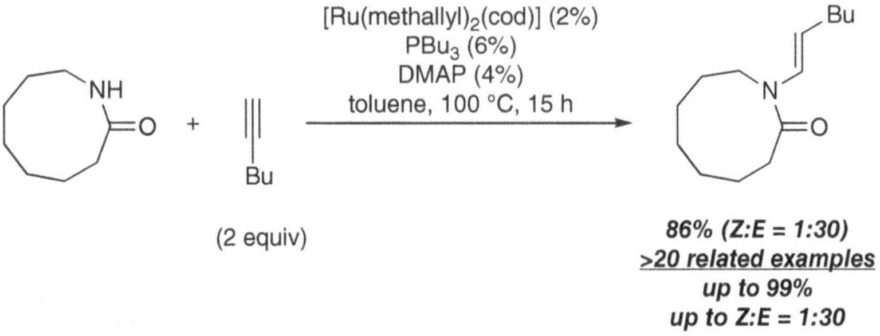

SCHEME 3.129   *E*-selective anti-Markovnikov hydroamination using ruthenium catalysts [140].

accomplished in two steps and entailed the gold-catalyzed addition of a primary amide to generate a hemiaminal as the first step. The catalyst system used in this step was comprised of a phosphine-ligated chlorogold complex bearing a bulky dialkylbiarylphosphine and silver(I) triflate along with several additives. The second step consisted of an acid-catalyzed Meyer–Schuster rearrangement. Curiously, the authors found that the use of a nonpolar solvent promoted the formation of the Z-stereoisomers, while analogous reactions using a polar solvent afforded a greater amount of the *E*-enamides. The authors also determined that the stereochemistry of the isolated Z-enamides could be reversed through the addition of acid in a polar solvent such as DMSO (Scheme 3.131).

The cesium hydroxide-promoted addition of benzimidazole to phenylacetylene proceeded with exclusive anti-Markovnikov regioselectivity and complete stereoselectivity for the formation of the Z-isomer (Scheme 3.132) [142]. Similar reactivity was obtained for pyrrole and imidazole; however, acyclic secondary amines such as diphenylamine or phenylmethylamine afforded mixtures of the stereoisomers. It should be noted that these additions were still regioselective for the anti-Markovnikov product. In related work, the base-assisted synthesis of enamines through the addition of indole derivatives to TMS-protected alkynes has been achieved using KOH as the promoter (Scheme 3.133) [143].

In addition to the methods described earlier, a number of additional approaches to the synthesis of enamines, enamides, and related compounds are known. One of these methods

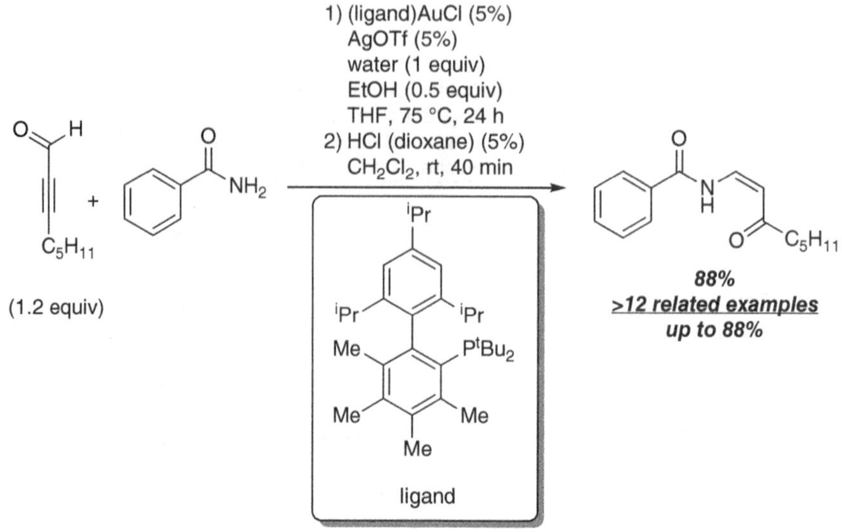

SCHEME 3.130 Gold-catalyzed synthesis of Z-enamides from propargyl aldehydes [141].

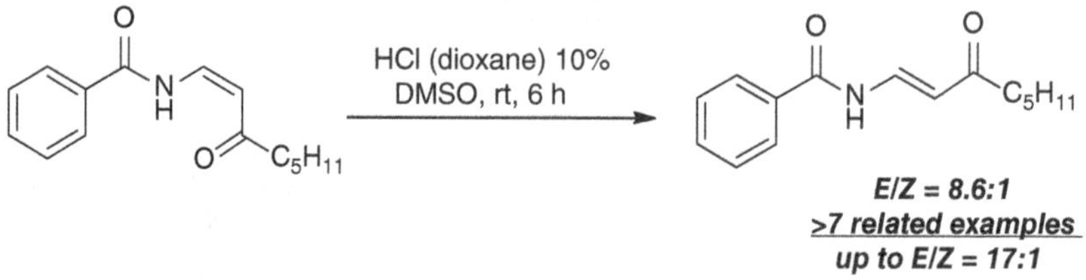

SCHEME 3.131 Acid-catalyzed enamide isomerization [141].

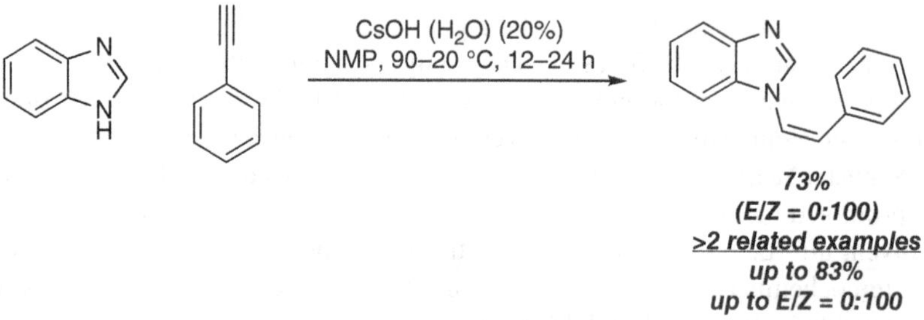

SCHEME 3.132 Base-assisted anti-Markovnikov hydroamination of phenylacetylene [142].

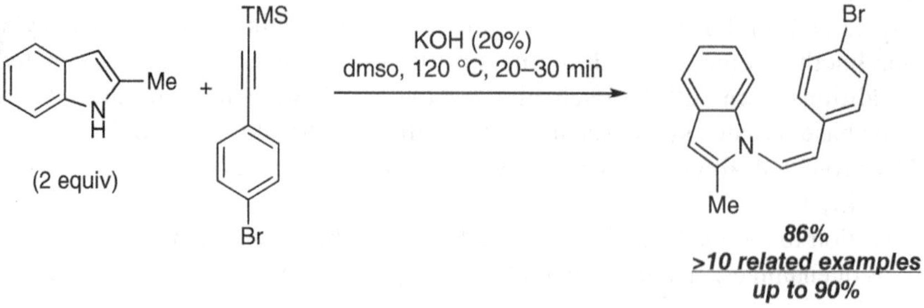

SCHEME 3.133 Synthesis of enamines with retention of an aryl bromide [143].

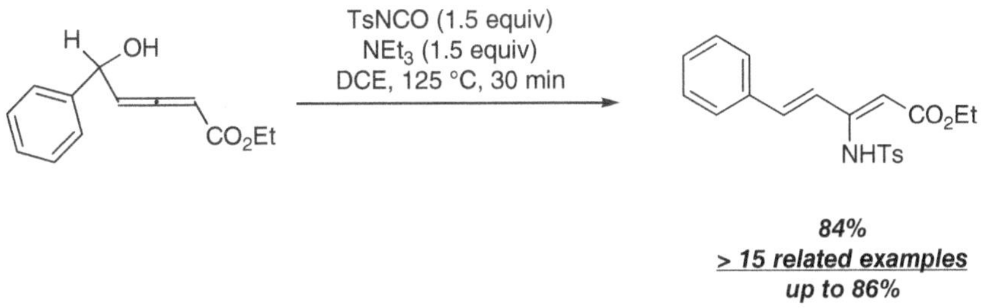

SCHEME 3.134   Preparation of *N*-tosyl-1,3-dien-2-yl amines from allenols [144].

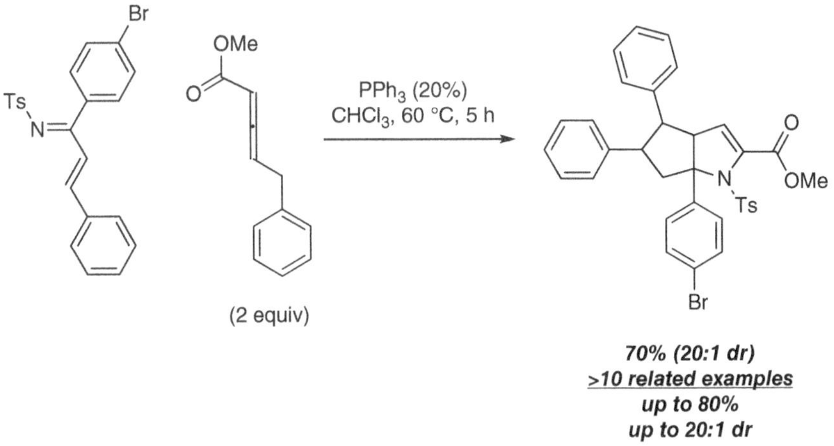

SCHEME 3.135   Triphenylphosphine-catalyzed annulation reactions [145].

entailed the synthesis of enamides through a decarboxylative amination reaction using alle-nols as substrates (Scheme 3.134) [144]. Although many decarboxylative approaches need transition metals to promote the reaction, the authors determined that no metals were needed for a successful reaction through a series of screening experiments.

The triphenylphosphine-catalyzed reaction between allenoates and unsaturated keti-mines affords aza-bicyclo[3,3,0]octane derivatives in good yield with excellent selectivity (Scheme 3.135 and Example 3.19) [145]. These reactions were quite practical since they were operationally trivial to set up, and an assortment of diverse heterocycles were readily accessed using this method.

In addition to the intermolecular hydroamination reactions, a significant number of intramolecular versions have been developed. As an example, the palladium-catalyzed cyclization of alkynylamides generated alkylidene lactams in moderate to good yield (Scheme 3.136) [146]. The reactions were quite rapid and afforded formal Markovnikov addition products.

The addition/cyclization of (pivaloyloxy)benzamides with a range of 1,6-enynes affords a range of heterocycles bearing both isoquinolone and cyclohexenone fragments (Scheme 3.137) [147]. The reaction was catalyzed by a crowded rhodium catalyst using

Example 3.19 Preparation of an aza-Bicyclo[3,3,0]octane Derivative [145].

Before starting this synthesis, carefully follow the safety instructions listed in Example 3.1.

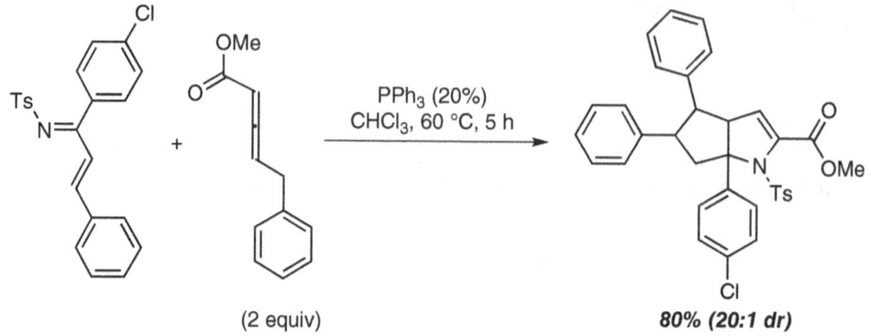

(2 equiv)   80% (20:1 dr)

Caution! Ethyl acetate and petroleum ether are highly flammable.

To a dry flask filled with nitrogen were added the ketimine (0.2 mmol) and allenoate (0.4 mmol) in 2 mL CHCl$_3$. PPh$_3$ (0.04 mmol) was added. This solution was stirred at 60 °C until the complete consumption of the starting material as monitored by TLC. After the removal of the solvent, the residue was subjected to chromatography on a silica gel (60–120 mesh) column using 20 : 1 petroleum ether/ethyl acetate solvent mixture as eluent to afford the aza-bicyclo[3,3,0]octane derivative.

*Adapted with permission from E. Li, P. Jia, L. Liang, Y. Huang, ACS Catal., **2014**, 4, 600–603. Copyright (2014) American Chemical Society.*

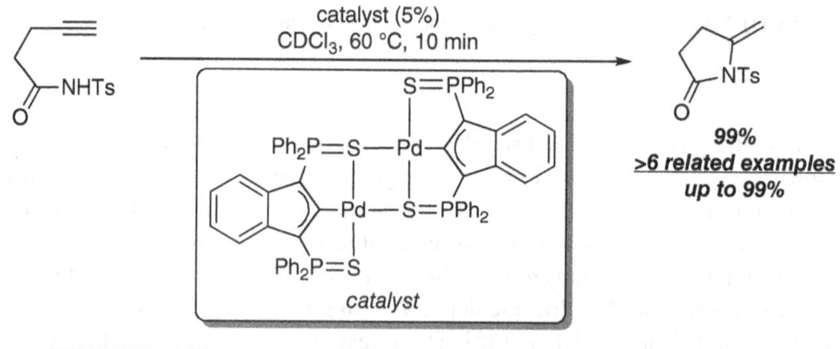

SCHEME 3.136 Palladium-catalyzed intramolecular hydroamination [146].

cesium acetate as a promoter and afforded good to excellent yields of the heterocycles. Although there was some substrate-specific reactivity noted by the authors, a range of electron-deficient and electron-rich benzamide derivatives were successfully used in this work.

Amino-N-vinylindoles have been generated through a palladium-catalyzed process (Scheme 3.138) [148]. The initial step in the process entailed the formal addition of brominated indoles to N-tosylhydrazones to generate a vinylindole species as an

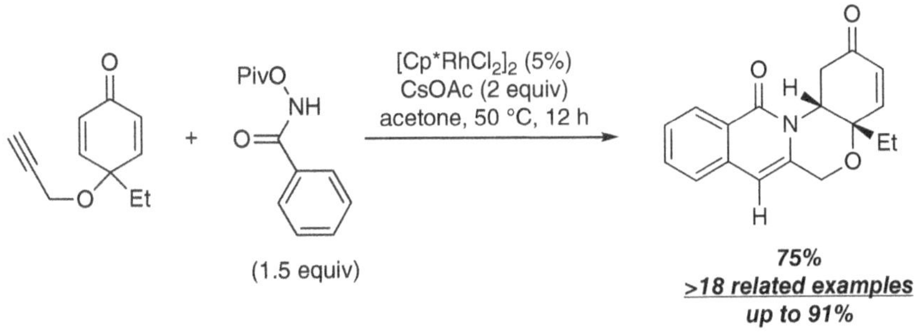

SCHEME 3.137   Rhodium-catalyzed preparation of *N*-heterocycles from benzamides [147].

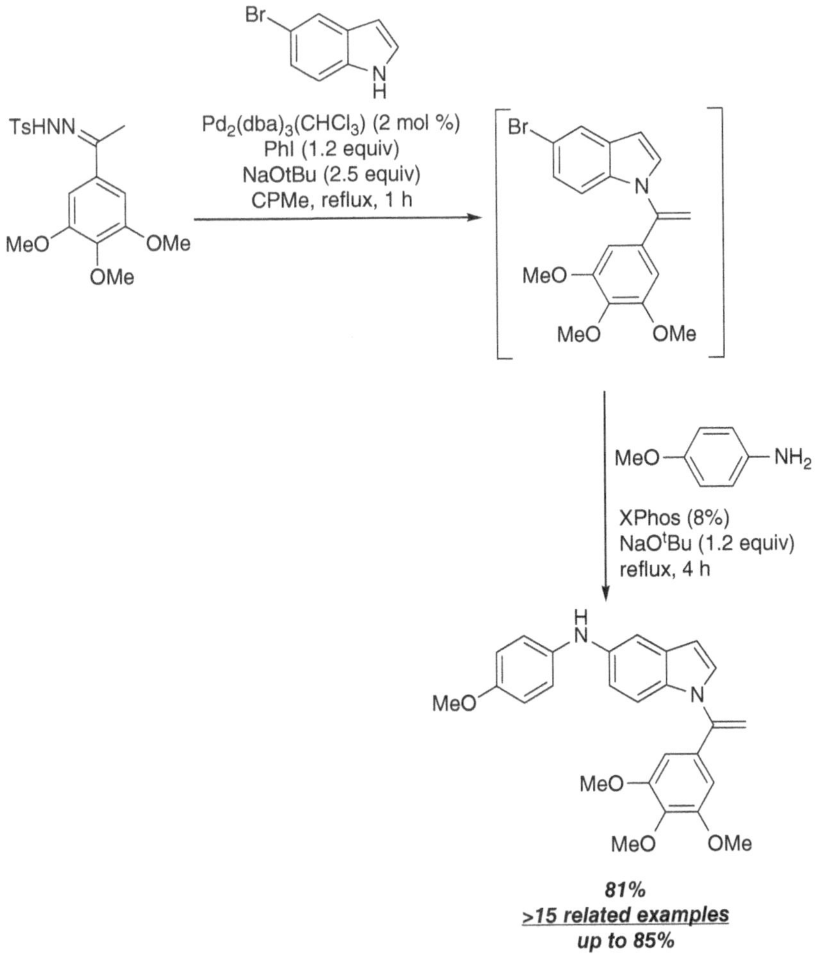

SCHEME 3.138   Palladium-catalyzed preparation of amino-*N*-vinylindoles [148].

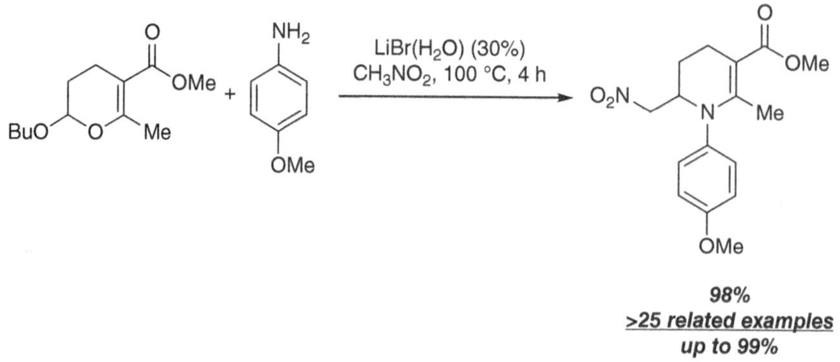

SCHEME 3.139   Preparation of tetrahydropyridines from dihydropyrans and anilines [149].

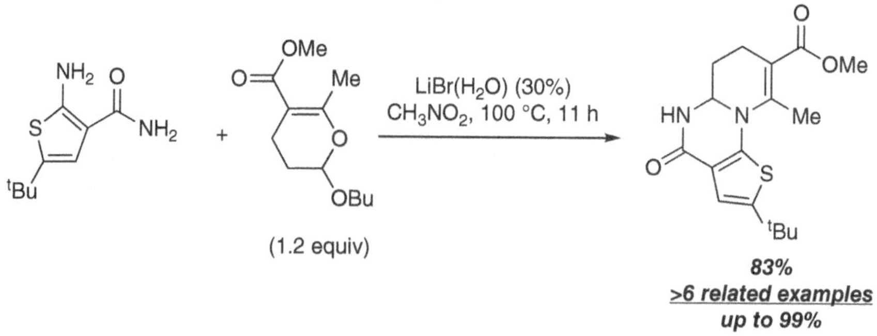

SCHEME 3.140   Synthesis of polycyclic systems from dihydropyrans [149].

intermediate. Treatment of this intermediate with an aniline derivative generated the amino-*N*-vinylindoles through a cross-coupling process.

A series of tetrahydropyridine derivatives have been generated through the treatment of dihydropyrans with aniline precursors (Scheme 3.139) [149]. This was an interesting reaction that resulted in the replacement of the oxygen in the dihydropyran precursor by the nitrogen from aniline. Moderate to good yields of the new tetrahydropyridines were obtained using an assortment of electron-rich and electron-poor anilines with the lowest yields obtained from electron-poor substrates. When substrates bearing an additional amino group were screened, a double cyclization occurred to generate a range of polycyclic systems (Scheme 3.140) [149].

A series of azomethine ylides have been generated through a rhodium-catalyzed reaction between sulfonyl triazoles and functionalized pyridines (Scheme 3.141) [150]. Using this approach, an array of sterically and electronically diverse azomethine ylides were generated by systematic modification of the precursors. In addition to the isolation of these intriguing species, they were also used as intermediates in the preparation of 1,4-diazepines through a one-pot multicomponent reaction involving an internal activated alkyne such as dimethyl acetylenedicarboxylate.

azomethine ylide synthesis

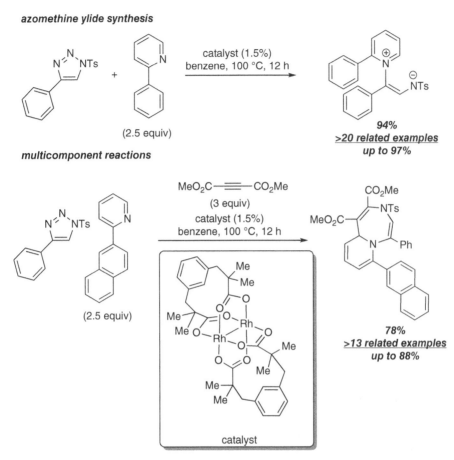

multicomponent reactions

SCHEME 3.141   Rhodium-catalyzed synthesis of azomethine ylides and their use in multicomponent reactions [150].

The copper-catalyzed addition of propioloylpyrazoles to azomethine imines has been achieved through the use of copper catalysts (Scheme 3.142) [151]. These enantioselective additions were carried out at subzero temperature and afforded the pyrazolines in good to outstanding yield. The enantioselectivity was also remarkably high and was fairly insensitive to the functional groups on the substrates.

Nitroalkenes are important compounds due to their ability to be readily converted into a wide assortment of derivatives through conjugate addition reactions. While a range of approaches have been developed for the synthesis of these compounds, the conversion of vinylboronic acids into nitroalkenes is particularly attractive (Scheme 3.143) [118]. The nitrating agent used in this chemistry was bismuth nitrate, and a moderate yield of the nitroalkene was obtained after heating for 12 h. While only a single example of a vinylboronic acid was screened, this chemistry has great potential for the extension to a vast array of nitroalkenes due to the high availability of vinylboronic acids.

The iron-catalyzed generation of imines using alcohols/amines with amines represents an attractive and practical approach to the preparation of imines. The protocol utilizes

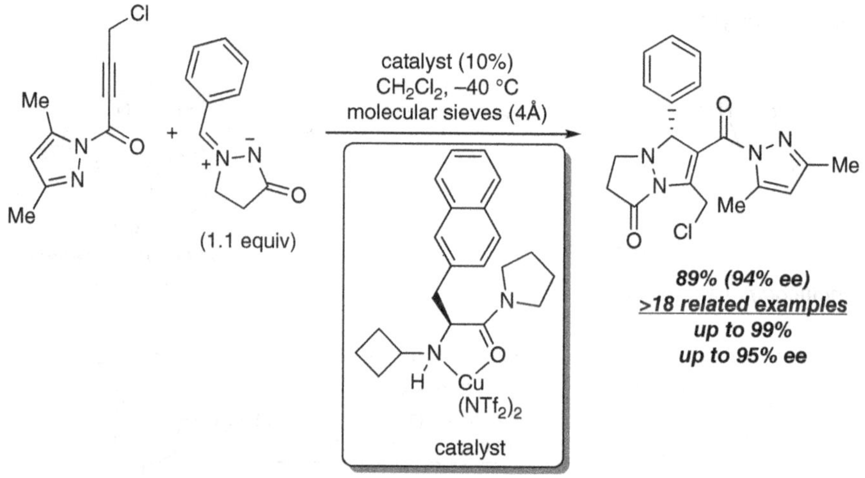

SCHEME 3.142   Preparation of pyrazolines through copper-catalyzed cycloaddition reactions [151].

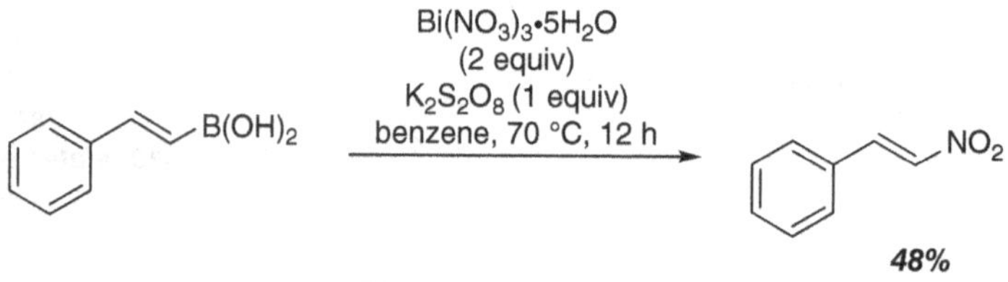

SCHEME 3.143   Nitration of a vinylboronic acid using bismuth nitrate [118].

inexpensive iron nitrate as the catalyst and air as the oxidant. The use of "air" as the oxygen source is a significant advantage of this protocol. While this approach generated moderate yields of the imines, the authors found that the addition of the catalytic amounts of TEMPO increased the yield of the reaction. The chemistry displayed wide substrate scope and was tolerant to a range of functional groups including preexisting halogens [152].

3.3.1   Troubleshooting the Synthesis of Vinylamines and Related Compounds

While it is not possible to predict all of the challenges and difficulties that will arise during the preparation of these compounds, the following suggestions will provide some direction for common issues and reaction requirements.

Problem/Issue/Goal	Possible Solution/Suggestion
Need a reliable method for the synthesis of an enamine	The palladium-catalyzed cross-coupling between vinyl bromides and secondary amines would be a reasonable approach [120]
Need a method for the preparation of enamides from vinyl halides	An assortment of copper-catalyzed reactions are known to promote this reaction [121, 123]
Can vinylboronates be converted into enamines?	Using copper compounds to promote the coupling of nitrogen nucleophiles with vinylboronates would be a good starting point [51, 125]

(Continued)

(Continued)

Problem/Issue/Goal	Possible Solution/Suggestion
Aryltrifluoroborate salts are significantly more stable than boronic acids or boronates. Will they still react in N-vinylation reactions?	A number of successful coupling reactions between aryltrifluoroborate salts with nitrogen nucleophiles have been reported [127–129]
Need to make an N-vinyl nitrone	This has been achieved using a copper-catalyzed coupling between vinylboronic acids and fluorenone oxime [126]
Need to carry out a Markovnikov hydroamination reaction between an aniline and a terminal alkyne	There are many catalysts that will promote this reaction with Markovnikov selectivity. Rhodium- and gold-catalyzed reactions were particularly high yielding and selective [135, 136]
Need to carry out an anti-Markovnikov hydroamination reaction (E-selective) between an aniline and a terminal alkyne	A ruthenium-catalyzed anti-Markovnikov addition of nitrogen nucleophiles to alkynes generated the E-isomers exclusively [137]
Need to carry out an anti-Markovnikov hydroamination reaction (Z-selective) between an aniline and a terminal alkyne	There are several approaches to this reaction. The cesium hydroxide-promoted version is operationally simple and exhibits high regioselectivity and stereoselectivity for a range of substrates [142]
Can internal activated alkynes be used in hydroamination reactions?	Nickel catalysts promoted the addition of nitrogen nucleophiles to internal alkynes [138]
Can TMS-protected alkynes be used in a base-assisted hydroamination reaction, or do they need to be deprotected first?	TMS-protected alkynes are excellent substrates for the base-assisted anti-Markovnikov selective hydroamination reaction [143]. No need to remove the protecting group prior to the hydroelementation reaction
Can dihydropyrans be converted into tetrahydropyridine derivatives?	Tetrahydropyridines have been generated through the treatment of dihydropyrans with aniline precursors [149]
The substrate is extremely crowded at the nucleophilic nitrogen. What would be a good approach for using it in a hydroamination reaction?	While there are several methods that could be successful, a rhodium-catalyzed hydroamination reaction has shown great tolerance to highly crowded substrates [139]
Can a vinylboronic acid be converted into a nitroalkene?	The direct conversion of vinylboronic acids into nitroalkenes has been accomplished using bismuth(III) nitrate [118]

3.4   SYNTHESIS OF YNAMIDES AND RELATED COMPOUNDS THROUGH NITROGEN–CARBON(SP) BOND-FORMING REACTIONS

While ynamines are highly reactive compounds that are quite challenging to generate and isolate, ynamides are significantly more stable and can be prepared in high yield through several methods [153–157]. One of the most straightforward approaches was initially reported by Stang [158] and entailed the preparation of ynamides through treatment of a secondary amide with a strong base followed by the addition of an alkynyliodonium salt (Scheme 3.144 and Example 3.20) [158–161]. The deprotonation step was commonly carried out at low temperatures, while the iodonium salt was added at ambient temperature.

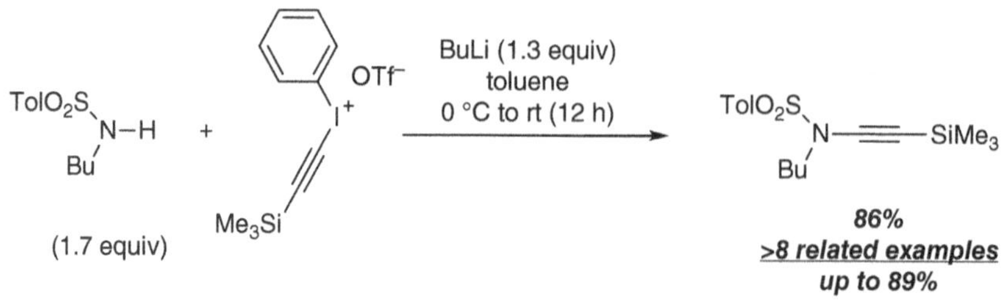

SCHEME 3.144 BuLi-promoted synthesis of ynamides [159].

Example 3.20 BuLi-Promoted Formation of Ynamides [159].

Before starting this synthesis, carefully follow the safety instructions listed in Example 3.1.

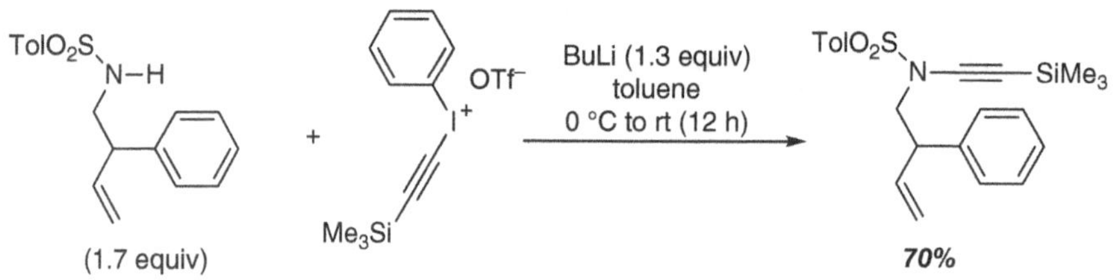

Caution! Petroleum ether and diethyl ether are highly flammable.

Caution! Organolithium reagents can react violently with moisture and must be stored and handled under an inert atmosphere.

n-BuLi (5.19 mmol, 3.25 mL of a 1.6 m solution in hexane) was added to a solution of the amide (1.3 g, 4.3 mmol) in absolute toluene (60 mL) under argon at 0 °C. After the mixture was allowed to warm to room temperature, the iodonium salt (1.13 g, 2.5 mmol) was added in small portions. The reaction mixture was stirred for 12 h and then filtered through a plug of silica gel. Purification by column chromatography (silica gel, petroleum ether/diethyl ether 9 : 1 (v/v)) gave analytically pure ynamide (1.19 g, 3.0 mmol, 70%).

*Adapted with permission from B. Witulski and T. Stengel, Angew. Chem. Int. Ed., **1998**, 37, 489–492. Copyright (1998) Wiley-VCH Verlag GmbH & Co. KGaA, Weinheim.*

The use of Grignard and organolithium reagents as the bases in this reaction typically worked quite well provided there are no other electrophiles in the substrates. While the direct product of this reaction is a TMS-protected ynamide, this protecting group was easily removed by TBAF without affecting the remainder of the molecule. In related work, ynamides were generated through a similar process using KHMDS as the base (Scheme 3.145) [162].

The flow of this reaction has been reversed, and the coupling of terminal alkynes with amide-functionalized iodonium reagents was developed (Scheme 3.146) [163]. This metal-free reaction was attractive since unfunctionalized alkynes could be used to generate the ynamides. The chemistry does require the preparation of the iodine(III) reagent, but once it was prepared, it could be used to functionalize an assortment of terminal acetylenes.

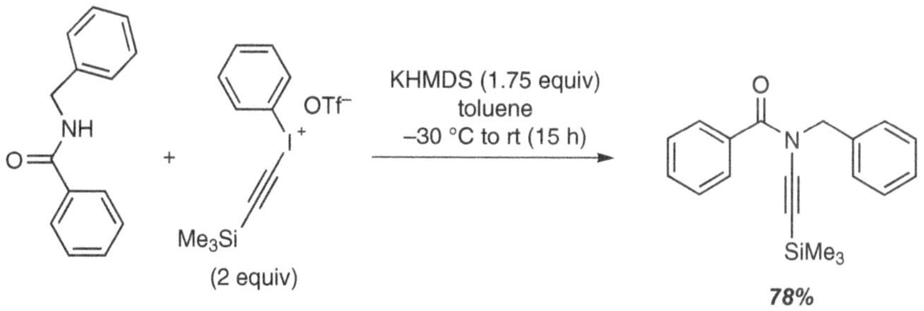

SCHEME 3.145   Synthesis of ynamides from phenyl(trimethylsilylethynyl)iodonium triflate [162].

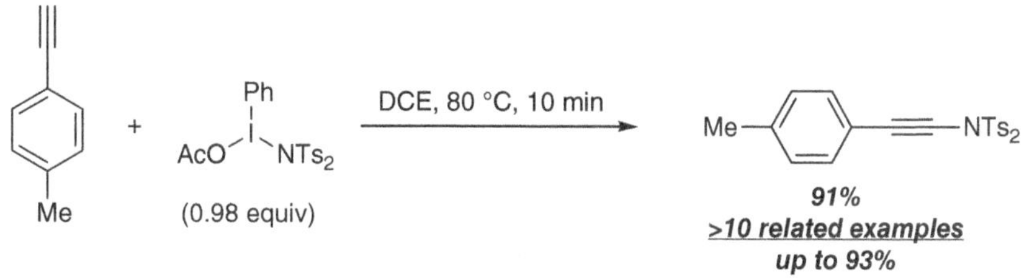

SCHEME 3.146   Metal-free synthesis of ynamides [163].

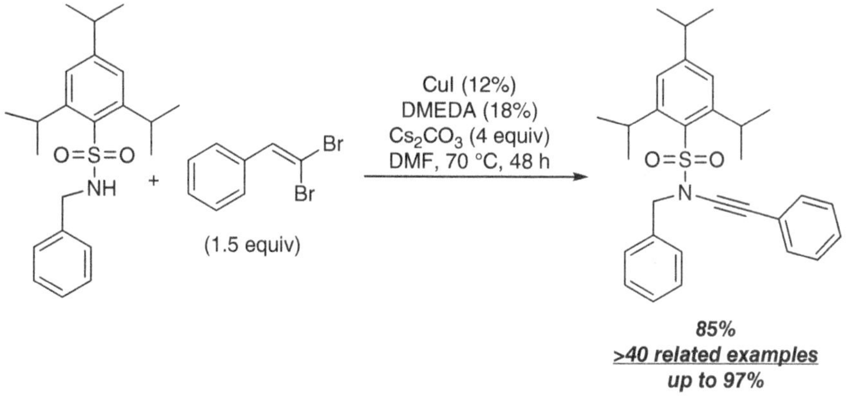

SCHEME 3.147   Synthesis of ynamides from *gem*-dibromoalkenes [164].

The authors noted that the synthesis was insensitive to the presence of electron-donating and electron-withdrawing groups, and the highest yields were observed with 3-ethynylanisole (93%).

In addition to these methods, several metal-catalyzed approaches are known to generate ynamides through nitrogen–carbon(sp) bond-forming reactions. One of these methods starts from *gem*-dibromoalkenes and uses copper complexes to promote the formation of ynamides (Scheme 3.147 and Example 3.21) [164]. Secondary amides and related compounds served as the nitrogen nucleophiles for this chemistry. The authors noted that the chemistry was quite sensitive to the base, and while the use of potassium carbonate and potassium phosphate either gave low conversions or significant amounts of secondary products, cesium carbonate cleanly afforded the ynamide. Building upon this observation,

Example 3.21   Copper-Catalyzed Synthesis of Ynamides from *gem*-Dibromoalkenes [164].

Before starting this synthesis, carefully follow the safety instructions listed in Example 3.1.

A 15 mL pressure tube was charged with sulfonamide (1.6 mmol), 1,1-dibromo-1-alkene (2.4 mmol), Cs$_2$CO$_3$ (2.1 g, 6.4 mmol), and copper(I) iodide (38 mg, 0.2 mmol). The tube was fitted with a rubber septum, evacuated under high vacuum, and backfilled with argon. Dry and degassed 1,4-dioxane or DMF (3 mL) and *N,N'*-dimethylethylenediamine (30 μL, 0.3 mmol) were next added, the rubber septa was replaced by a Teflon-coated screw cap, and the light blue-green suspension was heated at 70 °C for 48 h. The brownish suspension was cooled to room temperature. The crude reaction mixture was diluted with water and extracted with diethyl ether, and the combined organic layers were washed with brine, dried over MgSO$_4$, filtered, and concentrated. The crude residue was purified by flash chromatography over silica gel. The title compound was isolated as a white solid in 97% yield (637 mg).

Adapted with permission from A. Coste, G. Karthikeyan, F. Couty, *and* G. Evano, *Angew. Chem. Int. Ed.,* **2009**, *48,* 4381–4385. *Copyright (2009) Wiley-VCH Verlag GmbH & Co. KGaA, Weinheim.*

a vast array of ynamides were generated through systematic screening of various nitrogen precursors and *gem*-dibromoalkenes. The chemistry was insensitive to the presence of electron-rich or electron-poor arenes on the *gem*-dibromoalkene, and secondary tosyl-amides were among the most effective nitrogen nucleophiles. One of the more attractive aspects about this chemistry was that heteroaromatic examples as well as crowded substrates were also converted into ynamides in high yield. This is one of the most practical approaches to the preparation of ynamides due to the wide functional group tolerance and operational simplicity of the approach.

Ynamides have also been generated through an aerobic copper coupling between terminal alkynes and nitrogen nucleophiles (Scheme 3.148) [165]. The key advance with this chemistry was the ability to form the ynamides without prefunctionalization of the alkyne or secondary amide. The reaction needed copper to promote the coupling, and the authors found that copper(II) chloride was the most efficient catalyst for the reaction. Curiously, copper iodide was significantly less effective under the reaction conditions. The conversion was carried out under an atmosphere of oxygen, but high-pressure vessels were not needed since only 1 atm of oxygen was required for a successful conversion. Using the optimized system, an array of terminal alkynes were successfully coupled with an assortment of nitrogen

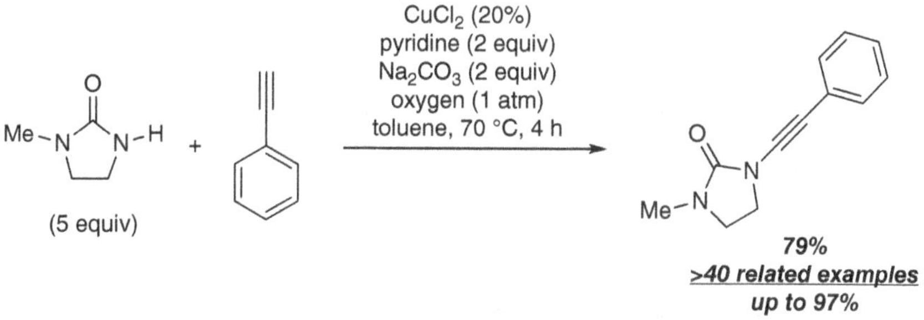

SCHEME 3.148   Copper-promoted coupling of terminal alkynes with secondary amides [165].

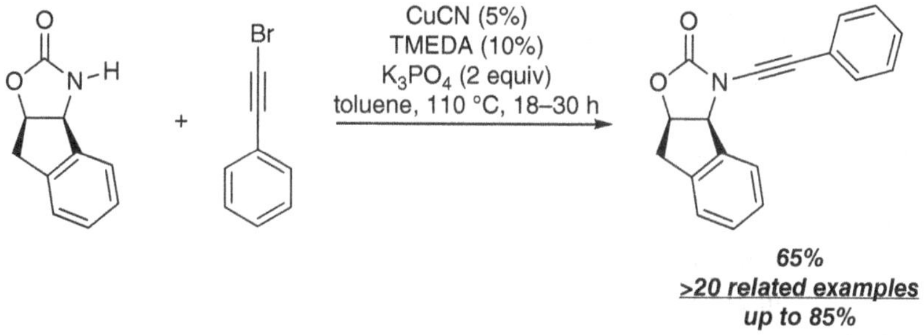

SCHEME 3.149   Copper-catalyzed synthesis of ynamides from haloalkynes [166].

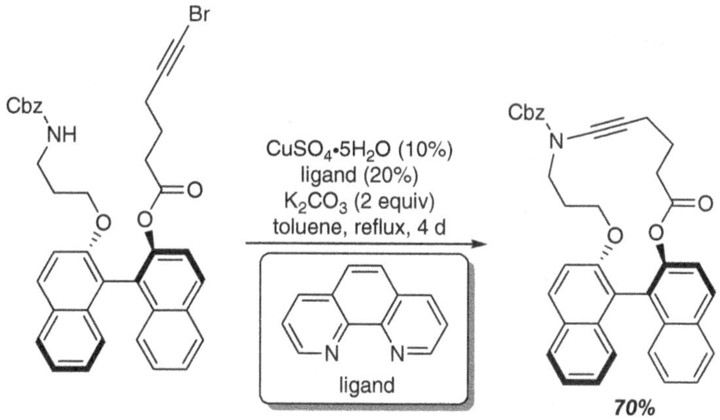

SCHEME 3.150   Synthesis of a macrocyclic ynamide [169].

nucleophiles to afford the ynamides in good to excellent yields. The authors noted an electronic effect on the nitrogen–carbon(sp) bond-forming reaction with electron-rich alkynes affording higher yields of the ynamides. It should be noted that if a haloalkyne can be generated, it can be coupled with a secondary amide to generate ynamides (Scheme 3.149) [166]. Several versions of this approach including the preparation of *N*-phosphoryl ynamides [167], N-alkynylated sulfoximines [168], an ynamide macrocycle [169] (Scheme 3.150), and a range of related compounds have been reported [170–175].

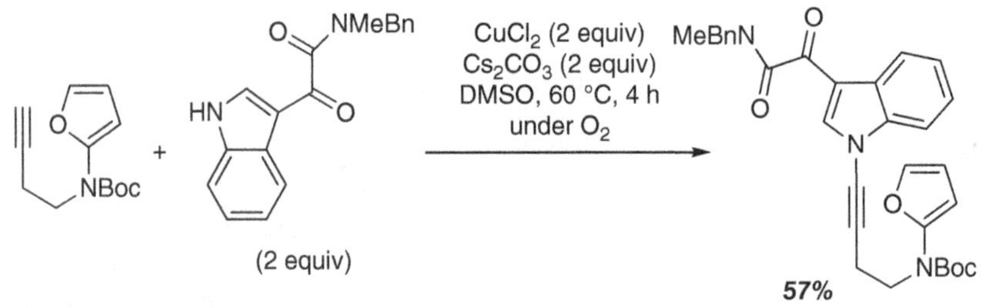

SCHEME 3.151   Copper-catalyzed synthesis of ynindoles [176].

Ynindoles have slowly been drawing the attention of the synthetic community due to the development of a few practical routes for their preparation. For the majority of examples, the isolated ynindoles were substituted with electron-withdrawing groups [165]. In an attempt to increase the scope of this chemistry and probe the reactivity of ynindoles in cycloaddition reactions, a range of indoles were screened for activity in an aerobic copper-catalyzed coupling reaction (Scheme 3.151) [176]. Copper(II) chloride was selected as the catalyst for the reaction and cesium carbonate served as the base. Heating indole derivatives bearing electron-withdrawing groups at the 3-position of the ring afforded moderate yields of the ynindoles (up to 57%). Indole and related compounds that were not substituted with electron-withdrawing groups were unresponsive under the reaction conditions only starting materials were isolated after heating. While the substrate scope of this reaction was limited, it is a reliable method for the preparation of ynindoles if the substrate is appropriately designed.

3.4.1   Troubleshooting the Synthesis of Ynamines and Ynamides

While it is not possible to predict all of the challenges and difficulties that will arise during the preparation of these compounds, the following suggestions will hopefully provide some direction for common issues and reaction requirements.

Problem/Issue/Goal	Possible Solution/Suggestion
Need a transition metal-free approach to the synthesis of ynamides	This has been accomplished through the treatment of a secondary amide with a strong base followed by the addition of an alkynyliodonium reagent [159, 162]
It is convenient to start the synthesis from a gem-dibromoalkene. Can these species be directly converted into ynamides?	A copper-catalyzed approach has been devised to convert gem-dibromoalkenes into ynamides [164]. This approach was remarkably tolerant to a wide variety of preexisting functional groups and bulky substrates
Starting material has a terminal alkyne. Can it be converted into an ynamide?	There are several approaches to consider. A metal-free coupling of terminal acetylenes with a functionalized iodine(III) reagent generated high yields of the ynamides [163]. An aerobic copper coupling reaction between terminal alkynes and secondary amides has also been successful [165]

(Continued)

(Continued)

Problem/Issue/Goal	Possible Solution/Suggestion
Would it be easier to generate an alkynyl bromide as the starting compound instead of the terminal alkyne?	This is perhaps the most popular starting point for the preparation of ynamides. This copper-catalyzed process typically uses a bromoalkyne and a secondary amide as substrates and mineral bases such as Cs_2CO_3. The process was attractive since it does not require the use of strong bases [166, 170–175]
Need a good method for the preparation of ynindoles	This intriguing class of compounds can be generated through a copper-catalyzed coupling reaction between terminal acetylenes and indoles [176]. It should be noted that there is often an electronic limitation with this chemistry and only indoles bearing electron-withdrawing groups are typically responsive to the chemistry

REFERENCES

[1] Huang, Y.-C.; Cao, C.; Tan, X.-L.; Li, X.; Liu, L. *Org. Chem. Front.* **2014**, *1*, 1050–1054.

[2] Ju, Y.; Varma, R. S. *Green Chem.* **2004**, *6*, 219–221.

[3] Enyong, A. B.; Moasser, B. *J. Org. Chem.* **2014**, *79*, 7553–7563.

[4] Wang, D.; Zhao, K.; Xu, C.; Miao, H.; Ding, Y. *ACS Catal.* **2014**, *4*, 3910–3918.

[5] Chavhan, S. W.; McAdam, C. A.; Cook, M. J. *J. Org. Chem.* **2014**, *79*, 11234–11240.

[6] Sorribes, I.; Junge, K.; Beller, M. *J. Am. Chem. Soc.* **2014**, *136*, 14314–14319.

[7] Do, H.-Q.; Bachman, S.; Bissember, A. C.; Peters, J. C.; Fu, G. C. *J. Am. Chem. Soc.* **2014**, *136*, 2162–2167.

[8] Mlynarski, S. N.; Karns, A. S.; Morken, J. P. *J. Am. Chem. Soc.* **2012**, *134*, 16449–16451.

[9] Xie, W.; Yang, J.; Wang, B.; Li, B. *J. Org. Chem.* **2014**, *79*, 8278–8287.

[10] Grigorjeva, L.; Daugulis, O. *Org. Lett.* **2014**, *16*, 4684–4687.

[11] Ma, W.; Ackermann, L. *ACS Catal.* **2015**, *5*, 2822–2825.

[12] Lu, Z.; Stahl, S. S. *Org. Lett.* **2012**, *14*, 1234–1237.

[13] Huang, D.; Wang, H.; Xue, F.; Guan, H.; Li, L.; Peng, X.; Shi, Y. *Org. Lett.* **2011**, *13*, 6350–6353.

[14] Zhu, C.; Ma, S. *Angew. Chem. Int. Ed.* **2014**, *53*, 13532–13535.

[15] Ghabraie, E.; Balalaie, S.; Mehrparvar, S.; Rominger, F. *J. Org. Chem.* **2014**, *79*, 7926–7934.

[16] Shi, Y.; Gulevich, A. V.; Gevorgyan, V. *Angew. Chem. Int. Ed.* **2014**, *53*, 14191–14195.

[17] Deshmukh, S. C.; Talukdar, P. *J. Org. Chem.* **2014**, *79*, 11215–11225.

[18] Trose, M.; Dell'Acqua, M.; Pedrazzini, T.; Pirovano, V.; Gallo, E.; Rossi, E.; Caselli, A.; Abbiati, G. *J. Org. Chem.* **2014**, *79*, 7311–7320.

[19] Huang, L.; Arndt, M.; GooBen, K.; Heydt, H.; GooBen, L. J. *Chem. Rev.* **2015**, *115*, 2596–2697.

[20] Zhu, S.; Buchwald, S. L. *J. Am. Chem. Soc.* **2014**, *136*, 15913–15916.

[21] Huehls, C. B.; Lin, A.; Yang, J. *Org. Lett.* **2014**, *16*, 3620–3623.

[22] Parra, A.; Amenos, L.; Guisan-Ceinos, M.; Lopez, A.; Garcia Ruano, J. L.; Tortosa, M. *J. Am. Chem. Soc.* **2014**, *136*, 15833–15836.

[23] Sevov, C. S.; Zhou, J.; Hartwig, J. F. *J. Am. Chem. Soc.* **2014**, *136*, 3200–3207.

[24] Schultz, E. E.; Lindsay, V. N. G.; Sarpong, R. *Angew. Chem. Int. Ed.* **2014**, *53*, 9904–9908.

[25] Musacchio, A. J.; Nguyen, L. Q.; Beard, G. H.; Knowles, R. R. *J. Am. Chem. Soc.* **2014**, *136*, 12217–12220.

[26] William, R.; Wang, S.; Ding, F.; Arviana, E. N.; Liu, X.-W. *Angew. Chem. Int. Ed.* **2014**, *53*, 10742–10746.

[27] Calder, E. D. D.; McGonagle, F. I.; Harkiss, A. H.; McGonagle, G. A.; Sutherland, A. *J. Org. Chem.* **2014**, *79*, 7633–7648.

[28] Beydoun, K.; Ghattas, G.; Thenert, K.; Klankermayer, J.; Leitner, W. *Angew. Chem. Int. Ed.* **2014**, *53*, 11010–11014.

[29] Sun, Z.; Winschel, G. A.; Zimmerman, P. M.; Nagorny, P. *Angew. Chem. Int. Ed.* **2014**, *53*, 11194–11198.

[30] Sarraf, D.; Richy, N.; Vidal, J. *J. Org. Chem.* **2014**, *79*, 10945–10955.

[31] Barber, D. M.; Duris, A.; Thompson, A. L.; Sanganee, H. J.; Dixon, D. J. *ACS Catal.* **2014**, *4*, 634–638.

[32] Bisai, V.; Suneja, A.; Singh, V. K. *Angew. Chem. Int. Ed.* **2014**, *53*, 10737–10741.

[33] Zhang, H.; Song, Y.; Zhao, J.; Zhang, J.; Zhang, Q. *Angew. Chem. Int. Ed.* **2014**, *53*, 11079–11083.

[34] Zhu, C.-L.; Zhang, F.-G.; Meng, W.; Nie, J.; Cahard, D.; Ma, J.-A. *Angew. Chem. Int. Ed.* **2011**, *50*, 5869–5872.

[35] Chen, X.-Y.; Gao, Z.-H.; Song, C.-Y.; Zhang, C.-L.; Wang, Z.-X.; Ye, S. *Angew. Chem. Int. Ed.* **2014**, *53*, 11611–11615.

[36] Cho, S. H.; Yoon, J.; Chang, S. *J. Am. Chem. Soc.* **2011**, *133*, 5996–6005.

[37] Sharma, U.; Kancherla, R.; Naveen, T.; Agasti, S.; Maiti, D. *Angew. Chem. Int. Ed.* **2014**, *53*, 11895–11899.

[38] Kiyokawa, K.; Yahata, S.; Kojima, T.; Minakata, S. *Org. Lett.* **2014**, *16*, 4646–4649.

[39] Banerjee, D.; Junge, K.; Beller, M. *Angew. Chem. Int. Ed.* **2014**, *53*, 13049–13053.

[40] Zhu, R.-Y.; Wang, C.-S.; Zheng, J.; Shi, F.; Tu, S.-J. *J. Org. Chem.* **2014**, *79*, 9305–9312.

[41] Fang, W.; Deng, Q.; Xu, M.; Tu, T. *Org. Lett.* **2013**, *15*, 3678–3681.

[42] Green, R. A.; Hartwig, J. F. *Org. Lett.* **2014**, *16*, 4388–4391.

[43] Peng, J.; Chen, M.; Xie, Z.; Luo, S.; Zhu, Q. *Org. Chem. Front.* **2014**, *1*, 777–781.

[44] Shin, Y.; Han, S.; De, U.; Park, J.; Sharma, S.; Mishra, N. K.; Lee, E.-K.; Lee, Y.; Kim, H. S.; Kim, I. S. *J. Org. Chem.* **2014**, *79*, 9262–9271.

[45] Peng, J.; Xie, Z.; Chen, M.; Wang, J.; Zhu, Q. *Org. Lett.* **2014**, *16*, 4702–4705.

[46] Wolfe, J. P.; Wagaw, S.; Marcoux, J.-F.; Buchwald, S. L. *Acc. Chem. Res.* **1998**, *31*, 805–818.

[47] Hartwig, J. F. *Angew. Chem. Int. Ed.* **1998**, *37*, 2046–2067.

[48] Surry, D. S.; Buchwald, S. L. *Chem. Sci.* **2011**, *2*, 27–50.

[49] Fors, B. P.; Watson, D. A.; Biscoe, M. R.; Buchwald, S. L. *J. Am. Chem. Soc.* **2008**, *130*, 13552–13554.

[50] Fors, B. P.; Davis, N. R.; Buchwald, S. L. *J. Am. Chem. Soc.* **2009**, *131*, 5766–5768.

[51] Jacobsen, M. F.; Knudsen, M. M.; Gothelf, K. V. *J. Org. Chem.* **2006**, *71*, 9183–9190.

[52] Chan, D. M. T.; Monaco, K. L.; Wang, R.-P.; Winters, M. P. *Tetrahedron Lett.* **1998**, *39*, 2933–2936.

[53] Lam, P. Y. S.; Clark, C. G.; Saubern, S.; Adams, J.; Winters, M. P.; Chan, D. M. T.; Combs, A. *Tetrahedron Lett.* **1998**, *39*, 2941–2944.

[54] Ma, D.; Cai, Q.; Zhang, H. *Org. Lett.* **2003**, *5*, 2453–2455.

[55] Shafir, A.; Buchwald, S. L. *J. Am. Chem. Soc.* **2006**, *128*, 8742–8743.

[56] Ma, D.; Cai, Q. *Acc. Chem. Res.* **2008**, *41*, 1450–1460.

[57] Kwong, F. Y.; Klapars, A.; Buchwald, S. L. *Org. Lett.* **2002**, *4*, 581–584.

[58] Gujadhur, R. K.; Bates, C. G.; Venkataraman, D. *Org. Lett.* **2001**, *3*, 4315–4317.

[59] He, C.; Chen, C.; Cheng, J.; Liu, C.; Liu, W.; Li, Q.; Lei, A. *Angew. Chem. Int. Ed.* **2008**, *47*, 6414–6417.

[60] Zhang, Z.; Zhang, W.; Li, J.; Liu, Q.; Liu, T.; Zhang, G. *J. Org. Chem.* **2014**, *79*, 11226–11233.

[61] Nallagonda, R.; Rehan, M.; Ghorai, P. *Org. Lett.* **2014**, *16*, 4786–4789.

[62] Zavesky, B. P.; Babij, N. R.; Wolfe, J. P. *Org. Lett.* **2014**, *16*, 4952–4955.

[63] Ueda, H.; Yamaguchi, M.; Kameya, H.; Sugimoto, K.; Tokuyama, H. *Org. Lett.* **2014**, *16*, 4948–4951.

[64] Song, J.; Fan, C.; Liu, G.; Qiu, G. *Org. Chem. Front.* **2014**, *1*, 1045–1049.

[65] Cao, H.; Liu, X.; Liao, J.; Huang, J.; Qiu, H.; Chen, Q.; Chen, Y. *J. Org. Chem.* **2014**, *79*, 11209–11214.

[66] Tu, X.-J.; Hao, W.-J.; Ye, Q.; Wang, S.-S.; Jiang, B.; Li, G.; Tu, S.-J. *J. Org. Chem.* **2014**, *79*, 11110–11118.

[67] Thirupathi, N.; Babu, M. H.; Dwivedi, V.; Kant, R.; Reddy, M. S. *Org. Lett.* **2014**, *16*, 2908–2911.

[68] Zhang, L.; Wang, H.; Yang, B.; Fan, R. *Org. Chem. Front.* **2014**, *1*, 1055–1057.

[69] Zardi, P.; Savoldelli, A.; Carminati, D. M.; Caselli, A.; Ragaini, F.; Gallo, E. *ACS Catal.* **2014**, *4*, 3820–3823.

[70] Jayakumar, J.; Parthasarathy, K.; Chen, Y.-H.; Lee, T.-H.; Chuang, S.-C.; Cheng, C.-H. *Angew. Chem. Int. Ed.* **2014**, *53*, 9889–9892.

[71] Wang, C.; Chen, C.; Zhang, J.; Han, J.; Wang, Q.; Guo, K.; Liu, P.; Guan, M.; Yao, Y.; Zhao, Y. *Angew. Chem. Int. Ed.* **2014**, *53*, 9884–9888.

[72] Chen, F.; Huang, X.; Li, X.; Shen, T.; Zou, M.; Jiao, N. *Angew. Chem. Int. Ed.* **2014**, *53*, 10495–10499.

[73] Nakamura, I.; Jo, T.; Zhang, D.; Terada, M. *Org. Chem. Front.* **2014**, *1*, 914–918.

[74] Matcha, K.; Antonchick, A. P. *Angew. Chem. Int. Ed.* **2014**, *53*, 11960–11964.

[75] Zheng, H.; Zhu, Y.; Shi, Y. *Angew. Chem. Int. Ed.* **2014**, *53*, 11280–11284.

[76] Muruganantham, R.; Mobin, S. M.; Namboothiri, I. N. N. *Org. Lett.* **2007**, *9*, 1125–1128.

[77] Pan, C.; Jin, N.; Zhang, H.; Han, J.; Zhu, C. *J. Org. Chem.* **2014**, *79*, 9427–9432.

[78] Yan, R.-L.; Luo, J.; Wang, C.-X.; Ma, C.-W.; Huang, G.-S.; Liang, Y.-M. *J. Org. Chem.* **2010**, *75*, 5395–5397.

[79] You, X.; Xie, X.; Sun, R.; Chen, H.; Li, S.; Liu, Y. *Org. Chem. Front.* **2014**, *1*, 940–946.

[80] John, A.; Nicholas, K. M. *J. Org. Chem.* **2011**, *76*, 4158–4162.

[81] Guimond, N.; Gouliaras, C.; Fagnou, K. *J. Am. Chem. Soc.* **2010**, *132*, 6908–6909.

[82] Vrijdag, J. L.; Delgado, F.; Alonso, N.; De Borggraeve, W. M.; Perez-Macias, N.; Alcazar, J. *Chem. Commun.* **2014**, *50*, 15094–15097.

[83] Chen, Z.-W.; Zhu, Y.-Z.; Ou, J.-W.; Wang, Y.-P.; Zheng, J.-Y. *J. Org. Chem.* **2014**, *79*, 10988–10998.

[84] Tomas-Mendivil, E.; Garcia-Alvarez, R.; Vidal, C.; Crochet, P.; Cadierno, V. *ACS Catal.* **2014**, *4*, 1901–1910.

[85] Grigorjeva, L.; Daugulis, O. *Angew. Chem. Int. Ed.* **2014**, *53*, 10209–10212.

[86] Hu, J.-D.; Cao, C.-P.; Lin, W.; Hu, M.-H.; Huang, Z.-B.; Shi, D.-Q. *J. Org. Chem.* **2014**, *79*, 7935–7944.

[87] Naresh, G.; Kant, R.; Narender, T. *Org. Lett.* **2014**, *16*, 4528–4531.

[88]   Lei, C.-H.; Zhao, L.; Wang, D.-X.; Zhu, J.; Wang, M.-X. *Org. Chem. Front.* **2014**, *1*, 909–913.

[89]   Grigorjeva, L.; Daugulis, O. *Org. Lett.* **2014**, *16*, 4688–4690.

[90]   Webb, N. J.; Marsden, S. P.; Raw, S. A. *Org. Lett.* **2014**, *16*, 4718–4721.

[91]   Shen, C.-H.; Li, L.; Zhang, W.; Liu, S.; Shu, C.; Xie, Y.-E.; Yu, Y.-F.; Ye, L.-W. *J. Org. Chem.* **2014**, *79*, 9313–9318.

[92]   Reddy, M. C.; Jeganmohan, M. *Org. Lett.* **2014**, *16*, 4866–4869.

[93]   Garzon, M.; Davies, P. W. *Org. Lett.* **2014**, *16*, 4850–4853.

[94]   Zhang, X.; Si, W.; Bao, M.; Asao, N.; Yamamoto, Y.; Jin, T. *Org. Lett.* **2014**, *16*, 4830–4833.

[95]   Xing, Y.; Cheng, B.; Wang, J.; Lu, P.; Wang, Y. *Org. Lett.* **2014**, *16*, 4814–4817.

[96]   Su, M.; Buchwald, S. L. *Angew. Chem. Int. Ed.* **2012**, *51*, 4710–4713.

[97]   Vijay Kumar, S.; Saraiah, B.; Parameshwarappa, G.; Ila, H.; Verma, G. K. *J. Org. Chem.* **2014**, *79*, 7961–7978.

[98]   Kim, J. H.; Choi, S. Y.; Bouffard, J.; Lee, S. *J. Org. Chem.* **2014**, *79*, 9253–9261.

[99]   Pooi, B.; Lee, J.; Choi, K.; Hirao, H.; Hong, S. H. *J. Org. Chem.* **2014**, *79*, 9231–9245.

[100]   Chen, Y.-F.; Hsieh, J.-C. *Org. Lett.* **2014**, *16*, 4642–4645.

[101]   Monir, K.; Bagdi, A. K.; Ghosh, M.; Hajra, A. *Org. Lett.* **2014**, *16*, 4630–4633.

[102]   Specklin, S.; Decuypere, E.; Plougastel, L.; Aliani, S.; Taran, F. *J. Org. Chem.* **2014**, *79*, 7772–7777.

[103]   Nageswar Rao, D.; Rasheed, S.; Vishwakarma, R. A.; Das, P. *Chem. Commun.* **2014**, *50*, 12911–12914.

[104]   Cao, J.-J.; Wang, X.; Wang, S.-Y.; Ji, S.-J. *Chem. Commun.* **2014**, *50*, 12892–12895.

[105]   Mamedov, V. A.; Zhukova, N. A.; Zamaletdinova, A. I.; Beschastnova, T. N.; Kadyrova, M. S.; Rizvanov, I. K.; Syakaev, V. V.; Latypov, S. K. *J. Org. Chem.* **2014**, *79*, 9161–9169.

[106]   Gupta, S.; Han, J.; Kim, Y.; Lee, S. W.; Rhee, Y. H.; Park, J. *J. Org. Chem.* **2014**, *79*, 9094–9103.

[107]   Gao, J.; Shao, Y.; Zhu, J.; Zhu, J.; Mao, H.; Wang, X.; Lv, X. *J. Org. Chem.* **2014**, *79*, 9000–9008.

[108]   Huang, F.; Wu, P.; Wang, L.; Chen, J.; Sun, C.; Yu, Z. *Chem. Commun.* **2014**, *50*, 12479–12481.

[109]   Kolb, H. C.; Finn, M. G.; Sharpless, K. B. *Angew. Chem. Int. Ed.* **2001**, *40*, 2004–2021.

[110]   Thirumurugan, P.; Matosiuk, D.; Jozwiak, K. *Chem. Rev.* **2013**, *113*, 4905–4979.

[111]   Moses, J. E.; Moorhouse, A. D. *Chem. Soc. Rev.* **2007**, *36*, 1249–1262.

[112]   Lim, D.; Brimble, M. A.; Kowalczyk, R.; Watson, A. J. A.; Fairbanks, A. J. *Angew. Chem. Int. Ed.* **2014**, *53*, 11907–11911.

[113]   Lal, S.; Rzepa, H. S.; Diez–Gonzalez, S. *ACS Catal.* **2014**, *4*, 2274–2287.

[114]   Li, W.; Wang, J. *Angew. Chem. Int. Ed.* **2014**, *53*, 14186–14190.

[115]   Lima, C. G. S.; Ali, A.; van Berkel, S. S.; Westermann, B.; Paixao, M. W. *Chem. Commun.* **2015**, *51*, 10784–10796.

[116]   Ramachary, D. B.; Shashank, A. B.; Karthik, S. *Angew. Chem. Int. Ed.* **2014**, *53*, 10420–10424.

[117]   Thomas, J.; John, J.; Parekh, N.; Dehaen, W. *Angew. Chem. Int. Ed.* **2014**, *53*, 10155–10159.

[118]   Manna, S.; Maity, S.; Rana, S.; Agasti, S.; Maiti, D. *Org. Lett.* **2012**, *14*, 1736–1739.

[119]   Kim, J.; Ok, J.; Kim, S.; Choi, W.; Lee, P. H. *Org. Lett.* **2014**, *16*, 4602–4605.

[120]   Barluenga, J.; Fernandez, M. A.; Aznar, F.; Valdes, C. *Chem. Commun.* **2002**, 2362–2363.

[121]   Shen, R.; Porco, J. A. Jr. *Org. Lett.* **2000**, *2*, 1333–1336.

[122]   Han, C.; Shen, R.; Su, S.; Porco, J. A. Jr. *Org. Lett.* **2004**, *6*, 27–30.

[123]   Jiang, L.; Job, G. E.; Klapars, A.; Buchwald, S. L. *Org. Lett.* **2003**, *5*, 3667–3669.

[124]   Pan, X.; Cai, Q.; Ma, D. *Org. Lett.* **2004**, *6*, 1809–1812.

[125]   Lam, P. Y. S.; Vincent, G.; Bonne, D.; Clark, C. G. *Tetrahedron Lett.* **2003**, *44*, 4927–4931.

[126]   Mo, D.-L.; Wink, D. A.; Anderson, L. L. *Org. Lett.* **2012**, *14*, 5180–5183.

[127]   Bolshan, Y.; Batey, R. A. *Angew. Chem.* **2008**, *120*, 2139–2142.

[128]   Bolshan, Y.; Batey, R. A. *Tetrahedron* **2010**, *66*, 5283–5294.

[129]   Bolshan, Y.; Batey, R. A. *Angew. Chem. Int. Ed.* **2008**, *47*, 2109–2112.

[130]   Brice, J. L.; Meerdink, J. E.; Stahl, S. S. *Org. Lett.* **2004**, *6*, 1845–1848.

[131]   Muller, T. E.; Hultzsch, K. C.; Yus, M.; Foubelo, F.; Tada, M. *Chem. Rev.* **2008**, *108*, 3795–3892.

[132]   Alonso, F.; Beletskaya, I. P.; Yus, M. *Chem. Rev.* **2004**, *104*, 3079–3160.

[133]   Evano, G.; Gaumont, A.-C.; Alayrac, C.; Wrona, I. E.; Giguere, J. R.; Delacroix, O.; Bayle, A.; Jouvin, K.; Theunissen, C.; Gatignol, J.; Silvanus, A. C. *Tetrahedron* **2014**, *70*, 1529–1616.

[134]   Yim, J. C.-H.; Schafer, L. L. *Eur. J. Org. Chem.* **2014**, 6825–6840.

[135]   Kumaran, E.; Leong, W. K. *Organometallics* **2012**, *31*, 1068–1072.

[136]   Malhotra, D.; Mashuta, M. S.; Hammond, G. B.; Xu, B. *Angew. Chem. Int. Ed.* **2014**, *53*, 4456–4459.

[137]   Das, U. K.; Bhattacharjee, M. *Chem Eur. J.* **2012**, *18*, 5180–5183.

[138]   Reyes-Sanchez, A.; Garcia-Ventura, I.; Garcia, J. J. *Dalton Trans.* **2014**, *43*, 1762–1768.

[139]   Alonso-Moreno, C.; Carrillo-Hermosilla, F.; Romero-Fernández, J.; Rodríguez, A. M.; Otero, A.; Antiñolo, A. *Adv. Synth. Catal.* **2009**, *351*, 881–890.

[140]   Gooßen, L. J.; Rauhaus, J. E.; Deng, G. *Angew. Chem. Int. Ed.* **2005**, *44*, 4042–4045.

[141]   Kim, S. M.; Lee, D.; Hong, S. H. *Org. Lett.* **2014**, *16*, 6168–6171.

[142]   Tzalis, D.; Koradin, C.; Knochel, P. *Tetrahedron Lett.* **1999**, *40*, 6193–6195.

[143]   Joshi, M.; Tiwari, R.; Verma, A. K. *Org. Lett.* **2012**, *14*, 1106–1109.

[144]   Wu, L.; Huang, H.; Liang, Y.; Cheng, P. *J. Org. Chem.* **2014**, *79*, 11264–11269.

[145]   Li, E.; Jia, P.; Liang, L.; Huang, Y. *ACS Catal.* **2014**, *4*, 600–603.

[146]   Espinosa-Jalapa, N. A.; Ke, D.; Nebra, N.; Le Goanvic, L.; Mallet-Ladeira, S.; Monot, J.; Martin-Vaca, B.; Bourissou, D. *ACS Catal.* **2014**, *4*, 3605–3611.

[147]   Fukui, Y.; Liu, P.; Liu, Q.; He, Z.-T.; Wu, N.-Y.; Tian, P.; Lin, G.-Q. *J. Am. Chem. Soc.* **2014**, *136*, 15607–15614.

[148]   Roche, M.; Bignon, J.; Brion, J.-D.; Hamze, A.; Alami, M. *J. Org. Chem.* **2014**, *79*, 7583–7592.

[149]   Sun, S.; Cheng, C.; Yang, J.; Taheri, A.; Jiang, D.; Zhang, B.; Gu, Y. *Org. Lett.* **2014**, *16*, 4520–4523.

[150]   Lee, D. J.; Han, H. S.; Shin, J.; Yoo, E. J. *J. Am. Chem. Soc.* **2014**, *136*, 11606–11609.

[151]   Hori, M.; Sakakura, A.; Ishihara, K. *J. Am. Chem. Soc.* **2014**, *136*, 13198–13201.

[152]   Zhang, E.; Tian, H.; Xu, S.; Yu, X.; Xu, Q. *Org. Lett.* **2013**, *15*, 2704–2707.

[153]   Ficini, J. *Tetrahedron* **1976**, *32*, 1449–1486.

[154]   Evano, G.; Coste, A.; Jouvin, K. *Angew. Chem. Int. Ed.* **2010**, *49*, 2840–2859.

[155]   Evano, G.; Jouvin, K.; Coste, A. *Synthesis* **2013**, *45*, 17–26.

[156]   DeKorver, K. A.; Li, H.; Lohse, A. G.; Hayashi, R.; Lu, Z.; Zhang, Y.; Hsung, R. P. *Chem. Rev.* **2010**, *110*, 5064–5106.

[157]   Feldman, K. S.; Bruendl, M. M.; Schildknegt, K.; Bohnstedt, A. C. *J. Org. Chem.* **1996**, *61*, 5440–5452.

[158] Zhdankin, V.; Stang, P. *Tetrahedron* **1998**, *54*, 10927–10966.

[159] Witulski, B.; Stengel, T. *Angew. Chem. Int. Ed.* **1998**, *37*, 489–492.

[160] Witulski, B.; Goessmann M. *Chem. Commun.* **1999**, 1879–1880.

[161] Martínez-Esperón, M. F.; Rodríguez, D.; Castedo, L.; Saá, C. *Tetrahedron* **2008**, *64*, 3674–3686.

[162] Tanaka, K.; Takeishi, K.; Noguchi, K. *J. Am. Chem. Soc.* **2006**, *128*, 4586–4587.

[163] Souto, J. A.; Becker, P.; Iglesias, A.; Muniz, K. *J. Am. Chem. Soc.* **2012**, *134*, 15505–15511.

[164] Coste, A.; Karthikeyan, G.; Couty, F.; Evano, G. *Angew. Chem. Int. Ed.* **2009**, *48*, 4381–4385.

[165] Hamada, T.; Ye, X.; Stahl, S. S. *J. Am. Chem. Soc.* **2008**, *130*, 833–835.

[166] Frederick, M. O.; Mulder, J. A.; Tracey, M. R.; Hsung, R. P.; Huang, J.; Kurtz, K. C. M.; Shen, L.; Douglas, C. J. *J. Am. Chem. Soc.* **2003**, *125*, 2368–2369.

[167] DeKorver, K. A.; Walton, M. C.; North, T. D.; Hsung, R. P. *Org. Lett.* **2011**, *13*, 4862–4865.

[168] Chen, X. Y.; Wang, L.; Frings, M.; Bolm, C. *Org. Lett.* **2014**, *16*, 3796–3799.

[169] Zhang, X.; Zhang, Y.; Huang, J.; Hsung, R. P.; Kurtz, K. C. M.; Oppenheimer, J.; Petersen, M. E.; Sagamanova, I. K.; Shen, L.; Tracey, M. R. *J. Org. Chem.* **2006**, *71*, 4170–4177.

[170] Mukherjee, A.; Dateer, R. B.; Chaudhuri, R.; Bhunia, S.; Karad, S. N.; Liu, R.-S. *J. Am. Chem. Soc.* **2011**, *133*, 15372–15375.

[171] Maity, P.; Klos, M. R.; Kazmaier, U. *Org. Lett.* **2013**, *15*, 6246–6249.

[172] Willumstad, T. P.; Haze, O.; Mak, X. Y.; Lam, T. Y.; Wang, Y.-P.; Danheiser, R. L. *J. Org. Chem.* **2013**, *78*, 11450–11469.

[173] Dunetz, J. R.; Danheiser, R. L. *Org. Lett.* **2003**, *5*, 4011–4014.

[174] Zhang, Y.; Hsung, R. P.; Tracey, M. R.; Kurtz, K. C. M.; Vera, E. L. *Org. Lett.* **2004**, *6*, 1151–1154.

[175] Istrate, F. M.; Buzas, A. K.; Jurberg, I. D.; Odabachian, Y.; Gagosz, F. *Org. Lett.* **2008**, *10*, 925–928.

[176] Chen, J.; Ferreira, A. J.; Beaudry, C. M. *Angew. Chem. Int. Ed.* **2014**, *53*, 11931–11934.

4

SYNTHESIS OF ORGANOPHOSPHINES, PHOSPHONATES, AND RELATED COMPOUNDS

4.1 INTRODUCTION TO THE SYNTHESIS OF ORGANOPHOSPHORUS COMPOUNDS GENERATED THROUGH THE FORMATION OF PHOSPHORUS–CARBON BONDS

Alkylphosphines and related compounds remain some of the most widely used organophosphorus reagents. As a result, a variety of synthetic methods have been developed that lead to the formation of these valuable compounds. For most syntheses, the air sensitivity of the products as well as the reagents is one of the most significant problems. To circumvent this issue, four general approaches have been designed (Figure 4.1). Pathway A is the most direct way to form the new organophosphorus compound since only the phosphorus–carbon(sp^3) bond-forming reaction is required along with excellent technique and rigorous exclusion of oxygen during the workup/isolation of the organophosphine. While this approach may not seem challenging if a glovebox is accessible, it is worth noting that all the manipulations including filtrations must be performed inside a glovebox. Furthermore, if the organophosphine requires purification by crystallization, this must be carried out inside the glovebox as well. Since it might take many days/weeks for the new organophosphine to crystallize, it could be challenging to maintain a completely clean and inert atmosphere inside the glovebox since other group members will likely need to use it. If it is not possible or unrealistic to maintain an oxygen-free atmosphere during the entire synthesis/isolation/purification of the organophosphine, pathway B has been a popular alternative. The initial step in this pathway is the same P–C bond-forming reaction from pathway A. However, rather than isolate the new organophosphine directly, this compound is protected prior to isolation through the formation of a phosphine oxide/sulfide or a phosphine-borane. These derivatives are often quite stable to oxygen and can often be isolated and purified without rigorous exclusion of oxygen. The downside to this approach is the need for a deboronation,

Practical Functional Group Synthesis, First Edition. Robert A. Stockland, Jr.
© 2016 John Wiley & Sons, Inc. Published 2016 by John Wiley & Sons, Inc.

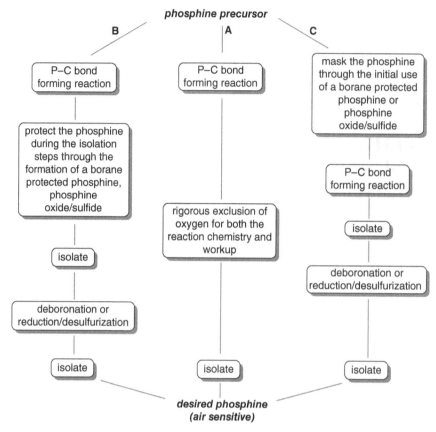

FIGURE 4.1   Common approaches to the synthesis of organophosphines.

reduction, or desulfurization step as well as an additional isolation/purification step. The success of this approach is evidenced by the vast number of organophosphines generated by following route B. The last general approach to the synthesis of air-sensitive organophosphines is summarized by pathway C. This method entails the use of a different phosphorus precursor (phosphine oxide/sulfide or phosphine-borane). This approach is attractive since the reagents and products are often stable to air. A significant issue with this approach is the potential for a different reactivity profile. For example, secondary phosphines can react very differently than secondary phosphine oxides in phosphorus–carbon(sp^3) bond-forming reactions. Similar to route B, the final steps in route C include deboronation, reduction, or desulfurization followed by purification/isolation.

4.2   SYNTHESIS OF ALKYLPHOSPHINES AND RELATED COMPOUNDS THROUGH THE FORMATION OF PHOSPHORUS–CARBON(SP^3) BONDS

Nucleophilic substitution chemistry is one of the classic approaches to the synthesis of phosphorus–carbon(sp^3) bonds. As expected for this type of reaction, primary and benzyl systems with good leaving groups work well. When starting from neutral phosphorus compounds, this is a popular route to the preparation of phosphorus ylides. Analogous reactions with anionic phosphorus reagents afford alkylphosphines. The typical way to generate an anionic phosphorus nucleophile is to treat a primary or secondary phosphine with a non-nucleophilic base. This is typically carried out in solution followed by the addition of the substrate to generate the alkylphosphine. An example of this chemistry was reported by Glueck and described the use of borane-protected primary phosphines as the phosphorus precursors (Example 4.1) [1]. Using sodium hydride as the base, the alkylphosphine was isolated in good yield after purification by column chromatography.

Example 4.1   Preparation of Alkylphosphines through Nucleophilic Substitution Chemistry [1].

Chemical Safety Instructions

Before starting this synthesis, all safety, health, and environmental concerns must be evaluated using the most recent information, and the appropriate safety protocols followed. All appropriate personal safety equipment must be used. All waste must be disposed of in accordance with all current local and government regulations.

The individual performing these procedures and techniques assumes all risks and is responsible for ensuring the safety of themselves and those around them. This chemistry is not intended for the novice and should only be attempted by professionals who have been well trained in synthetic organic chemistry. The authors and Wiley assume no risk and disclaim any liabilities for any damages or injuries associated in any way with the chemicals, procedures, and techniques described herein.

Caution! THF, diethyl ether, ethyl acetate, and hexane are highly flammable.

Caution! Hydrogen gas is highly flammable. These reactions should be carried out on small scales and in a well-ventilated fume hood.

Caution! Sodium hydride reacts violently with moisture and should be handled and stored under an inert atmosphere such as nitrogen or argon.

This reaction was carried out under an atmosphere of nitrogen. $PH_2Ph(BH_3)$ (1.09 g, 8.55 mmol) was dissolved in THF (10 mL) and added slowly to a stirred suspension of NaH (205 mg, 8.55 mmol) in THF (20 mL) at 0 °C. H_2 evolution was observed immediately, and by the end of the addition, a purple-gray solution formed. The solution was stirred at 0 °C for a further 45 min, and then a solution of 2-(2′-iodophenyl)ethyl iodide (3.06 g, 8.55 mmol) in THF (30 mL) was added dropwise over 10 min. The cooling bath was then removed and the reaction mixture was stirred for 18 h at room temperature. The pale green mixture was quenched by addition of H_2O (50 mL), transferred to a separating funnel, and Et_2O (60 mL) was added. The organic layer was then separated and the aqueous layer was washed with additional Et$_2$O (60 mL). The organic extracts were washed with brine (2×50 mL) and dried over MgSO$_4$. The solution was then filtered and all solvent was removed in vacuo to give a pale orange oil. The phosphine-borane (R_f=0.26, 10% ethyl acetate/hexane) was separated from a small amount of iodide starting material by dry-column flash chromatography on silica. A small amount of unreacted 2-(2'-iodophenyl)ethyliodide was eluted with hexane. Elution with 10% ethyl acetate/hexane yielded, after removal of solvent in vacuo, 2.33 g (6.58 mmol, 77%) of the product as a colorless cloudy oil.

Adapted with permission from T. J. Brunker, B. J. Anderson, N. F. Blank, D. S. Glueck, and A. L. Rheingold, *Org. Lett.*, **2007**, *9*, 1109–1112. *Copyright (2007) American Chemical Society.*

One of the oldest and most reliable approaches to the preparation of alkylphosphines is the reaction of chlorophosphines with Grignard and organolithium reagents [2, 3]. These reactions are often carried out at low temperatures to prevent undesired side reactions and typically generate excellent yields of the alkylphosphines. Although this process is often high yielding, the tolerance of this approach to many functional groups is poor. Despite this limitation, a myriad of alkylphosphines have been prepared and isolated using this method. The following sections will highlight representative examples of this approach.

One of the most common substrates for this approach is phosphorus trichloride. Addition of PCl$_3$ to a range of Grignard and organolithium reagents has generated a large number of alkylphosphine compounds. One of the more useful aspects of this chemistry is the ability to control the number of organic groups that will be added through the stoichiometry of the reaction. While there are a few substrate limitations to this approach, many reactions can be devised that will incorporate one, two, or three alkyl groups into the structure of the target compound. The remaining unreacted P—Cl groups can be functionalized further using different functional groups. The modular nature of this approach also makes it possible to build various frameworks with different organic fragments. This has been particularly useful when attempting to design a phosphine with a specific set of steric and electronic properties. Although PCl$_3$ and related chlorophosphines need to be handled and stored under an inert atmosphere such as nitrogen or argon, this remains the most popular approach to the synthesis of alkylphosphines.

In general, the operational difficulty of this reaction is high and the time needed to dry the glassware and reagents as well as purify solvents is moderate to high, since Grignard and organolithium reagents often react violently with moisture. Additionally, many of the phosphorus precursors, intermediates, and products are sensitive to oxygen and can ignite or simply oxidize upon exposure to air. As a result, rigorous exclusion of oxygen and moisture from the glassware, solvents, and reagents will aid in carrying out a successful alkylation reaction. As with many reactions, more time spent preparing to run the reaction will translate into cleaner and more repeatable reactions. Simply using reagent grade solvents could lead to secondary products and significant oxidation. This will require several additional purification steps following alkylation. Typically, the Grignard or organolithium reagent is generated in an inert solvent and added to a solution of the chlorophosphine at low temperature.

As a representative example of this chemistry, the synthesis of a dialkylchlorophosphine is shown in Scheme 4.1 [4]. The Grignard reagent was initially prepared from ^{t}BuCl and magnesium and slowly added to a solution of PCl$_3$. Monitoring the reaction by ^{31}P NMR confirmed the consumption of the starting material, and simply distilling the volatiles afforded the dialkylchlorophosphine. While this appears to be very straightforward, significant operational issues need to be addressed. As mentioned previously, the reaction needs to be carried out in flame-dried glassware with complete exclusion of oxygen and moisture, and all solvents need to be degassed prior to the reaction. Since the resulting chlorophosphine will likely be converted into another compound, it must also be stored under an inert atmosphere. An organolithium route to the preparation of this compound has also been described (Example 4.2) [5].

As mentioned previously, one of the biggest challenges encountered in the synthesis of alkylphosphines using Grignard and organolithium reagents is the reactivity of these reagents towards electrophiles. To circumvent these issues, Knochel generated the organolithium reagents at low temperature and converted them into organozinc compounds prior to the addition of the chlorophosphine [6]. This approach successfully avoided the classic issues encountered in these syntheses and enabled the preparation of a range of alkylphosphines bearing electrophiles including esters, nitriles, alkyl bromides, and chlorides. For ease of purification, the alkylphosphines were converted into the borane adducts following the phosphorus–carbon bond-forming reaction. Using this approach, moderate to excellent yields of the protected alkylphosphine were obtained (up to 86% for a substrate bearing a nitrile).

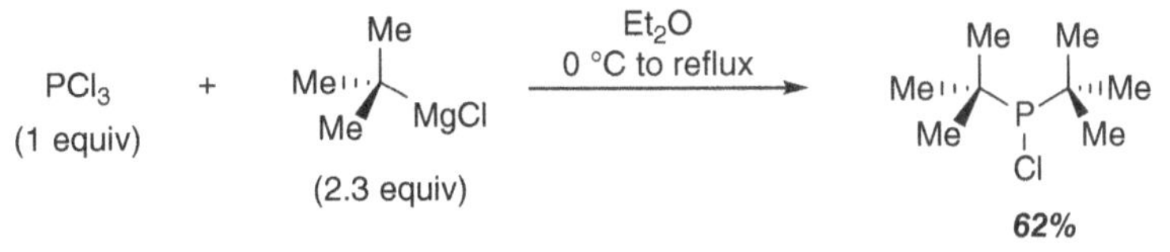

SCHEME 4.1   Preparation of $^tBu_2PCl$ through treatment of PCl_3 with a Grignard reagent [4].

Example 4.2   Preparation of $^tBu_2PCl$ Using an Organolithium Reagent [5].

Before starting this synthesis, carefully follow the chemical safety instructions listed in Example 4.1.

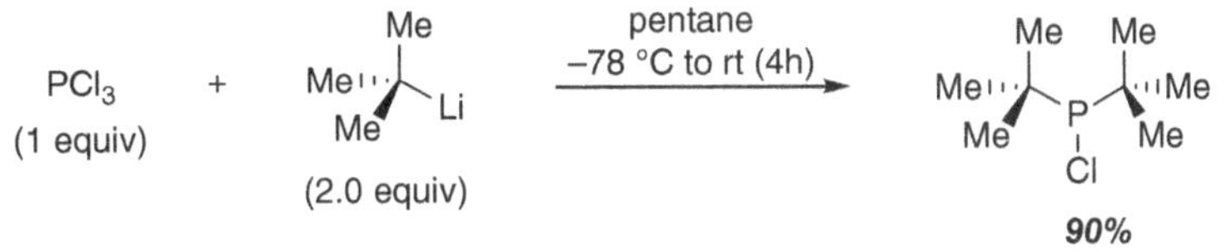

Caution! Organolithium reagents often react violently with moisture.

Caution! ^tBuLi is pyrophoric and must be stored and handled under inert atmosphere.

Caution! The organic solvents used in this synthesis are highly flammable.

Caution! Alkylphosphines and alkylchlorophosphines can react violently with oxygen/ moisture and must be stored and handled under an inert gas such as argon or nitrogen.
The chlorophosphine was prepared by slow addition of 20 mL (34 mmol) of ^tBuLi (1.7 M in pentane) to 17.0 mmol (2.33 g, 1.50 mL) of PCl_3 dissolved in 25 mL of dried, degassed pentane at −78 °C. The reaction was allowed to reach room temperature and then was stirred for an additional 4 h. Pentane was removed by distillation at ambient pressure, and the product was purified by distillation under 0.3 torr vacuum at 47–50 °C to obtain the chlorophosphine in 90% yield.

*Adapted with permission from A. A. Naiini, Y. Han, M. Akinc, J.G. Verkade, Inorg. Chem., **1993**, 32, 5394–5395. Copyright (1993) American Chemical Society.*

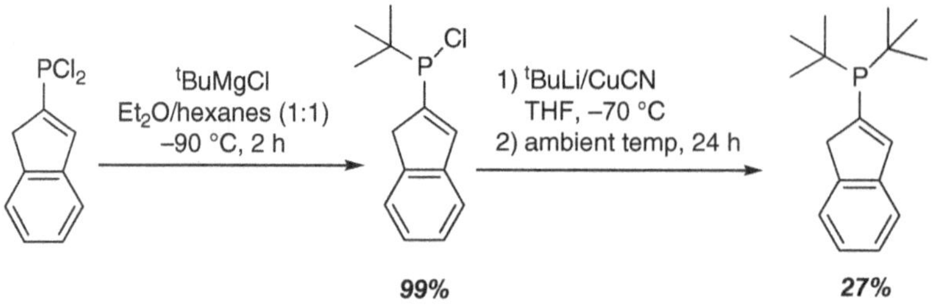

SCHEME 4.2 Modification of the Grignard approach for the preparation of alkylphosphines [7].

One of the common challenges encountered with this chemistry is incomplete consumption of the starting chlorophosphine. This is commonly encountered when bulky groups are present in the reagents. This was encountered during the synthesis of indene containing phosphine ligands (Scheme 4.2) [7]. It would have been reasonable to predict that two equivalents of the Grignard reagent would have added to the indenyldichlorophosphine reagent; however, significant amounts of secondary products were observed when this was attempted. Analysis of the reaction mixture suggested that metallation of the indene ring could have been occurring in addition to the P—C bond-forming reaction. The authors noted that the first equivalent of the Grignard reagent added smoothly to the chlorophosphine in outstanding yield. They modified the Grignard reagent through the formation of a copper complex and were able to obtain the alkylphosphine [8]. The addition of the copper compound facilitated the formation of the desired compound, but it introduced an operational challenge into the reaction. There are three main approaches to introducing solid reagents to an air-sensitive reaction that is already proceeding: (i) The reaction either needs to be carried out inside of a glovebox, where the top can be removed and the copper salt can be simply tipped into the flask. (ii) A high nitrogen or argon feed is introduced into the reaction vessel. The top is removed and the copper salt is tipped into the flask. For dense solids, this is often not problematic; however, for finely divided solids or less dense materials, the high nitrogen/argon purge can blow the solid out of the vessel before it reaches the reaction mixture. (iii) The solid material can be dissolved in a solvent and injected into the reaction vessel. The difficulty with this approach is that the solvent must be compatible with the reaction and rigorously dried and deoxygenated.

If there are small groups on either the Grignard reagent or the chlorophosphine, fully consuming the chlorophosphine is less problematic. A recent example of this entailed the use of a chlorophosphine with only methyl substituents and an adamantyl Grignard reagent (Example 4.3) [9]. Despite the bulky nature of the Grignard reagent, it was successfully added to the chlorophosphine in excellent yield. Although the reaction time was long (reflux overnight), the outstanding yield of this reaction makes this a reasonable approach to the synthesis of these compounds.

The synthesis of moderately bulky alkylphosphines through the use of small organo-lithium reagents has also been very successful (Example 4.4) [7]. Treatment of an indenyl-chlorophosphine with two equivalents of MeLi smoothly generated the alkylphosphine in excellent yield. As discussed previously, attempting to alkylate this chlorophosphine with bulky Grignard or organolithium reagents was unsuccessful due to significant amounts of side reactions. Using the smaller methyllithium reagent, the synthesis afforded the desired dialkylphosphine in excellent yield.

The synthesis of an extremely bulky Buchwald-type dialkylbiarylphosphine ligand was reported using a series of substitution reactions (Scheme 4.3) [10]. The first step in the

Example 4.3 Preparation of an Alkylphosphine Using a Grignard Approach [9].

Before starting this synthesis, carefully follow the chemical safety instructions listed in Example 4.1.

Caution! Grignard reagents often react violently with moisture.

Caution! Me$_2$PCl is pyrophoric and must be handled under inert atmosphere.

Caution! The organic solvents used in this synthesis are highly flammable.

Caution! Alkylphosphines and alkylchlorophosphines can react violently with oxygen/ moisture and must be stored and handled under an inert atmosphere.

This reaction was carried out under argon. A three-necked round-bottomed flask (500 mL) was equipped with a stir bar, a pressure-adjusted dropping funnel, and a condenser topped by an argon inlet. The apparatus was flame dried under argon. The reaction flask was charged with (1-adamantylmethyl) magnesium bromide (127 mL, 0.11 M, 14 mmol), and a solution of chlorodimethylphosphine (Caution! pyrophoric) (1.0 mL, 1.0 g, 10 mmol) in 20 mL of diethyl ether was placed in the addition funnel. The phosphine solution was added at −78 °C over a 45-min period. The reaction mixture (white precipitate, colorless solution) was allowed to warm slowly to room temperature and then heated under reflux overnight. The reaction mixture was cooled to 0 °C, and the reaction was quenched with 100 mL of degassed aqueous NH$_4$Cl [aqueous NH$_4$Cl in distilled water (l/l v/v)]. The ethereal layer was transferred by cannula to an argon-purged flask containing magnesium sulfate. The aqueous layer was extracted once with ether; the extract was combined with the ethereal supernatant. The solution was filtered through a medium porosity frit into an argon-purged flask. The magnesium sulfate was rinsed once with ether; the rinse was similarly filtered into the distillation flask. Solvent was removed under argon. Vacuum distillation afforded the phosphine as a colorless oil, bp 85 °C (0.15 torr), in 93% yield.

Adapted with permission from M. Hackett, G. M. Whitesides, *Organometallics,* **1987**, 6, 403–410. *Copyright (1987) American Chemical Society.*

synthesis entailed the single addition of an adamantyl group to ^{t}BuCl$_2$P using adamantylMgBr. Once this initial substitution was complete, a functionalized aryl Grignard was added to generate the bulky Buchwald-type ligand. Since the addition of the organolithium reagent was sluggish and the reaction needed to be heated to 140 °C for 24 h in a sealed tube to reach completion, there was a significant possibility that the organolithium reagent could metalate other areas of the substrate. To aid in the suppression of unwanted metallation, the authors added cuprous chloride to the reaction mixture to modify the reactivity of the organolithium reagent [8]. While the starting and intermediate chlorophosphines were air sensitive, the final product was not. This observation greatly simplified the purification of the phosphine ligand. It is also noteworthy to mention that this class of alkylphosphines is unusually stable to oxidation from exposure to air. While the overall reaction has a moderate to high level of operational difficulty due to the air sensitivity of the reagents and intermediates, the air stability of the products

Example 4.4 Synthesis of Alkylphosphines through Alkylation with Organolithium Reagents [7].

Before starting this synthesis, carefully follow the chemical safety instructions listed in Example 4.1.

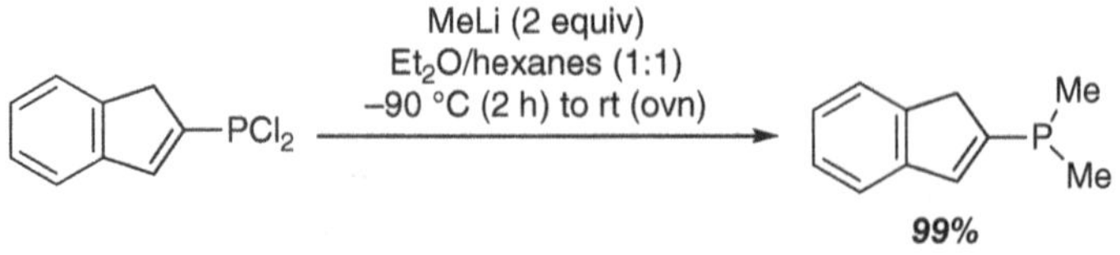

Caution! organolithium reagents often react violently with moisture.

Caution! Alkylphosphines and alkylchlorophosphines can react violently with oxygen/moisture.

Caution! The organic solvents used in this synthesis are highly flammable.

To a solution of the chlorophosphine (2.17 g, 10 mmol) in 60 mL of diethyl ether-hexanes (1 : 1, vol) was added dropwise MeLi (10.9 mL of 1.84 M solution in ether) in ether under vigorous stirring for 2 h at −90 °C. The reaction mixture was slowly warmed to ambient temperature, stirred overnight, and then filtered through a glass frit. The precipitate was additionally washed with 10 mL of ether. The combined extract was evaporated to dryness, and the residue was dried in a vacuum to afford the phosphine (1.76 g, 99%) as a colorless crystalline solid.

Adapted with permission from D. N. Kazulkin, A. N. Ryabov, V. V. Izmer, A. V. Churakov, I. P. Beletskaya, C. J. Burns, A. Z. Voskoboynikov, *Organometallics,* **2005**, *24*, 3024–3035. *Copyright (2005) American Chemical Society.*

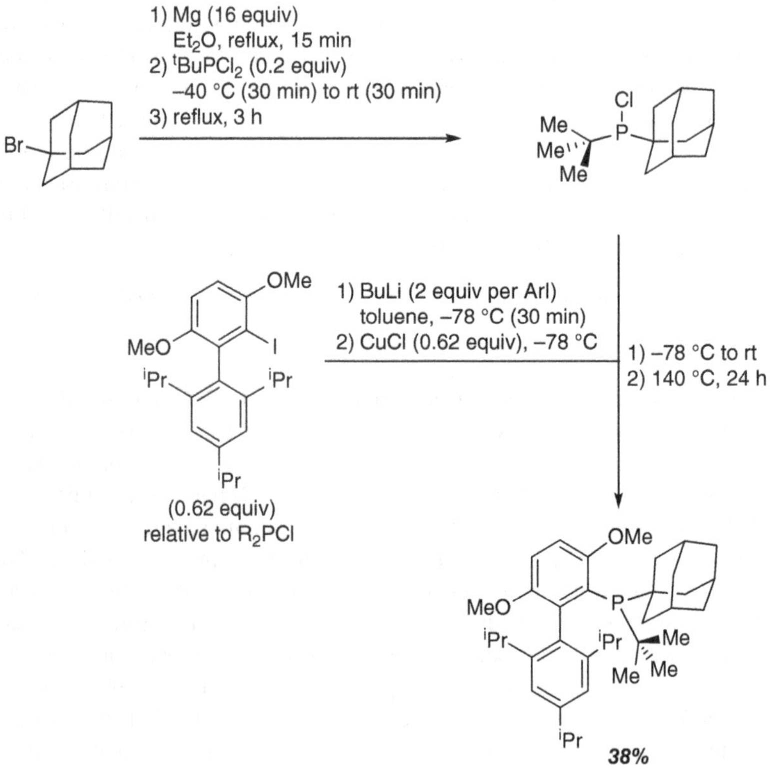

SCHEME 4.3 Preparation of a bulky alkylphosphine using a Grignard approach [10].

makes their isolation and use in catalysis quite practical. In related work, the preparation of a simple JohnPhos-type ligand has been described using a tandem approach (Example 4.5) [11].

In addition to the synthesis of achiral alkylphosphines, a number of approaches to the preparation of P-chiral phosphines have been explored using Grignard and organolithium approaches (Examples 4.6 and 4.7) [12, 13]. In general, the selectivity is higher when the reactions were carried out at low temperatures.

Example 4.5   Three-Step Approach to the Preparation of Methyl JohnPhos [11].

Before starting this synthesis, carefully follow the chemical safety instructions listed in Example 4.1.

Caution! Grignard reagents react violently with moisture.

Caution! Alkylphosphines can react violently with oxygen/moisture.

Caution! Several of the organic solvents used in this synthesis are highly flammable.

Step 1: A 250 mL round-bottomed flask fitted with a reflux condenser was charged with 2-bromobiphenyl (2.0225 g, 8.6 mmol), diisopropylethylamine (2.0700 g, 18 mmol), dimethylphosphonate (5.0800 g, 46 mmol), palladium(II) acetate (190.5 mg, 0.8 mmol), and 1,3-bis(diphenylphosphino)propane (696.3 mg, 1.7 mmol). The reaction was dissolved in toluene (100 mL) and heated to reflux for 36 h. The reaction was cooled to room temperature and diluted with EtOAc (100 mL). The organic layer was washed with saturated NaHCO$_3$ (3×100 mL) solution and then once with brine (100 mL). The organic layer was dried over Na$_2$SO$_4$, filtered, and the solvent was removed under vacuum to afford a yellow oil with residual white solids. The crude product was purified by flash column chromatography (60 mesh silica, EtOAc with 1% triethylamine, R_f=0.6), yielding a hygroscopic crystalline, white solid (1.5868 g, 71.1% yield).

Step 2: To a solution of dimethyl [1,1′-biphenyl]-2-ylphosphonate (95.1 mg, 1.8 mmol) and sodium trifluoromethanesulfonate (150.0 mg, 4.2 mmol) in THF (4 mL) at 0°C was added methylmagnesium bromide (0.32 mL, 3.0 M, 4.0 mmol) dropwise over 5 min. The reaction was then brought to reflux for 2 h. The reaction mixture was cooled to 0 °C and quenched with 0.1 M H_2SO_4 (4 mL). The reaction mixture was washed with dichloromethane (3×4 mL); then the organics were collected and dried over Na_2SO_4. The mixture was filtered and dried under vacuum to afford the phosphine oxide as a crystalline solid (72.6 mg, 87% yield).

Step 3: Under an inert atmosphere of N_2, [1,1′-biphenyl]-2-yldimethylphosphine oxide (368.0 mg, 1.6 mmol) was added to diisobutylaluminum hydride (1.2116 g, 8.0 mmol) and toluene (5 mL) in a heavy-walled unregulated pressure vessel. The reaction mixture was allowed to vent overnight before being sealed and heated to 150 °C for 12 h. The reaction mixture was allowed to cool and then (under an atmosphere of N_2) quenched with degassed 0.1 M HCl and diluted with diethyl ether (20 mL). The aqueous layer was washed with ether (3×20 mL). The collected organics were dried over Na_2SO_4 and then filtered and dried under vacuum, affording a very pure [1,1′-biphenyl]-2-yldimethylphosphine as a colorless, crystalline solid (288.3 mg, 84% yield).

Adapted with permission from A. J. Kendall, C. A. Salazar, P. F. Martino, D. R. Tyler, *Organometallics*, **2014**, *33*, 6171–6178. *Copyright (2014) American Chemical Society.*

Example 4.6   Preparation of a P-Chiral Bisphosphine through Alkylation of a Diphosphine [12].

Before starting this synthesis, carefully follow the chemical safety instructions listed in Example 4.1.

Caution! Grignard reagents often react violently with moisture.

Caution! Alkylphosphines can react violently with oxygen.

Caution! The organic solvents used in this synthesis are highly flammable.

A solution of enantioenriched (*S,S*)-benzodiphosphetane (25 mg, 0.10 mmol, er = 10:1) in 5 mL of CH_2Cl_2 was treated with methyl triflate (11 µL, 0.10 mmol). After 1 h, the solvent was removed in vacuo and the residue was redissolved in Et_2O (5 mL). The solution was cooled to −78 °C and treated with MeMgBr (34 µL of a 3.0 M solution in Et_2O, 0.10 mmol) and then allowed to warm to room temperature. After stirring for 2 h, a white precipitate had formed, and the ^{31}P NMR spectrum showed conversion to BenzP*. The solvent was removed in vacuo, then the residue was dissolved in Et_2O, and the solution was passed over a pipet silica plug. The solvent was removed under vacuum to give 13 mg (46%) of an oily white residue.

Adapted with permission from S. C. Reynolds, R. P. Hughes, D. S. Glueck, A. L. Rheingold, *Org. Lett.*, **2012**, *14*, 4238–4241. *Copyright (2012) American Chemical Society.*

Example 4.7   Preparation of JosiPhos Ligands through Substitution Chemistry [13].

Before starting this synthesis, carefully follow the chemical safety instructions listed in Example 4.1.

Step 1

74%

Caution! ^{t}BuLi is pyrophoric and must be handled under an inert atmosphere.

Caution! Pentane and hexane are highly flammable.

Caution! Chlorophosphines can react violently with moisture.

In an oven-dried Schlenk tube, (R)-[1-(dimethylamino)ethyl]ferrocene (Ugi's amine) (4.024 g, 15.652 mmol, 1 equiv) was dissolved in dry diethyl ether (to a concentration of 1–2 M), and this solution was cooled to −78 °C. To this solution, *tert*-butyllithium (12 mL, 1.6 M in pentane, 19.2 mmol, 1.2 equiv) was added dropwise with a syringe. The reaction mixture was stirred at this temperature for 30 min. Then the cooling bath was removed to allow the mixture to warm to room temperature. It was stirred for a further 60 min leading to the formation of a red-orange precipitate. The same amount of ether used before was added and the mixture was stirred for another 60 min giving a clear solution. The solution was cooled again to −78 °C and the corresponding secondary phosphane chloride (3 mL, 16.652 mmol, 1.1 equiv) was added dropwise with a syringe. The reaction was warmed to room temperature overnight. To the resulting orange suspension, a saturated solution of Na_2CO_3 was added slowly to quench the reaction. The biphasic mixture was separated in a separatory funnel, and the organic phase was washed with water and brine and dried with $MgSO_4$ and then filtered and concentrated to dryness in vacuo to give a sticky orange solid. The crude product recrystallized at −20°C from a saturated dichloromethane/hexane solution to afford the ferrocenylphosphane ligand as orange crystals (5.13 g, 74%).

Step 2

Before starting this synthesis, carefully follow the chemical safety instructions listed in Example 4.1.

96%

Caution! $HPCy_2$ is pyrophoric and must be handled under an inert atmosphere.

Caution! Allkylphosphines can react violently with air.

Caution! Glacial acetic acid must be used in a well ventilated fume hood.

In an oven-dried Schlenk tube, the ferrocenylphosphane prepared in step 1 (1.394 g, 3.16 mmol, 1 equiv) was dissolved in degassed glacial acetic acid (to a concentration of 0.5–1 M). To this solution, the dicyclohexylphosphine (0.7 mL, 3.46 mmol, 1.1 equiv) was added, and the mixture was stirred at 80 °C for 3 h. The solvent and volatiles were removed in vacuo to give an orange sticky solid. It was dissolved in dichloromethane and washed successively with a saturated Na_2CO_3 solution, brine, and water, dried with $MgSO_4$, and filtered. It was concentrated to dryness with a rotary evaporator to give the crude product. The crude product was purified by recrystallization from a saturated dichloromethane solution at −20°C to afford the JosiPhos ligand as orange crystals (1.813 g, 96%).

Adapted with permission from E. Mejía, R. Aardoom, A. Togni, *Eur. J. Inorg. Chem.*, **2012**, *5021–5032. Copyright (2012) Wiley-VCH Verlag GmbH & Co. KGaA, Weinheim.*

Example 4.8 Synthesis of P-Chiral Phosphines from a Borane-Protected Phosphinite [14].

Before starting this synthesis, carefully follow the chemical safety instructions listed in Example 4.1.

66% (92% ee)

Caution! THF, hexane, and ethyl acetate are highly flammable.

Caution! ^{t}BuLi is a pyrophoric compound and must be handled under an inert atmosphere such as nitrogen or argon.

This reaction was carried out under an atmosphere of argon. A 1.7 M hexane solution of *t*-BuLi (48.2 mL, 0.082 mol, 3 equiv vs. the phosphinite) was added slowly to a solution of the phosphinite (4.59 g, 0.027 mol) in THF (60 mL) cooled at −78 °C, and the resulting solution was then allowed to reach room temperature overnight. The reaction was quenched with H_2O (60 mL), THF was evaporated, and the product was extracted with CH_2Cl_2 and dried with $MgSO_4$. Evaporation of the solvent and flash chromatography (silica gel, hexane/ethyl acetate 95 : 5) gave the borane-protected phosphine as a white solid (46 g, 66%).

Adapted with permission from F. Maienza, F. Spindler, M. Thommen, B. Pugin, C. Malan, A. Mezzetti, *J. Org. Chem.*, **2002**, *67, 5239–5249. Copyright (2002) American Chemical Society.*

In addition to the use of chlorophosphines as substrates, phosphinates have also successfully been used to generate P—C(sp^3) bonds. Essentially, the RO group on the phosphinate was replaced by the organic fragment from the organolithium reagent. Mezzetti used this approach to generate a range of chiral phosphine ligands for use in asymmetric hydrogenation reactions (Example 4.8) [14]. The operational considerations were similar to the reactions described previously when chlorophosphines were used; however, the phosphinate reagents can be less sensitive than chlorophosphines. To ensure that no oxidation would occur, the authors generated the borane adduct prior to the alkylation reaction.

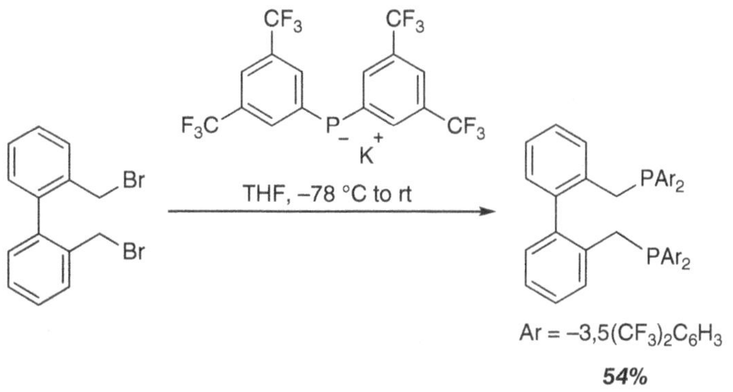

SCHEME 4.4   Synthesis of alkylphosphines using a potassium phosphide [16].

As noted by the examples listed previously, one of the most popular approaches to these reactions begins with an alkyl halide and proceeds by the addition of magnesium to generate a Grignard reagent followed by the chlorophosphine to generate the alkylphosphine. When the substrate is appropriately designed, an ortho-lithiation strategy for the generation of an organolithium reagent has also been successful. An example of this chemistry was recently reported by Beller and utilized 2,2′-dimethylbinaphthyl as a substrate [15]. Addition of two equivalents of BuLi along with TMEDA directly lithiated the methyl groups, and addition of one equivalent of a chlorophosphine generated the alkylphosphine. Isolated yields of the resulting chiral monodentate phosphines were moderate to good (60–83%).

The treatment of alkyl halides with group one metal phosphides can also be used to generate alkylphosphines (Scheme 4.4) [16–19]. This is the reverse of the common approach where the nucleophilic fragment is the Grignard or organolithium species. Generating and using a phosphide as the nucleophile has been successful in a number of cases. This reaction can be envisioned as a classic substitution reaction and will be most successful with substrates that are also susceptible to S_N2-type substitution. As a result, benzyl and primary halides are quite amenable to this approach and have been well studied [16, 17, 19]. As representative example, a group of pincer ligands were generated by initial deprotonation of a borane-protected secondary phosphine followed by addition of the alkyl halide (Example 4.9) [20]. The borane-protected secondary phosphine was used in the reaction to render the bisphosphine stable to air for workup. In this example, triflic acid/toluene followed by KOH/EtOH was used to remove the borane and generate the pincer ligand. In related work, *trans*-1,4-diphosphacyclohexane has been constructed from borane-protected 1,2-diphosphaethanes (Scheme 4.5) [21].

The metal-catalyzed alkylation of primary and secondary phosphines as well as chloro-phosphines remains a valuable approach to the synthesis of alkylphosphines. A range of transition metals have been used to promote these reactions. In addition to the preparation of achiral alkylphosphines, several P-chiral alkylphosphines have been synthesized through cross-coupling chemistry. A room-temperature copper-catalyzed alkylation of secondary phosphines has been reported by Glueck (Example 4.10) [22]. While many other copper-catalyzed reactions used a copper salt such as copper iodide along with a solubilizing ligand to generate a copper complex in solution, his approach opted for the use of a discrete copper species containing a large bite-angle bisphosphine (XantPhos) as the supporting ligand. One of the advantages to using the preformed copper complex was the ability to fully characterize the complex prior to addition to the catalytic reaction.

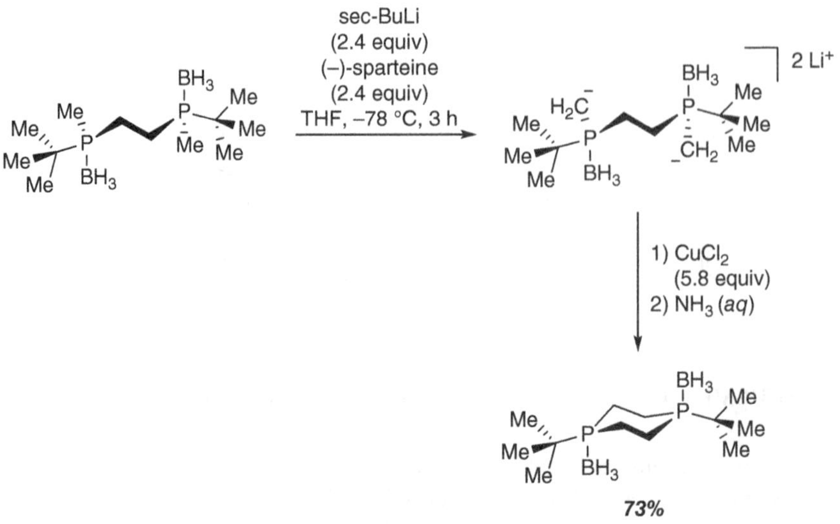

SCHEME 4.5   Synthesis of borane-protected bisphosphine [21].

Example 4.9   Using Lithium Phosphides to Prepare Alkylphosphines [20].

Before starting this synthesis, carefully follow the chemical safety instructions listed in Example 4.1.

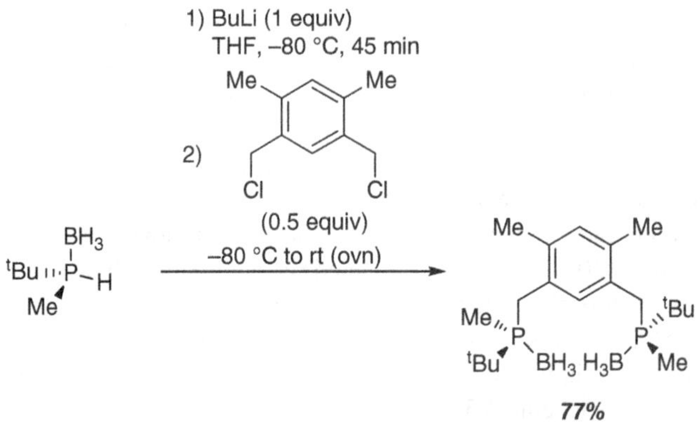

Caution! THF, hexane, and ethyl acetate are highly flammable.

Caution! Organolithium reagents can react violently with moisture and must be handled under an inert atmosphere such as nitrogen or argon.

This reaction was carried out under argon. To a stirred, cooled (−80 °C) solution of (*S*)-*tert*-butylmethylphosphine-borane (3.89 g, 33 mmol) in THF (33 mL) was slowly added *n*-BuLi (21.4 mL of 1.62 M hexane solution, 34.7 mmol) during 15 min under argon. After 30 min, a solution of 1,5-bis(chloromethyl)-2,4-dimethylbenzene (3.05 g, 15 mmol) in THF (15 mL) was added all at once, and the mixture was allowed to warm to room temperature. After stirring overnight, the reaction mixture was poured into a mixture of ethyl acetate (40 mL) and water (70 mL). The organic layer was separated and the aqueous layer was extracted with ethyl acetate (30 mL). The combined extracts were washed with a saturated NaCl solution, dried over Na_2SO_4, and evaporated. The residual pasty oil was triturated with *n*-hexane and the resulting solid material was collected on a glass filter and washed with *n*-hexane to give practically pure (*S*,*S*)-1,5-bis((boranato(*tert*-butyl)methylphosphino) methyl)-2,4-dimethylbenzene (4.24 g, 77%). Pure compound was obtained by recrystallization from hot ethyl acetate/*n*-hexane (1 : 3).

Adapted with permission from B. Ding, Z. Zhang, Y. Xu, Y. Liu, M. Sugiya, T. Imamoto, W. Zhang, *Org. Lett,* **2013**, *15*, 5476–5479. *Copyright (2013) American Chemical Society.*

Example 4.10   Copper-Catalyzed Alkylation of Secondary Phosphines [22].

Before starting this synthesis, carefully follow the chemical safety instructions listed in Example 4.1.

Caution! THF and petroleum ether are highly flammable.

Caution! Secondary phosphines can react violently with oxygen.

This reaction was carried out under an atmosphere of nitrogen. A slurry of [Cu(XantPhos) (NCMe)][PF$_6$] (22 mg, 0.027 mmol) in less than 1 mL of THF was treated with a solution of PHPh$_2$ (50 mg, 0.27 mmol) in 2 mL of THF and then with PhCH$_2$Br (32 μL, 0.27 mmol), resulting in a homogeneous solution. A solution of NaOSiMe$_3$ (30 mg, 0.27 mmol) in 2 mL of THF was added, resulting in a bright yellow solution and precipitation. Within a minute, the yellow color had subsided and the solution was clear with a large amount of precipitate. ^{31}P NMR spectroscopy showed full conversion to Ph$_2$PCH$_2$Ph in less than 15 min; the reaction mixture was pumped down under vacuum. The residue was dissolved in 10 mL of a 10% THF/petroleum ether solution, the solution was passed through a silica plug, and the resulting solution was pumped down again giving a white solid (73 mg, 0.26 mmol, 98%).

Adapted with permission from M. F. Cain, R. P. Hughes, D. S. Glueck, J. A. Golen, C. E. Moore, A. L. Rheingold, *Inorg. Chem.*, **2010**, *49*, 7650–7662. *Copyright (2010) American Chemical Society.*

The synthesis of the copper complex was straightforward and consisted of simply stirring [Cu(MeCN)$_4$]PF$_6$ with the commercially available bisphosphine ligand in THF. The coupling reaction proceeded in excellent yield using benzyl bromide as the substrate. In addition to the use of brominated substrates, benzyl chloride was also successfully used in the coupling reaction. The authors noted that when dichloromethane was used as the solvent, the copper complex activated the solvent towards HPPh$_2$, and Ph$_2$PCH$_2$Cl was generated. For this second reaction, the most active catalyst for this reaction was a triphos-ligated copper complex (Example 4.11). These approaches to the synthesis of alkylphosphines were quite fast at room temperature and generated near-quantitative yields of the alkylphosphines.

The synthesis of P-chiral phosphines has been achieved through the use of ruthenium complexes supported by chiral ligands (Schemes 4.6 and 4.7) [23]. The chemistry was carried out under mild conditions and afforded excellent yields and outstanding selectivities for both mono- and bisphosphines as their borane adducts. A critical parameter in this chemistry was the selection of the base for the chemistry. While common metal alkoxides such as NaOEt were unsuccessful, increasing the length of the carbon chain increased the solubility of the alkoxide in organic solvents and promoted the reaction.

Example 4.11   Copper-Catalyzed Phosphination of Dichloromethane [22].

Before starting this synthesis, carefully follow the chemical safety instructions listed in Example 4.1.

Caution! THF and petroleum ether are highly flammable.

Caution! Secondary phosphines can react violently with oxygen.

This reaction was carried out under an atmosphere of nitrogen. A solution of [Cu(triphos)(NCMe)][PF$_6$] (23 mg, 0.027 mmol, 10 mol%) and PHPh$_2$ (50 mg, 0.27 mmol) in 2 mL of CH$_2$Cl$_2$ was added to a stirring solution of NaOSiMe$_3$ (30 mg, 0.27 mmol) in 1 mL of CH$_2$Cl$_2$, resulting in a flash of yellow color, which dissipated within a minute. After 25 min, all of the PHPh$_2$ had been consumed, according to ^{31}P NMR monitoring. The solution was pumped down under vacuum, the residue was dissolved in 10 mL of 10% THF/petroleum ether, and this solution was passed through a silica plug. The filtrate was pumped down, giving a clear oil (54 mg, 0.23 mmol, 86%).

Adapted with permission from M. F. Cain, R. P. Hughes, D. S. Glueck, J. A. Golen, C. E. Moore, A. L. Rheingold, *Inorg. Chem.*, **2010**, *49*, 7650–7662. *Copyright (2010) American Chemical Society.*

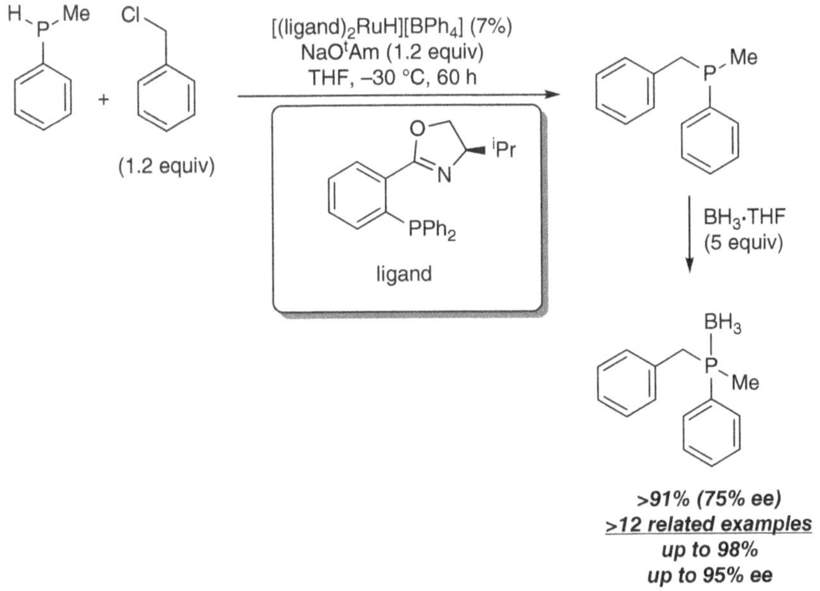

SCHEME 4.6   Ruthenium-catalyzed synthesis of P-chiral phosphines [23].

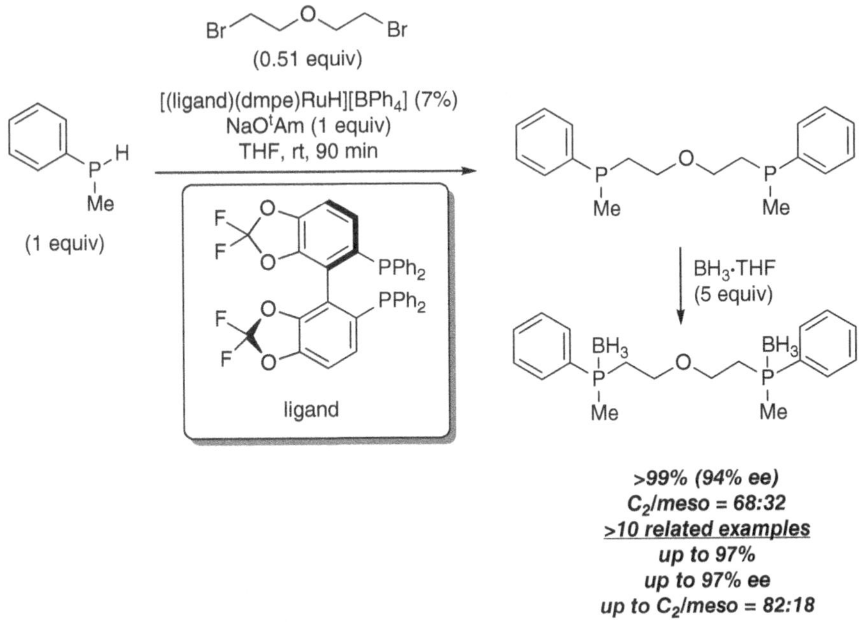

SCHEME 4.7   Ruthenium-catalyzed synthesis of bisphosphines [23].

While the vast majority of the phosphorus–carbon(sp^3) bond-forming reactions begin with hazardous HPR$_2$ precursors, Beller presented a clever solution to the challenges related to the initial storage and manipulation of the phosphine reagents (Scheme 4.8) [24]. The authors opted to use an air-stable and easily handled phosphine oxide in the reaction. Once the atmosphere was exchanged, they used a silane (TMDS) to reduce the phosphine oxide to the phosphine and a copper salt to promote the phosphorus–carbon bond-forming reaction. In addition to the preparation of monodentate phosphines, the preparation of bisphosphines was also achieved using this approach (Example 4.12) [24].

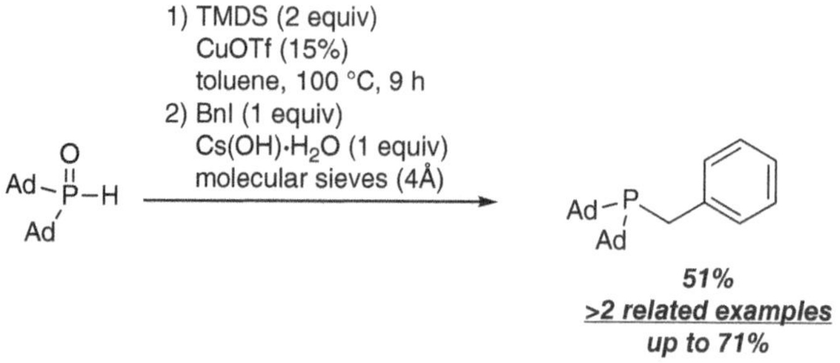

SCHEME 4.8   Synthesis of alkylphosphines from secondary phosphine oxides [24].

Example 4.12   Synthesis of Bisphosphines through Tandem Reduction/Coupling Reactions [24].

Before starting this synthesis, carefully follow the chemical safety instructions listed in Example 4.1.

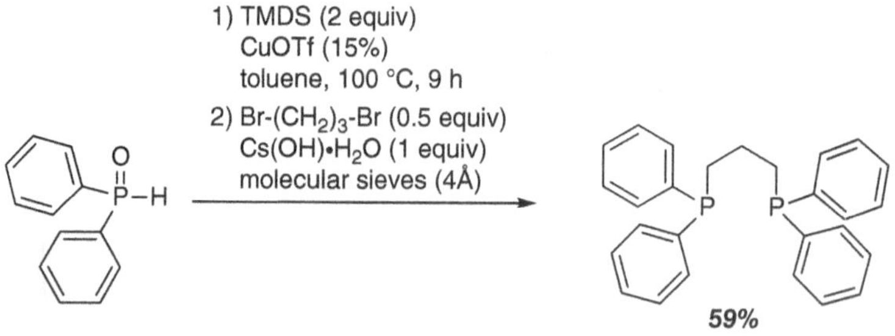

Caution! Alkylphosphines can react violently with oxygen.

Caution! Ethyl acetate is highly flammable.

A 10 mL dried Schlenk tube containing a stirring bar was charged with Cu(OTf)$_2$ (13.5 mg, 0.038 mmol) and diphenylphosphine oxide (50.0 mg, 0.25 mmol). Under argon flow, dry toluene (2 mL) and TMDS (90 μL, 0.5 mmol) were added, and then the mixture was stirred at 100 °C for 9 h. Then, the reaction mixture was cooled to room temperature, and N,N-dimethylformamide (degassed, 2 mL), Cs(OH).H$_2$O (43 mg, 0.25 mmol), and 4A molecular sieves (50 mg) were added under argon flow. After the suspension was stirred for 1 h at room temperature, 1,3-dibromopropane (12 μL, 0.125 mmol) was added under argon flow.

The suspension was heated to 100 °C and stirred for 24 h. Then, the reaction mixture was cooled to 0 °C and methanolic KOH (3 mL, 3 N) was added slowly. After stirring the mixture vigorously for 5 h at room temperature, water (3 mL) was added and the mixture was extracted by ethyl acetate (2×25 mL). The organic phase was combined and washed by HCl solution (aq., 3 mL, 1 N) and saturated NaHCO$_3$ solution (aq., 3 mL). Then, the organic phase was dried over Na$_2$SO$_4$ concentrated under vacuum. The residue was purified by column chromatography.

Adapted with permission from Y. Li, S. Das, S. Zhou, K. Junge, M. Beller, *J. Am. Chem. Soc.*, **2012**, *134*, 9727–9732. *Copyright (2012) American Chemical Society.*

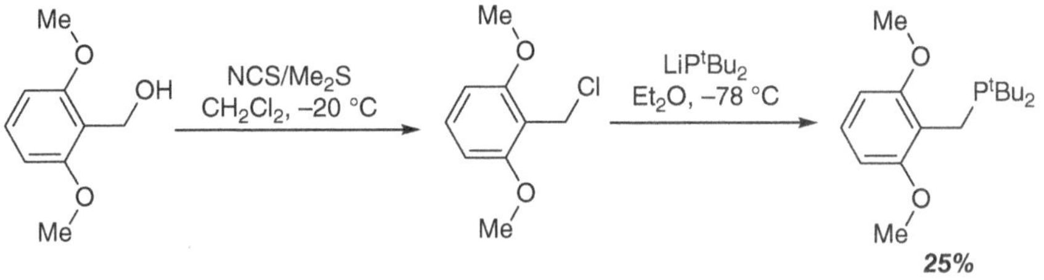

SCHEME 4.9 Synthesis of an alkylphosphine from a benzyl alcohol [25].

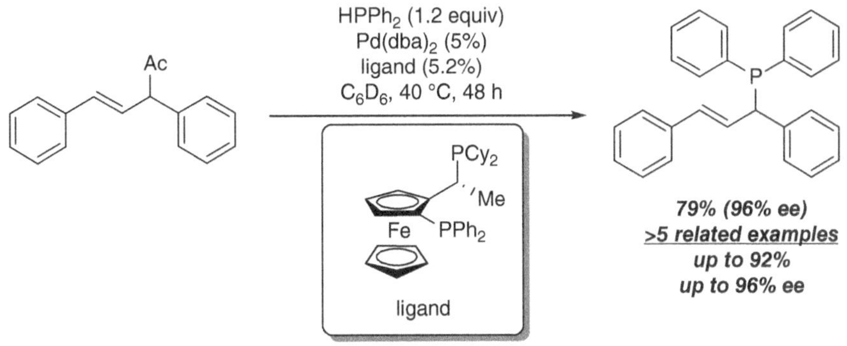

SCHEME 4.10 Palladium-catalyzed asymmetric synthesis of C-chiral monodentate phosphines [26].

Alkylphosphines can also be generated from benzyl alcohols through a two-step process (Scheme 4.9) [25]. The first step converted the benzyl alcohol into a benzyl chloride through the addition of NCS/Me$_2$S. This intermediate was converted into the alkylphosphine through the addition of a lithium phosphide.

The asymmetric allylic phosphination of allylic acetate derivatives proceeded under mild conditions to afford chiral phosphines in good to excellent yields (Scheme 4.10) [26]. The catalyst system was comprised of commercially available materials including a common palladium source and JosiPhos as the chiral ligand. The enantioselectivity of the process was outstanding (up to 96% ee), and the conditions were quite mild (40 °C). While only a single alkyl phosphine was screened, these results provided the proof of concept for the process, and extending the chemistry to more complex substrates and phosphorus compounds is likely to be straightforward due to the mild conditions as well as the high level of functional group tolerance commonly afforded by the use of palladium catalysts.

Elegant work by Glueck demonstrated the effectiveness of platinum and related catalysts in phosphorus–carbon bond-forming reactions [27–36]. As a representative example, the coupling of a secondary phosphine with benzyl halides was promoted by a platinum phosphide species bearing dupos as the supporting ligand (Scheme 4.11 and Example 4.13) [28]. The reaction time was a bit long (1 week for the anthracene derivative). In related work, the asymmetric phosphination of allylic acetates has been reported (Example 4.14) [26].

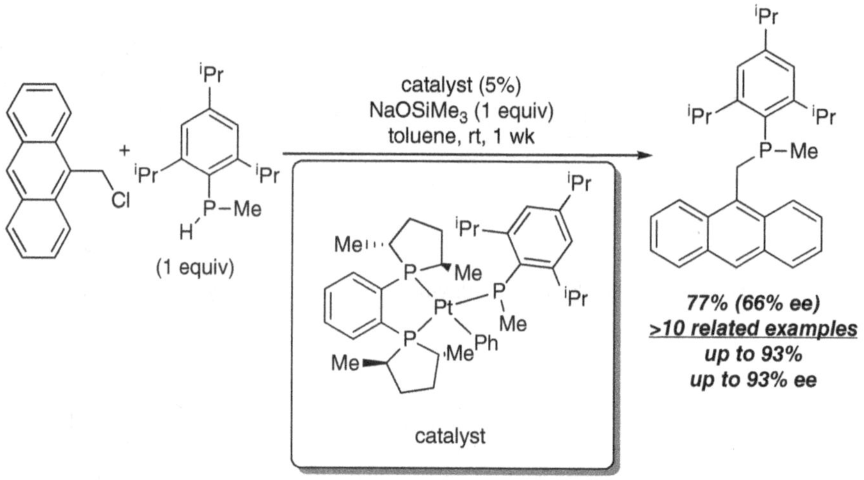

SCHEME 4.11   Platinum-catalyzed phosphorus–carbon bond-forming reactions [28].

Example 4.13   Preparation of an Enantioenriched Phosphine [28].

Before starting this synthesis, carefully follow the chemical safety instructions listed in Example 4.1.

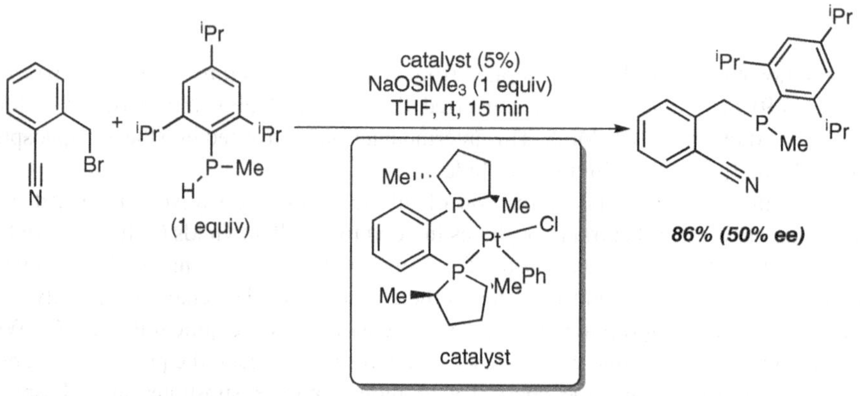

Caution! THF and petroleum ether are highly flammable.

To PHMe(Is) (25 mg, 0.1 mmol) in 0.1 mL of THF was added $NaOSiMe_3$ (11.2 mg, 0.1 mmol) suspended in 0.2 mL of THF. The mixture was added to Pt(Me-Duphos)(Ph)(Cl) (3.1 mg, 0.005 mmol) as the catalyst precursor (5 mol%) in 0.1 mL of THF and to 2-(bromomethyl)benzonitrile (19.6 mg, 0.1 mmol) in 0.1 mL of THF. The reaction mixture was transferred to an NMR tube and monitored by ^{31}P NMR spectroscopy. An increasing amount of white precipitate was observed as the reaction progressed. The reaction went to completion in approximately 15 min. The catalyst and NaBr were removed from the reaction mixture on a silica column (5 cm height, 0.6 cm diameter), using a 9:1 petroleum ether:THF mixture as eluent. The catalyst and NaBr did not elute. The solvent was removed under vacuum and 31 mg (86%) of colorless liquid (~50% ee) was isolated.

Adapted with permission from C. Scriban, D. S. Glueck, *J. Am. Chem. Soc.*, **2006**, *128*, 2788–2789. *Copyright (2006) American Chemical Society.*

Example 4.14   Asymmetric Phosphination of Allylic Acetates [26].

Before starting this synthesis, carefully follow the chemical safety instructions listed in Example 4.1.

Caution! Secondary phosphines can be pyrophoric.

Caution! Benzene is highly flammable.

1,3-Diphenylallyl acetate (29 mg, 0.115 mmol, 1.00 equiv) was added to a mixture of JosiPhos (4.8 mg, 8.046 mmol, 0.07 equiv) and [Pd-(dba)$_2$] (3.3 mg, 5.747 mmol, 0.05 equiv) in 1 mL of the chosen solvent, and the orange solution was stirred for 30 min. The secondary phosphine (0.115 mmol, 1.00 equiv) was then added and the mixture was heated to 40 °C in an oil bath. To remove the catalyst and by-products, the reaction solution was worked up in a glovebox by filtration through a short plug of silica gel (0.5 g) using toluene as an eluent. After removing the solvent in vacuo, the product was isolated as a colorless solid.

Adapted with permission from P. Butti, R. Rochat, A. D. Sadow, A. Togni, *Angew. Chem. Int. Ed.*, **2008**, *47*, 4878–4881. *Copyright (2008) Wiley-VCH Verlag GmbH & Co. KGaA, Weinheim.*

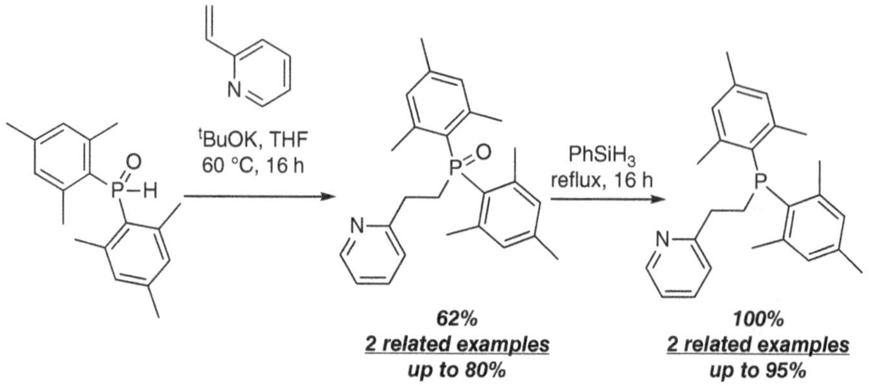

62%
2 related examples
up to 80%

100%
2 related examples
up to 95%

SCHEME 4.12   Alkoxide-mediated addition of secondary phosphine oxides to alkenes [55].

The addition of primary and secondary phosphines to alkenes is an atom-efficient approach to the formation of P—C(sp^3) bonds. Classically, this reaction has been promoted by radical initiators, bases, and metal catalysts. These approaches have been summarized in several reviews [37–44]. When activated alkenes such as acrylates are used as substrates, the reactions are typically referred to as *phospha*-Michael additions. Reactions that promote the addition to unactivated alkenes such has hexane are described as hydrophosphination reactions. The majority of the reactions that have been reported target the formation of neutral phosphines due to their value as ligands for transition metals. In addition, a range of addition reactions between tertiary phosphines and Michael acceptors leading to the formation of phosphonium salts has also been reported. The following sections will highlight representative reactions and example preparations of each class of reactions.

Due to the possibility of spontaneous ignition, great care must be taken when using primary and secondary phosphines in chemical reactions. While some of the substrates described later may not immediately ignite in air, virtually all of them will oxidize. Since phosphines are the targets of these reactions, this oxidation needs to be prevented unless a subsequent reduction step is planned.

None of the preparations listed herein will deal with the use of phosphine (PH$_3$) as a reagent. Although it has a rich history in the formation of phosphorus–carbon bonds [45–50], it remains an extremely dangerous reagent. With a boiling point below −50 °C and a propensity to spontaneously ignite in air, most common synthetic laboratories are not equipped to safely manipulate this compound.

The use of strong bases to promote the hydrophosphinylation of unactivated alkenes is one of the classic approaches to the synthesis of these compounds [39, 51–54]. Using this approach, the ^{t}BuOK-promoted hydrophosphinylation of vinylpyridines using secondary phosphine oxides was achieved (Scheme 4.12) [55]. Phosphine oxides were used as the substrate in this reaction to facilitate the isolation of an air-stable product that could be easily purified. The chemistry generated the anti-Markovnikov regioisomer, and reduction of the phosphine oxide with phenylsilane afforded the desired organophosphine in moderate to excellent yields. One of the most notable aspects of this chemistry was the ability to tolerate extremely bulky groups on the phosphorus reagent. In many hydrophosphinylation reactions, even incorporating a single ortho-substituent dramatically reduced the effectiveness of the reaction; however, using this

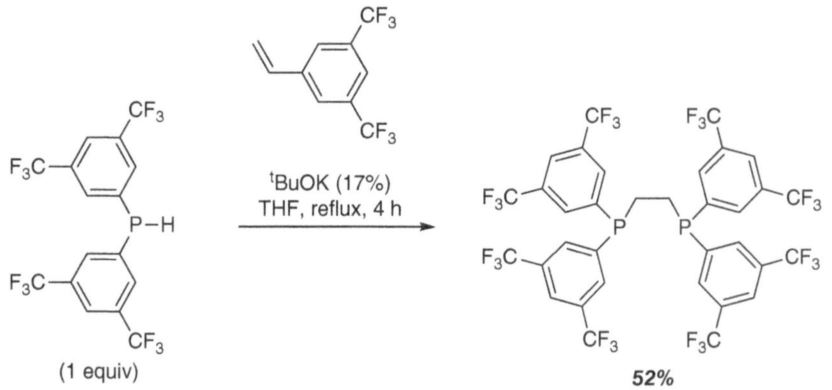

SCHEME 4.13 Synthesis of bisphosphines bearing electron-withdrawing groups [16].

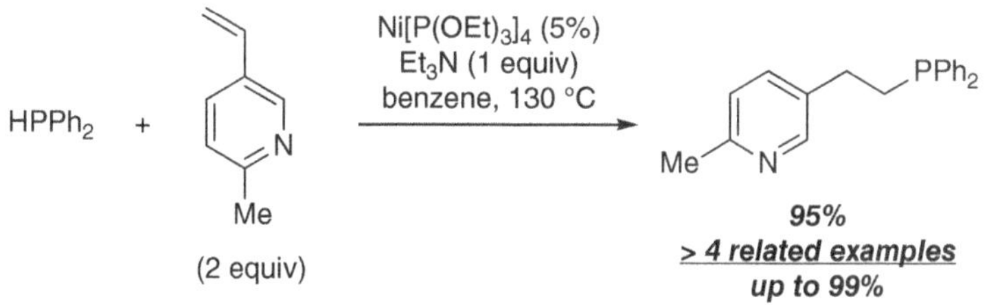

SCHEME 4.14 Addition of secondary phosphines to styrene derivatives [56].

base-assisted hydrophosphinylation approach, even bulky $Mes_2P(O)H$ was successfully added to vinyl pyridine (62% isolated).

The preparation of 1,2-bisphosphines bearing electron-withdrawing substituents was accomplished using a base-assisted hydrophosphination reaction (Scheme 4.13) [16]. While the yield of the reaction was moderate (52%), the strength of this reaction was the elimination of the normal requirement for the use of hazardous reagents such as $Li[3,5-C_6H_3(CF_3)_2]$ to generate the electron-deficient bisphosphine.

Beletskaya reported one of the first examples of the palladium- and nickel-catalyzed addition of secondary phosphines to styrene derivatives (Scheme 4.14 and Example 4.15) [56]. Common palladium and nickel catalysts such as $Pd(NCMe)_2Cl_2$ and $Ni[P(OEt)_3]_4$ promoted the addition reaction, and excellent yields of the addition products were obtained.

Platinum complexes are efficient catalysts for the addition of primary phosphines to alkenes. Higham described the synthesis of a tripodal ligand for chelation of transition metals generated through the double hydrophosphination of a primary phosphine (Scheme 4.15) [57]. One of the interesting aspects of this chemistry was the air stability of this primary phosphine. No oxidation products were observed following exposure of the solid phosphine or a solution ($CDCl_3$) to the atmosphere for several days. The working hypothesis was that primary phosphines with a high degree of conjugation are stable to oxidation without any additional stabilizing groups. This was an attractive approach as alternative methods of protecting primary phosphines focused on incorporating bulky

Example 4.15   Nickel-Catalyzed Hydrophosphination of Styrene Derivatives [56].

Before starting this synthesis, carefully follow the chemical safety instructions listed in Example 4.1.

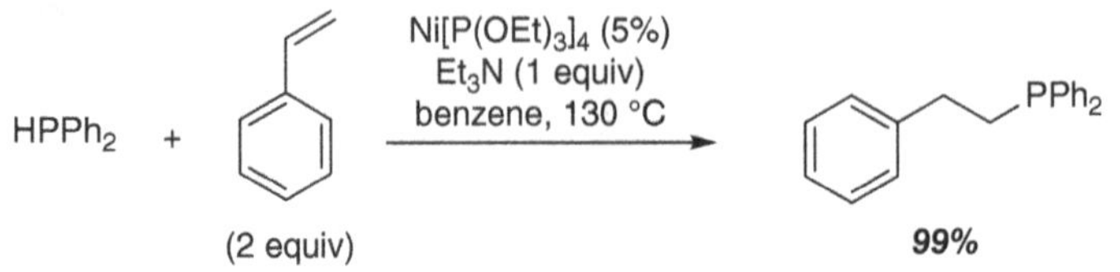

Caution! Benzene, THF, and hexane are highly flammable!

Caution! Secondary phosphines can spontaneously ignite in air and must be stored and handled under an inert atmosphere such as nitrogen or argon.

A mixture of 1.5 mmol of Ph_2PH, 3 mmol of alkene, 1.5 mmol of Et_3N, and 5 mol% of $Ni[P(OEt)_3]_4$ in 2 mL of benzene was placed in an NMR tube and sealed. The tube was heated in an oil bath at 130 °C, and the reaction was followed using ^{31}P NMR measurements. After completion of the reaction (disappearance of the signal of Ph_2PH at ~ −40.6 ppm), the solvents were evaporated under vacuum. The residue was recrystallized from a hexane/THF mixture and purified by column chromatography (hexane/benzene 3:1).

Adapted with permission from M. O. Shulyupin, M. A. Kazankova, I. P. Beletskaya, Org. Lett., 2002, 4, 761–763. Copyright (2002) American Chemical Society.

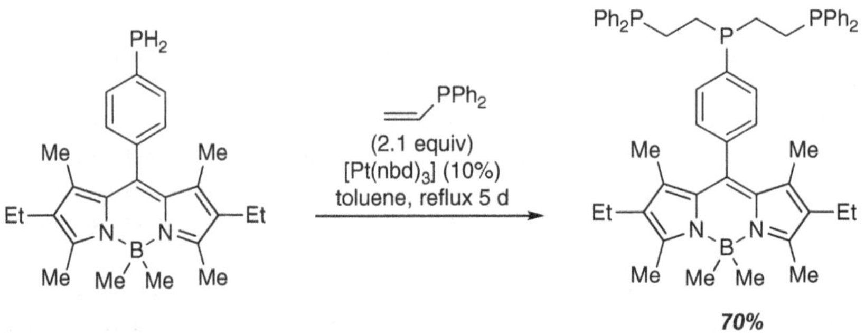

SCHEME 4.15   Hydrophosphination of an air-stable primary phosphine [57].

groups. Using the air-stable primary phosphine and a vinylphosphine as substrates, the authors carried out the addition reaction using a platinum catalyst. Although the reaction time was a bit long, the hydrophosphination reaction was high yielding and provided access to this tripodal phosphine.

An iron-catalyzed double hydrophosphination of alkynes generated 1,2-bisphosphines in excellent yields (Scheme 4.16 and Example 4.16) [58]. A range of iron compounds were screened for activity, and $CpFe(CO)_2Me$ was found to be the most effective. Simple iron salts such as ferrous or ferric chloride did not catalyze the reaction. The reaction was

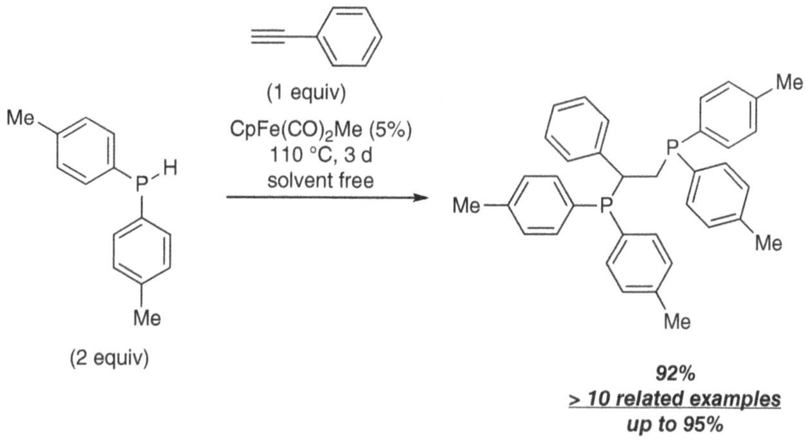

SCHEME 4.16   Synthesis of bisphosphines through a double hydrophosphination process [58].

Example 4.16   Iron-Catalyzed Double Hydrophosphination of Alkynes [58].

Before starting this synthesis, carefully follow the chemical safety instructions listed in Example 4.1.

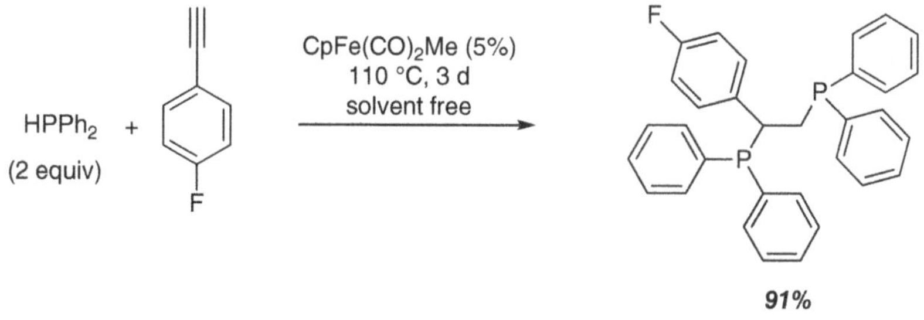

Caution! Hexane is highly flammable.

Caution! Secondary phosphines can spontaneously ignite in air and must be stored and handled under an inert atmosphere such as nitrogen or argon.

Secondary phosphine (2.5 mmol) was treated with acetylene (1.25 mmol) in the presence of iron catalyst (0.125 mmol, 5 mol% vs. phosphine) at 110 °C for 3 days. After removal of the volatile materials under reduced pressure, the residue was solved in CH_2Cl_2 (5 mL). The solution was passed through a celite column and the solvent (CH_2Cl_2) of the eluent was removed under reduced pressure. The residue was washed with hexane (2 mL, two times) and dried under vacuum to give the 1,2-bisphosphinoethane derivatives as a white powder.

Adapted with permission from M. Kamitani, M. Itazaki, C. Tamiya, H. Nakazawa, *J. Am. Chem. Soc.,* **2012,** *134,* 11932–11935. *Copyright (2012) American Chemical Society.*

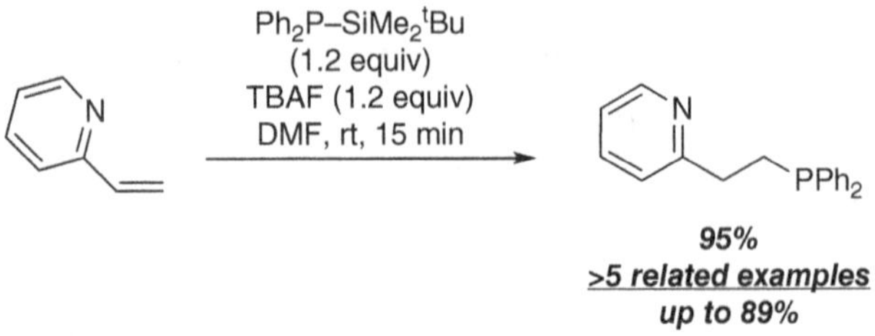

SCHEME 4.17   Metal-free addition of silylphosphines to vinyl pyridines [59].

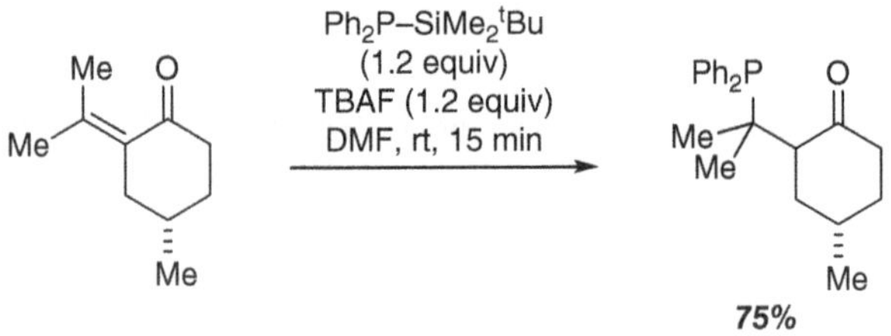

SCHEME 4.18   TBAF-promoted addition of silylphosphines to a tetrasubstituted alkene [59].

extremely sensitive to the conditions, and no addition product was observed when a range of organic solvents were used. Only when the reactions were carried out under solvent-free conditions was the addition product observed. Using this approach, several secondary phosphines and a range of terminal arylalkynes bearing either electron-donating or withdrawing groups were successfully used (up to 95% isolated yield). Even 4-ethynylaniline was successfully converted into the bisphosphine product (72%). While these compounds were successfully functionalized, none of the addition products were observed when dialkylphosphines or terminal alkylalkynes such as 1-hexyne were screened.

The transition metal-free addition of silylphosphines to styrenes, pyridines, and acrylates was promoted by a fluoride source (Scheme 4.17) [59]. The addition reaction occurred under extremely mild conditions and afforded moderate to excellent yields of the anti-Markovnikov addition product. The reaction was proposed to proceed through initial formation of a phosphide anion that reacted with the alkenes through a *phospha*-Michael-type addition. The proton source for the generation of the final adduct was proposed to be adventitious moisture in the commercially available TBAF solution. While activated alkenes were quite amenable to this approach, unactivated alkenes were unresponsive. 2-Vinylpyridine was particularly reactive and was cleanly converted into the alkylphosphine (95%). While most of the substrates were terminal Michael acceptors, a number of internal alkenes as well as a tetrasubstituted alkene were converted into the alkylphosphines in moderate yields (Scheme 4.18 and Example 4.17). The chemistry could be extended to terminal and internal alkynes, although a mixture of *E*- and *Z*-isomers were obtained with the *E/Z* > 1. Following this work, the authors were able take advantage of the need for another electrophile to complete the reaction and added an aldehyde to trap the intermediate carbanion [60]. This approach worked well and enabled the construction of gamma-hydroxyphosphonates in excellent yield (up to 88%).

Example 4.17   Fluoride-Promoted Addition of Silylphosphines to Alkenes [59].

Before starting this synthesis, carefully follow the chemical safety instructions listed in Example 4.1.

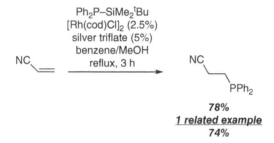

89%

Caution! Hexane and THF are highly flammable.
To a solution of (*tert*-butyldimethylsilyl)diphenylphosphine (72 mg, 0.24 mmol) and styrene (21 mg, 23 µL, 0.20 mmol) in DMF (1.5 ml) was added dropwise a 1 M TBAF solution in THF (240 µL, 0.24 mmol) at room temperature. After stirring for 15 min at room temperature, the solvent was removed under vacuum. The residue was purified by a column chromatography on silica gel ($CHCl_3$/Hex = 1 : 4, R_f = 0.4) to give the addition product (47 mg, 89%) as a colorless oil.

Reprinted from M. Hayashi, Y. Matsuura, Y. Watanabe, Fluoride-mediated phosphination of alkenes and alkynes by silylphosphines, Tetrahedron Letters, 45, 9167–9169. Copyright (2004), with permission from Elsevier.

SCHEME 4.19   Rhodium-catalyzed addition of silylphosphines to alkenes [61].

In addition of the use of TBAF to promote the addition of silylphosphines to alkenes, rhodium complexes also catalyzed the addition reaction (Scheme 4.19 and Example 4.18) [61]. While the activation of the Si—P bond in the TBAF-assisted reactions was proposed to proceed through the formation of diphenylphosphide, the mechanism proposed for the rhodium-catalyzed version entailed the initial oxidative addition of the silylphosphine to the rhodium(I) center to generate a rhodium(III) intermediate. No phosphide intermediates were proposed in the rhodium-catalyzed reaction. Although only a few examples were listed, the yields were moderate. As was the case with the TBAF-promoted reactions, only activated alkenes were successfully used in this reaction, and substrates such as 1-hexene were unreactive. Alkynes were successfully used in this chemistry and generated vinylphosphines in high yields with excellent regioselectivity for the β-adduct and high stereoselectivity for the *E*-isomer. As noted previously, the TBAF-promoted addition reaction generated a mixture of isomers.

Example 4.18 Addition of Silylphosphines to Activated Alkenes [61].

Before starting this synthesis, carefully follow the chemical safety instructions listed in Example 4.1.

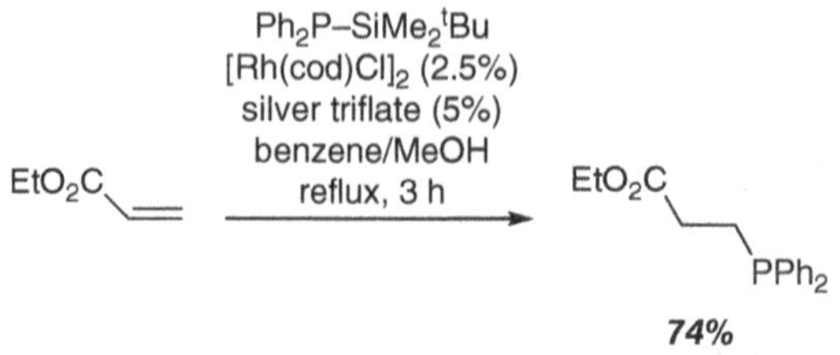

74%

Caution! Hexane and benzene are highly flammable.
To a solution of [Rh(cod)Cl]$_2$ (6 mg, 12.5 μmol) and MeOH (20 μL) in anhydrous benzene (2 mL) was added silver triflate (6 mg, 25 μmol) under argon. After the mixture was stirred for 20 min at room temperature, ethyl acrylate (0.30 mmol) and silylphosphine (0.6 mmol) were added successively. The mixture turned red with the addition of silylphosphine. The resulting mixture was heated under reflux until the color of the mixture turned from dark red to yellow (about 8–10 h). After the reaction mixture was concentrated, purification of the crude product by column chromatography on silica gel (eluent: CHCl$_3$/*n*-hexane = 1/1) afforded the corresponding alkylphosphine as a colorless oil (74%).

Adapted with permission from M. Hayashi, Y. Matsuura, and Y. Watanabe, *J. Org. Chem.*, **2006**, *71*, 9248–9251. *Copyright (2006) American Chemical Society.*

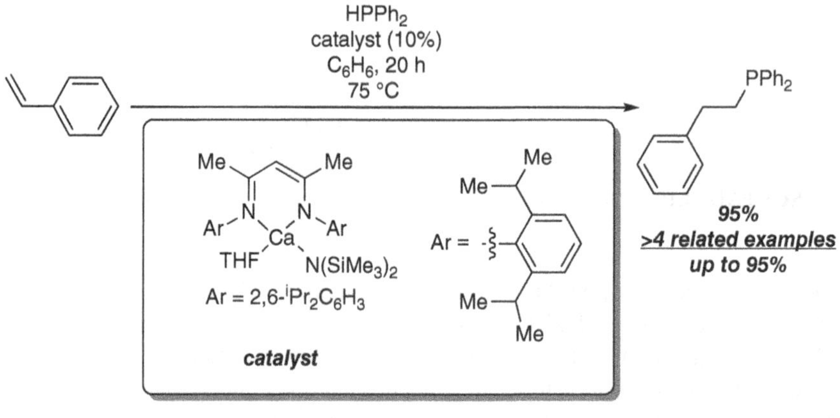

SCHEME 4.20 Calcium-catalyzed synthesis of alkylphosphines [62].

A calcium-catalyzed hydrophosphination reaction has been reported to functionalize unactivated alkenes such as isoprene and cyclohexadiene using a single-component calcium catalyst bearing a bulky diketiminate ligand (Scheme 4.20) [62]. This catalyst afforded alkylphosphines in excellent yields (up to 95%) with *anti*-Markovnikov regioselectivity. The diketinimate ligand was critical to the success of the reaction, and none of the addition products were observed using similar complexes such as [Ca(NSiMe$_3$)$_2$(THF)$_2$].

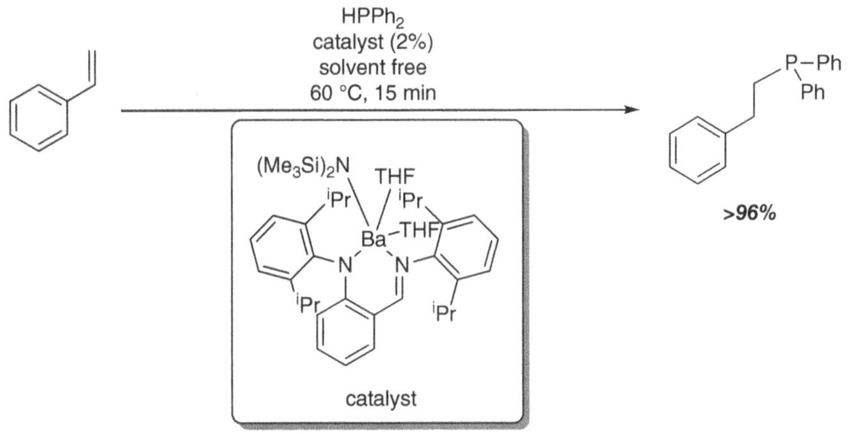

SCHEME 4.21   Barium-promoted addition of secondary phosphines to styrene [63].

Although the substrate scope was limited, it provided a transition metal-free approach to the synthesis of alkylphosphines from unactivated alkenes.

A barium-promoted hydrophosphination reaction has been reported (Scheme 4.21) [63, 64]. A bulky diketiminate was used as the supporting ligand to solubilize/stabilize the barium center. While addition reactions using HPPh$_2$ were high yielding and fast (96% after 15 min), analogous reactions with HPCy$_2$ were sluggish and incomplete reactions were observed. Even prolonged heating was unsuccessful and resulted in 42% conversion after 18.5 h at 60 °C. The investigation enabled the authors to compare the reactivity of the barium and calcium catalysts, and barium was found to be more active. A noteworthy aspect of this chemistry was that it was successful under solvent-free conditions with only 2% catalyst loading.

While the majority of hydrophosphination reactions were carried out in organic solvents, a copper-catalyzed hydrophosphination reaction was developed that could be carried out using pure water as the solvent (Scheme 4.22 and Example 4.19) [65]. The key to this chemistry was the addition of a surfactant to promote the formation of micelles. Once formed, the reaction was proposed to occur within the core of these aggregates. A considerable amount of substrate-specific reactivity was observed with this system with the highest yields obtained using electron-deficient substrates.

The iron-catalyzed addition of secondary phosphines to unactivated alkenes has been accomplished using an unusual oxo-bridged iron dimer as the catalyst (Scheme 4.23 and Example 4.20) [66]. The reactions were carried under very mild conditions and afforded moderate to good yields of the desired phosphines. The catalyst loading was quite low (0.5%), and an assortment of styrene derivatives bearing a range of electron-donating and electron-withdrawing groups was well tolerated. Additionally, vinyl pyridines and acrylates were successfully converted into tertiary phosphines using this approach. It should be noted that this iron catalyst was reported by the authors to be quite stable to oxygen.

Work by Leung demonstrated that palladacycles were effective catalysts for synthesis of alkyl phosphines through the addition of secondary phosphines to alkenes [67–74]. In addition, many of these reports described the asymmetric synthesis of phosphines using chiral palladium complexes. For example, the use of a chiral C, N-palladacycle for the asymmetric addition of secondary phosphines to enones was outlined (Scheme 4.24 and Example 4.21)

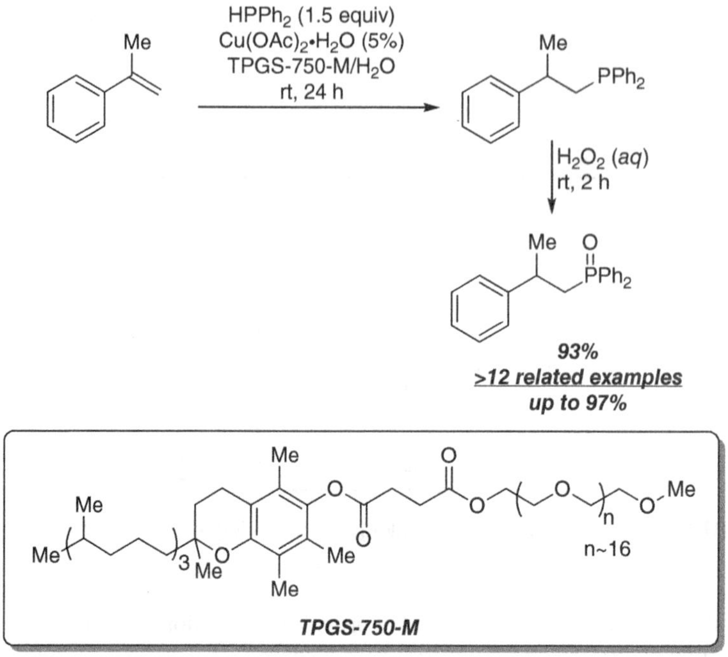

SCHEME 4.22 Copper-catalyzed synthesis of alkylphosphine oxides in water [65].

Example 4.19 Copper-Catalyzed Hydrophosphinations in Water [65].

Before starting this synthesis, carefully follow the chemical safety instructions listed in Example 4.1.

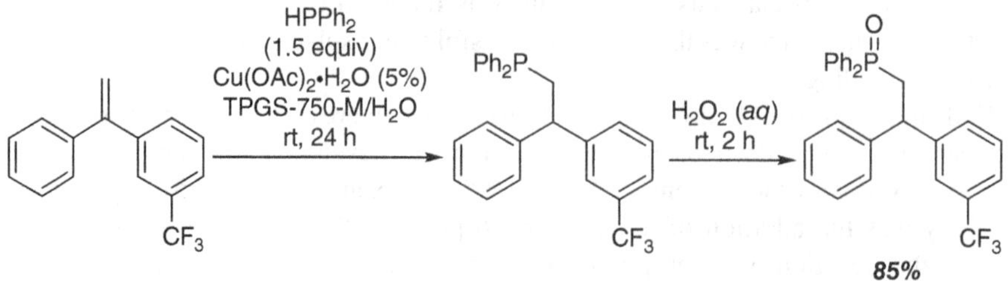

This reaction was carried out under an atmosphere of argon. The reaction was performed on a 0.25 mmol scale in a 5 mL microwave conical vial equipped with a septum and spin vane. First, Cu(OAc)$_2$·H$_2$O (2.5 mg, 0.0125 mmol) was added to the vial and a septum attached. The vial was purged of any oxygen with an argon flow followed by the addition of 2 wt% TPGS-750-M/H$_2$O (0.3 mL). Once the copper dissolved, HPPh$_2$ (65 μL, 70 mg, 0.375 mmol) was added. Upon the addition of HPPh$_2$, the solution became colored (red, brown, yellow, and orange). The color dissipated within a few to 10 min and the solution became cloudy and white. After the dissipation of color, the substrate (0.25 mmol) was added via syringe. The sides of the vial were subsequently washed with more 2 wt% TPGS-750-M/H$_2$O (0.2 mL). The reaction was vigorously stirred for 24 h. The mixture was cooled to 0°C and quenched with H$_2$O$_2$ (0.2 mL, 30% aq.) and stirred for 30 min. Next, Na$_2$S$_2$O$_3$ (0.3 mL, sat. aq.) was added and the mixture stirred at room temperature. The mixture was filtered through a pad of silica gel using 10% (v/v) MeOH–DCM. The mixture was then concentrated via rotary evaporation and purified by flash chromatography using MeOH–DCM to afford the phosphine oxide as a white solid (85%).

Adapted from N. A. Isley, R. T. H. Linstadt, E. D. Slack, B. H. Lipshutz, *Dalton Trans.,* **2014**, *43*, 13196–13200 *with permission of The Royal Society of Chemistry.*

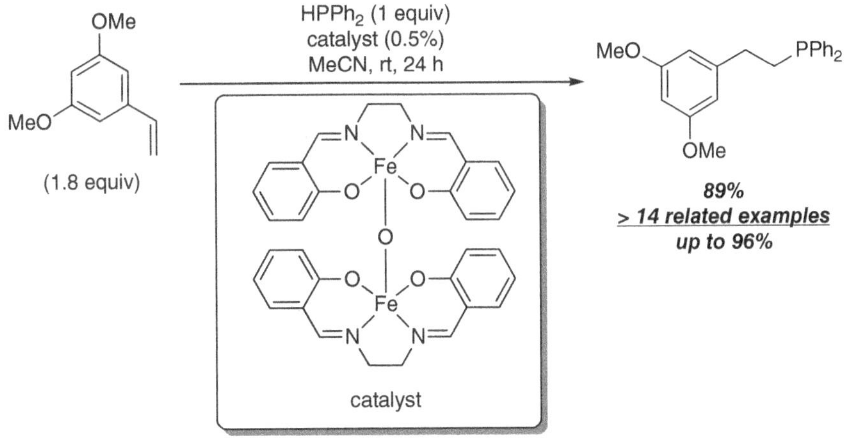

SCHEME 4.23 Iron-catalyzed addition of diphenylphosphine to alkenes [66].

Example 4.20 Hydrophosphination of Alkenes Using an Iron Catalyst [66].

Before starting this synthesis, carefully follow the chemical safety instructions listed in Example 4.1.

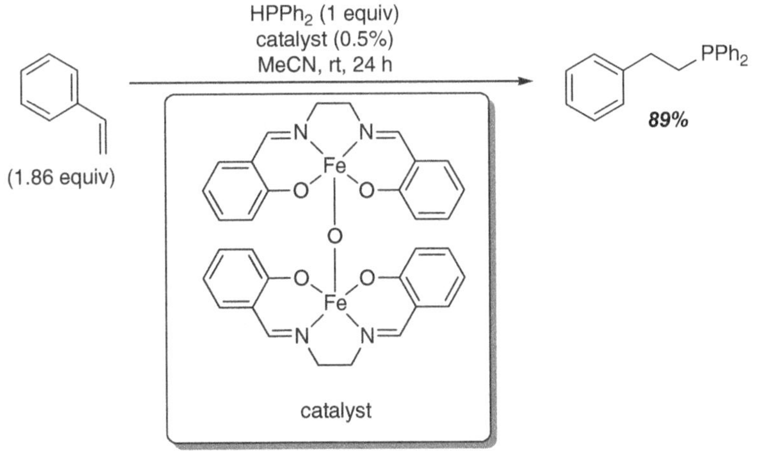

Caution! Secondary phosphines can ignite in the presence of air and must be handled under an inert atmosphere such as nitrogen or argon.

Caution! Pentane and ethyl acetate are highly flammable.

This reaction was carried out under an inert argon atmosphere in a glovebox. The iron catalyst (1.8 mg, 1 mol% Fe center) was weighed out into a Schlenk tube. CH_3CN (350 μL) was added to this (forming a deep red solution) followed by styrene (1.04 mmol, 1.86 equiv) and finally diphenylphosphine (100 μL, 0.57 mmol, 1 equiv). After stirring at room temperature for 24 h, the Schlenk tube was placed under vacuum, and the excess starting styrene and solvent were removed leaving an orange/red oil. For spectroscopic yields, reaction solutions were exposed to air and filtered through a silica gel plug using CH_2Cl_2. This removed the iron residue leaving a colorless oil. Solvent was removed by blowing nitrogen over the oil before addition of 45 μL 1,2-dichloroethane as an integration standard. $CDCl_3$ was used as the NMR solvent. The alkylphopshine was then isolated by column chromatography 1–5% EtOAc/pentane to afford a colorless oil (89%). The products oxidized slowly over time (this can be observed in some ^{31}P spectra as a peak at ~20 ppm).

Adapted from K. J. Gallagher and R. L. Webster, Chem. Commun., 2014, 50, 12109–12111 with permission of The Royal Society of Chemistry.

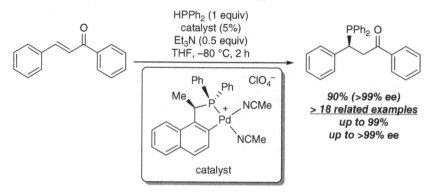

SCHEME 4.24 Enantioselective addition of secondary phosphines to alkenes [64].

Example 4.21 Palladium-Catalyzed Asymmetric Addition of Secondary Phosphines to Alkenes [67].

Before starting this synthesis, carefully follow the chemical safety instructions listed in Example 4.1.

Caution! THF and acetone are highly flammable.

Caution! Perchlorate salts can be explosive and must be handled with care and in small quantities.

Caution! Secondary phosphines can spontaneously ignite in air and must be handled and stored under an inert atmosphere such as nitrogen or argon.

This reaction was carried out under an atmosphere of nitrogen or argon. To a solution of Ph$_2$PH (55.9 mg, 0.30 mmol, 1.0 equiv) in THF (5 mL) is added the catalyst (9.4 mg, 0.015 mmol, 5 mol%) and the solution was cooled to −80 °C. Subsequently, the phenyl chalcone (0.30 mmol, 1.0 equiv) was added. Et$_3$N (15.2 mg, 0.15 mmol, 0.5 equiv) in THF (0.5 mL) was added dropwise. The solution was subsequently stirred at −80 °C. The reaction was monitored by ^{31}P NMR. After the reaction was completed, the mixture was warmed to room temperature and the solution was evaporated by vacuum pump to give the crude phosphine product (air sensitive) as a solid. The crude product was dissolved in 10 ml DCM and filtered through a short silica gel column using a pipette fixed on a two-neck Schlenk flask protected by nitrogen or argon. The solvent was removed by vacuum pump to give the product as solids. A single recrystallization from acetone gave enantiopure phosphine.

Adapted with permission from Y. Huang, S. A. Pullarkat, Y. Li, P.-H. Leung, *Inorg. Chem.,* **2012**, *51*, 2533–2540. *Copyright (2012) American Chemical Society.*

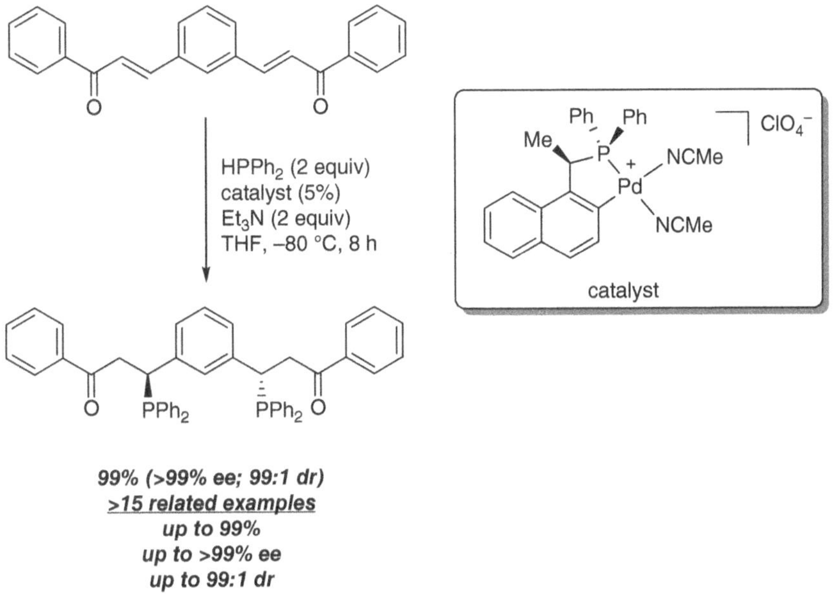

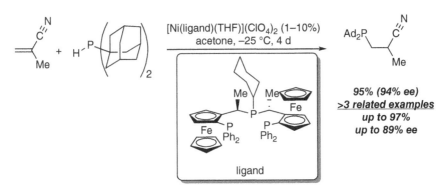

SCHEME 4.25 Palladium-catalyzed dual addition of secondary phosphines [75].

SCHEME 4.26 Nickel-catalyzed asymmetric synthesis of alkylphosphines through hydrophosphinylation [76, 77].

[67]. Using only a slight excess of the secondary phosphine, a wide range of enones bearing an assortment of electron-deficient and electron-rich aromatic substituents were successfully functionalized in excellent yields (up to 99%) with outstanding regioselectivity and high selectivity (up to 99% ee). The position of attachment was always beta to the carbonyl group. Although the reaction required cryogenic conditions, this requirement is mitigated by the excellent selectivity. In related work, the dual addition of secondary phosphines to dienones has been reported (Scheme 4.25) [75].

The asymmetric addition of secondary phosphines to methacrylonitrile has been reported by Togni (Scheme 4.26 and Example 4.22) [76, 77]. The reactions were carried out at low temperatures and required an extended stirring period; however, the regioselectivity and enantioselectivity were outstanding. The authors did observe a bit of

Example 4.22   Enantioselective Hydrophosphination of Activated Alkenes [76].

Before starting this synthesis, carefully follow the chemical safety instructions listed in Example 4.1.

Caution! Secondary phosphines can spontaneously ignite and must be handled under an inert atmosphere such as nitrogen or argon.

This reaction was carried out under argon. $[Ni(H_2O)_6](ClO_4)_2$ (0.0361 g, 0.0987 mmol) and Pigiphos (0.0898 g, 0.0988 mmol) were dissolved in 1 mL of methacrylonitrile to give a purple solution of $Ni(Pigiphos)(methacrylonitrile)](ClO_4)_2$. To this solution was added Cy_2PH (0.2 mL, 0.988 mmol), and the resulting mixture was stirred for 8 h. The volatile materials were removed under reduced pressure, and the mixture was extracted with pentane and filtered through Al_2O_3. Pentane was removed under reduced pressure, yielding 0.1873 g of $(C_6H_{11})_2PCH_2CH(CH_3)CN$ (0.0706 mmol, 71% yield).

Adapted with permission from A. D. Sadow, I. Haller, L. Fadini, A. Togni, J. Am. Chem. Soc., **2004**, *126,* 14704–14705. *Copyright (2004) American Chemical Society.*

---

substrate-specific reactivity, and while the use of electron-rich secondary alkylphosphines generated high yields of the hydrophosphination product, analogous reactions with diphenylphosphine oxide were less successful and the addition product was isolated in 10% yield with low ee (32%) For the right substrate, this is an effective approach to the preparation of alkylphosphines.

A chiral PCP pincer ligand has been used in the palladium-catalyzed addition of secondary phosphines to nitroalkenes (Scheme 4.27) [20]. The reactions were carried out at −40 °C in dichloromethane with a slight excess of the secondary phosphine and afforded excellent yields of the addition products. The process was highly regioselective for addition to the carbon beta to the nitro group. The enantioselectivity was less consistent and the authors noted some substrate dependence with the highest selectivity observed with para-substituted nitrostyrenes and the lowest with meta- and ortho-substituted substrates. The products were isolated as phosphine oxides in order to facilitate purification.

An uncatalyzed atom-efficient reaction is attractive since every atom in the starting materials and reagents is consumed in the process and converted into the desired compound

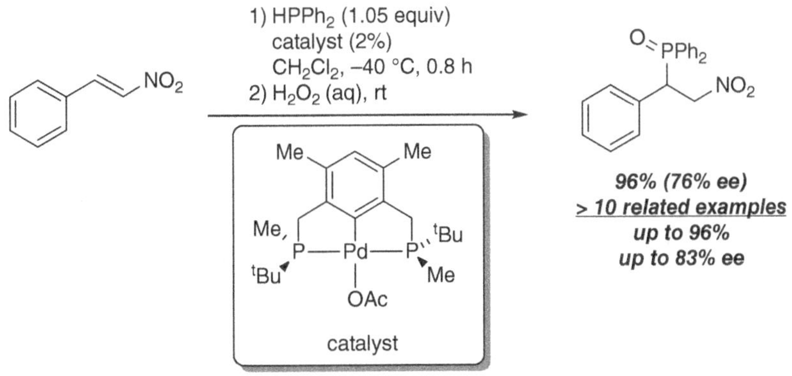

SCHEME 4.27 Use of a P-stereogenic PCP pincer ligand in asymmetric hydrophosphination reactions [20].

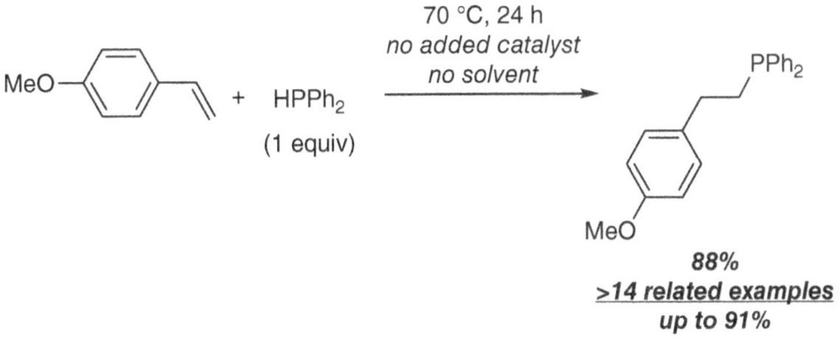

SCHEME 4.28 Uncatalyzed addition of secondary phosphine oxides to styrene derivatives [78].

with no waste generated. Ideally, this process would occur without the need for solvents or additives. To this end, the uncatalyzed addition of secondary phosphines to styrenes and activated alkenes under solvent-free conditions has been reported (Scheme 4.28) [78]. During the course of the investigation, the authors probed the effect of added solvents and found that a range of solvents inhibited the hydrophosphination reaction. While adding THF and DMF greatly reduced the conversion (2–19%), ethanol, dichloromethane, or acetonitrile completely inhibited the process (0%). Diphenylphosphine was the only secondary phosphine screened in the reaction, but a range of functionalized styrenes bearing both electron-donating and withdrawing groups were well tolerated by the reaction as well as vinyl ketones and acrylates. Although this reaction needed to be carried out under an inert atmosphere, the high yields and functional group tolerance make this a practical approach to the hydrophosphination of alkenes.

While the majority of the hydrophosphination reactions described previously led to the formation of products with almost complete anti-Markovnikov selectivity, devising an approach for the selective formation of the addition product with Markovnikov selectivity can be challenging [79]. To address this deficiency, Gaumont determined that a simple change to the catalyst structure switched the selectivity of the reaction (Scheme 4.29 and Example 4.23) [80]. The key observation was that using ferrous chloride as

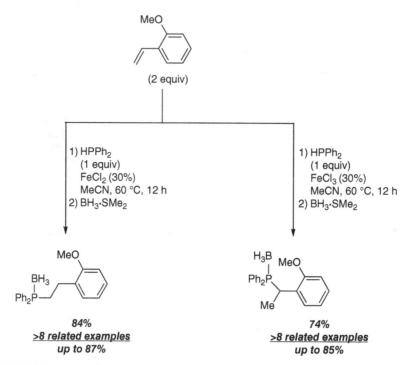

SCHEME 4.29 Hydrophosphination with Markovnikov and anti-Markovnikov selectivity [80].

Example 4.23 Markovnikov-Selective Hydrophosphination [80].

Before starting this synthesis, carefully follow the chemical safety instructions listed in Example 4.1.

Caution! Secondary phosphines can spontaneously ignite in air and must be stored under an inert atmosphere such as nitrogen or argon.

Caution! Pentane is highly flammable.

This reaction was carried out under an atmosphere of nitrogen. Styrene (120 μL, 1.04 mmol) and diphenylphosphine (100 μL, 0.57 mmol) were successively added to a solution of FeCl$_3$ (30 mg, 0.18 mmol) in acetonitrile (150 μL). The reaction mixture was then heated at 60 °C for 12 h. After cooling to 0 °C, BH$_3$·SMe$_2$ (63 μL, 0.7 mmol) was added dropwise and the reaction mixture stirred at RT for 30 min. Volatiles were eliminated under reduced pressure, and then the crude product was purified by silica gel chromatography with a CH$_2$Cl$_2$/pentane mixture (2:3) as eluent. Diphenyl-(1-phenyl-ethyl)phosphine-borane was isolated as a colorless oil (129 mg, 0.42 mmol) in 73% yield.

Adapted with permission from L. Routaboul, F. Toulgoat, J. Gatignol, J.-F. Lohier, B. Norah, O. Delacroix, C. Alayrac, M. Taillefer, A.-C. Gaumont, *Chem. Eur. J.,* **2013**, *19*, 8760–8764. *Copyright (2013) Wiley-VCH Verlag GmbH & Co. KGaA, Weinheim.*

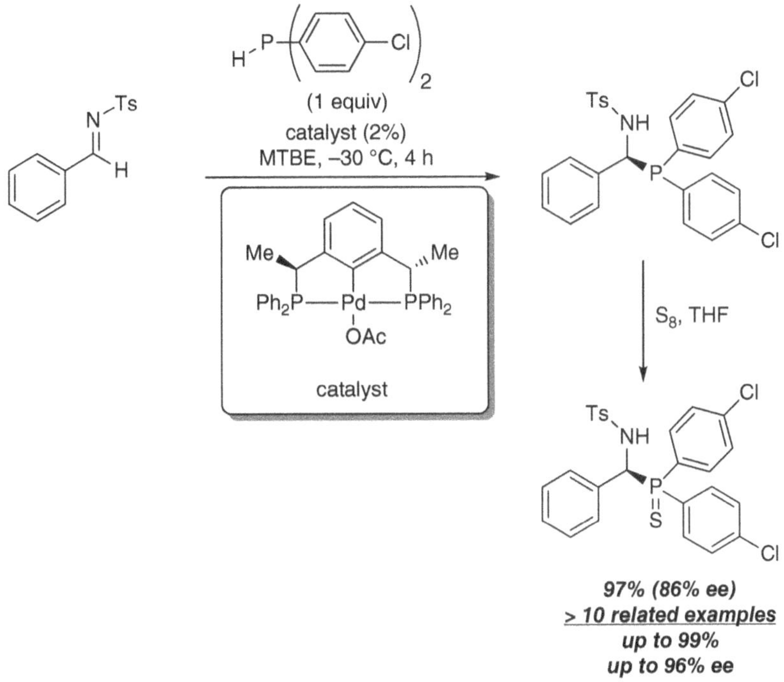

SCHEME 4.30   Addition of secondary phosphines to imines [81].

the catalyst generated the anti-Markovnikov addition product, while the use of ferric chloride afforded the Markovnikov product. Both of these iron salts are readily available from many commercial suppliers and inexpensive. The phosphines were isolated as borane adducts to facilitate purification by column chromatography. The authors also investigated the possibility that a contaminant in the commercially available iron compounds could be influencing the catalysis by adding various amount of common contaminants such as copper sources to the catalytic reactions as well as screening various grades of ferric chloride. They found that none of the common contaminants aided in the formation of the Markovnikov product. Furthermore, they found that several grades of ferric chloride were quite effective at promoting the addition reaction. While the precise mechanism and rationale for the differences between the iron compounds were not clear, this remains a rare example of an addition reaction that generated the Markovnikov addition product.

The asymmetric addition of phosphorus nucleophiles to imines provides direct access to phosphorus-containing chiral amines. While this chemistry is dominated by the use of phosphites and related compounds, a report by Duan outlines a successful protocol for the asymmetric addition of secondary phosphines to imines (Scheme 4.30) [81]. The reaction was catalyzed by a chiral palladium compound comprised of a palladium(II) center ligated by a resolved pincer compound. A range of aryl and heteroaryl imines as well as several secondary phosphines were successfully used in this reaction. To aid in the isolation of the phosphines, the authors added elemental sulfur at the end of the P—C bond-forming reaction. This chemistry provides access to a range of chiral N,P ligands for metal centers that were previously challenging to prepare and isolate.

Example 4.24   *Phospha*-Michael-Type Addition to Activated Alkenes [82].

Before starting this synthesis, carefully follow the chemical safety instructions listed in Example 4.1.

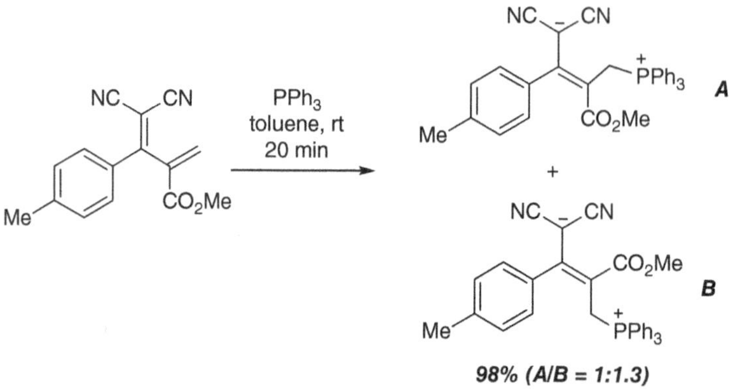

98% (A/B = 1:1.3)

Under argon atmosphere, to a solution of 4,4-dicyano-2-methylenebut-3-enoate 1a (26 mg, 0.1 mmol) in toluene (1.0 mL) was added PPh$_3$ (32 mg, 0.12 mmol) at room temperature. Then the resulting mixture was stirred at room temperature until the reaction was complete (monitoring by TLC). Then the solvent was removed under reduced pressure and the residue was purified by a flash column chromatography to afford the addition products.

Adapted with permission from X.-N. Zhang, G.-Q. Chen, X.-Y. Tang, Y. Wei, M. Shi, Angew. Chem. Int. Ed., **2014,** *53,* 10768–10773. *Copyright (2014) Wiley-VCH Verlag GmbH & Co. KgaA, Weinheim.*

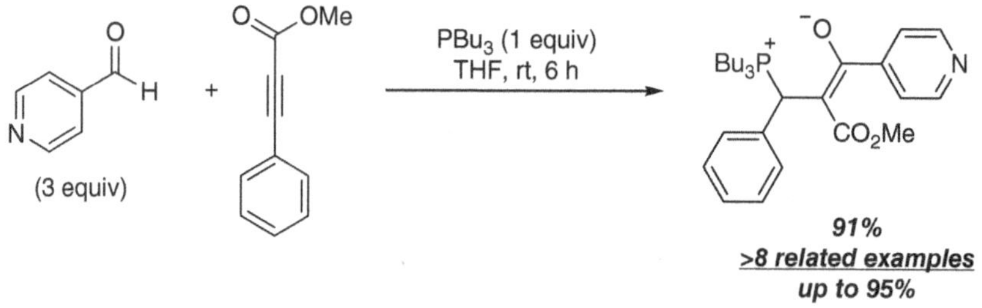

91%
≥8 related examples
up to 95%

SCHEME 4.31   Addition of phosphines to alkynes followed by trapping with an aldehyde to generate zwitterionic phosphonium enolates [83].

Tertiary phosphines also add to activated alkenes to generate zwitterionic compounds (Example 4.24) [82]. The reaction conditions were quite mild and high yields of the addition products were obtained. The reaction was selective for addition to the *beta*-position; however, a mixture of *E*- and *Z*-isomers of the final phosphonium species was obtained. While only a single example was reported for this reaction, it provides the foundation for extension to additional activated alkenes and phosphorus nucleophiles. In related work, tertiary phosphines were found to add to internal alkynes to generate a zwitterionic species containing a vinyl anion that was quenched by the addition of an aldehyde to generate stable tetravalent phosphonium enolate zwitterions (Scheme 4.31 and Example 4.25) [83].

Example 4.25   Preparation of a Zwitterionic Phosphonium Enolates [83].

Before starting this synthesis, carefully follow the chemical safety instructions listed in Example 4.1.

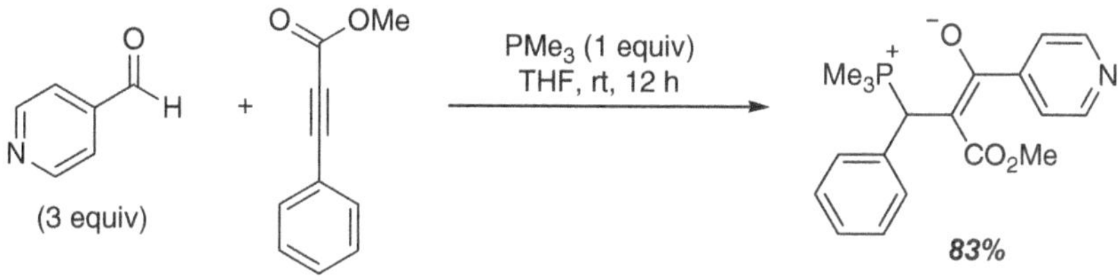

(3 equiv)

PMe₃ (1 equiv)
THF, rt, 12 h

83%

Caution! THF is flammable.

Caution! PMe₃ is a pyrophoric phosphine and must be handled under an inert atmosphere.
This reaction was carried out under an atmosphere of argon. THF (10 mL), 4-pyridine-carboxaldehyde (286 µL, 3.0 mmol), methyl phenylpropiolate (89 µL, 1.0 mmol), and trimethylphosphine (100 µL, 1.0 mmol) were added by syringe sequentially to a flame-dried 25 mL round-bottom flask under argon. Immediately after addition of the phosphine, the solution changed color to dark red. The reaction was stirred at room temperature for 12 h and monitored by TLC [R_f (EtOAc) = 0.0; R_f (10% MeOH in CH_2Cl_2) < 0.05]. The mixture was concentrated and subjected to FCC, first with EtOAc to elute 4-pyridinecarboxaldehyde and then with a gradient of MeOH (10–50%) in CH_2Cl_2 to elute the title compound, which was collected as a red oil (285 mg, 83%).

*Adapted with permission from X.-F. Zhu, C. E. Henry, O. Kwon, J. Am. Chem. Soc., **2007**, 129, 6722–6723. Copyright (2007) Wiley-VCH Verlag GmbH & Co. KGaA, Weinheim.*

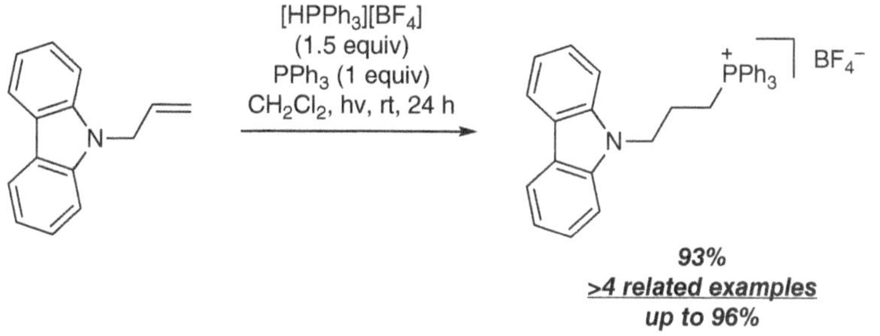

[HPPh₃][BF₄]
(1.5 equiv)
PPh₃ (1 equiv)
CH₂Cl₂, hv, rt, 24 h

93%
>4 related examples
up to 96%

SCHEME 4.32   Photochemical addition of phosphonium salts to alkenes [84].

While the synthesis of phosphonium salts from activated alkenes is readily achieved by the addition of triphenylphosphine and related substrates, unactivated alkenes are signifi-cantly less reactive in this type of reaction. To overcome this issue and provide a pathway to phosphonium salts from substrates such as 1-hexene, Grubbs recently reported the formal hydrophosphonation of these sluggish substrates (Scheme 4.32 and Example 4.26) [84]. The key substrate for the addition was the triphenylphosphonium salt [HPPh₃][BF₄]. The reaction was found to be promoted by both classic radical initiators such as AIBN and ACN. Additionally, UV light was found to be effective and was determined to be the

Example 4.26   Synthesis of Alkylphosphonium Salts through the Hydrophosphonation of Unactivated Alkenes [84].

Before starting this synthesis, carefully follow the chemical safety instructions listed in Example 4.1.

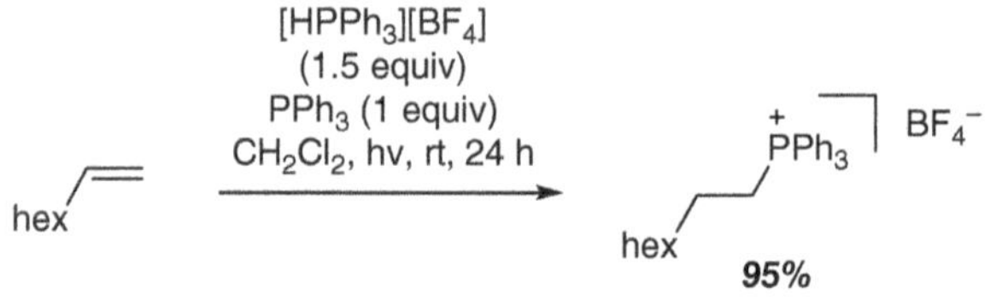

A Pyrex Schlenk tube equipped with a stir bar and PTFE closure was flame dried under vacuum. The tube was backfilled with Ar and charged with $[HPPh_3][BF_4]$ (525 mg, 1.5 mmol, 1.5 equiv) and PPh_3 (262 mg, 1 mmol, 1 equiv). The tube was evacuated and backfilled with Ar. The solids were dissolved in CH_2Cl_2 (10 ml). The substrate (1 mmol) was added in one portion. The Schlenk tube was sealed and placed in the photobox for 24 h. The solvent was removed from the reaction mixture via rotary evaporation to afford the phosphonium salt (95%).

Adapted with permission from C. S. Daeffler, R. H. Grubbs, Org. Lett., 2011, 13, 6429–6431. Copyright (2011) Wiley-VCH Verlag GmbH & Co. KGaA, Weinheim.

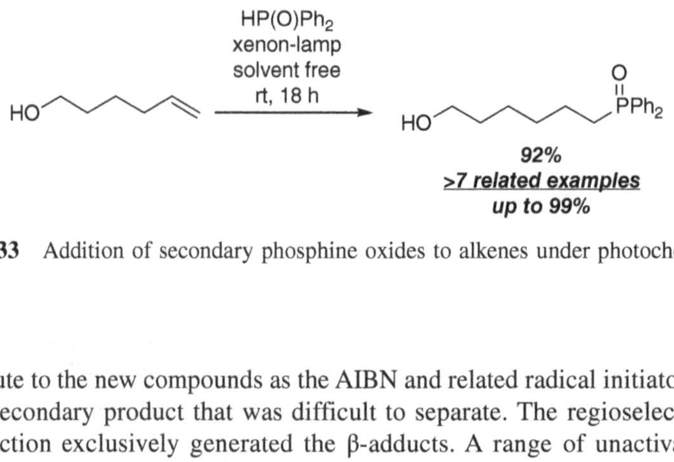

SCHEME 4.33   Addition of secondary phosphine oxides to alkenes under photochemical conditions [85].

preferred route to the new compounds as the AIBN and related radical initiators tended to generate a secondary product that was difficult to separate. The regioselectivity of the addition reaction exclusively generated the β-adducts. A range of unactivated alkenes were successfully functionalized using this approach, and the reaction was also scalable to 1 g samples. If the dichloromethane solvent was excluded, this approach is formally a halide-free route to phosphonium salts.

The addition of secondary phosphine oxides to unactivated alkenes has been accomplished using a xenon lamp (Scheme 4.33) [85]. The reaction proceeded under solvent-free conditions at room temperature and afforded moderate to excellent yields of the tertiary phosphine oxides. It should be noted that the authors determined that the addition of pyridine significantly accelerated the reaction. One of the practical advantages of this chemistry was that it was successful under solvent-free conditions.

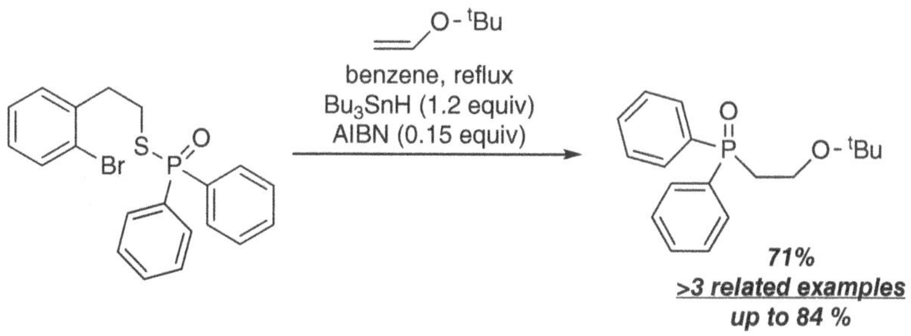

SCHEME 4.34   Addition of phosphinyl radicals to vinyl ethers [86].

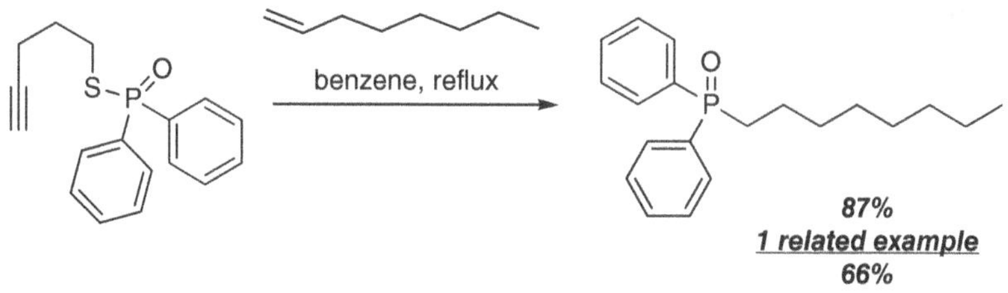

SCHEME 4.35   Tin-free addition of phosphinyl radicals to vinyl ethers [86].

Phosphinyl radicals have been generated from thiophosphine oxides and used in addition reactions involving activated and unactivated alkenes (Schemes 4.34 and 4.35) [86]. The initial part of the investigation utilized tributyltinhydride and AIBN as the radical initiating system. The key to generating the phosphinyl radicals was the design of a substrate that contained two key fragments: (i) a component that would react with the initiator to generate a radical intermediate and (ii) a component that would intramolecularly react with the intermediate radical to generate a phosphinyl radical that could be ejected from the precursor and added to an alkene. The substrate they designed included a strategically placed aryl bromide to generate the aryl radical and a tethered thiophosphonate to trap the intermediate radical and generate the phosphinyl radical through homolytic cleavage of the P—S bond. Once removed from the precursor, the phosphinyl radical added to the alkene. These reactions consistently generated the anti-Markovnikov products in moderate to high yields. An additional part of the investigation was focused on the removal of tin reagents from the system due to toxicity concerns. To this end, they devised a tin-free method for the addition reaction and used a tethered alkyne as a radical trap for the sulfur radical generated in the reaction. This approach was successful and moderate yields of several alkylphosphine oxides were obtained.

While several of the reactions discussed previously used UV light to promote the formation of phosphorus-centered radicals, there is significant interest in using visible light to promote these reactions. Kobayashi has devised an elegant solution to this problem through the use of a readily available photocatalyst (Scheme 4.36) [87]. Using this catalyst, he was able to promote the hydrophosphinylation of unactivated alkenes through the formation of phosphorus-centered radicals. The conditions of the reaction were quite mild and compatible with a range of functional groups. In addition to a range of internal and

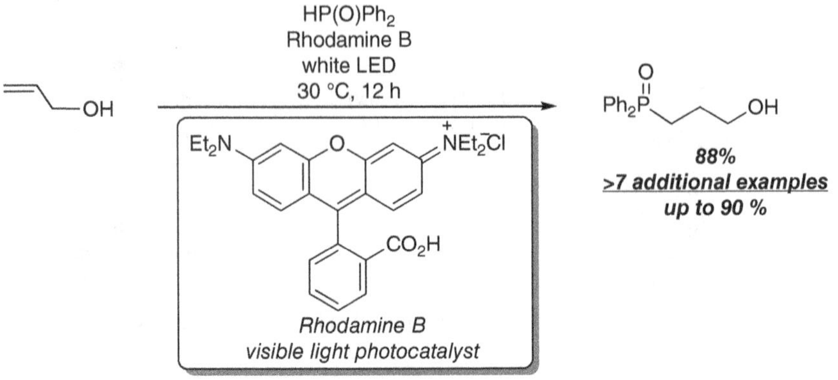

SCHEME 4.36 Addition of secondary phosphine oxides to an allylic alcohol using visible light catalysis [87].

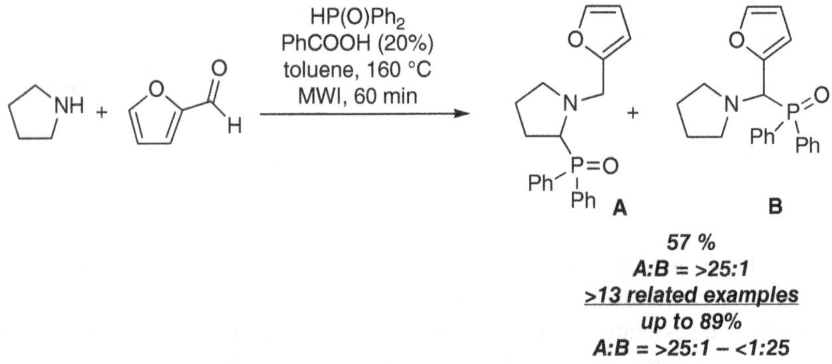

SCHEME 4.37 Benzoic acid promoted synthesis of α-aminophosphine oxides [88].

terminal alkenes, a substrate bearing a terminal alcohol was readily functionalized. Given the mild conditions and broad substrate scope, this is an attractive approach to the synthesis of alkylphosphine oxides through the hydrophosphinylation of alkenes.

The generation of α-aminophosphine oxides through a redox neutral process has been accomplished under metal-free conditions using only a weak acid as the catalyst (Scheme 4.37) [88]. The three-component reaction couples pyrrolidine, an aromatic aldehyde, and a secondary phosphine oxide to generate the α-functionalized tertiary amines. One of the issues with this chemistry was the formation of different regioisomers due to interception of the intermediate iminium ion by the secondary phosphine oxide. After some experimentation, the authors defined a catalyst system that would favor the formation of the functionalized pyrrolidine (isomer **A**). This chemistry was very sensitive to the reaction parameters, and even small modifications led to decreased regioselectivity. Furthermore, the strength of the acid was also found to play a role in the chemistry as increasing the strength of the acid (TFA) lead to reduced yields of the functionalized product. The chemistry was compatible with a range of electron-donating and electron-withdrawing groups and was also relatively insensitive to the incorporation of ortho-substituents on the aromatic aldehyde. Since the phosphonation

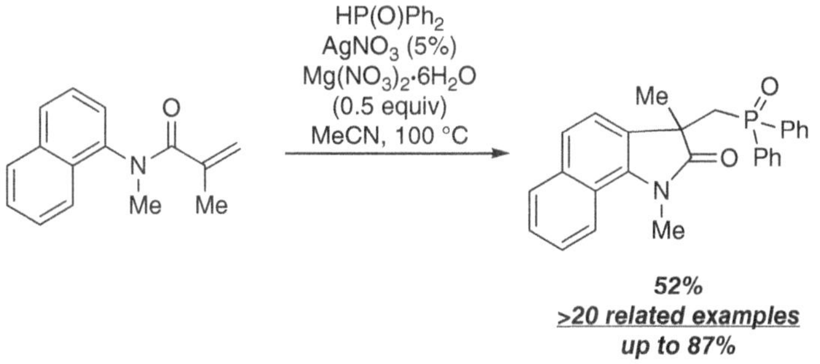

SCHEME 4.38   Synthesis of heterocycles bearing phosphine oxide pendant groups [89].

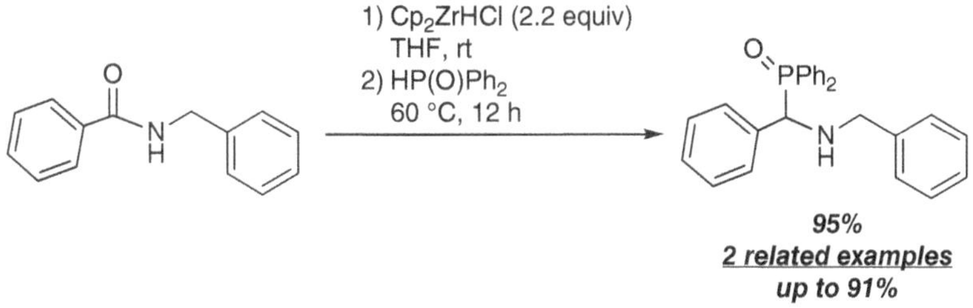

SCHEME 4.39   Synthesis of α-aminophosphine oxides [90].

of the alpha-position was coupled with a reductive alkylation at nitrogen, the process was overall redox neutral.

A cyclization reaction to generate heterocycles bearing phosphine oxides as pendant groups has been developed (Scheme 4.38) [89]. While several metal catalysts were active for this reaction, silver nitrate was selected for further study due to its high activity and low cost. A high catalyst loading (50%) was initially found to promote the reaction (79%); however, this catalyst loading was unattractive and modifications were made to the system in order to lower the amount of catalyst needed for the reaction. After some experimentation, the authors also found that the nitrate anion was critical to the success of the addition/cyclization reaction, but a high silver concentration was not. Thus, adding 0.5 equivalents of the magnesium nitrate along with a low silver nitrate loading (5%) promoted the addition reaction. Although the amount of silver could be lowered, it was still essential for a successful reaction. Control reactions carried out in the complete absence of silver led to no reaction. The chemistry was also compatible with a range of electron-withdrawing and donating groups on the aromatic ring. While it was insensitive to electronic effects, steric issues were discovered and incorporating substituents in the ortho-positions of the aromatic ring resulted in lower yields. Predictably, adding a substituent to the 3-position of the aromatic ring leads to mixtures of regioisomers. While the majority of the chemistry was focused on the synthesis of five-membered heterocyclic rings, the chemistry was extended to the formation of six-membered heterocycles as well as acyclic systems.

Tang has reported a reductive phosphinylation approach for the synthesis of α-amino-phosphine oxides (Scheme 4.39) [90]. The nitrogen-containing precursor for this reaction was an amide, while the phosphorus precursor was a secondary phosphine oxide. Amides

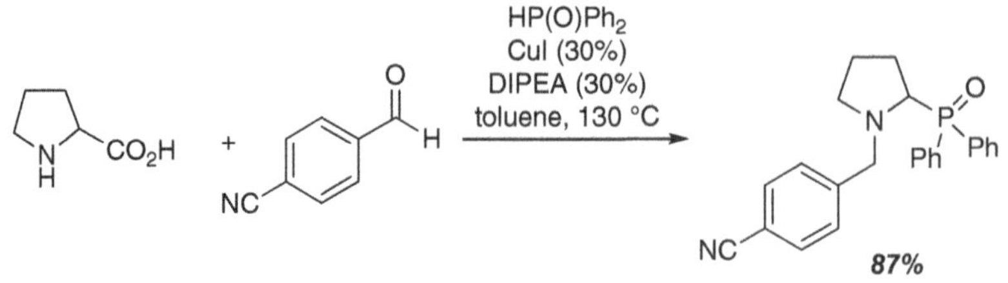

SCHEME 4.40   Decarboxylative approach to the synthesis of α-aminophosphine oxides [91].

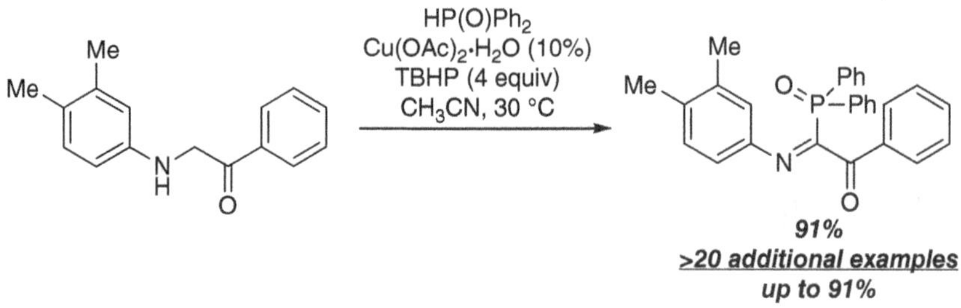

SCHEME 4.41   Synthesis of imidoylphosphine oxides [92].

are attractive starting materials since they are typically quite stable and readily available. Zirconocene hydrochloride (Schwartz's reagent) promoted the reaction and simply stirring the reagents for 12 h at 60 °C afforded high yields of the α-aminophosphine oxides. While only three examples were reported, this system has the potential for wide functional group compatibility. Although secondary amides needed 2.2 equivalents of the zirconocene catalyst for a successful reaction, tertiary amides only required 1.2 equivalents.

α-Aminophosphine oxides were prepared through a tandem decarboxylation coupling reaction between amino acids, aldehydes, and secondary phosphine oxides (Scheme 4.40) [91]. The reaction was catalyzed by copper salts and used DIPEA as a base. The ability to use carboxylic acids as substrates in these reactions is very attractive due to their stability and availability.

In related work, the preparation of imidoylphosphine oxides through successive oxidative dehydrogenation reactions has been achieved (Scheme 4.41) [92]. One of the challenges in this chemistry was inhibiting the competing formation of α-aminophosphine oxides. Many of the initial screenings generated 1 : 1 mixtures of the two products. Through the careful selection of reagent ratios, copper precatalysts, and oxidants, a successful system for the selective formation of the imidoylphosphine oxides was devised. The most effective catalyst system was comprised of copper acetate along with TBHP as oxidant and acetonitrile as the solvent. The chemistry displayed moderate functional group tolerance and was also successful with a number of heteroarenes. While the mechanism of this process was not fully elucidated, a control reaction in the presence of TEMPO was unsuccessful, which suggested that at least part of the process proceeded through a radical pathway.

The asymmetric addition of secondary phosphine oxides to imines for the synthesis of enantioenriched α-aminophosphine oxides has been reported (Scheme 4.42) [93].

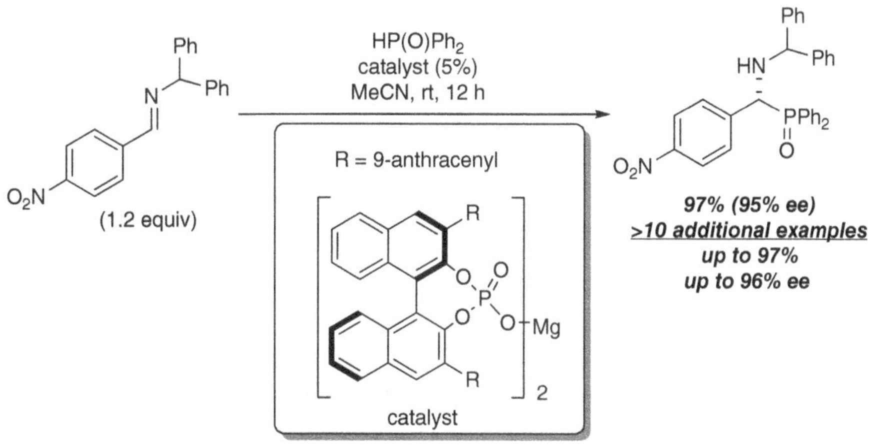

SCHEME 4.42   Addition of secondary phosphines to imines using a magnesium catalyst [93].

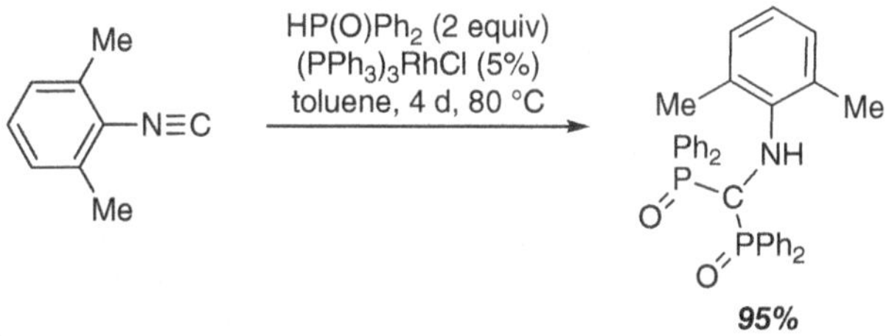

SCHEME 4.43   Addition of diphenylphosphine oxide to isocyanides [94].

The catalyst system was comprised of a magnesium salt of a BINOL-derived phosphate. The authors found that the calcium salt also displayed high activity; however, the sodium salt was considerably less active. Stirring a range of imines with diphenylphosphine oxide and this catalyst generated the α-aminophosphine oxides in high yields with moderate to excellent enantioselectivity. The enantioselectivity was dependent upon the organic fragment attached to the imine. Aromatic groups bearing either electron-donating or withdrawing groups afforded products with high stereoselectivity (ee's = 89–96%); however, the hydrophosphinylation of imines bearing alkyl groups such as n-butyl or n-propyl was less selective (ee's = 48–62%).

The synthesis of bisphosphinoylaminomethanes through the addition of secondary phosphine oxides to isocyanides has been reported by Han (Scheme 4.43) [94]. A range of metal catalysts were screened for activity towards the addition reaction, and rhodium(I) precursors were found to be the most effective. It was interesting that several group 10 catalysts (nickel and platinum) did not catalyze the process and palladium catalysts (Pd_2dba_3) promoted only a single addition of the secondary phosphine oxide that generated iminophosphine oxides. The determination of the scope of this approach was carried out through electronic variation of the groups on the diarylphosphine oxide. While the incorporation of either electron-donating or electron-withdrawing groups successfully generated the target compounds, the authors observed considerable substrate specific

SCHEME 4.44 Synthesis of functionalized alkylphosphine oxides [95].

reactivity. One drawback to this synthesis was the long reaction time (4 d). Furthermore, dialkylphosphine oxides such as $Bu_2P(O)H$ only generated trace amounts of the addition product.

An interesting approach to the synthesis of alkyl diarylphosphine oxides entailed the treatment of N-tosylhydrazones with secondary phosphine oxides (Scheme 4.44) [95]. The overall process can be described as a coupling reaction between a copper carbene species and the secondary phosphine oxide. After some experimentation, the authors found that copper(I) iodide was an effective catalyst for the reaction in the presence of a mineral base. The first step in the reaction was proposed to be the generation of a diazo compound through the reaction of the N-acylhydrazone with the mineral base. Addition of the copper to this intermediate afforded a copper carbene species along with concomitant formation of N_2. The final step involved reaction of the copper carbene with the secondary phosphine oxide followed by hydrogen transfer to generate the alkyl diarylphosphine oxides. It should also be noted that the vast majority of examples presented in this work were based upon N-tosylhydrazones derived from ketones. These are typically more challenging to functionalize than those originating from aldehydes. The chemistry was compatible with a range of electron-donating and electron-withdrawing groups. The reaction chemistry was also quite tolerant to the addition of one ortho-substituent on the aromatic ring. Only the N-tosylhydrazone derived from benzophenone was unreactive in the reaction. A cyclohexyl N-tosylhydrazone was converted into a cyclohexyl diphenylphosphine oxide using this approach, albeit in lower yields (53%). The authors also developed a one-pot reaction for the generation of the alkylphosphine oxides through the in situ generation of the N-tosylhydrazones. Care must be taken with this chemistry since nitrogen is liberated from the reaction, and the reaction vessels need to be properly vented to prevent explosions. Furthermore, a one-pot approach to the chemistry was developed.

The development of transition metal-free hydroelementation reactions has been a goal for the synthetic community for many years. As mentioned in the previous sections on Michael addition reactions, N-heterocyclic carbenes have been shown to be efficient catalysts for these addition reactions. One of the challenges that must be overcome when using these catalysts is the sensitivity of the catalysts to electrophiles. To circumvent this problem, carbon dioxide adducts have been synthesized and used as precatalysts for the phospha-Michael addition of secondary phosphine oxides to activated alkenes (Scheme 4.45) [96]. The success of this approach hinged on the observation that these adducts underwent cleavage in solution to generate the free carbene and carbon dioxide. The concentration of the free carbene was found to be dependent upon the solvent, and THF was found to be an effective solvent for the

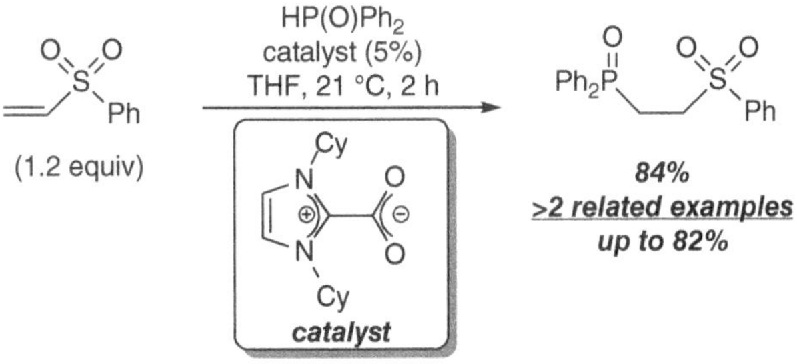

SCHEME 4.45   Use of a CO_2-trapped carbene as a precatalyst for the addition of secondary phosphine oxides to activated alkenes [96].

conversion of the precatalyst into the active species. A variety of carbene precursors were screened for activity, and the $ICy \cdot CO_2$ species was found to be a superior precatalyst. Using this system, simply stirring the reagents at room temperature for a couple of hours generated outstanding yields of the addition product. Diphenylphosphine oxide served as a representative P—H donor in this chemistry, and terminal alkenes bearing a range of electron-withdrawing groups were successfully functionalized. Acrylamide was the rare case that generated low yields of the *phospha*-Michael product. Overall, one of the most attractive aspects of this chemistry was the ability to successfully functionalize the alkenes under an atmosphere of air. The only part of the chemistry that required an inert atmosphere was the generation of the carbon dioxide adducts. Once they were isolated, all of the steps in the addition reactions were carried out under an atmosphere of air.

Diorgano alkylphosphonates are historically prepared by the treatment of a phosphite with an alkyl halide [97]. This transformation is commonly referred to as the Michaelis–Arbuzov reaction, and it has dominated the synthesis of phosphonates. The chemistry is often carried out under solvent-free conditions at elevated temperatures and affords moderate to good yields of the desired alkylphosphonate. It should be noted that a full equivalent of an undesired alkyl halide is released as part of the reaction. The alkylphosphonates generated from Arbuzov reactions are valuable compounds due to their popularity in synthesis. In much the same fashion that α-hydrogens on carbonyl compounds are more acidic than a typical hydrocarbon, the nature of the phosphonate fragment renders adjacent C—H bonds to be more acidic as well. As a result, organophosphonates have a rich history as intermediates in the synthesis of complex organic compounds. Popular reactions such as the Wittig olefination and Horner–Wadsworth–Emmons chemistry are only a few of the known transformations. The synthesis of lipoxygenase and microsomal prostaglandin E_2 synthase-1 inhibitors is an excellent example of this chemistry (Scheme 4.46) [98]. The authors needed to construct a precursor with an alkene bridging a hydrophobic aromatic fragment and a protected ester. The alkene needed to be generated regioselectively and required the *E*-stereochemistry. A further requirement of the synthesis was that the olefination reaction needed to ignore the ester fragment. This ester was critical since it would eventually be deprotected to generate a carboxylic acid that was essential to the binding and activity of the inhibitor. The key C—P bond-forming reaction occurred early in the synthesis and was achieved through an Arbuzov-type coupling reaction involving the treatment of an alkyl bromide with triethylphosphite. Once the phosphonate was incorporated, addition of a base (NaH) removed the acidic hydrogen on the carbon adjacent to the

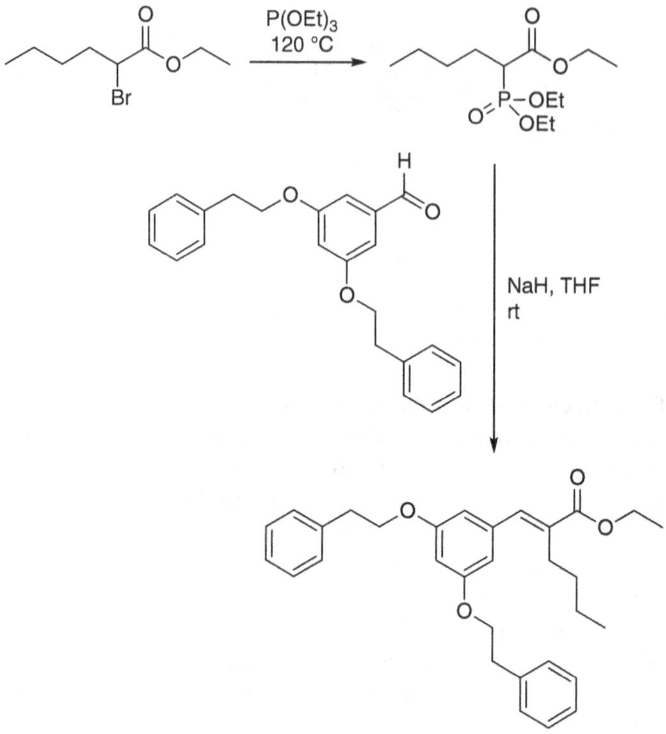

SCHEME 4.46 Tandem C—P bond-forming/olefination reactions [98].

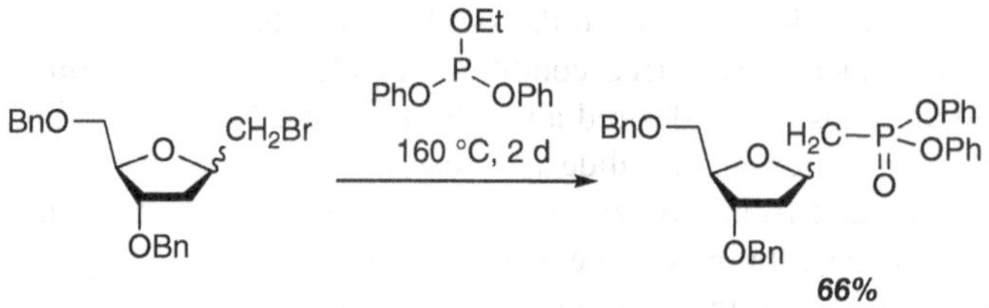

SCHEME 4.47 Preparation of a functionalized ribose through Arbuzov chemistry [99].

electron-withdrawing phosphonate fragment. The stabilized carbanion then added to the aldehyde fragment of the benzaldehyde derivative to generate the alkene along with the loss of phosphorus-containing functional group. Tandem C—P bond-forming/olefination reactions such as these remain some of the most popular approaches for the synthesis of functionalized alkenes.

Trialkylphosphites are not the only phosphorus compounds that are amenable to Arbuzov chemistry. As long as one alkoxy substituent is present on the phosphorus center, the classic Arbuzov chemistry is likely to be successful. This approach has been used to generate a vast array of complex organic structures. Quintiliani recently reported the synthesis of phosphonate and phosphonic acid derivatives of 2-deoxy-D-ribose (Scheme 4.47) [99]. The reaction required heating the reagents to 160 °C for 2 days, and even under these conditions, the phenyl groups were retained and the major reaction pathway was between the alkoxy group

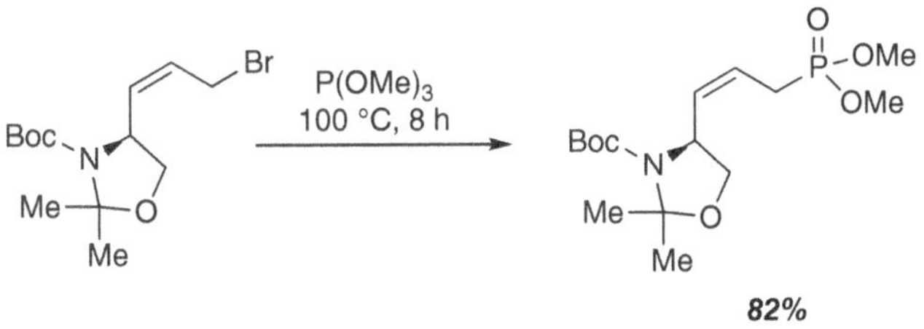

SCHEME 4.48   Synthesis of allylphosphonates using Arbuzov chemistry [100].

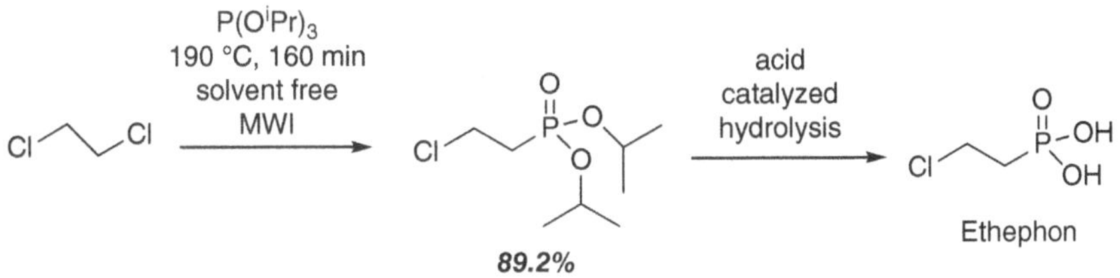

SCHEME 4.49   Synthesis of Ethephon through phosphorus-carbon(sp^3) bond formation/ hydrolysis [101].

on the phosphorus precursor and the alkyl bromide on the ribose. Naturally, these conditions are harsh and many substrates will not survive; however, if the substrate is robust, this is an effective way to prepare alkylphosphonates.

During the synthesis of a group of potential threonine synthase inhibitors, Zervosen studied the synthesis of allylic phosphonates through an Arbuzov-type reaction between allylic bromides and trialkylphosphites (Scheme 4.48) [100]. Trimethylphosphite was used as a representative phosphorus nucleophile in this chemistry. Similar to many of the Arbuzov-type reactions, the chemistry was operationally simple and proceeded under solvent-free conditions. Although the temperature of this process was lower than several of the systems described previously, there are still many substrates that will not survive heating to 100 °C for 8 h.

The microwave-assisted formation of 2-chloroethylphosphonates by irradiation of 1,2-dichloroethane and triethylphosphite mixtures has been investigated (Scheme 4.49) [101]. This reaction is of interest to the agricultural community since the resulting 2-chloroethyl dialkylphosphonate is a direct precursor to Ethephon. Despite the intense interest in this compound, the traditional synthetic procedures using conventional heating generated the target compound in low to moderate yields. In an effort to increase the yield of the reaction, the authors switched to a microwave-assisted approach to determine if the microwave reactor could increase the yield in the reaction. Due to the way that microwave radiation induced dielectric heating in the reaction mixture, it could heat the sample in a fashion that was not possible using conventional heating. Running the reaction in batch mode generated 89.2% of the desired phosphonate after irradiating a mixture of the phosphite and alkyl halide (1 : 10 ratio) for 160 min at 190 °C. Precise control of the temperature during this reaction was critical as increasing the temperature to 210 °C decreased the amount of the target and promoted the formation of secondary products. During the course of the investigation, the authors also probed the effect of solvents on the activity of the catalyst system. Curiously, the addition of one equivalent of toluene to the reaction mixture completely inhibited the formation of

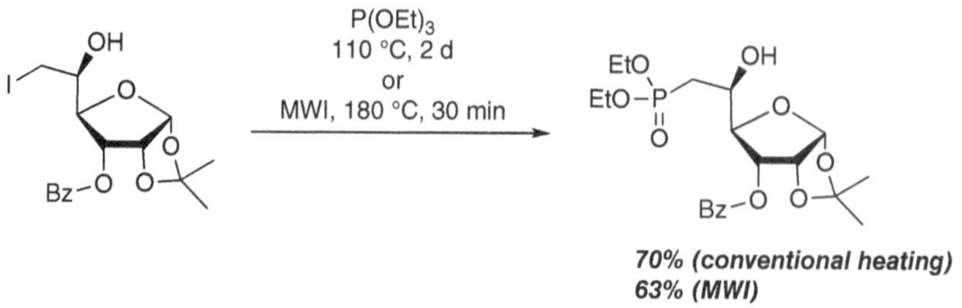

SCHEME 4.50 Functionalization of sugars through Arbuzov chemistry [102].

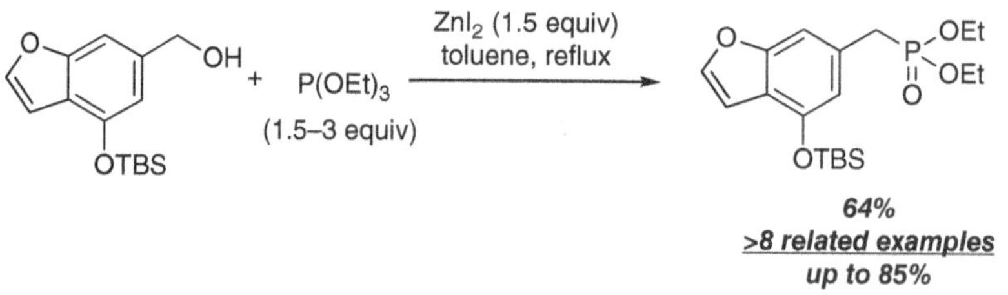

SCHEME 4.51 Lewis acid-assisted synthesis of benzylphosphonates [104].

Michaelis–Arbuzov product. The solvents acetonitrile and DMF were not as troublesome and the phosphonate precursor could still be isolated in moderate to good yields (MeCN=65%; DMF=70%). Additionally, the authors successfully adapted the chemistry to continuous flow while still using microwave heating to promote the reaction.

For the classic Arbuzov reaction between alkyl halides and trialkylphosphites, the order of reactivity is typically I>Br>Cl. Practically, this translates into the observation that reactions using alkyliodides will proceed at lower temperatures. An example of this entailed the functionalization of halogenated sugars with triethylphosphite (Scheme 4.50) [102, 103]. While most Arbuzov reactions with alkyl chlorides required prolonged heating at 190 °C or higher, the iodinated sugar was successfully phosphonated at 110 °C. It was also noteworthy that the presence of an unprotected secondary alcohol did not interfere with the chemistry. The use of a microwave reactor was also investigated, and moderate yields of the phosphonate were obtained after irradiating the sample for only 30 min. The yields obtained from the microwave-assisted reaction were slightly lower (63 vs. 70%) when compared with the conventional heating; however, the reduction in the reaction time could be worth the reduction in the yield. Despite their lower reactivity, alkyl chlorides are still practical electrophiles for this chemistry as they are considerably less expensive and readily available. It should be noted that if the alkyl chloride decomposes at the temperatures needed for a successful Arbuzov-type reaction, the chloride can be substituted by iodide using a Finkelstein reaction.

A significant amount of work has been focused on the synthesis of benzylic and allylic phosphonates. Historically, popular synthetic routes entailed the conversion of the alcohol into a good leaving group such as a mesylate or a halide followed by conversion to the organophosphorus compound. A more efficient synthesis would entail the direct conversion of the alcohol into the phosphonate. To this end, Wiemer described the use of zinc iodide as a Lewis acid catalyst for the direct conversion of benzylic and allylic alcohols into phosphonates (Schemes 4.51 and 4.52) [104]. Zinc iodide is inexpensive and

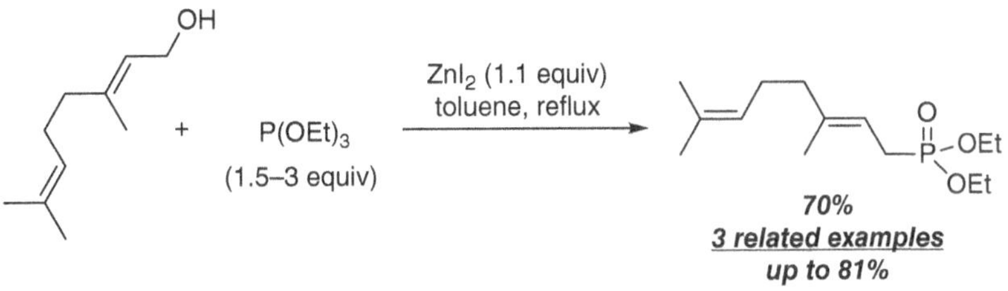

SCHEME 4.52   Lewis acid-assisted synthesis of allylphosphonates [104].

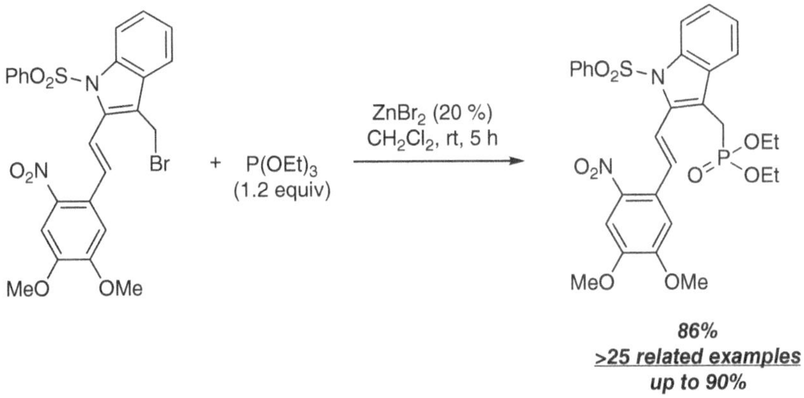

SCHEME 4.53   Lewis acid-assisted conversion of alkyl halides into alkylphosphonates [105].

readily available. Triethylphosphite served as the phosphorus nucleophile for this chemistry. A range of benzyl alcohols were successfully phosphonated using this approach with the highest yields obtained for substrates bearing electron-donating groups. Benzyl alcohols bearing electron-withdrawing groups could be functionalized, but the yields were typically fair to moderate. Once a successful system for the functionalization of benzyl alcohols was established, the chemistry was extended to allylic alcohols. High yields of the allylphosphonates were obtained for primary allylic alcohols, while tertiary allylic alcohols were sluggish and afforded the lowest conversions. While most benzylic and allylphosphonates were successfully functionalized using this catalyst system, analogous reactions with an alkyl alcohol were problematic and did not generate the target alkylphosphonate.

Within a few months following the submission of this chemistry by Weimer, an independent report on the use of Lewis acids to promote the conversion of benzyl halides/ alcohols into the corresponding phosphonates was submitted (Schemes 4.53 and 4.54) [105]. Similar to other Arbuzov-type reactions, the phosphorus nucleophile used in this chemistry was triethylphosphite. After screening a range of Lewis acids, the authors determined that zinc bromide was an effective catalyst for this reaction. For the benzyl bromides, the chemistry was operationally trivial and simply adding the zinc salt to a dichloromethane solution of the reagents followed by stirring at room temperature for 5 h generated moderate to excellent yields of the corresponding phosphonates. Predictably, benzyl chlorides were less reactive and some were unable to be functionalized using this approach. The chemistry was compatible with a range of electron-donating, and

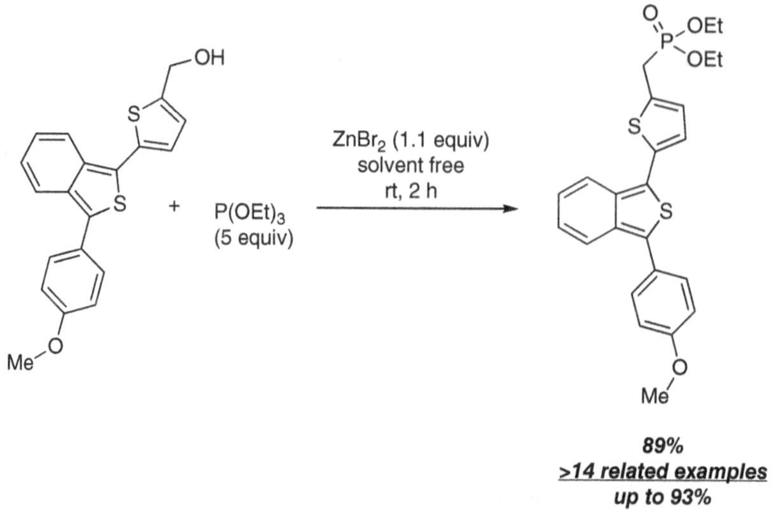

SCHEME 4.54   Lewis acid conversion of alcohols into alkylphosphonates [105].

electron-withdrawing groups as a host of functionalized benzyl halides were converted into benzylphosphonates. The authors also devised a catalyst system for the conversion of benzyl alcohols into benzylphosphonates. While the conversion of benzyl bromides used dichloromethane as the solvent, analogous reactions using benzyl alcohols were successful under solvent-free conditions. Solvent-free reactions are attractive from a sustainability perspective. For best results, the Lewis acid loading needed to be increased to 1.1 equivalents per benzyl alcohol. Similar to the benzyl halide chemistry, a range of functionalized substrates were readily converted into benzylphosphonates using this approach. While most of the examples listed contained a methylene spacer adjacent to the halide or alcohol, several secondary benzyl alcohols were also successfully functionalized. It should be noted that not every substrate was reactive towards the phosphite under the reaction conditions and simple alkyl halides such as butyl bromide failed to generate the target compounds. The broad substrate scope, the operational simplicity, and the high yields of the process makes this one of the most practical approaches to the synthesis of benzylphosphonates using Arbuzov chemistry.

The treatment of a secondary phosphite with a base followed by the addition of alkyl halide to generate an alkylphosphonate has been named the Michaelis–Becker reaction. This process is complementary to Arbuzov–Michaelis chemistry and has the practical advantage of generating organophosphonates at significantly lower temperatures than those typically needed for Arbuzov–Michaelis reactions. While the latter process typically requires heating to above 150 °C, Michaelis–Becker chemistry is often successful at room temperature or below (Scheme 4.55) [106]. One of the most popular bases for this chemistry is sodium hydride. The typical reaction entails stirring the secondary phosphite with the sodium hydride followed by the addition of the alkyl halide. The yields from this chemistry are typically good to excellent. The following sections will describe several versions of this reaction that have been reported.

The phosphonation of a 6-mesyloxymethyl-protected purine using the sodium salt of diethylphosphite has been reported (Scheme 4.56) [107]. Although it could be predicted that the 6-iodomethyl purine should have been able to be functionalized using standard

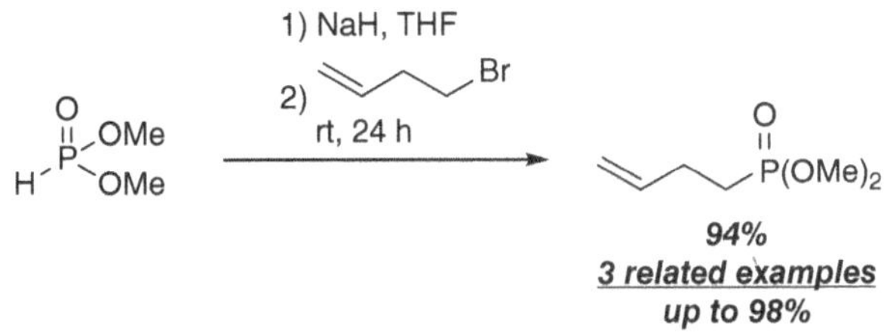

SCHEME 4.55 Synthesis of alkylphosphonates through base-assisted Michaelis–Becker reactions [106].

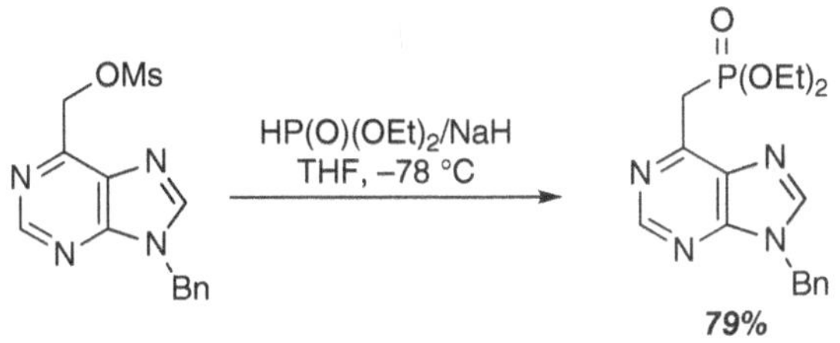

SCHEME 4.56 Michaelis–Becker synthesis of functionalized purines [107].

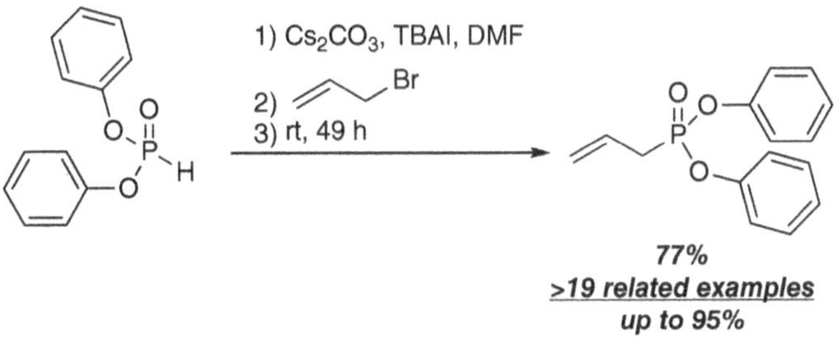

SCHEME 4.57 Cesium carbonate-promoted Michaelis–Becker reactions [108].

Arbuzov conditions, the authors obtained an intractable mixture using this approach. Moving to the mesylate analogue and using a Michaelis–Becker approach generated the phosphonated materials in high yield. While a number of similar reactions have been shown to be successful at room temperature and above, these reactions afforded the highest yields when they were performed at −78 °C. At higher temperatures, a side reaction with a second equivalent of the purine competed with the desired monofunctionalization reaction.

Sodium hydride remains one of the most popular bases for Michaelis–Becker reactions; however, the substrate scope is limited to compounds that do not contain electrophiles. While it might be predicted that the sodium hydride should be consumed by the time the alkyl halide has been added, this is practically difficult. Further, complicating this chemistry is the highly reactive nature of the sodium hydride. This results in difficulties when attempting to add precisely one equivalent to a reaction mixture. To identify a replacement for sodium hydride, the effectiveness of various bases on this reaction has been investigated (Scheme 4.57) [108]. A number of metal carbonates were screened due to their

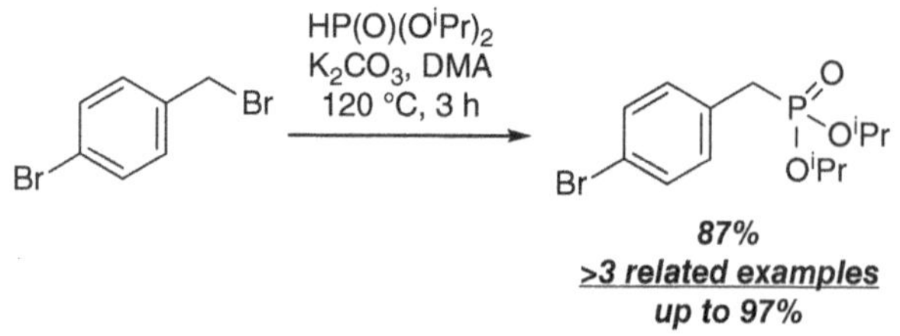

SCHEME 4.58   Potassium carbonate-promoted Michaelis–Becker reactions [108].

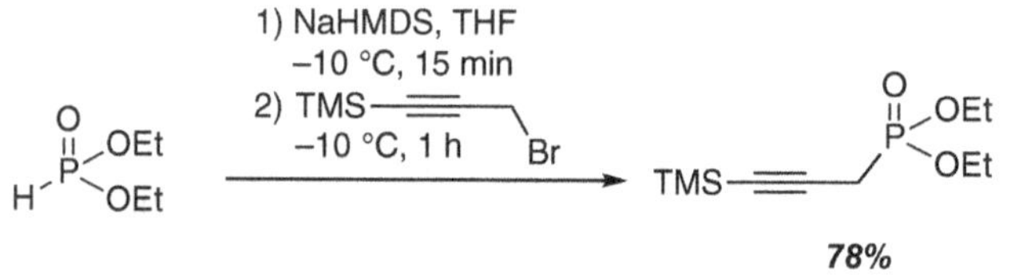

SCHEME 4.59   Using NaHMDS as the base in the synthesis of propargylphosphonates [110].

compatibility with a wide range of functional groups, low cost, and high availability. After a bit of experimentation, cesium carbonate was determined to be an effective base for the Michaelis–Becker reaction using DMF as the solvent along with tetrabutyl ammonium iodide as an additive. The effect of changing the organic fragments on the secondary phosphites was also investigated, and the authors found that both dialkyl- and diarylphosphites were successful in these reactions. Moderate to excellent yields of the alkylphosphonates were reported for a range of primary alkyl halides, while secondary alkyl halides such as bromocyclohexane were sluggish, and lower yields of the alkylphosphonates were obtained even with prolonged stirring. When compared with sodium hydride, cesium carbonate is significantly easier to store and handle. Due to these observations as well as the compatibility of cesium carbonate with a host of functional groups, this is a practical approach for the synthesis of these compounds. In related work, potassium carbonate has also been used to promote Michaelis–Becker reactions (Scheme 4.58) [109]. The temperature of this process was less than was typically needed for an Arbuzov-type reaction, but higher than when cesium carbonate was used.

Sodium bis(trimethylsilyl)amide is another base that has been successfully used in a Michaelis–Becker synthesis of alkylphosphonates (Scheme 4.59) [110]. Similar to the other reactions described previously, the reaction proceeds through initial generation of a sodium dialkylphosphonate followed by the addition of alkyl halide to generate the products. This base is attractive since it is relatively inexpensive and readily available as a solution. Overall, this type of base is significantly less hazardous to use than sodium metal or sodium hydride. The potassium salt of bis(trimethylsilyl)amide has also been used to promote the phosphonation of a benzyl bromide (Scheme 4.60) [111]. This chemistry was attractive as an aryl bromide was retained throughout the Michaelis–Becker chemistry and was available for further functionalization in subsequent reactions.

An alternative route to the synthesis of alkylphosphonates was recently developed during an investigation into modern methods of functionalizing nucleosides (Scheme 4.61) [112]. The authors approach started from nucleoside derivatives and used

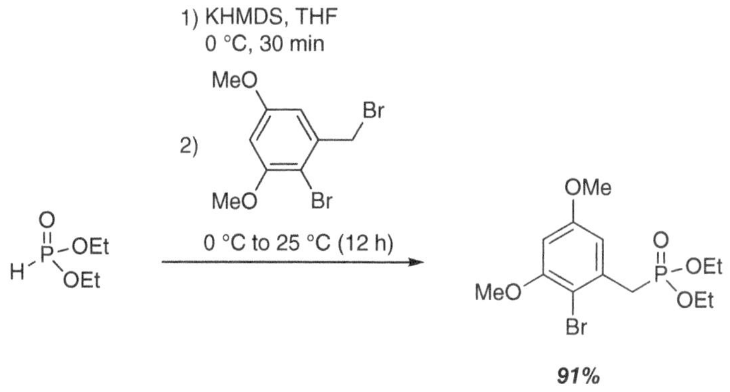

SCHEME 4.60   Using KHMDS in the synthesis of benzylphosphonates [111].

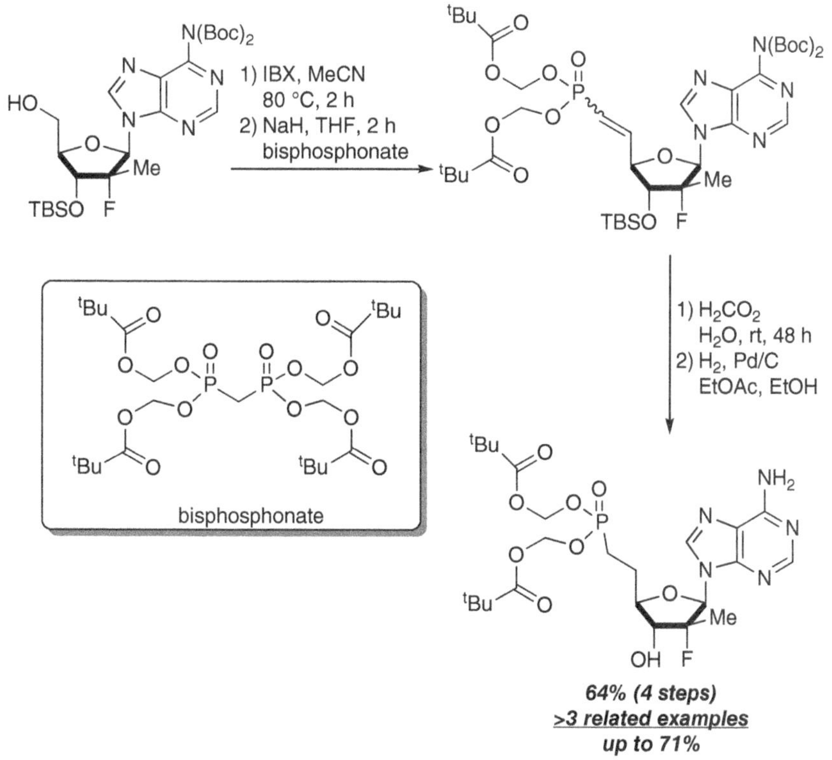

SCHEME 4.61   Preparation of nucleoside-derived alkylphosphonates [112].

Horner–Wadsworth–Emmons chemistry to generate an alkenylphosphonate intermediate that was reduced to generate the alkylphosphonate species. Normally, phosphonate fragments are introduced into the 5′-position using Michaelis–Arbuzov or Michaelis–Becker chemistry; however, this route is an attractive alternative since it avoided the high temperatures typically required for Michaelis–Arbuzov chemistry and it also eliminated the

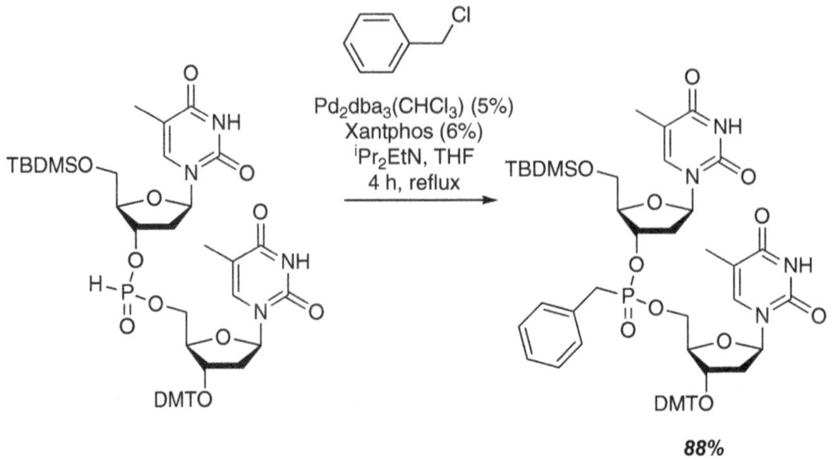

SCHEME 4.62 Palladium-catalyzed synthesis of nucleoside-derived benzylphosphonates [113, 114].

need to prepare an alkyl halide precursor. The first step in this chemistry transformed a primary alcohol into the corresponding aldehyde using 2-iodoxybenzoic acid (IBX) as the oxidizing agent. The authors also screened Dess–Martin periodinane as the oxidizing agent, but they obtained intractable mixtures in those reactions and concluded that the IBX oxidation was superior. Treatment of this intermediate aldehyde with an *in situ* generated bisphosphonate salt generated the alkenylphosphonate. Finally, reduction and deprotection afforded the target compounds in moderate yields. While most of the attention devoted to bisphosphonate compounds has been focused on their synthesis and biological activity, they are very practical precursors to alkenylphosphonate derivatives.

In addition to the methods described previously, Stawinski developed a palladium-catalyzed cross-coupling reaction for the formation alkylphosphonates (Scheme 4.62) [113, 114]. The key component of his approach was the use of a wide bite-angle diphosphine ligand (XantPhos). Such ligands have been shown to increase the rate of reductive elimination of alkylphosphonates from palladium centers [115, 116]. The cross-coupling chemistry displayed wide functional group compatibility, and range of aryl halides were readily incorporated. In addition to the ethers and esters screened in the initial studies, a substrate bearing an unprotected carboxylic acid was successfully coupled with diethylphosphite (99%). This was a rare example of such a reaction. Typically, these are challenging substrates to use in cross-coupling reactions. Once the catalyst system was defined, the authors moved to more complex substrates such as cholesterol and nucleoside derivatives. Even with these complex substrates, the chemistry was operational straightforward as the reagents and catalyst could be loaded in air, and the only precaution listed in the procedure was a nitrogen purge immediately before heating.

An interesting approach to the synthesis of alkylphosphonates from *N*-tosylhydrazones has been presented (Scheme 4.63) [95]. The overall reaction was essentially a coupling reaction between the *N*-tosylhydrazones and a secondary phosphite. The catalyst system for this chemistry consisted of copper iodide along with a mineral base. For the hydrazones derived from ketones, the most effective base was potassium carbonate; however, for analogous reactions involving hydrazones derived from aldehydes, the most effective

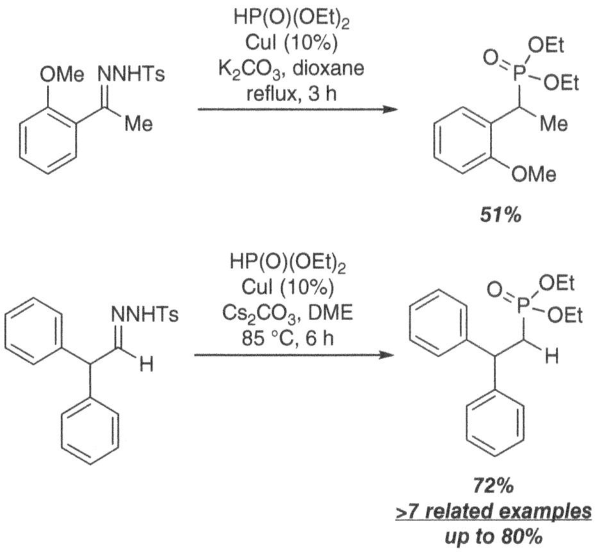

SCHEME 4.63   Conversion of *N*-tosylhydrazones into alkylphosphonates [95].

base was cesium carbonate. A proposed mechanism entailed initial conversion of the hydrazone into a diazo compound through treatment with the mineral base. Reaction of this intermediate with the copper(I) salt generated a copper carbene species and liberated N_2. Addition of the secondary phosphite to this species followed by hydrogen transfer generated the alkylphosphonates. While only one example of an *N*-tosylhydrazone derived from a ketone was presented, several examples of *N*-tosylhydrazones originating from aldehydes were listed. A range of electron-donating and electron-withdrawing groups on the aromatic groups were well tolerated. The addition of one ortho-substituent only slightly decreased the yield of the reaction. An alkyl *N*-tosylhydrazone was also successfully functionalized using this approach (65%). One drawback that could be envisioned was the need to prepare the *N*-tosylhydrazones prior to the coupling reaction. The authors addressed this issue through the *in situ* generation of the *N*-tosylhydrazones by treating aldehydes and ketones with tosylhydrazide prior to the addition of the secondary phosphite. Care must be taken with this chemistry since nitrogen is liberated from the reaction, and the reaction vessels need to be properly vented. The one-pot approach to the chemistry was attractive as only slight decreases in the overall yields were obtained.

The addition of secondary phosphites to alkenes is a powerful way to prepare functionalized alkylphosphonates. If the alkene contains electron-withdrawing substituents and is "activated" for 1,4-conjugate addition, the addition reaction is commonly referred to as a *phospha*-Michael addition. As with most Michael-type addition reactions, the reaction is typically regioselective and moderate to high yielding. The reaction is typically carried out under basic conditions, and a number of organic and mineral bases have been used to promote this reaction (Schemes 4.64–4.69) [117–121, 123]. When mild bases were used, the reactions were typically carried out in a one-pot fashion with the substrate, secondary phosphite, solvent, and base added at the same time. Reactions using stronger bases such as sodium hydride or lithium diisopropylamide were carried out in several steps. The process

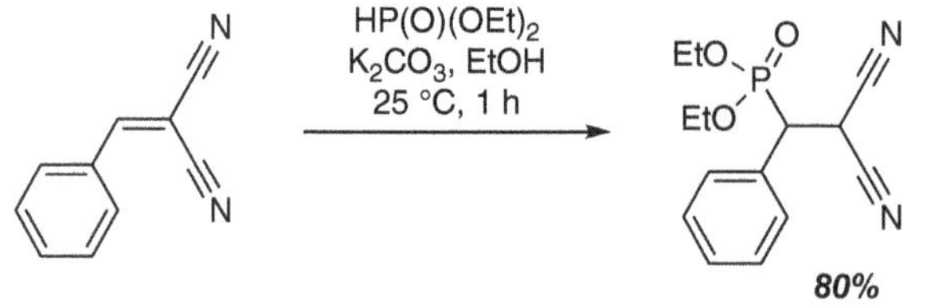

SCHEME 4.64 Potassium carbonate-promoted *phospha*-Michael reactions [117].

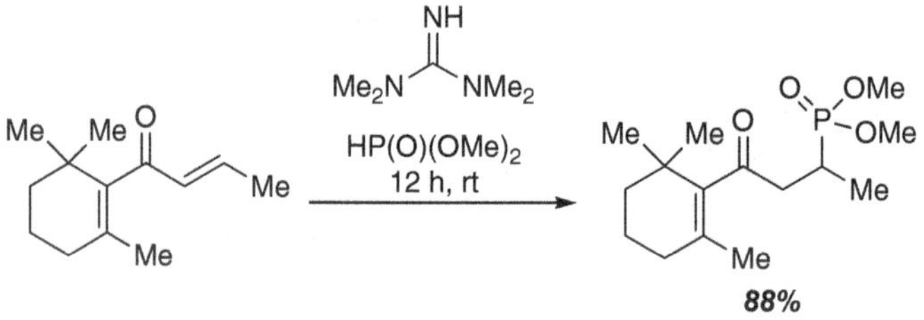

SCHEME 4.65 Tetramethylguanidine-promoted *phospha*-Michael reactions [118].

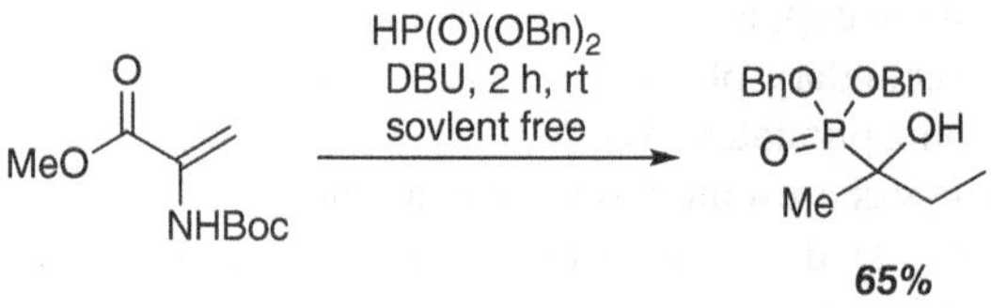

SCHEME 4.66 DBU-promoted *phospha*-Michael reactions [119].

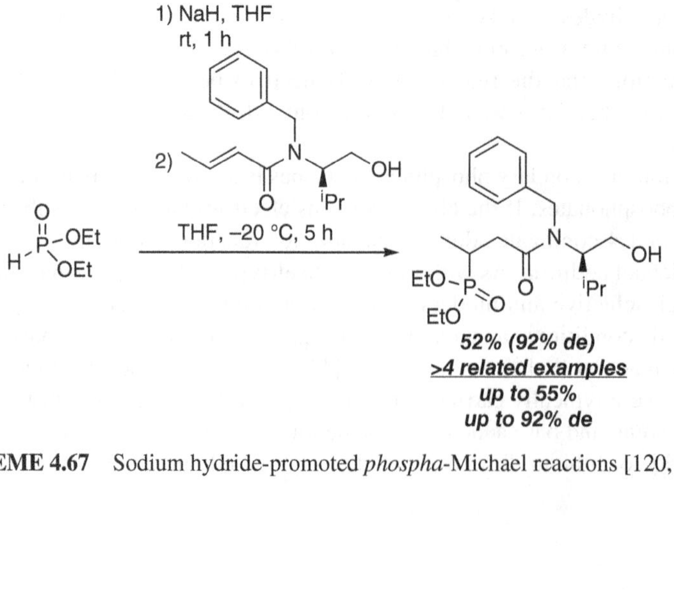

SCHEME 4.67 Sodium hydride-promoted *phospha*-Michael reactions [120, 121].

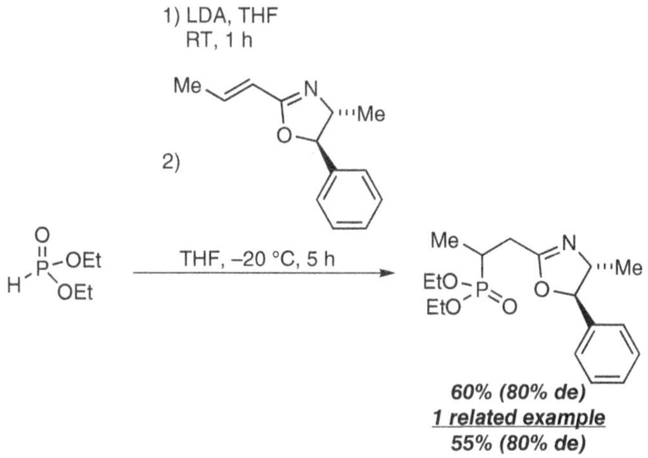

SCHEME 4.68 LDA-promoted *phospha*-Michael reactions [121].

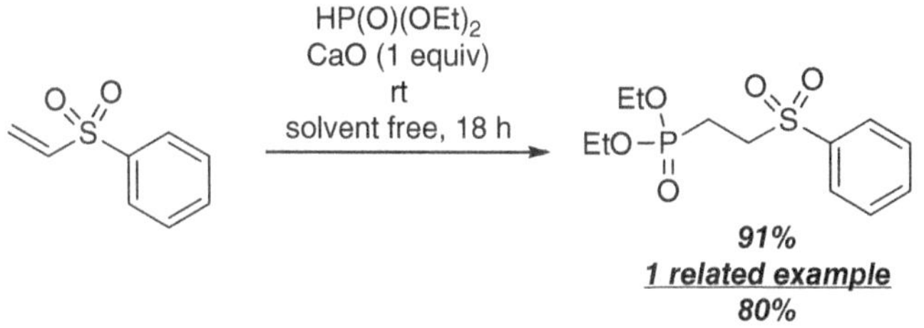

SCHEME 4.69 Calcium oxide-promoted *phospha*-Michael reactions [122].

typically involved initial treatment of the phosphite with the base to generate an MP(O)(OR)$_2$ salt followed by addition of the Michael acceptor to generate the 1,4-conjugate addition product. Provided that the substrate does not contain electrophiles that could intercept the phosphorus reagent, this is a reliable method for the preparation of a range of alkylphosphonate compounds. This chemistry has grown in popularity in recent years and has been reviewed [43].

For α,β-unsaturated ketones and related substrates, one of the challenges encountered when attempting to generate an alkylphosphonate through a 1,4-conjugate addition is competing 1,2-addition, which results in the formation of a hydroxyphosphonate. A practical solution to this problem has been reported and entailed the use of *N,O*-bis(trimethylsilyl) acetamide (BSA) in combination with a catalytic amount of trimethylsilyl triflate to generate a silyl enol ether intermediate (Scheme 4.70) [123]. Treatment of this compound with HCl afforded the β-ketophosphonate (1,4-addition product). After some optimization of the reaction conditions and catalyst loadings, the chemistry was extended to a range of cyclic enones. A variety of electronically varied substrates were successfully functionalized using this approach. It should be noted that substrates that generated a significant amount of the 1,2 adduct contained substituents in the β-position. For example, the use of 3-methyl cyclohexanone as the substrate generated 20% of desired 1,4-conjugate addition product and

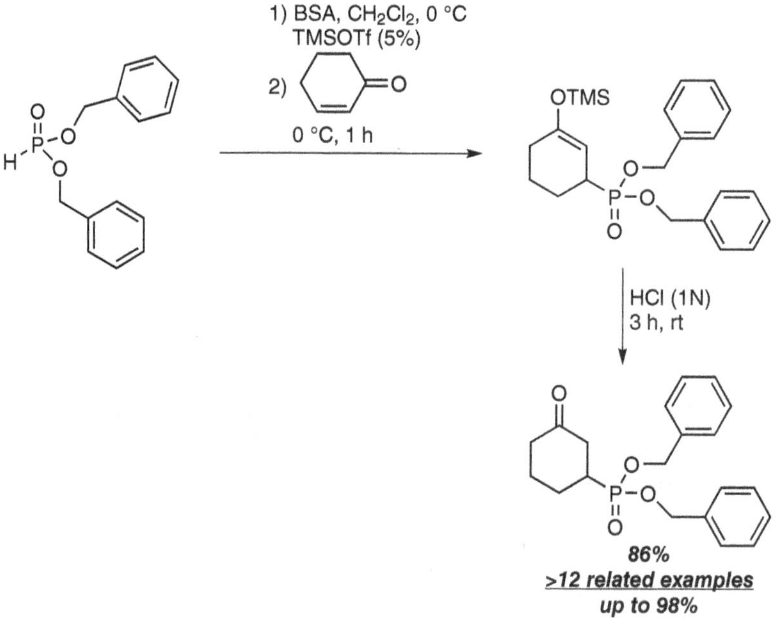

SCHEME 4.70 Selective addition of secondary phosphites to alkenes in the presence of carbonyl groups [123].

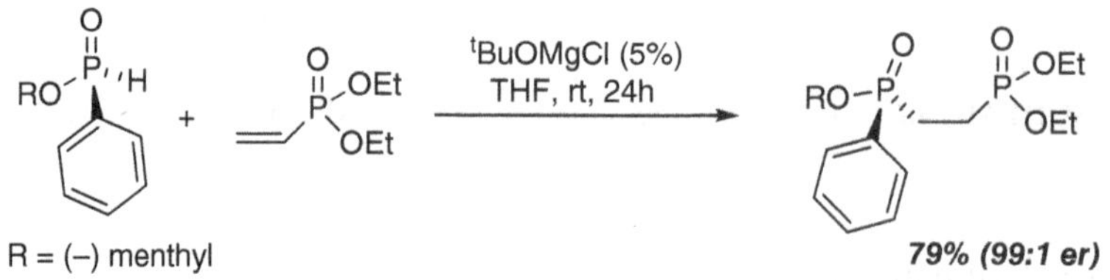

SCHEME 4.71 *Phospha*-Michael additions with retention of chirality [124].

68% of the hydroxyphosphonate. The use of five-membered cyclic enones also generated a small amount of the 1,2-addition product (10%). Despite these minor limitations, this represents a significant improvement in the selectivity of the *phospha*-Michael addition reaction under basic conditions.

The ability to preserve stereochemistry at phosphorus through a base-assisted *phospha*-Michael-type reaction has been investigated (Scheme 4.71) [124]. The resolved phosphorus precursor was a P-chiral hydrogen phosphinate bearing a (−) menthyl substituent. Several activated alkenes were screened for susceptibility towards the addition reaction along with a host of bases. Due to rapid epimerization of the phosphorus center in the presence of bases, the retention of stereochemistry was anticipated to be problematic. Indeed, using common bases such as sodium *tert*-butoxide to promote the reaction resulted in nearly complete scrambling of the stereochemistry (R:S=48:52). However, through a careful screening of several different bases, the authors discovered that a group of magnesium alkoxides promoted the *phospha*-Michael reaction with nearly complete

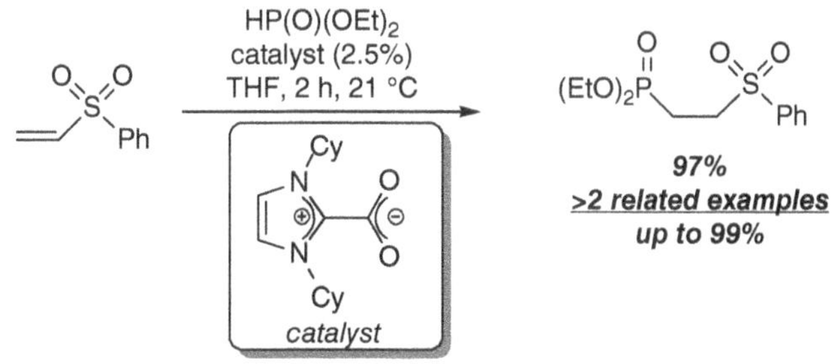

SCHEME 4.72   *Phospha*-Michael additions using carbene catalysts [96].

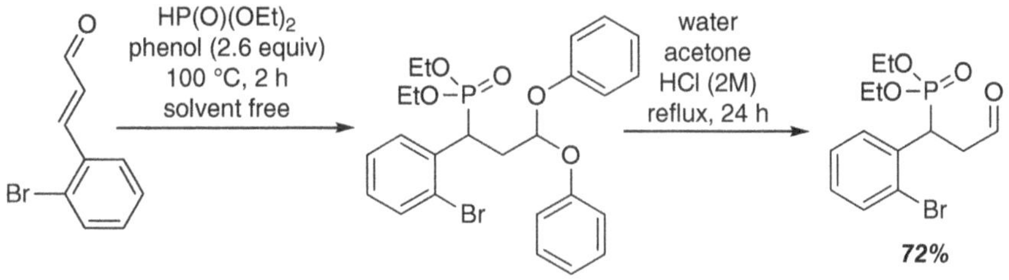

SCHEME 4.73   Synthesis of alkylphosphonates from aldehyde-containing alkenes [125–127].

preservation of the chirality at phosphorus (99:1 er). Several activated alkenes were screened for activity and moderate to excellent yields were observed. It was noteworthy that no scrambling of the —OR fragments on —CO_2R or —$P(O)(OR)_2$ groups was observed. This was a practical observation for the construction of complex organic molecules bearing P-chiral groups.

N-heterocyclic carbenes are efficient catalysts for the *phospha*-Michael addition of secondary phosphine oxides to activated alkenes [96]. These transition metal-free reactions employed a precatalyst comprised of the carbene attached to carbon dioxide. The carbene–CO_2 adducts were quite stable in the solid state; however, they underwent spontaneous cleavage in an organic solvent such as THF to generate the catalytically active carbene as well as carbon dioxide. This catalyst can also be used for the addition of hydrogen phosphonates to activated alkenes (Scheme 4.72) [96]. THF was an effective solvent for the reaction, and the $ICy \cdot CO_2$ adduct was found to be the most effective precatalyst. The reaction conditions for this addition reaction were quite mild and generated nearly quantitative yields of the addition products. Vinyl sulfones, ketones, and esters were all successfully transformed using this chemistry. Of the alkenes investigated in this reaction, the only alkene that did not generate the Michael product was acrylamide. Perhaps, the most practical aspect of this chemistry was the observation that all aspects of the addition reaction (charging, stirring, and product isolation) could be carried out under an atmosphere of air.

As mentioned previously, the use of aldehydes in *phospha*-Michael reactions can generate a range of 1,2-addition products (hydroxyphosphonates) or mixtures of 1,4- and 1,2-addition products. A solution to this problem has been developed that involved the conversion of the aldehyde into an acetal through the addition of excess phenol (Scheme 4.73) [125–127]. Once the aldehyde was masked, the addition of the phosphorus reagent to the alkene generated a β-diethylphosphonyl acetal. The addition was

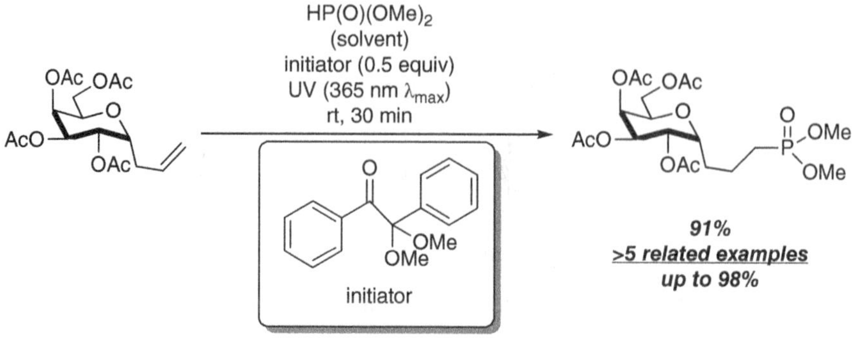

SCHEME 4.74   Radical addition of diethylphosphite to enopyranoses [128].

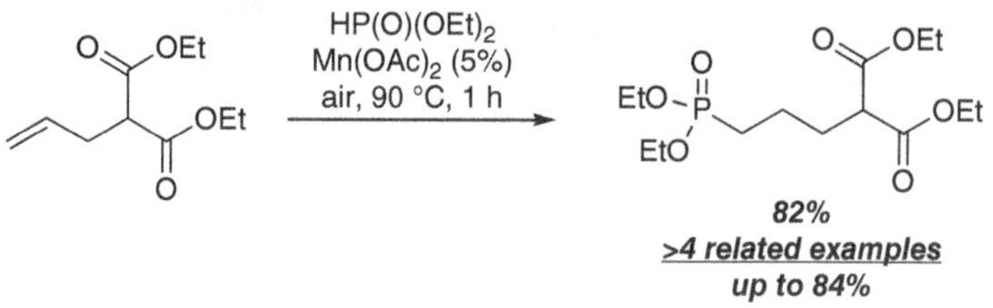

SCHEME 4.75   Manganese acetate-promoted addition of phosphites to alkenes [129].

regioselective and moderate yields of this intermediate were obtained. Acid-catalyzed cleavage of the acetal generated the β-diethylphosphonyl-aldehyde. The conditions for the collapse of the acetal were relatively mild, and the phosphonate diester unit remained intact through this step. Moderate to excellent yields of the addition product have been reported for the multistep synthesis when starting from cinnamaldehyde derivatives.

While considerable effort has been devoted to the development of base-assisted reactions, radical processes have also been devised. To this end, the radical hydrophosphonylation of enopyranoses has been achieved using a common initiator in conjunction with irradiation at 365 mn (Scheme 4.74) [128]. The goal of this work was to develop an addition reaction that would occur under mild conditions and not degrade sensitive biomolecules such as enopyranoses. While other reports outlined the use of acylphosphonates and more elaborate precursors for the generation of phosphoryl radicals, the authors investigated simple dialkylphosphites as substrates and found that they were active in the addition reaction. The authors selected a 2,2′-dimethoxy-acetophenone as the initiator, and after screening a range of conditions, the authors discovered that running the reaction in neat dimethylphosphite afforded the highest amount of the alkylphosphonate. A number of enopyranoses were functionalized using this approach with the lowest yields obtained when a methylene exo-glycal was used as a substrate (45%). Attempts at functionalizing a substrate that incorporated an endocyclic double bond were unsuccessful, and only starting material was observed after irradiation.

One of the most practical methods for the radical hydrophosphonylation of alkenes uses manganese (II) acetate to promote the addition reaction (Scheme 4.75) [129]. The chemistry was operationally trivial, and simply adding the reagents to a reaction vessel in air followed by stirring under air at 90 °C afforded fair to excellent yields of the alkylphosphonates. For terminal alkenes, the reaction selectively formed the anti-Markovnikov

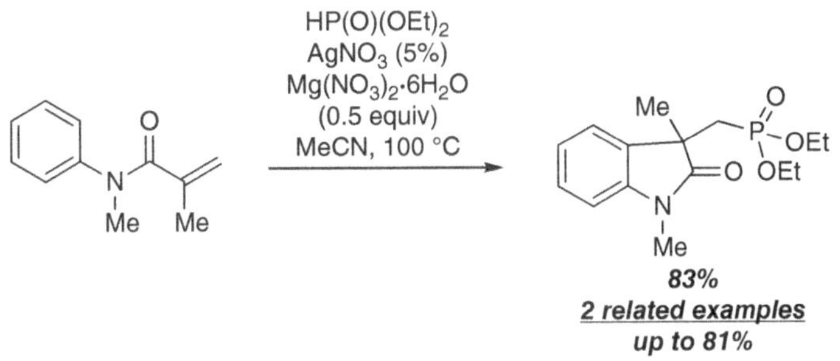

SCHEME 4.76   Synthesis of nitrogen heterocycles with tethered phosphonate groups [89].

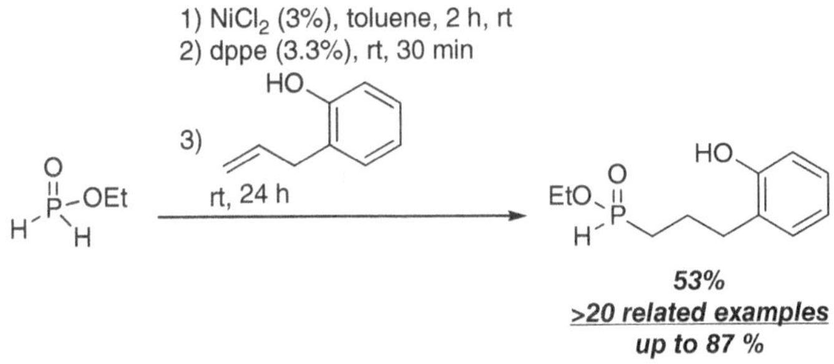

SCHEME 4.77   Nickel-catalyzed hydrophosphinylation [130].

addition product, while the use of internal alkenes generated mixtures. While the majority of examples were hydrocarbons, a β-diester was successfully transformed into the alkyl-phosphonate in excellent yield (82%) using this approach. The ability to promote the addition reaction in air makes this a very practical approach to the synthesis of alkylphosphonates.

Similar chemistry has been used in the preparation of nitrogen heterocycles bearing tethered phosphonate groups (Scheme 4.76) [89]. After screening a range of metal salts, the authors selected silver nitrate as the catalyst for the system due to its high activity and low cost. After some experimentation, the combination of silver nitrate and magnesium nitrate was determined to be an effective catalyst system for the addition/cyclization reaction. The rationale behind the two-component catalyst was a series of control reactions that revealed that a high nitrate concentration was essential, but a high silver concentration was not. Decreasing the silver concentration to 5% and adding 0.5 equivalents of the magnesium nitrate promoted the addition reaction. Although the amount of silver could be lowered, it was still essential for a successful reaction as its removal led to no reaction. While only a few examples were given for secondary phosphites, the chemistry should be extendable to a range of electronically varied substrates as analogous reactions with secondary phosphine oxides were insensitive to the electronic structure. Given the high isolated yields of the cyclized compounds as well as an inexpensive and readily available catalyst system, this approach is very attractive.

The nickel-catalyzed hydrophosphinylation of unactivated alkenes using phosphinates is an attractive approach for the synthesis of alkylphosphinates (Scheme 4.77) [130]. After screening a range of metal salts and supporting ligands, the combination of nickel (II) chloride

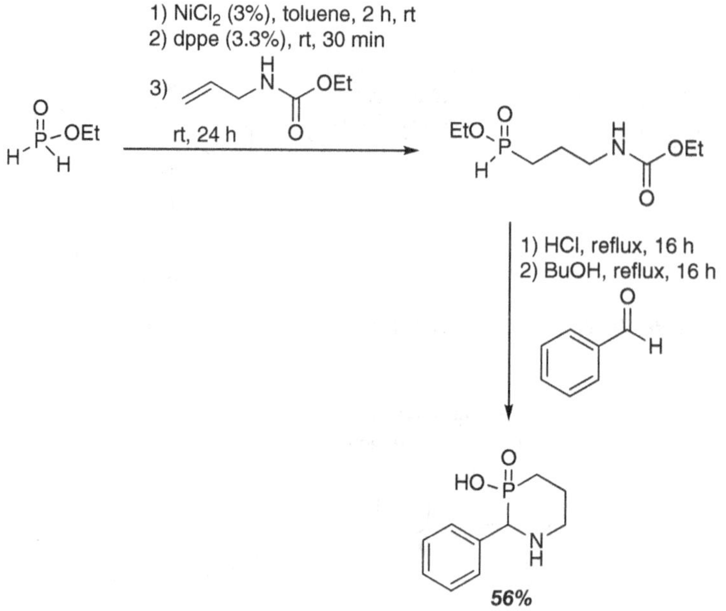

SCHEME 4.78 Nickel-catalyzed preparation of phosphorus-containing heterocycles [130].

and dppe was found to be the most effective. One of the competing reactions discovered during the course of this investigation was a transfer hydrogenation process. After some experimentation, this side reaction could be minimized if the nickel catalyst was preactivated. This step was operationally trivial as it involved stirring nickel chloride with the phosphinate at room temperature in toluene. After this preactivation, the diphosphine was added followed by the alkene. Once a suitable system was devised, it was extended to a range of alkenes including examples bearing ester, amide, TMS, and ketone containing groups. Perhaps, the most impressive was the ability to hydrophosphinylate an alkene bearing an unprotected phenol. Using a substrate that incorporated both alkene and phosphinate groups, the authors were able to use the nickel catalysis to promote the formation of a phosphorus-containing heterocycle (Scheme 4.78). This chemistry was sensitive to oxygen, and reactions exposed to air afforded lower yields of the target products. The chemistry was also scaled up (20 g) without a significant reduction in yield (79% isolated). Given that the chemistry proceeds in moderate to excellent yields at room temperature using both an inexpensive metal catalyst and supporting ligand, this is a very attractive process.

The simultaneous addition of multiple functional groups to unactivated alkenes is a powerful way to create heteroatom-dense compounds. To this end, Li investigated the use of common silver salts in combination with Selectfluor to promote the phosphonofluorination of alkenes (Scheme 4.79) [131]. The authors proposed a radical mechanism with the generation of phosphinyl radicals via a Ag(III)–Ag(II) reduction as a key step in the process. When terminal alkenes were used, analysis of the products revealed that the regioselectivity of the system favored the addition of the phosphonate fragment to the terminal position. The chemistry displayed broad functional group tolerance as evidenced by the large number of products containing alcohols, sulfonates, protected amines, amides, esters, ketone, and nitriles. While unactivated alkenes were readily functionalized using this approach, Michael acceptors such as methyl methacrylate did not generate phosphonofluorination products.

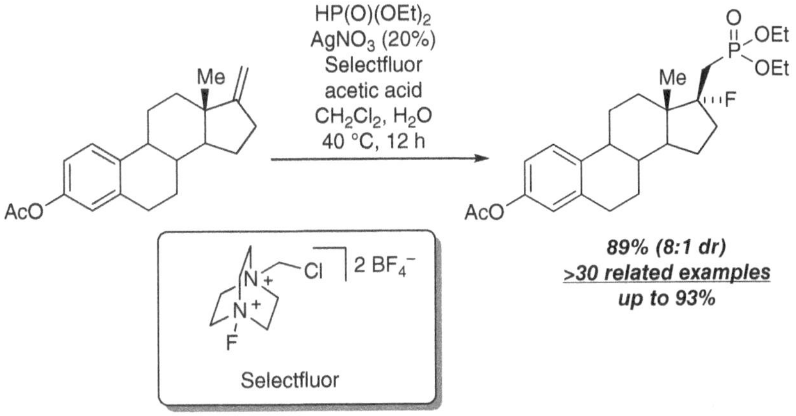

SCHEME 4.79 Radical phosphonofluorination of alkenes [131].

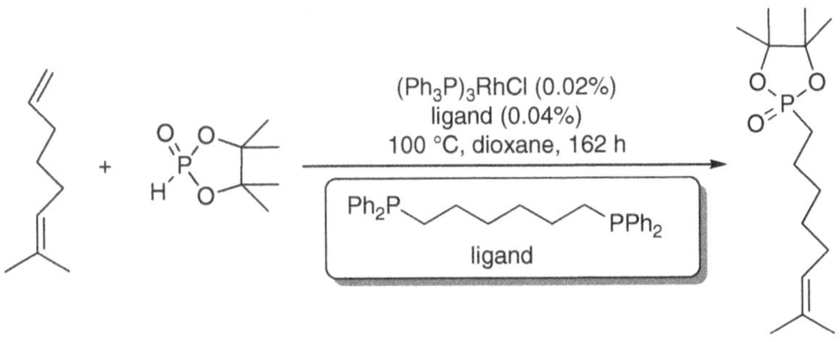

SCHEME 4.80 Rhodium-catalyzed addition of phosphites to unactivated alkenes [132].

The chemistry was also incompatible with functional groups that could be easily oxidized such as a phenol. Due to the mild conditions, as well as the high yields and readily available reagents, this is a very practical approach to the synthesis of β-fluoro-alkylphosphonates.

The rhodium-catalyzed addition of phosphites to unactivated alkenes has been reported (Scheme 4.80) [132]. The ability to lower the catalyst loading and still achieve high yields as well as the same product distribution is a critical concern for the development of sustainable reactions. Furthermore, lowering the amount of metal needed for a specific transformation is a practical concern from the standpoint of drug synthesis due to smaller amounts of metal that needs to be removed from an intermediate or prodrug. For this investigation, the phosphorus reagent was the pinacol-derived hydrogen phosphonate used in the studies described in previous sections, while the organic substrate was an unsymmetrical diene bearing both terminal and internal alkenes. When other secondary phosphites such as dimethylphosphite or diethylphosphite were screened, no addition products were observed. The authors started with Wilkinson's catalyst as the rhodium source. During the investigation, the authors noted that a small amount of excess phosphine significantly improved the yields of the reaction. To this end, they screened a range of monodentate and bidentate phosphine ligands and found that almost all phosphines could promote the addition reaction. However, only the bis(diphenylphosphino)hexane was able to promote

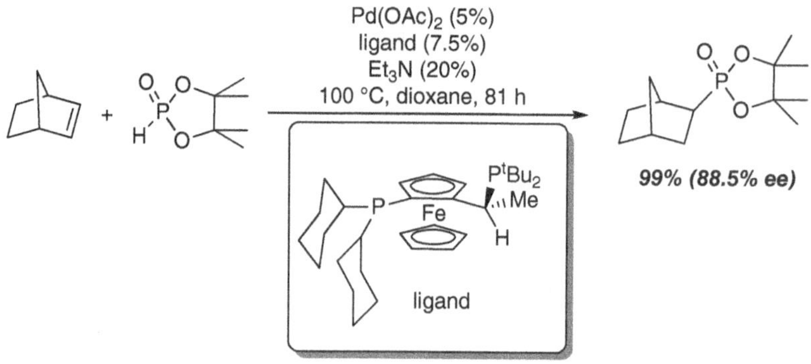

SCHEME 4.81   Asymmetric addition of phosphites to norbornene [133].

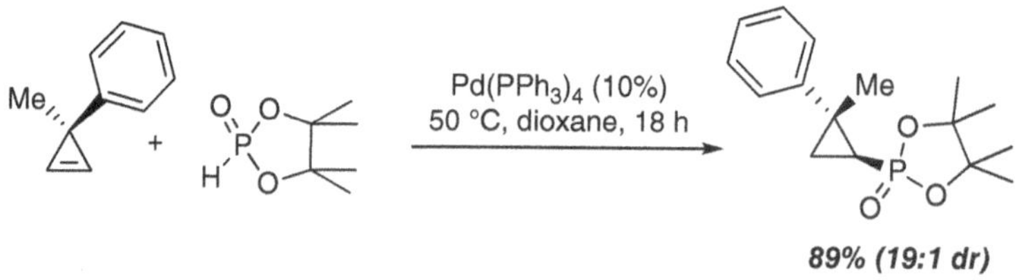

SCHEME 4.82   Hydrophosphonylation of unfunctionalized cyclopropenes [134].

the reaction at 0.02 mol% loading. The reaction was selective for the terminal alkene and also regioselective for the anti-Markovnikov product. For comparison, the authors also carried out a radical hydrophosphonylation reaction using these substrates and found that the radical process generated a significant amount of secondary products resulting from nonselective additions and cyclization. Thus, the metal-catalyzed process was significantly more controlled than the radical hydrophosphonylation reaction.

The asymmetric addition of the pinacol-derived hydrogen phosphonate to norbornene has been reported (Scheme 4.81) [133]. The catalyst system used for this transformation consisted of a palladium source as well as a resolved diphosphine ligand. After some experimentation, the authors determined that palladium acetate combined with a JosiPhos-type diphosphine ligand afforded high conversions with high selectivity. Dioxane was found to be the most effective solvent for the reaction. While a model reaction generated greater than 99% of the product with approximately 80% ee using conventional heating, microwave irradiation also promoted the formation of the alkylphosphonate in greater than 99% yield with approximately 80% ee; however, the reaction time was decreased from 17 h to 5 h using the microwave-assisted procedure. For the conventionally heated reactions, the addition of triethylamine (20 mol%) followed by stirring for 81 h resulted in an increase in the selectivity (88.5% ee). It was noteworthy that the yield and ee of the product from the model system were essentially unchanged when air (0.5 mL) was introduced into the reaction vessel. Overall, this was a practical addition to the growing body of reports on the asymmetric addition of hydrogen phosphonates to alkenes due to the ability of the catalyst system to tolerate contamination by air as well as the ability to use microwave irradiation to decrease the reaction time without sacrificing the yield or selectivity of the process.

The addition of phosphites to cyclopropenes offers a valuable strategy for the generation of cyclopropylphosphonates (Schemes 4.82 and 4.83) [134]. Similar to metal-catalyzed

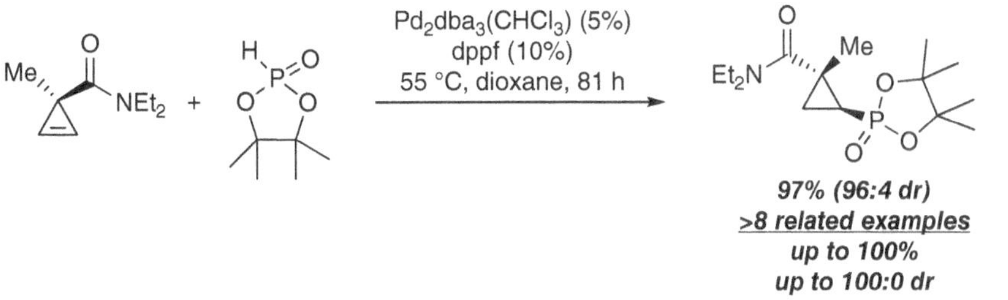

SCHEME 4.83   Hydrophosphonylation of amide-functionalized cyclopropenes [134].

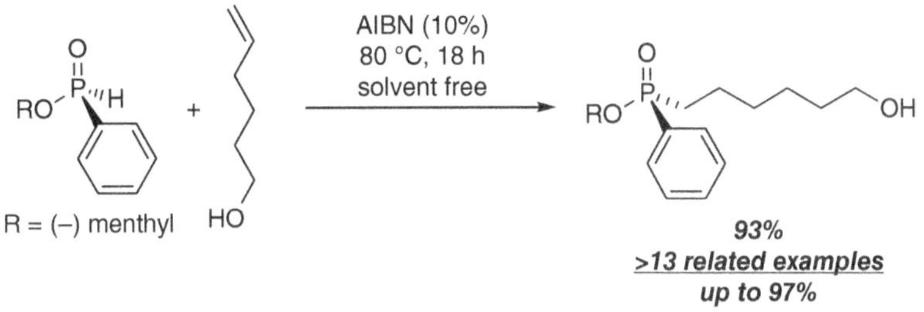

SCHEME 4.84   Radical addition to unactivated alkenes [124].

hydrophosphonylation reactions previously, the pinacol-derived hydrogen phosphonate was the most successful secondary phosphite. The authors screened a range of palladium sources and supporting ligands for activity towards the addition reaction. Due to the extreme ring strain of the system, another challenge that needed to be solved was the generation of secondary products due to ring-opening reactions. Indeed, a significant amount of the ring-opened product (40%) was obtained when a bulky phosphine ligand was screened (tris(2,4,6-trimethoxyphenyl)phosphine) along with palladium acetate as the catalyst. Furthermore, increasing the reaction temperature also increased the amount of this undesired allylphosphonate. After screening a range of palladium sources and supporting ligands, Pd(PPh$_3$)$_4$ was found to provide an attractive balance between reactivity and selectivity for hydrocarbon bearing cyclopropenes. After a successful catalyst system was found, the work was extended to a range of functionalized cyclopropenes. In contrast to the high selectivity observed in the screening of 3-phenyl-3-methylcyclopropene, the use of Pd(PPh$_3$)$_4$ as the catalyst was less selective, and an alternative catalyst system was devised. Moving to a two-component system comprised of Pd$_2$dba$_3$ (CHCl$_3$ solvate) along with dppf afforded high yields and selectivities for the reaction while minimizing the formation of ring-opening products. This chemistry illustrates the subtle interplay between metal catalysts and ligands when designing a successful catalytic reaction.

The addition of resolved P-chiral hydrogen phosphinates to alkenes has been accomplished without significant loss of the stereochemistry at phosphorus (Scheme 4.84) [124]. This reaction was achieved with the use of a radical initiator. One of the attractive aspects of this chemistry was that it proceeded under solvent-free conditions. The reaction was regioselective for the anti-Markovnikov product, and fair to moderate yields of the alkylphosphinate were obtained. Generally, high yields of the alkylphosphinate were obtained for a host of terminal alkenes as well as strained internal alkenes such as norbornene. Internal alkenes were significantly less reactive and satisfactory conversions were only

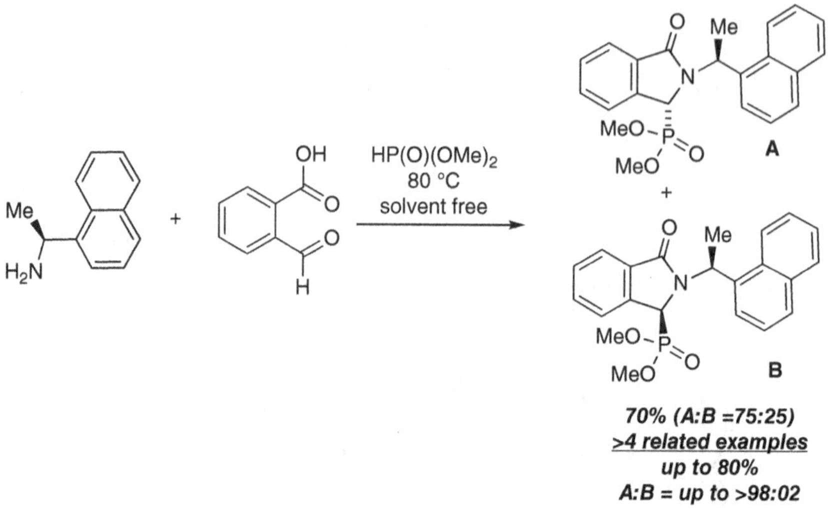

70% (A:B =75:25)
>4 related examples
up to 80%
A:B = up to >98:02

SCHEME 4.85   Diastereoselective synthesis of α-aminophosphonates [158].

observed after heating for 40 h. Even stannyl and acetoxyl olefins were successfully functionalized using this approach.

α-Aminophosphonates and their derivatives are a remarkably interesting class of molecules. They have applications in a host of different fields and are particularly attractive due to their biological activity [135–147]. The most common synthetic approach for the synthesis of α-aminophosphonates remains a three-component reaction between an amine, carbonyl compound, and secondary phosphite. Discovered by Fields [148] and Kabachnik [149] in the 1950s, this method still remains one of the most popular approaches to the synthesis of α-aminophosphonates [145, 147, 150, 151]. Following their pioneering work, a number of modifications to the reaction have been reported including the use of Lewis acids and bases [145, 147, 150–157].

A diastereoselective approach for the synthesis of α-aminophosphonates has been presented (Scheme 4.85) [158]. The key component for selectivity was the use of a resolved α-methylbenzylamine. Treatment of this precursor with 2-formylbenzoic acid followed by the addition of diethylphosphite generated the desired product in less than 1.5 h. The mechanism of the reaction was proposed to involve initial formation of a protonated imine from the reaction of the formylbenzoic acid with the amine. Attack of the phosphite on the imine followed by ring closure generated the cyclic α-aminophosphonates. The chiral induction was proposed to occur through selective attack of the phosphite on the *re* face of the protonated imine to generate one diastereomer. The selectivity of the addition reaction was controlled by the α-methylbenzylamine as evidenced by the ability of this reagent to dictate which diastereomer was formed in the reaction. Thus, beginning from the (*S*)-α-methylbenzylamine generated the (3*R*,1*S*) diastereomer, while the use of the amine in the (*R*)-configuration predominately afforded the (3*S*,1*R*) isomer. Operationally, the chemistry was attractive since it took place without the addition of any solvents. Only substrates bearing electron-neutral or electron-donating aryl groups were reported and generated moderate to excellent yields of the products. A single example of an alkylamine was screened and resulted in fair conversion (40%), but with

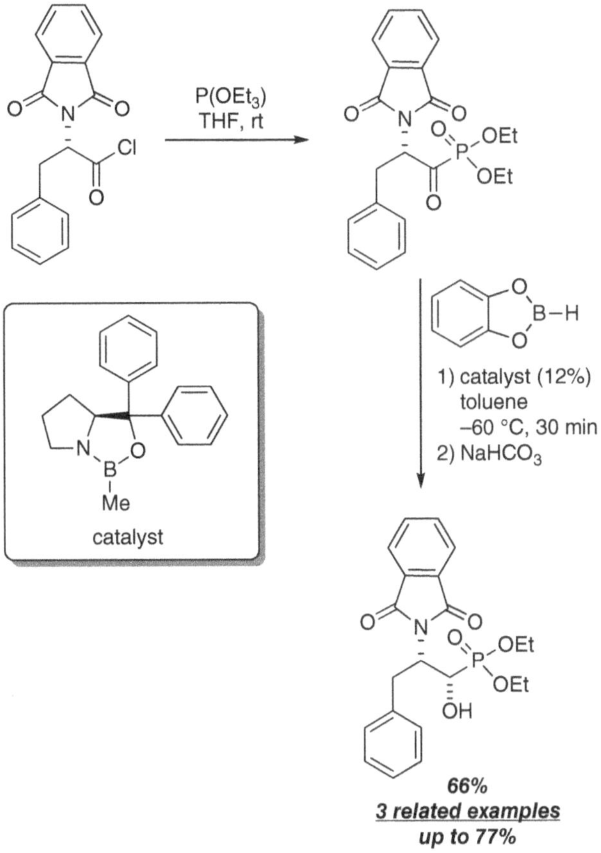

SCHEME 4.86   Synthesis of β-phthalimide-α-hydroxyphosphonates [159].

outstanding diastereoselectivity (>98 : 02). While only this single example was reported for alkylamines, the high diastereoselectivity was very impressive.

The diastereoselective synthesis of β-phthalimide-α-hydroxyphosphonates has been reported (Scheme 4.86) [159]. The authors used a resolved β-phthalimide acyl chloride as the starting point in the synthesis and converted it into an acylphosphonate through treatment with trimethylphosphite. The asymmetric reduction of the acylphosphonate by treatment with catecholborane in the presence of a catalytic amount of an oxazaborolidine as a catalyst generated the target compounds. Analogous reactions using $BH_3 \cdot SMe_2$ and the oxazaborolidine were not as selective and mixtures of diastereomers were generated. The phthalimide group was converted into an amino group through treatment with hydrazine in ethanol.

The hydrophosphonylation of imines bearing a variety of organic fragments has been reported under a range of conditions. Even crowded substrates are typically responsive to the addition reaction. As an example of this, an anthracene bearing imine has been successfully functionalized using secondary phosphites (Scheme 4.87) [160]. Simply refluxing the reaction mixture for 14 h afforded moderate yields of the α-aminophosphonates.

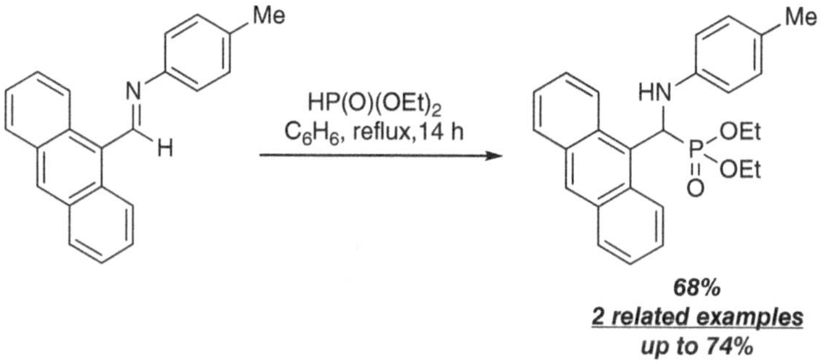

SCHEME 4.87 Uncatalyzed addition of secondary phosphites to imines [160].

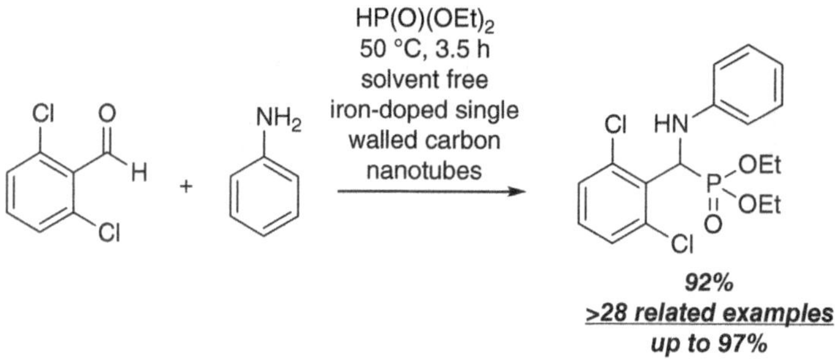

SCHEME 4.88 Use of an iron-based heterogeneous catalyst for the formation of α-aminophosphonates [161].

Heterogeneous catalysts have been used to promote the formation of α-aminophosphonates from aldehydes bearing bulky groups. One of the potential advantages of heterogeneous catalysts is the ability to recycle the catalyst from one reaction and use it to promote a subsequent reaction. To this end, Sharghi has reported the use of iron-doped single-walled nanotubes as a heterogeneous catalyst for the Kabachnik–Fields reaction (Scheme 4.88) [161]. The reaction conditions were mild and simply stirring the neat reagents under solvent-free conditions generated moderate to excellent yields of the aminophosphonates. In addition to a host of aldehydes, substrates bearing the bulky groups such as 2,6-dichlorobenzaldehyde were readily transformed using this catalyst system. The electronics of the system did not significantly affect the product distribution or yields. In terms of the recycling of the catalyst, the authors used the model system of aniline, 4-methoxybenzaldehyde, and diethylphosphite to probe the effectiveness of the heterogeneous catalyst on multiple reactions. The heterogeneous iron catalyst was recycled 10 times with no loss in activity. In addition to the use of bulky aldehydes, this catalyst system was able to promote the three-component reaction with several ketones. Dialkyl ketones such as cyclohexanone afforded high yields of the α-aminophosphonate (80%); however, alkyl aryl ketones such as acetophenone were sluggish under the reaction conditions, and only trace amounts of products were observed.

The synthesis of α-aminophosphonates bearing bulky groups has been achieved using chlorophosphites as the phosphorus precursors (Scheme 4.89) [162]. This chemistry was essentially a version of the Kabachnik–Fields reaction using a different phosphorus

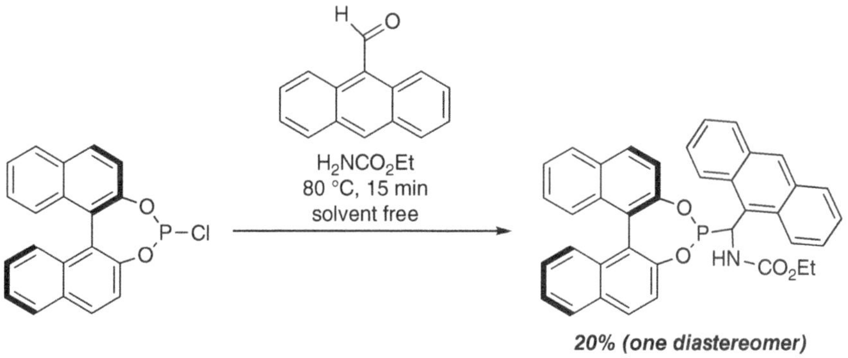

SCHEME 4.89   Conversion of chlorophosphites into α-aminophosphonates [162].

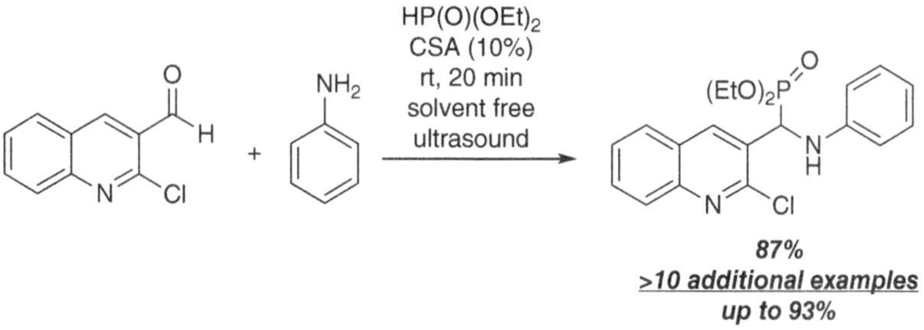

SCHEME 4.90   Acid-catalyzed ultrasound-promoted synthesis of α-aminophosphonates [163].

source. The chemistry was operationally trivial and simply stirring the reagents without the addition of solvent for 15 min at 80 °C afforded the α-aminophosphonate. Analysis of the phosphorus NMR spectrum revealed the presence of two diastereomers in a 1:1 ratio. One diastereomer was isolated from this mixture (20%).

The acid-catalyzed synthesis of α-aminophosphonates has been reported using an ultrasound bath to promote the reaction (Scheme 4.90) [163]. While a range of acid catalysts have shown activity towards this reaction, camphor sulfonic acid (CSA) was used in these studies. While the carbonyl fragment of the camphor group could have reacted with this phosphite under the acidic conditions, this side reaction was not observed. The ability to carry out the reaction under solvent-free conditions was a significant advantage of this chemistry. Using the ultrasound bath, the reaction times were decreased without decreasing the yields of the resulting hydroxyphosphonates. The scope of this reaction was broad and a range of functionalized aldehydes were successfully used.

Another example of the use of ultrasound to promote the synthesis of α-aminophosphonates has been reported and generated high yields of the functionalized materials using an uncatalyzed approach (Scheme 4.91) [164]. One of the advantages of these ultrasound-assisted reactions was the dramatic reduction in the reaction time. Most of the reactions were complete within 1 min. The authors also determined that removal of the solvent from these reactions was not detrimental. The chemistry displayed wide functional group compatibility, and a host of electron-donating and withdrawing groups were well tolerated. While aldehydes were readily functionalized using this approach, ketones such as acetone

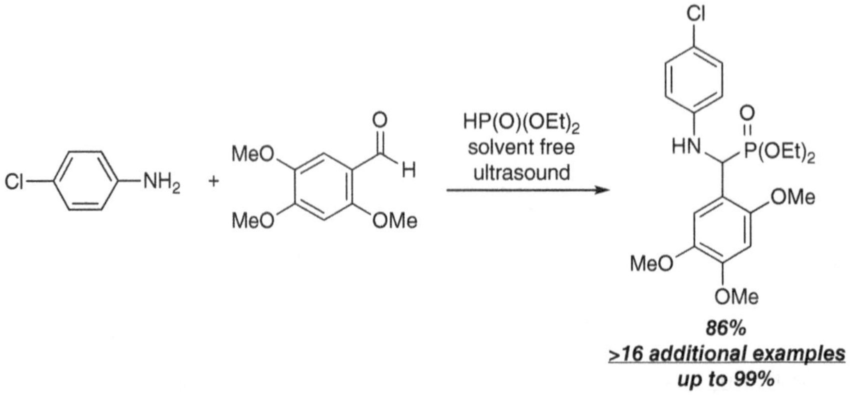

SCHEME 4.91 Ultrasound-assisted uncatalyzed synthesis of α-aminophosphonates [164].

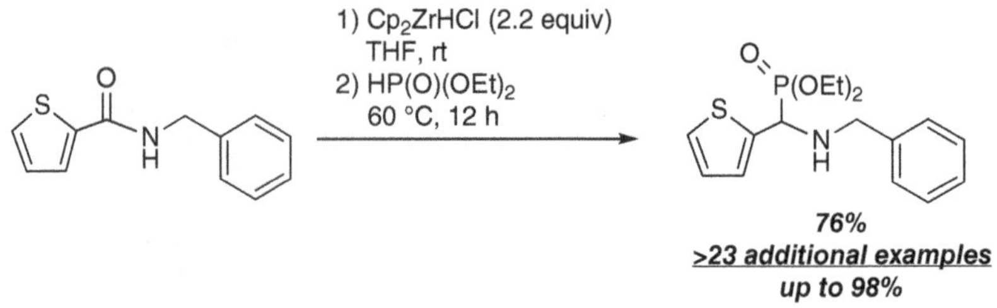

SCHEME 4.92 Preparation of α-aminophosphonates through reductive phosphonylation [90].

and cyclohexanone afforded only traces of the products. Despite the limitations, the ability to remove the solvent and the catalyst from the reactions makes this a very practical method for the synthesis of aminophosphonates.

A reductive phosphonylation approach has also been used to generate α-aminophosphonates (Scheme 4.92) [90]. This reaction can essentially be described as a direct conversion of amides into α-aminophosphonates. The ability to start with an amide is a significant advantage to this approach since they are robust and readily available. The catalyst used to promote this reaction was Schwartz's reagent, and simply stirring the reagents for 12 h at 60 °C afforded the target compounds in moderate to excellent yields. The substrate scope was broad and a range of amides were successfully transformed using this catalyst system. For secondary amides, 2.2 equivalents of the zirconocene catalyst was needed; however, for tertiary amides, only 1.2 equivalents was needed to promote the reaction. The use of bulky tertiary amides resulted in the slightly lower yields of the α-aminophosphonates. Given the mild conditions, wide substrate scope, and availability of the reagents and catalyst, this is a very attractive approach to the synthesis of these valuable compounds.

Although the Lewis acid-assisted Kabachik–Fields reaction is typically carried out in organic solvents, water is an attractive alternative, and several research groups have developed several versions of the classic reaction that can be carried out using water as the solvent. In addition to overcoming a potential catalyst poisoning issue with the use of water, the elimination of hazardous organic solvents used in these reactions is attractive for

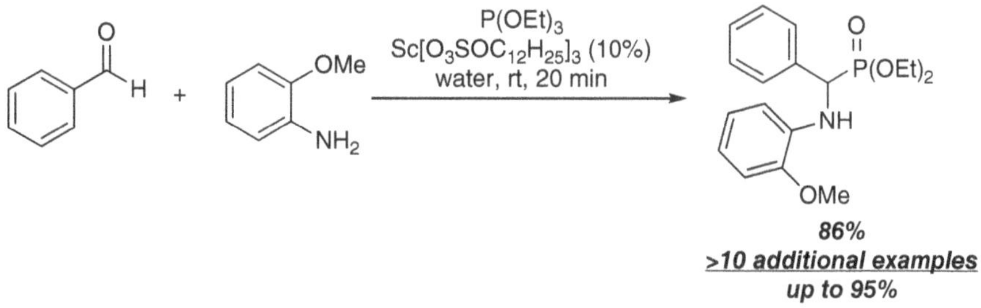

SCHEME 4.93   Synthesis of α-aminophosphonates in water [165].

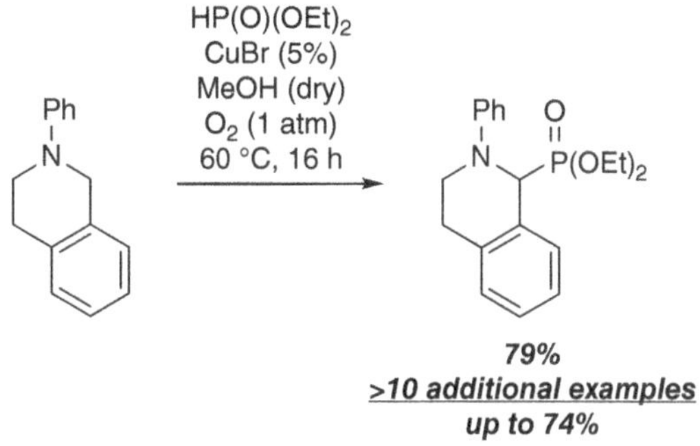

SCHEME 4.94   Synthesis of α-aminophosphonates through C—H activation [166].

sustainability issues. One of the first examples of this approach used a scandium-based catalyst bearing surfactant-like supporting ligands (Scheme 4.93) [165]. The ligand selected by the authors was dodecyl sulfate. This amphiphilic ligand was shown to be essential for the success of the three-component reaction as using a common scandium salt ($Sc(OTf)_3$) that had previously been shown to promote the reaction failed to generate any α-aminophosphonate when the reaction was carried out in water. The vast majority of approaches for the preparation of α-aminophosphonates used secondary phosphites as substrates; however, only traces of product were observed when these reagents were used. After some experimentation, the authors determined that triethylphosphite was an effective nucleophile and afforded high yields of the products. Using this system, a number of aldehydes including heterocycles were readily converted into α-aminophosphonates. This is a very practical and sustainable approach to the synthesis of these valuable compounds given the operational simplicity of this approach as well as the ability to carry out the three-component reaction in water.

The synthesis of α-aminophosphonates through an aerobic dehydrogenative cross-coupling approach has been reported (Scheme 4.94) [166]. This reaction substituted a hydrogen located adjacent to the nitrogen center in tetrahydroisoquinoline substrates with a secondary phosphite. The reaction was catalyzed by copper bromide in dry methanol and afforded moderate to high yields of the functionalized compounds. This chemistry was operationally straightforward as the reagents can be added to the reaction vessel under air, and the reaction itself was carried out under oxygen (1 atm). While a range of phosphites were successfully used in these reactions, only a few tetrahydroisoquinoline examples were

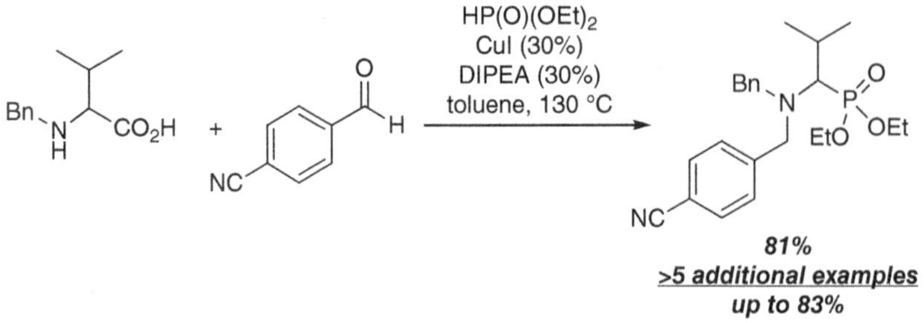

SCHEME 4.95 Copper-catalyzed synthesis of α-aminophosphonates [91].

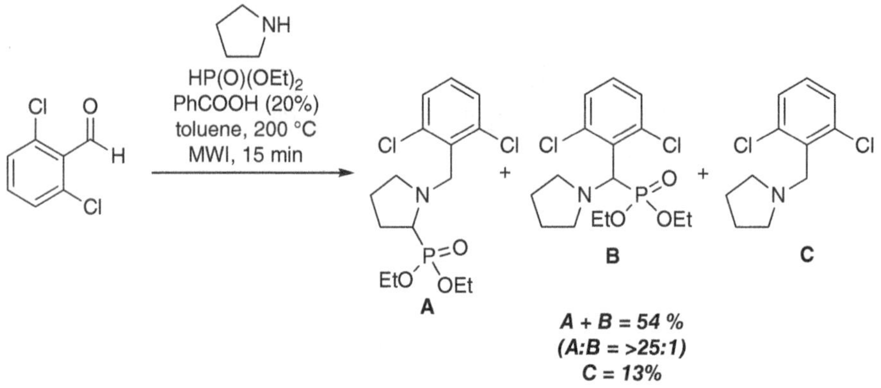

SCHEME 4.96 Benzoic acid-catalyzed synthesis of α-aminophosphonates [88].

screened. Thus, if the substrate has the correct architecture, this process will generate the α-aminophosphonates with excellent selectivity without prefunctionalization of the substrate. In related work, Prabhu reported the use of molybdenum trioxide to promote a similar reaction under aerobic conditions [167].

An interesting approach to the synthesis of α-aminophosphonates from amino acids and their derivatives has been reported (Scheme 4.95) [91]. The key to this chemistry was a copper-catalyzed tandem decarboxyation coupling reaction between amino acids, aldehydes, and secondary phosphites. The chemistry was compatible with a range of functional groups, and aldehydes bearing both electron-donating and electron-withdrawing groups were readily functionalized using this approach. The coupling reaction was sensitive to bulky groups, and incorporating ortho-substituents into the aldehyde resulted in sluggish reactions. The use of an *ortho*-nitro-substituted 4-cyanobenzaldehyde resulted in no conversion. Overall, this is a very attractive approach to the synthesis of α-aminophosphonates using readily available materials and catalysts.

A redox neutral process for the preparation of α-aminophosphonates has been reported (Scheme 4.96) [88]. One of the attractive aspects of this chemistry was that it was successful under metal-free conditions and only used a weak acid (benzoic acid) as the catalyst. This is formally a three-component reaction that couples a pyrrolidine, secondary phosphite, and an aromatic aldehyde to generate the α-aminophosphonates. One of the challenges with this type of reaction was the formation of different regioisomers. This arises due to a

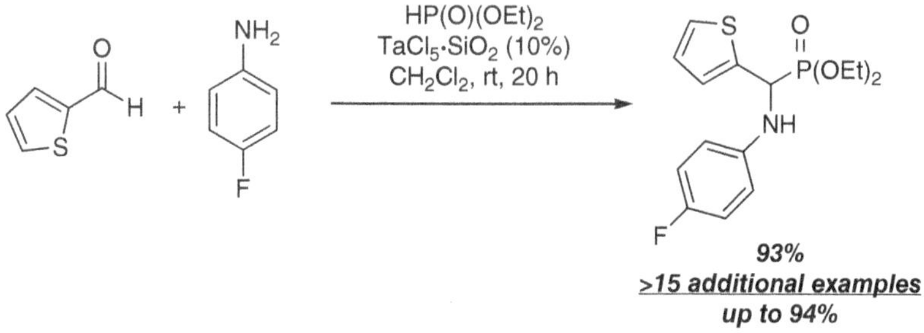

SCHEME 4.97   Synthesis of α-aminophosphonates using supported tantalum salts [156].

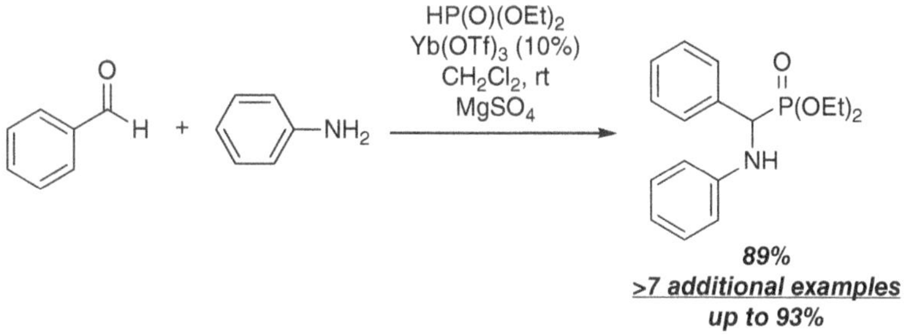

SCHEME 4.98   Preparation of α-aminophosphonates using Yb(OTf)$_3$ [168].

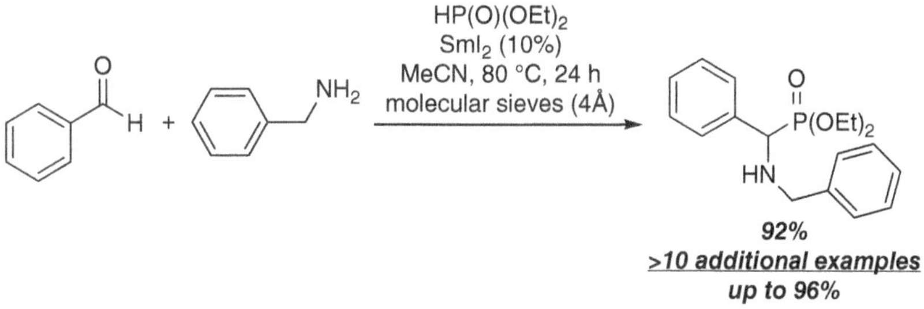

SCHEME 4.99   Samarium iodide-catalyzed synthesis of α-aminophosphonates [169].

competing secondary reaction between the phosphite and a proposed intermediate (iminium ion). Although only a few examples with phosphites were reported, the authors did find that the chemistry would favor addition of the phosphite to the pyrrolidine (isomer **A**). An additional product was observed in these reactions (**C**) and was presumably formed through a reductive amination process. While the scope of the reaction with regard to electron-donating and electron-withdrawing groups has yet to be determined, the chemistry was tolerant to the presence of ortho-substituents on the aromatic aldehyde. Overall, the process is described as redox neutral since formation of the α-aminophosphonate was coupled with a reductive N-alkylation.

In addition to these methods, a range of metal-catalyzed reactions have been developed for the synthesis of α-aminophosphonates. Several of these approaches are illustrated in Schemes 4.97–4.100).

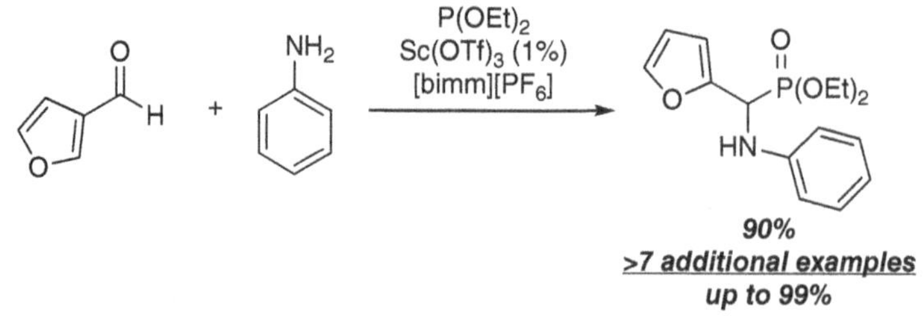

SCHEME 4.100 Scandium triflate-catalyzed synthesis of α-aminophosphonates [170].

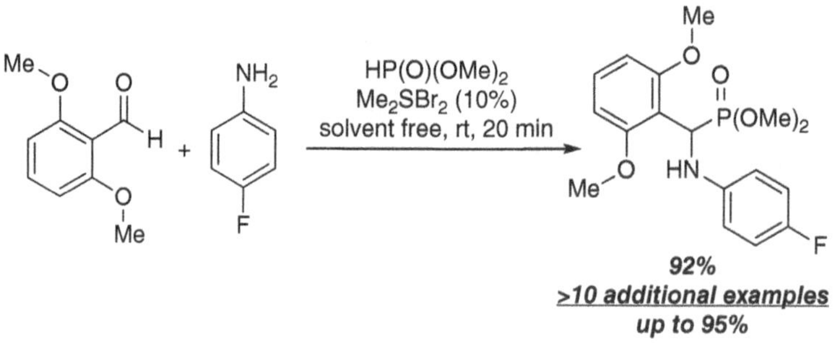

SCHEME 4.101 Preparation of α-aminophosphonate using (bromodimethyl)sulfonium bromide as the catalyst [171].

The use of (bromodimethyl)sulfonium bromide as an inexpensive and readily available catalyst for the preparation of α-aminophosphonates is an attractive approach (Scheme 4.101) [171]. The chemistry is operationally trivial as simply stirring the neat reagents and catalyst for 20 min at room temperature afforded excellent yields of the α-aminophosphonates. One of the impressive aspects of this chemistry was the ability to handle bulky substrates. For example 2,6-dimethoxylbenzaldehyde was converted into the α-aminophosphonate in 92% isolated yield. The chemistry was also tolerant of a range of electron-donating and electron-withdrawing groups. Overall, this represents a facile process for the synthesis of the α-aminophosphonates.

The sluggish reactivity of ketones remains one of the challenges with Kabachnik–Fields-type chemistry. One solution to this issue entailed the use of InCl₃ as a catalyst for the three-component reaction (Schemes 4.102 and 4.103) [172]. No additional cocatalysts were needed for this chemistry, and refluxing the reagents in THF for 25–30 h generated moderate to excellent yields of the α-aminophosphonates. The authors also screened a range of aldehydes for activity, and predictably found that the aldehydes were more reactive than the ketones. Indeed, the aldehydes could be successfully functionalized at room temperature. A range of functional groups including esters, nitro, and unprotected phenols were well tolerated by the process. The authors also used sonication to decrease the reaction time by roughly 50% with no significant change in the isolated yields of the products.

The indium chloride-mediated conversion of N-benzyloxycarbonylamino sulfones into α-aminophosphonates has been reported (Scheme 4.104) [173]. While N-benzyloxycarbonylamino sulfones might not be commonly used as a substrate, these compounds are typically stable solids that are readily generated from the corresponding aldehydes.

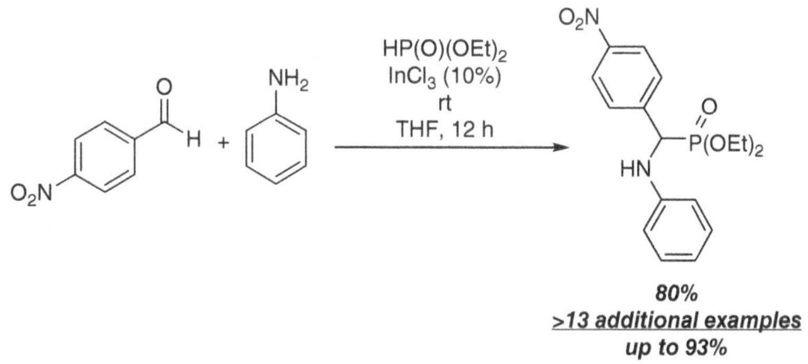

SCHEME 4.102   InCl$_3$-catalyzed conversion of aldehydes into α-aminophosphonates [172].

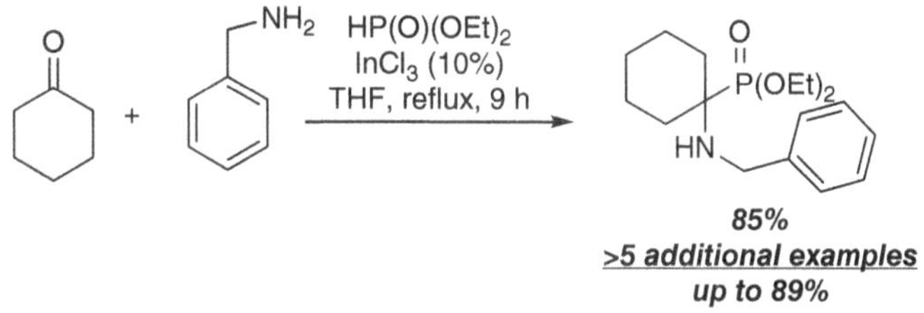

SCHEME 4.103   Conversion of ketones into α-aminophosphonates using InCl$_3$ as the catalyst [172].

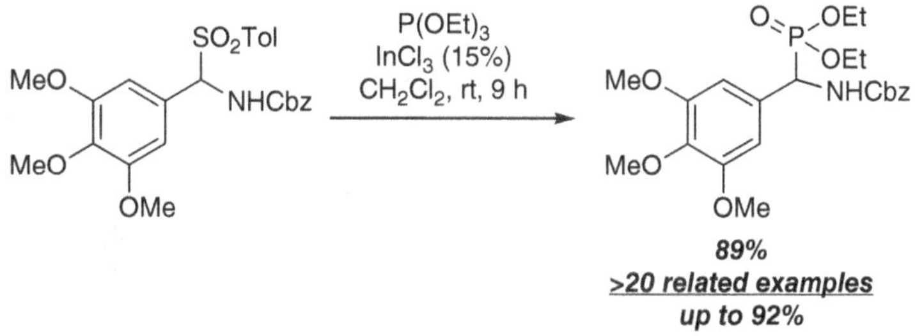

SCHEME 4.104   Synthesis of α-aminophosphonates from sulfones [173].

Treatment of these precursors with a Lewis acid was proposed to generate an *N*-acyliminium ion that reacted with the phosphorus nucleophile. Furthermore, while most of the methods for the synthesis of α-aminophosphonates used hydrogen phosphonates such as dimethylphosphite as the phosphorus nucleophile, low yields of the products were obtained in these reactions when these reagents were used. After some experimentation, the authors found that trialkylphosphites were superior and generated moderate to excellent yields of the products. Simply stirring the reagents in dichloromethane at room temperature for 8–20 h afforded the target compounds in excellent yields. The rate of the reaction was dependent upon the electronics of the system. Reactions that used *N*-benzyloxycarbonyl-amino sulfones bearing electron-donating groups on the aromatic ring reached completion faster than those bearing electron-withdrawing groups.

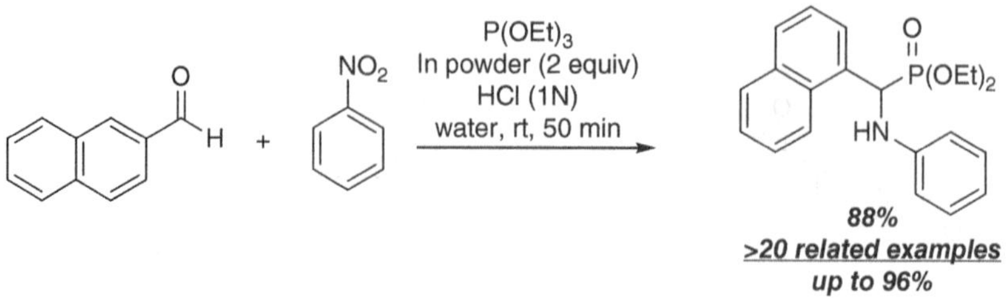

SCHEME 4.105   Synthesis of α-aminophosphonates using elemental indium as the catalyst precursor [174].

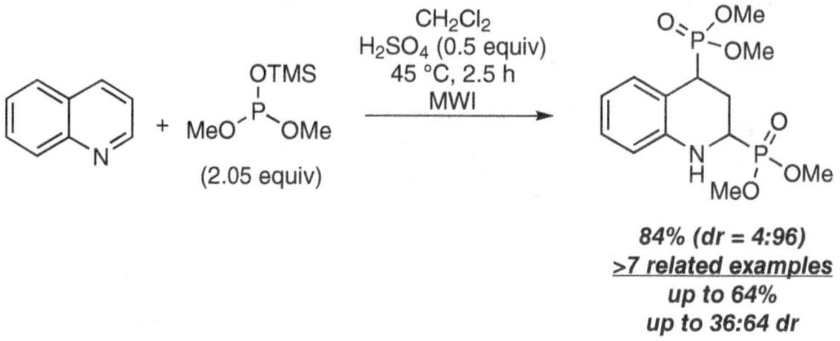

SCHEME 4.106   Sulfuric acid-promoted synthesis of α-aminophosphonates [175].

The use of nitroarenes in the synthesis of α-aminophosphonates has been reported (Scheme 4.105) [174]. After screening a range of metals for activity, indium was found to be the most effective for the reaction. The reaction could be carried out using water as the solvent. One issue with this chemistry was the need to use two equivalents of indium powder. However, the overall advantage of being able to use the elemental metal as the catalyst precursor might be worth it. Using this approach, a host of aldehydes were readily converted into α-aminophosphonates. The chemistry was compatible with a range of functional groups including halogens, amines, nitro, and ethers. This approach was successful with several ketones as well. The practical aspects of this approach include the ability to convert the functionalized nitro-containing aromatic compounds into α-amino-phosphonates through a one-pot reaction in water through the *in situ* generation of InCl$_3$ from elemental indium.

The synthesis of quinoline derivatives has been achieved using a sulfuric acid-promoted process (Scheme 4.106) [175]. A focused microwave reactor was used in these reactions and generally afforded higher conversions in shorter reaction times without a reduction in the diastereoselectivity. As anticipated, the addition of a methyl group to the 4-position resulted in sluggish conversion into the alkylphosphonate (2%). Furthermore, the double addition reaction was quite sensitive to the groups that were attached to the quinoline core. Incorporating the electron-donating methoxy group in the 6-position resulted in poor conversion (12%), while the addition of a nitro group in the same position afforded 52% of the product. In subsequent reactions, adding a methyl group to the same position generated 64% of the product, while the use of

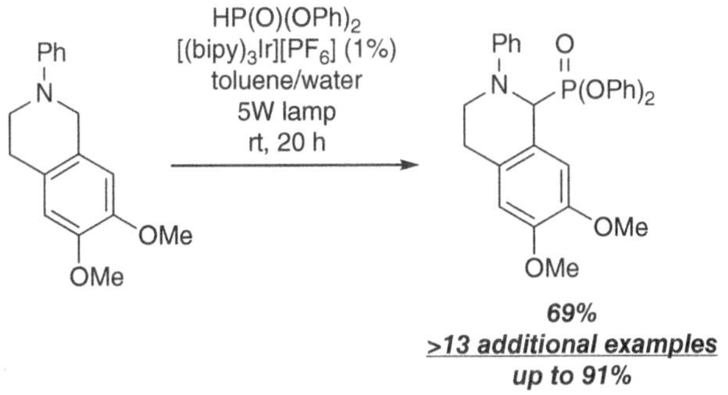

SCHEME 4.107 Synthesis of α-aminophosphonates through photoredox process [178].

6-fluoroquinoline was sluggish (19%). Thus, subtle differences in the composition of the reagents resulted in significantly different reactivity profiles.

While many organic transformations are promoted by UV light, there is considerable interest in the development of reactions that are promoted by visible light. One version of this chemistry used a photoredox catalyst to promote the reaction. To this end, a palladium (II) porphyrin was used to promote the synthesis of α-aminophosphonates using visible light at low catalyst loadings (0.05%) [176]. Rueping reported the use of a ZnO-based photocatalyst for similar reactions [177]. One example of a cross dehydrogenative coupling of amines and secondary phosphites generated α-aminophosphonates through a visible light photoredox process using a well-known transition metal catalyst (Scheme 4.107) [178]. The authors screened common iridium and ruthenium photoredox catalysts for activity and found that the iridium compounds generated the functionalized phosphonates in higher yields. Irradiating a biphasic solvent mixture (toluene/water) containing 1% of the photocatalyst, secondary phosphite, and tetrahydroisoquinoline afforded the title compounds. The chemistry was compatible with a range of functional groups and was successful at room temperature.

While several of the oxidative techniques listed previously utilize transition metal complexes as promoters, the use of a completely organic photocatalyst for the oxidative synthesis of α-aminophosphonates has been reported (Scheme 4.108 and Example 4.27) [179]. The organic photocatalyst the authors selected for this chemistry was eosin Y. This organic dye is attractive as it is inexpensive and readily available. Furthermore, the chemistry was operationally trivial as all of the reagents could be added under air, and rigorous exclusion of oxygen or moisture was not a concern. The conditions were mild, and excellent yields of the functionalized tetrahydroisoquinolines were obtained after only a few hours of irradiation. As mentioned previously, the only skeleton screened for this chemistry was the tetrahydroisoquinoline; however, if a system has the same or similar connectivity, the simplicity of this approach as well as the availability of the reagents makes this is an attractive method for the metal-free synthesis of α-aminophosphonates.

A metal-free dehydrogenative coupling approach was used by Wan for the generation of α-aminophosphonates using 2,3-dichloro-5,6-dicyanobenzoquinone (DDQ) as the oxidizing agent (Scheme 4.109) [180]. Similar to other studies, tetrahydroisoquinolines were selected by the authors as the representative substrates for the reaction along with several secondary phosphites. The conditions were mild, and moderate to excellent yields

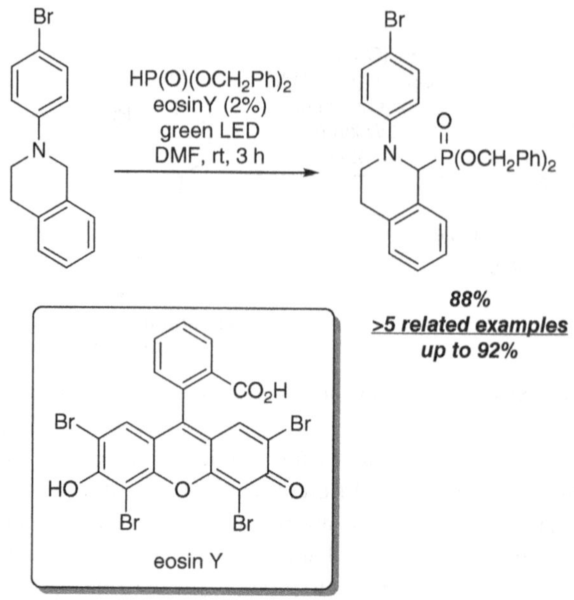

SCHEME 4.108 Preparation of alkylphosphonates promoted by visible light [179].

Example 4.27 Metal-Free Phosphorus–Carbon(sp³) Bond Formation Promoted by Visible Light [179].

Before starting this synthesis, carefully follow the chemical safety instructions listed in Example 4.1.

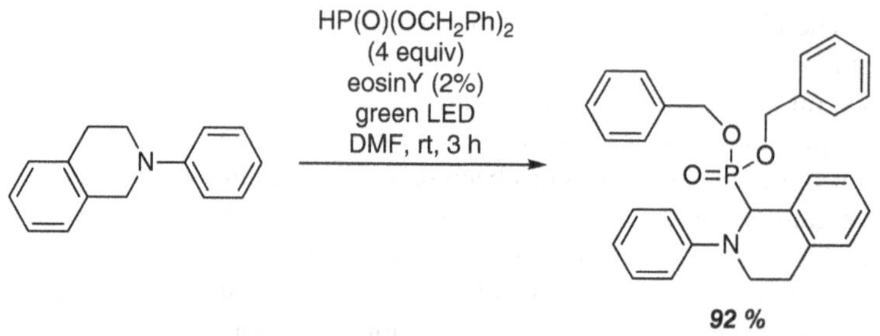

Caution! Ethyl acetate and hexanes are highly flammable.

In a 5 mL snap vial equipped with magnetic stirring bar, the tetrahydroisoquinoline derivative (1 equiv) and eosin Y (0.02 equiv) were dissolved in DMF (0.238 mmol/mL). Then dialkylphosphonate (4 equiv) was added, and the resulting mixture was irradiated through the vial's plane bottom side using green LEDs. After the reaction was completed (monitored by TLC), the mixture was transferred to the separating funnel, diluted with ethyl acetate, and washed with water. The aqueous phase was extracted three times with ethyl acetate. The combined organic layers were dried over MgSO$_4$, filtered, and concentrated in vacuum. Purification of the crude product was achieved by silica gel column chromatography using hexane/ethyl acetate as eluent that afforded the title compound (92%).

Adapted with permission from D. P. Hari, B. Konig, *Org. Lett.*, **2011**, *13*, 3852–3855. *Copyright (2011) American Chemical Society.*

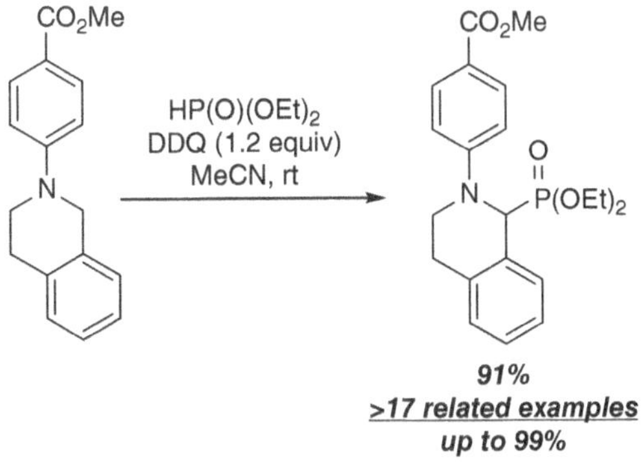

SCHEME 4.109   A dehydrogenative coupling approach to the synthesis of α-aminophosphonates [180].

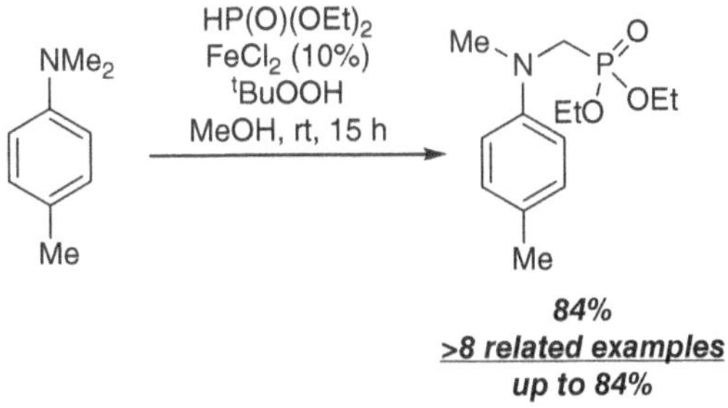

SCHEME 4.110   C—H phosphorylation of dimethyl anilines [182].

of the targets were obtained. This method is attractive due to the operational simplicity and the metal-free nature of the reaction. In related work, a metal-free cross dehydrogenative coupling reaction used molecular iodide and an oxygen atmosphere to promote the phosphonation of tetrahydroisoquinoline [181]. The reaction conditions were mild (room temperature, 24h), and moderate to excellent yields (45–75%) of the functionalized compounds were obtained.

While a number of reports have focused on the use of tetrahydroisoquinolines as substrates, Ofial has investigated the dehydrogenative phosphonation of *N,N*-dimethyl anilines (Scheme 4.110) [182]. The catalyst system the authors selected was comprised of iron(II) chloride and an oxidant (*tert*-butyl peroxide) with methanol as the solvent. While many of the substrates were functionalized at room temperature, aniline derivatives bearing electron-withdrawing groups required mild heating to achieve significant conversion. Overall, this method is attractive due to the ability to functionalize unactivated methyl groups under mild conditions.

One of the challenges with the metal-promoted Kabachnik–Fields synthesis of α-aminophosphonates was poisoning of the catalyst by the water generated in the reaction. In an effort to circumvent this issue, Reddy investigated a number of approaches that did not involve the use of metal catalysts (Scheme 4.111) [183]. After some experimentation, he determined that simply stirring the reagents in the presence of polyethylene glycol in water successfully promoted the synthesis of the α-aminophosphonates. The conditions

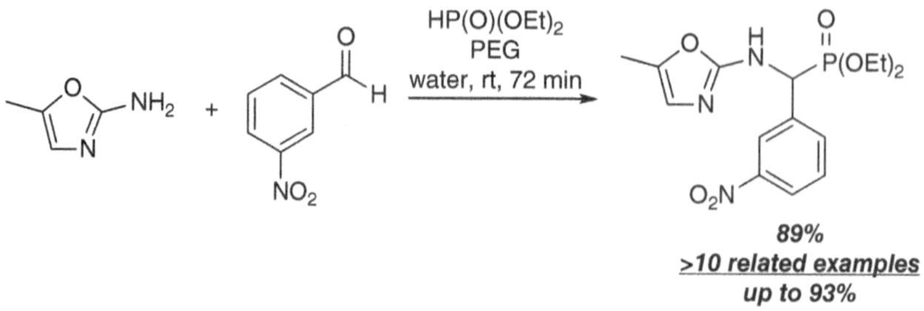

SCHEME 4.111   PEG-promoted synthesis of α-aminophosphonates [183].

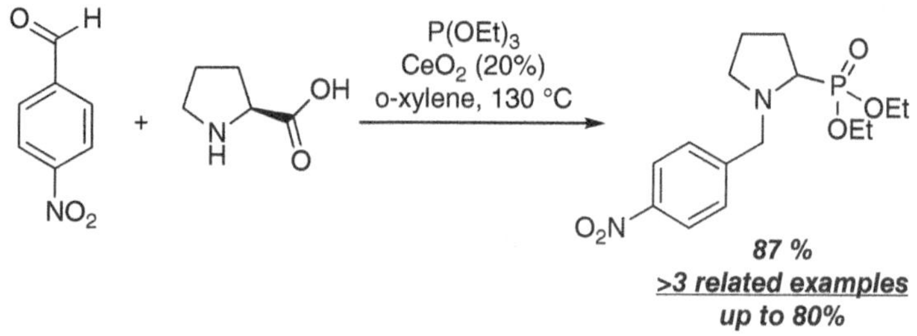

SCHEME 4.112   Decarboxylative approach to the synthesis of α-aminophosphonates [184].

for the reaction were quite mild, and moderate to excellent yields of the targets were obtained. A variety of functional groups including nitro, amino, and an unprotected phenol were well tolerated by this system. One limitation that could not be overcome was the sluggish reactivity of ketones. The authors also found that the PEG promoter could be recycled for several reactions. Due to the simplicity of the reaction and the ability to promote the reaction in water, this approach is one of the most practical syntheses of α-aminophosphonates.

The three-component reaction between proline, triethylphosphite, and aldehydes generated α-aminophosphonates through a decarboxylative coupling process (Scheme 4.112) [184]. A series of cerium catalysts were screened for activity, and CeO_2 was found to be the most active. Although $CeCl_3$ $(7H_2O)$ promoted similar reactions, it was not as active in these reactions. The solvent was also found to be a critical component and sluggish reactivity observed in reactions carried out in water or PEG. High yields of the products were obtained using nonpolar hydrocarbon solvents such as toluene. Since the reactions were typically carried out at 130 °C, o-xylene was substituted for toluene. The chemistry was compatible with a range of electron-donating and withdrawing groups on the aldehyde, but ketones and primary amino acids were unreactive.

In 2004, Jacobsen reported one of the first practical organocatalyzed asymmetric syntheses of α-aminophosphonate (Scheme 4.113) [185]. His approach entailed the use of thiourea-based organocatalyst to promote the hydrophosphonylation reaction. In addition to screening the imines for activity, the authors probed the effect of manipulating the organic fragments on the secondary phosphite. Although diethylphosphite was one of the most popular phosphorus reagents for this chemistry, its use only generated small

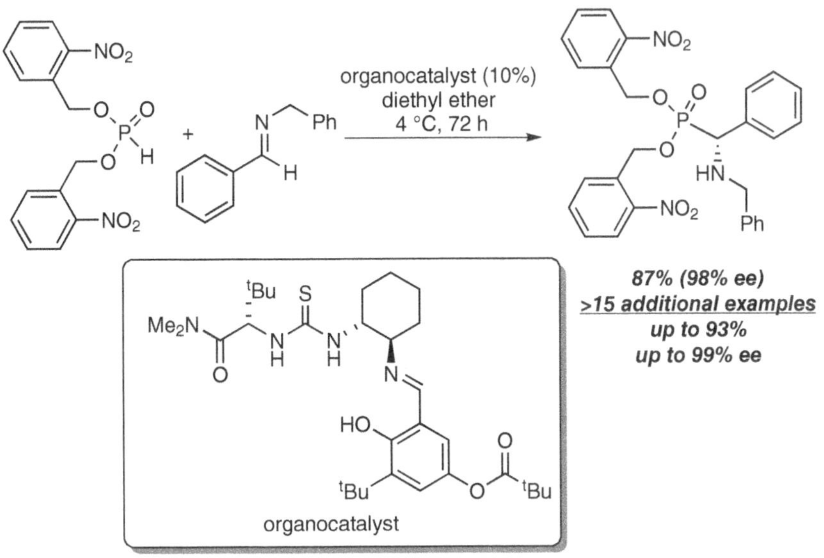

SCHEME 4.113   Asymmetric synthesis of α-aminophosphonates [185].

amounts (6%) of the target aminophosphonate using the organocatalyst. After some exper-imentation, the authors discovered that diarylphosphites bearing electron-withdrawing nitro groups afforded high conversions with excellent stereoselectivity. With a successful system in hand, the chemistry was extended to a host of functionalized imines. The authors demonstrated that these phosphonate diesters could be readily converted into the corresponding chiral α-aminophosphonic acids using asymmetric hydrogenation with no loss of stereochemistry. Due to the high yields and selectivity obtained using these thio-urea-based organocatalysts, this remains one of the most practical approaches for the asymmetric synthesis of α-aminophosphonates.

The use of chiral phosphoric acid derivatives as chiral auxiliaries for the asymmetric synthesis of α-aminophosphonates generated the functionalized compounds in excellent yields with moderate to excellent enantioselectivity (Scheme 4.114) [186]. The conditions for the reaction were quite mild, and simply stirring the reagents at room temperature afforded the α-aminophosphonates. The authors investigated the scope of the reaction and discovered that imines bearing unsubstituted arenes such as N-benzylidene p-anisidine afforded the product in high yield but with low enantioselectivity (52%). Changing the substrate to an α,β-unsaturated aldimine (cinnamaldehyde derivative) species bearing an ortho-substituent on the arene increased the enantioselectivity (82–90%). Additionally, the steric bulk of the dialkylphosphite was a critical component for obtaining a high enantioselectivity. While diethylphosphite generated the products with ee = 73%, changing to diisopropylphosphite increased the ee to 84%.

The asymmetric synthesis of α-aminophosphonates has been accomplished using quinine as an organocatalyst for the addition of phosphorus nucleophiles to imines (Scheme 4.115) [187]. The chemistry was operationally trivial, and simply stirring the reagents afforded the α-aminophosphonates. Reactions carried out at 20 °C generated the products in 2–3h with ee's ranging from 48 to 86%. Reducing the temperature to −20 °C increased the ee's to 88–94%. A range of functional groups were well tolerated by this

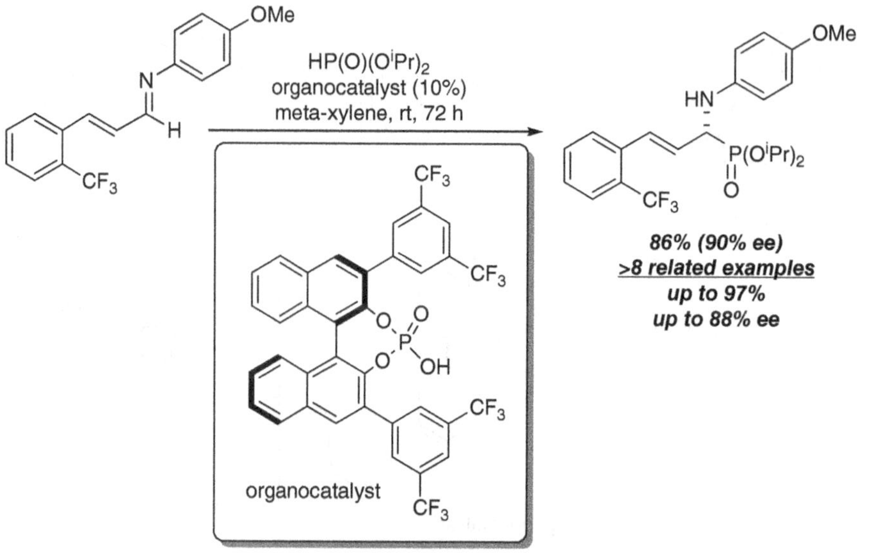

SCHEME 4.114   Phosphoric acid-catalyzed asymmetric synthesis of α-aminophosphonates [186].

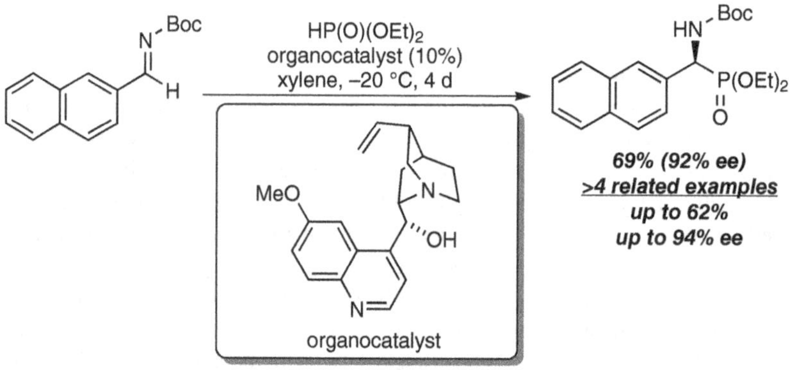

SCHEME 4.115   Organocatalyzed asymmetric addition of phosphites to imines [187].

chemistry with both electron-donating and electron-withdrawing groups generating the aminophosphonates with high ee's. The authors were also able to devise a quinine derivative that reversed the handedness of the aminophosphonates obtained from the addition reaction. This method is one of the most practical approaches for the asymmetric synthesis of these compounds due to the inexpensive and readily available organocatalyst as well as the mild conditions of the reaction.

One of the challenges encountered when designing a synthetic strategy for the preparation of chiral α-aminophosphonates containing a quaternary stereocenter adjacent to the phosphonate fragment is a lack of known procedures for this reaction [188–191]. Essentially, this would involve the asymmetric addition of phosphorus

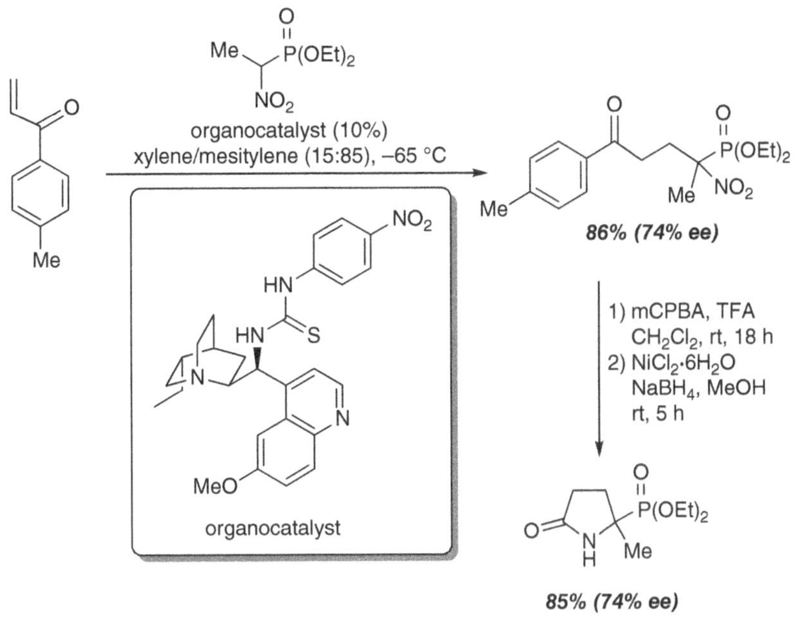

SCHEME 4.116 Preparation of chiral α-aminophosphonates containing a quaternary stereocenter adjacent to the phosphonate [194].

nucleophiles to ketimines. While high-yielding asymmetric reactions have been achieved using aldimines, extending this chemistry to ketimines has been challenging. Sluggish reactivity and difficulty in distinguishing the enantiotopic faces of the substrate have been proposed to explain the lack of existing protocols [192, 193]. An organocatalytic to this problem has recently been reported (Scheme 4.116) [194]. Their approach was comprised of an initial organocatalyzed asymmetric conjugate addition followed by cyclization/hydrogenation to generate the aminophosphonates. After screening several organocatalysts for efficacy towards the Michael addition, the authors selected an alkaloid-derived thiourea species and extended the optimized system to a range of aryl and alkyl vinyl ketones. One example was then further transformed into the α-aminophosphonate.

The asymmetric hydrophosphonylation of ketimines was reported by Nakamura and used either hydroquinine or hydroquinidine as the catalyst for the reaction (Scheme 4.117) [195]. One of the strengths of this method was that both enantiomers of the α-aminophosphonates could be selectively formed using the author's approach. The use of hydroquinine generated the S-configuration of products, while hydroquinidine promoted the formation of phosphonates with an R-configuration. The reactions were operationally straightforward, and simply adding the phosphite to a chilled solution (−20 °C) of the imine, base, and catalyst followed by stirring for several days afforded high yields with excellent selectivity. A range of bases were screened for activity, and sodium carbonate was found to generate the products with the highest stereoselectivity. Although the N-tosyl ketimines needed to be formed in a separate step, this preparation was high yielding.

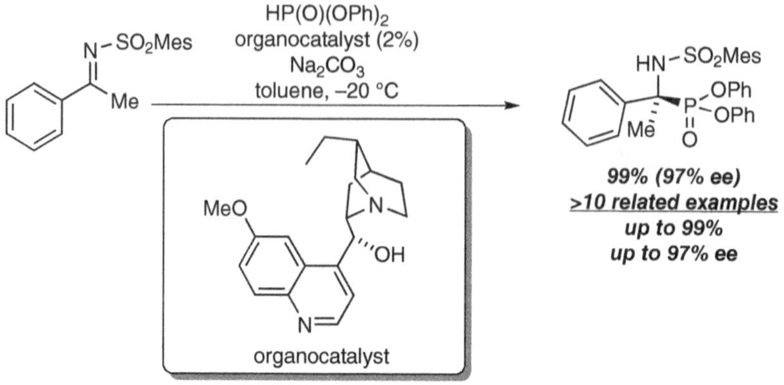

SCHEME 4.117   Organocatalyzed approach to the synthesis of chiral α-aminophosphonates bearing a quaternary stereocenter [195].

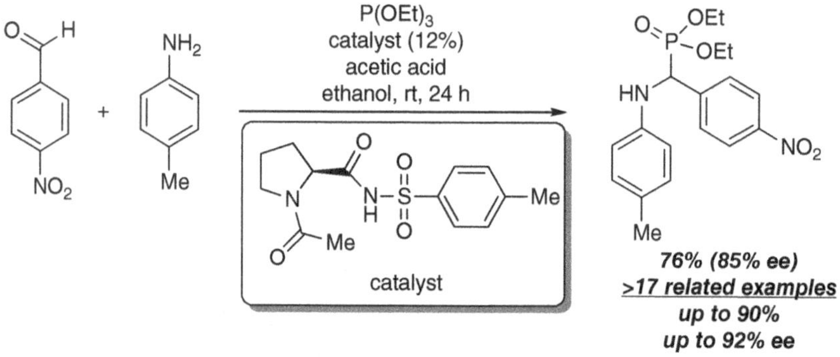

SCHEME 4.118   Organocatalyzed enantioselective synthesis of α-aminophosphonates [196].

Pyrrolidine derivatives catalyzed the enantioselective three-component reactions between aniline derivatives, aldehydes, and trialkylphosphites to generate α-aminophosphonates (Scheme 4.118) [196]. The reaction conditions were mild, and moderate to excellent yields of the products were obtained. Aldehydes bearing electron-donating substituents afforded moderate to excellent yields of the products; however, the selectivities were lower. In contrast, substrates bearing electron-withdrawing groups generated high yields of the α-aminophosphonates with excellent enantioselectivities. It should be noted that these effects were independent of the organic fragment on the aniline derivatives. A wide range of functional groups including halogen, nitro, ether, and unprotected phenols were well tolerated by this approach. Overall, this is an attractive approach for the preparation of α-aminophosphonates as the reactions are operationally simple and successful under metal-free conditions.

The asymmetric synthesis of α-aminophosphonate derivatives from resolved ketosulfinimines has been achieved (Scheme 4.119) [197]. These compounds have a chiral quaternary group adjacent to the phosphorus center and were generated by adding a chilled lithium phosphonate solution in THF to a cold solution of the ketosulfinimine. The reactions

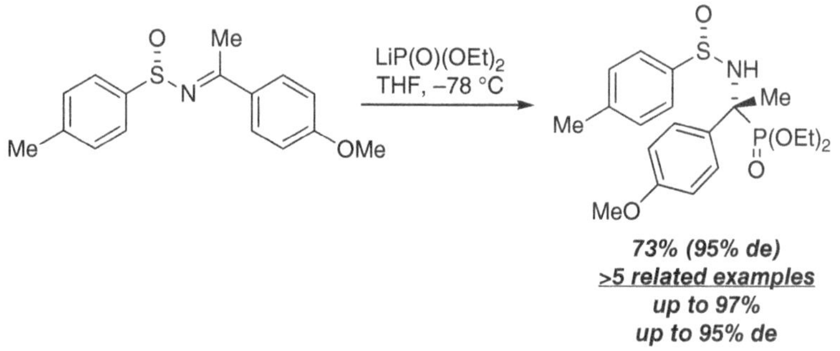

SCHEME 4.119   Asymmetric addition of a lithium phosphonate to ketosulfinimines [197].

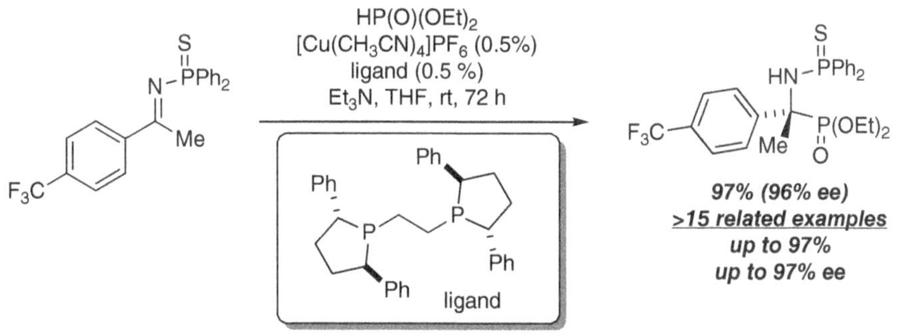

SCHEME 4.120   Copper-catalyzed enantioselective addition of phosphites to ketimines [198].

were highly diastereoselective and de's greater than 95% were commonly observed. While it might have been anticipated that the addition reaction would have generated a 50:50 mixture of the two diastereomers, only a single isomer was typically observed when the reaction was carried out at low temperatures. Furthermore, the authors ruled out the possibility of a reversible addition reaction under the reaction conditions using a control experiment.

The copper-catalyzed asymmetric hydrophosphonylation of ketimines has been reported (Scheme 4.120) [198]. Shibasaki used a readily available copper precursor in combination with a resolved bisphosphine to promote the addition reaction. For most of the examples, the authors selected ketimines that incorporated a small alkyl group on one side of the ketamine and an aromatic group on the other side. This substrate design may have contributed to the selectivity of the overall reaction [192, 193]. When they moved to substrates that reduced the steric bulk around the ketamine, the selectivity was reduced as well. The reactions occurred at mild conditions (room temperature) with a very low catalyst loading (0.5%). Due to the availability of the precursors, low catalyst loading, high conversions, high selectivity, and mild conditions, this is a very attractive catalyst system for the hydrophosphonylation of ketimines.

The use of salen-based aluminum catalysts for asymmetric hydrophosphonylation reactions has been reported; however, the selectivity tended to be low [199, 200]. Katsuki modified the structure of the salen ligand and developed a catalyst system for

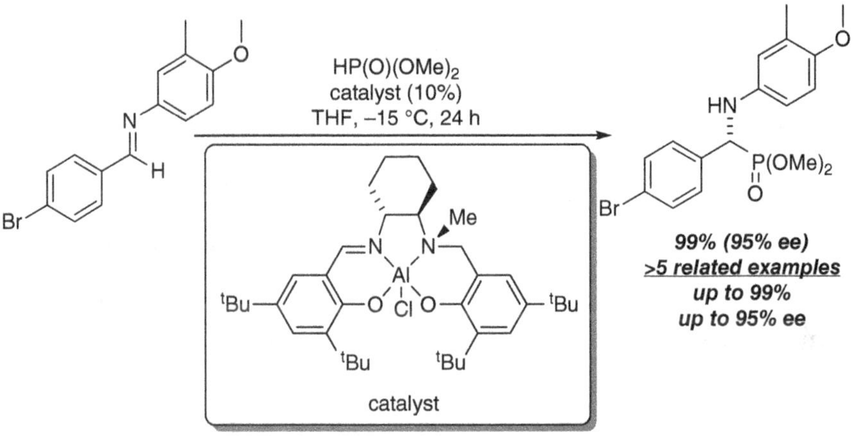

SCHEME 4.121 Enantioselective addition of phosphites to aldimines catalyzed by aluminum salen complex [201].

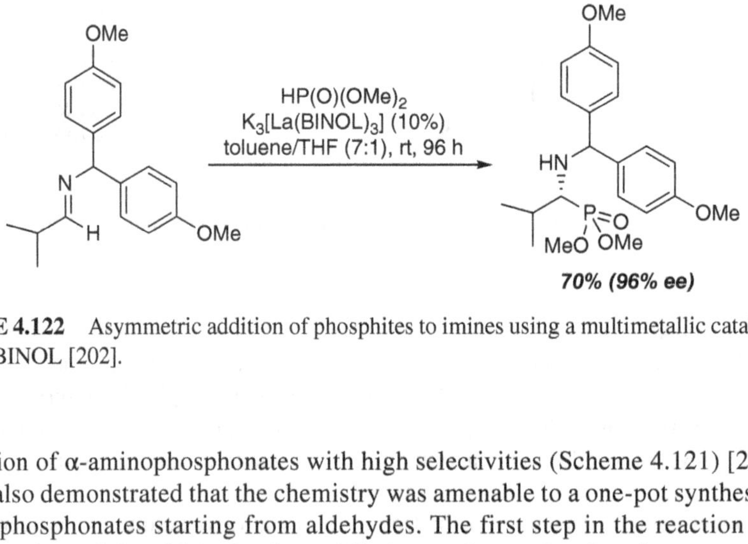

SCHEME 4.122 Asymmetric addition of phosphites to imines using a multimetallic catalyst along with (R)-BINOL [202].

preparation of α-aminophosphonates with high selectivities (Scheme 4.121) [201]. The authors also demonstrated that the chemistry was amenable to a one-pot synthesis of the α-aminophosphonates starting from aldehydes. The first step in the reaction entailed the conversion of the aldehydes into aldimines using molecular sieves to sequester the moisture generated in the reaction. Without purifying the aldimines, the reaction mixture was cooled to −15 °C and the aluminum catalyst and phosphite were added. Purification by column chromatography afforded the α-aminophosphonates in moderate to excellent yields.

A practical approach to the synthesis of chiral α-aminophosphonates was reported by Shibasaki (Scheme 4.122) [202]. His catalyst system was comprised of a multimetallic complex along with resolved BINOL as a ligand. After screening a host of different metal combinations, the most effective grouping was lanthanum and potassium. Using this system, a range of imines were successfully functionalized using dimethylphosphite as the source of the phosphonate fragment. The conditions for the reaction were quite mild (room temperature in toluene/THF), and moderate to excellent yields of the α-aminophosphonates were obtained with ee's as high as 96%. This multimetallic

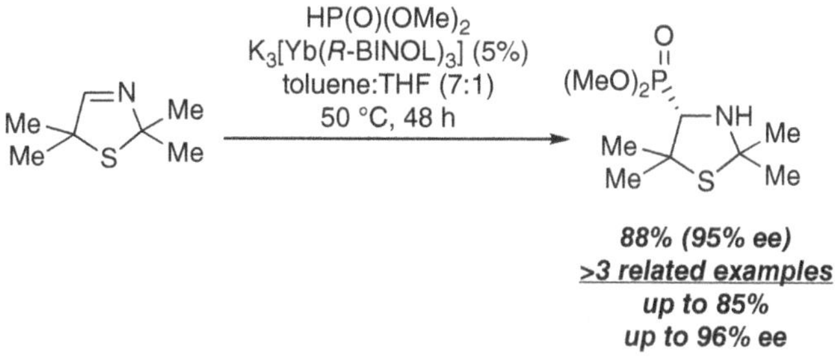

SCHEME 4.123 Asymmetric addition of phosphites to cyclic aldimines catalyzed by an ytterbium/potassium complex along with (R)-BINOL [203].

catalyst system was revolutionary and has become the standard by which other asymmetric approaches are evaluated.

Following this initial report, Shibasaki reported the asymmetric hydrophosphonylation of cyclic imines to generate α-aminophosphonates (Scheme 4.123) [203]. The authors screened a variety of lanthanide catalysts and found that ytterbium-based compounds displayed the most attractive balance between activity and selectivity. While a host of cyclic imines were available, the authors selected to screen several thiazolines as the resulting thiazolidinylphosphonates were essentially protected phosphonic acid versions of D-penicillamine. Using the catalyst formed from the combination of R-BINOL, KHMDS, and Yb(O^{i}Pr)$_3$, a number of thiazolines were selectively functionalized. In general, the yields and the enantioselectivity were excellent. Tetrasubstituted thiazolines were excellent substrates and afforded enantiomeric excesses between 92 and 96%. In the course of this investigation, titanium compounds were also screened for efficacy towards the hydrophosphonylation reaction, but they were not as effective as the lanthanide-containing catalysts. Following this work, the authors reported a refined method for the asymmetric synthesis of these thiazolidinylphosphonates based upon the concept of increased rigidity [204].

A series of lanthanide compounds were found to promote the hydrophosphonylation of imines at very low catalyst loadings (Scheme 4.124) [205]. While the addition to aldehydes was able to be carried out under solvent-free conditions, the analogous reactions with imines required a solvent (THF). In contrast to many of the reported hydrophosphonylation reactions that required a large excess of the phosphite, only 1.2 equivalents was needed in this work. The reaction conditions were quite mild, and simply stirring for 6 h at 40 °C afforded excellent yields for most substrates. While electron-donating groups on the C-terminal group of the aldimine afforded high yields of the hydrophosphonylation product, the incorporation of the electron-withdrawing nitro group resulted in a 42% isolated yield of the α-aminophosphonate. It was also possible that this lower yield could have been due to side reactions involving the nitro group and not just a simple electronic effect. One drawback to this chemistry was that the catalyst needed to be prepared in a separate step; however, the trade-off of preparing a catalyst that operates at such low loading may be worth the extra step. Due to the rapid reaction rates and high yields with low catalyst loading as well as the ability to carry out the reaction in the absence of solvent, this is a valuable method for the synthesis of α-aminophosphonates.

An asymmetric three-component reaction between an aldehyde, 2-aminophenol, and a secondary phosphite to generate a chiral α-aminophosphonate has been reported using

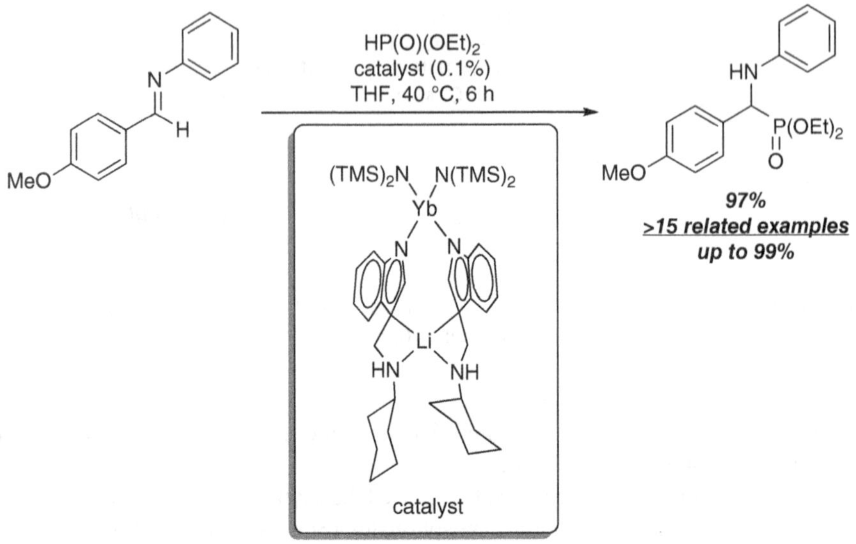

SCHEME 4.124   Asymmetric synthesis of α-aminophosphonates at low catalyst loadings [205].

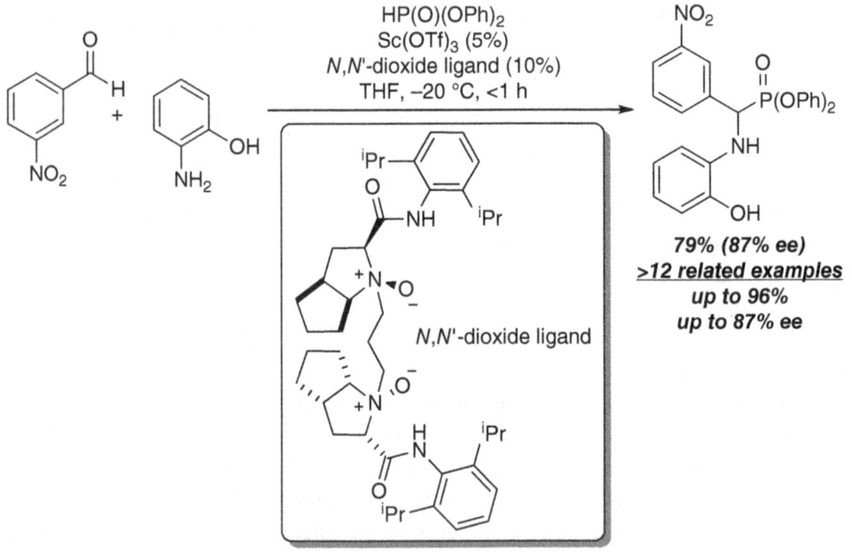

SCHEME 4.125   Asymmetric synthesis of α-aminophosphonates using scandium triflate along with a chiral supporting ligand [206].

scandium (III) triflate and an *N,N'*-dioxide ligand as the catalyst system (Scheme 4.125) [206]. The authors screened a range of metal triflates including yttrium, lanthanum, samarium, ytterbium, and scandium. Only scandium generated the products with the high selectivity. The ee's obtained from reactions using the other metal salts were less than 50%. Of particular note was the ability to use an unprotected phenol in the addition reaction. The reaction conditions were mild, and a range of functionalized aldehydes were successfully used.

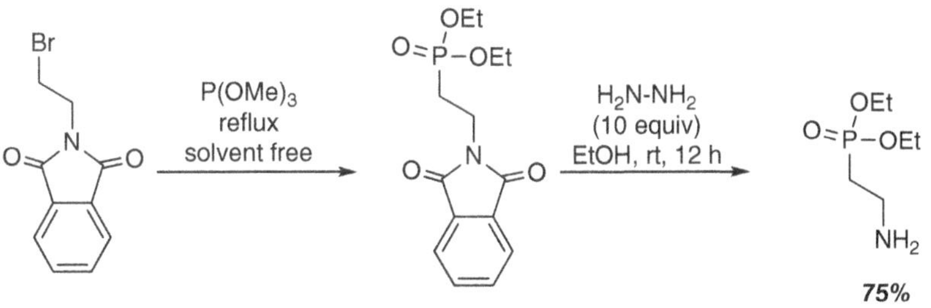

SCHEME 4.126 Synthesis of β-aminophosphonates through an Arbuzov process [207, 208].

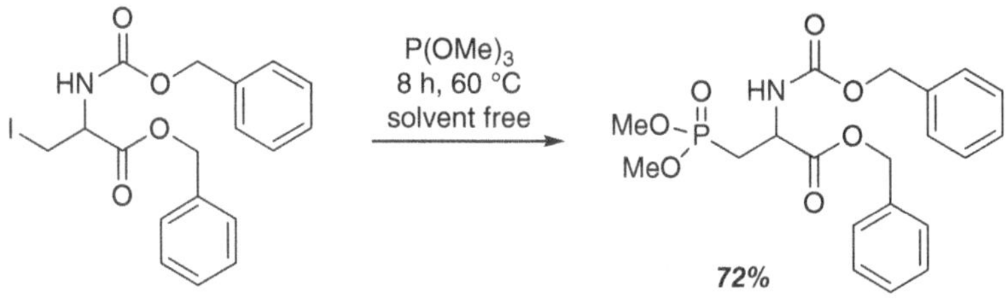

SCHEME 4.127 Preparation of β-aminophosphonates using an acyclic amide protecting group [209].

One of the classic approaches to the synthesis of β-aminophosphonates is the Arbuzov reaction between β-amino alkyl halides and trialkylphosphites. Due to the potential for competing SN$_2$ reactions between amine and alkyl halide, protected substrates are often used as starting points. Once the phosphonation process was completed, the protecting group was typically removed. An example of this process entailed the use of a phthalimide as an amine precursor (Scheme 4.126) [207, 208]. Treatment of a bromoethylphthalimide with triethylphosphite under solvent-free conditions generated the alkylphosphonate in excellent yields. Conversion of the phthalimide into an —NH$_2$ group was readily accomplished by the addition of hydrazine in ethanol at room temperature. With a moderate yield of 75% for two steps, this was a practical approach to the synthesis of β-aminophosphonates. In related work, an acyclic amide was used as the protecting group (Scheme 4.127) [209].

One of the most stereoselective approaches to the preparation of these compounds involved a ring opening of aziridines using secondary phosphites as nucleophiles (Scheme 4.128) [210]. While this approach might seem like an obvious route to the synthesis of these compounds, the use of phosphites has been limited due to sluggish reactivity. This desymmetrization reaction is promoted by the catalyst system comprised of cinchona alkaloid derivatives along with an equimolar amount of diethylzinc. The authors were also able to design a system for the selective synthesis of the other diastereomer based upon manipulation of the core of the chiral scaffold. While most of the substrates screened in this investigation were bicyclic aziridines, the authors did include several examples of nonbicyclic systems to demonstrate the scope of the chemistry. It should be noted that for several examples, the yields of the β-aminophosphonates were reduced due to a competing *phospha*-Brook rearrangement. This was primarily observed with a seven-membered bicyclic aziridine. It was noteworthy that although the yield was reduced in those reactions, the selectivity of the process was still very high.

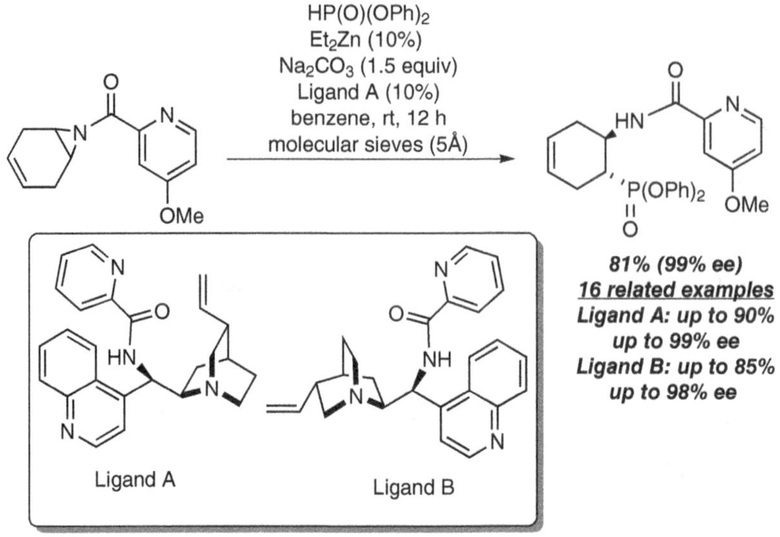

SCHEME 4.128   Asymmetric synthesis of β-aminophosphonates through the ring opening of aziridines [210].

As mentioned previously, a host of reports have focused on the preparation of substituted organophosphonates incorporating a single stereogenic center. The development of a selective approach to the synthesis of an organophosphonate with control over multiple chiral centers remains rare. To this end, Mukherjee developed a diastereoselective approach to the synthesis of dinitrophosphonates (Scheme 4.129) [211]. He used a tandem approach to the synthesis with the first step consisting of an Arbuzov reaction between an acyl chloride and triethylphosphite to generate the acylphosphonate. After converting this intermediate into a racemic α-nitrophosphonate, an asymmetric Michael addition to an nitro alkene generated the target compounds. The organocatalyst used to provide the stereocontrol in the reaction was a thiourea-based bifunctional derivative of quinine. After screening a range of conditions and solvents, dissolving the reagents in trifluorotoluene and cooling the reaction to −10 °C provided the optimal balance of activity and selectivity. The broad scope of the chemistry was demonstrated by the successful functionalization of a range of Michael acceptors incorporating both electron-donating and withdrawing groups.

Hydroxyphosphonates are valuable materials with applications in a number of fields including medicine [212–218] and organic synthesis [146, 219]. The pioneering work of Pudovik and Abramov outlined a synthetic approach that entailed the treatment of carbonyl compounds with secondary phosphites to afford the α-hydroxyphosphonates. These approaches remain the most popular routes to hydroxyphosphonates. Most of the recent work in this area has been focused on developing catalysts that exhibit increased substrate scope and are active under mild conditions. Given the biological activity of many examples, a significant amount of work has been devoted to designing catalysts for the asymmetric synthesis of α-hydroxyphosphonates [220–223].

The use of barium hydroxide to promote the addition of secondary phosphites to aldehydes has been reported (Scheme 4.130) [224]. One of the attractive aspects of this

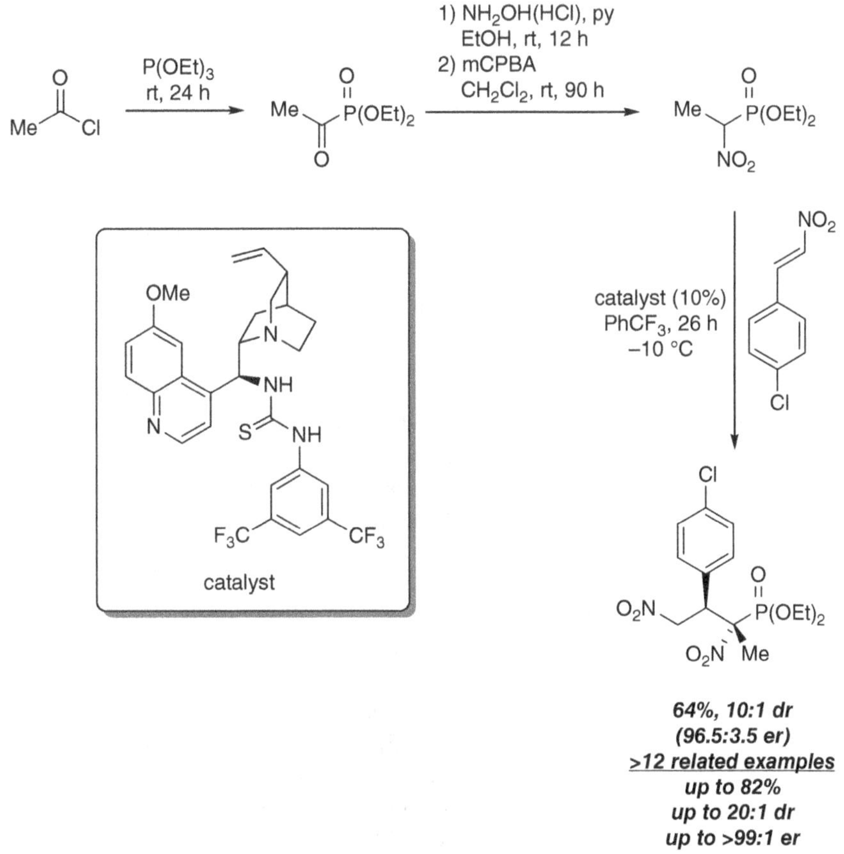

SCHEME 4.129 Organocatalytic approach to the synthesis of alkylphosphonates with multiple chiral centers [211].

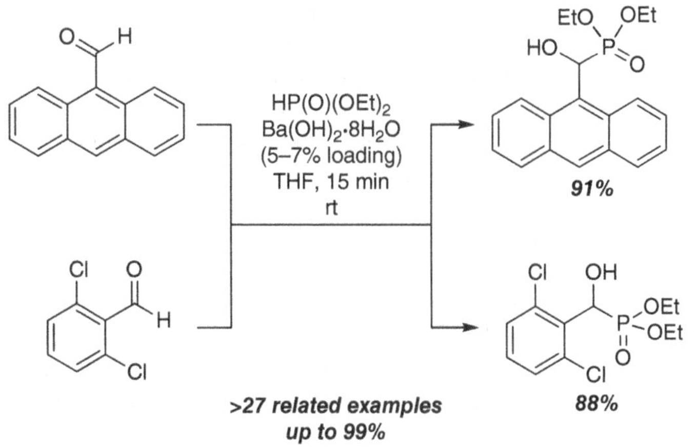

SCHEME 4.130 Barium hydroxide-promoted addition of phosphites to aldehydes [224].

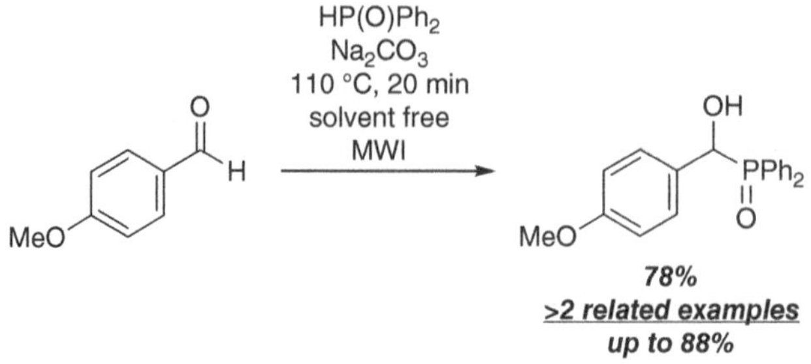

SCHEME 4.131   Microwave-assisted synthesis of α-hydroxyphosphine oxides [226].

chemistry was the use of an inexpensive catalyst that was stable to air and moisture. After screening a range of conditions, the authors devised a successful system based upon the use of the barium hydroxide (as the octahydrate) as the catalyst with THF as the solvent. The chemistry was quite rapid at room temperature and displayed wide functional group compatibility as a wide range of electron-donating and withdrawing groups were tolerated. One of the attractive aspects of this chemistry was its ability to handle bulky groups. Indeed, 9-anthracenecarboxaldehyde and 2,6-dichlorobenzaldehyde were readily functionalized using this approach. Overall, this approach is attractive due to operational simplicity of the method as well as the ability to handle a wide range of substrates with an inexpensive and robust catalyst.

A comparison of classic Pudovik reactions using ultrasound, solvent-free conditions, as well as radical initiation has been reported [225]. The conclusions from this investigation suggest that the selection of a successful approach for the preparation of an alkylphosphonate or hydroxyphosphonate depends largely on the substrate. Essentially, each substrate needs to be evaluated individually and assessed for the viability of each approach. While this can be a challenging goal for a complex substrate, the following sections will hopefully provide some direction regarding the most attractive approach with specific examples of successful reactions.

Keglevich has reported the synthesis of α-hydroxyphosphine oxides through the formal addition of secondary phosphine oxides to aldehydes (Scheme 4.131) [226]. The reactions were carried out using a focused microwave reactor under solvent-free conditions and generated the functionalized phosphine oxides in excellent yields. The authors found that a base was needed and found that sodium carbonate was as an inexpensive solution. The process was quite practical as the reagents were weighed out in the air and irradiated in open vessels. A number of functional groups including nitro, methoxy, and chloro were well tolerated, and the products were isolated by simply extracting the reaction reside with ethyl acetate followed by recrystallization.

The ring opening of epoxides using phosphorus nucleophiles typically generates β-hydroxyphosphonates. An example of this chemistry involved the formation of β-hydroxyphosphonates from functionalized nucleotides through the treatment of epoxides with secondary phosphites in the presence of N,O-bis(trimethyl)silyl acetamide followed by BF_3 (etherate) (Scheme 4.132) [227]. While only a few examples were presented as part of this work, there is great potential for the extension of this chemistry to more complex substrates bearing a range of functional groups.

During the synthesis of biologically active fosmidomycin derivatives, Pudovik chemistry was used to prepare α-hydroxyphosphonates (Scheme 4.133) [228]. The chemistry

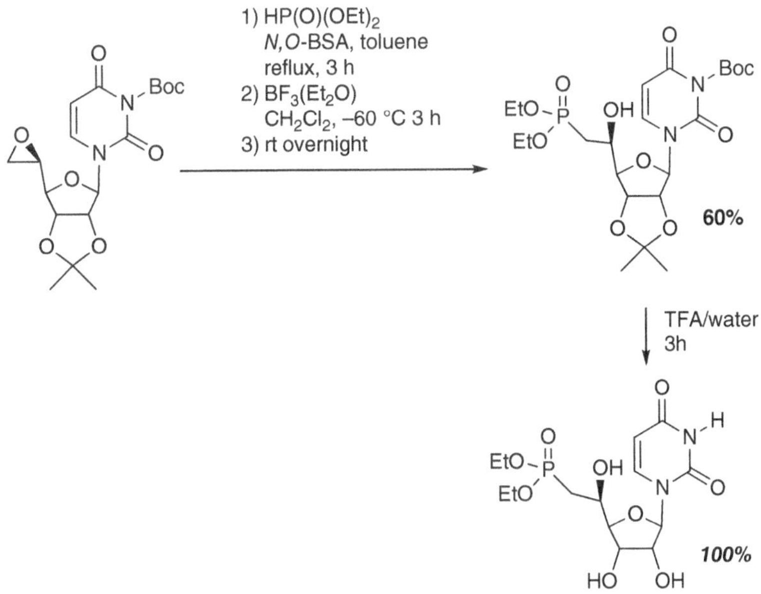

SCHEME 4.132   Synthesis of phosphorylated nucleosides [227].

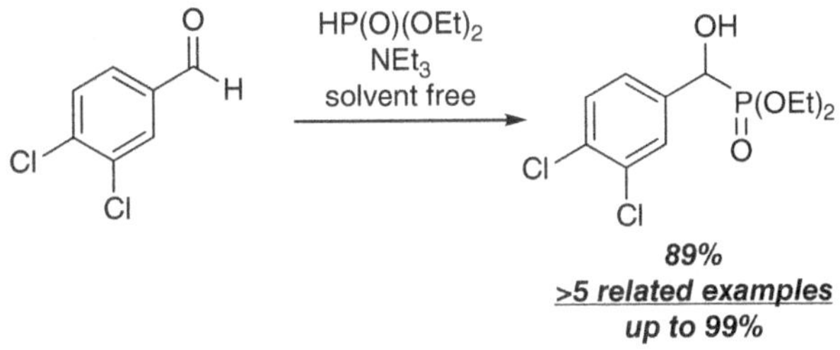

SCHEME 4.133   Synthesis of α-hydroxyphosphonates through Pudovick-type chemistry [228].

was operationally trivial, and simply adding the reagents together at 0 °C followed by gently warming to room temperature for 1 h generated the α-hydroxyphosphonates in high yields. This example demonstrates why this approach continues to be one of the most popular.

The solvent-free synthesis of hydroxyphosphonates has been achieved using a solid-supported phosphazene base (Scheme 4.134) [229]. The main goal of this chemistry was to develop a synthetic approach that had a low environmental factor (E-factor) by eliminating the organic solvent used for the reaction chemistry and lowering the amount of a base that was needed to promote the addition reaction. A range of supported bases bearing pendant tertiary amines, phosphazenes, pyridines, and triazabicyclic groups were screened for activity in this reaction. The phosphazene group was found to be significantly more active than the other bases and was used for the substrate screening. A range of electron-neutral and electron-rich aryl aldehydes as well as aliphatic aldehydes were compatible with this approach, and outstanding yields were observed for most products (94–99%).

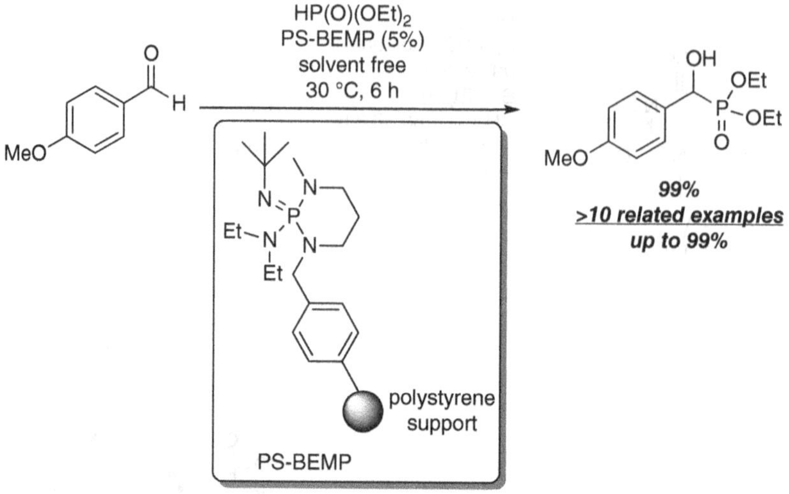

SCHEME 4.134 Solvent-free synthesis of α-hydroxyphosphonates [229].

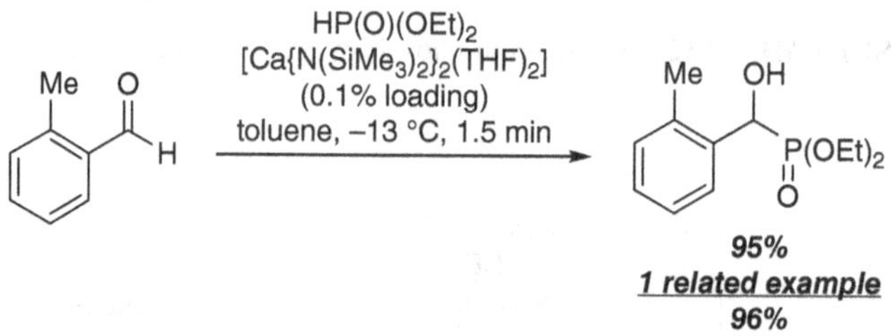

SCHEME 4.135 Calcium-catalyzed synthesis of α-hydroxyphosphonates from aldehydes [230].

It was interesting that no aldol products were observed when aldehydes containing enolizable hydrogens were screened. When an aldehyde bearing a 2-pyridyl or an electron-withdrawing aryl group was screened, a significant amount of the *phospha*-Brook rearrangement product was observed. Through the use of a flow system, the authors were able to significantly reduce the E-factors. Using several representative examples as case studies, they reported a reduction of E-factor by 92.7–96.3% using the flow system. For many aldehydes, this is one of the most sustainable approaches to the synthesis of α-hydroxyphosphonates.

While a host of reports have focused on the use of transition metals and lanthanides, the use of alkaline earth complexes as catalysts for the addition of secondary phosphine oxides to aldehydes and ketones is an attractive alternative due to the lower cost of the alkaline earths. To this end, a series of alkaline earth compounds have been prepared and screened for activity in hydrophosphonylation reactions (Schemes 4.135 and 4.136) [230]. Most of the work was focused on devising a system for problematic ketones, and after some experimentation, the authors discovered that a simple calcium-based complex promoted the addition reaction at low catalyst loadings (0.1%) The chemistry was operationally trivial, and simply stirring the neat reagents under solvent-free conditions generated moderate to excellent yields of the hydroxyphosphonates. The chemistry was sensitive to bulky groups on the aromatic ketone, and the incorporation of either a methyl group or a nitro group in the ortho-position of the aromatic group resulted in only trace amounts of the addition

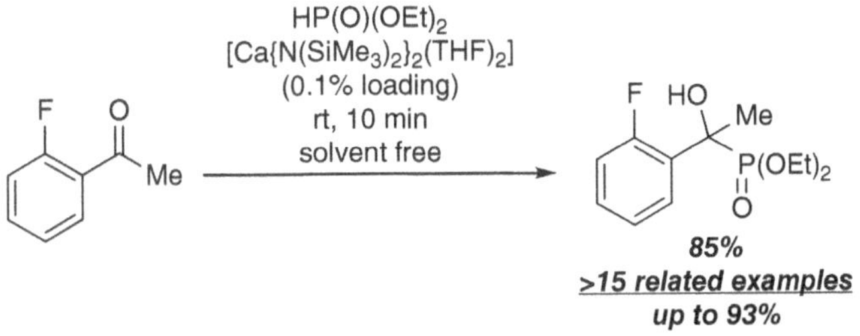

SCHEME 4.136   Calcium-catalyzed synthesis of α-hydroxyphosphonates from ketones [230].

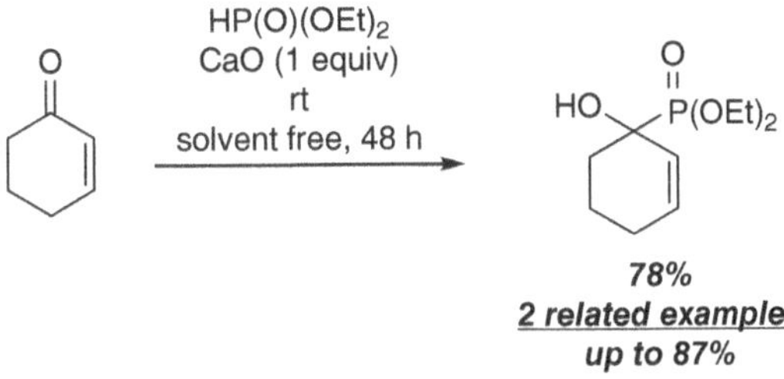

SCHEME 4.137   Calcium oxide-promoted synthesis of α-hydroxyphosphonates [122].

products. In addition to the work with ketones, the authors extended the chemistry to several aldehydes. Predictably, these substrates were more reactive and the hydrophosphonylation was successful at low temperature (−13 °C). Given the high yields of the hydroxyphosphonates, rapid reaction rates, and low catalyst loadings, this is one of the most attractive approaches for the synthesis of these compounds.

One of the challenges encountered in the synthesis of complex structures is the need to chemoselectively functionalize one part of the molecule while leaving another reactive site intact. The selective addition of secondary phosphites to an aldehyde or ketone while an activated alkene remains unchanged is an example of such a transformation. If the activated alkene is attached to the carbonyl fragment, the reaction is essentially a competition between a 1,2- and 1,4-addition of the phosphorus nucleophile. The selectivity of this reaction was probed using a readily available calcium base as the promoter and several cyclic and acyclic ketones as substrates (Scheme 4.137) [122]. The chemistry was operationally trivial and consisted of simply stirring the reagents in the presence of the inorganic base. For both types of ketones, the mode of addition of diethylphosphite was highly selective and favored the 1,2-hydrophosphonylation product. Although a stoichiometric amount of the base was used in this chemistry, it is inexpensive and readily available.

A rapid and high-yielding approach for the preparation of α-hydroxyphosphonates has been reported using $[(Me_3Si)_2N]_3La(\mu\text{-Cl})Li(THF)_3$ as the catalyst (Scheme 4.138) [231]. In addition to this lanthanum species, the authors screened a range of catalysts for activity towards the hydrophosphonylation reaction and discovered that simple salts such as $LaCl_3$ or $Yb(OTf)_3$ were not effective; whereas, the bimetallic system readily promoted the reaction. The authors selected dialkylphosphites as the phosphorus nucleophiles and a range of arylaldehydes as substrates. The chemistry was insensitive to electronic changes in the

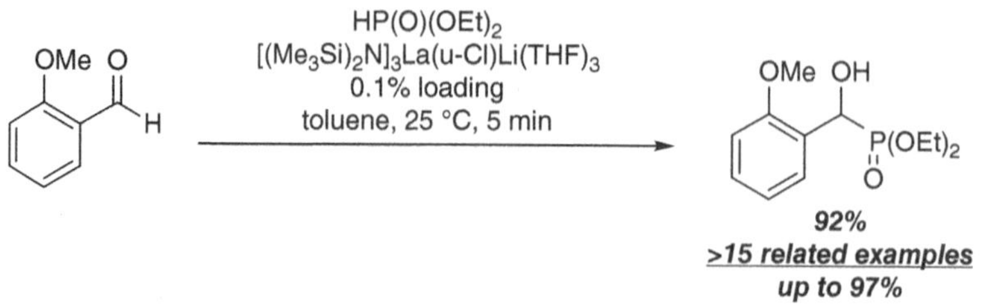

SCHEME 4.138   Preparation of α-hydroxyphosphonates using a lanthanum-based catalyst [231].

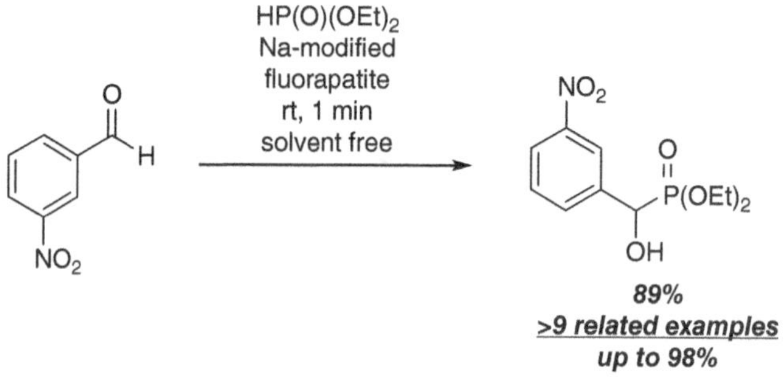

SCHEME 4.139   Synthesis of α-hydroxyphosphonates from aldehydes using sodium-modified fluorapatite [232].

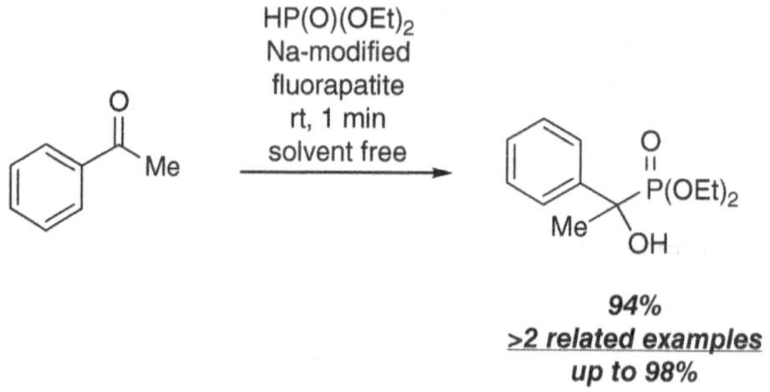

SCHEME 4.140   Synthesis of α-hydroxyphosphonates from ketones using sodium-modified fluorapatite [232].

aromatic groups as a range of arylaldehydes bearing electron-donating and withdrawing groups were successfully functionalized using this approach. As for the phosphites, even the bulky diisopropylphosphite was converted into the corresponding hydroxyphosphonate in high yield (94%). The practical advantages of this approach include a very low catalyst loading (0.1%) and fast reaction times (5 min).

The use of sodium-modified fluorapatite for the synthesis of α-hydroxyphosphonates has been reported (Schemes 4.139 and 4.140) [232]. For the reaction between secondary phosphites and aldehydes, the reaction conditions were quite mild, and excellent yields of the targets were obtained within a few minutes under solvent-free conditions using

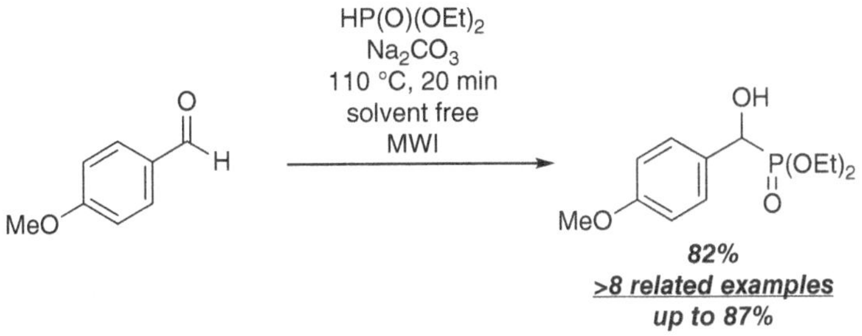

SCHEME 4.141   Microwave-assisted synthesis of α-hydroxyphosphonates [226].

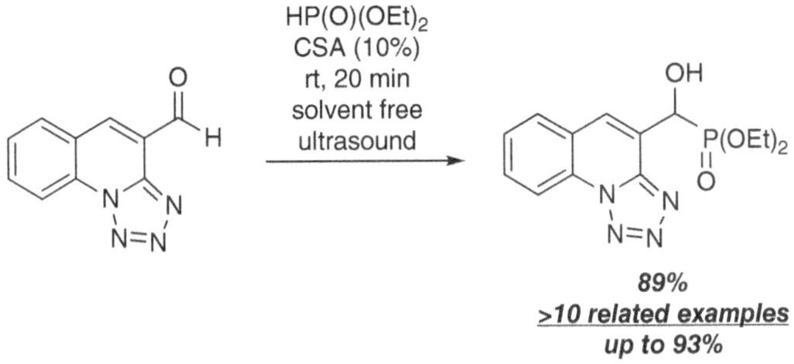

SCHEME 4.142   Ultrasound-promoted synthesis of α-hydroxyphosphonates [163].

this catalyst. The addition reactions were also operationally trivial as simply adding the reagents to a tube and stirring generated the α-hydroxyphosphonates. While a number of catalyst systems were unable to promote this reaction when ketones were used as substrates, the authors were pleased to find that the sodium-modified fluorapatite readily catalyzed the addition of phosphites to ketones under very mild conditions. One of the reasons why heterogeneous catalysts are so attractive is that they are easily separated from the reaction mixture and used for subsequent reactions. This was demonstrated using this sodium-modified fluorapatite. The catalyst was successfully recycled for four reactions before a significant decrease in reactivity was observed. Once the catalyst was washed with acetone and calcined at 500 °C, it was active for the addition reaction again. While long-term studies are needed, these results suggest that this catalyst system could be an attractive and sustainable route to the synthesis of hydroxyphosphonates from both aldehydes and ketones.

The microwave-assisted formation of α-hydroxyphosphonates has been reported by Keglevich (Scheme 4.141) [226]. In addition to rapidly forming the target compounds in excellent yields, the chemistry proceeded under solvent-free conditions using a readily available base. While many microwave-assisted reactions were carried out using closed reaction vials, these reactions were irradiated in open vessels. The incorporation of electron-donating and withdrawing groups into the aldehyde was well tolerated, and the products were isolated by simply extracting the reaction reside with ethyl acetate followed by recrystallization.

One of the most recent variations on the synthesis of α-hydroxyphosphonates entailed the use of ultrasound to promote the reaction (Scheme 4.142) [163]. The authors used a common ultrasound bath as well as CSA as the catalyst to promote the reaction.

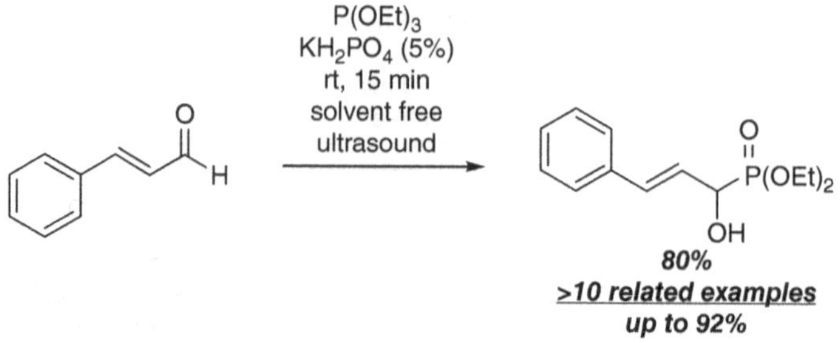

SCHEME 4.143   Preparation of α-hydroxyphosphonates using ultrasound and potassium dihydrogenphosphate [233].

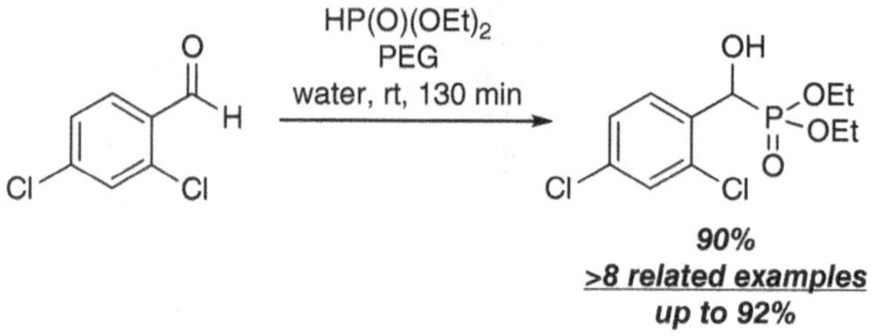

SCHEME 4.144   Synthesis of α-hydroxyphosphonates in water [183].

The reaction times were reduced without decreasing the yields of the resulting hydroxyphosphonates. In control reactions, the authors confirmed that the acid catalyst was still needed for a successful reaction as removing it resulted in no conversion. The scope of this reaction was broad, and a range of functionalized aldehydes were successfully used in this reaction. Of particular note was the successful use of an unprotected phenol in the synthesis.

The use of potassium dihydrogenphosphate as a catalyst for the ultrasound-assisted synthesis of α-hydroxyphosphonates has been reported (Scheme 4.143) [233]. The reactions were carried out under solvent-free conditions and generated moderate to excellent yields of the desired compounds after exposing the reaction mixtures to ultrasound for 5–45 min. A range of functional groups were well tolerated by this chemistry including electron-donating and withdrawing groups. Ketones such as acetophenone and cyclohexanone were unreactive even with extended reaction times.

The preparation of α-hydroxyphosphonates in a mixture of water and polyethylene glycol has been reported (Scheme 4.144) [183]. The chemistry was operationally trivial and involved simply stirring the reagents in the solvent mixture at room temperature. These aqueous reactions generated moderate to excellent yields of the compounds. The method was tolerant of a variety of functional groups including nitro, amino, and an unprotected phenol. Heteroarenes such as imidazole were also successfully functionalized. As with many related reactions, ketones were unreactive under the reaction conditions. This approach is one of the most practical syntheses of α-hydroxyphosphonates due to the simplicity of the reaction and the ability to promote the reaction in water at room temperature.

Building on the work of Spilling [234], Shibasaki [235], and Shibuya [236], a number of approaches to the asymmetric synthesis of hydroxyphosphonates have been developed.

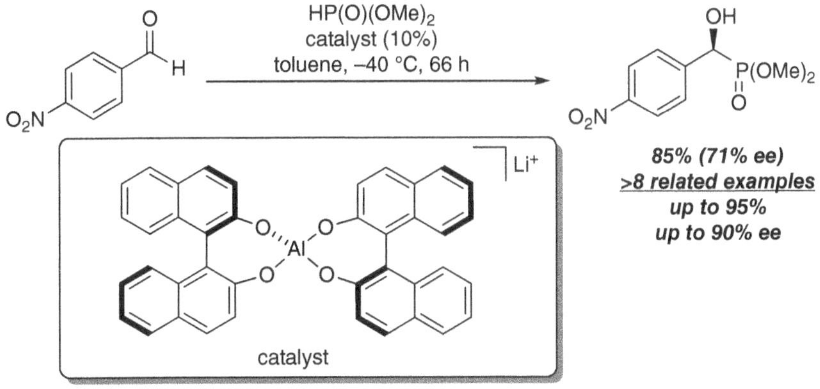

SCHEME 4.145 Asymmetric synthesis of α-hydroxyphosphonates using an aluminum-based catalyst [235].

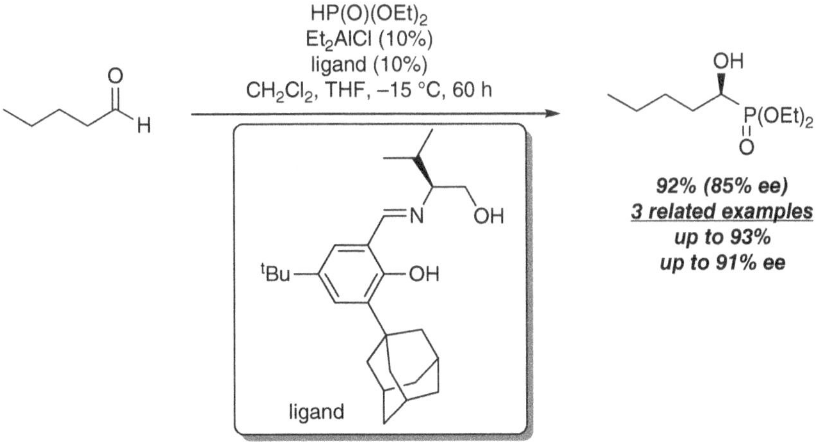

SCHEME 4.146 Asymmetric synthesis of α-hydroxyphosphonates from aliphatic aldehydes using an aluminum-based catalyst [237].

One of these approaches used a heterobimetallic catalyst consisting of two BINOL fragments attached to an aluminum center along with a lithium counterion (Scheme 4.145) [235]. Secondary phosphites were used as the phosphorus nucleophiles, and a wide range of aldehydes were readily functionalized using this approach. While moderate to high enantioselectivities were observed using arylaldehydes as substrates, aliphatic aldehydes were considerably less selective (ee's = 3–24%).

In terms of substrate scope and the ability to handle a wide variety of aldehydes, the catalyst system comprised of a Schiff base along with diethylaluminum chloride was a practical approach to the asymmetric synthesis of α-hydroxyphosphonates (Schemes 4.146 and 4.147) [237]. The catalyst preparation was relatively straightforward and could be carried out *in situ*. The strength of this transition metal-free approach was the ability to convert arylaldehydes bearing a wide range of electron-donating and withdrawing groups as well as ortho-substituents into α-hydroxyphosphonates with high

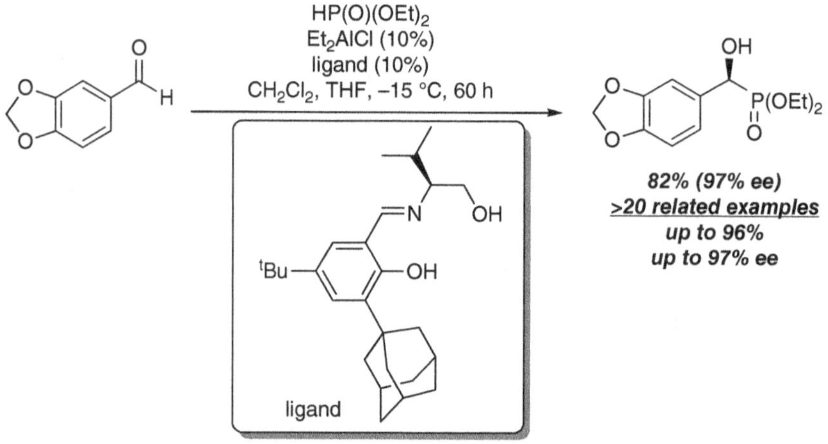

SCHEME 4.147 Asymmetric synthesis of α-hydroxyphosphonates from benzaldehydes using an aluminum-based catalyst [237].

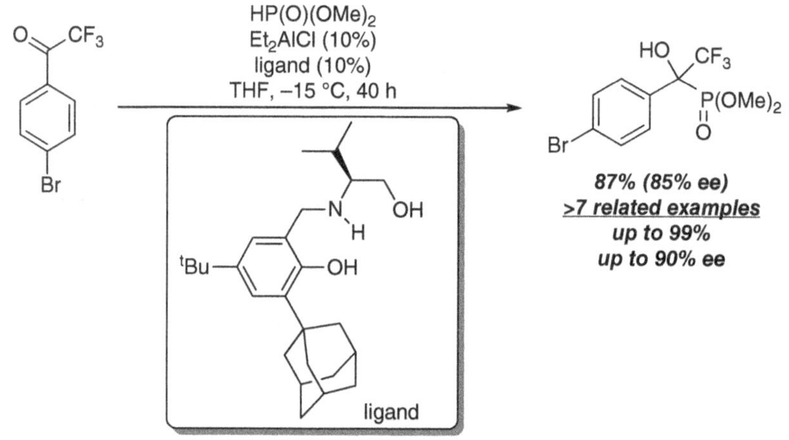

SCHEME 4.148 Enantioselective synthesis of α-hydroxyphosphonates bearing a trifluoromethyl group [238].

enantioselectivity. Heteroaromatic groups were also tolerated by this chemistry. Furthermore, the high activity of aliphatic aldehydes and the high ee's of the resulting organophosphonates were impressive. The mechanism of the reaction was investigated, and the authors proposed the active species in this chemistry to be an oligomeric/polymeric aluminum species.

Asymmetric hydrophosphonylation reactions were extended to trifluoromethyl ketones (Scheme 4.148) [238]. This approach was interesting since ketones often displayed sluggish reactivity in hydrophosphonylation reactions. While only a few examples were screened, several electron-donating and withdrawing groups were compatible with the catalyst system, and moderate to excellent yields of the hydroxyphosphonates were obtained. As mentioned in other sections of this text, the retention of the aryl bromide

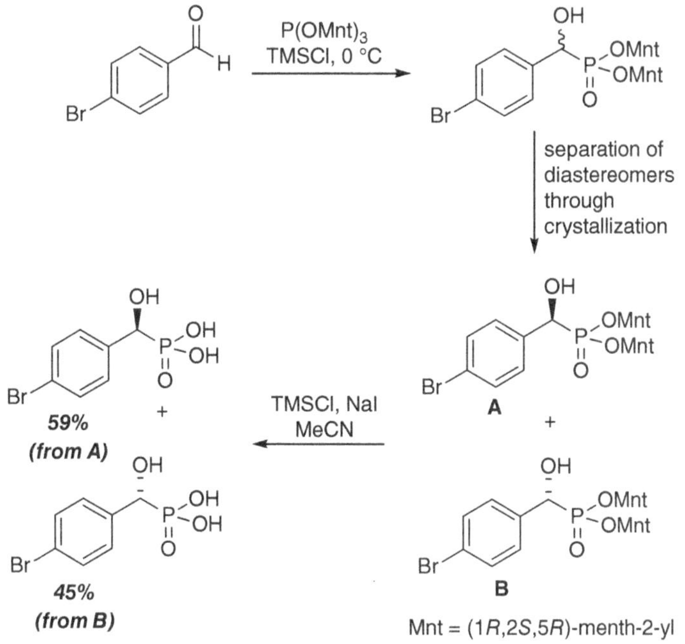

SCHEME 4.149   Asymmetric synthesis of hydroxyphosphonates using P(OMnt)$_3$ [212].

is advantageous for multistep syntheses as it can be readily transformed using cross-coupling chemistry. The authors carried out a mechanistic investigation of this chemistry in order to probe the nature of the active species. In contrast to the observations made when using alkyl and arylaldehydes, a monomeric aluminum species was suggested to be the active species. This chemistry provided a route for the incorporation of a valuable trifluoromethyl group into α-hydroxyphosphonates.

An asymmetric synthesis of α-hydroxyphosphonates has been reported by Ellman (Scheme 4.149) [212]. The overarching goal of this work was the development of new inhibitors for striatal-enriched protein tyrosine phosphatase (STEP) due to its high activity in disorders such as Alzheimer's disease. One of the most active compounds was an α-hydroxyphosphonic acid. The direct precursor to this was the corresponding phosphonate. Rather than approach the synthesis using an asymmetric protocol, the authors selected to generate diastereomers and separate them using classic techniques. Their approach entailed the use of tris[(1R,2S,5R)-menth-2-yl]phosphite (P(OMnt)$_3$) as the phosphorus nucleophile in the C—P bond-forming step. Once the mixture was generated, the diastereomers were separated by crystallization. Following the separation, the phosphonate esters were cleaved using TMSCl/NaI in acetonitrile to generate the enantiopure α-hydroxyphosphonic acids. The retention of the aryl bromide through the formation and separation/purification of the α-hydroxyphosphonate was critical since the compounds needed to be further functionalized using a range of coupling reactions. The compounds were further functionalized, screened for activity, and the α-hydroxyphosphonate with the R-configuration was significantly more inhibitory towards STEP.

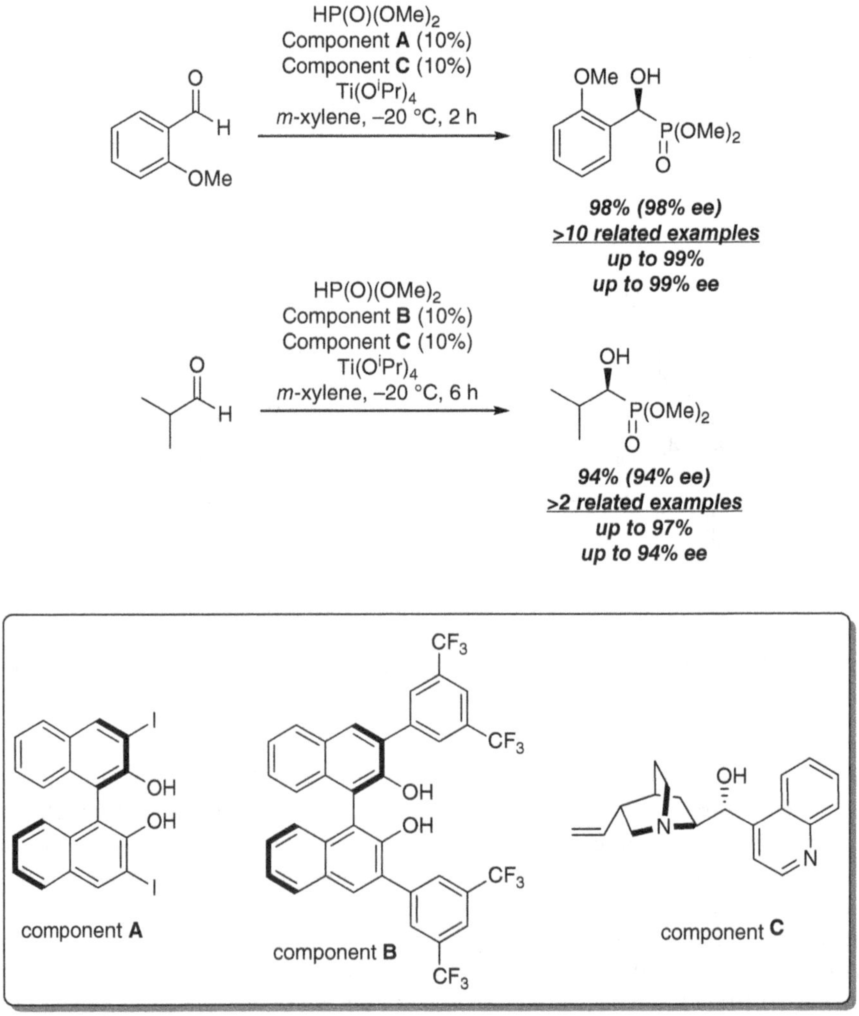

SCHEME 4.150   A multicomponent approach to the asymmetric synthesis of α-hydroxyl phosphonates [239].

   A multicomponent catalyst system has been used to promote the enantioselective hydrophosphonylation of aldehydes (Scheme 4.150) [239]. While a host of metal-based and organocatalysts promote asymmetric reactions by coordinating one part of the substrate while simultaneously activating/functionalizing another part of the molecule, the authors used a modular approach and separated the key components of the catalyst to allow for more flexibility in the active site. A range of BINOL ligands and cinchona alkaloids were screened for activity and selectivity in the titanium-catalyzed addition reaction, and the dual catalyst systems were found to be superior. For the troublesome aliphatic aldehydes, incorporating bulky fluorinated aryl groups into the BINOL structure afforded a reasonable balance of activity and selectivity for the reaction. In a

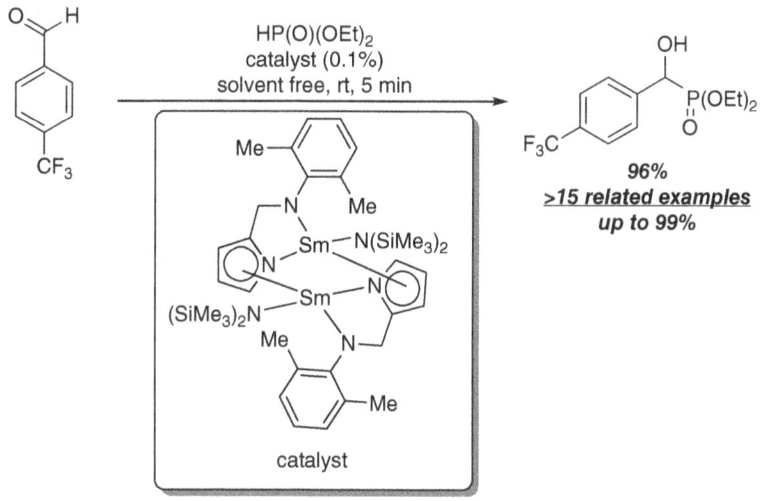

SCHEME 4.151   Asymmetric synthesis of α-hydroxyphosphonates using a samarium-based catalyst [240].

separate part of the investigation, the authors were able to decrease the loading of the titanium isopropoxide and ligands to 2.5%.

The enantioselective synthesis of α-hydroxyphosphonates has also been achieved using a samarium-based catalyst (Scheme 4.151) [240]. The authors used a bimetallic complex containing two samarium centers that were linked by pyrrolyl-based bridging ligands. The chemistry could be carried out under solvent-free conditions or in toluene solution. In either case, the products were formed in high yields within a few minutes at room temperature. No electronic preference was observed in chemistry, and a wide range of functionalized aldehydes were successfully used in the addition reaction. One of the most interesting aspects of this chemistry was the observation that ketones could also be functionalized using this catalyst. As mentioned previously, ketones have been challenging to functionalize in this manner due to sluggish reactivity. Although the catalyst needs to be prepared in a separate step, the rapid reaction rates, high yields, and low catalyst loadings make this a reasonable method for the preparation of α-hydroxyphosphonates.

The hydrophosphonylation of aldehydes and ketones bearing bulky groups can be quite challenging; however, several recent solutions have been reported. One such report entailed the use of a catalyst system comprised of iron (III) chloride ligated by Schiff bases (Scheme 4.152) [241]. An attractive aspect of this approach was that the catalyst system could be assembled *in situ*. There was no need to synthesize the iron complex prior to the hydrophosphonylation reaction. The phosphorus nucleophile used in this chemistry was a dialkylphosphite. After a series of screening experiments, the authors found that arylaldehydes bearing electron-withdrawing groups exhibited the fastest reaction rates. Substrates bearing electron-donating, heteroaromatic, or bulky groups were also successfully transformed into the α-hydroxyphosphonates at longer reaction times. The authors probed several mechanistic possibilities to explain the chemistry and suggested that the key

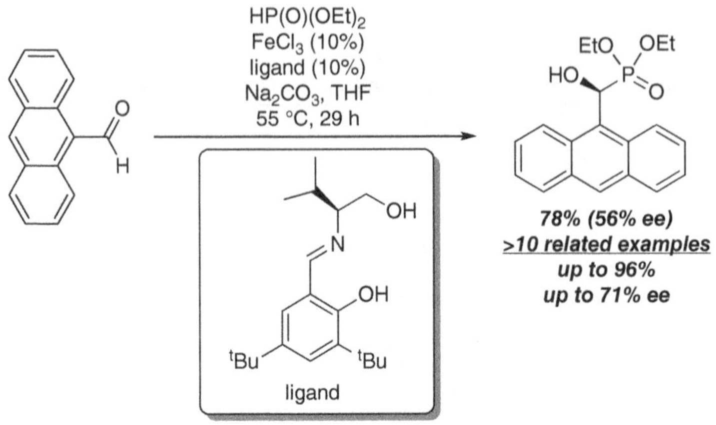

SCHEME 4.152 Preparation of resolved α-hydroxyphosphonates using an iron-based catalyst [241].

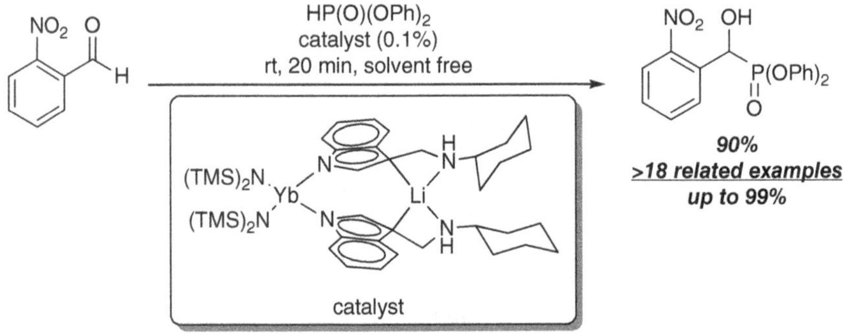

SCHEME 4.153 Asymmetric addition of phosphites to aldehydes catalyzed by an ytterbium-based catalyst [205].

intermediate was a dimeric iron complex containing both substrates coordinated to the same iron. While the secondary phosphite could have been coordinated through either the phosphorus or the oxygen, computational studies suggested that the attachment was through the oxygen. Nucleophilic attack of the coordinated phosphite on the carbonyl carbon was the key C—P bond-forming reaction. Using this mechanistic information as a guide, the authors added a base to the reaction in order to increase the concentration of the phosphite anion.

A series of ytterbium catalysts that promoted the addition of phosphites to imines were also active for the addition to aldehydes (Scheme 4.153) [205]. The catalyst was a well-defined bimetallic complex with a lanthanide and a lithium atom linked by amido-functionalized bis(indolyl) groups. After a bit of experimentation, the authors discovered that using ytterbium as the lanthanide center afforded the highest yields of the α-hydroxy-phosphonates. Furthermore, they determined that a solvent was not needed and high yields of the α-hydroxyphosphonates were obtained after simply stirring the reagents for 20 min. No significant electronic effect was observed in this chemistry as both electron-donating and electron-withdrawing groups were well tolerated.

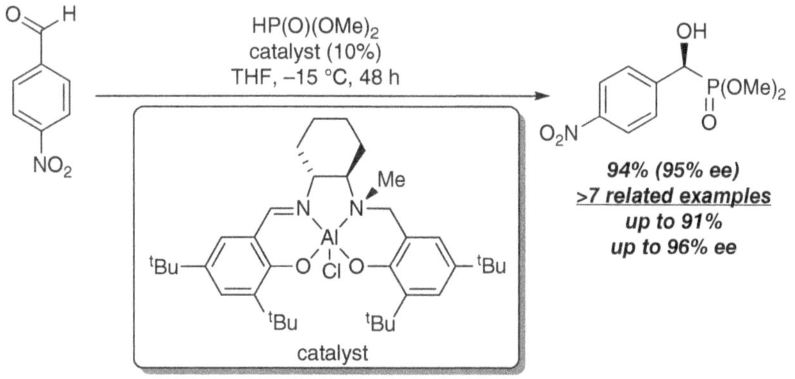

SCHEME 4.154   Enantioselective synthesis of α-hydroxyphosphonates [201, 242].

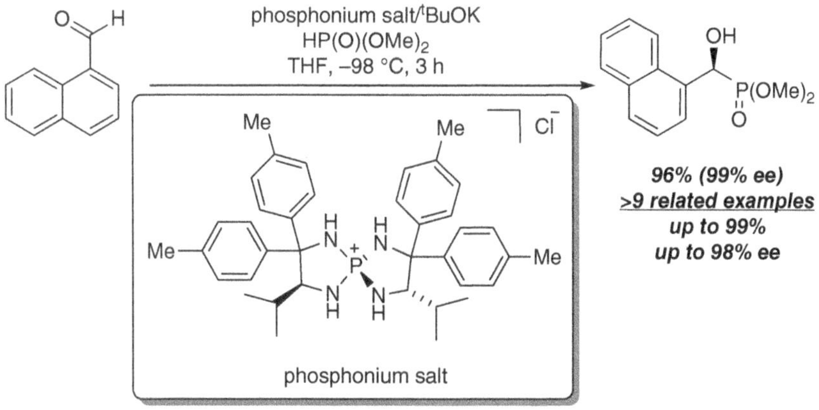

SCHEME 4.155   Enantioselective synthesis of α-hydroxyphosphonates from aldehydes using triaminoiminophosphorane-based catalysts [245].

The same aluminum catalyst that promoted the asymmetric synthesis of α-amino-phosphinates was also active for the asymmetric synthesis of α-hydroxyphosphonates (Scheme 4.154) [201, 242]. A range of electronically varied aromatic and aliphatic aldehydes were screened, and high enantioselectivities were obtained for both. In a subsequent report, the same group discovered that the addition of a base (K_2CO_3) increased the rate of the reaction and facilitated a reduction in the amount of catalyst that was needed to promote the addition reaction (1–4%) [243]. Thankfully, the enantioselectivity was retained in the updated version of the reaction (ee's = 93–98%). In addition, a computational study providing insight into the mechanism as well as the source of enantioselectivity has been reported [244].

The asymmetric hydrophosphonylation of aldehydes has been achieved using a catalyst based upon a triaminoiminophosphorane (Scheme 4.155) [245]. This was an interesting approach to this chemistry and generated the α-hydroxyphosphonates in excellent yields with high enantioselectivity. In addition to the hydrophosphonylation chemistry, the authors also investigated how readily common bases deprotonated phosphites in order to provide

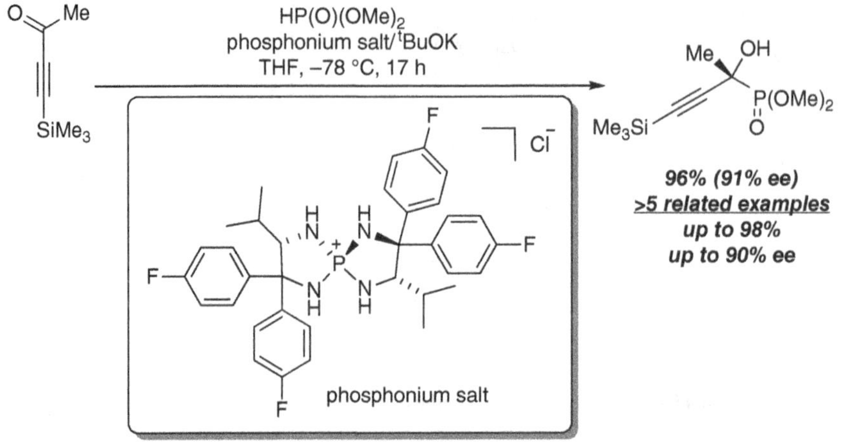

SCHEME 4.156 Enantioselective synthesis of α-hydroxyphosphonates from ynones [246].

insight into the reaction chemistry. Although the chemistry needed to be performed at very low temperatures, the isolated yields and selectivity of the process were impressive.

Following their work on the asymmetric synthesis of α-hydroxyphosphonates from arylaldehydes, the same group devised a route for the preparation of phosphorylated propargyl alcohols from ynones (Scheme 4.156) [246]. This structural motif was an attractive target since propargyl alcohols can be readily converted into a host of other functional groups. After screening a range of functionalized phosphonium salts, the authors discovered that a fluorinated phosphonium salt provided the best match of reactivity and selectivity. This was interesting since the most effective catalyst for this chemistry was electronically reversed from the most effective one from the arylaldehyde chemistry [245]. However, care must be taken with this, since moving to the electron-withdrawing trifluoromethyl group decreased the selectivity of the process. The authors proposed a mechanism for this reaction that was similar to the one they suggested for the arylaldehyde chemistry with the key step consisting of generation of a triaminoiminophosphorane through deprotonation of a tetraaminophosphonium salts with a powerful base. A range of ynones bearing alkyl groups were readily functionalized using this approach. Perhaps, one of the most attractive substrates screened in this chemistry was the trimethylsilyl group. This is a common protecting group for a terminal alkyne and is readily removed through treatment with a fluoride source. Phosphonylation of this valuable precursor proceeded in excellent yield with an impressive selectivity. Overall, this is a straightforward route to the preparation of phosphorylated propargyl alcohols.

The conversion of unactivated alkenes into β-hydroxyphosphonates in a one-pot reaction is a practical transformation as it incorporates both functionalities into the substrate (Scheme 4.157) [247]. This chemistry was accomplished using an iron-based catalyst system along with either pure oxygen or air as an oxidant. After screening a range of common iron compounds, the authors discovered that iron (II) phthalocyanine at 10 mol% loading was sufficient to promote the radical reaction. While acylphosphonates were used to generate phosphoryl radicals, the authors selected to use a dialkyl phosphorohydrazidate as a precursor to the phosphoryl radical species. A range of diorganophosphorohydrazidates were investigated for activity towards the addition reaction, and while substrates bearing linear or branched alkyl groups generated excellent yields of the β-hydroxy

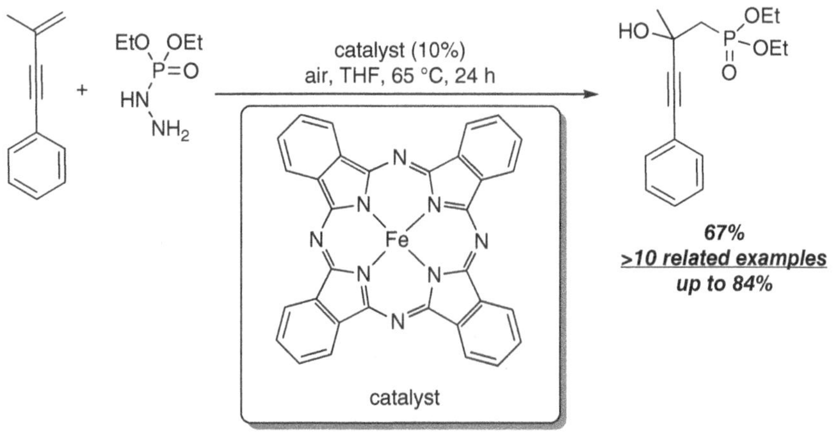

SCHEME 4.157 Iron-catalyzed synthesis of β-hydroxyphosphonates from alkenes [247].

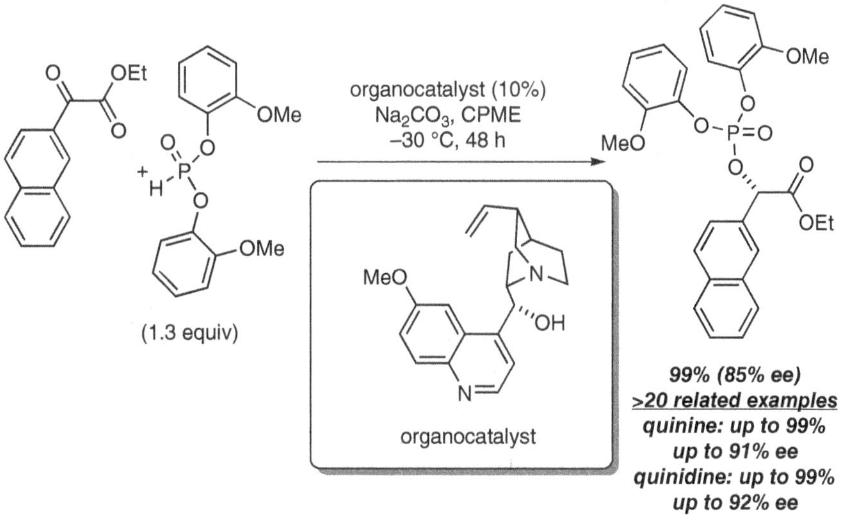

SCHEME 4.158 Organocatalytic additions of diarylphosphites to α-keto esters [248].

alkylphosphonates, analogous reactions with diphenylphosphorohydrazidate only generated a trace of the target product. As a result, the remaining alkenes were screened with diethyl phosphorohydrazidate. The chemistry was operationally trivial as simply adding the reagents together and refluxing under an atmosphere of air or oxygen afforded moderate fair to excellent yields of the target compounds. For the more reactive substrates, the concentration of oxygen in the atmosphere was sufficient to promote the reaction; however, for more troublesome substrates, pure oxygen was used. The reaction was regioselective and consistently generated the 1,2-addition product with the phosphonate fragment in the anti-Markovnikov position. The functional group compatibility was moderate, and a range of alkenes were successfully converted into the target products using this approach.

Although the *phospha*-Brook rearrangement can be a persistent and annoying problem when trying to promote Pudovik chemistry, it can be used to generate functionalized

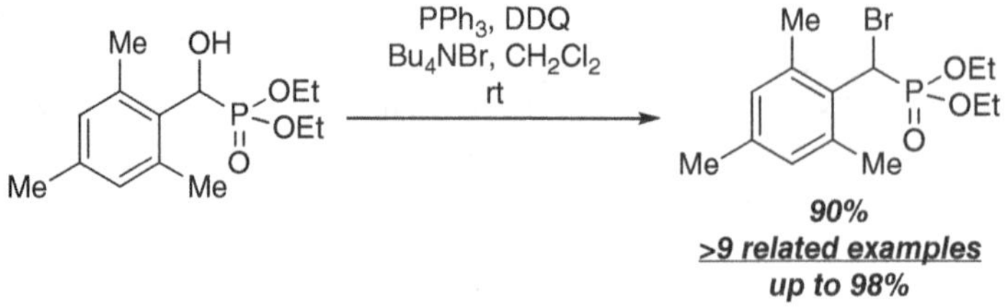

SCHEME 4.159   Conversion of α-hydroxyphosphonates into alkyl bromides [249].

SCHEME 4.160   Conversion of α-hydroxyphosphonates into alkyl azides [249].

phosphates. In a recent example of this, Nakamura reported the asymmetric synthesis of phosphate esters (Scheme 4.158) [248]. The reaction could be envisioned to proceed through multiple discrete steps including an initial Pudovik reaction between a diarylphosphite and an α-keto ester followed by a *phospha*-Brook rearrangement to generate an enolate. The final step in the process would consist of an enantioselective protonation of the enolate. The authors used quinine and related cinchona alkaloids as organocatalysts to simultaneously coordinate and activate the substrate during the protonation step. The phosphorus reagent used in these reactions was an orthofunctionalized diarylphosphite as it provided the desired products with high enantioselectivity. A range of α-keto esters were screened for activity in this reaction, and only minor electronic effects on the yields and enantioselectivity could be discerned. There was a steric issue with some of the compounds as those bearing ortho-substituents typically afforded products with lower ee's. Using quinine as the catalyst for this reaction resulted in the generation of the products with the R-configuration. The authors were pleased to find that the configuration could be reversed by using quinidine as the organocatalyst for this reaction. Due to the commercial availability of nearly all the reagents and catalysts, this is a very practical approach to the synthesis of phosphate esters.

Hydroxyphosphonates are valuable precursors to a range of α-functionalized phosphonate and phosphonic acid species. As an example, Firouzabadi and Iranpoor described the use of a triphenylphosphine/2,3-dichloro-5,6-dicyanobenzoquinone system to promote the synthesis of α-halogenated or α-azidophosphonates from α-hydroxyphosphonates (Schemes 4.159 and 4.160) [249]. The halogen source was tetrabutylammonium bromide/iodide, while sodium azide served as the azide source. The chemistry was operationally trivial, and simply stirring the reagents in a one-pot reaction generated moderate to excellent yields of the functionalized phosphonates. For the halogenation chemistry, the electronics of the system did not affect the success of the reaction, and the process was insensitive to bulky groups. For example, an α-hydroxyphosphonate

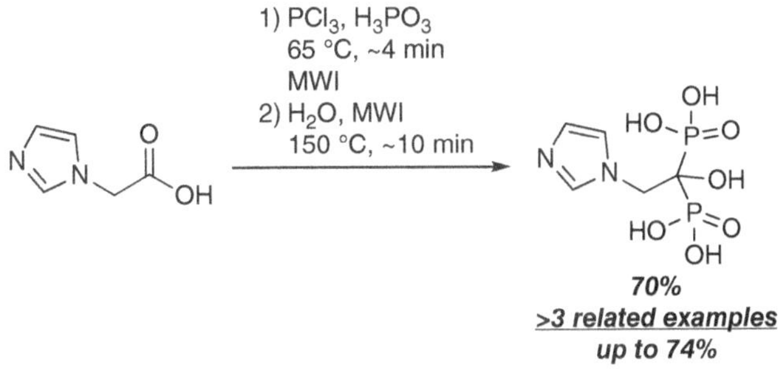

SCHEME 4.161   Microwave-assisted synthesis of bisphosphonic acid [250].

bearing a mesityl group was readily converted into the α-halogenated material (90%) after stirring at room temperature for 5.5 h. Curiously, attempts by the authors to chlorinate the hydroxyphosphonates using similar reactions were unsuccessful. The azidation reactions required refluxing in acetonitrile to promote the reaction, and the majority of substrates that were reported contained electron-withdrawing groups on the aromatic fragment.

Processes that result in the formal addition of two phosphonate fragments are valuable since the resulting compounds (bisphosphonates) are precursors to bisphosphonic acids. In general, there are two main approaches to the synthesis of the latter through phosphorus–carbon(sp^3) bond-forming reactions. One of the most popular methods entails the addition of phosphonic acid and trichlorophosphine to carboxylic acids. This method tends to be moderate to high yielding, but the conditions are harsh and acid-sensitive protecting groups will be unlikely to survive. A second approach starts from acyl chlorides and uses trialkylphosphites to generate bisphosphonates. These stable precursors can be stored until needed and converted into the bisphosphonic acids through simple cleavage of the esters. A microwave-assisted approach to the synthesis of bisphosphonic acids starting from carboxylic acids that significantly reduced the reaction time for the synthesis of several popular bisphosphonic acids was reported (Scheme 4.161) [250]. As with many microwave-assisted syntheses, careful control over the reaction time and conditions was critical to a successful reaction. The authors also compared the results from the microwave-assisted experiments with those obtained using conventional heating. In general, only minor differences in the yield and purity of the materials were observed between the two methods. However, the microwave-assisted procedure was significantly faster. However, as mentioned several times in this text, care must be taken when making assumptions about nonthermal microwave effects with reactions carried out in highly polar media. It is likely that the greater speed of the microwave-assisted procedure was due to the inability of conventional heating to accurately reproduce the thermal environment generated during microwave heating.

A key component in the synthesis of many complex organic molecules entails devising reactions and strategies that will preferentially transform one part of a compound while leaving another (sometimes very similar) part of the compound untouched. The synthesis of bisphosphonate species from a simple bifunctional substrate such as 4-chlorobutyryl chloride is an example of this challenge (Scheme 4.162) [251]. Both the primary alkyl chloride and the acyl chloride are known to react with trialkylphosphites to generate phosphonates. Through a series of screening reactions, the authors were able to determine

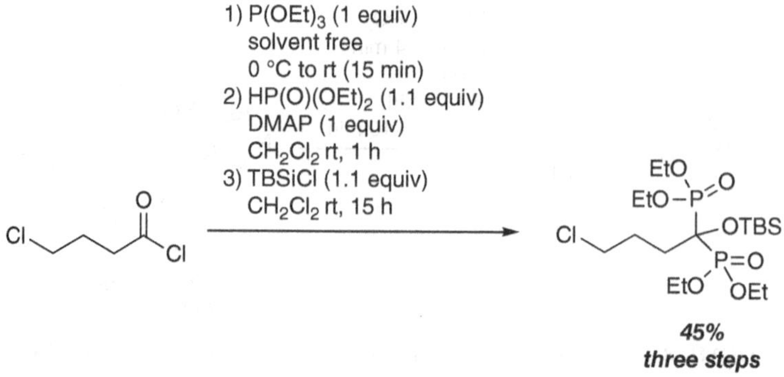

SCHEME 4.162   Chemoselective synthesis of bisphosphonates from acyl chlorides [251].

the conditions that would promote the reaction between the acyl chloride and the triethylphosphite while leaving the primary alkyl chloride untouched. The key was keeping the reaction temperature low. Arbuzov–Michaelis reactions with primary alkyl chlorides typically require heating to well above 100 °C. Once the intermediate acylphosphonate was generated, addition of a secondary phosphite and a base promoted the formation of the bisphosphonate. The final step in the synthesis of this intermediate was protection of the tertiary alcohol as the authors wanted to functionalize the alkyl chloride without interference from the newly formed alcohol.

The synthesis of tagged bioactive compounds remains a critical area of research as it enables researchers to follow where active species end up *in vivo*. Bisphosphonates have a high affinity for calcium as they are analogues of pyrophosphate, and if a bisphosphonate has been tagged with a dye, it could theoretically be monitored to provide the locations of high calcium concentration. Naturally, investigating bone adsorption might be the logical first thought for the value of this approach. Another potential use is the detection of microcalcifications in breast tissue. To this end, an NIR dye has been tagged with a bisphosphonate and investigated for this application (Scheme 4.163) [252]. The chemical synthesis of the tagged bisphosphonate was similar to that described previously using acyl chlorides. The initial step in the process entailed the conversion of a carboxylic acid into an acyl chloride through the addition of thionyl chloride. Generation of the bisphosphonate was accomplished through the next steps. The first phosphonate was incorporated using Arbuzov–Michaels chemistry, while the second was added via a hydrophosphonylation reaction. The authors were able to complete the synthesis in a one-pot fashion and reported an impressive yield of 91%. Finally, the tag was incorporated, and treatment with trimethylbromosilane followed by MeOH/H$_2$O cleaved the phosphodiesters and generated the target.

This same group reported a slight modification to this process that enabled the formation of bisphosphonic acids in a one-pot reaction (Scheme 4.164) [253]. Similar to the reactions previously, acyl halides were still the starting substrates for the initial formation of C—P bonds; however, the phosphorus nucleophile was tris(trimethylsilyl)phosphite. The idea behind the use of this reagent was to incorporate an organic fragment that could be cleaved under mild conditions to generate the phosphonic acid. This approach was successful and the simple addition of methanol afforded outstanding yields of the bisphosphonic acids for most substrates. A number of examples containing both aliphatic and aromatic groups

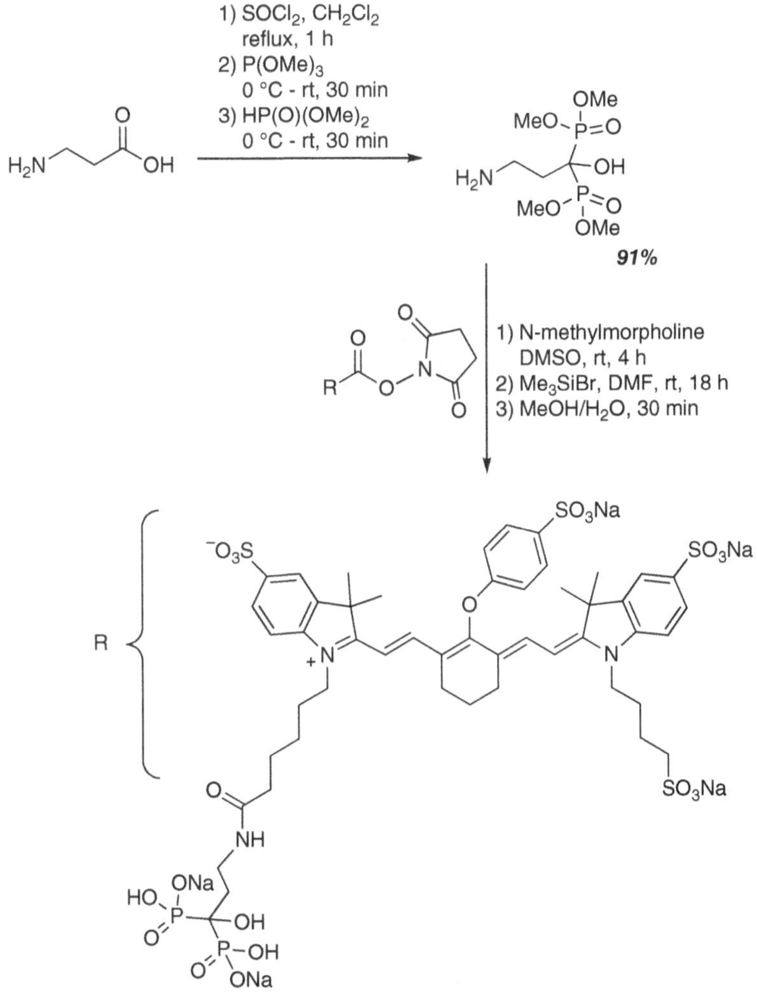

SCHEME 4.163   Synthesis of a tagged bisphosphonic acid derivative [252].

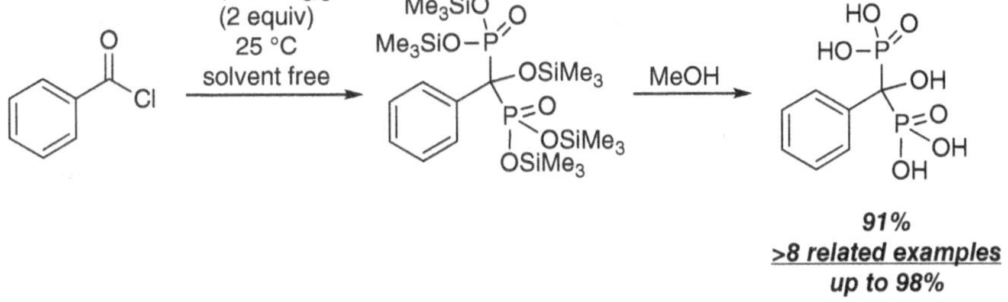

SCHEME 4.164   Synthesis of bisphosphonic acids from an acyl chloride [253].

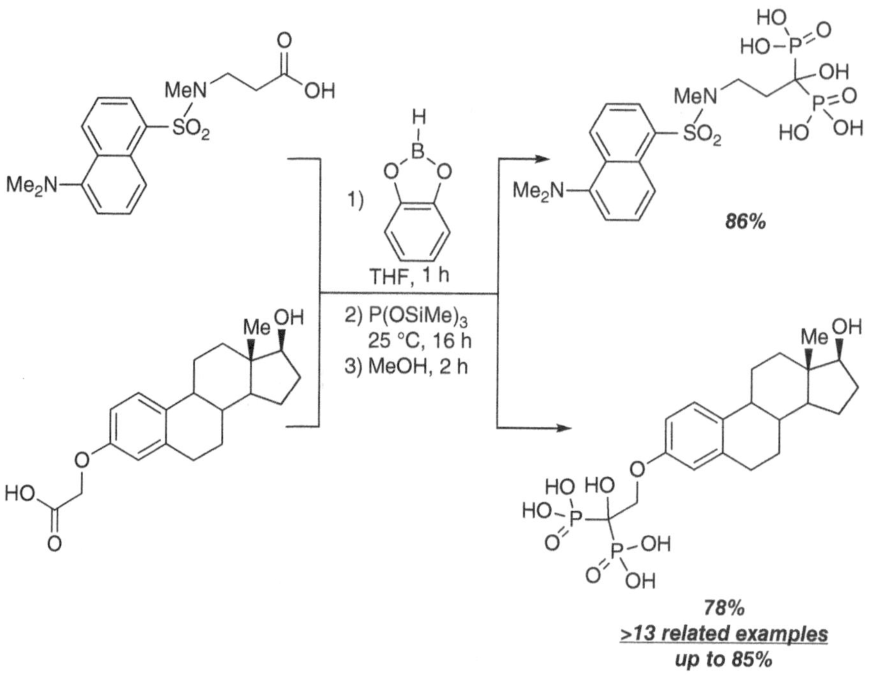

SCHEME 4.165   Synthesis of bisphosphonic acids from carboxylic acids [254].

(containing both electron-donating and electron-withdrawing groups) were readily converted using this approach. Another practical advantage of this chemistry was that several reactions proceeded under solvent-free conditions.

While the majority of the procedures described previously either begin from an acyl chloride or generate one *in situ*, an alternative approach was desired due to stability issues of acyl chlorides (Scheme 4.165) [254]. To this end, a route was devised that started from carboxylic acids and used a catecholborane to generate an intermediate phosphonate species. Treatment of this intermediate with tris(trimethylsilyl)phosphite at room temperature followed by hydrolysis generated the bisphosphonic acids in excellent yields. As mentioned previously, carboxylic acids are attractive substrates for this chemistry since they are readily available and less reactive than acyl chlorides. A range of functionalized substrates including examples bearing unprotected alcohols as well as primary or secondary amines were readily converted into bisphosphonic acids. Of particular interest was retention of a terminal alkyne through the process. This is a very attractive synthetic handle and can be used for further functionalization. In general, the ability to skip the generation of an acyl chloride intermediate makes this a practical choice for the synthesis of these valuable compounds.

Due to the biological activity of bisphosphonic acids, a vast array of derivatives and precursors have been generated including bile salts (Scheme 4.166) [255]. The overall process started with protection of the alcohols on the parent bile salt through the addition of formic acid. Once suitably protected, the addition of thionyl chloride converted the carboxylic acid into an acid chloride. Since these conversions are typically carried out using an excess of the thionyl chloride, the removal of unreacted material can be particularly troublesome. In these reactions, the removal of the excess thionyl chloride was achieved through repeated trituration with benzene. The final steps in the synthesis entailed

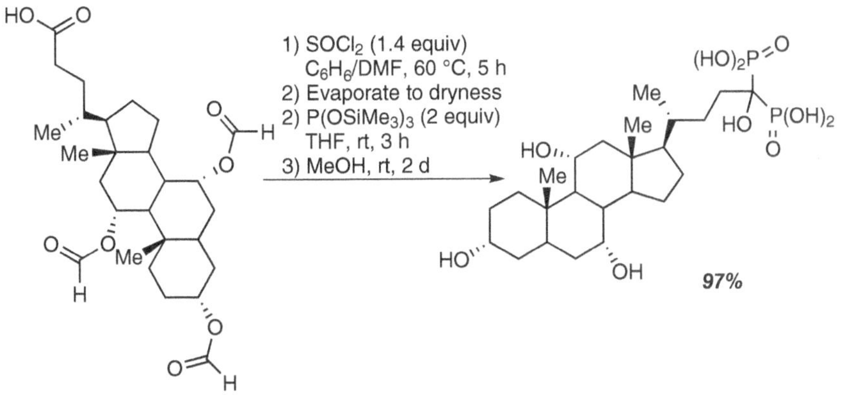

SCHEME 4.166   Preparation of a bile salt-functionalized bisphosphonic acid [255].

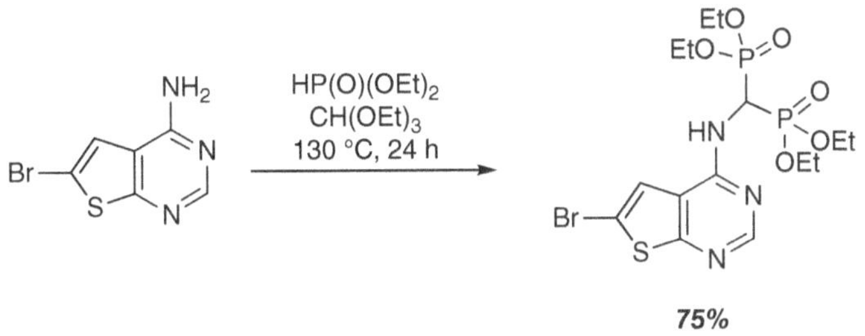

SCHEME 4.167   Preparation of an amine-linked bisphosphonate species [256].

the addition of tris(trimethylsilyl)phosphite followed by methanol to incorporate/deprotect the phosphonate fragments and regenerate the alcohols on the steroid. An outstanding yield of 97% was reported for this synthesis.

During the synthesis and investigation of potential inhibitors of human farnesyl pyrophosphate synthase, a series of bisphosphonates liked to a heterocycle through an amine spacer were prepared (Scheme 4.167) [256]. The synthetic method used for the generation of the α-aminobisphosphonate fragment entailed the treatment of the pyrimidine derivative with triethylorthoformate and diethylphosphite at elevated temperatures for 24 h. One of the attractive aspects of this chemistry was the retention of the vinyl bromide throughout the formation of the α-aminobisphosphonate. This bromine was subsequently exchanged for a series of aryl groups in a microwave-assisted cross-coupling reaction. This demonstrated the versatility of this method for the synthesis of these valuable compounds.

A microwave-assisted version of the aminophosphonate synthesis has been reported (Scheme 4.168) [257]. The chemistry was operationally trivial as simply mixing all of the reagents under neat conditions (no solvent) and irradiating generated moderate to excellent yields of the α-functionalized bisphosphonates. A range of electron-donating and electron-withdrawing groups were compatible with the chemistry. For the substrates screened in this chemistry, only 3-pyridyl- and benzylamines were problematic. These substrates generated a complex mixture of products under the standard reaction conditions developed during the investigation.

For the preparation of simple bisphosphonates, treatment of dichloromethane with a dialkylphosphite was moderately successful (Scheme 4.169) [258]. This reaction was

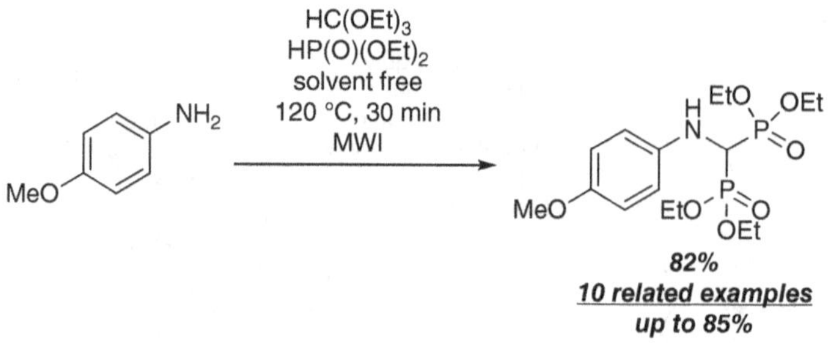

SCHEME 4.168   Preparation of bisphosphonates from aniline precursors [257].

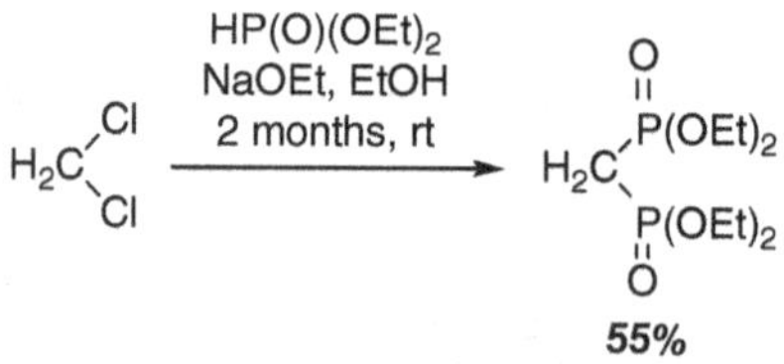

SCHEME 4.169   Preparation of bisphosphonates from dichloromethane [258].

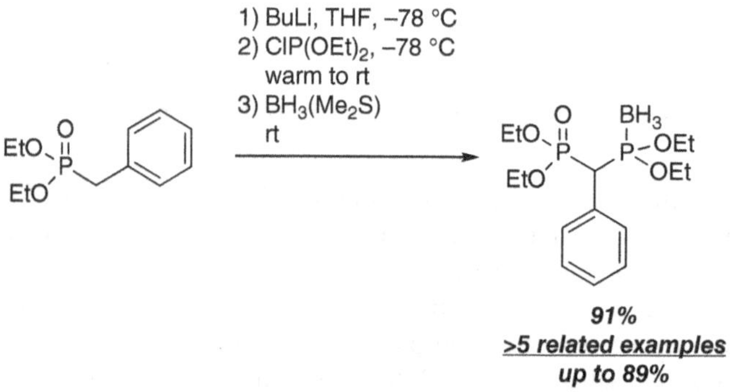

SCHEME 4.170   Synthesis of a mixed bisphosphorus species [259].

typically promoted by the addition of a base and was essentially a version of the classic Arbuzov reaction. The yields of the bisphosphonate compounds were typically moderate to good. This method remains one of the most popular approaches to the synthesis of this valuable class of compounds.

The synthesis of mixed bisphosphorus compounds has been reported by Montchamp (Scheme 4.170) [259]. His approach entailed the treatment of an alkylphosphonate precursor with a strong base at low temperatures followed by addition of a chlorophosphine or chlorophosphite to generate the bisphosphorus species. The final step in the process involved trapping the phosphine or phosphinate with BH$_3$ to prevent oxidation. The chemistry was successful with a range of substrates, and moderate to excellent yields of the bisphosphorus species were obtained. The resulting compounds are valuable intermediates in the construction of complex organic compounds, and this work provides a straightforward approach to their synthesis.

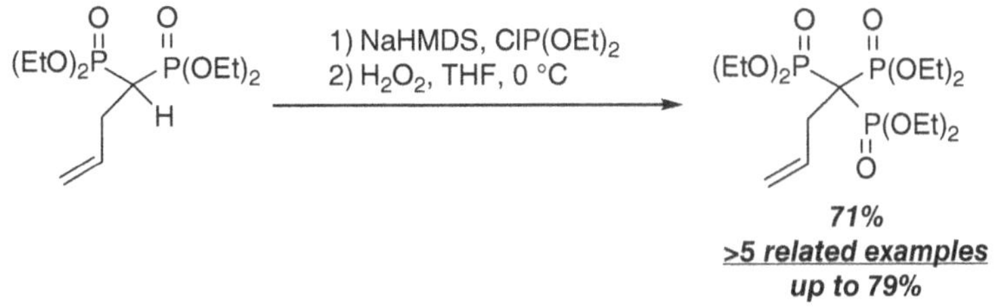

SCHEME 4.171 Conversion of a bisphosphonate into a trisphosphonate [260].

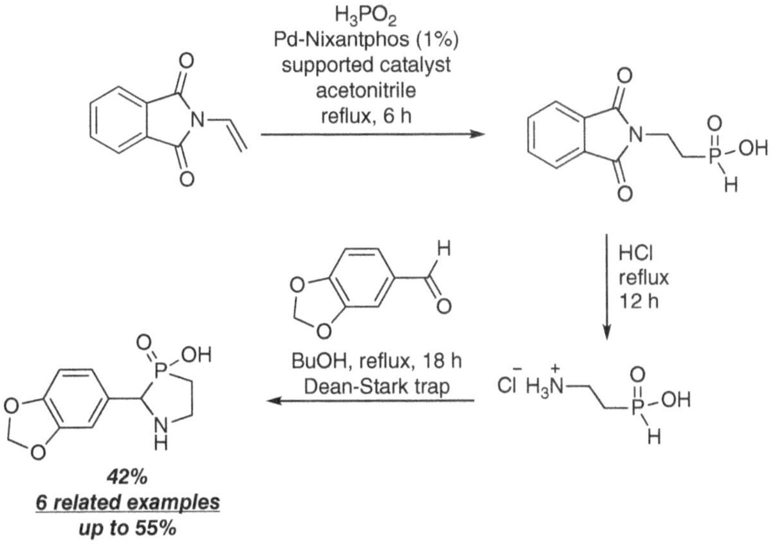

SCHEME 4.172 Preparation of the phosphorus-containing heterocycle [261].

A direct conversion of bisphosphonates into trisphosphonates through a base-assisted coupling reaction has been reported (Scheme 4.171) [260]. While attempts to generate a trisphosphate bearing additional functionality through the alkylation of precursors such as $HC(P(O)(OEt)_2)_3$ has been challenging, Wiemer reversed this approach and decided to investigate the coupling of a bisphosphonate with a chlorophosphinate followed by oxidation using hydrogen peroxide. An efficient base for this chemistry was NaHMDS and the coupling reaction proceeded under mild conditions to generate the products in moderate yields. Most of the examples contained hydrocarbon side groups; however, the retention of an alkene as well as an unprotected terminal alkyne through the preparation of the trisphosphonate was particularly attractive since these groups are synthetic "handles" that can be further functionalized in subsequent reactions.

The synthesis of phosphorus-containing heterocycles has been reported using a palladium-catalyzed approach (Scheme 4.172) [261]. The overall process involved two phosphorus–carbon(sp³) bond-forming reactions and started with a hydrophosphonylation reaction using a supported palladium catalyst to promote the addition. Conversion of the resulting phthalimide into a quaternary ammonium salt was accomplished by refluxing in concentrated hydrochloric acid. The final step in the procedure generated the second phosphorus–carbon(sp^3) bond and entailed refluxing the ammonium salt with an aldehyde in BuOH using a Dean–Stark trap. The overall synthesis generated the heterocyclic compounds in fair to moderate yields. The chemistry was also compatible with ketones and several examples were presented. In addition to the preparation of five-membered ring systems, the chemistry was extended to the synthesis of six-membered heterocycles. For sluggish substrates, the use of a focused microwave reactor promoted the cyclization reaction.

4.2.1 Troubleshooting the Synthesis of Alkylphosphines and Related Compounds

While it is impossible to anticipate the challenges of every new reaction/method that will be developed over the next few decades, there are some general trends that can be followed. While the section on solvents, additives, and general suggestions in Chapter 1 will be a good resource, the chart below will provide some direction when trying to select the most practical solution to the specific issues related to the synthesis of alkylphosphines and related compounds.

Problem/Issue/Goal	Possible Solution/Suggestion
A glovebox is not available	This will make the synthesis of many alkylphosphines very challenging. Unless the phosphorus precursor is packed in a septum-sealed bottle under an inert gas, even loading a reaction flask without introducing oxygen will be difficult. Furthermore, if purification by column chromatography is required, it will be difficult to maintain a completely inert atmosphere when collecting fractions
	It might be more practical to synthesize the protected phosphine (oxide, sulfide, borane) and convert it into the free phosphine following the P—C bond-forming step. Isolation of the target phosphine from the reduction or deboronation reaction is often significantly easier than from the initial P—C bond-forming reaction
Even with the use of a glovebox, partial oxidation of the phosphine was still observed following workup of the reaction	One solution is to rigorously purify and deoxygenate all of the reagents and solvents used during the workup procedure
While following a Grignard approach for the synthesis of a phosphine, other parts of the molecule are being metallated instead of the chlorophosphine	If the point of unwanted metalation contains a strong electrophile, the later will need to be protected or modified. In some cases, it might be possible to change the reactivity of the Grignard reagent through the addition of zinc or copper salts in order to minimize the formation of secondary products [6, 7]

(Continued)

(Continued)

Problem/Issue/Goal	Possible Solution/Suggestion
Highly functionalized substrates	This will depend upon the specific functional groups that are already present in the substrates, and each group must be evaluated for reactivity under the new set of conditions. Of the reactions described previously, the Grignard and organolithium approaches are the least tolerant. However, it should be noted that functionalized alkylzinc reagents (generated at low temperature from organolithium reagents) were successfully used in the synthesis of alkylphosphines [6]. Hydrophosphination and cross-coupling (Pd, Cu catalyzed) tend to be more tolerant
Need to make a phosphine with bulky groups on the phosphorus nucleophile	Several versions of the approaches described in the text are tolerant of bulky groups on the phosphorus center. In general, the Grignard and organolithium routes tend to be high yielding. Alternatively, ^{t}BuOK-promoted hydrophosphination of an alkene using a bulky secondary phosphine oxide such as $Mes_2P(O)H$ followed by reduction generated the alkylphosphine [55]
Normally, chlorophosphines are treated with Grignard reagents to generate alkylphosphines. Can the reagents be reversed?	The addition of alkyl halides to lithium and potassium phosphides has been used to generate alkylphosphines [16–19]
Need a green or sustainable version of the hydrophosphination reaction for the synthesis of an alkylphosphine	Try a solvent- and catalyst-free hydrophosphination [78]. This approach worked well with a range of styrenes and activated alkenes
Are there any alkylphosphines that are stable in air?	Dialkylbiarylphosphines tend to be very resistant to air oxidation. Even dimethylbiphenylphosphine can be generated and handled in air [11]
Are there any general approaches to the synthesis of chiral phosphines?	A number of methods are known for the synthesis of chiral phosphines [12, 14, 23, 64]. However, general approaches with broad substrate scope are rare. Each substrate should be evaluated against the known methods in order to design a reasonable synthesis
The starting material is a benzyl bromide derivative. Can the cross-coupling with a secondary phosphine be promoted over the formation of a phosphonium salt?	Try using a copper-catalyzed cross-coupling reaction [22]
Need a convenient route to a 1,2-bisphosphine	An attractive approach to this synthesis is the iron-catalyzed double hydrophosphination of an alkyne [58]
For air stability, a secondary phosphine oxide was used as the starting material; however, its reactivity was significantly different than the phosphine in cross-coupling reactions	If changing the catalyst, solvent, base, and other components of the coupling reaction does not generate the desired product, consider reducing the phosphine oxide in the reaction flask prior to addition of the catalyst and additives. This approach has been successfully implemented in the synthesis of alkylphosphines [24]

(Continued)

Problem/Issue/Goal	Possible Solution/Suggestion
Need a way to generate an alkylphosphine in water	This has been achieved through a copper-catalyzed process using specially designed surfactants [65]
Need a good approach to the synthesis of an alkylphosphine starting from an alkene	Hydrophosphination would be a good starting point. Base- [16] and transition metal-catalyzed routes are known [56]. An uncatalyzed addition reaction has also been reported [78]
The starting material is a secondary phosphine oxide. Can it be used in cross-coupling chemistry?	There are several cross-coupling approaches that have been successful using a range of metal catalysts [22, 24]
Need to generate an alkylphosphine oxide from an alkene	The addition of secondary phosphine oxides to alkenes would be a reasonable approach. In addition to a number of base- and transition metal-catalyzed reactions, visible light catalysis has also been used [87]
Need a reliable route to the synthesis of alkylphosphonates	Arbuzov chemistry remains one of the most popular approaches to the synthesis of these compounds [98–100]
Can allylic alcohols be used in Arbuzov-type chemistry?	The Lewis acid-promoted reaction between trialkylphosphites and allylic or benzylic alcohols has been achieved [104]
Need a general procedure for the preparation of α-aminophosphonates and α-hydroxyphosphonates	A vast array of reactions are known to generate these compounds. Potential routes should be evaluated against the substrates in order to design a viable approach
Are C—H phosphonation reactions known for the preparation of alkylphosphonates?	A variety of catalyst systems have been used to promote C—H phosphonation reactions [179–181]

4.3   SYNTHESIS OF ARYLPHOSPHINES AND RELATED COMPOUNDS THROUGH THE FORMATION OF PHOSPHORUS–CARBON(SP²) BONDS

While the formation of P—C(sp^3) bonds can be achieved through a range of substitution and addition reactions, practical methods leading to the formation of arylphosphines are generally limited to the treatment of chlorophosphines with Grignard/organolithium reagents or through cross-coupling. In a few cases, an appropriately functionalized aryl halide can directly be used in an uncatalyzed P—C(sp^2) bond-forming reaction. One of the most active areas of research entails the formation of arylphosphines through C—H activation processes. Overall, these synthetic routes have led to the preparation and isolation of thousands of arylphosphines. The following sections will highlight representative syntheses.

While many trialkylphosphines such as trimethylphosphine or triethylphosphine oxidize rapidly and sometimes violently in air, simple triarylphosphines such as triphenylphosphine have a measure of stability to air. With the former, they must be handled and stored under an inert atmosphere such as nitrogen or argon with rigorous exclusion of oxygen. In contrast, many triarylphosphines can survive limited exposure to air with minimal oxidation. Practically, this means that a number of triarylphosphines can typically be isolated by column chromatography or recrystallization without rigorous exclusion of oxygen. The level of air stability is substrate dependent, and best results are still often obtained with deoxygenated solvents and purged columns; however, gloveboxes are often not needed for

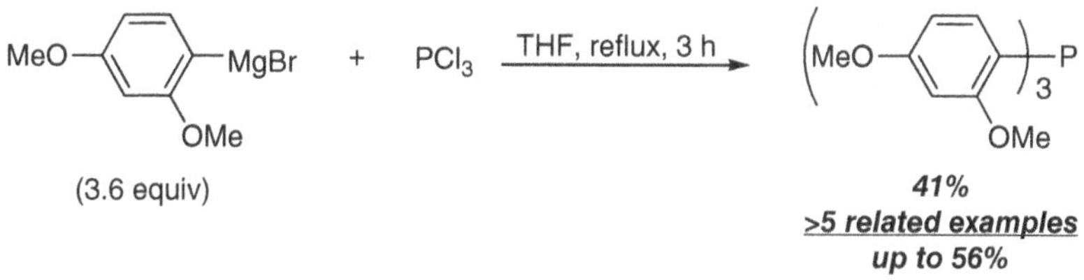

SCHEME 4.173   Synthesis of a triarylphosphine using a Grignard approach [263].

this class of compounds. Naturally, the phosphorus–carbon bond-forming reactions still need to be protected from oxygen due to the air sensitivity of the starting materials, but this is significantly easier than trying to isolate an air-sensitive compound by column chromatography without a glovebox.

If the arylphosphine is air sensitive and a glovebox is not available, alternative routes for the synthesis of the arylphosphine include the use of protected substrates such as phosphine oxides, sulfides, or phosphine-boranes. A general description of these approaches as they relate to general phosphorus–carbon bond-forming reactions is described in Section 4.1. In some cases, the reactivity of the phosphine oxide could be higher than the phosphine precursor, and proceeding through the phosphine oxide route is advantageous from a reactivity standpoint, not simply due to air sensitivity.

The treatment of chlorophosphines with aryl Grignard or organolithium reagents is one of the most popular ways to prepare arylphosphines. As mentioned previously, the operational difficulty with this approach is moderate to high due to the air sensitivity of the reagents and intermediates. Since the starting chlorophosphines are often commercially available, a glovebox is frequently not needed. A nitrogen/vacuum manifold will often suffice for setting up the reaction.

Symmetrical triarylphosphines are typically generated by treatment of trichlorophosphine with excess Grignard or organolithium reagent. This reaction is commonly quite tolerant to bulky groups on the aryl fragment, and moderate to high yields of triarylphosphine are common (Scheme 4.173 and Example 4.28) [263–266]. If the Grignard reagent is not commercially available, best results are typically obtained if the reagent is prepared immediately prior to use. Additionally, if a symmetrical triarylphosphine is the target, it is often more practical to use an excess of the Grignard reagent to completely convert PCl_3 into the triarylphosphine since it can be challenging to separate the secondary products resulting from incomplete conversion from the desired triarylphosphine.

One of the biggest drawbacks to the use of Grignard and organolithium reagents for the synthesis of organophosphorus compounds is the low functional group tolerance of this approach. Even poor electrophiles are commonly avoided and substrates bearing good electrophiles are generally thought of as synthetic "dead ends." Despite the problems with functional groups, a myriad of arylphosphines have been synthesized using this approach. To circumvent some of the issues with functional groups, Knochel reported a clever twist on the Grignard approach that enabled the use of arylhalides bearing reactive electrophiles (Scheme 4.174) [6]. His modification entailed initial formation of the organolithium reagent at low temperature (−90 °C). The next step was focused on modifying the reactivity of the organolithium reagent through the formation of an organozinc species. This was accomplished by simply adding zinc(II) bromide to the reaction mixture. Addition of PCl_3 to this reaction mixture followed by heating to 40 °C generated the functionalized triarylphosphine in moderate yield. The critical part of this synthesis was

Example 4.28   Preparation of a Symmetrical Triarylphosphine with Bulky Groups [263].

Before starting this synthesis, carefully follow the chemical safety instructions listed in Example 4.1.

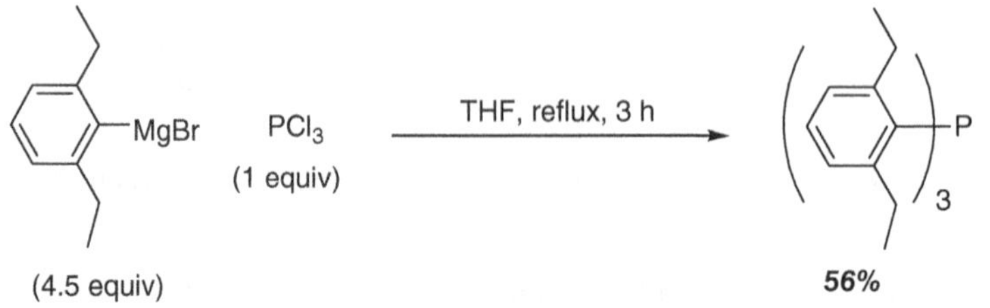

Caution! PCl₃ can react violently with moisture.

Caution! Grignard reagents can react violently with moisture.

Caution! THF, EtOAc, and ethanol are highly flammable.

In a 100 mL three-necked flask equipped with a dropping funnel and a reflux condenser was placed magnesium turnings (0.24 g, 10 mmol). In the dropping funnel was placed a solution of 2-bromo-1,3-diethylbenzene (1.52 mL, 9 mmol) in 20 mL of dry THF, and ca. 10% of the solution was added. The reaction was initiated by gentle heating of the mixture by a heat gun, followed by the addition of the rest of the aryl bromide solution over 10 min. After completion of the addition, the mixture was heated to reflux for 2 h and then cooled down to 0 °C. A solution of phosphorus trichloride (0.18 mL, 2 mmol) in 10 mL of dry THF was added dropwise over 5 min. The resulting suspension was heated to reflux for 3 h and then cooled to room temperature. The reaction was quenched with water (10 mL). The organic layer was separated, and the aqueous layer was extracted with EtOAc (20 mL × 2). The combined organic layer was washed with brine (50 mL × 2), dried over MgSO₄, and concentrated under reduced pressure. The crude product was recrystallized from hot ethanol to give the desired product as white solid (0.48 g, 56%).

Adapted with permission from W. Xu and N. Yoshikai, *Angew. Chem. Int. Ed.*, **2014**, *53*, 14166–14170. *Copyright (2014) Wiley-VCH Verlag GmbH & Co. KGaA, Weinheim.*

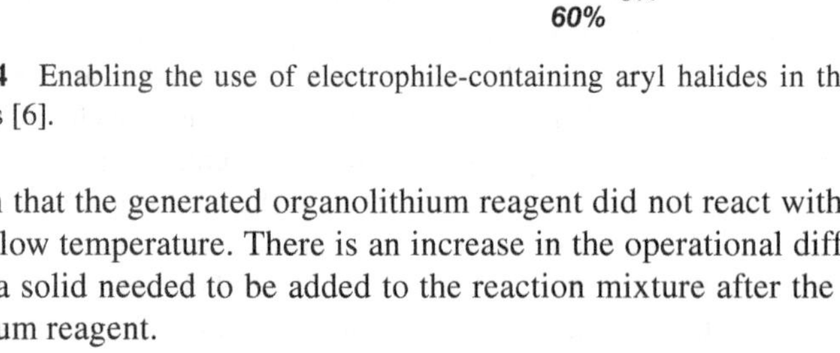

SCHEME 4.174   Enabling the use of electrophile-containing aryl halides in the synthesis of triarylphosphines [6].

the observation that the generated organolithium reagent did not react with the attached electrophile at low temperature. There is an increase in the operational difficulty of this reaction since a solid needed to be added to the reaction mixture after the formation of the organolithium reagent.

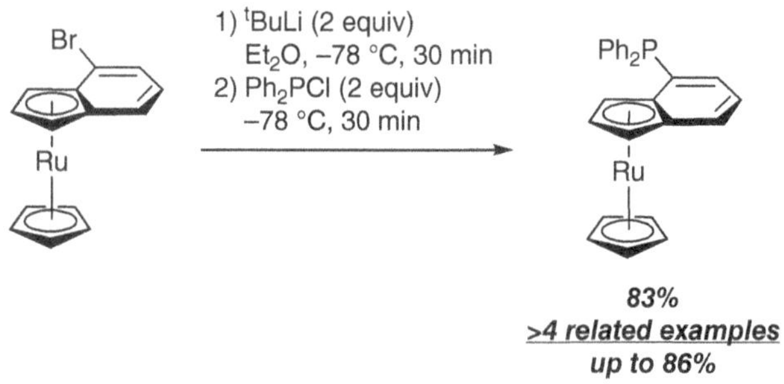

SCHEME 4.175 Phosphination of a brominated metallocene [271].

In addition to issues with the presence of organic electrophiles, metal centers can also react with the Grignard reagents or group 1 phosphides and complicate the phosphination reaction. One class of metal-containing substrates that are less reactive (at the metal center) with these powerful nucleophiles/bases are certain metallocenes. Naturally, not all metal-locene compounds will fit this category, but several have been successfully used in phosphination reactions. 1,1′-bis(diphenylphosphino)ferrocene and its derivatives are perhaps the most well-known examples [267–270]. In a typical synthesis, ferrocene is treated with BuLi in the presence of TMEDA at low temperatures followed by the addition of chlorodiphenylphosphine. This chemistry has been extended to a host of different metallocenes and chlorophosphines. An example of approach used substituted indenyl ruthenium complexes and their derivatives as substrates in the phosphination reaction (Scheme 4.175) [271]. The aryllithium reagent was generated by metal–halogen exchange at low temperatures, and the chlorophosphine was added without warming the reaction mixture. The chemistry afforded moderate to good yields of the arylphosphines and successfully used both aryl and alkyl-chlorophosphines. Since no organic electrophiles were present on the precursors substrates, this chemistry demonstrates that an appropriately designed metal complex can be used in phosphination reactions.

Using an organolithium approach, a highly modular and tunable synthesis of P-chiral arylphosphines has been reported (Example 4.29) [272]. The first step in the synthesis added a dichlorophosphine fragment to the cyclopentadienyl ring of Ugi's amine [273, 274] through lithiation and addition of PCl_3 [275]. The subsequent stepwise addition of different organolithium reagents selectively formed the P-chiral arylphosphine. While it might appear as through the addition of the RLi species would be random and scramble the stereochemistry at phosphorus, only a single diastereomer was generated in this reaction as determined by NMR spectroscopy. Furthermore, the chirality at phosphorus could be inverted by simply changing the order in which the different RLi species were added. Thus, the achieved stereoselectivity was not simply a steric issue with the more bulky organolithium reagents. The authors proposed that the pendant amine served as a directing group and interacted with the phosphorus center to form a rigid framework through the formation of a quaternary ammonium salt [276]. The geometry of this intermediate dictated the attack of the next RLi species. The chemistry might seem as though it would be operationally challenging given all of the additions at −78 °C; however, this is a one-pot reaction and does not require the isolation/purification of any of the intermediate species. It is simply a matter of warming and cooling the same flask. Given the modular nature of the synthesis, the high stereoselectivity, and the indifference to bulky groups, this is a practical approach to the synthesis of P-chiral arylphosphines.

Example 4.29   Preparation of P-Chiral Phosphines Using a Resolved Directing Group [272].

Before starting this synthesis, carefully follow the chemical safety instructions listed in Example 4.1.

49% (R_C,S_{FC},S_P)

Caution! Pentane and TBME are highly flammable.

Caution! Organolithium reagents can react violently with moisture and must be handled and stored under an inert atmosphere such as nitrogen or argon. ^tBuLi is particularly sensitive and must be handled with great care.

This reaction was carried out under an inert atmosphere with oven-dried glassware. To a solution of (R)-Ugi's amine (2.57 g, 10 mmol) in TBME (20 mL) was added 1.3 M ^tBuLi solution in n-hexane (8.5 mL, 11.05 mmol) at 0 °C. After the addition was complete, the mixture was warmed to room temperature and stirred for 1.5 h at room temperature. The mixture was then cooled to −78 °C and a solution of PCl₃ (1 mL, 11.46 mmol) in TBME was added slowly, and the mixture was warmed to room temperature and then stirred for 1.5 h. The mixture was then cooled to −78 °C again, and a suspension of PhLi (prepared from PhBr (1.75 g, 11.1 mmol) and 1.6 M n-BuLi solution in pentane (7.6 mL, 12.16 mmol) in TBME (40 mL) at −40 °C) was added slowly via a cannula. The mixture was warmed to room temperature and stirred for 1.5 h. The mixture was then cooled to −78 °C again, and a suspension of 3,5-Me₂C₆H₃Li (prepared from 3,5-Me₂C₆H₃Br (2.22 g, 12 mmol) and 1.6 M n-BuLi solution in pentane (8.4 mL, 13.44 mmol) in TBME (40 mL) at −40 °C) was added slowly via a cannula. The mixture was stirred overnight from −78 °C to room temperature and quenched by the addition of saturated NH₄Cl solution (40 mL). The organic layer was separated and dried over Na₂SO₄, and the solvent removed under reduced pressure. The residue was purified by chromatography to afford the title compound as a yellow solid (2.3 g, 49%).

Adapted with permission from H. Nie, L. Yao, B. Li, S. Zhang, and W. Chen, *Organometallics*, **2014**, *33*, 2109–2114. *Copyright (2014) American Chemical Society.*

While much of the attention in phosphine synthesis has been devoted to the use of aryl iodides, bromides, and chlorides, aryl fluorides can also be used. In addition to functionalizing aryl fluorides in superbasic conditions [277–279], Thiel reported the phosphorus–carbon bond-forming reaction using Me₃SiPPh₂ as the phosphinating agent and CsF as the

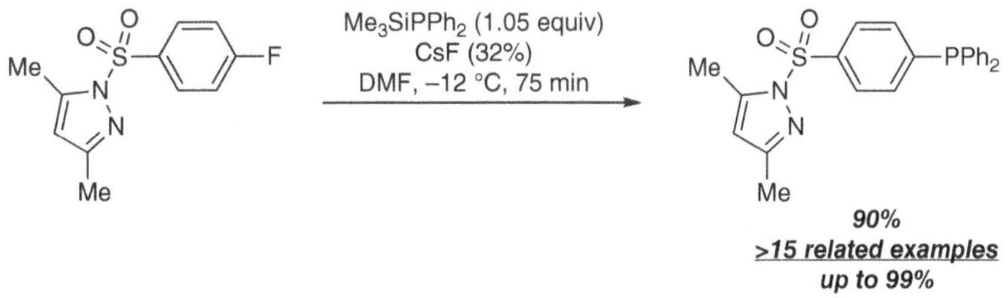

SCHEME 4.176   Fluoride-mediated synthesis of arylphosphines from aryl fluorides [280].

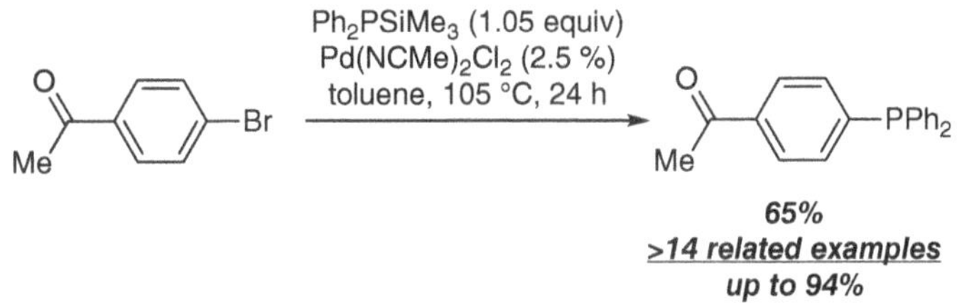

SCHEME 4.177   Palladium-catalyzed synthesis of arylphosphines [281].

catalyst for the reaction (Scheme 4.176) [280]. The chemistry followed classic S_NAr reactivity, and the key to this synthesis was the presence of electron-withdrawing groups on the aryl fluoride. As long as they were present, the synthesis generated good to excellent yields of the arylphosphines under very mild conditions. Some of the phosphination reactions occurred below room temperature. One of the key advantages of this system was broad scope of chemistry. Substrates bearing functional groups such as sulfones, ketones, esters, and nitriles were readily converted into arylphosphines using this approach. With few exceptions [6], these substrates are unusable using a Grignard approach. Thus, for the right system, this can be a very effective and transition metal-free route to the formation of arylphosphines.

One of the first examples of a palladium-catalyzed phosphine arylation was reported by Stille (Scheme 4.177) [281]. He screened both Ph_2PSiMe_3 and Ph_2PSnMe_3 and found that the former generated higher yields of the arylphosphine. One of the advantages of using palladium catalysts is their tolerance to functional groups. Substrates bearing nitriles, halogens, ethers, ketones, and aniline derivatives were converted into arylphosphines. Although normally robust substrates, alkyl aryl ethers are known to undergo dealkylation in the presence of phosphides to generate the alkylphosphine and phenoxide [282]. Although unactivated arenes such as bromotoluene were unreactive towards the P-arylation, activated substrates such as 4-bromoacetophenone were successfully used in the synthesis. One of the advantages of this approach was that a base was not required. Although the authors did not add a ligand to aid in solubilizing the palladium catalyst, the arylphosphine product of the reaction can function as a solubilizing ligand. Several extensions and modifications have been reported in order to increase the substrate scope or increase the operational simplicity [283, 284]. One such modification involved the generation of Ph_2PSnMe_3 in solution by the treatment of PPh_3 with sodium in liquid ammonia followed by the addition of Me_3SnCl [284]. The reported yields from this modification

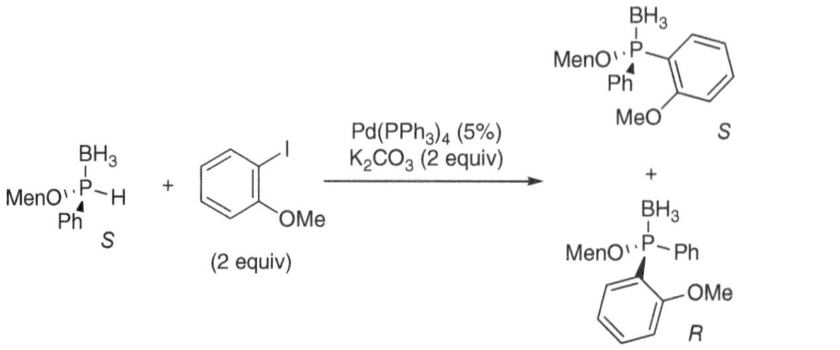

solvent	time	temp	yield	S:R
CH$_3$CN	16 h	50 °C	96%	100:0
THF	48 h	50 °C	76%	4:96

SCHEME 4.178   Effect of solvent on the stereochemical outcome of the phosphination reaction [286].

were generally excellent greater than 88%. As an example of how tolerant this chemistry was to electrophiles, 2-iodobenzoic acid was successfully phosphinated (69% isolated). Aryl triflates were also successfully functionalized using the one-pot approach [285]. The phosphorus coupling products were isolated as phosphine oxides by the addition of hydrogen peroxide.

The palladium-catalyzed coupling of iodoarenes with a resolved borane-protected phosphinate demonstrated that simple P—H donors could be utilized in cross-coupling reactions (Scheme 4.178) [286]. A soluble palladium(0) complex was used to catalyze the reaction. This approach required a base to consume the HI generated in the reaction, and after screening a range of mineral bases, the authors found that the identity of the base strongly influenced the stereochemistry of the diarylphosphinate formed in the reaction. Furthermore, they found that the solvent used in the reaction also played a significant role directing the stereochemistry. Using potassium carbonate and acetonitrile as the solvent exclusively generated the P$_S$ species as the sole product, while changing to the solvent to THF and using the same base was highly selective for the product with P$_R$ stereochemistry. While the precise role of the solvent was not elucidated, the cross-coupling chemistry was quite efficient under mild conditions, and the authors were able to devise a set of conditions for successfully retaining or inverting the stereochemistry at phosphorus. One of the key intermediates in the cross-coupling reaction was a palladium complex that contained the phosphide and the aryl fragment in a mutually cis arrangement on the metal center. Reductive elimination would generate the observed borane-protected arylphosphine. To provide insight into this and related reactions, Gaumont investigated the existence of this key intermediate [287]. A model system for the proposed structure was prepared and studied for its propensity to form the product. Upon monitoring a solution of this intermediate, the arylphosphine product was indeed generated. Furthermore, by changing the aryl group attached to palladium to —C$_6$F$_5$, the authors were able to slow the reductive elimination to a point that crystals suitable for X-ray diffraction could be grown. The X-ray structure clearly established the connectivity of the proposed intermediate.

Aryl bromides and iodides were the most popular aryl halides for these reactions since aryl chlorides and fluorides were typically sluggish in most of these reactions.

As mentioned in previous sections, aryl chlorides are attractive for coupling reactions since they are typically less expensive than aryl bromides or iodides. To address this issue, Buchwald developed a simple catalyst system consisting of a ferrocene-based ligand and Pd(OAc)$_2$ as the metal source (Scheme 4.179 and Example 4.30) [288]. While only a single aryl chloride was reported, it still demonstrated that aryl chlorides could be used in the

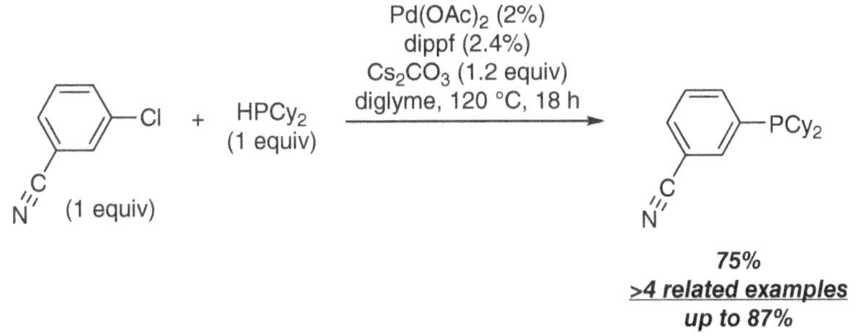

SCHEME 4.179 Palladium-catalyzed cross-coupling of aryl chlorides with secondary phosphines [288].

Example 4.30   Preparation of Arylphosphines through Cross-Coupling [288].

Before starting this synthesis, carefully follow the chemical safety instructions listed in Example 4.1.

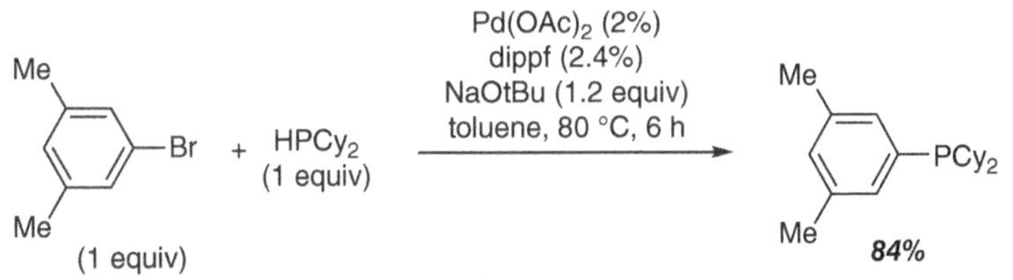

Caution! Secondary phosphines can spontaneously ignite in air and must be handled and stored under an inert atmosphere such as argon or nitrogen.

This reaction was carried out under an atmosphere of argon. Pd(OAc)$_2$ (4.5 mg, 0.020 mmol), dippf (10.0 mg, 0.024 mmol), base (1.2 mmol) were added to an oven-dried resealable Schlenk tube. The Schlenk tube was evacuated and backfilled with argon (3 cycles) and then charged with solvent (1.0 mL). The solution was stirred for 1 h at room temperature. Then the aryl halide (1.0 mmol) and the secondary phosphine (1.0 mmol) were added by syringe. The Schlenk tube was sealed with a Teflon valve, heated and stirred for 6 h. The reaction mixture was then allowed to reach room temperature. Ether (ca. 3 mL) was added and the aliquot was analyzed by GC. The reaction mixture was then filtered and concentrated. The crude product was purified by flash column chromatography on silica gel to afford the desired tertiary phosphine (0.225 g, 84%).

Reprinted from M. Murata and S. L. Buchwald, *A general and efficient method for the palladium-catalyzed cross-coupling of thiols and secondary phosphines, Tetrahedron Letters, 60,* 7397–7403. *Copyright (2004), with permission from Elsevier.*

Example 4.31 Microwave-Assisted Synthesis of Arylphosphines through Cross-Coupling [289].

Before starting this synthesis, carefully follow the chemical safety instructions listed in Example 4.1.

Caution! The metal catalyst could be reduced and plate on the sides of the reactor vial. This could lead to arching and a vessel failure.

This reaction was carried out under an atmosphere of nitrogen. 0.75 mmol (153 mg) of iodobenzene, 0.5 mmol (49 mg) of KOAc, and 2.5 mol% (3 mg) of $Pd(OAc)_2$ are filled into an appropriate nitrogen-flushed small Smith microwave process vial (ref. S1) and are admixed with NMP (1.0 mL). Subsequently, 0.5 mmol (93 mg) of Ph_2PH are added and the vial is sealed with a Teflon septum and placed into the SmithSynthesizer microwave cavity. After irradiation at 200 °C for 20 min and subsequent gas jet cooling (down to 40 °C), the mixture is diluted with water (5 mL) and allowed to stand in the refrigerator for approximately 1 h. The precipitate is filtered by suction, washed with water, and dried under vacuo, yielding 121 mg (92%) Ph_3P as an off-white crystalline powder.

Adapted with permission from A. Stadler and C. O. Kappe, *Organic Lett.*, **2002**, *4*, 3541–3543. *Copyright (2002) American Chemical Society.*

cross-coupling reactions. Using this approach, aryl bromides and iodides were successfully converted into the corresponding arylphosphines at 80–100 °C. Analogous reactions using 3-chlorobenzonitrile were sluggish; however, heating to 120 °C successfully promoted the coupling reaction. Due to the differences in the reactivity between the halogens (I > Br > Cl), the authors were able to develop a chemoselective phosphination reaction that afforded 2-halophenylphosphines in high yield (82–85%).

The microwave-assisted synthesis of arylphosphines through cross-coupling reactions has been reported (Example 4.31) [289]. Several homogeneous and heterogeneous palladium complexes were active towards the phosphination reaction, and moderate to excellent yields of the arylphosphines were obtained. Since unprotected secondary phosphines were used in this reaction, there is a bit of operational difficulty since the reagents needed to be loaded into the reaction vessel under an atmosphere of nitrogen. Due to the high absorbance of microwave radiation, the heterogeneous catalysts (Pd/charcoal) heated very rapidly and generated triarylphosphines in less than 3 min. Caution must be exercised in these reactions since the supported palladium catalysts can plate metal on the side of the flask that can result in arching and vessel failures/explosions. In addition to the palladium-catalyzed reactions, analogous reactions using nickel complexes were investigated. Using the model system comprised of phenyl triflate and diphenylphosphine, the authors found that heating a DMF solution of these reagents to 180 °C for 20 min

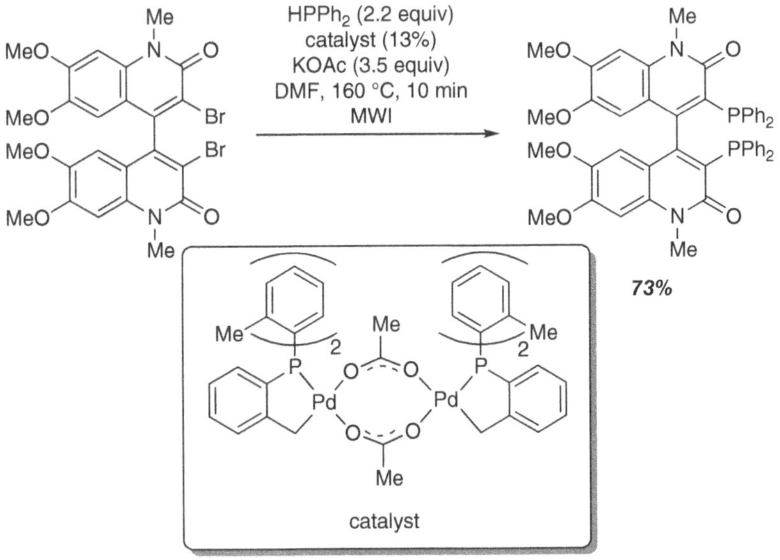

SCHEME 4.180 Microwave-assisted palladium-catalyzed synthesis of an az α-BINAP derivative [291].

generated triphenylphosphine (61% isolated) using $Ni(dppe)Cl_2$ (2%) as the catalyst. In related work, the synthesis of borane-protected arylphosphines based upon a helicene framework has been accomplished using a variation of this microwave-assisted protocol [290].

The synthesis of a BINAP derivative has been accomplished using a palladium-catalyzed microwave-assisted cross-coupling approach (Scheme 4.180) [291]. Using unprotected secondary phosphines and Herrmann's catalyst, the cross-coupling reaction generated moderate yields of the bisphosphine after 10 min of irradiation. It should be noted that the conditions and catalyst loadings were very specific. Decreasing the amount of the catalyst led to incomplete conversions and the generation of monosubstituted compounds, while increasing the temperature resulted in the formation of significant amount of secondary products. Dehalogenation was the most common side reaction for this chemistry.

Following this theme, a series of ruthenium metallocene complexes were phosphinated using a microwave-assisted approach (Scheme 4.181) [271]. The yield of the arylphosphine was moderate and the reactions were quite fast. While only a single set of substrates were screened in the microwave-assisted protocol, it does provide the proof of concept for the design of additional reactions. It should be noted that the temperature reached in this reaction is far above the boiling point for dichloromethane. Appropriate safety measures should be taken, and caution should be used when opening the reactor vials.

The coupling of dicyclohexylphosphine with arenediazonium tetrafluoroborates generates an aryldicyclohexylphosphine through a palladium-catalyzed cross-coupling reaction (Scheme 4.182) [292]. While only a single example was given, this chemistry has great potential due to the vast array of aniline derivatives that are readily available. Conversion of the anilines into the diazonium salts is typically straightforward and high yielding. As an example of the functional group tolerance of this chemistry, the presence of an ester on the arenediazonium salt did not alter the outcome of the cross-coupling reaction.

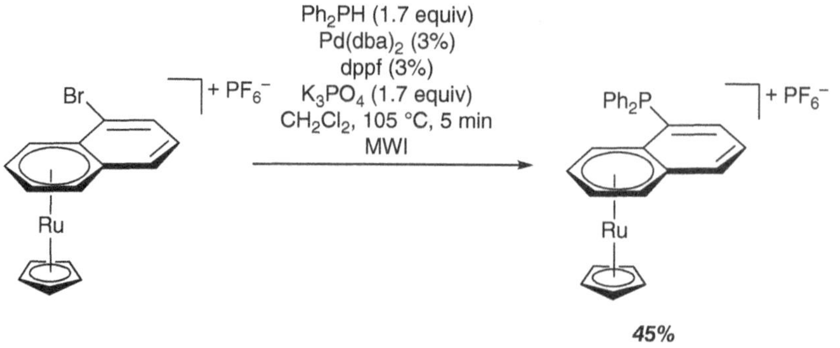

SCHEME 4.181   Microwave-assisted phosphination of a ruthenocene derivative [271].

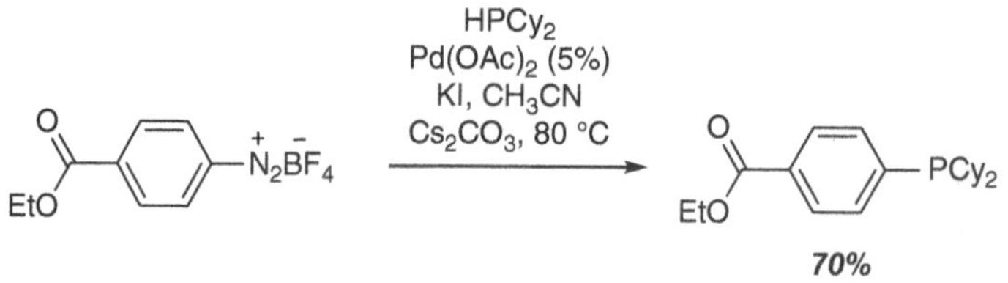

SCHEME 4.182   Using arenediazonium salts as substrates for the palladium-catalyzed synthesis of arylphosphines [292].

The synthesis of 2,2′-bis(diarylphosphino)biphenyls has attracted a significant amount of interest since the resulting compounds often chelate tightly to transition metals and serve as excellent ligands. During an investigation focused on the synthesis of a group of C_1-symmetric bisphosphines, the authors discovered some notable differences in the product distributions resulting from common methods for arylphosphonate synthesis [293]. The organolithium approach was the first method investigated by the authors. Starting from a range of 2,2′-dibromobiphenyls, metal–halogen exchange was carried out to generate the dilithiated species. Addition of ClPPh$_2$ (2 equiv) should have generated the bisphosphine species. Instead, significant amounts of a phosphafluorene species were generated. The authors tried changing the substituents on the biphenyl precursors to minimize formation of the phosphafluorene. While some of changes increased the amount of the desired bisphosphine relative to the phosphafluorene, yields still ranged from 10 to 47%. In an effort to increase the yields of the bisphosphine products and minimize the formation of the phosphafluorene side product, the authors turned to palladium-catalyzed cross-coupling chemistry (Scheme 4.183). They selected the common catalysts that were successful in other coupling reactions and attempted the phosphinylation of 2,2′-dibromobiphenyls. Disappointingly, only starting materials were recovered. Undeterred by this result, the authors pressed on and tried the reaction with 2,2′-diiodobiphenyls. This modification to the system was successful and moderate to excellent conversions were obtained. Perhaps most importantly, the relative amount of the phosphafluorene was significantly reduced using the palladium-catalyzed method. Once a successful method was devised, the authors extended the chemistry to a range of biphenyl systems and found that the chemistry was fairly

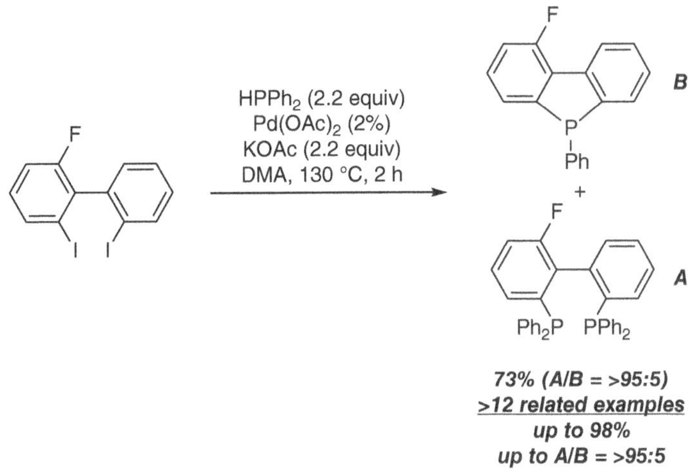

SCHEME 4.183 Preparation of the 2,2-bis(diphenylphosphino)biphenyls through cross-coupling [293].

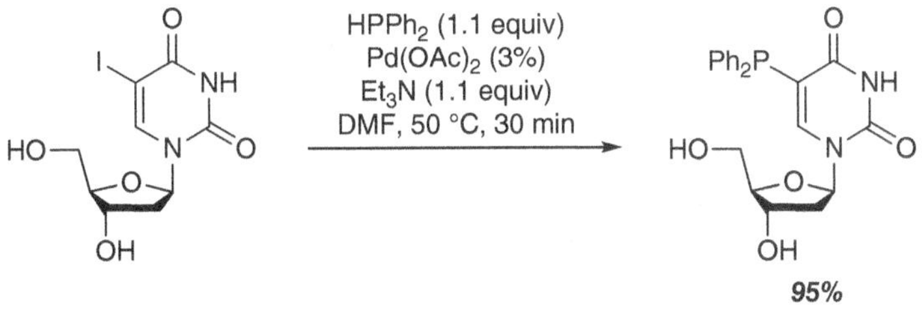

SCHEME 4.184 Phosphination of deoxyuridine using a cross-coupling approach [294].

tolerant to the presence of either electron-withdrawing or donating groups. Furthermore, several of the bisphosphinebiphenyls were not accessible following the organolithium pathway.

In addition to the use of mineral bases in the cross-coupling chemistry, organic bases such as triethylamine have also been successful. An example of this chemistry is the phosphination of deoxyuridine (Scheme 4.184) [294]. In addition to triethylamine, palladium acetate served as the second component of the catalyst system. No additional ligands were needed for this chemistry and the conditions were quite mild. The lack of a need for additional ligand was likely due to the possibility that both the secondary phosphine and the product could coordinate and stabilize the metal. It was noteworthy that this reaction chemistry tolerated the presence of a primary alcohol (5′-OH).

Arylnonaflates have also been successfully used as precursors in cross-coupling reactions that lead to the formation of P—C bonds (Scheme 4.185) [295]. In one of the initial reports of this chemistry, Lipshutz described the use of a readily available palladium(0) precatalyst along with a mineral base to promote the coupling reaction. The borane-protected phosphine was generated to inhibit unwanted oxidation during the reaction and workup. One of the most attractive aspects of this chemistry was the low reaction temperature. Simply stirring the reaction at 40 °C for a few hours generated a

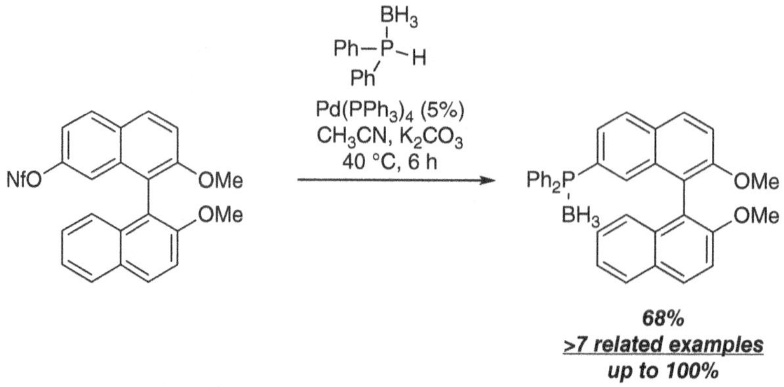

SCHEME 4.185   Palladium-catalyzed phosphination of arylnonaflates [295].

near-quantitative yield of a borane-protected arylphosphine. The highest yields were reported for substrates bearing electron-donating substituents, and this chemistry was remarkably tolerant of bulky groups on the arylnonaflate.

The phosphination of zinc porphyrins has been achieved using palladium catalysis (Example 4.32) [296]. The coupling reaction proceeded under mild conditions to generate the phosphination product in good yield. However, the phosphanylporphyrin was quite sensitive to air, and significant amounts of the oxide were formed upon exposure to the atmosphere. To avoid contaminating the target phosphine during the isolation step due to the formation of the oxide, the authors protected the reactive phosphine through the formation of an air-stable phosphine sulfide. This was readily achieved through the addition of elemental sulfur to the reaction mixture. After purification by column chromatography, the authors were able to generate the free phosphine through a desulfurization reaction using $P(NMe_2)_3$ in toluene. The removal of the sulfur was quite effective and the desired phosphanylporphyrin was isolated in outstanding yield by a simple recrystallization. Although the operational complexity of the overall process was moderate, it was well worth a little more work if the final compound could be isolated through a simple crystallization from the reaction mixture in excellent yield with no contamination from the oxide.

One of the attractive aspects of metal-catalyzed cross-coupling is the ability to control the selectivity of a reaction through careful choice of the supporting ligand on the metal center (Example 4.33) [297]. Due to the ability of arylphosphines to solubilize and stabilize palladium in several oxidation states, the design of either C-chiral or P-chiral arylphosphine ligands for asymmetric catalysis is of interest to those charged with synthesizing chiral molecules. To this end, a number of reported have outlined various approaches to the synthesis of chiral phosphines through metal-catalyzed coupling reactions.

In addition to the examples described previously, a tremendous variety of compounds have been prepared using palladium-catalyzed cross-coupling reactions due to the broad functional group tolerance. The combination of copper and palladium complexes have been used to promote the formation of borane-protected arylphosphines [298]. Palladium acetate catalyzed the coupling of halogenated benzoic acid derivatives leading to the formation of water-soluble phosphine ligands [299]. Aryl triflates have also been used in the phosphination reaction [300]. Functionalized tyrosine derivatives were converted

Example 4.32 Phosphination of a Zinc Porphyrin Using Palladium Catalysis and Protection/Deprotection of the Resulting Phosphine [296].

Before starting this synthesis, carefully follow the chemical safety instructions listed in Example 4.1.

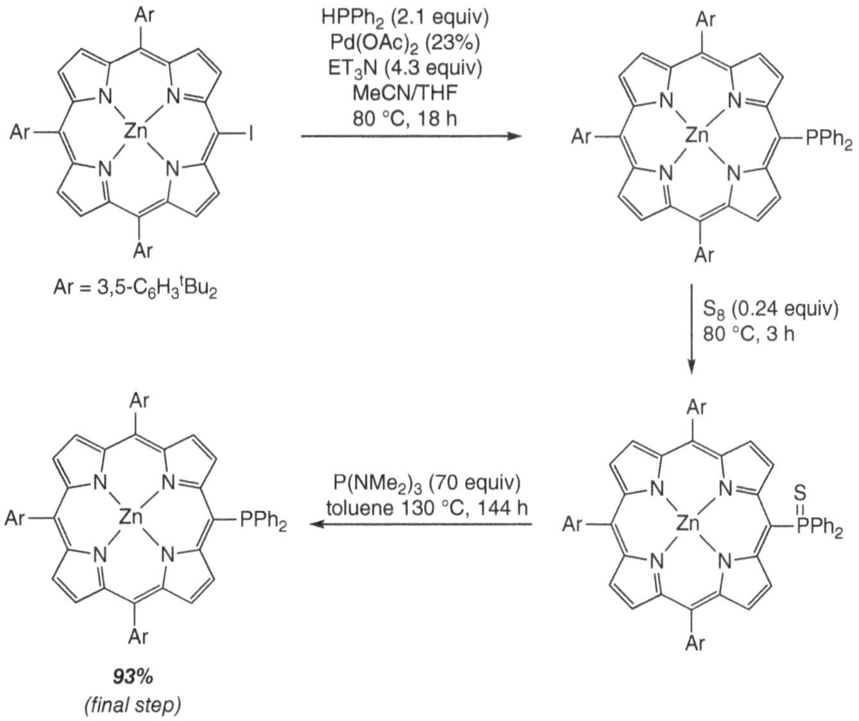

Ar = 3,5-C$_6$H$_3$^{t}Bu$_2$

HPPh$_2$ (2.1 equiv)
Pd(OAc)$_2$ (23%)
ET$_3$N (4.3 equiv)
MeCN/THF
80 °C, 18 h

S$_8$ (0.24 equiv)
80 °C, 3 h

P(NMe$_2$)$_3$ (70 equiv)
toluene 130 °C, 144 h

93%
(final step)

Caution! Secondary phosphines can react violently with air/moisture and must be stored and handled under an inert atmosphere such as nitrogen or argon.

Step 1: Preparation of the Phosphine Sulfide

This reaction was performed under an atmosphere of argon. A 30 mL flask containing the iodoporphyrin (222 mg, 0.209 mmol) and Pd(OAc)$_2$ (10.9 mg, 0.0486 mmol) was evacuated under vacuum and then filled with argon. The same manipulation was repeated three times. THF (11 mL), MeCN (8.8 mL), triethylamine (123 μL, 0.89 mmol), and diphenylphosphine (75 μL, 0.43 mmol) were added via syringes to the flask, and the resulting mixture was stirred at 80 °C (bath temperature) for 18 h. After checking for the consumption of the starting materials by TLC, S$_8$ (13 mg, 0.050 mmol) was added to the flask, and the mixture was stirred for an additional 3 h. The resulting mixture was filtered through a celite bed, and the filtrate was concentrated under reduced pressure to leave a solid, which was then chromatographed on silica gel using hexane and CH$_2$Cl$_2$ as eluents. The bluish-purple fraction (R_f=0.36; hexane/AcOEt = 5/1) was collected, concentrated, and reprecipitated from CH$_2$Cl$_2$/MeOH to give 4a as a purple solid (221 mg, 92%).

Step 2: Desulfurization Step

This reaction was performed under an atmosphere of argon. A 50 mL flask containing the phosphine sulfide (165 mg, 0.143 mmol) was evacuated and then filled with argon. The same manipulation was repeated three times. Toluene (20 mL) and P(NMe$_2$)$_3$ (1.82 mL, 10.0 mmol) were added via syringes to the flask, and the resulting mixture was stirred at 130 °C (bath temperature). After 144 h, the phosphine sulfide was consumed completely (checked by TLC). The solvent was concentrated under reduced pressure to leave a solid residue, which was reprecipitated from CH$_2$Cl$_2$/MeOH under argon atmosphere to give the phosphine as a purple solid (150 mg, 93%).

Adapted with permission from Y. Matano, K. Matsumoto, Y. Nakao, H. Uno, S. Sakaki, and H. Imahori, *J. Am. Chem. Soc.,* **2008**, *130*, 4588–4589. *Copyright (2008) American Chemical Society.*

Example 4.33   Asymmetric Synthesis of a Chiral Bisphosphine [297].

Before starting this synthesis, carefully follow the chemical safety instructions listed in Example 4.1.

92% (99% ee)

(4 equiv)

This reaction was carried out under inert atmosphere. To a solution of 32 mg of palladium bisacetate (0.144 mmol, 4 mol%) and 159 mg of diphenylphosphinoferrocene (0.287 mmol, 8 mol%) in distilled and degassed acetonitrile (5 mL), 3.1 mL of diisopropylethylamine (17.97 mmol, 5 equiv) and the aryl iodide (14.36 mmol, 4 equiv) were added, and the mixture was degassed three times and stirred at 80 °C under argon. Then, a solution of 1 g of the 6,6′-dimethoxy-biphenyl-2,2′-bisphosphine (3.59 mmol, 1 equiv) in distilled and degassed acetonitrile (15 mL) was added to the reaction mixture over 2 h using a syringe pump. After completion of the reaction (2–3 h), the mixture was concentrated and purified by silica gel flash chromatography (cyclohexane/EtOAc 70 : 30) to afford the bisphosphine in 92% yield.

Adapted with permission from L. Leseurre, F. Le Boucher d'Herouville, K. Puntener, M. Scalone, J.-P. Genet, V. Michelet, *Org. Lett.,* **2011**, *13*, 3250–3253. *Copyright (2011) American Chemical Society.*

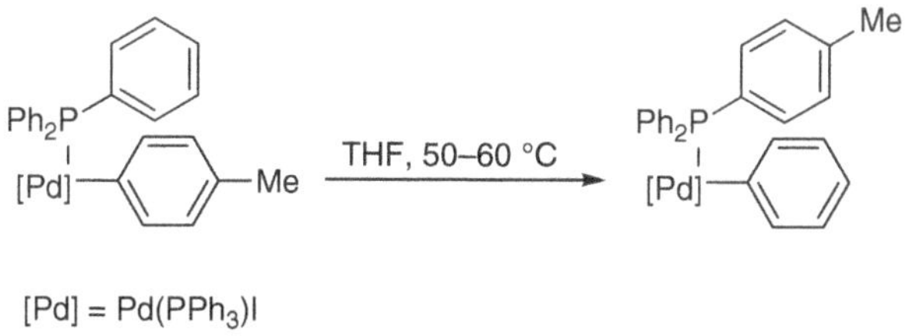

[Pd] = Pd(PPh₃)I

SCHEME 4.186   Aryl–aryl exchange between palladium and phosphorus [309].

into triflates that were successfully coupled with secondary phosphines to generate arylphosphine sulfides using a Pd(OAc)₂/dppp catalyst system followed by trapping the aryl-phosphine with elemental sulfur [301]. In related work, the cross-coupling of iodophenyl-analine with secondary phosphines was catalyzed by Pd(OAc)₂/PPh₃ [302]. Unprotected aniline derivatives have also been successfully used in the cross-coupling reaction [303]. Covalently attached polymeric phosphines were generated by coupling polymer chains bearing pendant iodophenyl groups and secondary phosphines in a postpolymerization modification [304, 305]. Pd₂(dba)₃ (CHCl₃ adduct) catalyzed the coupling of a xanthene derivative bearing two primary phosphines with 3-iodophenylguanidine to afford guani-dinium-modified bisphosphine ligands [306]. Substrates bearing multiple aryl halides have been phosphinated using palladium catalysts to generate oligomeric materials [307]. While most of the examples used organic solvents such as toluene, THF, or DMF, elegant work by Gaumont outlined the use of palladium catalysts in ionic liquids for the phosphorus–carbon bond-forming reaction [308]. An attractive aspect of this approach was that the palladium-laced ionic liquid could be reused for multiple subsequent reactions with minimal loss of activity.

During the course of an investigation into the stability and reactivity of arylpalladium complexes supported by two triphenylphosphine ligands, Cheng made an intriguing discovery (Scheme 4.186) [309]. The complex under scrutiny was a palladium II species bearing two triphenylphosphine groups as well as one tolyl and an iodide. Upon gently heating, the authors observed the aryl group directly attached to the palladium undergoing an exchange with one of the phenyl groups on the triphenylphosphine. Changing the system to ((C₆D₅)₃P)₂Pd(Ph)I revealed scrambling of the phenyl group on the palladium center with the deuterated groups on the phosphorus. Although this process was suggested in several previous reports, this detailed mechanistic work clearly demonstrated that this process was quite facile. The trick was adapting this remarkable observation to a catalytic process for the functionalization of arylphosphines.

The aryl–aryl exchange reaction was converted into a practical process for the prepara-tion of phosphorus–carbon bonds (Scheme 4.187) [310, 311]. The authors screened a range of catalysts for activity and found that palladium acetate was the most effective. No extra solubilizing ligand was added in these reactions. After analyzing the results from the aryla-tion reactions, an electronic effect was observed and substrates bearing electron-donating groups were found to undergo the exchange reaction at faster rates. The authors also inves-tigated the effect of changing the electronic composition of the triarylphosphine used as the phosphinating agent. Similar to the effect observed for substituents on the aryl triflate, incor-porating electron-rich aryl groups into the triarylphosphine resulted in faster reaction rates. The authors also screened trialkylphosphine groups in order to determine whether or not an alkyl group could participate in the exchange reaction. No exchange was observed when

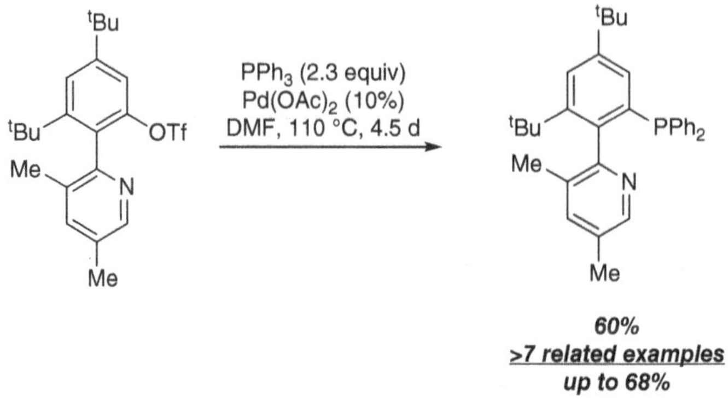

60%
≥7 related examples
up to 68%

SCHEME 4.187   Synthesis of arylphosphines using PPh₃ as the phosphinating agent [310, 311].

they used PCy₃ as a representative phosphinating agent. This chemoselectivity could be used to direct the substitution pattern of the resulting phosphine since an aryl group would be preferentially lost over the alkyl substituent. The success of the reaction was dependent upon the ability of the aryl triflate to add to the palladium center. Increasing the steric bulk about the triflate inhibited the reaction. It should be noted that the amount of triphenylphosphine had an effect on the reaction. While it might appear as through the 2.3 equivalents listed in Scheme 4.187 indicated that an excess was used, the true effect of the extra phosphine appears to be much more complex. Increasing to four equivalents of PPh₃ completely suppressed for the formation of the product in these reactions. Presumably, the larger excess of PPh₃ prevented the formation of a catalytically active species through coordination. One of the most attractive aspects of this chemistry was that a glovebox was not needed.

This chemistry was extended to a range of aryl halides and triflates [312–315]. In an effort to develop a more sustainable version of the reaction, a solvent-free modification was sought. After some experimentation, the authors found that simply removing the DMF from the experimental procedures and increasing the temperature slightly (115 °C) in the system they had already devised were all that was needed to convert the process into a solvent-free version. These new reactions were also faster than when DMF was used and were also quite tolerant to a range of functional groups including esters, ketones, nitriles, aldehydes, and ethers. As with many of the metal-mediated phosphination reactions, bulky substrates were problematic, and even adding a single ortho-substituent was enough to completely suppress the phosphination reaction when aryl bromides were used. Analogous reactions with aryl triflates were also sluggish and afforded low yields of the arylphosphines. Overall, this process was attractive since it did not require the use of a glovebox and could be carried out under solvent-free conditions.

The palladium-catalyzed synthesis of phosphole derivatives used a monophosphinated biphenyl as a substrate and generated the phosphole oxides in moderate to excellent yields (Scheme 4.188) [316]. Palladium acetate was the most effective metal source, and heating the reaction mixture in toluene generated the phospholes. Due to the sensitivity of the compounds to oxygen, they were isolated as phosphole oxides through the addition of hydrogen peroxide following the P—C bond-forming step. The authors did note that protection of the phosphorus could also be achieved through the formation of the borane-phosphine adduct. Similar to the other reports in this area, the cyclization reaction was quite tolerant to a number of functional groups, and substrates bearing ethers, amines, esters, nitriles, and halogens were successfully used. The retention of an aryl bromide was especially attractive

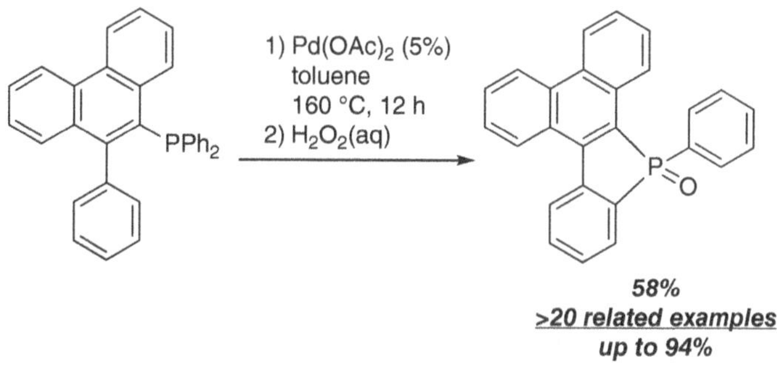

SCHEME 4.188 Synthesis of phospholes through an aryl–aryl exchange reaction [316].

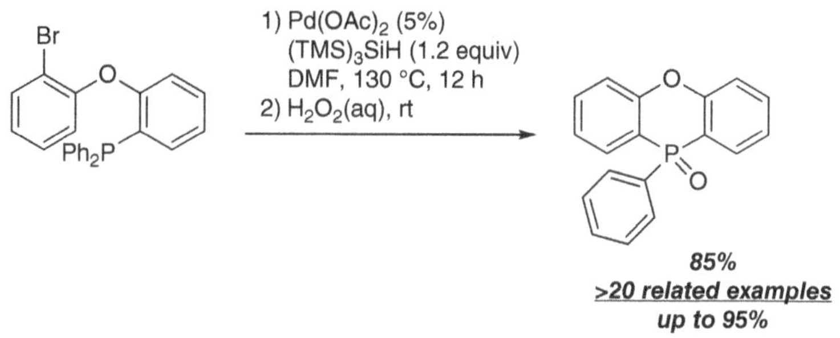

SCHEME 4.189 Synthesis of phosphacycles through a palladium-mediated exchange reaction [317].

since these functional groups are convenient starting points for further functionalization. The presence of ortho-substituents was also well tolerated, and the new P—C bond was established opposite the ortho-substituent. In addition to hydrocarbon-based biaryl substrates, heterobiaryl systems could also be functionalized using this approach.

This approach was adapted to the synthesis of six-membered phosphacycles (Scheme 4.189) [317]. Classically, these heterocycles were prepared by generation of a dilithiated species followed by trapping with a dichloroarylphosphine. Yields were typically low to moderate, and the presence of electrophiles was not tolerated due to the use of lithium reagents. Due to these limitations, a palladium-catalyzed route was investigated. Initially, the authors screened representative substrates for activity using the previously described cyclization reaction [316]. This approach was unsuccessful, and the authors proposed that the need to form a seven-membered ring as an intermediate could be problematic for the chemistry. After some experimentation, the authors determined that the substrate to be modified needed to include a bromide in the ortho-position and that a reducing agent was needed for an efficient process. Although zinc dust and Et_3SiH generated some of the desired compound (36–56%), bulky hydrosilanes such as $(TMS)_3SiH$ were much more effective and afforded high yields of the six-membered phosphacycle. Due to the air sensitivity of the resulting heterocycle, the products were converted into air-stable phosphine oxides to facilitate the isolation/purification. A range of functional groups including electrophiles were well tolerated by this approach, and moderate to excellent yields of the six-membered phosphacycles were obtained. When alkylarylphosphines were used as substrates, the cyclization reaction was selective for the retention of the alkyl substituent. This facilitated the development of a chemoselective synthesis for the cyclic phosphines.

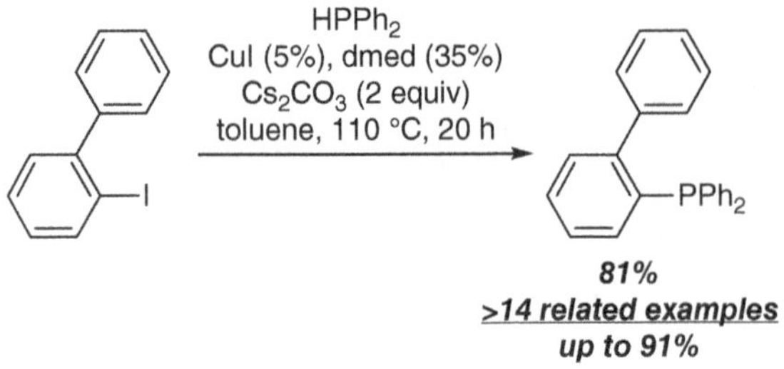

SCHEME 4.190   Copper-catalyzed synthesis of arylphosphines using a CuI/dmed catalyst system [318].

In addition to using the aryl–aryl exchange and related processes for the formation of new phosphorus–carbon bonds, the process can complicate what might be expected to be a straightforward synthesis. As described previously, the synthesis of 2,2′-bis(diphenylphosphino)biphenyls starting from 2,2′-diiodobiphenyls through a palladium-catalyzed cross-coupling reaction was hampered by the formation of a phosphafluorene (phosphole). An aryl–aryl exchange between the coordinated diphenylphosphide and the attached aryl ring of the 2,2′-dibromobiphenyls was proposed to be the mechanism behind the formation of this secondary product. The concomitant formation of triphenylphosphine was also observed. This process tends to be favored with electron-donating ligands, and the authors observed increased amounts of the phosphole in reactions using 2,2′-diiodobiphenyls bearing $-NH_2$ or $-OR$ functional groups. Furthermore, while it might seem like the Buchwald class of ligands would be susceptible to this chemistry, the vast majority of these arylphosphines have either two *tert*-butyl or cyclohexyl groups attached to the phosphorus center. These alkyl groups have been shown to be resistant to the exchange process.

Arylphosphines have also been prepared using copper-catalyzed cross-coupling reactions [318–320]. These approaches tended to use CuI as the copper source with (Scheme 4.190) [318, 320] or without a solubilizing ligand [319]. Cesium carbonate was found to be an effective base for this reaction and promoted the formation of a range of arylphosphines in moderate to excellent yields. An attractive aspect of this chemistry was the functional group tolerance. A wide range of aryl halides including an ester, unprotected aniline derivative, and aromatic heterocycles were successfully functionalized using this approach. While many metal-catalyzed C—P bond-forming reactions are slow when ortho-substituents are incorporated into the aryl halide, this approach generated high yields of bulky diarylbiarylphosphines. Even trimethyliodobenzene was successfully converted into the triarylphosphine in good yield (71%).

One approach to the synthesis of arylphosphines that avoids storing and handling highly air-sensitive HPR_2 species uses air-stable secondary phosphine oxides as substrates (Scheme 4.191 and Example 4.34) [24]. Instead of running the coupling reaction using the phosphine oxide and incorporating a reduction step at the end of the reaction to generate the phosphine, the authors opted to reduce the phosphorus species prior to the addition of the coupling reagents. While the common approach would have been to reduce the oxide after the coupling reaction, the differences in reactivity between the phosphine and phosphine oxides could lead to different product distributions. Their approach used tetramethyldisilazane (TMDS) as the reducing agent and copper triflate as the catalyst for the reaction. A number of substrates bearing existing functional groups were well tolerated by this chemistry, and moderate to excellent yields of the

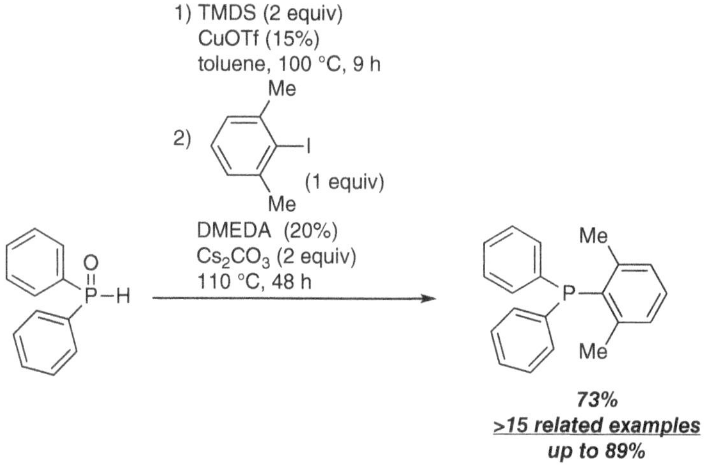

SCHEME 4.191 Copper-catalyzed synthesis of arylphosphines using reduction as the first step [24].

Example 4.34 Synthesis of Arylphosphines through Tandem Reduction/Coupling Reactions [24].

Before starting this synthesis, carefully follow the chemical safety instructions listed in Example 4.1.

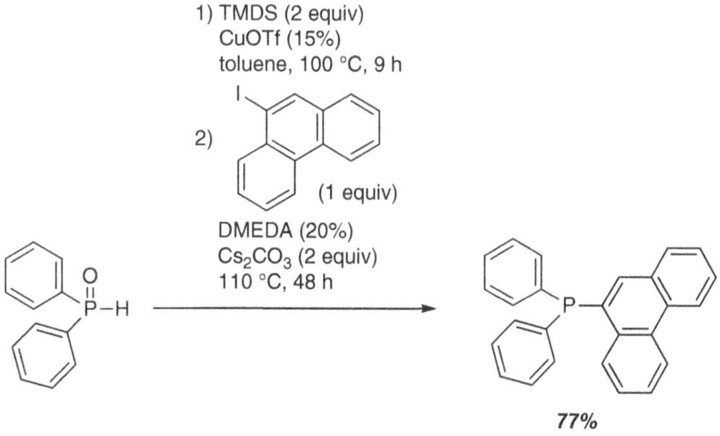

Caution! Ethyl acetate is highly flammable.

A 10 mL dried Schlenk tube containing a stirring bar was charged with Cu(OTf)$_2$ (27.0 mg, 0.075 mmol) and diphenylphosphine oxide (101.0 mg, 0.5 mmol). Under argon flow, dry toluene (2 mL) and TMDS (180 μL, 1.0 mmol) were added, and the mixture was stirred at 100 °C for 9 h. Then, the reaction mixture was cooled to room temperature, and *N,N'*-dimethyl-1,2-ethylenediamine (10.4 μL, 0.1 mmol), Cs$_2$CO$_3$ (326 mg, 1.0 mmol), and 9-iodophenanthrene (0.5 mmol) were added under argon flow. The suspension was heated to 110 °C and stirred for 48 h. Then, the reaction mixture was cooled to 0 °C and methanolic KOH (3 mL, 3 N) was added slowly. After stirring the mixture vigorously for 5 h at room temperature, water (3 mL) was added and the mixture was extracted by ethyl acetate (2×25 mL). The organic phase was combined and washed by HCl solution (aq., 5 mL, 1 N) and saturated NaHCO$_3$ solution (aq., 5 mL). Then, the organic phase was dried by Na$_2$SO$_4$ and concentrated under vacuum. The residue was purified by silica gel column chromatography.

Adapted with permission from Y. Li, S. Das, S. Zhou, K. Junge, M. Beller, *J. Am. Chem. Soc.*, **2012**, *134*, 9727–9732. *Copyright (2012) American Chemical Society.*

Example 4.35   Steroid-Derived BINAP-Type Bisphosphine [328].

Before starting this synthesis, carefully follow the chemical safety instructions listed in Example 4.1.

Caution! Secondary phosphines can react violently with air and must be handled and stored under an inert atmosphere such as nitrogen or argon.

This reaction was carried out under an atmosphere of argon. To a solution of $NiCl_2dppe$ (0.070 g, 0.13 mmol) in dimethyl acetamide (2 ml) was added diphenylphosphine (0.13 ml, 0.75 mmol) at room temperature, and the solution was heated to 100 °C. After 45 min, a solution of the bistriflate (1 g, 1.3 mmol) and 1,4-diazabicyclo[2.2.2]octane (0.62 g, 5.5 mmol) in dimethyl acetamide (4 ml) was added at once, the resulting green solution was kept at 100 °C, and three additional portions of diphenylphosphine (0.13 ml each) were added at 1, 3, and 5 h later. The reaction was kept at 100 °C for 6 days, and then the dark brown solution was diluted with MeOH. The desired product was filtered, and the filter cake was washed with MeOH and dried under vacuum. The crude product (0.820 g, 80%) was recrystallized from MeOH/Tol 10 : 1 to give 0.78 g pure phosphine.

Adapted with permission from V. Enev, Ch. L. J. Ewers, M. Harre, K. Nickisch, and J. T. Mohr, *J. Org. Chem.*, **1997**, *62*, 7092–7093. *Copyright (1997) American Chemical Society.*

arylphosphines were obtained. The ability of the system to handle bulky electrophiles was attractive as was the insensitivity to electron-donating and withdrawing groups. While monodentate phosphines are shown in Scheme 4.191, a series of bidentate bisphosphines were also generated using this approach.

In addition to the synthesis of achiral phosphines, cross-coupling chemistry has been extended to a wide array of chiral phosphines [321–327]. An example of this chemistry used common nickel salts to promote the conversion of a steroid-derived bis-triflate species into a bisphosphine (Example 4.35) [328].

A twist on the nickel-catalyzed reaction used elemental zinc as a reducing agent to aid in the regeneration of an active catalyst (Example 4.36) [329]. In addition to aiding the reduction of the nickel(II) species, the authors proposed that it also reacted with the chlorophosphine to generate Ph_2PZnCl, which functioned as a transmetallating agent in the reaction. While both organohalides and triflates were successfully functionalized by this approach (yields up to 95%), the highest yields were obtained with a naphthyl triflate

Example 4.36   Nickel-Catalyzed Phosphination Using Zn as a Reducing Agent [329].

Before starting this synthesis, carefully follow the chemical safety instructions listed in Example 4.1.

ClPPh$_2$ (1 equiv)
NiCl$_2$(dppe) (1.5%)
Zn (1.6 equiv)
DMF, 5–10 °C to 100 °C

89%

To a solution of the naphthyl triflate (10.6 g, 40 mmol), NiCl$_2$(dppe) (0.315 g, 0.60 mmol) and Ph$_2$PCl (8.85 g, 40 mmol) in anhydrous DMF (25 ml) was added Zn (4.10 g, 63 mmol) portionwise at 5–10 °C. The mixture was heated to 100–110 °C and monitored by GC. The reaction mixture was cooled to 80 °C, filtered, and rinsed with a minimal amount of DMF. The combined filtrate was cooled to 5 °C and the product crystallized overnight. The solids were filtered, rinsed with MeOH, and dried at room temperature under vacuum to give 11.21 g (89%) of the target phosphine as an off-white solid.

Adapted from D. J. Ager, M. B. East, A. Eisenstadt, and S. A. Laneman, *Chem. Commun.,* **1997**, 2359–2360 *with permission of The Royal Society of Chemistry.*

bearing an electron-withdrawing ester group adjacent to the triflate. The lowest yields were obtained using a vinyl bromide as the substrate.

Chiral benzophospholanes are an interesting class of compounds with potential applications as chiral ligands in enantioselective reactions. These reactions can be promoted by palladium complexes; however, the selectivity is moderate [1]. An alternate approach used an ephedrine-derived auxiliary (Scheme 4.192) [330]. One of the keys to the synthesis was the use of ether as the solvent. Other solvents such as THF were not as effective. The final step in the synthesis involved loss of the chiral auxiliary and formation of the borane-protected 1-phenyl-2-ox-α-1-phosphindane. The yield and selectivity of the process were outstanding (96%, 97.5% ee).

The incorporation of substituents in the 2- and 4-positions on a phosphole ring was one of the more challenging substitution patterns to achieve in phosphole synthesis. To this end, Jaroschik and Szymoniak designed an intriguing entry point to this rare class of compounds (Example 4.37) [331]. Their approach used a mixture of a zirconocene complex and lanthanum metal along with two equivalents of a terminal acetylene to generate a metallocycle that was trapped by the addition of a dichlorophosphine to generate a phosphole. A range of terminal acetylenes were successfully used in the synthesis, and moderate to good yields of the phospholes were obtained. The system was not restricted to aryl substituents, and using 1-hexyne as the substrate afforded the alkylphosphole in moderate yield (63%). The operational difficulty of this reaction is moderate; however, both steps were able to be completed in a single reaction vessel without purification of any intermediates.

The P-arylation of arynes is an interesting approach to the synthesis of arylphosphine oxides since it is typically a transition metal-free reaction that occurs at low temperatures (Scheme 4.193) [332]. In contrast to classic metal-mediated Arbuzov-type reactions between phosphinites and aryl halides that required high temperatures (150–200 °C), this

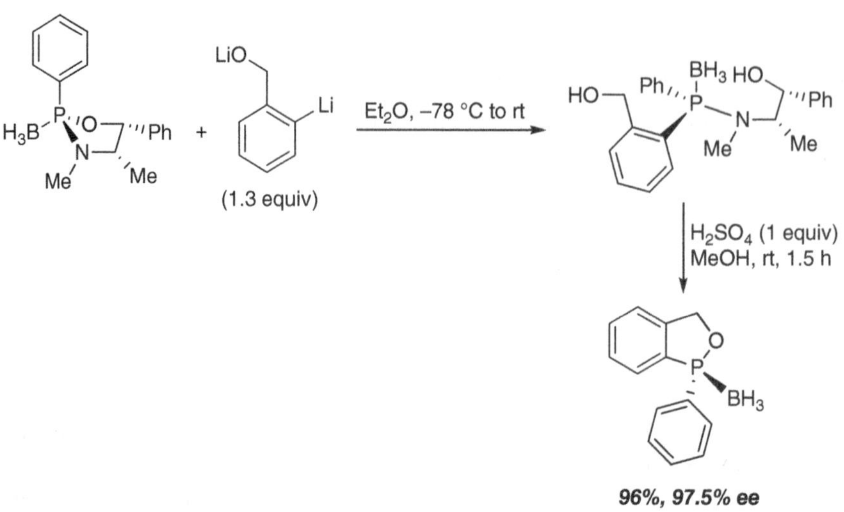

SCHEME 4.192 Synthesis of a borane-protected 1-phenyl-2-oxa-1-phosphindane [330].

Example 4.37 Synthesis of 2,4-Disubstituted Phospholes [331].

Before starting this synthesis, carefully follow the chemical safety instructions listed in Example 4.1.

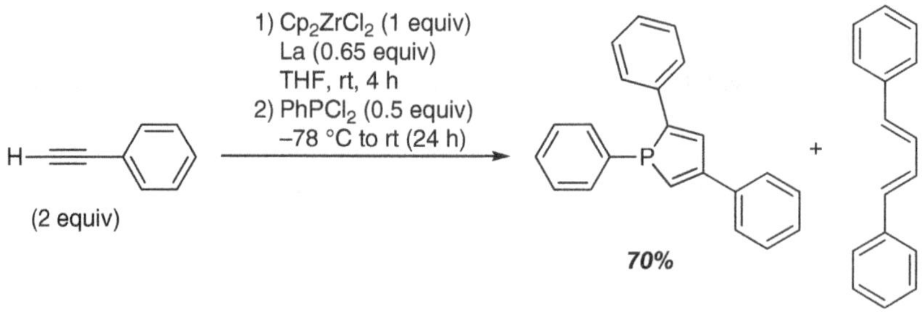

Caution! Ethyl acetate, THF, and petroleum ether are highly flammable.

Caution! Chlorophosphines can react violently with moisture and should be handled and stored under an inert atmosphere such as nitrogen or argon.

This reaction was carried out under an atmosphere of argon. A Schlenk tube was loaded with Cp$_2$ZrCl$_2$ (584 mg, 2.0 mmol), lanthanum (186 mg, 1.3 mmol), and THF (10 mL). The resulting mixture was stirred vigorously at room temperature until a deep red color appeared. At this stage, phenylacetylene (0.42 mL, 4.0 mmol) was added to the reaction mixture, and the stirring was continued until complete disappearance of the alkyne as shown by TLC. Then the optimized amount of dichlorophenylphosphine (0.14 mL, 1.0 mmol) was added at −78 °C. After slowly warming to room temperature, the reaction mixture was stirred for 18 h. After that time, petroleum ether (20 mL) was added to the brown solution, and the solution was filtered over a short column of basic aluminum oxide using petroleum ether–ethyl acetate 8 : 2 as the eluent. The solvent was evaporated, and the crude residue was purified by flash column chromatography on silica gel using petroleum ether to yield the phosphole (218 mg, 70%).

Adapted from G. Bousrez, F. Jaroschik, A. Martinez, D. Harakat, E. Nicolas, X. F. Le Goffb and J. Szymoniak, *Dalton Trans.*, **2013**, *42*, 10997–11004 *with permission of The Royal Society of Chemistry.*

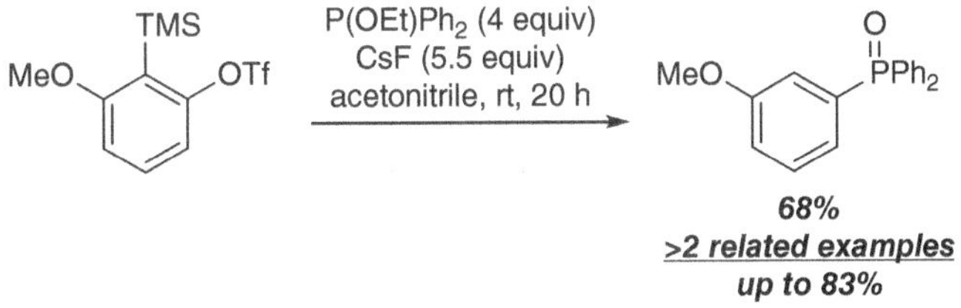

SCHEME 4.193 P-arylation of arynes [332].

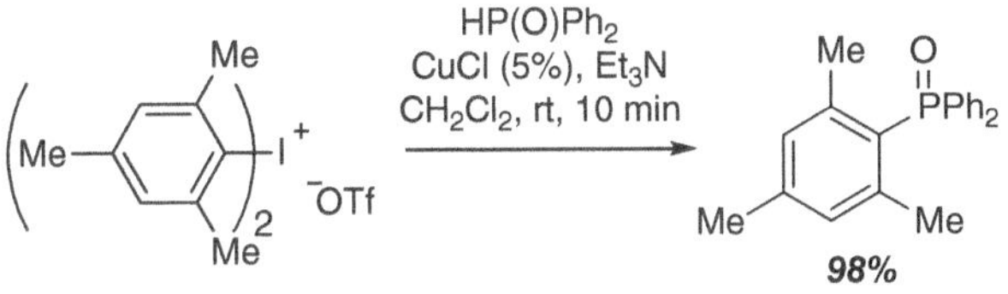

SCHEME 4.194 Synthesis of crowded phosphine oxides [333].

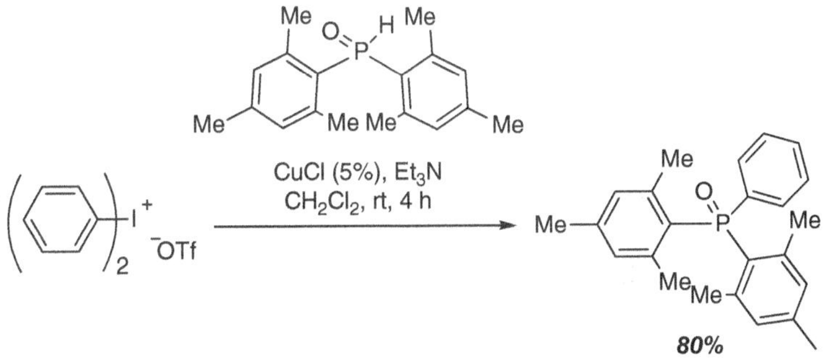

SCHEME 4.195 Reactivity of dimesitylphosphine oxide with diaryliodonium reagents [333].

chemistry occurred under mild conditions (room temperature) using readily available materials. The key component of the chemistry was the formation of an aryne intermediate. This was achieved through the use of a fluoride source that removed a TMS group that was adjacent to a triflate leaving group. The resulting aryne reacts with the phosphinite to generate the arylphosphine oxide. Although only four examples were given for this chemistry, it does represent a potentially valuable way to generate arylphosphine oxides under mild conditions without the addition of transition metals.

As with many cross-coupling reactions that result in the formation of C—P(O) bonds, the use of bulky electrophiles results in slow reactions. A potentially attractive solution to this problem has been reported by Tang (Schemes 4.194 and 4.195) [333]. He successfully coupled a range of diaryliodonium salts with secondary phosphine oxides using simple copper halides as catalysts. When unsymmetrical diaryliodonium salts were used, the bulkier group was preferentially transferred. Furthermore, the chemistry tolerated bulky groups on the secondary phosphine oxide. Another attractive aspect of this reaction was that it was rapid at room temperature. In addition to the synthesis of bulky phosphine oxides, the authors reported over two dozen additional examples (70–98% yields).

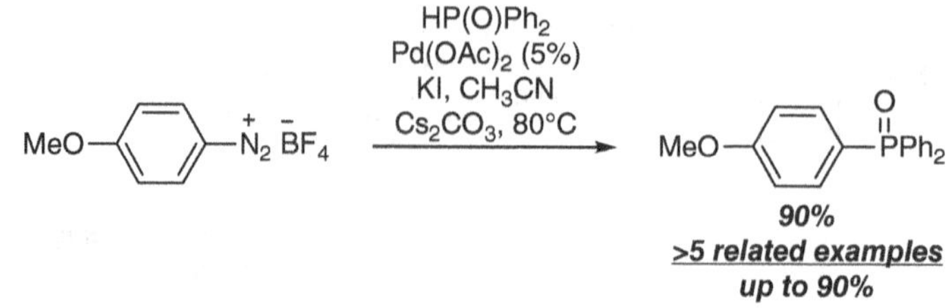

SCHEME 4.196   Palladium-catalyzed coupling of arenediazonium salts with secondary phosphine oxides [292].

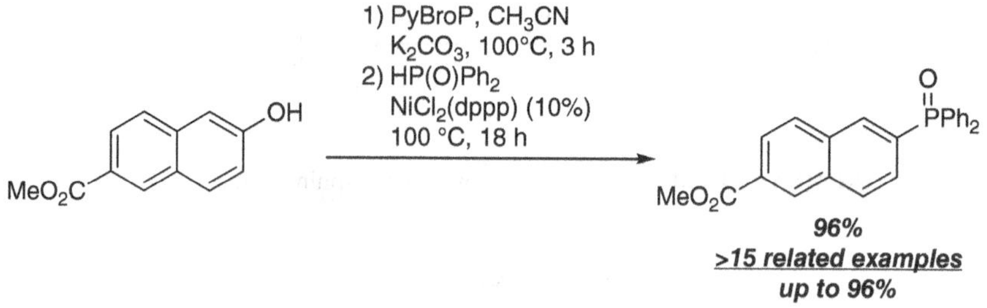

SCHEME 4.197   Nickel-catalyzed conversion of a phenol into a phosphine oxide using PyBroP as a promoter [334].

The palladium-catalyzed cross-coupling of arenediazonium tetrafluoroborates and secondary phosphine oxides generates arylphosphine oxides in moderate to excellent yields (Scheme 4.196) [292]. Palladium acetate was the most effective palladium precursor, and no additional ligand was needed for solubilization and stabilization of the palladium center. The chemistry was successful with arenes bearing esters, ethers, and halogens.

A nickel-catalyzed reaction that appeared to be a cross-coupling reaction between a phenol and a phosphine oxide fragment that resulted in an arylphosphine oxide has been reported (Scheme 4.197) [334]. Such a direct substitution reaction would be very curious; however, while this reaction appeared to be a direct substitution, it actually proceeded in tandem fashion. The first step entailed the conversion of the phenol into something that was susceptible to cross-coupling chemistry. After some experimentation, the authors found that bromotripyrrolidinophosphonium hexafluorophosphate (PyBroP) successfully converted the phenol into a phenolic phosphonium salt. The second part of the reaction involved the addition of a secondary phosphine oxide along with a preformed nickel catalyst and additives followed by heating to achieve the cross-coupling. The chemistry displayed wide functional group tolerance as substrates bearing esters, ethers, nitriles, pyridines, ketones, and naphthalene were successfully converted into phosphine oxides using this tandem approach. Due to the host of phenols that are readily available, this reaction is a very attractive approach for the synthesis of these valuable compounds.

One of the first reports of a microwave-assisted palladium-catalyzed reaction for the formation of an arylphosphine oxide described the synthesis of progesterone antagonists (Scheme 4.198) [335]. The chemistry was based upon functionalization of the aromatic group on the 11β-position of the steroid core. An aryl triflate and a secondary phosphine

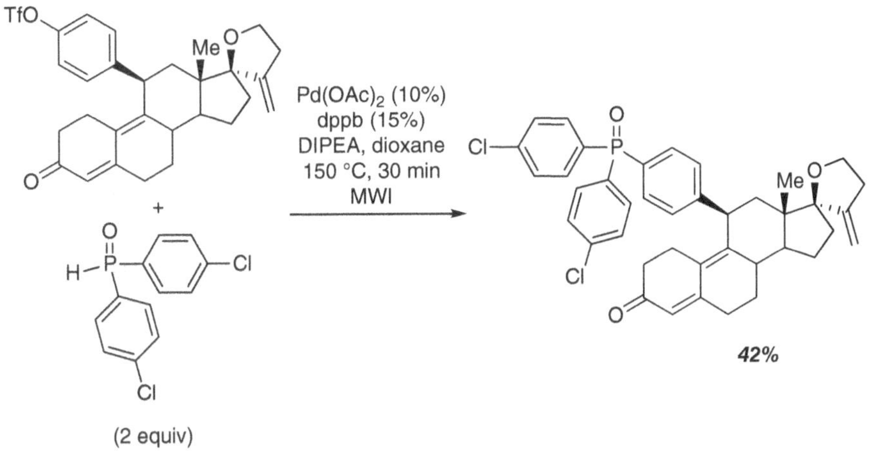

SCHEME 4.198 Microwave-assisted palladium-catalyzed coupling of secondary phosphine oxides with aryl triflates [335].

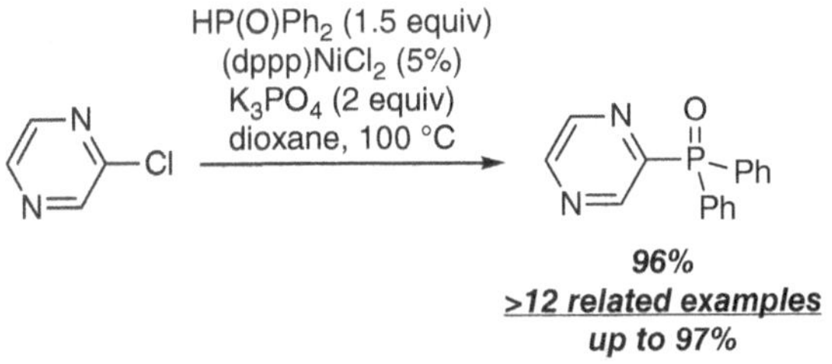

SCHEME 4.199 Nickel-catalyzed coupling of aryl bromides and chlorides with secondary phosphine oxides [336].

oxide served as coupling partners, and palladium acetate was an effective catalyst for this reaction. The authors found that the microwave-assisted version of the reaction consistently afforded higher yields of the cross-coupled product.

Han recently reported a nickel-catalyzed route to the formation of arylphosphine oxides starting from secondary phosphine oxides (Scheme 4.199) [336]. Using either aryl bromides or chlorides as coupling partners and a discrete nickel compound as a catalyst, a range of aryl halides were successfully converted into arylphosphine oxides in moderate to excellent yields. A significant electronic effect was observed in this chemistry, and the use of electron-donating groups on the aryl halide afforded lower yields of the arylphosphine oxides. For example, the use of electron-rich substrates such as 4-bromoanisole afforded 14% of the target, whereas the use of electron-deficient aryl bromides such as 4-bromobenzonitrile afforded 87% of the cross-coupled product. No examples containing ortho-substituents were reported, but the scope of the chemistry was relatively broad, and functional groups as ketones, esters, and nitriles were well tolerated. Heteroaryl substrates were also successfully used in this chemistry (79–97%).

Similar chemistry was reported by Yang (Schemes 4.200 and 4.201) [337]. Their approach used $(PPh_3)_2NiCl_2$ or $NiCl_2(DME)$ as the catalyst and either potassium carbonate or NaO^{t}Bu as the base along with DMF as the solvent. For some of the substrates,

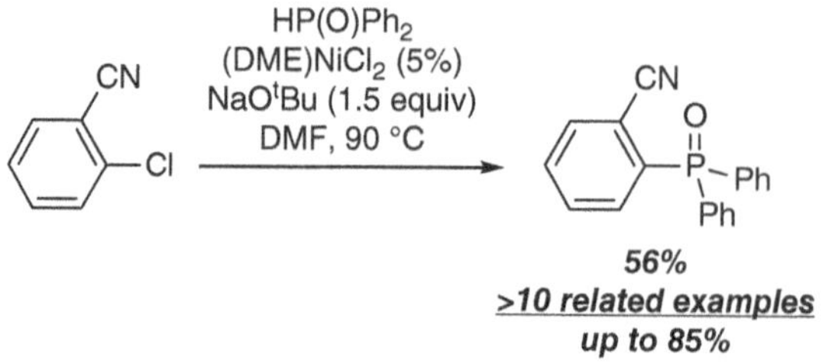

SCHEME 4.200   Synthesis of phosphine oxides through cross-coupling using (DME)NiCl$_2$ as the catalyst [337].

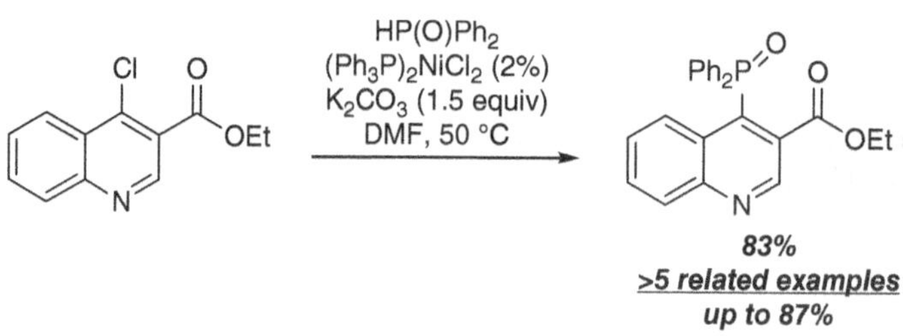

SCHEME 4.201   Synthesis of phosphine oxides through cross-coupling using (Ph$_3$P)NiCl$_2$ as the catalyst [337].

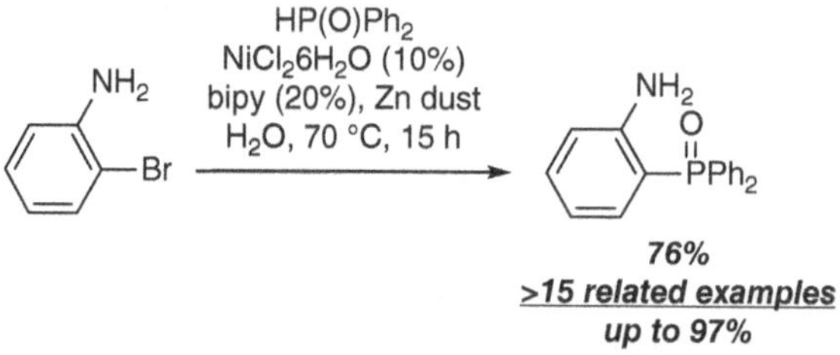

SCHEME 4.202   Preparation of phosphine oxides in water using nickel catalysts [338].

the catalyst loading was able to be decreased to 2%. A range of heteroaryl groups were successfully cross-coupled and yields of the arylphosphine oxides were moderate to good. The coupling reaction using heteroaryl groups was successful at 50 °C, whereas simple aryl chlorides required heating to 90 °C.

The nickel-catalyzed synthesis of arylphosphine oxides in water has been reported (Scheme 4.202) [338]. There is considerable interest in the use of water as a reaction solvent; however, water presents a number of challenges for transition metal catalysis. For this chemistry, a range of additives and supporting ligands for the nickel center were screened, and the most effective system used nickel chloride (as the hexahydrate), zinc powder, and bipy as a supporting ligand. The zinc powder was proposed to aid in the reduction step. The presence of a range of preexisting functional groups including amines, ketones, esters, and ethers did not reduce the effectiveness of this reaction, and excellent yields of the arylphosphine oxides were obtained.

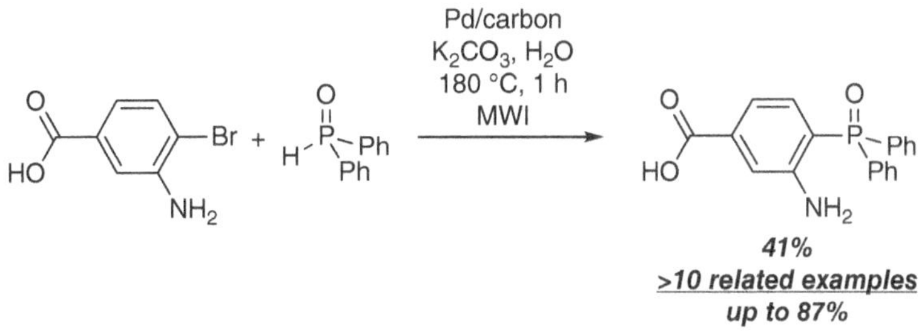

SCHEME 4.203   Palladium-catalyzed synthesis of arylphosphine oxides in water [339].

A palladium-catalyzed approach to the synthesis of arylphosphine oxides in water has been described (Scheme 4.203) [339]. One of the most interesting aspects of this chemistry was the ability to use halogenated benzoic acid derivatives as substrates. These electrophiles typically required protection/deprotection for their successful use in cross-coupling reactions due to undesired protonolysis reactions, catalyst poisoning, or decarboxylation. For this chemistry, the cross-coupling reaction was preferred. The authors also found that a focused microwave reactor was effecting at promoting the cross-coupling reaction in pure water with no added organic solvent. Furthermore, the authors reported that almost all the arylphosphine oxides could be isolated in greater than 95% purity without the use of any organic solvents. This was an attractive aspect of the chemistry since many organic reactions can be carried out in sustainable solvents or under solvent-free conditions, but liters of organic solvents are still needed for the isolation and purification steps. While the chemistry was successful for coupling reactions using bromobenzoic acid derivatives, iodobenzoic acids generated a significant amount of the dehydrohalogenation products and generated lower yields of the cross-coupling products. Chlorinated substrates were sluggish using this approach. The authors did note that a number of electron-donating groups were compatible with the cross-coupling reaction, and even a benzoic acid bearing an unprotected amine was successfully coupled with diphenylphosphine oxide. This system was attractive since it could be carried out in pure water, used a recyclable catalyst, and generated arylphosphine oxides bearing acidic functional groups in high yields.

The palladium-catalyzed cross-coupling of secondary phosphine oxides with aryl halides in water has also been reported using discrete palladacycles as catalysts (Scheme 4.204) [340]. No added ligands were needed for the metal center in this reaction although several additives were added to the reaction mixture including potassium fluoride and added TBAB. Aryl bromides, chlorides, and iodides were successfully converted into phosphine oxides using this approach, but aryl chlorides were sluggish.

Secondary phosphine oxides have been used in nickel-catalyzed cross-coupling reactions with aryl tosylates and mesylates for the preparation of arylphosphine oxides (Scheme 4.205) [341]. These substrates are typically more stable than aryl triflates and can be readily prepared from a wide range of phenols. In terms of the metal catalyst, the authors used a discrete species ((dppf)NiCl$_2$) and added extra supporting ligand (2 equiv per metal center) to prevent catalyst decomposition. The addition of zinc dust was essential to the success of the reaction, and no arylphosphine oxide was observed without it.

The C—H activation of pyridines has been coupled with a P—C bond-forming reaction to provide a selective method for the formation of arylphosphine oxides (Scheme 4.206) [342]. The catalyst system was comprised of a common palladium complex along with

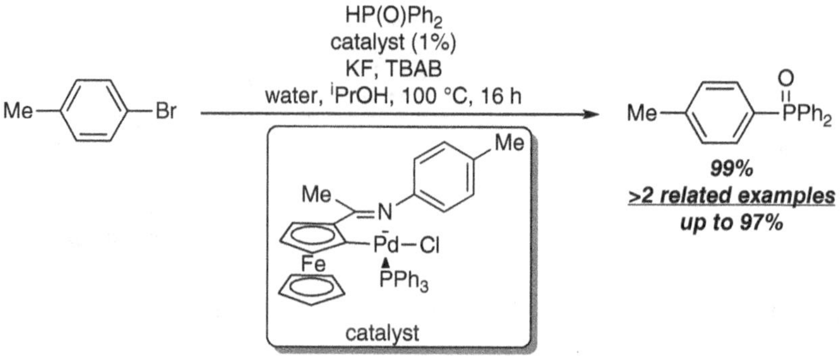

SCHEME 4.204   Using palladacycles to promote the cross-coupling reaction in water [340].

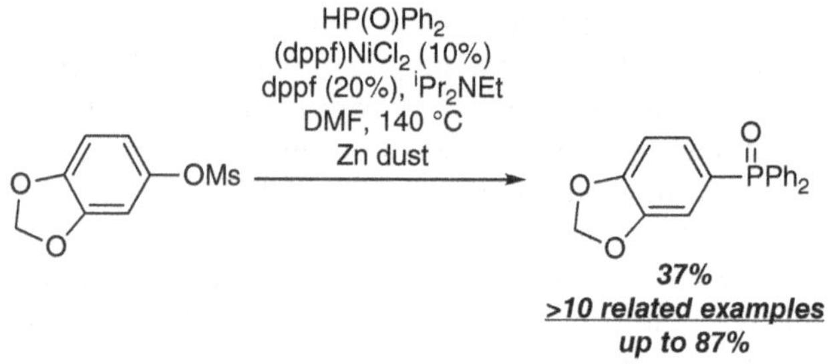

SCHEME 4.205   Nickel-catalyzed synthesis of arylphosphine oxides from aryl mesylates [341].

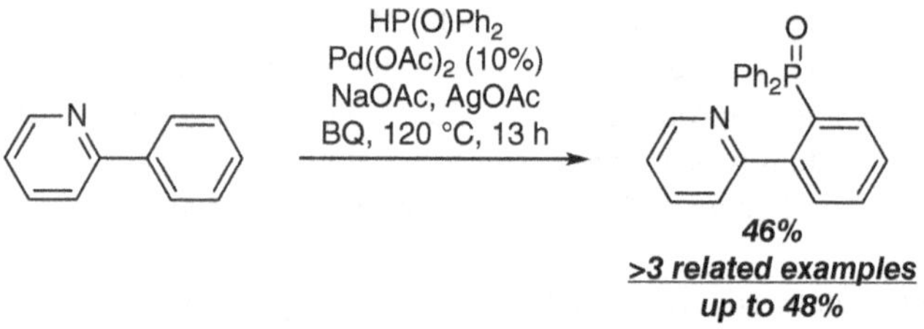

SCHEME 4.206   Palladium-catalyzed synthesis of arylphosphine oxides through C—H activation [342].

several acetate salts and benzoquinone. The latter was suggested to assist the formation of the arylphosphine oxides by increasing the rate of the reductive elimination step [343, 344]. While only a single pyridine derivative was screened, the authors did study a range of diarylphosphine oxides and found that the C—H activation/phosphonation process was insensitive to both electron-donating and withdrawing groups and fair to moderate yields of the arylphosphine oxides were obtained.

The palladium-catalyzed cross-coupling of halogenated pyrazoles with secondary phosphine oxides has been achieved using palladium acetate as the palladium source, triethylamine as the base, and XantPhos as the supporting ligand (Scheme 4.207) [345]. After some experimentation, the authors discovered that a preactivation step consisting of refluxing the catalyst components for 15 min prior to the introduction of the substrates afforded the highest conversions. A range of secondary phosphine oxides were functionalized

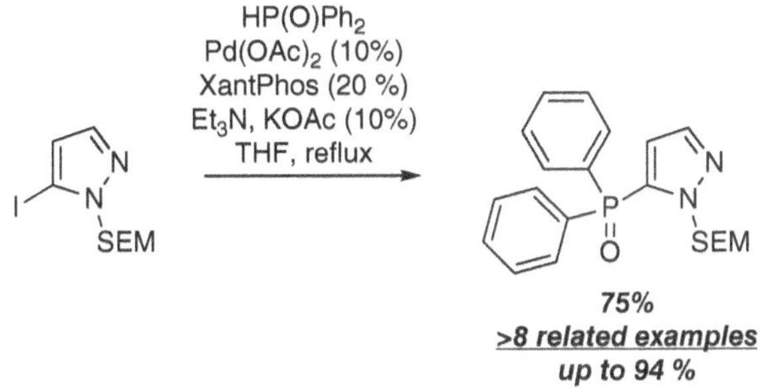

SCHEME 4.207 Palladium-catalyzed coupling of secondary phosphine oxides with iodinated heterocycles [345].

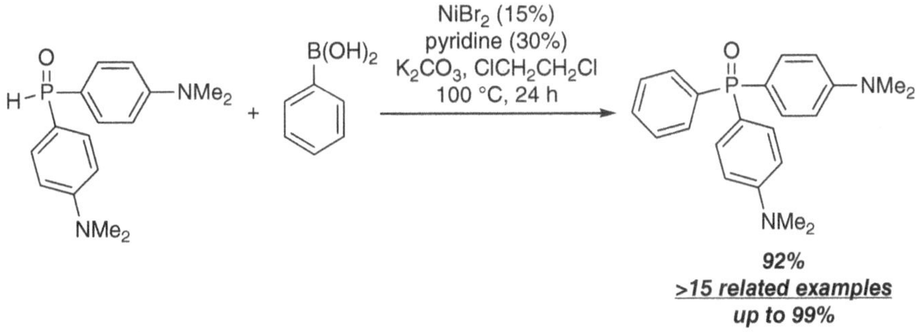

SCHEME 4.208 Nickel-catalyzed cross-coupling of arylboronic acids with secondary phosphine oxides [346].

and generated the desired products in low to excellent yields. This reaction was very substrate dependent and general trends were difficult to establish. Typically, the cross-coupling of iodinated pyrazoles generated higher yields of the phosphinylated pyrazoles; however, several examples where 5-bromopyrazole was superior were noted. The incorporation of bulky groups slowed the cross-coupling chemistry considerably and di-*tert*-butylphosphine oxide was unreactive using this catalyst system. While the catalyst loading was high (10%), this chemistry did demonstrate that 5-halogenated pyrazoles could be phosphinylated using a cross-coupling approach.

The nickel-catalyzed cross-coupling of boronic acids with secondary phosphine oxides is an attractive approach for the preparation of arylphosphine oxides (Scheme 4.208) [346]. After some experimentation, the authors found that nickel(II) bromide was the most effective nickel source. A mineral base was needed and potassium carbonate was effective. The substrate scope of this reaction was exceptionally high, and a range of functionalized boronic acids as well as secondary phosphine oxides were successfully cross-coupled. One drawback to this system was the use of 1,2-dichloroethane as the solvent for this reaction. It was noteworthy that the chemistry could be carried out under an atmosphere of air for some examples with only a minor reduction in the yields of the arylphosphine oxides. Thus, this reaction is very attractive as no glovebox or vacuum manifold was needed. This chemistry has the potential to generate a large number of different arylphosphine oxides due to the vast array of boronic acids that are readily available.

One of the challenges encountered when trying to add several different organic fragments to a phosphorus center using transition metal-catalyzed cross-coupling reactions is the selective activation of individual bonds. Essentially, a precursor needs to be constructed that will enable each new fragment to be added sequentially. This is quite a challenging task since many of the leaving groups typically encountered in organophosphorus chemistry are very similar and will react under similar conditions. A clever solution to this problem has been reported by Hayashi (Scheme 4.209) [347]. He used a series of selective additions to add three different groups to a phosphorus center. Following the generation of a precursor with suitably protected groups, a palladium-catalyzed cross-coupling between the precursor and an iodoarene selectively functionalized the hydroxymethyl fragment and formed the first C—P bond. The next step required the selective cleavage/functionalization of a single protecting group. The key finding was that the fragment bearing the silicon-based protecting group could be selectively removed using a fluoride source while leaving the remaining protecting group intact. The final cross-coupling reaction afforded the target in an impressive yield of 86% for all three steps. This is an interesting approach to the synthesis of unsymmetrically substituted organophosphorus compounds.

One of the oldest and most widely adopted methods for the synthesis of arylphosphonates involved the treatment of chlorophosphonates with organolithium or Grignard reagents. The chemistry was typically carried out at low temperatures in THF or a similar solvent. This chemistry was limited in scope due to the powerful nucleophiles/bases used in these reactions as well as the need to handle hazardous chlorophosphonates. Furthermore, while the organic fragments initially attached to the phosphorus center in phosphines and phosphine oxides are rarely displaced by the addition of excess organolithium or Grignard reagents, care must be taken when using chlorophosphonates as an excess of the organometallic reagent can result in the substitution of alkoxy groups. Overall, if the substrates do not contain substituents that are sensitive to powerful bases/nucleophiles, the treatment of chlorophosphonates with organolithium or Grignard reagents is an effective way to prepare arylphosphonates. A number of reviews on this approach to the synthesis of arylphosphonates have appeared [348–350]. An example of this approach was used to prepare components for dye-sensitized solar cells (Scheme 4.210) [351]. The chemistry consisted of initial deprotonation of terthiophene using BuLi at low temperature followed by addition of the chlorophosphonate. As mentioned in other sections, this metallation reaction is selective for the position adjacent to the sulfur.

The addition of secondary phosphites to arynes is an attractive procedure for the formation of arylphosphonates (Schemes 4.211 and 4.212) [332]. The resulting arylphosphonates were formed in moderate to excellent yields, and a range of substrates were compatible with the reaction chemistry with the provision that they possessed a TMS group adjacent to a triflate. In addition to the tertiary phosphites, diethyl phenylphosphonite was successfully added to the arynes to generate arylphosphinates. This approach is a potentially valuable way to generate arylphosphonates or phosphinates under mild conditions without the addition of transition metals.

As stated previously, arylhalides do not typically participate in Arbuzov-type chemistry due to their sluggish behavior in nucleophilic substitution reactions. Typically, metal salts are added to promote the reaction. However, suitably activated systems are known to participate in SNAr reactions using phosphorus nucleophiles. To this end, the pyrazinones have been phosphorylated through a metal-free Arbuzov reaction (Scheme 4.213) [352]. This approach to the functionalization of these heterocycles was attractive since it is operationally straightforward and was successful under solvent-free conditions. A focused microwave

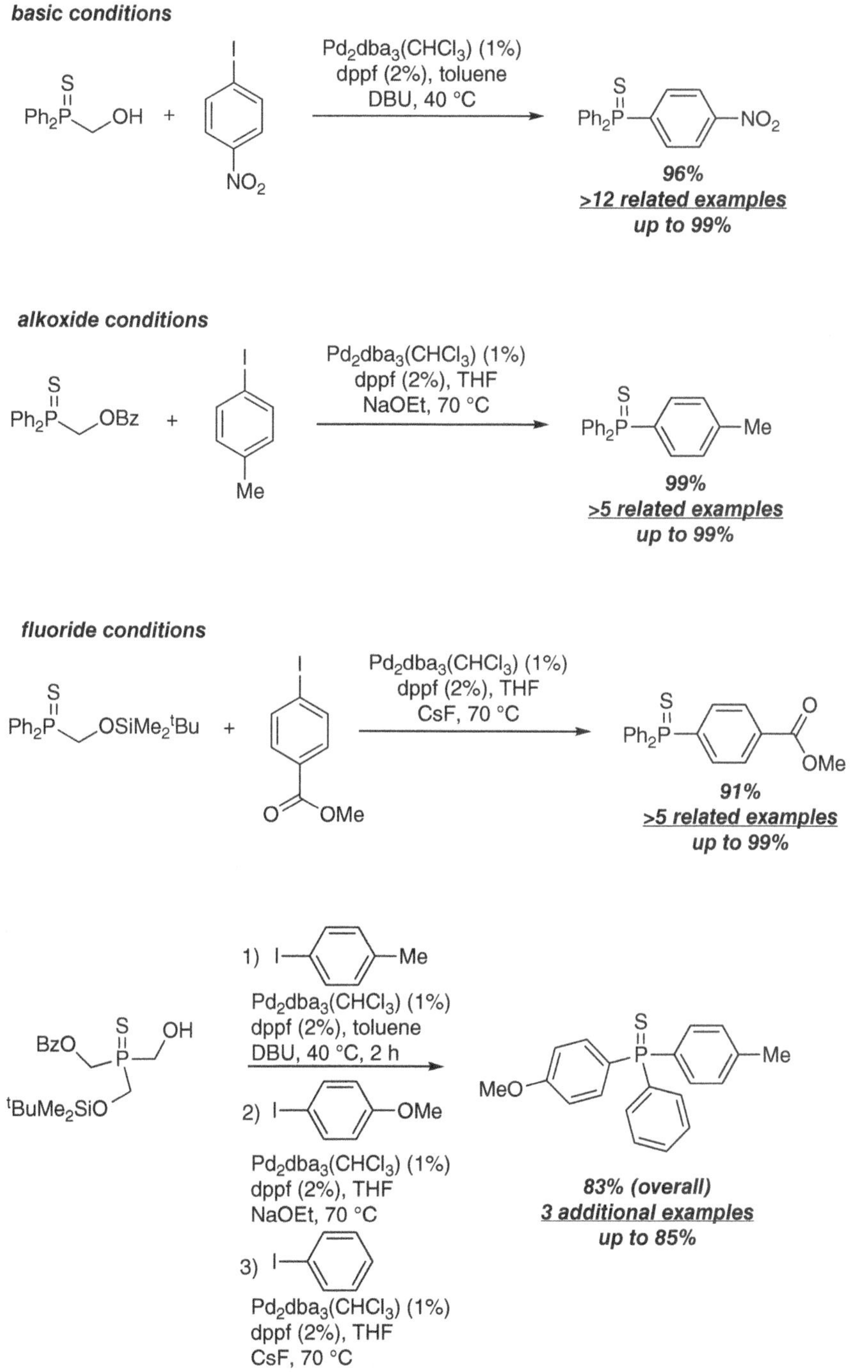

basic conditions

alkoxide conditions

fluoride conditions

SCHEME 4.209   Selective synthesis of unsymmetrical phosphine sulfides [347].

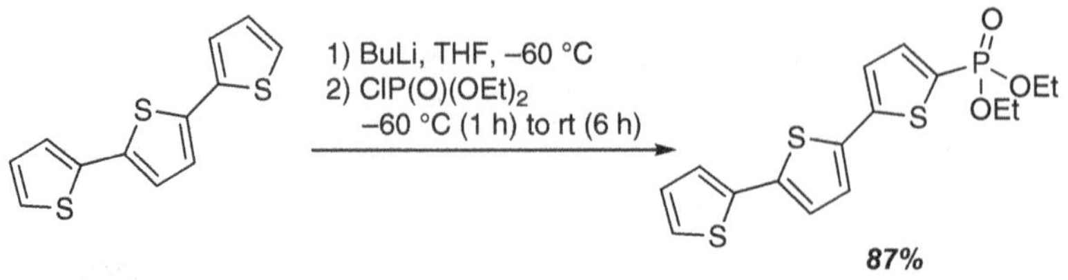

SCHEME 4.210   Synthesis of an arylphosphonate using an ortho-metallation approach [351].

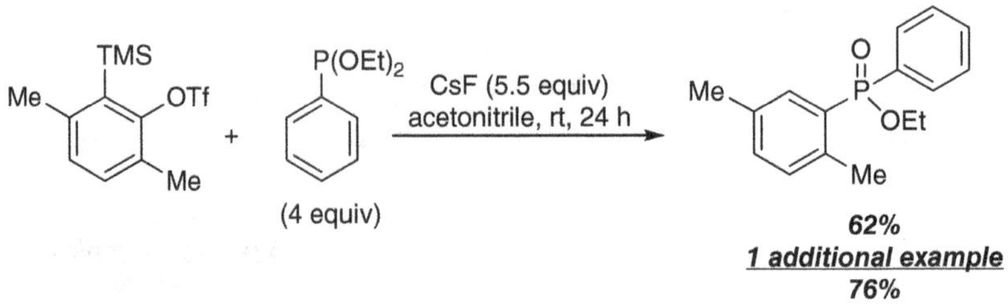

SCHEME 4.211   Synthesis of arylphosphinates through the addition of diethylphenylphosphite to arynes [332].

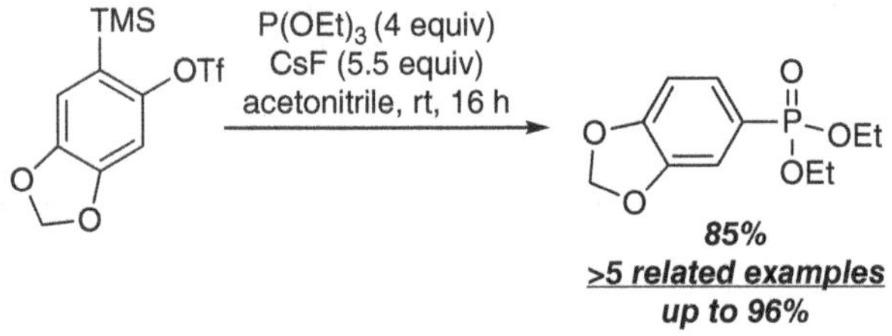

SCHEME 4.212   Synthesis of arylphosphonates through the addition of phosphites to arynes [332].

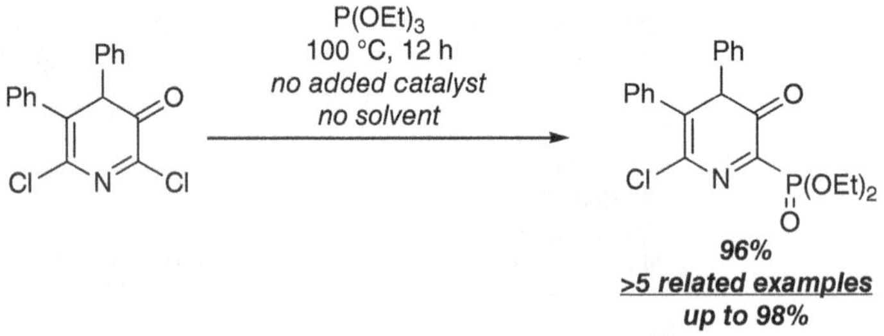

SCHEME 4.213   Metal-free microwave-assisted synthesis of arylphosphonates [352].

reactor was also screened and was able to generate 90% of the arylphosphonate after irradiating the reaction mixture for only 20 min.

Qu and Guo reported that specific positions on purine rings would undergo an Arbuzov-type reaction under solvent-free conditions using a focused microwave reactor (Scheme 4.214) [353]. This chemistry was tolerant of a range of groups at the N9 position

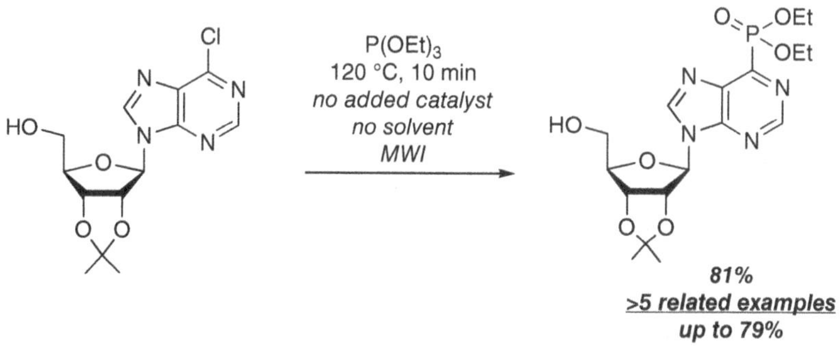

SCHEME 4.214   Metal-free phosphorylation of purines [353].

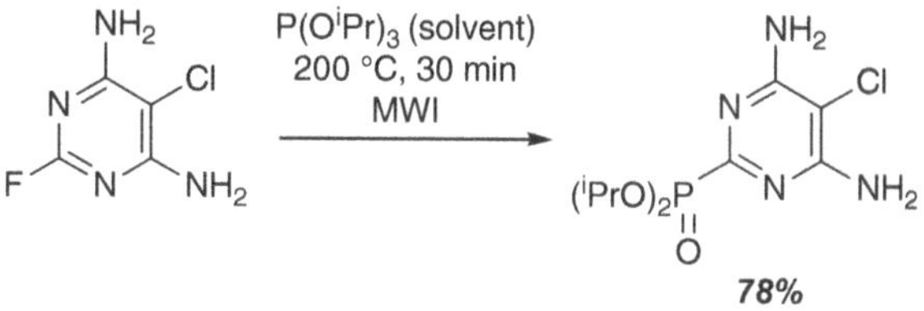

SCHEME 4.215   Chemoselective phosphorylation of mixed halide species [354].

and was even successful in the presence of an unprotected primary alcohol. No trans-esterification products were reported. The authors also screened several substrates that were chlorinated at both the 2- and 6-positions, and only position 6 was susceptible to phosphonylation using this chemistry.

The chemoselectivity of these metal-free Arbuzov-type reactions was investigated by Janeba (Scheme 4.215) [354]. Using a heterocycle that had both chloro and fluoro substituents, the authors found that simply heating the neat reagents using a focused microwave reactor without the addition of a metal catalyst afforded the phosphonylated material in moderate yields. Of interest was the observation that the fluoro group was replaced preferentially to the chloro. In addition to this example, a number of mono- and bisphosphonated compounds were prepared and characterized.

The synthesis of nucleoside-derived pyridylphosphonates under metal-free conditions was reported (Scheme 4.216) [355]. The chemistry was operationally straightforward and entailed treatment of a nucleoside-derived secondary phosphite with pyridine/trityl chloride and DBU to generate the arylphosphonates in excellent yields. It was noteworthy that the reaction was regioselective for substitution at the 4-position of the pyridine moiety. Based upon an investigation of potential intermediates, the mechanism was proposed to proceed through the initial generation of an N-tritylpyridinium ion. Reaction of this species with the secondary phosphite and DBU afforded a dihydropyridine intermediate that could be observed by ^{31}P NMR spectroscopy but could not be isolated as it underwent conversion into the arylphosphonate upon quenching/workup of the reaction. Following this report, a version of this approach was used to prepare a series of 4-pyridylphosphonothioates [356]. While it might have been predicted that the reaction chemistry would not change signifi-cantly when switching from the oxygen to sulfur, running the phosphonylation reaction under similar conditions afforded significant amounts of secondary products. After some

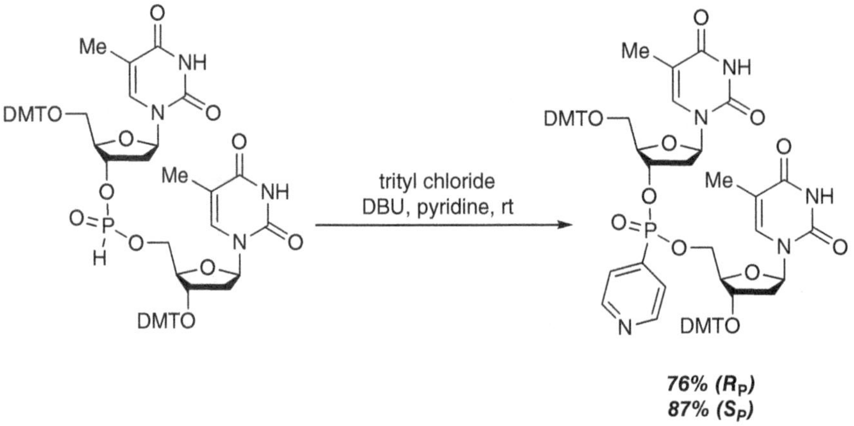

SCHEME 4.216   Metal-free synthesis of pyridylphosphonates [355].

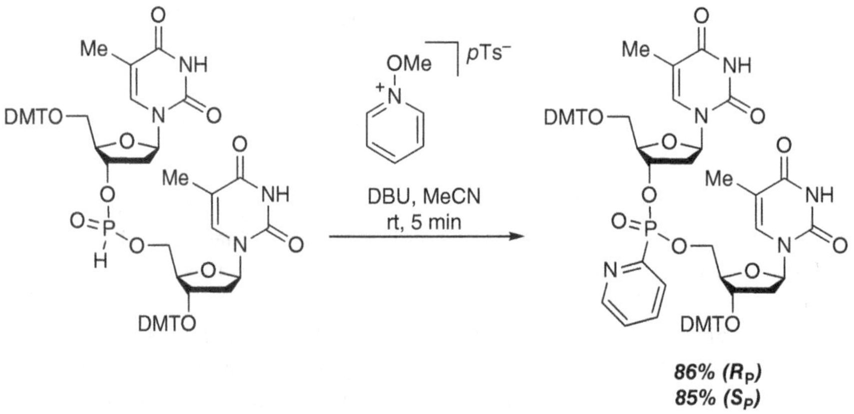

SCHEME 4.217   Preparation of pyridylphosphonates using pyridinium salts [357].

experimentation with the reaction conditions and amounts of reagents, they discovered that the use of a weaker base (Et_3N) generated the dihydropyridine intermediates in excellent yields. In contrast with the 4-pyridylphosphonates described previously, the sulfur analogue was significantly more stable and sluggishly transformed into the arylphosphonate. This reactivity issue was overcome through the use of elemental iodine to accelerate the reaction. Similar to the 4-pyridylphosphonate compounds, the stereochemistry of the starting secondary phosphites was retained through the arylation reaction, and the isolated yields were excellent (77–85%). Overall, this is a very practical and impressive approach for the preparation of pyridylphosphonates under metal-free conditions.

A system for the preparation of nucleoside-derived 2-pyridylphosphonates using a preformed pyridinium salt as the source of the pyridine fragment has been developed (Scheme 4.217) [357]. The base used for the reaction was DBU, and simply stirring the pyridinium salt with the secondary phosphite and DBU in acetonitrile at room temperature afforded the arylphosphonate in excellent yields. The phosphonylation was selective for the 2-position. The mechanism for this transformation was similar to the one proposed previously for the formation of 4-pyridylphosphonates. Initial attack of an *in situ* generated phosphite anion resulted in the formation of a dihydropyridine intermediate that underwent rearomatization along with loss of methoxide to generate the arylphosphonate. While the dihydropyridine could not be observed in these reactions, it was proposed as the likely intermediate. When diastereomerically pure secondary phosphites were used as substrates, the C—P bond-forming reaction occurred with retention of stereochemistry. Excellent yields of the functionalized nucleoside derivative were obtained after purification using standard column chromatography on silica gel. In a subsequent report, this research group was able to extend this chemistry to nucleoside-derived 2-pyridylphosphonothioates [356]. Starting from diastereomerically pure substrates, the arylphosphonate formation proceeded in high yields (79–84%) with retention of the stereochemistry at phosphorus. This metal-free process for the formation of pyridylphosphonates is attractive due to the rapid reactions and outstanding isolated yields of the diastereomerically pure compounds.

In addition to metal-catalyzed cross-coupling reactions, a range of metal-mediated Arbuzov-type reactions have been developed. Of the metals that have been used for this transformation, nickel is the most popular and is cost-effective relative to other group 10 metals such as palladium or platinum. The following sections will describe the advantages and limitations of this approach. One of the earliest accounts of a metal-mediated Arbuzov-type reaction entailed the use of commonly available nickel salts to promote the reaction between a trialkylphosphite and an aryl halide [358]. The reactions were carried out using neat reagents (no solvent) and consisted of simply adding nickel chloride or nickel bromide to the substrates followed by heating to 150–190 °C. These nickel salts were inexpensive, and moderate to excellent yields of the arylphosphonates were common. The typical issues with this chemistry included the excess of phosphite that was often needed for a successful reaction as well as the high reaction temperatures. Furthermore, the concomitant formation of the alkyl halide by-product was not desirable. However, if the substrates are robust, this method is a reliable procedure for the preparation of arylphosphonates and has been used to generate an assortment of arylphosphonates [42, 348–350, 359–363]. A microwave-assisted version of this reaction has been reported [364, 365].

One example of a microwave-assisted version of the nickel-catalyzed Arbuzov reaction was operationally trivial and simply involved irradiating the reagents under solvent-free conditions to generate the arylphosphonates (Scheme 4.218) [365]. The isolated yields of the products were good, and a number of number of fuctionalized compounds were screened.

Another example of a nickel-catalyzed Arbuzov reaction was reported by Teixeira (Scheme 4.219) [366]. The goal of their chemistry was to create new hybrid materials for

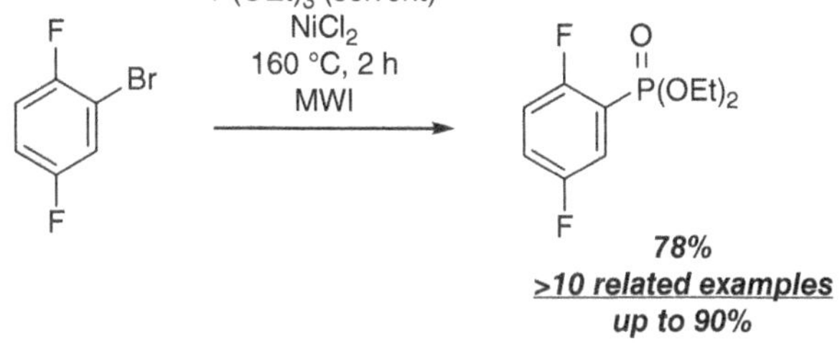

SCHEME 4.218   Nickel-catalyzed synthesis of arylphosphonates from trialkylphosphites [365].

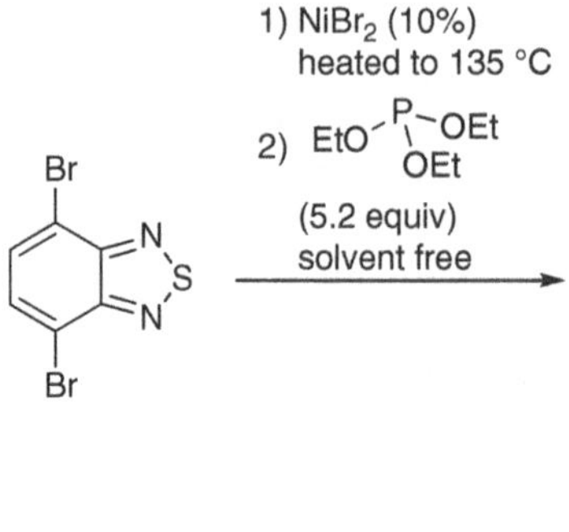

SCHEME 4.219   Nickel bromide-catalyzed synthesis of arylphosphonates [366].

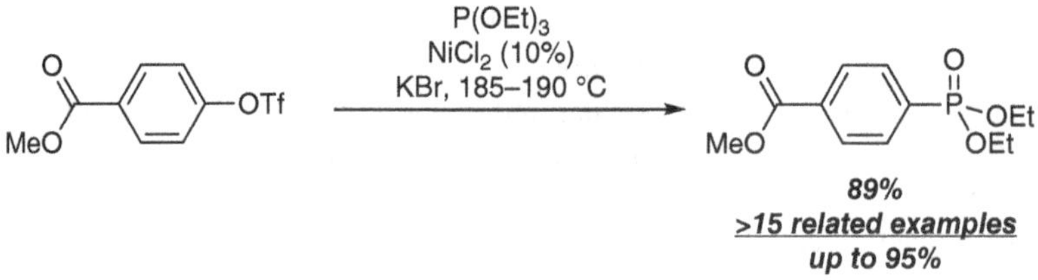

SCHEME 4.220   Nickel chloride-promoted synthesis of arylphosphonates using KBr as an additive [367].

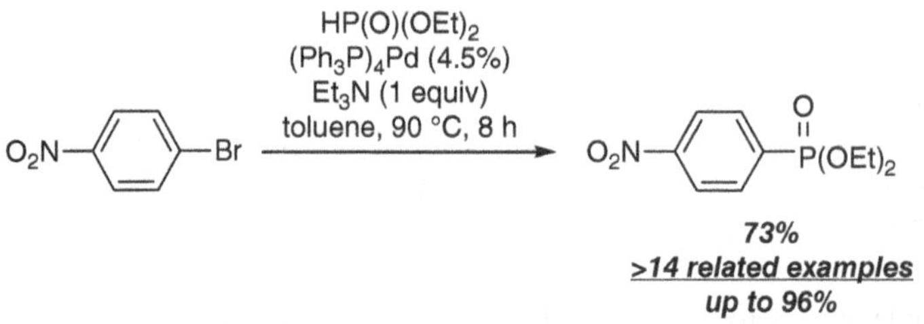

SCHEME 4.221   Classic Hirao coupling [368, 369].

full cells using a linked bisphosphonate. The authors screened a number of transition metals in order to generate the target species and found that simply adding an excess of triethylphosphite dropwise to a heated vessel containing the arylhalide and nickel bromide generated the target in excellent yield.

One of the first uses of aryl triflates in an metal-mediated Arbuzov-type reaction was reported by Zhang. He used nickel chloride as the metal catalyst and various additives to promote the formation of arylphosphonates from triethylphosphite (Scheme 4.220) [367]. While a range of additives such as LiBr and KI promoted the formation of the arylphosphonate, KBr was found to be superior. Most of the substrates screened in this work were hydrocarbon based; however, a number of examples contained functional groups such as an ester, acetate, and protected amines.

A significant breakthrough in the synthesis of arylphosphonates was developed by Hirao (Scheme 4.221) [368, 369]. He developed a palladium-catalyzed cross-coupling

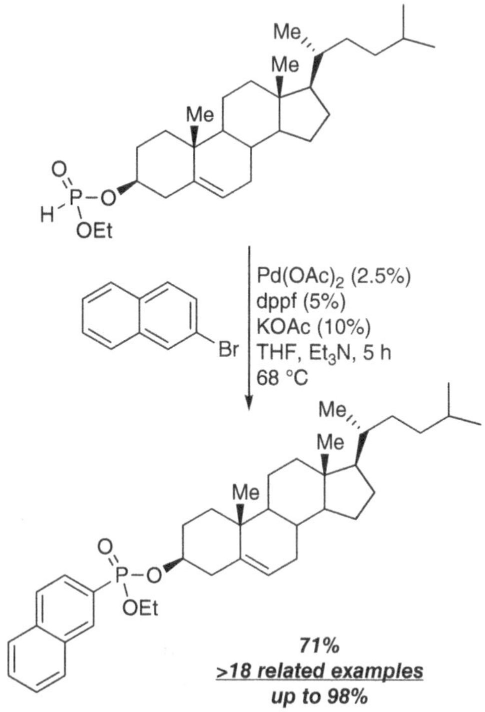

SCHEME 4.222   Palladium-catalyzed synthesis of arylphosphonates using potassium acetate as an additive [377].

reaction for the formation of arylphosphonates using a common palladium(0) catalyst $((Ph_3P)_4Pd)$ and triethylamine as the base. Instead of using trialkylphosphites (Arbuzov chemistry described previously) Hirao opted to use secondary phosphites for the coupling reaction. The temperatures required for a successful reaction were significantly lower than what was typically used in Arbuzov chemistry, and moderate to excellent yields of the targets were obtained. This chemistry displayed fairly broad functional group tolerance and was most effective with aryl bromides bearing electron-withdrawing groups. A host of mechanistic studies have appeared over the years describing refinements and advances to the classic palladium-catalyzed Hirao coupling reaction [370–376]. The overarching goals of these refinements were focused on (i) increasing the catalyst efficiency and lowering the catalyst loading, (ii) increasing the functional group compatibility, and (iii) lowering the reaction temperature.

A considerable amount of effort has been devoted to determining the effect of additives on metal-catalyzed cross-coupling reactions. Anionic species such as acetate have received a significant amount of interest due to observations that they were able to accelerate palladium-catalyzed versions of these reactions [371, 375]. To determine if this chemistry could be beneficial to the synthesis of arylphosphonates, Stawinski has investigated the effect of adding various acetate sources to palladium-catalyzed C—P bond-forming reactions (Scheme 4.222) [377, 378]. After screening a range of acetate sources, potassium acetate was selected for further studies. The catalyst system used in these studies was comprised of palladium acetate, dppf as the supporting ligand, and

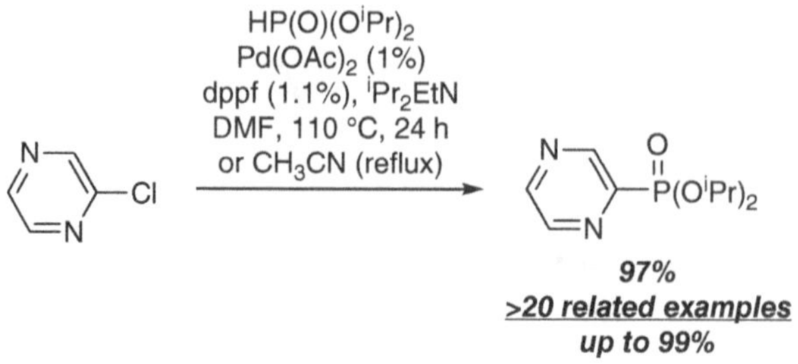

SCHEME 4.223 Synthesis of arylphosphonates from aryl chlorides using DMF as the solvent [379].

potassium acetate. During the course of the investigation, they also determined that the most active catalyst was generated by heating the components prior to addition of the secondary phosphite and arylhalide. Once an effective system was found, it was successfully extended to a range of substrates incorporating both electron-donating and withdrawing groups. While no significant electronic limitations were noted, the chemistry was sensitive to the presence of ortho-substituents on the aryl halide. Even with extended heating, incorporating a single methyl group into the ortho-position significantly reduced the amount of arylphosphonate formed in the reaction. In addition to common arylhalides, several nucleoside derivatives and a steroid were also functionalized using this approach. The authors carried out selective cross-coupling reactions using 4-bromoiodobenzene and related compounds as substrates and found that the iodo group was selectively replaced during the process. Through the screening of 2,4-dibromopyridine, they discovered that the bromine in the 2-position was preferentially functionalized. These selective coupling reactions are very attractive for the design of complex organic substrates as the remaining arylhalide can be transformed in subsequent reactions.

Montchamp has carried out a detailed analysis of palladium source, additives, solvents, and supporting ligands (Scheme 4.223) [379]. While the classic Hirao coupling reaction used $(Ph_3P)_4Pd$ and triethylamine as the base, Montchamp devised a superior system comprised of palladium acetate, dppf, and $^iPr_2EtN$. Palladium acetate is a readily available precursor for Pd(0) compounds, and dppf has been shown to increase the reductive elimination step (relative to triphenylphosphine) in related compounds. Using this system, he was able to decrease the palladium catalyst loading from 5% down to 1% and the supporting ligand from 4 equiv per palladium center down to 1.1 equiv per palladium center. Palladium is one of the more expensive metals, and reducing the amount needed to promote a reaction can be a very practical advantage when working with multigram scales. Heating the reaction mixture to 110 °C in DMF for 24 h or refluxing in acetonitrile afforded excellent yields of the arylphosphonates. In terms of the substrate scope, his system tolerated a wide range of functional groups including phenols, heteroaromatic compounds, and unprotected aniline derivatives. Perhaps, the most appealing aspect of the chemistry was the ability to couple troublesome aryl chlorides.

Montchamp followed up this work with the development of a catalyst system for the cross-coupling of hydrogen phosphinate esters with arylhalides (Scheme 4.224) [380]. While several aryl halides were screened in this chemistry, aryl chlorides were the focus of the investigation as they are typically sluggish in Hirao-type cross-coupling reactions.

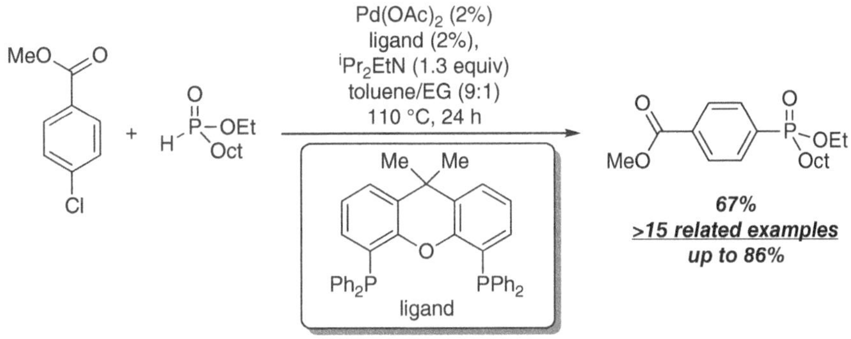

SCHEME 4.224 Palladium-catalyzed cross-coupling of aryl chlorides with hydrogen phosphinate esters [380].

Using the same catalyst system the authors devised for coupling reactions using hydrogen phosphonates, poor to moderate yields of the product was generated. After some experimentation, the authors discovered that increasing the catalyst loading to 2% and moving to a wide bite-angle diphosphine (XantPhos) successfully coupled a wide range of aryl chlorides with the HP(O)(OR)R species. The scope of the chemistry was broad, and a range of electron-donating/withdrawing groups as well as heteroaromatic substrates were well tolerated by the catalyst system. In an effort to probe the effectiveness of microwave irradiation on the reaction, screening reactions were performed, and the results mirrored the conventionally heated reactions. Given the greater abundance of aryl chlorides relative to the iodides and bromides, this is a very practical approach for the formation of arylphosphonates.

Following Montchamps cross-coupling work with aryl chlorides and phosphinates, Yang and Wu reported a similar reaction using a preformed palladium palladacyle as the catalyst and secondary phosphites as the phosphorus nucleophiles (Schemes 4.225 and 4.226) [109]. Electron-rich aryl chlorides tend to be sluggish in metal-mediated coupling reactions; however, using the palladacycle as the palladium source, these troublesome substrates were coupled with diisopropylphosphite in moderate to excellent yields in 3 h at 130 °C. In addition to aryl chlorides, aryl bromides were also screened and were predictably found to be more active in the cross-coupling reaction. As a result of the higher reactivity of the aryl bromides, the extra ligand was able to be eliminated from the reaction, and the reaction temperature and time were able to be further decreased. A range of functional groups including an unprotected carboxylic acid was successfully coupled using this palladacycle as the catalyst. Although this appears to be a highly active catalyst that is able to cross-couple problematic aryl chlorides, an issue with this chemistry is the requisite synthesis of the palladacycle. Following their report on the use of palladacycles as catalysts for C—P bond-forming reactions, the same authors found that they could use these catalysts to promote the formation of arylphosphonates in neat water [340]. Once they switched to water as the solvent, they needed to slightly modify the catalyst system in order to optimize the coupling reaction. They replaced the potassium carbonate base with potassium fluoride and added TBAB to the reaction mixture. One of the issues encountered when trying to use secondary phosphites in water is decomposition. The authors screened a range of additives and found that simply adding isopropanol to the reaction mixture

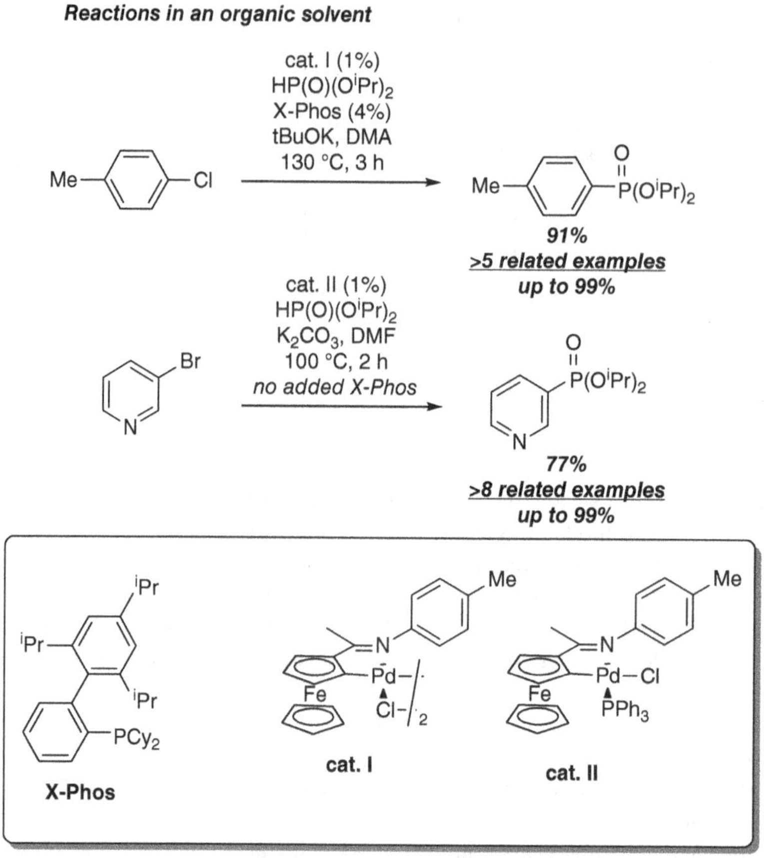

SCHEME 4.225 Synthesis of arylphosphonates in an organic solvent using a palladacycle-based catalyst [109].

effectively minimized the decomposition process and high yields of the cross-coupling products were obtained. Similar to their findings in organic solvents, the palladium-catalyzed process was tolerant of a range of functional groups on the aryl chloride and was even successful using substrates with ortho-substituents. Aryl bromides and iodides could also be successfully phosphonated in water using a similar catalyst system.

A microwave-assisted version of the Hirao coupling has been described (Schemes 4.227 and 4.228) [381]. Using a robust and readily available palladium catalyst, Stawinski coupled a range of arylhalides with secondary phosphites to generate arylphosphonates in high yields. He found that simply irradiating the reaction mixture for 10 min at 120 °C afforded excellent yields of the arylphosphonates. He also reported a similar example of the phosphonation chemistry using conventional heating and found that they needed to reflux the reaction mixture in THF for 4–5 h in order to achieve significant conversions [382]. Once the microwave-assisted procedure was screened with common aryl halides, the authors successfully extended the chemistry to include P(O)–H donors bearing steroidal and nucleoside fragments. They found that the use of a Cs_2CO_3 as the base for the coupling of simple arenes substrates afforded higher yields of the arylphosphonate; however, when the nucleoside-containing hydrogen phosphonate was used, triethylamine

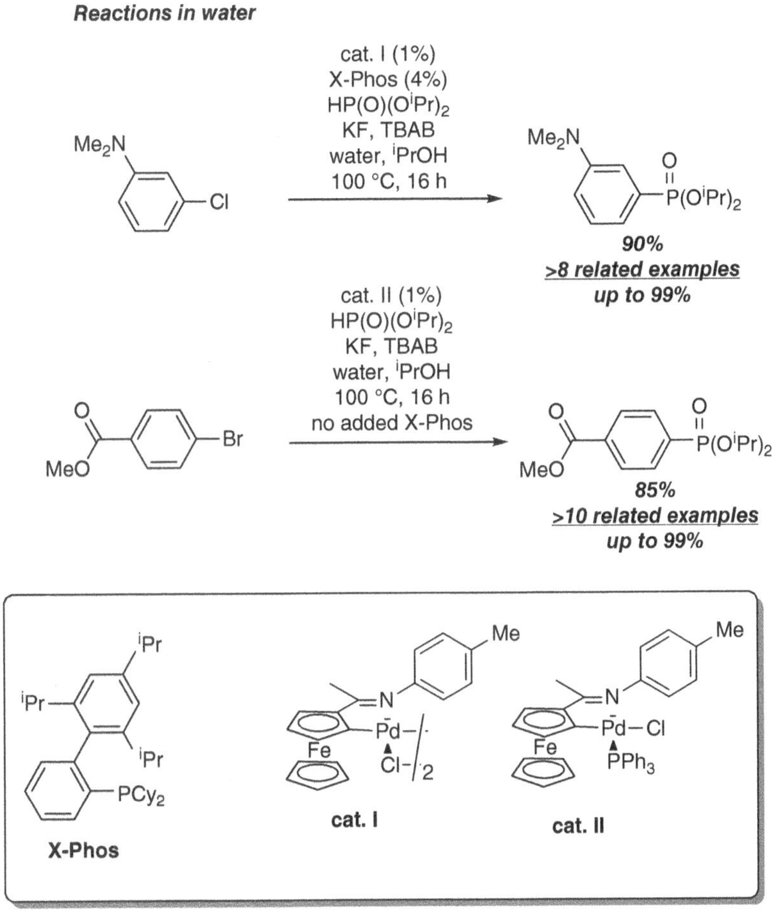

Reactions in water

SCHEME 4.226   Synthesis of arylphosphonates in water using a palladacycle-based catalyst [340].

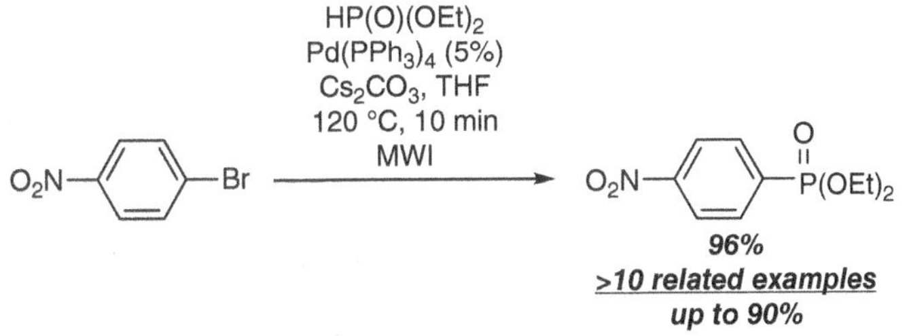

SCHEME 4.227   Microwave-assisted palladium-catalyzed synthesis of arylphosphonates [381].

was a superior base. Furthermore, the stereochemistry of the nucleoside derivative was not scrambled during the functionalization process. This racemization was often observed in similar reactions using conventional heating [382–384]. Given the speed and high yields of this reaction, it could be the synthetic method of choice for the synthesis of

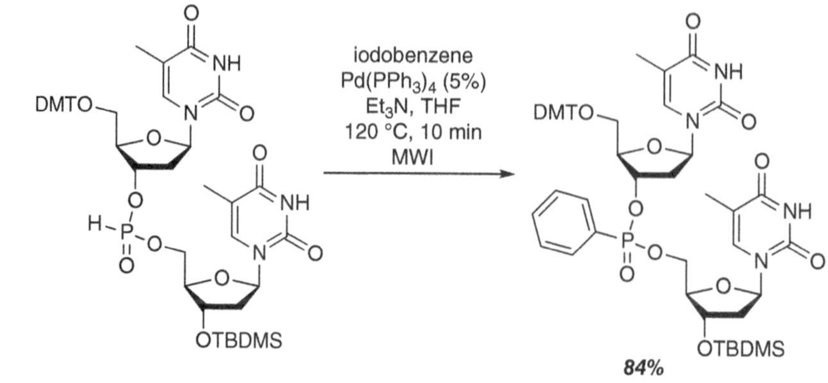

SCHEME 4.228   Preparation of a nucleoside-derived arylphosphonate [381].

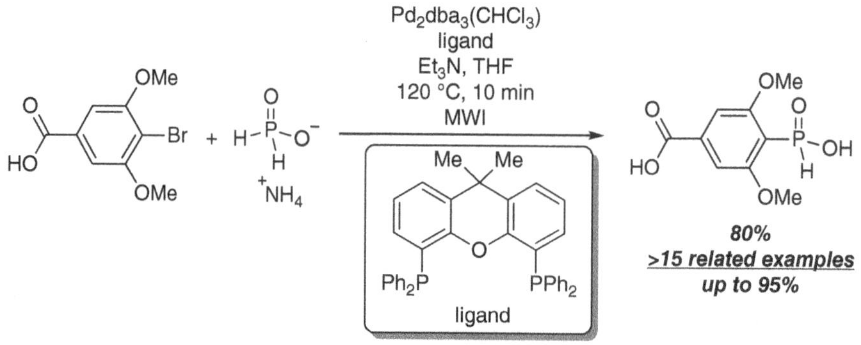

SCHEME 4.229   Synthesis of crowded arylphosphonates [387].

arylphosphonates. Following this report, a microwave-assisted synthesis of acyclic nucleoside phosphonates was reported [385]. A similar method (conventional heating) was used to generate dinucleoside pyridylphosphonates [382, 386].

The microwave-assisted synthesis of arylphosphinic acids has been described (Schemes 4.229 and 4.230) [387]. This palladium-catalyzed cross-coupling reaction used a Pd(0) precursor along with XantPhos as the supporting ligand. One of the attractive aspects of this chemistry was the low palladium loading. Most of the coupling reactions used 0.1% of the catalyst and only one equivalent of the supporting ligand per metal center. A wide range of preexisting functional groups were well tolerated by this approach. Furthermore, bulky aryl halides including 2,6-disubstituted aryl halides were successfully phosphonated. This approach could also be adapted for the synthesis of unsymmetrical diarylphosphinic acids. By performing the cross-coupling reaction twice, two different aryl groups could be added to the phosphorus center to generate the unsymmetrical compound. Although it would be attractive to carry out the addition of both groups in a one-pot reaction, best results were obtained when one of the aryl groups was added followed by an ethyl acetate extraction. The second group was subsequently added. Due to the simplicity of this approach, wide steric, electronic, and functional group compatibility, this is one of the most attractive and practical syntheses of symmetrical and unsymmetrical diarylphosphinic acids.

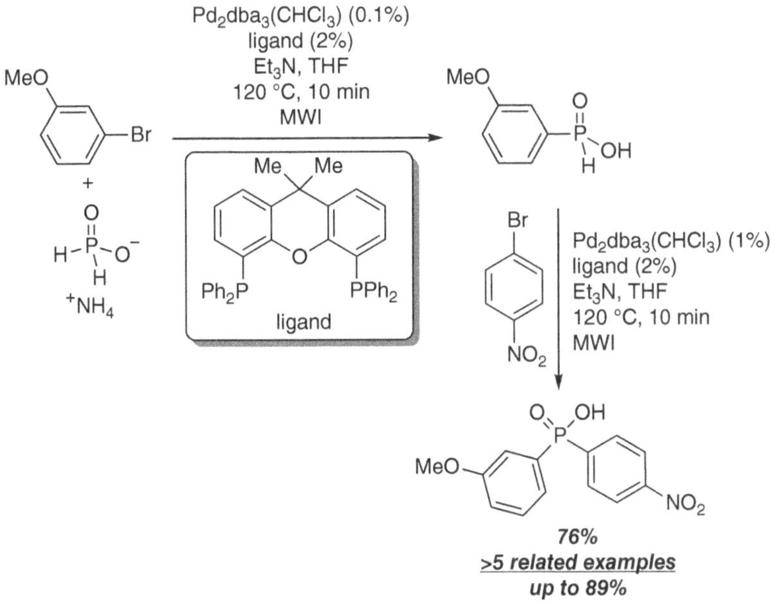

SCHEME 4.230   Synthesis of multiple aryl groups through sequential functionalizations [387].

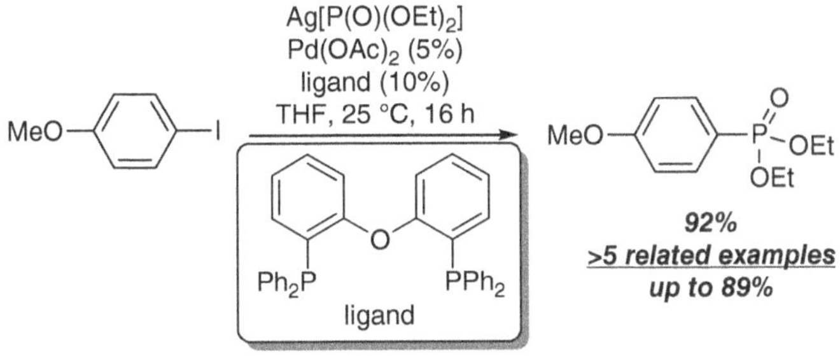

SCHEME 4.231   Synthesis of arylphosphonates at low temperatures [388].

A base was typically needed for the vast majority of Hirao-type coupling reactions. Thus, using this chemistry to add phosphonate fragments was challenging for substrates containing base-sensitive groups. These groups could be protected/deprotected, but this adds several steps to the overall process. The development of a base-free coupling reaction would remove these extra steps. There have been few solutions to this problem. One of the first examples of a base-free process entailed the use of a silver phosphonate (Scheme 4.231) [388]. This reagent does not require a base to deprotonate the a $HP(O)(OR)_2$ reagent since the phosphonate group was transferred to palladium through transmetallation from the silver salt. A range of functionalized aryl iodides including a phenol served as substrates for the phosphorylation chemistry, and moderate to excellent yields of the arylphosphonates were reported. While a range of electron-donating and withdrawing groups were well tolerated by this chemistry, there were steric limits on this reaction. While adding a single ortho-substituent did not significantly affect the yield of this reaction, adding two groups

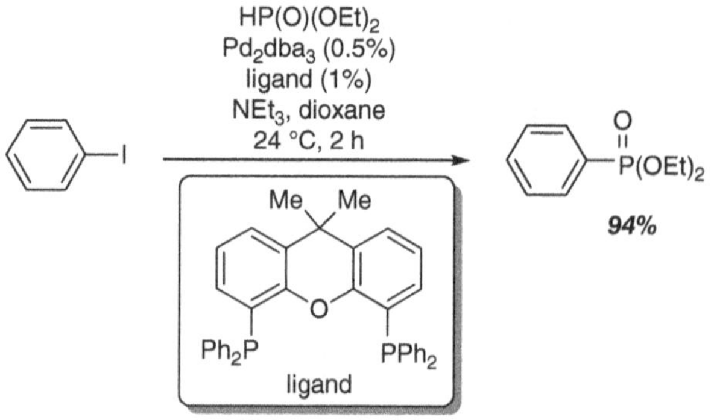

SCHEME 4.232 Low-temperature coupling of aryl halides with secondary phosphites [389].

completely suppressed the cross-coupling reaction. One drawback to this chemistry was that the silver salts used to transfer the phosphonate fragment were not thermally stable. Best results were obtained when these reagents were freshly prepared. This chemistry also addressed the question of whether or not the free phosphite typically found in Hirao-type couplings could inhibit the reaction by coordinating to the metal center and poisoning the catalyst. In this example, there is no free phosphite present, and the reaction was one of the first examples of a room-temperature Hirao-type coupling reaction. Historically, most palladium-catalyzed P—C bond-forming reactions were carried out at high temperatures. Thus, the silver phosphonate chemistry was amenable for the use of base-sensitive as well as temperature-sensitive substrates.

As mentioned previously, the vast majority of Hirao and related reactions were carried out at high temperatures. Not only can this cause substrate decomposition, but it can racemize chiral species. Due to the plethora of temperature-sensitive substrates used in modern synthesis, the development of a low-temperature version of this valuable reaction would be of great interest to the synthetic community. As mentioned previously, one of the first examples of such reaction entailed the use of a silver phosphonate in place of the secondary phosphites; however, this chemistry required the synthesis and isolation of the silver salt. Another solution to this issue was reported by Herzon (Scheme 4.232) [389]. Using Pd$_2$dba$_3$ as the palladium source and the wide bite-angle diphosphine XantPhos as the supporting ligand, they were able to develop a rapid room-temperature phosphonation of iodoarenes. While only a single example of an arylphosphonate was given, this catalyst system has the potential to have great functional group tolerance due to the broad functional group compatibility typically displayed in related Hirao coupling reactions. Furthermore, the authors were able to decrease the catalyst loading to 1% palladium with one equivalent of the supporting ligand, making this even more attractive for large-scale preparations. Similar to the Pd/silver phosphonate system discussed previously, this room-temperature coupling reaction is limited to iodoarenes.

A version of the previous palladium-catalyzed approach to the synthesis of arylphosphonates has been reported using aryl imidazolylsulfonates and secondary phosphites as substrates (Scheme 4.233) [390]. While it might not be immediately thought of as a very practical approach, the imidazolylsulfonates are easy to prepare and are relatively stable. The authors screened a variety of supporting ligands and found that dppp was the most effective. Curiously, the catalyst system comprised of XantPhos and palladium acetate

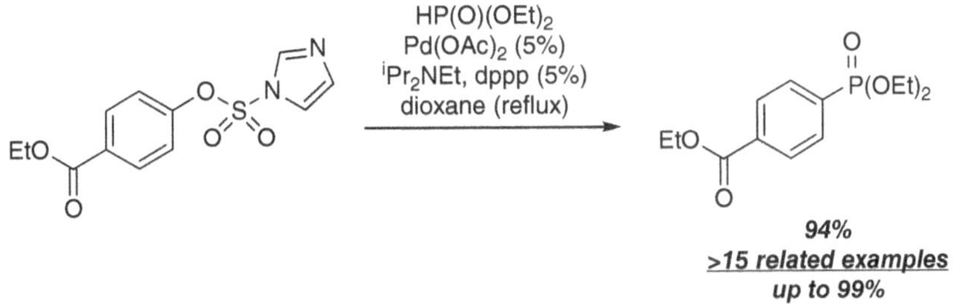

SCHEME 4.233   Synthesis of arylphosphonates from aryl imidazolylsulfonates [390].

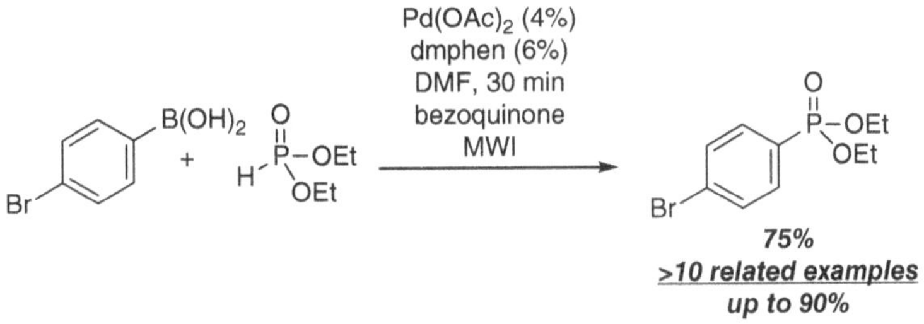

SCHEME 4.234   Palladium-catalyzed coupling of arylboronic acids with secondary phosphites [391].

only afforded trace amounts of the arylphosphonate. Several mineral bases such as Cs_2CO_3 afforded poor to fair yields of the target molecules, while a sterically hindered amine consistently generated higher yields of the arylphosphonates. Using the optimized system, a range of functionalized aryl imidazolylsulfonates bearing esters, ketones, ortho-substituents, and ethers were readily transformed into arylphosphonates in excellent yields.

While several microwave-assisted routes to the formation of arylphosphonates have been discussed previously, a modification of this approach was reported by Larhed (Scheme 4.234) [391]. His catalyst system consisted of a common palladium salt, supporting ligands, and additives. While this might not initially read like something different, the authors reported a catalytic cycle where an oxidizing agent was proposed to assist in the formation of palladium phosphonate adducts. The majority of the palladium-catalyzed coupling reactions use phosphine ligands; however, for reactions carried out under an atmosphere of air and oxidizing conditions, phosphine oxidation remains a significant problem. In order to circumvent this issue, the authors opted to use 2,9-dimethyl-1,10-phenanthroline (dmphen) as the ligand for the palladium center. One of the arguments that the authors used as support for the unusual palladium catalysis was the observation that a base was not needed for a successful reaction. In addition to being a mechanistic probe, the development of a base-free reaction will enable the use of base-sensitive substrates. A range of functional groups were compatible with the system including aryl halides, esters, and acetates. The reaction was operationally trivial as well. All the reagents were loaded into the reaction vial under an atmosphere of air, and the reaction vessel was capped under air. No glovebox or inert atmosphere manifold was needed

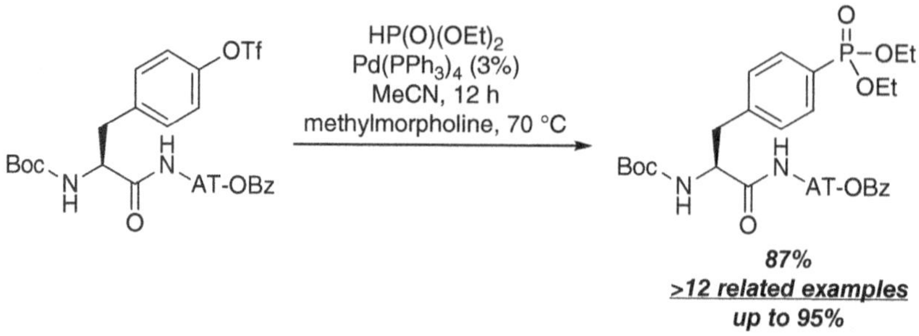

SCHEME 4.235   Palladium-catalyzed synthesis of arylphosphonates from aryl triflates [392].

for this chemistry. One prerequisite to this chemistry is that none of the substrates could contain functional groups that were sensitive to benzoquinone oxidation. The high yields of the arylphosphonates, the speed of the reactions, as well as the ability to carry the reaction out under an atmosphere of air makes this an extremely practical synthesis of arylphosphonates.

While most of the examples discussed previously used aryl halides as coupling partners for the preparation of arylphosphonates, aryl triflates were also effective substrates for C—P bond-forming reactions (Scheme 4.235) [392]. During an investigation into the design and synthesis of protein kinase inhibitors, Petrakis developed a protocol for the conversion of aryl triflates into arylphosphonates. Using a single-component palladium catalyst and methylmorpholine as a base, a range of aryl halides as well several small peptides were successfully functionalized. The chemistry was compatible with both electron-donating and electron-withdrawing groups, and while many of the metal-catalyzed approaches to the synthesis of arylphosphonates were shown to be less effective with bulky substrates, the use of aryl triflates bearing a single ortho-substituent was successfully phosphonated.

In related work, the palladium-catalyzed coupling of a progesterone derivative with secondary phosphites generated the arylphosphonate in moderate yield (Scheme 4.236) [335]. Palladium acetate was an effective catalyst for the reaction, and dppb served as an effective supporting ligand for the palladium center. Using this catalyst system, moderate yields of the arylphosphonate were obtained. While heating the reaction mixture using conventional heating did generate the desired compound, a microwave-assisted version of the reaction was higher yielding and cleaner.

An interesting variation of the classic Hirao coupling reaction entailed the use of ethanol as the solvent and dicyclohexylmethylamine as the base (Scheme 4.237) [393]. Although flammable, ethanol is an attractive solvent for the development of environmentally benign synthesis. By matching the ethanol with the organic fragments on the secondary phosphite, the authors were able to mitigate any transesterification reactions. The catalyst loading was relatively low in these reactions, and the chemistry displayed fairly broad functional group compatibility.

The use of arenediazonium tetrafluoroborates as substrates for palladium-catalyzed cross-coupling reactions generates arylphosphonates in moderate to excellent yields (Schemes 4.238 and 4.239) [292]. While most palladium-catalyzed C—P bond-forming reactions use secondary phosphites as substrates, this approach uses trialkylphosphites. The chemistry displayed excellent functional group compatibility as arenes bearing esters,

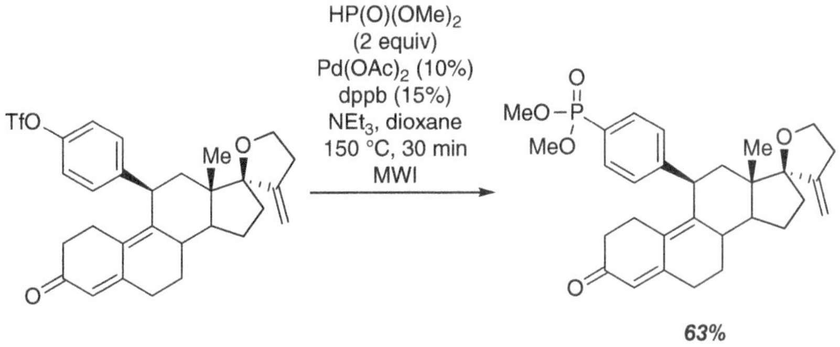

SCHEME 4.236 Microwave-assisted synthesis of a progesterone-derived arylphosphonate [335].

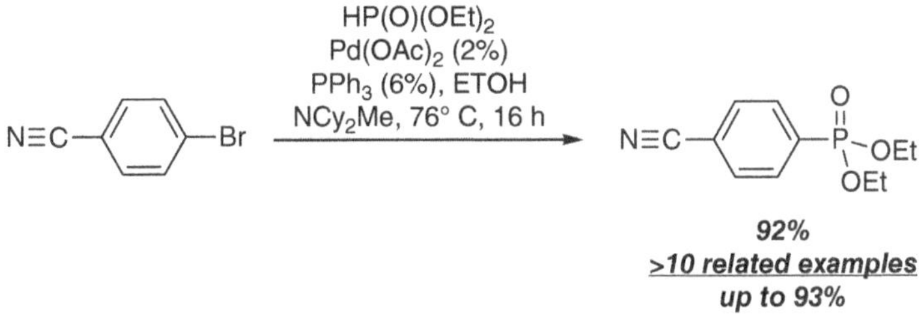

SCHEME 4.237 Palladium-catalyzed synthesis of arylphosphonates in ethanol [393].

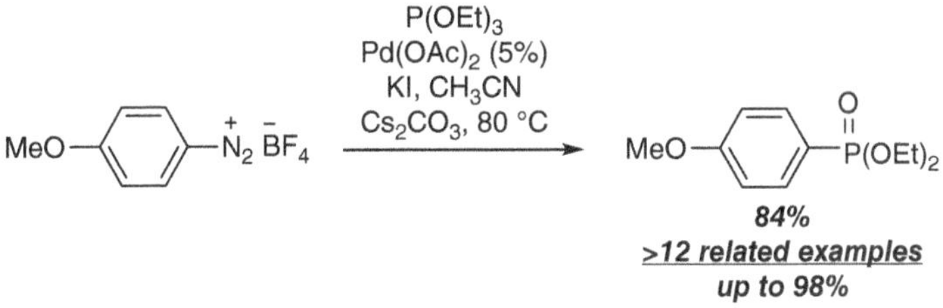

SCHEME 4.238 Conversion of arenediazonium salts into arylphosphonates [292].

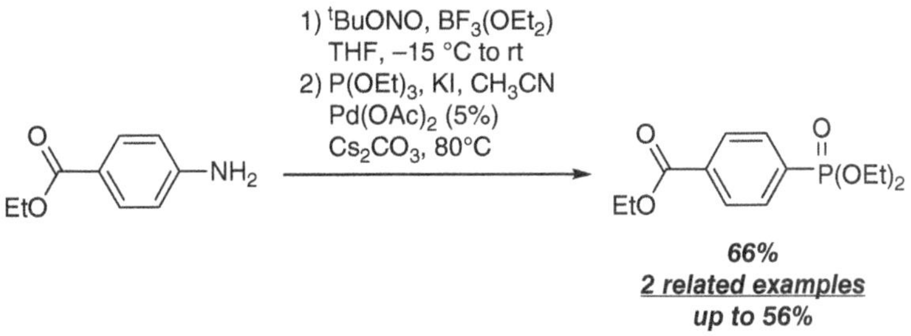

SCHEME 4.239 One-pot conversion of anilines into arylphosphonates through the formation of arenediazonium salts [292].

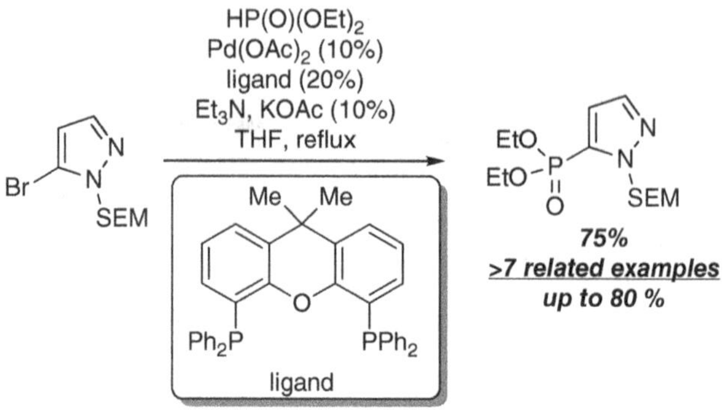

SCHEME 4.240   Palladium-catalyzed synthesis of phosphonylated heterocycles [345].

ethers, halogens, nitriles, and nitro were successfully cross-coupled. No additional supporting ligand was needed for the generation of an active catalyst. This was presumably due to the ability of the phosphite to solubilize/stabilize to the palladium center. The authors presented an elegant one-pot synthesis of arylphosphonates from anilines using an *in situ* generation of the diazonium species followed by addition of the phosphite and catalyst. Given the host of aniline derivatives that are readily available, this has the potential to be a very practical solution to the synthesis of arylphosphonates.

A cross-coupling approach to the synthesis of phosphonylated heteroaromatic compounds has been reported (Scheme 4.240) [345]. The phosphorus components were secondary phosphites, and 5-iodo and bromopyrazoles served as the heterocycle reagents. After some experimentation, the authors settled on a palladium-catalyzed Hirao-type coupling reaction. While several acyclic and cyclic secondary phosphites were successfully added to the pyrazole, diallylphosphite did not generate the cross-coupling product. The yields of the coupling reaction were substrate dependent, and some examples were challenging to rationalize. While the focus of the work was devoted to challenging 5-halogenated pyrazoles, the authors also screened 3- and 4-iodopyrazoles for activity. These compounds were more active and afforded excellent yields of the arylphosphonates (83–87%).

In terms of operational simplicity, one of the most significant advances entailed the use of copper catalysts for the synthesis of arylphosphonates. Many copper catalysts do not need to be stored in gloveboxes, and the common supporting ligands used for the solubilization of the copper sources tend to be air stable. A number of reports have detailed the use of copper catalysts to promote the formation of C—P bonds [394–395].

A copper-catalyzed cross-coupling reaction of aryl halides and hydrogen phosphonates has been used to generate arylphosphonates (Scheme 4.241) [318]. The catalyst system comprised of CuI and *N,N*-dimethylethylenediamine afforded the highest yields of the arylphosphonates. The most effective base for this chemistry was found to be cesium carbonate. This approach has the potential to be significantly more tolerant to functional groups than the lithiation/phosphonation approach, and indeed, the authors reported a successful cross-coupling between dibutylphosphite and 2-iodoaniline without protection of the amine. This approach was also successful with ortho-substituted aryl halides. Care must be taken when using arylhalides bearing ketone-based functional groups. A significant amount of the hydroxyphosphonate product can be formed due to addition of the phosphorus nucleophile to the carbonyl group.

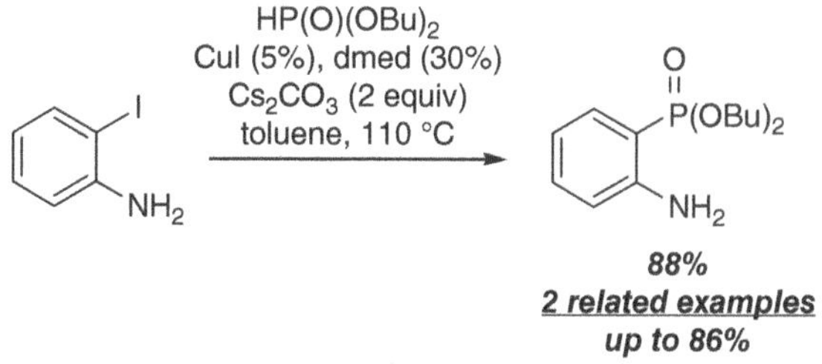

SCHEME 4.241 Copper-catalyzed synthesis of arylphosphonates [318].

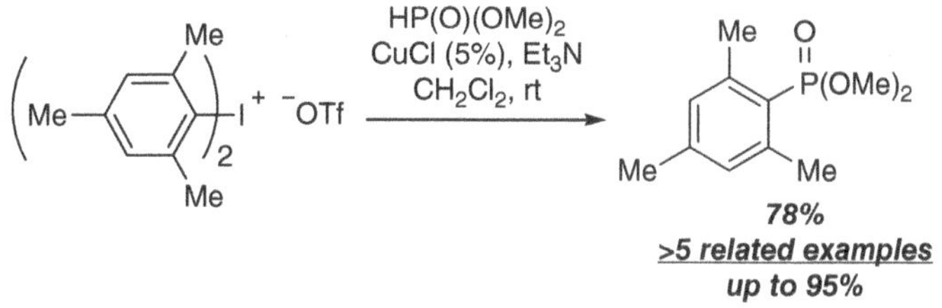

SCHEME 4.242 Using iodonium salts in the copper-catalyzed synthesis of arylphosphonates [333].

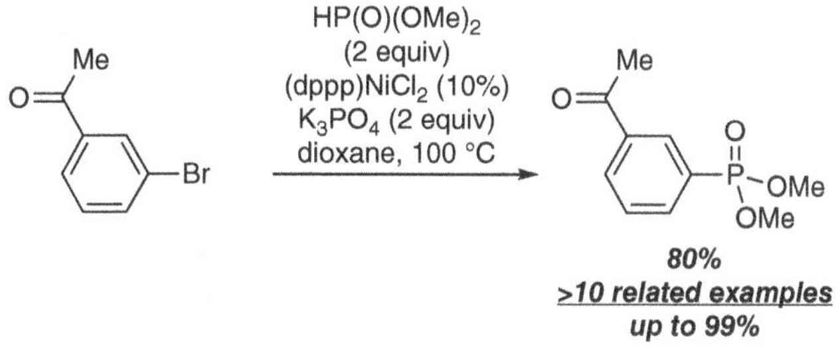

SCHEME 4.243 Nickel-catalyzed coupling of secondary phosphites with aryl halides [336].

While most of the palladium- and copper-catalyzed routes to arylphosphonates were sluggish with crowded electrophiles, the use of iodonium salts has circumvented this limitation (Scheme 4.242) [333]. The diaryliodonium salts were easily prepared, and excellent yields of the arylphosphonates were reported using copper chloride as the catalyst for the coupling reaction. In contrast to other reports of copper-catalyzed coupling reactions, no solubilizing ligands were needed to promote a successful reaction. The speed of this approach was another practical advantage (10 min at room temperature). In addition to the hydrogen phosphonates screened in this reaction, a single example of a phosphinate (HP(O)(OEt)Ph was reported (94% conversion).

A nickel-catalyzed route to the formation of arylphosphonates starting from aryl halides and secondary phosphites has been reported (Scheme 4.243) [336]. While some of the methods described previously added nickel salts such as $NiCl_2$ to the reaction mixture along with a ligand to solubilize/stabilize the metal center, the authors opted to use a preformed nickel complex. A significant electronic effect was observed in this

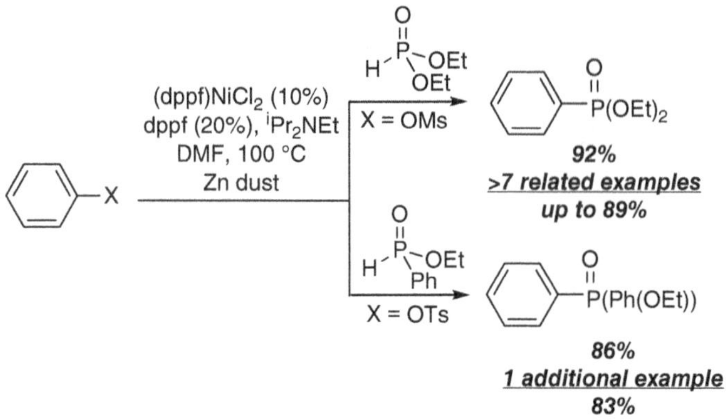

SCHEME 4.244   Synthesis of arylphosphonates through nickel-catalyzed cross-coupling [341].

chemistry and lower yields of the arylphosphonates were obtained when electron-rich aryl halides were used. There was also a steric issue with the cross-coupling reaction, and substrates bearing a single ortho-substituent were sluggish. No examples with multiple ortho-substituents were reported. The scope of the chemistry was relatively broad and functional groups such as ketones, esters, and nitriles were well tolerated. Aryl chlorides were also successfully coupled with the secondary phosphite in moderate to excellent yields (95%).

Aryl tosylates and mesylates have also been used as substrates in the nickel-catalyzed synthesis of arylphosphonates (Scheme 4.244) [341]. These substrates are typically more stable than aryl triflates and are easily prepared. Given that the immediate precursors of aryl tosylates are phenols, this chemistry has the possibility of having a wide scope due to the plethora of phenols that are known. The catalyst system used in this work was comprised of a preformed nickel complex along with zinc dust as a reducing agent. The zinc dust was essential to the synthesis, and no consumption of the starting aryl mesylates was observed in its absence. Through a series of screening experiments, the authors determined that a small amount of added ligand significantly promoted the reaction. In related work, phenols were converted into arylphosphonates through a nickel-catalyzed cross-coupling reaction using bromotripyrrolidinophosphonium hexafluorophosphate (PyBroP) as an activating agent (Schemes 4.245) [334].

The functionalization of C—H bonds remains one of the most difficult challenges in synthetic chemistry, and while a number of catalysts have been developed to promote specific reactions, general approaches are quite rare. Many of these reactions were proposed to occur through radical processes; however, the precise mechanism for many reactions remains unknown. Several versions of this chemistry have been used to generate arylphosphonates, and the nature and reactivity of phosphorus-based radicals have been reviewed [396]. Some of the earliest reports on this chemistry were based upon using anionic oxidation [397], peroxides [398], cerium nitrate [399], and manganese(III) acetate [400]. The following sections will outline specific examples of successful C—H phosphonations.

The formation of arylphosphonates using a Mn(II)/Co(II)/O$_2$ catalyst system occurs under mild conditions (Scheme 4.246) [401]. Although the use of monosubstituted arenes formed a mixture of ortho-, meta-, and para-phosphonated compounds, one very practical

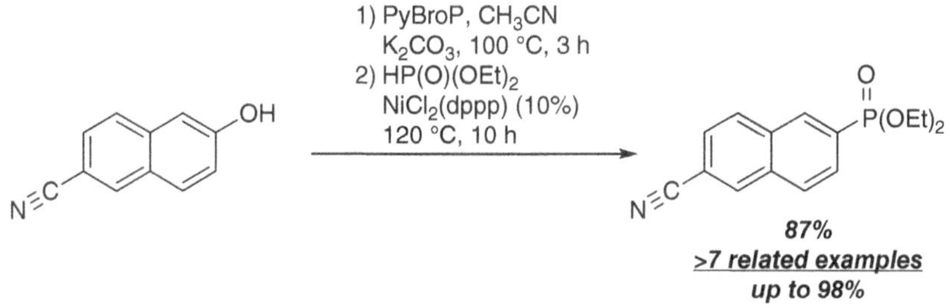

SCHEME 4.245   Nickel-catalyzed synthesis of arylphosphonates from phenols [334].

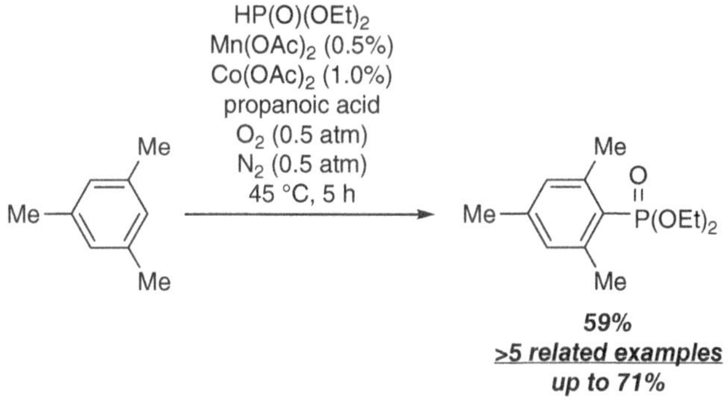

SCHEME 4.246   C–H phosphorylation of mesitylene [401].

advantage to this chemistry was the successful synthesis of crowded arylphosphonates from bulky arenes. As an example, the phosphonation of mesitylene using diethylphosphite generated the monofunctionalized product in moderate yields (59%) at 45 °C. This was attractive since many of the low-temperature Hirao-type phosphonation reactions were sluggish or completely suppressed when bulky substrates were used. The authors used 2,6-di-*tert*-butyl-4-methylphenol (BHT) to probe the possibility of a radical process. Indeed the addition of this radical trap to the reaction mixture fully inhibited the formation of the arylphosphonate. The chemistry was operationally trivial as the reagents were charged under an atmosphere of air and the reactions were carried out under $O_2 : N_2$ (1 : 1). A related system using only manganese (III) acetate (3 equiv) to promote the phosphonation improved the selectivity for addition to the ortho-position of monosubstituted arenes [402].

The regioselective phosphonation of heteroaromatic compounds also used the manganese(III) acetate as a promoter (Scheme 4.247) [403]. The authors screened a number of solvents, conditions, and reagent ratios to find an efficient system that afforded high yields of the heteroarylphosphonates. Although they opted to use acetic acid for the majority of the reactions, the authors did report a successful phosphonation (76%) under solvent-free conditions. For many of the substrates, the addition of the phosphonate fragment occurred regioselectively. When both positions 4 and 5 were available for substitution, the phosphonate fragment was selectively added to position 5. However, if both positions 2 and 5 were available, functionalization occurred in both positions. The

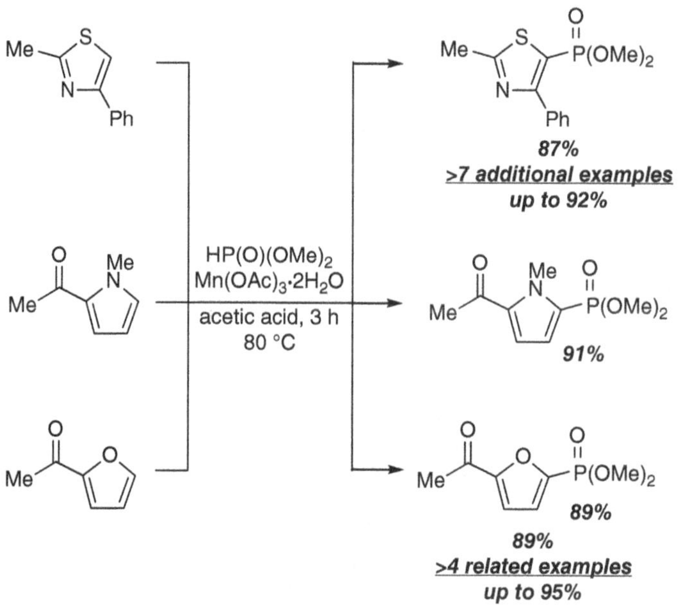

SCHEME 4.247   Manganese(III) acetate-promoted phosphonation of heteroarenes [403].

distribution of products was dependent upon the groups that were initially attached to the thiazole. The authors successfully extended the chemistry to a range of furans and a single example of a pyrrole.

A catalyst system for the C—H phosphonation of arenes bearing electron-withdrawing substituents such as amides and sulfonamides has been developed (Schemes 4.248 and 4.249) [404]. The combination of silver(I) sulfate and the potassium salt of the peroxodisulfate dianion successfully phosphonated a range of arenes bearing electron-withdrawing groups. An acetonitrile/water mixture was used as the solvent for this chemistry. The addition was highly regioselective and added the phosphonate fragment adjacent to the electron-withdrawing group. This chemistry was also operationally trivial as simply adding the reagents in air and heating afforded the arylphosphonates in moderate to excellent yields.

An effective C—H phosphonation of pyrimidylphosphonates has been reported (Scheme 4.250) [405]. Manganese (III) acetate served as the promoter for this chemistry, and simply heating to 50 or 80 °C in acetic acid successfully phosphonated the C-5 position on the pyrimidine. Curiously, the cerium nitrate [399] was an ineffective catalyst for this transformation. An attractive aspect of this chemistry was that the primary and secondary alcohols on the sugar did not need to be protected prior to the phosphonation reaction. One drawback to this approach was that three equivalents of the manganese reagent and four equivalents of the secondary phosphite were needed for high conversions. However, these issues are overshadowed by the ability to functionalize a C—H bond under such mild conditions. Overall, this is a very attractive process for the synthesis of pyrimidylphosphonates.

A dehydrogenative cross-coupling between pyridines or five-membered heteroarenes with secondary phosphites has recently been developed (Schemes 4.251 and 4.252) [406]. The approach used silver nitrate as the promoter and $K_2S_2O_8$ as an oxidant. When the

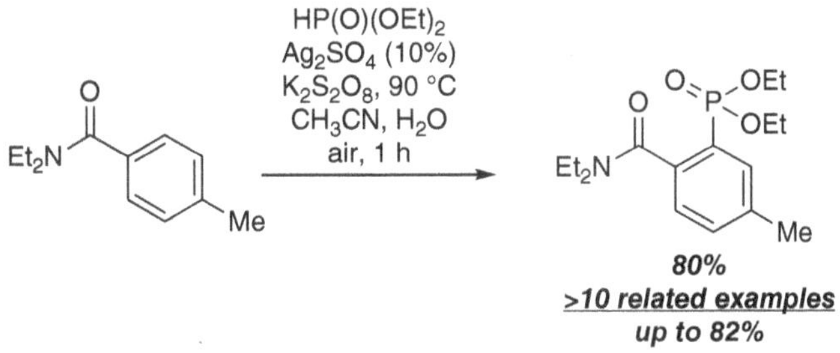

SCHEME 4.248   Regioselective formation of arylphosphonates [404].

SCHEME 4.249   Synthesis of arylphosphonates in air [404].

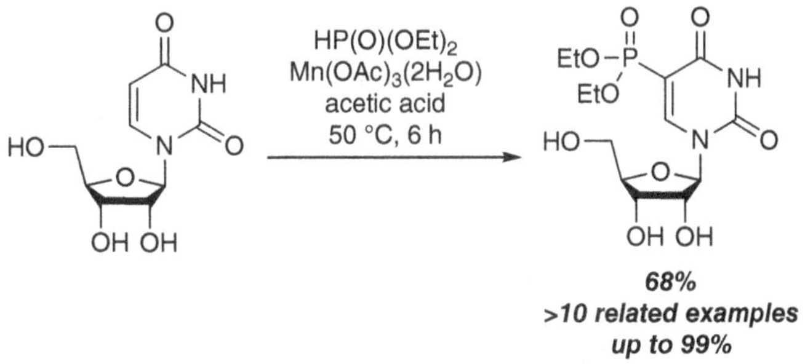

SCHEME 4.250   Preparation of pyrimidylphosphonates [405].

five-membered heteroaryl substrates were screened, moderate to excellent conversions were observed after simply stirring the reagents in 1:1 mixture of water and dichloromethane at room temperature. A range of substituted furans, pyrroles, thiophenes, and a thiazole were all regioselectively phosphonated using this catalyst system. While the full mechanism of the reaction was not elucidated, the addition of TEMPO to the reaction mixture suppressed the cross-coupling reaction suggesting that a component of the pathway was radical in nature. When pyridine derivatives were screened, the initial yields of the reaction were lower and analysis of the reaction mixture revealed the presence of N-oxides. The addition of a reducing agent to the reaction mixture after the phosphonylation reaction was complete circumvented this problem, and high yields of the phosphonylated pyridines were

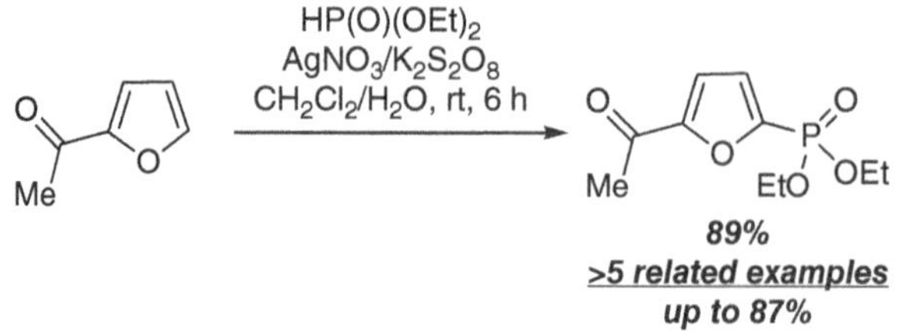

89%
>5 related examples
up to 87%

SCHEME 4.251   Synthesis of arylphosphonates from heteroarenes [406].

70%
>5 related examples
up to 81%

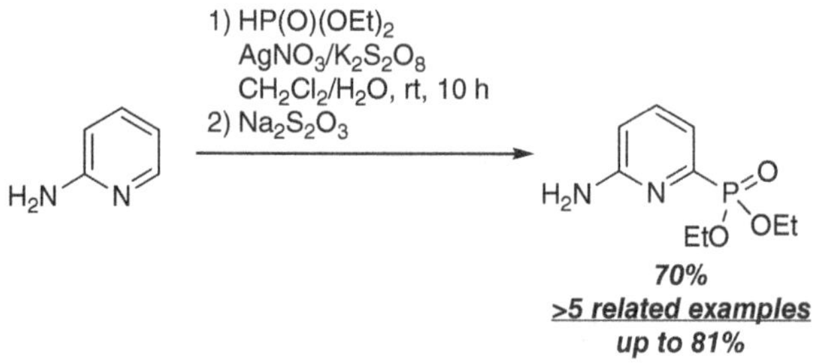

SCHEME 4.252   Synthesis of pyridylphosphonates from heteroarenes [406].

obtained. Similar to the results obtained using five-membered heteroarenes, the substitution pattern was regioselective for the position adjacent to the heteroatom. The ability to perform the reaction at room temperature without removal of the atmosphere or moisture was a significant practical advantage of this chemistry.

Li recently reported the synthesis of arylphosphonates through the palladium-catalyzed phosphonylation of C—H bonds (Scheme 4.253) [407]. The catalyst system was comprised of palladium acetate along with proline as the supporting ligand. One of the key components of the reaction was the use of $K_2S_2O_8$ as an oxidant. The authors proposed that the addition of this oxidizing agent promoted the formation of a palladium (IV) intermediate. The authors probed the possibility of a radical process through the addition of TEMPO as a radical scavenger. The yield of the arylphosphonate was essentially the same with the addition of TEMPO suggesting that this may not be a radical process. While a range of heterocycles were successfully phosphonated using this system, N-methylbenzimidazole and N-methylindole were resistant to functionalization and no phosphonated products were observed.

The direct phosphonylation of indoles has been reported using readily available secondary phosphites as the phosphorus precursors (Scheme 4.254) [408]. The chemistry was operationally trivial, and simply adding the reagents together and heating afforded the functionalized materials. After investigating the efficacy of a range of organic and metal-based promoters, silver(I) acetate was selected for the screening reactions. In general, substrates bearing either electron-donating or electron-withdrawing groups were successfully functionalized with higher yields of the arylphosphonates observed for the former. For most substrates, the chemistry was regioselective and afforded moderate yields of the indoles functionalized at C2; however, the regioselectivity was reduced in N-alkylated examples. The mechanism for these reactions was probed, and evidence for the possibility of a radical pathway was presented including the complete suppression of the reaction

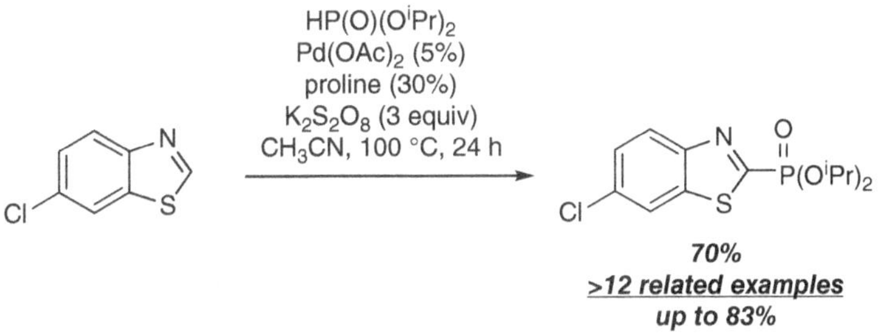

SCHEME 4.253   Synthesis of arylphosphonates through C—H phosphonylation [407].

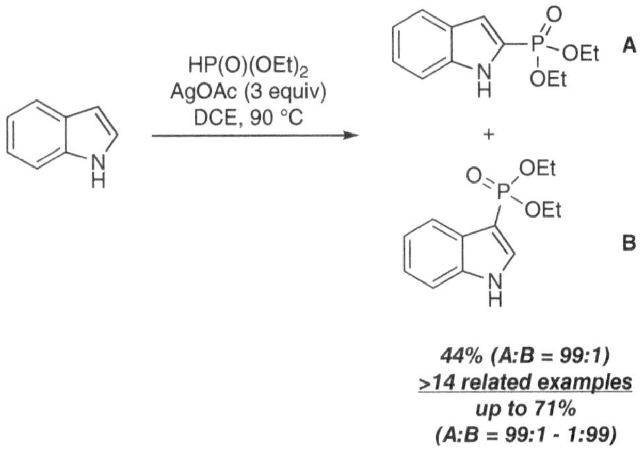

SCHEME 4.254   Selective phosphonylation of unfunctionalized indoles [408].

when BHT was added to the reaction mixture. One drawback to this system was the high loading of the silver salt. However, this chemistry offered a way to generate the phosphonated compounds without prefunctionalization of the indoles.

The selective synthesis of pyridylphosphonates through C—H phosphonylation has been reported (Scheme 4.255) [342]. This palladium-catalyzed process attached the —P(O) $(OR)_2$ fragment to the ortho-position of an attached aromatic ring. Benzoquinone was a key component of the catalyst system and was proposed to promote the reaction through the acceleration of the reduction step [343, 344]. In addition to pyridine derivatives, a host of different heterocycles were successfully functionalized using this chemistry. This chemistry displayed moderate functional group compatibility with electron-withdrawing groups on the pyridine derivative affording the lowest yields of the arylphosphonate. The authors also investigated different sources of the phosphonate fragment and found that diisopropylphosphite was the most active in the C—H activation reaction. It was also critical to add the phosphite very slowly to prevent deactivation of the catalyst. Although a number of requirements need to be met for this approach to be successful, it is an attractive way to prepare arylphosphonates as it results from C—H activation and is selective for the ortho-position.

Following this report, the palladium-catalyzed synthesis of heteroarylphosphonates using a masked phosphonate precursor was presented (Scheme 4.256) [409]. While the

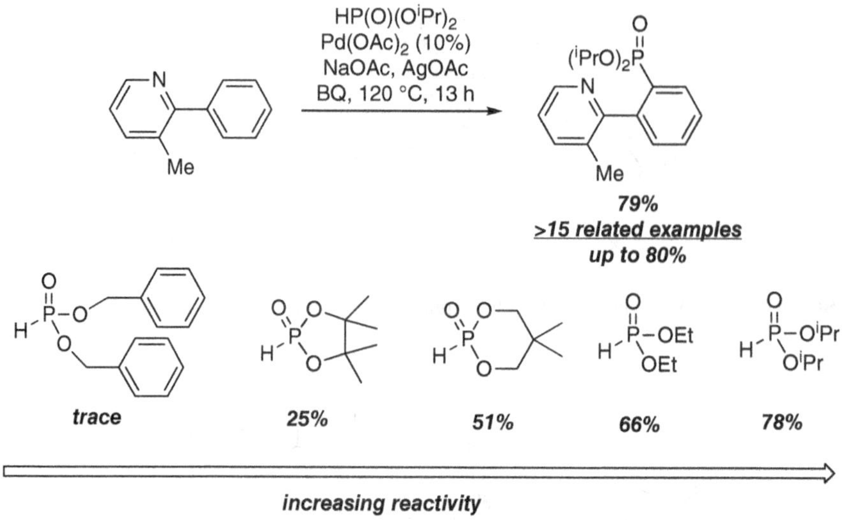

SCHEME 4.255 Pyridylphosphonate synthesis through palladium-catalyzed C—H activation [342].

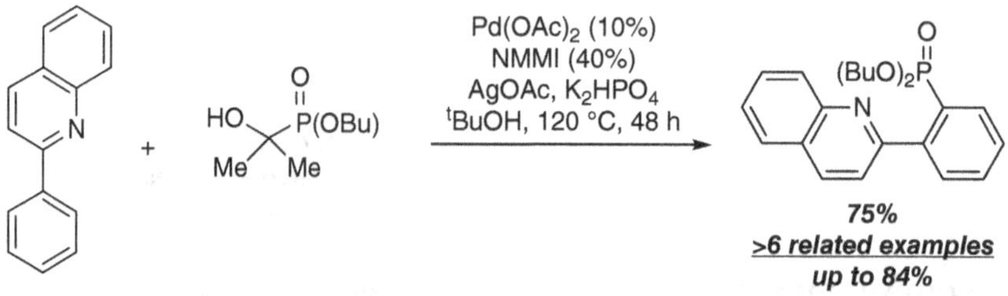

SCHEME 4.256 Heteroarylphosphonate synthesis through palladium-catalyzed C—H activation [409].

use of a syringe pump to slowly deliver reagents to the reaction mixture has been popular, this work investigated reagents that would slowly form small amounts of the phosphorus reagent in solution. They used an α-hydroxyphosphonate that would slowly convert into the secondary phosphite under the reaction conditions. These compounds are attractive substrates since they are readily prepared by the treatment of the hydrogen phosphonates with ketones. Essentially, the reverse of this reaction would generate the hydrogen phosphonate and the ketone. Through mechanistic studies, the authors determined that this reverse reaction was possible under the reaction conditions. The chemistry was regioselective and the phosphonate group was consistently added at the ortho-position. The use of α-hydroxyphosphonates as precursors to secondary phosphites could be of use in other reactions where the presence of high concentrations of secondary phosphites would be detrimental.

The synthesis of arylphosphonates has been accomplished using a copper acetate-promoted C—H phosphonylation process (Scheme 4.257) [410]. The reaction was carried out in DMSO under mild conditions and afforded moderate to good yields of the arylphosphonates. The chemistry was relatively insensitive to the electronic composition of the arene, and an assortment of arylphosphonates were prepared. Due to the presence

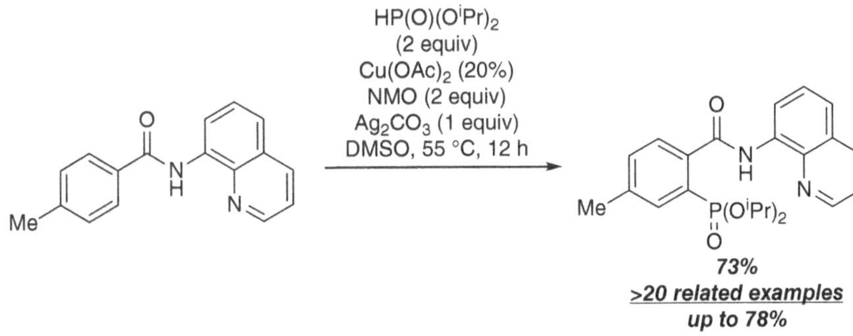

SCHEME 4.257   Synthesis of arylphosphonates through C—H activation [410].

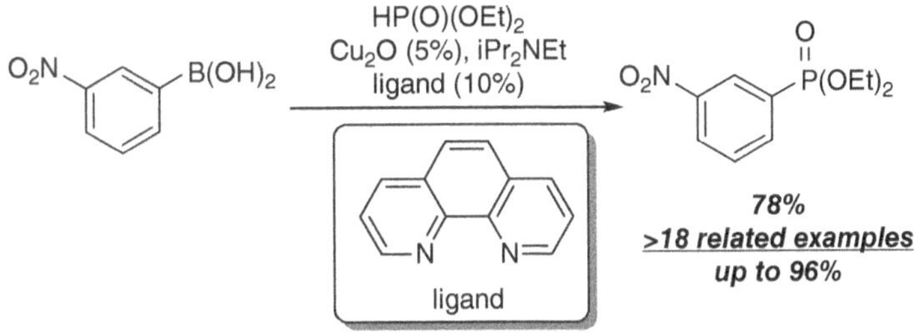

SCHEME 4.258   Copper-catalyzed synthesis of arylphosphonates using arylboronic acids in air [411].

of the ortho-directing group, the addition occurred selectively at the position adjacent to the directing group.

The copper-catalyzed coupling of boronic acids with hydrogen phosphonates is an attractive approach to the synthesis of arylphosphonates because the reaction can be carried out under an atmosphere of air at room temperature (Scheme 4.258) [411]. A wide variety of preexisting functional groups were well tolerated by this approach, and the isolated yields were excellent. The reaction time was a bit long (24 h); however, the ability to eliminate the use of a glovebox or even a vacuum line makes the reaction very practical.

The synthesis of arylphosphonates through a nickel-catalyzed cross-coupling of secondary phosphites with boronic acids has been reported (Scheme 4.259) [346]. The catalyst system used for this chemistry was comprised of nickel(II) bromide along with 2,2′-bipyridine as the supporting ligand. A number of bases were screened, and the authors found that a mineral base generated excellent yields of the arylphosphonates. It was noteworthy that the chemistry could be carried out under an atmosphere of air. As a result, this reaction was quite practical as no glovebox or vacuum manifold was needed. While only one example was reported, this approach to the synthesis of arylphosphonates has great potential due to the vast array boronic acids that are readily available.

An interesting approach to the synthesis of chiral phosphine and phosphine oxides including Buchwald-type biarylphosphines has recently been reported (Scheme 4.260) [412]. The approach was similar to the chemistry reported by Juge's that entailed initial formation of a P-, N-, O-heterocycle followed by selective bond cleavage using organometallic reagents to generate chiral phosphines. While Juge chose to protect the phosphorus center through the initial formation of a phosphine-borane, the authors selected to generate a phosphine oxide as the protecting group. One of the challenges encountered when developing a process that will

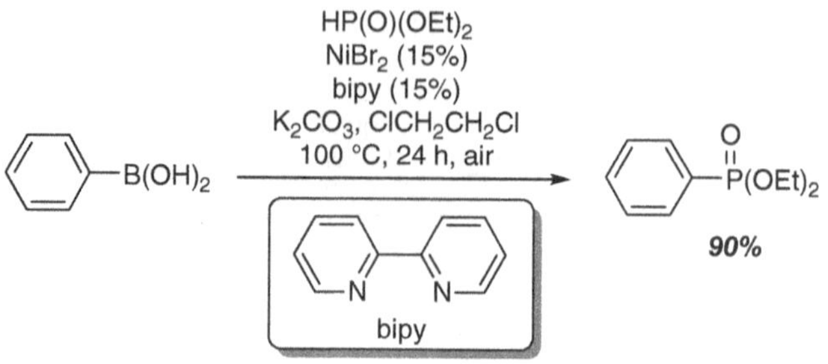

SCHEME 4.259 Nickel-catalyzed coupling of arylboronic acids with secondary phosphites in air [346].

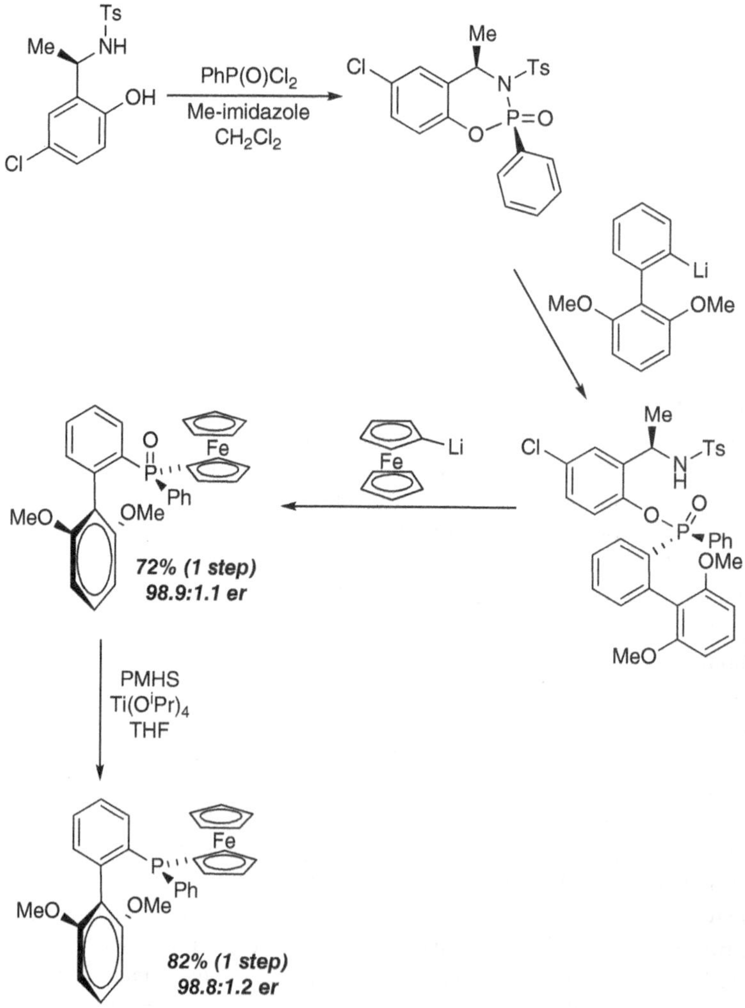

SCHEME 4.260 Synthesis of a chiral phosphine [412].

selectively cleave one bond while leaving the remainder intact is the partial reactivity of the other bonds. This challenge was addressed through the attachment of a tosyl group to the nitrogen of an amino alcohol. Once attached to the phosphorus center, the -NTs group was readily and selectively cleaved by the addition of the lithiated biaryl fragment. Treatment of the resulting species with feroccenyl lithium generated the chiral phosphine oxide in excellent yield with high selectivity. Reduction using polymethylhydrosiloxane and titanium isopropoxide afforded the phosphine with retention of the chirality. In addition to the synthesis of Buchwald-type biarylphosphines, the authors reported the preparation of over a dozen additional examples. Several of these examples were successfully scaled up to greater than 100 g.

4.3.1 Troubleshooting the Synthesis of Arylphosphines and Related Compounds

While it is impossible to anticipate the challenges of every new reaction/method that will be developed over the next few decades, there are some general trends that can be followed. While the section on solvents, additives, and general suggestions in Chapter 1 will be a good resource, the chart below will provide some direction when trying to select the most practical solution to the specific issues related to the synthesis of arylphosphines and related compounds.

Reaction Requirement/Problem	Possible Solution
Partial oxidation of the phosphine during workup	One solution is to rigorously purify and deoxygenate all of the reagents and solvents used during the workup procedure
	Alternatively, it might be more practical to fully convert the phosphine into a phosphine oxide, sulfide, or borane adduct prior to isolation. Isolation of the target phosphine from the reduction reaction is often significantly easier than from the initial P—C bond-forming reaction. Furthermore, the secondary phosphine oxide might be more reactive in phosphination reactions [300, 389, 413]
The starting material contains bulky groups	If there are no other electrophiles present in the compound, treatment of a chlorophosphine with a Grignard/organolithium reagent has been an effective strategy [265]. Additionally, a copper-catalyzed coupling reaction has successfully used bulky aryl halides [24]
Can arylphosphines and related compounds be generated through cross-coupling reactions?	This is a popular approach to the synthesis of these compounds. A number of metal complexes have served as catalysts for these reactions [281, 286, 288, 318]. Palladium-catalyzed reactions versions have shown great tolerance to preexisting functional groups [294]
Are microwave-assisted methods known for the cross-coupling methods?	This reaction is quite amenable to microwave heating [289, 291]
The starting material can be converted into an aryl triflate. Can these be used in a cross-coupling reaction leading to the formation of an arylphosphine?	A variety of metal-catalyzed coupling reactions have successfully used aryl triflates for the synthesis of arylphosphines [310, 329, 413]
Can arylboronic acids be used in cross-coupling reactions leading to the formation of arylphosphine oxides?	Nickel bromide is an attractive catalyst for the coupling reaction between arylboronic acids and secondary phosphine oxides [346]

(Continued)

(Continued)

Reaction Requirement/Problem	Possible Solution
Are there transition metal-free routes to the synthesis of arylphosphine oxides and phosphonates?	For the proper system, the addition of phosphorus reagent to arynes is an attractive approach [332]. Activated aryl halides also can be functionalized using metal-free syntheses [352–354]
The cross-coupling chemistry used for the synthesis of arylphosphine oxides or phosphonates is sluggish and low yielding	For some palladium-catalyzed reactions, the addition of acetate additives has been beneficial for increasing the reaction rates [377, 378]
Can arenediazonium salts be used in the synthesis of arylphosphine oxides and phosphonates?	This has been achieved through a palladium-catalyzed cross-coupling reaction [292]
Are solvent-free processes known for the synthesis of arylphosphonates?	A number of reports describe solvent-free approaches to the synthesis of arylphosphonates [366]
Can protic solvents be used in the palladium-catalyzed cross-coupling of secondary phosphites with aryl bromides?	Ethanol was successfully used as the solvent in such a cross-coupling reaction [393]
Can diaryliodonium salts be used in cross-coupling reactions that lead to the formation of arylphosphine oxides and phosphonates?	This has been achieved, and a copper-catalyzed process is particularly attractive due to the mild conditions and broad substrate scope [333]
Need to use water as solvent	This has been achieved using a palladium-catalyzed approach [340]
Are low-temperature preparations of arylphosphine oxides and phosphonates known?	Low-temperature reactions are possible through a number of palladium-catalyzed Hirao-type cross-coupling reactions [388, 389]
Need a base-free approach to the synthesis of arylphosphonates since the substrate contains base-sensitive groups	A number of approaches have been developed for the base-free synthesis of arylphosphonates [388, 391]
Can arylphosphonates be prepared through C—H phosphonylation reactions?	A number of approaches are known to promote this reaction. Each substrate should be evaluated against these conditions in order to design a successful coupling reaction [403–406, 408]

4.4   SYNTHESIS OF VINYLPHOSPHINES AND RELATED COMPOUNDS THROUGH THE FORMATION OF PHOSPHORUS–CARBON(SP²) BONDS

Vinylphosphines are a valuable class of compounds with a myriad of applications in organic synthesis and catalysis. Vinylphosphines also have a rich history in coordination chemistry due to the geometry and reactivity of this fragment with transition metal centers [414–420]. The proximity and flexibility of the alkene have been used to generate a host of coordination compounds in which the vinylphosphine is coordinated to the metal center through the phosphorus as well as through the alkene. In addition to providing a unique architecture for discrete coordination compounds, vinylphosphines are also valuable linkers in the formation of 3D arrays and coordination polymers [421–424].

Treating chlorophosphines with nucleophilic vinyl fragments is one of the oldest ways to generate vinylphosphines. While some of the early work in this area used organolead and mercury reagents [425], the vast majority of procedures used Grignard and vinyllithium species. Using this approach, moderate yields of the organophosphines were commonly obtained. The drawback to this methodology was the lack of functional group tolerance due to the use of powerful nucleophiles/bases. The troublesome oxidation of the phosphine precursors as well as the vinylphosphine products was another hurdle that needed to be circumvented in these syntheses. However, if the target substrate was not sensitive to these reagents and the oxidation issues were resolved, this approach remains a popular route to the preparation of vinylphosphines.

One of the simplest vinylphosphines, trivinylphosphine, was prepared through the treatment of PCl_3 with several equivalents of vinylmagnesium bromide [426]. Great care was taken to ensure that both the phosphorus precursor and the Grignard reagent were introduced slowly and simultaneously. The trivinylphosphine was isolated as a liquid after distillation under vacuum, albeit in low yield (18%). They dealt with the oxidation issue by performing all of the manipulations under an atmosphere of nitrogen. In related work, a fluorinated Grignard was also reported to react with PBr_3 to generate a completely fluorinated derivative (Scheme 4.261) [427]. The compound was isolated in 9.2% yield following bulb-to-bulb distillation. While many of the Grignard-based approaches to the synthesis of vinylphosphines employ chlorophosphines, the authors found that the reaction of the Grignard reagent with PCl_3 did not generate the desired trivinylphosphine. One of the challenges with this chemistry was reported to be oligomerization and polymerization.

In an effort to increase the isolated yields of the vinylphosphines, one approach that has been used with some success involved protection of the reactive phosphine center through the formation of a phosphine-borane (Scheme 4.262) [8]. Rather than add the borane at the end of the reaction, the authors generated the protected phosphine prior to the addition of the Grignard reagent. Significantly higher yields of the vinylphosphine were formed using this approach. It was noteworthy that the P—B bond survived treatment with such a potent nucleophile. Given the ease of formation of the phosphine-boranes, this is a practical synthesis of unsubstituted vinylphosphines.

While high yields were obtained in the preceding example due to the protection of the chlorophosphine through the formation of the borane adduct, changing the phosphorus

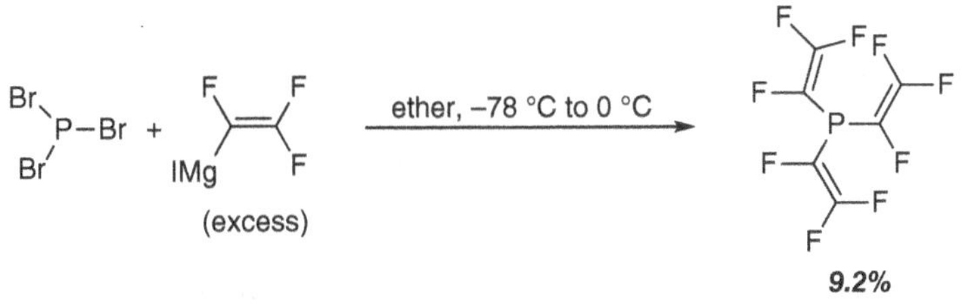

SCHEME 4.261   Synthesis of a fluorinated trivinylphosphine [427].

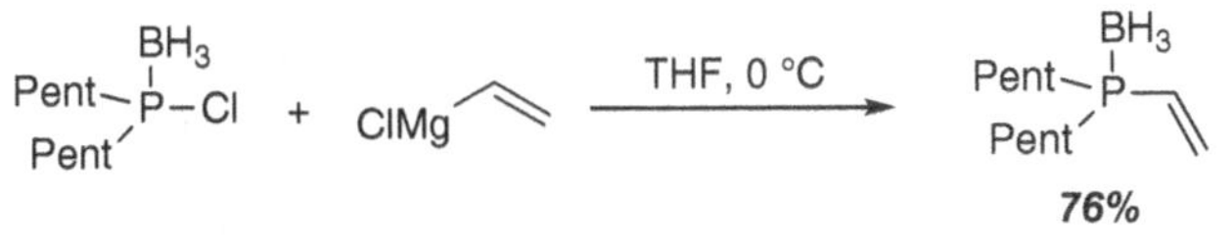

SCHEME 4.262   Synthesis of vinylphosphines using protected chlorophosphines [8].

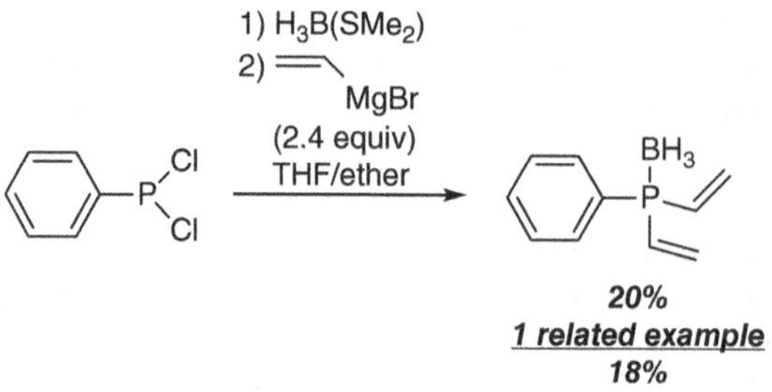

SCHEME 4.263   Synthesis of divinyl phenylphosphine [428].

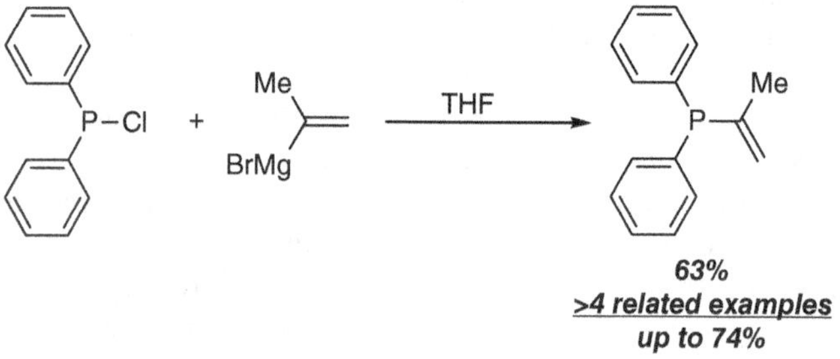

SCHEME 4.264   Synthesis of vinylphosphines using an α-substituted vinyl Grignard reagent [262].

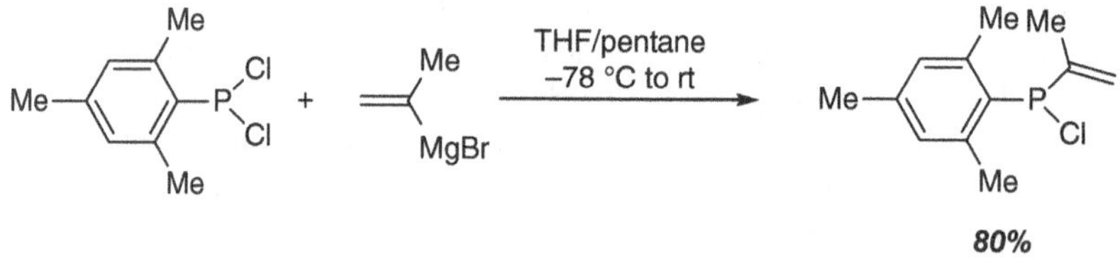

SCHEME 4.265   Synthesis of a crowded vinylchlorophosphine using a Grignard approach [430].

substrate to phenyldichlorophosphine resulted in a significant reduction in the yield of the divinylphosphine (Scheme 4.263) [428].

In contrast to the low yields typically observed using unsubstituted vinyl Grignard reagents, higher yields were reported when groups were attached to the vinyl fragment (Scheme 4.264) [262]. Mixing several substituted vinyl bromides with magnesium turnings followed by the addition of diphenylchlorophosphine afforded the vinyldiphenylphosphines in moderate yields. It was noteworthy that this process was quite tolerant of substitution in the α-position. As with the trivinylphosphines discussed previously, the new compounds were isolated by distillation.

This approach to the synthesis of vinylphosphines also tolerates bulky groups on the phosphorus center (Schemes 4.265 and 4.266) [429–431]. Several reports outlined successful additions of Grignard reagents to bulky chlorophosphines at room temperature and above. The yields of the vinylphosphines from these reactions were typically good relative to when unencumbered chlorophosphines were functionalized. Naturally, these reactions still need to be rigorously protected from moisture, and electrophiles cannot be present on either substrate; however, the ability to handle bulky groups remains one of the attractive

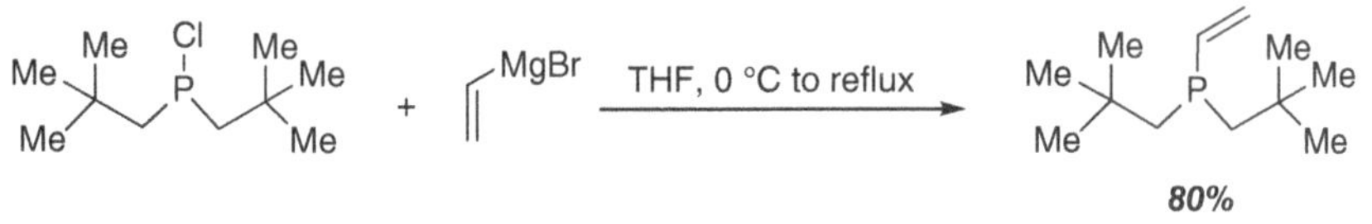

SCHEME 4.266 Synthesis of a crowded vinylphosphine [429].

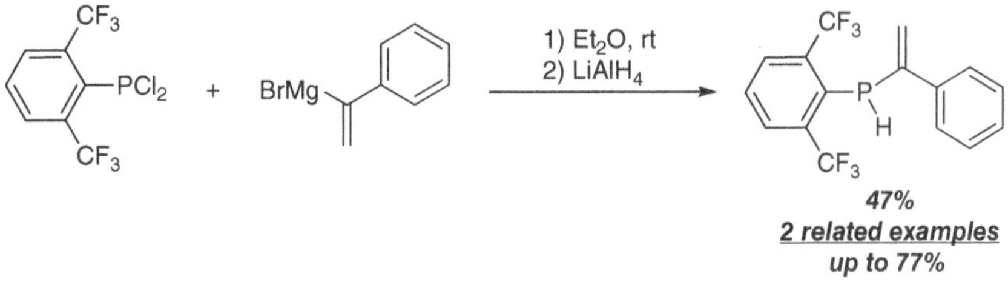

SCHEME 4.267 Synthesis of a crowded secondary phosphine [432].

aspects of this approach. As will be noted in later parts of this section, this is not always the case with other methods.

In addition to successfully functionalizing bulky chlorophosphines, this chemistry was also tolerant of electron-withdrawing groups on the chlorophosphine precursors (Scheme 4.267) [432]. Treatment of $(2,6-(CF_3)_2C_6H_3)PCl_2$ with a substituted vinyl Grignard reagent followed by reduction using lithium aluminum hydride afforded secondary phosphines in moderate yields. The authors were targeting these secondary vinylphosphines as they were using them as precursors for the synthesis of phosphaalkenes through a 1,3-hydrogen migration process. Once isolated, the conversion was trivial as simply adding catalytic amounts of a base (DBU) converted the vinylphosphines into phosphalkenes. Although the vinylation reaction was successful, it was sensitive to the groups on the Grignard reagent. While the use of 2-methylpropenylmagnesium bromide afforded 77% of the vinylphosphine after distillation, only 47% of the target phosphine was isolated from analogous reactions with the bulky 2-phenylpropenylmagnesium bromide. The vinylphosphines could be stored for long periods of time under an inert atmosphere.

This approach to the synthesis of vinylphosphines has been extended to chlorophosphines bearing —C_6F_5 groups (Scheme 4.268) [433]. The chemistry was similar to the other reactions in that the Grignard reagent was added at low temperatures to a solution of the chlorophosphine followed by warming to room temperature and stirring for a couple of hours. The presence of the electron-withdrawing fluorines did not impede the reaction, and a moderate yield of the target phosphine was obtained. These vinylphosphines were of subject of the investigation since they were used as substrates in the synthesis of "frustrated" Lewis acid–base complexes.

As mentioned previously, one of the biggest drawbacks to this chemistry was the intolerance to the presence of electrophiles. To circumvent this limitation, Knochel reported a twist on the Grignard chemistry that significantly expanded the scope of this reaction (Scheme 4.269) [8]. Through the formation of organocopper species, the reactivity of the nucleophile was altered such that moderate electrophiles could be tolerated. The authors also protected the phosphorus center of the chlorophosphine through the formation of

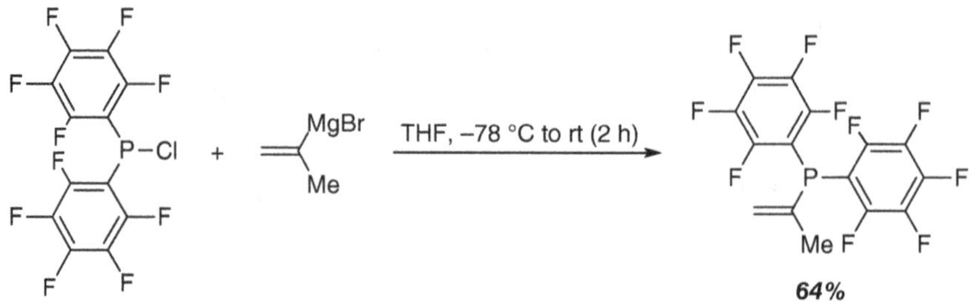

SCHEME 4.268   Synthesis of a fluorinated vinylphosphine [433].

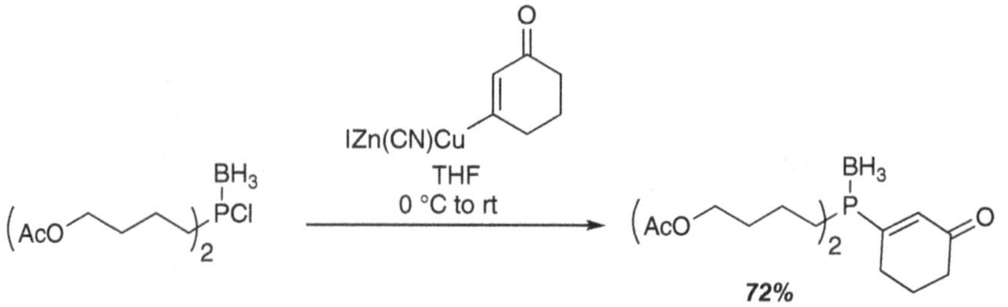

SCHEME 4.269   Modification of the reactivity of vinyl Grignard reagents through the formation of organocopper species [8].

borane adduct. Using this approach, electrophiles on both the chlorophosphine and the nucleophile were well tolerated, and the borane-protected vinylphosphine was isolated in excellent yields.

Reversing the roles of the substrates in this chemistry has also been used to generate vinylphosphines (Scheme 4.270) [434]. While the carbon fragment was the nucleophile in the reactions described previously, the phosphorus precursor can also be used in this role. These phosphides were commonly generated in solution through the treatment of the R_2PH species with BuLi. In contrast to the reactions discussed previously, the yields of the vinylphosphines were lower when the roles were reversed due to the reduced susceptibility of vinyl halides to undergo secondary reactions. This method also necessitates the handling of potentially hazardous secondary phosphines. Although the yields from these reactions were quite low, the reagents are readily available, and it might be more practical to simply accept the lower yield and move on to the next step in a complex synthesis.

A metal-free approach to the synthesis of phosphorus–carbon bonds has been reported (Scheme 4.271) [7, 435]. The authors treated PCl_5 with indene under mild conditions and generated an intermediate phosphonium salt. This compound was converted into the dichloroindenylphosphine upon treatment with trichlorosilane. This reaction was facile, and the authors report quantitative conversion into the target phosphine. An alternative procedure for the conversion of the phosphonium salt into the vinylphosphine target involved treatment with diethylphosphite and triethylamine [435]. While this scope of this approach has not been thoroughly investigated, it does represent a metal-free approach to the synthesis of vinylphosphines.

One of the biggest challenges for the modern synthetic chemist entails the functionalization of substrates that already contain a number of functional groups. To this end, the

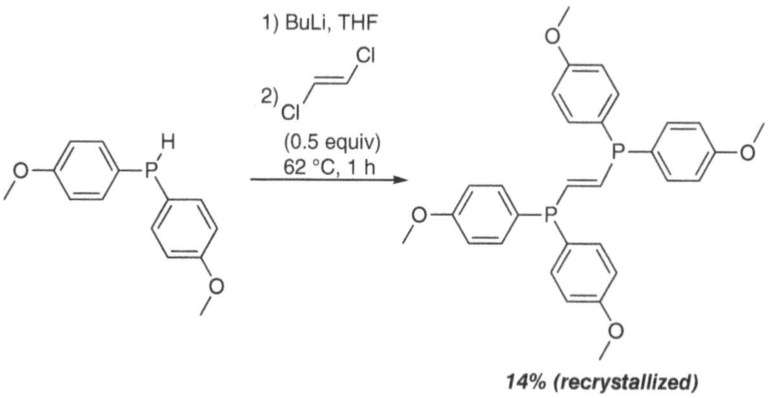

SCHEME 4.270 Synthesis of vinylphosphines using lithium phosphides [434].

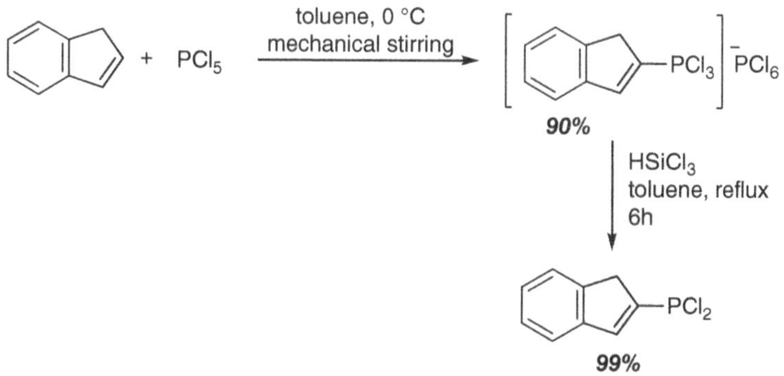

SCHEME 4.271 Metal-free synthesis of vinylphosphines [7].

metal-free phosphination of enamines is a difficult transformation. While attempts to incorporate a phosphine fragment into simple enamines generated highly reactive compounds with labile P—C bonds that were challenging to isolate, the incorporation of a nitrile or ester fragment in the β-position of enamines led to isolable compounds (Schemes 4.272 and 4.273) [436]. The scope and substrate dependence of this reaction was investigated using several electron-withdrawing groups and enamines. For nitrile-containing substrates, the phosphination occurred in the β-position of the enamine resulting in the formation of vinylphosphines. This chemistry was quite sensitive to the structure of the starting enamine, and multiple products were commonly observed. Due to the sensitivity to air, the vinylphosphines were converted into phosphonium salts or phosphine oxides/sulfides prior to isolation and purification. Exchanging the nitrile for an ester fragment altered the product distribution, and phosphination at the beta carbon was no longer the dominant product. Instead, the phosphorus nucleophile added to the methyl group of the enamine to generate an allylphosphine. As with the vinylphosphines, the authors converted the phosphines into air-stable species through the formation of phosphonium salts or phosphine oxides/sulfides. While the product distribution was very sensitive to the composition of the starting material, it does present an interesting and metal-free route to the formation of functionalized vinylphosphines.

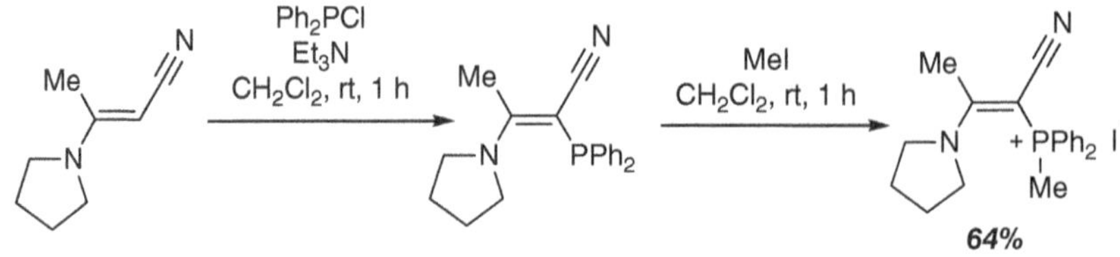

SCHEME 4.272   Synthesis of vinylphosphines and subsequent reaction with methyl iodide [436].

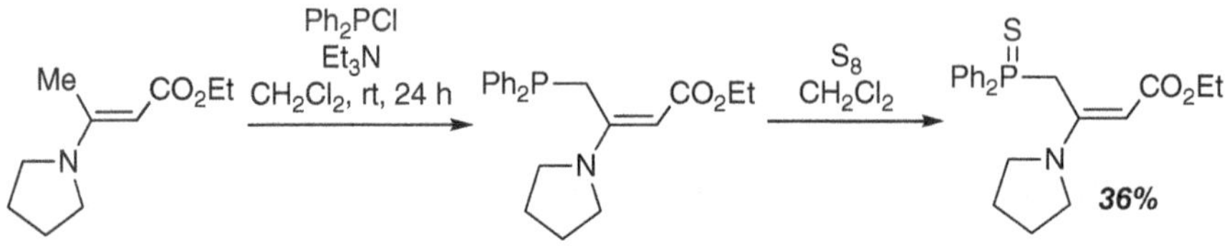

SCHEME 4.273   Synthesis of allylphosphines and subsequent reaction with sulfur [436].

The synthesis of vinylphosphines can also be accomplished by metal-catalyzed cross-coupling chemistry. The practical advantages to this chemistry include the ability to tolerate the presence of a wide range of functional groups, the vinylphosphonates are typically isolated in excellent yields, and the reaction conditions can be quite mild. One of the most significant advantages of the cross-coupling chemistry was that the substitution is almost always regio- and stereoselective. The phosphorus-containing fragment was almost always incorporated into the substrate in the same position occupied by the leaving group. The following sections will highlight the use of palladium, copper, ruthenium, as well as other metal catalysts for the formation of vinylphosphines.

One of the earliest examples of a palladium-catalyzed cross-coupling reaction that led to the formation of vinylphosphines was reported by Beletskaya (Scheme 4.274) [437, 438]. Using a palladium(II) precatalyst, secondary phosphines were coupled with vinyl halides under mild conditions, and the resulting vinylphosphines were isolated in moderate to excellent yields. A range of vinyl bromides and chlorides were successfully used. This stands in contrast to hydrophosphination reactions discussed later that result in mixtures of regioisomers. The authors handled the air sensitivity of the products through rigorous exclusion of oxygen from the reactions. It is worth noting for this chemistry that the catalyst was not air sensitive. While many of the vinyl halides could be functionalized at 20 °C, several substrates required heating (70–120 °C) for longer reaction times. Despite the elevated temperatures, excellent yields of the vinylphosphines were reported for a wide range of substrates. The authors also successfully used trimethylsilyl-protected diarylphosphines in the synthesis of the vinylphosphines. Following this report, most of the efforts in this area have been focused on increasing the substrate scope, developing more active catalysts, and the utilization of leaving groups other than halogens.

While the vast majority of reports on the synthesis of vinylphosphines used diarylphosphine precursors, similar reactions with dialkylphosphine-based compounds have also been described (Scheme 4.275) [7]. These substrates as well as the vinylphosphine products are much more oxygen sensitive than their analogous diarylphosphine counterparts. However, the authors were still able to isolate the sensitive product by only using good technique and rigorous exclusion of oxygen. Even after recrystallization, the target

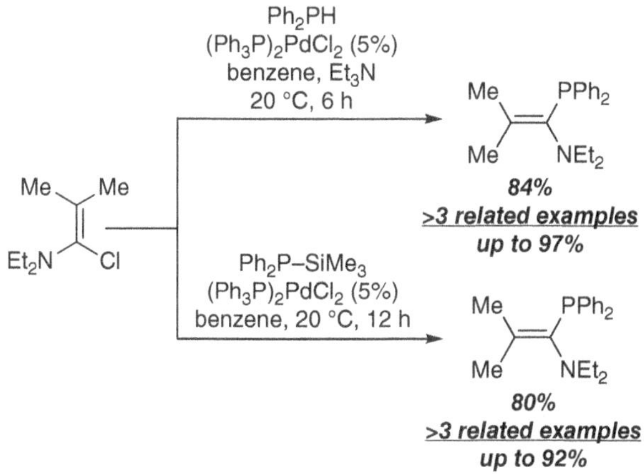

SCHEME 4.274 Synthesis of vinylphosphines through palladium-catalyzed cross-coupling chemistry [437].

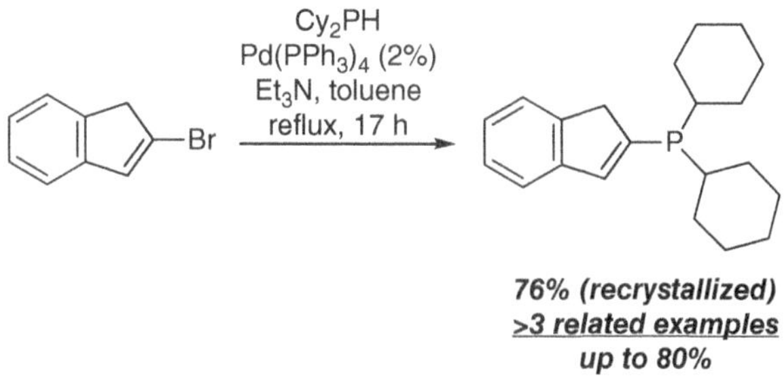

SCHEME 4.275 Synthesis of vinylphosphines through cross-coupling [7].

vinylphosphine was isolated in 76% yield. This chemistry was quite substrate dependent, and the lowest yields were observed for the bulky ^{t}Bu$_2$PH secondary phosphine (5%).

Vinyl triflates have also been successfully used in cross-coupling reactions to generate vinylphosphines (Scheme 4.276) [295]. These precursors are attractive substrates for cross-coupling reactions since they are readily prepared. One of the earliest reports of this chemistry utilized Pd(PPh$_3$)$_4$ as the catalyst for the reaction along with potassium carbonate as the base. To prevent the unwanted oxidation of the phosphorus center, the authors protected the secondary phosphine precursors through the formation of phosphine-boranes. This approach to the phosphorus–carbon bond-forming reaction was also attractive because the conditions were quite mild. Simply stirring the reaction at 40 °C for a few hours generated a near-quantitative conversion into the borane-protected vinylphosphine. This reaction also serves as an example of how the palladium-catalyzed cross-coupling chemistry is tolerant of functional groups already present on the substrates.

Crowded vinyl triflates could also be used in the cross-coupling reactions (Scheme 4.277) [439, 440]. This chemistry was attractive since the vinyl triflates could be readily generated from ketones. With the triflates in hand, the catalyst system comprised of palladium acetate along with a flexible diphosphine ligand and an amine as the base was used to generate the vinylphosphines. This palladium salt has been used as an air-stable precursor to a

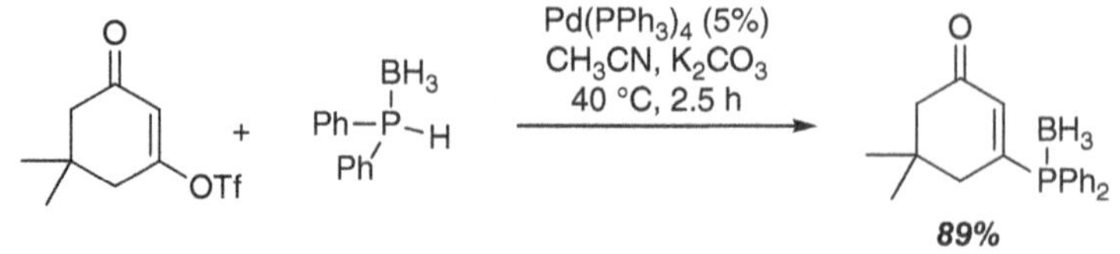

SCHEME 4.276   Palladium-catalyzed synthesis of protected vinylphosphines from vinyl triflates [295].

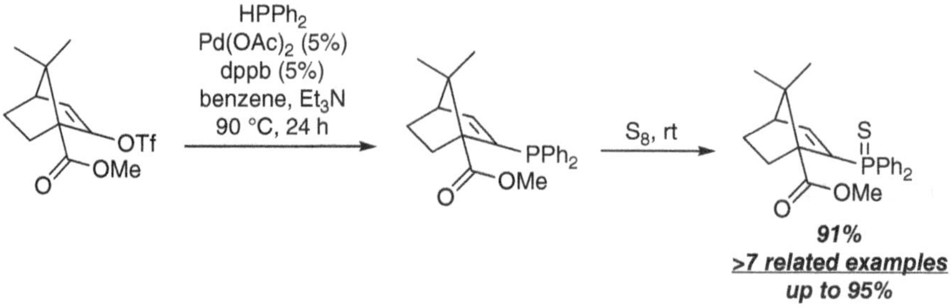

SCHEME 4.277   Use of vinyl triflates as substrates in the palladium-catalyzed synthesis of vinylphosphines [439, 440].

catalytically active palladium(0) species. These authors opted to protect the phosphine at the end of the reaction rather than generate the phosphine-borane prior to the cross-coupling. Furthermore, while the cross-coupling reactions generated the vinylphosphines in moderate to excellent yields at 40 °C, the reactions involving ketopinic acid derivatives required higher temperatures to achieve similar conversions.

Borane-protected secondary phosphines were used in the cross-coupling reaction with vinyl triflates (Scheme 4.278) [441, 442]. The authors screened a range of metal compounds and found that a palladium(II) catalyst bearing a flexible diphosphine ligand was effective. The bulky and unactivated nature of the vinyl triflate required elevated temperatures. Similar to the preceding example, potassium carbonate was used as the base in these reactions. During the course of the investigation, the authors noticed that some of the borane was lost from the phosphorus center during the reaction with some of the vinyl triflates. To maintain the air stability of the product, a second equivalent of the borane precursor was added following the P—C bond-forming reaction. The authors also screened chiral vinyl triflates in the reaction. The cross-coupling chemistry proceeded cleanly to generate the borane-protected vinylphosphine with no loss of stereochemistry.

A cross-coupling approach has been used to functionalize chiral cycloalkenyl substrates in order to prepare enantiopure chiral phosphine precursors (Scheme 4.279) [443]. The catalyst system used cesium carbonate as the base and afforded excellent yields of the vinylphosphine oxide after oxidation with hydrogen peroxide. No scrambling of the neighboring chiral centers was reported.

Cross-coupling chemistry has also been used to promote P—C bond-forming reactions between norbornadienyl triflates with secondary phosphines (Scheme 4.280) [444]. Palladium acetate was used as the palladium source in this reaction along with BINAP as the supporting ligand. The reaction conditions were mild (45 °C), and the authors oxidized

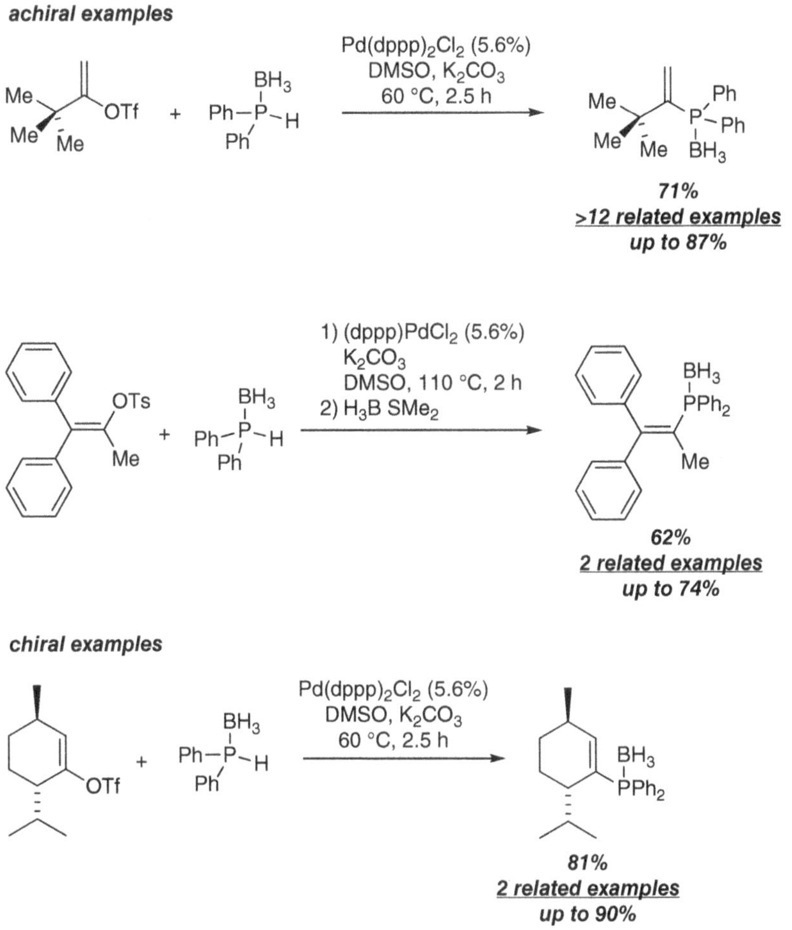

achiral examples

71%
>12 related examples
up to 87%

62%
2 related examples
up to 74%

chiral examples

81%
2 related examples
up to 90%

SCHEME 4.278   Using vinyl triflates and related compounds in the palladium-catalyzed synthesis of vinylphosphines [441, 442].

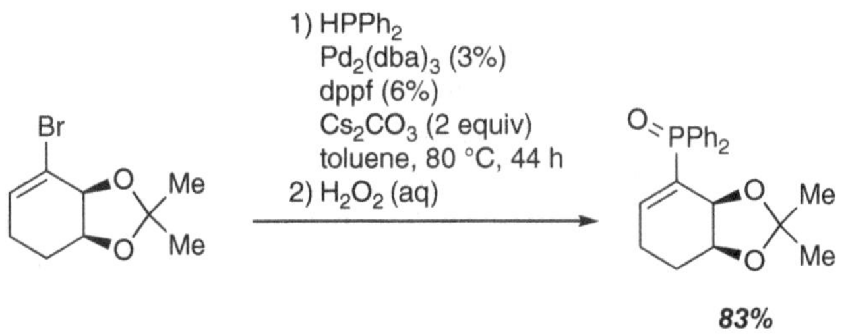

83%

SCHEME 4.279   Synthesis of functionalized vinylphosphine oxides through cross-coupling [443].

the vinylphosphine for air stability prior to isolation. In addition to providing air stability, phosphine oxides were typically stable under standard hydrogenation conditions. Since part of the synthetic strategy was to reduce the alkene using hydrogenation following the P—C bond-forming reaction, the oxide was preferred over the phosphine-borane protecting

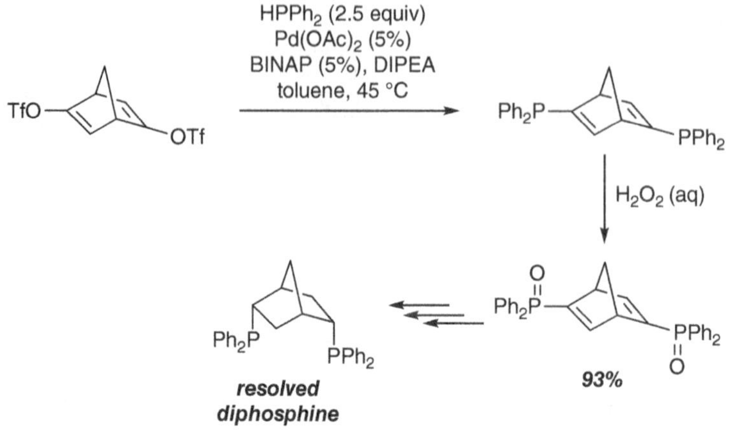

SCHEME 4.280   Synthesis of a chiral phosphine from a vinylphosphine oxide [444].

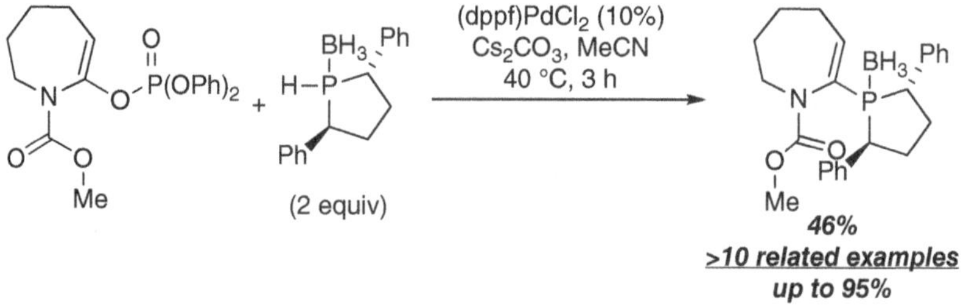

SCHEME 4.281   Synthesis of protected vinylphosphines from enol phosphates [447].

group as the latter tends to poison hydrogenation catalysts [445]. After several steps, the phosphine oxide was reduced using the $Ti(O^iPr)_4/HSi(OEt)_3$ [446] system to generate the chiral diphosphine.

A variation on the cross-coupling approach to the synthesis of vinylphosphines through the use of enol phosphates as substrates has been reported (Scheme 4.281) [447]. The biggest difference in this cross-coupling chemistry relative to what was described previously was the use of a phosphate leaving group. Although not as popular as halides or triflates, these enol phosphates can be readily generated. The palladium-catalyzed coupling reactions were rapid and the conditions were mild. Furthermore, the addition of a resolved phosphine-borane precursor resulted in retention of the chirality at phosphorus. While the overwhelming majority of vinylphosphines contain only aryl fragments in addition to the vinyl group, this report demonstrated that protected dialkylphosphines can be successfully used in these reactions. While almost all the substrates screened were successfully converted into the borane-protected vinylphosphine, reactions involving substrates bearing a *tert*-butyl ester fragment were sluggish and afforded no vinylphosphine in several cases.

One of the most general approaches to the formation of vinylphosphines is the nickel-catalyzed cross-coupling of vinyl halides with secondary phosphines

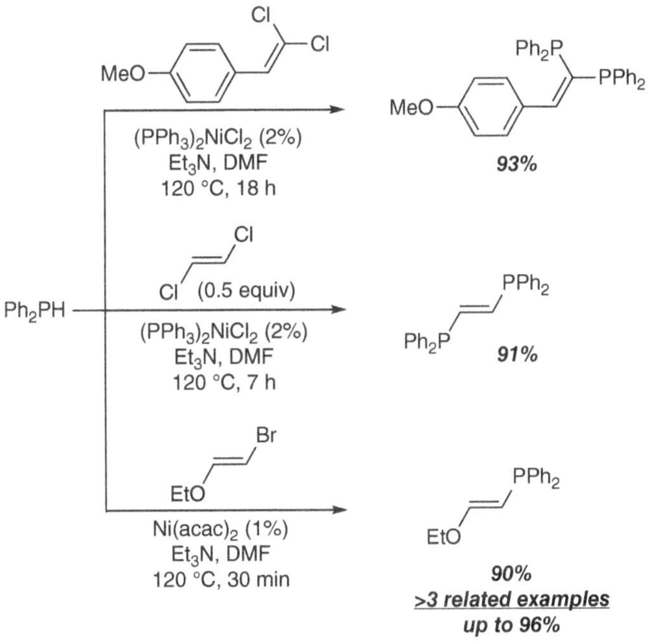

SCHEME 4.282   Synthesis of vinylphosphines through nickel-catalyzed cross-coupling [448].

(Scheme 4.282) [448]. The authors used a relatively stable nickel(II) precatalyst along with triethylamine as the base to promote the coupling reactions. For some substrates, this nickel-catalyzed approach to the synthesis was significantly higher yielding than the Grignard method. An impressive aspect of this chemistry was the ability to use vinyl chlorides in the cross-coupling reaction. These substrates have greater availability than vinyl bromides; however, they tend to be less reactive in cross-coupling reactions. Although the vinyl chlorides could be used, it should be noted that vinyl bromides were more reactive and excellent yields of the target vinylphosphines were obtained after significantly shorter reaction times.

Although the palladium- and nickel-catalyzed synthesis of vinylphosphines through cross-coupling reactions is a popular approach, copper complexes can also be used to promote these reactions (Scheme 4.283) [318]. Buchwald selected a common vinyl iodide and diphenylphosphine to serve as representative substrates for the synthesis of vinylphosphines. The most active catalyst was comprised of CuI and *N,N*-dimethylethyl-enediamine, and the most effective base for this chemistry was found to be cesium carbonate. After heating in toluene for 9 h, the vinylphosphine was isolated in excellent yield. The authors dealt with the oxygen sensitivity of the reagents and products by performing the reaction and workup under rigorous removal of oxygen. While only a single example using a secondary phosphine was reported, this approach has the potential to be significantly more tolerant to functional groups than the lithiation/phosphination approach.

While the majority of the cross-coupling reactions discussed in this section used a reactive P—H donor as the phosphorus source, chlorophosphines have been used in Ni/Zn promoted reactions (Scheme 4.284) [329]. The authors proposed that a nickel compound

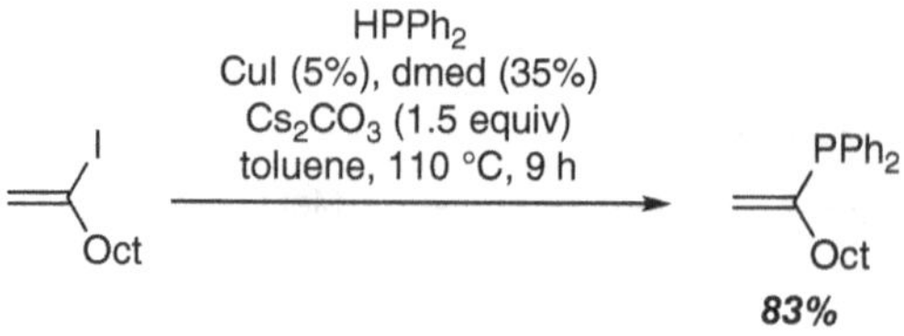

HPPh$_2$
CuI (5%), dmed (35%)
Cs$_2$CO$_3$ (1.5 equiv)
toluene, 110 °C, 9 h

83%

SCHEME 4.283 Synthesis of vinylphosphines through copper-catalyzed cross-coupling [318].

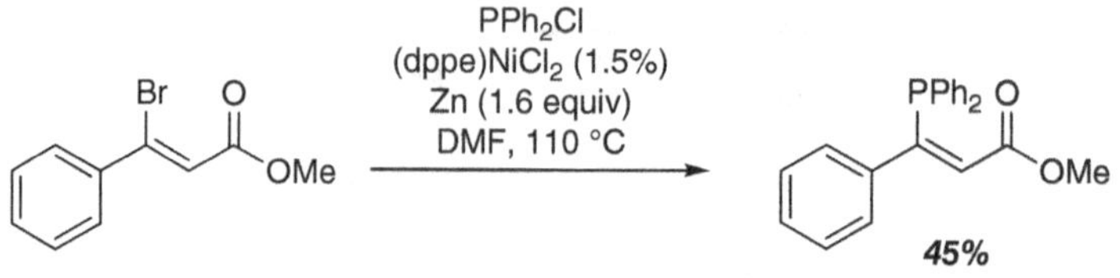

PPh$_2$Cl
(dppe)NiCl$_2$ (1.5%)
Zn (1.6 equiv)
DMF, 110 °C

45%

SCHEME 4.284 Synthesis of vinylphosphines using a nickel/zinc catalyst system [329].

was the catalyst for the reaction, while the zinc served as a reducing agent and generated a zinc phosphide that was able to transfer the PPh$_2^-$ fragment to the nickel center. One of the biggest advantages to this chemistry was the elimination of the potentially pyrophoric secondary phosphine. The authors demonstrated that this chemistry successfully functionalized a number of compounds bearing a range of functional groups.

The synthesis of vinylphosphines through the addition of P—H and related bonds to alkynes is an attractive approach. The chemistry is typically atom efficient and high yielding. The biggest challenge with this chemistry is designing a catalyst system that provides a high level of regio- and stereocontrol to the addition reaction. While these issues can often be mitigated through the judicious selection of reaction conditions, metal catalyst, and supporting ligand, general processes with broad substrate scope and high selectivity remain rare. These addition reactions are typically promoted by simply heating or through the addition of bases, radical initiators, or transition metal catalysts.

The base-catalyzed hydrophosphination of internal alkynylbisphosphines using secondary phosphines generated the tetrasubstituted vinylphosphine in moderate yields (Scheme 4.285) [449]. The authors dealt with the oxygen sensitivity through rigorous removal of oxygen from the synthesis and purification steps.

The transition metal-free addition of sodium phosphides to protected alkynols proceeds under mild conditions to selectively form the 1,1-addition products (Scheme 4.286) [73]. The alcohol was deprotonated prior to the addition of the anionic phosphorus reagent to ensure the selective addition to the alkyne. The lithium alkoxide was hydrolyzed after the P—C bond-forming reaction was obtained. The conditions were extremely mild, and moderate yields of the addition product were obtained.

The addition of borane-protected secondary phosphines to terminal alkynes proceeds under microwave-assisted conditions without the addition of a transition metal, base, or radical initiator (Scheme 4.287) [450]. Furthermore, the authors discovered that the addition reaction did not need to be carried out under an atmosphere of nitrogen. The chemistry predominately generated the Z-isomer, although the formation of the E-isomer was never completely prevented. It was noteworthy that a terminal alkyne containing a primary alcohol was successfully functionalized. Clearly, the

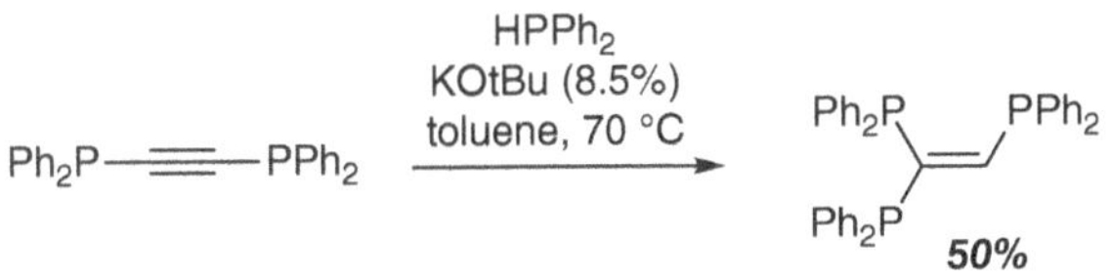

SCHEME 4.285   Synthesis of vinylphosphines through a base-assisted addition reaction [449].

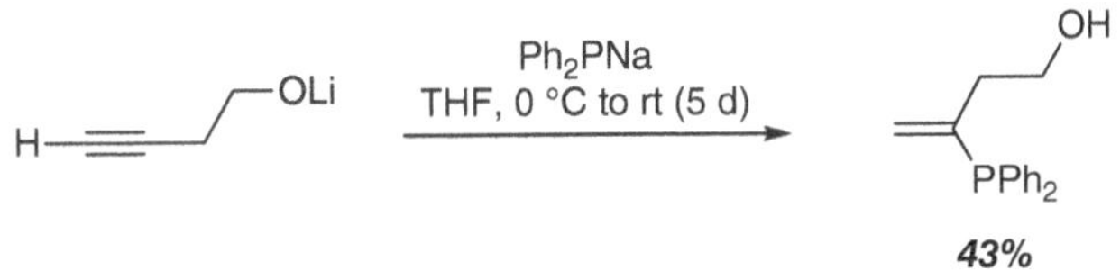

SCHEME 4.286   Addition of sodium phosphide to an alkyne [73].

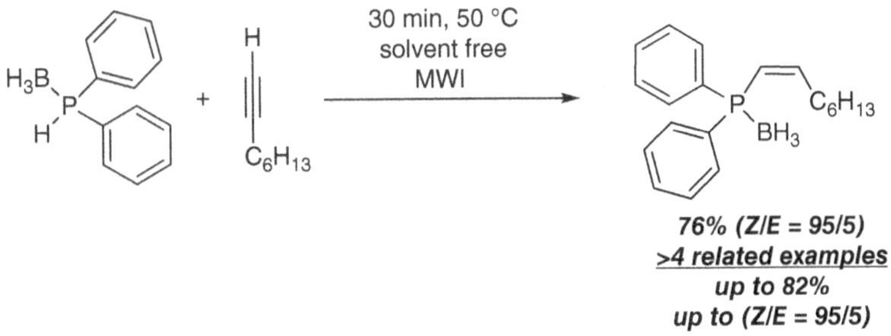

76% (Z/E = 95/5)
≥4 related examples
up to 82%
up to (Z/E = 95/5)

SCHEME 4.287   Addition of a protected secondary phosphine to an alkyne [450].

hydrophosphination process was preferred over the hydroalkoxylation reaction. In contrast to the reactivity of the alkyl-substituted alkynes, none of the hydrophosphination addition product was observed when phenylacetylene was screened. Analysis of the reaction mixture revealed that the phenylacetylene had polymerized rather than take part in the addition reaction.

The addition of primary phosphines to activated alkynes proceeds in the presence of KOH (Scheme 4.288) [451]. Representative substrates included several primary phosphines and a number of cyanoacetylenes. For the vast majority of examples, the conditions were mild, and moderate to excellent yields of the addition products were obtained. The addition reaction was regioselective for the addition of beta to the cyano group and stereoselective for the formation of the Z-isomer. While the addition of secondary phosphines to activated alkynes proceeded without the addition of a catalyst, no reaction was observed with the primary phosphines without the addition of the potassium hydroxide. The authors expanded upon this observation to investigate the addition of secondary phosphines to activated propargyl alcohols. In the absence of added bases, the addition reaction proceeded to generate the vinylphosphine in excellent yield with retention of the alcohol moiety. The regioselectivity and stereochemistry were the same in these reactions and resulted in the formation of Z-isomers. The authors have also reported the addition of

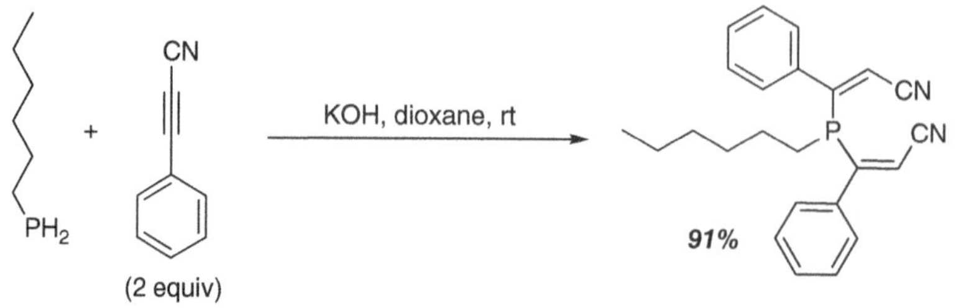

SCHEME 4.288 Base-catalyzed addition of a primary phosphine to an activated alkyne [451].

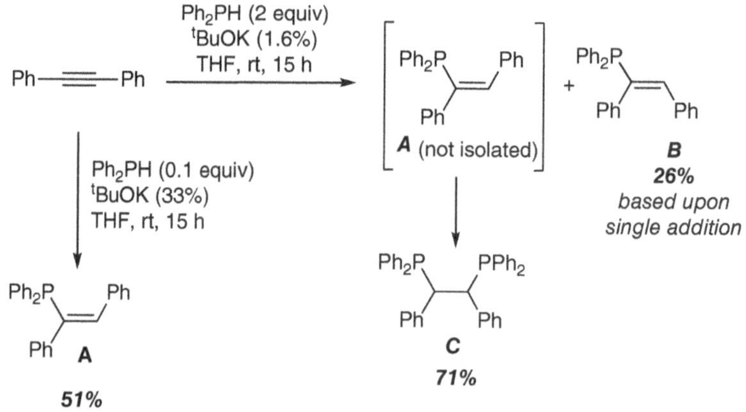

SCHEME 4.289 Addition of secondary phosphines to internal alkynes [453, 454].

primary phosphines to terminal alkynes such as phenylacetylene [452]. *Operationally, extreme care must be used when handling and manipulating these pyrophoric phosphines.*

The base-catalyzed addition of secondary phosphines to diphenylacetylene has been investigated (Scheme 4.289) [453, 454]. Using KOtBu as the base, mixtures of products were generated. When two equivalents of the secondary phosphine along with a catalytic amount of the base were used, a vinylphosphine intermediate was formed and further transformed into the diphosphine through addition of a second equivalent of the phosphine. A small amount of the vinylphosphine with E-stereochemistry was formed in this reaction as well. For comparison, the authors generated the Z-vinylphosphine through the use of an excess of diphenylacetylene along with a higher concentration of the base. While only one example was screened, this work outlines several of the challenges encountered with base-catalyzed hydrophosphinations involving internal alkynes.

In contrast to this chemistry, the hydrophosphination of borane-protected secondary phosphines to internal alkynes cleanly generates vinylphosphines (Schemes 4.290 and 4.291) [455, 456]. This chemistry was attractive since internal alkynes are significantly more challenging to functionalize relative to terminal alkynes. The author's approach involved the use of a stoichiometric amount of sodium hydride to promote the addition reaction. The chemistry was selective for the formation of the Z-isomers, and the protected vinylphosphines were formed in moderate to excellent yields. A range of internal alkynes bearing a

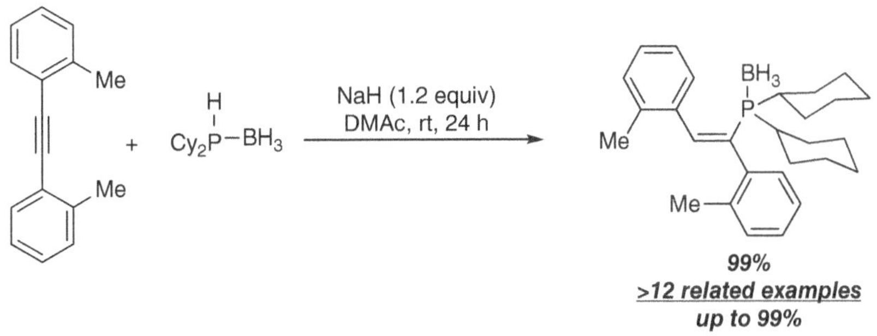

SCHEME 4.290   Addition of protected secondary phosphines to internal alkynes [455, 456].

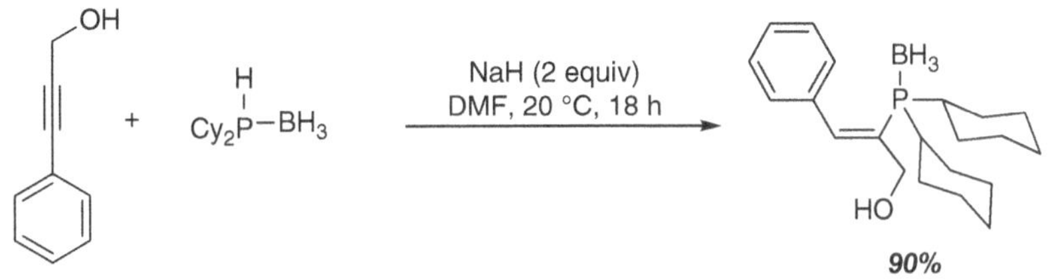

SCHEME 4.291   Addition of protected secondary phosphines to propargyl alcohols [455, 456].

range of electron-donating and withdrawing groups were readily functionalized using this system. Of note was the ability of the system to functionalize internal alkynes bearing primary alcohols. The authors discovered that the addition of a second equivalent of sodium hydride deprotonated the alcohol and generated the alkoxide. This alkoxide did not inter-fere with the P—C bond-forming reaction and was reprotonated during the workup. Another attractive aspect of this chemistry was its tolerance to moderately bulky groups on the internal alkyne. Furthermore, the effect of changing the groups attached to phosphorus on the addition reaction was investigated, and the addition reaction was found to be tolerant to a wide range of alkyl and aryl groups. Even the bulky borane-protected di-*tert*-butyl phosphine was successfully added to an internal alkyne in 88% yield. When unsymmetrical internal arylalkynes were screened, the phosphorus fragment was predominantly incorpo-rated beta to the aromatic group. Given the ability to tolerate a wide range of electron-donating and withdrawing groups as well as bulky groups on the phosphorus center as well as the internal alkyne, this approach is one of the most practical synthetic methods for the preparation of vinylphosphines.

The hydrophosphination of alkynes under radical conditions has been reported [457, 458]. One example of this chemistry used AIBN to promote the addition of diphe-nylphosphine to terminal and internal alkynes (Scheme 4.292) [458]. The reactions were carried out under argon and without the addition of a solvent. After stirring for 10–156 h, moderate to excellent yields of the addition products were isolated. It should be noted that the authors successfully carried out the addition reaction using a mercury lamp to promote the radical addition reaction. For those versions of the chemistry, typical reaction times were 3–5 days and similar isolated yields were reported. While the chemistry was highly regioselective and generally formed the anti-Markovnikov addition products, only moderate

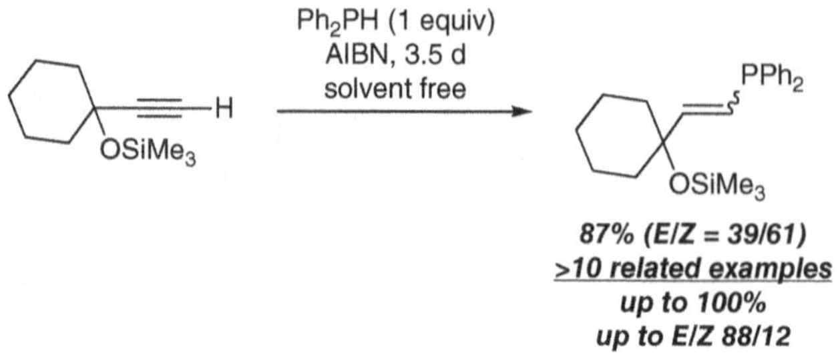

87% (E/Z = 39/61)
>10 related examples
up to 100%
up to E/Z 88/12

SCHEME 4.292   Radical addition of secondary phosphines to alkynes [458].

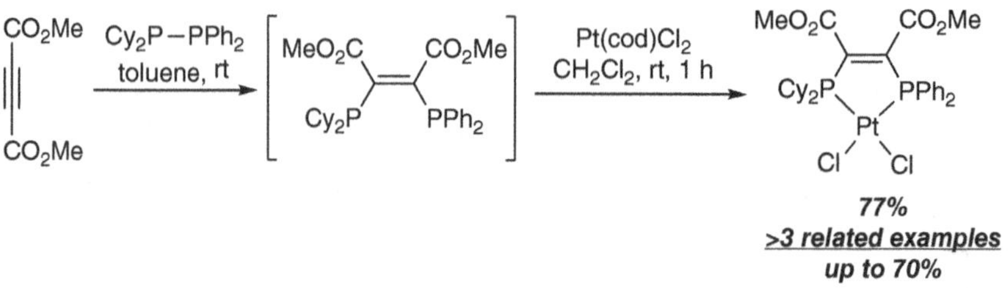

77%
>3 related examples
up to 70%

SCHEME 4.293   Addition of tetraorganodiphosphines to alkynes [459].

levels of stereoselectivity were observed and mixtures of the E- and Z-isomers were obtained for almost all substrates.

The addition of symmetrical and unsymmetrical tetraorganodiphosphine species to activated alkynes occurred without the addition of a catalyst or promoter (Scheme 4.293) [459]. This chemistry generated the cis-isomers and was operationally trivial. Simply adding the diphosphine to the alkyne at room temperature in toluene solution afforded the vinyldiphosphines. When unsymmetrical R_4P_2 precursors were used, the reaction generated the unsymmetrical vinyldiphosphines. However, when an activated terminal alkyne was screened, the addition reaction was not selective and generated a mixture of regioisomers. Although this process was not selective with terminal alkynes, it did demonstrate that only a single activating group was needed to promote the addition of the diphosphine to the alkyne. Rather than attempt to isolate and purify these compounds, the authors generated the vinylphosphines in situ and used them to prepare several palladium and platinum compounds for use as catalysts in organic transformations. The yields of the metallated bisphosphine compounds were moderate to good.

The preparation of benzophospholes was achieved through P—H activation followed by annulation (Scheme 4.294) [460]. Treatment of an internal alkyne with a secondary phosphine oxide using silver oxide as a promoter generated moderate to good yields of the benzophospholes. In one study, the authors initially screened a series of group 9 and 10 salts for activity towards the addition reaction, but discovered that the functionalization reaction could be promoted by silver oxide or silver acetate alone [460]. After some experimentation, the chemistry functionalized a range of symmetrical and unsymmetrical alkynes with the highest yields obtained with diarylacetylenes (>90%). The lowest yields were obtained using an unsymmetrical internal alkyne (3-phenylpropynenitrile, 43%). Unsymmetrical secondary phosphine oxides were also well tolerated

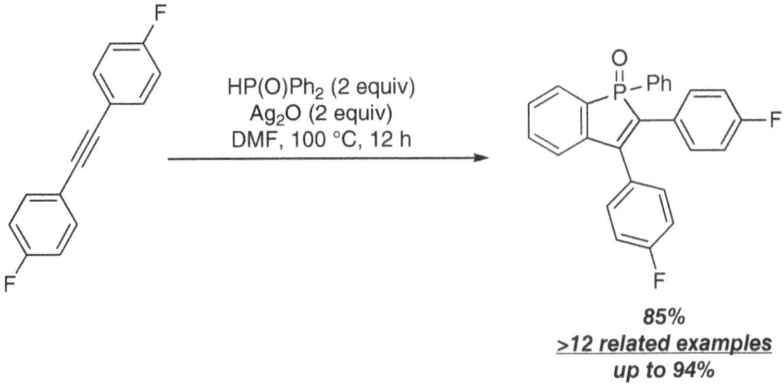

SCHEME 4.294   Silver-promoted annulation reaction between secondary phosphine oxides and internal alkynes [460].

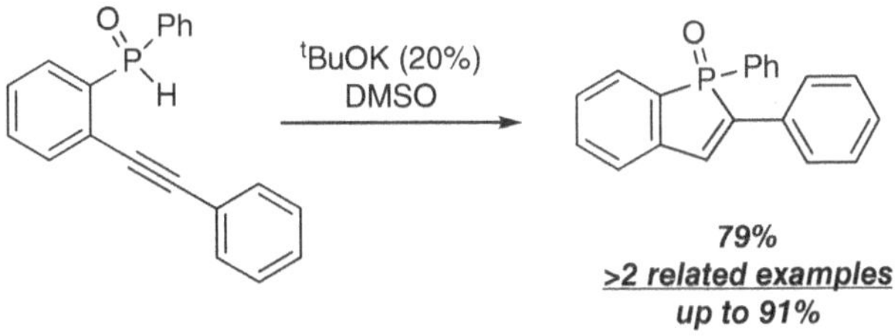

SCHEME 4.295   Base-catalyzed intramolecular hydrophosphinylation [461].

by the chemistry, and moderate to high yields of the cyclization products were obtained. It was noteworthy that the system could effectively couple bulky phosphine oxides. In related work, the base-assisted synthesis of benzophospholes has been reported (Scheme 4.295) [461].

Gem-dibromoalkenes can be converted into vinylphosphine oxides (Scheme 4.296) [462]. Nickel bromide was used as the catalyst in this reaction along with 2,2′-bipyridine as the solubilizing/stabilizing ligand. As with many of the other systems described in this section, the chemistry was highly regioselective; however, the process consistently generated mixtures of stereoisomers. The chemistry predominately formed the *E*-isomer; however, the *Z*-isomer was still formed in most cases.

The synthesis of 1,2-diphosphinoethenes through the radical addition of tetraphenyl-diphosphines to alkynes offers the synthetic chemist a direct route to the formation of 1,2-diphosphinoethenes. However, the realization of this process as a practical route has not occurred due to low reaction yields. Furthermore, although tetraorganodiphosphines are attractive sources of the organophosphorus fragment in P—C bond-forming reactions, they have rarely been selected as viable precursors as they can be quite sensitive to oxidation. A recent solution to this problem entailed the generation of the tetraorganodi-phosphine reagent in solution (Scheme 4.297) [463]. The addition of alkynes to this phosphorus precursor readily converted the terminal alkynes into vinyldiphosphines. For the selective formation of the diphosphine without contamination due to the monoad-duct, an excess of the tetraphenyldiphosphine was needed. For terminal alkynes, the

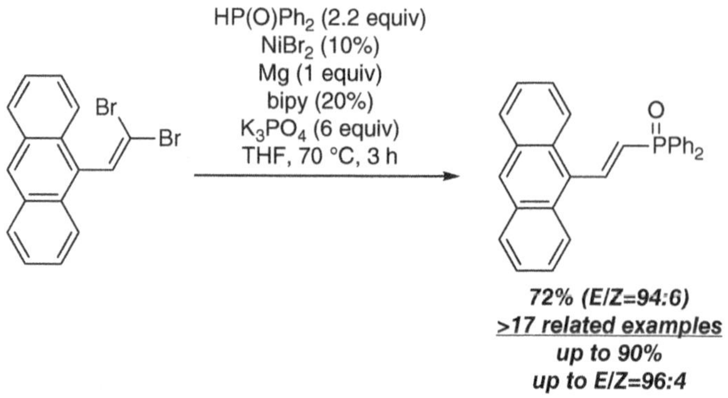

72% (E/Z=94:6)
>17 related examples
up to 90%
up to E/Z=96:4

SCHEME 4.296 Conversion of 1,1-dibromoalkenes into vinylphosphine oxides [462].

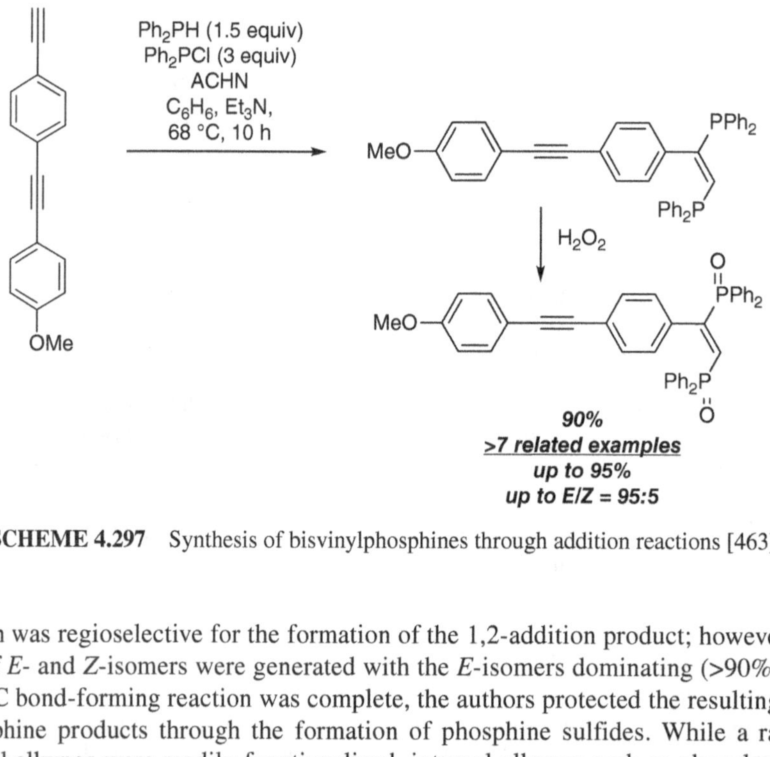

90%
>7 related examples
up to 95%
up to E/Z = 95:5

SCHEME 4.297 Synthesis of bisvinylphosphines through addition reactions [463].

reaction was regioselective for the formation of the 1,2-addition product; however, mixtures of E- and Z-isomers were generated with the E-isomers dominating (>90%). Once the P—C bond-forming reaction was complete, the authors protected the resulting vinyldiphosphine products through the formation of phosphine sulfides. While a range of terminal alkynes were readily functionalized, internal alkynes such as phenylacetylene were unreactive under the reaction conditions. Of note was the ability to selectively functionalize a terminal alkyne bearing a chlorinated hydrocarbon chain. Although alkyl chlorides are well known to readily add to phosphines to generate phosphonium salts, this secondary reaction was not observed in this chemistry. The ability to generate the tetraorganodiphosphine reagent *in situ* makes this a practical approach to the formation of vinyldiphosphines. Furthermore, a more general synthesis of diorganodiphosphines has recently been reported [464]. The chemistry was tolerant to a range of electron-donating and withdrawing groups, and over 10 examples were reported with isolated yields reaching up to 99%.

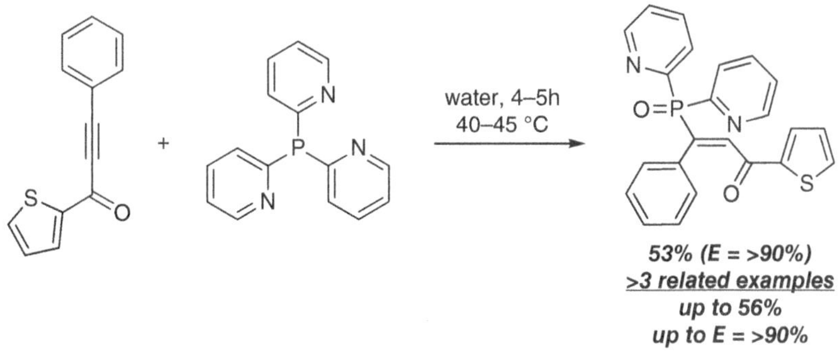

SCHEME 4.298   Using trispyridylphosphine in addition reactions with activated alkynes [465].

While the majority of the examples in this section have used diarylphosphines as the phosphorus nucleophile, trispyridylphosphine has successfully been used in the addition reaction (Scheme 4.298) [465]. One of the key aspects of this chemistry was the cleavage of a phosphorus–carbon bond in the triarylphosphine along with the formation of a phosphorus–carbon bond between the phosphorus center and the alkyne. The overall reaction was highly regioselective and incorporated the phosphorus fragment on the carbon bearing the aryl group. The reaction was also highly stereoselective for the formation of the *E*-isomers (>90%) with the exception of cyanophenylacetylene. For this substrate, the stereochemistry was reversed and the *Z*-isomer dominated (>90%). The authors proposed a mechanism for this reaction that proceeded through a vinylphosphonium salt followed by reaction with water and conversion into the vinylphosphine oxide along with concomitant loss of pyridine. While the majority of synthetic strategies are solely focused on the formation of phosphorus–carbon bonds, approaches that contain a sacrificial organic fragment could serve as alternative methods.

Calcium-based catalysts have also shown promise in hydrophosphination reactions (Scheme 4.299) [62]. Treatment of phenylacetylene with diphenylphosphine in the presence of a β-diketiminate-ligated calcium complex afforded an excellent yield of the vinylphosphine. Furthermore, the reaction was highly selective for the formation of the *E*-stereoisomer. This chemistry provided a transition metal-free route to the synthesis of these valuable targets.

A slightly different calcium catalyst was also shown to promote the hydrophosphination of internal alkynes (Scheme 4.300) [466, 467]. Instead of incorporating a solubilizing ligand into the structure of the catalyst, the authors opted to use a slightly more polar solvent (THF) to increase the amount of dissolved calcium catalyst. While the calcium compound loaded into the flask contained 4-coordinated THF molecules, calculations suggested that the active species in this chemistry was a calcium phosphide species with three-coordinated THF molecules. Using this catalyst system, they were able to generate high yields of the addition product at room temperature in a relatively short period of time. When unsymmetrical alkylarylalkynes were screened, the reaction was found to be both regioselective for addition to the carbon bearing the alkyl group and stereoselective for the *Z*-isomer. The authors also noted that no further addition of the diphenylphosphine was observed once the vinylphosphine was generated. When the authors carried out analogous reactions using 1,4-diphenyl-1,3-butadiene, the phosphorus fragment was selectively added on the carbons bearing the phenyl substituents in the *EE*-configuration. Impressively,

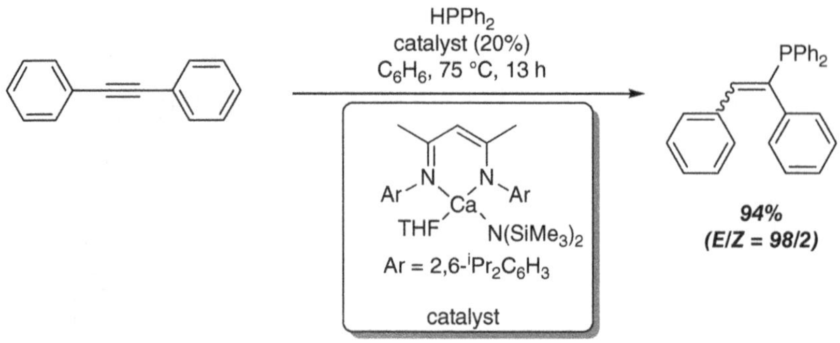

SCHEME 4.299   Synthesis of vinylphosphines using calcium catalysts [62].

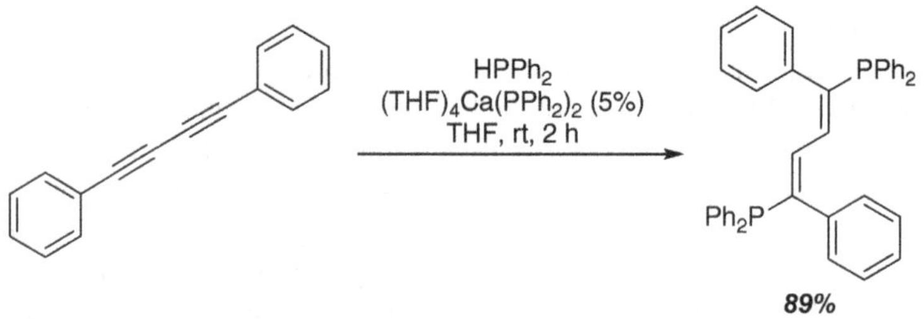

SCHEME 4.300   Double addition of diphenylphosphine to conjugated diynes [466, 467].

this addition reaction was also successful with 1,4-dimesityl-1,3-butadiene. This crowded system was successfully hydrophosphinated using the calcium catalyst system in 76% as a 45 : 10 : 45 mixture of isomers (*E,E*; *E,Z*; *Z,Z*) from which the *Z,Z*-isomer was selectively crystallized.

A comparison between the activity and selectivity of calcium and ytterbium catalysts towards hydrophosphination reactions has been reported (Scheme 4.301) [468]. The ligand system for both catalysts was a tridentate amidinate ligand system. Terminal alkynes and phenylacetylene derivatives were successfully converted into vinylphosphines; however, analogous reactions with 4-hexyne failed to generate the addition product. With few exceptions, both catalysts successfully promoted the addition and generated the vinyl-phosphines in excellent yields. Both catalyst systems were regioselective for addition of the phosphorus fragment to the terminal carbon when terminal alkynes were screened and to the carbon beta to the phenyl ring in internal arylacetylenes. Although the regioselectivity was same, the stereochemistry was reversed for most substrates. The calcium catalyst generated vinylphosphines with predominately *Z*-configuration. Although there were deviations, the ytterbium catalyst generally reversed the selectivity and generated the *E*-isomers. Curiously, both catalyst systems generated the same stereochemistry when 2-pyridylacetylene was screened.

The use of copper catalysts to promote the hydrophosphination of alkynylphosphines was reported (Scheme 4.302) [469]. The authors screened a range of catalysts for these reactions and found CuI to be effective for the addition reactions. The catalyst system was comprised of a low loading of this copper salt along with a mineral base. The reaction

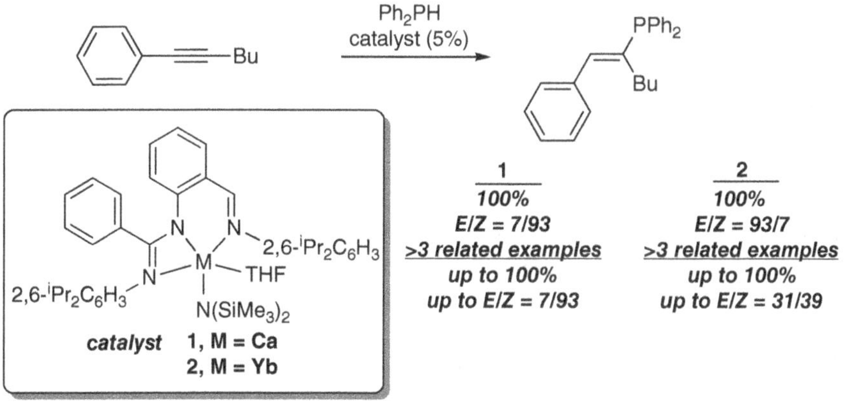

SCHEME 4.301 Using calcium and ytterbium compounds as catalysts for hydrophosphination reactions [468].

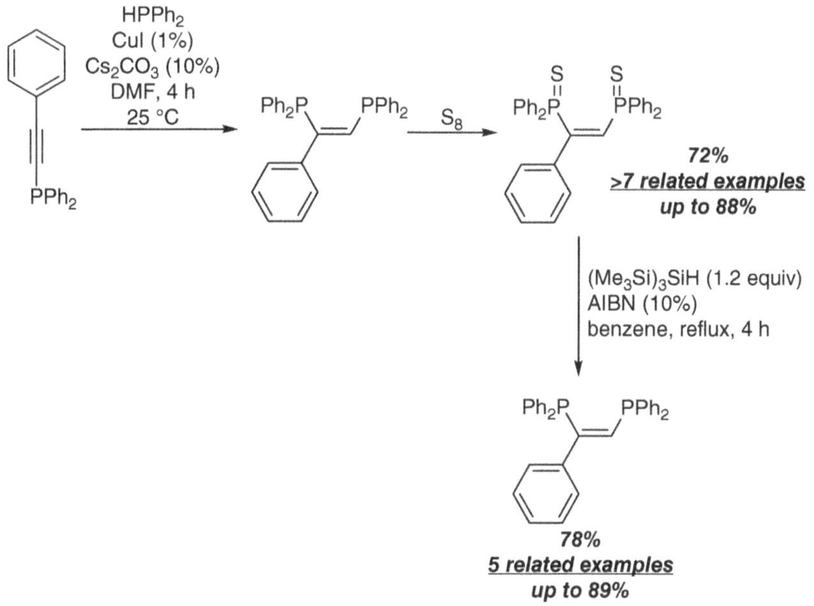

SCHEME 4.302 Copper-catalyzed synthesis of vinylphosphines from secondary phosphines [469].

conditions were quite mild, and moderate to excellent yields of the hydrophosphination products were obtained after stirring for a few hours at room temperature. Following the P—C bond-forming reaction, the authors protected the phosphorus centers from uncontrolled oxidation through the formation of phosphine sulfides. Once isolated, desulfurization was carried out using $(Me_3Si)_3SiH/AIBN$ to generate the target bisphosphines. The reaction was highly regioselective and generated the 1,2-diphosphinoethenes. The process was also stereoselective for the formation of the Z-isomer. As a result, these new bisphosphines were attractive ligands for transition metals. A range of functional groups were tolerated by the reaction chemistry including a propargyl alcohol. Given the mild conditions and low catalyst loadings, this was a very practical approach to the synthesis diphosphinoethenes.

In addition to promoting cross-coupling reactions between secondary phosphines and vinyl halides, nickel acetylacetonate is an effective catalyst for the hydrophosphination of internal alkynes (Scheme 4.303) [470]. In the course of discovering an effective catalyst for the addition reaction, the authors screened a range of metal salts and a range of terminal alkynes. While most terminal alkynes generated mixtures of regio- or stereoisomers with several different metal catalysts, diphenylacetylene cleanly formed the vinylphosphine with *E*-stereochemistry in quantitative yield. Given the relative low cost of nickel catalysts relative to palladium or rhodium as well as the outstanding yield of the addition reaction, this is a very practical approach to the hydrophosphination of internal alkynes.

The palladium-catalyzed hydrophosphination of terminal alkynes using borane-protected phosphines proceeded under mild conditions to regioselectively generate the 1,1-addition product (Scheme 4.304) [450]. The most effective palladium catalyst was palladium acetate along with a bisphosphine as the supporting ligand. One of the noteworthy aspects of this chemistry was the tolerance to the presence of a primary alcohol. It should be noted that the addition still occurred when the reaction was carried out in the absence of the palladium catalyst; however, the 1,2-addition product was predominately formed. The regioselectivity of this reaction along with the tolerance to alcohols renders this to be a very practical approach to synthesis of these compounds.

The synthesis of chiral vinylphosphines based upon the palladium-catalyzed hydrophosphinylation of alkynes has been described (Scheme 4.305) [471]. As with many asymmetric palladium-catalyzed reactions, the key to this chemistry was the use of a chiral

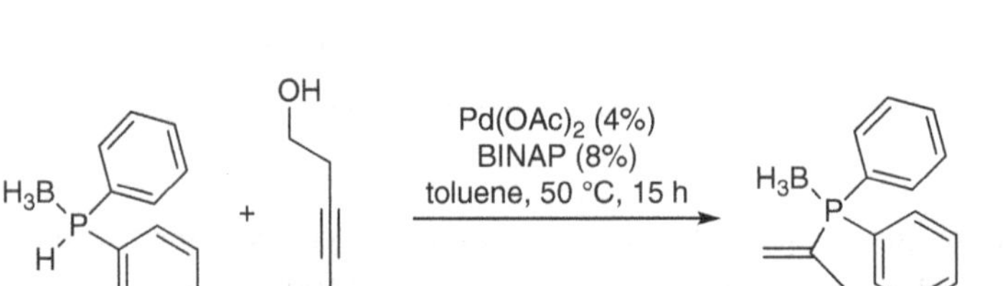

SCHEME 4.303   Selective addition of a secondary phosphine to diphenylacetylene [470].

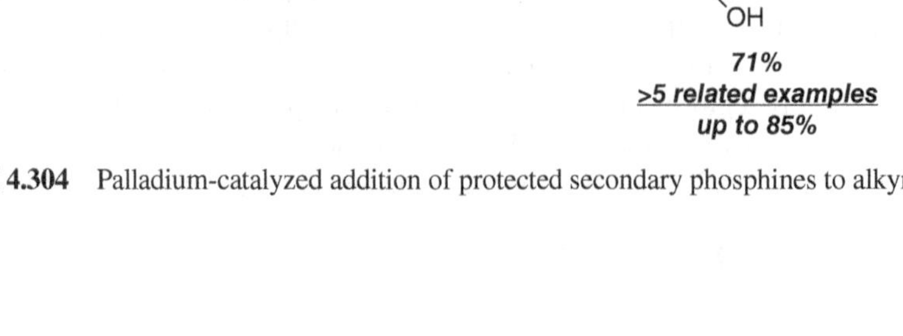

SCHEME 4.304   Palladium-catalyzed addition of protected secondary phosphines to alkynes [450].

environment around the metal center. However, the prediction of which ligand architecture will provide the highest enantiomeric excess for a specific transformation is a challenging task. It has often much more practical to select a number of possible ligands and screen them for activity. To this end, the authors investigated a range of phosphine and nitrogen compounds and discovered that a DUPHOS ligand combined with palladium acetate provided the best match of reactivity and selectivity for the addition reaction. While only a single substrate was screened and moderate ee's were obtained, it provides the proof of concept that the chemistry is attainable.

The use of cobalt compounds as catalysts for the addition of secondary phosphines to both internal and terminal alkynes has been described (Scheme 4.306) [472]. After screening a range of metal salts and additives, the authors found that the combination of cobalt(II) acetylacetonate and BuLi generated an active catalyst for the addition reaction. With terminal alkynes, a mixture of regioisomers was formed with the exception of alkynes bearing bulky groups. *Tert*-butylacetylene and related compounds exclusively generated the 1,2 adducts with *E*-stereochemistry. Separation of the regioisomers was achieved by column chromatography or crystallization. Internal alkynes were cleanly converted into vinylphosphines with *E*-stereochemistry. The vinylphosphines were converted into phosphine sulfides to generate air-stable products.

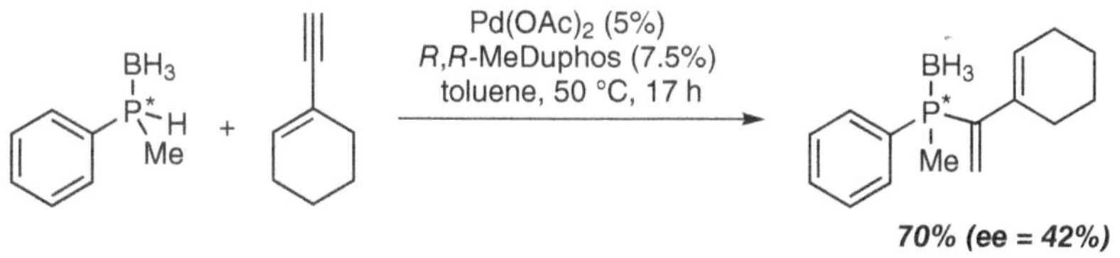

SCHEME 4.305   Synthesis of chiral-borane protected vinylphosphines [471].

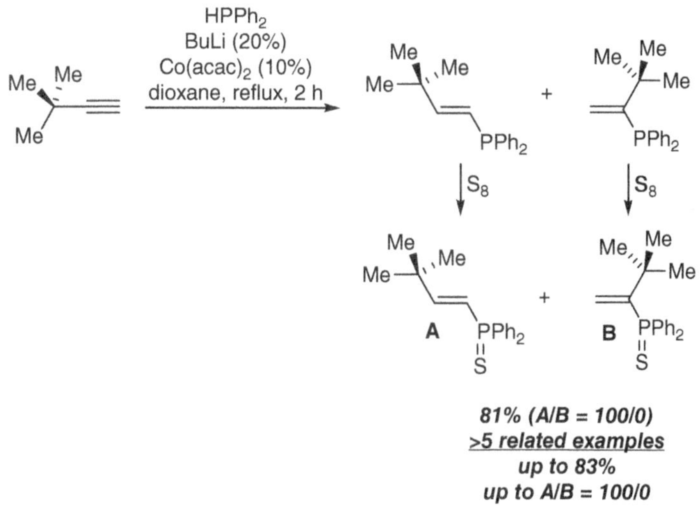

SCHEME 4.306   Regio- and stereoselective synthesis of vinylphosphines [472].

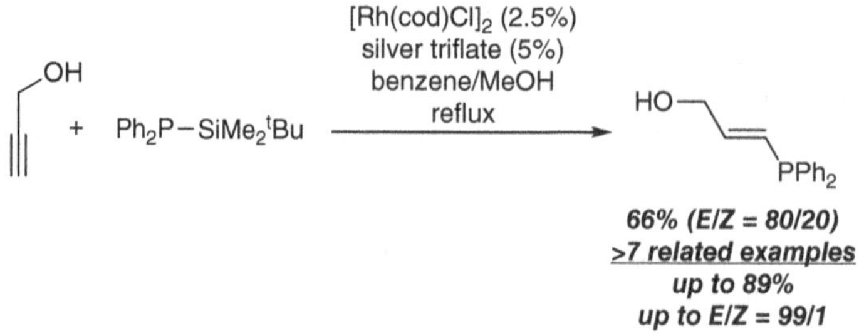

SCHEME 4.307 Rhodium-catalyzed addition of silylphosphines to propargyl alcohols [61].

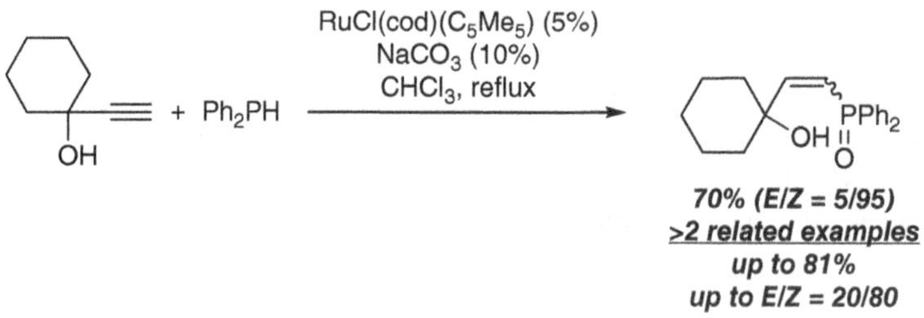

SCHEME 4.308 Synthesis of vinylphosphine oxides through a ruthenium-catalyzed process [473].

Rhodium compounds were also found to catalyze the formation of vinylphosphines (Scheme 4.307) [61]. Instead of using secondary phosphines as the phosphorus precursors, the authors opted to use silylphosphines. After screening a range of group 9 and 10 complexes and additives for activity towards the addition reaction, [Rh(cod)Cl]$_2$ along with silver triflate was found to be an effective system for the addition reaction. Using this catalyst system, the hydrophosphination of terminal alkynes readily afforded the vinylphosphines with high regioselectivity. The stereochemistry of the vinylphosphines was predominately anti with almost all of the examples exhibiting E/Z greater than 92:8. Of note was the ability to functionalize a propargyl alcohol without protection of the alcohol. Symmetrical internal alkynes such as 6-dodecyne were readily converted into vinylphosphines with high selectivity for the formation of the E-isomers (>99:1). When unsymmetrical substrates such as alkyl or arylpropiolates were screened, the phosphorus fragment was incorporated into the compound beta to the ester fragment. While the regioselectivity was consistent across a range of internal alkynes, the stereochemistry was found to be predominately anti with the lowest E/Z ratios (80/20) found when ethyl phenylpropiolate was screened.

The ruthenium-catalyzed hydrophosphination of propargyl alcohols has been reported (Scheme 4.308) [473]. Similar to the rhodium-catalyzed reaction described previously, this reaction was attractive as it facilitated the construction of these valuable fragments without protection/deprotection of the alcohol. During their initial studies, the authors discovered that the composition of the ruthenium catalyst was critical to the selectivity of the addition reaction. While many common ruthenium compounds promoted the reaction, most of them exhibited poor selectivity. The ruthenium compound that displayed the best combination of activity and selectivity was RuCl(cod)(C$_5$Me$_5$). Using this catalyst, the addition reaction strongly favored the Z-isomers, and E/Z ratios were reported to

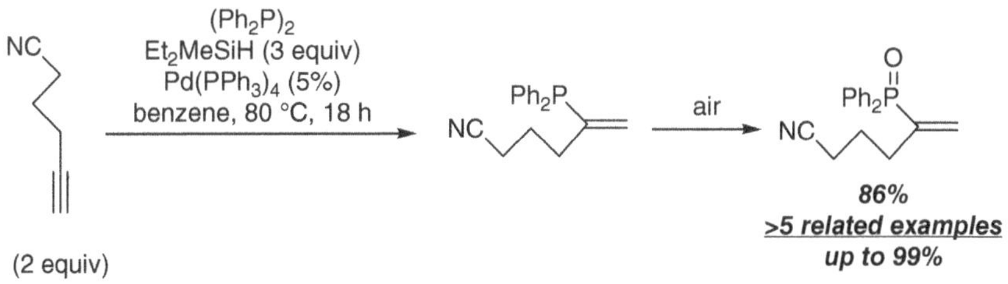

SCHEME 4.309 Palladium-catalyzed addition of tetraorganophosphines to terminal alkynes [474, 475].

be as high as 5:95. The authors carried out a mechanistic investigation of this reaction and proposed a vinylidene intermediate.

Palladium compounds were discovered to catalyze the addition of tetra-organodiphosphines to alkynes (Scheme 4.309) [474, 475]. While the radical addition reaction generated diphosphinoethene derivatives, the palladium-catalyzed version of this reaction only incorporated a single phosphorus fragment. After screening a range of palladium precatalysts for activity towards the addition reaction, the authors selected a common palladium(0) species as an effective catalyst. Furthermore, while the catalyst system generated some of the product under rigorously deoxygenated conditions, the authors discovered that small amounts of oxygen were needed to generate a highly active catalyst. Using this approach, a series of terminal alkynes were readily functionalized. As with many of the palladium-catalyzed processes described herein, the addition reaction was highly regioselective and generated the 1,1-addition products. Of note was the ability to selectively functionalize the alkyne in the presence of a nitrile or alkyl chloride.

1,1-Diphosphinoethenes are challenging to synthesize in high yields. While the treatment of vinylidene chlorides with Grignard reagents or organolithium species is low yielding, nickel-catalyzed cross-coupling of 1,1-dichloroethenes and secondary phosphines has been used with some success [448], although the substrate scope is limited. An interesting approach to the synthesis of these compounds has been reported using the formation of a vinylcuprate intermediate as a key step (Scheme 4.310) [476]. The authors generated air-stable phosphine sulfides to facilitate the isolation of the target species. Once isolated and purified, the desulfurization of the target compounds was achieved through the addition of trimethylaminophosphine and heating in toluene. Given the high yields of this approach as well as the historical challenges associated with the syntheses of these compounds, this is a very practical approach.

The thiophosphination of alkynes incorporates both a diorganophosphino group and a phenylthio fragment into the substrate and converts the alkyne into an alkene (Scheme 4.311) [477]. The chemistry was straightforward and irradiation of the reagents under neat conditions generated the functionalized materials. A range of terminal alkynes and an internal alkyne were readily functionalized using this approach. When terminal alkynes were screened, the addition reaction was regioselective for the addition of the phosphorus fragment to the internal alkyne carbon, while the arylsulfide was attached to the terminal position. In addition, the thiophosphination generated predominately *E*-isomers. Curiously, a double phosphination or thiolation was not observed in these reactions. Analogous reactions with internal alkynes were also regioselective; however, the stereoselectivity

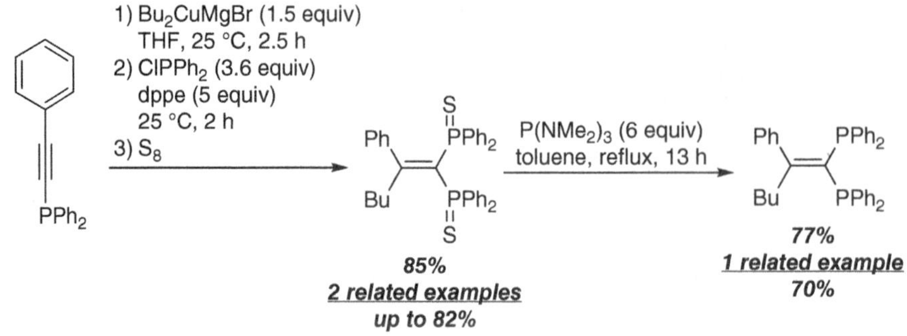

SCHEME 4.310   Synthesis of 1,1-diphosphinoethenes [476].

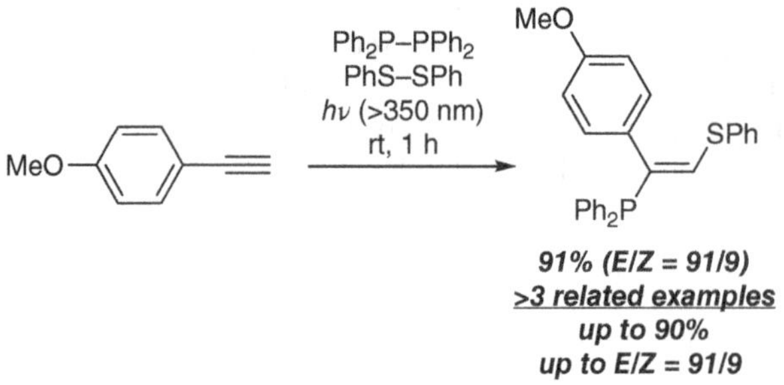

SCHEME 4.311   Thiophosphination of a terminal alkyne [477].

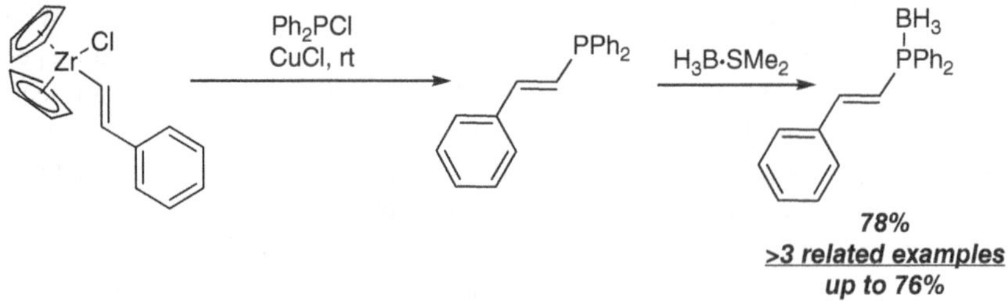

SCHEME 4.312   Copper-promoted conversion of vinylzirconocenes into protected vinylphosphines [478].

for the *E*-isomer was not as high. Given the high efficiency of this reaction, this is a very attractive approach for the synthesis of these compounds.

The copper-promoted reaction between vinylzirconocenes and chlorophosphines generated vinylphosphines in moderate to excellent yields (Scheme 4.312 and 4.313) [478]. While this might seem like a cumbersome approach to the synthesis of vinylphosphines, the vinylzirconocene precursors were readily prepared from Cp$_2$ZrHCl. Once generated, the vinylphosphines were protected from unwanted air oxidation through the formation of phosphine-boranes. Using this protocol, a host of vinylzirconocenes were readily

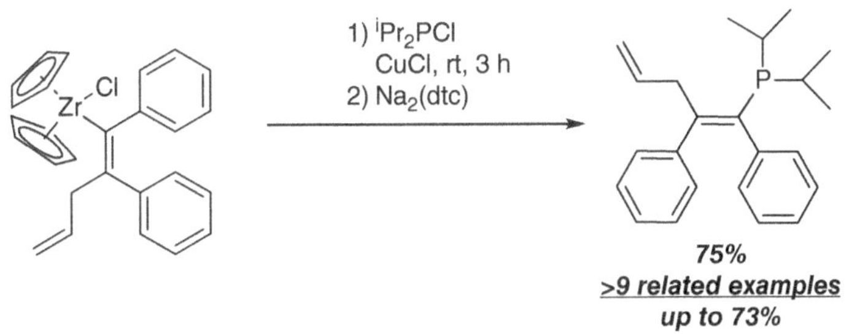

SCHEME 4.313 Copper-promoted conversion of α-substituted vinyl zirconocenes into crowded vinylphosphines through the addition of Na$_2$(dtc) [478].

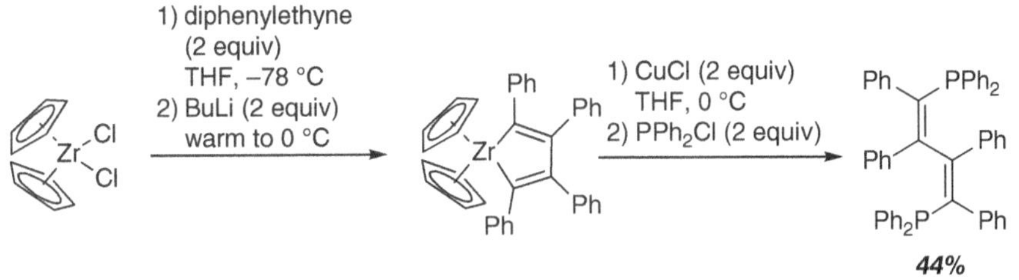

SCHEME 4.314 Zirconocene-mediated synthesis of vinylphosphines [416].

transformed into vinylphosphines. Bulky vinylzirconocenes as well as crowded chlorophosphines were well tolerated by the chemistry. Given the high yields, broad substrate scope, and straightforward reaction chemistry, this is quite a versatile approach to the synthesis of vinylphosphines.

The synthesis of vinylphosphines has been accomplished using a complimentary process (Scheme 4.314) [416]. The first part of the chemistry consisted of metallocycle formation by treatment of zirconocene dichloride with two equivlents of BuLi followed by the addition of two equivalents of an internal alkyne such as diphenylacetylene. Treatment of this intermediate with a chlorophosphine should have generated vinylphosphines. However, the authors found that this reaction was low yielding. After some experimentation, they discovered that the conversion of the zirconium metallocycle into a vinylcopper species through the addition of copper chloride followed by addition of the chlorophosphine was an effective route to the formation of the vinylphosphines.

This chemistry has recently been extended to the formation of a C$_2$-symmetric bisphosphine by functionalizing a tethered diyne (Scheme 4.315) [479]. Although this system has the potential to generate chiral diphosphines through atropisomeric chirality, the prerequisite for this is the conformational rigidity of the system. Unfortunately the diphosphine lost its optical purity upon simply stirring in octane at 40 °C. However, the authors found that oxidation to the phosphine oxide imparted a significant amount of stability to the system and no loss of chirality was observed upon heating the bisphosphine oxide to over 100 °C.

Acylphosphines can be prepared through the metal-free treatment of acyl chlorides with secondary phosphines (Scheme 4.316) [480]. This approach is high yielding

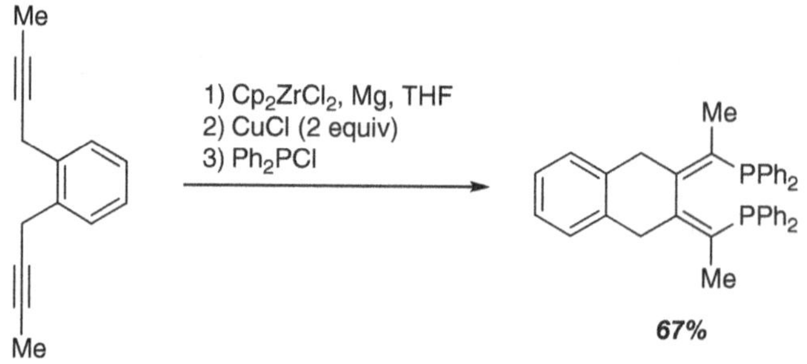

SCHEME 4.315   Synthesis of a C_2-symmetric bisphosphine [479].

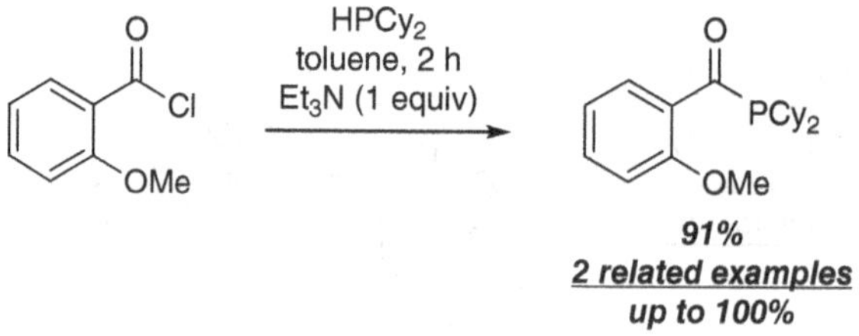

SCHEME 4.316   Synthesis of acylphosphines [480].

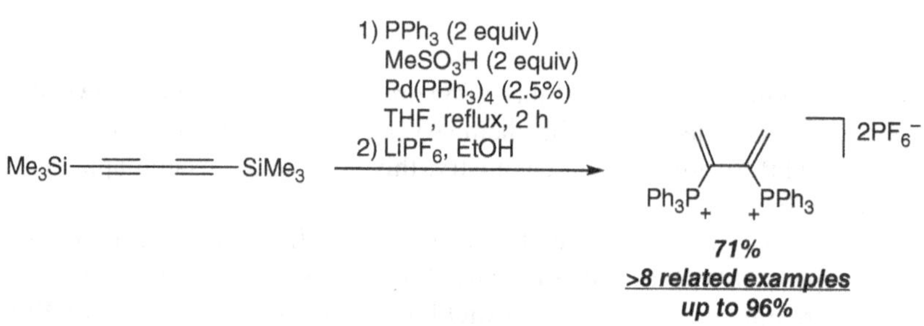

SCHEME 4.317   Palladium-catalyzed double addition of triphenylphosphine to diynes [481].

and uses an organic base to consume the equivalent of HCl generated in the process. Yields of the acylphosphines were outstanding and the chemistry occurred under mild conditions.

The palladium-catalyzed double addition of triphenylphosphine to unactivated diynes proceeds in moderate to high yields (Schemes 4.317 and 4.318) [481]. The authors found that several palladium(0) complexes successfully promoted the addition, while $(Ph_3P)_2PdCl_2$ did not. Thus, using $Pd(PPh_3)_4$ as the precatalyst, a number of terminal alkynes were readily transformed into vinylphosphonium salts. The reaction was highly regioselective and most substrates added the phosphine to the internal carbon. Even 4-octyne was readily functionalized using this system. Once the P—C bond-forming reaction was complete, the authors exchanged the anion for PF_6^- through the addition of $LiPF_6$. While palladium catalysts promoted addition to the internal carbon, the regioselectivity of the rhodium-catalyzed reaction was reversed, and the phosphine added to

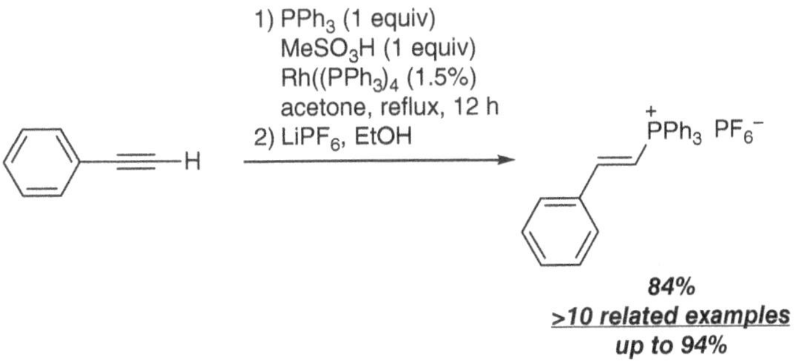

SCHEME 4.318   Rhodium-catalyzed addition of triphenylphosphine to alkynes [481].

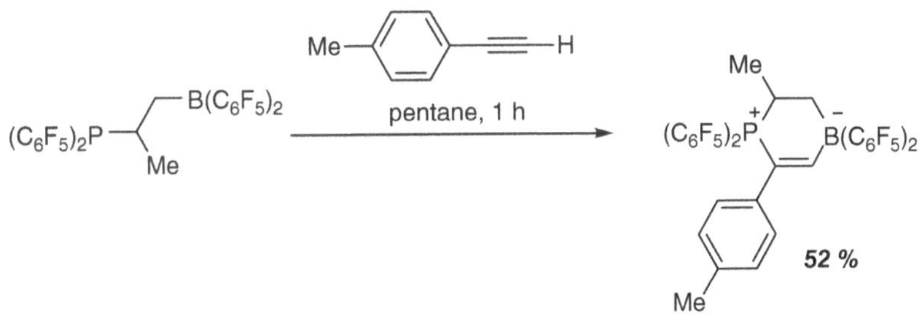

SCHEME 4.319   Synthesis of frustrated Lewis acid–base pairs [433].

the terminal carbon. Similar to the palladium chemistry, the counterion was exchanged for PF_6^- following the P—C bond-forming reaction.

One of the more novel synthetic approaches to the synthesis of vinylphosphines entailed the addition of alkynes to a functionalized tertiary phosphine (Scheme 4.319) [433]. These compounds are part of a special class of compounds that contain an intramolecular frustrated Lewis acid–base pair (FLP). These FLPs have been used to react with small molecules such as hydrogen and carbon dioxide. The P—C bond-forming reaction was operationally trivial and consisted of mixing the alkyne with the specialized precursor in a hydrocarbon solvent for 1 h. The product precipitated and was separated by simple filtration. Although the substrate scope for this reaction is limited at present, this approach does provide a pathway to an intriguing class of vinylphosphines.

While phosphines have received a great deal of attention as organocatalysts for organic transformations, phosphine oxides have been growing in popularity over the last decade. One of the most significant practical advantages of the latter group of compounds is their air stability. While phosphines are often quite sensitive to air oxidation, phosphine oxides can typically be stored for long periods of time under air with no further oxidation. A recent example of the use of phosphine oxides as organocatalysts for organic transformations entails the asymmetric addition of aldehydes to allylsilanes (Scheme 4.320) [479]. The authors initially generated the bisphosphine precursor through a zirconium-promoted cyclization/phosphination reaction and generated the bisphosphine oxide through oxidation (hydrogen peroxide). This bisphosphine oxide catalyst was proposed to promote the reaction by acting as a chiral Lewis base and coordinating the chlorosilane to

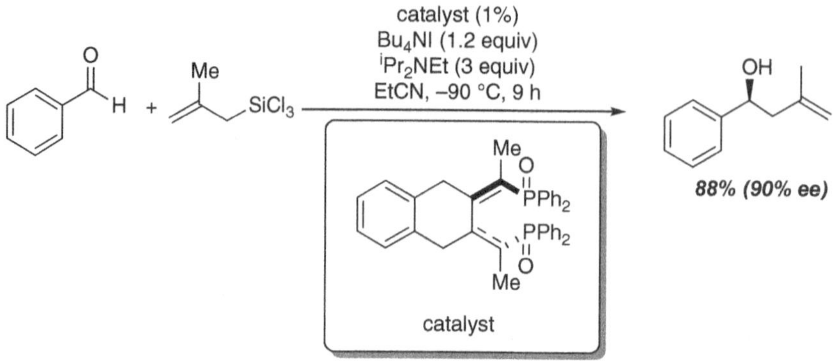

SCHEME 4.320   Application of a bisphosphine oxide as an organocatalyst [479].

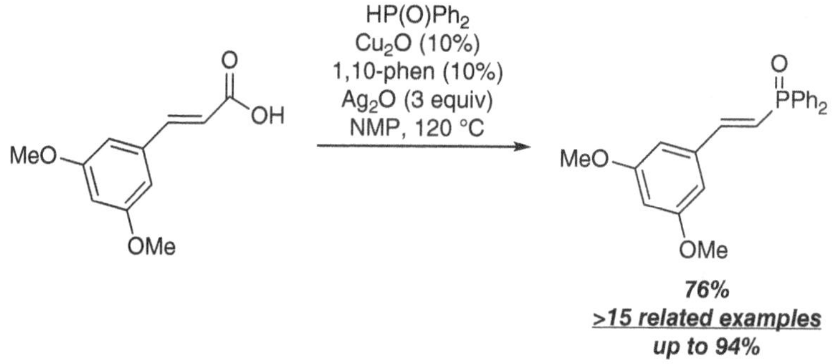

SCHEME 4.321   Copper oxide-catalyzed decarboxylative C–P cross-coupling [482].

provide preorganization/activation of the system prior to addition of the aldehyde. It is well known that diphosphines and diphosphine oxides can exhibit atropisomeric chirality through restricted rotation and while the diphosphine lost its optical purity upon simply stirring in a hydrocarbon solvent at room temperature, the analogous bisphosphine oxide was conformationally rigid at temperatures up to 135 °C. The authors used this bisphosphine oxide to promote the enantioselective allylation reaction at low temperatures (–90 °C). An attractive aspect of this chemistry was the observation that the bisphosphine oxide could be recovered using column chromatography (silica gel) with no loss of optical purity.

Recently, a practical synthesis of vinylphosphine oxides was reported by Liang and Yang (Scheme 4.321) [482]. Their approach started with a secondary phosphine oxide and an alkenylpropiolic acid. Secondary phosphine oxides are attractive starting materials and are readily prepared by several routes. After screening a range of copper catalysts, copper(II) oxide was selected for the screening runs. The chemistry is operationally simple as all of the reagents and catalysts were loaded in air. The only precaution for this chemistry was exchanging the atmosphere for argon prior to heating. In related work, copper chloride was used to promote a decarboxylative C–P coupling (Scheme 4.322) [483].

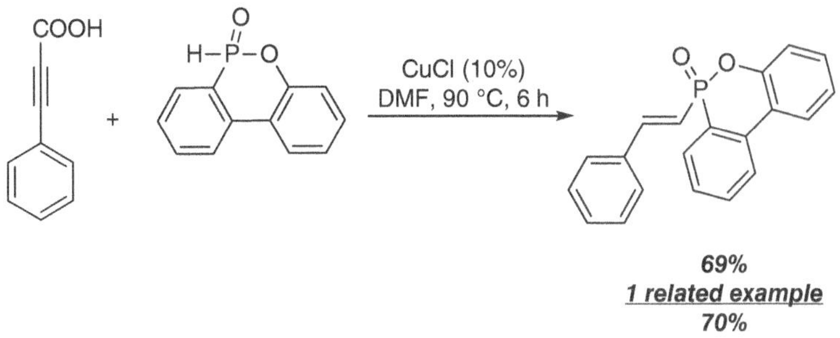

SCHEME 4.322   Copper chloride-catalyzed decarboxylative C—P cross-coupling [483].

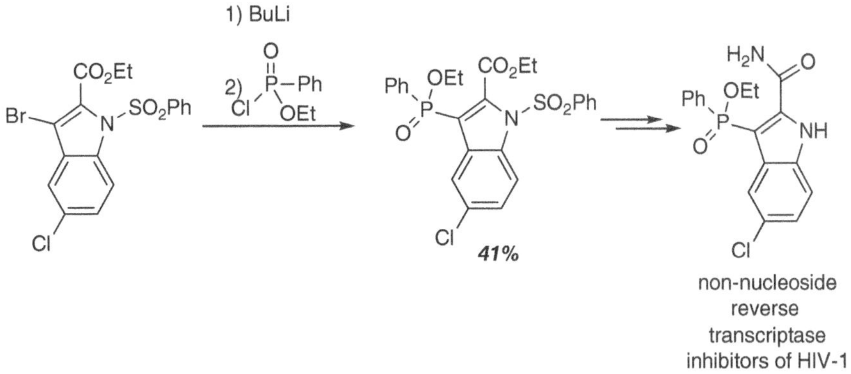

SCHEME 4.323   Synthesis of a highly functionalized vinylphosphinate [484].

While Grignard and organolithium reagents are commonly avoided when substrates contain electrophiles, there are examples where the metalation chemistry can compete with the secondary reactions between the organometallic reagent and the other electrophiles. Recently, Alexandre reported the formation of a vinylphosphinate using BuLi (Scheme 4.323) [484]. The interesting aspect of this reaction was that the starting vinyl halide contained both sulfone and ester functional groups. These groups are known to react with organolithium/Grignard reagents and complicate the desired metalation reaction. The phosphinates were intermediates in an overall synthesis of phosphinylated indoles. The latter displayed significant activity as reverse transcriptase inhibitors of HIV-1.

The addition of secondary phosphine oxides to alkynes remains one of the more popular routes to the synthesis of vinylphosphine oxides. An array of metal catalysts are known to promote this reaction, and despite the attention this area has received, general processes that are regio- and stereoselective and are tolerant to a wide range of preexisting functional groups are rare. Essentially, each set of substrates needs to be evaluated against the strengths and weaknesses of each approach in order to design a successful synthesis. Several of these methods are outlined in Schemes 4.324 through 4.330.

The synthesis of iminophosphine oxides through the addition of secondary phosphine oxides to isocyanides has been reported by Han (Scheme 4.331) [94]. A range of metal catalysts were screened for activity towards the addition reaction, and Pd$_2$dba$_3$ was found

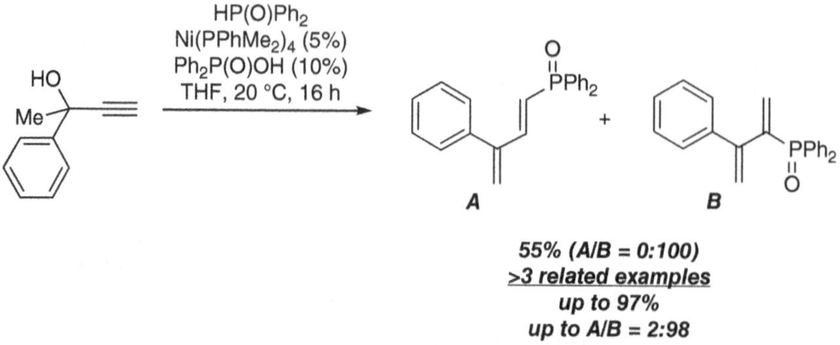

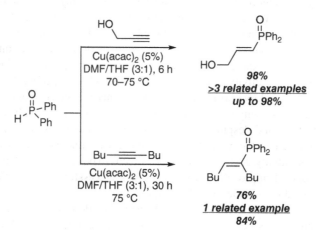

55% (A/B = 0:100)
>3 related examples
up to 97%
up to A/B = 2:98

SCHEME 4.324   Nickel-catalyzed addition of secondary phosphine oxides to propargyl alcohols leading to the formation of 1,3-butadienes [485].

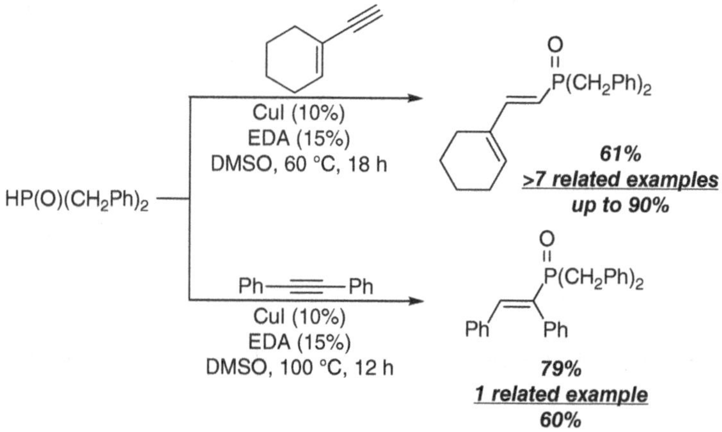

61%
>7 related examples
up to 90%

79%
1 related example
60%

SCHEME 4.325   Copper iodide-catalyzed addition of secondary phosphine oxides to internal and terminal alkynes [486].

98%
>3 related examples
up to 98%

76%
1 related example
84%

SCHEME 4.326   Copper acetylacetonate-catalyzed addition of secondary phosphine oxides to propargyl alcohols and internal alkynes [487].

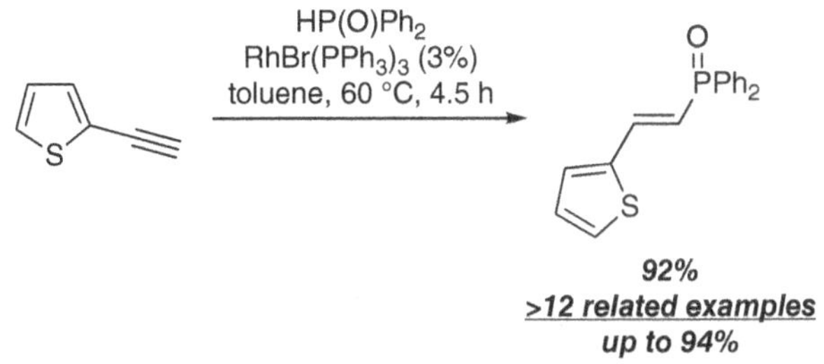

SCHEME 4.327   Rhodium-catalyzed addition of secondary phosphine oxides to terminal alkynes [488].

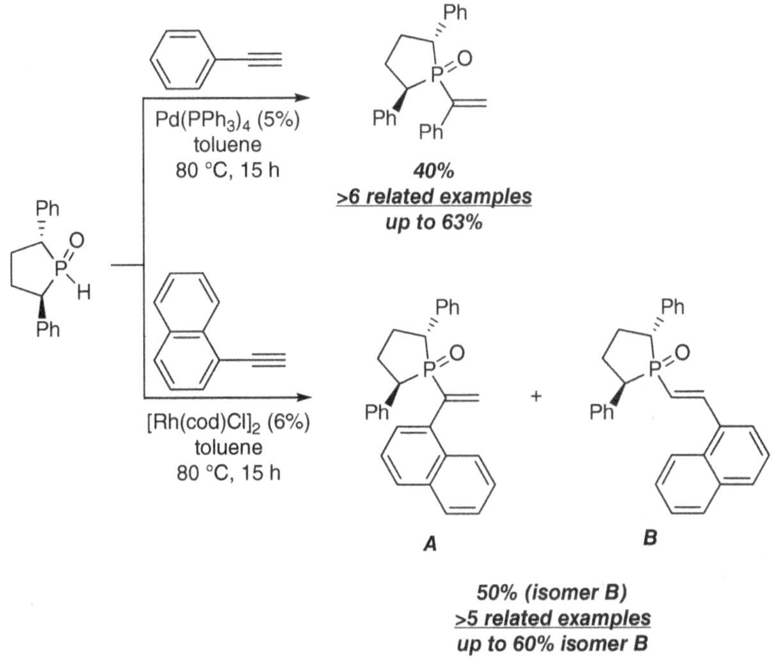

SCHEME 4.328   Rhodium-catalyzed addition of chiral secondary phosphine oxides to terminal alkynes [489].

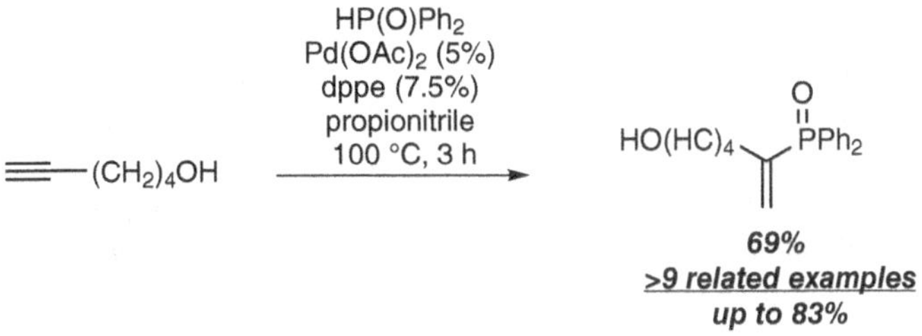

SCHEME 4.329   Palladium-catalyzed addition of secondary phosphine oxides to terminal alkynes [490].

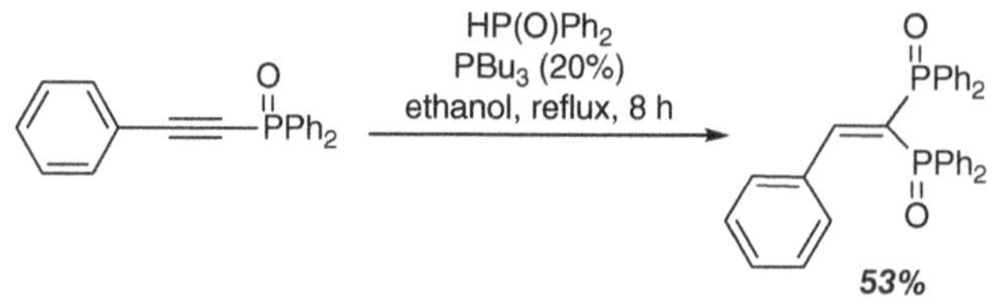

SCHEME 4.330   Phosphine-catalyzed addition of secondary phosphine oxides to activated alkynes [491].

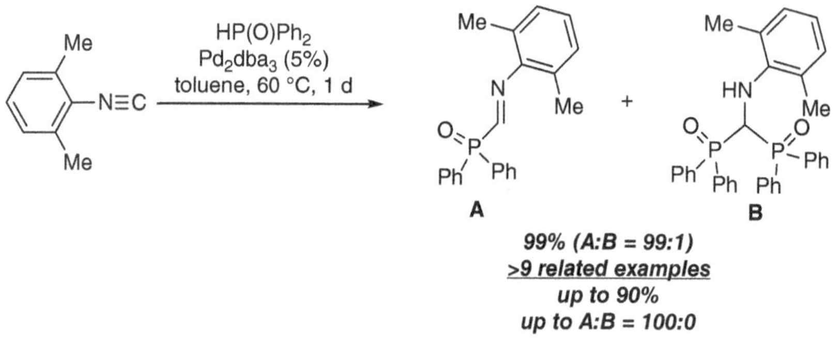

SCHEME 4.331   Palladium-catalyzed addition of secondary phosphine oxides to isocyanides [94].

to be the most effective. It was interesting that several nickel and platinum salts did not promote the reaction while rhodium compounds promoted a double addition reaction leading to the formation of bisphosphinoylaminomethanes. Curiously, while most palladium-catalyzed processes require the use of stabilizing ligands such as phosphines or carbenes, this process did not require the use of added ligands. Presumably, the secondary phosphine oxide used as the reagent in this chemistry was able to solubilize and stabilize the palladium species generated in the reaction. The chemistry generated predominately the single addition product, although small amounts of the side product resulting from double addition could not be completely eliminated for most of the examples (1–5%). Curiously, complete elimination of the latter was observed when the secondary phosphine oxide contained at least one alkyl group. Thus, the addition of dibutylphosphine oxide to the isocyanide resulted in formation of the iminophosphine oxide with no contamination from the double addition species. In addition to screening several isocyanides for activity in this reaction, the authors probed the effectiveness of electronic variation of the diarylphosphine oxide. It was impressive that the incorporation of either electron-donating or electron-withdrawing groups on the diarylphosphine oxide generated high yields of the target compounds. While bulky isocyanides were well tolerated, the conversion was influenced by adding bulky groups to the phosphine oxide. The use of bulky substrates such as dicyclohexylphosphine oxides resulted in lower yields of the target compounds.

   The development of a palladium-catalyzed cross-coupling process for the preparation of vinylphosphonates by Hirao was one of the turning points for the preparation of vinylphosphonates [368]. His chemistry was operationally simple and utilized a readily available palladium(0) compound as the catalyst and triethylamine as the base. The temperatures required for his reaction were mild compared to Arbuzov chemistry, and the phosphorus–carbon bond-forming reaction was compatible with a wide range of functional groups.

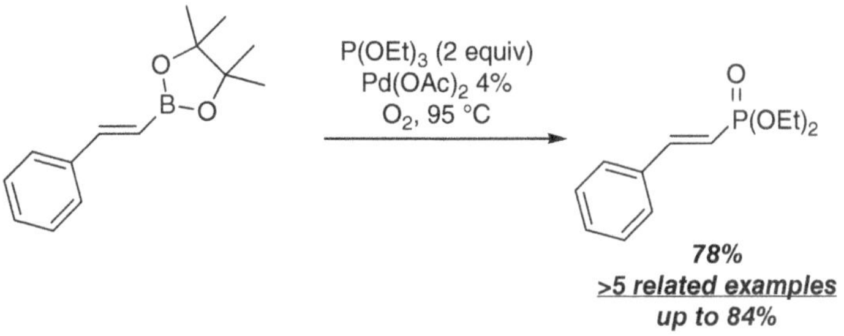

SCHEME 4.332   Synthesis of vinylphosphonates through palladium-catalyzed cross-coupling [493].

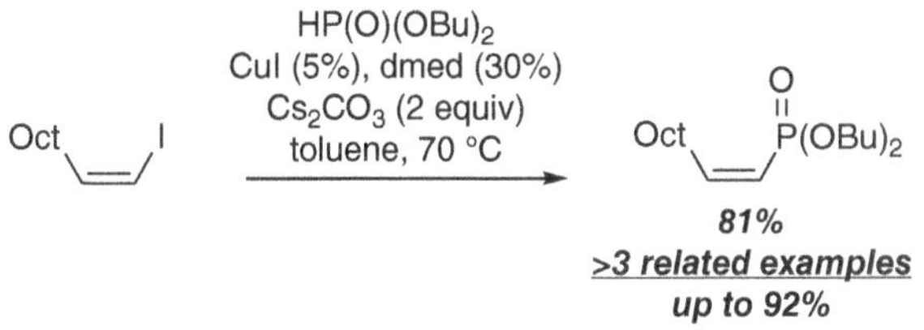

SCHEME 4.333   Synthesis of vinylphosphonates through copper iodide-catalyzed cross-coupling [318].

Perhaps, one of the most attractive aspects of this and related cross-coupling approaches for the formation of vinylphosphonates is the retention of the stereochemistry of the alkene. Following his pioneering work, a host of related transition metal-mediated reactions were developed.

The palladium-catalyzed cross-coupling of vinylboronates with phosphites generates vinylphosphonates in moderate to excellent yields. Since most of the H—P addition reactions generate E-vinylphosphonates, this cross-coupling reaction is attractive as it can be tuned to generate either E- or Z-stereochemistry (Scheme 4.332) [493]. This stereocontrol arose from the use of E- or Z-vinylboronates. Essentially, if the process started with a Z-vinylboronates, Z-vinylphosphonates were generate. This approach was attractive as the E- and Z-vinylboronates can be readily prepared [494–496]. The authors screened a number of group 10 catalysts to determine which catalyst was the most effective for the coupling reaction and found that palladium acetate provided the highest yield of the alkenylphosphonates without scrambling the stereochemistry. This reaction is operationally simple as all reagents were loaded in air and the reaction was carried out under an atmosphere of oxygen. This chemistry can also be carried out without the addition of a solvent.

A copper-catalyzed cross-coupling approach to the synthesis of vinylphosphonates has been reported and uses CuI as the copper source and N,N-dimethylethylenediamine as the supporting ligand (Scheme 4.333) [318]. An attractive aspect of this chemistry is the retention of the stereochemistry of the starting alkene. This is an advantage over the hydrophosphonylation methodology that tends to form E-alkenes or a mixture of the E- and Z-isomers. This approach provided a pathway for the preparation of pure Z-vinylphosphonates. Due to the mild reaction conditions, the chemistry was also successful with functionalized alkenes including vinyl esters. In related work, Stawinski reported a microwave-assisted

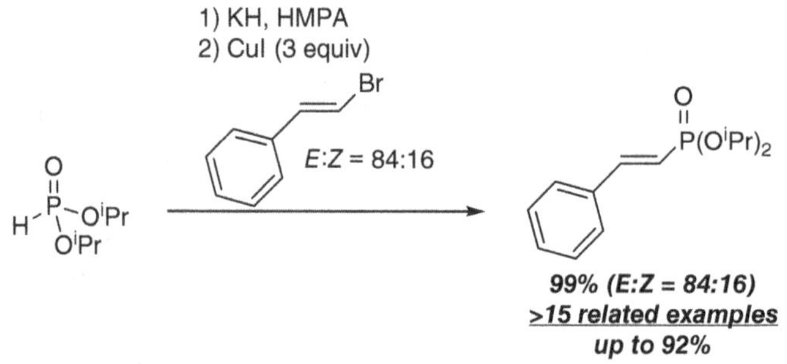

SCHEME 4.334   Copper-catalyzed coupling of secondary phosphites with vinyl bromides [497].

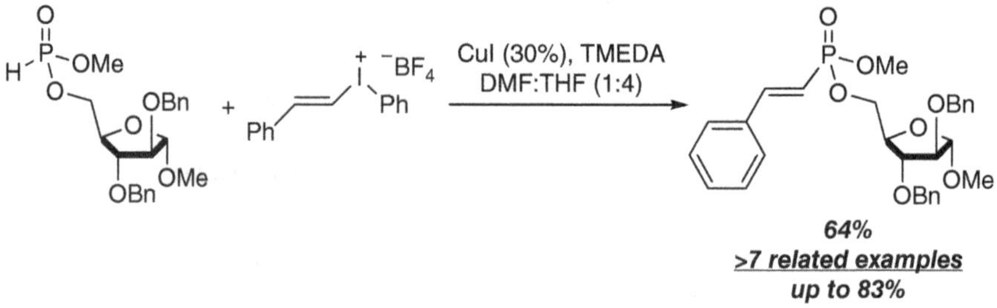

SCHEME 4.335   Copper-catalyzed coupling of secondary phosphites with iodonium salts [498].

cross-coupling reaction for the preparation of vinylphosphonates [381]. The stereochemistry was retained in this reaction as well (i.e., Z-vinyl bromides afforded Z-vinylphosphonates). Given the speed and high yields of this reaction as well as the availability of the substrates, it could be the synthetic method of choice for the synthesis of vinylphosphonates.

The copper-catalyzed cross-coupling of vinyl bromides with secondary phosphites has been reported (Scheme 4.334) [497]. Copper iodide was an effective catalyst for this reaction; however, three equivalents of CuI were needed for each vinyl bromide. The authors were pleased to find that the stereochemistry of the vinylphosphonate mirrored the starting vinyl bromide. Although HMPA was a superior solvent for the reaction chemistry, the authors successfully used other solvents such as DMF and tetramethylurea due to the potential carcinogenicity of HMPA. The overall reaction was carried out in two stages with the phosphonate anion generated in the first step, while the cross-coupling reaction was carried out in the second step.

The copper-catalyzed cross-coupling of hydrogen phosphonates with vinyliodonium tetrafluoroborates generated vinylphosphonates in moderate to excellent yield (Scheme 4.335) [498]. The authors screened a number of catalyst systems and found that copper(I) iodide along with TMEDA successfully promoted the C—P bond-forming reaction. While the catalyst loading was a little high (30%), the reaction proceeded at room temperature. This is a significant advantage to this approach. The initial stereochemistry was retained through the formation of the vinylphosphonate. While the base majority of transition metal-catalyzed approaches for the formation of vinylphosphonates used either an organic or mineral base, the authors found that this chemistry proceeded in the absence of an added base.

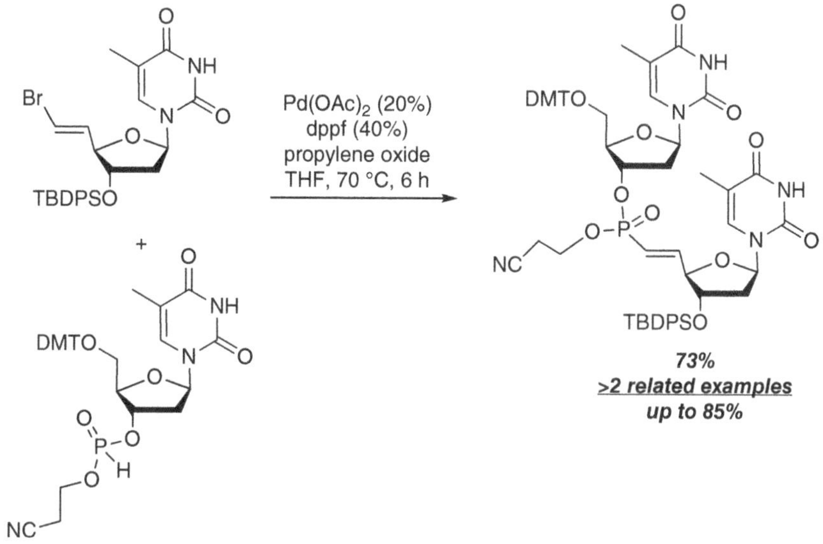

SCHEME 4.336   Palladium-catalyzed vinylphosphonate-linked nucleosides [499].

The preparation of vinylphosphonate-linked oligonucleic acids using palladium-catalyzed cross-coupling chemistry has been reported by Hayes (Scheme 4.336) [499]. His catalyst system was comprised of palladium acetate and dppf along with propylene oxide dissolved in THF. The key step in this chemistry was the observation that the requisite hydrogen phosphonates could be prepared directly from commercially available 5'-DMT-cyanoethyl phosphoramidites. The conversion was accomplished by simply treating these precursors with tetrazole for 10 min followed by the addition of water. Furthermore, this reaction proceeds in near-quantitative yields, and the hydrogen phosphonates could be successfully used in the palladium-catalyzed reaction after simply drying the reaction mixture with magnesium sulfate and removal of the volatiles. This was a very practical advancement in the preparation of these precursors as many related protocols resulted in low yields and time-consuming separations. Cross-coupling these substrates with a thymidine-derived vinyl bromide generated vinylphosphonate-linked dinucleosides in moderate to high yields.

A decarboxylative coupling between a carboxylic acid and a hydrogen phosphonate has the potential to be a very popular approach to the synthesis of vinylphosphonates (Scheme 4.337) [482]. Both the carboxylic acid and the hydrogen phosphonate are stable materials that rarely need special storage and handling. A range of copper catalysts successfully promoted the reaction, and copper (I) oxide was selected for the screening reactions. A phosphinate version of the reaction was also reported. The chemistry is operationally simple as all of the reagents and catalysts were loaded in air. The only precaution for this chemistry was exchanging the atmosphere for argon prior to heating.

The development of transformations that can be carried out under air instead and do not require the use of gloveboxes or vacuum manifolds is extremely practical due to their operational simplicity. To this end, a potentially valuable microwave-assisted formation of vinylphosphonates was reported by Larhed (Scheme 4.338) [391]. Using palladium acetate as the catalyst and 2,9-dimethyl-1,10-phenanthroline (dmphen) as the supporting

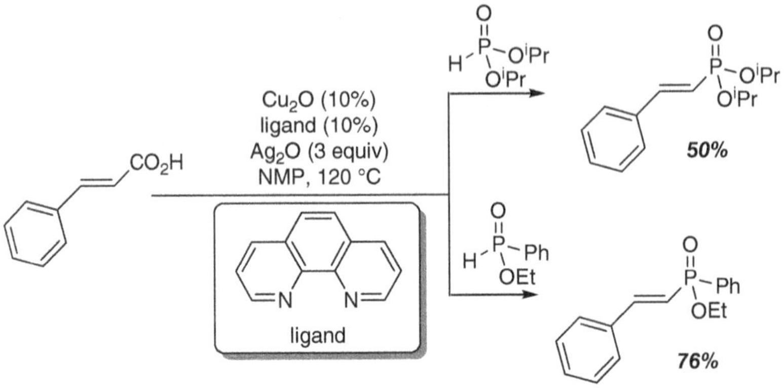

SCHEME 4.337 Copper-catalyzed decarboxylative synthesis of vinylphosphonates and phosphinates [482].

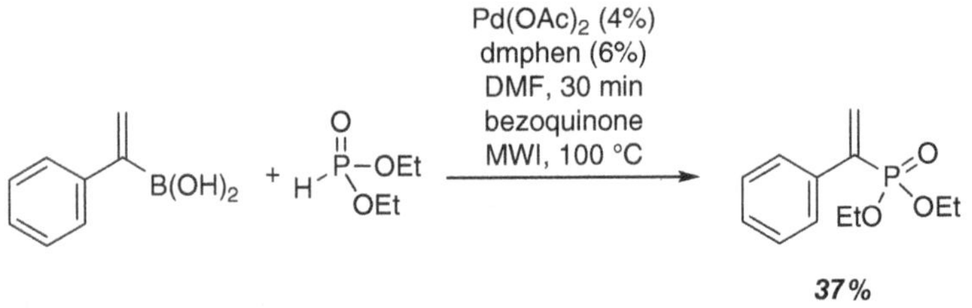

SCHEME 4.338 Palladium-catalyzed coupling of secondary phosphites with boronic acids [391].

ligand, a hydrogen phosphonate was successfully coupled with a vinylboronic acid to generate the vinylphosphonate under an atmosphere of air. The authors needed to use a robust supporting ligand since phosphine oxidation can be problematic when using phosphine-based ancillary ligands. While this might seem like a classic palladium-catalyzed cross-coupling reaction, the authors observed that no added base was needed for the coupling reaction. While providing evidence for a different type of catalytic cycle, the removal of the base from the reaction will enable the use of base-sensitive substrates. While the reaction generated the cross-coupling product simply under an atmosphere of air, adding benzoquinone increased the conversion and decreased the reaction time. The reaction was operationally trivial as well. All the reagents were loaded into the reaction vial under an atmosphere of air and purging with an inert gas prior to irradiation was not needed. One issue with this chemistry is that functional groups that are sensitive to benzoquinone oxidation could present challenges. However, this issue is relatively minor given the practical advantage of being able to carry the reaction out under air. While only a single example of a vinylphosphonate was listed, there is significant potential for further development.

Similar to the sections previously on hydroelementation reactions, the addition of secondary phosphites to alkynes is a popular approach to the synthesis of vinylphosphonates. Many of these reactions are problematic due to the formation of mixtures containing

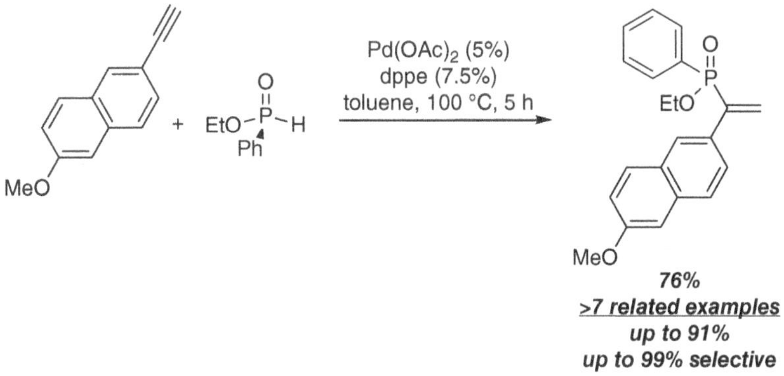

SCHEME 4.339 Palladium-catalyzed P—H addition reactions [500].

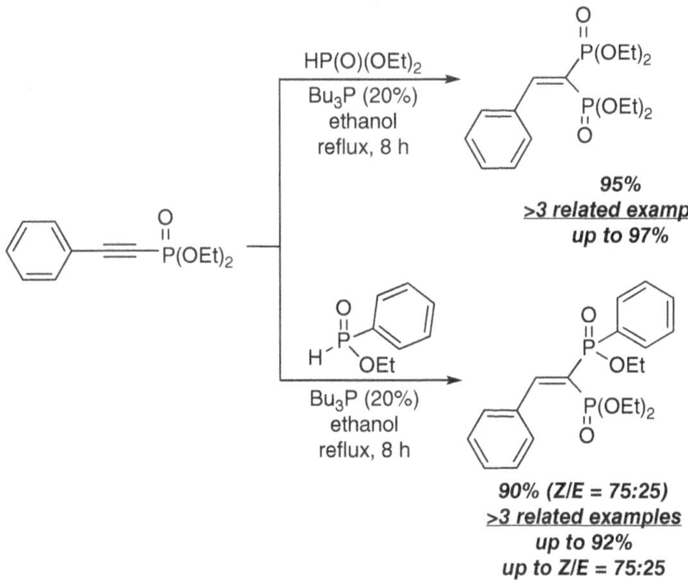

SCHEME 4.340 Phosphine-catalyzed P—H addition reactions [491].

regio- and stereoisomers. A number of the more selective approaches to this chemistry are summarized in Schemes 4.339–4.343.

The treatment of acyl chlorides with trialkylphosphites is one of the oldest known methods for the preparation of acylphosphonates (Scheme 4.344) [503, 504]. The reaction was typically carried out by simply adding the phosphite to the acyl chloride with stirring. The C—P bond-forming process was exothermic, and the phosphite needed to be added slowly to maintain a low temperature. This was critical since secondary products appeared at higher reaction temperatures. One of the attractive aspects of this chemistry was that no solvents were used. The scope of the reaction chemistry was broad, and a host of acyl

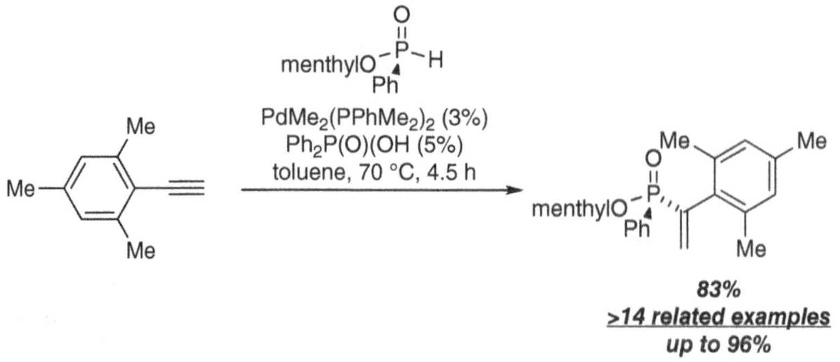

SCHEME 4.341   Palladium-catalyzed P—H addition reactions using a resolved phosphinate [501].

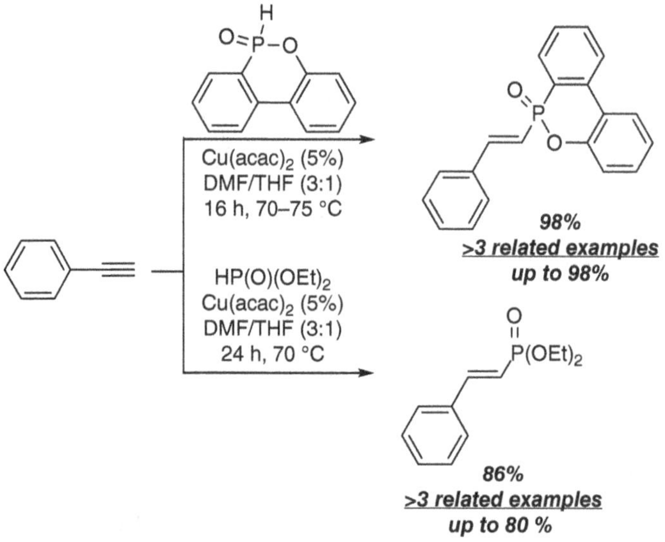

SCHEME 4.342   Copper-catalyzed P—H addition reactions using phenylacetylene [487].

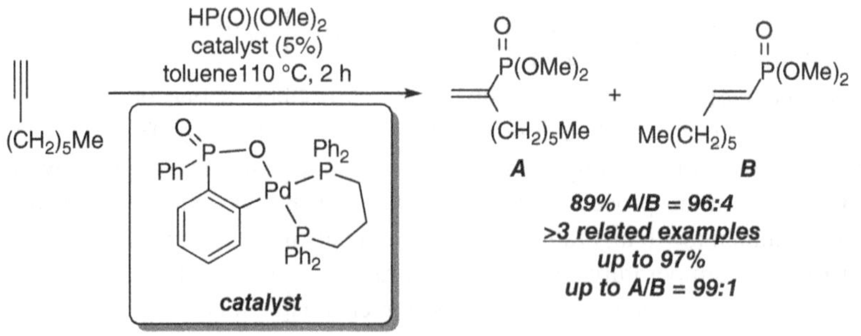

SCHEME 4.343   Palladium-catalyzed Markovnikov-selective P—H addition reactions [502].

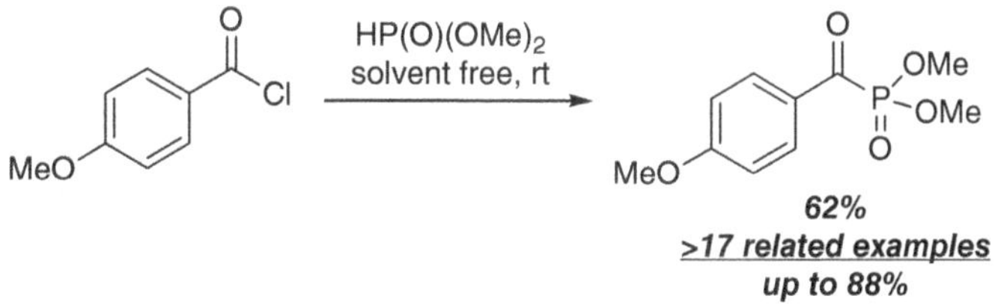

SCHEME 4.344   Synthesis of an acylphosphonate [503, 504].

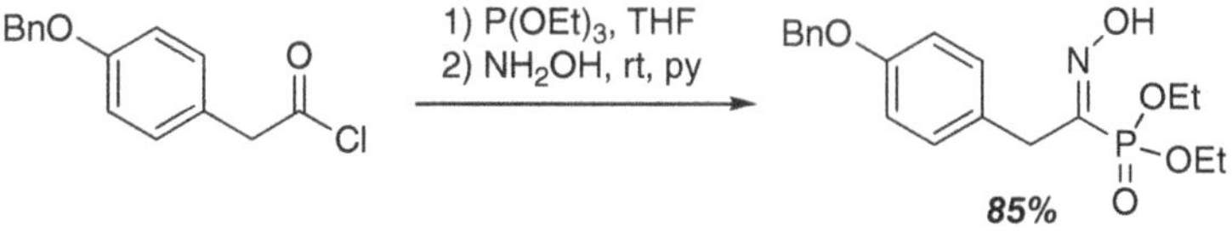

SCHEME 4.345   Synthesis of 1-hydroxyiminophosphonates [505].

halides bearing alkyl and aryl groups were successfully functionalized using this approach. Care must be taken to ensure that anhydrous conditions were maintained as the acylphosphonates decomposed upon exposure to moisture. As with the other Arbuzov-type reactions, one of the drawbacks of this chemistry is the generation of a full equivalent of alkyl halide that must be dealt with through recycling or disposal. Although this reaction was developed in the 1960s, it remains the most popular method for the preparation of acylphosphonates.

The synthesis of 1-hydroxyiminophosphonates has been accomplished in a one-pot reaction (Scheme 4.345) [505]. Initial treatment of the functionalized acyl chloride with triethylphosphite generated the acylphosphonate through an Arbuzov reaction. Treatment of this compound with hydroxylamine and pyridine at room temperature afforded the 1-hydroxyiminophosphonates. While only a few examples were listed, the high yields of the target compounds were attractive. These interesting compounds have been used as substrates for a range of reactions. Of particular interest is the ability to convert them into α-aminophosphonates through a nickel-catalyzed reduction using sodium borohydride in methanol [506].

4.4.1   Troubleshooting the Synthesis of Vinylphosphines and Related Compounds

While it is impossible to anticipate the challenges of every new reaction/method that will be developed over the next few decades, there are some general trends that can be followed. The chart below will provide some direction when trying to select the most practical solution to the specific issues related to the synthesis of vinylphosphines and related compounds.

Problem/Issue/Goal	Possible Solution/Suggestion
Partial oxidation of the phosphine during workup	One solution is to rigorously purify and deoxygenate all of the reagents and solvents used during the workup procedure
	Alternatively, it might be more practical to fully convert the phosphine into a phosphine oxide, sulfide, or borane adduct prior to isolation

(Continued)

(Continued)

Problem/Issue/Goal	Possible Solution/Suggestion
Can vinyl Grignard reagents be used to generate vinylphosphines and phosphonates?	This is a classic approach to the synthesis and typically involves the addition of the vinyl Grignard to a solution of a chlorophosphine or chlorophosphonate. As long as the substrates do not possess electrophiles, this is an effective approach to the synthesis of these compounds [262, 429, 430, 432]
The starting vinyl bromide has a carbonyl group. Will this eliminate a Grignard approach?	Perhaps not. Knochel has developed an approach that converts Grignard/organolithium reagents into organozinc reagents that tolerate the presence of some electrophiles [8]
The Grignard approach typically uses chlorophosphines as substrates. Can this chemistry be reversed through the addition of lithium phosphides to vinyl halides?	This approach is not as popular as the classic approach, but it has been successfully used to make vinylphosphines [434]
Can vinylphosphines and phosphonates be generated through cross-coupling chemistry?	This is a popular approach due to the functional group tolerance of palladium- and copper-catalyzed reactions [7, 295, 318, 437, 440]
The starting material has a terminal alkyne. Can it be converted into a vinylphosphine?	This addition process is typically promoted by the addition of bases, radical initiators, and transition metal catalysts. General approaches with wide functional group tolerance are rare due to varying levels of control over the regio- and stereochemistry [449, 450, 457]
Can a vinylphosphonate be generated from a propargyl alcohol with retention of the alcohol?	The ruthenium-catalyzed hydrophosphonylation of propargyl alcohols would be a reasonable approach [473]
The starting secondary phosphine was converted into a phosphine oxide to increase the air stability. Are these secondary phosphine oxides reactive in hydrophosphinylation reactions?	Secondary phosphine oxides are quite active in hydrophosphinylation reactions. These reactions were typically promoted by the addition of bases, transition metals, or radical initiators [485–488, 507, 508]
Would like a green or sustainable version of the hydrophosphination	Try a solvent- and catalyst-free hydrophosphination [78]. This approach worked well with a range of styrenes and activated alkenes
Are there solvent- and catalyst-free procedures for the addition of secondary phosphines to alkynes?	The addition of diphenylphosphine to phenylacetylene in the absence of solvents and catalysts has been achieved [78]
Can a vinyl triflate be used in a cross-coupling reaction to generate a vinylphosphine?	A reasonable approach would be to use a cross-coupling reaction between a borane-protected secondary phosphine and a vinyl triflate [441]
The starting material can be easily converted into a 1,1-dibromoalkene. Can this transformed into a vinylphosphine oxide?	This conversion has been accomplished using a nickel-catalyzed process [462].
Can vinylphosphonates be generated through cross-coupling chemistry?	This is a classic approach to these compounds and is typically promoted by palladium and copper catalysts [318, 368, 493]
Can a vinylcarboxylic acid be used as a starting material for the synthesis of a vinylphosphonate?	The copper-catalyzed decarboxylative phosphonylation of carboxylic acids has been achieved [482]

4.5   SYNTHESIS OF ALKYNYLPHOSPHINES AND RELATED COMPOUNDS THROUGH THE FORMATION OF PHOSPHORUS–CARBON(sp) BONDS

Alkynylphosphines are valuable compounds with applications in a number of fields. They provide a unique geometry for the preparation of macromolecular architectures [509, 510], have been incorporated into potential metalloenediyne therapeutics [511], offer a range of coordination modes [509, 510, 512], and have served as substrates in a range of reactions [513–516]. Recently, several excellent reviews have been published on the typical routes to synthesis of alkynylphosphines [517, 518]. The following sections will summarize the traditional and current synthetic methods and highlight the practical aspects of each approach. Traditionally, alkynylphosphines were prepared by treating an organometallic acetylide with a chlorophosphine [517–519]. Although it is possible, most procedures do not isolate the acetylide reagent. It was generated in solution through the treatment of a terminal acetylene with a Grignard reagent or organolithium species at low temperatures. Organolithium reagents were some of the most popular reagents for this chemistry, and a host of alkynylphosphines were created using this approach. The chemistry remains one of the most popular and successful approach for the synthesis of these molecules provided there are no electrophiles in either substrate. The preparation of alkynylphosphines has also been accomplished by treatment of an alkynylphosphonite with two equivalents of a Grignard reagent [520]. The alkynyl fragment was initially incorporated into the substrate by treatment of a chlorophosphite with a lithium acetylide, and subsequent treatment of the alkynylphosphonite with the Grignard reagent generated the alkynylphosphine. While the alkoxy groups were displaced by the Grignard reagents, this chemistry does demonstrate the lability of the chloride relative to the alkoxy groups for the preparation of organophosphorus compounds.

One solution to the problem of using strongly nucleophilic organometallic reagents was developed by Beletskaya [521]. She developed a nickel-catalyzed procedure that cross-coupled a terminal alkyne with a chlorophosphine using triethylamine as the base. Isolated yields were good to excellent with minimal formation of the phosphine oxide after heating (toluene, 80 °C) for only 10–15 min. The most problematic substrate was the bulky di-*tert*-butylchlorophosphine. Refluxing with phenylacetylene in toluene for 2 weeks afforded 25% of the alkynylphosphine. Demonstrating the tolerance of this chemistry to electrophiles, this methodology was recently used for the preparation of alkynylphosphines containing esters [469]. A series of C_3-symmetric trisphosphines were also generated using this approach (Scheme 4.346) [522].

While most efforts in this area have been focused on developing a catalytic system that will tolerate a wide range of functionalized alkynes, an alternative approach to the

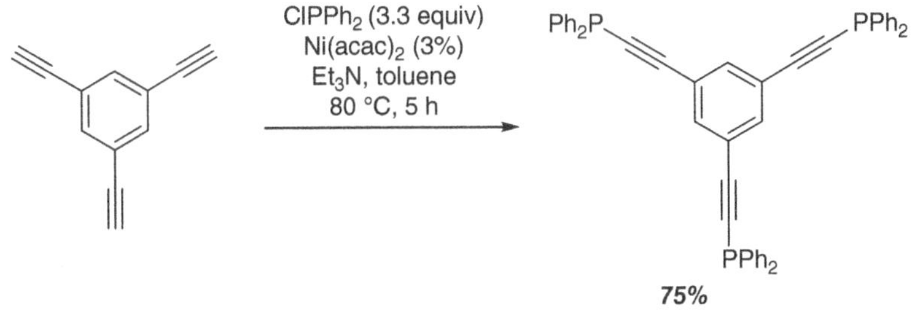

SCHEME 4.346   Nickel-catalyzed coupling of chlorophosphines with terminal alkynes [522].

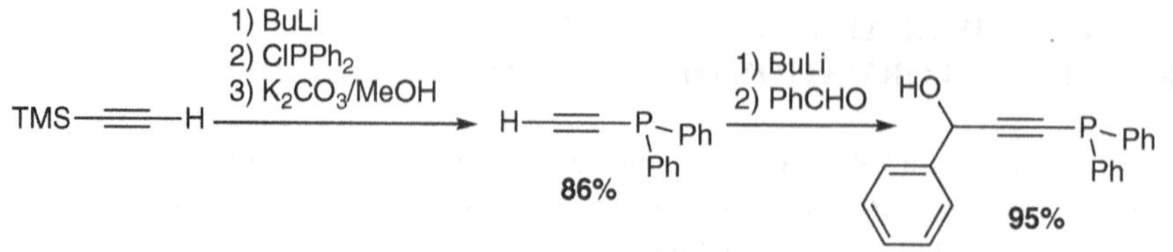

SCHEME 4.347   Stepwise approach to the synthesis of a phosphinated propargyl alcohol [469].

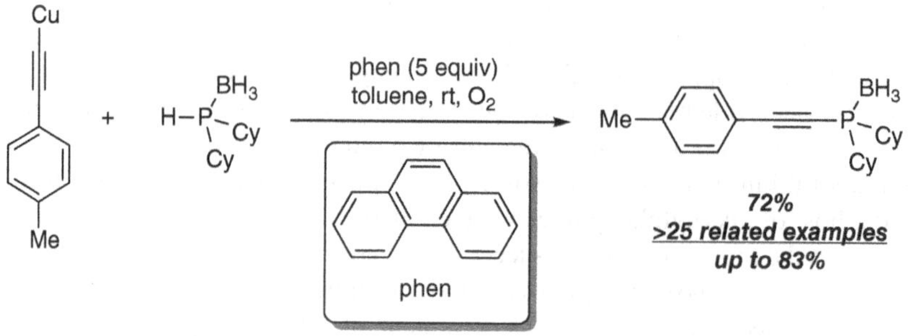

SCHEME 4.348   Oxidative alkynylation with borane protection of the resulting alkynylphosphines [523].

preparation of functionalized alkynylphosphines involves addition of the functional group after initial synthesis of the alkynylphosphine. This can be readily accomplished through the incorporation of a terminal acetylene. Deprotonation followed by the addition of an electrophile will afford the new alkynylphosphine with the functional group across the alkyne from the phosphine. Oshima and Yorimitsu demonstrated this using benzaldehyde as the electrophile (Scheme 4.347) [469]. The trick with this chemistry was protection of the acetylene during the formation of the alkynylphosphine. The TMS group is one of the most well-established protecting groups for alkynes, and it is easily removed.

As with most synthetic methods for the preparation of phosphines, rigorous exclusion of oxygen is typically required for a high-yielding process. This air sensitivity can present challenges for the isolation and purification of these compounds. Even a small amount of oxygen can convert a significant amount of the desired phosphine into a phosphine oxide. Naturally, the undesired phosphine oxide can often be removed using chromatography or crystallization. However, these extra steps cost more time and present further opportunities for the unwanted oxidation to occur. One approach that has been successful at circumventing the oxidation problem is to bind the phosphorus atom to BH_3 during the final step in the synthesis. These borane adducts are often significantly more stable towards oxidation than the free phosphine and can be readily purified with minimal oxidation. Once purified, treatment of the phosphine-boranes with DABCO readily generated the alkynylphosphines.

The oxidative alkynylation of phosphine-boranes using copper acetylides as substrates has been reported (Schemes 4.348 and 4.349) [523]. While many organometallic acetylides such as lithium phenylacetylide are extremely reactive towards moisture/electrophiles and cannot typically be handled in air, copper acetylides are relatively stable and are a very practical source of the acetylide fragment. As a direct result of their stability towards electrophiles, a

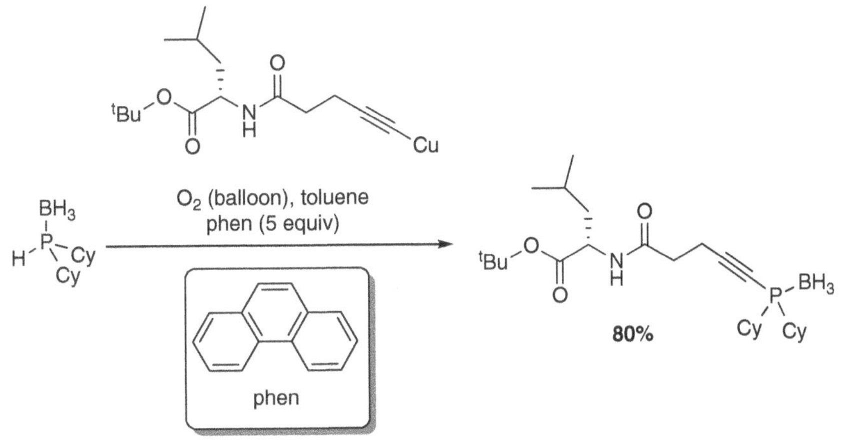

SCHEME 4.349 Oxidative alkynylation of borane-phosphines using a functionalized organocopper reagent [523].

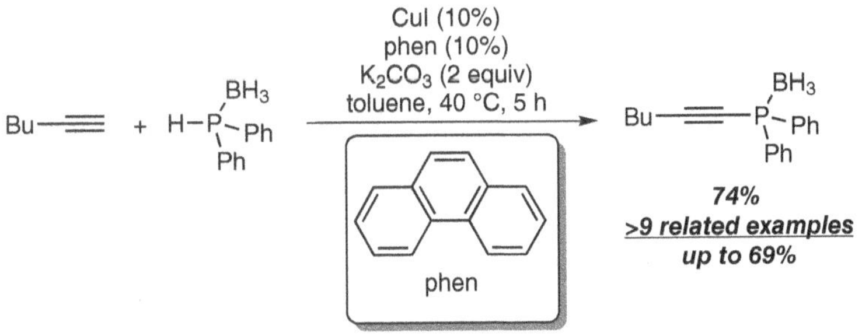

SCHEME 4.350 Copper-catalyzed coupling of protected secondary phosphines with terminal alkynes [524].

host of copper acetylides bearing a range of functional groups are known. Since the alkynylation reactions were carried out under an atmosphere of oxygen, neither a glovebox nor a nitrogen line were needed for this chemistry. While this is advantageous for small-scale reactions, the use of molecular oxygen during scale-up could introduce a hazard. The authors were pleased to find that the electronic limitation on the part of the alkyne that was discovered during their initial studies was no longer an issue and a wide range of functional groups were tolerated by the chemistry. However, they discovered an interesting electronic effect during the substrate screening. While moderate to excellent yields of the target phosphines were obtained using dialkylphosphine-boranes, the incorporation of a single aromatic group completely shut down the cross-coupling chemistry. Despite this electronic limitation, this approach remains one of the most practical methods for the preparation of alkynylphosphines.

While many of the methods described in previous sections utilized an electrophilic source of phosphorus for the synthesis of alkynylphosphines, a recent report by Gaumont summarized their investigations into devising a copper-catalyzed cross-coupling approach using nucleophilic phosphines (Scheme 4.350) [524]. While it may initially not seem attractive to use potentially pyrophoric secondary phosphines for the synthesis of

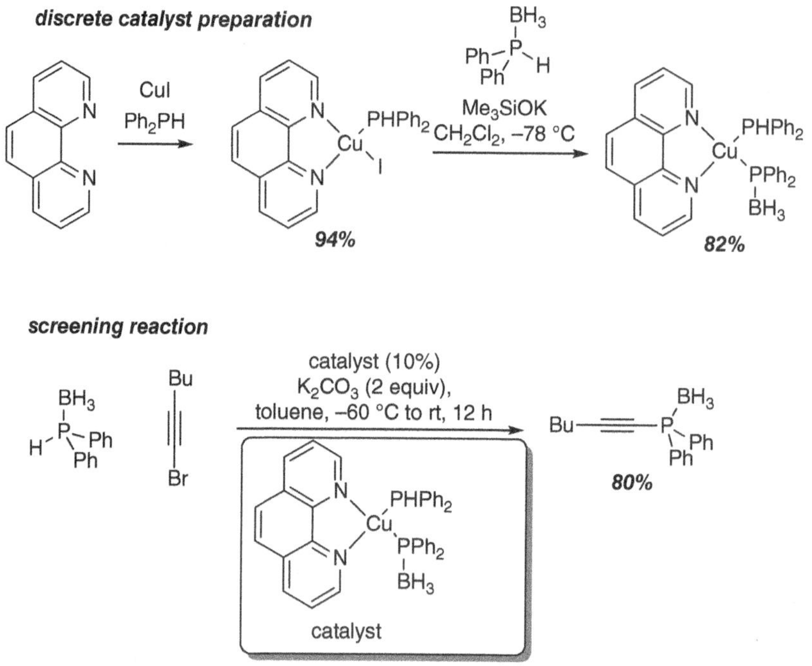

SCHEME 4.351 Copper-catalyzed synthesis of protected alkynylphosphines using a preformed copper catalyst [525].

triorganophosphines, the authors circumvented this issue by using BH_3 to stabilize the secondary phosphines prior to the cross-coupling reaction. Using readily available copper iodide and 1,10-phenanthroline as an ancillary ligand for the copper, moderate to excellent yields of the phosphine-boranes were obtained in 5–38 h at 60 °C. Most 1-bromoalkynes were successfully used; however, terminal alkynes containing electron-withdrawing groups did not react under the reaction conditions. In related work, the use of a preformed copper species promoted the P—C coupling reaction (Scheme 4.351) [525].

One of the classic ways to generate alkynylphosphine oxides through the formation of P—C bonds entails the coupling of chlorophosphine oxides and related compounds with metal acetylides [526–530]. This method suffers from functional group limitations due to the powerful bases (LDA, RLi, and Grignard reagents) used to promote the metalation reaction. However, if no electrophiles are present, this is a reliable method for the formation of alkynylphosphine oxides. While there are a number of reactions that generate alkynylphosphine oxides from chlorophosphines and secondary phosphine oxides, those methods are still not as popular as simply oxidizing alkynylphosphines. The most popular oxidant for this reaction is hydrogen peroxide, and the yields of the oxidized product are typically very high. If the phosphine is commercially available, the oxidation is typically trivial. A variation on this approach consists of synthesizing an alkynylphosphine through metallation/phosphination or cross-coupling followed by oxidation [531–536]. The following sections will highlight examples using these classic synthetic procedures as well as recent advances for the synthesis of alkynylphosphine oxides and related compounds.

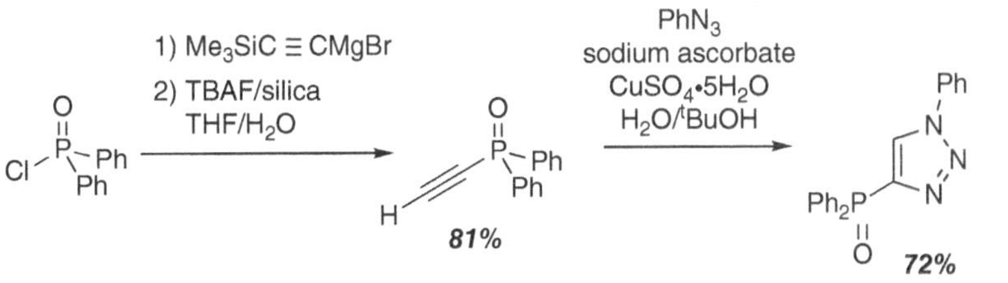

SCHEME 4.352   Synthesis of an alkynylphosphine oxide followed by conversion into a triazole [526].

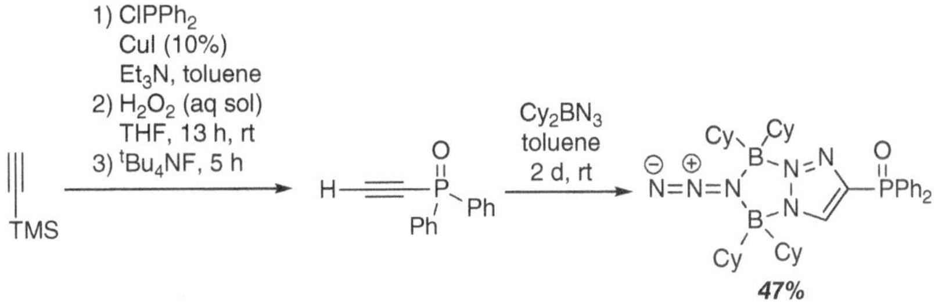

SCHEME 4.353   Synthesis and application of an alkynylphosphine oxide [537].

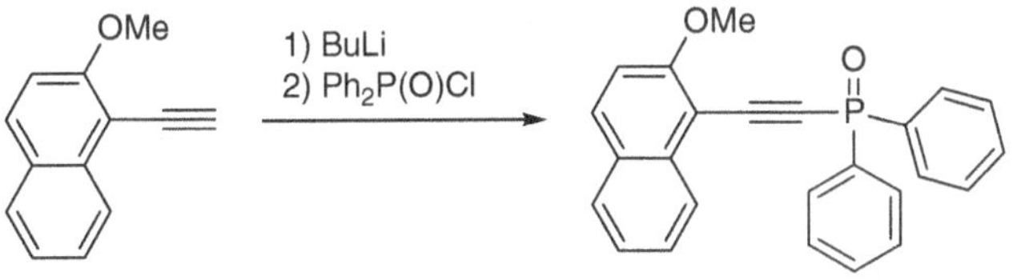

SCHEME 4.354   Synthesis of a crowded alkynylphosphine oxide [529].

A version of this approach was used to generate the precursors needed for the preparation of *phospha*-substituted triazoles (Scheme 4.352) [526]. The alkynylation reaction proceeded smoothly and afforded the ethynylphosphine oxide in excellent yield. Treatment of this compound with phenyl azide under cycloaddition conditions generates the 1,4-disubstituted triazoles in high yield.

A similar approach has been used for the formation of a $C_2N_4B_2$ bicyclic framework (Scheme 4.353) [537]. The heterocycle was constructed by simply stirring Cy_2BN_3 with an ethynylphosphine oxide. The latter was prepared using a copper-catalyzed P—C bond-forming reaction followed by oxidation using hydrogen peroxide and cleavage of the TMS protecting group. The authors found that terminal alkynes bearing electron-neutral or donating groups such as phenylacetylene, 4-*tert*-butylphenylacetylene, and trimethylsilylacetylene were unsuccessful. The cycloaddition reaction was only successful when electron-withdrawing groups were present on the alkyne.

The reaction of chlorophosphine oxides with metal acetylides is typically tolerant of bulky groups on the alkyne (Scheme 4.354) [529]. A recent example of this entails the synthesis of an alkynylphosphine oxide from chlorodiphenylphosphine oxide and 1-ethynyl-2-methoxynaphthalene. Treatment of the alkyne with BuLi and subsequent addition of the

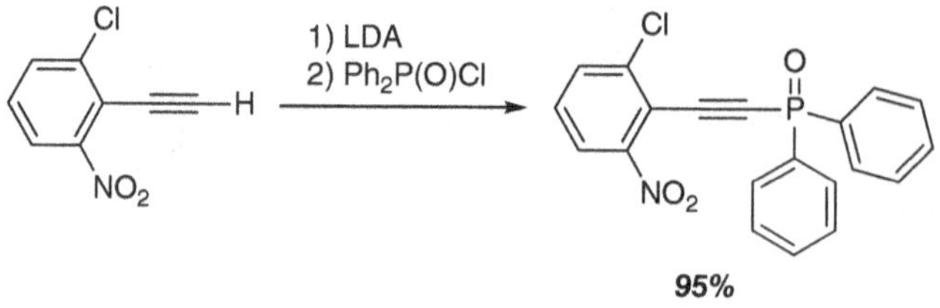

SCHEME 4.355   Successful phosphinylation of a nitro-substituted arylalkyne [528].

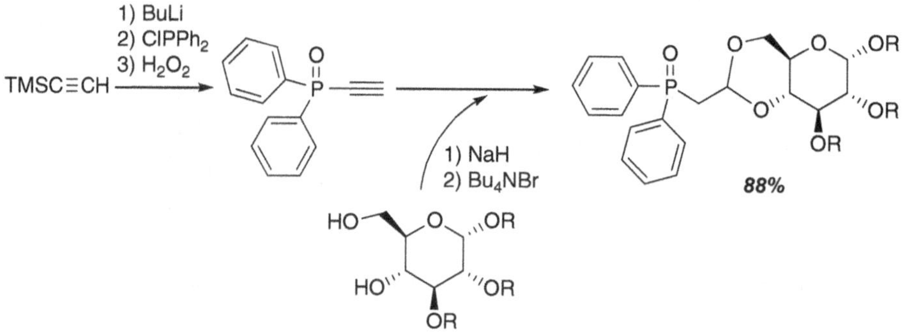

SCHEME 4.356   Synthesis and application of an ethynylphosphine oxide [536].

chlorophosphine afforded the target as a pale yellow solid. Unfortunately, the isolated yield was not reported for this reaction.

An interesting and potentially valuable modification to this approach has recently been reported by Carter (Scheme 4.355) [528]. His group recently demonstrated the successful coupling of a phenylacetylene derivative bearing a nitro group with chlorodiphenylphosphine oxide. Nitro groups are often problematic when using Grignard reagents. Furthermore, the incorporation of two ortho-substituents on the alkyne does not reduce the effectiveness of this reaction, and excellent yields of the alkynylphosphine oxide were obtained.

An interesting example of how the alkynylphosphine oxide group has been used in glyco-chemistry entailed initial formation of the alkynylphosphine by lithiation of the TMS-protected acetylene followed by addition of the chlorophosphine (Scheme 4.356) [536]. Oxidation using hydrogen peroxide afforded the alkynyphosphine oxide. Treatment of this compound with the glucose derivative afforded the functionalized phosphine oxides in high yields.

One of the most practical syntheses of alkynylphosphine oxides starts with a secondary phosphine oxide and an aryl propiolic acid (Scheme 4.357) [482]. The majority of synthetic approaches for the formation of alkynylphosphine oxides used hazardous reagents such as chlorophosphine oxides. Secondary phosphine oxides are more attractive starting materials and are readily prepared by several routes. After some experimentation, the authors discovered that the addition of a palladium catalyst and a supporting ligand to the reaction mixture promoted the decarboxylation reaction and generated the alkynylphosphine oxides in fair to moderate yields. The chemistry is operationally simple as all of the reagents and catalysts were loaded in air. The only precaution for this chemistry was exchanging the atmosphere for argon prior to heating.

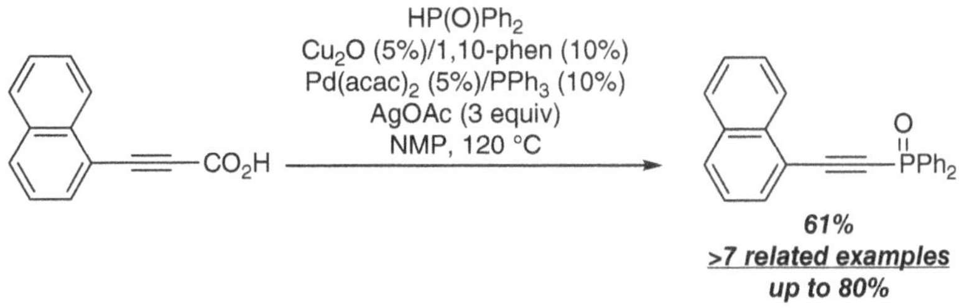

SCHEME 4.357 Synthesis of alkynylphosphine oxides through a decarboxylative coupling reaction [482].

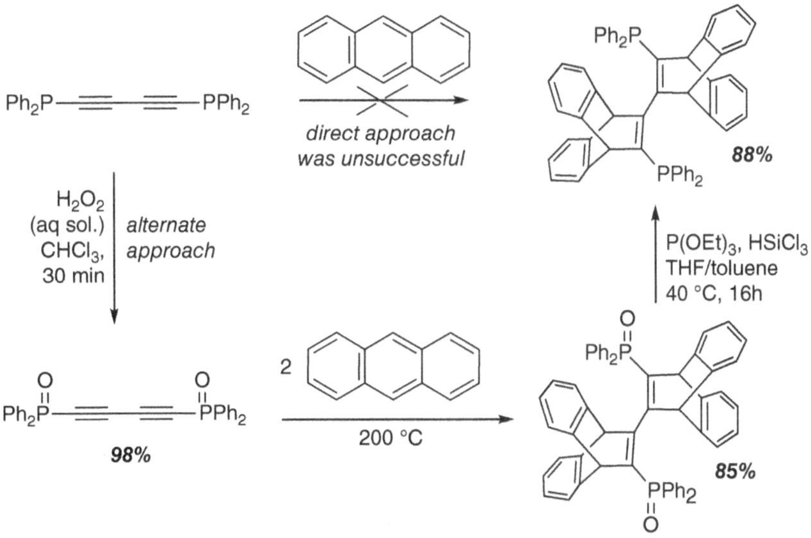

SCHEME 4.358 Synthesis of a CATPHOS-type ligand from an alkynylphosphine oxide [538].

As mentioned previously, one of the reasons for the preparation of an alkynylphosphine oxide was so that the phosphorus center could be protected while other parts of the molecule were altered. Once the changes to the target were complete, the phosphine oxide could be reduced back to the phosphine. This approach was recently used to generate a class of CATPHOS-type ligands for transition metals (Scheme 4.358) [538]. The authors initially attempted to generate the diphosphine ligand by simply heating anthracenes and bis(diphenylphosphinyl)buta-1,3-diyne together. However, this generated an intractable mixture of products. The solution to this problem was protection of the phosphine through the formation of the phosphine oxide. Once the phosphine was suitably protected, the double cycloaddition reaction proceeded smoothly and generated moderate yields of the bisphosphine oxide. Reduction afforded the desired diphosphine ligand.

While a host of reports outline the use of organophosphines as organocatalysts, the use of phosphine oxides is less common. A recent report by Tanaka provides an elegant example of the full synthesis of an axially chiral phosphine oxide as well as its

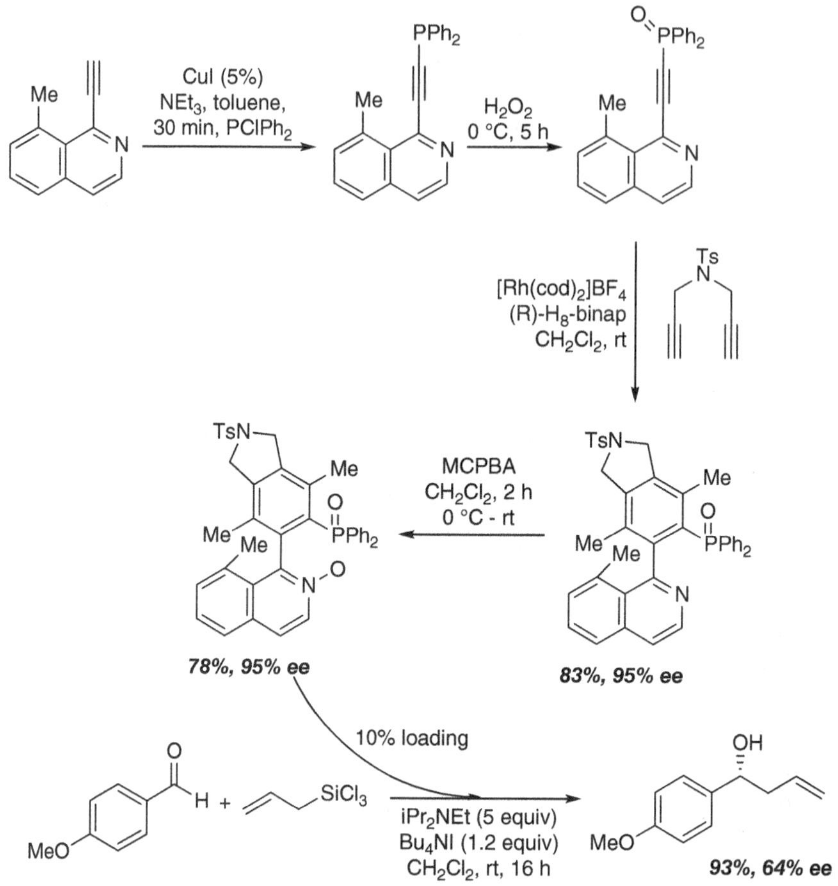

SCHEME 4.359   Synthesis and application of an alkynylphosphine oxide as an organocatalyst [539].

application to the asymmetric allylation of benzaldehyde derivatives (Scheme 4.359) [539]. The initial steps in the synthesis entailed the copper-catalyzed phosphination of an alkynylisoquinoline derivative followed by oxidation using hydrogen peroxide to generate the alkynylphosphine oxide derivative. A rhodium/binap-catalyzed [2+2+2] cycloaddition using a tosylamide-linked 1,6-diyne generated the axially chiral phosphine oxide in excellent yields with high enantiomeric excess. The authors found that the removal of the 8-methyl group from the isoquinoline starting materials decreased the selectivity of the reaction. Oxidation using MCPBA generated the *N*-oxide in excellent yield with retention of the chirality. Using this compound in the asymmetric allylation of benzaldehyde afforded the secondary alcohol in high yield with moderate selectivity. A similar approach was used to generate NU-BIPHEP-type phosphine and phosphine oxide compounds [540].

The alkynylation of secondary phosphine oxides using iodonium salts is an effective route to the synthesis of alkynylphosphine oxides (Scheme 4.360) [541]. The chemistry was operationally trivial, and good to excellent yields of the alkynylphosphine oxides were

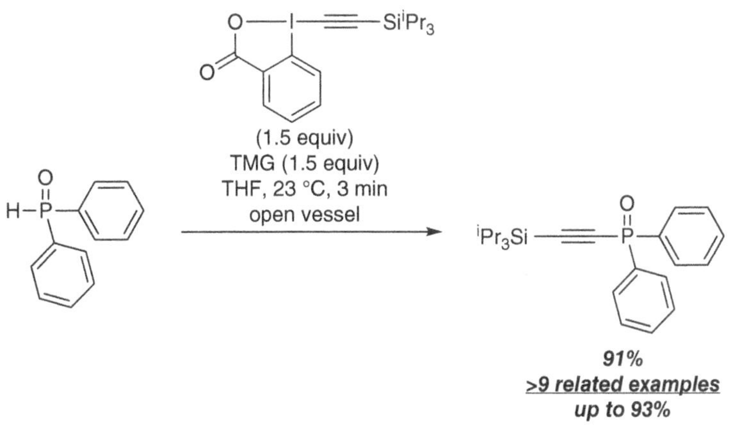

SCHEME 4.360 Alkynylation of secondary phosphine oxides using iodonium species [541].

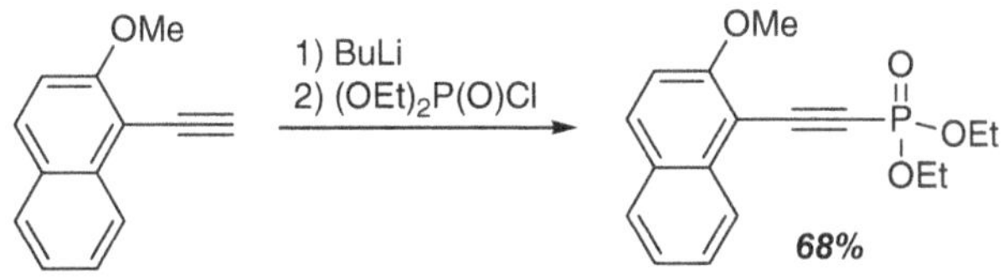

SCHEME 4.361 Synthesis of a crowded alkynylphosphonate [529].

obtained. One of the most practical aspects of the chemistry was that it could be carried out in an open vessel under air.

1-Alkynylphosphonates are valuable compounds with applications in organic synthesis, inorganic chemistry, and materials science. The electron-withdrawing phosphonate group renders the alkyne fragment susceptible to attack by nucleophiles, and a host of functionalized derivatives have been prepared using this approach [542, 543]. Classic methods for the preparation of alkynylphosphonates have been summarized in a number of excellent reviews [349, 542]. Some of the earliest reports of alkynylphosphonate synthesis involved an Arbuzov-type reaction between haloalkynes and trialkylphosphites. This chemistry was sensitive to the organic fragment on the alkyne; however, for the substrates that worked, the reaction was often successful at low temperatures. A number of reports outlined the preparation of alkynylphosphonates by treating chlorophosphonates with Grignard or organolithium reagents. While this chemistry does generate alkynylphosphonates, it was severely limited in scope due to necessity of powerful nucleophiles/bases. Furthermore, it required handling of the hazardous chlorophosphonates. The following sections will summarize the historical and current synthetic methods for the preparation of these interesting compounds.

A recent example of the tolerance of this chemistry to steric bulk was reported by Tanaka (Scheme 4.361) [529]. Treatment of 1-ethynyl-2-methoxynaphthalene with BuLi at low temperatures followed by addition of the chlorophosphonate afforded the alkynylphosphonate in moderate yields (68%) after purification by column chromatography.

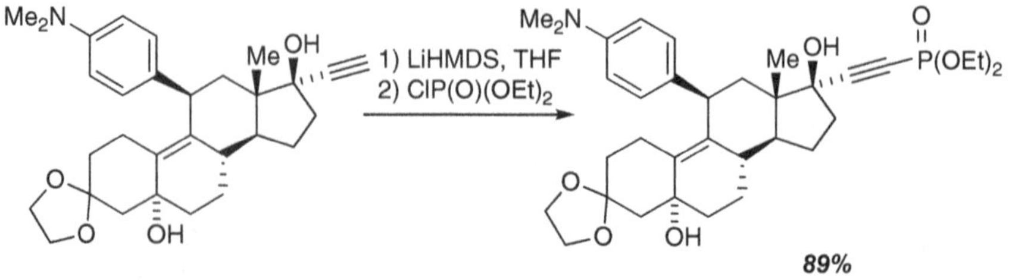

SCHEME 4.362   Synthesis of a mifepristone-derived alkynylphosphonate [544].

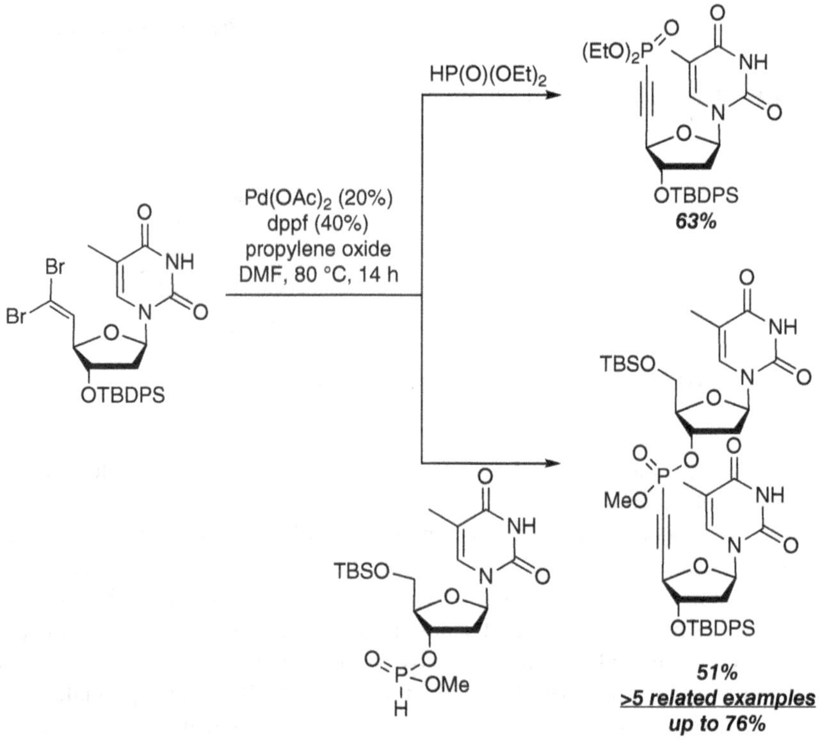

SCHEME 4.363   Synthesis of alkynylphosphonates from 1,1-dibromoalkenes [545].

A group of mifepristone derivatives containing an alkynylphosphonate in the 17-position have been generated following a metallation/phosphonylation approach (Scheme 4.362) [544]. The authors screened a range of bases for activity in this reaction including BuLi, NaH, MeMgBr, and LiHMDS. Only the latter generated significant amounts of the phosphorylated material. It should be noted that protection of the 17-OH and the 5-OH groups was not needed.

An interesting synthesis of an alkynylphosphonate was discovered by accident while trying to selectively monophosphorylate a 1,1-dibromoalkene (Scheme 4.363) [545]. Instead of generating the desired 1-bromo-1-phosphinoyl alkene, an alkynylphosphonate

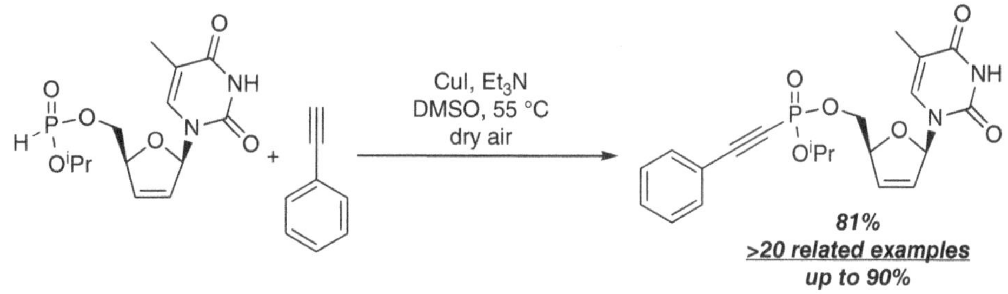

SCHEME 4.364   Copper-catalyzed oxidative coupling of terminal alkynes with secondary phosphites [517].

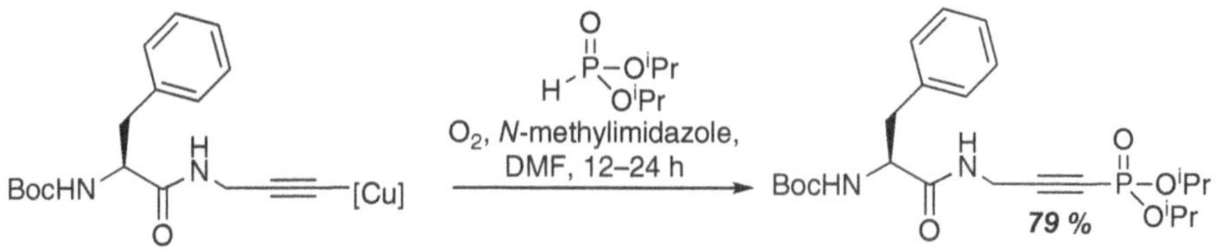

SCHEME 4.365   Synthesis of an alkynylphosphonate through an oxidative coupling reaction [550].

was the sole product. After some experimentation, a catalyst system comprised of palladium acetate and dppf was determined to be the most effective. The chemistry was attractive since the starting 1,1-dibromides can be readily prepared from aldehydes [546]. This chemistry displayed wide functional group tolerance, and despite the high palladium loading (20%), was a very effective way to generate alkynylphosphonates. Following this initial report, a number of nucleoside phosphonates have been prepared and characterized [547, 548].

One of the most promising approaches to the synthesis of alkynylphosphonates was recently reported by Han and coworkers (Scheme 4.364) [517]. He adapted an oxidative copper-catalyzed cross-coupling reaction to generate alkynylphosphonates through the use of terminal alkynes and hydrogen phosphonates as substrates. Since the reaction occured under an atmosphere of air using air-stable copper salts, the chemistry was operationally trivial. In contrast with the classic methods of alkynylphosphonate synthesis described previously, this methodology was tolerant of a wide range of functional groups including cyano, chloro, hydroxyl, alkoxy, carboxyl, carbonyl, amino, sulfonamide, and carbamoyl. A solid-supported copper catalyst has also been shown to promote this reaction [549]. This approach represents one of the most practical approaches to the synthesis of alkynylphosphonates.

The copper-catalyzed oxidative coupling approach for the synthesis of alkynylphosphonates has also been reported using discrete copper acetylides and hydrogen phosphonates (Scheme 4.365) [550]. One of the most attractive aspects of this chemistry was its tolerance to a wide range of functional groups. The conditions were quite mild (room temperature), and the isolated yields of the alkynylphosphonates were moderate to excellent with the exception of a substrate containing a primary alcohol.

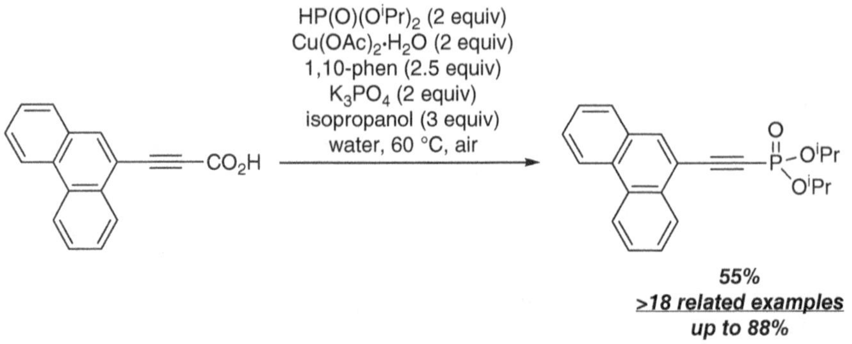

SCHEME 4.366 Decarboxylative synthesis of an alkynylphosphonate [551].

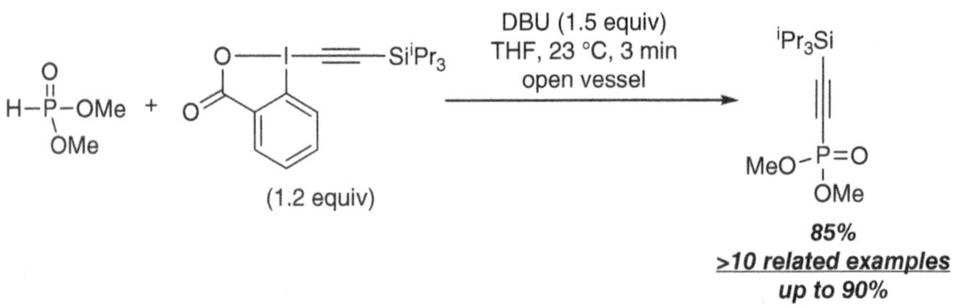

SCHEME 4.367 Transition metal-free synthesis of alkynylphosphonates in open vessels [541].

While the decarboxylation reactions described previously typically used a polar organic solvent such as DMSO or DMF, a successful version of this reaction in water has recently been reported (Scheme 4.366) [551]. Historically, water has been a troublesome solvent for reactions involving dialkyl hydrogen phosphonates due to facile decomposition into undesired secondary reaction products. The authors circumvented this issue through the addition of isopropanol to the reaction mixture. The authors had previously discovered that this additive suppressed the decomposition reaction and facilitated the use of these versatile reagents in water. Using hydrated copper (II) acetate along with a chelating ligand (phen) as the catalyst system, a range of arylpropiolic acids were successfully coupled with hydrogen phosphonates to afford moderate to excellent yields of the alkynylphosphonates. Only three equivalents of the isopropanol additive were needed for a successful coupling reaction. Similar to the other dehydrogenative processes, the reaction was carried out under an atmosphere of air. In terms of the functional group tolerance, a wide range of electron-donating and withdrawing groups were tolerated by the chemistry with the highest yields generally observed with electron-neutral and withdrawing groups.

As mentioned in the previous section, the treatment of secondary phosphine oxides with alkynyl iodonium salts is a reliable method for the preparation of phosphorus–carbon(sp) bonds. This chemistry can be extended to secondary phosphites for the preparation of alkynylphosphonates (Scheme 4.367) [541]. The reactions were carried out under mild conditions in reaction vessels that were open to the atmosphere.

4.5.1   Troubleshooting the Synthesis of Alkynylphosphines and Related Compounds

While it is impossible to anticipate the challenges of every new reaction/method that will be developed over the next few decades, there are some general trends that can be followed. While the section on solvents, additives, and general suggestions in Chapter 1 will be a good resource, the chart below will provide some direction when trying to select the most practical solution to the specific issues related to the synthesis of alkynylphosphines and related compounds.

Problem/Issue/Goal	Possible Solution/Suggestion
A glovebox is not available	This will make the synthesis of many alkynylphosphines very challenging. Unless the phosphorus precursor is packed in a septum-sealed bottle under an inert gas, even loading a reaction flask without introducing oxygen will be difficult. Furthermore, if purification by column chromatography is required, it will be difficult to maintain a completely inert atmosphere when collecting fractions
	It might be more practical to synthesize the protected phosphine (oxide, sulfide, borane) and convert it into a phosphine following the P—C bond-forming step. Isolation of the target phosphine from the reduction or deboronation reaction is often significantly easier than from the initial P—C bond-forming reaction
Partial oxidation of the phosphine during workup	One solution is to rigorously purify and deoxygenate all of the reagents and solvents used during the workup procedure
The starting material is a terminal acetylene. What is a reliable method for converting it into an alkynylphosphine?	Deprotonation of the acetylene followed by addition of a chlorophosphine is a popular approach [509, 519]. This method is likely to be successful as long as electrophiles are not present on either substrate. If electrophiles are present, a nickel-catalyzed cross-coupling has been successful [521, 522]
What is a reliable method for converting a phosphine-borane into an alkynylphosphine?	Several methods are known to promote this conversion. The copper-catalyzed approach has been successful with a variety of substrates [524, 525]
Can alkynylphosphine oxides be prepared through the treatment of acetylides with chlorophosphine oxides?	This is a popular route to alkynylphosphine oxides provided there are no electrophiles present on either substrate [526, 528, 529]
Need a transition metal-free route to the synthesis of alkynylphosphine oxides	Iodonium salts have been used to promote this reaction [541]
Are there decarboxylative approaches to the synthesis of alkylphosphine oxides that start from propiolic acids?	Using copper salts to promote the decarboxylative cross-coupling between the propiolic acids and secondary phosphine oxides would be a reasonable approach to this chemistry [482]
Can alkynylphosphonates be generated through Arbuzov-type chemistry?	This reaction is possible and uses trialkylphosphites and haloalkynes as substrates. Yields tend to be highly variable
The starting material is a terminal alkyne. Need a reliable method of making the alkynylphosphonate	Deprotonation of the alkyne followed by addition of the chlorophosphonate would be a reasonable approach [528, 544]

(*Continued*)

(Continued)

Problem/Issue/Goal	Possible Solution/Suggestion
Can 1,1-dibromoalkenes be converted into alkynylphosphonates?	Treatment of the 1,1-dibromoalkene with a secondary phosphite in the presence of palladium catalysts generated the alkynylphosphonate [545, 546]
Can alkynylphosphonates be generated through a decarboxylative phosphonation?	In much the same manner that alkynylphosphine oxides could be prepared through a metal-catalyzed decarboxylative coupling, alkynylphosphonates can also be generated using this method [551]
The transition metal-free synthesis of alkynylphosphine oxides can be accomplished using iodonium salts. Can this chemistry be extended to alkynylphosphonates?	This has been accomplished by simply substituting a secondary phosphite for the phosphine oxide in the metal-free coupling reaction [541]

REFERENCES

[1] Brunker, T. J.; Anderson, B. J.; Blank, N. F.; Glueck, D. S.; Rheingold, A. L. *Org. Lett.* **2007**, *9*, 1109–1112.

[2] Gilheany, D. G.; Mitchell, C. M. In *The Chemistry of Organophosphorus Compounds*; Hartley, F. R., Ed.; John Wiley & Sons, Ltd: Chichester, **1990**; Vol. 1, pp 151–190.

[3] Engel, R.; Cohen, J. I. In *Synthesis of Carbon-Phosphorus Bonds*; Engel, R., Ed.; CRC Press: Boca Raton, FL, **2003**.

[4] Meiners, J.; Friedrich, A.; Herdtweck, E.; Schneider, S. *Organometallics* **2009**, *28*, 6331–6338.

[5] Naiini, A. A.; Han, Y.; Akinc, M.; Verkade, J. G. *Inorg. Chem.* **1993**, *32*, 5394–5395.

[6] Langer, F.; Knochel, P. *Tetrahedron Lett.* **1995**, *36*, 4591–4594.

[7] Kazul'kin, D. N.; Ryabov, A. N.; Izmer, V. V.; Churakov, A. V.; Beletskaya, I. P.; Burns, C. J.; Voskoboynikov, A. Z. *Organometallics* **2005**, *24*, 3024–3035.

[8] Longeau, A.; Langer, F.; Knochel, P. *Tetrahedron Lett.* **1996**, *37*, 2209–2212.

[9] Hackett, M.; Whitesides, G. M. *Organometallics* **1987**, *6*, 403–410.

[10] Su, M.; Buchwald, S. L. *Angew. Chem. Int. Ed.* **2012**, *51*, 4710–4713.

[11] Kendall, A. J.; Salazar, C. A.; Martino, P. F.; Tyler, D. R. *Organometallics* **2014**, *33*, 6171–6178.

[12] Reynolds, S. C.; Hughes, R. P.; Glueck, D. S.; Rheingold, A. L. *Org. Lett.* **2012**, *14*, 4238–4241.

[13] Mejía, E.; Aardoom, R.; Togni, A. *Eur. J. Inorg. Chem.* **2012**, *2012*, 5021–5032.

[14] Maienza, F.; Spindler, F.; Thommen, M.; Pugin, B.; Malan, C.; Mezzetti, A. *J. Org. Chem.* **2002**, *67*, 5239–5249.

[15] Junge, K.; Oehme, G.; Monsees, A.; Riermeier, T.; Dingerdissen, U.; Beller, M. *Tetrahedron Lett.* **2002**, *43*, 4977–4980.

[16] Casey, C. P.; Paulsen, E. L.; Beuttenmueller, E. W.; Proft, B. R.; Petrovich, L. M.; Matter, B. A.; Powell, D. R. *J. Am. Chem. Soc.* **1997**, *119*, 11817–11825.

[17] McKinstry, L.; Livinghouse, T. *Tetrahedron Lett.* **1994**, *35*, 9319–9322.

[18] Sues, P. E.; Lough, A. J.; Morris, R. H. *Organometallics* **2011**, *30*, 4418–4431.

[19] Victoria Jimenez, M.; Perez-Torrente, J. J.; Isabel Bartolome, M.; Oro, L. A. *Synthesis* **2009**, 1916–1922.

[20] Ding, B.; Zhang, Z.; Xu, Y.; Liu, Y.; Sugiya, M.; Imamoto, T.; Zhang, W. *Org. Lett.* **2013**, *15*, 5476–5479.

[21] Morisaki, Y.; Imoto, H.; Ouchi, Y.; Nagata, Y.; Chujo, Y. *Org. Lett.* **2008**, *10*, 1489–1492.

[22] Cain, M. F.; Hughes, R. P.; Glueck, D. S.; Golen, J. A.; Moore, C. E.; Rheingold, A. L. *Inorg. Chem.* **2010**, *49*, 7650–7662.

[23] Chan, V. S.; Chiu, M.; Bergman, R. G.; Toste, F. D. *J. Am. Chem. Soc.* **2009**, *131*, 6021–6032.

[24] Li, Y.; Das, S.; Zhou, S.; Junge, K.; Beller, M. *J. Am. Chem. Soc.* **2012**, *134*, 9727–9732.

[25] Weissman, H.; Shimon, L. J. W.; Milstein, D. *Organometallics* **2004**, *23*, 3931–3940.

[26] Butti, P.; Rochat, R.; Sadow, A.; Togni, A. D. *Angew. Chem. Int. Ed.* **2008**, *47*, 4878–4881.

[27] Chapp, T. W.; Glueck, D. S.; Golen, J. A.; Moore, C. E.; Rheingold, A. L. *Organometallics* **2010**, *29*, 378–388.

[28] Scriban, C.; Glueck, D. S. *J. Am. Chem. Soc.* **2006**, *128*, 2788–2789.

[29] Anderson, B. J.; Guino-o, M. A.; Glueck, D. S.; Golen, J. A. DiPasquale, A. G.; Liable-Sands, L. M.; Rheingold, A. L. *Org. Lett.* **2008**, *10*, 4425.

[30] Scriban, C.; Glueck, D. S.; Golen, J. A.; Rheingold, A. L. *Organometallics* **2007**, *26*, 1788–1800.

[31] Anderson, B. J.; Glueck, D. S.; DiPasquale, A. G.; Rheingold, A. L. *Organometallics* **2008**, *27*, 4992–5001.

[32] Chapp, T. W.; Schoenfeld, A. J.; Glueck, D. S. *Organometallics* **2010**, *29*, 2465–2473.

[33] Tang, W.; Zhang, X. *Chem. Rev.* **2003**, *103*, 3029–3070.

[34] Wicht, D. K.; Kourkine, I. V.; Lew, B. M.; Nthenge, J. M.; Glueck, D. S. *J. Am. Chem. Soc.* **1997**, *119*, 5039–5040.

[35] Wicht, D. K.; Kourkine, I. V.; Kovacik, I.; Glueck, D. S.; Concolino, T. E.; Yap, G. P. A.; Incarvito, C. D.; Rheingold, A. L. *Organometallics* **1999**, *18*, 5381–5394.

[36] Scriban, C.; Glueck, D. S.; Zakharov, L. N.; Kassel, W. S.; DiPasquale, A. G.; Golen, J. A.; Rheingold, A. L. *Organometallics* **2006**, *25*, 5757–5767.

[37] Greenhalgh, M. D.; Jones, A. S.; Thomas, S. P. *ChemCatChem* **2015**, *7*, 190–222.

[38] Koshti, V.; Gaikwad, S.; Chikkali, S. H. *Coord. Chem. Rev.* **2014**, *265*, 52–73.

[39] Delacroix, O.; Gaumont, A. C., *Curr. Org. Chem.* **2005**, *9*, 1851–1882.

[40] Glueck, D. S. *Top. Organomet. Chem.* **2010**, *31*, 65–100.

[41] Tanaka, M. *Top. Curr. Chem.* **2004**, *232*, 25–54.

[42] Beletskaya, I. P.; Kazankova, M. A. *Russ. J. Org. Chem.* **2002**, *38*, 1391–1430.

[43] Enders, D.; Saint-Dizier, A.; Lannou, M. I.; Lenzen, A. *Eur. J. Org. Chem.* **2006**, 29–49.

[44] Beletskaya, I. P.; Ananikov, V. P.; Khemchyan, L. L. In *Synthesis of Phosphorus Compounds via Metal-Catalyzed Addition of P-H Bond to Unsaturated Organic Molecules*; Peruzzini, M. G. L., Ed.; Catalysis by Metal Complexes: **2011**; Vol. *37*, pp 213–264.

[45] Ellis, J. W.; Harrison, K. N.; Hoye, P. A. T.; Orpen, A. G.; Pringle, P. G.; Smith, M. B. *Inorg. Chem.* **1992**, *31*, 3026–3033.

[46] Pringle, P. G.; Smith, M. B. *J. Chem. Soc. Chem. Comm.* **1990**, 1701–1702.

[47] Costa, E.; Pringle, P. G.; Smith, M. B.; Worboys, K. *J. Chem. Soc. Dalton Trans.* **1997**, 4277–4282.

[48] Hoye, P. A. T.; Pringle, P. G.; Smith, M. B.; Worboys, K. *J. Chem. Soc. Dalton Trans.* **1993**, 269–274.

[49] Costa, E.; Pringle, P. G.; Worboys, K. *Chem. Commun.* **1998**, 49–50.

[50] Robertson, A.; Bradaric, C.; Frampton, C. S.; McNulty, J.; Capretta, A. *Tetrahedron Lett.* **2001**, *42*, 2609–2612.

[51] Bunlaksananusorn, T.; Knochel, P. *Tetrahedron Lett.* **2002**, *43*, 5817–5819.

[52] Trofimov, B. A.; Brandsma, L.; Arbuzova, S. N.; Malysheva, S. F.; Gusarova, N. K. *Tetrahedron Lett.* **1994**, *35*, 7647–7650.

[53] Casey, C. P.; Paulsen, E. L.; Beuttenmueller, E. W.; Proft, B. R.; Matter, B. A.; Powell, D. R. *J. Am. Chem. Soc.* **1999**, *121*, 63–70.

[54] Chou, H.-H.; Raines, R. T. *J. Am. Chem. Soc.* **2013**, *135*, 14936–14939.

[55] Flapper, J.; Kooijman, H.; Lutz, M.; Spek, A. L.; van Leeuwen, P. W. N. M.; Elsevier, C. J.; Kamer, P. C. J. *Organometallics* **2009**, *28*, 3272–3281.

[56] Shulyupin, M. O.; Kazankova, M. A.; Beletskaya, I. P. *Org. Lett.* **2002**, *4*, 761–763.

[57] Davies, L. H.; Stewart, B.; Harrington, R. W.; Clegg, W.; Higham, L. J. *Angew. Chem. Int. Ed.* **2012**, *51*, 4921–4924.

[58] Kamitani, M.; Itazaki, M.; Tamiya, C.; Nakazawa, H. *J. Am. Chem. Soc.* **2012**, *134*, 11932–11935.

[59] Hayashi, M.; Matsuura, Y.; Watanabe, Y. *Tetrahedron Lett.* **2004**, *45*, 9167–9169.

[60] Hayashi, M.; Matsuura, Y.; Watanabe, Y. *Tetrahedron Lett.* **2005**, *46*, 5135–5138.

[61] Hayashi, M.; Matsuura, Y.; Watanabe, Y. *J. Org. Chem.* **2006**, *71*, 9248–9251.

[62] Crimmin, M. R.; Barrett, A. G. M.; Hill, M. S.; Hitchcock, P. B.; Procopiou, P. A. *Organometallics* **2007**, *26*, 2953–2956.

[63] Liu, B.; Roisnel, T.; Carpentier, J.-F.; Sarazin, Y. *Angew. Chem. Int. Ed.* **2012**, *51*, 4943–4946.

[64] Rodriguez-Ruiz, V.; Carlino, R.; Bezzenine-Lafollee, S.; Gil, R.; Prim, D.; Schulz, E.; Hannedouche, J. *Dalton Trans.*, **2015**, *44*, 12029–12059.

[65] Isley, N. A.; Linstadt, R. T. H.; Slack, E. D.; Lipshutz, B. H. *Dalton Trans.* **2014**, *43*, 13196–13200.

[66] Gallagher, K. J.; Webster, R. L. *Chem. Commun.* **2014**, *50*, 12109–12111.

[67] Huang, Y.; Pullarkat, S. A.; Li, Y.; Leung, P. *Inorg. Chem.* **2012**, *51*, 2533–2540.

[68] Yuan, M.; Zhang, N.; Pullarkat, S. A.; Li, Y.; Liu, F.; Pham, P.; Leung, P.-H. *Inorg. Chem.* **2010**, *49*, 989–996.

[69] Chen, S.; Ng, J. K.-P.; Pullarkat, S. A.; Liu, F.; Li, Y.; Leung, P.-H. *Organometallics* **2010**, *29*, 3374–3386.

[70] Huang, Y.; Pullarkat, S. A.; Li, Y.; Leung, P.-H. *Chem. Commun.* **2010**, *46*, 6950–6952.

[71] Xu, C.; Kennard, G. J. H.; Hennersdorf, F.; Li, Y.; Pullarkat, S. A.; Leung, P.-H. *Organometallics* **2012**, *31*, 3022–3026.

[72] Huang, Y.; Chew, R. J.; Pullarkat, S. A.; Li, Y.; Leung, P.-H. *J. Org. Chem.* **2012**, *77*, 6849–6854.

[73] Pullarkat, S. A.; Yi, D.; Li, Y.; Tan, G.-K.; Leung, P.-H. *Inorg. Chem.* **2006**, *45*, 7455–7463.

[74] Chew, R. J.; Huang, Y.; Li, Y.; Pullarkat, S. A.; Leung, P. *Adv. Synth. Cat.* **2013**, *355*, 1403–1408.

[75] Huang, Y.; Chew, R. J.; Li, Y.; Pullarkat, S. A.; Leung, P.-H. *Org. Lett.* **2011**, *13*, 5862–5865.

[76] Sadow, A. D.; Haller, I.; Fadini, L.; Togni, A. *J. Am. Chem. Soc.* **2004**, *126*, 14704–14705.

[77] Sadow, A. D.; Togni, A. *J. Am. Chem. Soc.* **2005**, *127*, 17012–17024.

[78] Alonso, F.; Moglie, Y.; Radivoy, G.; Yus, M. *Green Chem.* **2012**, *14*, 2699–2702.

[79] Douglass, M. R.; Marks, T. J. *J. Am. Chem. Soc.* **2000**, *122*, 1824–1825.

[80] Routaboul, L.; Toulgoat, F.; Gatignol, J.; Lohier, J.-F.; Norah, B.; Delacroix, O.; Alayrac, C.; Taillefer, M.; Gaumont, A.-C. *Chem. Eur. J.* **2013**, *19*, 8760–8764.

[81] Huang, M.; Li, C.; Huang, J.; Duan, W.-L.; Xu, S. *Chem. Commun.* **2012**, *48*, 11148–11150.

[82] Zhang, X.-N.; Chen, G.-Q.; Tang, X.-Y.; Wei, Y.; Shi, M. *Angew. Chem. Int. Ed.* **2014**, *53*, 10768–10773.

[83] Zhu, X.-F.; Henry, C. E.; Kwon, O. *J. Am. Chem. Soc.* **2007**, *129*, 6722–6723.

[84] Daeffler, C. S.; Grubbs, R. H. *Org. Lett.* **2011**, *13*, 6429–6431.

[85] Kawaguchi, S.; Nomoto, A.; Sonoda, M.; Ogawa, A. *Tetrahedron Lett.* **2009**, *50*, 624–626.

[86] Carta, P.; Puljic, N.; Robert, C.; Dhimane, A.-L.; Fensterbank, L.; Lacote, E.; Malacria, M. *Org. Lett.* **2007**, *9*, 1061–1063.

[87] Yoo, W.-J.; Kobayashi, S. *Green Chem.* **2013**, *15*, 1844–1848.

[88] Das, D.; Seidel, D. *Org. Lett.* **2013**, *15*, 4358–4361.

[89] Li, Y.-L.; Sun, M.; Wang, H.-L.; Tian, Q.-P.; Yang, S.-D. *Angew. Chem. Int. Ed.* **2013**, *52*, 3972–3976.

[90] Gao, Y.; Huang, Z.; Zhuang, R.; Xu, J.; Zhang, P.; Tang, G.; Zhao, Y. *Org. Lett.* **2013**, *15*, 4214–4217.

[91] Yang, D.; Zhao, D.; Mao, L.; Wang, L.; Wang, R. *J. Org. Chem.* **2011**, *76*, 6426–6431.

[92] Yang, B.; Yang, T.-T.; Li, X.-A.; Wang, J.-J.; Yang, S.-D. *Org. Lett.* **2013**, *15*, 5024–5027.

[93] Ingle, G. K.; Liang, Y.; Mormino, M. G.; Li, G.; Fronczek, F. R.; Antilla, J. C. *Org. Lett.* **2011**, *13*, 2054–2057.

[94] Hirai, T.; Han, L.-B. *J. Am. Chem. Soc.* **2006**, *128*, 7422–7423.

[95] Chen, Z.-C.; Zhou, Z.-C.; Hua, H.-L.; Duan, X.-H.; Luo, J.-Y.; Wang, J.; Zhou, P.-X.; Liang, Y.-M. *Tetrahedron* **2013**, *69*, 1065–1068.

[96] Hans, M.; Delaude, L.; Rodriguez, J.; Coquerel, Y. *J. Org. Chem.* **2014**, *79*, 2758–2764.

[97] Bhattacharya, A. K.; Thyagarajan, G. *Chem. Rev.* **1981**, *81*, 415–430.

[98] Hieke, M.; Greiner, C.; Dittrich, M.; Reisen, F.; Schneider, G.; Schubert-Zsilavecz, M.; Werz, O. *J. Med. Chem.* **2011**, *54*, 4490–4507.

[99] Quintiliani, M.; Balzarini, J.; McGuigan, C. *Tetrahedron* **2013**, *69*, 9111–9119.

[100] Gahungu, M.; Arguelles-Arias, A.; Fickers, P.; Zervosen, A.; Joris, B.; Damblon, C.; Luxen, A. *Bioorg. Med. Chem.* **2013**, *21*, 4958–4967.

[101] Jansa, P.; Holy, A.; Dracinsky, M.; Baszczynski, O.; Cesnek, M.; Janeba, Z. *Green Chem.* **2011**, *13*, 882–888.

[102] Gallier, F.; Peyrottes, S.; Périgaud, C. *Eur. J. Org. Chem* **2007**, 925–933.

[103] Peyrottes, S.; Gallier, F.; Béjaud, J.; Périgaud, C. *Tetrahedron Lett.* **2006**, *47*, 7719–7721.

[104] Barney, R. J.; Richardson, R. M.; Wiemer, D. F. *J. Org. Chem.* **2011**, *76*, 2875–2879.

[105] Rajeshwaran, G. G.; Nandakumar, M.; Sureshbabu, R.; Mohanakrishnan, A. K. *Org. Lett.* **2011**, *13*, 1270–1273.

[106] André, V.; Lahrache, H.; Robin, S.; Rousseau, G. *Tetrahedron* **2007**, *63*, 10059–10066.

[107] Hasník, Z.; Pohl, R.; Hocek, M. *Tetrahedron Lett.* **2010**, *51*, 2464–2466.

[108] Cohen, R. J.; Fox, D. L.; Eubank, J. F.; Salvatore, R. N. *Tetrahedron Lett.* **2003**, *44*, 8617–8621.

[109] Xu, K.; Hu, H.; Yang, F.; Wu, Y. *Eur. J. Org. Chem.* **2013**, 319–325.

[110] van Summeren, R. P.; Feringa, B. L.; Minnaard, A. J. *Org. Biomol. Chem.* **2005**, *3*, 2524–2533.

[111] Snyder, S. A.; Breazzano, S. P.; Ross, A. G.; Lin, Y.; Zografos, A. L. *J. Am. Chem. Soc.* **2009**, *131*, 1753–1765.

[112] Pradere, U.; Amblard, F.; Coats, S. J.; Schinazi, R. F. *Org. Lett.* **2012**, *14*, 4426–4429.

[113] Laven, G.; Stawinski, J. *Synlett* **2009**, 225–228.

[114] Laven, G.; Kalek, M.; Jezowska, M.; Stawinski, J. *New J. Chem.* **2010**, *34*, 967–975.

[115] Stockland, R. A. Jr.; Levine, A. M.; Giovine, M. T.; Guzei, I. A.; Cannistra, J. C. *Organometallics* **2004**, *23*, 647–656.

[116] Levine, A. M.; Stockland, R. A. Jr.; Clark, R.; Guzei, I. *Organometallics* **2002**, *21*, 3278–3284.

[117] Platonov, A.; Sivakov, A. A.; Chistokletov, V. N.; Maiorova, E. D. *Russ. Chem. Bull.* **1999**, *48*, 367–370.

[118] Simoni, D.; Invidiata, F. P.; Manferdini, M.; Lampronti, I.; Rondanin, R.; Roberti, M.; Pollini, G. P. *Tetrahedron Lett.* **1998**, *39*, 7615–7618.

[119] Sperandio, D.; Gangloff, A. R.; Litvak, J.; Goldsmith, R.; Hataye, J. M.; Wang, V. R.; Shelton, E. J.; Elrod, K.; Janc, J. W.; Clark, J. M.; Rice, K.; Weinheimer, S.; Yeung, K.-S.; Meanwell, N. A.; Hernandez, D.; Staab, A. J.; Venables, B. L.; Spencer, J. R. *Bio. Org. Med. Chem. Lett.* **2002**, *12*, 3129–3133.

[120] Castelot-Deliencourt, G.; Roger, E.; Pannecoucke, X.; Quirion, J.-C. *Eur. J. Org. Chem.* **2001**, 3031–3038.

[121] Castelot-Deliencourt, G.; Pannecoucke, X.; Quirion, J.-C. *Tetrahedron Lett.* **2001**, *42*, 1025–1028.

[122] Martinez-Castro, E.; Lopez, O.; Maya, I.; Fernandez-Bolanos, J. G.; Petrini, M. *Green Chem.* **2010**, *12*, 1171–1174.

[123] Mori, I.; Kimura, Y.; Nakano, T.; Matsunaga, S.; Iwasaki, G.; Ogawa, A.; Hayakawa, K. *Tetrahedron Lett.* **1997**, *38*, 3543–3546.

[124] Han, L.-B.; Zhao, C.-Q. *J. Org. Chem.* **2005**, *70*, 10121–10123.

[125] Andaloussi, M.; Henriksson, L. M.; Wieckowska, A.; Lindh, M.; Bjorkelid, C.; Larsson, A. M.; Suresh, S.; Iyer, H.; Srinivasa, B. R.; Bergfors, T.; Unge, T.; Mowbray, S. L.; Larhed, M.; Jones, T. A.; Karlen, A. *J. Med. Chem.* **2011**, *54*, 4964–4976.

[126] Jansson, A. M.; Wieckowska, A.; Bjorkelid, C.; Yahiaoui, S.; Sooriyaarachchi, S.; Lindh, M.; Bergfors, T.; Dharavath, S.; Desroses, M.; Suresh, S.; Andaloussi, M.; Nikhil, R.; Sreevalli, S.; Srinivasa, B. R.; Larhed, M.; Jones, T. A.; Karlen, A.; Mowbray, S. L. *J. Med. Chem.* **2013**, *56*, 6190–6199.

[127] Haemers, T.; Wiesner, J.; Poecke, S. V.; Goeman, J.; Henschker, D.; Beck, E.; Jomaa, H.; Calenbergh, S. V. *Bioorg. Med. Chem. Lett.* **2006**, *16*, 1888–1891.

[128] Dondoni, A.; Staderini, S.; Marra, A. *Eur. J. Org. Chem.* **2013**, *2013*, 5370–5375.

[129] Tayama, O.; Nakano, A.; Iwahama, T.; Sakaguchi, S.; Ishii, Y. *J. Org. Chem.* **2004**, *69*, 5494–5496.

[130] Ortial, S.; Fisher, H. C.; Montchamp, J.-L. *J. Org. Chem.* **2013**, *78*, 6599–6608.

[131] Zhang, C.; Li, Z.; Zhu, L.; Yu, L.; Wang, Z.; Li, C. *J. Am. Chem. Soc.* **2013**, *135*, 14082–14085.

[132] Ajellal, N.; Thomas, C. M.; Carpentier, J.-F. *Adv. Synth. Catal.* **2006**, *348*, 1093–1100.

[133] Xu, Q.; Han, L. B. *Org. Lett.* **2006**, *8*, 2099–2101.

[134] Alnasleh, B. K.; Sherrill, W. M.; Rubin, M. *Org. Lett.* **2008**, *10*, 3231–3234.

[135] Moonen, K.; Laureyn, I.; Stevens, C. V. *Chem. Rev.* **2004**, *104*, 6177–6215.

[136] Atherton, F. R.; Hassall, C. H.; Lambert, R. W. *J. Med. Chem.* **1986**, *29*, 29–40.

[137] Allen, M. C.; Fuhrer, W.; Tuck, B.; Wade, R.; Wood, J. M. *J. Med. Chem.* **1989**, *32*, 1652–1661.

[138] Giannousis, P. P.; Bartlett, P. A. *J. Med. Chem.* **1987**, *30*, 1603–1609.

[139] Strater, N.; Lipscomb, W. N. *Biochemistry* **1995**, *34*, 9200–9210.

[140] Mucha, A.; Kafarski, P.; Berlicki, L. *J. Med. Chem.* **2011**, *54*, 5955–5980.

[141] Kafarski, P.; Lejczak, B. *Phosphorus, Sulfur Silicon Relat. Elem.* **1991**, *63*, 193–215.

[142] Collinsova, M.; Jiracek, J. *Curr. Med. Chem.* **2000**, *7*, 629–647.

[143] Hecker, S. J.; Erion, M. D. *J. Med. Chem.* **2008**, *51*, 2328–2345.

[144] Berlicki, L.; Kafarski, P. *Curr. Org. Chem.* **2005**, *9*, 1829–1850.

[145] Orsini, F.; Sello, G.; Sisti, M. *Curr. Med. Chem.* **2010**, *17*, 264–289.

[146] Fields, S. C. *Tetrahedron* **1999**, *55*, 12237–12273.

[147] Naydenova, E. D.; Todorov, P. T.; Troev, K. D. *Amino Acids* **2010**, *38*, 23–30.

[148] Fields, E. K. *J. Am. Chem. Soc.* **1952**, *74*, 1528–1531.

[149] Kabachnik, M. I.; Medved, T. Y. *Dokl. Akad. Nauk SSSR* **1952**, *83*, 689–692.

[150] Keglevich, G.; Balint, E. *Molecules* **2012**, *17*, 12821–12835.

[151] Bhadury, P. S.; Song, B.; Yang, S.; Zhang, Y.; Zhang, S. *Curr. Org. Chem.* **2008**, *5*, 134–150.

[152] Yokomatsu, T.; Yoshida, Y.; Shibuya, S. *J. Org. Chem.* **1994**, *59*, 7930–7933.

[153] Yadav, J. S.; Reddy, B. V. S.; Raj, K. S.; Reddy, K. B.; Prasad, A. R. *Synthesis* **2001**, *15*, 2277–2280.

[154] Ha, H. J.; Nam, G. S. *Synth. Commun.* **1992**, *22*, 1143–1148.

[155] Bhattacharya, A. K.; Raut, D. S.; Rana, K. C.; Polanki, I. K.; Khan, M. S.; Iram, S. *Eur. J. Med. Chem.* **2013**, *66*, 146–152.

[156] Chandrasekhar, S.; Prakash, S. J.; Jagadeshwar, V.; Narsihmulu, C. *Tetrahedron Lett.* **2001**, *42*, 5561–5563.

[157] Reddy, B. V. S.; Krishna, A. S.; Ganesh, A. V.; Kumar, G. G. K. S. N. *Tetrahedron Lett.* **2011**, *52*, 1359–1362.

[158] Viveros-Ceballos, J. V.; Cativiela, C.; Ordonez, M. *Tetrahedron: Asymmetry* **2011**, *22*, 1479–1484.

[159] Barco, A.; Benetti, S.; Bergamini, P.; De Risi, C.; Marchetti, P.; Pollini, G. P.; Zanirato, V. *Tetrahedron Lett.* **1999**, *40*, 7705–7708.

[160] Kraicheva, I.; Tscheva, I.; Vodenicharova, E.; Tashev, E.; Tosheva, T.; Kril, A.; Topashka-Ancheva, M.; Iliev, I.; Gerasimova, T.; Troev, K. *Bioorg. Med. Chem.* **2012**, *20*, 117–124.

[161] Sharghi, H.; Ebrahimpourmoghaddam, S.; Doroodmand, M. M. *Tetrahedron* **2013**, *69*, 4708–4724.

[162] Kumara Swamy, K. C.; Kumaraswamy, S.; Senthil Kumar, K.; Muthiah, C. *Tetrahedron Lett.* **2005**, *46*, 3347–3351.

[163] Shinde, P. V.; Kategaonkar, A. H.; Shingate, B. B.; Shingare, M. S. *Tetrahedron Lett.* **2011**, *52*, 2889–2892.

[164] Dar, B.; Singh, A.; Sahu, A.; Patidar, P.; Chakraborty, A.; Sharma, M.; Singh, B. *Tetrahedron Lett.* **2012**, *53*, 5497–5502.

[165] Manabe, K.; Kobayashi, S. *Chem. Commun.* **2000**, 669–670.

[166] Basle, O.; Li, C.-J. *Chem. Commun.* **2009**, 4124–4126.

[167] Alagiri, K.; Devadig, P.; Prabhu, K. R. *Tetrahedron Lett.* **2012**, *53*, 1456–1459.

[168] Qian, C.; Huang, T. *J. Org. Chem.* **1998**, *63*, 4125–4128.

[169] Xu, F.; Luo, Y.; Deng, M.; Shen, Q. *Eur. J. Org. Chem.* **2003**, 4728–4730.

[170] Lee, S.; Park, J. H.; Kang, J.; Lee, J. K. *Chem. Commun.* **2001**, 1698–1699.

[171] Kudrimoti, S.; Bommena, V. R. *Tetrahedron Lett.* **2005**, *46*, 1209–1210.

[172] Ranu, B. C; Hajra, A.; Jana, U. *Org. Lett.* **1999**, *1*, 1141–1143.

[173] Das, B.; Damodar, K.; Bhunia, N. *J. Org. Chem.* **2009**, *74*, 5607–5609.

[174] Das, B.; Satyalakshmi, G.; Suneel, K.; Damodar, K. *J. Org. Chem.* **2009**, *74*, 8400–8402.

[175] De Blieck, A.; Masschelein, K. G. R.; Dhaene, F.; Rozycka-Sokolowska, E.; Marciniak, B.; Drabowicz, J.; Stevens, C. V. *Chem. Commun.* **2010**, *46*, 258–260.

[176] To, W.-P.; Liu, Y.; Lau, T.-C.; Che, C.-M. *Chem. Eur. J.* **2013**, *19*, 5654–5664.

[177] Rueping, M.; Zoller, J.; Fabry, D. C.; Poscharny, K.; Koenigs, R. M.; Weirich, T. E.; Mayer, J. *Chem. Eur. J.* **2012**, *18*, 3478–3481.

[178] Rueping, M.; Zhu, S.; Koenigs, R. M. *Chem. Commun.* **2011**, *47*, 8679–8681.

[179] Hari, D. P.; Konig, B. *Org. Lett.* **2011**, *13*, 3852–3855.

[180] Wang, H.; Li, X.; Wu, F.; Wan, B. *Tetrahedron Lett.* **2012**, *53*, 681–683.

[181] Dhineshkumar, J.; Lamani, M.; Alagiri, K.; Prabhu, K. R. *Org. Lett.* **2013**, *15*, 1092–1095.

[182] Han, W.; Ofial, A. R. *Chem. Commun.* **2009**, 6023–6025.

[183] Rao, K. U. M.; Jayaprakash, S. H.; Nayak, S. K.; Reddy, C. S. *Catal. Sci. Technol.* **2011**, *1*, 1665–1670.

[184] Firouzabadi, H.; Iranpoor, N.; Ghaderi, A.; Ghavami, M. *Tetrahedron Lett.* **2012**, *53*, 5515–5518.

[185] Joly, G. D.; Jacobsen, E. N. *J. Am. Chem. Soc.* **2004**, *126*, 4102–4103.

[186] Akiyama, T.; Morita, H.; Itoh, J.; Fuchibe, K. *Org. Lett.* **2005**, *7*, 2583–2585.

[187] Pettersen, D.; Marcolini, M.; Bernardi, L.; Fini, F.; Herrera, R. P.; Sgarzani, V.; Ricci, A. *J. Org. Chem.* **2006**, *71*, 6269–6272.

[188] Bernardi, L.; Zhuang, W.; Jorgensen, K. A. *J. Am. Chem. Soc.* **2005**, *127*, 5772–5773.

[189] Kim, S. M.; Kim, H. R.; Kim, D. Y. *Org. Lett.* **2005**, *7*, 2309–2311.

[190] Kuwano, R.; Nishio, R.; Ito, Y. *Org. Lett.* **1999**, *1*, 837–839.

[191] Kiyohara, H.; Matsubara, R.; Kobayashi, S. *Org. Lett.* **2006**, *8*, 5333–5335.

[192] Shibasaki, M.; Kanai, M. *Chem. Rev.* **2008**, *108*, 2853–2873.

[193] Bella, M.; Gasperi, T. *Synthesis* **2009**, 1583–1614.

[194] Bera, K.; Namboothiri, I. N. N. *Org. Lett.* **2012**, *14*, 980–983.

[195] Nakamura, S.; Hayashi, M.; Hiramatsu, Y.; Shibata, N.; Funahashi, Y.; Toru, T. *J. Am. Chem. Soc.* **2009**, *131*, 18240–18241.

[196] Thorat, P. B.; Goswami, S. V.; Magar, R. L.; Patil, B. R.; Bhusare, S. R. *Eur. J. Org. Chem* **2013**, *2013*, 5509–5516.

[197] Davis, F. A.; Lee, S.; Yan, H.; Titus, D. D. *Org. Lett.* **2001**, *3*, 1757–1760.

[198] Yin, L.; Bao, Y.; Kumagai, N.; Shibasaki, M. *J. Am. Chem. Soc.* **2013**, *135*, 10338–10341.

[199] Duxbury, J. P.; Cawley, A.; Thornton-Pett, M.; Wantz, L.; Warne, J. N. D.; Greatrex, R.; Brown, D.; Kee, T. P. *Tetrahedron Lett.* **1999**, *40*, 4403–4406.

[200] Ward, C. V.; Jiang, M.; Kee, T. P. *Tetrahedron Lett.* **2000**, *41*, 6181–6184.

[201] Saito, B.; Egami, H.; Katsuki, T. *J. Am. Chem. Soc.* **2007**, *129*, 1978–1986.

[202] Sasai, H.; Arai, S.; Tahara, Y.; Shibasaki, M. *J. Org. Chem.* **1995**, *60*, 6656–6657.

[203] Groger, H.; Saida, Y.; Sasai, H.; Yamaguchi, K.; Martens, J.; Shibasaki, M. *J. Am. Chem. Soc.* **1998**, *120*, 3089–3103.

[204] Schlemminger, I.; Saida, Y.; Groger, H.; Maison, W.; Durot, N.; Sasai, H.; Shibasaki, M.; Martens, J. *J. Org. Chem.* **2000**, *65*, 4818–4825.

[205] Zhu, X.; Wang, S.; Zhou, S.; Wei, Y.; Zhang, L.; Wang, F.; Feng, Z.; Guo, L.; Mu, X. *Inorg. Chem.* **2012**, *51*, 7134–7143.

[206] Zhou, X.; Shang, D.; Zhang, Q.; Lin, L.; Liu, X.; Feng, X. *Org. Lett.* **2009**, *11*, 1401–1404.

[207] Gali, H.; Prabhu, K. R.; Karra, S. R.; Katti, K. V. *J. Org. Chem.* **2000**, *65*, 676–680.

[208] Kosolapoff, G. M. *J. Am. Chem. Soc.* **1947**, *69*, 2112–2113.

[209] Brachwitz, H.; Olke, M.; Bergmann, J.; Langen, P. *Bioorg. Med. Chem. Lett.* **1997**, *7*, 1739–1742.

[210] Hayashi, M.; Shiomi, N.; Funahashi, Y.; Nakamura, S. *J. Am. Chem. Soc.* **2012**, *134*, 19366–19369.

[211] Tripathi, C. B.; Kayal, S.; Mukherjee, S. *Org. Lett.* **2012**, *14*, 3296–3299.

[212] Baguley, T. D.; Xu, H.-C.; Chatterjee, M.; Nairn, A. C.; Lombroso, P. J.; Ellman, J. A. *J. Med. Chem.* **2013**, *56*, 7636–7650.

[213] Snoeck, R.; Holy, A.; Dewolf-Peeters, C.; Van Den Oord, J.; De Clercq, E.; Andrei, G. *Antimicrob. Agents Chemother.* **2002**, *46*, 3356–3361.

[214] Szymanska, A.; Szymczak, M.; Boryski, J.; Stawinski, J.; Kraszewski, A.; Collu, G.; Sanna, G.; Giliberti, G.; Loddo, R.; La Colla, P. *Bioorg. Med. Chem.* **2006**, *14*, 1924–1934.

[215] Lee, M. V.; Fong, E. M.; Singer, F. R.; Guenette, R. S. *Cancer Res.* **2001**, *61*, 2602–2608.

[216] Patel, D. V.; Rielly-Gauvin, K.; Ryono, D. E.; Free, C. A.; Rogers, W. L.; Smith, S. A.; DeForrest, J. M.; Oehl, R. S.; Petrillo, E. W. Jr. *J. Med. Chem.* **1995**, *38*, 4557–4569.

[217] Tao, M.; Bihovsky, R.; Wells, G. J.; Mallamo, J. P. *J. Med. Chem.* **1998**, *41*, 3912–3916.

[218] Dellaria, J. F.; Maki, R. G.; Stein, H. H.; Cohen, J.; Whittern, D.; Marsh, K.; Hoffman, D. J.; Plattner, J. J.; Perun, T. J. *J. Med. Chem.* **1990**, *33*, 534–542.

[219] Sobhani, S.; Tashrifi, Z. *Tetrahedron* **2010**, *66*, 1429–1439.

[220] Merino, P.; Marques-Lopez, E.; Herrera, R. P. *Adv. Synth. Catal.* **2008**, *350*, 1195–1208.

[221] Kolodiazhnyi, O. I. *Tetrahedron* **2005**, *16*, 3295–3340.

[222] Albrecht, L.; Albrecht, A.; Krawczyk, H.; Jørgensen, K. A. *Chem. Eur. J*, **2010**, *16*, 28–48.

[223] Odinets, I. L.; Matveeva, E. V. *Russ. Chem. Rev.* **2012**, *81*, 221–238.

[224] Pandi, M.; Chanani, P. K.; Govindasamy, S. *Appl. Catal. A* **2012**, *441–442*, 119–123.

[225] Semenzin, D.; Etemad-Moghadam, G.; Albouy, D.; Diallo, O.; Koenig, M. *J. Org. Chem.* **1997**, *62*, 2414–2422.

[226] Keglevich, G.; Toth, V. R.; Drahos, L. *Heteroat. Chem.* **2011**, *22*, 15–17.

[227] Hospital, A.; Meurillon, M.; Peyrottes, S.; Perigaud, C. *Org. Lett.* **2013**, *15*, 4778–4781.

[228] Brucher, K.; Illarionov, B.; Held, J.; Tschan, S.; Kunfermann, A.; Pein, M. K.; Bacher, A.; Grawert, T.; Maes, L.; Mordmuller, B.; Fischer, M.; Kurz, T. *J. Med. Chem.* **2012**, *55*, 6566–6575.

[229] Angelini, T.; Bonollo, S.; Lanari, D.; Pizzo, F.; Vaccaro, L. *Org. Biomol. Chem.* **2013**, *11*, 5042–5046.

[230] Liu, B.; Carpentier, J.-F.; Sarazin, Y. *Chem. Eur. J.* **2012**, *18*, 13259–13264.

[231] Wu, Q.; Zhou, J.; Yao, Z.; Xu, F.; Shen, Q. *J. Org. Chem.* **2010**, *75*, 7498–7501.

[232] Ramananarivo, H. R.; Solhy, A.; Sebti, J.; Smahi, A.; Zahouily, M.; Clark, J.; Sebti, S. *ACS Sustain. Chem. Eng.* **2013**, *1*, 403–409.

[233] Mandhane, P. G.; Joshi, R. S.; Nagargoje, D. R.; Gill, C. H. *Tetrahedron Lett.* **2010**, *51*, 1490–1492.

[234] Groaning, M. D.; Rowe, B. J.; Spilling, C. D. *Tetrahedron Lett.* **1998**, *39*, 5485–5488.

[235] Arai, T.; Bougauchi, M.; Sasai, H.; Shibasaki, M. *J. Org. Chem.* **1996**, *61*, 2926–2927.

[236] Yokomatsu, T.; Yamagishi, T.; Shibuya, S. *Tetrahedron* **1993**, *4*, 1783–1784.

[237] Zhou, X.; Liu, X.; Yang, X.; Shang, D.; Xin, J.; Feng, X. *Angew. Chem. Int. Ed.* **2008**, *47*, 392–394.

[238] Zhou, X.; Zhang, Q.; Hui, Y.; Chen, W.; Jiang, J.; Lin, L.; Liu, X.; Feng, X. *Org. Lett.* **2010**, *12*, 4296–4299.

[239] Yang, F.; Zhao, D.; Lan, J.; Xi, P.; Yang, L.; Xiang, S.; You, J. *Angew. Chem. Int. Ed.* **2008**, *47*, 5646–5649.

[240] Zhou, S.; Wu, Z.; Rong, J.; Wang, S.; Yang, G.; Zhu, X.; Zhang, L. *Chem. Eur. J.* **2012**, *18*, 2653–2659.

[241] Muthupandi, P.; Sekar, G. *Org. Biomol. Chem.* **2012**, *10*, 5347–5352.

[242] Saito, B.; Katsuki, T. *Angew. Chem. Int. Ed.* **2005**, *44*, 4600–4602.

[243] Suyama, K.; Sakai, Y.; Matsumoto, K.; Saito, B.; Katsuki, T. *Angew. Chem. Int. Ed.* **2010**, *49*, 797–799.

[244] Li, W.; Qin, S.; Su, Z.; Yang, H.; Hu, C. *Organometallics* **2011**, *30*, 2095–2104.

[245] Uraguchi, D.; Ito, T.; Ooi, T. *J. Am. Chem. Soc.* **2009**, *131*, 3836–3837.

[246] Uraguchi, D.; Ito, T.; Nakamura, S.; Ooi, T. *Chem. Sci.* **2010**, *1*, 488–490.

[247] Taniguchi, T.; Idota, A.; Yokoyama, S.; Ishibashi, H. *Tetrahedron Lett.* **2011**, *52*, 4768–4770.

[248] Hayashi, M.; Nakamura, S. *Angew. Chem. Int. Ed.* **2011**, *50*, 2249–2252.

[249] Firouzabadi, H.; Iranpoor, N.; Sobhani, S. *Tetrahedron* **2004**, *60*, 203–210.

[250] Mustafa, D. A.; Kashemirov, B. A.; McKenna, C. E. *Tetrahedron Lett.* **2011**, *52*, 2285–2287.

[251] Vachal, P.; Hale, J. J.; Lu, Z.; Streckfuss, E. C.; Mills, S. G.; MacCoss, M.; Yin, D. H.; Algayer, K.; Manser, K.; Kesisoglou, F.; Ghosh, S.; Alani, L. L. *J. Med. Chem.* **2006**, *49*, 3060–3063.

[252] Bhushan, K. R.; Tanaka, E.; Frangioni, J. V. *Angew. Chem. Int. Ed.* **2007**, *46*, 7969–7971.

[253] Lecouvey, M.; Mallard, I.; Bailly, T.; Burgada, R.; Leroux, Y. *Tetrahedron Lett.* **2001**, *42*, 8475–8478.

[254] Egorov, M.; Aoun, S.; Padrines, M.; Redini, F.; Heymann, D.; Lebreton, J.; Mathe-Allainmat, M. *Eur. J. Org. Chem.* **2011**, 7148–7154.

[255] Bortolini, O.; Fantin, G.; Fogagnolo, M.; Rossetti, S.; Maiuolo, L.; Di Pompo, G.; Avnet, S.; Granchi, D. *Eur. J. Med. Chem.* **2012**, *52*, 221–229.

[256] Leung, C. Y.; Park, J.; De Schutter, J. W.; Sebag, M.; Berghuis, A. M.; Tsantrizos, Y. S. *J. Med. Chem.* **2013**, *56*, 7939–7950.

[257] Kaboudin, B.; Alipour, S. *Tetrahedron Lett.* **2009**, *50*, 4243–4245.

[258] Bortolini, O.; Mulani, I.; De Nino, A.; Maiuolo, L.; Nardi, M.; Russo, B.; Avnet, S. *Tetrahedron* **2011**, *67*, 5635–5641.

[259] Ortial, S.; Thompson, D. A.; Montchamp, J.-L. *J. Org. Chem.* **2010**, *75*, 8166–8179.

[260] Smits, J. P.; Wiemer, D. F. *J. Org. Chem.* **2011**, *76*, 8807–8813.

[261] Queffelec, C.; Ribiere, P.; Montchamp, J.-L. *J. Org. Chem.* **2008**, *73*, 8987–8991.

[262] Julienne, D.; Delacroix, O. Gaumont, A.-C. *Curr. Org. Chem.* **2010**, *14*, 457–482.

[263] Xu, W.; Yoshikai, N. *Angew. Chem. Int. Ed.* **2014**, *53*, 14166–14170.

[264] Horner, L.; Simons, G. *Phosphorus Sulfur Silicon Relat. Elem.* **1983**, *15*, 165–175.

[265] Sasaki, S.; Yoshifuji, M. *Curr. Org. Chem.* **2007**, *11*, 17–31.

[266] Butler, I. R.; Gerner, P. *Inorg. Chim. Acta* **2013**, *396*, 1–5.

[267] Togni, A.; Hayashi, T. In *Ferrocenes: Homogeneous Catalysis, Organic Synthesis, Materials Science;* VCH Publishers: Weinheim; New York, **1995**, pp 3–104.

[268] Phillips, E. S., *Ferrocenes Compounds, Properties and Applications*, Nova Science Pub Inc., New York, **2011**.

[269] Štěpnička, P. *Ferrocenes Ligands,Materials and Biomolecules*, John Wiley & Sons, Inc., West Sussex, **2008**.

[270] Bandoli, G.; Dolmella, A. *Coord. Chem. Rev.* **2000**, *209*, 161–196.

[271] Buchgraber, P.; Mercier, A.; Yeo, W. C.; Besnard, C.; Kundig, E. P. *Organometallics* **2011**, *30*, 6303–6315.

[272] Nie, H.; Yao, L.; Li, B.; Zhang, S.; Chen, W. *Organometallics* **2014**, *33*, 2109–2114.

[273] Marquarding, D.; Klusacek, H.; Gokel, G.; Hoffmann, P.; Ugi, I. *J. Am. Chem. Soc.* **1970**, *92*, 5389–5393.

[274] Battelle, L. F.; Bau, R.; Gokel, G. W.; Oyakawa, R. T.; Ugi, I. K. *J. Am. Chem. Soc.* **1973**, *95*, 482–486.

[275] Tschirschwitz, S.; Lonnecke, P.; Hey-Hawkins, E. *Dalton Trans.* **2007**, 1377–1382.

[276] Chen, W.; Mbafor, W.; Roberts, S. M.; Whittall, J. *J. Am. Chem. Soc.* **2006**, *128*, 3922–3923.

[277] Bitterer, F.; Herd, O.; Hessler, A.; Kuhnel, M.; Rettig, K.; Stelzer, O.; Sheldrick, W. S.; Nagel, S.; Rosch, N. *Inorg. Chem.* **1996**, *35*, 4103–4113.

[278] Herd, O.; Langhans, K. P.; Stelzer, O.; Weferling, N.; Sheldrick, W. S. *Angew. Chem. Int. Ed.* **1993**, *32*, 1058–1059.

[279] Herd, O.; Hessler, A.; Langhans, K. P.; Stelzer, O.; Sheldrick, W. S.; Weferling, N. *J. Organomet. Chem.* **1994**, *475*, 99–111.

[280] Reis, A.; Dehe, D.; Farsadpour, S.; Munstein, I.; Sun, Y.; Thiel, W. R. *New J. Chem.* **2011**, *35*, 2488–2495.

[281] Tunney, S. E.; Stille, J. K. *J. Org. Chem.* **1987**, *52*, 748–753.

[282] Mann, F. G.; Pragnell, M. J. *J. Chem. Soc.* **1965**, 4120–4127.

[283] Le Floch, P.; Carmichael, D.; Ricard, L.; Mathey, F. *J. Am. Chem. Soc.* **1993**, *115*, 10665–10670.

[284] Martín, S. E.; Bonaterra, M.; Rossi, R. A. *J. Organomet. Chem.* **2002**, *664*, 223–227.

[285] Bonaterra, M.; Rossi, R. A.; Martin, S. E. *Organometallics* **2009**, *28*, 933–936.

[286] Oshiki, T.; Imamoto, T. *J. Am. Chem. Soc.* **1992**, *114*, 3975–3977.

[287] Gaumont, A.-C.; Brown, J. M.; Hursthouse, M. B.; Coles, S. J. *Chem. Commun.* **1999**, 63–64.

[288] Murata, M.; Buchwald, S. L. *Tetrahedron* **2004**, *60*, 7397–7403.

[289] Stadler, A.; Kappe, C. O. *Org. Lett.* **2002**, *4*, 3541–3543.

[290] Zadny, J.; Velisek, P.; Jakubec, M.; Sykora, J.; Církva, V.; Storch, J. *Tetrahedron* **2013**, *69*, 6213–6218.

[291] Arshad, N.; Hashim, J.; Kappe, C. O. *J. Org. Chem.* **2008**, *73*, 4755–4758.

[292] Berrino, R.; Cacchi, S.; Fabrizi, G.; Goggiamani, A.; Stabile, P. *Org. Biomol. Chem.* **2010**, *8*, 4518–4520.

[293] Bonnafoux, L.; Gramage-Doria, R.; Colobert, F.; Leroux, F. R. *Chem. Eur. J.* **2011**, *17*, 11008–11016.

[294] Ropartz, L.; Meeuwenoord, N. J.; van der Marel, G. A.; van Leeuwen, P. W. N. M.; Slawin, A. M. Z.; Kamer, P. C. J. *Chem. Commun.* **2007**, 1556–1558.

[295] Lipshutz, B. H.; Buzard, D. J.; Yun, C. S. *Tetrahedron Lett.* **1999**, *40*, 201–204.

[296] Matano, Y.; Matsumoto, K.; Nakao, Y.; Uno, H.; Sakaki, S.; Imahori, H. *J. Am. Chem. Soc.* **2008**, *130*, 4588–4589.

[297] Leseurre, L.; Le Boucher d'Herouville, F.; Puntener, K.; Scalone, M.; Genet, J.-P.; Michelet, V. *Org. Lett.* **2011**, *13*, 3250–3253.

[298] Al-Masum, M.; Livinghouse, T. *Tetrahedron Lett.* **1999**, *40*, 7731–7734.

[299] Herd, O.; Heßler, A.; Hingst, M.; Tepper, M.; Stelzer, O. *J. Organomet. Chem* **1996**, *522*, 69–76.

[300] Martorell, G.; Garcias, X.; Janura, M.; Saa, J. M. *J. Org. Chem.* **1998**, *63*, 3463–3467.

[301] Gilbertson, S. R.; Starkey, G. W. *J. Org. Chem.* **1996**, *61*, 2922–2923.

[302] Kraatz, H.-B.; Pletsch, A. *Tetrahedron* **2000**, *11*, 1617–1621.

[303] Hessler, A.; Stelzer, O.; Dibowski, H.; Worm, K.; Schmidtchen, F. P. *J. Org. Chem.* **1997**, *62*, 2362–2369.

[304] Bergbreiter, D. E.; Liu, Y.-S.; Furyk, S.; Case, B. L. *Tetrahedron Lett.* **1998**, *39*, 8799–8802.

[305] Persigehl, P.; Jordan, R.; Nuyken, O. *Macromolecules* **2000**, *33*, 6977–6981.

[306] Wasserscheid, P.; Waffenschmidt, H.; Machnitzki, P.; Kottsieper, K. W.; Stelzer, O. *Chem. Commun.* **2001**, 451–452.

[307] Ziessel, R. F.; Charbonniere, L. J.; Mameri, S.; Camerel, F. *J. Org. Chem.* **2005**, *70*, 9835–9840.

[308] Vallette, H.; Pican, S.; Boudou, C.; Levillain, J.; Plaquevent, J.-C.; Gaumont, A.-C. *Tetrahedron Lett.* **2006**, *47*, 5191–5193.

[309] Kong, K. C.; Cheng, C. H. *J. Am. Chem. Soc.* **1991**, *113*, 6313–6315.

[310] Kwong, F. Y.; Chan, K. S. *Organometallics* **2000**, *19*, 2058–2060.

[311] Kwong, F. Y.; Chan, K. S. *Organometallics* **2001**, *20*, 2570–2578.

[312] Kwong, F. Y.; Lai, C. W.; Chan, K. S. *Tetrahedron Lett.* **2002**, *43*, 3537–3539.

[313] Kwong, F. Y.; Lai, C. W.; Yu, M.; Chan, K. S. *Tetrahedron* **2004**, *60*, 5635–5645.

[314] Kwong, F. Y.; Lai, C. W.; Yu, M.; Tian, Y.; Chan, K. S. *Tetrahedron* **2003**, *59*, 10295–10305.

[315] Kwong, F. Y.; Chan, K. S. *Chem. Commun.* **2000**, 1069–1070.

[316] Baba, K.; Tobisu, M.; Chatani, N. *Angew. Chem. Int. Ed.* **2013**, *52*, 11892–11895.

[317] Baba, K.; Tobisu, M.; Chatani, N. *Org. Lett.* **2015**, *17*, 70–73.

[318] Gelman, D.; Jiang, L.; Buchwald, S. L. *Org. Lett.* **2003**, *5*, 2315–2318.

[319] Allen, D. V.; Venkataraman, D. *J. Org. Chem.* **2003**, *68*, 4590–4593.

[320] Tani, K.; Behenna, D. C.; McFadden, R. M.; Stoltz, B. M. *Org. Lett.* **2007**, *9*, 2529–2531.

[321] Birdsall, D. J.; Hope, E. G.; Stuart, A. M.; Chen, W.; Hu, Y.; Xiao, J. *Tetrahedron Lett.* **2001**, *42*, 8551–8553.

[322] Nakamura, Y.; Takeuchi, S.; Zhang, S.; Okumura, K.; Ohgo, Y. *Tetrahedron Lett.* **2002**, *43*, 3053–3056.

[323] Kerrigan, N. J.; Dunne, E. C.; Cunningham, D.; McArdle, P.; Gilligan, K.; Gilheany, D. G. *Tetrahedron Lett.* **2003**, *44*, 8461–8465.

[324] Knöpfel, T. F.; Aschwanden, P.; Ichikawa, T.; Watanabe, T.; Carreira, E. M. *Angew. Chem. Int. Ed.* **2004**, *43*, 5971–5973.

[325] Clarke, M. L.; Orpen, A. G.; Pringle, P. G.; Turley, E. *Dalton Trans.* **2003**, 4393–4394.

[326] McCarthy, M.; Guiry, P. J. *Tetrahedron* **1999**, *55*, 3061–3070.

[327] Lacey, P. M.; McDonnell, C. M.; Guiry, P. J. *Tetrahedron Lett.* **2000**, *41*, 2475–2478.

[328] Enev, V.; Ewers, C. L. J.; Harre, M.; Nickisch, K.; Mohr, J. T. *J. Org. Chem.* **1997**, *62*, 7092–7093.

[329] Ager, D. J.; East, M. B.; Eisenstadt, A.; Laneman, S. A. *J. Chem. Soc. Chem. Commun.* **1997**, 2359–2360.

[330] Rast, S.; Mohar, B.; Stephan, M. *Org. Lett.* **2014**, *16*, 2688–2691.

[331] Bousrez, G.; Jaroschik, F.; Martinez, A.; Harakat, D.; Nicolas, E.; Le Goff, X. F.; Szymoniak, J. *Dalton Trans.* **2013**, *42*, 10997–11004.

[332] Dhokale, R. A.; Mhaske, S. B. *Org. Lett.* **2013**, *15*, 2218–2221.

[333] Xu, J.; Zhang, P.; Gao, Y.; Chen, Y.; Tang, G.; Zhao, Y. *J. Org. Chem.* **2013**, *78*, 8176–8183.

[334] Zhao, Y.-L.; Wu, G.-J.; Han, F.-S. *Chem. Commun.* **2012**, *48*, 5868–5870.

[335] Jiang, W.; Allan, G.; Fiordeliso, J. J.; Linton, O.; Tannenbaum, P.; Xu, J.; Zhu, P.; Gunnet, J.; Demarest, K.; Lundeen, S.; Sui, Z. *Bioorg. Med. Chem.* **2006**, *14*, 6726–6732.

[336] Zhao, Y.-L.; Wu, G.-S.; Li, Y.; Gao, L.-X.; Han, F.-J. *Chem. Eur. J.* **2012**, *18*, 9622–9627.

[337] Zhang, H.-Y.; Sun, M.; Ma, Y.-N.; Tian, Q.-P.; Yang, S. *Org. Biomol. Chem.* **2012**, *10*, 9627–9633.

[338] Zhang, X.; Liu, H.; Hu, X.; Tang, G.; Zhu, J.; Zhao, Y. *Org. Lett.* **2011**, *13*, 3478–3481.

[339] Rummelt, S. M.; Ranocchiari, M.; van Bokhoven, J. A. *Org. Lett.* **2012**, *14*, 2188–2190.

[340] Xu, K.; Yang, F.; Zhang, G.; Wu, Y. *Green Chem.* **2013**, *15*, 1055–1060.

[341] Shen, C.; Yang, G.; Zhang, W. *Org. Biomol. Chem.* **2012**, *10*, 3500–3505.

[342] Feng, C. G.; Ye, M.; Xiao, K.-J.; Li, S.; Yu, J.-Q. *J. Am. Chem. Soc.* **2013**, *135*, 9322–9325.

[343] Chen, X.; Li, J.-J.; Hao, X.-S.; Goodhue, C. E.; Yu, J.-Q. *J. Am. Chem. Soc.* **2006**, *128*, 78–79.

[344] Chen, X.; Goodhue, C. E.; Yu, J.-Q. *J. Am. Chem. Soc.* **2006**, *128*, 12634–12635.

[345] Tran, G.; Gomez Pardo, D.; Tsuchiya, T.; Hillebrand, S.; Vors, J.-P.; Cossy, J. *Org. Lett.* **2013**, *15*, 5550–5553.

[346] Hu, G.; Chen, W.; Fu, T.; Peng, Z.; Qiao, H.; Gao, Y.; Zhao, Y. *Org. Lett.* **2013**, *15*, 5362–5365.

[347] Hayashi, M.; Matsuura, T.; Tanaka, I.; Ohta, H.; Watanabe, Y. *Org. Lett.* **2013**, *15*, 628–631.

[348] Ivonin, S. P.; Tolmachev, A. A.; Pinchuk, A. M. *Curr. Org. Chem.* **2008**, *12*, 25–38.

[349] Demmer, C. S.; Krogsgaard-Larsen, N.; Bunch, L. *Chem. Rev.* **2011**, *111*, 7981–8006.

[350] Van der Jeught, S.; Stevens, C. V. *Chem. Rev.* **2009**, *109*, 2672–2702.

[351] Liao, K.; Anwar, H.; Hill, I. G.; Vertelov, G. K.; Schwartz, J. *ACS Appl. Mater. Interfaces* **2012**, *4*, 6735–6746.

[352] Alen, J.; Dobrzanska, L.; DeBorggraeve, W. M.; Compernolle, F. *J. Org. Chem.* **2007**, *72*, 1055–1057.

[353] Qu, G.-R.; Xia, R.; Yang, X.-N.; Li, J.-G.; Wang, D.-C.; Guo, H. M. *J. Org. Chem.* **2008**, *73*, 2416–2419.

[354] Jansa, P.; Hradil, O.; Baszczyński, O.; Dračínský, M.; Klepetářová, B.; Holý, A.; Balzarini, J.; Janeba, Z. *Tetrahedron* **2012**, *68*, 865–871.

[355] Kers, A.; Stawiński, J. *Tetrahedron Lett.* **1999**, *40*, 4263–4266.

[356] Johansson, T.; Stawinski, J. *Tetrahedron* **2004**, *60*, 389–395.

[357] Johansson, T.; Kers, A.; Stawinski, J. *Tetrahedron Lett.* **2001**, *42*, 2217–2220.

[358] Tavs, P. *Chem. Ber.* **1970**, *103*, 2428.

[359] Baillie, C.; Xiao, J. L. *Curr. Org. Chem.* **2003**, *7*, 477–514.

[360] Tappe, F. M. J.; Trepohl, V. T.; Oestreich, M. *Synthesis* **2010**, 3037–3062.

[361] Balthazor, T. M.; Miles, J. A.; Stults, B. R. *J. Org. Chem.* **1978**, *43*, 4538–4540.

[362] Balthazor, T. M. *J. Org. Chem.* **1980**, *45*, 2519–2522.

[363] Balthazor, T. M.; Grabiak, R. C. *J. Org. Chem.* **1980**, *45*, 5425–5426.

[364] Villemin, D.; Elbilali, A.; Simeon, F.; Jaffres, P.-A.; Maheut, G.; Mosaddak, M.; Hakiki, A. *J. Chem. Res. Synop.* **2003**, 436–437.

[365] Keglevich, G.; Grün, A.; Bölcskei, A.; Drahos, L.; Kraszni, M.; Balogh, G. T. *Heteroat. Chem.* **2012**, *23*, 574–582.

[366] Teixeira, F. C.; Rangel, C. M.; Teixeira, A. P. S. *New J. Chem.* **2013**, *37*, 3084–3091.

[367] Yang, G.; Shen, C.; Zhang, L.; Zhang, W. *Tetrahedron Lett.* **2011**, *52*, 5032–5035.

[368] Hirao, T.; Masunaga, T.; Yamada, N.; Ohshiro, Y.; Agawa, T. *Bull. Chem. Soc. Jpn.* **1982**, *55*, 909–913.

[369] Hirao, T.; Masunaga, T.; Ohshiro, Y.; Agawa, T. *Synthesis* **1981**, 56–57.

[370] Kalek, M.; Stawinski, J. *Organometallics* **2007**, *26*, 5840–5847.

[371] Amatore, C.; Jutand, A. *Acc. Chem. Res.* **2000**, *33*, 314–321.

[372] Kozuch, S.; Shaik, S.; Jutand, A.; Amatore, C. *Chem. Eur. J.* **2004**, *10*, 3072–3080.

[373] Kohler, M. C.; Grimes, T. V.; Wang, X.; Cundari, T. R.; Stockland, R. A. Jr. *Organometallics* **2009**, *28*, 1193–1201.

[374] Kohler, M. C.; Stockland, R. A. Jr.; Rath, N. P. *Organometallics* **2006**, *25*, 5746–5756.

[375] Amatore, C.; Jutand, A. *J. Organomet. Chem.* **1999**, *576*, 254–278.

[376] Defacqz, N.; de Bueger, B.; Touillaux, R.; Cordi, A.; Marchand-Brynaert, J. *Synthesis* **1999**, 1368–1372.

[377] Kalek, M.; Jezowska, M.; Stawinski, J. *Adv. Synth. Catal.* **2009**, *351*, 3207–3216.

[378] Kalek, M.; Stawinski, J. *Organometallics* **2008**, *27*, 5876–5888.

[379] Belabassi, Y.; Alzghari, S.; Montchamp, J.-L. *J. Organomet. Chem.* **2008**, *693*, 3171–3178.

[380] Deal, E. L.; Petit, C.; Montchamp, J.-L. *Org. Lett.* **2011**, *13*, 3270–3273.

[381] Kalek, M.; Ziadi, A.; Stawinski, J. *Org. Lett.* **2008**, *10*, 4637–4640.

[382] Johansson, T.; Stawinski, J. *Chem. Commun.* **2001**, 2564–2565.

[383] Xu, Y.; Zhang, J. *Chem. Commun.* **1986**, 1606.

[384] Zhang, J.; Xu, Y.; Huang, G.; Guo, H. *Tetrahedron Lett.* **1988**, *29*, 1955–1958.

[385] Hockova, D.; Dracinsky, M.; Holy, A. *Eur. J. Org. Chem.* **2010**, *2010*, 2885–2892.

[386] Zmudzka, K.; Johansson, T.; Wojcik, M.; Janicka, M.; Nowak, M.; Stawinski, J.; Nawrot, B. *New J. Chem.* **2003**, *27*, 1698–1705.

[387] Kalek, M.; Stawinski, J. *Tetrahedron* **2009**, *65*, 10406–10412.

[388] Kohler, M. C.; Sokol, J. G.; Stockland Jr., R. A. *Tetrahedron Lett.* **2009**, *50*, 457–459.

[389] Bloomfield, A. J.; Herzon, S. B. *Org. Lett.* **2012**, *14*, 4370–4373.

[390] Luo, Y.; Wu, J. *Organometallics* **2009**, *28*, 6823–6826.

[391] Andaloussi, M.; Lindh, J.; Sävmarker, J.; Sjöberg, P. J. R.; Larhed, M. *Chem. Eur. J* **2009**, *15*, 13069–13074.

[392] Petrakis, K. S.; Nagabhushan, T. L. *J. Am. Chem. Soc.* **1987**, *109*, 2831–2833.

[393] Goossen, L. J.; Dezfuli, M. K. *Synlett* **2005**, 445–448.

[394] Huang, C.; Tang, X.; Fu, H.; Jiang, Y.; Zhao, Y. *J. Org. Chem.* **2006**, *71*, 5020–5022.

[395] Rao, H.; Jin, Y.; Fu, H.; Jiang, Y.; Zhao, Y. *Chem. Eur. J.* **2006**, *12*, 3636–3646.

[396] Marque, S.; Tordo, P. *Top. Curr. Chem.* **2005**, *250*, 43–76.

[397] Ohmori, H.; Nakai, S.; Masumi, M. *J. Chem. Soc. Perkin I* **1979**, 2023–2026.

[398] Jason, E. F.; Fields, E. K. *J. Org. Chem.* **1962**, *27*, 1402–1405.

[399] Kottmann, H.; Skarzewski, J.; Effenberger, F. *Synthesis* **1987**, 797–801.

[400] Wang, G.-W.; Wang, C.-Z.; Zou, J.-P. *J. Org. Chem.* **2011**, *76*, 6088–6094.

[401] Kagayama, T.; Nakano, A.; Sakaguchi, S.; Ishii, Y. *Org. Lett.* **2006**, *8*, 407–409.

[402] Xu, W.; Zou, J.-P.; Zhang, W. *Tetrahedron Lett.* **2010**, *51*, 2639–2643.

[403] Mu, X.-J.; Zou, J.-P.; Qian, Q.-F.; Zhang, W. *Org. Lett.* **2006**, *8*, 5291–5293.

[404] Mao, X.; Ma, X.; Zhang, S.; Hu, H.; Zhu, C.; Cheng, Y. *Eur. J. Org. Chem.* **2013**, *2013*, 4245–4248.

[405] Kim, S. H.; Kim, S. H.; Lim, C. H.; Kim, J. N. *Tetrahedron Lett.* **2013**, *54*, 1697–1699.

[406] Xiang, C.-B.; Bian, Y.-J.; Mao, X.-R.; Huang, Z.-Z. *J. Org. Chem.* **2012**, *77*, 7706–7710.

[407] Hou, C.; Ren, Y.; Lang, R.; Hu, X.; Xia, C.; Li, F. *Chem. Commun.* **2012**, *48*, 5181–5183.

[408] Wang, H.; Li, X.; Wu, F.; Wan, B. *Synthesis* **2012**, *44*, 941–945.

[409] Li, C.; Yano, T.; Ishida, N.; Murakami, M. *Angew. Chem. Int. Ed.* **2013**, *52*, 9801–9804.

[410] Wang, S.; Guo, R.; Wang, G.; Chen, S.-Y.; Yu, X.-Q. *Chem. Commun.* **2014**, *50*, 12718–12721.

[411] Zhuang, R.; Xu, J.; Cai, Z.; Tang, G.; Fang, M.; Zhao, Y. *Org. Lett.* **2011**, *13*, 2110–2113.

[412] Han, Z. S.; Goyal, N.; Herbage, M. A.; Sieber, J. D.; Qu, B.; Xu, Y.; Li, Z.; Reeves, J. T.; Desrosiers, J.-N.; Ma, S.; Grinberg, N.; Lee, H.; Mangunuru, H. P. R.; Zhang, Y.; Krishnamurthy, D.; Lu, B. Z.; Song, J. J.; Wang, G.; Senanayake, C. H. *J. Am. Chem. Soc.* **2013**, *135*, 2474–2477.

[413] Cai, D.; Payack, J. F.; Bender, D. R.; Hughes, D. L.; Verhoeven, T. R.; Reider, P. J. *J. Org. Chem.* **1994**, *59*, 7180–7181.

[414] Baya, M.; Buil, M. L.; Esteruelas, M. A.; Onate, E. *Organometallics* **2005**, *24*, 2030–2038.

[415] Kenward, A. L.; Piers, W. E. *Angew. Chem. Int. Ed.* **2008**, *47*, 38–41.

[416] Doherty, S.; Knight, J. G.; Robins, E. G.; Scanlan, T. H.; Champkin, P. A.; Clegg, W. *J. Am. Chem. Soc.* **2001**, *123*, 5110–5111.

[417] Pietsch, J.; Braunstein, P.; Chauvin, Y. *New J. Chem.* **1998**, *22*, 467–472.

[418] Hao, P.; Zhang, S.; Yi, J.; Sun, W.-H. *J. Mol. Cat. A. Chem.* **2009**, *302*, 1–6.

[419] Yu, B.; Yan, X.; Wang, S.; Tang, N.; Xi, C. *Organometallics* **2010**, *29*, 3222–3226.

[420] Zhou, Y.; Chen, C.; Yan, X.; Xi, C. *Organometallics* **2014**, *33*, 844–846.

[421] Brandys, M.-C.; Puddephatt, R. J. *J. Am. Chem. Soc.* **2001**, *123*, 4839–4840.

[422] Brandys, M.-C.; Puddephatt, R. J. *J. Am. Chem. Soc.* **2002**, *124*, 3946–3950.

[423] DelNegro, A. S.; Woessner, S. M.; Sullivan, B. P.; Dattelbaum, D. M.; Schoonover, J. R. *Inorg. Chem.* **2001**, *40*, 5056–5057.

[424] Hunks, W. J.; Lapierre, J.; Jenkins, H. A.; Puddephatt, R. J. *J. Chem. Soc. Dalton Trans.* **2002**, 2885–2889.

[425] Kaesz, H. D.; Stone, F. G. A. *J. Org. Chem.* **1959**, *24*, 635–637.

[426] Maier, L.; Seyferth, D.; Stone, F. G. A.; Rochow, E. G. *J. Am. Chem. Soc.* **1957**, *79*, 5884–5889.

[427] Cowley, A. H.; Taylor, M. W. *J. Am. Chem. Soc.* **1969**, *91*, 1929–1933.

[428] Dunne, K. S.; Lee, S. E.; Gouverneur, V. *J. Organomet. Chem.* **2006**, *691*, 5246–5259.

[429] King, R. B.; Cloyd, J. C.-Jr.; Reimann, R. H. *J. Org. Chem.* **1976**, *41*, 972–977.

[430] Mercier, F.; Hugel-Le Goff, C.; Mathey, F. *Tetrahedron Lett.* **1989**, *30*, 2397–2398.

[431] Mercier, F.; Mathey, F. *Tetrahedron Lett.* **1989**, *30*, 5269–5270.

[432] Yam, M.; Tsang, C.-W.; Gates, D. P. *Inorg. Chem.* **2004**, *43*, 3719–3723.

[433] Stute, A.; Kehr, G.; Daniliuc, C. G.; Frohlich, R.; Erker, G. *Dalton Trans.* **2013**, *42*, 4487–4499.

[434] Fessler, M.; Czermak, G.; Eller, S.; Trettenbrein, B.; Bruggeller, P.; Bettucci, L.; Bianchini, C.; Meli, A.; Ienco, A.; Oberhauser, W. *Dalton Trans.* **2009**, 1859–1869.

[435] Timokhin, B.; Dmitriev, V.; Vengelnikova, V.; Donskikh, V.; Kalabina, A. *Zh. Obshch. Khim.* **1983**, *53*, 291–294.

[436] Kostyuk, A. N.; Svyaschenko, Y. V.; Volochnyuk, D. M.; Lysenko, N. V.; Tolmachev, A. A.; Pinchuk, A. M. *Tetrahedron Lett.* **2003**, *44*, 6487–6491.

[437] Kazankova, M. A.; Chirkov, E. A.; Kochetkov, A. N.; Efimova, I. V.; Beletskaya, I. P. *Tetrahedron Lett.* **1999**, *40*, 573–576.

[438] Veits, Y. A.; Karlstedt, N. B.; Beletskaya, I. P. *Tetrahedron Lett.* **1995**, *36*, 4121–4124.

[439] Gilbertson, S. R.; Fu, Z.; Starkey, G. W. *Tetrahedron Lett.* **1999**, *40*, 8509–8512.

[440] Gilbertson, S. R.; Fu, Z. *Org. Lett.* **2001**, *3*, 161–164.

[441] Julienne, D.; Lohier, J.-F.; Delacroix, O.; Gaumont, A.-C. *J. Org. Chem.* **2007**, *72*, 2247–2250.

[442] Julienne, D.; Delacroix, O.; Gaumont, A.-C. *Phosphorus, Sulfur Silicon Relat. Elem.* **2009**, *184*, 846–856.

[443] Boyd, D. R.; Sharma, N. D.; Kaik, M.; Bell, M.; Berberian, M. V.; McIntyre, P. B. A.; Kelly, B.; Hardacre, C.; Stevenson, P. J.; Allen, C. C. R. *Adv. Synth. Catal.* **2011**, *353*, 2455–2465.

[444] Vandyck, K.; Matthys, B.; Willen, M.; Robeyns, K.; Van Meervelt, L.; van der Eycken, J. *Org. Lett.* **2006**, *8*, 363–366.

[445] Börner, A.; Ward, J.; Ruth, W.; Holz, J.; Kless, A.; Heller, D.; Kagan, H. B. *Tetrahedron* **1994**, *50*, 10419–10430.

[446] Coumbe, T.; Lawrence, N. J.; Muhammad, F. *Tetrahedron Lett.* **1994**, *35*, 625–628.

[447] Cieslikiewicz, M.; Bouet, A.; Jugé, S.; Toffano, M.; Bayardon, J.; West, C.; Lewinski, K.; Gillaizeau, I. *Eur. J. Org. Chem.* **2012**, *2012*, 1101–1106.

[448] Shulyupin, M. O.; Chirkov, E. A.; Kazankova, M. A.; Beletskaya, I. P. *Synlett* **2005**, 658–660.

[449] Schmidbaur, H.; Paschalidis, C.; Reber, G.; Müller, G. *Chem. Ber.* **1988**, *121*, 1241–1245.

[450] Mimeau, D.; Gaumont, A. C. *J. Org. Chem.* **2003**, *68*, 7016–7022.

[451] Gusarova, N. K.; Shaikhudinova, S. I.; Arbuzova, S. N.; Vakul'skaya, T. I.; Sukhov, B. G.; Sinegovskaya, L. M.; Nikitin, M. V.; Mal'kina, A. G.; Chernysheva, N. A.; Trofimov, B. A. *Tetrahedron* **2003**, *59*, 4789–4794.

[452] Malysheva, S. F.; Sukhov, B. G.; Larina, L. I.; Belogorova, N. A.; Gusarova, N. K.; Trofimov, B. A. *Russ. J. Gen. Chem.* **2001**, *71*, 1907–1911.

[453] Bookham, J. L.; McFarlane, W.; Thornton-Pett, M.; Jones, S. *J. Chem. Soc. Dalton Trans.* **1990**, 3621–3627.

[454] Bookham, J. L.; Smithies, D. M.; Wright, A.; Thornton-Pett, M.; McFarlane, W. *J. Chem. Soc. Dalton Trans.* **1998**, 811–818.

[455] Busacca, C. A.; Farber, E.; DeYoung, J.; Campbell, S.; Gonnella, N. C.; Grinberg, N.; Haddad, N.; Lee, H.; Ma, S.; Reeves, D.; Shen, S.; Senanayake, C. H. *Org. Lett.* **2009**, *11*, 5594–5597.

[456] Busacca, C. A.; Qu, B.; Farber, E.; Haddad, N.; Grět, N.; Saha, A. K.; Eriksson, M. C.; Wu, J.-P.; Fandrick, K. R.; Han, S.; Grinberg, N.; Ma, S.; Lee, H.; Li, Z.; Spinelli, M.; Gold, A.; Wang, G.; Wipf, P.; Senanayake, C. H. *Org. Lett.* **2013**, *15*, 1132–1135.

[457] Guillemin, J.-C.; Janati, T.; Lassalle, L. *Adv. Space Res.* **1995**, *16*, 85–92.

[458] Mitchell, T. N.; Heesche, K. *J. Organomet. Chem.* **1991**, *409*, 163–170.

[459] Dodds, D. L.; Haddow, M. F.; Orpen, A. G.; Pringle, P. G.; Woodward, G. *Organometallics* **2006**, *25*, 5937–5945.

[460] Chen, Y.-R.; Duan, W.-L. *J. Am. Chem. Soc.* **2013**, *135*, 16754–16757.

[461] Sanji, T.; Shiraishi, K.; Kashiwabara, T.; Tanaka, M. *Org. Lett.* **2008**, *10*, 2689–2692.

[462] Liu, L.; Wang, Y.; Zeng, Z.; Xu, P.; Gao, Y.; Yin, Y.; Zhao, Y. *Adv. Synth. Catal.* **2013**, *355*, 659–666.

[463] Sato, A.; Yorimitsu, H.; Oshima, K. *Angew. Chem. Int. Ed.* **2005**, *44*, 1694–1696.

[464] Molitor, S.; Becker, J.; Gessner, V. H. *J. Am. Chem. Soc.* **2014**, *136*, 15517–15520.

[465] Arbuzova, S. N.; Gusarova, N. K.; Glotova, T. E.; Ushakov, I. A.; Verkhoturova, S. I.; Korocheva, A. O.; Trofimov, B. A. *Eur. J. Org. Chem.* **2014**, *2014*, 639–643.

[466] Al-Shboul, T. M. A.; Görls, H.; Westerhausen, M. *Inorg. Chem. Commun.* **2008**, *11*, 1419–1421.

[467] Al-Shboul, T. M. A.; Pálfi, V. K.; Yu, L.; Kretschmer, R.; Wimmer, K.; Fischer, R.; Görls, H.; Reiher, M.; Westerhausen, M. *J. Organomet. Chem.* **2011**, *696*, 216–227.

[468] Hu, H.; Cui, C. *Organometallics* **2012**, *31*, 1208–1211.

[469] Kondoh, A.; Yorimitsu, H.; Oshima, K. *J. Am. Chem. Soc.* **2007**, *129*, 4099–4104.

[470] Kazankova, M. A.; Efimova, I. V.; Kochetkov, A. N.; Afanas'ev, V. V.; Beletskaya, I. P. *Russ. J. Org. Chem.* **2002**, *38*, 1465–1474.

[471] Join, B.; Mimeau, D.; Delacroix, O.; Gaumont, A.-C. *Chem. Commun.* **2006**, 3249–3251.

[472] Ohmiya, H.; Yorimitsu, H.; Oshima, K. *Angew. Chem. Int. Ed. Engl.* **2005**, *44*, 2368–2370.

[473] Jerome, F.; Monnier, F.; Lawicka, H.; Derien, S.; Dixneuf, P. H. *Chem. Commun.* **2003**, 696–697.

[474] Nagata, S.; Kawaguchi, S.; Matsumoto, M.; Kamiya, I.; Nomoto, A.; Sonoda, M.; Ogawa, A. *Tetrahedron Lett.*, **2007**, *48*, 6637–6640.

[475] Kawaguchi, S.; Nagata, S.; Nomoto, A.; Sonoda, M.; Ogawa, A. *J. Org. Chem.* **2008**, *73*, 7928–7933.

[476] Kanemura, S.; Kondoh, A.; Yorimitsu, H.; Oshima, K. *Org. Lett.* **2007**, *9*, 2031–2033.

[477] Shirai, T.; Kawaguchi, S.; Nomoto, A.; Ogawa, A. *Tetrahedron Lett.* **2008**, *49*, 4043–4046.

[478] Miyaji, T.; Xi, Z.; Ogasawara, M.; Nakajima, K.; Takahashi, T. *J. Org. Chem.* **2007**, *72*, 8737–8740.

[479] Ogasawara, M.; Kotani, S.; Nakajima, H.; Furusho, H.; Miyasaka, M.; Shimoda, Y.; Wu, W.-Y.; Sugiura, M.; Takahashi, T.; Nakajima, M. *Angew. Chem. Int. Ed.* **2013**, *52*, 13798–13802.

[480] Angharad Baber, R.; Clarke, M. L.; Guy Orpen, A.; Ratcliffe, D. A. *J. Organomet. Chem.* **2003**, *667*, 112–119.

[481] Arisawa, M.; Yamaguchi, M. *J. Am. Chem. Soc.* **2000**, *122*, 2387–2388.

[482] Hu, J.; Zhao, N.; Yang, B.; Wang, G.; Guo, L.-N.; Liang, Y.-M.; Yang, S.-D. *Chem. Eur. J.* **2011**, *17*, 5516–5521.

[483] Hu, G.; Gao, Y.; Zhao, Y. *Org. Lett.* **2014**, *16*, 4464–4467.

[484] Alexandre, F.-R.; Amador, A.; Bot, S.; Caillet, C.; Convard, T.; Jakubik, J.; Musiu, C.; Poddesu, B.; Vargiu, L.; Liuzzi, M.; Roland, A.; Seifer, M.; Standring, D.; Storer, R.; Dousson, C. B. *J. Med. Chem.* **2011**, *54*, 392–395.

[485] Han, L. B.; Ono, Y.; Yazawa, H. *Org. Lett.* **2005**, *7*, 2909–2911.

[486] Niu, M.; Fu, H.; Jiang, Y.; Zhao, Y. *Chem. Commun.* **2007**, 272–274.

[487] Trostyanskaya, I. G.; Beletskaya, I. P. *Tetrahedron* **2014**, *70*, 2556–2562.

[488] Han, L.-B.; Zhao, C.-Q.; Tanaka, M. *J. Org. Chem.* **2001**, *66*, 5929–5932.

[489] Duraud, A.; Toffano, M.; Fiaud, J.-C. *Eur. J. Org. Chem.* **2009**, *2009*, 4400–4403.

[490] Dobashi, N.; Fuse, K.; Hoshino, T.; Kanada, J.; Kashiwabara, T.; Kobata, C.; Nune, S. K.; Tanaka, M. *Tetrahedron Lett.* **2007**, *48*, 4669–4673.

[491] Lecercle, D.; Sawicki, M.; Taran, F. *Org. Lett.* **2006**, *8*, 4283–4285.

[492] Salin, A. V.; Ilin, A. V.; Shamsutdinova, F. G.; Fatkhutdinov, A. R.; Galkin, V. I.; Islamov, D. R.; Kataeva, O. N. *Tetrahedron Lett.* **2015**, *56*, 6282–6286.

[493] Kabalka, G. W.; Guchhait, S. K. *Org. Lett.* **2003**, *5*, 729–731.

[494] Brown, H. C.; Gupta, S. K. *J. Am. Chem. Soc.* **1975**, *97*, 5249–5255.

[495] Deloux, L.; Srebnik, M. *J. Org. Chem.* **1994**, *59*, 6871–6873.

[496] Srebnik, M.; Bhat, N. G.; Brown, H. C. *Tetrahedron Lett.* **1988**, *29*, 2635–2638.

[497] Ogawa, T.; Usuki, N.; Ono, N. *J. Chem. Soc. Perkin Trans. 1* **1998**, 2953–2958.

[498] Thielges, S.; Bisseret, P.; Eustache, J. *Org. Lett.* **2005**, *7*, 681–684.

[499] Abbas, S.; Bertram, R. D.; Hayes, C. J. *Org. Lett.* **2001**, *3*, 3365–3367.

[500] Kumar Nune, S.; Tanaka, M. *Chem. Commun.* **2007**, 2858–2860.

[501] Han, L. B.; Zhao, C. Q.; Onozawa, S. Y.; Goto, M.; Tanaka, M. *J. Am. Chem. Soc.* **2002**, *124*, 3842–3843.

[502] Xu, Q.; Shen, R.; Ono, Y.; Nagahata, R.; Shimada, S.; Goto, M.; Han, L.-B. *Chem. Commun.* **2011**, *47*, 2333–2335.

[503] Berlin, K. D.; Taylor, H. A. *J. Am. Chem. Soc.* **1964**, *86*, 3862–3866.

[504] Berlin, K. D.; Hellwege, D. M.; Nagabhushanam, M. *J. Org. Chem.* **1965**, *30*, 1265–1267.

[505] Boyer, S. H.; Jiang, H.; Jacintho, J. D.; Reddy, M. V.; Li, H.; Li, W.; Godwin, J. L.; Schulz, W. G.; Cable, E. E.; Hou, J.; Wu, R.; Fujitaki, J. M.; Hecker, S. J.; Erion, M. D. *J. Med. Chem.* **2008**, *51*, 7075–7093.

[506] Demir, A. S.; Tanyeli, C.; Şeşenoğlu, Ö.; Demiç, Ş.; Evin, Ö. Ö. *Tetrahedron Lett.* **1996**, *37*, 407–410.

[507] Trofimov, B. A.; Malysheva, S. F.; Gusarova, N. K.; Belogorlova, N. A.; Vasilevsky, S. F.; Kobychev, V. B.; Sukhov, B. G.; Ushakov, I. A. *Mendeleev Commun.* **2007**, *17*, 181–182.

[508] Yamamoto, Y.; Fukatsu, K.; Nishiyama, H. *Chem. Commun.* **2012**, *48*, 7985–7987.

[509] Low, P. J. *J. Clust. Sci.* **2008**, *19*, 5–46.

[510] Schauer, P. A.; Low, P. J. *Eur. J. Inorg. Chem.* **2012**, 390–411.

[511] Lindahl, S. E.; Park, H.; Pink, M.; Zaleski, J. M. *J. Am. Chem. Soc.* **2013**, *135*, 3826–3833.

[512] Miller, A. D.; Johnson, S. A.; Tupper, K. A.; McBee, J. L.; Tilley, T. D. *Organometallics* **2009**, *28*, 1252–1262.

[513] Imamoto, T.; Saitoh, Y.; Koide, A.; Ogura, T.; Yoshida, K. *Angew. Chem. Int. Ed.* **2007**, *46*, 8636–8639.

[514] Ochida, A.; Ito, H.; Sawamura, M. *J. Am. Chem. Soc.* **2006**, *128*, 16486–16487.

[515] Ito, H.; Ohmiya, H.; Sawamura, M. *Org. Lett.* **2010**, *12*, 4380–4383.

[516] Ortial, S.; Montchamp, J.-L. *Org. Lett.* **2011**, *13*, 3134–3137.

[517] Gao, Y.; Wang, G.; Chen, L.; Xu, P.; Zhao, Y.; Zhou, Y.; Han, L.-B. *J. Am. Chem. Soc.* **2009**, *131*, 7956–7957.

[518] Bernoud, E.; Veillard, R.; Alayrac, C.; Gaumont, A.-C. *Molecules* **2012**, *17*, 14573–14587.

[519] Shiue, T.-W.; Yeh, W.-Y.; Lee, G.-H.; Peng, S.-M. *Organometallics* **2006**, *25*, 4150–4154.

[520] Aguiar, A. M.; Irelan, J. R. S.; Morrow, C. J.; John, J. P.; Prejean, G. *J. Org. Chem.* **1969**, *34*, 2684–2686.

[521] Beletskaya, I. P.; Afanasiev, V. V.; Kazankova, M. A.; Efimova, I. V. *Org. Lett.* **2003**, *5*, 4309–4311.

[522] Di Credico, B.; Fabrizi de Biani, F.; Gonsalvi, L.; Guerri, A.; Ienco, A.; Laschi, F.; Peruzzini, M.; Reginato, G.; Rossin, A.; Zanello, P. *Chem. Eur. J.* **2009**, *15*, 11985–11998.

[523] Jouvin, K.; Veillard, R.; Theunissen, C.; Alayrac, C.; Gaumont, A.-C.; Evano, G. *Org. Lett.* **2013**, *15*, 4592–4595.

[524] Bernoud, E.; Alayrac, C.; Delacroix, O.; Gaumont, A.-C. *Chem. Commun.* **2011**, *47*, 3239–3241.

[525] Abdellah, I.; Bernoud, E.; Lohier, J.-F.; Alayrac, C.; Toupet, L.; Lepetit, C.; Gaumont, A.-C. *Chem. Commun.* **2012**, *48*, 4088–4090.

[526] van Assema, S. G. A.; Tazelaar, C. G. J.; de Jong, G. B.; van Maarseveen, J. H.; Schakel, M.; Lutz, M.; Spek, A. L.; Slootweg, J. C.; Lammertsma, K. *Organometallics* **2008**, *27*, 3210–3215.

[527] Slowinski, F.; Aubert, C.; Malacria, M. *J. Org. Chem.* **2003**, *68*, 378–386.

[528] Ashburn, B. O.; Carter, R. G.; Zakharov, L. N. *J. Am. Chem. Soc.* **2007**, *129*, 9109–9116.

[529] Nishida, G.; Noguchi, K.; Hirano, M.; Tanaka, K. *Angew. Chem. Int. Ed.* **2007**, *46*, 3951–3954.

[530] Nishida, G.; Noguchi, K.; Hirano, M.; Tanaka, K. *Angew. Chem. Int. Ed.* **2008**, *47*, 3410–3413.

[531] Bushuk, S. B.; Carre, F. H.; Guy, D. M. H.; Douglas, W. E.; Kalvinkovskya, Y. A.; Klapshina, L. G.; Rubinov, A. N.; Stupak, A. P.; Bushuk, B. A. *Polyhedron* **2004**, *23*, 2615–2623.

[532] Heller, B.; Gutnov, A.; Fischer, C.; Drexler, H.-J.; Spannenberg, A.; Redkin, D.; Sundermann, C.; Sundermann, B. *Chem. Eur. J* **2007**, *13*, 1117–1128.

[533] Charrier, C.; Chodkiewicz, W.; Cadiot, P. *Bull. Soc. Chim. Fr.* **1966**, 1002–1011.

[534] Doherty, S.; Knight, J. G.; Hashmi, A. S. K.; Smyth, C. H.; Ward, N. A. B.; Robson, K. J.; Tweedley, S.; Harrington, R. W.; Clegg, W. *Organometallics* **2010**, *29*, 4139–4147.

[535] Yang, X.; Matsuo, D.; Suzuma, Y.; Fang, J.-K.; Xu, F.; Orita, A.; Otera, J.; Kajiyama, S.; Koumura, N.; Hara, K. *Synlett* **2011**, 2402–2406.

[536] Pellizzaro, L.; Tatibouët, A.; Fabris, F.; Rollin, P.; Lucchi, O. D. *Tetrahedron Lett.* **2009**, *50*, 101–103.

[537] Melen, R. L.; Stephan, D. W. *Dalton Trans.* **2013**, *42*, 4795–4798.

[538] Doherty, S.; Knight, J. G.; Smyth, C. H.; Harrington, R. W.; Clegg, W. *Organometallics* **2008**, *27*, 1679–1682.

[539] Sakiyama, N.; Hojo, D.; Noguchi, K.; Tanaka, K. *Chem. Eur. J.* **2011**, *17*, 1428–1432.

[540] Doherty, S.; Knight, J. G.; Smyth, C. H.; Harrington, R. W.; Clegg, W. *Org. Lett.* **2007**, *9*, 4925–4928.

[541] Chen, C. C.; Waser, J. *Chem. Commun.* **2014**, *50*, 12923–12926.

[542] Iorga, B.; Eymery, F.; Carmichael, D.; Savignac, P. *Eur. J. Org. Chem.* **2000**, 3103–3115.

[543] Mo, J.; Kang, D.; Eom, D.; Kim, S. H.; Lee, P. H. *Org. Lett.* **2013**, *15*, 26–29.

[544] Jiang, W.; Allan, G.; Chen, X.; Fiordeliso, J. J.; Linton, O.; Tannenbaum, P.; Xu, J.; Zhu, P.; Gunnet, J.; Demarest, K.; Lundeen, S.; Sui, Z. *Steroids* **2006**, *71*, 949–954.

[545] Lera, M.; Hayes, C. J. *Org. Lett.* **2000**, *2*, 3873–3875.

[546] Desai, N. B.; McKelvie, N.; Ramirez, F. *J. Chem. Soc.* **1962**, *84*, 1745–1747.

[547] Gallier, F.; Alexandre, J. A. C.; El Amri, C.; Deville-Bonne, D.; Peyrottes, S.; Perigaud, C. *Chem. Med. Chem.* **2011**, *6*, 1094–1106.

[548] Meurillon, M.; Gallier, F.; Peyrottes, S.; Perigaud, C. *Tetrahedron* **2009**, *65*, 6039–6046.

[549] Liu, P.; Yang, J.; Li, P.; Wang, L. *Appl. Organomet. Chem.* **2011**, *25*, 830–835.

[550] Jouvin, K.; Heimburger, J.; Evano, G. *Chem. Sci.* **2012**, *3*, 756–760.

[551] Li, X.; Yang, F.; Wu, Y.; Wu, Y. *Org. Lett.* **2014** *16*, 992–995.

5

SYNTHESIS OF THIOETHERS, SULFONES, AND RELATED COMPOUNDS

5.1 SYNTHESIS OF THIOETHERS AND RELATED COMPOUNDS THROUGH THE FORMATION OF SULFUR–CARBON(sp³) BONDS

Sulfur–carbon(sp³) bonds are commonly generated using nucleophilic substitution chemistry. These classic reactions are typically robust and tolerant to a wide variety of solvents, substrates, and reactions conditions. In addition to this methodology, a host of metal and organocatalyzed approaches have been devised and implemented in the syntheses of sulfur-containing compounds. The following sections will highlight some of the developments in this area.

The synthesis of alkyl aryl sulfides has been achieved using sodium thiosulfate as the sulfur source (Scheme 5.1) [1]. This was a valuable approach since many of the classic methods for the synthesis of alkyl sulfides required the use of a thiol. Using this approach, an array of aryl halides and alkyl chlorides were converted into alkyl aryl sulfides. While aryl halides bearing electron-withdrawing groups as well as heteroaromatic halides were successfully transformed using this approach, electron-rich aryl halides were sluggish. Overall, this is an attractive approach to the synthesis of these compounds.

The palladium-catalyzed hydrothiolation of vinyl ethers generated alkyl sulfides with Markovnikov selectivity (Scheme 5.2) [2]. A range of palladium catalysts were screened as part of the investigation, and palladium acetate was found to be the most effective. No additional stabilizing/solubilizing ligand was needed for a successful addition. Curiously, Pd(PPh₃)₄ was a poor catalyst for this chemistry. In addition to vinyl ethers, *N*-vinyl lactams could also be used as substrates in the addition reaction. Similar to the vinyl ethers, the addition chemistry generated predominantly the Markovnikov products.

Although there are a host of methods for the incorporation of a trifluoromethylthio fragment into organic compounds, the copper-catalyzed trifluoromethylthiolation of

Practical Functional Group Synthesis, First Edition. Robert A. Stockland, Jr.
© 2016 John Wiley & Sons, Inc. Published 2016 by John Wiley & Sons, Inc.

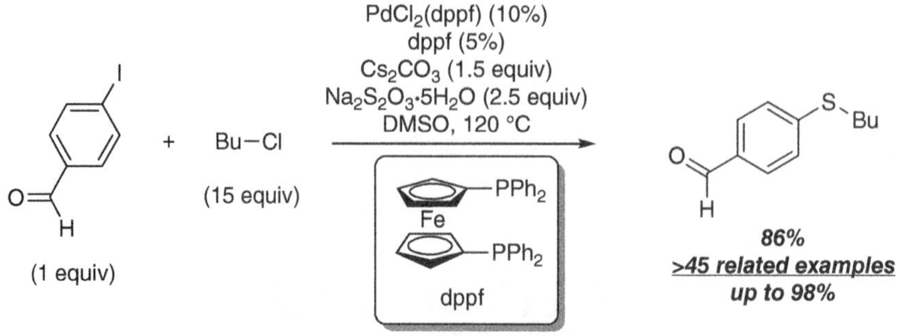

SCHEME 5.1   Synthesis of alkyl aryl sulfides using $Na_2S_2O_3$ as a sulfurating agent [1].

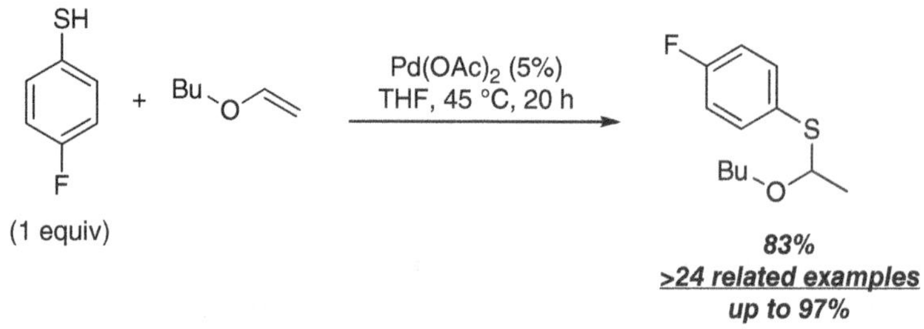

SCHEME 5.2   Palladium-catalyzed hydrothiolation of vinyl ethers [2].

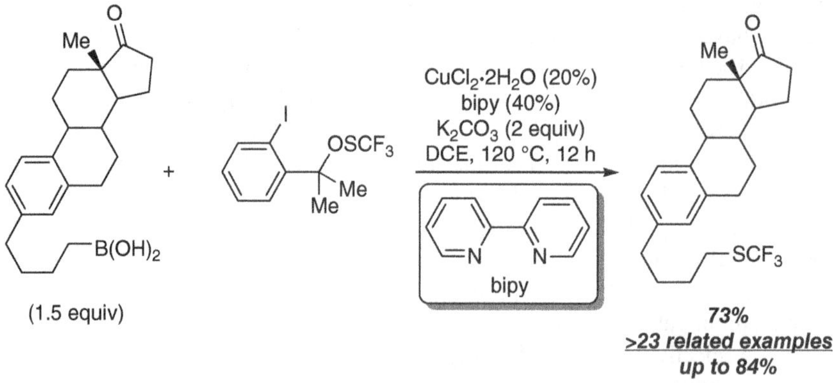

SCHEME 5.3   Trifluoromethylthiolation of alkylboronic acids [3].

alkylboronic acids was a convenient approach (Scheme 5.3) [3]. The key components in this chemistry were an electrophilic trifluoromethylthiolating agent along with a copper catalyst supported by a dipyridyl ligand. While some sulfur–carbon bond-forming reactions did not require the addition of an added ligand due to the possibility that the starting materials or products could coordinate/stabilize the metal center, bipy was needed in this work to generate high yields of the target compounds. Both primary and secondary alkylboronic acids were well tolerated by this chemistry, and moderate to high yields were obtained for most substrates.

The addition of sulfur nucleophiles to activated alkenes is a popular and valuable approach to the synthesis of sulfur–carbon(sp^3) bonds. Due to the importance of the products, a vast array of catalysts and conditions have been developed to promote this reaction. One of the most direct approaches to these thia-Michael additions entailed simply stirring a thiol with the activated alkene in water (Scheme 5.4) [4]. This remarkably straightforward approach was also fast, and outstanding yields of a range of sulfides were obtained in less than 15 min. It should be noted that water was critical for the success of this reaction. Removing the water and performing the reaction under neat conditions was less successful.

In addition to the chemistry described previously, a variety of enantioselective thia-Michael reactions have been developed [5]. An example of one of these approaches involved the addition of allylic thiols to highly activated ketones using a thiourea organocatalyst (Scheme 5.5) [6]. This thia-Michael addition was carried out in a cascade process with another Michael addition to generate trisubstituted tetrahydrothiophenes under mild conditions. The reaction was highly enantioselective and high yielding. In related work,

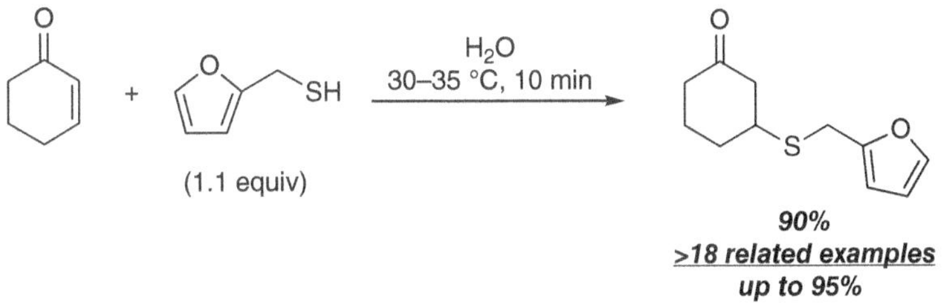

SCHEME 5.4 Thia-Michael reaction in water [4].

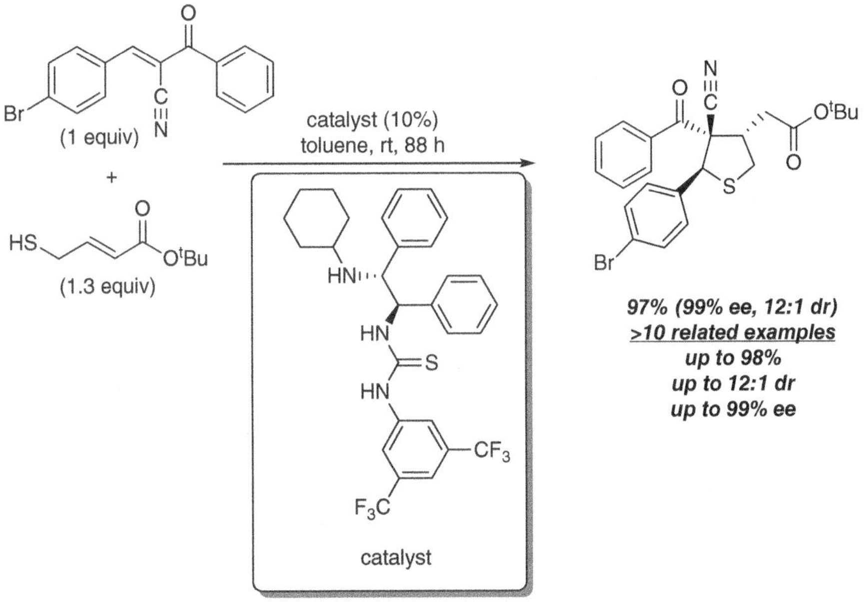

SCHEME 5.5 Enantioselective thia-Michael reactions [6].

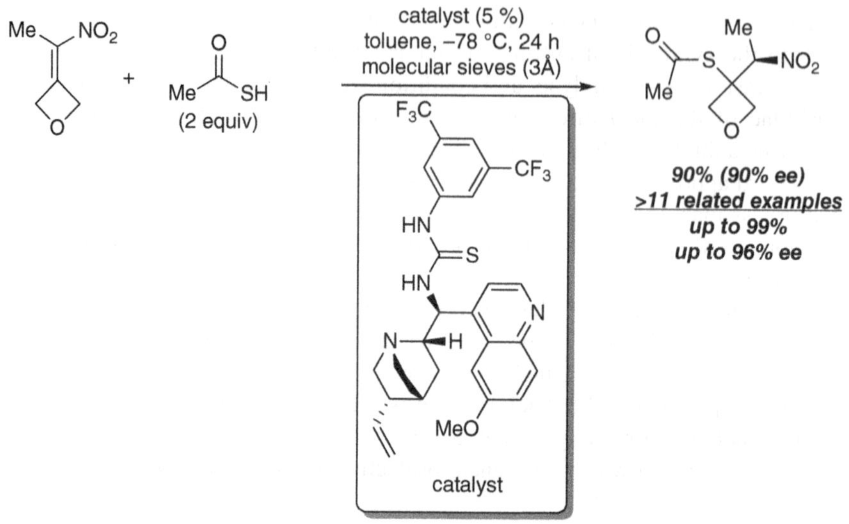

SCHEME 5.6   Enantioselective addition of thioacetic acid to activated alkenes [7].

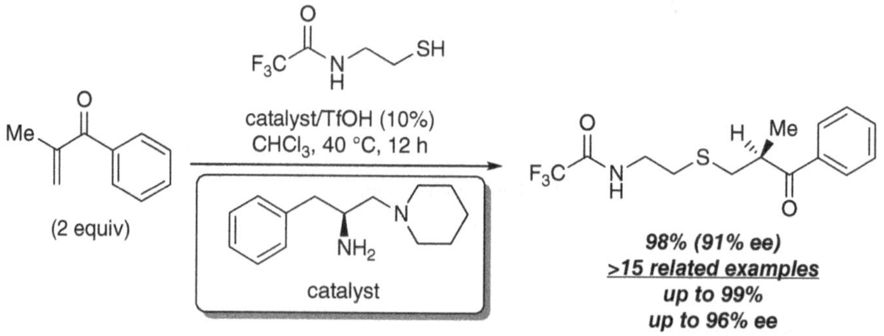

SCHEME 5.7   Organocatalyzed asymmetric thia-Michael addition reactions [8].

the organocatalyzed addition of thioacids to activated alkenes has been achieved (Scheme 5.6) [7]. Additionally, chiral primary amines were found to promote the enantioselective thia-Michael addition of thiols to α-substituted enones (Scheme 5.7) [8].

While most of the methods described previously employed thiols as the thiolating agents, saccharin-based compounds have also shown promise and have been used in asymmetric sulfenylations (Scheme 5.8) [9]. Treating a variety of silyl enol ethers with these saccharin reagents generated α-thiolated ketones in good to excellent yields with outstanding enantioselectivities. A chiral selenophosphoramide served as the catalyst for the reaction and was effective at relatively low loadings. Overall, this is an attractive organocatalyzed approach to the synthesis of thiolated ketones.

In addition to the thia-Michael chemistry described previously that was successful in water, a range of metal-catalyzed approaches were devised that used organic solvents. These approaches are important if the substrate undergoes secondary reactions in the presence of water. One of the metal-catalyzed approaches used an NHC-supported iron catalyst to promote a thia-Michael addition (Scheme 5.9) [10]. The organic solvent played a critical role in

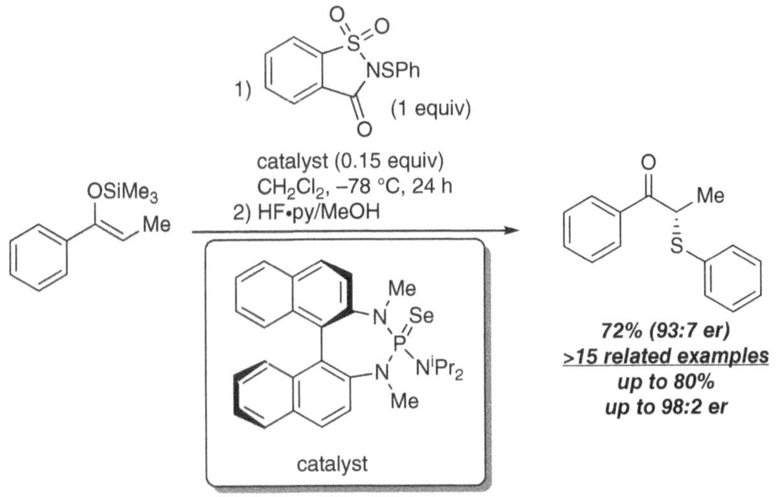

SCHEME 5.8   Asymmetric sulfenylation of silyl enol ethers [9].

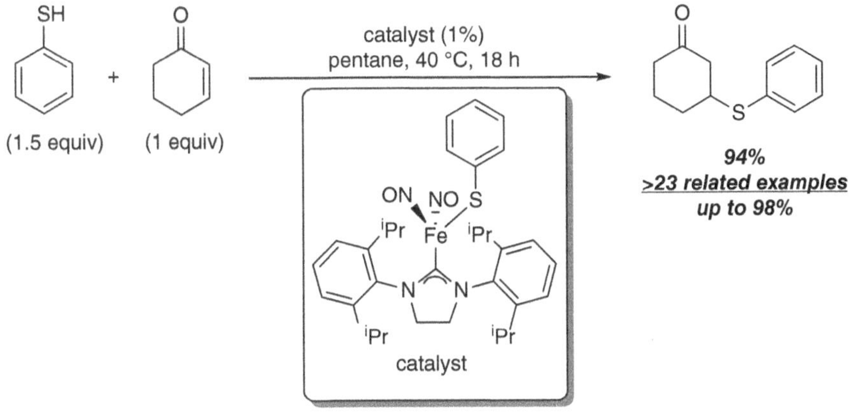

SCHEME 5.9   Iron-catalyzed thia-Michael additions [10].

the chemistry. The use of polar solvents such as THF and acetonitrile led to poor yields; however, changing to pentane afforded moderate to excellent yields of the thia-Michael product. A variety of enones were successfully functionalized with a range of thiols using this approach. The high yields and excellent functional group tolerance make this a valuable approach.

As mentioned in the phosphorus chapter, *N*-heterocyclic carbenes are efficient catalysts for the *phospha*-Michael additions of secondary phosphine oxides to activated alkenes. Instead of using the sensitive free carbenes directly, masked carbenes were used as precatalysts for the addition reaction. These precatalysts were comprised of the carbene bound to carbon dioxide and were synthesized by simply bubbling CO_2 through a solution of the carbene. In solid form, these adducts were quite stable and could be handled in air. These adducts were used to promote addition reactions involving thiophenols (Scheme 5.10) [11]. After a series of screening experiments, THF was selected as the solvent for the reactions, and the most active precatalyst

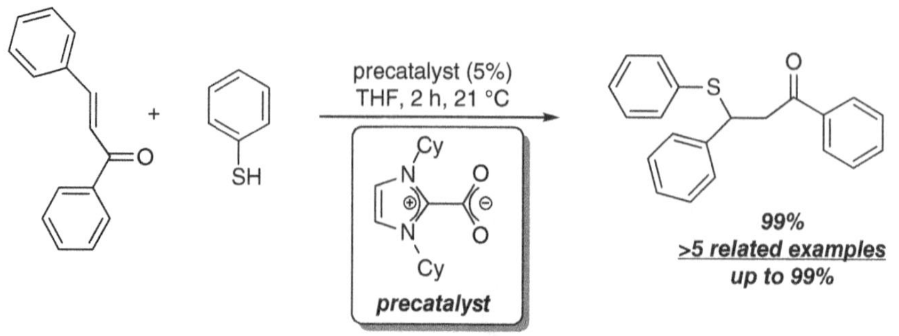

SCHEME 5.10   Carbene-promoted thia-Michael addition reactions [11].

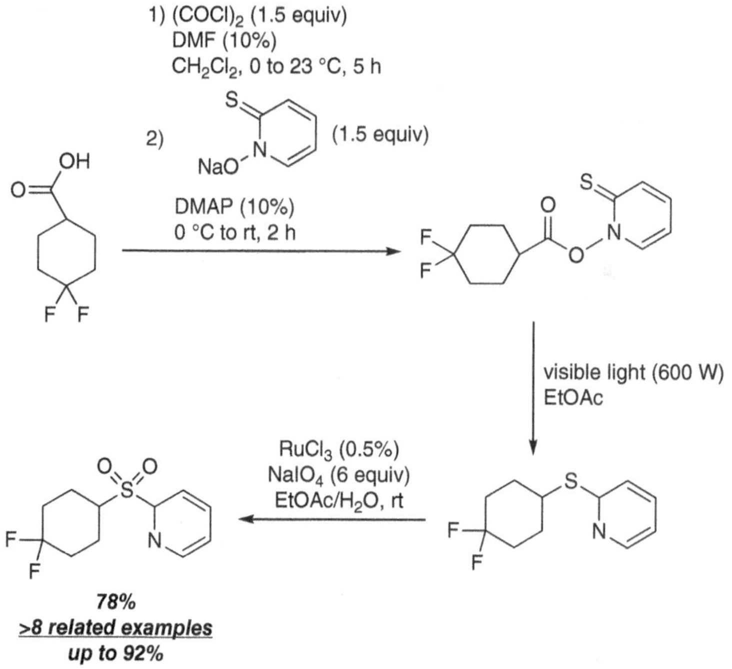

SCHEME 5.11   Conversion of carboxylic acids into sulfones [12].

was found to be ICy·CO$_2$. The reaction conditions were quite mild and nearly quantitative yields of the thioethers were generated. Substrates bearing vinyl ketones, sulfones, esters, acrylonitrile, and acrylamide were successfully functionalized. The high reactivity of acrylamide in these thia-Michael reactions stands in contrast to the sluggish behavior of this substrate in the *phospha*-Michael additions. It could be argued that the most practical aspect of this transition metal-free chemistry was the observation that the all aspects of the thia-Michael addition reaction were performed without any concern for the exclusion of air.

The conversion of carboxylic acids into sulfones has been achieved through a multistep reaction (Scheme 5.11) [12]. This approach was used to generate a number of new aryl and alkyl sulfones that were generated in good to excellent yield. The functional group tolerance of chemistry was outstanding, and a host of alkyl and heteroaromatic substrates were functionalized.

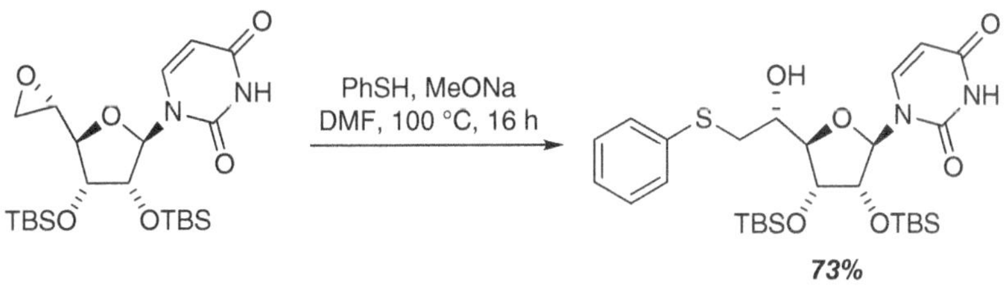

SCHEME 5.12   Ring opening of epoxides with thiolate anions [13].

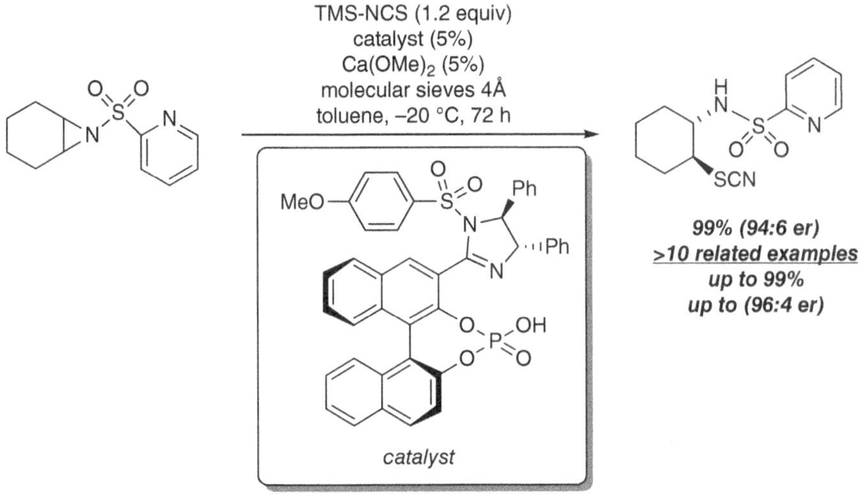

SCHEME 5.13   Imidazoline-phosphoric acid-catalyzed opening of aziridines [14].

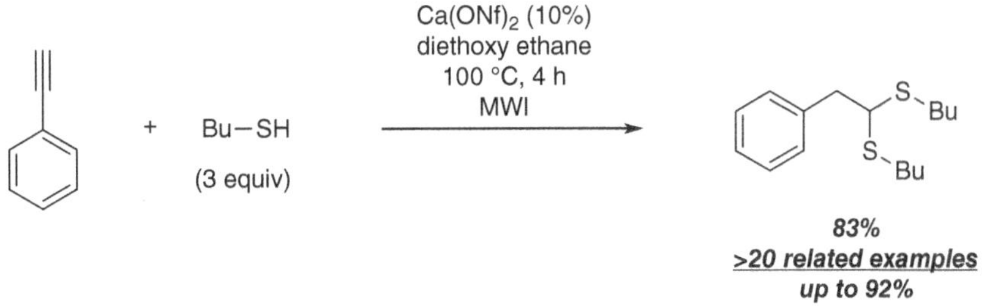

SCHEME 5.14   Synthesis of dithioacetals from alkynes [15].

The ring opening of epoxides is a powerful approach to the synthesis of carbon–heteroelement bonds. This chemistry has been extended to a uridine-bearing epoxide (Scheme 5.12) [13]. Treatment of this precursor with a thiophenolate generated the corresponding thioether in good yield. While only a single example of this chemistry was reported, it provided the foundation for expansion of this work to additional substrates. In related work, the ring opening of aziridines has been reported using a chiral phosphoric acid catalyst (Scheme 5.13) [14].

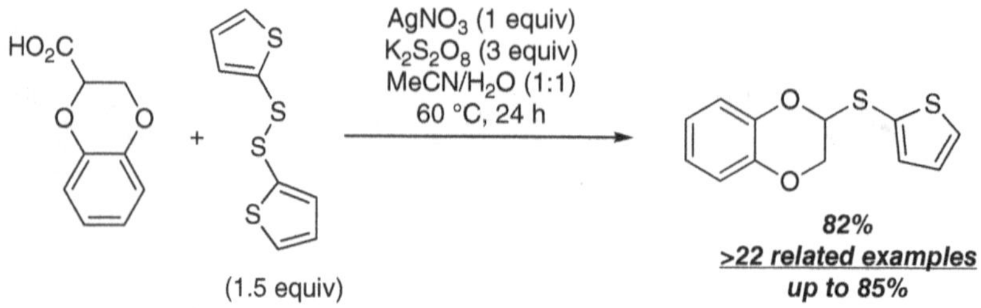

SCHEME 5.15   Decarboxylative thiolation of alkyl carboxylic acids [16].

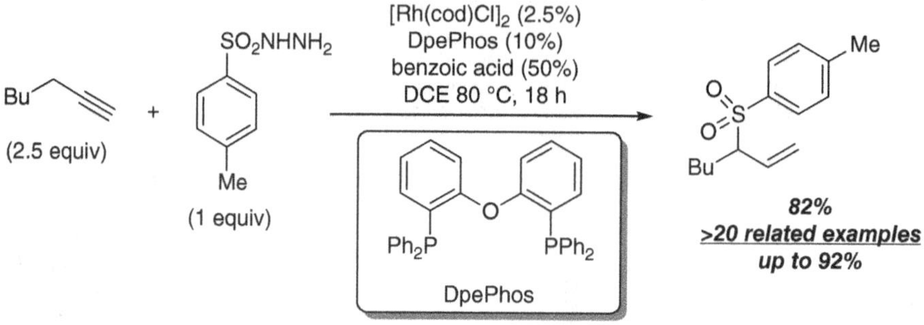

SCHEME 5.16   Rhodium-catalyzed synthesis of allylic sulfones [17].

The microwave-assisted calcium-catalyzed double addition of thiols to alkynes generated dithioacetals (Scheme 5.14) [15]. The authors screened a range of calcium catalysts and found that $Ca(ONf)_2$ was the most effective under the reaction conditions. Due to the long reaction times needed using conventional heating, the chemistry was carried out in a microwave reactor. The reaction times were significantly decreased, and the yields of the reaction were good to excellent (up to 92%). Using this approach, a range of alkynes was selectively converted into the anti-Markovnikov dithioacetals.

Due to the stability and availability of carboxylic acids, they remain some of the most attractive starting points for the synthesis of complex organic compounds. Using these compounds as substrates, the decarboxylative thiolation of carboxylic acids has been achieved using silver nitrate as a promoter (Scheme 5.15) [16]. Disulfides served as the sulfur source for this chemistry, and moderate to good yields of the thioethers from a host of alkyl carboxylic acids.

The preparation of allylic sulfones has been accomplished through a rhodium-catalyzed addition of sulfonyl hydrazides to terminal alkynes (Scheme 5.16) [17]. The catalyst system was comprised of a common rhodium complex along with a large bite-angle bisphosphine (DpePhos). This supporting ligand is more flexible than XantPhos and can twist to accommodate the needs of the metal center. The authors also found that benzoic acid was needed for a successful transformation. Eliminating or reducing the amount of this additive led to lower yields of the allylic sulfone. While the role of the benzoic acid was not fully elucidated, it was a critical component. The hydrosulfination reaction was moderate to high yielding and was highly selective for the formation of the allylic sulfone. The main side products were vinyl sulfones, and the amount of these secondary products varied with substrate. Many examples contained less than 10% of these contaminants.

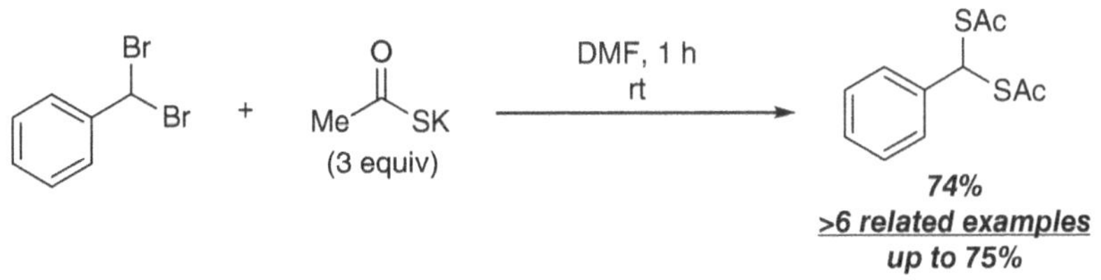

SCHEME 5.17 Synthesis of *gem*-dithioacetates through substitution chemistry [18].

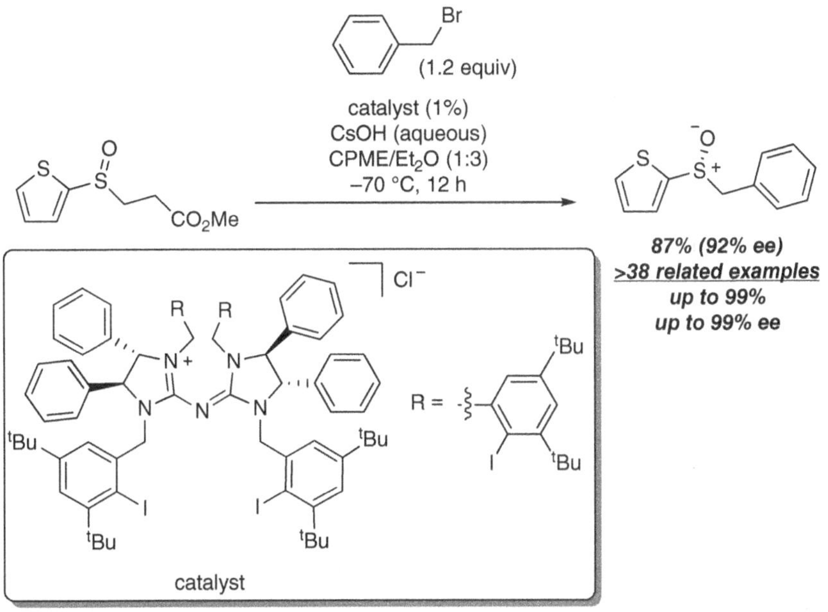

SCHEME 5.18 Enantioselective synthesis of sulfoxides using a chiral organocatalyst [19].

Due to the issues regarding safe handling of H_2S, researchers have sought safer sources of this reagent. Furthermore, easily handled alternatives that are able to deliver a precise amount of H_2S to a reaction mixture are highly valuable. A group of these H_2S sources are based upon a *gem*-dithioacetate fragment. These compounds contain two thioacetate groups that were attached to the core of the molecule through substitution chemistry using *gem*-dibromo species as substrates. This substitution chemistry rapidly generated the *gem*-dithioacetates in moderate to good yields under mild conditions (Scheme 5.17). Since the starting materials for the *gem*-dibromo compounds were benzaldehyde derivatives, this chemistry has broad substrate scope due to the large number of benzaldehydes that are readily available. It should be noted that these *gem*-dithioacetates were stable in water, and minimal amounts of hydrolysis was observed. The stability of these compounds to water was a considerable advantage to these hydrogen sulfide precursors.

The enantioselective benzylation of sulfoxides using a chiral organocatalyst has been achieved (Scheme 5.18) [19]. The chemistry required cryogenic conditions; however, the product yields and enantioselectivity were outstanding. A range of benzyl halides bearing

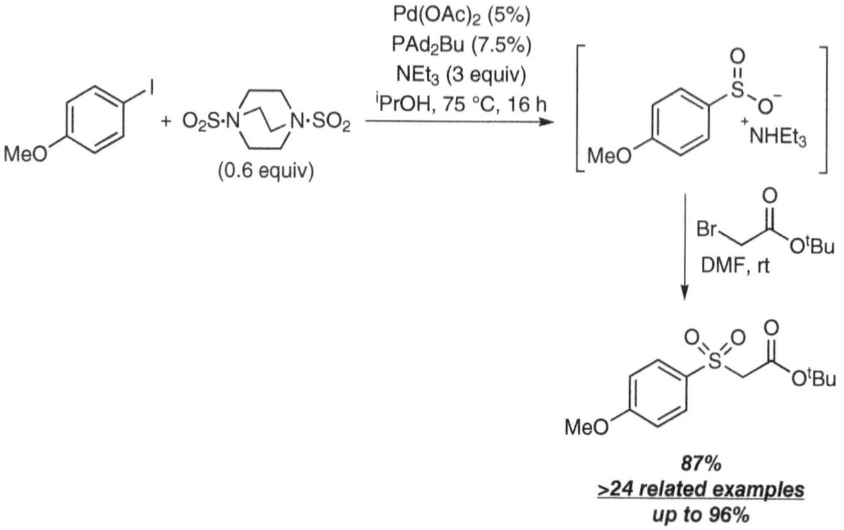

SCHEME 5.19   Synthesis of β-keto aryl sulfones using a nongaseous source of SO$_2$ [20].

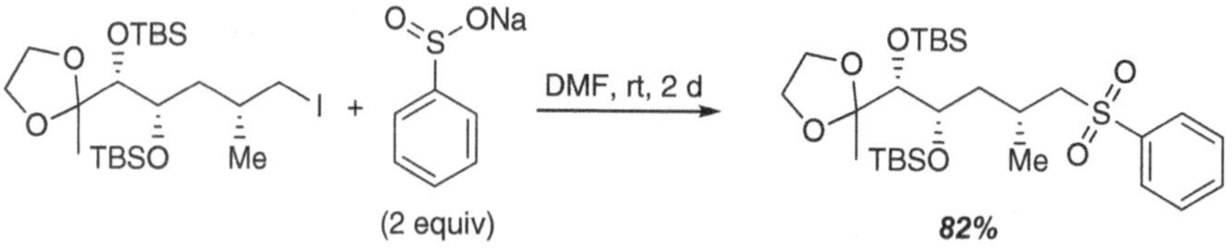

SCHEME 5.20   Nucleophilic substitution chemistry with sodium phenylsulfinate [21].

electron-donating and withdrawing groups were successfully functionalized using this approach along with a variety of sulfoxides.

In an effort to generate a nongaseous source of SO$_2$ that could be delivered directly into a reaction mixture, a DABCO-(SO$_2$)$_2$ species was prepared and evaluated in the synthesis of sulfones from aryl halides (Scheme 5.19) [20]. The catalyst system consisted of palladium acetate along with a bulky trialkylphosphine and an excess of an organic base. Heating the reaction mixture for 16 h afforded good yields of an intermediate ammonium sulfinate species that could be alkylated through the addition of benzyl bromide. In related work, the addition of aryliodonium salts generated diaryl sulfones.

In addition to well-known nucleophilic behavior of thiolate anions, sulfinates are also active in substitution reactions. To this end, the conversion of a primary alkyl iodide into an alkyl sulfone has been accomplished by simply stirring the sodium sulfinate salt with the alkyl halide in DMF at room temperature (Scheme 5.20) [21]. While only a few examples were described, this work demonstrates the ability of sulfinate salts to participate in nucleophilic substitution reactions.

5.1.1   Troubleshooting the Synthesis of Thioethers and Related Compounds

While it is impossible to predict every challenge that will arise during the synthesis of sulfur–carbon(sp^3) bonds, the following suggestions will provide some direction when designing a synthesis.

Reaction Requirement/Substrate	Possible Solution/Suggestion
Need to start from an activated alkene	Try simply stirring a thiol with the substrate in water [4]. Reaction times were typically less than 15 min for greater than 90% conversion
Need to perform an asymmetric thia-Michael	This is a challenging reaction and general systems with wide substrate scope are rare. Many of the current approaches have been reviewed [5]
Need to make a group of alkyl sulfides and would rather not handle a thiol due to the odor	A palladium-catalyzed "thiol-free" approach to the synthesis of alkyl aryl sulfides has been developed and uses sodium thiosulfate as the sulfur source [1]
Need to use water as solvent	If the area of the substrate that needs to be thiolated can be configured into an activated alkene, a thia-Michael reaction has been shown to be successful in pure water [4]
Need to make a dithioacetal and an alkyne is a convenient entry point	The calcium-catalyzed double addition of thiols to alkynes generated dithioacetals using a microwave-assisted approach [15]
Need a thia-Michael addition that can tolerate air (oxygen)	The metal-free carbene-catalyzed thia-Michael reaction can be carried out with a host of substrates without any precautions for the removal of oxygen [11].
Need a source of electrophilic sulfur	A range of these compounds have been described and used. Saccharin-derived electrophilic sources of sulfur have been used in asymmetric sulfenylation reactions [9]
Starting from a vinyl lactam and need Markovnikov selectivity	The palladium-catalyzed addition of thiols to vinyl ethers and lactams would be a reasonable place to start [2]
Need an inexpensive metal to promote a thia-Michael addition reaction	The iron-catalyzed thia-Michael addition to enones has been reported [10].
Need to make several sulfones, and it would be convenient to start from carboxylic acids	A reasonable entry point to the chemistry would be to convert the carboxylic acids into sulfinate salts [12]. Alternatively, a silver-catalyzed decarboxylative thiolation will generate sulfone precursors [16]
Would like a practical nongaseous source of SO_2	A DABCO-SO_2 species has been generated and used in the conversion of aryl halides into sulfones [20]
Would like a practical nongaseous source of H_2S that is relatively stable to water	Compounds containing *gem*-dithioacetates would be a possible solution [18]

5.2 SYNTHESIS OF ARYL THIOETHERS AND RELATED COMPOUNDS THROUGH THE FORMATION OF SULFUR–CARBON(SP²) BONDS

Due to sluggish reactivity of aryl and vinyl halides in nucleophilic substitution reactions, the formation of sulfur–carbon(sp²) bonds is typically carried out using transition metal catalysis [22–27]. While the field is dominated by the use of palladium, copper, and nickel catalysts, considerable advances have been made using more abundant metal catalysts such as iron. Additionally, a number of transition metal-free approaches have been developed for the formation of sulfur–carbon(sp²) bonds. The following sections will highlight representative examples of C—S bond forming reactions.

Some of the first palladium-catalyzed C—S bond-forming reactions generated diaryl sulfides through the coupling of arylthiols with aryl halides [28, 29]. In order to improve the catalyst system and broaden the substrate range, a variety of palladium sources and ancillary ligands have been screened for activity towards the coupling reaction. The catalyst system comprised of palladium acetate and dippf was particularly effective and promoted the coupling of aryl bromides and chlorides with both alkyl- and arylthiols (Scheme 5.21 and Example 5.1) [30]. A range of preexisting groups including esters, ethers, and nitriles were tolerated by this chemistry along with bulky aryl halides such as 2,6-dimethylchlorobenzene (95% yield of the diaryl sulfide). In related work, $Pd_2(dba)_3$/DpePhos was able to promote the coupling of aryl iodides with aryl sulfides [31], and secondary phosphine oxides were used as supporting ligands in carbon–sulfur

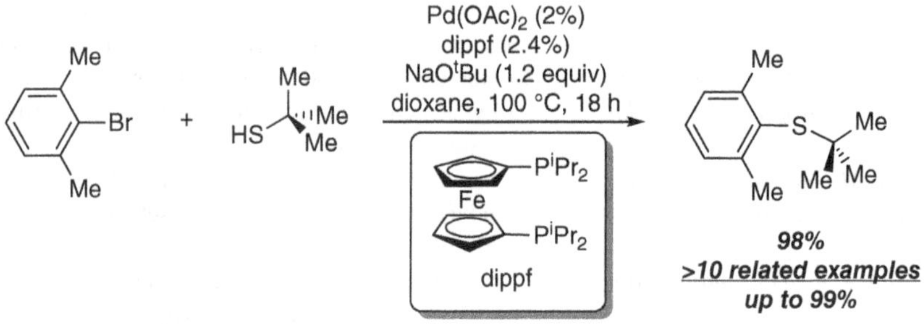

SCHEME 5.21   Palladium-catalyzed synthesis of alkyl aryl sulfides [30].

Example 5.1   Palladium-Catalyzed Synthesis of Aryl Sulfides [30].

Chemical Safety Instructions

Before starting this synthesis, all safety, health, and environmental concerns must be evaluated using the most recent information, and the appropriate safety protocols followed. All appropriate personal safety equipment must be used. All waste must be disposed of in accordance with all current local and government regulations.

The individual performing these procedures and techniques assumes all risks and is responsible for ensuring the safety of themselves and those around them. This chemistry is not intended for the novice and should only be attempted by professionals who have been well trained in synthetic organic chemistry. The authors and Wiley assume no risk and disclaim any liabilities for any damages or injuries associated in any way with the chemicals, procedures, and techniques described herein.

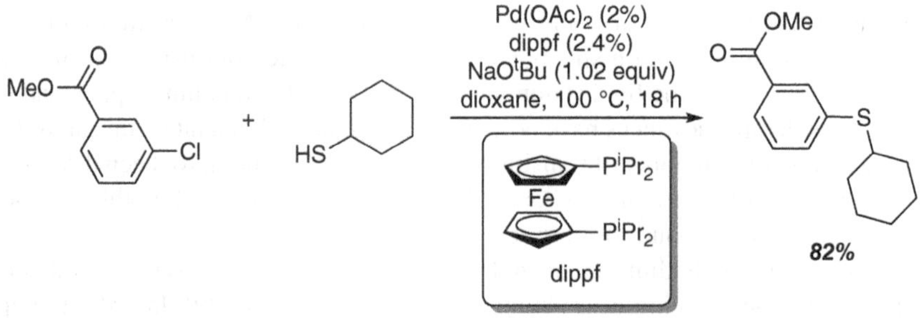

Caution! Dioxane is highly flammable.

Caution! Volatile thiols that have a potent odor are best used in a well-ventilated fume hood.

This reaction was carried out under an atmosphere of argon. $Pd(OAc)_2$ (0.020 mmol), dippf (0.024 mmol), NaO^{t}Bu (1.02 mmol), and the aryl halide (1.0 mmol) (if a solid) were added to an oven-dried resealable Schlenk tube. The tube was evacuated and backfilled with argon (three cycles) and then charged with dioxane (2.0 mL). The solution was stirred for 1 h at room temperature. Then aryl halide (1.0 mmol) (if a liquid) and the cyclohexanethiol (1.0 mmol) was added by syringe. The Schlenk tube was sealed with a Teflon valve, heated to 100 °C, and stirred for 18 h. The reaction mixture was then allowed to reach room temperature. Ether (ca. 3 mL) was added and an aliquot was removed and analyzed by GC. The reaction mixture was then filtered and concentrated. The crude product was purified by flash column chromatography on silica gel to afford the desired thioether (82%) as a yellow oil.

Reprinted from M. Murata and S. L. Buchwald, Tetrahedron Lett., 60, 7397–7403. Copyright (2004), with permission from Elsevier.

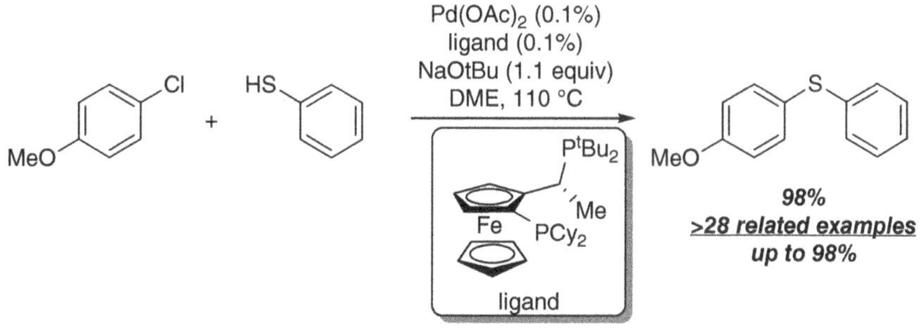

SCHEME 5.22 Palladium-catalyzed synthesis of diaryl sulfides [34].

bond-forming reactions [32]. Furthermore, the synthesis of aryl trifluoromethyl sulfides has been achieved using a palladium-catalyzed approach with BrettPhos as the supporting ligand and $AgSCF_3$ as the sulfur source [33].

Hartwig used palladium acetate as the metal precursor and a chelating bisphosphine (JosiPhos) as the supporting ligand for the synthesis of a number of diaryl sulfides (Scheme 5.22) [34]. One of the most attractive aspects of this chemistry was the observation that high yields could be obtained using 0.1% of the palladium catalyst. In subsequent reports, the authors were able to refine the aryl chloride coupling chemistry [35] and lower the catalyst loading even further (0.01% for selected substrates). They were also able to increase the scope of the reaction by changing the substrate to an aryl bromide and iodide [36]. Overall, this is a very attractive approach to the synthesis of these valuable compounds.

The synthesis of phenothiazine derivatives has been achieved through a palladium-catalyzed three-component reaction (Scheme 5.23) [37]. The catalyst system comprised of $Pd_2(dba)_3$ and dppf was the most effective combination for the reaction. Some of the other phosphine ligands screened such as XPhos and BINAP generated significant amounts of an acyclic amine product. The reaction was typically carried out in two stages with the first

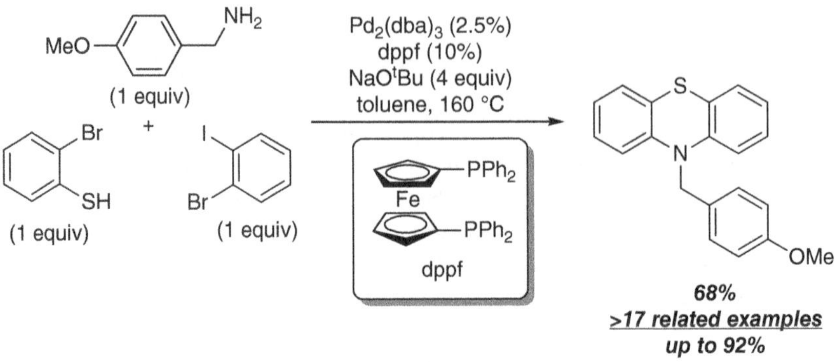

SCHEME 5.23   Synthesis of phenothiazine derivatives based upon a three-component coupling reaction [37].

step consisting of gently heating the reagents and catalyst at 60 °C for 20 min, while the second step involved heating to 160 °C for several hours or overnight. Catalyst deactivation was an issue with this chemistry, and the authors solved this problem by thoroughly mixing the reagents (prior to initial heating period). The reaction was also successfully carried out using microwave heating. Due to the poor absorption of microwave radiation by toluene, best results were obtained using toluene/DMF as the solvent system.

NHC-ligated palladium complexes have also been used in the synthesis of sulfur–carbon (sp^2) bonds [38]. A range of aryl bromides and chlorides were successfully coupled with organothiols using a low catalyst loading (0.2%) to generate organosulfides in moderate to excellent yields. In addition to the low Pd loading, the ability to transform both aryl- and alkylthiols into organosulfides was an attractive aspect of this chemistry. Furthermore, the catalyst system was particularly effective at coupling bulky substrates to afford crowded diaryl sulfides (Scheme 5.24). In related work, *N*-amido imidazolium species have been used as supporting ligands in the palladium-catalyzed synthesis of diaryl sulfides [39].

A detailed investigation of the palladium-catalyzed synthesis of diaryl ethers from aryl benzyl sulfides has been reported by Walsh (Scheme 5.25) [40]. This debenzylative coupling reaction was catalyzed by palladium and proceeded with gentle heating to afford the diaryl sulfides in moderate to high yield. The selection of base was critical to the success of the reaction, and while common metal alkoxides such as NaO^{t}Bu were ineffective, NaN(SiMe$_3$)$_2$ successfully promoted the reaction. The chemistry was also extended to heteroaryl bromides (Example 5.2). A practical advantage of this chemistry was the circumvention of the need for thiols to be used as coupling partners. Overall, this is a valuable approach to the synthesis of diaryl sulfides.

The palladium-catalyzed directed arylthiolation of arenes has been achieved using common palladium complexes as catalysts (Scheme 5.26) [41]. The most effective palladium source was Pd(OAc)$_2$, and no added ligand was needed to solubilize/stabilize the metal center. Given the propensity of sulfur donors to bind to palladium, it is likely that a sulfur-containing species present in the reaction mixture was bound to the metal center to prevent irreversible generation of metallic palladium. In addition to the metal catalyst, the other key reagent in the system was the succinimide-based thiolating agent, generated from succinimide and various thiols. The C—H thiolation chemistry proceeded under acidic conditions at room temperature and generated a host of the diorganosulfides in moderate to

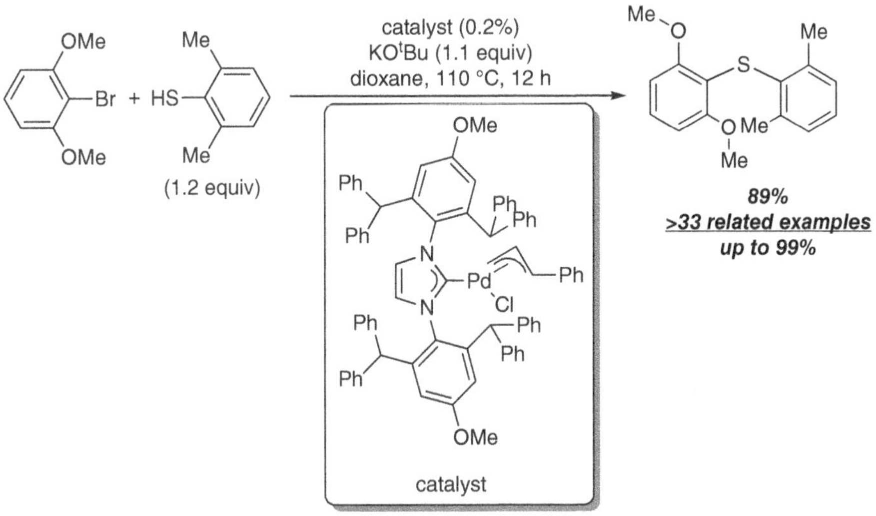

SCHEME 5.24   Palladium-catalyzed synthesis of crowded diaryl sulfides [38].

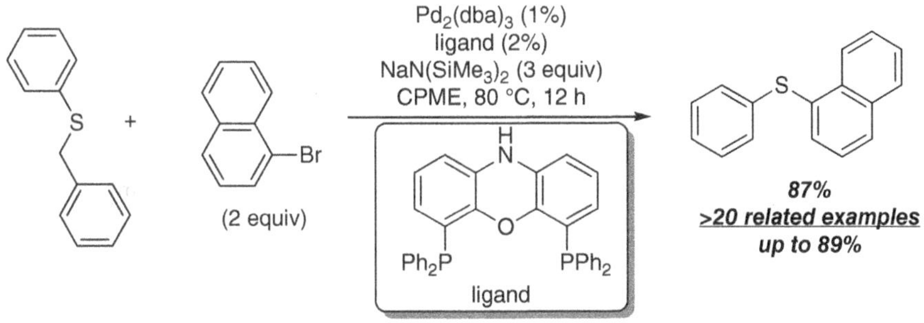

SCHEME 5.25   Preparation of diaryl sulfides from aryl benzyl sulfides.

excellent yield. Furthermore, most reactions were complete within 45 min. Although the arylthiolating reagent needed to be generated, this is an attractive approach to the synthesis of a range of diorganosulfides. In related work, an iron-catalyzed version of this reaction has been reported [42].

In addition to palladium-catalyzed coupling reactions, analogous processes have also been carried out using copper complexes as catalysts [24]. One of the more practical versions of these reactions was reported using a catalyst system comprised of a common copper(I) salt along with ethylene glycol and potassium carbonate (Scheme 5.27) [43]. This system was able to promote the coupling of a range of aryl iodides with both alkyl- and arylthiols to generate diorganosulfides in good to excellent yields. The presence of a range of functional groups including esters, carboxylic acids, aldehydes, ketones, nitro groups, alcohols, and halogens was well tolerated by this approach. Of particular note was the ability to perform the chemistry using reagent-grade solvents. Extensive degassing was not needed, although it should be noted that reactions carried out under an atmosphere of air resulted in varying amounts of diaryl disulfides.

Example 5.2   Preparation of 1-Methyl-5-(Phenylthio)-1*H*-Indole from Phenyl Benzyl Sulfide [40].

Before starting this synthesis, carefully follow the chemical safety instructions listed in Example 5.1.

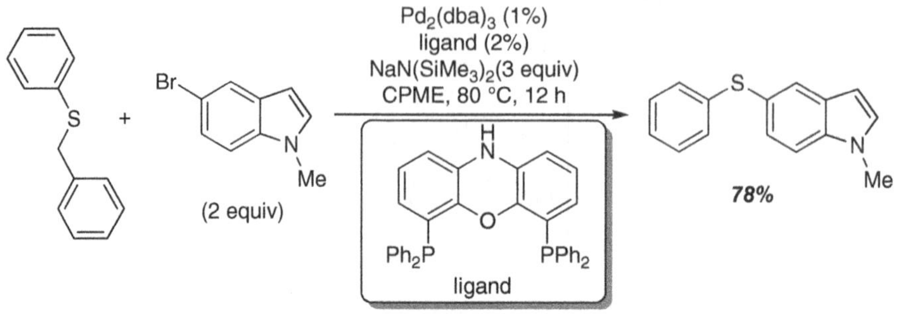

Caution! Hexanes, cyclopentyl methyl ether, and ethyl acetate are highly flammable.

Caution! NaN(SiMe₃)₂ can react violently with moisture and must be stored and handled under an inert atmosphere.

To an oven-dried microwave vial equipped with a stir bar was added Pd(dba)$_2$ (1.15 mg, 0.002 mmol) and NiXantPhos (2.21 mg, 0.004 mmol) under nitrogen atmosphere inside a glovebox. Next, 2.0 mL dry CPME via syringe was added. After the catalyst/ligand solution was stirred for 2 h at 24 °C inside the glovebox, benzyl phenyl sulfide (40.6 mg, 0.20 mmol, 1.0 equiv) was added to the reaction vial followed by NaN(SiMe$_3$)$_2$ (110.4 mg, 0.60 mmol, 3 equiv). The microwave vial was sealed, removed from the glovebox, and the bromoindole (69.2 μL, 0.40 mmol, 2.0 equiv) was added by syringe under a nitrogen atmosphere. Note that if the aryl bromide was a solid, it was added to the reaction vial before the NaN(SiMe$_3$)$_2$. The reaction mixture was heated to 80 °C in an oil bath and stirred for 12 h. The sealed vial was cooled to room temperature, five drops water was added via syringe, and then the vial opened to air. The reaction mixture was passed through a short pad of silica gel and rinsed with 5 mL 10:1 ethyl acetate:methanol. The solvent was removed by rotary evaporator. The crude product was purified by flash chromatography on silica gel (eluted with hexanes:EtOAc = 30:1) to give the product (37.2 mg, 78% yield) as a solid.

Adapted with permission from J. Mao, T. Jia, G. Frensch, and P. J. Walsh, Org. Lett., **2014,** *16,* 5304. *Copyright (2014) American Chemical Society.*

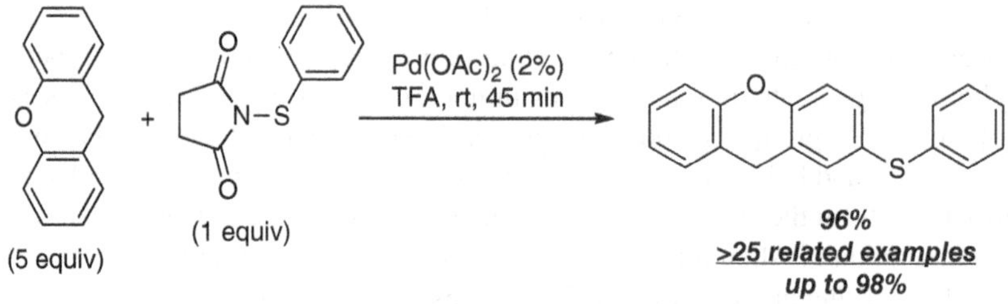

SCHEME 5.26   Palladium-catalyzed C—H thiolation of arenes [41].

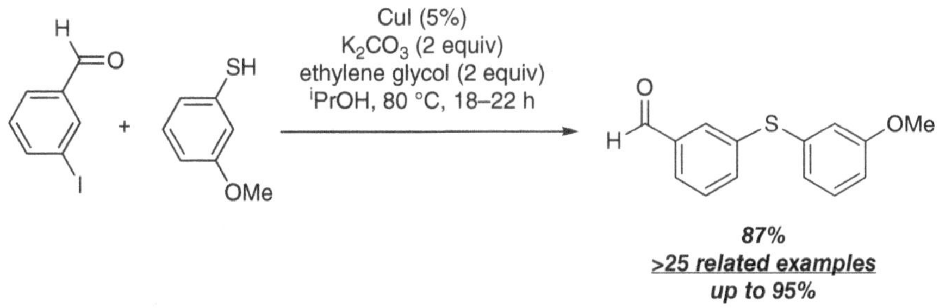

SCHEME 5.27   Copper-catalyzed diaryl sulfide synthesis [43].

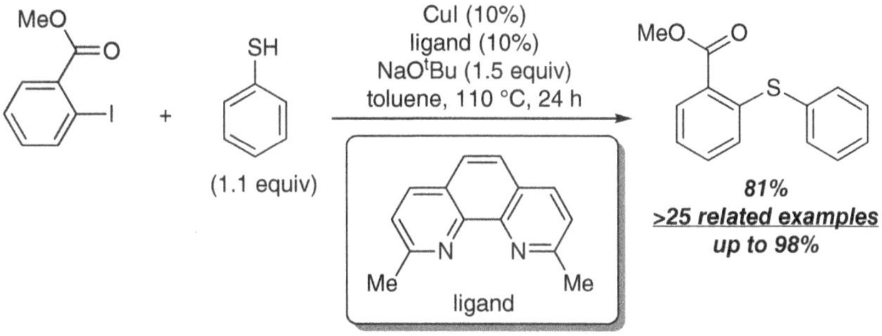

SCHEME 5.28   CuI/neocuproine-catalyzed synthesis of diaryl sulfides [44].

While the previous example generated the copper catalyst in solution using ethylene glycol as the supporting ligand, neocuproine can also be used to stabilize/solubilize copper salts and promote the synthesis of diaryl sulfides (Scheme 5.28) [44]. To this end, Venkataraman reported the synthesis of diorganosulfides through the coupling of aryl iodides with alkyl- or arylthiols. The chemistry proceeded under mild conditions and afforded good to excellent yields of the target compounds. The functional group tolerance of this chemistry was high, and substrates bearing a range of electron-donating and withdrawing groups were successfully converted into diorganosulfides.

The vast majority of diorganosulfide syntheses require the use of an alkyl- or arylthiol. A notable exception to this chemistry is Walsh's debenzylative coupling chemistry described previously [40]. An alternative approach using carbon disulfide as the sulfur source has been reported using a ligand-free copper coupling reaction (Scheme 5.29) [45]. The functional group tolerance of this reaction was outstanding as a range of functionalized aryl iodides were successfully employed in this work. Even iodophenols were used with minimal complications. The competing reaction observed using iodophenols was due to the coupling of the alcohol fragment of the substrate with the aryl iodide fragment of another to generate diaryl ethers. This was an attractive approach to the synthesis of diaryl sulfides chemistry since it significantly broadened the scope of the "thiol-free" coupling reactions. In addition to the intermolecular coupling reaction, an intramolecular version of the reaction was reported and successfully generated several heterocycles through a cyclization reaction (Scheme 5.30). In related work, a microwave-assisted thiol-free synthesis of diorganosulfides has been developed using sulfonyl hydrazides as the sulfur-containing

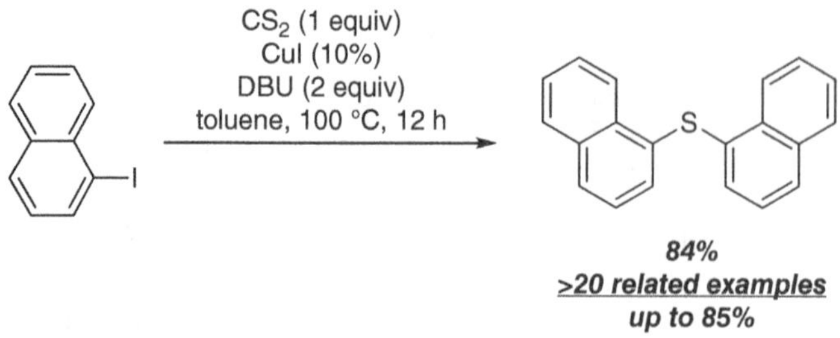

SCHEME 5.29   Diaryl sulfide synthesis using CS_2 as the sulfur source [45].

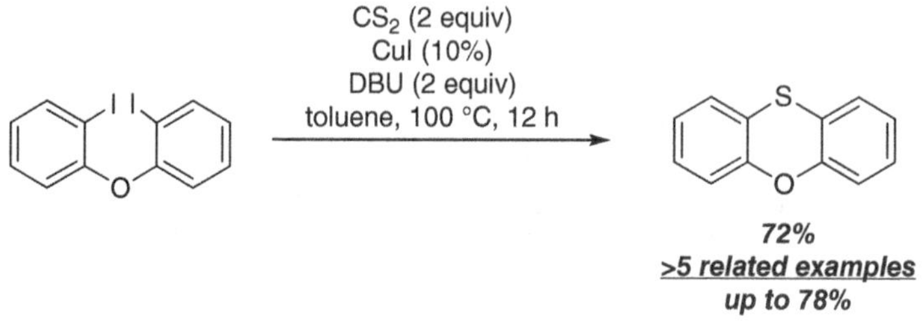

SCHEME 5.30   Heterocycle synthesis using CS_2 as the sulfur source [45].

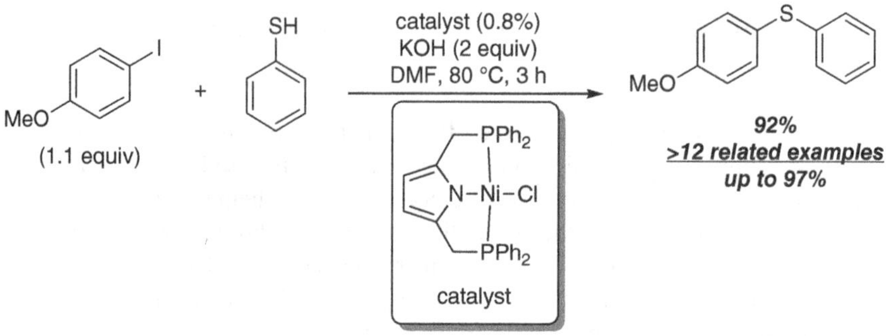

SCHEME 5.31   Nickel-catalyzed synthesis of unsymmetrical diaryl sulfides [47].

precursors [46]. Several of these methods are synthetically valuable since they enabled the synthesis of unsymmetrical diaryl sulfides.

One of the reports of a nickel-catalyzed synthesis of diaryl sulfides used $NiCl_2(dppf)$ as the metal source and elemental zinc as a reducing agent to promote the coupling of phenyl mesylate with sodium benzenethiolate [48]. Other reports of nickel-catalyzed diaryl sulfide synthesis were focused on modifying the classic system to increase the scope and functional group tolerance of the process [49–51]. An attractive version of the nickel-catalyzed synthesis of unsymmetrical diaryl sulfides involved the use of a pincer nickel complex as the catalyst for the reaction (Scheme 5.31) [47]. Treating a range of aryl iodides and bromides with alkyl- and arylthiols in the presence of a low loading of the nickel catalyst afforded excellent yields of the unsymmetrical diorganosulfides. In related work, the nickel-catalyzed synthesis of aryl trifluoromethyl sulfides has been achieved using $Ni(cod)_2$ as the

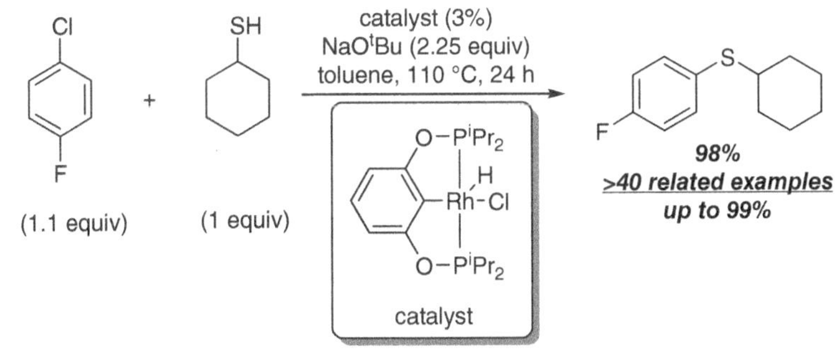

SCHEME 5.32   Rhodium-catalyzed coupling of aryl chlorides with thiols [53].

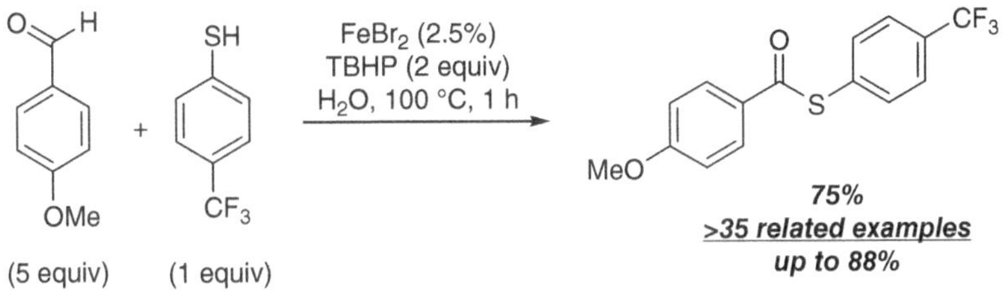

SCHEME 5.33   Synthesis of thioesters in water [54].

nickel source and a chelating bipyridine as the supporting ligand [52]. This thiol-free approach used $[NMe_4][SCF_3]$ as the sulfur source and generated moderate to good yields of the aryl trifluoromethyl sulfides at room temperature.

The rhodium-catalyzed synthesis of a range of diorganosulfides employed both alkyl bromides and chlorides as coupling partners along with a range of alkyl- and arylthiols (Scheme 5.32) [53]. The catalyst was comprised of a rhodium center supported by a PCP pincer ligand. A host of aryl halides bearing electron-donating and electron-withdrawing groups were successfully coupled with a wide range of alkyl- and arylthiols to generate an array of diorgano-sulfides. Over 40 examples were described with isolated yields of the organosulfides as high as 99%. While this chemistry does require the preparation of the rhodium catalyst, the outstanding yields along with the broad scope of the chemistry make this an attractive approach.

The synthesis of thioesters through the coupling aldehydes with thiols has been achieved using common iron salts as catalysts (Scheme 5.33) [54]. The reactions required an oxidant, and *tert*-butyl hydroperoxide was found to be effective under the reaction conditions. The reactions were carried out in pure water and afforded moderate to excellent yields of the thioesters. The scope of the chemistry was quite broad as both alkyl and aryl aldehydes as well as alkyl- and arylthiols were successfully coupled as well as several heteroaromatic aldehydes. Overall, this is a practical approach to the preparation of thioesters.

The transition metal-free coupling of diphenyl disulfide with unfunctionalized arenes proceeds through a C—H activation process to afford the diaryl sulfides in moderate to excellent yields (Scheme 5.34) [55]. A vast array of arenes were successfully used in this chemistry including bromomesitylene. Instead of forming the new carbon–sulfur bond at the brominated position, the substrate underwent a C—H activation to generate a bromi-nated diaryl sulfide. The retention of the bromine was attractive since it could be further

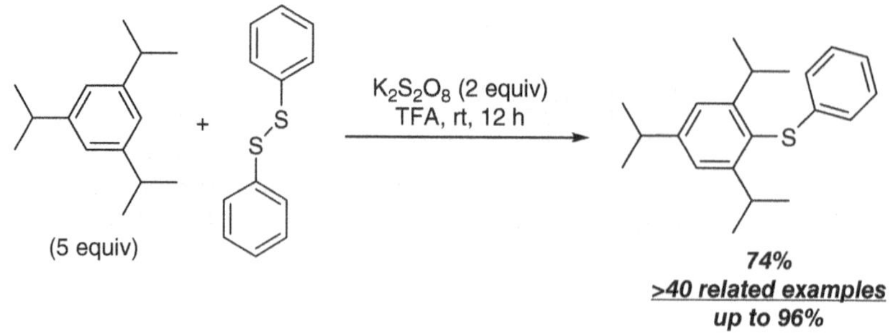

SCHEME 5.34   Coupling of arenes with diphenyl disulfide [55].

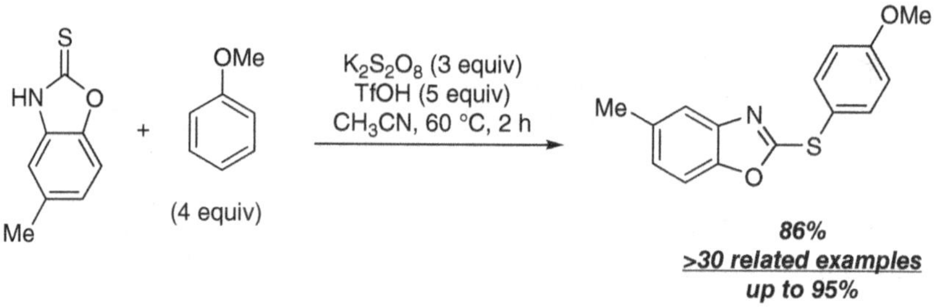

SCHEME 5.35   C—H thiolation of heterocycles [56].

functionalized in subsequent reactions. Furthermore, while phenols can be problematic in syntheses leading to the formation of diorganosulfides, this chemistry efficiently function-alized several substrates with retention of the phenol. Similar to the brominated example described previously, the retention of the phenol was attractive for further functionaliza-tion. In related work, the C—H thiolation of heterocycles was achieved using a $K_2S_2O_8$/ Bronsted acid system (Scheme 5.35) [56].

N-chlorosuccinimide promoted the sulfenylation of azaheterocycles under very mild conditions without the addition of transition metal catalysts (Scheme 5.36) [57]. The new sulfur–carbon bond was formed through a regioselective C—H activation pro-cess. The overall transformation was accomplished in two steps and was completed within a few hours at room temperature. A host of arylthiols and heterocycles were successfully coupled in this chemistry, and isolated yields of the diaryl sulfides were as high as 97%. The strength of this approach was the speed of the reaction as well as the availability of the reagents. In related work, the NCS-promoted coupling of thiols with organozinc [58] or Grignard reagents generated a variety of unsymmetrical diorganosul-fides (Scheme 5.37) [59].

The reaction of alkyl aryl sulfides with diaryliodonium salts generated unsymmetrical diaryl sulfides in moderate to excellent yields through cleavage of the sulfur–carbon(sp^3) bond and formation of a new sulfur–carbon(sp^2) bond (Scheme 5.38) [60]. The synthesis was transition metal free and tolerated a host of different electron-withdrawing and donating groups. Another noteworthy aspect of the chemistry was that it was successful in an acidic environment. This renders the work complementary to the transition metal-catalyzed routes that are typically carried out under basic conditions. In related work, the

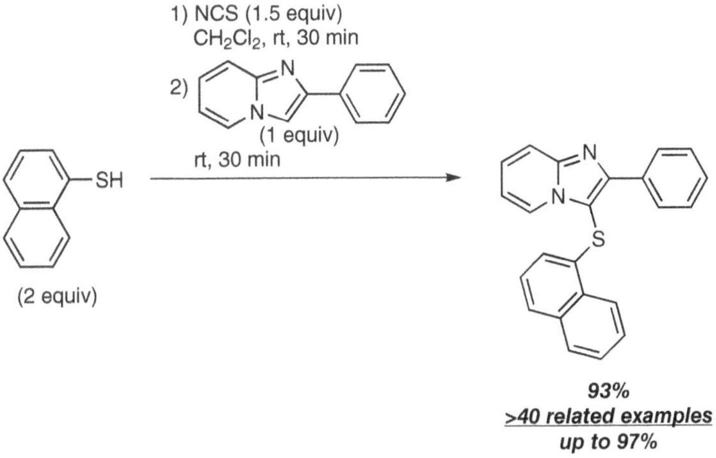

SCHEME 5.36 C—H sulfenylation of heterocycles [57].

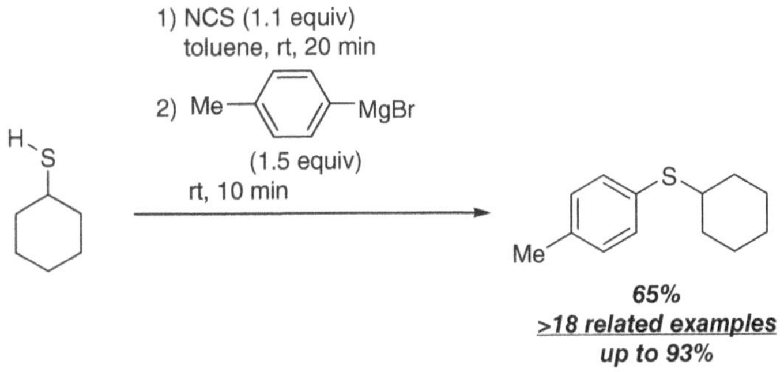

SCHEME 5.37 Synthesis of unsymmetrical diorganosulfides through sulfenylchloride intermediates [59].

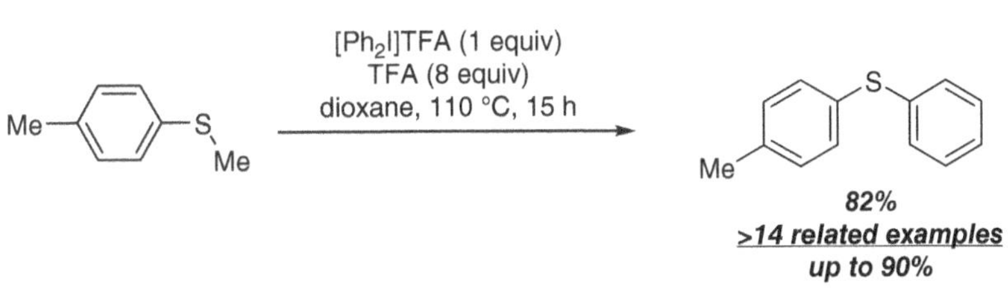

SCHEME 5.38 Transition metal-free synthesis of unsymmetrical diaryl sulfides [60].

metal-free synthesis of sulfur-containing heterocycles was accomplished using a tandem coupling/cyclization reaction (Scheme 5.39) [61].

Benzothiazoles have been generated by treating 2-halonitroarenes with elemental sulfur and benzylamines (Scheme 5.40 and Example 5.3) [62]. This three-component reaction generated moderate to high yields of the benzothiazoles from a range of prefunctionalized

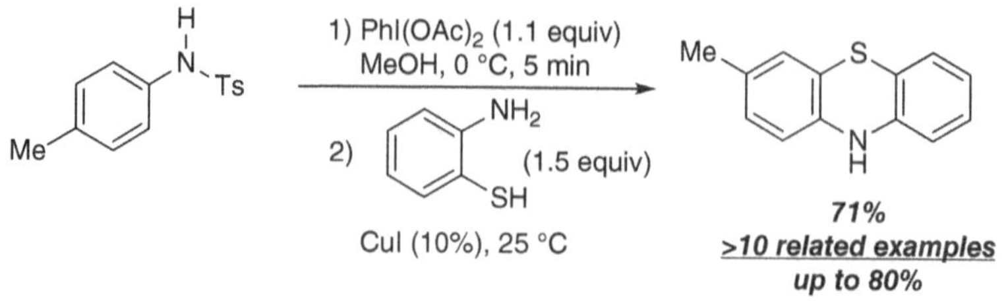

SCHEME 5.39   Synthesis of phenothiazine derivatives [61].

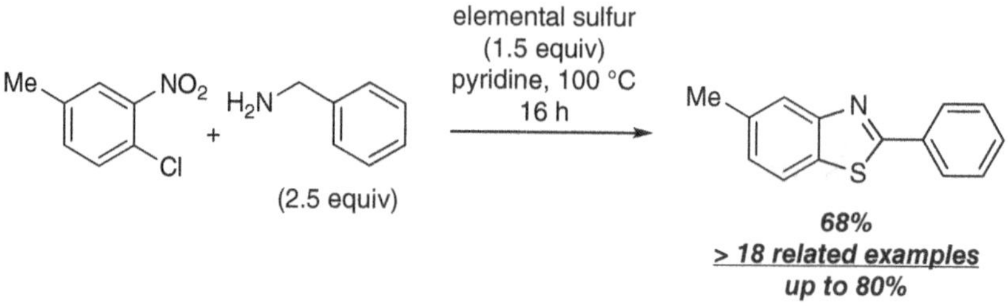

SCHEME 5.40   A three-component reaction leading to benzothiazoles [62].

Example 5.3   Three-Component Reaction Leading to the Formation of Benzothiazoles [62].

Before starting this synthesis, carefully follow the chemical safety instructions listed in Example 5.1.

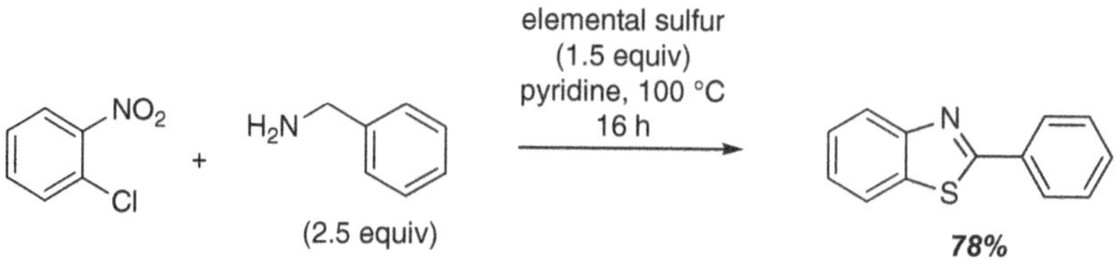

Caution! Heptane and ethyl acetate are highly flammable.

A mixture of *o*-chloronitrobenzene (5 mmol), sulfur (240 mg, 7.5 mmol), and amine (12.5 mmol) in pyridine (1 mL) was stirred under an argon atmosphere in a 25 mL test tube with pressure equalization at 100 °C for 16 h. After being cooled to room temperature, the volatiles were removed in vacuo. The crude reaction mixture was triturated with CH_2Cl_2 (5×2 mL). The combined CH_2Cl_2 layers were concentrated and purified by silica gel column chromatography (heptane/EtOAc; 95/5) to afford the desired benzothiazole as a pale yellow powder (78%).

Adapted with permission from T. B. Nguyen, L. Ermolenko, P. Retailleau, A. Al-Mourabit, *Angew. Chem. Int. Ed.,* **2014**, *53*, 13808–13812. *Copyright (2014) Wiley-VCH Verlag GmbH & Co. KGaA, Weinheim.*

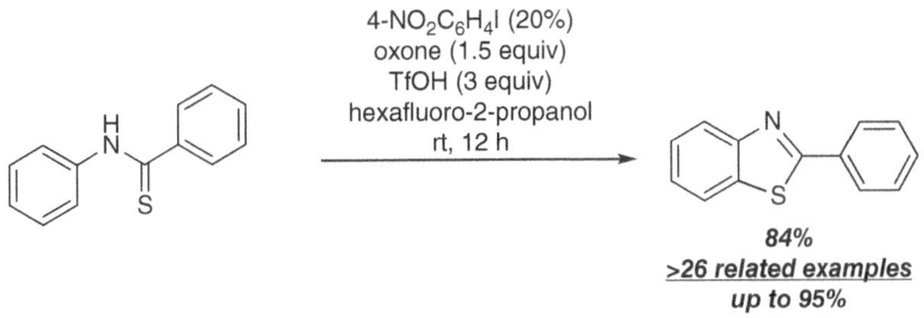

SCHEME 5.41   Benzothiazole synthesis through C–H functionalization [63].

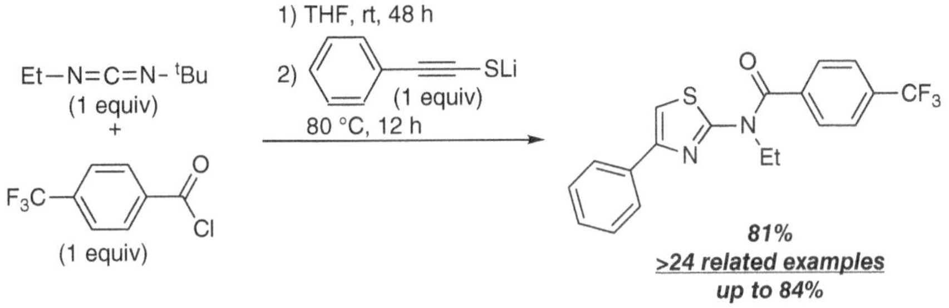

SCHEME 5.42   Preparation of 2-aminothiazoles [64].

substrates including heteroaromatic examples. Electron-donating and electron-withdrawing groups were well tolerated using this approach. Although the system needs to be set up for the cyclization, this remains an efficient metal-free route to the synthesis of these compounds. In related work, benzothiazoles were generated from aryl iodides and oxone in a C–H functionalization reaction (Scheme 5.41) [63].

The synthesis of 2-aminothiazoles has been achieved through a multicomponent reaction that cleaved carbodiimides as part of the sulfur–carbon bond-forming reaction (Scheme 5.42 and Example 5.4) [64]. The overall reaction was broken into two steps, with the initial step consisting of treatment of an acyl chloride with the carbodiimide. Addition of the lithium alkynethiolate to this mixture generated the 2-aminothiazoles in moderate to high yields. The functional group tolerance of this process was excellent, and the presence of a range of electron-withdrawing or donating substituents on the acyl chloride was well tolerated. The authors did note that the selectivity of the process was dependent upon the sterics of the system. Overall, if the system is well designed, this is an efficient method for the incorporation of 2-aminothiazoles into a host of organic structures.

Functionalized benzothiophenes have been generated through an iodocyclization reaction (Scheme 5.43) [65]. This transition metal-free approach to the synthesis of these compounds proceeded under very mild conditions and generated outstanding yields of the cyclization products in a few hours at room temperature. Even substrates bearing primary alcohols were successfully cyclized.

Example 5.4   Preparation of 2-Aminothiazoles [64].

Before starting this synthesis, carefully follow the chemical safety instructions listed in Example 5.1.

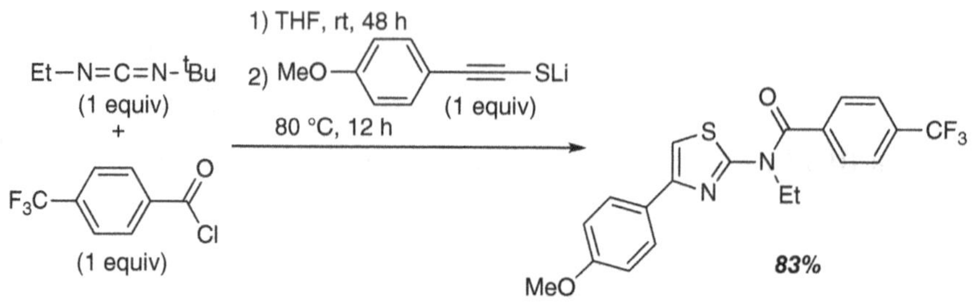

Caution! The organic solvents used in this preparation are highly flammable.

This procedure was carried out in a glovebox or in vessels with a nitrogen atmosphere. In a 25 mL flask, the acid chloride (1 mmol) was added to the carbodiimide (1 mmol) in THF (10 mL), and the mixture was stirred at room temperature for 48 h. In another 25 mL flask, *n*-BuLi (1 mmol, 1.6 M in hexane) was added dropwise at −78 °C to a stirred solution of terminal alkyne (1 mmol) in THF (5 mL). After stirring at −78 °C for 0.5 h, sulfur (1 mmol, 32 mg) was added and the reaction mixture was allowed to warm to room temperature for 2 h. The two reaction solutions were mixed into one flask, which was heated to 80 °C for 12 h. The volatiles were removed under vacuum, and the residue purified by chromatography to afford the 2-aminothiazole (337 mg, 83%).

Adapted with permission from Y. Wang, F. Zhao, Y. Chi, W.-X. Zhang, Z. Xi, J. Org. Chem., **2014**, *79,* 11146–11154. *Copyright (2014) American Chemical Society.*

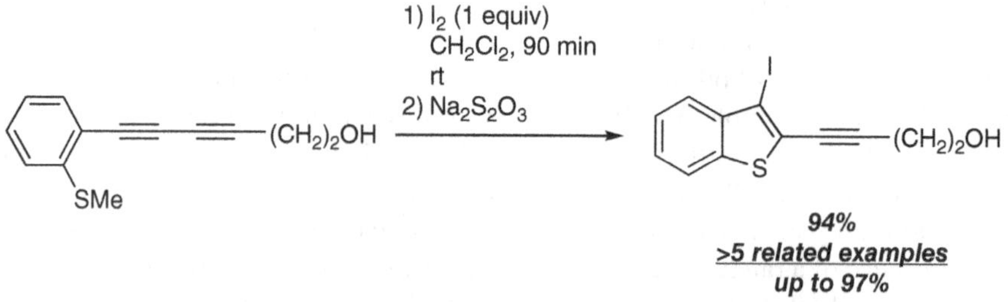

SCHEME 5.43   Synthesis of functionalized benzothiophenes through iodocyclization [65].

Aryl pyrazolone thioethers have been generated through an iodine-catalyzed reaction promoted by the addition of TsOH (Scheme 5.44) [66]. The chemistry was tolerant of a wide variety of electron-withdrawing and donating groups, and substrates bearing ethers, nitriles, halogens, trifluoromethyl groups, and even primary alcohols were successfully transformed into the target compounds.

While much of the chemistry described previously was focused on the synthesis of sulfides and related compounds, a range of diaryl sulfones can also be generated through

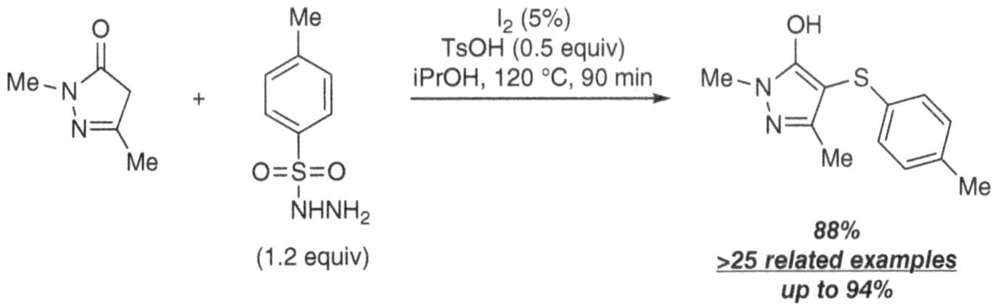

SCHEME 5.44   C—H thiolation of pyrazolones to afford aryl pyrazolone thioethers [66].

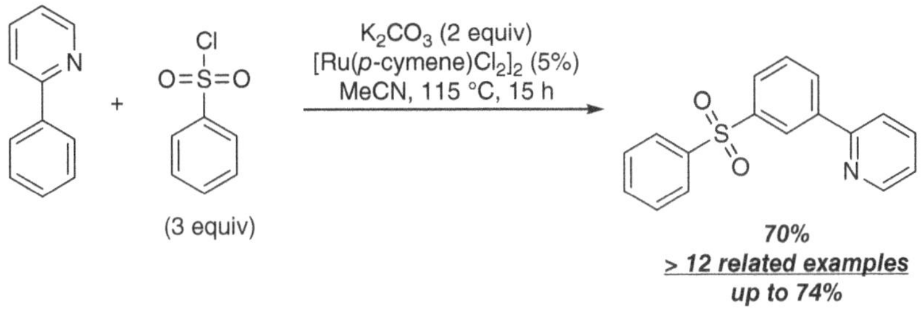

SCHEME 5.45   Ruthenium-catalyzed synthesis of diaryl sulfones [67].

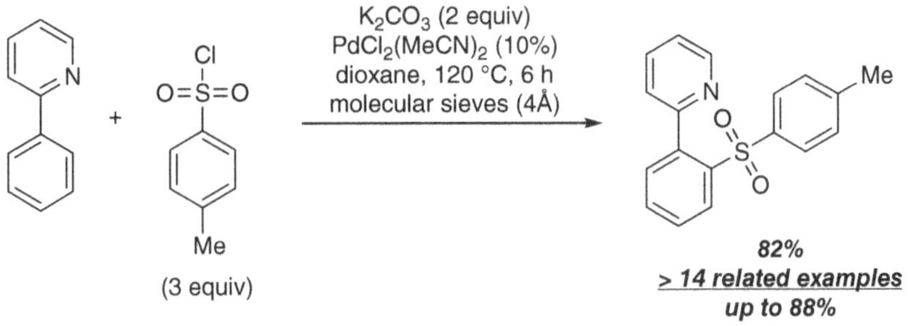

SCHEME 5.46   Palladium-catalyzed ortho-sulfonation of 2-phenylpyridine [68].

sulfur–carbon(sp²) bond-forming reactions. A number of metal complexes are known to promote this reaction including palladium and ruthenium. The ruthenium-catalyzed version of this reaction was cost effective while still maintaining a high level of reactivity (Scheme 5.45) [67]. Additionally, the authors discovered that one of the most active ruthenium catalysts was also simple to generate. The overall sulfur–carbon bond-forming reaction was a C—H sulfonation of 2-phenylpyridine using an arylsulfonyl chloride as the sulfur source. This reaction proceeded in moderate yields and selectively generated the meta-substituted diaryl sulfone. This chemistry stands in contrast to a very similar palladium-catalyzed process that generated the ortho-substituted product (Scheme 5.46) [68].

The palladium-catalyzed coupling of aryl halides with a sodium aryl sulfinate afforded symmetrical and unsymmetrical diaryl sulfones in good to excellent yield (Scheme 5.47) [69]. A common palladium(0) compound was used as the metal source, and XantPhos served as the supporting/stabilizing ligand. A base was also needed for this chemistry, and

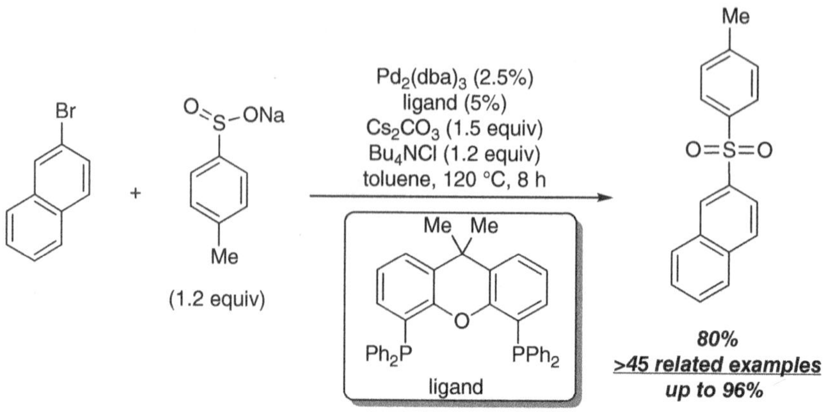

SCHEME 5.47   Palladium-catalyzed synthesis of diaryl sulfones [69].

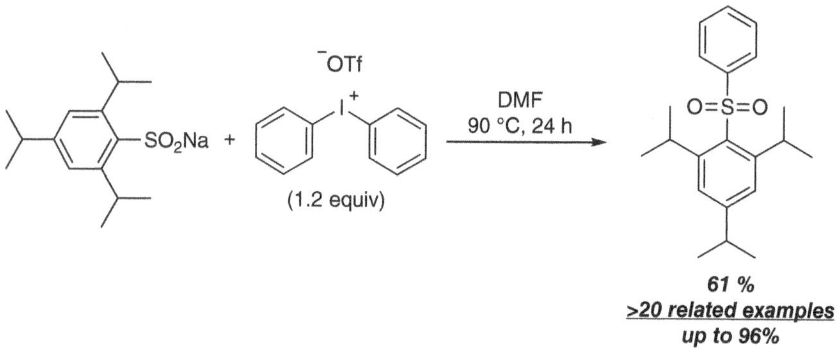

SCHEME 5.48   Transition metal-free synthesis of diaryl sulfones [70].

a mineral base (Cs_2CO_3) was found to be the most effective. Aryl halides as well as triflates were successfully functionalized through this approach.

A transition metal-free approach to the synthesis of diaryl sulfones has been devised using diaryliodonium salts (Scheme 5.48) [70]. The reactions were carried out in DMF at 90 °C and afforded high yields of the diaryl sulfones. A vast array of electron-donating and electron-withdrawing groups were well tolerated by this approach as well as heteroaromatic compounds. One of the strengths of this approach was the ability to functionalize extremely bulky substrates. As an example, sodium 2,4,6-triisopropylphenyl sulfinate was arylated in 61% yield. If a metal-free synthesis is needed for the synthesis of an unsymmetrical diaryl sulfone, this chemistry would be a reasonable approach.

5.2.1   Troubleshooting the Synthesis of Thioethers and Related Compounds

While it is impossible to predict every challenge that will arise during the synthesis of sulfur–carbon(sp²) bonds, the following suggestions will provide some direction when designing a synthesis.

Reaction Requirement	Possible Solution
Need a general route to the preparation of sulfur–carbon(sp²) bonds	Metal-catalyzed cross-coupling of aryl halides with thiols remains one of the more popular routes to the synthesis of sulfur–carbon(sp^2) bonds [23–25, 71]
Need to couple bulky aliphatic thiols and bulky aryl halides.	A palladium-catalyzed approach has shown great tolerance to bulky groups [30]
Need to generate bulky diaryl sulfides	Using an NHC-ligated palladium complex to catalyze the coupling of two bulky arylthiols would be a good place to start [38]
Need a copper-catalyzed approach to the synthesis of diaryl sulfides	A catalyst system comprised of copper iodide and ethylene glycol would be an excellent place to start [43]. Discrete copper catalysts have also shown activity towards the coupling reaction [44]
It would be convenient to have a nickel-catalyzed approach to the synthesis of sulfur–carbon bonds	Several nickel compounds have shown activity towards this coupling reaction. A nickel catalyst supported by a pincer ligand was particularly active for this transformation [47]
It would be convenient to use an aryl chloride in the cross-coupling reaction	A rhodium-catalyzed process has been shown to be quite effective at coupling aryl chlorides [53]. A palladium-catalyzed approach has also been successful [34]
Need a thiol-free synthesis of diaryl sulfides	Cross-coupling aryl benzyl sulfides with aryl iodides would be a good place to start [40]. An interesting approach using CS$_2$ as the sulfur source has been used in the copper-catalyzed synthesis of diaryl sulfides [45]
Need a transition metal-free route to the preparation of diaryl sulfides.	The conversion of alkyl aryl sulfides or thiophenols into diaryl sulfides was achieved using diaryliodonium salts [60, 61]. In related work, the sulfenylation of heteroaromatic compounds by thiols was promoted by NCS [57, 59]
Need to directly arylthiolate unactivated arenes	A palladium-catalyzed C—H arylthiolation would be a reasonable place to start [41]
Need a general route to the preparation of thioesters	A reasonable entry point to this chemistry would be the use of an iron-catalyzed process in water [54]
Can diaryl sulfides be generated from disulfides?	An effective approach for this reaction has been reported [55]
Need a good synthesis for benzothiazoles	Several methods are known for this transformation. A three-component coupling reaction between elemental sulfur, chlorinated nitroarenes, and benzylamines would be a reasonable approach [62]
Need a general method for the synthesis of aryl sulfones	The palladium-catalyzed coupling of sodium aryl sulfinates with aryl halides would be a reasonable approach [69]
Need a metal-free synthesis of aryl sulfones	Diaryliodonium salts have been successfully used to promote this reaction [70]
Can aryl sulfones be generated through C—H activation processes?	These reactions are typically promoted by metal catalysts. While ortho-directing groups have promoted the formation of ortho-substituted arenes using palladium catalysts [68], ruthenium complexes have been shown to generate the meta-substituted aryl sulfone using the same ortho-directing group [67]

5.3 SYNTHESIS OF VINYL THIOETHERS AND RELATED COMPOUNDS THROUGH THE FORMATION OF SULFUR–CARBON(sp^2) BONDS

The copper-catalyzed coupling of organothiols with vinyl halides remains one of the most efficient approaches to the synthesis of vinyl sulfides (Scheme 5.49) [72]. The scope of the reaction was broad and a score of electronically varied substrates were successfully coupled using this approach. Bulky substrates such as 2,6-dimethylthiophenol and heterocycles were also well tolerated by this chemistry. As with many other cross-coupling reactions, the stereochemistry of the starting vinyl halide was retained in the product. As a result, this chemistry was a valuable approach to the synthesis of these compounds when a specific stereochemistry was required. In related work, the temperature of the coupling reaction was able to be reduced and the electronic scope of the reaction was increased by generating the copper catalyst in solution through the combination of 1,2-cyclohexanediol and copper iodide (Scheme 5.50) [73]. The preparation of vinyl sulfides has also been achieved using a secondary phosphine oxide-ligated palladium catalyst [74].

The metal-free synthesis of vinyl sulfides from *gem*-dibromoalkenes proceeds under mild conditions (Scheme 5.51) [75]. A variety of thiols served as the source of the

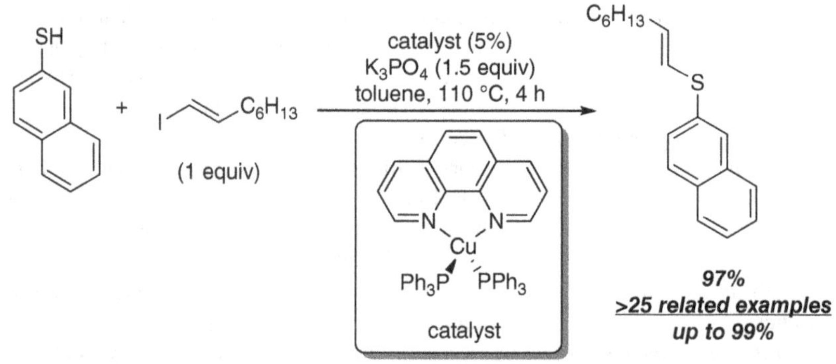

SCHEME 5.49 Synthesis of vinyl sulfides through copper-catalyzed cross-coupling [72].

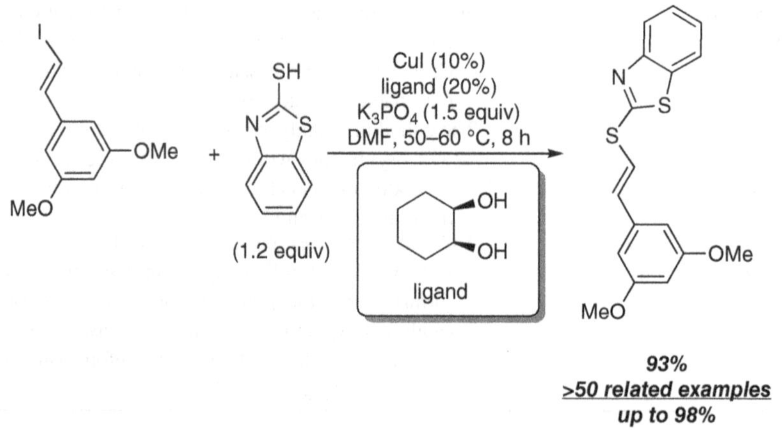

SCHEME 5.50 Preparation of vinyl sulfides using a copper catalyst generated in situ [73].

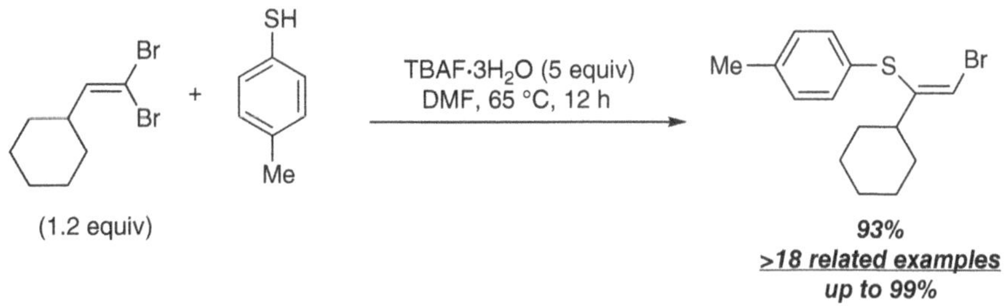

SCHEME 5.51 Conversion of 1,1-dibromoalkenes into vinyl sulfides [75].

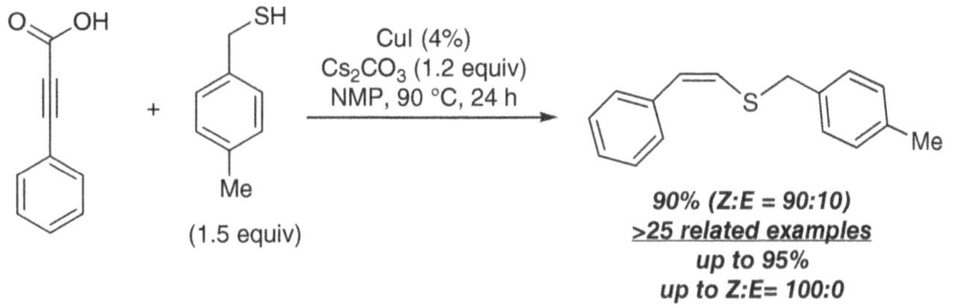

SCHEME 5.52 Vinyl sulfide synthesis through decarboxylation of carboxylic acids [76].

sulfur-containing fragment, and simply stirring the reaction mixture with gentle heating afforded moderate to excellent yields of the vinyl sulfides for most substrates. The chemistry was regioselective for attachment of the sulfur to the carbon bearing the organic fragment and stereoselective for the formation of the Z-isomer. The TBAF was proposed to serve as a base for the generation of the thiolate nucleophiles as well as for conversion of the *gem*-dibromo species into a 1-bromoalkyne. Nucleophilic attack of the thiolate anion on the haloalkyne followed by trapping by another equivalent of thiol generated the observed compound. Overall, this is a valuable and selective route for the preparation of vinyl sulfides.

The formation of sulfur–carbon(sp^2) bonds through a copper-catalyzed decarboxylation process combined with a cross-coupling reaction has been achieved (Scheme 5.52) [76]. The most effective catalyst was simply copper iodide without any added ligand. It should be noted that the starting thiol as well as the vinyl sulfide product could coordinate to the copper center and serve as a stabilizing ligand. The process was regioselective and predominantly generated the Z-vinyl sulfides. This stereoselectivity was highly dependent upon the substrate, and while several carboxylic acid/thiol combinations generated exclusively the Z-isomer, ~50 : 50 ratios were also obtained. Overall, if the system is well designed, this is a very attractive and selective approach to the preparation of vinyl sulfides.

The addition of sulfur–hydrogen bonds to alkynes (hydrothiolation) is a valuable and popular approach to the synthesis of vinyl sulfides [23]. Some of the earliest reports of this chemistry involved the use of molybdenum [77] and palladium complexes [78]. While discussions of the challenges encountered when trying to design an effective cross-coupling reactions often focus on concerns about reactivity and substrate scope, hydroelementation reactions introduce further complications due to regio- and stereoselectivity issues. In many

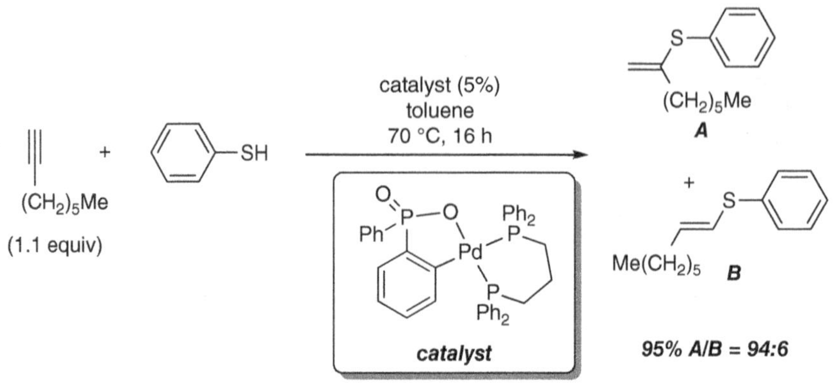

SCHEME 5.53   *E*-selective palladium-catalyzed hydrothiolation of alkynes [79].

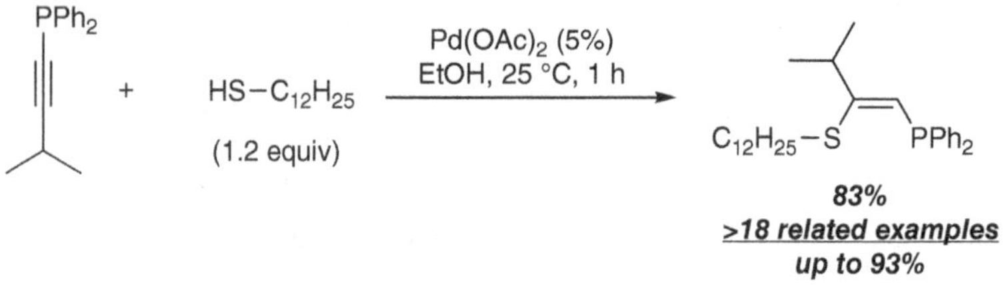

SCHEME 5.54   Hydrothiolation of alkynylphosphines [80].

cases, reactions are highly regioselective, but they generate mixtures of *E/Z*-isomers. To this end, considerable effort has been devoted to the design and investigation of new catalysts and conditions that will selectively promote the formation of one isomer.

The palladium-catalyzed hydrothiolation of terminal alkynes has been achieved using a metallocycle catalyst that was generated through the treatment of palladium acetate with phosphinic acids (Scheme 5.53) [79]. Using this catalyst system, benzenethiol was added to 1-octyne in moderate yield with high selectivity for the Markovnikov-isomer. While only a single hydrothiolation example was reported, this chemistry provides the foundation for the design of additional palladium-catalyzed reactions.

One of the strengths of palladium-catalyzed cross-coupling and hydroelementation reactions is the tolerance to preexisting functional groups. This was exemplified in the hydrothiolation of 1-alkynylphosphines (Scheme 5.54) [80]. Palladium acetate was used as the catalyst for these reactions and no added stabilizing/solubilizing ligand was needed for a successful hydrothiolation reaction. Treatment of these internal alkynes with thiols generated vinyl sulfides in moderate to good yields after only 1 h of stirring under very mild conditions. The overall process was remarkably regio- and stereoselective and afforded (*Z*)-1-phosphino-2-thio-1-alkenes.

The conversion of propargyl alcohols into vinyl sulfides has been achieved using nickel catalysts under solvent free conditions (Scheme 5.55) [81]. The reactions were quite rapid and proceeded with only gentle heating. For most substrates screened, the process was highly selective for the Markovnikov addition, and only traces of the anti-Markovnikov product were observed.

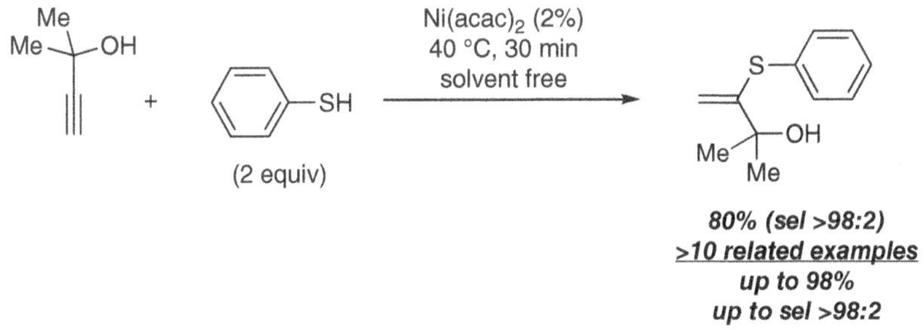

SCHEME 5.55 Nickel-catalyzed hydrothiolation of alkynes [81].

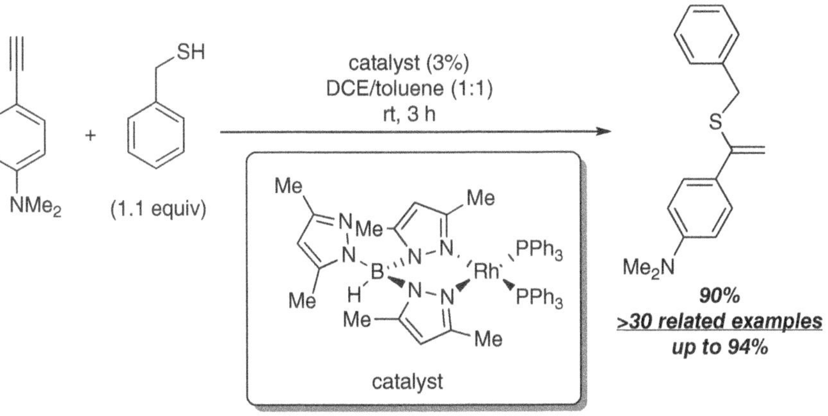

SCHEME 5.56 Rhodium-catalyzed hydrothiolation of alkynes [82].

The rhodium-catalyzed addition of sulfur–hydrogen bonds to alkynes has been extensively investigated by Love (Scheme 5.56) [82–84]. One of the most selective rhodium catalysts was ligated by two triphenylphosphines and tris-3,5-dimethyl pyrazolylborate. The hydrothiolation reactions selectively generated the Markovnikov product after stirring for only a few hours. The functional group tolerance of this approach was outstanding, and a range of alkyl- and arylthiols were successfully used. On the alkyne side, excellent yields were obtained with alkylalkynes and most arylalkynes. Electron-deficient arylalkynes were sluggish and afforded lower yields of the vinyl sulfides. Although the rhodium catalyst needed to be prepared in this work, the selectivity and ability to use a wide assortment of thiols and alkynes make this one of the most attractive routes for the preparation of vinyl sulfides. In related work, an NHC-ligated rhodium complex was found to direct the mode of thiol addition through modification of the catalyst system (Scheme 5.57) [85, 86]. The use of organozirconium complexes for the Markovnikov addition of thiols to alkynes has also been reported [87].

The overall anti-addition of S—H bonds to alkynes has been achieved using cesium carbonate in DMSO (Scheme 5.58) [88]. A key component of this chemistry was the addition of a radical inhibitor (TEMPO) to suppress a background radical reaction. While the radical process generated vinyl sulfides, the stereoselectivity was poor. When terminal alkynes or unsymmetrical internal alkynes were used, the chemistry regioselectively

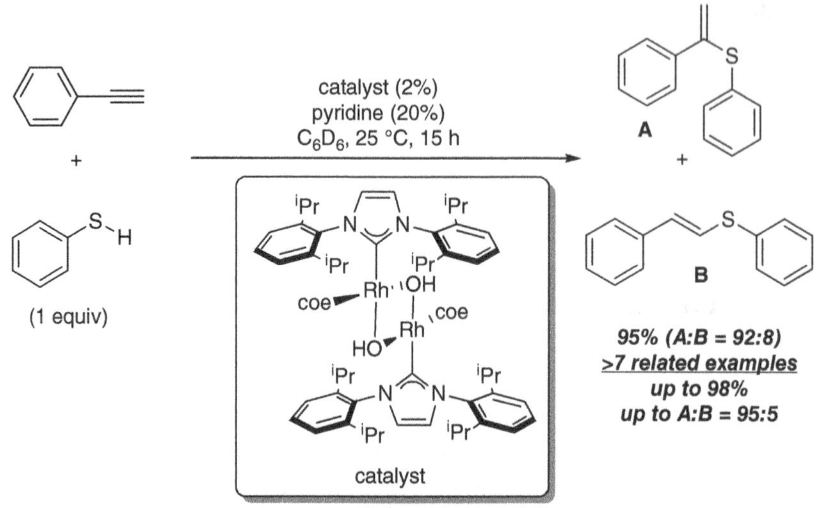

SCHEME 5.57   Addition of thiols to alkynes using an NHC-ligated rhodium catalyst [85].

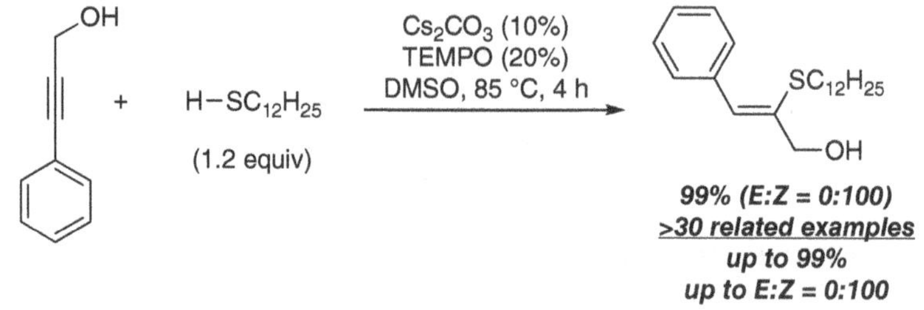

SCHEME 5.58   Overall Z-selective hydrothiolation of internal and terminal alkynes [88].

generated the anti-Markovnikov product with good to outstanding stereoselectivity for the Z-isomer. The functional group tolerance was outstanding, and even unprotected amines or propargyl alcohols were successfully transformed into vinyl sulfides. In related work, the addition of anthracene-based thiolates to a range of 3-aryl-2-propynenitriles has been achieved (Scheme 5.59) [89].

The preparation of vinyl sulfones has been achieved through a decarboxylative sulfonation of vinyl carboxylic acids (Scheme 5.60) [90]. This copper-catalyzed reaction proceeded in moderate to excellent yields and displayed a high degree of tolerance to a wide range of preexisting functional groups including heteroaromatic substrates. While the chemistry was not sensitive to electronic effects, increasing the steric bulk around the alkene was detrimental as evidenced by the sluggish reactivity of α-methylcinnamic acid (0% conversion). In addition to the retention of the stereochemistry from the starting carboxylic acid, this chemistry is particularly attractive since it could be carried out under an atmosphere of air. No glovebox or vacuum manifold was needed for this chemistry. In related work, the transition metal-free regioselective synthesis of vinyl sulfides from terminal alkynes has been achieved using sulfonyl hydrazides [91]. As with many similar

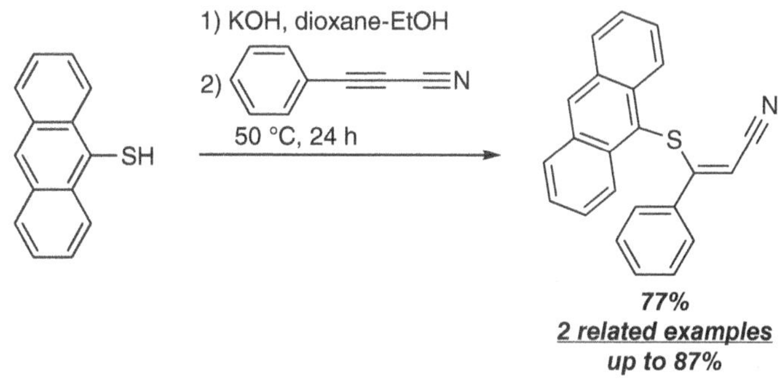

SCHEME 5.59   Hydrothiolation of 3-aryl-2-propynenitriles [89].

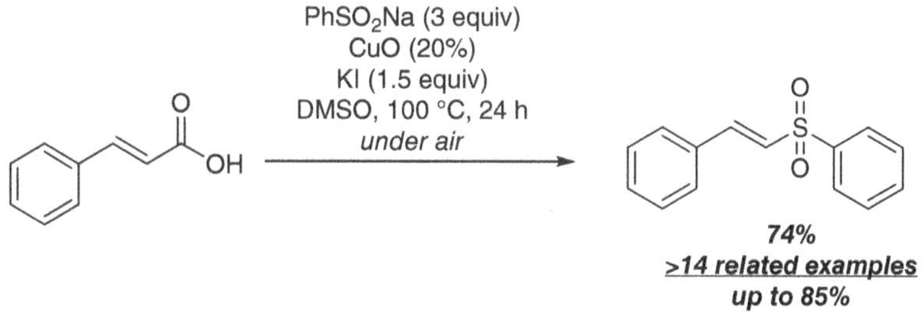

SCHEME 5.60   Preparation of vinyl sulfones through a copper-promoted decarboxylation reaction [90].

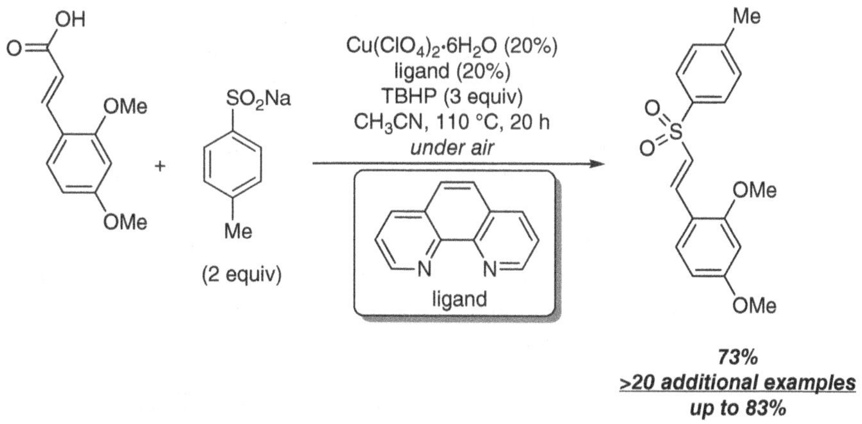

SCHEME 5.61   Decarboxylative synthesis of vinyl sulfones [92].

reactions, while the regiochemistry was defined, the stereochemistry was less controlled and mixtures were often obtained. It should be noted that several substrates generated a single isomer. In related work, $Cu(ClO_4)_2$ has also been used to effect a decarboxylative conversion of carboxylic acids into vinyl sulfones (Scheme 5.61) [92].

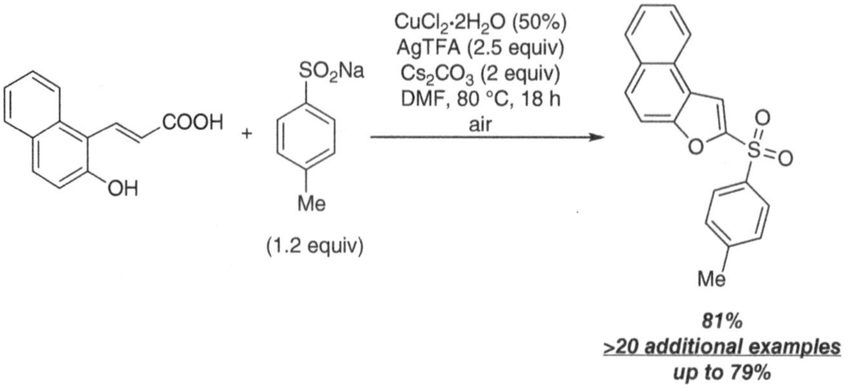

SCHEME 5.62   Conversion of cinnamic acid derivatives into sulfonated benzofurans [94].

The generation of 2-sulfonylbenzofurans from cinnamic acid derivatives has been reported (Scheme 5.62) [94]. While many of the classic protocols for the preparation of these compounds required multistep syntheses or the use of air-sensitive materials, the authors were able to devise a synthesis for the target compounds using a cascade reaction comprised of a protodecarboxylation, an oxidative sulfonylation, and an intramolecular cyclization. Both copper and silver salts were used to promote the transformation, and the absence of either metal salt reduced the conversion. One of the most attractive aspects of this chemistry was the ability to load the reactor vial and heat the reaction under an atmosphere of air. No glovebox or nitrogen manifold were needed for this chemistry. Air also functioned as an oxidant in this chemistry. The addition reaction was influenced by the electronic nature of the substituents as incorporating electron-donating substituents resulted in high yields of the sulfones while substrates bearing electron-withdrawing groups were sluggish or unreactive under similar conditions. In related work, the use of a carbon dioxide atmosphere promoted the formation of Z-isomers in a copper-catalyzed hydrothiolation of alkynes [93].

A series of vinyl sulfones have been prepared through palladium-catalyzed cross-coupling (Scheme 5.63) [69]. A palladium(0) precursor along with a large bite-angle bisphosphine ligand was used to promote the coupling of a variety of vinyl halides and triflates with sulfinic acid salts to generate vinyl sulfones. The most effective base for this chemistry was cesium carbonate, and good to excellent yields of the target sulfones were obtained after stirring for 9 h with gentle heating. In related work, the preparation of halogenated vinyl sulfones has been reported using an iron-mediated halosulfonation of terminal alkynes [95].

An interesting approach to the preparation of vinyl sulfones entailed the use of iridium complexes to catalyze an allylic substitution/isomerization reaction starting from allyl sulfinates (Scheme 5.64) [96]. A common iridium complex was used as the metal source, and a chiral phosphoramidite served as the supporting ligand for this chemistry. The yields were good to excellent, and a range of electron-donating and withdrawing groups were well tolerated. The overall reaction was highly regio- and stereoselective for the formation of the E-isomer. If an allylic sulfinate can be generated as the starting point, this is an effective means of generating the vinyl sulfide.

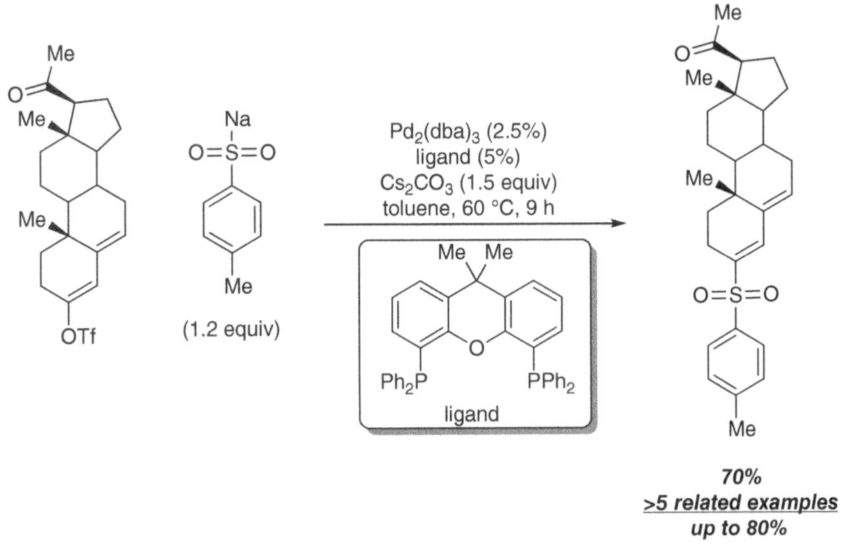

SCHEME 5.63   Preparation of vinyl sulfones through palladium-catalyzed cross-coupling [69].

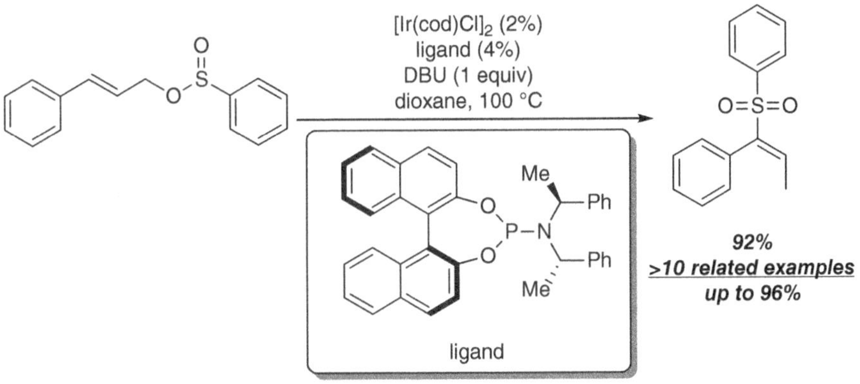

SCHEME 5.64   Iridium-catalyzed preparation of vinyl sulfones [96].

In addition to reactions that form a single sulfur–carbon(sp^2) bond, a range of reactions have been reported that generate several S—C bonds [23]. An example of such a reaction used common palladium catalysts to promote the addition of disulfides to terminal alkynes [97]. The two sulfur-containing fragments were added mutually cis, and moderate to excellent yields of the bissulfides were obtained for a range of functionalized alkynes including examples containing propargyl alcohols. In related work, the addition of dialkyl disulfides to terminal alkynes was promoted by a rhodium complex and TFA [98]. Similar to the previous case, the two sulfur-containing groups were selectively added such that a Z-vinyl bissulfide was obtained, and the chemistry displayed a wide functional group tolerance.

5.3.1 Troubleshooting the Synthesis of Vinyl Sulfides and Related Compounds

While it is impossible to predict every challenge that will arise during the synthesis of sulfur–carbon(sp^2) bonds for the synthesis of vinyl sulfides and related compounds, the following suggestions will provide some direction when designing a synthesis.

Reaction Requirement	Suggestion/Possible Solution
Must exclusively generate an E- *or* Z-*isomer of a vinyl sulfide*	Using a copper-catalyzed cross-coupling would be a reasonable place to start since the coupling reaction will retain the geometry of the starting vinyl halide [72]. The in situ generation of the catalytically active species is particularly attractive [73]
Need a metal-free approach to the synthesis of vinyl sulfides	The conversion of 1,1-dibromoalkenes into vinyl sulfides would be a good entry point into this chemistry [75]
It would be convenient to start the synthesis from a carboxylic acid	Copper catalysts have been shown to catalyze a tandem decarboxylative coupling of alkynoic acids and thiols [76]
Need a method for the formation of Z-*vinyl sulfides*	In addition to a cross-coupling approach between Z-vinyl halides and thiols, a copper-catalyzed coupling process starting from an alkynoic acid predominately generates the Z-isomer [76]. A transition metal-free approach has also been devised [88]
Need a method for the Markovnikov addition of thiols to alkynes	Several catalysts will generate the Markovnikov product. A palladium-catalyzed approach is particularly attractive [79]. The rhodium-catalyzed reaction has been well studied [82]
Need to effect a hydrothiolation on a propargyl alcohol with retention of the alcohol group	This can be a challenging reaction; however, a simple nickel catalyst has shown considerable reactivity and selectivity (Markovnikov product) in this reaction [81]. A transition metal-free approach has also been devised [88]
Can internal alkynes be used in hydrothiolation reactions?	While some substrate-specific reactivity was observed, a range of internal alkynes has been successfully converted into vinyl sulfides. The metal-free addition of bulky thiolates to propynenitriles serves as a representative example [89]
Need a good way to make a vinyl sulfone.	An attractive approach would be the decarboxylative sulfonylation of alkenyl carboxylic acids [90, 92]
Can vinyl sulfones be generated through cross-coupling?	Palladium-catalyzed reactions are attractive entry points for this chemistry [69]

5.4 SYNTHESIS OF THIOETHERS AND RELATED COMPOUNDS THROUGH THE FORMATION OF SULFUR–CARBON(SP) BONDS

One of the classic routes to the synthesis of alkynyl aryl sulfides involved the treatment of bromoalkynes with sodium thiophenolate and related compounds [100]. Using this approach, moderate to good yields of the alkynyl aryl sulfides were obtained under very mild conditions. Following this work, a considerable amount of effort has been focused on increasing the scope and functional group tolerance of the chemistry. One modification entailed the generation of an acetylide anion through the treatment of a terminal alkyne with BuLi followed by the addition of a disulfide [101, 102]. Although this approach was successful at generating the alkynyl sulfide, a significant amount of a bissulfide side product was formed as a result of thiolate anion generated in the first step of the reaction attacking the desired alkynyl sulfide. This issue was solved through the addition of methyl iodide. While this additive was successful in preventing this secondary reaction, the resulting methylated products were sometimes challenging to remove from the desired alkynyl sulfides. Exchanging the methyl iodide for 4-nitrobenzyl bromide successfully trapped with thiolate anion, and the resulting sulfur-containing species were readily separable from the target alkynyl sulfides (Scheme 5.65) [101].

The transition metal-free synthesis of alkynyl aryl sulfides from thiophenols has been achieved using alkynyliodonium salts as the alkynylating agents (Scheme 5.66 and Example 5.5) [103]. The alkynylation reactions were remarkably fast, and outstanding yields of a range of alkynyl aryl sulfides were obtained after stirring for only 5 min. The presence of a range of preexisting electron-donating and withdrawing groups was well tolerated in this chemistry and halogens, esters, heterocycles, ethers, and glycosides were successfully converted in this work. Even substrates bearing carboxylic acids were functionalized. Dialkynyl sulfides could also be generated starting from sodium hydrogen sulfide. In addition to the broad functional group tolerance, the observation that the reactions could be carried out in an open vessel with no protection from oxygen or moisture was one of the

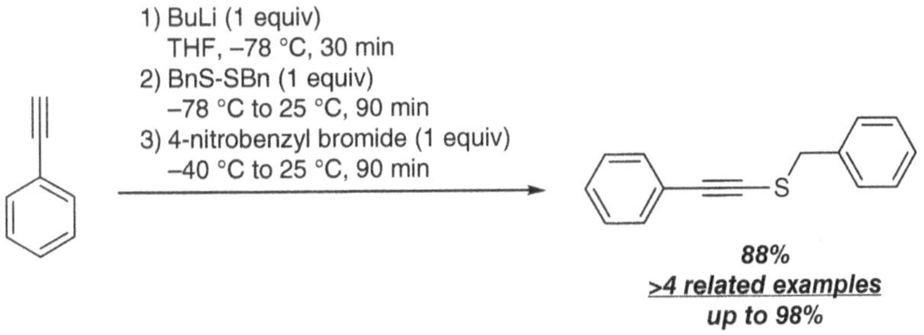

1) BuLi (1 equiv)
   THF, −78 °C, 30 min
2) BnS-SBn (1 equiv)
   −78 °C to 25 °C, 90 min
3) 4-nitrobenzyl bromide (1 equiv)
   −40 °C to 25 °C, 90 min

88%
>4 related examples
up to 98%

SCHEME 5.65   Conversion of terminal alkynes into alkynyl sulfides [101].

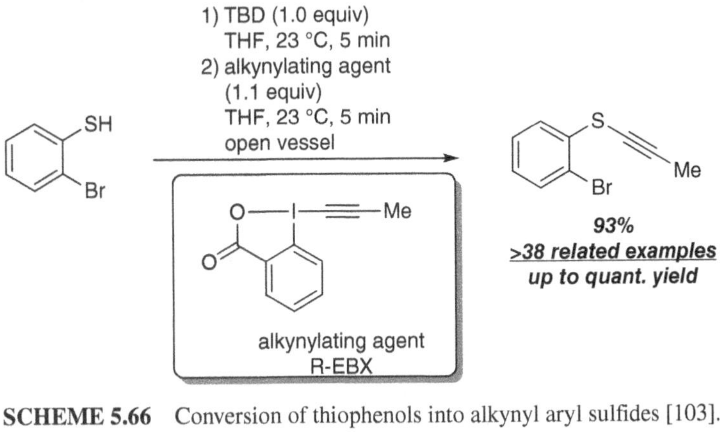

1) TBD (1.0 equiv)
   THF, 23 °C, 5 min
2) alkynylating agent
   (1.1 equiv)
   THF, 23 °C, 5 min
   open vessel

93%
>38 related examples
up to quant. yield

alkynylating agent
R-EBX

SCHEME 5.66   Conversion of thiophenols into alkynyl aryl sulfides [103].

Example 5.5   Preparation of an Alkynyl Alkyl Sulfide [103].

Before starting this synthesis, carefully follow the chemical safety instructions listed in Example 5.1.

Caution! Ethyl acetate and pentane are highly flammable.

This procedure was carried out in an open vessel. A round-bottom flask (25 mL) was charged with a magnetic stirring bar, TrpCys dipeptide (0.094 g, 0.200 mmol), TMG (30.0 μL, 0.240 mmol), and THF (5.0 mL). After stirring the resulting solution for 5 min at room temperature, the alkynylating agent (R-EBX) (0.220 mmol) was added as a solid in one portion. The resulting reaction mixture was stirred for 5 min at room temperature. Next, the mixture was diluted with water (10 mL) and extracted with EtOAc (3 × 15 mL). The combined organic layers were dried over MgSO$_4$, filtered, and concentrated in vacuo. The crude reaction mixture was purified by flash chromatography using EtOAc/pentane as the eluent to afford the title compound as a white solid (0.098 g, 80%).

Adapted with permission from R. Frei, M. D. Wodrich, D. P. Hari, P.-A. Borin, C. Chauvier, J. Waser, *J. Am. Chem. Soc.*, **2014**, *136*, 16563–16573. *Copyright (2014) American Chemical Society.*

most practical aspects of the chemistry. In related work, the preparation of alkynyl sulfones was achieved using these and related alkynylating agents [104, 105].

The trifluoromethyl group is an attractive fragment to incorporate into organic structures due to the unique properties it imparts. To this end, a clever approach was developed to generate alkynyl trifluoromethyl sulfides from terminal alkynes (Scheme 5.67) [106]. While a range of metal-catalyzed processes are known to incorporate a trifluoromethyl group, combining this process with sulfurization step has been challenging. The authors solved this issue through the combination of elemental sulfur with CF$_3$SiMe$_3$. In combination with KF, this duo was able to promote the trifluoromethylthiolation of terminal alkynes in

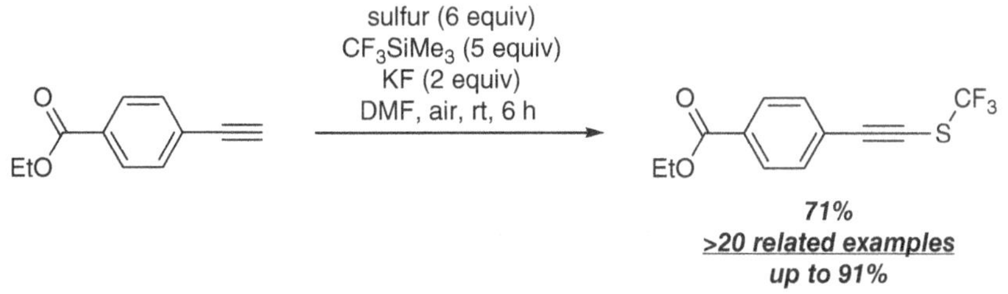

SCHEME 5.67   Trifluoromethylthiolation of alkynes in air [106].

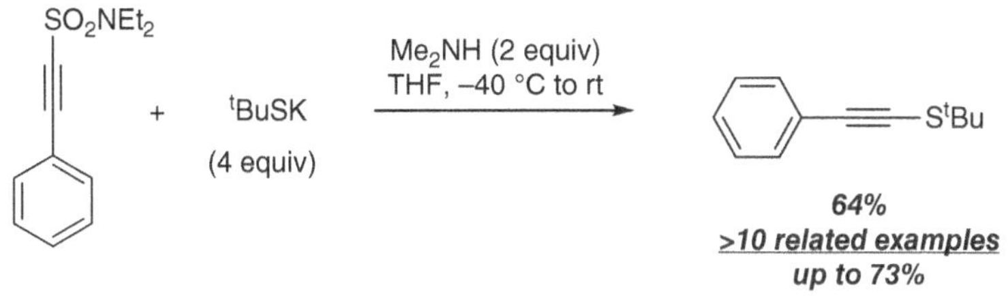

SCHEME 5.68   Synthesis of alkynyl sulfides from sulfonamides [107].

moderate to high yields. The functional group tolerance of this approach was outstanding, and a range of electron-donating and withdrawing groups were successfully transformed into the target compounds. One of the more practical aspects of this chemistry was the observation that it could be carried out under an atmosphere of air.

The synthesis of alkynyl sulfides from sulfonamides using potassium thiolate salts generated moderate yields of the alkynyl sulfides after only a few minutes (Scheme 5.68) [107]. The synthesis was carried out in two stages. The first step entailed the synthesis of thiolate anions through the addition of potassium hydride to solutions of the thiol. The addition of the sulfonamide along with dimethylamine comprised the second step. While the role of the amine was not fully elucidated, it was needed for a successful conversion as its absence led to poor conversions. There is a level of operational complexity to this approach since potassium hydride needed to be handled in the first step of the synthesis.

5.4.1   Troubleshooting the Synthesis of Alkynyl Thioethers and Related Compounds

While it is impossible to predict every challenge that will arise during the synthesis of sulfur–carbon(sp) bonds, the following suggestions will provide some direction when designing a synthesis.

Reaction Requirement/Challenge	Suggestion/Possible Solution
Need to start from a disulfide	Treating a disulfide with a lithium acetylide followed by a thiolate trapping agent would be a reasonable approach provided the disulfide does not contain electrophiles [101]
Need to alkynylate a thiol	Direct alkynylation has been reported using alkynyliodonium salts [103]

(Continued)

(Continued)

Reaction Requirement/Challenge	Suggestion/Possible Solution
Do not have a glovebox or nitrogen line and need to prepare an alkynyl aryl sulfide from a thiophenol	Alkynyliodonium salts have been shown to alkynylate a range of thiophenols in an open flask [103]
Do not have a glovebox or nitrogen line and need to generate an alkynyl trifluoromethyl sulfide	A trifluoromethylthiolation of terminal alkynes has been reported that it does not require a glovebox and can be carried out under air [106]
Need to generate a dialkynyl sulfide	The double alkynylation of sodium hydrogen sulfide using alkynyliodonium salts would be a good place to start [103]

REFERENCES

[1] Qiao, Z.; Wei, J.; Jiang, X. *Org. Lett.* **2014**, *16*, 1212–1215.

[2] Tamai, T.; Ogawa, A. *J. Org. Chem.* **2014**, *79*, 5028–5035.

[3] Shao, X.; Liu, T.; Lu, L.; Shen, Q. *Org. Lett.* **2014**, *16*, 4738–4741.

[4] Khatik, G. L.; Kumar, R.; Chakraborti, A. K. *Org. Lett.* **2006**, *8*, 2433–2436.

[5] Chauhan, P.; Mahajan, S.; Enders, D. *Chem. Rev.* **2014**, *114*, 8807–8864.

[6] Meninno, S.; Croce, G.; Lattanzi, A. *Org. Lett.* **2013**, *15*, 3436–3439.

[7] Phelan, J. P.; Patel, E. J.; Ellman, J. A. *Angew. Chem. Int. Ed.* **2014**, *53*, 11329–11332.

[8] Fu, N.; Zhang, L.; Luo, S.; Cheng, J.-P. *Org. Lett.* **2014**, *16*, 4626–4629.

[9] Denmark, S. E.; Rossi, S.; Webster, M. P.; Wang, H. *J. Am. Chem. Soc.* **2014**, *136*, 13016–13028.

[10] Alt, I.; Rohse, P.; Plietker, B. *ACS Catal.* **2013**, *3*, 3002–3005.

[11] Hans, M.; Delaude, L.; Rodriguez, J.; Coquerel, Y. *J. Org. Chem.* **2014**, *79*, 2758–2764.

[12] Gianatassio, R.; Kawamura, S.; Eprile, C. L.; Foo, K.; Ge, J.; Burns, A. C.; Collins, M. R.; Baran, P. S. *Angew. Chem. Int. Ed.* **2014**, *53*, 9851–9855.

[13] Fer, M. J.; Doan, P.; Prange, T.; Calvet-Vitale, S.; Gravier-Pelletier, C. *J. Org. Chem.* **2014**, *79*, 7758–7765.

[14] Nakamura, S.; Ohara, M.; Koyari, M.; Hayashi, M.; Hyodo, K.; Nabisaheb, N. R.; Funahashi, Y. *Org. Lett.* **2014**, *16*, 4452–4455.

[15] Hutka, M.; Tsubogo, T.; Kobayashi, S. *Organometallics* **2014**, *33*, 5626–5629.

[16] Wang, P.-F.; Wang, X.-Q.; Dai, J.-J.; Feng, Y.-S.; Xu, H. *Org. Lett.* **2014**, *16*, 4586–4589.

[17] Xu, K.; Khakyzadeh, V.; Bury, T.; Breit, B. *J. Am. Chem. Soc.* **2014**, *136*, 16124–16127.

[18] Zhao, Y.; Kang, J.; Park, C.-M.; Bagdon, P. E.; Peng, B.; Xian, M. *Org. Lett.* **2014**, *16*, 4536–4539.

[19] Zong, L.; Ban, X.; Kee, C. W.; Tan, C.-H. *Angew. Chem. Int. Ed.* **2014**, *53*, 11849–11853.

[20] Emmett, E. J.; Hayter, B. R.; Willis, M. C. *Angew. Chem. Int. Ed.* **2014**, *53*, 10204–10208.

[21] Lu, H.; Zhang, F.-M.; Pan, J.-L.; Chen, T.; Li, Y.-F. *J. Org. Chem.* **2014**, *79*, 546–558.

[22] Beletskaya, I.; Cheprakov, A. *Coord. Chem. Rev.* **2004**, *248*, 2337–2364.

[23] Beletskaya, I. P.; Ananikov, A. V. *Chem. Rev.* **2011**, *111*, 1596–1636.

[24] Ley, S. V.; Thomas, A. W. *Angew. Chem. Int. Ed.* **2003**, *42*, 5400–5449.

[25] Kondo, T.; Mitsudo, T. *Chem. Rev.* **2000**, *100*, 3205–3220.

[26] Hartwig, J. F. *Acc. Chem. Res.* **2008**, *41*, 1534–1544.

[27] Monnier, F.; Taillefer, M. *Angew. Chem. Int. Ed.* **2009**, *48*, 6954–6971.

[28] Migita, T.; Shimizu, T.; Asami, Y.; Shiobara, J.; Kato, Y.; Kosugi, M. *Bull. Chem. Soc. Jpn.* **1980**, *53*, 1385–1389.

[29] Kosugi, M.; Shimizu, T.; Migita, T. *Chem. Lett.* **1978**, 13–14.

[30] Murata, M.; Buchwald, S. L. *Tetrahedron* **2004**, *60*, 7397–7403.

[31] Schopfer, U.; Schlapbach, A. *Tetrahedron* **2001**, *57*, 3069–3073.

[32] Li, G. Y. *Angew. Chem. Int. Ed.* **2001**, *40*, 1513–1516.

[33] Teverovskiy, G.; Surry, D. S.; Buchwald, S. L. *Angew. Chem. Int. Ed.* **2011**, *50*, 7312–7314.

[34] Fernandez-Rodriguez, M. A.; Shen, Q.; Hartwig, J. F. *J. Am. Chem. Soc.* **2006**, *128*, 2180–2181.

[35] Fernández-Rodríguez, M. A.; Shen, Q.; Hartwig, J. F. *Chem. Eur. J.* **2006**, *12*, 7782–7796.

[36] Fernandez-Rodriguez, M. A.; Hartwig, J. F. *J. Org. Chem.* **2009**, *74*, 1663–1672.

[37] Dahl, T.; Tornøe, C. W.; Bang-Andersen, B.; Nielsen, P.; Jørgensen, M. *Angew. Chem. Int. Ed.* **2008**, *47*, 1726–1728.

[38] Bastug, G.; Nolan, S. P. *J. Org. Chem.* **2013**, *78*, 9303–9308.

[39] Byeun, A.; Baek, K.; Han, M. S.; Lee, S. *Tetrahedron Lett.* **2013**, *54*, 6712–6715.

[40] Mao, J.; Jia, T.; Frensch, G.; Walsh, P. J. *Org. Lett.* **2014**, *16*, 5304–5307.

[41] Saravanan, P.; Anbarasan, P. *Org. Lett.* **2014**, *16*, 848–851.

[42] Tian, H.; Zhu, C.; Yang, H.; Fu, H. *Chem. Commun.* **2014**, *50*, 8875–8877.

[43] Kwong, F. Y.; Buchwald, S. L. *Org. Lett.* **2002**, *4*, 3517–3520.

[44] Bates, C. G.; Gujadhur, R. K.; Venkataraman, D. *Org. Lett.* **2002**, *4*, 2803–2806.

[45] Zhao, P.; Yin, H.; Gao, H.; Xi, C. *J. Org. Chem.* **2013**, *78*, 5001–5006.

[46] Singh, N.; Singh, R.; Raghuvanshi, D. S.; Singh, K. N. *Org. Lett.* **2013**, *15*, 5874–5877.

[47] Venkanna, G. T.; Arman, H. D.; Tonzetich, Z. J. *ACS Catal.* **2014**, *4*, 2941–2950.

[48] Percec, V.; Bae, J.-Y.; Hill, D. H. *J. Org. Chem.* **1995**, *60*, 6895–6903.

[49] Zhang, Y.; Ngeow, K. C.; Ying, J. Y. *Org. Lett.* **2007**, *9*, 3495–3498.

[50] Taniguchi, N. *J. Org. Chem.* **2004**, *69*, 6904–6906.

[51] Xu, X.-B.; Liu, J.; Zhang, J.-J.; Wang, Y.-W.; Peng, Y. *Org. Lett.* **2013**, *15*, 550–553.

[52] Zhang, C.-P.; Vicic, D. A. *J. Am. Chem. Soc.* **2012**, *134*, 183–185.

[53] Timpa, S. D.; Pell, C. J.; Ozerov, O. V. *J. Am. Chem. Soc.* **2014**, *136*, 14772–14779.

[54] Huang, Y.-J.; Lu, S.-Y.; Yi, C.-L.; Lee, C.-F. *J. Org. Chem.* **2014**, *79*, 4561–4568.

[55] Prasad, C. D.; Balkrishna, S. J.; Kumar, A.; Bhakuni, B. S.; Shrimali, K.; Biswas, S.; Kumar, S. *J. Org. Chem.* **2013**, *78*, 1434–1443.

[56] Varun, B. V.; Prabhu, K. R. *J. Org. Chem.* **2014**, *79*, 9655–9668.

[57] Ravi, C.; Chandra Mohan, D.; Adimurthy, S. *Org. Lett.* **2014**, *16*, 2978–2981.

[58] Yonova, I. M.; Osborne, C. A.; Morrissette, N. S.; Jarvo, E. R. *J. Org. Chem.* **2014**, *79*, 1947–1953.

[59] Cheng, J.-H.; Ramesh, C.; Kao, H.-L.; Wang, Y.-J.; Chan, C.-C.; Lee, C.-F. *J. Org. Chem.* **2012**, *77*, 10369–10374.

[60] Wagner, A. M.; Sanford, M. S. *J. Org. Chem.* **2014**, *79*, 2263–2267.

[61] Zhang, L.; Wang, H.; Yang, B.; Fan, R. *Org. Chem. Front.* **2014**, *1*, 1055–1057.

[62] Nguyen, T. B.; Ermolenko, L.; Retailleau, P.; Al-Mourabit, A. *Angew. Chem. Int. Ed.* **2014**, *53*, 13808–13812.

[63] Alla, S. K.; Sadhu, P.; Punniyamurthy, T. *J. Org. Chem.* **2014**, *79*, 7502–7511.

[64] Wang, Y.; Zhao, F.; Chi, Y.; Zhang, W.-X.; Xi, Z. *J. Org. Chem.* **2014**, *79*, 11146–11154.

[65] Danilkina, N. A.; Kulyashova, A. E.; Khlebnikov, A. F.; Brase, S.; Balova, I. A. *J. Org. Chem.* **2014**, *79*, 9018–9045.

[66] Zhao, X.; Zhang, L.; Li, T.; Liu, G.; Wang, H.; Lu, K. *Chem. Commun.* **2014**, *50*, 13121–13123.

[67] Saidi, O.; Marafie, J.; Ledger, A. E. W.; Liu, P. M.; Mahon, M. F.; Kociok-Kohn, G.; Whittlesey, M. K.; Frost, C. G. *J. Am. Chem. Soc.* **2011**, *133*, 19298–19301.

[68] Zhao, X.; Dimitrijevic, E.; Dong, V. M. *J. Am. Chem. Soc.* **2009**, *131*, 3466–3467.

[69] Cacchi, S.; Fabrizi, G.; Goggiamani, A.; Parisi, L. M.; Bernini, R. *J. Org. Chem.* **2004**, *69*, 5608–5614.

[70] Umierski, N.; Manolikakes, G. *Org. Lett.* **2013**, *15*, 188–191.

[71] Okauchi, T.; Kuramoto, K.; Kitamura, M.; *Synlett*, **2010**, 2891–2894.

[72] Bates, C. G.; Saejueng, P.; Doherty, M. Q.; Venkataraman, D. *Org. Lett.* **2004**, *6*, 5005–5008.

[73] Kabir, M. S.; Lorenz, M.; Van Linn, M. L.; Namjoshi, O. A.; Ara, S.; Cook, J. M. *J. Org. Chem.* **2010**, *75*, 3626–3643.

[74] Li, G. Y. *J. Org. Chem.* **2002**, *67*, 3643–3650.

[75] Xu, H.; Gu, S.; Chen, W.; Li, D.; Dou, J. *J. Org. Chem.* **2011**, *76*, 2448–2458.

[76] Ranjit, S.; Duan, Z.; Zhang, P.; Liu, X. *Org. Lett.* **2010**, *12*, 4134–4136.

[77] McDonald, J. W.; Corbin, J. L.; Newton, W. E. *Inorg. Chem.* **1976**, *15*, 2056–2061.

[78] Kuniyasu, H.; Ogawa, A.; Sato, K.-I.; Ryu, I.; Kambe, N.; Sonoda, N. *J. Am. Chem. Soc.* **1992**, *114*, 5902–5903.

[79] Xu, Q.; Shen, R.; Ono, Y.; Nagahata, R.; Shimada, S.; Goto, M.; Han, L.-B. *Chem. Commun.* **2011**, *47*, 2333–2335.

[80] Kondoh, A.; Yorimitsu, H.; Oshima, K. *Org. Lett.* **2007**, *9*, 1383–1385.

[81] Ananikov, V. P.; Orlov, N. V.; Beletskaya, I. P. *Organometallics* **2006**, *25*, 1970–1977.

[82] Yang, J.; Sabarre, A.; Fraser, L. R.; Patrick, B. O.; Love, J. A. *J. Org. Chem.* **2009**, *74*, 182–187.

[83] Cao, C.; Fraser, L. R.; Love, J. A. *J. Am. Chem. Soc.* **2005**, *127*, 17614–17615.

[84] Sabarre, A.; Love, J. *Org. Lett.* **2008**, *10*, 3941–3944.

[85] Palacios, L.; Artigas, M. J.; Polo, V.; Lahoz, F. J.; Castarlenas, R.; Perez-Torrente, J. J.; Oro, L. A. *ACS Catal.* **2013**, *3*, 2910–2919.

[86] Di Giuseppe, A.; Castarlenas, R.; Perez-Torrente, J. J.; Crucianelli, M.; Polo, V.; Sancho, R.; Lahoz, F. J.; Oro, L. A. *J. Am. Chem. Soc.* **2012**, *134*, 8171–8183.

[87] Weiss, C. J.; Marks, T. J. *J. Am. Chem. Soc.* **2010**, *132*, 10533–10546.

[88] Kondoh, A.; Takami, K.; Yorimitsu, H.; Oshima, K. *J. Org. Chem.* **2005**, *70*, 6468–6473.

[89] Ishii, A.; Aoki, Y.; Nakata, N. *J. Org. Chem.* **2014**, *79*, 7951–7960.

[90] Jiang, Q.; Xu, B.; Jia, J.; Zhao, A.; Zhao, Y.-R.; Li, Y.-Y.; He, N.-N.; Guo, C.-C. *J. Org. Chem.* **2014**, *79*, 7372–7379.

[91] Singh, R.; Raghuvanshi, D. S.; Singh, K. N. *Org. Lett.* **2013**, *15*, 4202–4205.

[92] Rokade, B. V.; Prabhu, K. R. *J. Org. Chem.* **2014**, *79*, 8110–8117.

[93] Nurhanna Riduan, S.; Ying, J. Y.; Zhang, Y. *Org. Lett.* **2012**, *14*, 1780–1783.

[94] Li, H.-S.; Liu, G. *J. Org. Chem.* **2014**, *79*, 509–516.

[95] Li, X.; Shi, X.; Fang, M.; Xu, X. *J. Org. Chem.* **2013**, *78*, 9499–9504.

[96] Xu, Q.-L.; Dai, L.-X.; You, S.-L. *Org. Lett.* **2010**, *12*, 800–803.

[97] Kuniyasu, H.; Ogawa, A.; Miyazaki, S.-I.; Ryu, I.; Kambe, N.; Sonoda, N. *J. Am. Chem. Soc.* **1991**, *113*, 9796–9803.

[98] Arisawa, M.; Yamaguchi, M. *Org. Lett.* **2001**, *3*, 763–764.

[99] Witulski, B.; Alayrac, C. *Sci. Synth.* **2005**, *24*, 797–819.

[100] Ziegler, G. R.; Welch, C. A.; Orzech, C. E.; Kikkawa, S.; Miller, S. I. *J. Am. Chem. Soc.* **1963**, *85*, 1648–1651.

[101] Riddell, N.; Tam, W. *J. Org. Chem.* **2006**, *71*, 1934–1937.

[102] Kabanyane, S. T.; MaGee, D. I. *Can. J. Chem.* **1992**, *70*, 2758–2763.

[103] Frei, R.; Wodrich, M. D.; Hari, D. P.; Borin, P.-A.; Chauvier, C.; Waser, J. *J. Am. Chem. Soc.* **2014**, *136*, 16563–16573.

[104] Tykwinski, R. R.; Williamson, B. L.; Fischer, D. R.; Stang, P. J.; Arif, A. M. *J. Org. Chem.* **1993**, *58*, 5235–5237.

[105] Chen, C. C.; Waser, J. *Org. Lett.* **2015**, *17*, 736–739.

[106] Chen, C.; Chu, L.; Qing, F.-L. *J. Am. Chem. Soc.* **2012**, *134*, 12454–12457.

[107] Gray, V. J.; Cuthbertson, J.; Wilden, J. D. *J. Org. Chem.* **2014**, *79*, 5869–5874.

6

SYNTHESIS OF ORGANOBORONIC ACIDS, ORGANOBORONATES, AND RELATED COMPOUNDS

6.1 SYNTHESIS OF ALKYLBORONATES AND RELATED COMPOUNDS THROUGH THE FORMATION OF BORON–CARBON(SP³) BONDS

One of the earliest approaches to the preparation of alkylboronates entailed the treatment of trialkylborates with organolithium reagents [1]. The reactions were typically carried out at low temperatures and afforded outstanding yields (up to 98%) of the alkylboronates. This chemistry was quite successful for unfunctionalized organolithium reagents such as MeLi and BuLi; however, this approach was limited in scope due to the propensity of organolithium reagents to react with preexisting functional groups. In related work, the direct synthesis of alkylboronates from pinacolborane has been achieved through the simple addition of alkyl Grignard reagents (Scheme 6.1) [2]. The chemistry was carried out at room temperature and afforded moderate to excellent yields of the alkylboronates. Primary, secondary, and tertiary Grignard reagents were all transformed into alkylboronates. Naturally, this approach was generally not successful with substrates bearing electrophiles. It should be noted that the chemistry was redesigned into a one-pot reaction (for a single alkyl bromide) where the alkyl halide, elemental magnesium, and pinacolborane were stirred together (Scheme 6.2). This approach afforded the alkylboronate in 90% yield. An advantage to this approach was the ability to generate allylboronates using a one-pot reaction, which eliminated the need to generate the Grignard reagent in a separate step. Using this method, several allyl bromides could be cleanly converted into the allylboronates without significant formation of secondary products due to Wurtz-type coupling.

The transition metal-free synthesis of an alkylboronate has been achieved using a PhMe₂Si-Bpin as the borylating agent and an alkyl bromide as the substrate (Scheme 6.3) [3]. One of the issues with this chemistry was a competing silylation reaction, which

Practical Functional Group Synthesis, First Edition. Robert A. Stockland, Jr.
© 2016 John Wiley & Sons, Inc. Published 2016 by John Wiley & Sons, Inc.

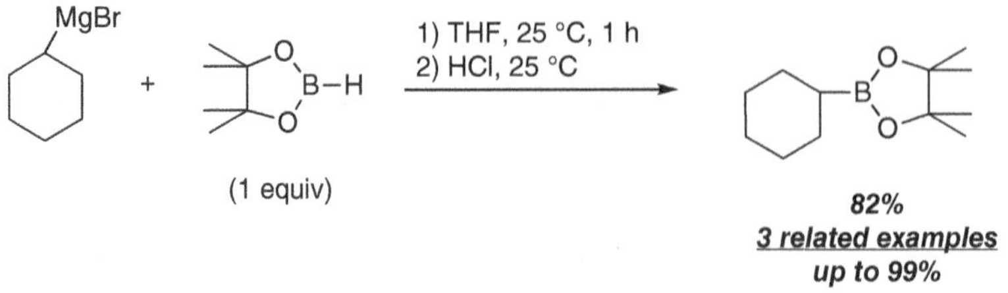

SCHEME 6.1   Alkylboronate synthesis through the treatment of pinacolborane with Grignard reagents [2].

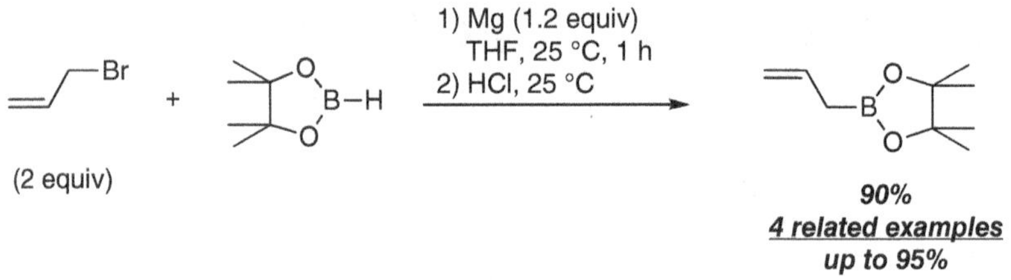

SCHEME 6.2   Synthesis of allylborane [2].

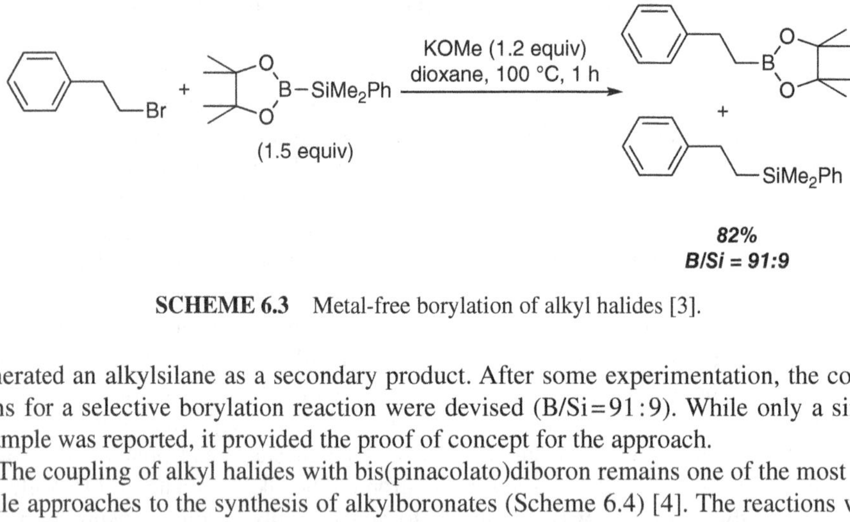

SCHEME 6.3   Metal-free borylation of alkyl halides [3].

generated an alkylsilane as a secondary product. After some experimentation, the conditions for a selective borylation reaction were devised ($B/Si = 91:9$). While only a single example was reported, it provided the proof of concept for the approach.

The coupling of alkyl halides with bis(pinacolato)diboron remains one of the most versatile approaches to the synthesis of alkylboronates (Scheme 6.4) [4]. The reactions were catalyzed by a triphenylphosphine-stabilized copper complex and occurred under mild conditions using a common base. A wide range of functional groups including esters, nitriles, ketones, amides, and vinyl groups were all successfully transformed into alkylboronates. Primary and secondary alkyl halides as well as cyclic and acyclic substrates were susceptible to the copper chemistry. This is an attractive approach to the synthesis of alkylboronates given the broad substrate scope, mild conditions, and remarkable functional group tolerance. In related work, a XantPhos-supported copper complex catalyzed the borylation of unactivated alkyl halides at room temperature [5] as well as the borylative cyclization of unactivated vinyl halides [6]. In addition to the copper catalysts, a nickel complex promoted the borylation of alkyl halides under mild conditions [7]. This nickel-catalyzed process is particularly useful since it successfully borylated tertiary alkyl halides. It should also be noted that there are several reports on the direct borylation of alkanes through metal-mediated C—H borylation reactions [8–10].

A zinc-catalyzed borylation of unactivated alkyl halides has been achieved (Scheme 6.5) [11]. Simple zinc(II) chloride was used as the metal precursor, while a moderately bulky NHC served as the supporting ligand. This chemistry was carried out at room temperature and afforded moderate to excellent yields of the alkylboronates. Similar to the copper chemistry described earlier, the zinc-catalyzed chemistry successfully functionalized a range of substrates including cyclic and acyclic alkyl halides. This system was also able to borylate several tertiary alkyl halides, although there was some substrate specificity noted by the authors. It is also noteworthy that the authors carried out the borylation reaction under an atmosphere of air in a sealed vial and still obtained a significant amount of the alkylboronate.

The addition of B_2pin_2 to activated alkenes is an efficient way to generate alkylboronates. One version of this reaction entailed the use of a platinum complex to catalyze the addition, which generated remarkably high yields of the boronates when heated (Scheme 6.6) [12, 13]. While there was some substrate specificity, this was an effective approach to the synthesis of alkylboronate compounds from Michael acceptors. It should be noted that a chiral imidazolinium salt catalyzed the enantioselective borylation of acyclic enones with excellent selectivity (up to er = 98 : 2 [14]). The copper-catalyzed borylation of cyclopropenes has also been reported using bisphosphine as the chiral ligand (Schemes 6.7 and 6.8) [15, 16] in

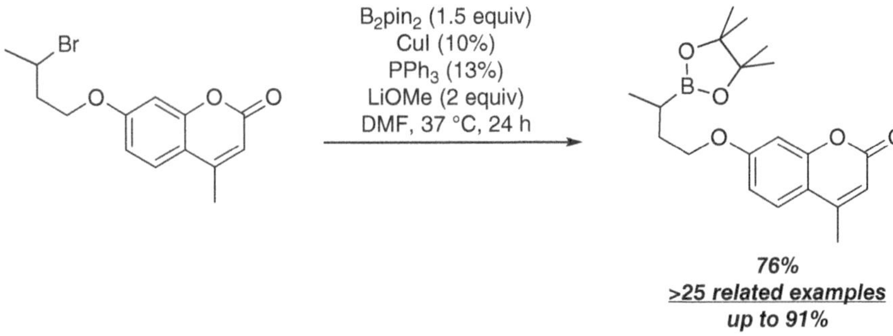

SCHEME 6.4 Copper-catalyzed borylation of alkyl halides [4].

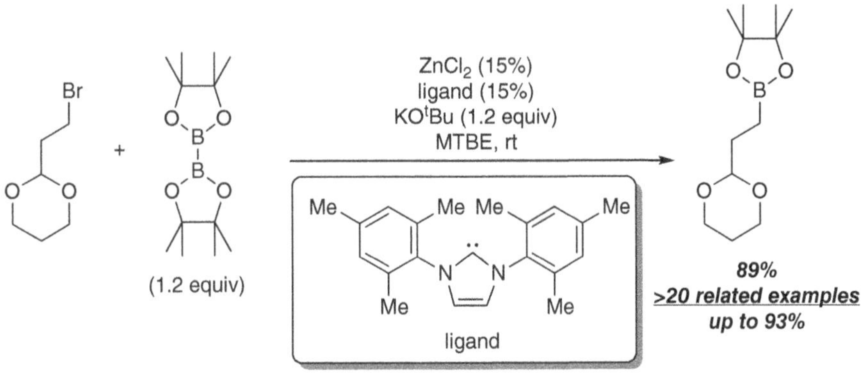

SCHEME 6.5 Zinc-catalyzed conversion of alkyl halides into alkylboronates [11].

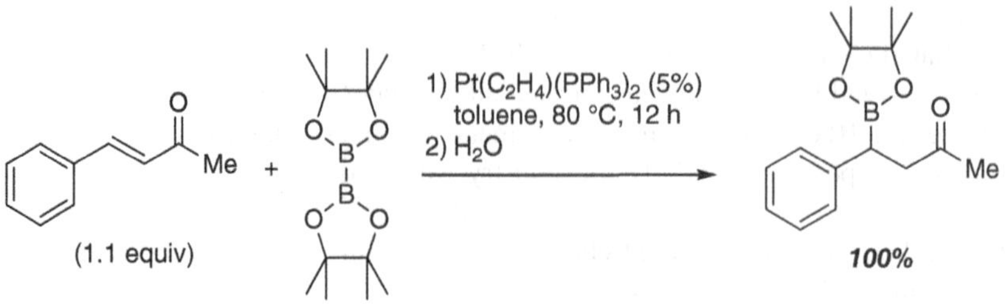

SCHEME 6.6   Platinum-catalyzed borylation of activated alkenes [12, 13].

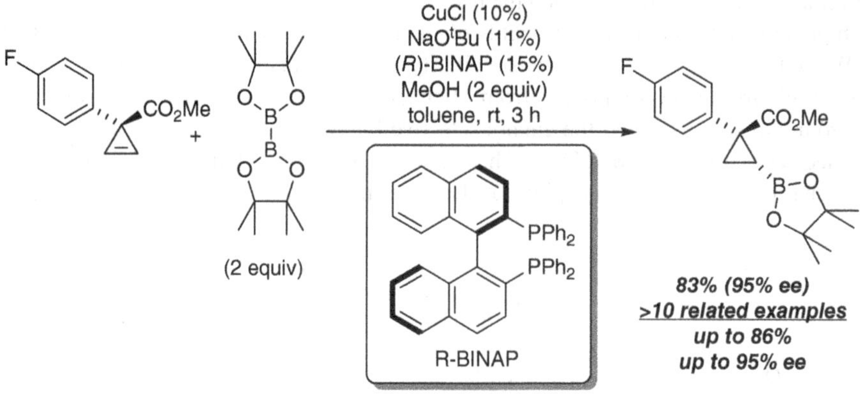

SCHEME 6.7   Copper-catalyzed borylation of functionalized cyclopropenes [15].

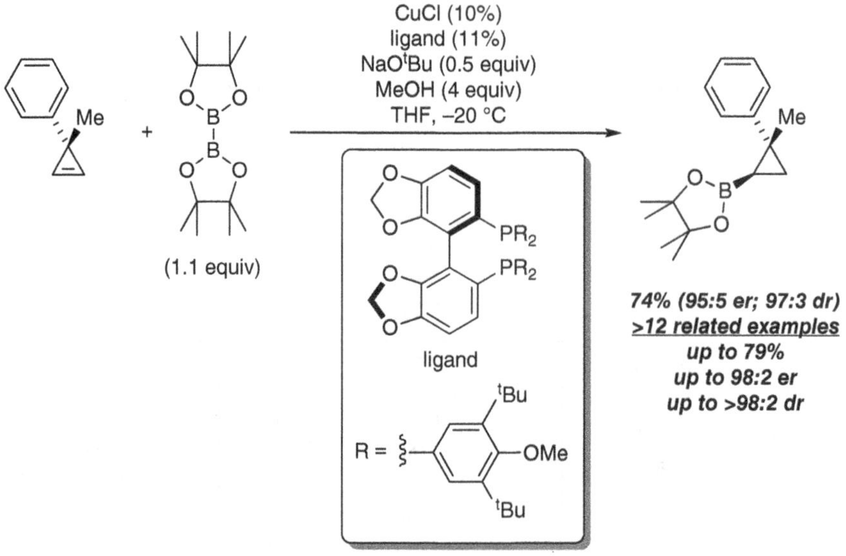

SCHEME 6.8   Asymmetric borylation of unactivated cyclopropenes [16].

addition to the enantioselective borylation of cinnamonitrile and related esters [17]. This is an ongoing area of research, and several excellent reviews have appeared [18, 19].

The 1,4-hydroboration of conjugated dienes is an attractive approach to the preparation of allylboronates. These reactions are often catalyzed by transition metal complexes [20] and can be tuned to generate high yields of the desired compounds under very mild conditions while minimizing the competing 1,2-additions. In addition to palladium- and rhodium-promoted reactions, a nickel-catalyzed process has been reported [21]. Ni(cod)$_2$ was used as the nickel source and tricyclohexylphosphine served as the solubilizing/stabilizing ligand. Treating a range of conjugated dienes with pinacolborane at room temperature for a few hours afforded excellent yields of the allylboronates. Instead of isolating the organoboron compounds, the authors converted them into allylic alcohols. One of the practical aspects of this approach was the ability to selectively generate the Z-isomers while minimizing the formation of any 1,2-hydroboration products.

Several iron-catalyzed versions of this reaction have been reported [22, 23]. One of the earliest reports used an iminopyridine-supported iron catalyst to promote the addition of pinacolborane to conjugated dienes (Scheme 6.9) [23]. The reaction conditions were extremely mild (room temperature), and a relatively low iron loading was needed (4%) to effect the transformation. The addition reaction was highly selective for the formation of the linear isomer and was stereoselective for the generation of the E-isomer. In addition to the hydroboration chemistry, the authors also designed a one-pot approach for the synthesis of allylic alcohols by combining this B–H addition chemistry with an oxidation step. In related work, a slight modification to the supporting ligand and reaction composition enabled the iron-catalyzed synthesis of secondary allylboronates from conjugated dienes (Scheme 6.10) [22].

The palladium-catalyzed conversion of allylic chlorides into allylboronates has been achieved using ligand-free palladium catalysts (Scheme 6.11) [24]. Developing such procedures was attractive from a practical perspective since there are no added ligands that

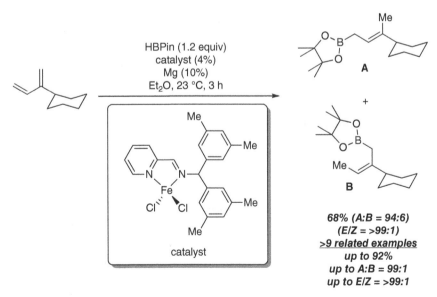

SCHEME 6.9   Iron-catalyzed hydroboration of conjugated dienes [23].

need to be removed during the purification steps. The reaction conditions were mild, and a range of allylic chlorides were converted into allylboronates in good yield. Of particular note was the ability to transform allylic chlorides bearing vinyl chloride groups without significant formation of side products resulting from oxidative addition of the vinyl chloride to the palladium center. In related work, the nickel-catalyzed conversion of allylic acetates into allylboronates was achieved using a catalyst system comprised of Ni(cod)$_2$ and tricyclohexylphosphine (Scheme 6.12) [24].

The formation of branched allylboronates followed by a 1,3-borotropic shift generated a range of linear allylboronates (Scheme 6.13) [25]. The two-step reaction was highly selective

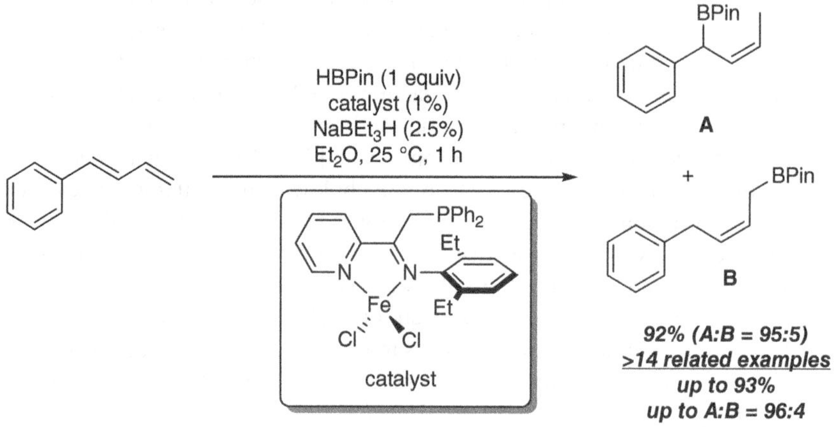

SCHEME 6.10 Iron-catalyzed synthesis of allylboronates [22].

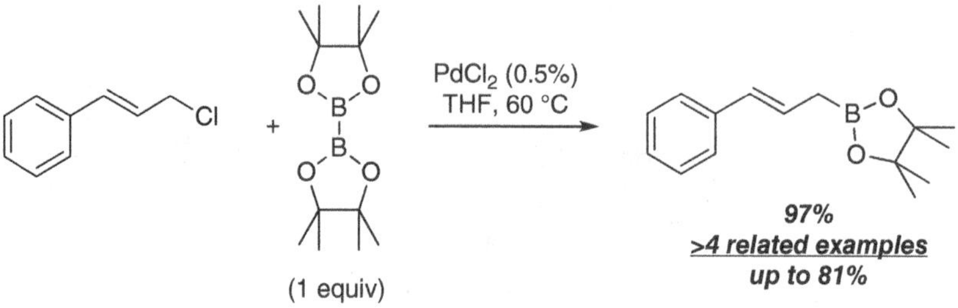

SCHEME 6.11 Palladium-catalyzed borylation of allylic chlorides [24].

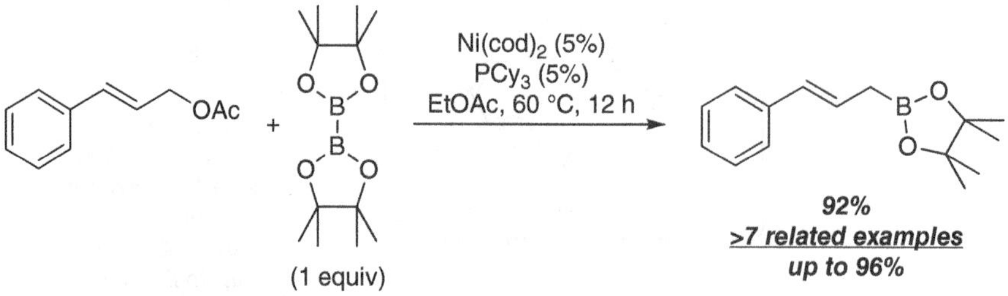

SCHEME 6.12 Nickel-catalyzed borylation of allylic acetates [24].

for the formation of the *E*-isomer (up to >95 : 5) and generated moderate to high yields of the allylboronates. It should be noted that an ester was successfully borylated using this approach.

The enantioselective synthesis of chiral allylboronates has been achieved through a copper-catalyzed reaction (Scheme 6.14) [26]. The catalyst system was comprised of a simple copper salt along with a resolved bisphosphine ligand. After screening a range of chiral bisphosphine ligands, the (*R,R*)-BenzP* (shown in Scheme 6.14) was the most effective. Curiously, Duphos and Segphos (popular bisphosphine ligands) were considerably less effective for this chemistry. Using the optimized catalyst system, a range of allylic acetals

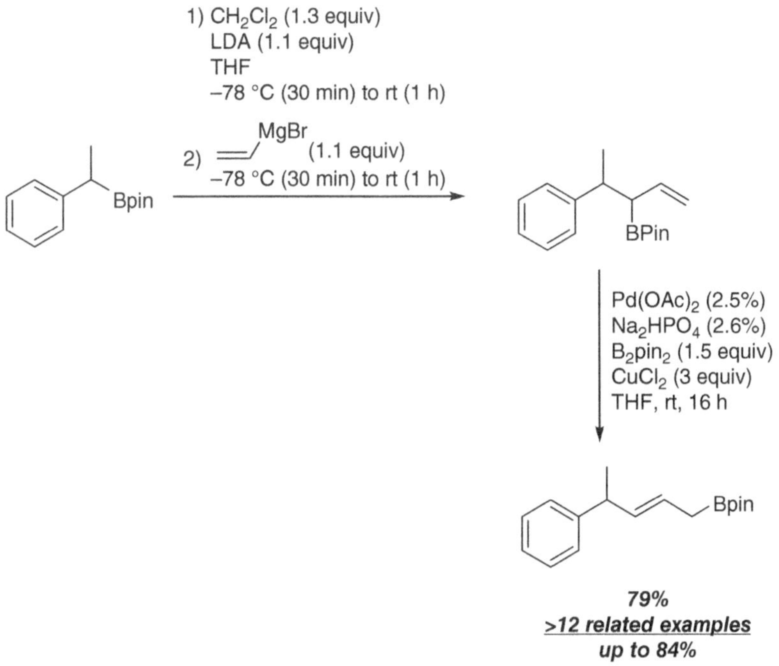

SCHEME 6.13   Synthesis and rearrangement of allylboronates [25].

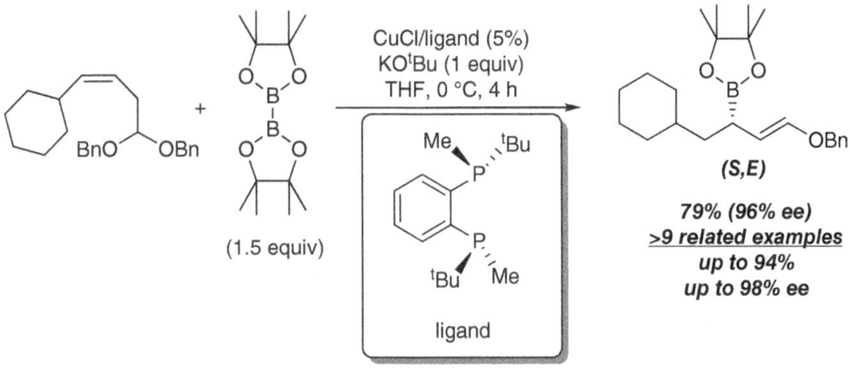

SCHEME 6.14   One-pot synthesis of allylboronates [2].

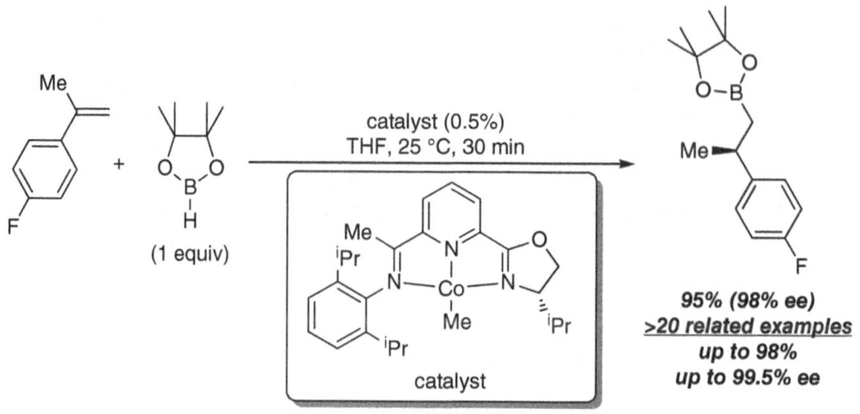

SCHEME 6.15   Cobalt-catalyzed asymmetric synthesis of alkylboronates [27].

were borylated under very mild conditions to generate the chiral allylboronates in excellent yields with outstanding selectivity.

The addition of B–H bonds across unactivated alkenes is an effective approach to the synthesis of carbon–boron bonds. A cobalt-catalyzed version of this reaction was reported using pinacolborane as the boron precursor (Scheme 6.15) [27]. While the majority of examples were based upon a styrene core, several unconjugated alkenes were successfully borylated using this approach. The atom-efficient reaction occurred under mild conditions using a low loading of the cobalt catalyst and generally afforded outstanding yields of the allylboronates with high selectivities. Given the interest in developing cobalt-based catalysts for organic transformations, as well as the high selectivity of this process, this should be a valuable approach to the synthesis of these compounds. It should be noted that a platinum-catalyzed enantioselective diboronation of vinylboronates has also been achieved using Pt(dba)$_3$ along with a chiral phosphonate as the supporting ligand [28].

MIDA boronates are an attractive class of compounds due to their stability and base-promoted reactivity. With some variations, the synthesis of these valuable compounds typically involved the addition of MIDA to a solution of the boronic acid. A tandem approach to the synthesis has been reported using the hydroboration of alkenes as the first step in the synthesis followed by the conversion of the crude alkylboronic acid into the MIDA boronate (Scheme 6.16) [29, 30]. The yields of this chemistry were typically good to excellent, and a range of functional groups were well tolerated.

Benzylnitriles can also be converted into benzylboronates (Scheme 6.17) [31]. This was a transition metal catalyzed process. A chloride-bridged rhodium dimer was used as the precursor to the active species, and a bisphosphine ligand with a large bite angle served as the supporting ligand. Although a benzylnitrile could be converted into the benzylboronate under the reaction conditions, analogous reactions using simple alkylnitriles were less successful.

A stereoselective boronate rearrangement (Aggarwal's rearrangement) is a powerful way to generate alkylboronates. The classic version of this reaction uses sec-BuLi at low temperatures (−78 °C) to promote the process; however, a modification to this procedure enabled the development of a rearrangement process that occurs without the need for such

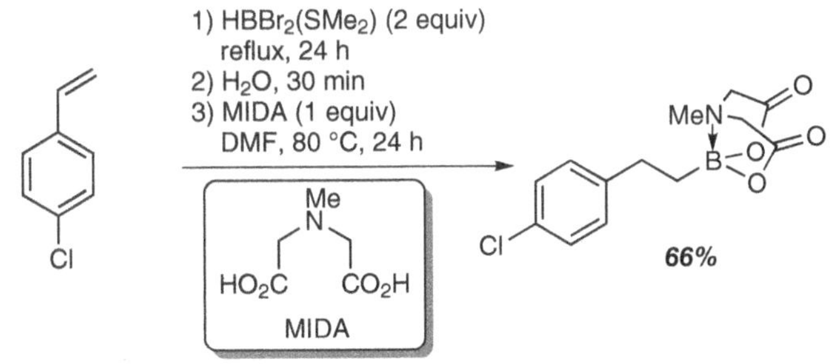

SCHEME 6.16   Synthesis of alkyl MIDA boronates [29].

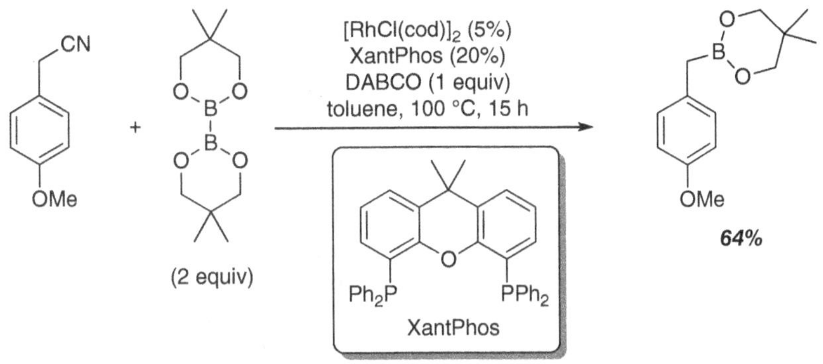

SCHEME 6.17   Rhodium-catalyzed preparation of alkylboronates from alkylnitriles [31].

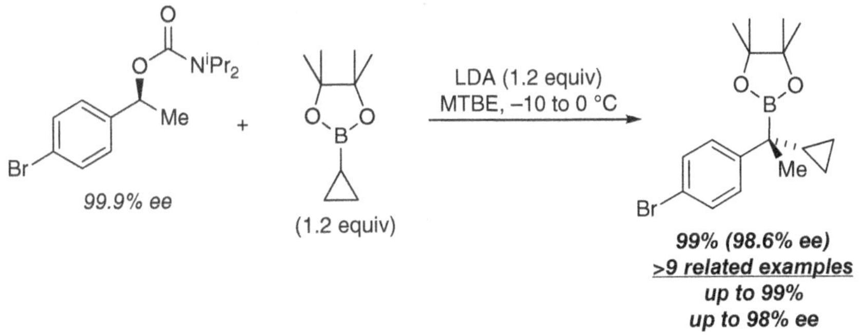

SCHEME 6.18   LDA-promoted boronate rearrangement reaction [32].

low temperatures (Schemes 6.18) [32]. The most significant modification to the classic system was the substitution of LDA for sec-BuLi. Additionally, MTBE was used as the solvent for the reaction instead of diethyl ether. Using this new system, the rearrangement reaction only needed slight cooling (−10 °C) for a successful conversion.

A series of bistrifluoromethylated alkylboronates have been generated through the reaction of arylboroxines with an excess of CF_3CHN_2 (Scheme 6.19) [33]. The key starting materials for this chemistry were arylboroxines. These substrates were relatively easy to generate from arylboronic acids through an azeotropic removal of water while heating. A range of arylboroxines were successfully used in this chemistry although ortho-substituted

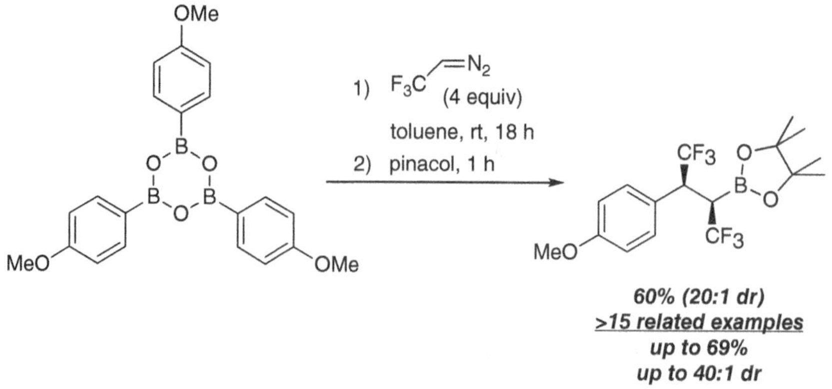

SCHEME 6.19 Application of arylboroxines to the synthesis of alkylboronates [33].

Example 6.1 Metal-Free Preparation of Trifluoromethylated Alkylboronates [33].

Chemical Safety Instructions

<u>**Before starting this synthesis, all safety, health, and environmental concerns must be evaluated using the most recent information, and the appropriate safety protocols followed. All appropriate personal safety equipment must be used. All waste must be disposed of in accordance with all current local and government regulations.**</u>

The individual performing these procedures and techniques assumes all risks and is responsible for ensuring the safety of themselves and those around them. This chemistry is not intended for the novice and should only be attempted by professionals who have been well trained in synthetic organic chemistry. The authors and Wiley assume no risk and disclaim any liabilities for any damages or injuries associated in any way with the chemicals, procedures, and techniques described herein.

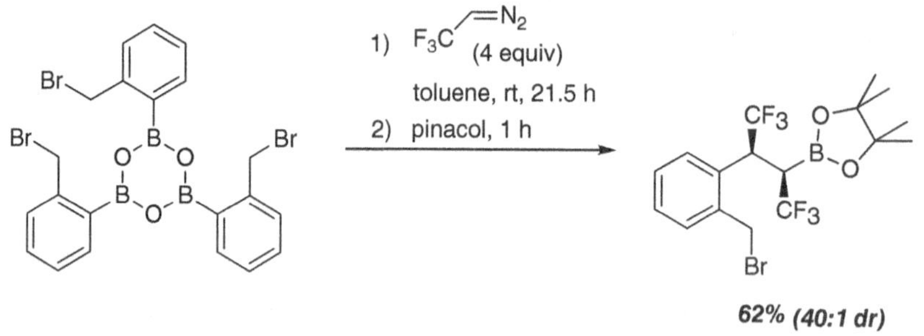

A cyclic boroxine (0.33 mmol) was added to a 20 mL Biotage microwave vial equipped with a stir bar. The vial was sealed and purged with argon three times. A solution of CF$_3$CHN$_2$ in toluene (ca. 0.5 m, ca. 4 mmol, 8 mL) was added under an argon atmosphere, and the reaction mixture was stirred at room temperature for a specified amount of time. Pinacol (124 mg, 1.05 mmol) in CH$_2$Cl$_2$ (2 mL) was then added under argon, and the reaction mixture was stirred for a further 1 h. The crude product mixture was passed through a plug of Celite, washed with CH$_2$Cl$_2$ (3×3 mL), and concentrated under vacuum. The desired product was recrystallized from a solution of the crude product in CH$_2$Cl$_2$/hexanes.

Adapted with permission from G. A. Molander, D. Ryu, *Angew. Chem. Int. Ed.,* **2014**, *53*, 14181–14185. *Copyright (2014) Wiley-VCH Verlag GmbH & Co. KGaA, Weinheim.*

substrates were sluggish. It should be noted that a bromomethyl substituent of the arylboroxines was well tolerated (Example 6.1) [33]. A number of preexisting functional groups including esters, ethers, heteroaromatics, and alkenes were tolerated, and moderate to excellent yields of the alkylboronates were obtained.

Alkyltrifluoroborates are an important class of compounds and are generally prepared by treating a solution of the alkylboronic acid with KHF_2. They can also be generated in solution by the treatment of alkylboronates with KHF_2 [34–36].

6.1.1 Troubleshooting the Synthesis of Alkylboronates and Related Compounds

While it is impossible to anticipate every issue that will arise during the preparation of alkylboronates and related compounds, the following guide will provide some suggestions in order to circumvent many common issues.

Problem/Issue/Goal	Possible Solution/Suggestion
Need a transition metal-free procedure for the synthesis of an alkylboronate	If the substrate does not contain electrophiles, try a one-pot magnesiation/borylation of aryl halides using pinacolborane [2]
Need to borylate a secondary alkyl bromide bearing a range of functional groups	The copper-catalyzed coupling of B_2pin_2 with alkyl halides has shown great functional group tolerance
Are transition metal-free syntheses of allylboronates known?	Similar to the earlier examples, if the substrate does not have electrophiles present, a noncryogenic treatment of magnesium with allyl bromide in the presence of pinacolborane is an effective route to these compounds [2]
Synthesis of allylboronates through metal-catalyzed cross-coupling	Palladium, nickel, and copper compounds are all known to promote coupling reactions for the synthesis of allylboronates [24, 26]
Can inexpensive metals be used for the cross-coupling reaction?	A zinc-catalyzed cross-coupling is known and was able to promote the borylation of a range of alkyl bromides [11]
Need to synthesize an alkylboronate from an activated alkene	Several catalysts and approaches would work for this reaction. A platinum-catalyzed version of this reaction using B_2pin_2 would be a reasonable starting point [12, 13]
Need to generate an allylboronate from a diene	An iron-catalyzed process will convert 1,3-dienes into allylboronates [22, 23]
Are asymmetric hydroboration reactions known?	Using cobalt catalysts to promote asymmetric hydroboration reactions would be a reasonable starting point [27]
Need to make an alkyl MIDA boronate	Several approaches to this reaction have been reported including a convenient one-pot hydroboration of alkenes followed by the addition of MIDA to generate the MIDA boronate [29]
Need to generate an aryltrifluoroborate salt	A convenient precursor for the direct synthesis is not readily available. Try generating an alkylboronic acid or alkylboronate and treat it with a solution of KHF_2 [37, 38]. For arylboronates, a 4 : 1 THF/water mixture was successful [38]

6.2 SYNTHESIS OF ARYLBORONATES AND RELATED COMPOUNDS THROUGH THE FORMATION OF BORON–CARBON(sp²) BONDS

One of the classic approaches to the synthesis of arylboronic acids involves the generation of an aryl Grignard or aryllithium reagent at low temperature followed by the addition of a trialkylborate [39, 40]. Although a host of arylboronates and related compounds have been generated using this chemistry, the presence of electrophiles on the precursors to the Grignard or RLi species can lead to significant amounts of unwanted secondary products. While several approaches have been devised to circumvent this issue [39], most of these methods require the operator to maintain very low temperatures. Although most synthetic laboratories are well equipped to cool reactions using dry ice/solvent baths, it would significantly increase the operational simplicity of the approach if the reactions could be carried out near room temperature. To this end, the combination of I/MgBr exchange along with the addition of a chelating ligand (bis[2-(N,N-dimethylaminoethyl)] ether) was found to change the reactivity of the aryl Grignard to such an extent that it could successfully be used near room temperature to generate arylboronic acids from substrates bearing nitriles and esters (Scheme 6.20 and Example 6.2) [41]. The chemistry was fairly rapid and generated moderate yields of the functionalized boronic acids. Even brominated heteroarenes were successfully converted into arylboronic acids using this approach. Another method for the preparation of arylboronic acids was derived from Knochel's approach to the synthesis of pinacolylboronate esters from diiodoarenes [42] and entailed the use of a noncryogenic halogen/MgBr exchange reaction along with LiCl modified Grignard reagents [43]. In addition to increased functional group tolerance, this approach also converted bulky aryl bromides into boronic acids. For example, mesitylboronic acid was generated from bromomesitylene in 91% yield. Additionally, the use of (iPrO)Bpin was found to be effective in the synthesis of functionalized arylboronate esters containing aldehydes, esters, halogens, and secondary alcohols [44].

Selective borylation of aryl versus vinyl iodides has been accomplished by metalation of o-iodo-2-(2-iodovinyl)benzene with iPrMgCl·LiCl followed by the addition of iPrOB-Pin (Scheme 6.21) [45]. This chemistry was valuable since the borylation of the aryl iodides with retention of the vinyl iodide leaves the latter for further functionalization. While only a few substrates were screened in this study, it provided the proof of concept for this line of research.

A modification to the classic preparation of arylboronates using a Grignard-type approach entailed the use of elemental magnesium to generate the organometallic reagent in solution from the aryl halide in the presence of pinacolborane (Scheme 6.22) [2]. This

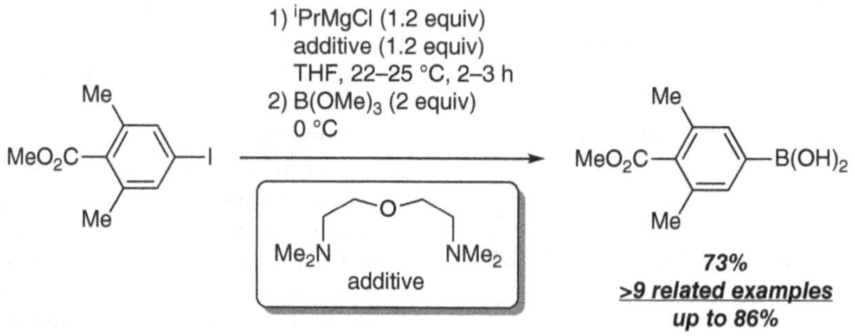

SCHEME 6.20 Synthesis of arylboronic acids using an I/MgBr exchange reaction [41].

Example 6.2   Borylation of Functionalized Aryl Iodides Using Grignard Reagents [41].

Before starting this synthesis, carefully follow the safety instructions listed in Example 6.1.

Caution! THF, hexane, and ethyl acetate are highly flammable.

Caution! Grignard reagents can react violently with moisture and should be handled and stored under an inert atmosphere.

To a solution of bis[2-(*N*,*N*-dimethylamino)ethyl] ether (0.22 mL, 1.2 mmol) in THF (5 mL) was added isopropylmagnesium chloride (0.60 mL, 1.2 mmol, 2 M solution in THF) at 15 °C. The mixture was stirred at this temperature for 20 min. Methyl 3-methoxy-4-iodobenzoate (290 mg, 1.0 mmol) was added. After the resulting mixture was stirred at 22–25 °C for 10 min, trimethylborate (0.23 mL, 2.0 mmol) was added at 0 °C. The mixture was then quenched with 0.1 N HCl and extracted with EtOAc. The extract was dried over $MgSO_4$ and concentrated. The residue was purified by crystallization in hexane to give the arylboronic acid (187 mg, 89%).

*Adapted with permission from X.-J. Wang, X. Sun, L. Zhang, Y. Xu, D. Krishnamurthy, and C. H. Senanayake, Org. Lett., **2006**, 8, 305–307. Copyright (2006) American Chemical Society.*

one-pot approach avoided the use of cryogenic conditions and was carried out at room temperature. The isolated yields were good to outstanding, and no transition metal catalyst was needed. It should also be noted that the borylation reaction could be carried out using preformed Grignard reagents as well. As anticipated, the functional group tolerance was moderate, and only a few functional groups such as ethers and halogens were shown to be successfully functionalized. However, this is a convenient approach to the synthesis of arylboronates for an appropriately designed substrate.

When sensitive functional groups were present, a successful borylation reaction can still be carried out provided the sensitive groups were protected prior to addition of the organolithium reagent. To this end, the borylation of *N*-methyl-4-bromooxindole was accomplished in a four-step process (Scheme 6.23) [46]. In the first step, the oxindole was converted into a silyl ketene aminal. The second and third steps were focused on the metalation of the intermediate through the addition of BuLi followed by borylation. The final step was regeneration of the oxindole with retention of the arylboronate fragment.

In addition to the Grignard/RLi approach to the preparation of arylboronates, transition metal-catalyzed coupling and C—H borylation reactions have grown in popularity due in part to their ability to tolerate a wide variety of functional groups that are often incompatible

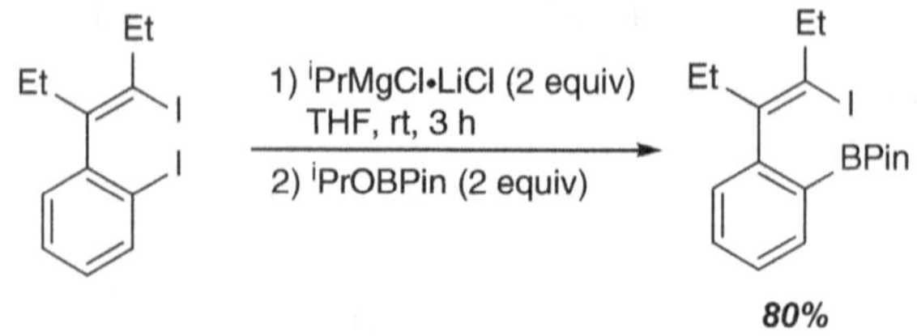

SCHEME 6.21   Chemoselective metalation/borylation [45].

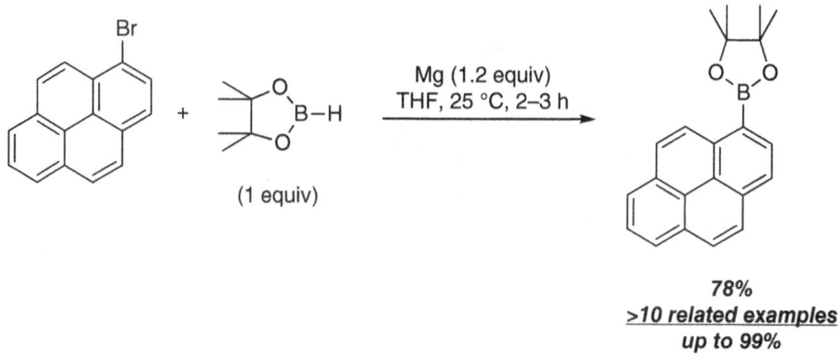

SCHEME 6.22   One-pot approach to the synthesis of arylboronates using elemental magnesium [2].

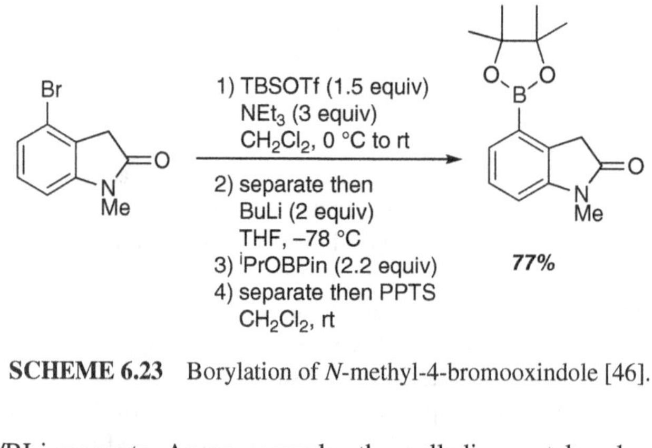

SCHEME 6.23   Borylation of *N*-methyl-4-bromooxindole [46].

with Grignard/RLi reagents. As an example, the palladium-catalyzed coupling of aryl halides with reagents such as B$_2$pin$_2$ was developed and remains one of the most popular approaches to the synthesis of arylboronates [47]. In this work, PdCl$_2$(dppf) was used as the palladium precursor, and a range of aryl halides served as the substrates. For most aryl halides screened in this study, the reaction conditions were mild, and isolated yields were typically good to excellent after heating for a few hours (1–4 h). The majority of the arylboronates were generated from aryl bromides; however, if the reactions were sluggish, aryl iodides were used. Crowded aryl halides were also successfully functionalized. As an example, iodomesitylene was successfully borylated using the catalyst system (80 °C, 24 h)

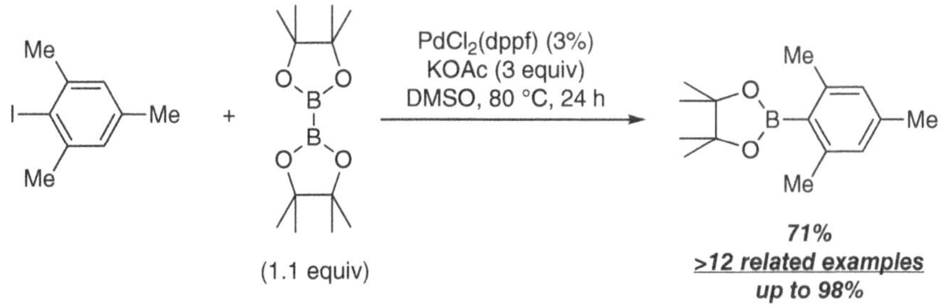

SCHEME 6.24   Palladium-catalyzed coupling of B$_2$pin$_2$ with aryl halides [47].

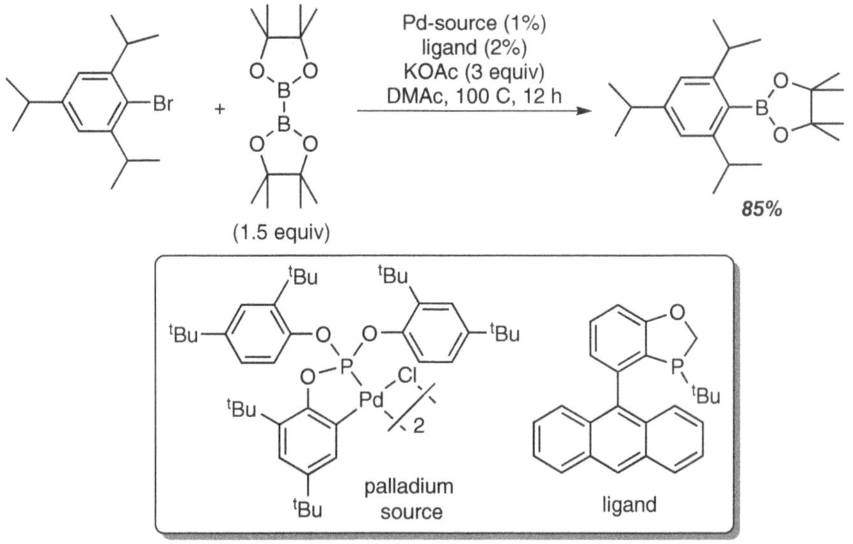

SCHEME 6.25   Palladium-catalyzed coupling of bulky aryl halides with B$_2$pin$_2$ [48].

(Scheme 6.24) [47]. Another report used a naphthalene-containing phosphine ligand and a bulky phosphite-ligated palladium precursor to promote the coupling of bulky aryl bromides such as bromo-2,4,6-triisopropylbenzene with B$_2$pin$_2$ (Scheme 6.25) [48].

The presence of strongly electron-withdrawing groups on the aryl halide is often well tolerated by palladium-catalyzed coupling chemistry. An example of this chemistry was reported using PdCl$_2$(dppf) and bis(glycolato)diboron as the boron precursor (Scheme 6.26) [49]. After heating at 100 °C for 1 h, the arylboronate was isolated in excellent yield with retention of the nitro group. In many cases, the borylation was carried out as the first step in a tandem reaction in order to generate a specific functional group [50].

In addition to the use of B$_2$pin$_2$ as the boron precursor, pinacolborane has also been used in the coupling reaction (Scheme 6.27) [51]. Pinacolborane is attractive for this chemistry since this reagent is less expensive and more atom economical; however, it has been challenging to develop practical reactions using this borylating agent. To address this issue, Buchwald developed a palladium-catalyzed route using an air-stable palladium precursor ligated by a bulky dialkylbiarylphosphine ligand. A wide variety of aryl halides bearing electron-donating and electron-withdrawing groups were well tolerated by this chemistry, and isolated yields

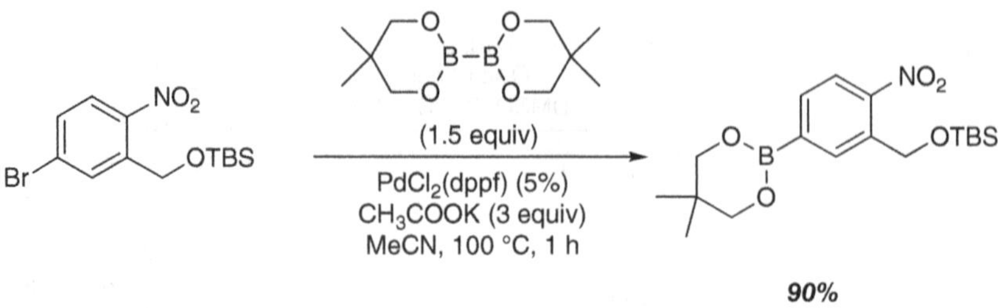

SCHEME 6.26   Palladium-catalyzed synthesis of arylboronates [49].

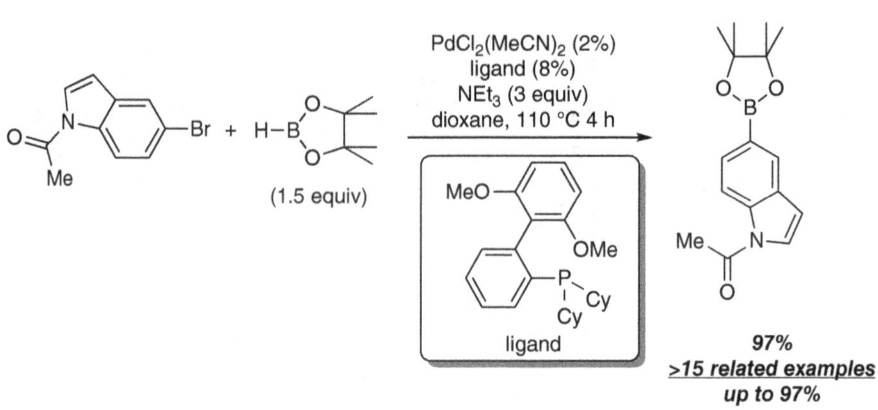

SCHEME 6.27   Palladium-catalyzed synthesis of arylboronates using pinacolborane [51].

were typically good to excellent (up to 97%). This catalyst system was also quite efficient with bulky substrates, and bromomesitylene was converted into the arylboronate (90%). Overall, this is an attractive route to the preparation of arylboronates. In addition to the palladium-catalyzed reaction, the coupling of aryl halides with pinacolborane and related borylating agents has been reported using copper [52] and nickel catalysts [53–55].

The synthesis of arylboronates from tetrahydroxydiboron ($B_2(OH)_4$) has been reported (Scheme 6.28) [56]. The first step in the coupling reaction was catalyzed by an XPhos-ligated palladium species, while the addition of a diol or MIDA comprised the second step in the process. If a boronic acid was desired, it could be isolated in high yield by a simple aqueous workup followed by a hexane wash. Even a diol bearing an unprotected carboxylic acid was successfully used with retention of the carboxyl group. It should be noted that aryl chlorides generated high yields of the arylboronate species, while aryl bromides were unable to be used in the chemistry due to undesired coupling reactions. In addition, the intermediate boronate species could be readily converted into the aryltrifluoroborate salt through the addition of aqueous KHF_2 [57]. This chemistry has also been extended to the synthesis of related trifluoroborates using a nickel-catalyzed coupling reaction [58] and has been used in tandem with cross-coupling chemistry to generate a number of functionalized biaryl and heterobiaryl compounds [59].

While palladium-catalyzed routes to the synthesis of arylboronates remain some of the most popular approaches, several zinc-catalyzed reactions have been reported [60, 61]. These reactions are valued since zinc is significantly less expensive than palladium and can

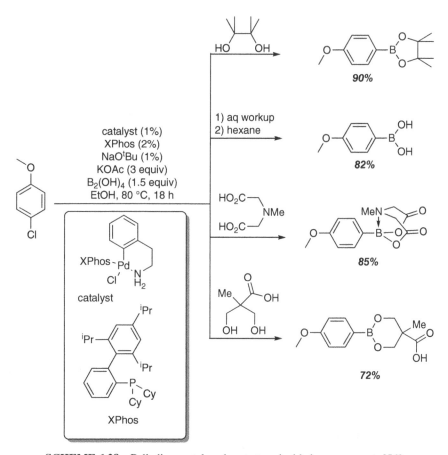

SCHEME 6.28 Palladium-catalyzed route to valuable boron reagents [56].

generate fewer health concerns. One of the catalyst systems used diethylzinc as the catalyst for the borylation reaction [60], and although successful, this reagent can be challenging to handle and store since it tends to react violently with moisture. If the laboratory is not equipped to handle this reagent, the use of zinc(II) bromide as the catalyst for the reaction might be a viable alternative (Scheme 6.29) [61]. This inorganic salt is much easier to store and handle than diethylzinc. Using $ZnBr_2$, the borylation reaction was successful under very mild conditions and tolerated a wide range of functional groups including halogens, nitriles, esters, and heterocycles. When alcohols and aldehydes were used, only small amounts of the target arylboronates were observed.

In addition to the transition metal-catalyzed conversion of aryl halides and free arenes into arylboronates, nitriles can also be used as precursors (Scheme 6.30) [31]. The reactions were catalyzed by the combination of a common rhodium salt with a large bite angle ligand. As described earlier, alkylnitriles were generally unreactive under the reaction conditions, although a benzylnitrile was successfully converted in a benzylboronate. While this was the only example of an alkylnitrile that could be functionalized, a host of arylnitriles were transformed into the corresponding arylboronates in good to high yield. A host of electron-withdrawing and electron-donating groups were well tolerated by this chemistry, and yields of the arylboronates were moderate to good.

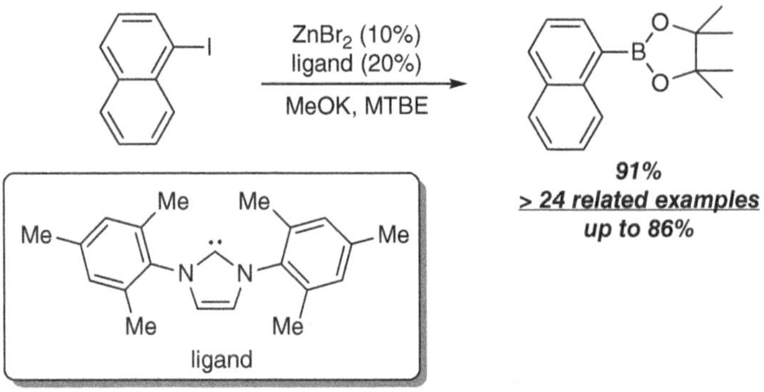

SCHEME 6.29   Zinc-catalyzed synthesis of arylboronates [61].

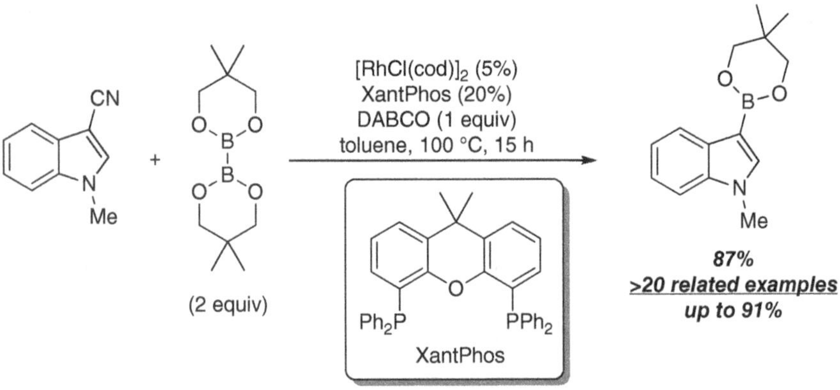

SCHEME 6.30   Conversion of arylnitriles into arylboronates [31].

The rhodium-catalyzed conversion of aryl pyridyl ethers into arylboronates has been achieved using an NHC-supported rhodium catalyst (Scheme 6.31) [62]. The main theme of this work was the use of rhodium complexes to promote the cleavage of the pyridyl ether fragment and borylation of the arene. The reaction was carried out at elevated temperatures and afforded moderate to good yields of the arylboronates. One of the most impressive aspects of this chemistry was its tolerance to a wide range of functional groups. Heteroaryl ethers as well as substrates bearing esters, amides, and even a free amine were successfully converted into arylboronates. If the substrate is appropriately functionalized, this would be a reasonable approach to the formation of arylboronates.

A metal-free route to the synthesis of arylboronates has been reported through the deaminoborylation of anilines (Scheme 6.32) [63]. This chemistry used B_2pin_2 as the source of the boron fragment and employed Sandmeyer-type chemistry to promote the carbon–boron bond-forming reaction. The functional group tolerance was outstanding, and a wide range of preexisting groups were tolerated by the borylation reaction. The reaction yields were moderate, but if a metal-free route is needed for a specific application, this is a reasonable choice. In related work, this chemistry was extended to a range of nitroarenes [64].

Ito designed a metal-free approach to the borylation of aryl halides using a silylborane as the borylating agent (Scheme 6.33) [3]. The reaction conditions were quite mild, and a host of functionalized aryl halides were converted into arylboronates. Substrates bearing esters, halogens, amides, and unprotected amines were well tolerated. The chemistry also converted bulky aryl halides into arylboronates. For example, bromo-2,4,6-triisopropylbenzene

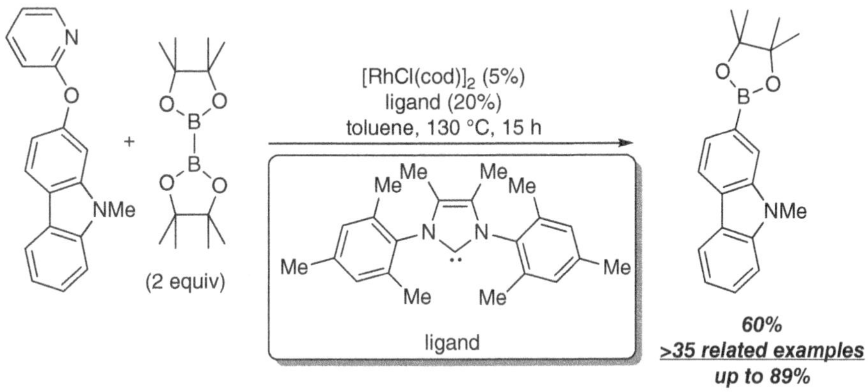

SCHEME 6.31   Rhodium-catalyzed borylation through C—O bond cleavage [62].

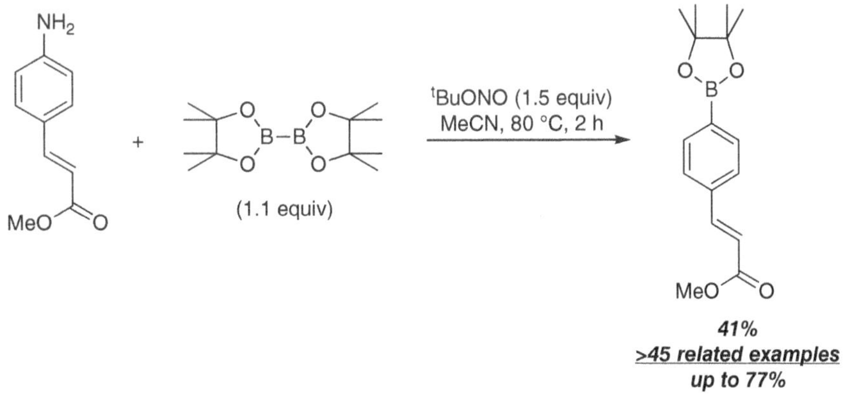

SCHEME 6.32   Metal-free deaminoborylation of arylamines [63].

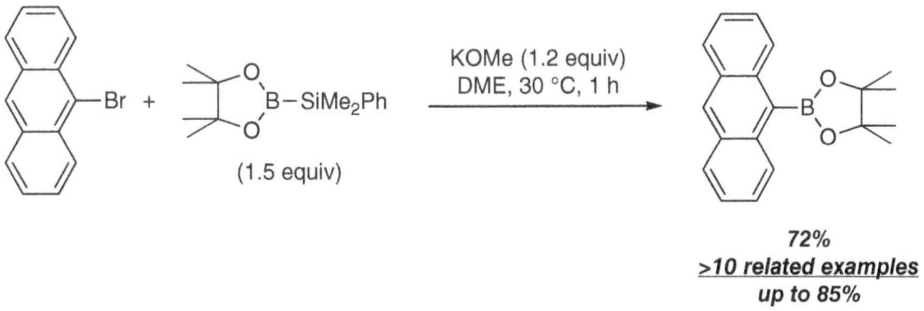

SCHEME 6.33   Metal-free synthesis of arylboronates [3].

was borylated in 82% isolated yield. It should be noted that the reaction was quite sensitive to the solvent and base. Altering the catalyst system resulted in the generation of varying amounts of vinylsilanes. For example, one of the screening reactions generated a 50 : 50 mixture of borylated and silylated products using dichloromethane as the solvent and KOMe as the base. If the general procedure was followed, this approach successfully generated the arylboronates in moderate to excellent yields with minimal contamination due to the silylated compounds.

While a number of transition metal catalysts promoted an overall C—H borylation of arenes, most examples suffered from harsh reaction conditions and poor activity. Smith and Hartwig have developed approaches to this reaction using iridium complexes as catalysts (Example 6.3) [38, 65–71]. During this study, a low catalyst loading (3%) of an air-stable iridium complex successfully borylated a range of arenes. An interesting aspect of this chemistry was the ability to borylate arenes bearing halogens. This chemistry stands in contrast to the common reactivity pattern exhibited by aryl bromides in cross-coupling chemistry. To this end, the ability to promote a C—H borylation with retention of the halogen is valuable since the resulting brominated arylboronate could then be subsequently functionalized to generate a host of derivatives. A C—H borylation process leading to the formation of arylboronates can also be achieved using palladium catalysts [72], and related C—H borylation reactions have been developed using amine directing groups [73]. It is worth noting that this iridium-catalyzed chemistry works so well that it has been combined with other transformations to provide functionalized derivatives through tandem reactions [74].

One of the advantages of the metal-catalyzed approach to the synthesis of arylboronates is the tolerance to a wide range of functional groups. This often eliminated the need to protect/deprotect sensitive functional groups during a synthesis. An example of this was recently reported using an iridium-catalyzed C—H borylation of indoles (Scheme 6.34 and Example 6.4) [75]. Even a primary alcohol and an amide were successfully retained through the boron–carbon(sp^2) bond-forming reaction.

The synthesis of diarylborinic acids and their derivatives was accomplished using a Grignard approach (Scheme 6.35 and Example 6.5) [76]. In some cases, the diarylborinic

Example 6.3   Solvent-Free Iridium-Catalyzed C—H Borylation [68].

Before starting this synthesis, carefully follow the safety instructions listed in Example 6.1.

Caution! Benzene is highly flammable.

A 25 mL flask assembled with a magnetic stirring bar, a septum inlet, and a condenser was charged with [IrCl(COD)]$_2$ (10.1 mg, 0.015 mmol), 2,2′-bipyridine (4.7 mg, 0.03 mmol), and bis(pinacolato)diboron (254.0 mg, 1.0 mmol) and then flushed with nitrogen. The arene (60 mmol) was added, and the mixture was stirred at 80 °C for 16 h. The reaction mixture was analyzed by GC and GC/mass spectroscopy. The product was extracted with benzene, washed with brine, and dried over MgSO$_4$. Kugelrohr distillation gave analytically pure samples (73%).

Adapted with permission from T. Ishiyama, J. Takagi, K. Ishida, N. Miyaura, N. R. Anastasi, *and* J. F. Hartwig, *J. Am. Chem. Soc.*, **2002**, *124*, 390–391. *Copyright (2002) American Chemical Society.*

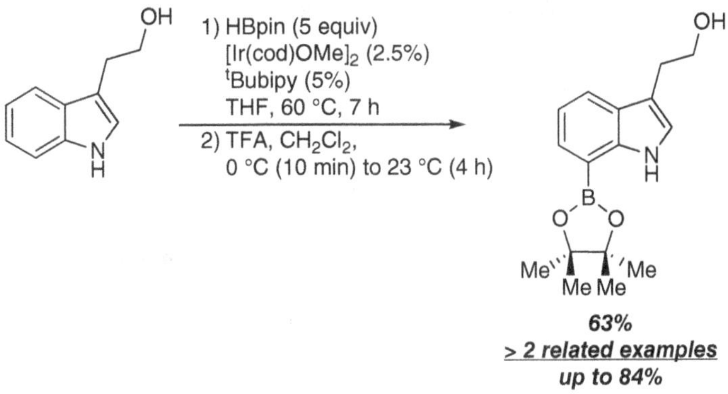

SCHEME 6.34 Iridium-catalyzed C—H borylation of indoles [75].

Example 6.4 C7-Selective Boronation of Alkylindoles [75].

Before starting this synthesis, carefully follow the safety instructions listed in Example 6.1.

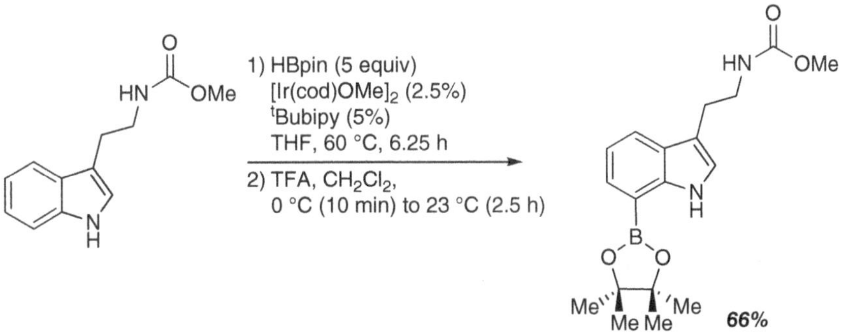

Caution! THF, hexanes, and acetone are highly flammable.

This reaction was carried out under an atmosphere of argon. A pressure tube was charged sequentially with (1,5-cyclooctadiene)(methoxy)iridium(I) dimer (5.0 mg, 7.5 μmol, 2.5 mol%), 4,4′-di-tert-butyl-2,2′-bipyridine (4.0 mg, 15 μmol, 5.0 mol%), and tryptamine methyl carbamate (67.0 mg, 307 μmol, 1 equiv). The contents of the reaction vessel were kept under an argon atmosphere. Freshly distilled anhydrous tetrahydrofuran (2 mL) was introduced to the flask via a gastight syringe to afford a dark-brown solution. Pinacolborane (218 μL, 1.50 mmol, 4.86 equiv) was added in a single portion via a gastight syringe, at which point the solution turned bright red. The pressure tube was sealed and the reaction mixture stirred at 60 °C for 6 h and 15 min and then cooled to 0 °C, whereupon anhydrous dichloromethane (3 mL) was added under an argon atmosphere. Trifluoroacetic acid (2 mL) was then added dropwise via a gastight syringe to afford an orange solution. The solution was stirred at 0 °C for 10 min and then warmed to 23 °C and stirred for 2 h and 30 min. The solution was diluted with dichloromethane (50 mL) and washed with saturated aqueous sodium bicarbonate solution (50 mL). The organic layer was dried over anhydrous sodium sulfate, filtered, and concentrated under reduced pressure. The resulting brown residue was purified by flash column chromatography on silica gel (eluent: 10% acetone, 20% dichloromethane, 70% hexanes) to provide the boronic ester (70.0 mg, 66.2%) as a white waxy solid.

Adapted with permission from R. P. Loach, O. S. Fenton, K. Amaike, D. S. Siegel, E. Ozkal, *and* M. Movassaghi, *J. Org. Chem.*, **2014**, *79*, 11254–11263. *Copyright (2014) American Chemical Society.*

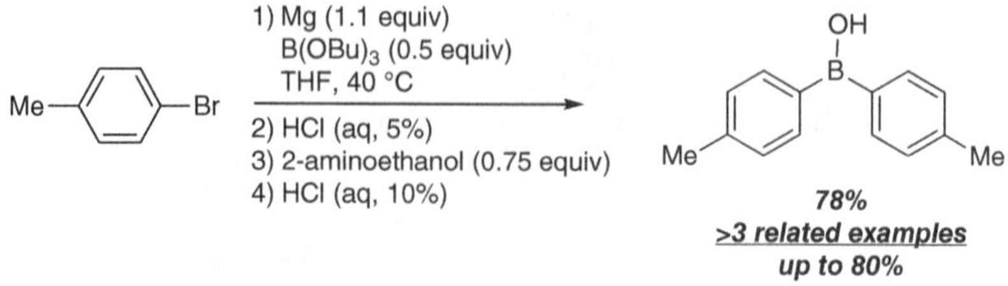

SCHEME 6.35 Preparation of diphenylborinic acid [76].

Example 6.5 Preparation of Diphenylborinic Anhydride [76].

Before starting this synthesis, carefully follow the safety instructions listed in Example 6.1.

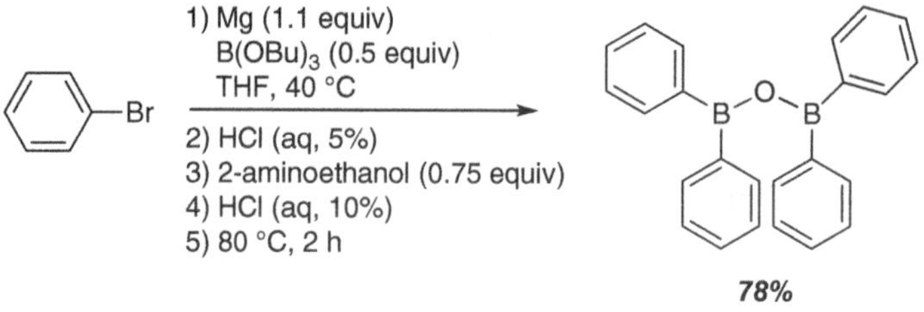

Caution! THF and ethyl acetate are highly flammable.

Caution! Grignard reagents can react violently with moisture and must be handled and stored under an inert atmosphere.

Under an N_2 atmosphere, a mixture of $B(OBu)_3$ (11.51 g, 50 mmol) and bromobenzene (15.70 g, 100 mmol) in 50 mL of THF was added dropwise to a stirred mixture of magnesium turnings (2.64 g, 110 mmol) and a small crystal of I_2 in THF (50 mL) at 40 °C over a period of 30 min. The reaction was maintained at 40 °C for an additional 2 h and then hydrolyzed by the addition of 100 mL of 5% HCl (aq) after being cooled to room temperature. The mixture was extracted with EtOAc and then concentrated to 20 mL before 2-ethanolamine (4.58 g, 75 mmol) was added. The resulting solution was stirred at room temperature for 2 h and then washed with water. The organic layer was concentrated under vacuum to obtain the crude product of 2-aminoethoxydiphenyl borate, which was recrystallized in ethanol and then acidified with 50 mL of 10% HCl (aq). The mixture was extracted with EtOAc, washed with brine, dried over Na_2SO_4, filtered, and concentrated under vacuum to afford diphenylborinic acid, which was further converted into anhydride by heating at 80 °C for 2 h under vacuum.

Adapted with permission from X. Chen, H. Ke, Y. Chen, C. Guan, *and* G. Zou, *J. Org. Chem.,* **2012**, *77*, *7572–7578. Copyright (2012) American Chemical Society.*

acids were challenging to purify, and they were converted into more stable species through the addition of 2-aminoethanol. This was a convenient protecting group as it was easily removed by the addition of HCl. Once isolated, they could be converted into diphenyl boronic anhydrides by simply heating at 80 °C under vacuum. The anhydrides and boronic acids were successfully used in palladium-catalyzed cross-coupling reactions.

Aryltrifluoroborates are extremely valuable compounds for synthetic applications, and the vast majority of aryltrifluoroborates have been prepared by treatment of the arylboronic acids with KHF_2 [37]. This method typically generated good to excellent yields of the aryltrifluoroborate salts and was tolerant to a wide range of functional groups. Even arylboronates were effectively converted into the $ArBF_3K$ species upon treatment with KHF_2 in a biphasic solvent system [38].

6.2.1 Troubleshooting the Synthesis of Arylboronates and Related Compounds

While it is impossible to anticipate every issue that will arise during the preparation of arylboronates and related compounds, the following guide will provide some suggestions in order to circumvent many common issues.

Problem/Issue/Goal	Possible Solution/Suggestion
Need a transition metal-free synthesis of an arylboronic acid	If the substrate does not contain electrophiles, the classic approach would entail the treatment of trialkylborates with Grignard/RLi reagents followed by hydrolysis [39, 40]
Need a transition metal-free synthesis of an arylboronate	The conversion of aryl bromides into arylboronates has been achieved using a silylboronate [3]
	If the substrate does not contain electrophiles, try a one-pot magnesiation/borylation of aryl halides using pinacolborane [2]. If the substrates do contain sensitive functional groups, try adding a reagent that will modify the reactivity of the Grignard reagent [41, 42]
The substrate contains both vinyl and aryl iodides. The aryl iodide needs to be preferentially borylated with retention of the vinyl iodide	This has been achieved using a modified Grignard reagent [45]
Can a Grignard approach to the synthesis of arylboronates be used under noncryogenic conditions?	This has been achieved by generating the Grignard reagent in the presence of the pinacolborane [2, 41]
Can metal-catalyzed cross-coupling be used to generate arylboronates?	Several approaches to this synthesis are known. The palladium-catalyzed version of this reaction is especially attractive [47, 48] and is able to borylate very bulky aryl halides [48]
Can tetrahydroxydiboron be used as a precursor?	The palladium-catalyzed coupling of this reagent with a range of aryl halides is an extremely efficient approach to the synthesis of arylboronates and related compounds [56]. Additionally, MIDA boronates can be generated using this approach
Can other metals be used in the cross-coupling reaction?	A zinc-catalyzed process has been reported and generates good to excellent yields of the arylboronates [61]
Is it possible to design a system for the conversion of ethers into arylboronates?	Although a few successful systems are known [62], general processes are still quite rare
The synthesis needs to start from an aniline derivative	An interesting one-pot conversion of anilines into arylboronates has been developed [63]
Can an unfunctionalized arene be borylated?	Several C—H borylation processes have been developed. One of the most practical is an iridium-catalyzed reaction [65, 68]
Is the C—H borylation reaction tolerant to existing aryl halides?	The iridium-catalyzed reaction enables the borylation of arenes with retention of existing aryl bromides that can be functionalized in subsequent reactions [68]

(Continued)

(Continued)

Problem/Issue/Goal	Possible Solution/Suggestion
Can the iridium chemistry tolerate substrates with primary alcohols?	Yes, a version of the iridium-catalyzed C—H borylation is tolerant to the presence of primary alcohols [75]
Need to generate an aryltrifluoroborate salt	A convenient precursor for the direct synthesis is not readily available. Try generating an arylboronic acid or arylboronate, and treat it with a solution of KHF_2 [37, 38]. For arylboronates, a 4:1 THF/water mixture was successful [38]

6.3 SYNTHESIS OF VINYLBORONATES AND RELATED COMPOUNDS THROUGH THE FORMATION OF BORON–CARBON(SP²) BONDS

The conversion of vinyl triflates into vinylboronates has been achieved using a palladium-catalyzed process (Scheme 6.36) [77]. Bis(pinacolato)diboron was used as the boron precursor, and the catalyst was a triphenylphosphine-ligated palladium complex. The reaction conditions were mild, and gently heating vinyl triflates with the catalyst system and a slight excess of B_2pin_2 afforded good to excellent yields of the vinylboronates. The tolerance of the chemistry to bulky substrates was also very good, and vinyl halides could also be used as substrates in this chemistry.

The palladium-catalyzed synthesis of vinylboronates was reported using pinacolborane as the borylating agent (Scheme 6.37) [51]. The catalyst system was comprised of $PdCl_2(MeCN)_2$ along with a bulky dialkylbiarylphosphine as the solubilizing/stabilizing ligand. While only a single substrate was screened in this study, it provides the proof of concept that the coupling chemistry commonly used for the formation of arylboronates could be extended to the preparation of vinylboronates.

The zinc-catalyzed borylation of vinyl bromides has been achieved using a catalyst comprised of zinc(II) bromide and a carbene ligand (Scheme 6.38) [61]. The chemistry occurred under mild conditions and was moderately stereoselective.

While most of the hydroboration methods resulted in the formation of β-substituted species, a rare example of an addition reaction that formed the α-substituted vinylboronate has been reported (Example 6.6) [78]. The overall reaction was a two-step process that proceeds through the initial formation of a vinyl aluminate that was quenched by the addition of methoxy(pinacolato)borane. The reaction was highly regioselective, and less than 2% of the β-substituted product was formed. In addition to the functionalization of styrene, other arylalkynes as well as alkylalkynes were converted into α-substituted vinylboronates in high yield with outstanding selectivity.

The addition of B–H bonds across alkynes has the potential to generate a range of products due to scrambling of the regiochemistry and stereochemistry of the addition. Considerable advances have been made concerning the ability to direct the addition reaction and selectively generate a single isomer. While some catalysts and additives were selective for a single mode of addition, it was often challenging to modify the catalyst system in order to generate a different regioisomer. A rare example of this chemistry involved the development of a catalyst that could switch between a Markovnikov and anti-Markovnikov hydroboration of an alkyne through a modification of the reaction conditions (Example 6.7) [79]. Additionally, the modification was trivial and consisted of whether or not methanol was present in the reaction mixture. In the presence of methanol, the reaction

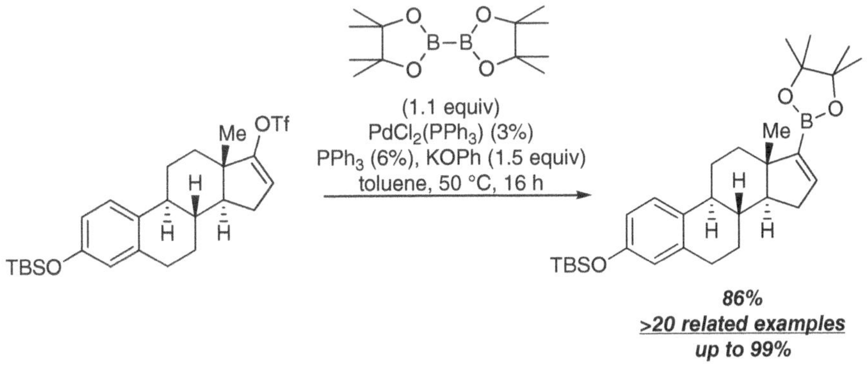

SCHEME 6.36 Palladium-catalyzed synthesis of vinylboronates from vinyl triflates [77].

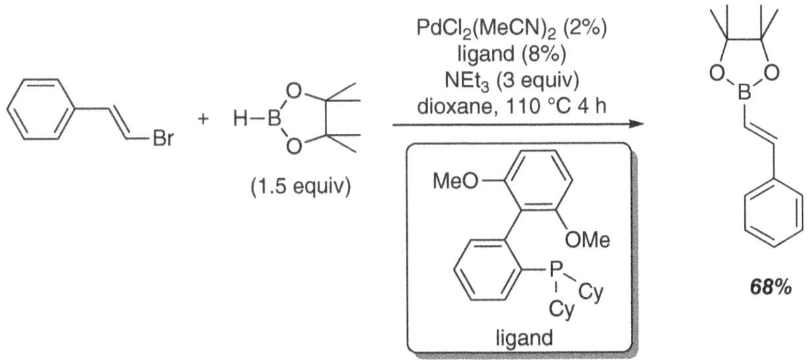

SCHEME 6.37 Palladium-catalyzed coupling of vinyl halides with pinacolborane [51].

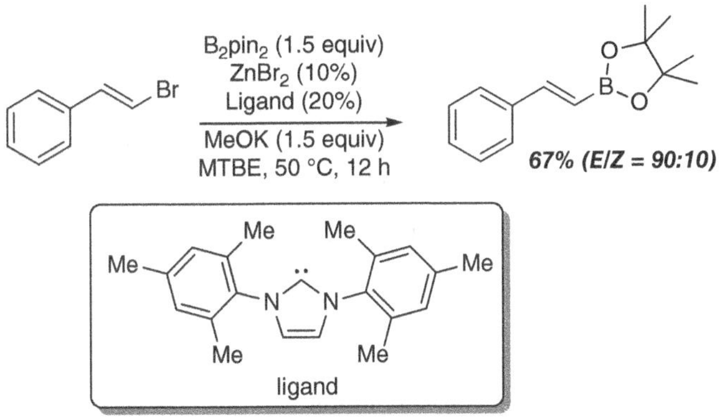

SCHEME 6.38 Zinc-catalyzed borylation of bromostyrene [61].

was selective for the formation of the β-substituted vinylboronate (anti-Markovnikov). If methanol was never added, the reaction generated the α-substituted isomer (Markovnikov). In addition to being able to control the regiochemistry, the hydroboration reaction can be combined with an alkylation step to achieve a one-pot methylboronation of alkynes. The ability to perform this reaction under air makes this process even more attractive.

Example 6.6   Nickel-Catalyzed Synthesis of α-Substituted Vinylboronates [78].

Before starting this synthesis, carefully follow the safety instructions listed in Example 6.1.

75% (A/B = >98:2)
5 related examples
up to 94%
up to A/B = >98:2

Caution! THF, ethyl acetate, and hexanes are highly flammable.

Caution! Dibal-H can react violently with moisture and must be handled and stored under an inert atmosphere.
Commercial grade 1,3-bis(diphenylphosphino)propane nickel(II) chloride (Ni(dppp)Cl$_2$, 16.3 mg, 0.0300 mmol) is placed in a flame-dried 10 mL round-bottom flask equipped with a stir bar and a refluxing condenser. The apparatus is sealed with a septum and purged with N$_2$ for approximately 10 min. Tetrahydrofuran (3.0 mL) is added through a syringe, followed by dropwise addition of dibal-H (232 μL, 1.30 mmol) at 22 °C (gas evolution occurs as dibal-H is added). The resulting black solution is allowed to cool to 0 °C (ice bath) before phenylacetylene (110 μL, 1.00 mmol) is added slowly over 5 min (reaction is exothermic). The resulting black solution is allowed to warm to 22 °C and stir for an additional 2 h. After 2 h, 2-methoxy-4,4,5,5-tetramethyl-1,3,2-dioxaborolane (MeO-Bpin; 492 μL, 3.00 mmol) is added dropwise through a syringe into the reaction solution at 0 °C (ice bath). The resulting solution is allowed to be heated to 80 °C and stir for 24 h before the reaction is quenched by dropwise addition of water (3.0 mL) at 0 °C (ice bath). The mixture is allowed to warm to 22 °C and stir for one additional hour before it is washed with Et$_2$O (5.0 mL×3). The combined organic layers are passed through a plug of anhydrous MgSO$_4$ and concentrated under vacuum to afford yellow oil, which is purified by silica gel chromatography (40/1 hexanes/ethyl acetate) to afford the desired product as colorless oil (173.0 mg, 0.752 mmol, 75% yield).

Adapted with permission from F. Gao, *and* A. H. Hoveyda, *J. Am. Chem. Soc.*, **2010**, *132*, 10961–10963. *Copyright (2010) American Chemical Society.*

It should be noted that a similar carboboration reaction has been carried out using a XantPhos-supported copper catalyst [80].

Negishi reported the regioselective and stereoselective addition of boron–halogen bonds across terminal alkynes (Scheme 6.39) [81]. This work utilized propyne as a prototypical terminal alkyne and BBr$_3$ as the boron source. The chemistry took place at low temperatures and generated the 1,2-adduct with Z-stereochemistry. Trapping of this compound through the addition of pinacol afforded the vinylboronate.

In addition to the transition metal-catalyzed conversion of arylnitriles into arylboronates, vinylnitriles can also be used as precursors (Scheme 6.40) [31]. Similar to the

Example 6.7 Directed Hydroboration and Methylboronation of Alkynes [79].

Before starting any of the following syntheses, carefully follow the safety instructions listed in Example 6.1.

General procedure for α-hydroboration of internal alkynes in air

alpha-seletive hydroboration

catalyst

A vial was charged in air with [Cu(Cl)-(IMes)], NaOH (2.5 mg, 12 mol%), the alkyne (0.50 mmol), and CPME (0.6 mL). The mixture was stirred for 5 min at rt and cooled to −30 °C. Pinacolborane [HB(pin)] was added slowly (0.75 mmol, 0.11 mL). The vial was closed with a screw cap, and the reaction mixture was stirred at 80 °C for 20 h. The conversion was determined by GC analysis. The volatiles were removed in vacuo, and the product was purified by column chromatography (SiO$_2$). The regioselectivity and the α/β ratio of the products were determined by ^{1}H NMR.

General procedure for the β-hydroboration of internal alkynes in air

beta-selective hydroboration

A vial was charged in air with [Cu(Cl)(IMes)] (appropriate amount using a stock solution of 0.005 mmol Cu in 3.3 mL of CH$_2$Cl$_2$; the latter was evaporated in vacuo before adding other reagents), NaOH (0.025 mmol, 1 mg), bis(pinacolato)diboron (0.55 mmol, 140 mg), the alkyne (0.5 mmol), CPME (0.6 mL), and MeOH (1.0 mmol, 0.05 mL). The vial was closed with a screw cap, and the reaction mixture was stirred at rt for 16 h. The conversion was determined by GC analysis. The volatiles were removed in vacuo, and the product was purified by column chromatography (SiO$_2$).

General procedure for the carboboration of internal alkynes in air

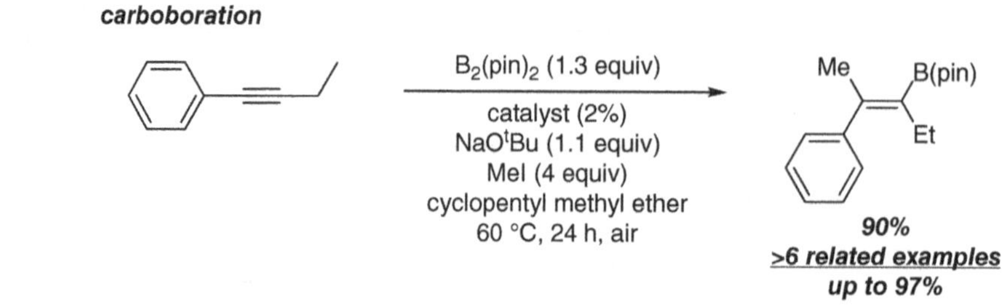

carboration

A vial was charged in air with [Cu(Cl)(IMes)] (4.1 mg, 2 mol%), NaOtBu (0.55 mmol, 53 mg), bis(pinacolato)diboron (0.65 mmol, 165 mg), alkyne (0.5 mmol), CPME (1.4 mL), and the electrophile (3–4 equiv). The vial was closed with a screw cap, and the reaction mixture was stirred at 60 °C for 24 h. The conversion was determined by GC analysis. The volatiles were removed in vacuo, and the product was purified by column chromatography (SiO$_2$).

Adapted with permission from Y. D. Bidal, F. Lazreg, and C. S. J. Cazin, ACS Catal., 2014, 4, 1564–1569. Copyright (2014) American Chemical Society.

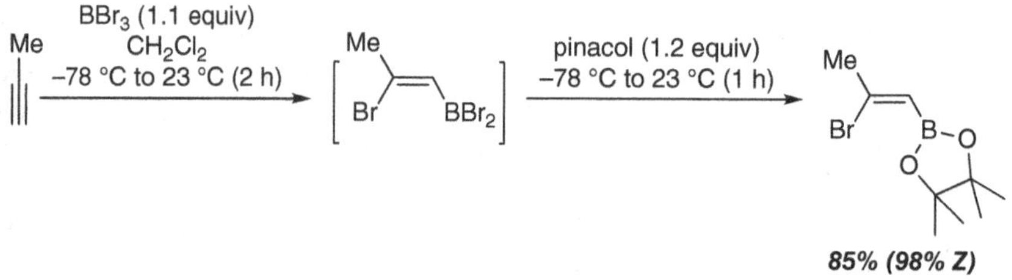

SCHEME 6.39 Addition of B–Br bonds to alkynes and conversion into a vinylboronate [81].

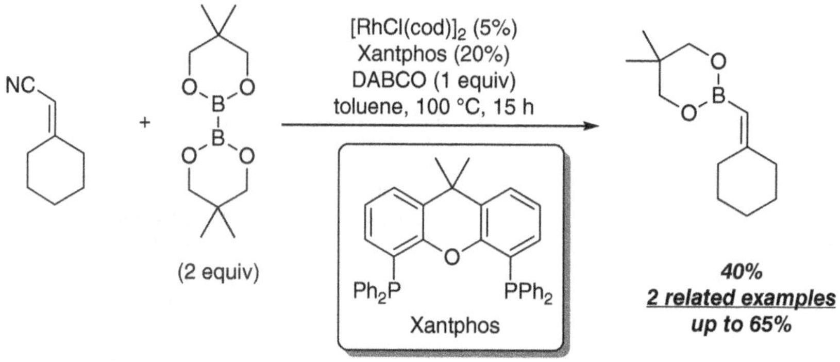

SCHEME 6.40 Synthesis of vinylboronates through cleavage of C–C bonds [31].

reactions using arylnitriles, these reactions were catalyzed by rhodium complexes using XantPhos as the supporting ligand. While only a few examples were listed, this chemistry provided the proof of concept for the synthesis of vinylboronates from vinylnitriles.

A vinylboronate was prepared through a metal-free borylation reaction starting from a trisubstituted vinyl bromide (Scheme 6.41) [3]. The key in this chemistry was the use of

PhMe$_2$Si-Bpin as the borylating agent and KOMe as the base. The solvent was also a critical choice for this chemistry as significant amounts of a silane were formed when the dichloromethane was used as the solvent in related arylboronate syntheses.

The stereoselective synthesis of Z-vinylboronates is one of the more challenging goals to reach using hydroelementation chemistry. Conceivably, cross-coupling chemistry could accomplish this goal provided that a suitable vinyl precursor was available with Z-stereo-chemistry already established. The ability to selectively convert alkynes into Z-vinylboro-nates was achieved through the use of ruthenium catalysts (Scheme 6.42) [82]. Treatment of terminal alkynes with pinacolborane in the presence of a ruthenium complex afforded good to excellent yields of the target compounds with a high degree of selectivity for the Z-isomer. Internal alkynes as well as terminal alkenes were unresponsive under the reaction conditions.

A clever transition metal-free synthesis of Z-vinylboronates was reported by Molander (Scheme 6.43) [83]. Instead of focusing on trying to devise a ligand architecture for a

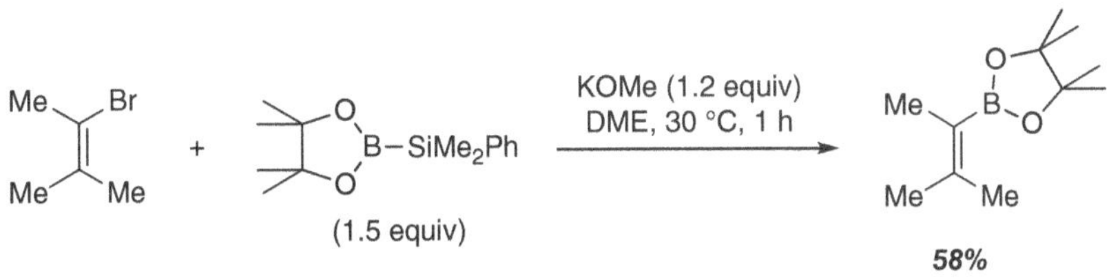

SCHEME 6.41   Metal-free synthesis of a vinylboronate [3].

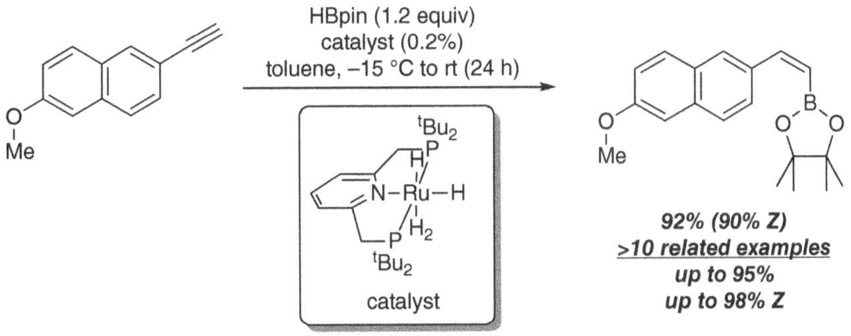

SCHEME 6.42   Ruthenium-catalyzed preparation of Z-vinylboronates [82].

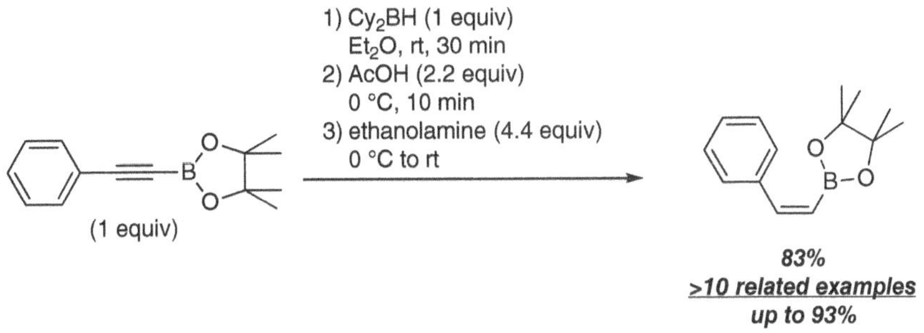

SCHEME 6.43   Stereoselective synthesis of Z-vinylboronates using hydroboration/protode-boronation [83].

transition metal catalyst that would promote a specific mode of addition, he used a hydroboration/protodeboronation sequence to selectively generate the Z-vinylboronate. While a range of preexisting functional groups were well tolerated, substrates bearing nitro groups and nitriles were unable to be converted due to competing secondary reactions. It should be noted that these arylboronate compounds could be readily converted into the vinyltrifluoroborate salts with retention of the Z-stereochemistry upon treatment of the vinylboronate with an aqueous solution of KHF_2. Overall, the chemistry was transition metal-free and formed the target compounds in good to excellent yield.

The formal hydrochlorination of alkynylboronates can be achieved through the use of $Cp_2Zr(H)Cl$ followed by the addition of NCS (Scheme 6.44) [84]. The chemistry was regioselective for the formation of the E-isomer, and the yields of the chlorinated vinylboronates were fair to moderate. The incorporation of the chlorine was valuable since it provided a "synthetic handle" for further functionalization. A variation of this approach was used in the stereoselective hydroboration of terminal alkynes (Scheme 6.45) [85].

The silylborylation of alkynylboronates is an effective route to the generation of gem-diborylated alkenes (Scheme 6.46) [86]. The chemistry was catalyzed by a common palladium complex without an added supporting ligand. The chemistry was highly regioselective and generated the target compound in moderate yield. Given the high regioselectivity of this process, it is an attractive route to the synthesis of vinylboronates.

The addition of B_2pin_2 to terminal alkynes has been achieved using nanoporous gold catalysts (Scheme 6.47) [87]. The addition was remarkably selective for incorporation of the pinacolborane fragments on the same side of the resulting alkene. A host of terminal

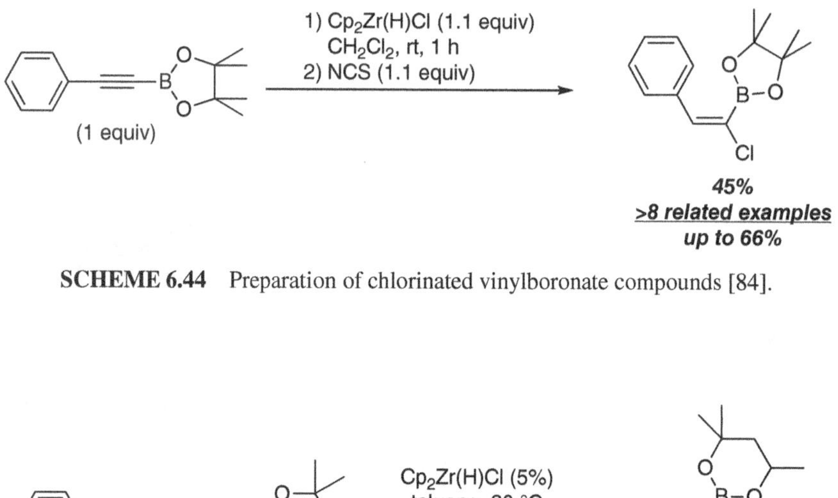

SCHEME 6.44   Preparation of chlorinated vinylboronate compounds [84].

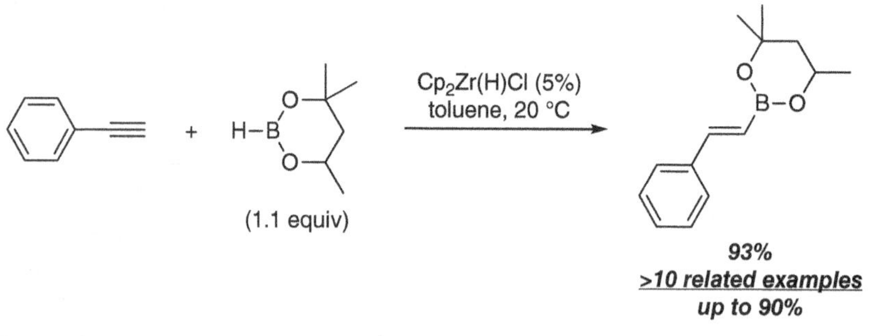

SCHEME 6.45   Hydroboration of terminal alkynes using Schwartz's reagent as the catalyst [85].

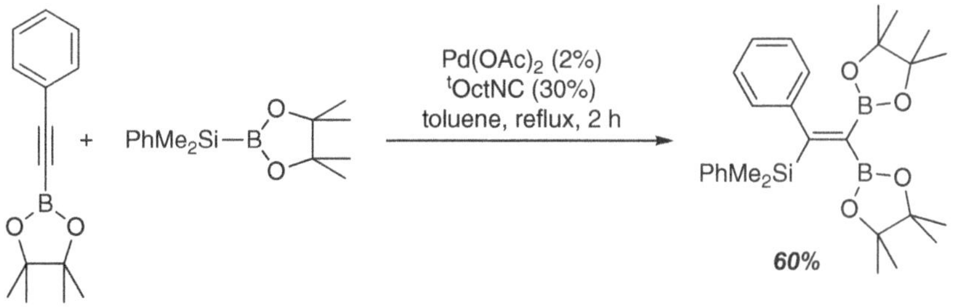

SCHEME 6.46 Palladium-catalyzed silylborylation of alkynylboronates [86].

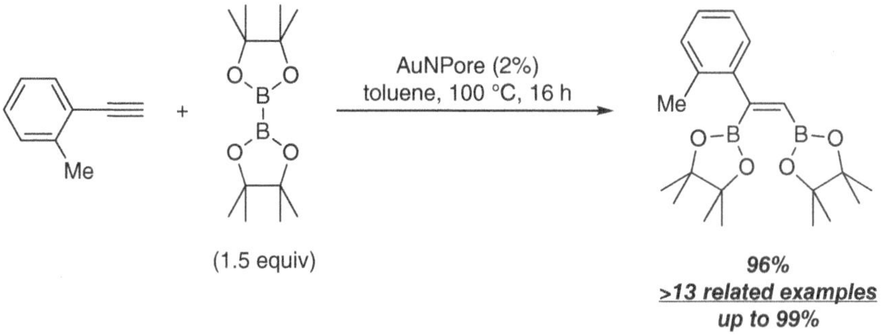

(1.5 equiv)

96%
>13 related examples
up to 99%

SCHEME 6.47 Gold-catalyzed addition of B_2pin_2 to alkynes [87].

and internal alkenes were successfully borylated using this approach with the lowest yields obtained using unfunctionalized internal alkynes. In related work, the enantioselective diborylation of allenes has been achieved using a palladium catalyst [88–90].

As described earlier, a variety of transition metal catalysts have been used to promote the addition of B–H bonds across alkynes. The development of metal-free approaches has been challenging to design due to the unique reactivity profile afforded by transition metal catalysts. To this end, a metal-free catalyst system has been devised for the addition of pinacolborane to internal alkynes (Scheme 6.48) [91]. Surprisingly, only a carboxylic acid additive was needed to effect the hydroboration reaction. In addition to common alkynes, an alkynylboronate was successfully borylated using the carboxylic acid catalyst. The functional group tolerance was remarkably broad, and a host of alkynes were successfully functionalized. Given the wide substrate scope, availability of the precursors and catalysts, and the metal-free nature of the chemistry, this is a reasonable approach to selective hydroboration reactions.

An intriguing approach to the synthesis of vinylboronates entailed the formal alkynylboronation of internal alkynes (Scheme 6.49) [92]. The reaction was catalyzed by nickel complexes using a basic monodentate phosphine as a supporting ligand for the metal center. A range of internal alkynes were screened for activity toward the addition reaction, and moderate to high selectivities for the *cis*-isomer (up to 99:1) were obtained. When unsymmetrical internal alkynes were screened, the chemistry was largely regioselective as well (up to 94:6).

The synthesis of allenylboronates has been achieved using a copper catalyst supported by a large bite angle bisphosphine ligand (Scheme 6.50) [93]. The reactions were carried

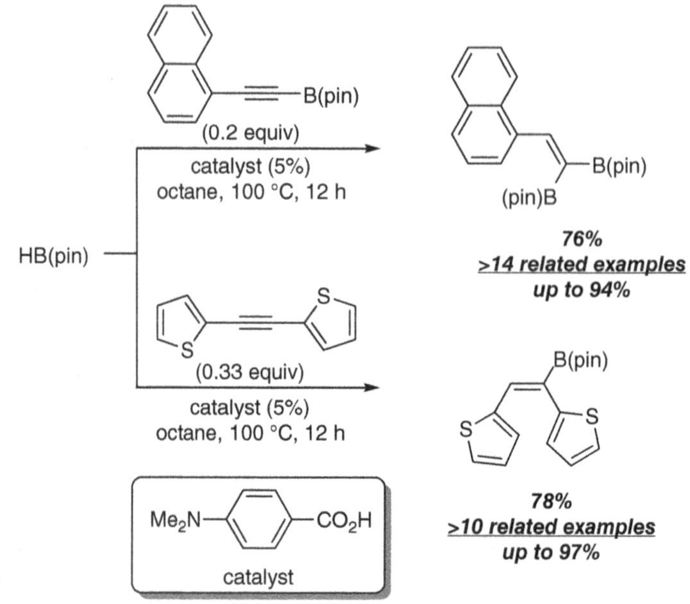

SCHEME 6.48   Addition of pinacolborane to alkynes catalyzed by carboxylic acids [91].

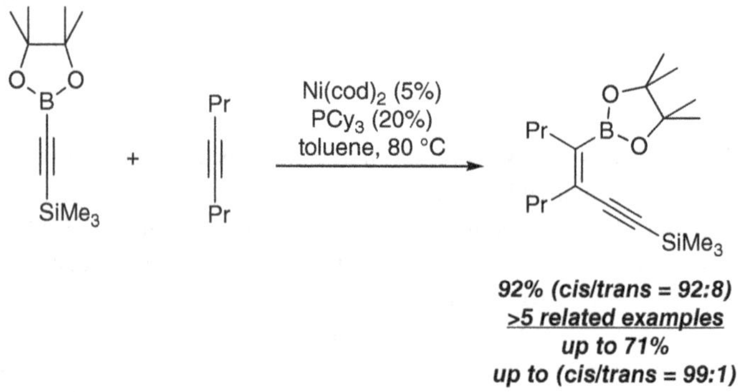

SCHEME 6.49   Borylation of internal alkynes [92].

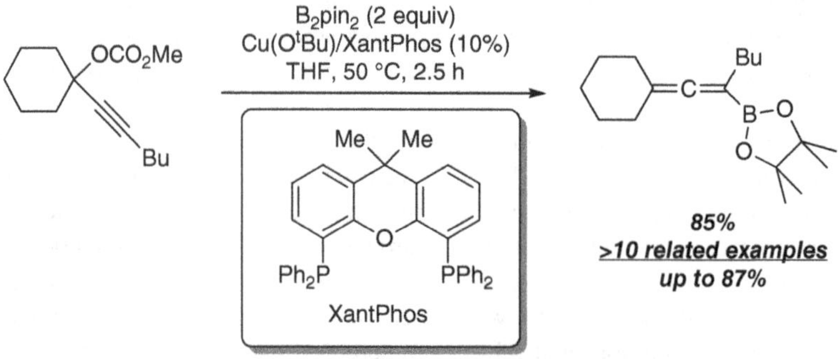

SCHEME 6.50   Synthesis of allenylboronates [93].

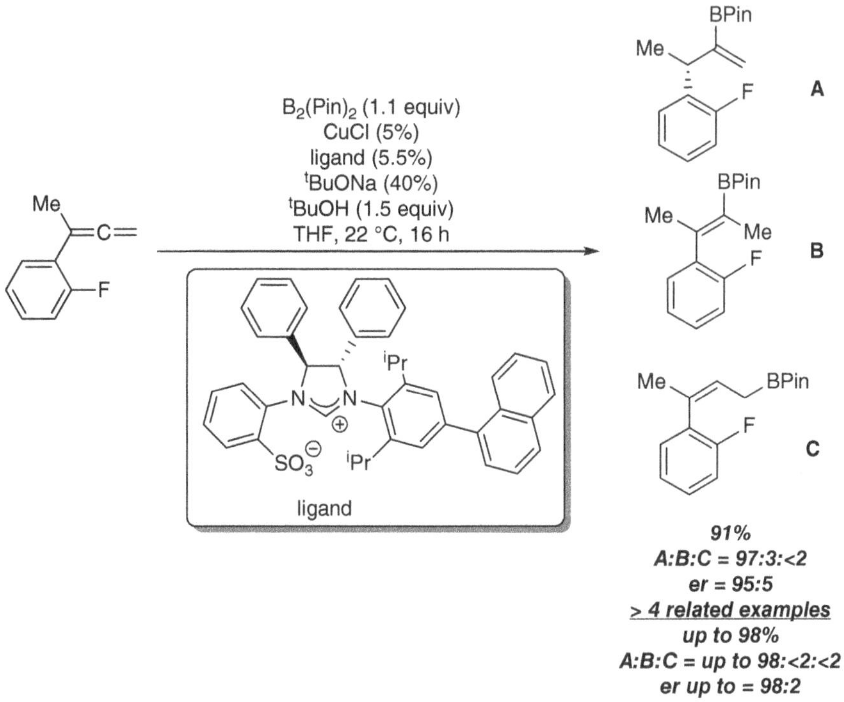

SCHEME 6.51 Copper-catalyzed borylation of allenes [95].

out with gentle heating and generated moderate to excellent yields of the allenylboronates and were tolerant to several functional groups. Similar work was carried out using a combination of copper and palladium catalysts [94].

The copper-catalyzed reaction of allenes with B_2pin_2 could generate a range of regio- and stereoisomers. However, through careful selection of supporting ligand and catalyst, the regioselective formation of vinylboronates was achieved (Scheme 6.51) [95]. Additionally, the use of a chiral supporting ligand for the copper center enabled the development of an enantioselective borylation reaction. After screening a range of ligands, a chiral N-heterocyclic carbene ligand was selected for the borylation reaction. The selectivity was generally excellent, and enantiomeric ratios of 96:4 were obtained for several substrates. The competing formation of allyl and additional vinylboronates was minimized through careful screening of the reaction conditions.

6.3.1   Troubleshooting the Synthesis of Vinylboronates and Related Compounds

While it is impossible to anticipate every issue that will arise during the preparation of vinylboronates and related compounds, the following guide will provide some suggestions in order to circumvent many common issues.

Problem/Issue/Goal	Possible Solution/Suggestion
Need a transition metal-free synthesis of a vinylboronate	If the substrate does not contain electrophiles, try a one-pot magnesiation/borylation of aryl halides using pinacolborane [2]. An alternative approach uses a silylboronate as the borylating agent [3]
Can a vinyl triflate be converted into a vinylboronate?	Palladium-catalyzed coupling between B$_2$pin$_2$ and vinyl triflates will generate vinylboronates [77]
Starting material is a vinyl bromide	Vinyl bromides are readily converted into vinylboronates through palladium-catalyzed cross-coupling reactions [51]. A zinc-catalyzed approach has been reported and is attractive due to the reduced cost of zinc [61]
Need to promote a Markovnikov hydroboration using HBpin	Several approaches have been used for this reaction. Both nickel- and copper-catalyzed processes have been used to generate vinylboronates [78, 79]
Need to generate vinylboronate from an alkyne with anti-Markovnikov selectivity and E-*stereochemistry*	A copper-catalyzed hydroboration would be a reasonable place to start [79]. Additionally, a zirconium-catalyzed approach has been successful [85]
Need to selectively generate a Z-*vinylboronate or related compound*	Consider starting from an alkynylboronate [84] and selectively reducing it to the Z-vinylboronate [82, 83]
It would be convenient to effect a carboboration on an internal alkyne	This reaction has been achieved through a borylation reaction in the presence of an alkyl halide [79]
Can a vinylnitrile be converted into a vinylboronate?	This intriguing reaction is catalyzed by rhodium complexes bearing large bite angle bisphosphines [31]
Need to start from an alkyne and generate a disubstituted vinylboronate with Z-*stereochemistry*	This synthesis can be challenging to accomplish using hydroboration chemistry. A ruthenium-catalyzed hydroboration reaction has been shown to generate the Z-vinylboronate [82]. A clever transition metal-free approach by Molander entailed initial formation of the alkynylboronate followed by hydroboration/protodeboronation to generate the Z-vinylboronate [83]
Need to generate a 1,1-diboronate from an alkyne	An interesting metal-free approach has been developed starting from an alkynylboronate. A carboxylic acid-catalyzed hydroboration afforded 1,1-diboronates in moderate to excellent yields [91]
Need to generate a syn-1,2-diboronate from an alkyne	The use of gold nanoparticles has been reported to promote the diborylation of terminal alkynes with Z-stereochemistry [87]
Need to make an allenylboronate	The copper-catalyzed conversion of propargylic carbonates into allenylboronates would be a convenient entry point to this chemistry [93]

Problem/Issue/Goal	Possible Solution/Suggestion
Need to generate an vinyltrifluoroborate salt	A convenient precursor for the direct synthesis is not readily available. Try generating an arylboronic acid or arylboronate, and treat it with a solution of KHF_2 [37, 38]. For arylboronates, a 4:1 THF/water mixture was successful [38, 83]
Is there a practical source of ethenylboronic acid?	While ethenylboronic acid has been reported to be unstable, its cyclodehydration product (a cyclotriboroxane) was quite stable as its pyridine adduct. It successfully served as a vinylboronic acid equivalent in coupling reactions [96]

6.4   SYNTHESIS OF ALKYNYLBORONATES AND RELATED COMPOUNDS THROUGH THE FORMATION OF BORON–CARBON(sp) BONDS

A classic approach to the synthesis of alkynylboron compounds involved the addition of lithium acetylides to trialkylborates [97]. This methodology was extended to pinacolborane systems, and a range of alkynylboronates were generated following this approach (Scheme 6.52) [84, 91, 98–102]. In addition to the use of trialkylborates, pinacolborane has also been used in the synthesis of alkynylboronates [103].

The synthesis of alkynyltrifluoroborate salts begins with the generation of lithium acetylides at low temperature followed the addition of a trialkylborate (Scheme 6.53) [104]. Once the intermediate alkynylboronate was generated, aqueous KHF_2 was added to generate the target compounds. The resulting alkynyltrifluoroborates were isolated in moderate to excellent yields. Both alkyl- and arylalkynes were well tolerated by this chemistry.

The synthesis of MIDA ethynylboronate was achieved by treating ethynylmagnesium bromide with trimethylborate followed by the addition of MIDA [105, 106]. The yield of this alkynylboronate was moderate; however, this compound was a valuable precursor for the synthesis of a host of internal alkynylboronates. Additional examples were generated through a palladium-catalyzed reaction using the ethynyl MIDA boronate as a substrate along with a variety of aryl bromides and iodides.

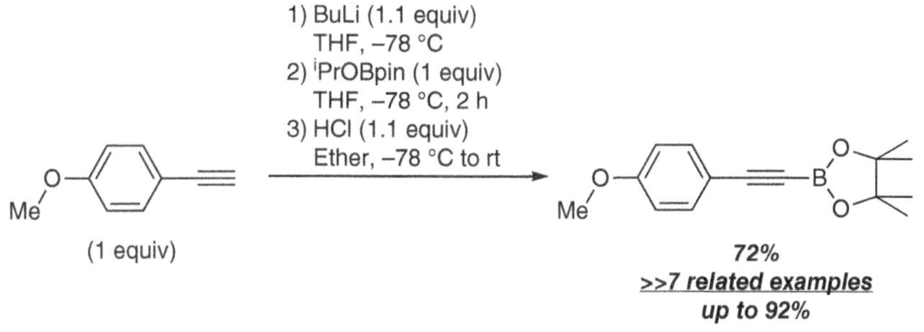

SCHEME 6.52   Synthesis of alkynylboronates using an organolithium approach [84].

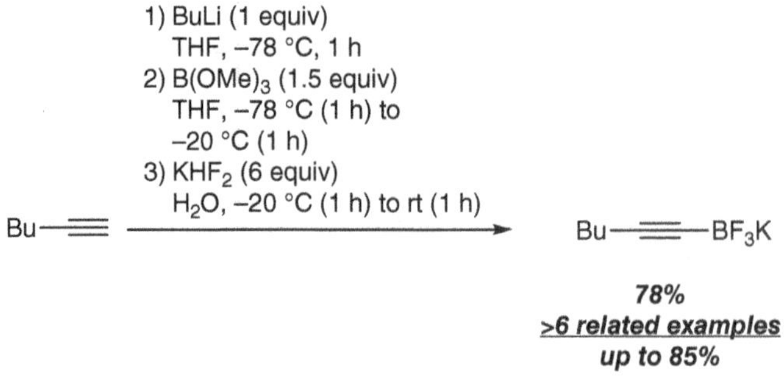

SCHEME 6.53 Synthesis of alkynyltrifluoroborate salts [104].

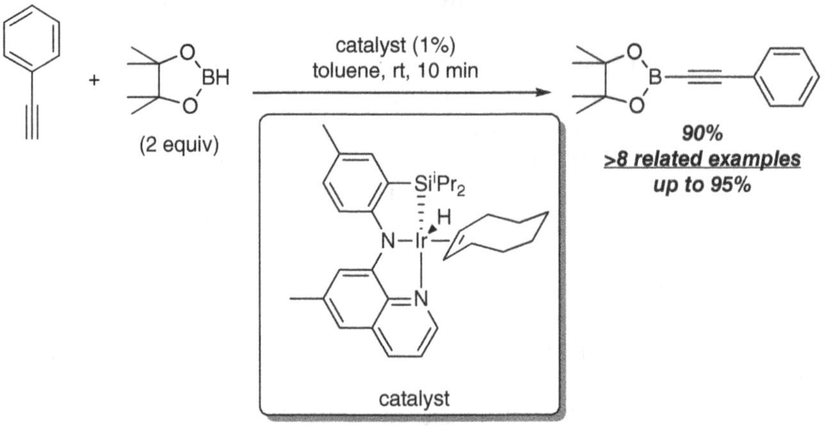

SCHEME 6.54 Iridium-catalyzed synthesis of alkynylboronates [107].

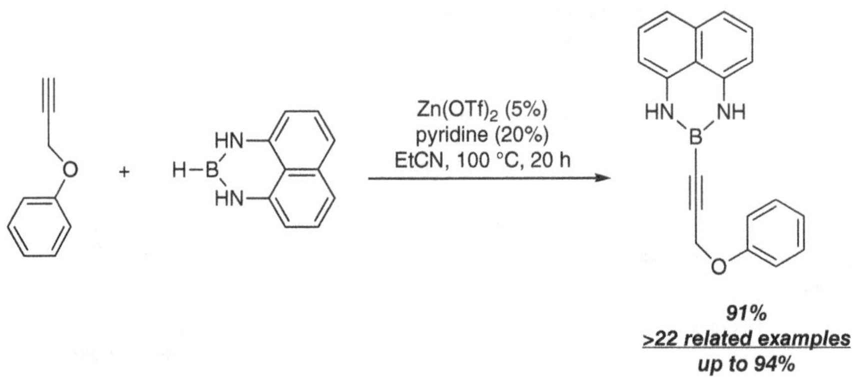

SCHEME 6.55 Zinc-catalyzed coupling of terminal alkynes with 1,8-naphthalenediaminatoborane [108].

A coupling reaction between terminal alkynes and pinacolboranes would be an attractive approach to the synthesis of alkynylboronates. Elegant work by Ozerov made this challenging reaction a reality using an iridium-catalyzed dehydrogenative borylation process (Scheme 6.54) [107]. The reaction occurred under mild conditions, and outstanding yields of the alkynylboronates were obtained using only 1% catalyst loading. Caution! A full equivalent of highly flammable hydrogen gas is released from this reaction. Given the wide assortment of terminal alkynes that are readily available, this is an attractive approach to the synthesis of functionalized alkynylboronates.

The coupling of terminal alkynes with 1,8-naphthalenediaminatoborane has been achieved using a Lewis acid-catalyzed approach (Scheme 6.55) [108]. The authors screened a range of B–H donors including pinacolborane, but only the 1,8-naphthalenediaminatoborane was effective in this reaction. A host of terminal alkynes were successfully functionalized including a range of alkylalkyne and arylalkynes bearing a range of electron-donating and electron-withdrawing groups. Heteroaromatic substrates were also well tolerated by the chemistry. Even an alkylalkyne bearing a primary alcohol was transformed into the alkynylboronate with retention of the alcohol. This is a very practical approach to the synthesis of alkynylboronates given the wide scope of the chemistry and availability of the materials.

6.4.1 Troubleshooting the Synthesis of Alkynylboronates and Related Compounds

While it is impossible to anticipate every issue that will arise during the preparation of alkynylboronates and related compounds, the following guide will provide some suggestions in order to circumvent many common issues.

Problem/Issue/Goal	Possible Solution/Suggestion
Need to use a functionalized alkyne bearing electrophiles	While it might appear that a Grignard reaction would be out of the question, a reasonable approach might involve synthesis of an ethynylboronate using a Grignard approach and then incorporate the functional groups using cross-coupling chemistry [105]
Need to generate an alkynyltrifluoroborate salt	A convenient precursor for the direct synthesis is not readily available. Try generating an arylboronic acid or arylboronate, and treat it with a solution of KHF$_2$ [37, 38].
Can an alkynyl MIDA boronate be prepared?	Initial treatment of an alkynyl Grignard reagent with trimethylborate followed by the addition of MIDA generated the MIDA alkynylboronates [106]
Need a high-yielding synthesis for an alkynylboronate that does not generate significant amounts of salt by-products	The iridium-catalyzed dehydrogenative coupling of terminal alkynes with HBpin is an attractive solution [107]
Need to generate an alkynylboron species from an alkyne bearing a primary alcohol	This would be challenging for several of the approaches discussed in this section; however, the successful use of 1,8-naphthalenediaminatoborane in place of HBpin for the synthesis of alkynylboronates has been reported [108]

REFERENCES

[1] Brown, H. C.; Cole, T. E. *Organometallics* **1983**, *2*, 1316–1319.

[2] Clary, J. W.; Rettenmaier, T. J.; Snelling, R.; Bryks, W.; Banwell, J.; Wipke, W. T.; Singaram, B. *J. Org. Chem.* **2011**, *76*, 9602–9610.

[3] Yamamoto, E.; Izumi, K.; Horita, Y.; Ito, H. *J. Am. Chem. Soc.* **2012**, *134*, 19997–20000.

[4] Yang, C.-T.; Zhang, Z.-Q.; Tajuddin, H.; Wu, C.-C.; Liang, J.; Liu, J.-H.; Fu, Y.; Czyzewska, M.; Steel, P. G.; Marder, T. B.; Liu, L. *Angew. Chem. Int. Ed.* **2012**, *51*, 528–532.

[5] Ito, H.; Kubota, K. *Org. Lett.* **2012**, *14*, 890–893.

[6] Kubota, K.; Yamamoto, E.; Ito, H. *J. Am. Chem. Soc.* **2013**, *135*, 2635–2640.

[7] Dudnik, A. S.; Fu, G. C. *J. Am. Chem. Soc.* **2012**, *134*, 10693–10697.

[8] Hartwig, J. F. *Acc. Chem. Res.* **2012**, *45*, 864–873.

[9] Shimada, S.; Batsanov, A. S.; Howard, J. A. K.; Marder, T. B. *Angew. Chem. Int. Ed.* **2001**, *40*, 2168–2171.

[10] Waltz, K. M.; Hartwig, J. F. *Science* **1997**, *277*, 211–213.

[11] Bose, S. K.; Fucke, K.; Liu, L.; Steel, P. G.; Marder, T. B. *Angew. Chem. Int. Ed.* **2014**, *53*, 1799–1803.

[12] Lawson, Y. G.; Lesley, G. M. J.; Norman, N. C.; Rice, C. R.; Marder, T. B. *Chem. Commun.* **1997**, 2051–2052.

[13] Bell, N. J.; Cox, A. J.; Cameron, N. R.; Evans, J. S. O.; Marder, T. B.; Duin, M. A.; Elsevier, C. J.; Baucherel, X.; Tulloch, A. A. D.; Tooze, R. P. *Chem. Commun.* **2004**, 1854–1855.

[14] Wu, H.; Radomkit, S.; O'Brien, J. M.; Hoveyda, A. H. *J. Am. Chem. Soc.* **2012**, *134*, 8277–8285.

[15] Tian, B.; Liu, Q.; Tong, X.; Tian, P.; Lin, G. Q. *Org. Chem. Front.* **2014**, *1*, 1116–1122.

[16] Parra, A.; Amenos, L.; Guisan-Ceinos, M.; Lopez, A.; Garcia Ruano, J. L.; Tortosa, M. *J. Am. Chem. Soc.* **2014**, *136*, 15833–15836.

[17] Lee, J. E.; Yun, J. *Angew. Chem. Int. Ed.* **2008**, *47*, 145–147.

[18] Cid, J.; Gulyas, H.; Carbo, J. J.; Fernandez, E. *Chem. Soc. Rev.* **2012**, *41*, 3558–3570.

[19] Hartmann, E.; Vyas, D. J.; Oestreich, M. *Chem. Commun.* **2011**, *47*, 7917–7932.

[20] Satoh, M.; Nomoto, Y.; Miyaura, N.; Suzuki, A. *Tetrahedron Lett.* **1989**, *30*, 3789–3792.

[21] Ely, R. J.; Morken, J. P. *J. Am. Chem. Soc.* **2010**, *132*, 2534–2535.

[22] Cao, Y.; Zhang, Y.; Zhang, L.; Zhang, D.; Leng, X.; Huang, Z. *Org. Chem. Front.* **2014**, *1*, 1101–1106.

[23] Wu, J. Y.; Moreau, B.; Ritter, T. *J. Am. Chem. Soc.* **2009**, *131*, 12915–12917.

[24] Zhang, P.; Roundtree, I. A.; Morken, J. P. *Org. Lett.* **2012**, *14*, 1416–1419.

[25] Unsworth, P. J.; Leonori, D.; Aggarwal, V. K. *Angew. Chem. Int. Ed.* **2014**, *53*, 9846–9850.

[26] Yamamoto, E.; Takenouchi, Y.; Ozaki, T.; Miya, T.; Ito, H. *J. Am. Chem. Soc.* **2014**, *136*, 16515–16521.

[27] Zhang, L.; Zuo, Z.; Wan, X.; Huang, Z. *J. Am. Chem. Soc.* **2014**, *136*, 15501–15504.

[28] Coombs, J. R.; Zhang, L.; Morken, J. P. *J. Am. Chem. Soc.* **2014**, *136*, 16140–16143.

[29] St. Denis, J. D.; Scully, C. C. G.; Lee, C. F.; Yudin, A. K. *Org. Lett.* **2014**, *16*, 1338–1341.

[30] Grob, J. E.; Nunez, J.; Dechantsreiter, M. A.; Hamann, L. G. *J. Org. Chem.* **2011**, *76*, 4930–4940.

[31] Tobisu, M.; Kinuta, H.; Kita, Y.; Remond, E.; Chatani, N. *J. Am. Chem. Soc.* **2012**, *134*, 115–118.

[32] Fandrick, K. R.; Patel, N. D.; Mulder, J. A.; Gao, J.; Konrad, M.; Archer, E.; Buono, F. G.; Duran, A.; Schmid, R.; Daeubler, J.; Fandrick, D. R.; Ma, S.; Grinberg, N.; Lee, H.; Busacca, C. A.; Song, J. J.; Yee, N. K.; Senanayake, C. H. *Org. Lett.* **2014**, *16*, 4360–4363.

[33] Molander, G. A.; Ryu, D. *Angew. Chem. Int. Ed.* **2014**, *53*, 14181–14185.

[34] Molander, G. A.; Figueroa, R. *Org. Lett.* **2006**, *8*, 75–78.

[35] Molander, G. A.; Ham, J. *Org. Lett.* **2006**, *8*, 2031–2034.

[36] Molander, G. A.; Petrillo, D. E. *Org. Lett.* **2008**, *10*, 1795–1798.

[37] Vedejs, E.; Chapman, R. W.; Fields, S. C.; Lin, S.; Schrimpf, M. R. *J. Org. Chem.* **1995**, *60*, 3020–3027.

[38] Murphy, J. M.; Tzschucke, C. C.; Hartwig, J. F. *Org. Lett.* **2007**, *9*, 757–760.

[39] Mkhalid, I. A. I.; Barnard, J. H.; Marder, T. B.; Murphy, J. M.; Hartwig, J. F. *Chem. Rev.* **2010**, *110*, 890–931.

[40] Kato, S.; Shimizu, S.; Kobayashi, A.; Yoshihara, T.; Tobita, S.; Nakamura, Y. *J. Org. Chem.* **2014**, *79*, 618–629.

[41] Wang, X.-J.; Sun, X.; Zhang, L.; Xu, Y.; Krishnamurthy, D.; Senanayake, C. H. *Org. Lett.* **2006**, *8*, 305–307.

[42] Baron, O.; Knochel, P. *Angew. Chem. Int. Ed.* **2005**, *44*, 3133–3135.

[43] Leermann, T.; Leroux, F. R.; Colobert, F. *Org. Lett.* **2011**, *13*, 4479–4481.

[44] Jiang, Q.; Ryan, M.; Zhichkin, P. *J. Org. Chem.* **2007**, *72*, 6618–6620.

[45] Wei, J.; Zhang, Y.; Zhang, W.-X.; Xi, Z. *Org. Chem. Front.* **2014**, *1*, 983–987.

[46] Cleary, L.; Pitzen, J.; Brailsford, J. A.; Shea, K. J. *Org. Lett.* **2014**, *16*, 4460–4463.

[47] Ishiyama, T.; Murata, M.; Miyaura, N. *J. Org. Chem.* **1995**, *60*, 7508–7510.

[48] Tang, W.; Keshipeddy, S.; Zhang, Y.; Wei, X.; Savoie, J.; Patel, N. D.; Yee, N. K.; Senanayake, C. H. *Org. Lett.* **2011**, *13*, 1366–1369.

[49] Boinapally, S.; Huang, B.; Abe, M.; Katan, C.; Noguchi, J.; Watanabe, S.; Kasai, H.; Xue, B.; Kobayashi, T. *J. Org. Chem.* **2014**, *79*, 7822–7830.

[50] Zhang, Y.; Lu, B. Z.; Li, G.; Rodriguez, S.; Tan, J.; Wei, H.-X.; Liu, J.; Roschangar, F.; Ding, F.; Zhao, W.; Qu, B.; Reeves, D.; Grinberg, N.; Lee, H.; Heckmann, G.; Niemeier, O.; Brenner, M.; Tsantrizos, Y.; Beaulieu, P. L.; Hossain, A.; Yee, N.; Farina, V.; Senanayake, C. H. *Org. Lett.* **2014**, *16*, 4558–4561.

[51] Billingsley, K. L.; Buchwald, S. L. *J. Org. Chem.* **2008**, *73*, 5589–5591.

[52] Zhu, W.; Ma, D. *Org. Lett.* **2006**, *8*, 261–263.

[53] Leowanawat, P.; Resmerita, A.-M.; Moldoveanu, C.; Liu, C.; Zhang, N.; Wilson, D. A.; Hoang, L. M.; Rosen, B. M.; Percec, V. *J. Org. Chem.* **2010**, *75*, 7822–7828.

[54] Wilson, D. A.; Wilson, C. J.; Moldoveanu, C.; Resmerita, A.-M.; Corcoran, P.; Hoang, L. M.; Rosen, B. M.; Percec, V. *J. Am. Chem. Soc.* **2010**, *132*, 1800–1801.

[55] Moldoveanu, C.; Wilson, D. A.; Wilson, C. J.; Leowanawat, P.; Resmerita, A.-M.; Liu, C.; Rosen, B. M.; Percec, V. *J. Org. Chem.* **2010**, *75*, 5438–5452.

[56] Molander, G. A.; Trice, S. L. J.; Dreher, S. D. *J. Am. Chem. Soc.* **2010**, *132*, 17701–17703.

[57] Molander, G. A.; Trice, S. L. J.; Kennedy, S. M.; Dreher, S. D.; Tudge, M. T. *J. Am. Chem. Soc.* **2012**, *134*, 11667–11673.

[58] Molander, G. A.; Cavalcanti, L. N.; Garcia-Garcia, C. *J. Org. Chem.* **2013**, *78*, 6427–6439.

[59] Molander, G. A.; Trice, S. L. J.; Kennedy, S. M. *J. Org. Chem.* **2012**, *77*, 8678–8688.

[60] Nagashima, Y.; Takita, R.; Yoshida, K.; Hirano, K.; Uchiyama, M. *J. Am. Chem. Soc.* **2013**, *135*, 18730–18733.

[61] Bose, S. K.; Marder, T. B. *Org. Lett.* **2014**, *16*, 4562–4565.

[62] Kinuta, H.; Tobisu, M.; Chatani, N. *J. Am. Chem. Soc.* **2015**, *137*, 1593–1600.

[63] Qiu, D.; Jin, L.; Zheng, Z.; Meng, H.; Mo, F.; Wang, X.; Zhang, Y.; Wang, J. *J. Org. Chem.* **2013**, *78*, 1923–1933.

[64] Qiu, D.; Wang, S.; Tang, S.; Meng, H.; Jin, L.; Mo, F.; Zhang, Y.; Wang, J. *J. Org. Chem.* **2014**, *79*, 1979–1988.

[65] Cho, J.-Y.; Tse, M. K.; Holmes, D.; Maleczka, R. E. Jr.; Smith, M. R. III. *Science* **2002**, *295*, 305–308.

[66] Boebel, T. A.; Hartwig, J. F. *J. Am. Chem. Soc.* **2008**, *130*, 7534–7535.

[67] Robbins, D. W.; Boebel, T. A.; Hartwig, J. F. *J. Am. Chem. Soc.* **2010**, *132*, 4068–4069.

[68] Ishiyama, T.; Takagi, J.; Ishida, K.; Miyaura, N.; Anastasi, N. R.; Hartwig, J. F. *J. Am. Chem. Soc.* **2002**, *124*, 390–391.

[69] Larsen, M. A.; Hartwig, J. F. *J. Am. Chem. Soc.* **2014**, *136*, 4287–4299.

[70] Preshlock, S. M.; Ghaffari, B.; Maligres, P. E.; Krska, S. W.; Maleczka, R. E. Jr.; Smith, M. R. III. *J. Am. Chem. Soc.* **2013**, *135*, 7572–7582.

[71] Ghaffari, B.; Preshlock, S. M.; Plattner, D. L.; Staples, R. J.; Maligres, P. E.; Krska, S. W.; Maleczka, R. E. Jr.; Smith, M. R. III. *J. Am. Chem. Soc.* **2014**, *136*, 14345–14348.

[72] Dai, H.-X.; Yu, J.-Q. *J. Am. Chem. Soc.* **2012**, *134*, 134–137.

[73] Roering, A. J.; Hale, L. V. A.; Squier, P. A.; Ringgold, M. A.; Wiederspan, E. R.; Clark, T. B. *Org. Lett.* **2012**, *14*, 3558–3561.

[74] Partridge, B. M.; Hartwig, J. F. *Org. Lett.* **2013**, *15*, 140–143.

[75] Loach, R. P.; Fenton, O. S.; Amaike, K.; Siegel, D. S.; Ozkal, E.; Movassaghi, M. *J. Org. Chem.* **2014**, *79*, 11254–11263.

[76] Chen, X.; Ke, H.; Chen, Y.; Guan, C.; Zou, G. *J. Org. Chem.* **2012**, *77*, 7572–7578.

[77] Takagi, J.; Takahashi, K.; Ishiyama, T.; Miyaura, N. *J. Am. Chem. Soc.* **2002**, *124*, 8001–8006.

[78] Gao, F.; Hoveyda, A. H. *J. Am. Chem. Soc.* **2010**, *132*, 10961–10963.

[79] Bidal, Y. D.; Lazreg, F.; Cazin, C. S. J. *ACS Catal.* **2014**, *4*, 1564–1569.

[80] Alfaro, R.; Parra, A.; Aleman, J.; Garcia Ruano, J. L.; Tortosa, M. *J. Am. Chem. Soc.* **2012**, *134*, 15165–15168.

[81] Wang, C.; Tobrman, T.; Xu, Z.; Negishi, E. *Org. Lett.* **2009**, *11*, 4092–4095.

[82] Gunanathan, C.; Holscher, M.; Pan, F.; Leitner, W. *J. Am. Chem. Soc.* **2012**, *134*, 14349–14352.

[83] Molander, G. A.; Ellis, N. M. *J. Org. Chem.* **2008**, *73*, 6841–6844.

[84] Gazic-Smilovi, I.; Casas-Arce, E.; Roseblade, S. J.; Nettekoven, U.; Zanotti-Gerosa, A.; Kovacevic, M.; Casar, Z. *Angew. Chem. Int. Ed.* **2012**, *51*, 1014–1018.

[85] PraveenGanesh, N.; d'Hondt, S.; Chavant, P. Y. *J. Org. Chem.* **2007**, *72*, 4510–4514.

[86] Jiao, J.; Nakajima, K.; Nishihara, Y. *Org. Lett.* **2013**, *15*, 3294–3297.

[87] Chen, Q.; Zhao, J.; Ishikawa, Y.; Asao, N.; Yamamoto, Y.; Jin, T. *Org. Lett.* **2013**, *15*, 5766–5769.

[88] Sieber, J. D.; Morken, J. P. *J. Am. Chem. Soc.* **2006**, *128*, 74–75.

[89] Burks, H. E.; Liu, S.; Morken, J. P. *J. Am. Chem. Soc.* **2007**, *129*, 8766–8773.

[90] Pelz, N. F.; Woodward, A. R.; Burks, H. E.; Sieber, J. D.; Morken, J. P. *J. Am. Chem. Soc.* **2004**, *126*, 16328–16329.

[91] Ho, H. E.; Asao, N.; Yamamoto, Y.; Jin, T. *Org. Lett.* **2014**, *16*, 4670–4673.

[92] Suginome, M.; Shirakura, M.; Yamamoto, A. *J. Am. Chem. Soc.* **2006**, *128*, 14438–14439.

[93] Ito, H.; Sasaki, Y.; Sawamura, M. *J. Am. Chem. Soc.* **2008**, *130*, 15774–15775.

[94] Zhao, T. S. N.; Yang, Y.; Lessing, T.; Szabo, K. J. *J. Am. Chem. Soc.* **2014**, *136*, 7563–7566.

[95] Jang, H.; Jung, B.; Hoveyda, A. H. *Org. Lett.* **2014**, *16*, 4658–4661.

[96] McKinley, N. F.; O'Shea, D. F. *J. Org. Chem.* **2004**, *69*, 5087–5092.

[97] Brown, H. C.; Bhat, N. G.; Srebnik, M. *Tetrahedron Lett.* **1988**, *29*, 2631–2634.

[98] Ahammed, S.; Kundu, D.; Ranu, B. C. *J. Org. Chem.* **2014**, *79*, 7391–7398.

[99] Deloux, L.; Skrzypczak-Jankun, E.; Cheesman, B. V.; Srebnik, M.; Sabat, M. *J. Am. Chem. Soc.* **1994**, *116*, 10302–10303.

[100] Renaud, J.; Graf, C.-D.; Oberer, L. *Angew. Chem. Int. Ed.* **2000**, *39*, 3101–3104.

[101] Hansen, E. C.; Lee, D. *J. Am. Chem. Soc.* **2005**, *127*, 3252–3253.

[102] Buttner, M. W.; Natscher, J. B.; Burschka, C.; Tacke, R. *Organometallics* **2007**, *26*, 4835–4838.

[103] Janetzko, J.; Batey, R. A. *J. Org. Chem.* **2014**, *79*, 7415–7424.

[104] Molander, G. A.; Katona, B. W.; Machrouhi, F. *J. Org. Chem.* **2002**, *67*, 8416–8423.

[105] Struble, J. R.; Lee, S. J.; Burke, M. D. *Tetrahedron* **2010**, *66*, 4710–4718.

[106] Lee, S. J.; Anderson, T. M.; Burke, M. D. *Angew. Chem. Int. Ed.* **2010**, *49*, 8860–8863.

[107] Lee, C.; Zhou, J.; Ozerov, O. V. *J. Am. Chem. Soc.* **2013**, *135*, 3560–3566.

[108] Tsuchimoto, T.; Utsugi, H.; Sugiura, T.; Horio, S. *Adv. Synth. Cat.* **2015**, *357*, 77–82.

7

SYNTHESIS OF ORGANOHALIDES

7.1 SYNTHESIS OF ALKYL HALIDES THROUGH THE FORMATION OF HALOGEN–CARBON(SP³) BONDS

Alkyl fluorides are valuable compounds for study as well as convenient precursors for the construction of complex organic architectures with a high degree of functionalization. A range of routes have been devised for the synthesis of these compounds, and a number of reviews on the synthesis of alkyl fluorides have appeared [1–6]. One of the most direct routes to the preparation of an alkyl fluoride would be the deoxyfluorination of alcohols. Fundamentally, this would be analogous to a fluorine version of the Appel reaction, and while the Appel reaction is one of the most successful approaches to the synthesis of alkyl chlorides and bromides [7], it has been challenging to extend the chemistry to the preparation of alkyl fluorides [6].

In one version, primary alcohols were converted into alkyl fluorides through the initial formation of a triflate (Scheme 7.1) [8]. Once formed, the triflate was converted into an alkyl fluoride through a copper-catalyzed reaction using KF as a nucleophilic source of the fluorine. For most substrates, this reaction is remarkably fast, and good to excellent yields were obtained in less than 10 min. The chemistry was a bit sensitive to moisture, and the authors noted that the KF needed to be dried prior to use. It was also noteworthy that none of the common elimination products were observed with this chemistry. Additionally, a range of functional groups including nitro, nitrile, ester, ether, halogen, and tosyl were well tolerated by this chemistry. Given the broad functional group tolerance and the large number of alcohols that are readily available, this is one of the more practical routes to alkyl fluorides.

The zinc-catalyzed asymmetric fluorination of malonates used a bisoxazoline ligand to achieve high selectivity (Scheme 7.2) [9]. While a number of Lewis acids showed activity

Practical Functional Group Synthesis, First Edition. Robert A. Stockland, Jr.
© 2016 John Wiley & Sons, Inc. Published 2016 by John Wiley & Sons, Inc.

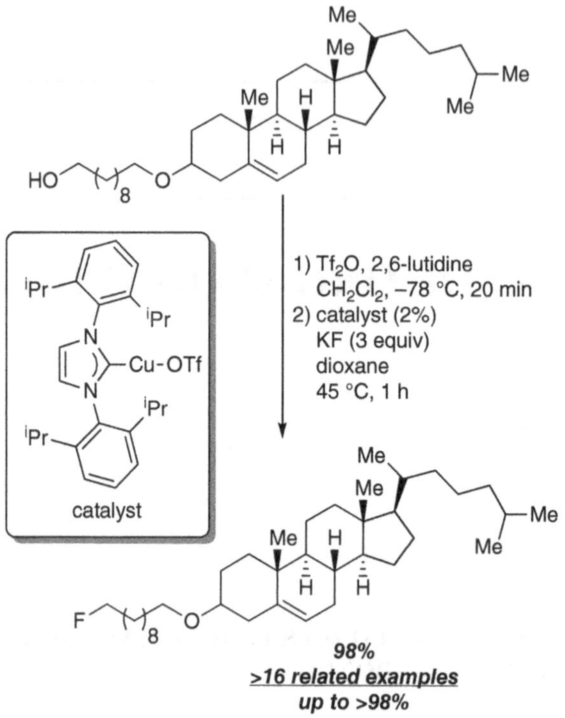

SCHEME 7.1 Copper-catalyzed fluorination of alkyl triflates [8].

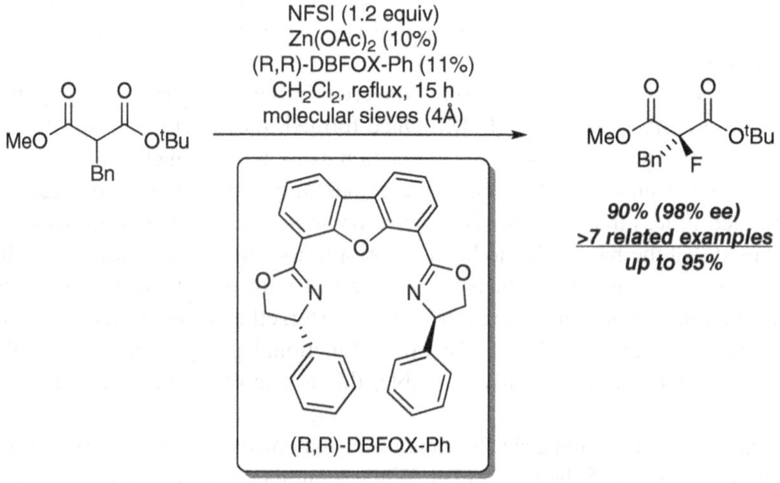

SCHEME 7.2 Zinc-catalyzed asymmetric fluorination of malonates [9].

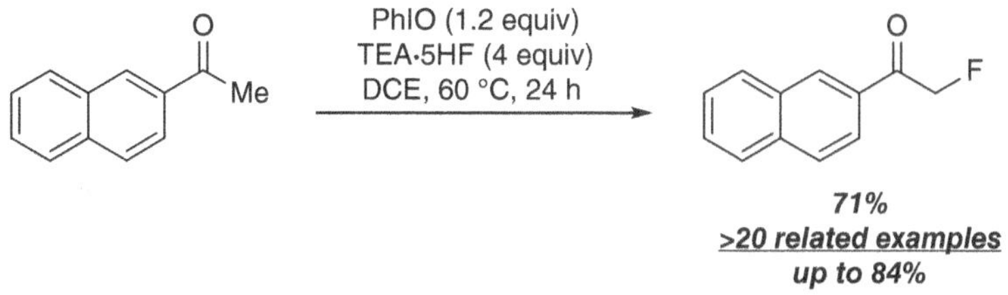

SCHEME 7.3   Transition metal-free fluorination of methyl ketones [10].

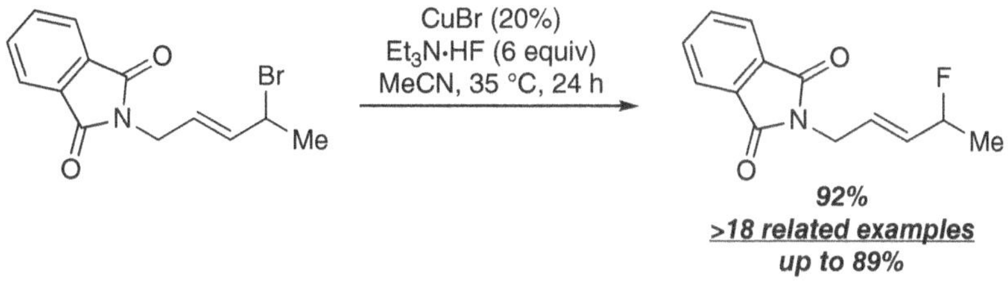

SCHEME 7.4   Copper-catalyzed fluorination of allylic bromides [11].

towards the reaction, zinc(II) acetate was the most effective. While only a few substrates were screened, this work provided the proof of concept for the chemistry and the potential for further development. In related work, the α-fluorination of methyl ketones has been achieved using a Lewis base adduct of HF and PhIO as an oxidizing agent (Scheme 7.3) [10].

The copper-catalyzed fluorination of allylic bromides and chlorides has been achieved using common copper salts such as CuBr as catalysts (Scheme 7.4) [11]. Several other first-row metal compounds (FeBr$_2$ and CoCl$_2$) were screened; however, only the copper salts were successful at promoting the fluorination reaction. The reaction occurred under mild conditions using triethylamine-HF adduct as the fluorine source. Yields were generally moderate to good, and a wide array of functional groups was well tolerated by the chemistry. It is worth noting that compounds bearing donor heteroatoms such as oxygen and nitrogen were generally fluorinated, while an allylic bromide on a pure hydrocarbon chain was unreactive under the reaction conditions.

The fluorination of unfunctionalized C—H bonds is a challenging reaction; however, it is a highly valuable transformation. To this end, the palladium-catalyzed fluorination of benzyl C—H bonds was developed (Example 7.1) [12]. The chemistry used a common palladium salt to catalyze the process, and N-fluoropyridinium tetrafluoroborate served as a source of electrophilic fluorine. While only a few substrates were screened in this work, it provided the proof of concept for the approach. Following this report, a host of benzyl C—H fluorination reactions have been reported. Several of these reactions are summarized in Schemes 7.5 through 7.9 and Examples 7.1 through 7.3.

"Chiral anion phase-transfer catalysis" has been extended to the asymmetric synthesis of allylic fluorides (Scheme 7.10) [20]. The key with this chemistry was judicious selection of the electrophilic fluorine source. Selectfluor displays poor solubility in nonpolar solvents such as cyclohexane; however, combining this F$^+$ source with a chiral phosphate salt generated afforded a soluble ion pair that promoted the chemistry. A wide range of

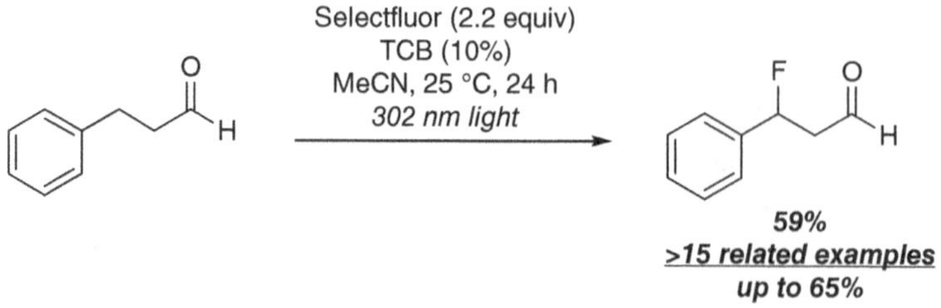

SCHEME 7.5 Photocatalyzed benzyl C—H fluorination [13].

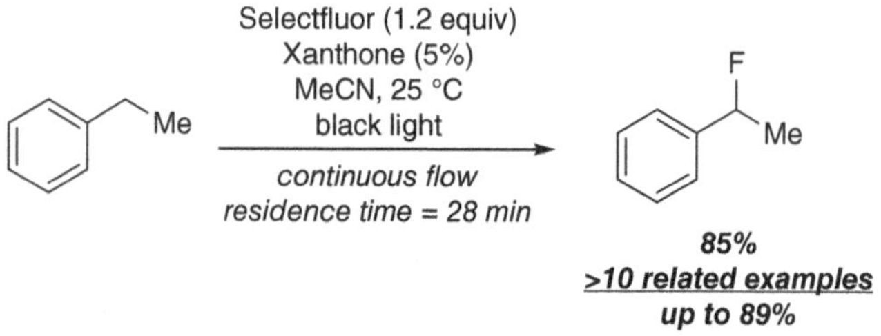

SCHEME 7.6 Photocatalyzed benzyl C—H fluorination under continuous flow [14].

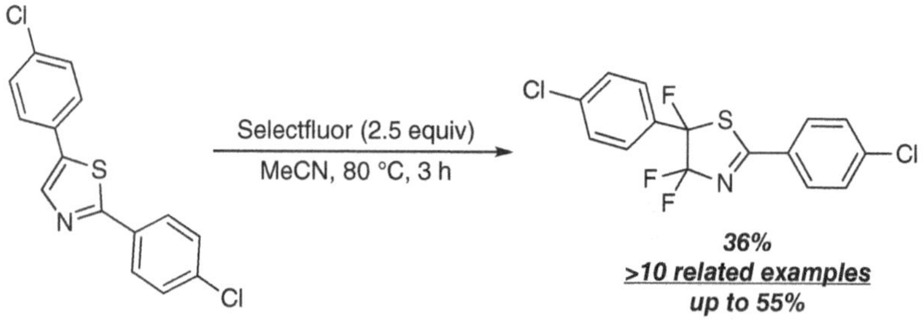

SCHEME 7.7 Fluorination of C—H bonds [15].

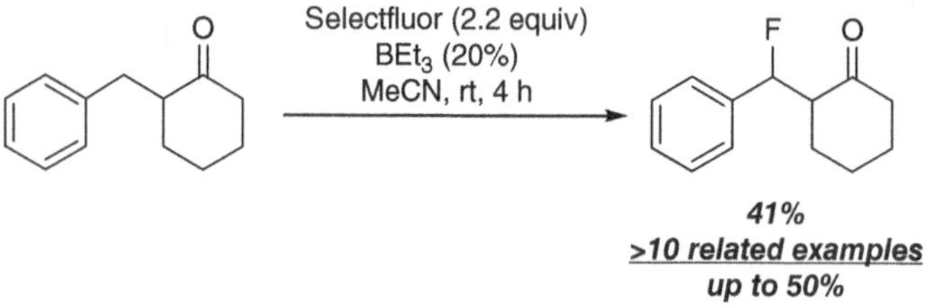

SCHEME 7.8 Radical benzyl C—H fluorination using triethylborane [16].

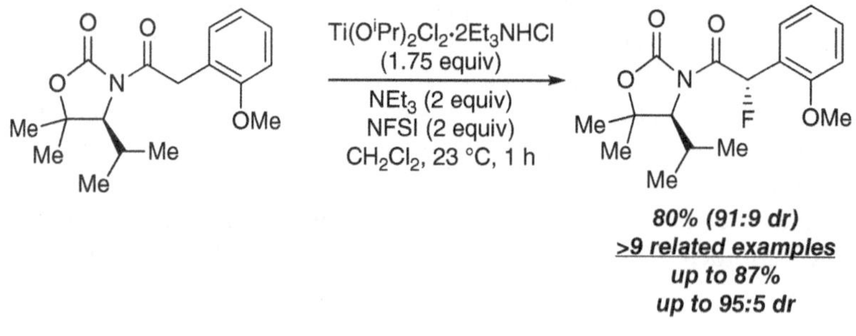

SCHEME 7.9 Asymmetric benzyl C—H fluorination using titanium alkoxide [17].

Example 7.1   Palladium-Catalyzed Fluorination through C—H Activation [12].

Chemical Safety Instructions

Before starting this synthesis, all safety, health, and environmental concerns must be evaluated using the most recent information, and the appropriate safety protocols followed. All appropriate personal safety equipment must be used. All waste must be disposed of in accordance with all current local and government regulations.

The individual performing these procedures and techniques assumes all risks and is responsible for ensuring the safety of themselves and those around them. This chemistry is not intended for the novice and should only be attempted by professionals who have been well trained in synthetic organic chemistry. The authors and Wiley assume no risk and disclaim any liabilities for any damages or injuries associated in any way with the chemicals, procedures, and techniques described herein.

(1.6 equiv)   **57%**

Caution! Benzene and hexanes are highly flammable.

Caution! The health effects of benzene have not been fully established. All current information needs to be reviewed prior to use this solvent.

5,8-Dimethylquinoline (105.7 mg, 0.669 mmol, 1 equiv), N-fluoro-2,4,6-trimethylpyridinium tetrafluoroborate (237.9 mg, 1.05 mmol, 1.6 equiv), and Pd(OAc)$_2$ (15.6 mg, 0.069 mmol, 10 mol %) were combined in benzene (4.5 mL) in an 80 mL microwave reaction vessel equipped with a Teflon stir bar. The reaction mixture was heated in the microwave at 110 °C for 1 h (250 W, 110 °C, 1 h, 10 min ramp). The resulting yellow suspension was cooled to room temperature, naphthalene was added as an internal standard, and the reaction was analyzed by gas chromatography (which showed ~57% GC yield). The reaction was then diluted with acetonitrile (10 mL), filtered through a plug of Na$_2$CO$_3$ (2.5 mg) and Celite (0.5 mg), and the plug was washed with copious acetonitrile (3 × 50 mL). The resulting solution was evaporated to dryness, and the brown solid was purified by chromatography on silica gel (R_f = 0.07 in 90% toluene/10% hexanes). The fluorinated product was obtained as a yellow oil (67 mg, 57% yield). Note: GC–MS analysis of the crude reaction mixture also showed unreacted material (~4% GC yield), as well as traces of the corresponding acetoxylated product (~5% GC yield), the corresponding phenylated product (~2% GC yield), and the corresponding difluorinated product (~4% GC yield).

Adapted with permission from K. L. Hull, W. Q. Anani, and M. S. Sanford, J. Am. Chem. Soc., 2006, 128, 7134–7135. Copyright (2006) American Chemical Society.

compounds were selectively functionalized with the highest selectivity obtained using unsubstituted or para-substituted arenes. Substrates bearing ortho-substituents were not as selective (74–77% ee). Overall, this is a very attractive route to the synthesis of chiral allylic fluorides.

Example 7.2 Asymmetric Fluorination of Branched Cyclohexanones [18].

Before starting this synthesis, carefully follow the chemical safety instructions listed in Example 7.1.

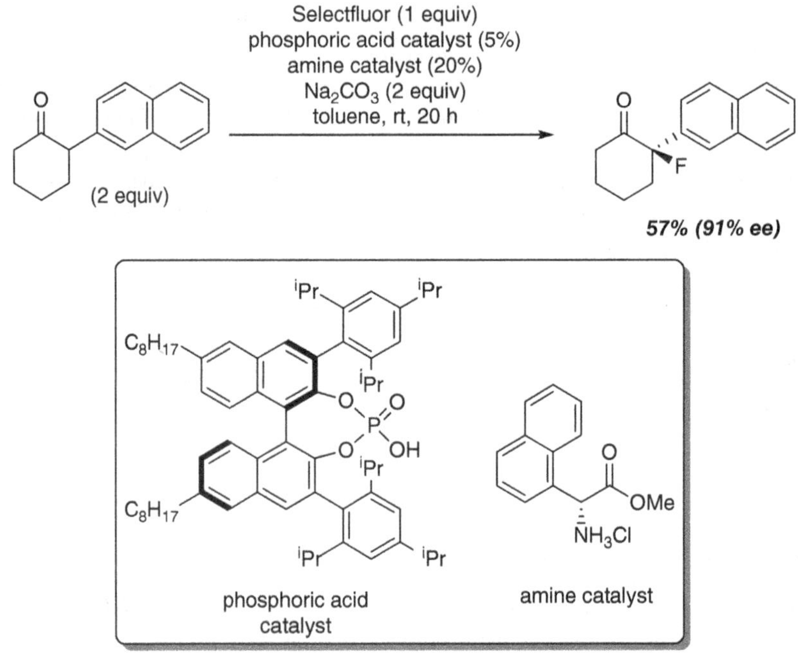
Selectfluor (1 equiv)
phosphoric acid catalyst (5%)
amine catalyst (20%)
Na_2CO_3 (2 equiv)
toluene, rt, 20 h
(2 equiv)
57% (91% ee)
phosphoric acid catalyst
amine catalyst

Caution! Hexane and ethyl acetate are highly flammable.

To the substrate (0.20 mmol) in a 1 dram (15 × 45 mm) vial equipped with an 8 mm magnetic stirrer bar was added toluene (1.0 mL). Subsequently, $Na_2CO_3{\cdot}H_2O$ (24.8 mg, 0.20 mmol), Selectfluor (95%, 36.0 mg, 0.10 mmol), (D)-amino acid methyl ester hydrochloride salt (0.02 mmol), and (*R*)-C_8-TRIP (5.0 mg, 0.005 mmol) were added. The vial was capped with a screw cap and stirred rapidly for 40 h at room temperature, the vial standing on the stirrer plate. During this time, the vial was shaken every 10 h to agitate material adhered to the sides of the vial. After this time, the reaction was diluted with ethyl acetate and poured into satd. $NaHCO_3$ solution. After extraction, the aqueous layer was extracted with further EtOAc, and the combined organics were, dried (Na_2SO_4), and evaporated in vacuo. The crude residue was purified by flash column chromatography (7 : 3 hexane/DCM) to give the title compound as an oil (13.8 mg, 0.0570 mmol, 29% based on ketone, 57% based on Selectfluor). It is notable that fast and efficient stirring should be maintained in order to achieve reliable results.

Adapted with permission from X. Yang, R. J. Phipps, and F. D. Toste, *J. Am. Chem. Soc.*, **2014**, *136*, 5225–5228. *Copyright (2014) American Chemical Society.*

Example 7.3   C—H Fluorination with Nucleophilic Fluoride [19].

Before starting this synthesis, carefully follow the chemical safety instructions listed in Example 7.1.

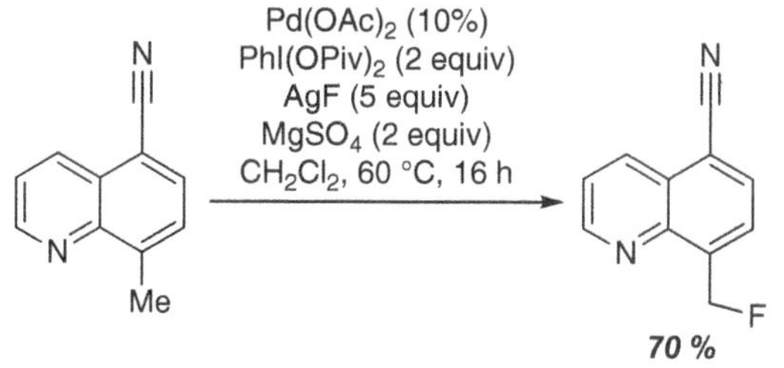

Pd(OAc)$_2$ (10%)
PhI(OPiv)$_2$ (2 equiv)
AgF (5 equiv)
MgSO$_4$ (2 equiv)
CH$_2$Cl$_2$, 60 °C, 16 h

70 %

Caution! Ethyl acetate is highly flammable.

Caution! The reaction is carried out in a sealed vessel above the boiling point of the solvent. Care must be taken to cool the vessel to room temperature before opening.

On the benchtop, PhI(OPiv)$_2$ (650 mg, 1.60 mmol, 2.0 equiv) and AgF (508 mg, 4.00 mmol, 5.0 equiv) were weighed into a 20 mL vial, and CH$_2$Cl$_2$ (8.0 mL) was added. The vial was sealed with a Teflon-lined cap (with Teflon tape covering the threads of the vial), and this mixture was heated at 60 °C for 1 h. The vial was removed from the heat and allowed to cool to rt (~10 min). Then, Pd(OAc)$_2$ (18 mg, 0.08 mmol, 0.10 equiv), MgSO$_4$ (193 mg, 1.60 mmol, 2.0 equiv), and 8-methylquinoline-5-carbonitrile (135 mg, 0.80 mmol, 1.0 equiv) were added. The vial was sealed with a Teflon-lined cap and heated at 60 °C for an additional 16 h. A GC yield of 67% was obtained by analysis of the crude reaction mixture using hexadecane as a standard. The reaction mixture was then filtered through a plug of Celite, the plug was washed with CH$_2$Cl$_2$ (100 mL), and the filtrate was concentrated under vacuum. The resulting crude product was purified by chromatography on silica gel with a gradient of 0–4% EtOAc in CH$_2$Cl$_2$. The fluorinated product was obtained as a tan solid (105 mg, 70%).

Adapted with permission from K. B. McMurtrey, J. M. Racowski, and M. S. Sanford, Org. Lett., 2012, 14, 4094–4097. Copyright (2012) American Chemical Society.

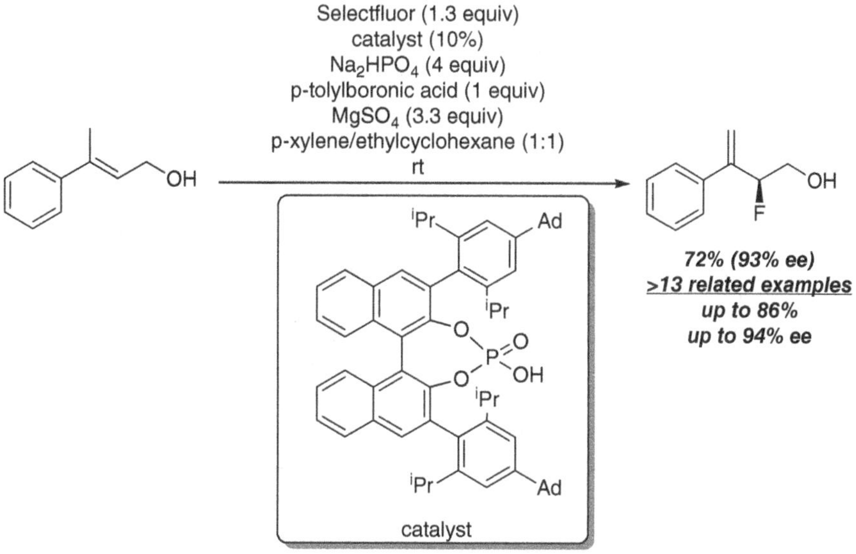

Selectfluor (1.3 equiv)
catalyst (10%)
Na$_2$HPO$_4$ (4 equiv)
p-tolylboronic acid (1 equiv)
MgSO$_4$ (3.3 equiv)
p-xylene/ethylcyclohexane (1:1)
rt

72% (93% ee)
>13 related examples
up to 86%
up to 94% ee

catalyst

SCHEME 7.10   Enantioselective synthesis of allylic fluorides [20].

One of the most efficient methods of incorporating multiple functional groups into the same compound is through a halofunctionalization reaction [3]. To this end, an iron-catalyzed aminofluorination reaction has been developed using hydroxylamines as substrates (Example 7.4) [21]. Control reactions revealed that both triethylamine-HF and XtalFluor-E were needed to promote the chemistry and the latter was proposed to suppress a competing

Example 7.4   Iron-Catalyzed Aminofluorination [21].

Before starting this synthesis, carefully follow the chemical safety instructions listed in Example 7.1.

58% (>20:1 dr)

Caution! Hexane and ethyl acetate are highly flammable.

Caution! Even though $Et_3N \cdot 3HF$ is considerably safer to handle and manipulate than HF, it must still be treated with great care, and all current safety procedures for its use and storage must be reviewed and followed.

To a flame-dried 5 mL vial (vial A) with freshly activated 4Å molecular sieves (equal weight to starting material) and a magnetic stir bar were added $Fe(BF_4)_2 \cdot 6H_2O$ (0.01 mmol, 3.4 mg) and ligand (0.01 mmol, 2.4 mg). After the vial was evacuated and backfilled with argon for three times, anhydrous and degassed DCM (1.0 mL) was added. The mixture was stirred vigorously at room temperature for 20 min. During this time, to another flame-dried vial (vial B) was added XtalFluor-E (0.12 mmol, 27.5 mg). After the vial was evacuated and backfilled with argon for three times, anhydrous and degassed DCM (1.0 mL) was added. $Et_3N \cdot 3HF$ (0.14 mmol, 23 µL) was subsequently added into vial B, and the solution was stirred for 3 min till it became clear. The vial A was cooled in an ice-water bath and the solution in vial B was added into it dropwise. The acyloxyl carbamate (0.1 mmol, dissolved in 1 mL of DCM) was then added to vial A dropwise at the same temperature. The mixture was stirred at 0 °C for another 10 min and then kept stirring at room temperature until the acyloxyl carbamate was consumed monitored by TLC. After the reaction was quenched by saturated aqueous sodium bicarbonate solution (1 mL), the aqueous layer was separated and extracted by DCM (2 mL × 3). The combined organic layer was dried over anhydrous Na_2SO_4 and concentrated in vacuo. The residue was purified by flash columns with 1 : 1 hexane and ethyl acetate as the eluent to afford fluorooxazolidinone (58%, >20 : 1 d.r.).

Adapted with permission from D.-F. Lu, G.-S. Liu, C.-L. Zhu, B. Yuan, and H. Xu, Org. Lett., 2014, 16, 2912–2915. Copyright (2014) American Chemical Society.

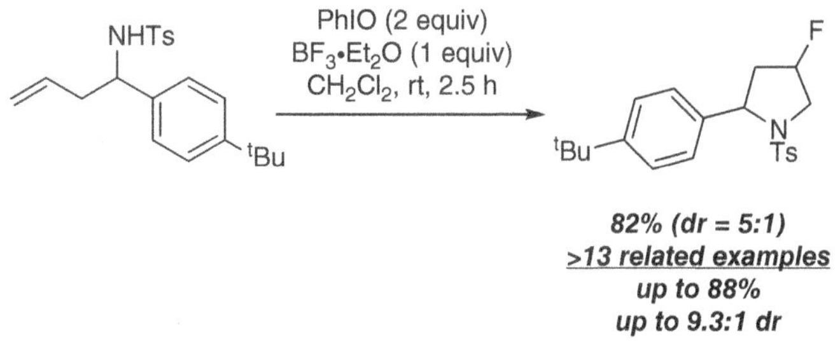

SCHEME 7.11   Fluorination using BF_3 etherate as F source [22].

aminohydroxylation pathway. The triethylamine-HF adduct was also critical for the chemistry since decomposition of the starting material was observed in its absence. The selectivity of the reaction was moderate to high, and diastereomeric ratios of greater than 20:1 were found for several substrates. It should also be noted that BF_3 etherate has been found to function as a fluorine source in iodosobenzene-promoted aminofluorination reactions (Scheme 7.11) [22].

While the metal-catalyzed fluorination of benzyl C—H bonds has generated a host of valuable compounds, extending this chemistry to saturated hydrocarbons remains a current and significant challenge. To address this issue, a successful approach for the copper-catalyzed fluorination of unfunctionalized hydrocarbons has been devised (Example 7.5) [23, 24]. The catalyst for this reaction was an bisimine copper complex, and Selectfluor was used as an electrophilic source of fluorine. The copper complex was essential for the reaction as no fluorination was observed in its absence. A host of cyclic and acyclic saturated hydrocarbons were successfully fluorinated using this approach, and moderate to good yields of the alkyl fluorides were obtained. For substrates such as ethylbenzene and dihydrocoumarin, fluorination of the benzylic position was preferred.

In related work, the development of a process for the fluorination of unactivated C—H bonds has been achieved using visible light catalysis (Scheme 7.12) [25]. The process used a decatungstate species as the photocatalyst and NFSI as the fluorine source and was attractive as the photocatalyst was trivial to prepare. The reaction conditions were mild, and moderate to excellent yields of the fluorinated compounds were obtained. This chemistry tolerated the presence of a range of preexisting functional groups including esters, acetates, and heterocycles. It was also noteworthy that acyl fluorides could be generated from aldehydes using this chemistry.

The palladium-catalyzed fluoroarylation of styrene derivatives is an attractive reaction since it extends the carbon framework of the substrate while simultaneously incorporating the halogen (Scheme 7.13) [26]. A version of this three-component reaction used palladium acetate as the metal precursor along with a bipyridine derivative to stabilize/solubilize the metal. While some palladium-catalyzed reactions were successful without the addition of a coordinating ligand, the bis*tert*-butylbipyridine was critical to the success of this chemistry since its absence resulted in exclusive generation of Heck products. Additionally, Selectfluor was the most active fluorine source for this chemistry. The authors used amide groups to direct the point of fluorination and proposed that the process involved a Pd(II)/Pd(IV) cycle. It was also noteworthy that the reaction chemistry was not only stable to moisture, but the reaction rate was increased when small amounts of water were added to the system. Furthermore, the addition of a phosphate bearing long side chains functioned as a phase-transfer agent and promoted the reaction. Once the system was optimized, it was extended to a range of substrates bearing a broad spectrum of functional groups including halogens, ethers, esters, protected amines, and sulfones. Even the bulky 2,6-dimethylphenylboronic acid was successfully used in the reaction (45%). The chemistry could be extended to the asymmetric fluoroarylation reactions by incorporating a chiral ligand on the metal center (Example 7.6).

Example 7.5   Copper-Catalyzed Fluorination of Alkanes [23].

Before starting this synthesis, carefully follow the chemical safety instructions listed in Example 7.1.

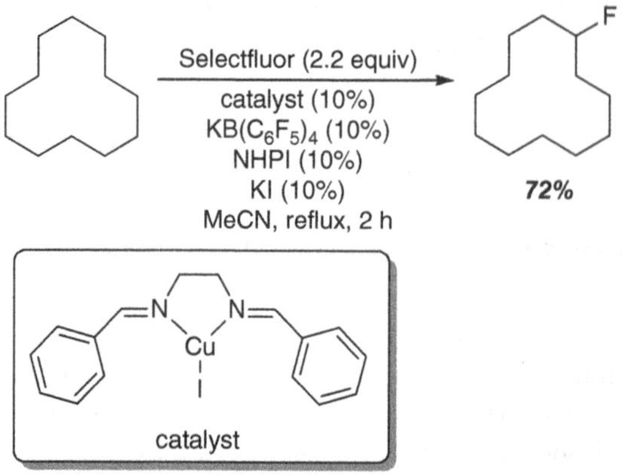

An oven-dried, 10 mL round-bottom flask equipped with a stir bar was placed under an atmosphere of N_2. $KB(C_6F_5)_4$ (22.0 mg, 0.025 mmol, 0.1 equiv), Selectfluor (195 mg, 0.55 mmol, 2.2 equiv), copper(I) iodide (5.0 mg, 0.025 mmol, 0.1 equiv), N,N-bis(phenylmethylene)-1,2-ethanediamine (6.0 mg, 0.025 mmol, 0.1 equiv), hydroxyphthalimide (4.0 mg, 0.025 mmol, 0.1 equiv), and potassium iodide (4.2 mg, 0.025 mmol, 0.1 equiv) were then added, followed by degassed MeCN (3.0 mL). The reaction mixture was then stirred for 10 min. Under a stream on N_2, the cycloalkane (0.25 mmol, 1.0 equiv) was added neat and the mixture heated to reflux for 2 h. The reaction was monitored by ^{19}F NMR at 30 min intervals. Final yields were determined either by ^{19}F NMR spectroscopy using 3-chlorobenzotrifluoride as an internal standard or from isolated material (column chromatography on silica).

Adapted with permission from S. Bloom, C. R. Pitts, D. C. Miller, N. Haselton, M. G. Holl, E. Urheim, and T. Lectka, *Angew. Chem. Int. Ed.*, **2012**, *51*, 10580–10583. *Copyright (2012) Wiley-VCH Verlag GmbH & Co. KGaA, Weinheim.*

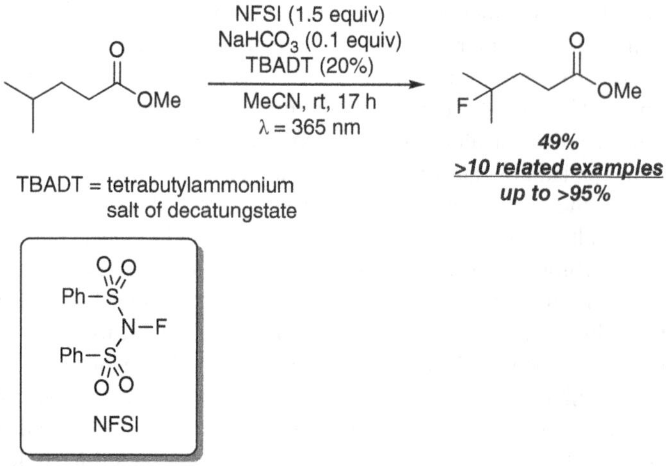

SCHEME 7.12   Light-promoted fluorination of C—H bonds [25].

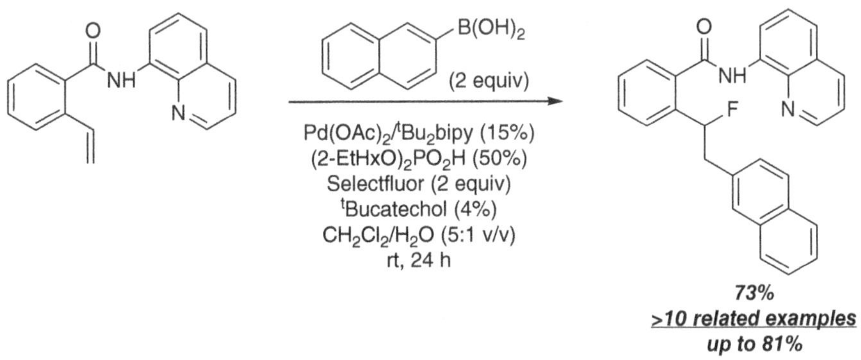

SCHEME 7.13   Palladium-catalyzed fluoroarylation of styrene derivatives [26].

Example 7.6   Asymmetric Fluoroarylation of Styrenes [26].

Before starting this synthesis, carefully follow the chemical safety instructions listed in Example 7.1.

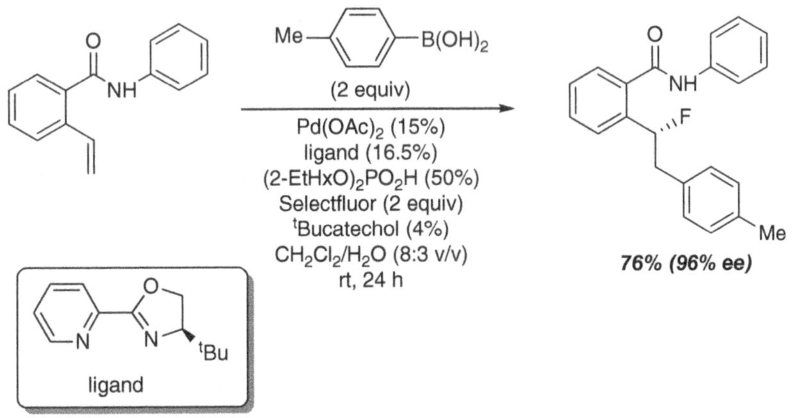

Caution! Ethyl acetate and isohexanes are highly flammable.

Part 1: Preparation of the Catalyst.   Pd(OAc)$_2$ (3.3 mg, 0. 01 mmol, 15 mol %) was added to a solution of (*S*)-2-(*tert*-butyl)-4-(pyridin-2-yl)-2,5-dihydrooxazole (3.3 mg, 0.0165 mmol) in CH$_2$Cl$_2$ (0.2 mL), and the reaction mixture was stirred for 30 min.

Part 2: Asymmetric Synthesis of the Benzyl Fluoride.   The solution containing the catalyst was then added to a solution of the styrene derivative (0.1 mmol), arylboronic acid (0.2 mmol), bis(2-ethylhexyl) hydrogen phosphate (16 mg, 0.05 mmol), Selectfluor (106 mg, 0.2 mmol), and *tert*-butyl catechol (0.6 mg, 0.004 mmol) in CH$_2$Cl$_2$ (0.8 mL)/ water (0.3 mL). The reaction mixture was then degassed with N$_2$ and vigorously stirred for 15 h. The reaction was passed through Celite, concentrated under reduce pressure, and purified by column chromatography (0–10% EtOAc in isohexanes) to give the fluorocarboxamide (25.5 mg, 76%) as a white solid.

Adapted with permission from E. P. A. Talbot, T. de A. Fernandes, J. M. McKenna, and F. D. Toste, *J. Am. Chem. Soc.*, **2014**, *136*, 4101–4104. *Copyright (2014) American Chemical Society.*

Example 7.7   One-Pot Anti-Markovnikov Hydrofluorination of Unactivated Alkenes [27].

Before starting this synthesis, carefully follow the chemical safety instructions listed in Example 7.1.

Caution! Ethyl acetate and hexane are highly flammable.

Rh(PPh$_3$)$_3$Cl (3.7 mg, 0.004 mmol) and 2-(but-3-en-1-yl)isoindoline-1,3-dione (40.2 mg, 0.2 mmol) were placed in a Schlenk tube. The vessel was evacuated and filled with nitrogen for three times. Anhydrous CH$_2$Cl$_2$ (0.2 mL) and pinacolborane (32 μL, 0.24 mmol) were then added successively at room temperature. The reaction mixture was stirred for another 3 h at room temperature. After completion, AgNO$_3$ (6.8 mg, 0.04 mmol) and Selectfluor (212 mg, 0.6 mmol) were added into the reaction mixture, followed by CH$_2$Cl$_2$ (0.4 mL), TFA (0.4 mL), H$_3$PO$_4$ (0.1 mL), and water (0.9 mL). The reaction mixture was then stirred at 50 °C for another 8 h. The resulting mixture was extracted with CH$_2$Cl$_2$ (15 mL × 3). The combined organic phases were dried over anhydrous Na$_2$SO$_4$. After the removal of solvent under reduced pressure, the crude product was purified by column chromatography on silica gel with hexane/ethyl acetate (5 : 1, v : v) as the eluent to give the pure product 2-(4-fluorobutyl)isoindoline-1,3-dione as a yellow oil (29.6 mg, 67%).

Adapted with permission from Z. Li, Z. Wang, L. Zhu, X. Tan, and C. Li, J. Am. Chem. Soc., 2014, 136, 16439–16443. Copyright (2014) American Chemical Society.

The formal addition of HF to alkenes would be an attractive approach to the synthesis of alkyl fluorides. To circumvent the issues surrounding the use of hazardous HF for this reaction, a two-step approach has been developed (Example 7.7) [27]. The first step in the process consisted of a rhodium-catalyzed anti-Markovnikov hydroboration of an alkene in order to generate an intermediate alkylboronate. The second step entailed the use of a silver salt to promote a fluorodeboronation of the intermediate. The result of this two-step process was an overall hydrofluorination of an unactivated alkene. In related work, the regioselectivity of the hydrofluorination reaction was reversed using a cobalt-catalyzed process (Markovnikov selective) (Scheme 7.14) [28].

The synthesis of α-fluoroketones is an valuable transformation due to the importance of these compounds in medicine and organic synthesis. The conversion of styrene derivatives into α-fluoroketones would be an attractive reaction due to the abundance of available styrene derivatives. To achieve this goal, an oxyfluorination reaction has been developed under metal-free conditions using IBX and molecular oxygen as oxidizing agents (Scheme 7.15) [29]. While a number of the common sources of electrophilic fluorine were active in the reaction, Selectfluor was the most effective. Yields of the fluorinated ketones were moderate to good, and a wide range of functional groups were well tolerated by this fluorination reaction.

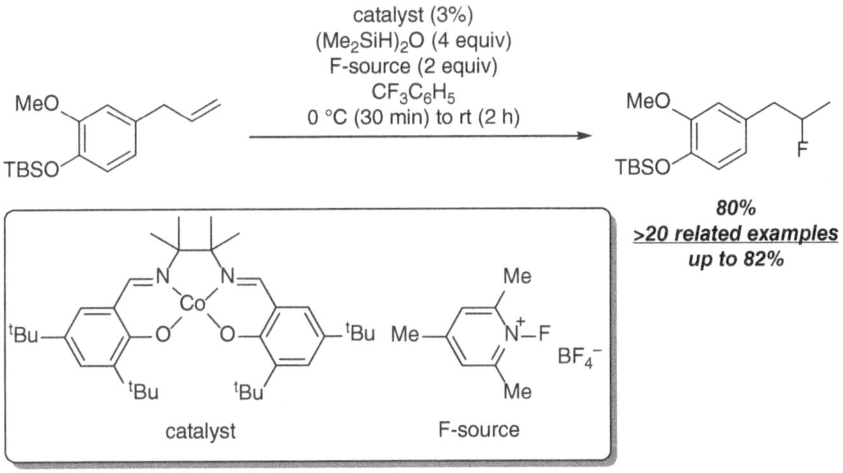

SCHEME 7.14   Cobalt-catalyzed hydrofluorination of unactivated alkenes [28].

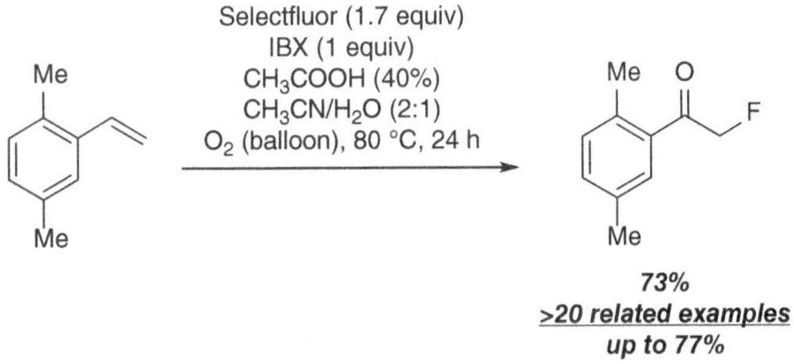

SCHEME 7.15   Transition metal-free oxyfluorination of styrene derivatives [29].

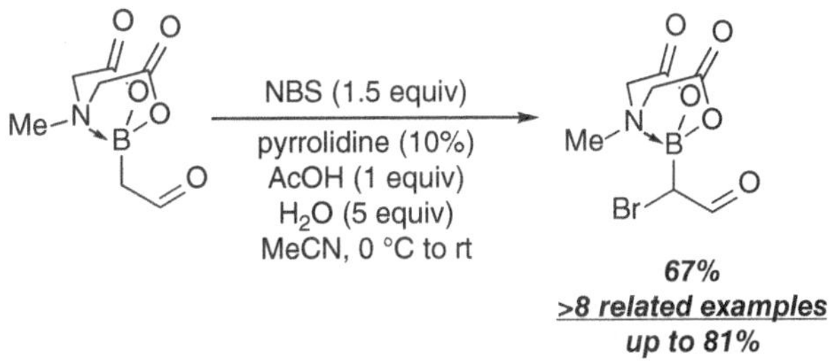

SCHEME 7.16   Bromination of MIDA boronates [30].

While a great deal of attention has been given to MIDA boronates due to their ability to transmetallate palladium in cross-coupling reactions, they can also serve as substrates for halogenation reactions (Scheme 7.16) [30]. This chemistry was catalyzed by pyrrolidine and used NBS as a bromine source. It is noteworthy to point out that the chemistry was optimized using a stoichiometric amount of acetic acid and a slight excess of water. This solvent mixture aided in reducing the amount of dibrominated products that were generated in the reaction. A range of MIDA boronates were successfully monohalogenated using this approach in moderate to good yields. The chemistry could be extended to the synthesis of chlorinated derivatives by exchanging NBS for NCS (Example 7.8). Curiously, the yields of the α-chlorinated compounds were typically higher (up to 95%) than the brominated analogues.

An asymmetric chlorination/reduction reaction for the synthesis of β-chloroalcohols has been developed using MacMillan's resolved catalyst and NCS as the chlorine source (Scheme 7.17) [31]. The reaction was highly selective and enantiomeric ratios of >95:5 were common.

Example 7.8   Chlorination of MIDA Boronates [30].

Before starting this synthesis, carefully follow the chemical safety instructions listed in Example 7.1.

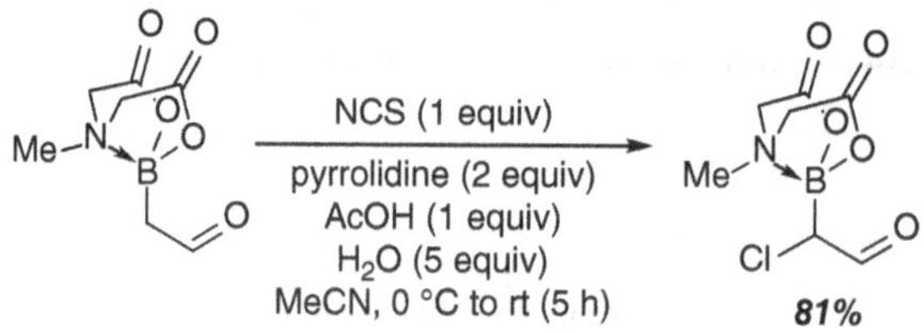

Caution! Hexane and ethyl acetate are highly flammable.

To a solution of the parent boryl aldehyde 1 (1.5 mmol, 1.0 equiv) in CH$_3$CN (15 mL) was added pyrrolidine (3.0 mmol, 2.0 equiv), AcOH (1.5 mmol, 1.0 equiv), and H$_2$O (7.5 mmol, 5.0 equiv). The mixture was stirred for 5 min. *N*-chlorosuccinimide (1.5 mmol, 1 equiv) was added in one portion, and the reaction mixture stirred at room temperature for 5 h. Afterwards, the solvent was removed in vacuo to afford the crude chlorinated enamine. The crude residue was passed through a silica gel column (hexanes/EtOAc 8:2 to pure EtOAc) to afford the chloroaldehyde.

*Adapted with permission from J. D. St. Denis, A. Zajdlik, J. Tan, P. Trinchera, C. F. Lee, Z. He, S. Adachi, and A. K. Yudin, J. Am. Chem. Soc., **2014**, 136, 17669–17673. Copyright (2014) American Chemical Society.*

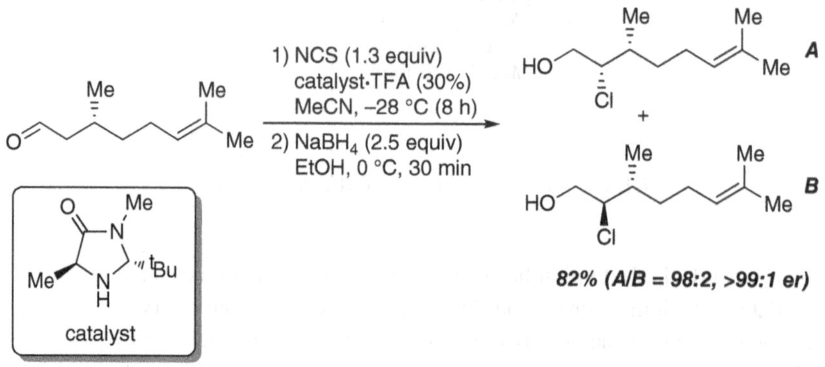

SCHEME 7.17   One-pot chlorination/reduction of resolved citronellal [31].

A halogenative semipinacol rearrangement has been used to generate a range of alkyl fluorides, bromides, and iodides. These reactions are often promoted through the use of a resolved catalyst along with an electrophilic source of the halogen. This area has been the subject of intense research, and representative approaches are illustrated in Schemes 7.18 through 7.20 and Example 7.9.

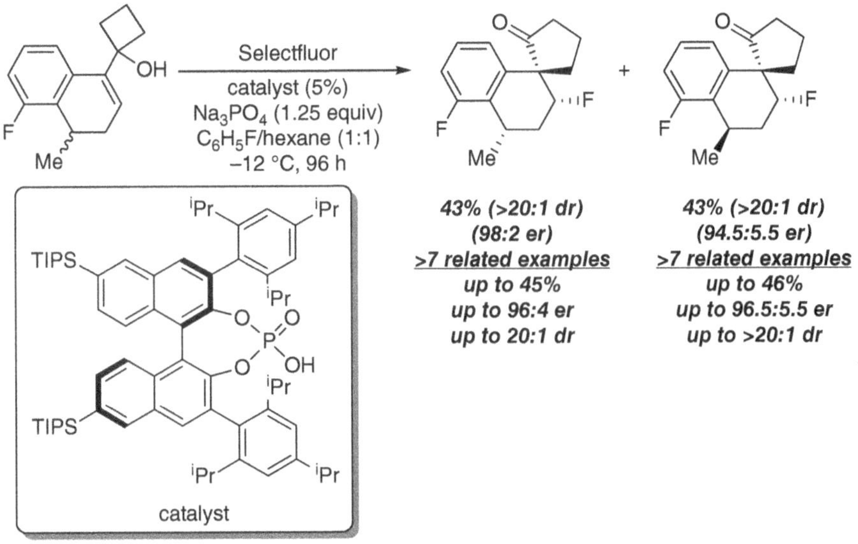

SCHEME 7.18   Fluorinative stereodivergent semipinacol rearrangement [32].

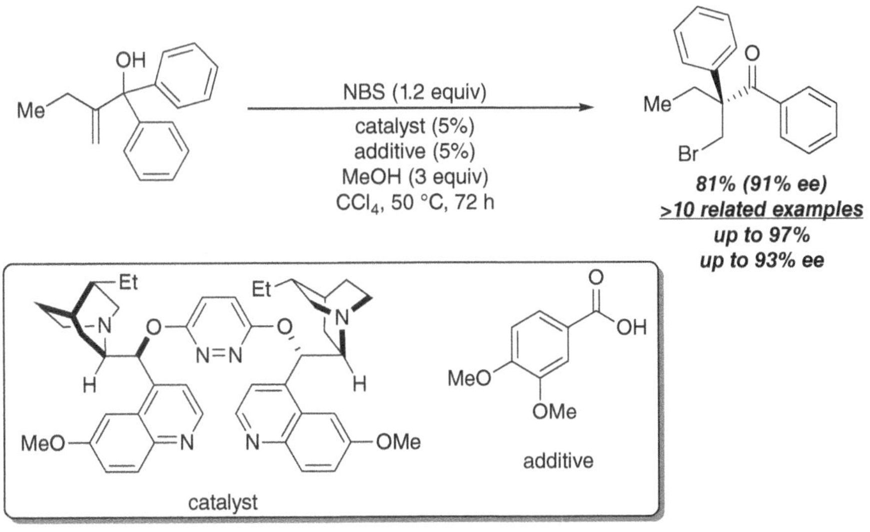

SCHEME 7.19   Asymmetric semipinacol rearrangement using a DHQD-based organocatalyst [33].

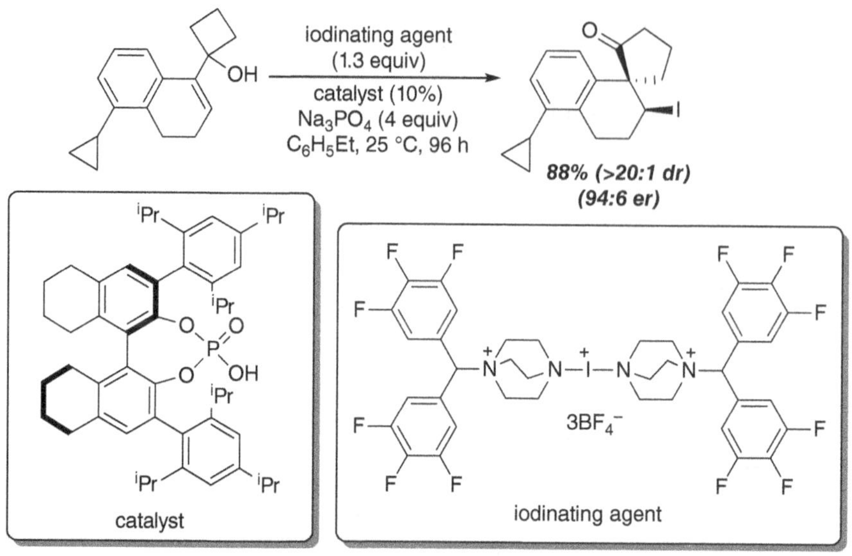

SCHEME 7.20 Asymmetric bromination–semipinacol rearrangement [34].

Example 7.9 Iodination–Semipinacol Rearrangement Using a Phosphoric Acid Catalyst [34].

Before starting this synthesis, carefully follow the chemical safety instructions listed in Example 7.1.

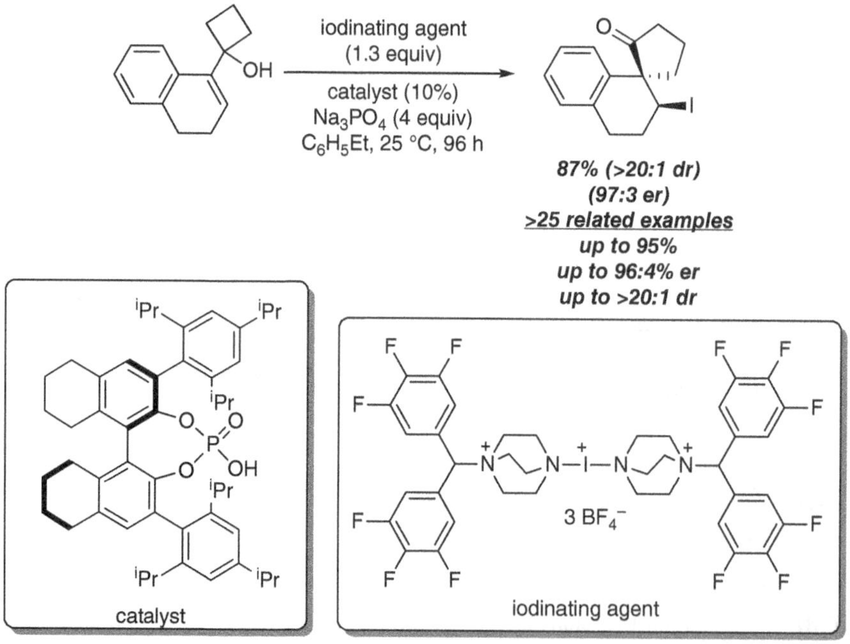

*Caution! Methyl **tert**-butyl ether, hexane, ethylbenzene, and diethyl ether are highly flammable.*

To a well-stirred solution of 1-(3,4-dihydronaphthalen-1-yl)cyclobutanol (40 mg, 0.20 mmol, 1.0 equiv) and chiral phosphoric acid (15 mg, 0.02 mmol, 10 mol %) in anhydrous ethylbenzene (4.0 mL, 0.05 M) were added iodinating reagent (302 mg, 0.26 mmol, 1.3 equiv) and powdered anhydrous Na_3PO_4 (131 mg, 0.80 mmol, 4.0 equiv). The resultant heterogeneous mixture was stirred at ambient temperature for 48–72 h. Saturated aqueous $Na_2S_2O_3$ was then added to quench the reaction. The layers were separated, and the aqueous layer was extracted with methyl *tert*-butyl ether. The combined organic extracts were dried over anhydrous Na_2SO_4, filtered and concentrated in vacuo. Conversions and diastereomeric ratios (d.r.) were determined by ^1H NMR analysis of the crude compounds. Pure material was obtained after purification by flash chromatography on silica gel (*n*-hexane/Et_2O 99 : 1 to 95 : 5) as a white amorphous solid (57 mg, 87%). Enantiomeric ratios (e.r.) were determined by chiral HPLC or chiral SFC analysis of purified compounds.

Adapted with permission from F. Romanov-Michailidis, L. Guenee, and A. Alexakis, *Org. Lett.*, **2013**, *15*, 5890–5893. *Copyright (2013) American Chemical Society.*

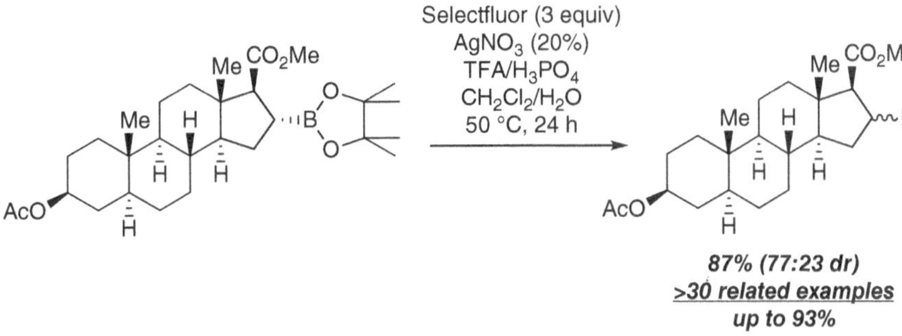

87% (77:23 dr)
>30 related examples
up to 93%

SCHEME 7.21   Silver-promoted fluorination of alkylboronates using Selectfluor [27].

A catalytic amount of silver nitrate was used to promote the fluorination of alkylboronates (Scheme 7.21) [27]. The silver was critical to the reaction as no fluorination occurred upon its absence. Selectfluor served as the source of fluorine in this reaction, and cyclic alkylboronates as well as primary, secondary, and tertiary substrates were all successfully converted into alkyl fluorides using this approach. It should be noted that the chemistry was not only successful in the presence of water, but it also tolerated the presence of acid.

Copper salts were found to catalyze the aminofluorination of styrene derivatives (Scheme 7.22) [35]. NFSI served as the fluorine source in these reactions, and the addition chemistry was regioselective for the formation of the α-fluorination product. Mechanistic studies suggested that NFSI served as a source of radical nitrogen in addition to acting as a source of fluorine. The reaction typically generated good to excellent yields of the aminofluorination products and was insensitive to a number of preexisting functional groups such as ethers and halogens.

Common silver compounds such as silver nitrate catalyzed the decarboxylative fluorination of carboxylic acids (Scheme 7.23 and Example 7.10) [36]. Selectfluor served as the fluorine source for this chemistry, and simply refluxing the reagents in acetone/water mixtures generated good to excellent yields of the fluorinated compounds. Not only was the reaction successful in the presence of water, but also water was a critical component of the

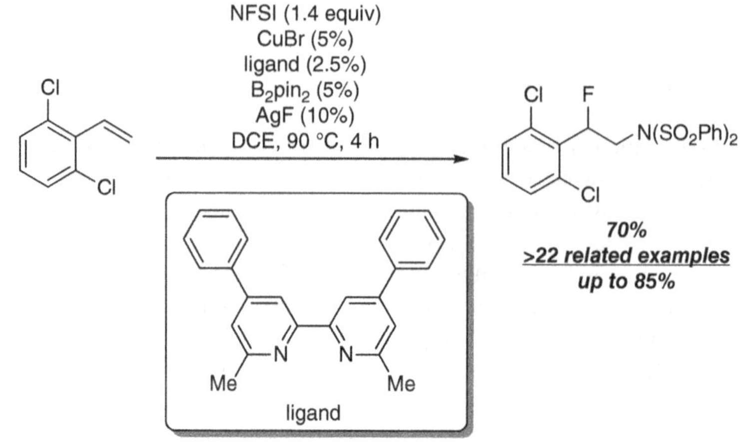

SCHEME 7.22 Copper-catalyzed aminofluorination of styrene derivatives [35].

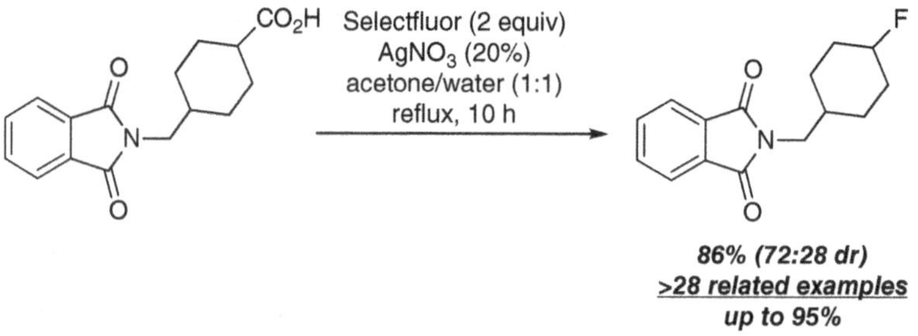

SCHEME 7.23 Silver-catalyzed decarboxylative fluorination [36].

Example 7.10 Silver-Catalyzed Decarboxylative Fluorination [36].

Before starting this synthesis, carefully follow the chemical safety instructions listed in Example 7.1.

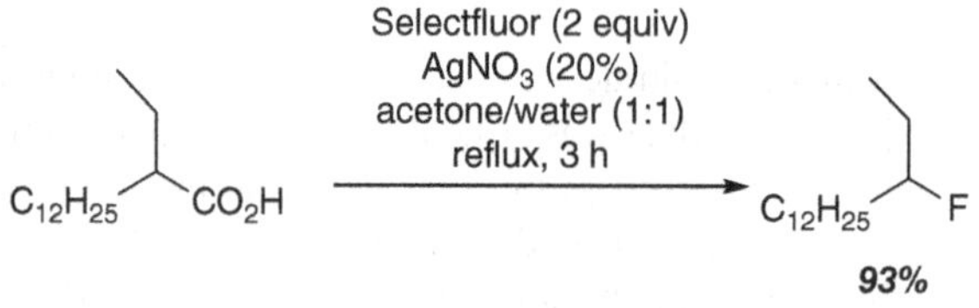

Caution! Hexane is highly flammable.

2-Ethyltetradecanoic acid (51.2 mg, 0.20 mmol), AgNO₃ (6.8 mg, 0.04 mmol), and Selectfluor (141.6 mg, 0.4 mmol) were placed in a Schlenk tube. The reaction vessel was evacuated and filled with nitrogen. Acetone (2 mL) and water (2 mL) were then added. The reaction mixture was then stirred at refluxing temperature for 10 h. Upon completion of the reaction, the resulting mixture was cooled down to room temperature and extracted with dichloromethane (15 mL × 3). The combined organic phase was dried over anhydrous Na₂SO₄. After the removal of solvent under reduced pressure, the crude product was purified by column chromatography on silica gel with hexane as the eluent to give the pure product 3-fluoropentadecane as a colorless oil (42.7 mg, 93%).

Adapted with permission from F. Yin, Z. Wang, Z. Li, and C. Li, *J. Am. Chem. Soc.*, **2012**, *134*, 10401–10404. Copyright (2012) American Chemical Society.

reaction. Analogous reactions carried out under anhydrous conditions resulted in no conversion. A number of primary, secondary, and tertiary carboxylic acids bearing a range of preexisting functional groups such as acetate, esters, ethers, halogens, and ketones were all successfully fluorinated using this approach. Given the vast array of carboxylic acids that are available, this is quite an attractive approach to the synthesis of alkyl fluorides. The chemistry has been extended to the synthesis of di- and trifluoromethyl arenes [37].

For the synthesis of alkyl bromides and chlorides, the Appel reaction remains one of the most widely adopted and successful approaches [7]. The classic reaction uses PPh$_3$ and carbon tetrachloride to convert alcohols into alkyl chlorides. The reaction is operationally trivial, often high yielding, and has been extended to bromination reactions using CBr$_4$. Furthermore, an extension of this classic reaction enabled the conversion of alcohols into alkyl iodides (Scheme 7.24 and Example 7.11) [38, 39]. In related work, visible light catalysis has been used to convert alcohols into alkyl halides [40]. The following examples will highlight additional advancements that have been made to this classic reaction.

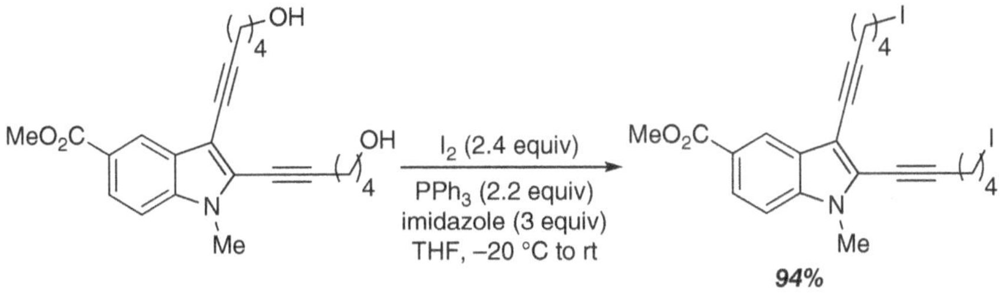

SCHEME 7.24   Conversion of alcohols into alkyl iodides using Appel-type chemistry [38].

Example 7.11   Conversion of a Primary Alcohol into an Alkyl Iodide [39].

Before starting this synthesis, carefully follow the chemical safety instructions listed in Example 7.1.

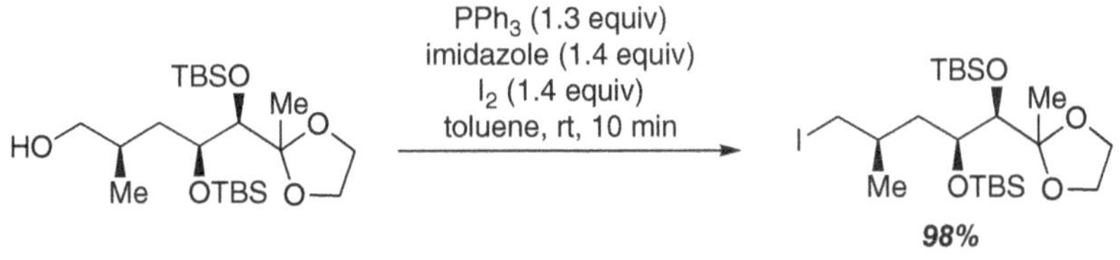

Caution! Ether, petroleum ether, and ethyl acetate are highly flammable.
To a solution of the alcohol (467 mg, 1.04 mmol) in toluene (5.0 mL) under an argon atmosphere were added successively Ph$_3$P (354 mg, 1.35 mmol), imidazole (99 mg, 1.46 mmol), and I$_2$ (371 mg, 1.46 mmol), and the mixture was stirred at rt for 10 min. After being quenched by the addition of the saturated aqueous Na$_2$S$_2$O$_3$ at 0 °C, the organic phase was separated, and the aqueous layer was extracted with ether (3 × 20 mL). The combined organic layers were washed successively with saturated NaHCO$_3$ solution and brine, dried over Na$_2$SO$_4$, and concentrated in vacuo. The residue was purified by flash column chromatography on silica gel (petroleum ether/EtOAc 50 : 1) to afford the alkyl iodide (360 mg, 98%) as an oil.

Adapted with permission from H. Lu, F.-M. Zhang, J.-L. Pan, T. Chen, and Y.-F. Li, *J. Org. Chem.*, **2014**, *79*, 546–548. Copyright (2014) American Chemical Society.

The generation of alkyl chlorides from alcohols has been reported using cyanuric chloride [41]. The chemistry was carried out in neat alcohol with heating, and moderate to good yields of the alkyl chlorides were obtained. To tolerate a wider range of substrates, the cyanuric chloride was pretreated with DMF prior to the introduction of the alkyl chloride (Scheme 7.25) [42]. This also facilitated the development of a room temperature version of this reaction. Most reactions were complete within a few minutes to a few hours, and a host of substrates bearing esters, protected amines, alkenes, ethers, and thioethers were successfully chlorinated. In addition, the monochlorination of diols was achieved using this system, and the approach was extended to the synthesis of benzyl bromides by adding sodium bromide to the reaction mixture (Example 7.12).

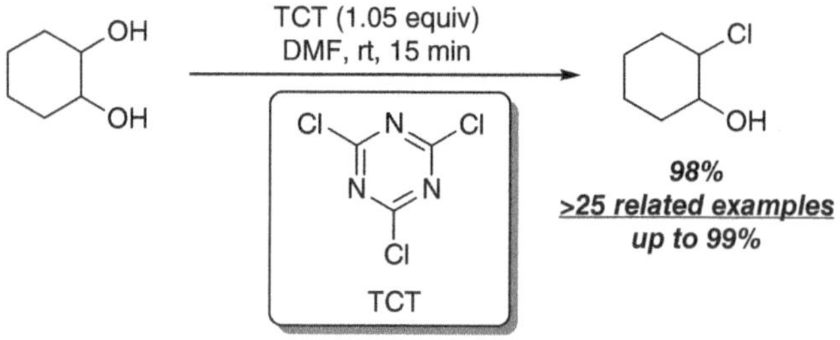

SCHEME 7.25   Selective monochlorination of diols using TCT [42].

Example 7.12   Room Temperature Conversion of Benzyl Alcohol to Benzyl Bromide [42].

Before starting this synthesis, carefully follow the chemical safety instructions listed in Example 7.1.

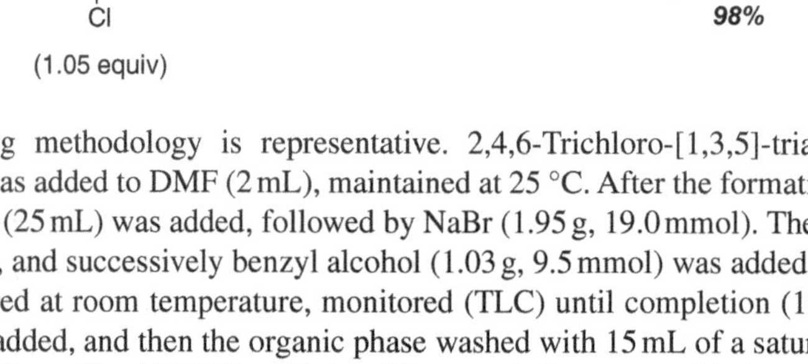

The following methodology is representative. 2,4,6-Trichloro-[1,3,5]-triazine (1.83 g, 10.0 mmol) was added to DMF (2 mL), maintained at 25 °C. After the formation of a white solid, CH_2Cl_2 (25 mL) was added, followed by NaBr (1.95 g, 19.0 mmol). The mixture was stirred for 8 h, and successively benzyl alcohol (1.03 g, 9.5 mmol) was added. The mixture was then stirred at room temperature, monitored (TLC) until completion (15 min). Water (20 mL) was added, and then the organic phase washed with 15 mL of a saturated solution of Na_2CO_3, followed by 1 N HCl and brine. The organic layer were dried (Na_2SO_4) and the solvent evaporated to yield benzyl bromide, b.p. 198 °C (1.59 g, 98%).

Adapted with permission from L. De Luca, G. Giacomelli, and A. Porcheddu, *Org. Lett.*, **2002**, *4*, 553–555. *Copyright (2002) American Chemical Society.*

Another alternative to the PPh$_3$/CX$_4$ approach to the synthesis of alkyl halides used a nucleophilic source of the halogen along with an XtalFluor-E (Scheme 7.26) [43]. While the precise mechanism of this reaction was not determined, the XtalFluor-E was proposed to initially react with the nucleophilic Br- (from TEAB) to generate diethylaminosulfur bromodifluoride. An attractive aspect of this chemistry was the reduction in the amount of organic side products resulting from the synthesis. For this chemistry, only water-soluble secondary products were generated. This facilitated the purification and isolation of the target compounds. A wide range of alcohols were successfully functionalized using this approach, and the chemistry was extended to several chlorinations and iodinations. Given the ease of separating unwanted organic products from the reaction mixture as well as the commercial availability of all the reagents, this chemistry is likely to be a popular approach to the synthesis of these compounds.

As noted earlier, triphenylphosphine oxide is typically a by-product of the Appel reaction. A significant amount of this by-product is generated from a classic Appel process, and it has been difficult to remove from the product in some cases. A clever approach to the synthesis entailed the use of catalytic amounts of this by-product to promote the reaction (Scheme 7.27) [44]. The key to this chemistry was the use of oxalyl chloride to generate chlorophosphonium salts that were the active species in the chlorodeoxygenation reaction. The reaction conditions were quite mild, and a host of alcohols were successful converted into alkyl chlorides.

In addition to the Appel reaction, the conversion of carboxylic acids into alkyl bromides and chlorides is an attractive approach to the synthesis of these compounds due in part to

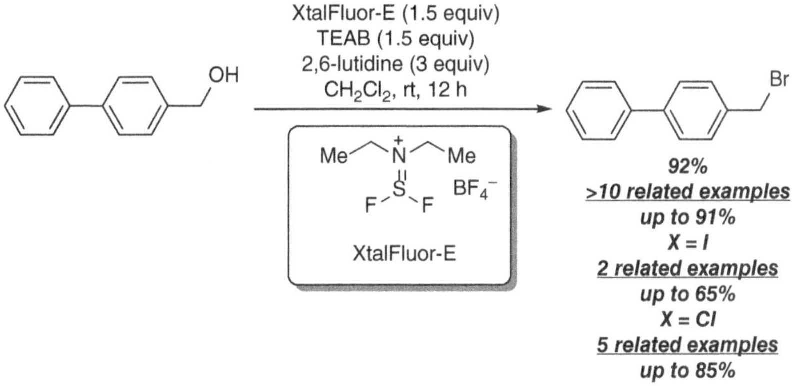

SCHEME 7.26   Conversion of alcohols into alkyl halides [43].

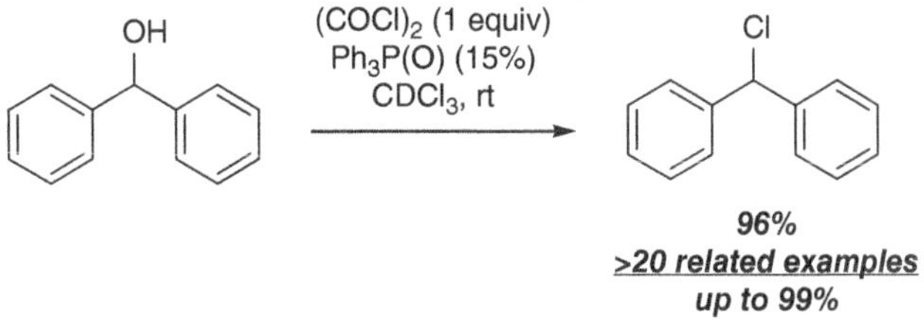

SCHEME 7.27   Triphenylphosphine oxide-catalyzed Appel reaction [44].

the widespread availability and stability of the carboxylic acids. One of the earliest examples of this chemistry was reported by Hunsdiecker and entailed treating silver carboxylates with elemental bromine to afford alkyl bromides [45]. Following this work, a number of modifications were made to the system including the use of lead(IV) acetate/LiX combinations (Kochi reaction) [46] and the initial conversion of the carboxylic acid into a thiohydroxamate ester, followed by treatment with $XCCl_3$ and a radical initiator (Barton decarboxylation) [47]. Additionally, alkyl iodides can be generated through the addition of $PhI(OAc)_2$ to the carboxylic acid, followed by elemental iodine under UV light (Suárez modification) [48].

It would be advantageous to develop a chlorodecarboxylation reaction that only required catalytic amounts of the heavy metal. To address this issue, a method that was catalytic with respect to the silver salt was devised (Scheme 7.28) [49]. The key to the chemistry was the use of *tert*-butyl hypochlorite as the chlorine source. The redesigned reactions proceeded under very mild conditions using only 5% of a silver complex. Primary, secondary, and tertiary alcohols as well as bulky substrates were all successfully chlorinated using this approach.

The reductive bromination of carboxylic acids was achieved using a catalytic amount of indium(III) bromide along with a siloxane and a bromine source (Scheme 7.29) [50]. After some experimentation, the authors discovered that the combination of TMDS and TMSBr was an effective combination for the bromination reaction. Elemental bromine, NBS, and LiBr were less effective under the reaction conditions. A range of electron-donating and electron-withdrawing groups were well tolerated by this system, and even a functionalized phenol was converted into an alkyl bromide in high yield (89%). A few substrates were problematic including an electron-rich benzoic acid as well as a nitroarene. Overall, the chemistry occurred under very mild conditions, displayed a broad substrate scope, and was

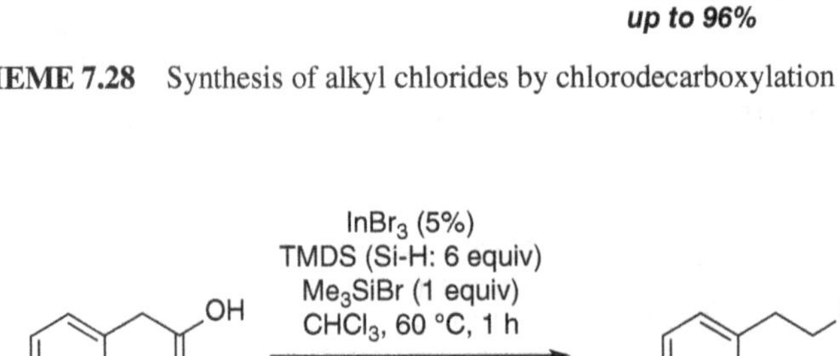

SCHEME 7.28   Synthesis of alkyl chlorides by chlorodecarboxylation [49].

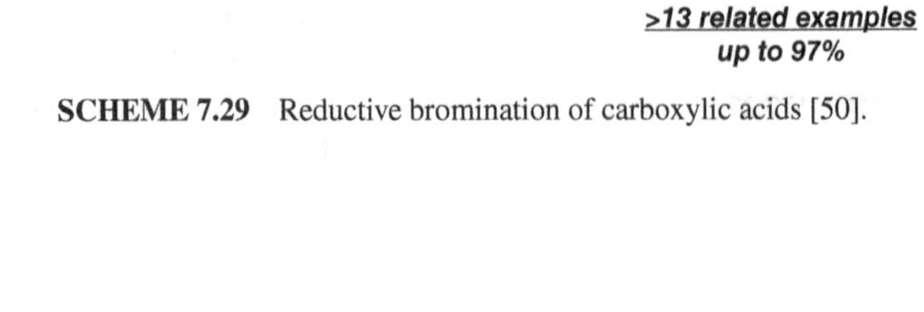

SCHEME 7.29   Reductive bromination of carboxylic acids [50].

operationally straightforward. *Caution! Although it was not part of their optimized system, the authors did encounter an explosion during one of the screening experiments using elemental bromine. One possible explanation for this observation was the generation of an explosive silane during the reaction* [51]. *Thus, care must be taken using this approach as well as during attempts to modifying the system.* In related work, the gallium(III) chloride-catalyzed reductive chlorination of aliphatic and aromatic carboxylic acids has been achieved (Scheme 7.30) [52].

One of the most practical approaches to the synthesis of alkyl iodides is through a Finkelstein reaction [53]. The reactions are operationally trivial and are often not sensitive to air. The classic version of this reaction entails the treatment of an alkyl chloride or bromide with sodium iodide in acetone. It is an equilibrium process and is driven by the precipitation of sodium chloride or sodium bromide. The moderate solubility of sodium iodide in acetone is a key component to this reaction. The process is remarkably facile and high yielding and typically only requires a filtration and removal of the volatiles. This chemistry can be combined with other approaches for the synthesis of a wide assortment of alkyl iodides. For example, a tandem approach to the synthesis of alkyl iodides used the conversion of carboxylic acids into alky chlorides through a Hunsdiecker reaction as the first step in the process. The alkyl chlorides were transformed into alkyl iodides through a Finkelstein reaction during the second step of the process. Modern versions of this reaction have used mesylates as leaving groups in the presence of KX (Scheme 7.31) [54].

The copper-catalyzed enantioselective chlorination of β-keto esters has been achieved using resolved pyridyl spirooxazoline ligands (Scheme 7.32) [55]. NCS served as the source of the halogen, and an effective copper catalyst was Cu(OTf)$_2$. In contrast with many asymmetric reactions that needed to be carried out at subzero temperatures, high

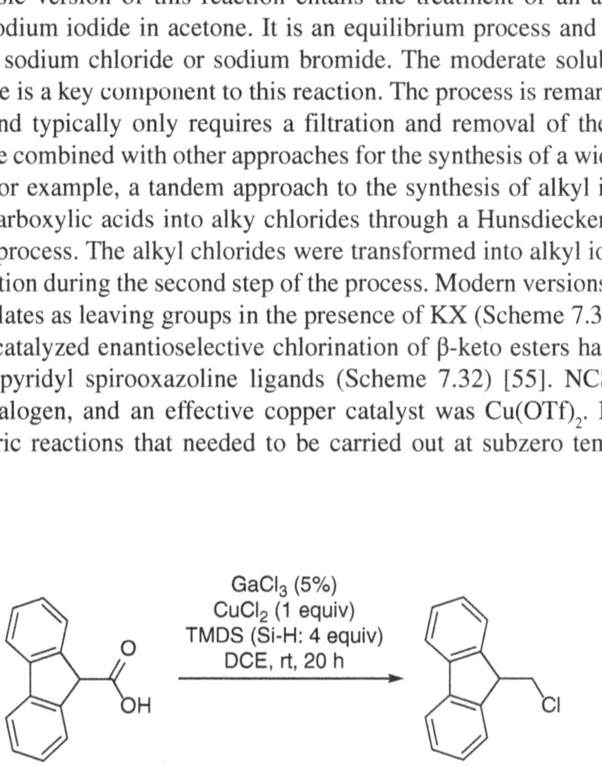

GaCl$_3$ (5%)
CuCl$_2$ (1 equiv)
TMDS (Si-H: 4 equiv)
DCE, rt, 20 h

72%
>15 related examples
up to 92%

SCHEME 7.30 Reductive chlorination of carboxylic acids using GaCl$_3$ [52].

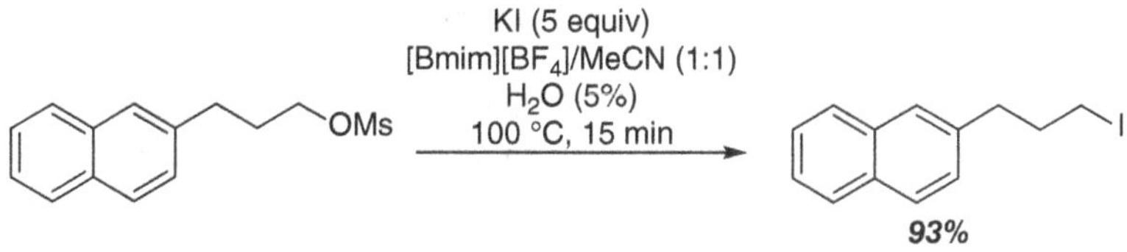

KI (5 equiv)
[Bmim][BF$_4$]/MeCN (1:1)
H$_2$O (5%)
100 °C, 15 min

93%

SCHEME 7.31 Using alkyl mesylates in Finkelstein reactions [54].

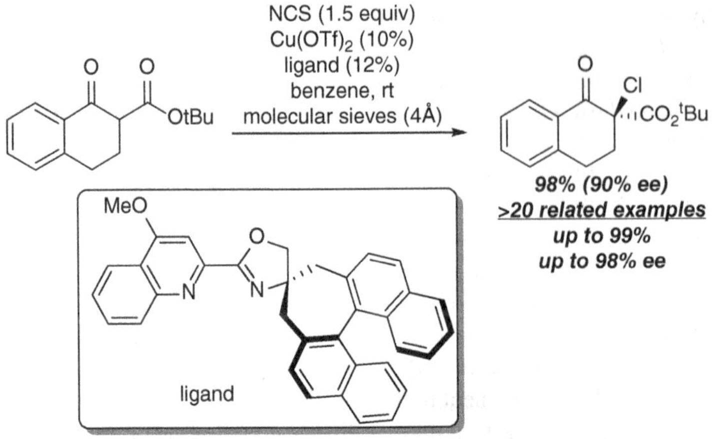

SCHEME 7.32   Enantioselective chlorination of keto esters [55].

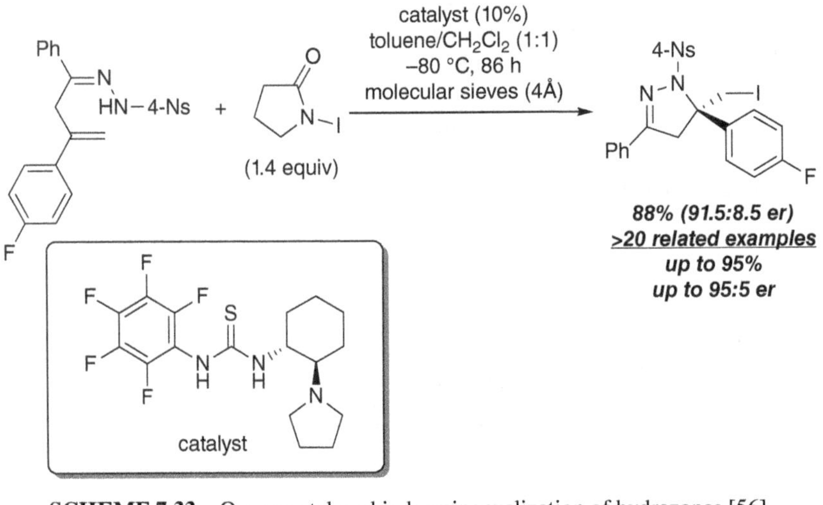

SCHEME 7.33   Organocatalyzed iodoaminocyclization of hydrazones [56].

selectivity was obtained with this system at room temperature. The yields of the reaction were outstanding, and the selectivity was very high (up to 98% ee).

The iodoaminocyclization of hydrazones proceeded under very mild conditions to generate pyrazolines with high selectivity (Scheme 7.33 and Example 7.13) [56]. The organocatalytic reaction was promoted by the use of a bifunctional thiourea catalyst. Using *N*-iodopyrrolidinone as the source of electrophilic iodine leads to the highest selectivity for the reaction. Control reactions were carried out, and the authors noted that strict control of the temperature was needed to prevent a background reaction. Additionally, including molecular sieves in the reaction mixture led to an increase in the selectivity.

Halocyclization reactions are valuable transformations since they often impart chirality while incorporating the halogen. A host of these reactions have been reported, and new

Example 7.13 Alkene *trans*-Heterodifunctionalization Using *N*-Iodopyrrolidinone [56].

Before starting this synthesis, carefully follow the chemical safety instructions listed in Example 7.1.

95% (92.5:7.5 er)

Caution! Petroleum ether and ethyl acetate are highly flammable.
In an oven-dried 25 mL round-bottom flask fitted with a long neck, preactivated 4 Å molecular sieves (100 mg) were added. Molecular sieves were activated by heating under high vacuum at 150 °C for 2 h, followed by cooling to room temperature under argon. To the flask, *N*-iodopyrrolidinone (14.2 mg, 0.067 mmol, 1.4 equiv) was added. The reaction flask was degassed in vacuo and cooled to −80 °C under argon. After 30 min., a solution of the catalyst shown in Scheme 7.33 (2.4 mg, 0.0048 mmol, 0.1 equiv) in 1 : 1 toluene/CH$_2$Cl$_2$ (0.8 mL) was slowly added. The resulting mixture was stirred at −80 °C for 30 min., and a solution of the hydrazone (0.048 mmol, 1.0 equiv) in 1 : 1 toluene/CH$_2$Cl$_2$ (4.0 mL) was added over 1 h by using a syringe pump. Reaction mixture was stirred at −80 °C for 86 h. Reaction was monitored by TLC and quenched by adding saturated aqueous Na$_2$S$_2$O$_3$ solution (5 mL) at −80 °C. The aqueous layer was extracted with CH$_2$Cl$_2$ (3×8 mL). The combined organic layers were dried over anhydrous Na$_2$SO$_4$ and concentrated under reduced pressure. The crude product was purified by column chromatography on silica gel (230–400 mesh) using 93 : 7 petroleum ether/EtOAc to afford the title compound as a white solid (25 mg, 95%).

Note: The temperature must be maintained at −80 °C to suppress the background reaction. No side products were formed as long as the temperature was maintained. Furthermore, the reaction must be quenched prior to warming.

Adapted with permission from C. B. Tripathi and S. Mukherjee, *Org. Lett.*, **2014**, *16*, 3368–3371. Copyright (2014) American Chemical Society.

developments are occurring rapidly. A number of advancements in this area are shown in Schemes 7.34 through 7.38 and Example 7.14.

The site-selective C—H bromination of hydrocarbons is an attractive reaction since it converts classically unusable compounds into reactive substrates. To this end, the bromination of a series of cycloalkanes has been accomplished using *N*-bromoamides [62]. Furthermore, this reaction was promoted by visible light. Once an effective system was devised, it was extended to (+)-sclareolide (Example 7.15).

The gold-catalyzed dihydrofluorination of internal and terminal alkynes was promoted by a gold compound bearing a dialkylbiarylphosphine as a solubilizing/stabilizing ligand (Scheme 7.39) [63]. The chemistry proceeded under very mild conditions and afforded outstanding yields of the 1,1-difluorinated compounds with a high degree of selectivity for the Markovnikov product. A range of electron-withdrawing and electron-donating groups

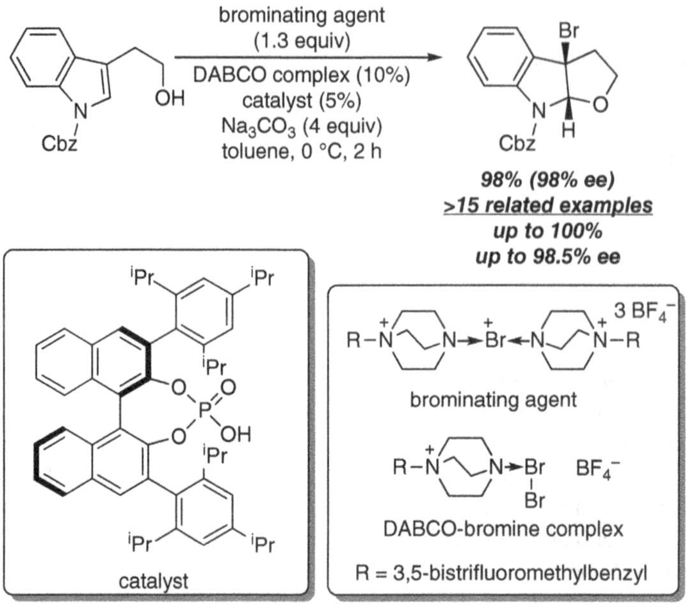

SCHEME 7.34   Asymmetric bromocyclization using a DABCO-bromine adduct [57].

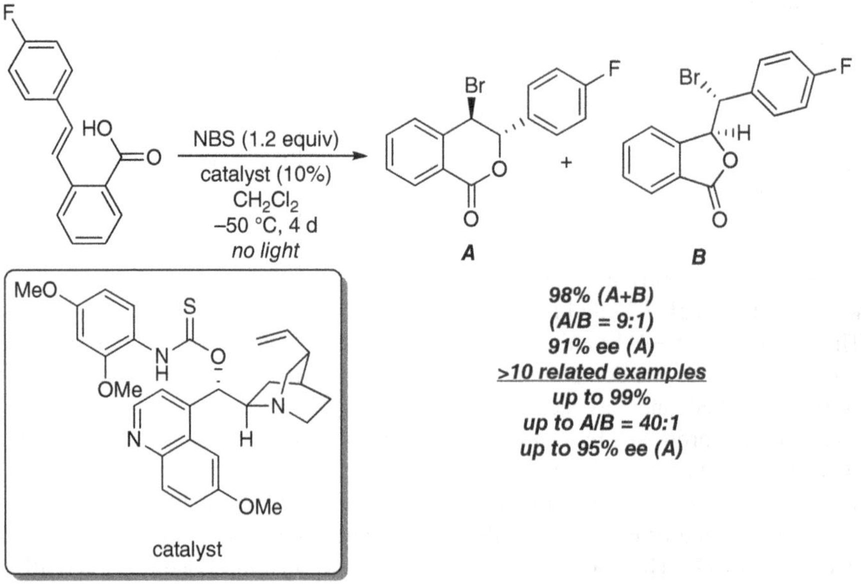

SCHEME 7.35   Enantioselective bromocyclization of styrenyl carboxylic acids [58].

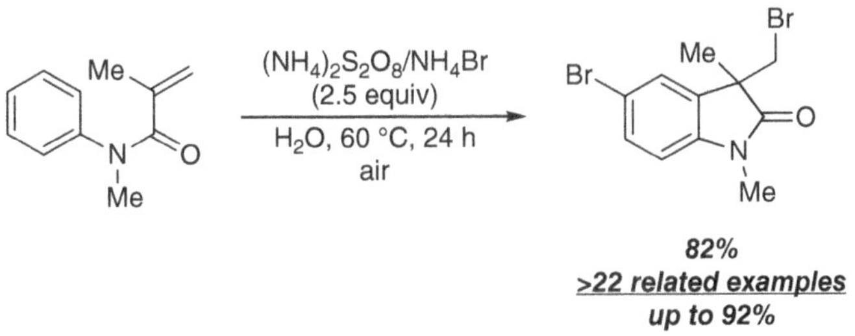

SCHEME 7.36 Bromocyclization in air [59].

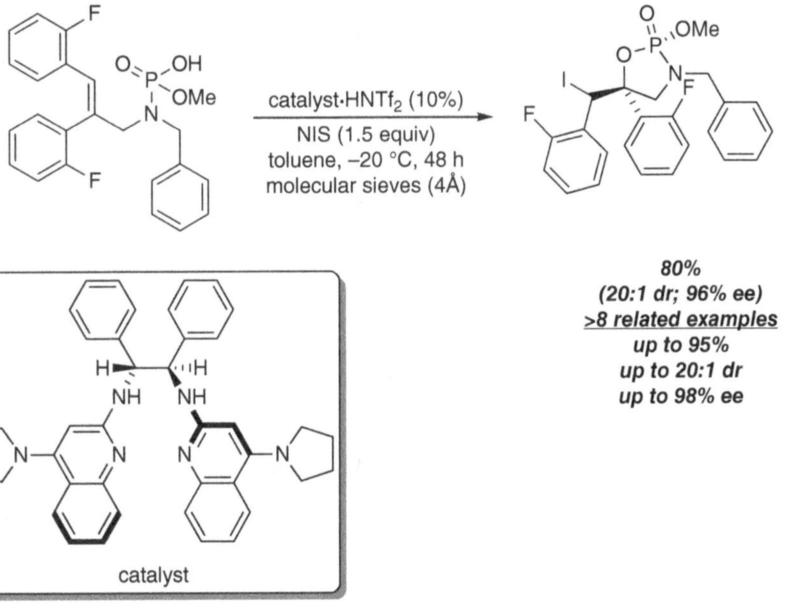

SCHEME 7.37 Asymmetric phosphoramidate synthesis through iodocyclization [60].

as well as heteroaryl bearing alkynes were successfully used in this work. Even the notoriously unreactive internal alkyne, diphenylacetylene, was fluorinated (79%). Due to the hazardous nature of HF, the authors opted to use an adduct (DMPU/HF) in the reactions. In addition to being safer to manipulate, an advantage of this reagent is the ability to add a precise amount of "HF" to a reaction.

The tandem diiodohydration of propargyl alcohols has been achieved using gold catalysts (Scheme 7.40) [64]. This is a valuable transformation that has been challenging to accomplish using conventional methodology. The substrates evaluated included propargyl alcohols bearing aromatic groups as well as an ynol ether. Most were converted into the desired compounds in moderate to excellent yields, although it should be noted that tertiary alcohols were sluggish under the reaction conditions. This chemistry was attractive since it was able to be carried out in the presence of water. Indeed, an acetonitrile/water mixture

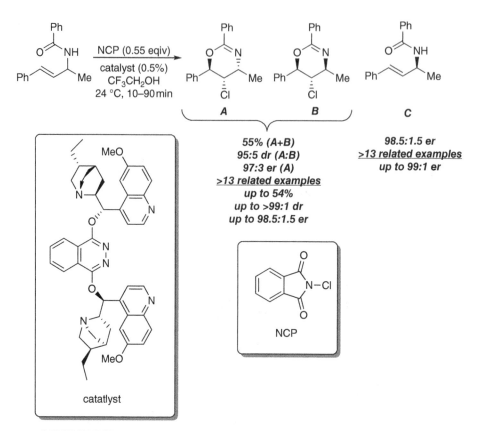

SCHEME 7.38 Kinetic resolution through organocatalyzed chlorocyclization [61].

Example 7.14   Preparation of Oxindoles in Water and Air [59].

Before starting this synthesis, carefully follow the chemical safety instructions listed in Example 7.1.

To a 25 mL Schlenk tube were added $(NH_4)_2S_2O_8$ (6.0 equiv), NH_4Cl (3.0 equiv), N-aryl-lacrylamide (0.2 mmol), and H_2O (1.0 mL). Then, the tube was stirred at 60 °C for the indicated time until complete consumption of the starting material (as detected by TLC). After the reaction was finished, the reaction mixture was cooled to room temperature, and ethyl acetate (20 mL) and water (10 mL) were added. The organic layer was separated, and the aqueous phase was extracted with ethyl acetate. The combined organic layers were washed with brine, dried over Na_2SO_4, filtered, and concentrated in vacuum. The oxindole product was obtained after purification by flash column chromatography (petroleum ether/ethyl acetate 8 : 1) as a light yellow solid (36.2 mg, 65%).

Adapted with permission from M.-Z. Zhang, W.-B. Sheng, Q. Jiang, M. Tian, Y. Yin, C.-C. Guo. J. Org. Chem. **2014**, *79*, 10829. *Copyright (2014) American Chemical Society.*

Example 7.15 Site-Selective Bromination of Hydrocarbons Using Visible Light [62].

Before starting this synthesis, carefully follow the chemical safety instructions listed in Example 7.1.

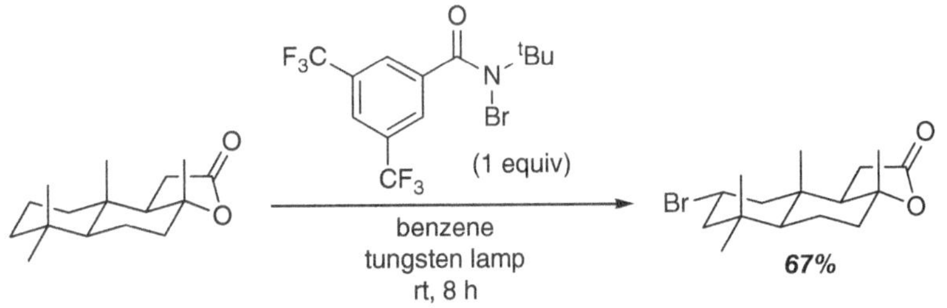

Caution! Benzene, ethyl acetate, pentane, and hexanes are highly flammable.
A flame-dried, 1 dram vial was charged with a stir bar and fitted with a PTFE-lined screw cap. Bromoamide (152.4 mg, 0.390 mmol) was added to the vial in the absence of ambient light, and the reaction was taken into a glovebox and dissolved in 1.2 mL of dry, freeze–pump–thawed benzene. Sclareolide (27.9 mg, 0.111 mmol) was then added. The reaction was then sealed with Teflon tape and taken out of the glovebox and placed in a circulating cooling bath (roughly room temperature) and irradiated with two 100 W tungsten filament lightbulbs for 8 h. When the reaction was complete, it was concentrated under reduced pressure and dissolved in pentanes. The resulting suspension was filtered through cotton and concentrated a second time. The crude reaction mixture was purified using column chromatography eluting with 5–10% EtOAc in hexanes to give the product (24.4 mg, 67%) as a white solid.

Adapted with permission from V. A. Schmidt, R. K. Quinn, A. T. Brusoe, and E. J. Alexanian, J. Am. Chem. Soc., **2014**, *136*, 14389–14392. *Copyright (2014) American Chemical Society.*

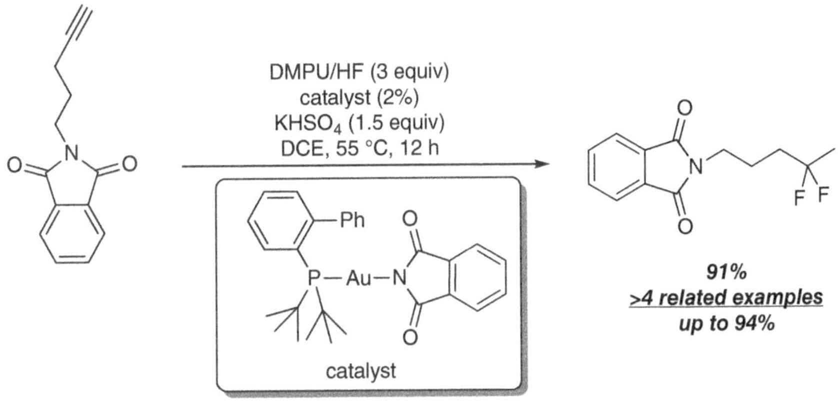

SCHEME 7.39 Gold-catalyzed regioselective double addition of HF to alkynes [63].

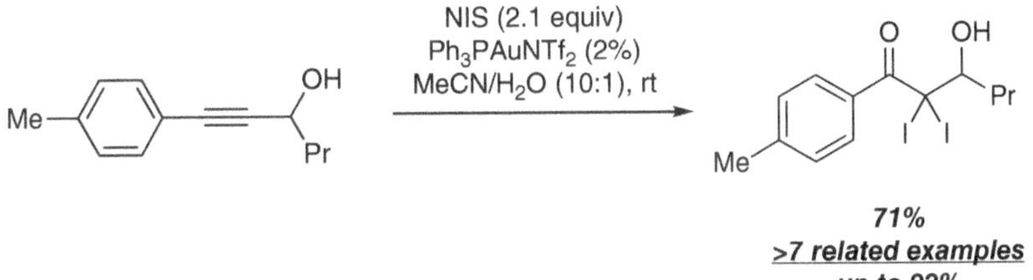

SCHEME 7.40 Gold-catalyzed conversion of propargyl alcohols into diiodo-β-hydroxyketones [64].

was among the most effective solvent combinations screened. In related work, the generation of dibromotoluene was accomplished by treating benzaldehyde with a slight excess of BBr$_3$ in dichloromethane (DCM) (Scheme 7.41) [65], and the palladium-catalyzed asymmetric iodination of unactivated C(sp^3)—H bonds was reported [66].

The chloroformyloxylation of alkenes has been accomplished using PhICl$_2$ [67]. The key to this chemistry was actually to use wet DMF and avoid anhydrous conditions. When the reaction was carried out using dry solvents, none of the chlorinated product was formed. The reactions were moderate to high yielding under very mild conditions and were tolerant of a broad range of functional groups. While most of the examples incorporated an activating group on the alkene (Michael acceptors), even *trans*-stilbene and styrene were successfully functionalized. Furthermore, these chlorination reactions could be carried out under air. Naturally, this is a significant advantage since gloveboxes and nitrogen lines were not needed. Curiously, when a more electron-donating alkene was used, the direct chlorination of the alkene was observed (Schemes 7.42 and 7.43). Further experimentation revealed that the direction of the reaction (chloroformyloxylation vs. direct halogenation) depended upon a subtle

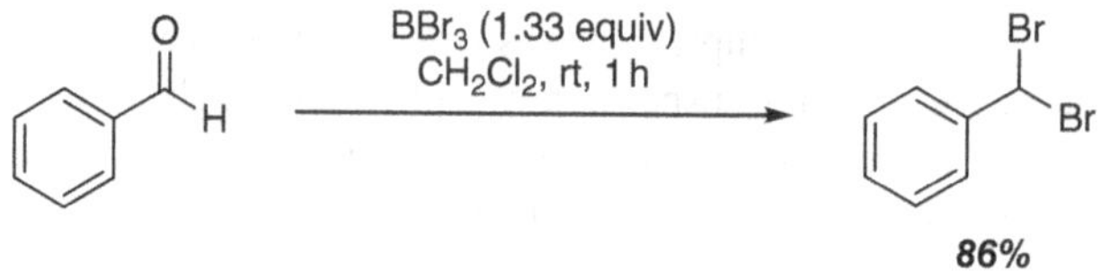

SCHEME 7.41   Conversion of benzaldehyde into dibromotoluene [65].

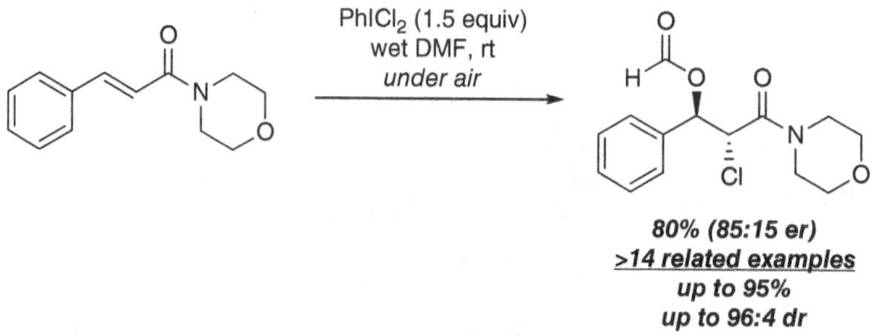

SCHEME 7.42   Chloroformyloxylation of alkenes [67].

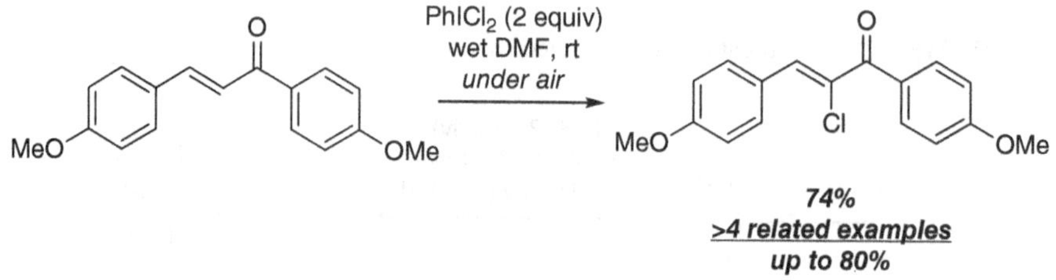

SCHEME 7.43   Direct α-chlorination of alkenes [67].

balance between the electron-withdrawing group and electron-donating nature of the remaining substituents. If the system was carefully designed, it could be directed to generate a single product.

Trichloroisocyanuric acid (TCCA) promoted the conversion of alcohols into α-chloro-aldehydes and ketones (Scheme 7.44) [68]. In the course of the investigation, two separate procedures were developed in order to increase the scope of the approach. When primary alcohols were screened, the authors discovered that an oxidation catalyst was needed to generate high yields of the chlorinated aldehyde and determined that TEMPO was an effective choice. When secondary alcohols were used as substrates, the chlorination reaction was high yielding for selected substrates without TEMPO as long as methanol was present (2 equiv). Through the course of the investigation, a wide variety of alcohols were successfully functionalized. In related work, if the primary or secondary alcohol was substituted for a propargyl alcohol, an α,α-dichloroketone was generated with retention of the alcohol (Scheme 7.45) [64].

An interesting reaction was observed when THF was treated with triphenylphosphine and CBr$_4$. After prolonged stirring in chloroform, a dibromination reaction was observed along with ring opening of the THF (Example 7.16) [69]. This intriguing reaction takes place under mild conditions affording the acyclic 1,4-dibromobutane in outstanding yield.

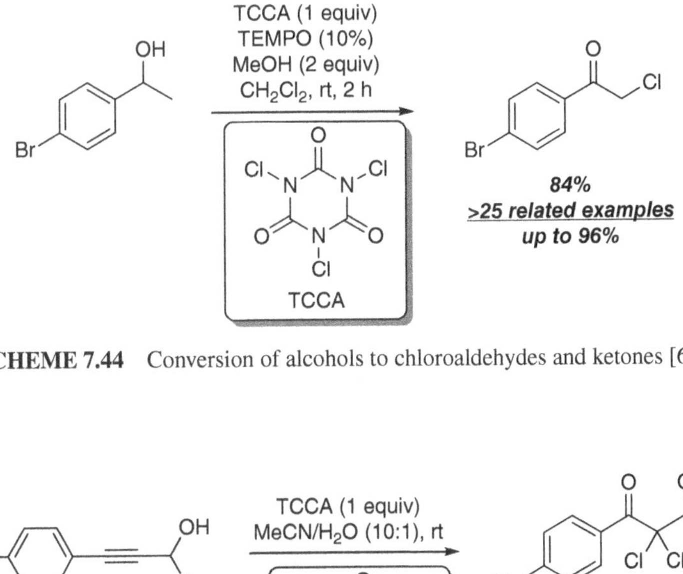

SCHEME 7.44 Conversion of alcohols to chloroaldehydes and ketones [68].

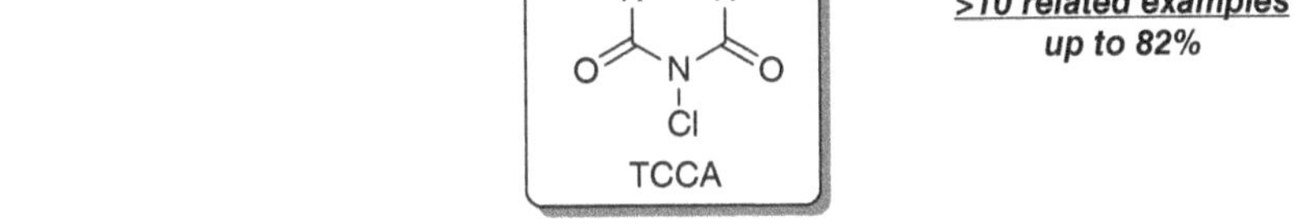

SCHEME 7.45 Chlorination of a propargyl alcohol [64].

Example 7.16   Ring opening and dibromination of cyclic ethers [69].

Before starting this synthesis, carefully follow the chemical safety instructions listed in Example 7.1.

THF (596 mg, 8.27 mmol) was dissolved in 15 mL of $CHCl_3$, and 6.12 g (18.45 mmol) of CBr_4 was added. Then, 8.29 g (31.61 mmol) of Ph_3P dissolved in 5 mL of $CHCl_3$ was added dropwise to the stirred mixture over 60 min. After stirring for 10 d at room temperature, the reaction mixture was quenched with water, transferred into a separatory funnel, and the organic layer was washed once with water. After drying over $MgSO_4$, filtration, and rotary evaporating of the solvent, the crude product was Kugelrohr distilled (9 Torr, oven temperature ca. 90 °C) to afford the dibromo species (1.66 g, 93%).

Adapted with permission from P. Billing and U. H. Brinker, J. Org. Chem., 2012, 77, 11227–11231. Copyright (2012) American Chemical Society.

7.1.1   Troubleshooting the Synthesis of Alkyl Halides and Related Compounds

While it will not be possible to anticipate all of the needs of the synthetic chemist over the next few decades, this list will provide some insight into issues commonly encountered when trying to prepare alkyl halides. This chemistry can be particularly confusing as several protocols will work for a couple of the halogens, but not for all of them. Care must be taken to make sure that the chemistry will be successful for the substrate under investigation.

Problem/Issue/Goal	Possible Solution/Suggestion
Is it possible to generate an alkyl fluoride through a Finkelstein reaction?	The direct conversion has been challenging, and general approaches are rare, but some substrate-specific reactions have been developed. For example, the copper-catalyzed fluorination of allylic bromides has been developed and tends to be high yielding [11]
A process for the fluorination of benzyl C—H bonds would be convenient	This is a challenging reaction; however, benzylic and α-hydrogens of ketones (and related substrates) are quite susceptible to fluorination, and number of successful approaches have been reported using a range of catalyst systems [1, 12–16]
Need to fluorinate a saturated hydrocarbon	Practical approaches to the fluorinative C—H activation of saturated hydrocarbons remain rare. Recent advances using copper catalysts and visible light catalysis have been reported [23, 25]
Need to convert an alcohol into a fluoride	This would be a fluorinative Appel reaction. While this reaction has been very successful with chlorides and bromides, its extension to fluorine has been difficult [6]. A reasonable approach would be to convert the alcohol into a triflate followed by a fluorination using copper catalysts [8]

(Continued)

(Continued)

Problem/Issue/Goal	Possible Solution/Suggestion
Substrate needs water present to dissolve	Several approaches are quite tolerant to the presence of moisture [26, 27, 36]. If the starting material contains or can be converted into a carboxylic acid, the silver-catalyzed decarboxylative fluorination in acetone/water (1 : 1) would be a reasonable place to start [36]
Starting from an unactivated alkene and need to hydrofluorinate without handling HF	A clever two-step approach has been reported for the anti-Markovnikov addition [27], while a single-step cobalt-catalyzed reaction has been shown to generate the Markovnikov product [28]
Need an enantioselective fluorination process	General approaches with wide substrate scope are rare. These reviews and reports on selective and non-selective fluorination would be reasonable places to consult prior to designing a synthesis [1–6]
Need to make a chiral allylic fluoride	A reasonable place to start would be to use a combination of Selectfluor and a chiral phosphoric acid catalyst (chiral anion phase-transfer catalysis) [20]
Can an alkylboronate be fluorinated?	A silver-catalyzed approach to the conversion of functionalized alkylboronates into alkyl fluorides has been reported [27]
Does a fluorinative decarboxylation reaction exist?	The decarboxylative fluorination of a range of carboxylic acids has been achieved using silver salts as catalysts [36]
Starting from an alcohol and need to generate alkyl chloride or bromide	Appel chemistry would be effective. Catalytic versions using $PPh_3P(O)$ as the catalyst are also known [44]
Need an asymmetric brominative semipinacol rearrangement	While a number of approaches to this reaction are known, an organocatalyzed version is attractive [33]
Need to effect a halodecarboxylation reaction (Hunsdiecker-type chemistry)	Several catalysts and approaches have been reported for this chemistry [45, 47, 48]. The silver-catalyzed version of this reaction is attractive [49]
Is reductive halogenation of a carboxylic acid possible?	Several versions of this reaction are known [50, 52]
It would be convenient to be able to promote a brominative C—H activation	This is a challenging reaction. Visible light catalysis has been shown to brominate unactivated C—H bonds [62]
Can a mesylate be converted into an alkyl iodide?	This reaction works quite well using KI as the iodinating agent [54]
Need to monochlorinate a diol	This reaction has been achieved using TCT as the chlorinating agent [42]
Need a method for the triphenylphosphine-free bromination/chlorination of a benzyl alcohol	Several approaches to this reaction have been reported [42–44]
Need to convert an alcohol into an alkyl iodide	A classic Appel reaction would be a good place to start [38, 39]
Starting from an alkyl chloride or bromide: need to make an alkyl iodide	This is a classic setup for a Finkelstein reaction. Treat the alkyl chloride (bromide) with sodium iodide in acetone. The equilibrium will shift to the alkyl iodide side since the sodium bromide and chloride are not soluble in acetone

(Continued)

(Continued)

Problem/Issue/Goal	Possible Solution/Suggestion
Need an asymmetric iodinative semipinacol rearrangement	While a number of approaches are known for this reaction, an organocatalyzed version is attractive [34]
Need to make a 1,1-difluorinated alkyl fragment	Starting from the alkyne is an attractive entry point to this chemistry as treating terminal alkynes with DMPU/ HF has been shown to generate 1,1-difluorinated compounds [63]
Need to make a 1,1-dichlorinated alkyl fragment	There are a number of approaches to this chemistry, and the conversion of propargyl alcohols using TCCA as the chlorinating agent in methanol/water is a practical method [64]
Need to make a 1,1-dibrominated alkyl fragment	Treatment of aldehydes with BBr_3 would be a good place to start [65]
Need to make a 1,1-diiodinated alkyl fragment	There are a number of approaches for this reaction. For propargyl alcohols, a gold-catalyzed iodination with NIS generates the 1,1-diiodinated compounds [64]

7.2   SYNTHESIS OF ARYL HALIDES THROUGH THE FORMATION OF HALOGEN–CARBON(SP^2) BONDS

Historically, aryl fluorides were prepared by either Balz–Schiemann-type reactions or though the Halex process [70–75]. Although a number of these procedures were quite successful, many of them required the use of toxic reagents or harsh conditions. To address these issues and provide practical routes to aryl fluorides, a host of alternative protocols have been developed [6, 76–78]. In addition to this work, a number of methods for the synthesis of aryl halides have been developed. The following sections will highlight several practical versions.

The preparation of aryl fluorides through the treatment of aryl Grignard reagents with a source of F^+ has been reported by Knochel and Beller (Schemes 7.46 and 7.47) [79–81]. Knochel generated aryl Grignards at low temperature and treated them with NFSI. The choice of solvent was critical to the success of the fluorination step, and the authors found that CH_2Cl_2/perfluorodecalin mixtures (4:1) afforded higher yields of the aryl fluorides.

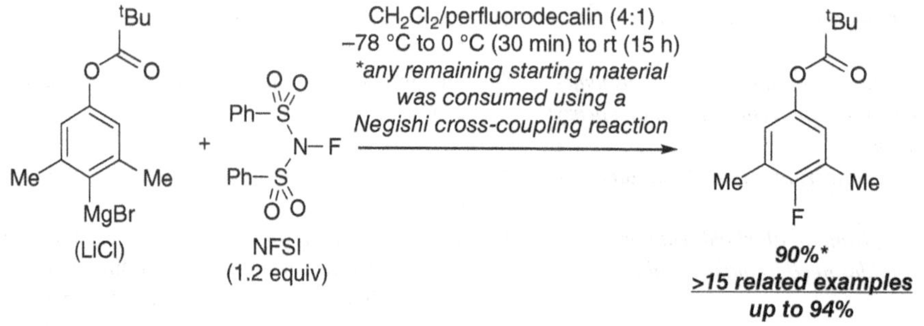

SCHEME 7.46   Synthesis of functionalized aryl fluorides by treatment of aryl Grignard reagents with NFSI [80].

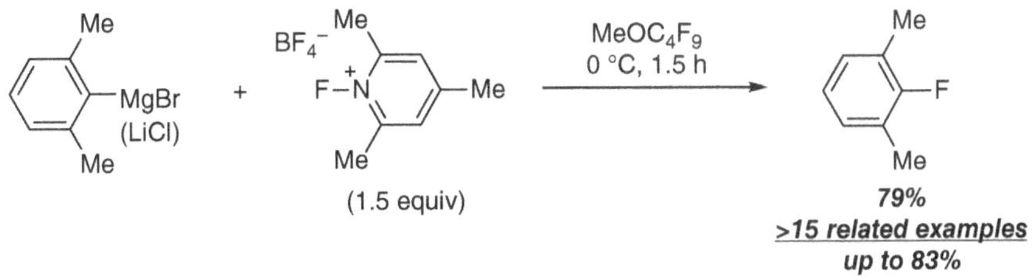

SCHEME 7.47   Synthesis of aryl fluorides by treatment of aryl Grignard reagents with F+ [79].

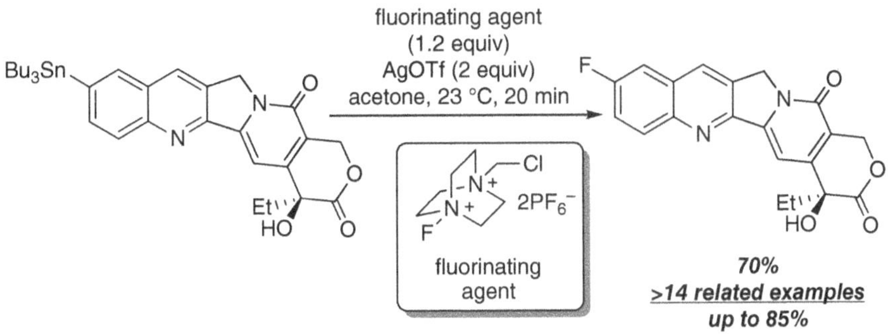

SCHEME 7.48   Fluorination of arylstannanes [82, 83].

The operational complexity was moderate due to the need to prepare the moisture sensitive Grignard reagent; however, this was offset by the observation that the entire reaction could be carried out in a single reaction vessel. Despite the use of Grignard reagents, a range of amides, ethers, and acetates were successfully fluorinated. One of the challenges with this type of reaction was the separation of the hydrolysis products as well as unreacted starting materials from the products. When this was problematic, the authors devised a clever solution to this issue and converted any of the unreacted starting material into a biaryl species through a Negishi coupling reaction prior to attempting to purify the products. As shown in Scheme 7.46, very bulky Grignard reagents were successfully employed in this reaction. Beller's approach was also highly dependent upon the solvent, and he found that the highest yields of the aryl fluorides were obtained using methoxyperfluorobutane or hexane along with N-fluoro-2,4,6-trimethylpyridinium tetrafluoroborate as the source of electrophilic fluorine. Similar to the NFSI/CH$_2$Cl$_2$/perfluorodecalin system used by Knochel, the chemistry was tolerant to the presence of several functional groups and fluorinated bulky aryl Grignards (Scheme 7.47). The authors did report that the chemistry was sensitive to the order of addition and best results were obtained if the aryl Grignard was added to a solution of the fluorinating agent. Overall, if the initial aryl halide is accessible, this is a high yielding one-pot method for the preparation of aryl fluorides.

The silver-promoted/catalyzed treatment of arylstannanes with electrophilic fluorine was effective method for the synthesis of aryl fluorides [82, 83]. Some of the early reports used an excess of the silver reagents to promote the reactions (Scheme 7.48). F-TEDA-PF$_6$ was an effective F+ source for these reactions, and the conditions were remarkably mild. One of the most attractive aspects of this chemistry was the speed of the reactions.

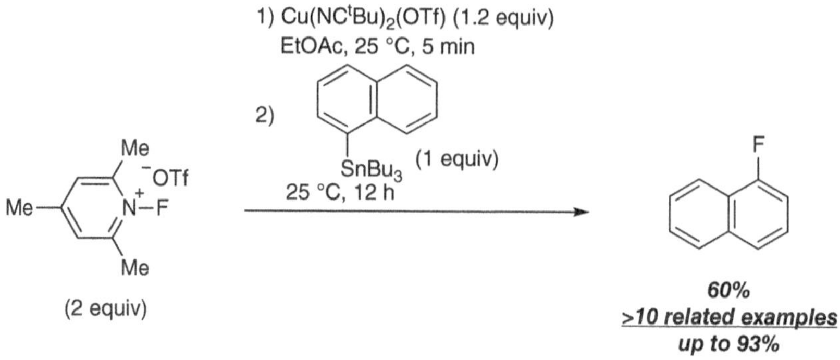

SCHEME 7.49   Copper-promoted fluorination of arylstannanes [84].

Most were complete within 20 min at 23 °C. This avoided some functional group issues encountered with nucleophilic fluoride chemistry and facilitated the incorporation of an fluorine into highly functionalized compounds at a very late stage in the total synthesis. In a subsequent report, the same authors were able to devise a system for catalytic fluorination [83]. Using only 5% of a silver oxide as the metal catalyst, a host of arylstannanes were successfully functionalized. Higher temperatures were typically needed to promote these reactions, and outstanding yields were obtained for a range of biologically active compounds including estrone and DOPA.

The copper-promoted fluorination of arylstannanes has also been reported (Scheme 7.49) [84]. The copper source was (Cu(NC^{t}Bu)$_2$(OTf)), and N-fluoro-2,4,6-trimethylpyridinium tetrafluoroborate was found to be the most effective source of electrophilic fluorine. During the initial stages of the investigation, all of the reagents and additives were added to the reaction vessel at the same time; however, this resulted in low yields of the aryl fluoride (6%). The yield of the fluorination reaction was significantly higher if the copper precursor and F$^+$ source were mixed for 5 min prior to the introduction of the arylstannane. The authors proposed that this prestirring period facilitated the formation of a copper(III)–F species that reacted with the arylstannanes to generate the aryl fluoride product. The fluorination reaction was successful at room temperature, and a variety of substrates bearing electron-donating and electron-withdrawing substituents were successfully functionalized using this approach along with bulky substrates.

The palladium-catalyzed coupling of aryl triflates with nucleophilic F$^-$ sources generated excellent yields of aryl fluorides (Scheme 7.50) [85, 86]. The catalyst system was comprised of a Pd(II) source along with a bulky dialkylbiarylphosphine ligand. Using cesium fluoride as the source of nucleophilic fluoride, moderate to excellent yields of the aryl fluorides were obtained in 12 h at 110 °C. The substrate scope for this palladium-catalyzed process was broad, and the incorporation of either electron-donating or electron-withdrawing groups did little to slow the process. A few systems were problematic including substrates bearing potential donors for the metal center (nitrogen, carbonyl) adjacent to the triflate, as well as a few heteroaryl groups and highly electron-rich arenes. For some of these systems, the sluggish reactivity could be due to a reduction in the activity of the palladium center due to coordination of donor groups to the metal center. Overall, this remains one of the most practical approaches to the synthesis of aryl fluorides. An improved version of this initial system used a diadamantylbiarylphosphine ligand to promote the

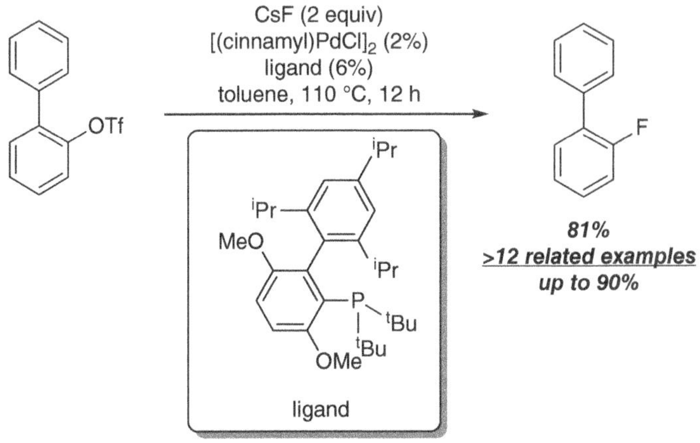

SCHEME 7.50   Palladium-catalyzed fluorination of aryl triflates using CsF [85].

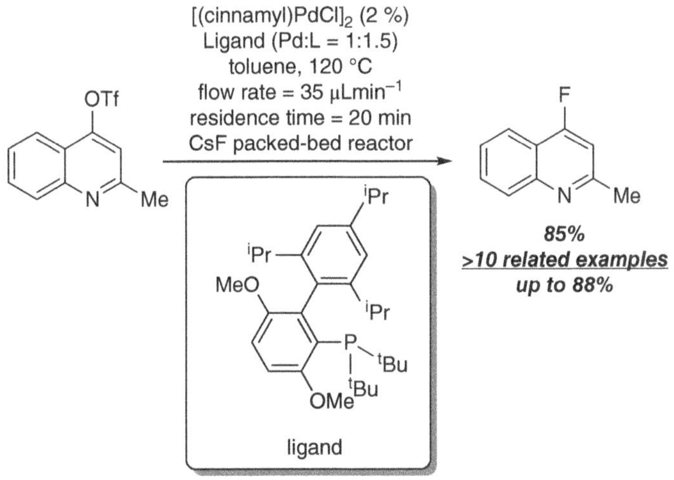

SCHEME 7.51   Palladium-catalyzed fluorination of aryl triflates under continuous flow [88].

fluorination of the more challenging electron-rich aryl triflates [87]. Furthermore, a continuous flow version of this reaction has been developed and used a CsF packed reactor tube (Scheme 7.51) [88]. Excellent yields of aryl and heteroaryl fluorides were obtained at 120 °C (residence time of 20 min).

   Given the abundance of aryl bromides, the development of a reaction that would convert them into aryl fluorides would be a valuable transformation. To address this goal, a metal-catalyzed example of this reaction was developed using a catalyst system comprised of a palladium complex bearing a bulky dialkylbiarylphosphine ligand, silver fluoride, and KF (Scheme 7.52) [89]. A host of brominated heteroarenes were successfully fluorinated using this system, and moderate to excellent yields were obtained for most substrates. Bromopyrimidine was one of the more challenging substrates and was fluorinated in 51% yield after heating at 150 °C. A significant advantage of this system is the virtual

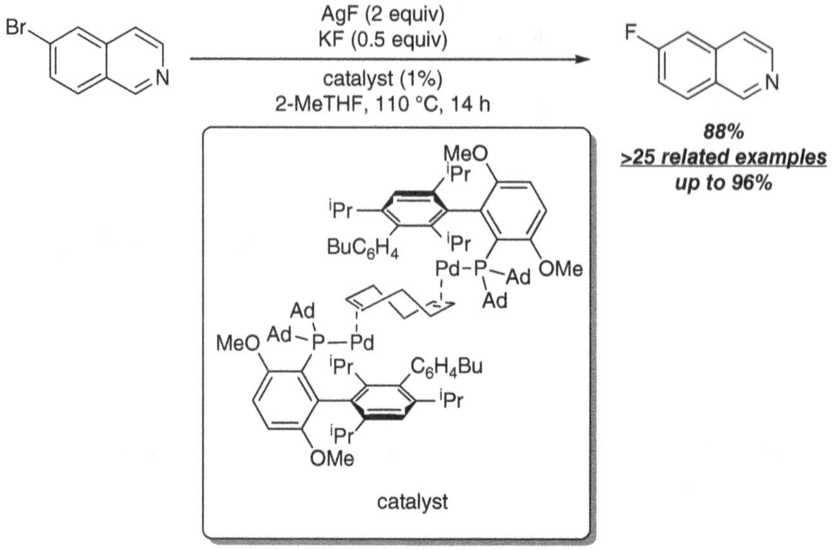

SCHEME 7.52 Palladium-catalyzed conversion of heteroaryl bromides into heteroaryl fluorides [89].

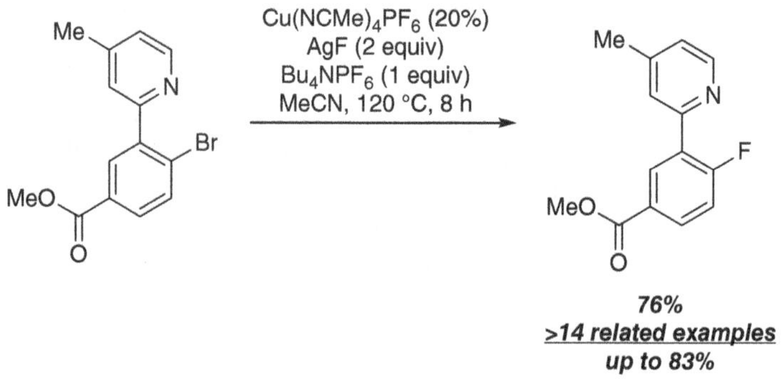

SCHEME 7.53 Copper-catalyzed conversion of aryl bromides into aryl fluorides [90].

elimination of the reduction side products that normally plague similar reactions. These secondary products can be challenging to separate from the desired aryl fluorides.

A copper-catalyzed version of this reaction has been developed (Scheme 7.53) [90]. A readily available copper complex was used as the catalyst, and silver fluoride served as the source of nucleophilic fluoride. After some experimentation, the authors discovered that a pyridine substituent was also needed for a successful fluorination reaction. It is noteworthy to mention that no reaction was observed upon removal of this pyridyl group. Simple substrates such as iodobenzene were unreactive under the conditions studied. The authors proposed that the pyridyl group could have promoted the reaction by stabilizing the active copper catalyst and preventing the formation of catalytically inactive CuF_2. Electron-rich and electron-poor aryl bromides were both successfully converted into aryl fluorides with nitro and -OCF_3 groups affording the lowest yields. A number of functional groups including ethers, ketones, and halides were tolerated by the chemistry.

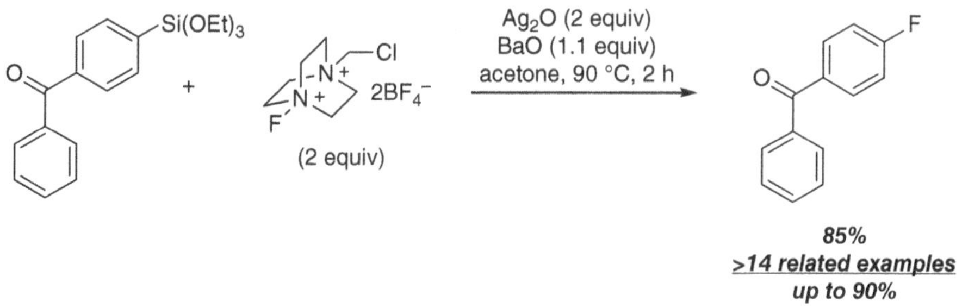

SCHEME 7.54   Fluorination of aryltrialkoxysilanes using Selectfluor [91].

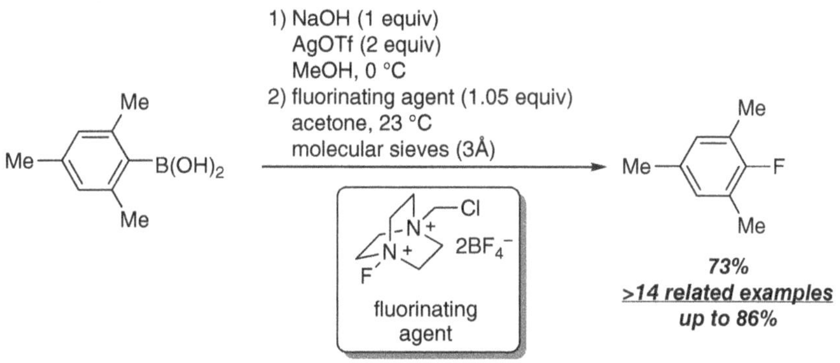

SCHEME 7.55   Silver-mediated fluorination of arylboronic acids [92].

The silver-promoted fluorination of aryltrialkoxysilanes is an attractive approach to the preparation of aryl fluorides (Scheme 7.54) [91]. One of the key aspects of this reaction was the addition of a base to trap any acidic species generated from the fluorination reaction. After screening a series of mineral and organic bases, BaO was found to be the most effective. The chemistry exhibited broad substrate scope, and a range of esters, halogens, ketones, and amides were successfully fluorinated in moderate to excellent yield. The authors noted that the silver salt can be isolated by a simple filtration at the end of the reaction and recycled.

Arylboronic acids are attractive precursors for the preparation of aryl fluorides due to their low toxicity and availability. Using these substrates, a silver-promoted fluorination of arylboronic acids was developed (Scheme 7.55) [92]. Silver(I) triflate was used as the silver source, and Selectfluor served as the source of electrophilic fluorine. The operational complexity of this chemistry was moderate since the overall reaction needed two steps with a solvent removal step in the middle, but this is offset by fact that everything occurs in the same flask. The first step in the reaction entailed the formation of an arylsilver species resulting from base-promoted transmetallation between the boronic acid and a silver precursor. Treatment of this intermediate with the fluorinating agent generated the aryl fluorides in good to excellent yields. One of the complications typically associated with the synthesis of aryl fluorides using electrophilic fluorine is the generation of small amounts of arenes. These contaminants can be challenging to separate from the desired aryl fluoride. Thankfully, the two-step approach employed by the authors was successful at mitigating this issue. It should be noted that the fluorination chemistry is sensitive to moisture and

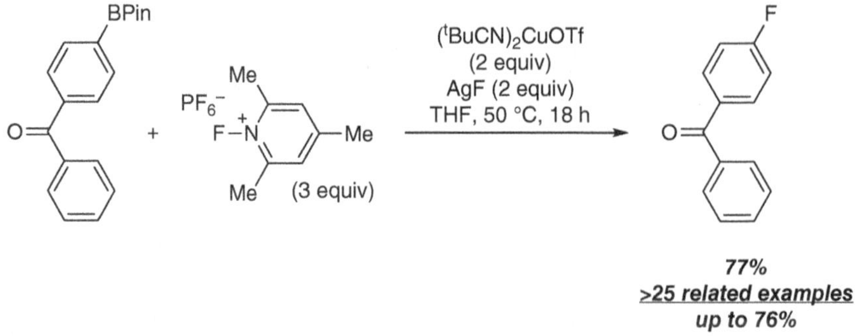

SCHEME 7.56   Copper-promoted conversion of arylboronates into aryl fluorides [93].

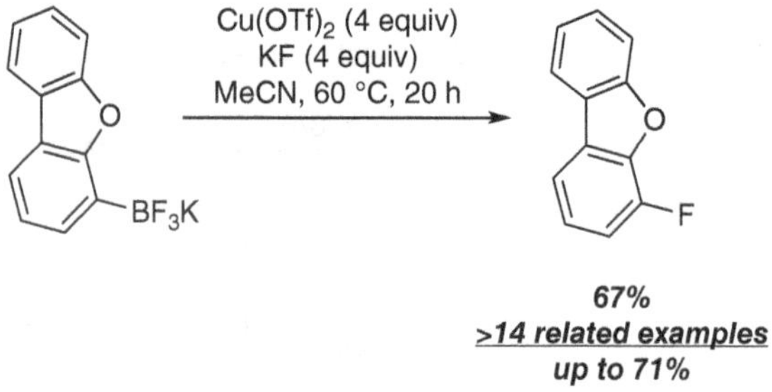

SCHEME 7.57   Fluorination of aryltrifluoroborate salts using KF [94].

significant amounts of phenols were generated if water was present. This issue was circumvented through the addition of molecular sieves. The authors also successfully scaled up the chemistry to gram-scale reactions.

In addition to arylboronic acids, arylboronates have also been successfully used in fluorination reactions (Scheme 7.56) [93]. These boron compounds are attractive substrates since they are typically more robust than other boron species and can often be stored for long periods of time. The catalyst system for this reaction was a copper(I) triflate species along with 2 equiv of silver fluoride. While several sources of electrophilic fluorine generated the aryl fluorides, N-fluoro-2,4,6-trimethylpyridinium hexafluorophosphate was the most effective. The chemistry displayed broad functional group tolerance with the lowest yields obtained with heteroarylboronates. The authors were also able to devise a one-pot borylation–fluorination reaction starting from the parent arene. The arene was converted into an arylboronate through an iridium-catalyzed borylation reaction in the first step of the reaction, while fluorination was achieved during the second step. This is particularly attractive since it facilitates the conversion of unfunctionalized substrates into aryl fluorides.

Aryltrifluoroborate salts can also be converted into aryl fluorides. One version of this process involved the use of nucleophilic fluorinating sources such as potassium fluoride (Scheme 7.57) [94]. A common copper compound along with an excess of KF promoted the fluorination reaction under mild conditions. Moderate to excellent yields of the aryl fluorides were obtained along with minimal amounts of the protodeboronation products. A range of aryltrifluoroborates including examples bearing electron-donating and electron-withdrawing

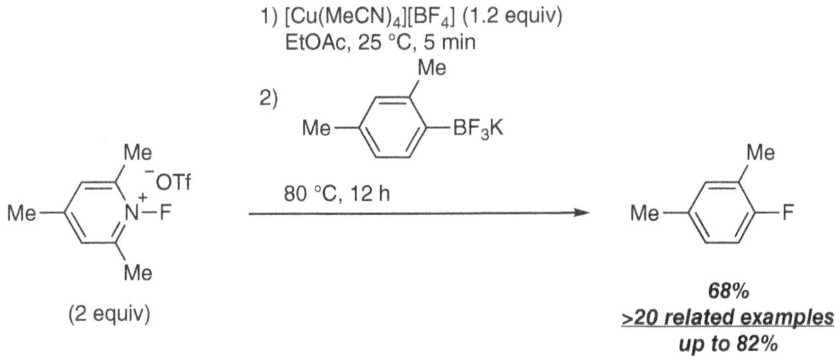

SCHEME 7.58   Copper-promoted fluorination of aryltrifluoroborate salts [84].

groups as well as heteroaryltrifluoroborates were successfully fluorinated using this approach. It should be noted that the fluorination chemistry was fairly tolerant to the presence of moisture and oxygen. While the overall yields of the aryl fluoride were lower in the presence of these contaminants, a significant amount of the target compound was still formed.

Another approach to the conversion of aryltrifluoroborate salts into aryl fluorides also used a common copper salt to promote the reactions; however, this chemistry used an electrophilic source of fluorine (Scheme 7.58) [84]. The most effective source of F+ was found to be N-fluoro-2,4,6-trimethylpyridinium triflate. Similar to the authors' previous work with arylstannanes, the key to the synthesis was prestirring the F+ source and the copper complex prior to the introduction of the aryltrifluoroborate salts. Using this system, a wide range of aryltrifluoroborate salts bearing ethers, halogens, aldehydes, ketones, amides, and esters were converted into aryl fluorides.

In addition to the use of copper catalysts to promote the fluorination of aryltrifluoroborates, palladium complexes can also catalyze these reactions (Example 7.17) [95]. The authors used a terpy-supported palladium catalyst along with Selectfluor to convert a range of aryltrifluoroborates into aryl fluorides. The reaction conditions were quite mild, and a range of functionalized borates bearing amides, ketones, ethers, carboxylic acids, and nitriles were well tolerated by this chemistry. Even a primary alcohol was ignored by the fluorination reaction. Yields were typically good to excellent (up to 99%), and bulky substrates such as 2,4,6-triisopropylphenyltrifluoroborate were successfully fluorinated (63%). Although this chemistry required the synthesis of the palladium catalyst, the synthesis of this compound is trivial and high yielding (98%). The broad scope of this reaction along with the availability of the aryltrifluoroborate salts makes this a very attractive approach to the preparation of aryl fluorides.

The nucleophilic fluorination of unsymmetrical diaryliodonium salts has been achieved (Scheme 7.59) [96]. This chemistry was promoted by common copper compounds and used potassium fluoride as the nucleophilic fluoride source. Conceivably, the fluorination reaction could take place on either side of the diaryliodonium salt. Under the reaction conditions, the authors noted that the copper compound promoted the fluorination of the smaller group. Curiously, if the copper was removed, the mesityl side was preferentially fluorinated. The chemistry was tolerant to a wide range of functional groups, and a host of electron-withdrawing and electron-donating groups were successfully functionalized. Heteroaryl substrates were also fluorinated using the copper/KF system in high yields with excellent selectivity. This chemistry has been extended to ^{18}F-labeled compounds [97].

Example 7.17 Palladium-Catalyzed Fluorination of Aryltrifluoroborate Salts [95].

Before starting this synthesis, carefully follow the chemical safety instructions listed in Example 7.1.

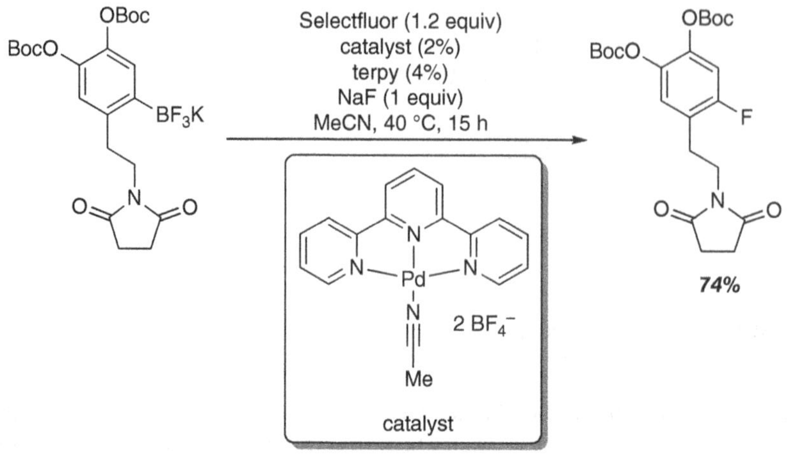

Caution! Ethyl acetate, hexanes, and acetonitrile are highly flammable.
To a mixture of palladium catalyst (2.2 mg, 4.0 µmol, 0.020 equiv), terpy (1.9 mg, 8.0 µmol, 0.040 equiv), aryltrifluoroborate (108 mg, 200 µmol, 1.00 equiv), Selectfluor (85.0 mg, 240 µmol, 1.20 equiv), and sodium fluoride (8.40 mg, 200 µmol, 1.00 equiv) was added acetonitrile (2.0 mL, 0.1 M) at 23 °C. The reaction mixture was stirred for 15 h at 40 °C, allowed to cool to 23 °C, and then transferred to a separatory funnel. Dichloromethane (20 mL) was added, and the organic layer was washed with water (20 mL). The aqueous layer was extracted with dichloromethane (4 × 20 mL). The combined organic layers were washed with brine (20 mL), dried over sodium sulfate, filtered, and concentrated in vacuo to afford a yellow solid. The solid was purified by chromatography on silica gel eluting with a solvent mixture of hexanes/ethyl acetate (1 : 1 (v/v)) to afford the title compound (67.1 mg, 148 µmol, 74% yield) as a colorless crystalline solid.

Adapted with permission from A. R. Mazzotti, M. G. Campbell, P. Tang, J. M. Murphy, and T. Ritter, *J. Am. Chem. Soc.*, **2013**, *135*, 14012–14015. *Copyright (2013) American Chemical Society.*

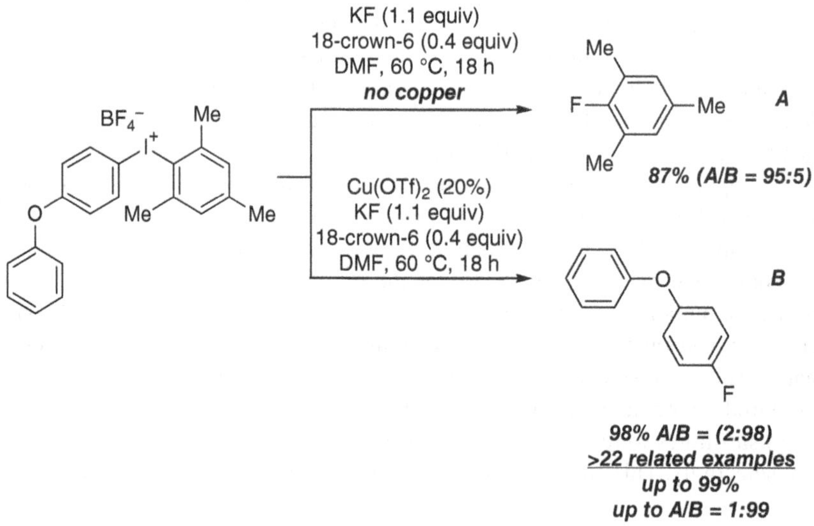

SCHEME 7.59 Copper-promoted fluorination of diaryliodonium salts [96].

The conversion of phenols into aryl fluorides (deoxyfluorination) has been achieved using a fluorinating agent derived from readily available imidazolium salts (PhenoFluor; Scheme 7.60) [98, 99]. As part of the investigation, the authors compared the reactivity of this fluorinating agent with common sources, and while many of these reagents were not active in the deoxyfluorination reaction, PhenoFluor successfully fluorinated a wide range of phenols. The chemistry was sensitive to the electronic composition of the phenol, and substrates bearing electron-withdrawing groups such as nitro and $-CF_3$ were fluorinated in 3 h at 80 °C, while analogous reactions with phenols bearing electron-donating substituents such as methoxy or methyl required longer reaction times (18–20 h) and elevated temperatures (110 °C). It was also noteworthy that secondary products resulting from deoxyprotonation reactions were not observed. This chemistry required the preparation of the fluorinating agent, but the deoxyfluorination is straightforward and high yielding.

The ortho-fluorination of arenes has been achieved through the incorporation of directing groups on the aromatic ring. The most common substituents used for this application are nitrogen-containing groups such as pyridine. One of the early examples of this chemistry entailed the use of N-fluoropyridinium tetrafluoroborate as the source of electrophilic fluorine [12]. A common palladium complex was used as the catalyst for the reaction, and moderate yields of the ortho-fluorinated arenes were isolated. It was noteworthy that competing arylation and acetoxylation reactions were not observed in this chemistry. This was a microwave-assisted reaction, and reactions were complete within a few hours. While monofluorination was the primary focus of the work, difluorination was also possible by altering the stoichiometry of the reaction (Scheme 7.61).

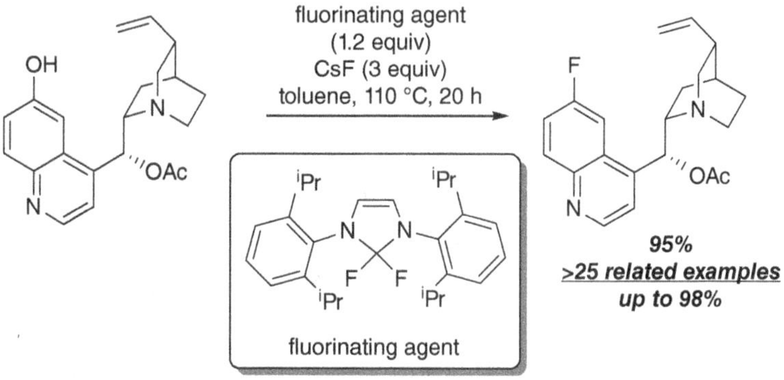

SCHEME 7.60   Deoxyfluorination of phenols [98].

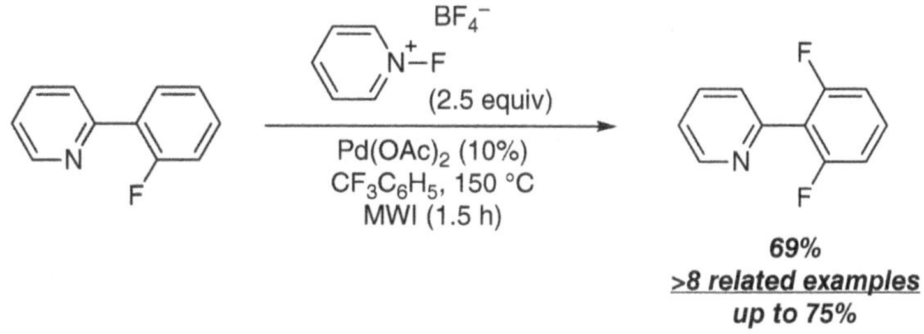

SCHEME 7.61   Microwave-assisted fluorination [12].

Another example of this approach used an amide derivative to direct the fluorination of functionalized arenes (Scheme 7.62) [100]. This palladium-catalyzed process was a formal C—H fluorination, and a range of F⁺ sources were screened for activity. These screening experiments revealed that *N*-fluoro-2,4,6-trimethylpyridinium triflate was the most effective. Curiously, the use of Selectfluor or NFSI only generated trace amounts of the product. Another critical component for the reaction was NMP, and although its precise role was not determined, its presence (and amount) was critical to the success of the reaction. While many of the ortho-directing groups are an integral part of the structure of the products, one of the strengths of this approach was the ability to readily remove the trifylamide and generate additional classes of fluorinated compounds.

As mentioned earlier, the attachment of Lewis bases to specific positions on the framework of aromatic ring is an effective means by which the position of fluorination can be directed. Another example of this chemistry used a quinoxaline-derived ring system as the directing group and palladium acetate as the catalyst (Scheme 7.63) [101]. The chemistry required an electrophilic source of fluorine, and the authors determined that the combination of NFSI in TFA afforded the desired fluorination products in moderate to good yields. One of the attractive aspects of this chemistry was the propensity of the chemistry to generate the monofluorinated compounds. When a second fluorination site is available, difluorination can be a challenging problem to overcome since the difluorinated compounds are often difficult to separate from the desired monosubstituted species. If the difluorinated compounds were needed, the authors also noted that these compounds could be generated using this catalyst system simply through modification of the reaction stoichiometry. Furthermore, the fluorination reactions could be carried out in air. This observation was a

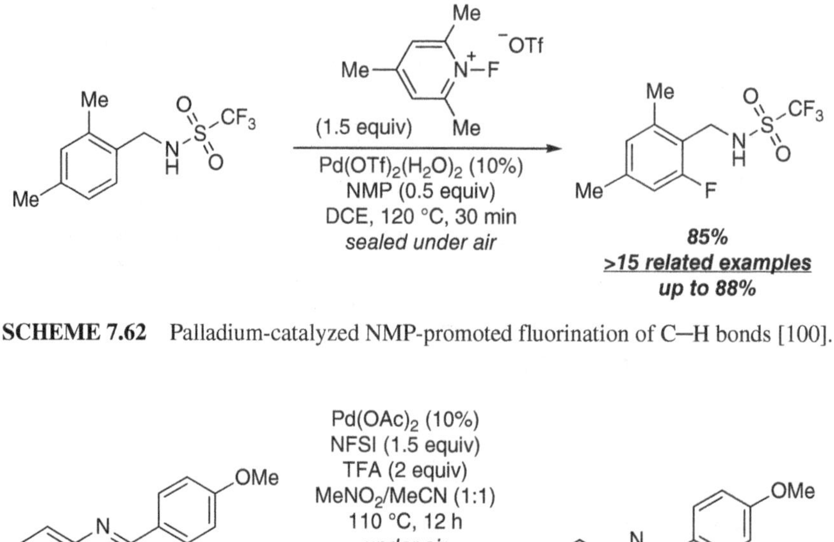

SCHEME 7.62   Palladium-catalyzed NMP-promoted fluorination of C—H bonds [100].

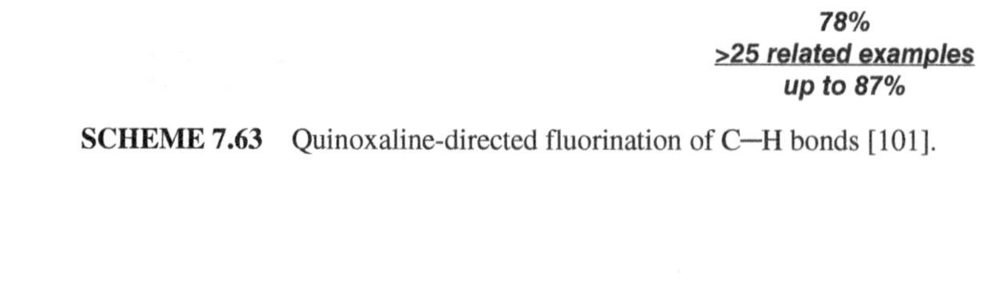

SCHEME 7.63   Quinoxaline-directed fluorination of C—H bonds [101].

significant advancement since the chemistry could be carried out without the need for a glovebox or nitrogen manifold.

The use of *O*-methyl oxime ethers as directing groups for palladium-catalyzed fluorination reactions is one of the most effective routes to the targeted synthesis of aryl fluorides (Scheme 7.64) [102–106]. The fluorination reaction was highly selective for the formation of the monofluorinated product, and only small amounts of difluorination were observed. One of the challenges with this chemistry was the competing formation of an ortho-acetoxylation product. After screening a range of palladium catalysts, the authors found that Pd$_2$dba$_3$ successfully promoted the reaction while minimizing the formation of this side product. Another advantage to this approach was the observation that the commonly used silver salts could be replaced by potassium nitrate with no loss of activity. In fact, the conditions of the optimized reaction were some of the mildest that have been reported for the C—H fluorination reaction. The fluorination chemistry effectively passed over ketones, esters, sulfones, nitriles, nitro, and a range of other functional groups and generated the aryl fluorides in good to excellent yield. Substrates bearing electron-withdrawing groups were a little sluggish and required gentle heating to promote the reaction.

An amide was also successfully used as a directing group for the ortho-fluorination of arenes (Scheme 7.65) [107]. The investigation started by screening a range of functionalized compounds for activity in the fluorination reaction. Simple compounds such as benzoic

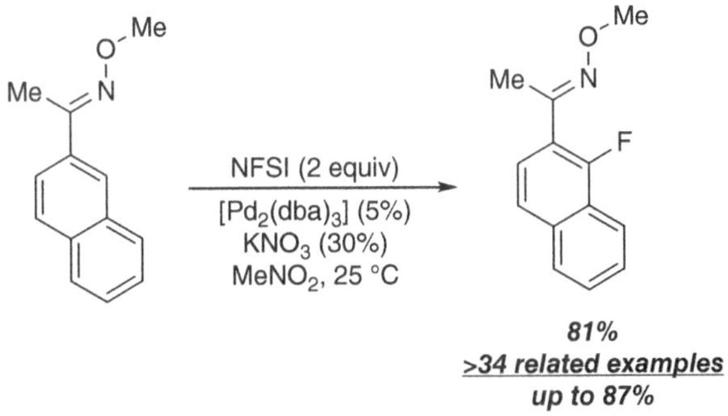

81%
≥34 related examples
up to 87%

SCHEME 7.64   Nitrate-promoted palladium-catalyzed C—H fluorination using *O*-methyl oxime ethers as directing groups [102].

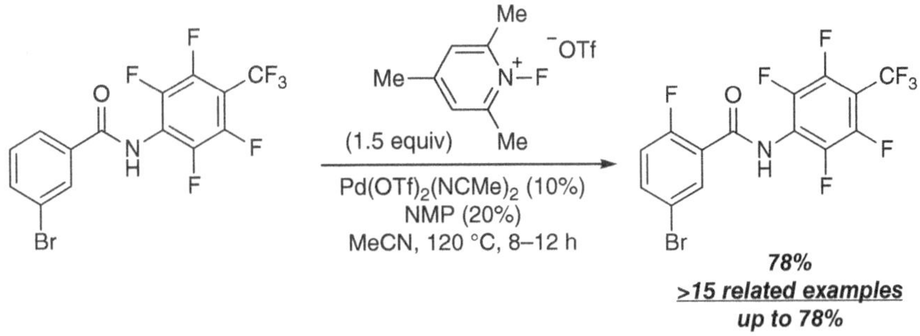

78%
≥15 related examples
up to 78%

SCHEME 7.65   Palladium-catalyzed ortho-directed fluorination of *N*-phenylbenzamide derivatives [107].

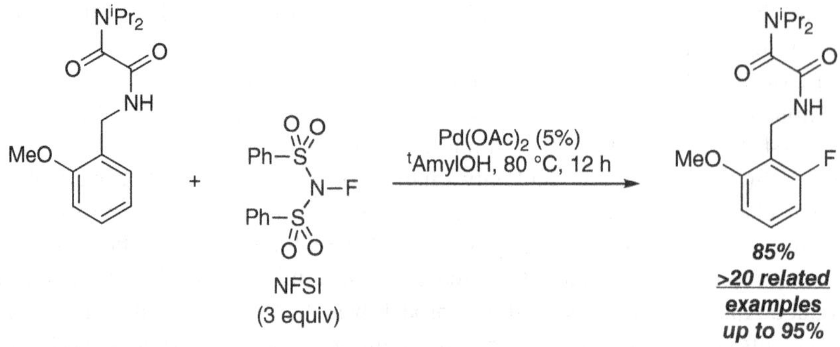

SCHEME 7.66 Palladium-catalyzed ortho-directed fluorination using NFSI [108].

acid and *N*-phenylbenzamide were completely unreactive in the electrophilic fluorination reaction using *N*-fluoro-2,4,6-trimethylpyridinium triflate. When the authors screened an extremely electron-withdrawing *N*-phenylbenzamide derivative, fluorination of the arene occurred. The yield of this reaction was moderate (84%), but the reaction was largely selective for the formation of the monosubstituted product. Only small amounts of the difluorinated species were observed (4%). Similar to other studies, the difluorinated product could be obtained in good to excellent yields by simple modification of the reaction conditions and stoichiometry. It is worth noting that the electron-withdrawing *N*-phenyl-benzamide used to direct the fluorination reaction could be readily converted into a carboxylic acid upon treatment with KOH/ethylene glycol. This would enable the synthesis of a host of fluorinated carboxylic acid derivatives.

The fluorination of oxalyl amide bearing benzylamine derivatives has been achieved (Scheme 7.66) [108]. This chemistry used a common palladium complex as the catalyst and NFSI as the source of electrophilic fluorine. It should be noted that several sources of F⁺ were active in this reaction and NFSI was selected due to its relative low cost. The functional group tolerance was fairly broad, and a range of functionalized oxalyl amide bearing benzylamine derivatives were fluorinated in moderate to excellent yields using this approach. One of the advantages of this directing group was the ability to readily cleave it and generate additional classes of compounds. Ns-protected amines were generated through the addition of NaH/NsCl, and benzylamines were formed by the addition of sodium hydroxide. This chemistry is attractive since the directing group could be added and removed fairly easily.

Although NFSI was successful for the monofluorinations, it was not as active in the difluorination reactions and tended to form mixtures of the mono- and difluorinated compounds. After some experimentation, *N*-fluoro-2,4,6-trimethylpyridinium triflate was found to be more active in the difluorination reaction and generated moderate to excellent yields of the functionalized compounds (Example 7.18).

Given the value of aryl chlorides, bromides, and iodides as substrates in organic transformations, a host of protocols have been devised for their synthesis. One of the classic methods for the synthesis of aryl halides were versions of the halodediazoniation reactions [109–111]. For a range of electron-rich arenes as well as many common aromatic compounds, bromination is commonly achieved simply through the addition of NBS to the arene in DMF [112]. Yields are good to excellent, and no transition metal catalyst was needed. In addition to NBS, elemental bromine has also been successfully used for the bromination

Example 7.18   Difluorination of Functionalized Arenes [108].

Before starting this synthesis, carefully follow the chemical safety instructions listed in Example 7.1.

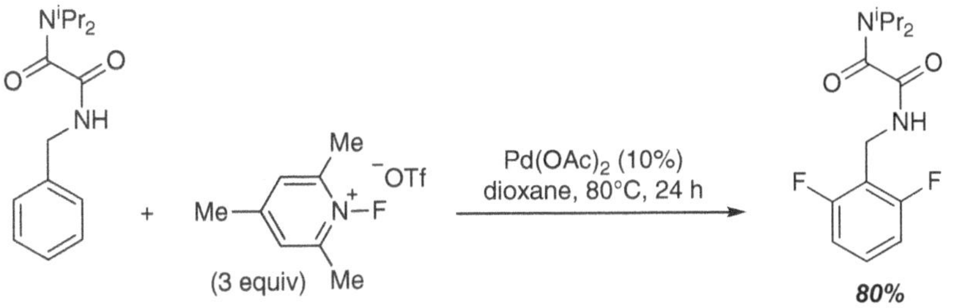

Caution! Dioxane is highly flammable.
A mixture of oxalamide (0.2 mmol, 1.0 equiv), Pd(OAc)$_2$ (44 mg, 0.10 equiv), N-fluoro-2,4,6-trimethylpyridinium triflate (0.6 mmol, 3.0 equiv), and 1,4-dioxane (2 mL) in a 25 mL glass vial (sealed with PTFE cap) was heated at 80 °C for 24 h. The reaction mixture was cooled to room temperature and concentrated in vacuo. The resulting residue was purified by column chromatography on silica gel to give the product as a pale yellow oil (80%).

Adapted with permission from C. Chen, C. Wang, J. Zhang, and Y. Zhao, *J. Org. Chem.*, **2015**, *80*, 942–949. *Copyright (2015) American Chemical Society.*

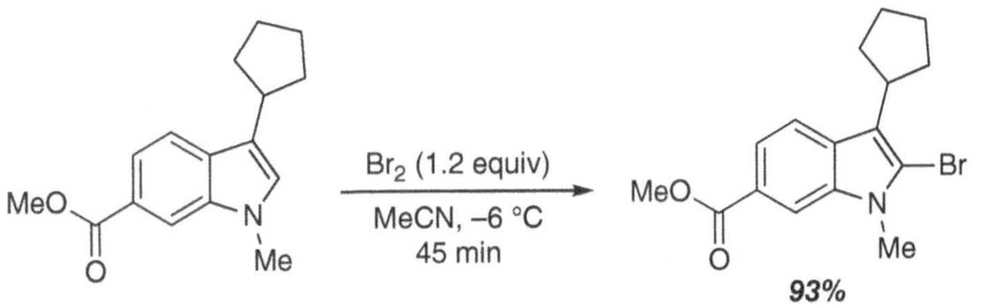

SCHEME 7.67   Bromination of a C—H bond using elemental bromine [114].

of activated arenes [113] and heteroarenes (Scheme 7.67) [114]. Analogous reactions using NIS in an acidic medium were also successful [115]. Heteroaryl substrates such as pyrroles were brominated using a combination of elemental bromine in carbon tetrachloride [116]. Aryl chlorides were generated using TMSCl as the chlorinating agent [117]. The following sections will highlight some of the recent contributions to this area.

The dibromination of bisquinolones proceeds under mild conditions using NBS as the halogen source (Example 7.19) [118]. Simply stirring the arene with an excess of NBS at room temperature generated excellent yields of the brominated species. Using a reduced amount of NBS (2.5 equiv) resulted in a mixture of the mono- and disubstituted compounds. If desired, the monosubstituted product was readily separated from the reaction mixture in moderate yield (62%).

The combination of potassium iodide and *tert*-butyl hydrogen peroxide (TBHP) is an alternative approach to the halogenation of highly activated arenes (Scheme 7.68) [119].

Example 7.19   Mono- and Dibromo Bisquinolone Derivatives Using NBS/DMF [118].

Before starting this synthesis, carefully follow the chemical safety instructions listed in Example 7.1.

Caution! Ethyl acetate is highly flammable.

Preparation of the monobrominated bisquinolone. A mixture of 109 mg (0.25 mmol) of 4,4′-bis(6,7-dimethoxy-1-methylquinolin-2(1H)-one), 111.2 mg (0.625 mmol, 2.5 equiv) of N-bromosuccinimide (NBS), and 0.6 mL of DMF in a vial equipped with a magnetic stir bar was stirred for 35 min at room temperature. The reaction was quenched by addition of ice water and was filtered. The resulting crude product mixture was purified by automated flash chromatography using EtOAc/CH₃CN as eluent to give 80 mg (62%) of the monobrominated material as a yellow solid.

Preparation of the dibrominated bisquinolone. A mixture of 109 mg (0.25 mmol) of 4,4′-bis(6,7-dimethoxy-1-methylquinolin-2(1H)-one), 222.4 mg (1.25 mmol, 5.0 equiv) of N-bromosuccinimide (NBS), and 0.6 mL of DMF in a vial equipped with a magnetic stir bar was stirred for 50 min at room temperature. The reaction was quenched by addition of ice water and was filtered. The resulting crude product mixture was purified by automated flash chromatography using EtOAc/CH₃CN as eluent to give 122 mg (82%) of the dibrominated material as a yellow solid.

Adapted with permission from N. Arshad, J. Hashim, and C. O. Kappe, J. Org. Chem., 2008, 73, 4755–4758. Copyright (2008) American Chemical Society.

No metal catalyst was needed for this iodination reaction, and moderate to excellent yields of the aryl iodides were obtained.

One of the classic approaches to the synthesis of aryl iodides starts from aryl bromides. Treatment of the ArBr species with an organolithium reagent such as BuLi generated the

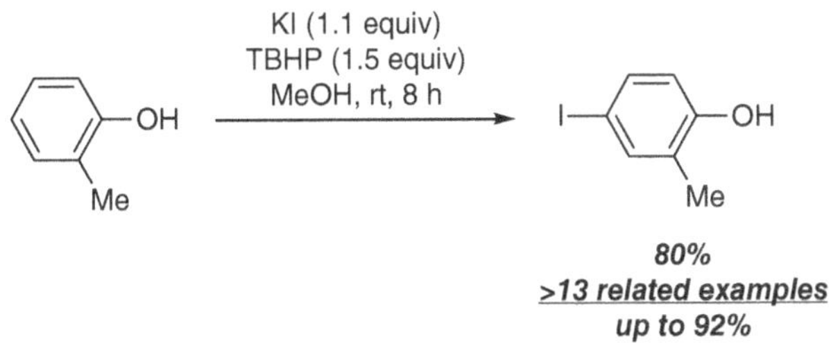
KI (1.1 equiv)
TBHP (1.5 equiv)
MeOH, rt, 8 h
OH
Me
I
OH
Me
80%
>13 related examples
up to 92%

SCHEME 7.68 Iodination of arenes using KI/TBHP [119].

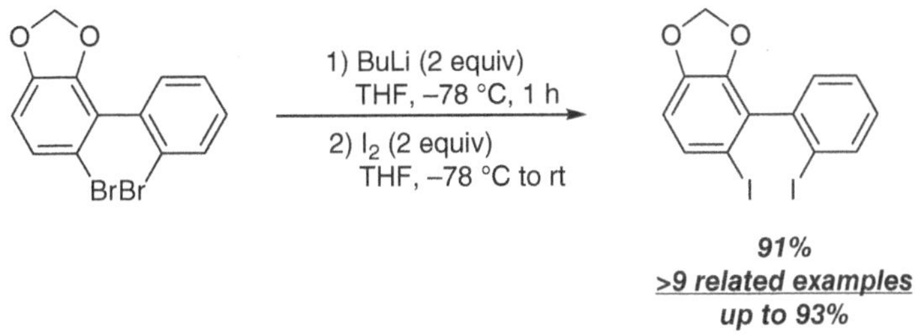
O
O
BrBr
1) BuLi (2 equiv)
THF, –78 °C, 1 h
2) I2 (2 equiv)
THF, –78 °C to rt
O
O
I I
91%
>9 related examples
up to 93%

SCHEME 7.69 Halogen exchange using BuLi/X_2 [120].

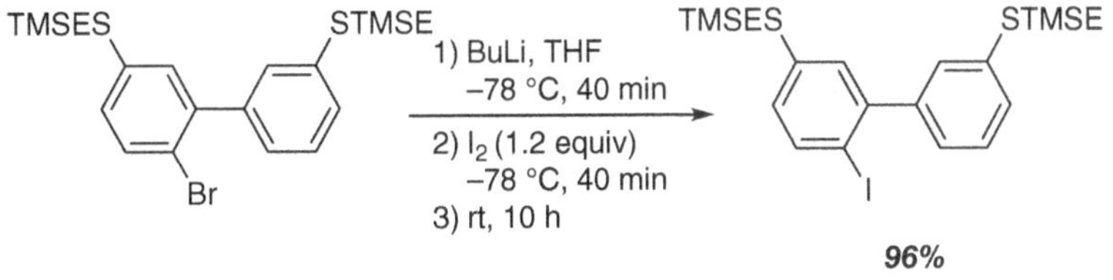
TMSES
STMSE
Br
1) BuLi, THF
–78 °C, 40 min
2) I2 (1.2 equiv)
–78 °C, 40 min
3) rt, 10 h
TMSES
STMSE
I
96%

SCHEME 7.70 Halogen exchange using a metallation/halogenation approach [121].

aryllithium reagent, and quenching at low temperature with elemental iodine generated the iodinated compounds (Schemes 7.69 and 7.70) [120, 121]. This approach tended to be high yielding and often worked quite well. One of the limitations with this chemistry was the general lack of tolerance for substrates with good electrophiles.

This approach is often used in tandem with an orthometallation reaction to generate halogenated arenes. A variety of modifications to the metallation step have been devised to manipulate the reactivity of the metallated species including the addition of metals such as zinc or copper. This is often done to facilitate the use of substrates bearing moderate electrophiles. Examples of this chemistry are shown in Examples 7.20 and 7.21 [90, 122]. Typically, the nitrile would be expected to react with the lithium reagent; however, the authors converted the organolithium reagent into an organozinc reagent through the addition of zinc(II) chloride to the metallated arene, followed by the addition of elemental bromine to generate the aryl bromide. Typically, this approach was carried out at low temperatures to prevent side reactions from occurring.

Electron-poor arenes can be challenging to directly halogenate using common approaches. An effective system consisted of tribromoisocyanuric acid (TBCA) as an electrophilic source of bromine along with TFA as the solvent (Scheme 7.71) [123].

Example 7.20   Zinc-Assisted Synthesis of 1,2-Dibromoarenes [122].

Before starting this synthesis, carefully follow the chemical safety instructions listed in Example 7.1.

87%

Caution! Hexanes, THF, and tert-butyl methyl ether are highly flammable.

Caution! Organolithium salts can react violently with moisture and must be handled and stored over an inert atmosphere such as nitrogen or argon.

To n-butyllithium in hexanes (1.95 mL, 1.55 M, 3.02 mmol) at −20 °C was added a solution of 2,2,6,6-tetramethylpiperidine (0.51 mL, 3.02 mmol) in THF (4.75 mL). After aging for 30 min, the solution was cooled to −78 °C, and a solution of 3-bromobenzonitrile (0.5 g, 2.7 mmol) in THF (2.8 mL) was added dropwise so as to maintain the temperature at less than −70 °C. After 2 h, a 1.0 M solution of ZnCl$_2$ in THF (2.75 mL, 2.75 mmol) was added dropwise and allowed to stir for an additional 30 min. Bromine (0.21 mL, 4.12 mmol) was added dropwise at a rate to keep the temperature below −50 °C. The reaction was monitored by HPLC (typical reaction time <30 min), and the completed reaction was allowed to warm to 25 °C and quenched with H$_2$O. The product was extracted two times into *tert*-butyl methyl ether, and the combined organic layers were washed with 1 M HCl (aq) followed by H$_2$O. The organic layer was dried over Na$_2$SO$_4$, filtered, and concentrated. The residue was purified by flash column chromatography with hexanes to afford 0.62 g of 2,3-dibromobenzonitrile as slightly yellow crystals (2.38 mol, 87%).

Adapted with permission from K. Menzel, E. L. Fisher, L. DiMichele, D. E. Frantz, T. D. Nelson, and M. H. Kress, *J. Org. Chem.*, **2006**, *71*, 2188–2191. *Copyright (2006) American Chemical Society.*

Example 7.21   Bromination of a Pyridine Ring [90].

Before starting this synthesis, carefully follow the chemical safety instructions listed in Example 7.1.

25%

Caution! Hexane, ethyl acetate, petroleum ether, and THF are highly flammable.

Caution! Organolithium reagents can react violently with moisture and need to be handled and stored under an inert atmosphere such as nitrogen or argon.

To a 250 mL three-necked bottle, the mixture of DMAE (2.7580 g, 32 mmol) and anhydrous hexane (40 mL) was cooled to −5 °C. n-BuLi (25.6 mL, 2.5 M) was added slowly via dropping funnel. The resultant orange solution was maintained −5 °C for 15 min, then 4-*tert* pyridine

(2.1663 g, 16 mmol) was added, and the mixture was continued to stir for 1 h. The red mixture was cooled again to −78 °C, and the solution of CBr_4 (13.25 g, 40 mmol) in THF was added. The mixture was maintained at −78 °C for 1 h. The dark brown mixture was allowed to warm to room temperature and continued stirring for additional 1 h. The mixture was quenched at 0 °C by adding 40 mL H_2O. The mixture was extracted by ethyl acetate. The combined organic phase was dried over anhydrous $MgSO_4$, and the residue was purified by flash chromatography (petroleum ether/ethyl acetate = 12 : 1) to provided 2-bromo-4-*t*-butylpyridine (0.88 g, 25% yield).

Adapted from X. Mu, H. Zhang, P. Chen and G. Liu, *Chem. Sci.*, **2014**, *5*, 275–280 *with permission of The Royal Society of Chemistry.*

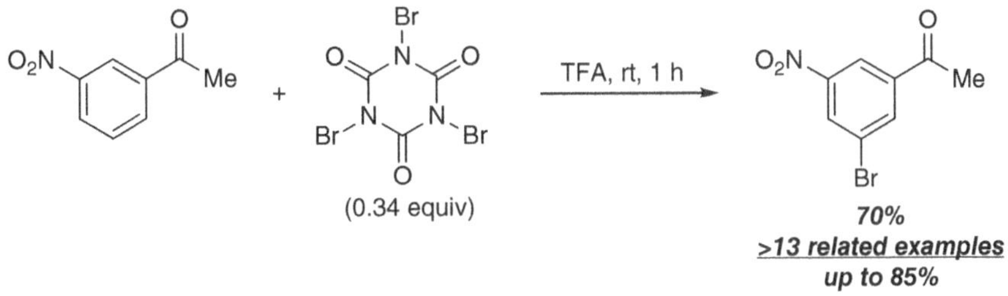

SCHEME 7.71   Bromination of deactivated arenes [123].

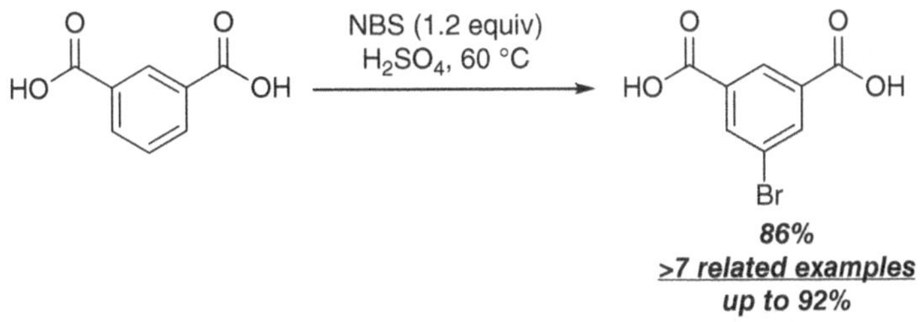

SCHEME 7.72   Bromination of deactivated arenes using NBS in sulfuric acid [124].

This transition metal-free approach proceeded under very mild conditions and was typically regioselective for the formation of the meta-substituted product. It was noted that the amount of halogenation was dependent upon the acid used in the reaction. Changing from TFA to acetic acid resulted in sluggish reactivity, while sulfuric acid resulted in moderate conversion into the monobrominated product along with the formation of the dibrominated species (10%). In related work, the use of NBS in concentrated sulfuric acid effectively brominated a series of deactivated arenes [124]. One difference between the two approaches was that the latter was a bit more effective for the bromination of benzoic acid derivatives (Scheme 7.72). These approaches are complementary and are effective solutions for moderately deactivated systems, provided that the substrates do not contain functional groups that are sensitive to the acidic conditions.

Elemental bromine is a popular brominating agent. However, it has significant health concerns and should always be handled with great care. An alternative brominating agent is pyridinium tribromide. While bromine is a liquid, pyridinium tribromide is a solid and

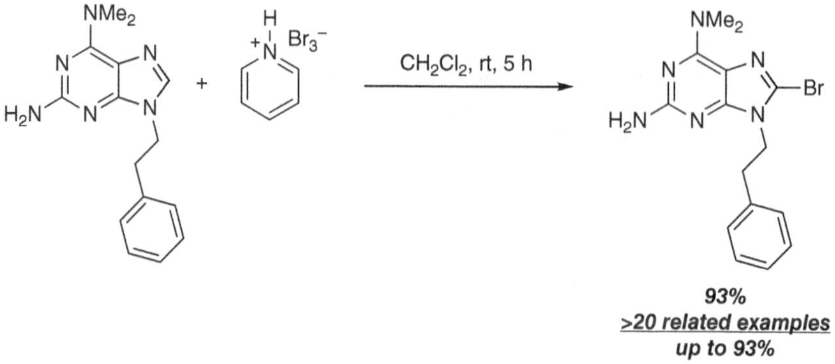

93%
>20 related examples
up to 93%

SCHEME 7.73   Bromination of purine derivatives using pyridinium bromide [125].

is much easier to work with. The bromination of a series of purine derivatives serves as an example of this (Scheme 7.73) [125]. The authors simply dissolved the purine derivatives in DCM and added the pyridinium tribromide to afford the brominated compounds. After screening a range of purines, the authors discovered that the bromination reaction was only successful if the purine was substituted with electron-donating groups. Substrates such as 6-chloropurine were unreactive under the reaction conditions.

Another approach to the halogenation of electron-deficient arenes used a powerful Lewis acid and NXS as a source of the halogen (Scheme 7.74 and Example 7.22) [126]. This chemistry was successful for chlorination, bromination, and iodination, and one of the attractive aspects of this chemistry was that no additional solvents were needed for the reaction. The conditions were mild, and good to excellent yields of the aryl halides were obtained.

The copper chloride-promoted chlorination of arylboronic acids (chlorodeboronation) is an effective method for the preparation of 3-substituted aryl chlorides (Scheme 7.75) [127]. The chemistry used a stoichiometric amount of NCS and a catalytic amount of copper chloride to generate the aryl chlorides in 18–24 h at 80 °C. The isolated yields were outstanding, and the chemistry tolerated both electron-donating and electron-withdrawing groups. This was particularly noteworthy given the challenges commonly associated with the use of electron-deficient arylboronic acids. The authors also investigated the effect of adding a stoichiometric amount of the copper chloride and found that the reaction time was decreased to only a few hours for most substrates.

The conversion of aryltrifluoroborate salts into aryl chlorides has been achieved using sodium hypochlorite (Scheme 7.76) [128]. This metal-free chlorodeboronation reaction was operationally straightforward, and simply adding an aqueous bleach solution to the aryltrifluoroborate salts in water/ethyl acetate afforded excellent yields of the aryl chlorides. The chemistry was limited to electron-rich aryltrifluoroborate salts, and little conversion was observed when electron-poor aryltrifluoroborate salts were screened. After some experimentation, the authors discovered that exchanging the sodium hypochlorite for TCICA enabled the chlorodeboronation of the electron-poor aryltrifluoroborates. One of the more practical advantages of this chemistry was the observation that the reactions could be carried out in an open vessel. The bromination of aryltrifluoroborate salts was also achieved through the use of NaBr/chloramine-T mixtures [129]. Yields were typically good, and bulky substrates were brominated in high yeild (2,6-dimethylphenyltrifluoroborate, 83%).

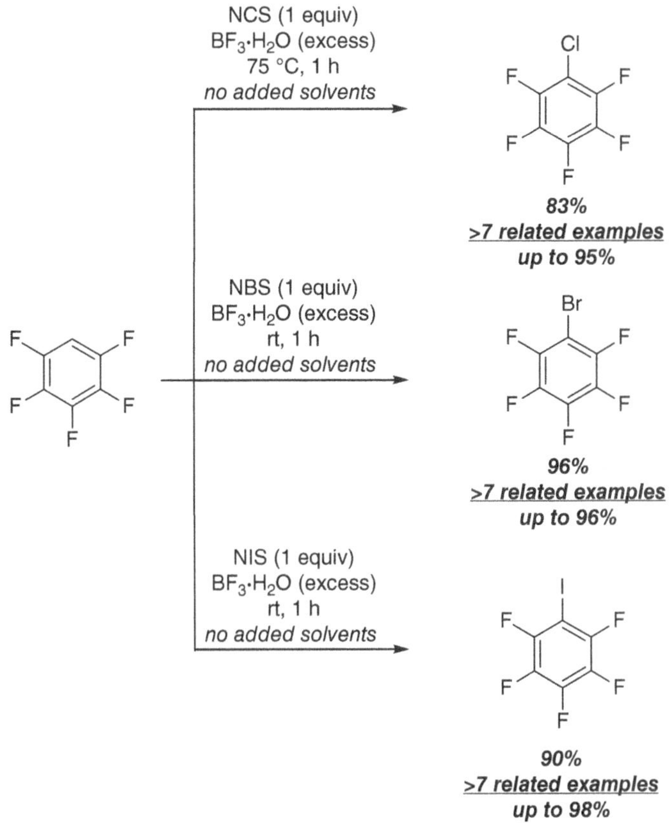

SCHEME 7.74   Halogenation of electron-deficient arenes [126].

Example 7.22   Halogenation of Electron-Deficient Arenes Using BF_3–H_2O [126].

Before starting this synthesis, carefully follow the chemical safety instructions listed in Example 7.1.

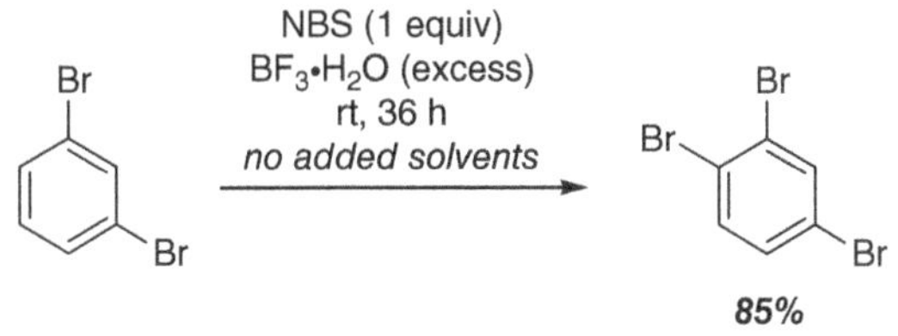

Caution! BF₃ adducts can be highly reactive and must be handled with care.

Caution! Petroleum ether is highly flammable.

BF_3–H_2O (4 mL) was added to a mixture of the deactivated aromatic compound (10 mmol) and *N*-bromosuccinimide (10 mmol) in a 30 mL Nalgene bottle. The bottle was closed, and the mixture was stirred at the required temperature (room temperature for most reactions) for 36 h.

The progress of the reaction was monitored by GC–MS analysis. The reaction mixture was then completely transferred to a separatory funnel using water (30 mL) and dichloromethane (2×30 mL) and thoroughly shaken to extract the product into the organic layer. The organic layer was washed with saturated solution of sodium bisulfite or sodium thiosulfate (15 mL) to remove any free halogen present, followed by saturated sodium bicarbonate solution (30 mL) and brine (30 mL). It was then washed with water (2×30 mL) and dried over anhydrous sodium sulfate. The solvent was removed under reduced pressure, and the product was purified by fractional distillation or recrystallization from petroleum ether.

Adapted with permission from G. K. S. Prakash, T. Mathew, D. Hoole, P. M. Esteves, Q. Wang, G. Rasul, and G. A. Olah, J. Am. Chem. Soc., 2004, 126, 15770–15776. Copyright (2004) American Chemical Society.

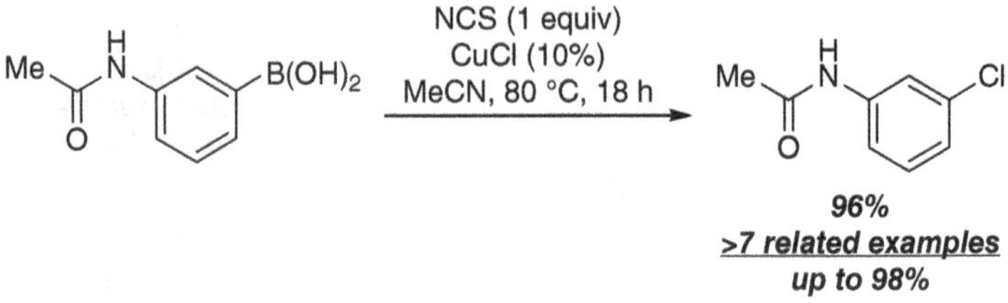

SCHEME 7.75   Copper-promoted chlorodeboronation reactions [127].

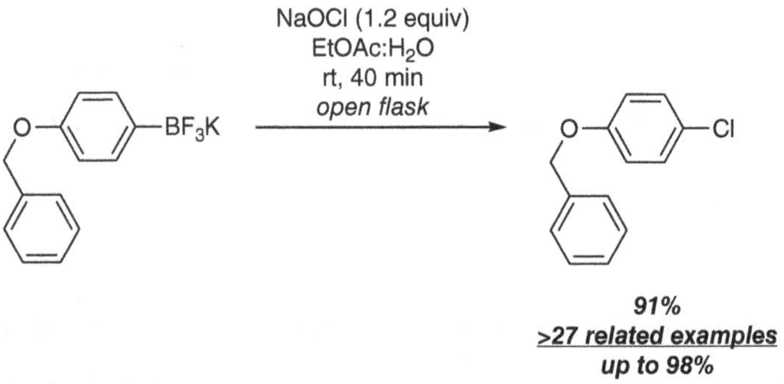

SCHEME 7.76   Chlorodeboronation of electron-rich aryltrifluoroborates [128].

While the chlorodeboronation reactions using NCS required a copper catalyst to generate reasonable amounts of the aryl chlorides, analogous reactions with NBS or NIS were successful without the addition of a catalyst. Simply stirring the boronic acid with the halosuccinimide generated moderate to good yields of the aryl bromides and iodides (Scheme 7.77) [130]. The conditions were quite mild, and a wide range of electron-donating and electron-withdrawing groups were well tolerated. For the iododeboronation reactions, bulky arylboronic acids were also successfully used. The generation of NIS in solution through the treatment of NCS with NaI was also a successful method (Scheme 7.78). Given the wide range of arylboronic acids that are available, this metal-free synthesis of aryl halides is an attractive approach.

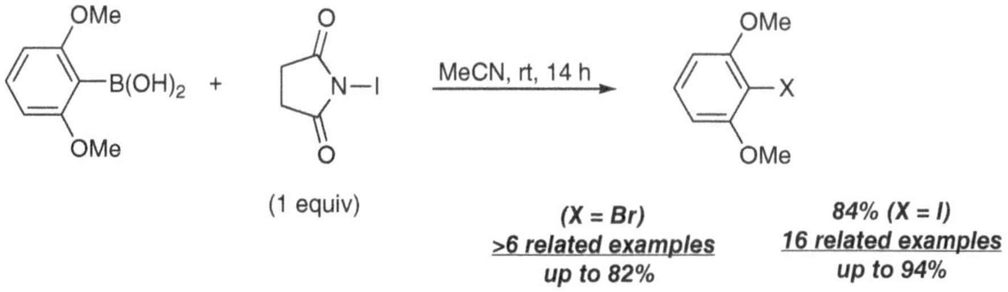

SCHEME 7.77 Metal-free halodeboronation reactions [130].

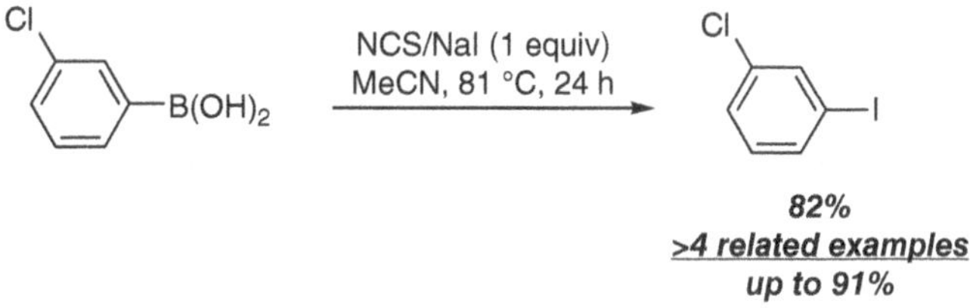

SCHEME 7.78 Iododeboronation of arylboronic acids using NIS generated in solution [130].

A sequential borylation/bromination one-pot reaction selectively converted arenes into aryl bromides (Example 7.23) [131]. Borylation was the first step in the process and was achieved using an iridium catalyst and B_2Pin_2. Once this step was complete, the solvent was removed under vacuum and replaced with a methanol/water mixture, and an excess of copper(II) bromide was added. The chemistry displayed broad substrate scope and effectively converted a range of functionalized arenes into aryl bromides. The chemistry was extended to aryl chlorides, and over 10 arenes were successfully chlorinated in moderate to good yields.

While the iridium-catalyzed sequential borylation/halogenation sequence successfully generated aryl bromides and chlorides, extending this chemistry to aryl iodides was initially unsuccessful. The first step (borylation) was not the problem, and another approach to the halogenation step was sought. After some experimentation, a solution was devised that entailed the use of a catalytic amount of copper along with a solubilizing/stabilizing ligand and a nucleophilic source of iodine (Scheme 7.79) [132]. The authors also discovered that this second step could be carried out in air. While the goal of their work was the development of the sequential reaction, the ability to iodinate the arylboronate ester in air using commercially available materials was quite an attractive approach to the synthesis of aryl iodides. This chemistry provides an elegant approach to the synthesis of aryl iodides.

While the previous sections have been focused on the replacement of a boronic acid, trifluoroborate, or pinacol-functionalized boronate ester with a halogen, it would be highly valuable to halogenate the arene with retention of the boron-containing fragment. The reactivity of the boron species could then be utilized in subsequent reactions. An example of this chemoselective halogenation reaction has been reported (Scheme 7.80) [133]. Using a gold catalyst and a slight excess of a halosuccinimide, a series of arylboronates were successfully halogenated with retention of the boronate ester. The conditions of the reaction were mild, and good to excellent yields of the halogenated compounds were obtained. This chemistry displayed broad scope, and a host of functionalized arylboronate esters

Example 7.23 Iridium-Catalyzed Conversion of Arenes into Aryl Bromides [131].

Before starting this synthesis, carefully follow the chemical safety instructions listed in Example 7.1.

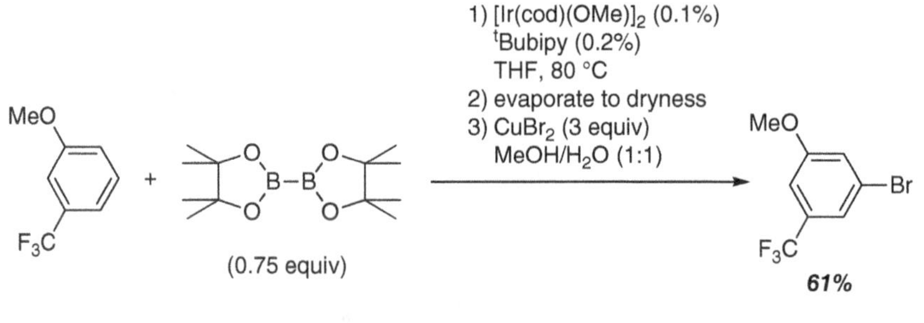

Caution! THF, diethyl ether, and pentane are highly flammable.

[Ir(COD)(OMe)]$_2$ (3.4 mg, 0.005 mmol), dtbpy (2.9 mg, 0.016 mmol), and B$_2$pin$_2$ (370.5 mg, 1.46 mmol) were combined, and the bomb was evacuated and refilled with nitrogen three times. Under a positive flow of nitrogen, dry THF (4.0 mL) and arene (2.00 mmol) were added. The bomb was then sealed and heated overnight in an 80 °C oil bath. After heating overnight, the volatiles were removed under vacuum, and 25.0 mL of MeOH and 25.0 mL of a CuBr$_2$ (1.34 g, 6.0 mmol) solution were added. The bomb was then resealed and heated in an 80 °C oil bath for 4 h. Reaction progress was monitored by gas chromatography. After heating, the reaction mixture was cooled to room temperature and was then extracted with Et$_2$O (3×40 mL). The product was then purified by column chromatography using silica gel and an eluent mixture of 2% CH$_2$Cl$_2$/98% pentane to afford a clear oil (61%).

Adapted with permission from J. M. Murphy, X. Liao, and J. F. Hartwig, J. Am. Chem. Soc., 2007, 129, 15434–15435. Copyright (2007) American Chemical Society.

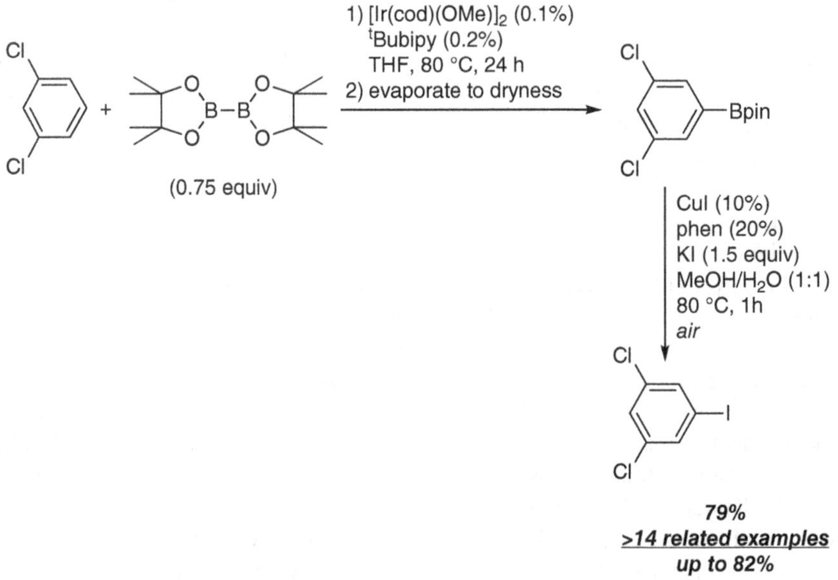

SCHEME 7.79 Conversion arenes into aryl iodides using a borylation/iodination approach [132].

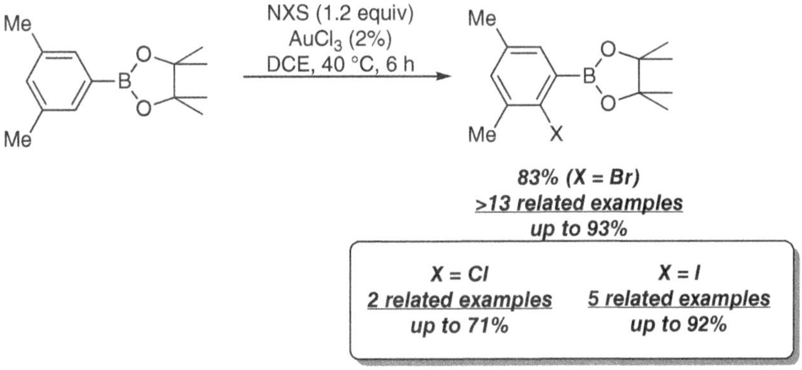

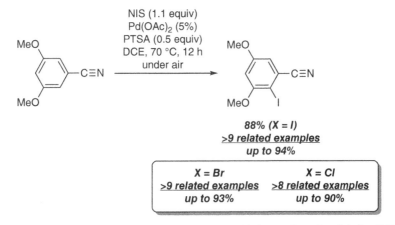

SCHEME 7.80   Gold-catalyzed halogenation of arylboronate esters [133].

SCHEME 7.81   Palladium-catalyzed ortho-directed halogenation of arylnitriles [139].

were successfully halogenated. Even a phenol-containing substrate was halogenated in high yield. It should be noted that the chemistry was highly regioselective and the typical point of halogenation was adjacent to the boronate ester.

One of the more popular approaches to the regioselective halogenation of arenes entails incorporating a directing group into the substrate. To this end, a variety of these directing groups have been used for the ortho-halogenation of arenes [134–138].

The ortho-selective halogenation of arenes using a nitrile as the directing group has been developed using common palladium complexes to catalyze the reaction (Scheme 7.81) [139]. The authors found that NXS reagents were effective sources of the halogen needed for the reaction and that DCE was as superior solvent. Other common solvents used in halogenation reactions such as acetonitrile and DMF afforded lower yields of the aryl halide. After a bit of experimentation, it was discovered that an acid needed to be added to the reaction in order to generate high yields of the aryl halide. The authors proposed that it served to generate a more potent X^+ source through interaction with the NXS. The reaction conditions were fairly mild, and a range of electron-donating and electron-withdrawing groups did not interfere with the halogenation reaction, and good to excellent yields of the aryl halides were obtained. One of the attractive aspects about this chemistry was that the reactions could be carried out under air. This removes the need for a glovebox or vacuum/N_2 manifold. A similar approach used a tetrazole as the ortho-directing group (Scheme 7.82) [140].

Esters also effectively direct the halogenation of arenes [141]. The palladium-catalyzed halogenation reaction occurred adjacent to the benzoate, and a range of electron-donating and electron-withdrawing groups were well tolerated by this approach. Heteroaryl esters

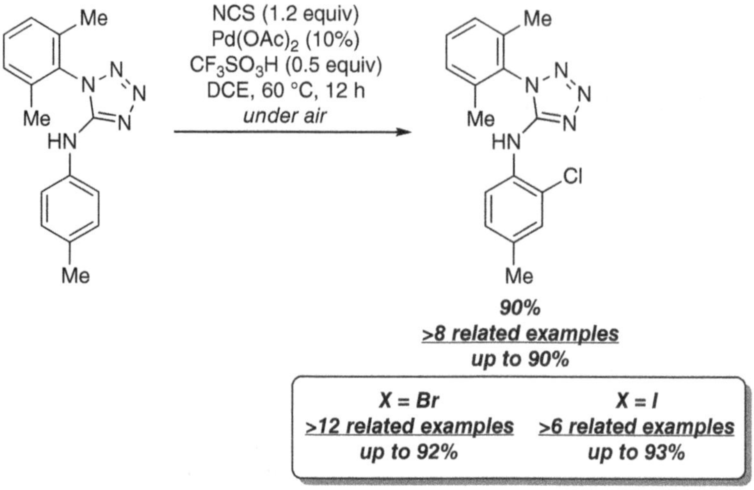

SCHEME 7.82   Palladium-catalyzed halogenation of aminotetrazoles [140].

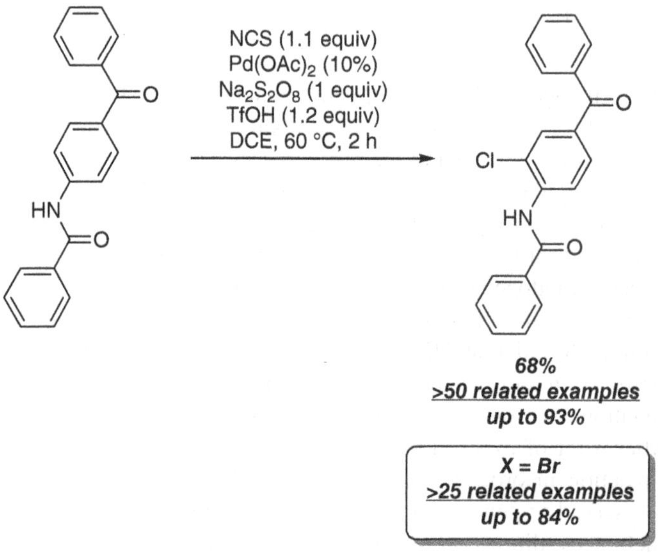

SCHEME 7.83   A competition experiment to probe the directing ability of various functional groups [141].

were also successfully halogenated albeit in lower yield, and for the chlorination of a sulfur-heterocycle serves as an example (39%). It was also noteworthy that increasing the concentration of the NXS reagent generated dichlorination and dibromination products. The substitution was regioselective for both of the positions adjacent to the ester. In order to probe the directing ability of different functional groups, a series of competition experiments were carried out using substrates bearing a range of common directing groups (Scheme 7.83) [141]. While there was some substrate specificity in the reactions, a general order of directing ability could be devised (Figure 7.1).

One of the first examples of the ortho-directed palladium-catalyzed halogenation reaction used common palladium catalysts and NXS as the sources of the halogen (Example 7.24) [142]. Acetic acid and acetonitrile were used as the solvents for this chemistry, and several different nitrogen-containing groups served as the ortho-directors. Moderate yields of the aryl halides were obtained after heating to 100–120°C for 12 h.

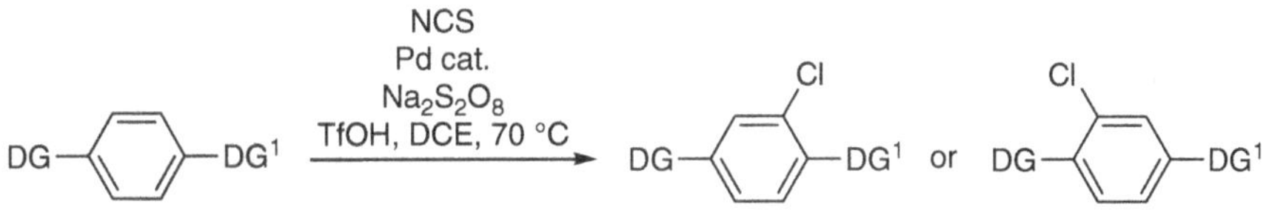

DG priority: NHAc > CONHR > C=O > SO₂NHR > CO₂Et > CONR₂ > SO₂NR₂

FIGURE 7.1 A general order of directing ability for different groups [141].

Example 7.24 Palladium-Catalyzed Monohalogenation Reactions [142].

Before starting this synthesis, carefully follow the chemical safety instructions listed in Example 7.1.

Chlorination

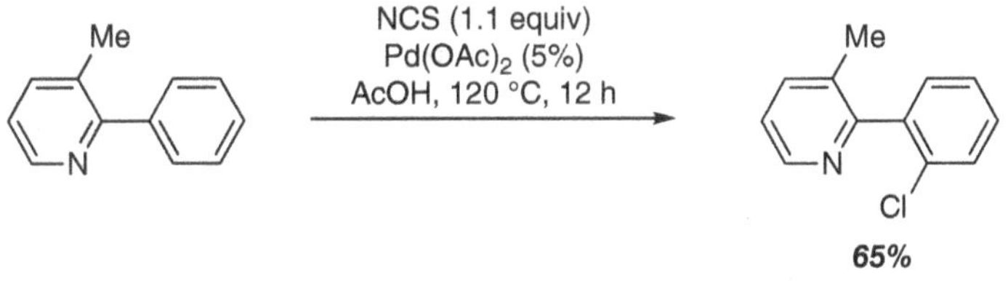

Caution! Hexanes and ethyl acetate are highly flammable!
3-Methyl-2-phenylpyridine (100 mg, 0.591 mmol, 1 equiv), NCS (86.8 mg, 0.650 mmol, 1.10 equiv), and Pd(OAc)$_2$ (6.6 mg, 0.029 mmol, 5 mol %) were combined in a 20 mL vial. AcOH (4.9 mL) was added, the vial was sealed with a Teflon-lined cap, and the mixture was heated at 120 °C for 12 h. The solvent was removed under vacuum. The crude product was dissolved in CH$_2$Cl$_2$, filtered through a pad of silica, and washed with a 50/50 mixture of hexanes/ethyl acetate (300 mL). The filtrate was concentrated, and the resulting oil was purified by chromatography on silica gel (R_f=0.3 in 5% EtOAc/95% CH$_2$Cl$_2$). The chlorinated product was isolated as a clear oil (78 mg, 65% yield).

Bromination

56%

3-Methyl-2-phenylpyridine (100 mg, 0.591 mmol, 1 equiv), NBS (126 mg, 0.709 mmol, 1.20 equiv), and Pd(OAc)$_2$ (6.6 mg, 0.029 mmol, 5 mol %) were combined in a 20 mL vial. AcOH (4.9 mL) was added, the vial was sealed with a Teflon-lined cap, and the mixture was heated at 120 °C for 12 h. The solvent was removed under vacuum. The crude product was dissolved in CH$_2$Cl$_2$, filtered through a pad of silica, and washed with a 50/50 mixture of hexanes/ethyl acetate (300 mL). The filtrate was concentrated, and the resulting oil was purified by chromatography on silica gel ($R_f = 0.3$ in 5% EtOAc/95% CH$_2$Cl$_2$). The brominated product was isolated as a clear oil (82 mg, 56% yield).

Iodination

79%

3-Methyl-2-phenylpyridine (50.0 mg, 0.295 mmol, 1 equiv), NIS (79.8 mg, 0.354 mmol, 1.20 equiv), and Pd(OAc)$_2$ (3.3 mg, 0.015 mmol, 5 mol %) were combined in a 20 mL vial. CH$_3$CN (2.5 mL) was added, the vial was sealed with a Teflon-lined cap, and the mixture was heated at 100 °C for 12 h. The solvent was removed under vacuum. The crude product was dissolved in CH$_2$Cl$_2$, filtered through a pad of silica, and washed with a 50/50 mixture of hexanes/ethyl acetate (150 mL). The filtrate was concentrated, and the resulting oil was purified by chromatography on silica gel ($R_f = 0.3$ in 5% EtOAc/95% CH$_2$Cl$_2$). The iodinated product was isolated as a clear oil (69 mg, 79% yield).

Adapted with permission from D. Kalyani, A. R. Dick, W. Q. Anani, and M. S. Sanford, *Org. Lett.*, **2006**, *8*, 2523–2526. *Copyright (2006) American Chemical Society.*

While a number of ortho-directed halogenations have been promoted by palladium and nickel, a rhodium-catalyzed process has also been used (Scheme 7.84) [143]. An amide was used as the directing group in this chemistry, and NBS and NIS were used as halogen sources. The conditions were fairly mild, and high yields of the iodinated and brominated arenes were isolated.

Due to the high cost of rhodium, it would be advantageous to develop a process that utilized less expensive metals. To this end, a cobalt-catalyzed ortho-directed halogenation was devised (Scheme 7.85 and Example 7.25) [144]. Similar to the rhodium-catalyzed reactions, the process was successful with NBS and NIS. The cobalt chemistry required elevated temperatures and longer reactions times at a higher catalyst loading (10% vs. 1% for Rh). The yields were not as high as those obtained with rhodium catalysts, and the substrate scope of the reaction was not as broad.

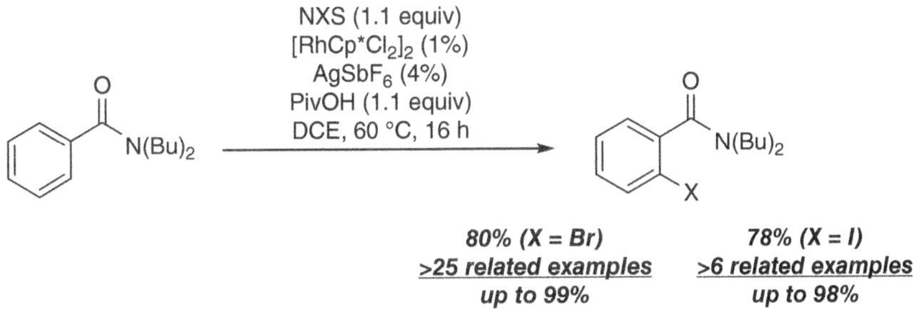

SCHEME 7.84 Rhodium-catalyzed halogenation of tertiary benzamides [143].

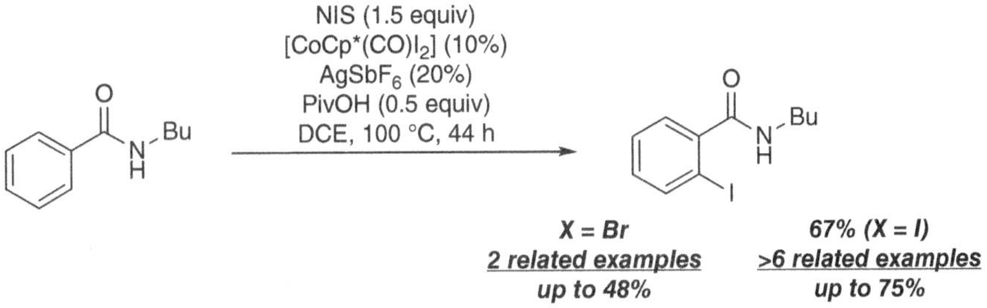

SCHEME 7.85 Cobalt-catalyzed ortho-halogenation of benzamides [144].

Example 7.25 Cobalt-Catalyzed Monobromination of a Phenylpyridine Derivative [144].

Before starting this synthesis, carefully follow the chemical safety instructions listed in Example 7.1.

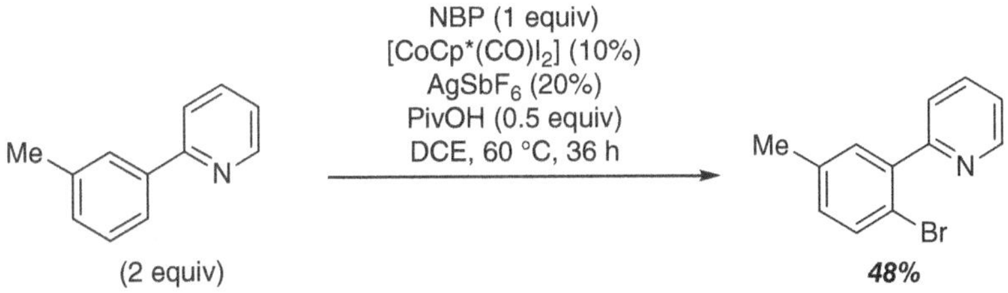

Caution! Ethyl acetate and pentane are highly flammable.
In a 10 mL flame-dried Schlenk tube with a stir bar, the phenylpyridine derivative (0.80 mmol, 2.0 equiv), *N*-bromophthalimide (NBP) (0.40 mmol, 1.0 equiv), and [Cp*Co(CO)I$_2$] (10 mol %) were added under argon. Then, the tube was transferred into the glovebox, and AgSbF$_6$ (20 mol %) was added. After moving out of the glovebox, a solution of PivOH (50 mol %) in DCE (2 mL) was injected into the tube under argon. Then, the tube was sealed and the mixture was stirred at 60 °C for 36 h. After cooling down, the volatiles were removed, and the analytically pure brominated pyridine derivative was obtained by flash chromatography (silica; gradient of *n*-pentane/EtOAc).

Adapted with permission from D.-G. Yu, T. Gensch, F. de Azambuja, S. Vásquez-Céspedes, and F. Glorius, *J. Am. Chem. Soc.*, **2014**, *136*, 17722–17725. *Copyright (2014) American Chemical Society.*

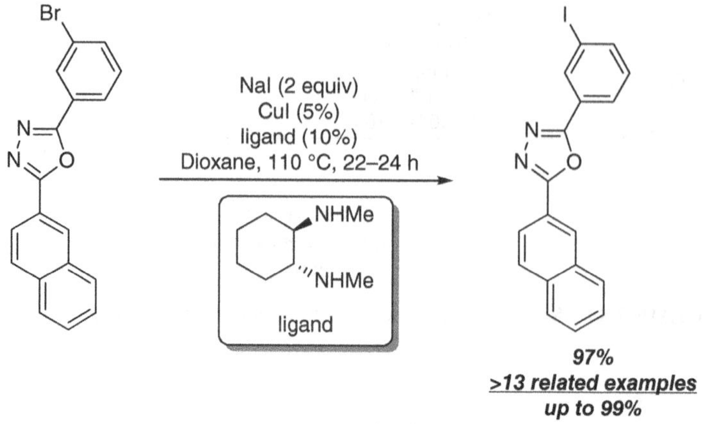

SCHEME 7.86   A copper-catalyzed aromatic Finkelstein reaction [146].

The classic Finkelstein reaction is commonly carried out with alkyl halides and a source of nucleophilic halogen. Naturally, examples of this chemistry are dominated by primary alkyl halides. The reaction is typically carried out in a solvent that fully dissolves the source of the nucleophilic halogen. Ideally, the precipitated salt of the displaced halide is insoluble in the reaction medium and is removed by a simple filtration. An aromatic version of this reaction that affords aryl iodides is attractive due to the higher reactivity of aryl iodides in transition metal-catalyzed cross-coupling reactions [145]. A copper-catalyzed example of this chemistry used sodium iodide as the nucleophilic iodine and dioxane as the solvent (Scheme 7.86) [146]. The reaction times were a bit long, but the isolated yields of the aryl iodides were generally outstanding. The overall halogen exchange reaction can also be achieved through the treatment of aryl bromides with BuLi followed by elemental iodine. However, there are significant limitations with the use of organolithium reagents, and this approach is significantly more operationally complex. Given the simplicity of this approach and the outstanding yields of the aryl iodides, this is a practical way to generate aryl iodides from bromides.

This aromatic Finkelstein reaction has been modified for continuous flow conditions (Scheme 7.87) [147]. The copper-catalyzed reaction used a NaI packed-bed reactor as the source of the nucleophilic iodine and dioxane as the solvent. A range of substrates bearing electron-donating and electron-withdrawing groups as well as heteroaromatic examples were successfully halogenated using this approach. The chemistry was largely regioselective with the iodine occupying the position vacated by the bromine.

A nickel-catalyzed aromatic halogen exchange reaction used nickel(II) bromide as the catalyst and sodium iodide as the source of the nucleophilic iodide (Scheme 7.88) [148]. One of the key components of this reaction was 2.5 equivalents of Bu_3P per metal center. Both activated and deactivated aromatic systems were iodinated in moderate to good yields. Although the conditions were a bit harsh, it is an alternative to the more expensive palladium-catalyzed reaction.

The palladium-catalyzed iodination of benzoic acid derivatives has been reported [149]. The metal catalyst in this chemistry was palladium acetate, and IOAc was the source of electrophilic iodine (Scheme 7.89). The IOAc was conveniently generated in solution through the reaction of elemental iodine and $PhI(OAc)_2$. Despite the electron-deficient

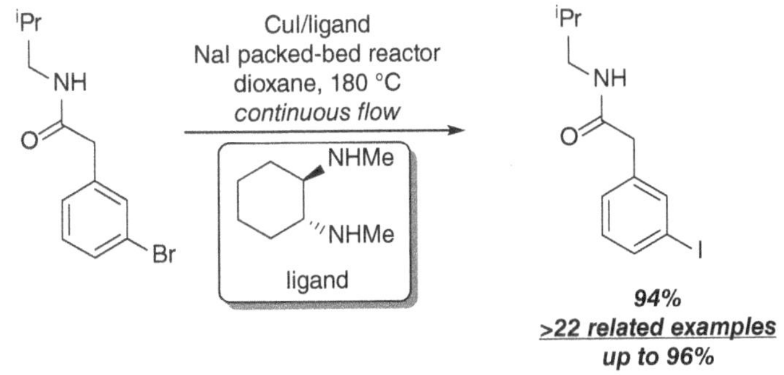

SCHEME 7.87   An aromatic Finkelstein reaction under continuous flow [147].

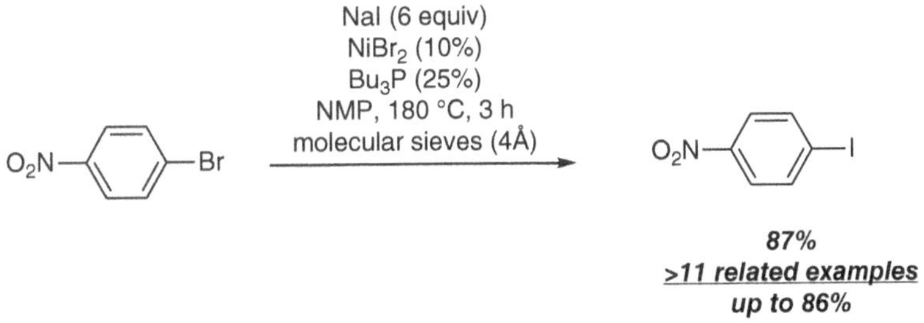

SCHEME 7.88   Nickel-catalyzed aromatic Finkelstein reaction [148].

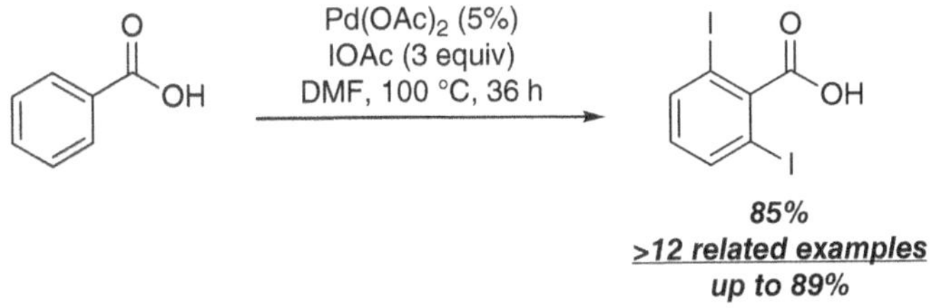

SCHEME 7.89   Palladium-catalyzed double iodination of benzoic acid [149].

ring, diiodination was achieved after heating in DMF for 36 h. The yields of the aryl diiodides were excellent. Monoiodination was more challenging. After a bit of experimentation, the authors discovered that the addition of a source of nucleophilic iodide successfully aided the selective formation of the monoiodinated product in moderate yields (Scheme 7.90).

The palladium-catalyzed halogenation of aryl triflates used a common source of palladium(0) and a bulky dialkylbiarylphosphine ligand to promote the reaction (Scheme 7.91) [150]. The process needed a nucleophilic source of the halogen, and both KBr and KCl were found to be suitable. The yields of this process were good to excellent, and a range of functionalized substrates were successfully halogenated including substrates bearing

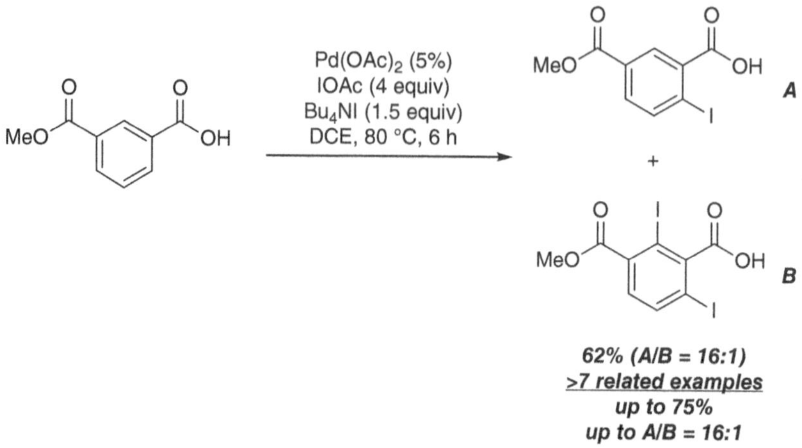

SCHEME 7.90   Palladium-catalyzed monoiodination of benzoic acid derivatives [149].

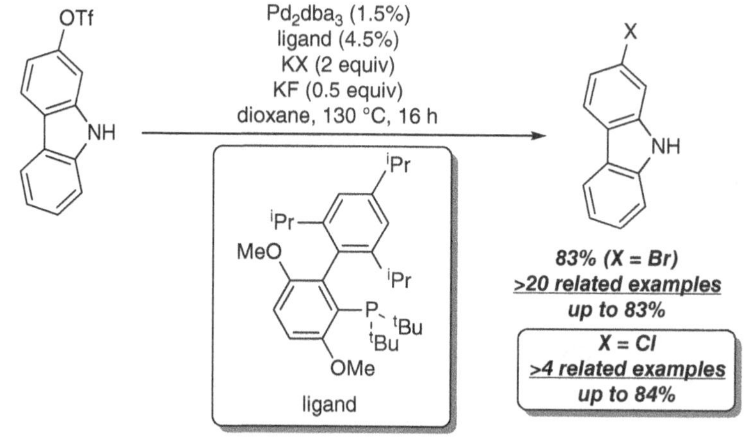

SCHEME 7.91   Palladium-catalyzed halogenation of aryl triflates [150].

electron-donating or electron-withdrawing groups. The chemistry was also successful with bulky or heteroaryl substrates. It was also noteworthy that reagents bearing unprotected —NH groups could be halogenated. One of the key elements was the addition of KF as a promoter for the chemistry. The fluorine did not end up in the final product; however, it was critical to the success of the reaction.

The ruthenium-catalyzed bromination of aryl triflates has been achieved (Scheme 7.92) [151]. Similar to the palladium example described earlier, the chemistry needed a source of nucleophilic bromide, and a group I metal salt was used. The solvent was critical in this chemistry, and DMI was found to be the most effective for the halogenation reaction. The substrate scope was broad, and substrates bearing electron-donating and electron-withdrawing groups as well as heteroaryl systems were well tolerated by this approach. Isolated yields were typically good to outstanding.

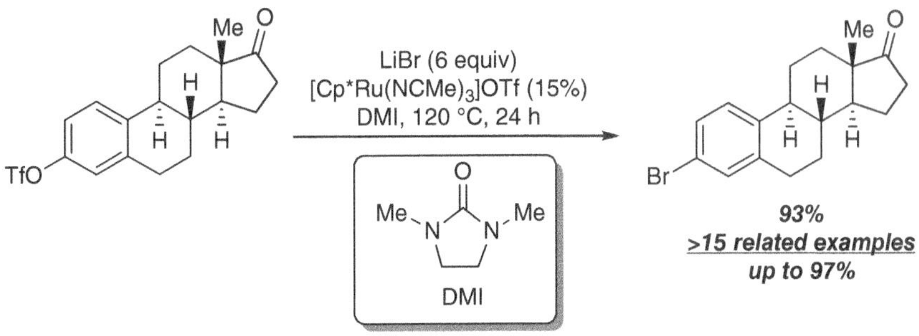

SCHEME 7.92 Ruthenium-catalyzed halogenation of aryl triflates [151].

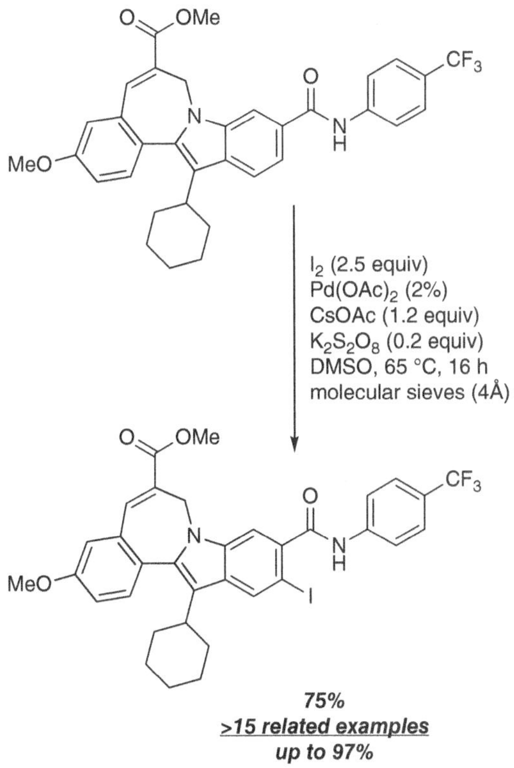

SCHEME 7.93 Palladium-catalyzed iodination using elemental iodine [152].

The palladium-catalyzed iodination of complex organic structures bearing a host of functional groups has been developed using elemental iodine (Scheme 7.93) [152]. Both mono- and diiodination were achieved using this approach by simply altering the stoichiometry of the reaction. One of the strengths of this approach was the ability to functionalize a host of heteroaromatic compounds in high yield. It should also be noted that the chemistry could be scaled up to gram-scale reactions. Furthermore, the palladium could be lowered to 0.5% loading.

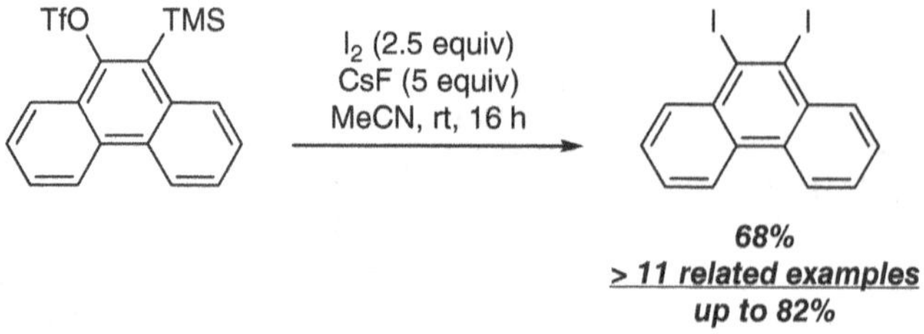

SCHEME 7.94   Synthesis of 1,2-diiodo arenes by trapping benzynes with I$_2$ [153].

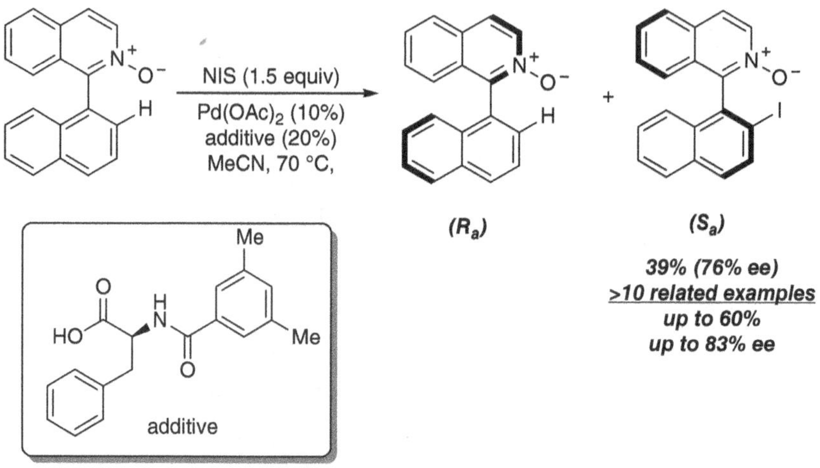

SCHEME 7.95   Palladium-catalyzed asymmetric synthesis of aryl iodides [154].

An intriguing approach to the synthesis of 1,2-diiodoarenes was proposed to proceed through a benzyne intermediate that was trapped by elemental iodine to generate a diiodo species (Scheme 7.94) [153]. The chemistry occurred under mild conditions, and moderate yields of the target compounds were obtained. The method appeared to be a general approach to these compounds, and the biggest limitation could be the availability of the substrates.

An asymmetric example of the ortho-directed halogenation reaction used a common palladium catalyst and NIS as an iodine source iodine to promote the selective formation of a chiral aryl iodide through kinetic resolution C—H halogenation reaction (Scheme 7.95) [154]. Naturally, the key to the stereoenrichment was the use of an appropriate chiral ligand for the metal center. After screening a range of ligands, a phenylalanine derivative was found to generate aryl iodides with the highest enantioselectivity. The yields of this reaction were fair, but moderate to good ee's were obtained.

The iodination of thiophene rings attached to a borane has been achieved using only NIS as the iodinating agent (Scheme 7.96) [155]. The chemistry was carried out under mild conditions in a mixture of acetic acid and chloroform. The reaction was regioselective for the position adjacent to the sulfur center, and simply stirring at room temperature for a few hours generated moderate yields of the diiodinated compound.

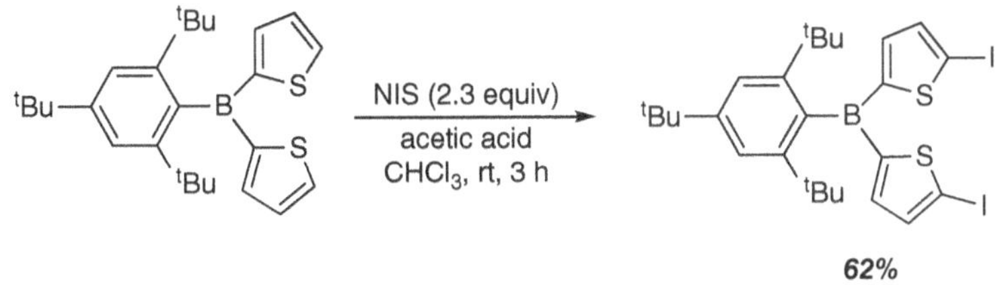

SCHEME 7.96   Iodination of a borane using NIS [155].

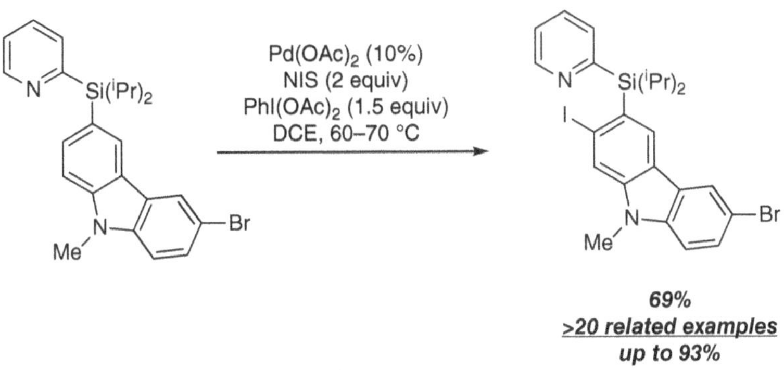

69%
>20 related examples
up to 93%

SCHEME 7.97   Use of a silicon-based directing group in a palladium-catalyzed iodination reaction [156].

The use of a silicon containing ortho-directing group has also been successful (Scheme 7.97) [156]. The attractiveness of this approach was due to the ease in which the silicon-containing fragment could be cleaved. This chemistry stands in contrast to the majority of the directing groups that are strongly bound to the arene and unable to be easily removed. Indeed, the authors converted the halogenated compounds into a number of valuable derivatives including a phenol and an arylboronate diester.

One of the more practical approaches to the synthesis of heteroaryl halides is through a one-pot process that combined several sequential reactions (Example 7.26) [157]. The first step in the process was a palladium-catalyzed coupling reaction between an acid chloride and a terminal alkyne to generate an ynone. This intermediate was not isolated, but was converted into a 4-iodopyrrole through an addition–cyclocondensation reaction. A range of electron-donating and electron-withdrawing aryl and alkyl acid chlorides were successfully functionalized using this approach. Yields were moderate to good, and the entire sequence was completed in a few hours. The one-pot nature of this chemistry makes it an attractive approach to the synthesis of 4-iodopyrroles. In related work, the generation of aryl halides through oxidative bromination of arenes using molecular oxygen as the oxidant and a palladium–polyoxometalate species as the catalyst has been reported [158].

The use of diaryliodonium salts in organic synthesis has grown significantly due to their unique reactivity profiles [159]. The ability to tune the electronic and steric components of the diaryliodonium salt to suit the needs of specific reaction is driving this chemistry.

Example 7.26   Multicomponent Synthesis of Iodinated Pyrroles [157].

Before starting this synthesis, carefully follow the chemical safety instructions listed in Example 7.1.

Caution! THF, petroleum ether, and ethyl acetate are highly flammable.

PdCl$_2$(PPh$_3$)$_2$ (70 mg, 0.10 mmol) and CuI (39 mg, 0.20 mmol) were placed under argon in a screw-cap vessel, which was then dried with a heat gun and cooled to the room temperature. Then, 25 mL of dry THF was added, and the mixture was degassed with argon. Dry triethylamine (0.69 mL, 5.00 mmol), 5.00 mmol of the acid chloride, and *tert*-butyl prop-2-ynylcarbamate (776 mg, 5.00 mmol) were successively added to the mixture, which was stirred at room temperature until the conversion was complete (monitored by TLC). Then, sodium iodide (3.79 g, 25.0 mmol), toluene-4-sulfonic acid monohydrate (1.94 g, 10.0 mmol), and 5 mL of *tert*-butanol were successively added to the mixture, which was stirred at room temperature for 1 h (monitored by TLC). The reaction mixture was diluted with 50 mL brine, the phases were separated, and the aqueous phase was extracted with dichloromethane (3×25 mL). The combined organic layers were dried with anhydrous sodium sulfate. After removal of the solvents in vacuo, the residue was absorbed onto Celite and chromatographed on silica gel with petroleum ether (boiling range 40–60 °C)/ ethyl acetate to give the iodinated pyrrole (1.40 g, 73%).

Adapted with permission from E. Merkul, C. Boersch, W. Frank, and T. J. J. Muller, *Org. Lett*, **2009**, *11*, 2269–2272. *Copyright (2009) American Chemical Society.*

The following sections will highlight a few of the common methods for the formation of these valuable reagents.

The preparation of diaryliodonium triflates has been achieved by adding the oxidizing agent directly to a mixture of the aryl iodide and free arene [160, 161]. This chemistry was promoted by the presence of triflic acid and was quite fast under mild conditions. While many of the reactions were complete in 10 min, the bulkier arenes required slightly longer reaction times (Scheme 7.98). One of the attractive aspects of the chemistry was the ability to create electronically and sterically diverse unsymmetrical diaryliodonium triflates through minor modifications to the system. For symmetrical diaryliodonium triflates, it was convenient to simply mix iodine with the arene along with the oxidizing agent and triflic acid (Scheme 7.99).

The preparation of unsymmetrical diaryliodonium salts from readily available arylboronic acids and aryl halides has been reported (Scheme 7.100) [162]. MCPBA was

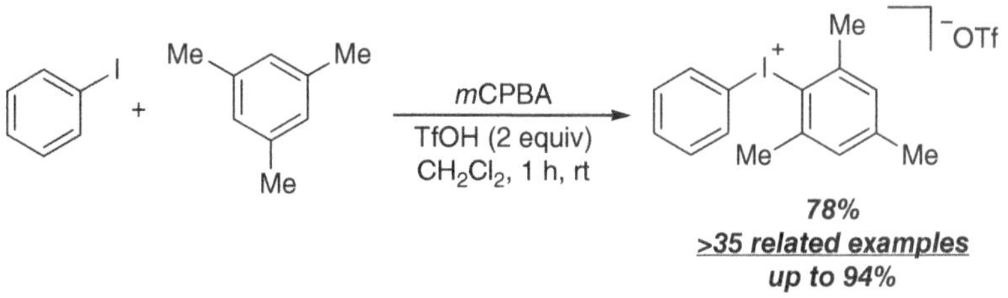

SCHEME 7.98   Synthesis of unsymmetrical diaryliodonium triflates [160, 161].

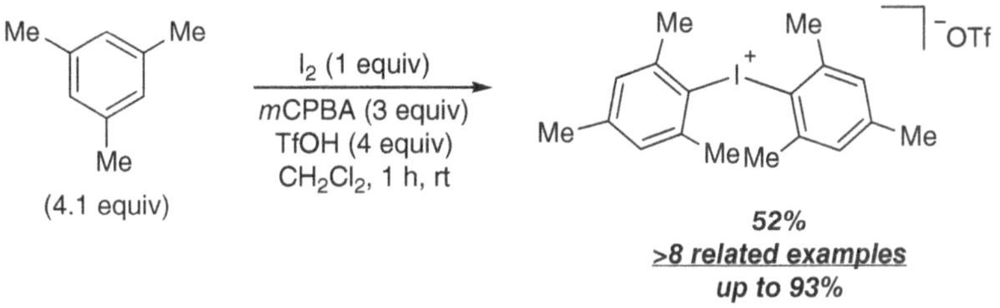

SCHEME 7.99   Synthesis of bulky symmetrical diaryliodonium triflates [160, 161].

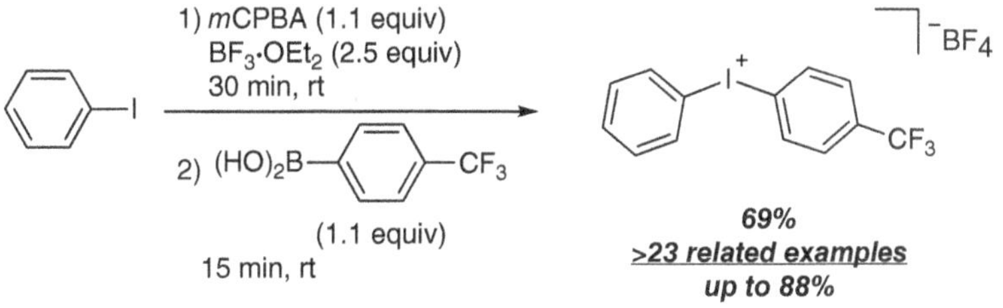

SCHEME 7.100   Synthesis of diaryliodonium salts from arylboronic acids [162].

used as the oxidizing agent in this chemistry, and BF_3 (etherate) served as a convenient Lewis acid. An added benefit was that the BF_3 was converted into BF_4^- during the course of the reaction and served as the counterion for the diaryliodonium cation. However, the initial screens were unsuccessful due to a competing reaction between the boronic acid and the oxidizing agent. The authors solved this problem by allowing the aryl halide to stir with the oxidizing agent for 30 min prior to the addition of the boronic acid. Using this approach, a host of unsymmetrical diaryliodonium salts were prepared. Both electron-rich and electron-deficient arenes were converted into the diaryliodonium salts with only minor modifications to the system. The operational simplicity of the approach coupled with the availability of the reagents makes this an attractive approach to the synthesis of both symmetrical and unsymmetrical diaryliodonium salts. A version of this approach was used to create an estrone-derived diaryliodonium tetrafluoroborate (Example 7.27) [96].

Example 7.27   Synthesis of an Estrone-Derived Diaryliodonium Tetrafluoroborate [96].

Before starting this synthesis, carefully follow the chemical safety instructions listed in Example 7.1.

1) BF$_3$·OEt$_2$ (1.1 equiv)
   CH$_2$Cl$_2$, 0 °C, 10 min
2) MesI(OAc)$_2$ (1.05 equiv)
   rt, 2 h
3) NaBF$_4$(aq), 30 min

73%
>23 related examples
up to 100%

Caution! Diethyl ether is highly flammable.

The estrone-derived arylboronic acid (1.0 equiv) and CH$_2$Cl$_2$ (0.075 M) were combined in an oven-dried round-bottom flask equipped with a stir bar. The mixture was cooled to 0 °C, BF$_3$·OEt$_2$ (1.10 equiv) was added, and the mixture was stirred for 10 min. 2-(Diacetoxyiodo) mesitylene (1.05 equiv) was then added as a solution in CH$_2$Cl$_2$ (0.33 M), and the mixture was warmed to room temperature and stirred for 2 h. The reaction was quenched by the addition of saturated aqueous NaBF$_4$. After 30 min of vigorous stirring, the aqueous layer was extracted with CH$_2$Cl$_2$ (2×). The combined organic layers were dried over MgSO$_4$, filtered, and concentrated under vacuum. Et$_2$O was added to the residual solid, and the diaryliodonium tetrafluoroborate (73%) was collected via filtration, washed with Et$_2$O, dried under vacuum overnight, and stored in a dry box under N$_2$ until use.

Adapted with permission from N. Ichiishi, A. J. Canty, B. F. Yates, and M. S. Sanford, *Org. Lett.*, **2013**, *15*, 5134–5137. *Copyright (2013) American Chemical Society.*

7.2.1   Troubleshooting the Synthesis of Aryl Halides and Related Compounds

While it will not be possible to anticipate all of the needs of the synthetic chemist over the next few decades, this list will provide some insight into issues commonly encountered when trying to prepare aryl halides. This chemistry can be particularly confusing as several protocols will work for several of the halogens, but not for all of them. Care must be taken to make sure that the chemistry will be successful for the substrate under investigation.

Problem/Issue/Goal	Possible Solution/Suggestion
Can a Grignard reagent be converted into an aryl fluoride?	Yes, several approaches to this reaction have been reported using electrophilic fluorinating agents [79–81]
Can aryl triflates be converted into aryl fluorides through metal-catalyzed cross-coupling chemistry?	Palladium-catalyzed versions of this reaction have been reported [85]. This chemistry has been adapted for continuous flow [88]

(Continued)

(Continued)

Problem/Issue/Goal	Possible Solution/Suggestion
Can aryl bromides be converted into aryl fluorides through metal-catalyzed fluorination chemistry?	Both palladium [89] and copper catalysts [90] are known to promote this reaction
Need to generate a very bulky aryl halide	The silver-promoted fluorination of bulky arylboronic acids would be a good place to start. The chemistry generates bulky aryl fluorides in moderate to high yields [92]
Need a transition metal-free route to bulky aryl fluorides	The selective cleavage of unsymmetrical diaryliodonium salts using KF selectively generates the more hindered aryl fluoride [96]
Can organoboronates and related compounds be converted into aryl fluorides?	A range of organoboron compounds can be readily fluorinated using copper and palladium catalysts [84, 93–95]
Need to start from a phenol; can it be converted into an aryl fluoride?	The transition metal-free deoxyfluorination of phenols has been achieved using PhenoFluor and CsF [98]
It would be convenient to be able to selectively fluorinate aryl C—H bonds	General approaches to this reaction are rare with the exception of the ortho-position. The palladium-catalyzed ortho-directed fluorination of aryl C—H bonds has been effective [12, 100, 101, 108, 163]
The starting compound already has an aryl bromide. Can the C—H fluorination reaction be designed to retain the bromine?	This has been achieved using a palladium-catalyzed fluorination reaction [107]
Need an alternative to elemental bromine that has similar reactivity	A reasonable replacement is pyridinium tribromide [125]
Starting material is an electron-rich arene and need to brominate	These reactions are typically facile using NBS and do not need a catalyst [112, 118]
Need to regioselectively brominate a arene in a position adjacent to a nitrogen	There are many successful approaches to this reaction. As long as the substrate does not have additional electrophiles, orthometallation with an organolithium reagent such as BuLi followed by Br_2 has been successful [122]. CBr_4 has also been used as the brominating agent in these reactions [90]
Need a meta-bromination of an electron-deficient arene	A reasonable starting point would be to use TBCA in TFA [123] or NBS in H_2SO_4 [124]
Need to halogenate an electron-poor arene	The combination of NXS and a powerful Lewis acid such as boron trifluoride have promoted this transformation [126]
Can an arylboronic acid be converted into an aryl chloride?	This reaction has been achieved using copper catalysts and NCS as the chlorine source [127]
It would be convenient to be able to halogenate an arylboronate with retention of the boron-containing fragment	A gold catalyst has been able to promote this reaction using NXS as the halogen source [133]
It would be convenient to brominate an arene in a position not directly adjacent to the directing group	This has been achieved using a two-step approach using initial boronation, followed by halogenation with $CuBr_2$ [131, 132]

(Continued)

(Continued)

Problem/Issue/Goal	Possible Solution/Suggestion
Starting material is an electron-rich arene. What is a good iodination approach?	These reactions are typically facile using NIS [115]. Ki/TBHP has also been successful [119]
Need to generate an aryl iodide from an aryl bromide using a substrate that does not have any electrophiles present	As long as the substrate does not have any electrophiles, a high-yielding approach would be to metallate the substrate using BuLi and quench with elemental iodine [120, 121]
Need to generate a bulky aryl iodide or bromide	The iodination/bromination of boronic acids using NIS in acetonitrile would be a good place to start [130]
NIS tends to degrade. Is there a better way to store/use NIS?	Instead of struggling with storage issues, consider the generation of NIS in solution through the treatment of NCS with NaI [130]
Need to halogenate an arene ortho *to a nitrogen- or oxygen-containing functional group*	This is readily achieved by a host of catalysts and conditions [134–144]
Can an aryl bromide be converted into an aryl iodide using an aromatic Finkelstein reaction?	This has been achieved using several catalyst systems including nickel [148], copper [146], and palladium. A microwave-assisted approach to this reaction has also been reported [164]. The copper system has been adapted for continuous flow [147]
Substrate is a carboxylic acid. Does the carboxylic acid need to be protected prior to iodination?	Depends—a palladium-catalyzed iodination of benzoic acid and related compounds generates moderate to excellent yields without protection of the carboxylic acid [149]
Need to convert an aryl triflate into an aryl bromide	There are several reasonable approaches to this challenge. Both palladium and ruthenium catalysts have been used to effect this transformation [150, 151]
It would be convenient to halogenate aryl C—H bonds of arenes bearing electron-withdrawing groups or heteroaromatics	Using a palladium-catalyzed approach with elemental iodine or NIS as the halogen source would be a good place to start [152]. For some thiophene-containing compounds, NIS was able to iodinate ring without the addition of a catalyst [155]
Need to make a diaryliodonium salt	Classic approaches to this chemistry start from aryl halides and an oxidizing agent such as MCPBA [160, 161]. If an arylboronic acid is available, it is also a convenient entry point to the chemistry [96]

7.3 SYNTHESIS OF VINYL HALIDES THROUGH THE FORMATION OF HALOGEN–CARBON(SP²) BONDS

Vinyl fluorides can be generated from vinylboronic acids following a two-step protocol using a mixture of sodium hydroxide and silver triflate, followed by a the addition of a fluorinating agent (Scheme 7.101) [92]. One of the attractive aspects of this reaction

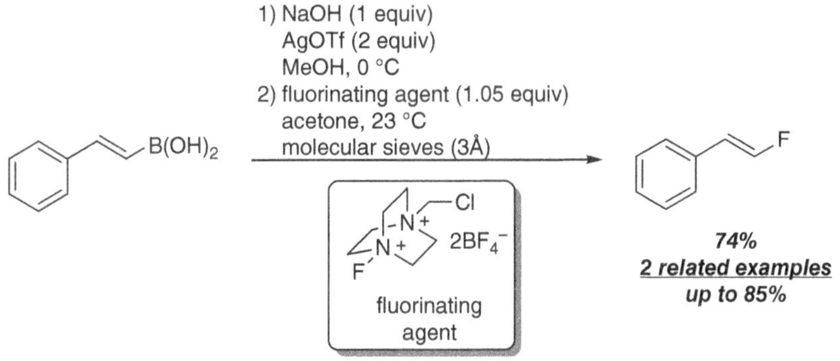

SCHEME 7.101   Synthesis of vinyl fluorides from vinylboronic acids [92].

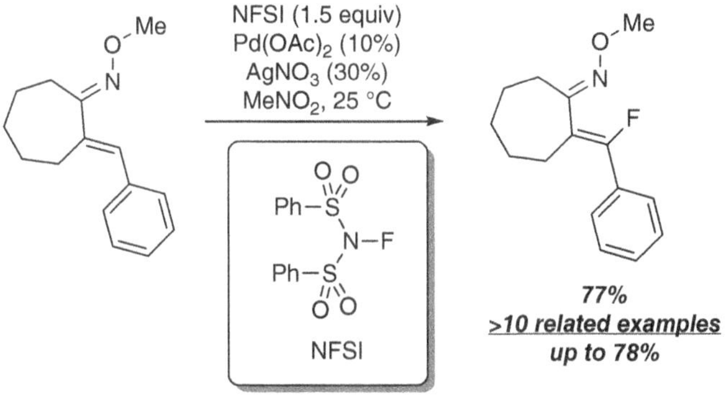

SCHEME 7.102   Palladium-catalyzed addition fluorination of C—H bonds [102].

was the ability to telescope the borylation step with a fluorodeboronation reaction to generate vinyl fluorides from terminal acetylenes in a one-pot reaction.

One of the most direct routes to the generation of vinyl fluorides involves the design of a process for the fluorination of vinyl C—H bonds (Scheme 7.102) [102]. This has been achieved using a silver nitrate-promoted palladium-catalyzed C—H activation reaction. The yields were moderate to good under very mild conditions using NFSI as an electrophilic fluorinating agent.

Gold complexes bearing a bulky dialkylbiarylphosphine ligand promoted the hydrofluorination of a wide range of internal and terminal alkynes (Scheme 7.103) [63]. Instead of using hazardous HF in the reaction, the authors used an organic base to bind the HF so it can be readily handled [72, 165]. Olah's reagent is an excellent example of this approach and he used pyridine as the organic base [165]. The authors decided to use DMPU instead of pyridine since it has an increased ability to hydrogen bond and is less basic than pyridine. Using this DMPU/HF reagent, a host of internal and terminal alkynes were successfully functionalized. Even the normally sluggish diphenylacetylene was successfully converted into the corresponding vinyl fluoride. Dihydrofluorination was also possible and generated *gem*-difluoromethylene compounds. In related work, vinyl fluorides were generated through a Shapiro fluorination reaction using protected trishydrazones as the substrates, BuLi as the base, and NFSI as a source of electrophilic fluorine [166]. This chemistry

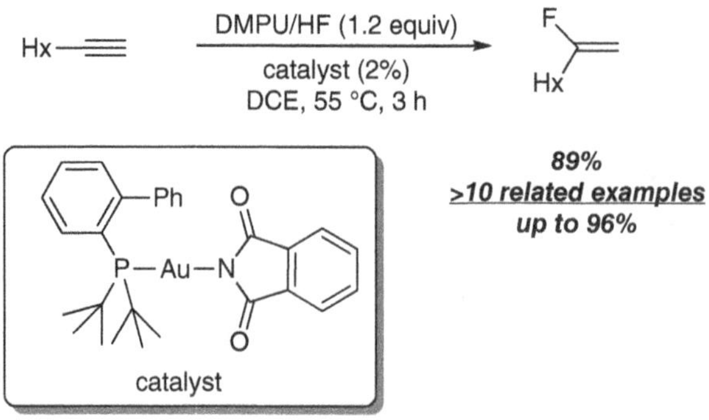

SCHEME 7.103   Gold-catalyzed hydrofluorination of terminal alkynes [63].

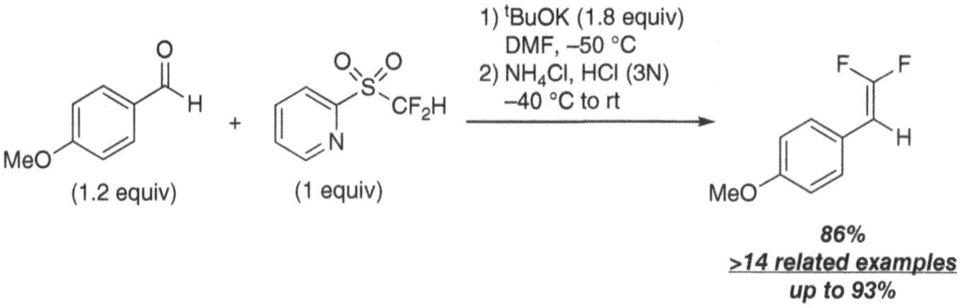

SCHEME 7.104   *gem*-Difluoroolefination of aldehydes [167].

provides a valuable route to vinyl fluorides from ketones since the hydrazones can be generated from them.

Aldehydes were converted into *gem*-difluoroalkenes using difluoromethyl 2-pyridyl sulfone as a difluoroolefination agent (Scheme 7.104) [167]. The chemistry was carried out at low temperatures and afforded moderate to excellent yields of the difluoroalkenes. A range of substrates were screened, and ethers, halides, and even a secondary alcohol were tolerated by the halogenation reaction. In contrast with previous reports, this approach is operationally straightforward.

The conversion of diazo compounds into *gem*-difluoroalkenes is an attractive route to the synthesis of these compounds (Scheme 7.105) [168]. The reactions are catalyzed by a catalytic amount of copper salt with an equimolar amount of cesium fluoride as an additive. It is noteworthy to point out the copper loading in this example is significantly lower than other copper-catalyzed fluorinations. In general, diaryl diazomethanes bearing electron-donating groups afforded lower yields than electron-deficient arenes.

There has been considerable interest in the development of reliable procedures for the preparation of acyl fluorides. To this end, several routes to these compounds have been devised including the treatment of acyl chlorides with anhydrous TBAF in THF [169] and the combination of aminodifluorosulfinium salts with TEA·3HF for the conversion of carboxylic acids into acyl fluorides [170]. In addition to these methods, fluorocarbonylation has been used to generate acyl bromides. Typically, this approach has high operational complexity due

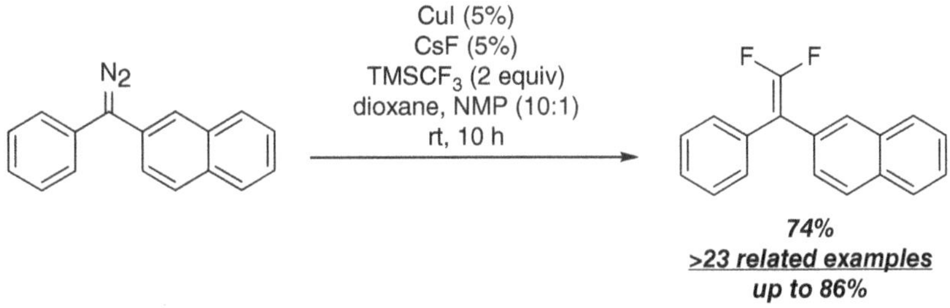

SCHEME 7.105   Conversion of diazo compounds into *gem*-difluoroalkenes [168].

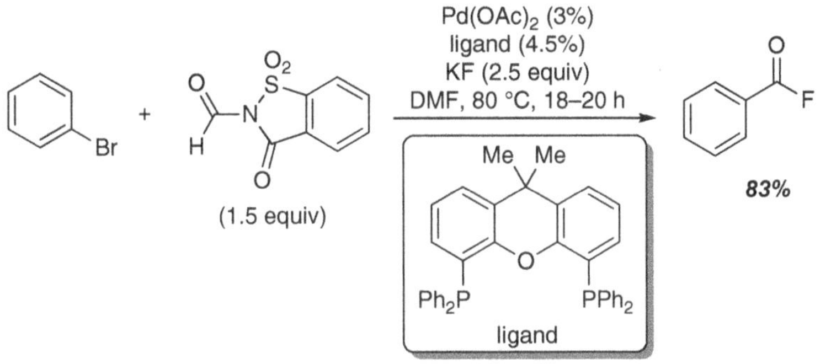

SCHEME 7.106   Acyl fluoride synthesis from aryl bromides [171].

to the use of gaseous carbon monoxide; however, a recent report used *N*-formylsaccharin as the CO source for the synthesis of acyl fluorides (Scheme 7.106) [171]. This reagent is attractive since it is a crystalline solid and can be weighed into the reaction vessel. As a result, the stoichiometry between the aryl halide and the CO source is very precise. To further the scope of the chemistry, the authors generated the acyl fluoride in solution and used it without isolation in a number of subsequent reactions. In addition to these methods, acyl fluorides could also be generated directly from aldehydes using a decatungstate photocatalyst and black light (365 nm) [25].

One of the more operationally simple approaches to the preparation of vinyl halides is the Hunsdiecker decarboxylative halogenation of alkenyl carboxylic acids. While a number of modifications to this reaction have been reported, one of the most practical versions used triethylamine and NXS to promote the reaction (Scheme 7.107) [172, 173]. The reaction was remarkably fast and high yielding. Given the ability to generate vinyl chlorides, bromides, and iodides from readily available alkenyl carboxylic acids, this is a very attractive approach. In related work, the combination of Selectfluor and KBr converted a series of cinnamic acid derivatives into vinyl bromides. The reaction time was longer (5 h), and the reaction was selective for the formation of the β-(*E*)-isomer [174].

Vinyl bromides can be converted into vinyl iodides using a Finkelstein-type reaction (Example 7.28) [146]. The process was copper catalyzed and used sodium iodide as a source of nucleophilic iodine. While only a single example was listed, this procedure has the potential to be very attractive to the synthetic community due to operational simplicity.

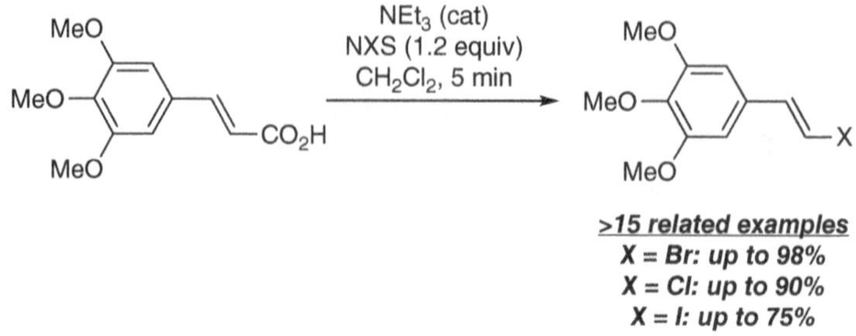

>15 related examples
X = Br: up to 98%
X = Cl: up to 90%
X = I: up to 75%

SCHEME 7.107   Triethylamine-catalyzed Hunsdiecker decarboxylative halogenation [172].

Example 7.28   A Vinyl Finkelstein Reaction [146].

Before starting this synthesis, carefully follow the chemical safety instructions listed in Example 7.1.

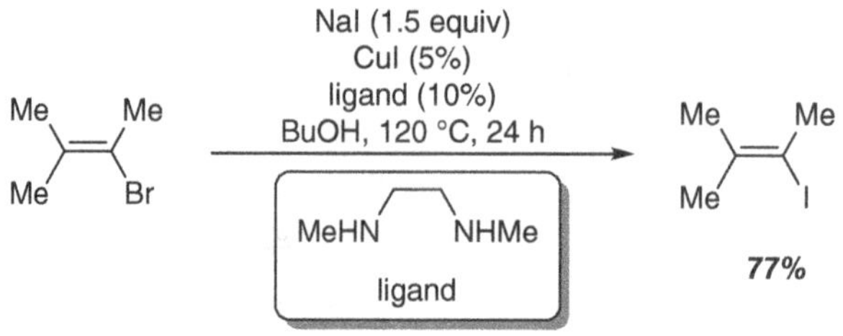

Caution! Pentane is highly flammable!

A 50 mL Schlenk tube was charged with CuI (382 mg, 2.01 mmol, 5.0 mol %), NaI (9.00 g, 60.0 mmol), evacuated, and backfilled with argon. *N,N′*-Dimethylethylenediamine (426 μL, 4.00 mmol, 10 mol %), 2-bromo-3-methyl-2-butene (4.65 mL, 40.1 mmol), and *n*-butanol (20 mL) were added under argon. The Schlenk tube was sealed with a Teflon valve, and the reaction mixture was stirred at 120 °C for 24 h. The resulting tan suspension was allowed to reach room temperature, poured into pentane (200 mL), and washed with a solution of 30% aq ammonia (10 mL) in water (200 mL), followed by water (3 × 200 mL). The organic phase was dried (MgSO₄) and concentrated to approximately 10 mL volume. The residue was distilled collecting the fraction boiling at 120–140 °C to give 6.08 g (77% yield) of 2-iodo-3-methyl-2-butene as a colorless liquid (>95% pure).

Adapted with permission from A. Klapars and S. L. Buchwald, J. Am. Chem. Soc., 2002, 124, 14844–14845. Copyright (2002) American Chemical Society.

The conversion of vinyl triflates into vinyl chlorides and bromides employed a common palladium(0) precatalyst and a bulky dialkylbiarylphosphine ligand to catalyze the process (Scheme 7.108) [150]. KBr and KCl were used as the halogen sources, and a substoichiometric amount of KF was needed to generate high yields of the vinyl halides. The precise role of the fluoride reagent was not determined. While only a few cyclic vinyl triflates were screened, good to excellent yields were obtained.

The conversion of aryl triflates into vinyl bromides used a ruthenium complex bearing a bulky cyclopentadienyl ligand to catalyze the reaction (Scheme 7.109) [151]. A slight

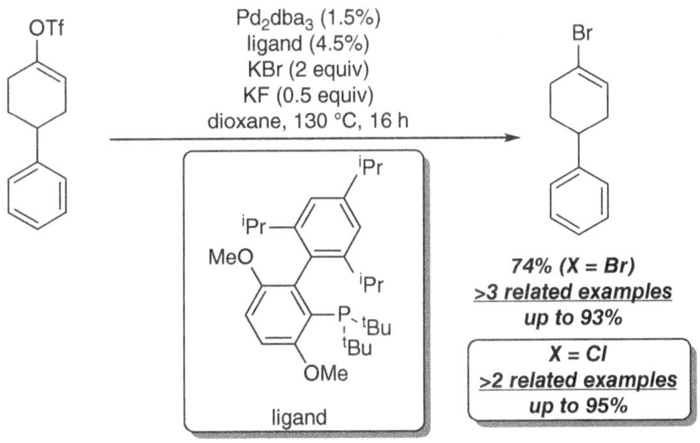

SCHEME 7.108 Palladium-catalyzed conversion of vinyl triflates into vinyl halides [150].

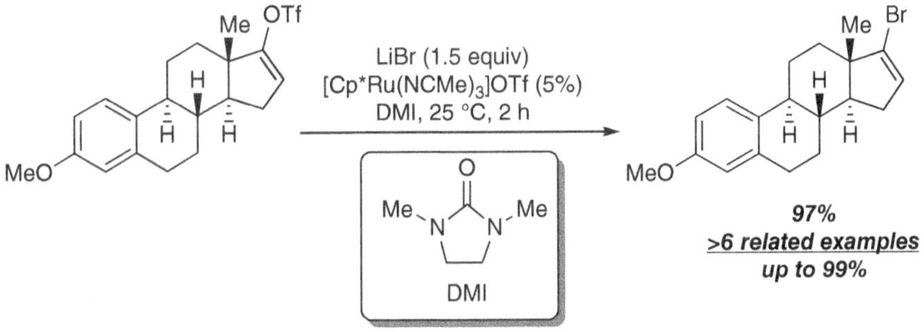

SCHEME 7.109 Ruthenium-catalyzed conversion of vinyl triflates into vinyl bromides [151].

excess of LiBr served as source of nucleophilic bromide, and DMI was the solvent. The reactions were remarkably fast for most substrates (10 min) and afforded good to outstanding yields of the vinyl bromides. While only a few substrates were screened, this chemistry has the potential for widespread adoption due to its speed and simplicity. It should be noted that vinylstannanes and related compounds can also be transformed into vinyl bromides upon treatment with elemental iodine [175].

The transition metal-free conversion of vinyltrifluoroborate salts into vinyl bromides has been achieved (Scheme 7.110) [129]. The reaction was operationally straightforward, and simply stirring all of the reagents together for 10 min afforded excellent yields of the vinyl bromides. The chemistry was not even sensitive to water. It should be noted that the chloramine-T is particularly hazardous and the most recent safety bulletins should be consulted before using this reagent.

One of the more practical approaches to the synthesis of vinyl chlorides involves the treatment of vinyltrifluoroborates with TCCA as the chlorinating agent (Scheme 7.111) [128]. The reaction was trivial to setup as simply adding the reagent to a flask followed by stirring at room temperature afforded excellent yields of the vinyl chloride. The halogen occupied the position vacated by the trifluoroborate, and significant scrambling was not

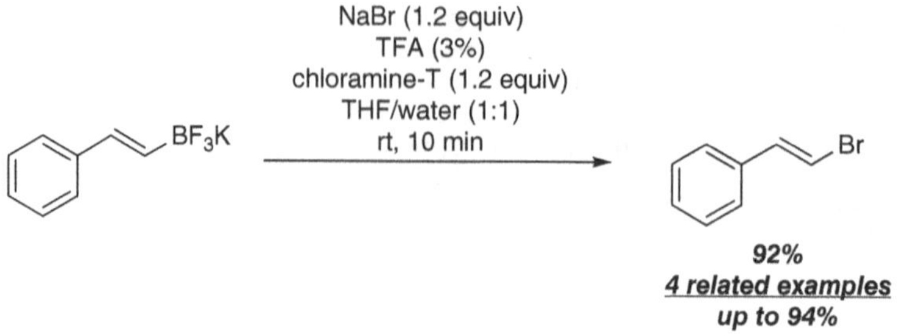

SCHEME 7.110 Conversion of vinyltrifluoroborate salts into vinyl bromides [129].

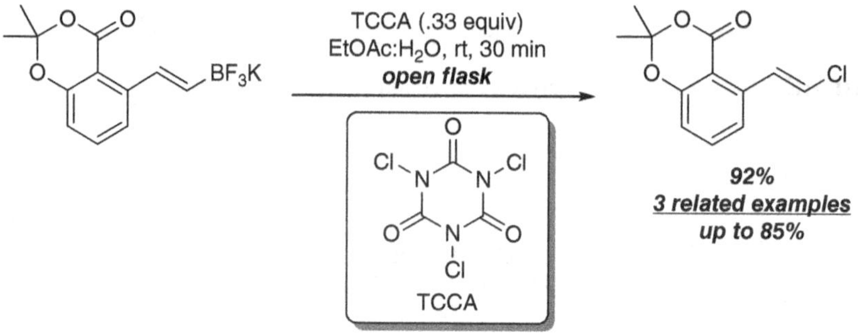

SCHEME 7.111 Conversion of vinyltrifluoroborates into vinyl chlorides [128].

reported. The aspect of the chemistry that makes it practical was the observation that the reactions can be carried out in an open vessel. While only a few examples were reported, this chemistry has great potential for the synthesis of a host of vinyl chlorides.

Elegant work by Negishi demonstrated that aryl bromides and iodides could be regioselectively and stereoselectively generated from a single alkyne through a series of functionalization reactions (Scheme 7.112) [177]. Using propyne as a representative substrate, initial bromoboration generated an intermediate that was trapped by the addition of pinacol to generate an isolable vinylboronate ester. The chemistry was highly regioselective and stereoselective for the formation of the (Z)-α-bromo isomer. Alkylation of this species through a Negishi coupling exchanged the bromide for an alkyl group, and subsequent treatment of the resulting boronate with elemental iodide generated a vinyl iodide in outstanding yield.

The iodination and bromination of vinylamides have been achieved using rhodium catalysts and NIS (Scheme 7.113) [178]. The reaction occurred under mild conditions and generally afforded excellent yields of the iodinated and brominated alkenes. The process was highly regioselective for the formation of the β-substitution product. Control reactions revealed that the rhodium catalyst was critical to the success of the reaction as its removal leads to poor yields and mixtures of products.

The conversion of aryl allylic alcohols into halogenated vinyl ketones has been achieved (Scheme 7.114) [179]. The process occurred under very mild conditions using oxalyl halides and was generally regioselective for the formation of the α-substitution product. A number of electron-donating and electron-withdrawing groups were well tolerated by the chemistry, provided they were attached to the arene in specific locations. Best results

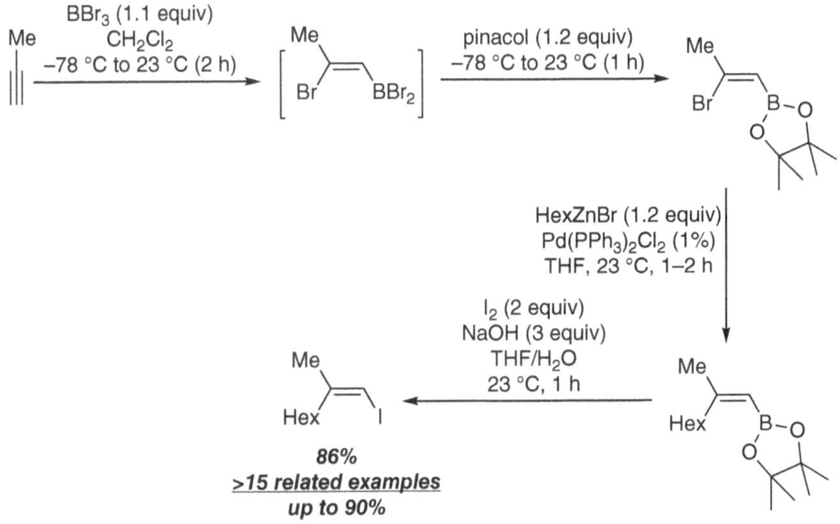

SCHEME 7.112 Synthesis of vinyl iodides and bromides based upon initial bromoboration of an alkyne [177].

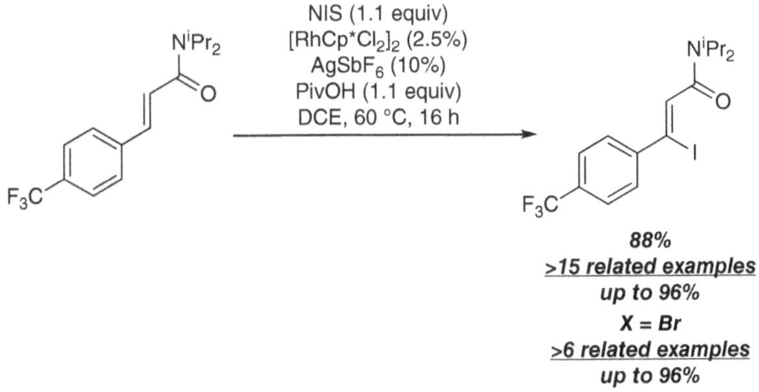

SCHEME 7.113 Rhodium-catalyzed halogenation of vinylamides [178].

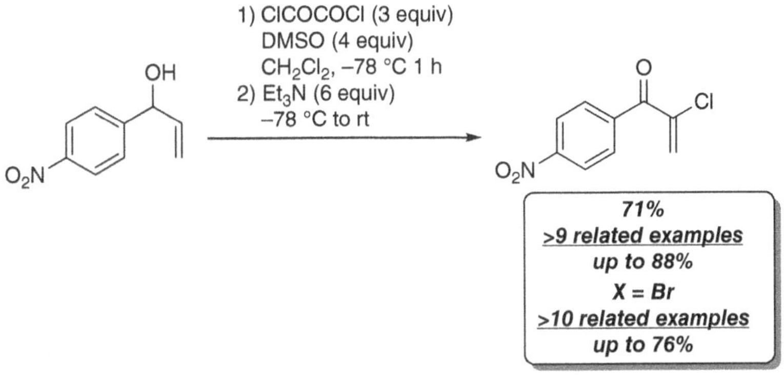

SCHEME 7.114 Conversion of allylic alcohols to halogenated vinyl ketones [179].

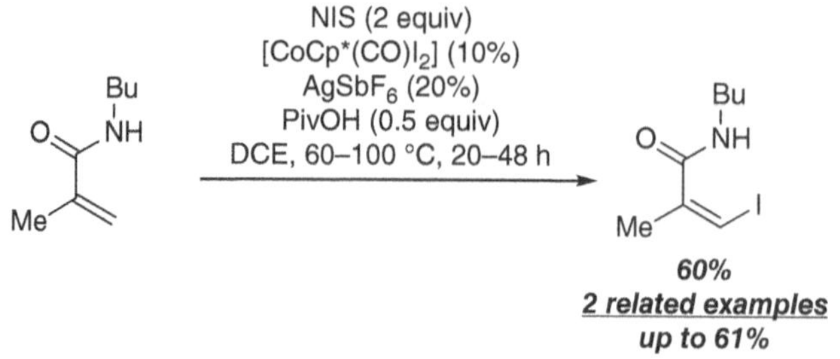

SCHEME 7.115   Cobalt-catalyzed iodination of Michael acceptors [144].

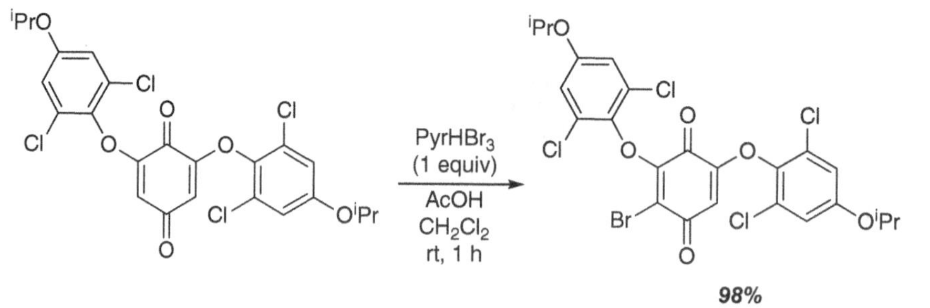

SCHEME 7.116   Bromination of a dichloroquinone derivative using pyridinium tribromide [180].

were obtained with the electron-withdrawing groups occupying ortho- and para-positions, while the electron-donating groups needed to be attached to the meta-position. Analogous reactions carried out using substrates bearing electron-donating groups in the ortho- and para-positions resulted in the formation of allylic chlorides when oxalyl chloride was used and intractable mixtures when oxalyl bromide was used.

A considerable amount of effort has been devoted to the development of catalytic reactions that use less expensive transition metal catalysts. As an example, the cobalt-catalyzed iodination of activated alkenes has been reported (Scheme 7.115) [144]. While only a few examples with Michael-type acceptors were shown, this method has the potential for significant expansion.

The bromination of a dichloroquinone derivative was achieved using pyridinium tribromide as the brominating agent (Scheme 7.116) [180]. The reaction conditions were extremely mild, and the yield of the brominated product was outstanding (98%). The fact that pyridinium tribromide is a solid and can be weighed out in air remains one of the biggest advantages of this reagent.

The regioselective synthesis of vinyl iodides from alkynes used a combination of elemental iodine and a secondary phosphine oxide to achieve a formal hydroiodination of an alkyne (Scheme 7.117) [181]. Control reactions revealed that the phosphine oxide was critical to the success of the reaction and that changing the phosphorus reagent to a hydrogen phosphonate or a phosphine resulted in lower yields. The authors proposed that I_2 reacted with 1 equiv of $HP(O)Ph_2$ to generate HI along with $IP(O)Ph_2$. Once formed, HI was quickly sequestered by a second equivalent of secondary phosphine oxide. It is this adduct that was proposed to be active in the hydroiodination reaction. A range of functionalized alkyl and arylalkynes were successfully transformed using this approach. The ability to iodinate both

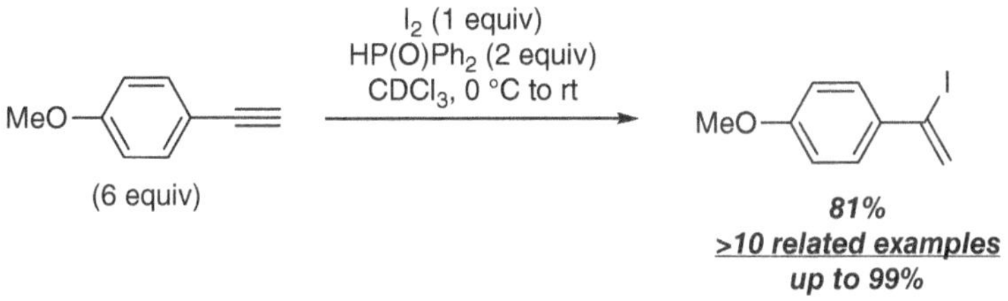

SCHEME 7.117 Formal hydroiodination of alkyne using the I_2/HP(O)Ph$_2$ system [181].

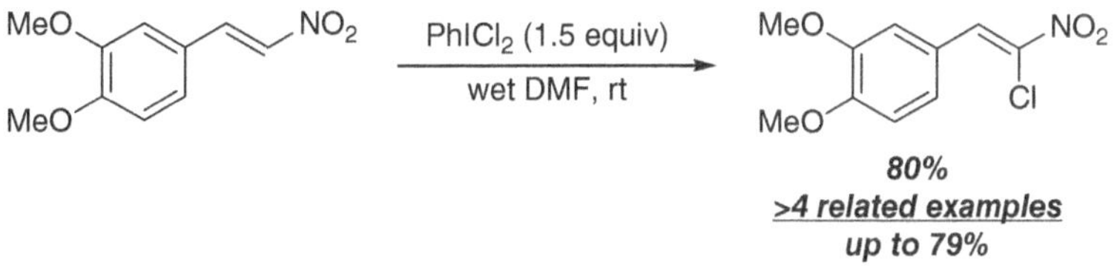

SCHEME 7.118 α-Chlorination of activated alkenes using iodobenzene dichloride [67].

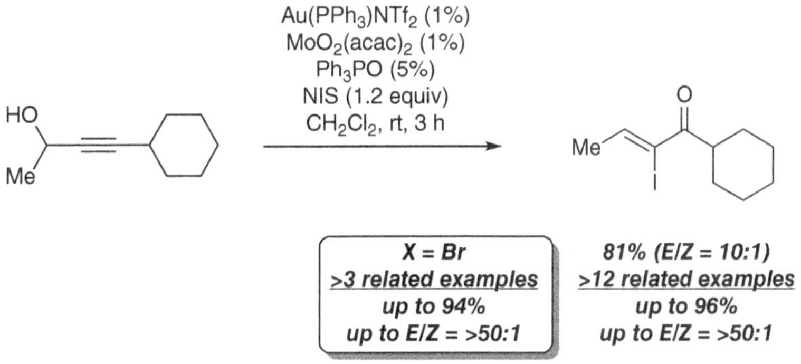

SCHEME 7.119 Preparation of vinyl iodides and bromides from propargyl alcohols [183].

electron-deficient and electron-rich aryl alkynes was exemplified by the observation that both 4-ethynylanisole and 4-ethynyl-α,α,α-trifluorotoluene gave the same isolated yields of the vinyl iodides (81%). Given the ease in which this chemistry was carried out should make it an attractive method for the preparation of α-iodinated alkenes.

Chlorinated styrene derivatives were generated using iodobenzene dichloride (Scheme 7.118) [67]. The key to this chemistry was the use of wet DMF as the solvent. The conditions of the reaction were quite mild, and yields of the halogenated compounds were good. The chemistry was regioselective for the incorporation of the chlorine adjacent to the activating group. In addition to these examples, activated internal alkynes were converted into (E)-1-chloro-2-iodoalkenes through the addition of tetrabutylammonium iodide in DCE at 80 °C [182]. The chemistry was regioselective and stereoselective and incorporated the iodine adjacent to the activating group.

Propargyl alcohols can be converted into vinyl iodides and bromides using a dual catalyst approach (Scheme 7.119) [183]. The combination of a common gold complex with a molybdenum species generated a highly active catalyst for this reaction. The precise

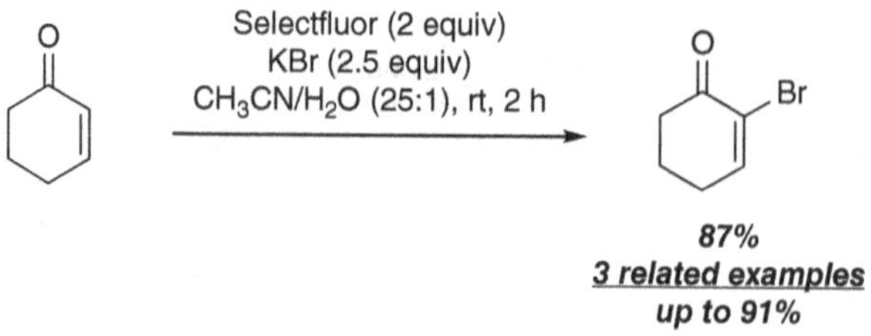

SCHEME 7.120   α-Bromination of activated alkenes [174].

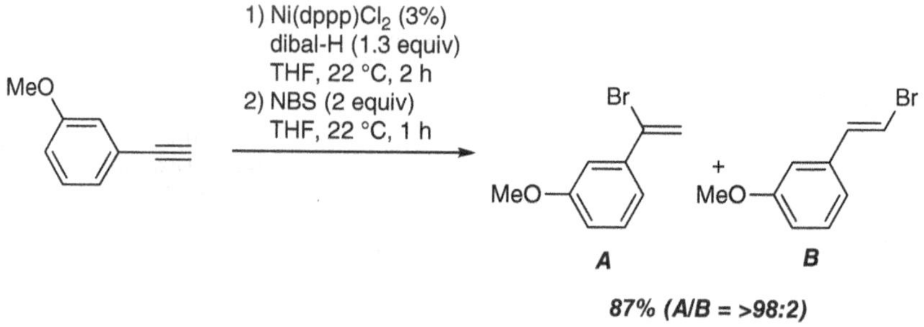

SCHEME 7.121   Nickel-catalyzed α-selective formal hydrobromination of alkynes [184].

interaction of the two catalysts was not determined (the catalytically active species); however, lower yields were obtained if either metal was absent from the reaction mixture. NIS or NBS served as the halogen source along with PPh$_3$O as an additive. The function of the triphenylphosphine oxide was not clear, although the authors speculated that it acted upon the Mo center and not the gold. Its presence was important since it inhibited the formation of undesired secondary products (enones). Given the mild conditions, relatively fast reaction rate, and high commercial availability of the reagents, this is an attractive approach for the synthesis of vinyl iodides and bromides.

Selectfluor has been used to brominate the α-position of both cyclic and acyclic activated alkenes (Scheme 7.120) [174]. The reaction conditions were remarkably mild, and the chemistry could tolerate the presence of water. The bromination reaction was selective for the formation of the α-substitution product, and moderate to excellent yields were obtained. While only a few examples were described in this report, the operational simplicity and the widespread availability of a host of Michael acceptors make this an attractive approach.

The ability of a catalyst to direct the stereoselectivity and regioselectivity of a chemical reaction remains an integral part of why transition metal catalysts are so attractive for organic synthesis. An intriguing example of this outlined the use of nickel catalysts to promote the regioselective synthesis of vinyl bromides [184]. While the overall reaction was a halogenation, the first step in the reaction was a hydroalumination process using dibal-H. Quenching this highly reactive intermediate with either NBS or Br$_2$ afforded the vinyl bromide (Scheme 7.121). The chemistry occurred under mild conditions and was highly selective for α-substitution. When the authors switched to monodentate phosphine ligands, the selectivity was reversed, and the reaction favored the formation of the β-substituted vinyl bromide (Scheme 7.122). This was an attractive approach for the synthesis of these compounds since a slightly different catalyst structure completely reversed the regioselectivity of the reaction.

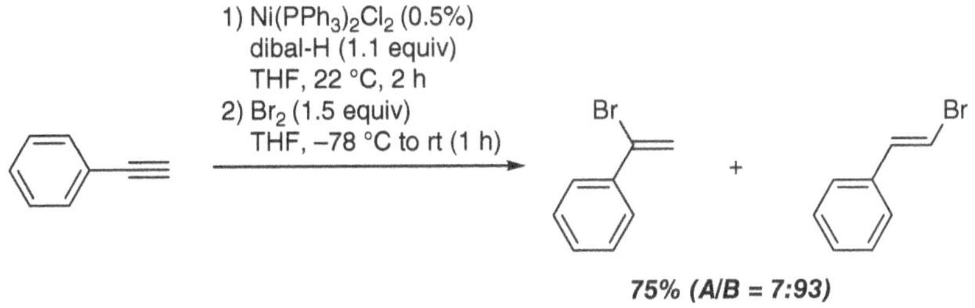

SCHEME 7.122   Nickel-catalyzed β-selective formal hydrobromination of alkynes [184].

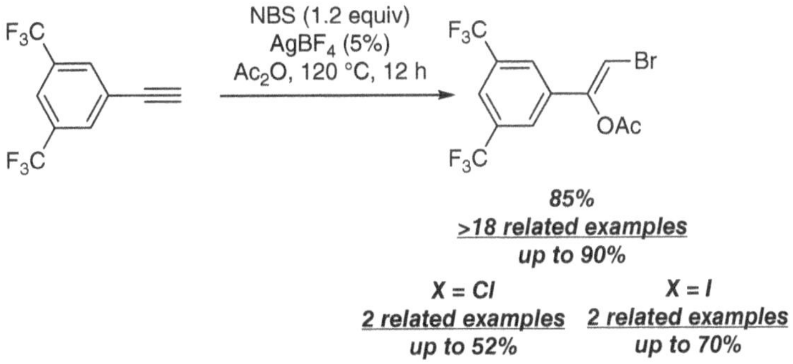

SCHEME 7.123   Silver-promoted synthesis of β-haloenol acetates [185].

The difunctionalization of terminal alkynes using NXS reagents and acetic anhydride generated halogenated acetates (Scheme 7.123) [185]. The reaction was promoted by silver, and it is worth noting that the amount of silver needed for a successful functionalization of the alkyne was quite low (5%). Common palladium and copper catalysts such palladium(II) acetate and copper(I) iodide were ineffective. The chemistry was highly regioselective and stereoselective for the formation of the β-halogenated Z-isomer. This was a very attractive approach given the operational simplicity of the method and the wide assortment of terminal alkynes that are readily available. In related work, the halosulfonylation of terminal alkynes has been achieved using iron catalysts [186].

Alkoxides and terminal alkynes have been coupled through a titanium(IV) bromide-mediated reaction (Scheme 7.124) [187]. This was an attractive approach to the synthesis since alkoxides can be readily generated and a wide range of terminal alkynes are commercially available. The coupling reaction took place under very mild conditions and afforded moderate to good yields of the vinyl bromide. The chemistry was also regioselective and highly stereoselective for the E-isomer. It should be noted that titanium(IV) bromide can react violently with moisture and should be handled and stored under an inert atmosphere such as nitrogen or argon.

The synthesis of vinyl iodides and bromides starting from unactivated styrenes has been achieved using a silylative coupling reaction, followed by trapping with NXS (Scheme 7.125) [188]. The authors focused on the preparation of vinyl iodides and bromides due to their popularity in transition metal-catalyzed cross-coupling reactions. The first step in the synthesis was a ruthenium-catalyzed silylative coupling using the vinyl silane. The intermediate vinyl

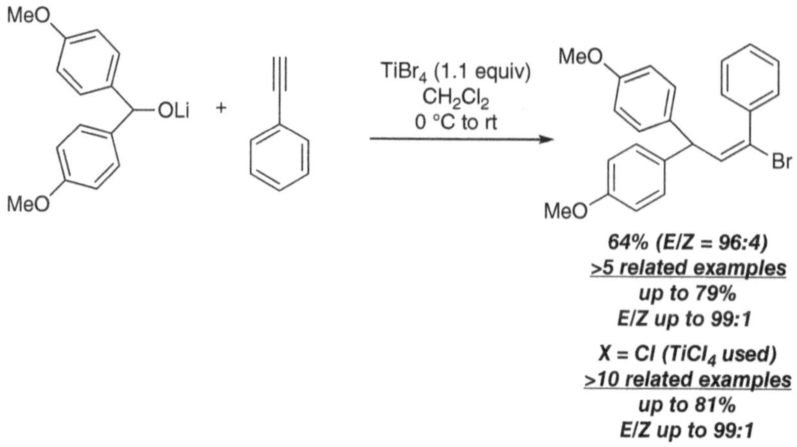

SCHEME 7.124 Titanium-mediated synthesis of vinyl bromides and chlorides [187].

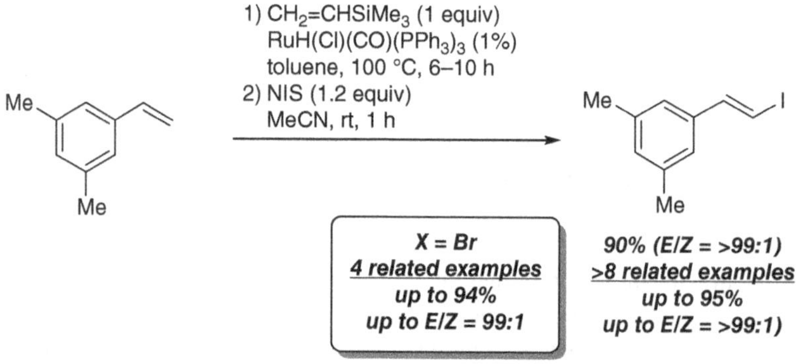

SCHEME 7.125 Ruthenium-catalyzed halogenation of unfunctionalized styrenes [188].

silane was trapped in the second step by NXS and converted into the vinyl bromide through a halodesilylation reaction. The overall process was highly regioselective for the formation of the β-substitution product and highly stereoselective for the *E*-isomer. This reaction is attractive due to the low cost and widespread availability of the styrene derivatives as well as the operational simplicity of the approach.

In addition to these examples, a formal insertion of acetylene into the C—Cl bond of acyl chlorides generated a host of vinyl chlorides. The chemistry was largely selective for the formation of the *E*-isomer, and moderate to good yields were obtained [189]. In related work, a curious solvent effect was observed [191]. When DMF was used as the solvent, clean formation of the dihydrooxazoles was observed. However, when analogous reactions were carried out using DCM as the solvent, functionalized oxazolidines bearing a quaternary carbon center were formed. As long as DMF was the solvent, high-yielding conversions were obtained.

The intramolecular iodocyclization of ynamides bearing an ethoxyether has been achieved using an electrophilic source of iodine (Scheme 7.126) [190]. The substrate was designed such that the ethoxyether group was adjacent to the ynamide on the aromatic ring. As a result, the ethoxyether was in a good position to act as a leaving group during the cyclization reaction. While elemental iodine and NIS afforded some of the cyclization product, the activity was low for some screening reactions. Other preliminary reactions generated a significant amount of an undesired diiodinated enamide. Bis(2,4,6-collidine) iodonium hexafluorophosphate generated moderate yields of the cyclization product without significant formation of the secondary product. It should be noted that this reaction was remarkably fast and outstanding yields of the cyclized compounds were obtained in a few seconds at room temperature. In addition to these examples, a host of halocyclizations and related reactions have been developed. A sampling of these reactions is presented in Schemes 7.127 through 7.138 and Examples 7.29 and 7.30.

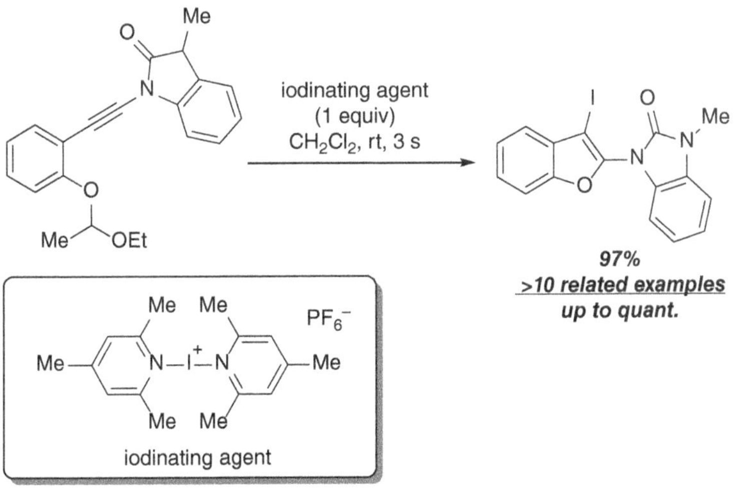

SCHEME 7.126   Iodocyclization of substituted ynamides [190].

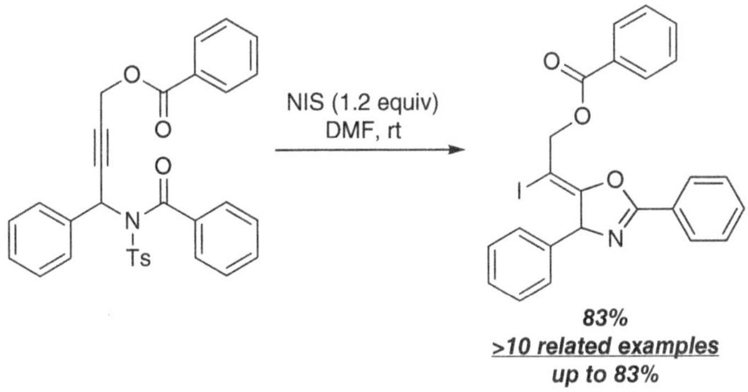

SCHEME 7.127   NIS-promoted synthesis of dihydrooxazoles in DMF [191].

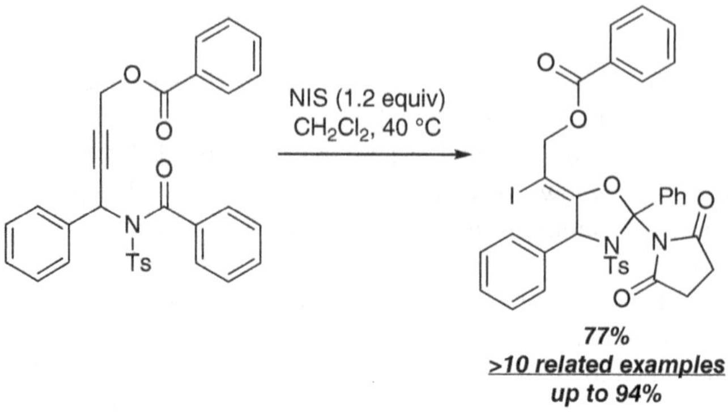

SCHEME 7.128 NIS-promoted synthesis of oxazolidine in dichloromethane [191].

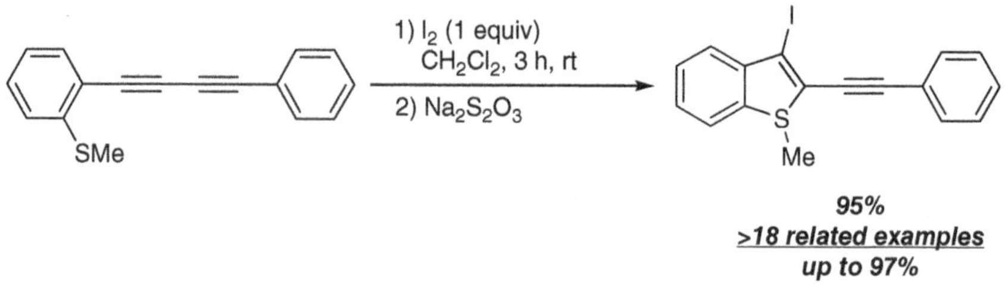

SCHEME 7.129 Eelectrophilic cyclization of diynes [38].

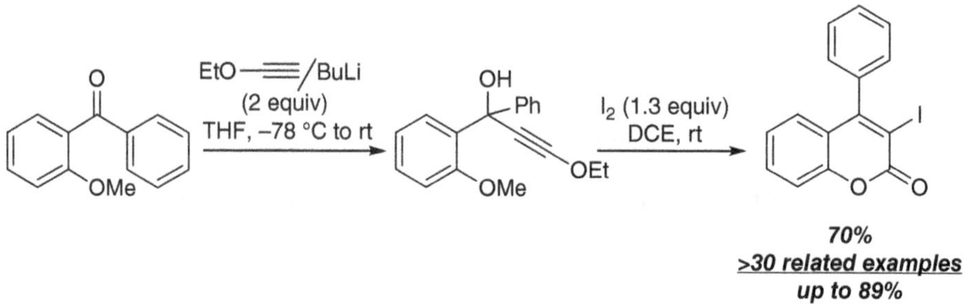

SCHEME 7.130 Iodocyclization of propargyl alcohols [192].

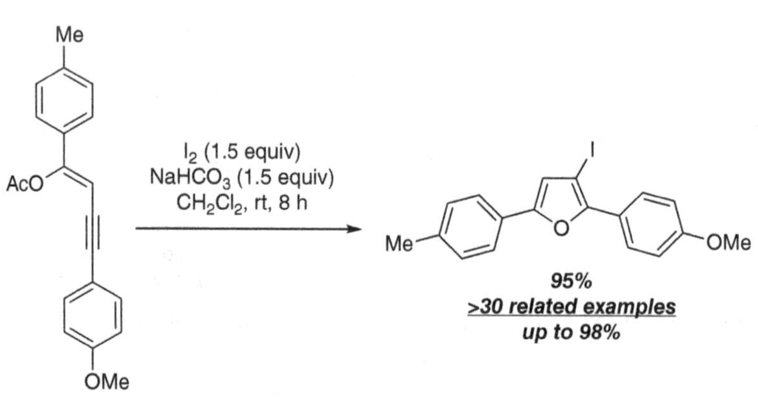

SCHEME 7.131 Synthesis of 3-iodofurans through iodocyclization [193].

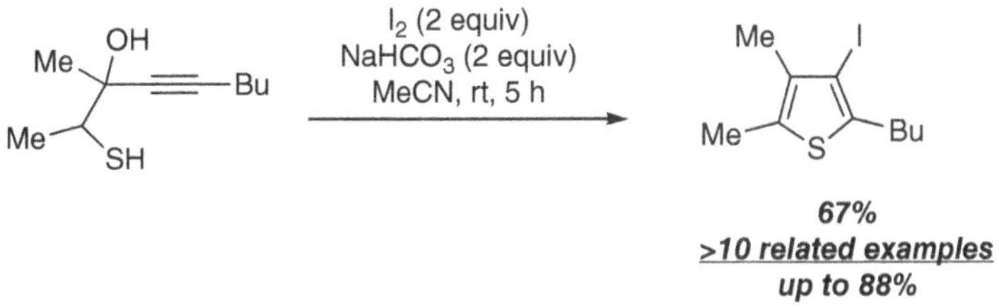

SCHEME 7.132 Synthesis of 3-iodothiofurans through iodocyclization [194].

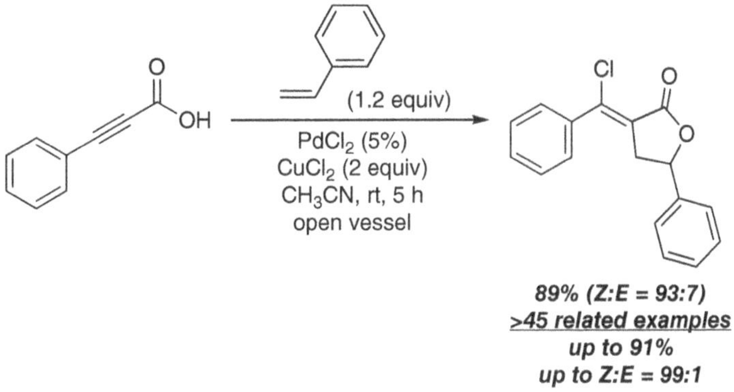

SCHEME 7.133 Palladium-catalyzed synthesis of vinyl chlorides [195].

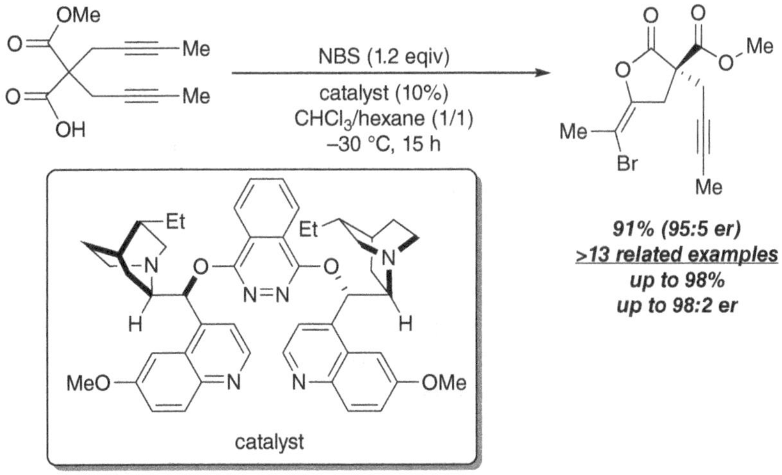

SCHEME 7.134 Brominative cyclization of a diyne [196].

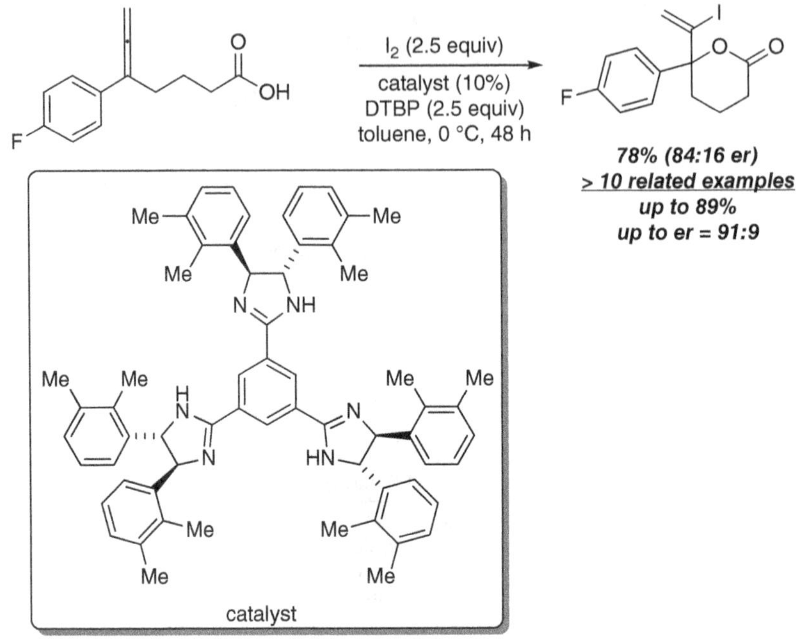

SCHEME 7.135   Asymmetric synthesis of vinyl iodides through iodolactonization [197].

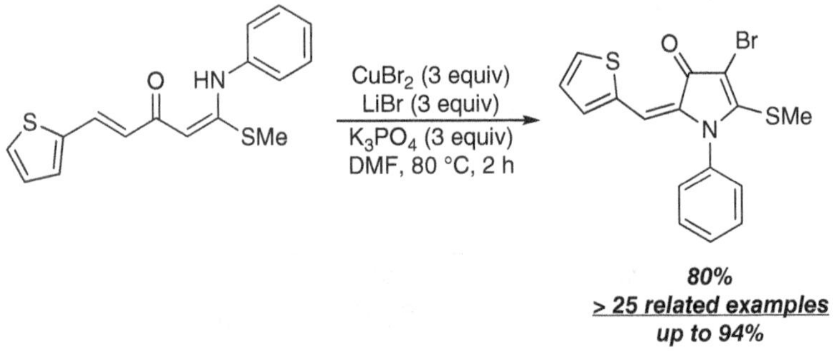

SCHEME 7.136   Copper-promoted synthesis of vinyl bromides through cyclization [198].

The conversion of aldehydes into vinyl iodides through the formation and subsequent iodination of hydrazones is known as the Barton vinyl iodide synthesis [202]. This synthesis can be quite challenging, and a practical approach to this chemistry has been developed using *N-tert*-butyldimethylsilylhydrazones as intermediates (Scheme 7.139) [203]. The initial formation of the *N*-silylhydrazones was catalyzed by scandium(III) triflate, and the iodination step was aided by the addition of 2-*tert*-butyl-1,1,3,3-tetramethylguanidine (BTMG). In addition to the generation of vinyl iodides, the chemistry was also extended to the synthesis of vinyl bromides.

In related work, the halogenation of tosylhydrazones has also been used to generate vinyl halides (Scheme 7.140) [204]. Instead of DBU or similar base, a mineral base

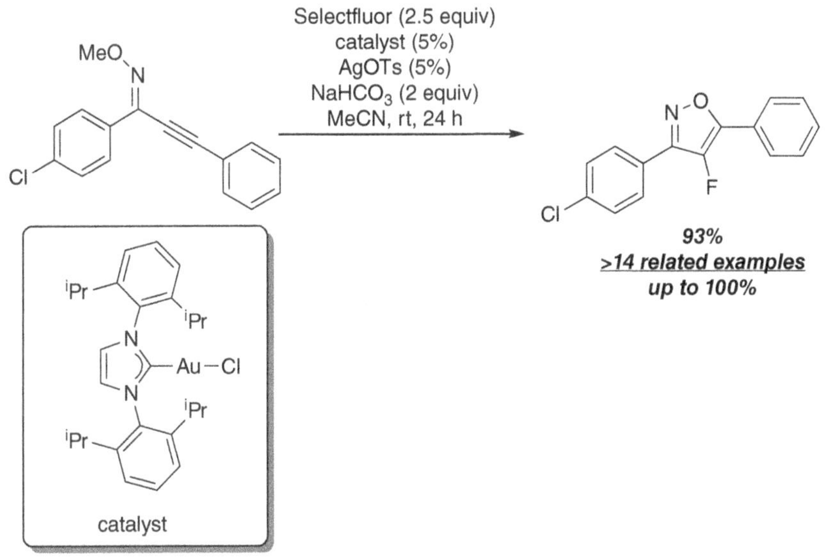

SCHEME 7.137 Gold-catalyzed synthesis of heterocycles [199].

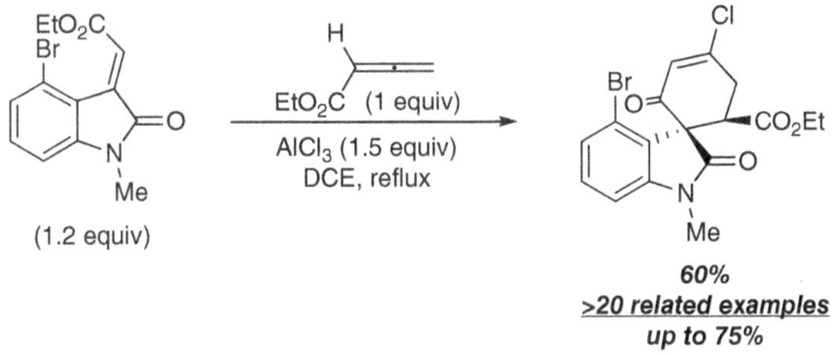

SCHEME 7.138 Cyclization of allenoates using aluminum chloride [200].

Example 7.29 Synthesis of Vinyl Bromides through an Intramolecular Cyclization [201].

Before starting this synthesis, carefully follow the chemical safety instructions listed in Example 7.1.

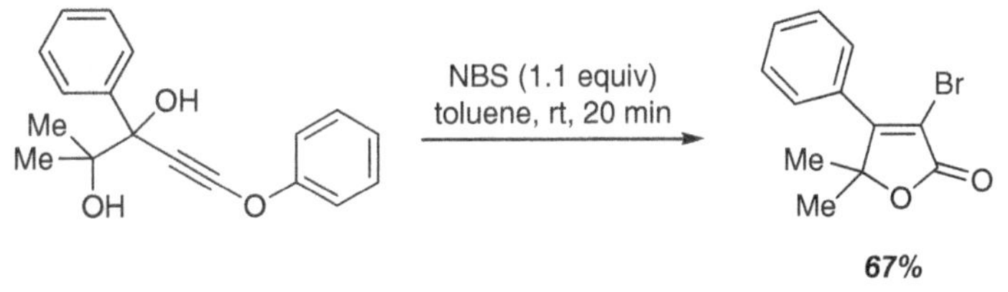

Under a nitrogen atmosphere, NBS (36.1 mg, 0.20 mmol) was added to a solution of propargyl alcohol (52.0 mg, 0.18 mmol) in toluene (0.46 mL) at room temperature. After stirring for 20 min, the reaction mixture was filtered through a pad of Celite. The filtrate was evaporated in vacuo. The residue was purified by column chromatography (CH_2Cl_2) to give the vinyl bromide (32.8 mg, 67%) as a white solid.

Adapted with permission from M. Egi, Y. Ota, Y. Nishimura, K. Shimizu, K. Azechi, and S. Akai, *Org. Lett.,* **2013**, *15*, 4150–4153. *Copyright (2013) American Chemical Society.*

Example 7.30   Palladium-Catalyzed Intermolecular Oxyvinylcyclization [195].

Before starting this synthesis, carefully follow the chemical safety instructions listed in Example 7.1.

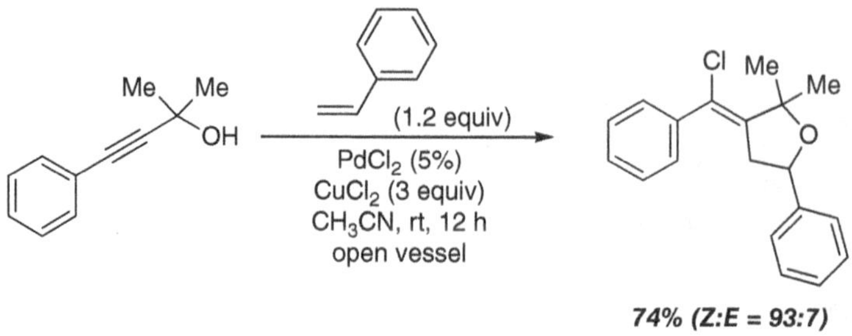

74% (Z:E = 93:7)

A mixture of the alkynol (0.5 mmol), styrene (0.60 mmol), PdCl$_2$ (5 mol %), CuCl$_2$ or CuCl$_2 \cdot$2H$_2$O (1.0 or 1.5 mmol), and MeCN (1 mL) was placed in a test tube (10 mL) equipped with a magnetic stir bar. The mixture was stirred at room temperature under air. After the reaction was completed, purification of the mixture on a preparative TLC afforded the lactone as a colorless oil (110.3 mg, 74%).

Adapted with permission from Z. Zhang, L. Ouyang, W. Wu, J. Li, Z. Zhang, H. Jiang, *J. Org. Chem.*, **2014**, *79*, 10734–10742. *Copyright (2014) American Chemical Society.*

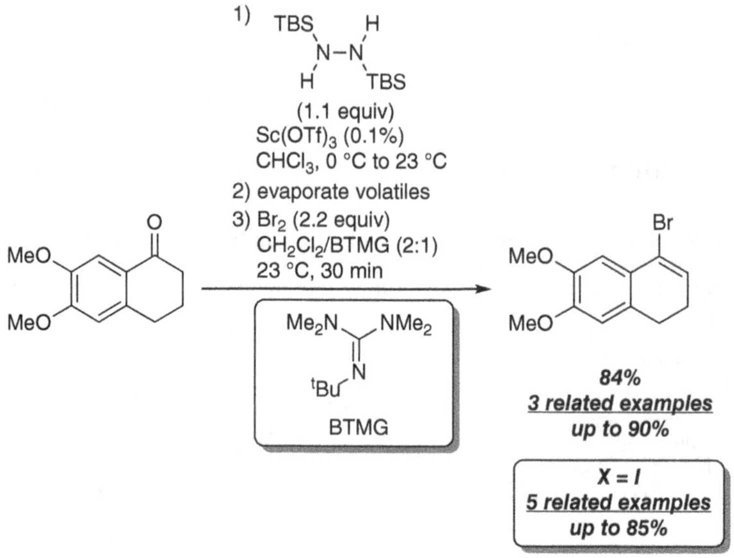

SCHEME 7.139   Conversion of aldehydes into vinyl bromides [203].

successfully promoted the halogenation reaction. After some experimentation, the authors also discovered that both NBS and TBAB were needed for a high-yielding conversion. It was noteworthy that the reaction was highly regioselective for the formation of the α-substituted vinyl bromide.

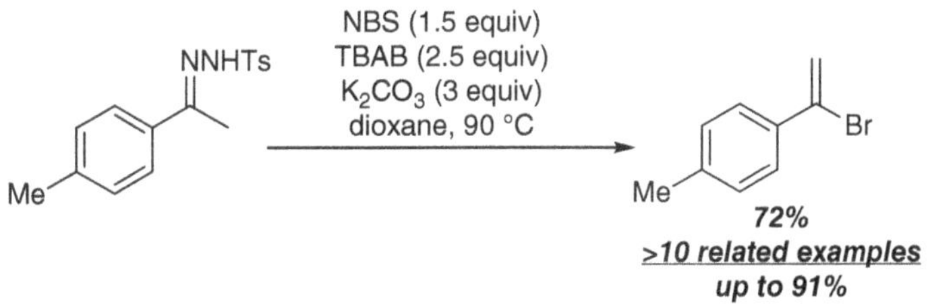

SCHEME 7.140 Regioselective conversion of tosylhydrazones into vinyl halides [204].

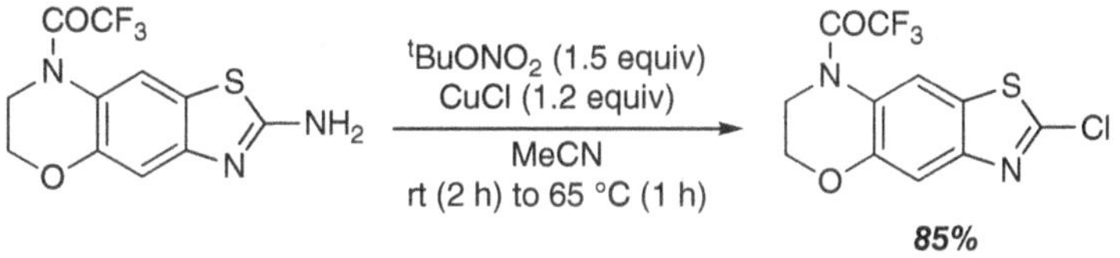

SCHEME 7.141 Chlorodeamination of heterocycles [205].

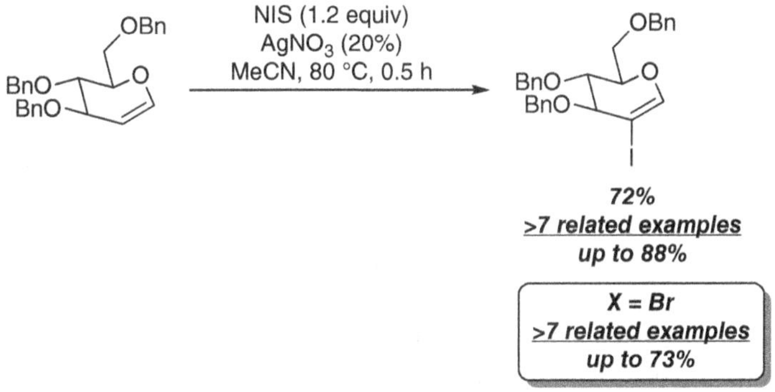

SCHEME 7.142 Synthesis of halogenated glycals [206].

The chlorodeamination of heterocycles has been reported using *tert*-butyl nitrite and copper chloride as promoters (Scheme 7.141) [205]. The conditions were quite mild, and an excellent yield of the chlorinated compound was obtained.

The iodination of glycals has been achieved using NXS reagents along with a catalytic amount of silver nitrate (Scheme 7.142) [206]. The chemistry was high yielding, occurred under fairly mild conditions, and was successful for the synthesis of both iodinated and brominated compounds. It was noteworthy that the halogenation reaction regioselectively produced the 2-haloglycal.

A metal-free approach to the stereoselective synthesis of (*E*)-β-halogenated styrenes has been reported (Scheme 7.143) [207]. For bromination reactions, the chemistry needed an excess of dibromomethane along with a potent base to achieve high conversion. Presumably, the initial step in the reaction was the formation of a *gem*-dibromo species that underwent a stereospecific elimination reaction to afford the *E*-vinyl bromide. Vinyl iodides and chlorides were also constructed using this approach. The ability to generate high yields of

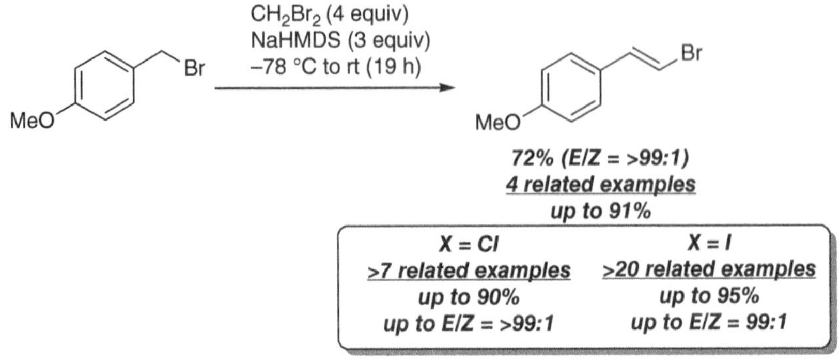

SCHEME 7.143   Synthesis of halogenated styrenes from benzyl halides [207].

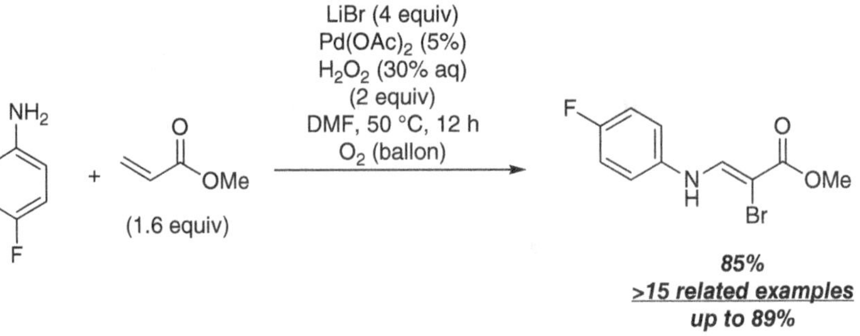

SCHEME 7.144   Dehydrogenative aminohalogenation of alkenes [208].

the vinyl iodides was attractive since these compounds typically display higher reactivities in functionalization reactions.

Amino-functionalized vinyl bromides have been generated through the dehydrogenative aminohalogenation of Michael acceptors (Scheme 7.144) [208]. The multicomponent reaction was catalyzed by common palladium salts and used hydrogen peroxide and elemental oxygen as the oxidizing agents. Although a nucleophilic source of bromine was added to the reaction, the authors proposed that the active bromine species was actually Br^+ generated through oxidation by H_2O_2. The solvent was a critical component for this reaction, and while the use of DMF cleanly generated the vinyl bromides, analogous reactions carried out in THF formed nonhalogenated enamines. Given the cost of many of the electrophilic sources of halogens, the ability to generate X^+ from common X^- salts using hydrogen peroxide is an attractive alternative.

gem-Dihalogenated compounds are an important class of vinyl halides with applications in a number of fields [209]. A representative sampling of the synthetic methods used to generate these valuable compounds is described in the following sections. Some of these approaches do not generate the two carbon–halogen bonds directly; however, these are some of the most practical routes to the preparation of the *gem*-dihalogenated compounds.

In addition to a Wittig-type approach to the synthesis of *gem*-dichloroalkenes, this class of compounds has also been generated through a three-step process (Example 7.31) [210]. While it may seem as though this was a operationally complex procedure, it was somewhat mitigated by the observation that it all proceeds in the same reaction vessel. The authors

Example 7.31   Synthesis of a *gem*-Dichloroalkene through a Non-Wittig Approach [210].

Before starting this synthesis, carefully follow the chemical safety instructions listed in Example 7.1.

1) CCl_3COOH (1.5 equiv)
   CCl_3COONa (1.5 equiv)
   DMF, 25–35 °C, 2 h
2) Ac_2O, 5 °C to 25 °C (1 h)
3) Zn (2 equiv)
   AcOH, 0 °C to 60 °C (1 h)

88%

Caution! Hexane and ethyl acetate are highly flammable.

Caution! This reaction evolves CO_2, and the additions must be made carefully and slowly to prevent overflow of the reaction vessel.

To a stirred solution of trichloroacetic acid (105 g, 0.642 mol) and cyclopropylcarboxaldehyde (30 g, 0.428 mol) in DMF (300 mL) at 25 °C was added sodium trichloroacetate (119 g, 0.642 mol) in portions. The internal temperature was kept below 35 °C by addition control. After addition was completed, the mixture was stirred at room temperature for 4 h with continuous evolution of CO_2. The solution was cooled to 5 °C, and acetic anhydride (80.77 mL, 0.856 mol) was carefully added. Strong CO_2 evolution was observed. The mixture was allowed to warm to room temperature and stirred for an additional hour. The reaction mixture was diluted with acetic acid (400 mL) and cooled to 0 °C. To the solution, the zinc powder (55.9 g, 0.856 mol) was added in one portion. The solution was stirred for 1 h at 60 °C and then was cooled to room temperature. To the solution, water was added and then extracted with hexanes. The combined organic phases were washed with water and saturated aqueous solution of sodium chloride. The organic phase was dried over $MgSO_4$, filtered, and concentrated by rotary evaporation. The crude 1,1-dichloro-2-cyclopropylethylene was obtained in a relative good purity. Both flash chromatography (hexane/EtOAc, 9 : 1) and distillation (bp = 47–5 °C/2 Torr) methods for the purification of this compound had been applied. Purification by distillation yielded 44.07 g (88%) of the desired compound.

Reprinted from Z. Wang, S. Campagna, G. Xu, M. E. Pierce, J. M. Fortunak and P. N. Confalone, *Tetrahedron Lett.*, *41*, 4007–4009. *Copyright (2000), with permission from Elsevier.*

noted that careful addition of the acetic anhydride and sodium trichloroacetate was needed to prevent rapid evolution of carbon dioxide. Nine aldehydes were converted into vinylic dichlorides with excellent isolated yields (up to 95%). In related work, oxone has been used to promote the conversion of terminal alkynes into dihalogenated ketones [211].

One of the oldest and most reliable routes to the preparation of *gem*-dibromoalkenes is also one of the most operationally straightforward approaches (Scheme 7.145) [212]. Simply adding triphenylphosphine to a DCM solution of carbon tetrabromide followed by addition of the aldehyde generated good yields of the dibromostyrene. Using this approach, a host of dibromoalkenes have been generated [209, 213].

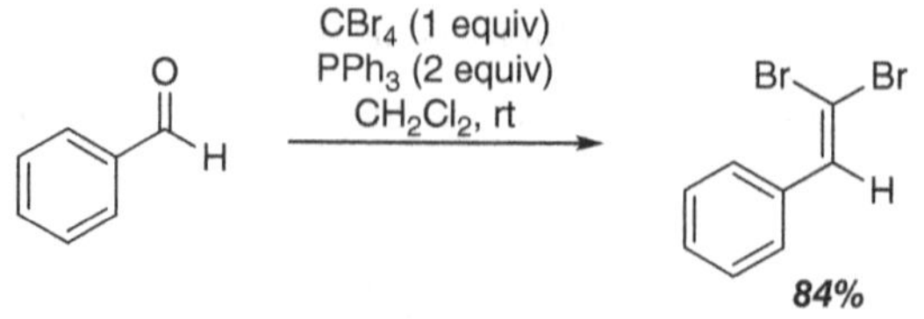

SCHEME 7.145   Synthesis of dibromoalkenes through a Wittig-type approach [212].

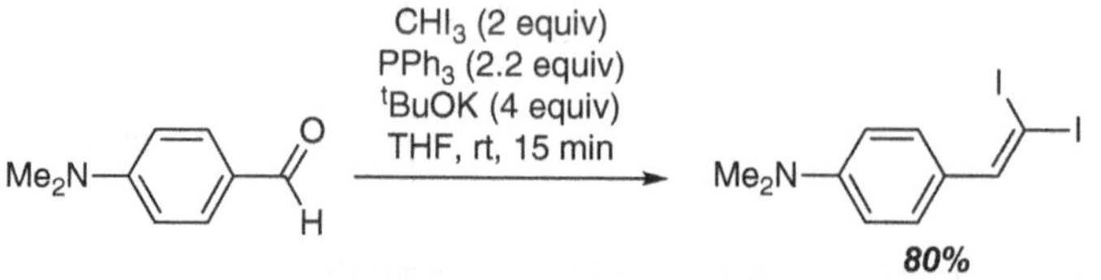

SCHEME 7.146   Synthesis of *gem*-diiodoalkene from aldehydes [214].

Similar to the syntheses described earlier, one of the classic approaches to the synthesis of *gem*-diiodoalkynes involved the addition of an aldehyde to a solution of a Wittig-type reagent generated from treating triphenylphosphine with iodoform in the presence of a strong base (Scheme 7.146) [214]. Moderate yields of the *gem*-diiodide were formed, and the authors did not observe any Ph_3PI_2 in this reaction.

A wide range of approaches have been used to generate *gem*-diiodoalkenes (Example 7.32) [209, 215]. One group that has been challenging to functionalize is benzaldehydes due to the propensity for generation of an iodoalkyne instead of the *gem*-diiodo species [216].

The dibromination of alkynes was accomplished using copper(II) bromide (Scheme 7.147) [217]. No other reagents were needed for the chemistry, and simply stirring the alkyne with the copper salt in acetonitrile generated the target compounds in high yield. Molecular sieves were used to inhibit the formation of ketone-containing products. The chemistry was highly regioselective and stereoselective for the formation of the 1,2-(*E*)-dibromoalkenes. Both terminal and internal alkynes and a host of substrates bearing alkyl, aryl, heterocycles and protected amines were converted into the 1,2-(*E*)-dibromoalkenes. Even a terminal alkyne bearing a primary alcohol was successfully brominated using this approach. Given the broad scope of this reaction and the tolerance to a range of functional groups, this approach is likely to be a popular route. NBS has also been found to promote the dihalogenation of alkenes [218], and elemental iodine has been shown to add to terminal and internal alkynes under solvent- and catalyst-free conditions at room temperature (Scheme 7.148) [219].

Most mixed halogenation reactions are not selective and scramble the arrangement of the halogens. This can be problematic when trying to generate compounds with a specific arrangement of halogens. To address this issue, a practical stepwise approach to the synthesis of Z-1-iodo-2-bromoalkenes has been reported (Scheme 7.149) [220]. The critical component in this chemistry was to attach the bromine to the substrate prior to the addition of the iodide source. Once the alkynyl bromide was isolated, addition of a nucleophilic source of iodine along with acetic anhydride selectively formed the mixed product. No scrambling of the halogens was observed in this chemistry. A wide range of alkyl and

Example 7.32   Generation of an α-Aryl *gem*-Diiodo Species [215].

Before starting this synthesis, carefully follow the chemical safety instructions listed in Example 7.1.

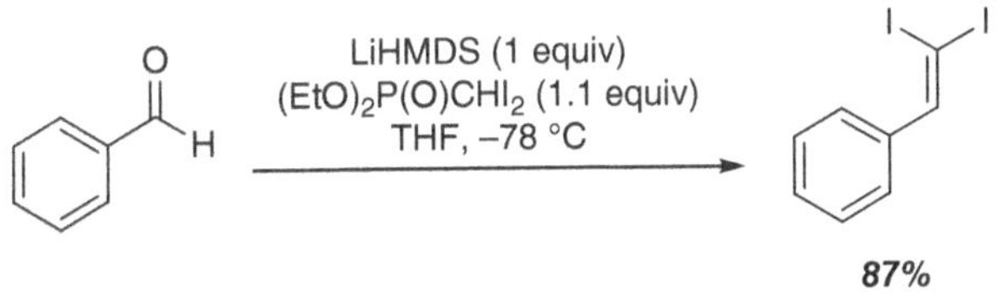

LiHMDS (1 equiv)
(EtO)$_2$P(O)CHI$_2$ (1.1 equiv)
THF, −78 °C

87%

Caution! THF and ethyl acetate are highly flammable.

At room temperature and under an argon atmosphere, a flame-dried round-bottom flask equipped with a magnetic stirrer and fitted with a septum was charged with THF (15 mL), distilled HMDS (0.22 mL, 1.0 mmol, 1.0 equiv). The solution was stirred and cooled to 0 °C. *n*-BuLi (1.0 mmol, 1.0 equiv) was added slowly. The solution was stirred at 0 °C for 5 min and then cooled to −78 °C. A solution of diiodomethyldiethylphosphonate (444 mg, 1.1 mmol, 1.1 equiv) in 5 mL of THF was added slowly via cannula. The orange anion was stirred for 10 min at that temperature. A solution of benzaldehyde (106 mg, 1.0 mmol, 1.0 equiv) in 4.0 mL of THF was added slowly via cannula. The mixture was stirred for 10 additional minutes. Water was added, and the mixture was allowed to warm to room temperature and diluted with AcOEt. Aqueous layer was extracted twice with AcOEt, and combined organic layers were washed with brine, dried over MgSO$_4$. The solvent was removed in vacuo. The remaining oil was submitted to flash chromatography on silica gel. The gem-diiodoalkene (309 mg, 0.87 mmol) was isolated as a colorless compound, or as a pale pink compound, in 87% yield.

Adapted with permission from J.-M. Cloarec and A. B. Charette, *Org. Lett.*, **2004**, *6*, 4731–4734. *Copyright (2004) American Chemical Society.*

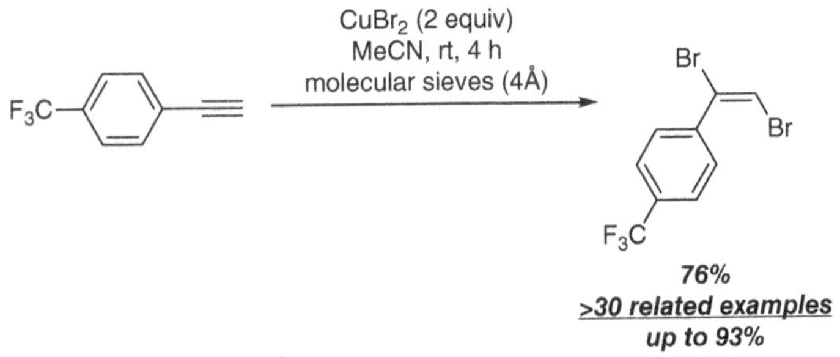

CuBr$_2$ (2 equiv)
MeCN, rt, 4 h
molecular sieves (4Å)

76%
>30 related examples
up to 93%

SCHEME 7.147   Regio- and stereoselective dibromination of alkynes [217].

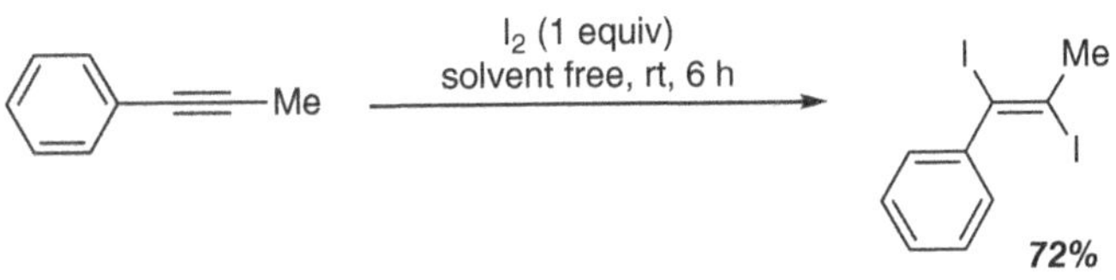

I$_2$ (1 equiv)
solvent free, rt, 6 h

72%

SCHEME 7.148   Solvent- and catalyst-free halogenation of phenylacetylene derivatives [219].

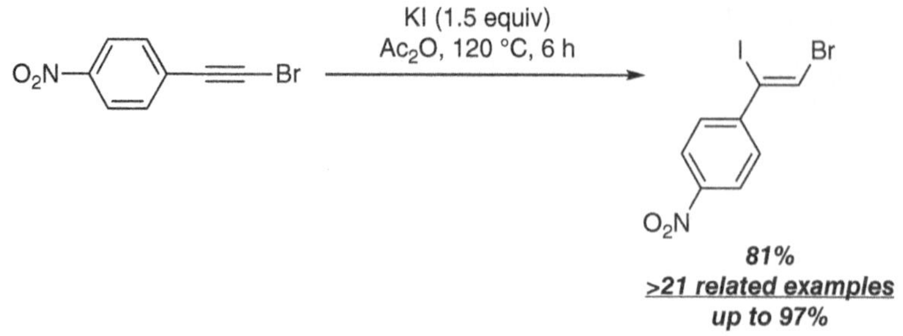

SCHEME 7.149   Regio- and stereoselective synthesis of Z-1-iodo-2-bromoalkenes [220].

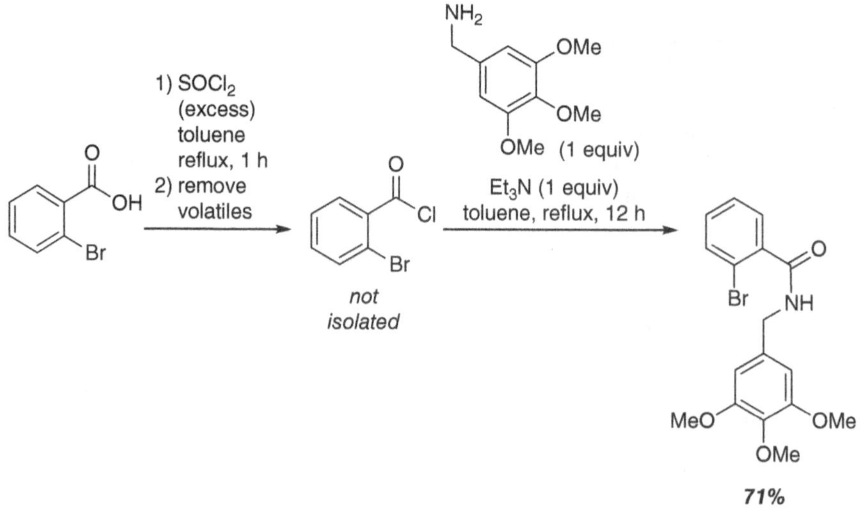

SCHEME 7.150   Generation and use of acyl chlorides in solution [221].

arylalkynes bearing either electron-donating or electron-withdrawing groups were success-fully converted into the mixed species.

Acyl halides are some of the most versatile reagents in organic synthesis. Acyl chlorides remain arguably the most common examples of these compounds, and one of the most popular approaches to the synthesis of these valuable compounds entails the treatment of carboxylic acids with thionyl chloride. The conversion is typically achieved by the addition of thionyl chloride to a solution of the carboxylic acid in a nonpolar solvent along with a mild base to consume the concomitant formation of HCl. The operational complexity of this chemistry is typically low to moderate due to the sensitivity of the thionyl chloride and resulting acyl chloride to moisture. Some of the sensitivity can be mitigated by generating the acyl chlorides in solution. This is an attractive approach to the chemistry and has been used to generate a range of functionalized amides (Scheme 7.150) [221].

While a host of acyl chlorides have been generated and used in the synthesis of new com-pounds, acyl bromides and iodides have received less attention. Since the halide is lost in most chemical reactions involving acyl halides, the heavier congeners are often not needed since the reactivity of acyl chlorides is typically sufficient for most reactions. If needed for a specific application, acyl bromides and iodides can be generated through a halogen exchange by treating acyl chlorides with TMSBr or TMSI [222, 223]. Acyl bromides can be generated directly by the addition of NBS and triphenylphosphine to carboxylic acids [224], the addition of tribromoacetate/triphenylphosphine to carboxylic acids under acid-free conditions [225], or other brominating agents [226].

7.3.1   Troubleshooting the Synthesis of Vinyl Halides and Related Compounds

While it will not be possible to anticipate all of the needs of the synthetic chemist over the next few decades, this list will provide some insight into issues commonly encountered when trying to prepare vinyl halides. This chemistry can be particularly confusing as several protocols will work for several of the halogens, but not for all of them. Care must be taken to make sure that the chemistry will be successful for the substrate under investigation.

Problem/Issue/Goal	Possible Solution/Suggestion
Need to generate a vinyl fluoride	One of the most convenient solutions to this problem is to use a silver-promoted conversion of vinylboronic acids into vinyl fluorides. The conditions are quite mild, and yields of the vinyl fluoride were good [92]
It would be convenient to start from an alkene and generate the vinyl fluoride through C—H activation	This is a challenging reaction, and general approaches are rare. Many reactions tend to be substrate specific. For example, methyl oxime ethers can be converted into vinyl fluorides under very mild conditions using silver salts and NFSI as the fluorine source [102]
Need to convert an alkyne into a vinyl fluoride	There are several approaches to this conversion. The direct addition of HF using DMPU/HF would be a good place to start [63]. Additionally, it might be more practical to convert the alkyne into a vinylboronate species and then to the vinyl fluoride [92]
What is a convenient substitute for HF?	A host of reagents have been developed to serve as HF substitutes. Olah's pyridine adduct is perhaps the most well known [165], and other examples with different organic bases have also been used [63]
Need to generate a gem-*difluoroalkene*	A good entry point for this chemistry would be the conversion of diazo compounds into *gem*-difluoroalkenes [168]
Can an acyl chloride be converted into an acyl fluoride?	This is a valuable reaction and has been accomplished using several routes including simply treating the acyl chloride with TBAF in THF [169]
Need to prepare an acyl fluoride from an aryl bromide	This fluorocarbonylation reaction is known, and a recent modification to the palladium-catalyzed version of this reaction used *N*-formylsaccharin as the carbon monoxide source to avoid the need for high pressures of gaseous CO [171]
Need to start from a carboxylic acid and generate a vinyl halide	The conversion of carboxylic acids to vinyl chlorides, bromides, and iodides can be achieved in less than 5 min using NXS and a catalytic amount of triethylamine [172]

(Continued)

Problem/Issue/Goal	Possible Solution/Suggestion
Can a Finkelstein reaction be used with a vinyl halide?	A copper-catalyzed vinyl Finkelstein has been developed and would be a good entry point to this chemistry [146]
Substrate needs to have water present in order to dissolve	Several vinyl halide syntheses are tolerant to water [129]
Need to start from a vinyl triflate and make a vinyl chloride or bromide	Palladium complexes have been shown to promote this reaction using nucleophilic sources of the bromide or chloride [150]. Ruthenium complexes have also been active towards this reaction [151]
Can a vinyltrifluoroborate salt be converted into a vinyl bromide or chloride?	This transition metal-free reaction is typically high yielding. Bromination was promoted by the chloramine-T along with a nucleophilic bromine source [129], while chlorination used TCCA [128]
Need to generate a vinyl iodide from an alkyne	Treatment of the alkyne with elemental iodine and a secondary phosphine oxide is a particularly attractive approach to the synthesis of vinyl iodides with Markovnikov selectivity [181]. An attractive entry point for this chemistry would be to convert the alkyne into a vinylboronate followed by the addition of iodine under basic conditions to afford the vinyl iodide [177]
The starting material is a vinylamide, and it needs to be iodinated or brominated with retention of the amide	This has been accomplished using a rhodium-catalyzed reaction with NIS or NBS serving as the halogen source [178]
Need to make a halogenated vinyl ketone bearing an electron-deficient arene	A good starting point for this chemistry would be a allylic alcohol. Conversion of this allylic alcohol into a chlorinated or brominated vinyl ketone can be accomplished using oxalyl halides [179]. A gold-catalyzed process that affords the iodinated vinyl ketones was successful at room temperature [183]
Need to iodinate/brominate an activated alkene with retention of the alkene— essentially a C—H halogenation	While there are several approaches to this chemistry, a cobalt-catalyzed process was successful and selective for the anti-Markovnikov addition product with Z-stereochemistry [144]. A transition metal-free version of this reaction has been reported using a nucleophilic bromine source [174]
What is a convenient source of elemental bromine?	Pyridinium tribromide is a solid that can be weighed out in air and functions as a source of elemental bromine [180]
Need to selectively iodinate or brominate an unactivated styrene with retention of the alkene	A ruthenium-catalyzed process is an attractive approach for this reaction and uses NXS as the halogen source [188]
Need to chlorinate a activated alkene with retention of the alkene	PhICl$_2$ has been used to effect this transformation in wet DMF [67]
Need to effect a formal hydrobromination of an alkyne with Markovnikov selectivity	A nickel-catalyzed reaction using a chelating bisphosphine ligand converted terminal alkynes into vinyl bromides with the desired selectivity [184]
Need to effect a formal hydrobromination of an alkyne with anti-Markovnikov selectivity	The anti-Markovnikov product can be generated using a nickel catalyst ligated by monodentate phosphine ligands [184]

(Continued)

(Continued)

Problem/Issue/Goal	Possible Solution/Suggestion
Can an ynamide be used in halocyclization reactions?	There are many examples of successful halocyclization reactions using ynamides [190]
Can a benzyl halide be converted into a vinyl halide?	Treatment of benzyl halides with dihalomethane and NaHMDS affords vinyl halides with anti-Markovnikov selectivity and *E*-stereochemistry [207]
Starting from an aldehyde, need to generate a 1,1-dihaloalkene	This conversion tends to work quite well following a Wittig-type process using CX_4 or CHX_3 [212, 214, 215]. Care must be exercised with this chemistry due to a propensity to generate the haloalkyne instead of the *gem*-dihalo species. Non-Wittig approaches have also been devised [210]
Need to generate an E-1,2-dihaloalkene starting from an alkyne	For bromination, this chemistry has been developed using copper bromide in acetonitrile [217]. Preparation of the iodinated version was trivial and entailed simply stirring the alkyne with elemental iodine [219]. It should be noted that treating an alkynyl bromide with KI and acetic anhydride afforded the *Z*-1-iodo-2-bromoalkene [220]

7.4 SYNTHESIS OF ALKYNYL HALIDES THROUGH THE FORMATION OF HALOGEN–CARBON(SP) BONDS

Mono- and dihalogenated acetylenes are the simplest members of this group; however, they have rarely been used as reagents in organic synthesis since they have been reported to be explosive in contact with oxygen, upon heating, or mechanical shock [227, 228]. While few reports have outlined the use of chloro- and bromoacetylene, iodoacetylene has been prepared and used in solution. For example, treatment of tributyl(ethynyl)tin with iodine in THF solution afforded the desired monohalogenated compound [229]. The authors reported that the THF solution of iodoacetylene was stable up to 160 °C. Recently, a synthesis of diiodoacetylene in DMF solution was reported [230]. The authors used TMS-protected acetylene as the substrate and NIS as the halogen source along with silver nitrate. Once formed, the diiodoacetylene was stored as a solution in DCM.

In contrast, 1-haloalkynes of the type XCCR (X = Br, Cl, F; R = hydrocarbon) are often isolable and stable species that have been used in a wide variety of organic transformations [231]. The following sections will focus on the various methods for the synthesis of haloalkynes and highlight the strengths and weaknesses of each approach.

1-Fluoroalkynes remain the rarest of the 1-haloalkynes. Although several routes have been devised for the synthesis of these compounds [232, 233], they have received scant attention from the synthetic community. Part of this lack of interest has been proposed to be due to the potential for explosions [232] as well as undefined toxicity.

One of the few examples of a synthetic approach involved the deprotonation of a terminal acetylene, followed by the addition of a source of electrophilic fluorine (Example 7.33) [234]. After some experimentation, the authors found that the most effective base for the reaction was BuLi and NFSI proved to be the most effective iodinating agent. Selectfluor was less effective for this chemistry (only 10% conversion). The yields of the fluoroalkynes were moderate to good (up to 80%), and several existing functional groups including

Example 7.33   Preparation of a 1-Fluoroalkyne [234].

Before starting this synthesis, carefully follow the chemical safety instructions listed in Example 7.1.

Caution! THF and hexane are highly flammable.

Caution! Organolithium reagents can react violently with moisture and should be handled under an inert atmosphere such as nitrogen or argon.

This reaction was carried out under an atmosphere of argon. A solution of phenylacetylene (0.20 mmol, 20.4 mg) in THF (5.0 mL) was added dropwise with *n*-BuLi (2.40 M, 92 µL, 0.22 mmol, 1.1 equiv) over 5 min at −78 °C under an argon atmosphere. The resulting suspension was stirred at this temperature for 30 min, and a solution of NFSI (98%, 0.24 mmol, 77.1 mg, 1.2 equiv) in THF (3.0 mL) was added. After stirring at this temperature for 1 h, the cooling bath was removed, and the reaction was allowed to slowly warm to room temperature and stirred for 18 h. The reaction was quenched with saturated ammonium chloride solution, and the organic layers were separated. The aqueous layer was washed twice with diethyl ether, and the combined organic layer was washed with brine, dried over MgSO$_4$, filtered, and then concentrated in vacuo. The residue was purified on silica using hexane as solvent system to afford the fluorinated product (15.1 mg, 63% yield).

Note: Very little is known regarding the hazards of this class of compounds. Similar compounds are known to be explosive. The possibility for toxicity is also very high. Until the health and safety issues are fully defined, 1-fluoroalkynes should only be prepared in very small quantities and only if they are critical to the investigation. These compounds should be handled with great care in any amounts.

Adapted with permission from C. Liu, H. Ma, J. Nie, J. Ma, Chin. J. Chem., **2012**, *30,* 47–52. *Copyright (2012) Wiley-VCH Verlag GmbH & Co. KGaA, Weinheim.*

halogens and ethers were tolerated by this approach. The fluorination chemistry was also successfully extended to diynes (up to 90% yield). It should be noted that very little is known about the stability and toxicity of this class of compounds and great care should be exercised when preparing and handling them.

In contrast with the scarcity of fluoroalkynes, numerous examples of chloroalkynes are known, and several reliable methods for their preparation and isolation have been developed. Some of the first examples of chloroalkynes were prepared from alkynylmercury complexes [235], and although this method was successful, the use of mercury salts is not

Example 7.34   Synthesis of a Chloroalkyne Using BuLi/NCS [238].

Before starting this synthesis, carefully follow the chemical safety instructions listed in Example 7.1.

1) BuLi (1.1 equiv)
   THF, −78 °C, 30 min

2) NCS (1.1 equiv)
   −78 °C (1 h) to rt (18 h)

73%

Caution! Diethyl ether, THF, and hexanes are highly flammable.

Caution! Organolithium reagents can react violently with moisture and must be handled and stored under an inert atmosphere such as nitrogen or argon.

A solution of phenylacetylene (2.0 g, 19.6 mmol) in THF (40 mL) was treated dropwise with *n*-butyllithium (8.0 mL, 2.5 M in hexanes, 21.5 mmol) over 5 min at −78 °C under an N_2 atmosphere. The resulting suspension was stirred at this temperature for 30 min, at which time a solution of *N*-chlorosuccinimide (2.87 g, 21.5 mmol) in THF (20 mL) was added rapidly through a cannula. After stirring at this temperature for 1 h, the cooling bath was removed, and the reaction was allowed to slowly warm to room temperature and stirred there for 18 h. The reaction was quenched with saturated NH_4Cl (25 mL), and the aqueous layer was separated and extracted with Et_2O (3 × 50 mL). The combined organic layers were dried with Na_2SO_4, filtered, and evaporated to dryness in vacuo. Purification by flash chromatography (SiO_2, hexanes) yielded 1.94 g (73%) of 1-(2-chloroethynyl)benzene as a clear oil.

Adapted with permission from D. Sud, T. J. Wigglesworth, and N. R. Branda, *Angew. Chem. Int. Ed.*, **2007**, *46*, 8017–8019. *Copyright (2007) Wiley-VCH Verlag GmbH & Co. KGaA, Weinheim.*

attractive due to health concerns and waste disposal. A classic approach to the synthesis of chloroalkynes involved deprotonation of a terminal acetylene, followed by the addition of a source of electrophilic chlorine. Several versions of this approach have been reported and used to generate a number of chloroalkynes. The combination of NaOH and sodium hypochlorite was used to generate a range of chloroalkynes (Strauss method) [236]. Two of the more interesting practical aspects of this reaction were that it tolerated the presence of a propargyl alcohol and could be carried out using mostly water as the solvent. Another version of this reaction used BuLi as the base and CCl_4 as the halogen source [237]. The yields from this reaction were moderate to good; however, this reaction cannot tolerate the presence of moisture, and there are significant issues due to the use of CCl_4. Furthermore, the scope of this reaction is limited due to the use of Grignard and organolithium reagents.

One of the modifications to this reaction entailed replacing carbon tetrachloride with NCS as the source of electrophilic chlorine (Example 7.34) [238]. This version of the reaction offered greater control over the amount of reactive chlorine that was introduced into the reaction. Although the yield of the chlorination product was increased, the process still suffered from poor functional group tolerance due to the use of organolithium reagents. If no electrophiles are present on the substrate, this is a reliable method for the synthesis of chloroalkynes. One of the valuable substrates successfully prepared using this approach is (chloroethynyl)triethylsilane [239].

A metal-free approach to the synthesis of chloroalkynes has been developed (Scheme 7.151) [240]. Starting from terminal alkynes and using *N*-chlorophthalimide

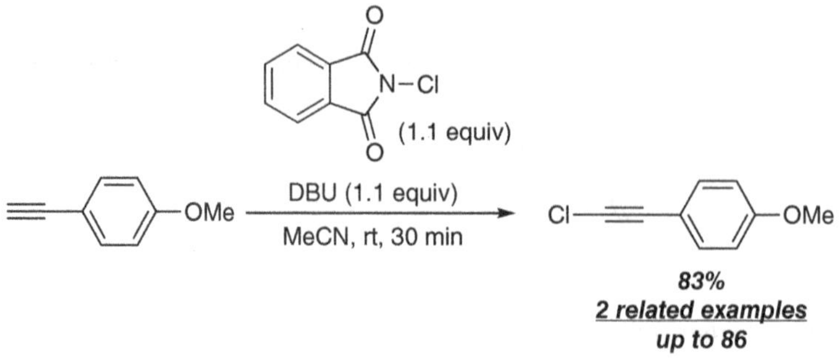

SCHEME 7.151   Synthesis of alkynyl chlorides using *N*-chlorophthalimide [240].

(NCP) as the source of electrophilic chlorine and DBU as the base, excellent yields of the chlorinated alkynes were obtained (up to 86%). The conditions were quite mild and simply required stirring at room temperature in acetonitrile. Similar to other reports, analogous reactions using NCS were unsuccessful [241]. While only three examples were listed, the high yields and mild conditions were attractive.

A remarkably flexible system for the preparation of chloroalkynes using silane-protected alkynes was reported by Szafert [242] (Example 7.35). These silane-protected substrates often have a longer shelf life than terminal alkynes; thus, a procedure to convert these species into chloroalkynes would be attractive to the synthetic community. The authors borrowed a page from the synthesis of bromoalkynes and attempted to convert silane-protected alkynes into chloroalkynes using silver salts and NCS. While quite successful with NBS (for the preparation of bromoalkynes), analogous reactions with NCS were completely unsuccessful, and none of the chlorinated species was observed. Undeterred by this result, they discovered that the addition of a fluoride source to the reaction promoted the chlorination. The reaction conditions were extremely mild (room temperature), and the functional group tolerance was outstanding. The presence of halogens, ethers, alcohols, nitro groups, nitriles, aldehydes, and ketones did not significantly lower the yield of the reaction (>20 examples; up to 98% yield). Only minor electronic effects were observed as substrates bearing either electron-withdrawing or electron-donating groups were success-fully chlorinated. Many of these substrates were challenging to functionalize using the BuLi/Cl$^+$ systems. Terminal alkynes were also able to be chlorinated under the same reaction conditions (Scheme 7.152). Although it would be reasonable to presume that the TBAF was critical for the desilylation reaction and that it would not be needed if the starting compound was a terminal alkyne, the authors determined that none of the chloroalkyne was formed when the TBAF was removed. These results were similar to those mentioned earlier. Essentially very little or no conversion was observed using the AgX/NCS system. When they added the TBAF to these screening reactions, excellent conversions were obtained. While the precise role of the TBAF was not elucidated, it clearly is serving both as a desilylating agent and as a promoter for the chlorination reaction. The system was also able to convert bulky alkynylsilanes into chloroalkynes. Indeed, using 4-ethynylbenzoni-trile protected by the bulky TIPS group as the substrate afforded the chloroalkyne in out-standing yield (96%) after stirring for 4 h at room temperature. The authors also report that the order of addition was critical. The TBAF needed be added last. If the NCS was added after the TBAB, secondary products were observed and the target chloroalkyne was not observed. The versatility of this reaction as well as the ability to carry out the process

Example 7.35   Conversion of a Silane-Protected Alkyne into a Chloroalkyne [242].

Before starting this synthesis, carefully follow the chemical safety instructions listed in Example 7.1.

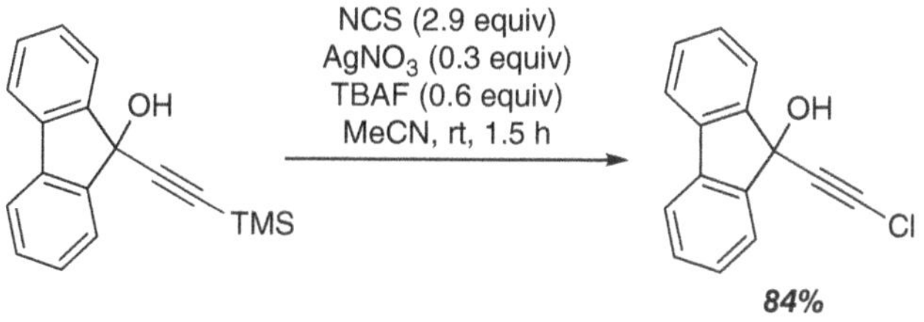

Caution! Hexane, ether, and acetonitrile are highly flammable.
A sample of ((trimethylsilyl)ethynyl)-9*H*-fluoren-9-ol (0.193 g, 0.693 mmol) was dissolved in acetonitrile (10 mL). NCS (0.271 g, 1.99 mmol), AgNO₃ (0.035 g, 0.21 mmol), and TBAF (0.42 mL, 0.42 mmol, 1 M in THF) were added, and the mixture was stirred for 1.5 h. After this time, solvent was removed under reduced pressure, and crude product was purified by filtration trough a short silica gel plug (hexane/Et₂O, 1 : 1, v/v) to afford the chloroalkyne (0.140 g, 0.582 mmol, 84%) as a yellow oil.

Adapted with permission from N. Gulia, B. Pigulski, M. Charewicz, and S. Szafert, *Chem. Eur. J.*, **2014**, *20*, 2746–2749. *Copyright (2014) Wiley-VCH Verlag GmbH & Co. KGaA, Weinheim.*

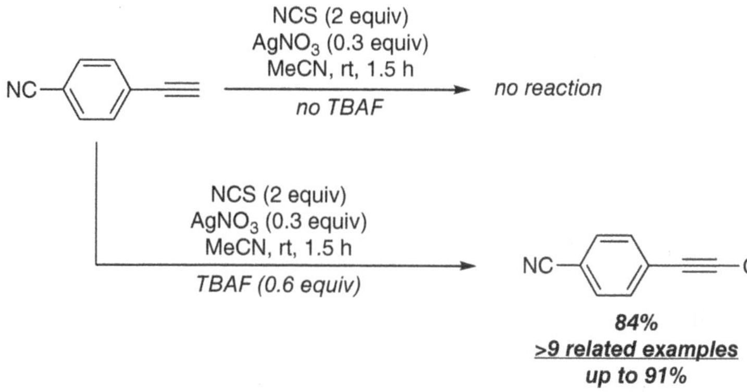

SCHEME 7.152   Dependence of the chlorination reaction on TBAF [242].

at room temperature with minimal operational complexity makes this a very attractive approach for the synthesis of these compounds.

The use of bromoalkynes as substrates and intermediates in organic transformations is significantly more widespread than chloroalkynes. Similar to the chloroalkynes, some of the earliest examples of bromoalkynes were generated using alkynylmercury intermediates [235]. Naturally, this process still suffers from toxicity and waste issues and has not been a popular choice for the synthesis of these compounds. The Strauss method and its modifications [243] have been considerably more widely adopted, and a number of bromoalkynes have been prepared by versions of this approach [236]. This chemistry is remarkably high yielding (>90%)

Example 7.36   Preparation of Bromoalkynes Using KOH/Br$_2$ in Water [244].

Before starting this synthesis, carefully follow the chemical safety instructions listed in Example 7.1.

91%
>2 related examples
up to 95%

Caution! Diethyl ether is highly flammable.

Caution! Br$_2$ is a hazardous liquid that must be handled with extreme care.
To a cold (−5 to 0 °C) solution of KOH (111.5 g, 1.75 mol, 7 equiv) in water (440 mL) was added pure bromine (42.4 g, 250 mmol, 1 equiv) slowly in a hood. After 15 min, propargyl alcohol (19.5 mL, 335 mmol, 1.34 equiv) was added dropwise via an addition funnel to the light orange solution while maintaining the temperature between 0 and 5 °C. After 30 min of stirring at 0 °C, the reaction was allowed to reach room temperature, and the product was extracted several times with diethyl ether (4 × 100 mL). Solvent removal afforded a pure product in 91% yield.

Adapted with permission from J. P. Marino and H. N. Nguyen, *J. Org. Chem.*, **2002**, *67*, 6841–6844. *Copyright (2002) American Chemical Society.*

when the substrate contains polar functional groups such as alcohols or dimethylamines (Example 7.36) [244]. BuLi has also been used as the base in this reaction [245].

While most of the protocols leading to the preparation of bromoalkynes begin with a terminal alkyne or a silane-protected alkyne, propiolic acids can also be used in these reactions (Hunsdiecker reaction) [172]. Several versions of this decarboxylative bromination reaction have been reported under various conditions. One version of this reaction used tetrabutylammonium trifluoroacetate (TBAFTFA) along with NBS to generate the bromoalkynes (Scheme 7.153) [246]. The conditions of the reaction were remarkably mild, and moderate to good yields of the target compounds were obtained after simply stirring the reagents at room temperature. Curiously, attempts to prepare chloroalkynes using this approach were unsuccessful, and only starting materials were recovered. The same group later reported that catalytic amounts of triethylamine in combination with NBS served as an effective approach to the synthesis of bromoalkynes (Scheme 7.154) [172]. One of the attractive aspects of this approach was that it was tolerant to small amounts of water. This increased the operational simplicity of the reaction since anhydrous solvents were not required.

Arguably, the most widely adopted approach for the synthesis of bromoalkynes uses silver nitrate in combination with NBS to convert terminal alkynes into bromoalkynes (Scheme 7.155) [241]. The operational simplicity of this approach is extremely high since simply adding everything to the flask and stirring for less than 1 h at room temperature often generate high yields of the brominated product. Common solvents are acetone and acetonitrile. The functional group tolerance of this approach is outstanding, and even substrates bearing unprotected alcohols as well as esters have been successfully converted into bromoalkynes [241, 247–250]. Substrates bearing these groups would have been

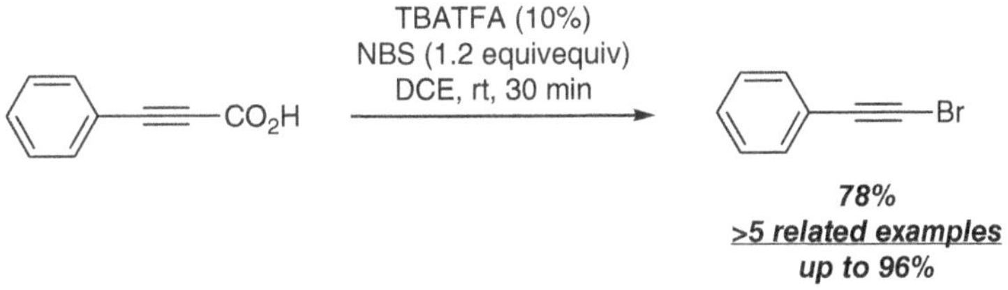

SCHEME 7.153   Hunsdiecker synthesis of bromoalkynes using TBAFTFA/NBS [246].

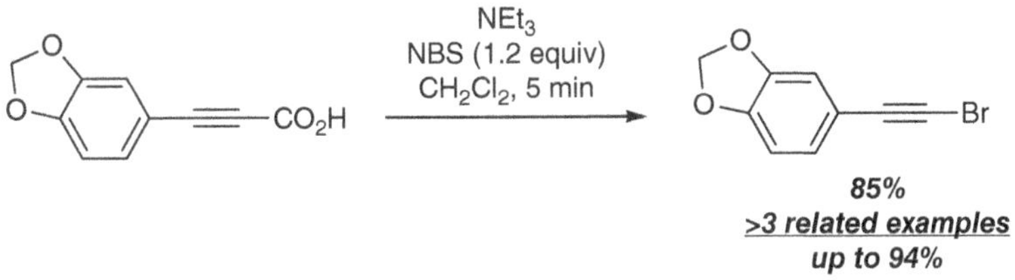

SCHEME 7.154   Triethylamine-promoted Hunsdiecker synthesis of bromoalkynes [172].

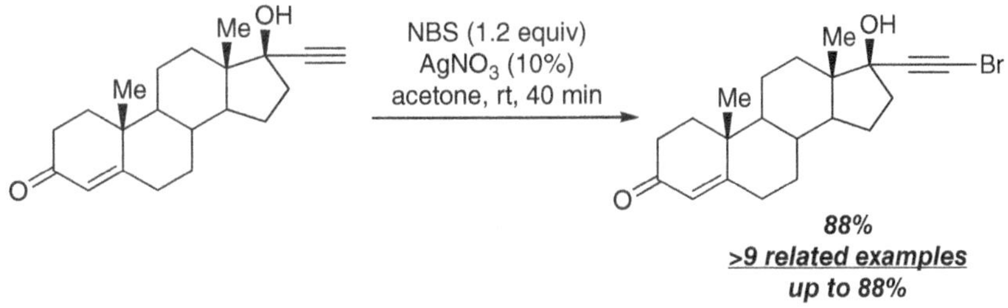

SCHEME 7.155   Preparation of bromoalkynes using the NBS/AgNO$_3$ [241].

challenging to functionalize using the BuLi/NBS approach described earlier. A polar group is not required for this chemistry, and pure hydrocarbons such as 1-hexyne have been successfully converted into bromoalkynes using this approach [251]. Given the broad scope of this reaction as well as the availability of many precursors, this approach might be the most effective synthesis of bromoalkynes.

In addition to the silver-promoted chemistry, gold compounds have shown a propensity to catalyze the conversion of terminal alkynes into bromoalkynes (Scheme 7.156) [252]. The authors screened a range of catalysts for activity towards the halogenation reaction and found that ^{t}Bu$_3$PAuNTf$_2$ was the most active. The conditions for this reaction were mild (room temperature), and the yields were outstanding (up to 95%).

A metal-free approach using DBU as a base and NBS as the source of bromine has been reported and generated excellent yields of the bromoalkyne (up to 99%; Example 7.37) [240]. The chemistry was remarkably fast at room temperature (1 min) in acetonitrile. Other solvents such as CH$_2$Cl$_2$ and DMF were effective as well; however, the use of more nonpolar solvents such as toluene and THF afforded lower yields of the bromoalkyne. This chemistry was also tolerant to the presence of a number of functional

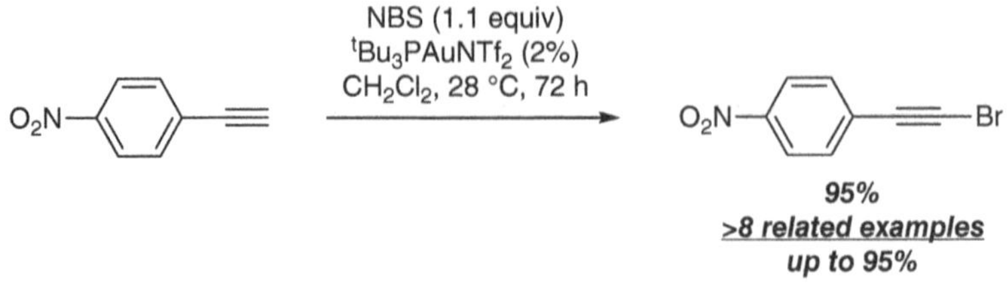

SCHEME 7.156   Gold-catalyzed bromination of terminal alkynes [252].

Example 7.37   Preparation of Bromoalkynes Using DBU/NBS [240].

Before starting this synthesis, carefully follow the chemical safety instructions listed in Example 7.1.

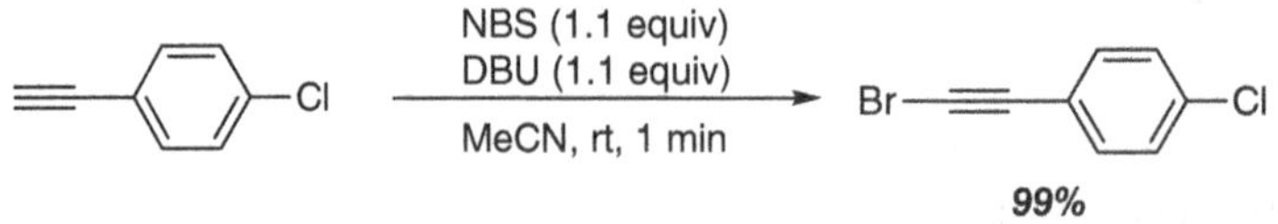

Caution! Petroleum ether and acetonitrile are highly flammable!
To a solution of 1-chloro-4-ethynylbenzene (136.6 mg, 1.0 mmol) in MeCN (2.0 mL) were added NBS (195.8 mg, 1.1 mmol) and DBU (0.159 mL, 1.1 mmol). The mixture was stirred at room temperature for 1 min. The reaction mixture was poured into water and then extracted with CH_2Cl_2 (3×10 mL). The combined organic phase was washed with water (3×10 mL), filtered, and concentrated under reduced pressure. The crude product was purified by flash chromatography (silica gel, petroleum ether as eluent) to give the bromoalkyne (213 mg, 99%) as a white solid.

Adapted from M. Li, Y. Li, B. Zhao, F. Liang, L.-Y. Jin, *RSC Adv.*, **2014**, *4*, 30046–30049 *with permission of The Royal Society of Chemistry.*

groups such as halogens, ethers, and heteroaromatic substrates. This reaction is trivial to setup and carryout, metal-free, and generates the target compounds quickly.

In addition to the use of terminal alkynes and propiolic acids as substrates for the synthesis of haloalkynes, silane-protected alkenes can also be converted into bromoalkynes in a one-pot reaction (Scheme 7.157 and Example 7.38) [253, 254]. This desilylation/bromination reaction uses silver(I) fluoride to aid in the removal of the silicon protecting group and NBS as a source of bromine. A range of functional groups were well tolerated by this chemistry, and the reaction conditions were mild. While the obvious choice might be to start from a terminal alkyne, these silane-protected compounds typically have a long shelf life and are more tolerant to chemistry performed on other parts of the substrate. The ability to convert them into the halogenated compounds in a one-pot reaction remains the biggest practical advantage of this approach.

The conversion of alkynyltrifluoroborate salts into alkynyl bromides has been achieved using chloramine-T and sodium bromide in aqueous THF (Scheme 7.158) [129]. In addition to the alkynyltrifluoroborate salts derived from arylalkynes, nonaromatic alkynyltrifluoroborate substrates were also successfully functionalized in high yield. Simply stirring at room temperature for 20 min afforded excellent yields of the alkynyl bromide.

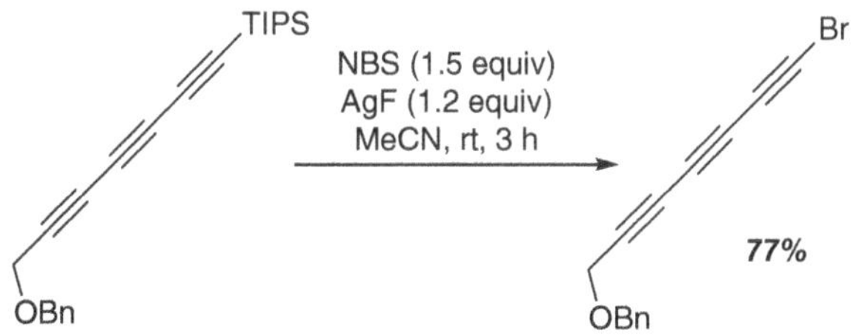

SCHEME 7.157   Desilylation/bromination using AgF [253].

Example 7.38   Synthesis of a Bromoalkyne from a TIPS-Protected Alkyne [254].

Before starting this synthesis, carefully follow the chemical safety instructions listed in Example 7.1.

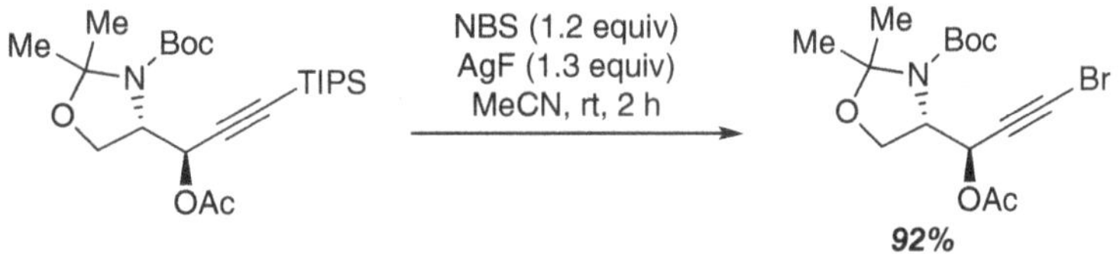

Caution! Hexane, acetonitrile and ethyl acetate are highly flammable.
To a solution of the TIPS-protected alkyne (1.80 g, 3.97 mmol) in acetonitrile (40 mL) were added NBS (848 mg, 4.76 mmol) and AgF (640 mg, 5.04 mmol) in the dark. The reaction mixture was stirred at room temperature for 2 h and then filtered through a pad of Celite. The filtrate was diluted with EtOAc, washed with water, dried over MgSO$_4$, filtered, and concentrated in vacuo. The resulting residue was purified by silica gel column chromatography (hexane/EtOAc, 8 : 1) to give the bromoalkyne (1.37 g, 92%) as a waxy solid.
Note: The author's comment on adding the NBS and the silver salt in the dark should be strictly heeded for best results.

Adapted with permission from S. Kim, Lee, Y. M.; Kang, H. R.; Cho, J.; Lee, T.; and Kim, D. *Org. Lett.*, **2007**, *9*, 2127–2130. *Copyright (2007) American Chemical Society.*

Given their widespread applications in organic synthesis, a wide range of approaches for the preparation of iodoalkynes have been reported [172, 176, 209, 255–259]. Similar to the preparations of bromoalkynes, one of the classic approaches to the synthesis of iodoalkynes involves the treatment of a terminal alkyne with a base, followed by the addition of a source of iodine [255]. The bases typically used for this reaction are commonly Grignard or organolithium reagents. Typically, a terminal alkyne is dissolved in a solvent such as THF, cooled to −78 °C and treated with the Grignard reagent followed by elemental iodine. After warming to room temperature and purification, moderate to excellent yields are typically obtained. Normally the use of such powerful bases precludes the presence of electrophiles in the substrates. However, it should be noted that a number of electrophiles are tolerant to a select group of bases at low temperature. In these examples, as long as the stoichiometry of the base is carefully controlled and any excess is quenched prior to warming, the synthesis should be successful.

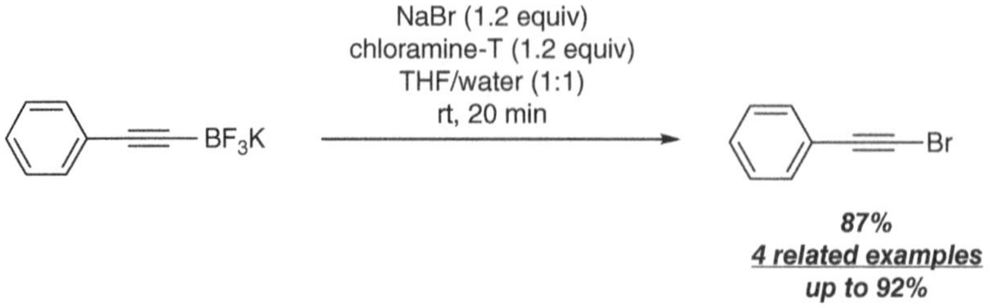

SCHEME 7.158   Conversion of alkynyltrifluoroborate salts into alkynyl bromides [129].

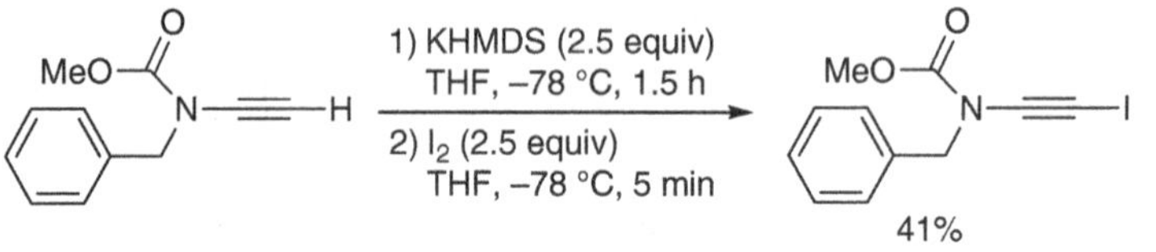

SCHEME 7.159   Iodination of ynamides using KHMDS/I$_2$ [260].

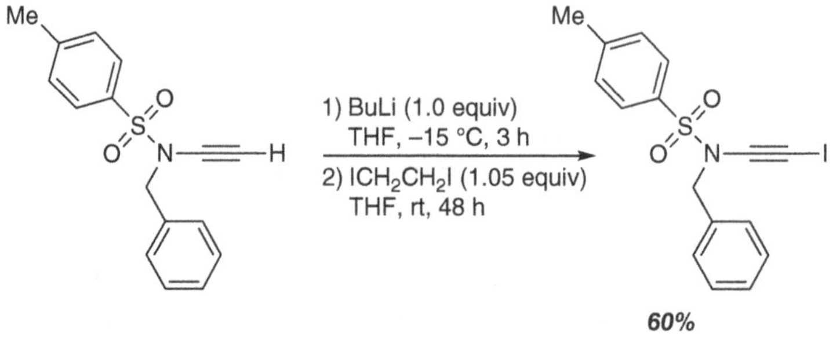

SCHEME 7.160   Iodination of ynamides using BuLi/I$_2$ [260].

An example of this chemistry was focused on the iodination of ynamides (Schemes 7.159 and 7.160) [260]. Although ynamides are typically presumed to be intolerant to organolithium or Grignard reagents, the authors were able to successfully deprotonate the alkyne while leaving the ynamide fragment intact by keeping the temperature at −78 °C for the deprotonation step as well as for the addition of elemental iodine. The authors also used a bulky base (KHMDS) that has poor nucleophilic character for the deprotonation step. It should be noted that this hindered base was not the only factor in the development of a successful process as the much more nucleophilic BuLi was also successfully used to deprotonate a sulfone containing ynamide without significant secondary reactions. Furthermore, the BuLi chemistry was quite successful despite only lowering the temperature to −15 °C. It should be noted that the authors initially attempted the synthesis of these iodinated ynamides using silver nitrate and NIS. This chemistry was unsuccessful and prompted the search for a different route. Once isolated, the resulting iodinated ynamides could be stored in DCM for several weeks and could be purified by column chromatography (silica gel), provided that the silica was pretreated with NEt$_3$.

The Hunsdiecker decarboxylative iodination is also a successful approach for the synthesis of iodoalkynes (Scheme 7.161 and Example 7.39) [172, 246]. The chemistry is

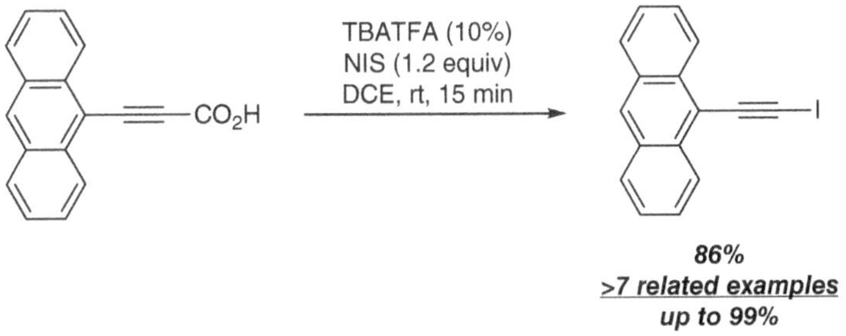

SCHEME 7.161 Synthesis of iodoalkynes using TBATFA-promoted decarboxylation [246].

Example 7.39 Triethylamine-Assisted Hunsdiecker Reaction for the Synthesis of Iodoalkynes [172].

Before starting this synthesis, carefully follow the chemical safety instructions listed in Example 7.1.

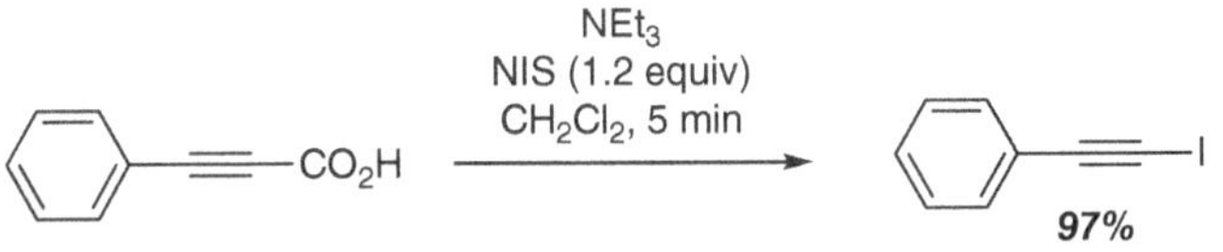

Caution! Ethyl acetate and hexane are highly flammable.
Triethylamine (28 μL) was added to a solution of substituted propiolic acid (1 mM) in 3 mL dichloromethane. After the mixture was stirred for 5 min at room temperature, N-iodosuccinimide (1.2 mM) was added. After 5 min, solvent was removed under reduced pressure, and the mixture was subjected to column chromatography (silica gel 60–120 mesh, eluent 1% ethyl acetate/hexane) to afford 1-iodoalkynes.

Adapted with permission from J. P. Das and S. Roy, J. Org. Chem., 2002, 67, 7861–7864. Copyright (2002) American Chemical Society.

typically promoted by either TBATFA or triethylamine and affords excellent yields of the iodoalkynes in less than 20 min at room temperature. If a carboxylic acid is a convenient starting point, this is a very effective approach to the synthesis.

One of the transition metal-free approaches to the preparation of iodoalkynes uses *tert*-butyl hydroperoxide in combination with potassium iodide as the oxidizing–iodinating system (Scheme 7.162) [119]. There is no need for prefunctionalization of the alkyne since terminal alkynes worked well. Although the presence of an oxidizing agent could be problematic for some substrates, primary alcohols were successfully functionalized using this reagent with minimal oxidation. Another advantage of this system was the atom economic nature of the addition with respect to the iodide. Only 1 equivalent of "I" was needed for this transformation. Some of the other routes to these compounds lose equivalents of iodide either in the reaction itself or in the preparation of iodinating agents.

One of the most practical approaches to the synthesis of iodoalkynes is a copper-catalyzed iodination of terminal alkynes using N-iodomorpholine-hydrogen iodide (Example 7.40) [261, 262]. The reaction conditions were extremely mild, and no Grignard reagents are needed. As a result, the functional group tolerance was very high. The yields

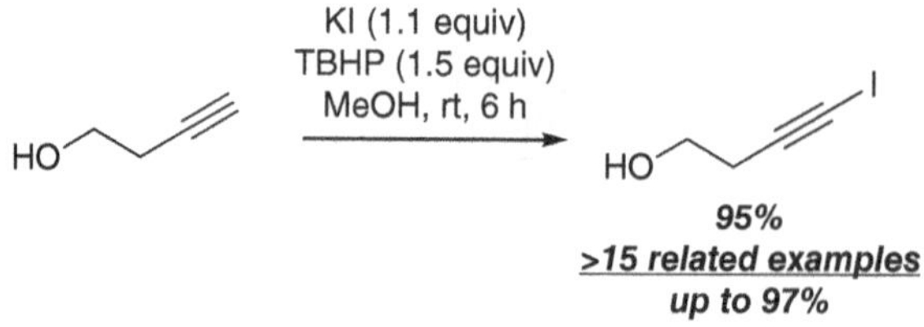

SCHEME 7.162   Synthesis of iodoalkynes using potassium iodide as the halogen source [119].

Example 7.40   Preparation of Iodoalkynes Using *N*-Iodomorpholine-Hydrogen Iodide [261].

Before starting this synthesis, carefully follow the chemical safety instructions listed in Example 7.1.

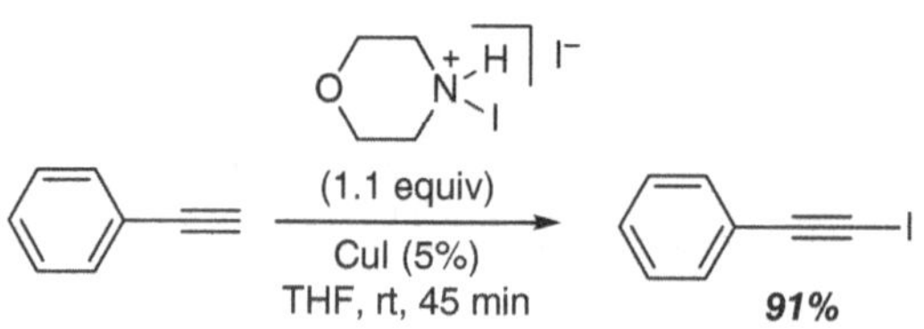

Caution! THF is highly flammable.

Step 1: Preparation of *N*-Iodomorpholine-Hydrogen Iodide

A solution of iodine (25.40 g, 0.10 mol) in MeOH (400 mL) was treated dropwise with morpholine (8.71 mL, 0.10 mol). On addition, the solution rapidly changed from dark purple brown to light orange, and a fine orange precipitate was generated. The solution was stirred for approximately 45 min, and then solid was isolated by filtration. The solid was transferred to a round-bottom flask and dried under vacuum. Once the material reached a free-flowing consistency, it was placed in a plastic bottle and stored in the fridge. This procedure gave *N*-iodomorpholine-hydrogen iodide as an orange crystalline powder (30.34 g, 0.09 mol, 89%), which was used without further purification or characterization.

Step 2: Preparation of 1-Iodo-Phenylacetylene

Phenylacetylene (8.17 g, 80.00 mmol) was dissolved in THF (200 mL) and treated with CuI (0.762 g, 4.00 mmol) and *N*-iodomorpholine (30.00 g, 88.00 mmol). The reaction mixture was stirred at room temperature for 45 min, after which a fine white precipitate had formed. The suspension was poured onto a pad of neutral alumina (400 mL), and the filtrate was collected under vacuum. The solid phase was washed with DCM (4 × 100ml), and the combined organic fractions were pooled and concentrated by evaporation, giving 1-iodo-phenylacetylene (16.61 g, 72.82 mmol, 91%) as a yellow oil.

Adapted with permission from J. E. Hein, J. C. Tripp, L. B. Krasnova, K. B. Sharpless, and V. V. Fokin, *Angew. Chem. Int. Ed.*, **2009**, *48*, 8018–8021. *Copyright (2009) Wiley-VCH Verlag GmbH & Co. KGaA, Weinheim.*

of the iodoalkynes were good to excellent (>10 examples, up to 96%). Although this chemistry required the synthesis of the *N*-iodomorpholine-hydrogen iodide, the preparation of this reagent was trivial. Furthermore, the operational simplicity of this approach was very high. Simply stirring the reagents at room temperature followed by a filtration and removal of the volatiles afforded the iodoalkynes. Even with the requisite synthesis of the iodinating agent, this is one of the most attractive approaches to the synthesis of iodoalkynes.

Terminal alkynes remain one of the most common starting materials for the preparation of iodoalkynes. To increase the variety of starting materials that could be used for the preparation of iodoalkynes, a one-pot conversion of benzylic or allylic halides into iodoalkynes was developed (Example 7.41) [263]. These are attractive starting materials since a wide array of them are commercially available. The number of steps needed for the synthesis

Example 7.41   Conversion of a Benzyl Halide into an Iodoalkyne [263].

Before starting this synthesis, carefully follow the chemical safety instructions listed in Example 7.1.

Caution! THF, diethyl ether, 2-propanol, and hexanes are highly flammable.
Note: The authors report that the iodoalkynes are sensitive to light and need to be stored in the dark and should be stored at −20 °C in the dark.

To a flame-dried and argon-flushed 25 mL round-bottom flask equipped with a septum and a magnetic stir bar was added solid NaHMDS (550.1 mg, 3.0 mmol, 3.0 equiv). The flask was covered with aluminum foil, and NaHMDS was solubilized with 2.2 mL of anhydrous THF and 1.0 mL of anhydrous Et$_2$O. The flask was then cooled to −78 °C using an acetone/dry ice cooling bath and stirred for 2 min. To another separate flame-dried and argon-flushed 10 mL round-bottom flask equipped with a septum and a magnetic stir bar was added iodoform (590.5 mg, 1.5 mmol, 1.5 equiv), and it was solubilized in 1.0 mL of anhydrous THF. The yellow solution was then added dropwise via syringe to the cooled solution of NaHMDS (upon addition of a few drops of the iodoform solution, the reaction solution changed drastically from a slightly yellow solution to a black suspension). Upon completion of the addition, the iodoform flask was rinsed with THF (200 μL), and the solution was transferred to the reaction. The reaction was then slowly warmed to −20 °C using a 1 : 1 *i*-PrOH : H$_2$O/dry ice cooling bath and stirred for 1 h at −20 °C. To another separate flame-dried and argon-flushed 10 mL round-bottom flask equipped with a septum and a magnetic stir bar was added the benzyl bromide (1.0 mmol, 1.0 equiv). The bromide was solubilized in 1.0 mL of anhydrous THF and stirred at room temperature for 2 min.

The solution of bromide was added dropwise to the reaction using a syringe. Upon completion of the addition, the bromide flask was rinsed with THF (~200 μL), and the solution was transferred to the reaction. The reaction was stirred at −20 °C for 5 h. Then, KOtBu (112.2 mg, 1.0 mmol, 1.0 equiv) was added to the reaction, and the reaction was stirred from −20 °C to room temperature overnight. The reaction was diluted with 8 mL of a 1 : 1 mixture of Et$_2$O/hexanes, and the black/brown suspension was filtered on a pad of diatomaceous earth with a sintered funnel under water-mediated vacuum. The pad was rinsed with another 16 mL of a 1 : 1 mixture of Et$_2$O/hexanes. The crude filtrate was then evaporated to dryness under reduced pressure and purified by column chromatography (using a gradient of 100% hexanes to 5% EtOAc/hexanes) to afford the iodoalkyne (217.0 mg, 69%).

Adapted with permission from G. Pelletier, S. Lie, J. J. Mousseau, and A. B. Charette, *Org. Lett.*, **2012**, *14*, 5464–5467. *Copyright (2012) American Chemical Society.*

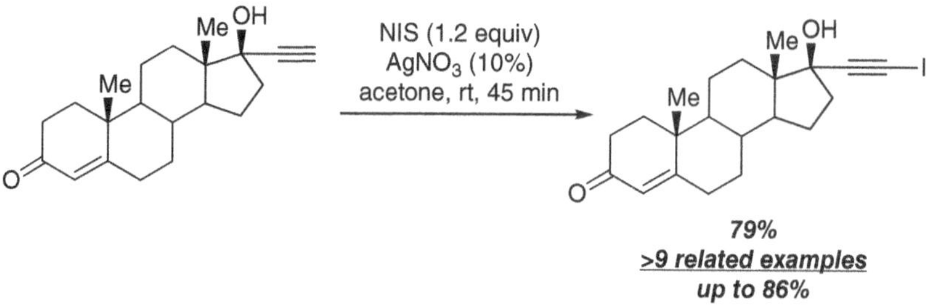

SCHEME 7.163   Synthesis of iodoalkynes using AgNO$_3$/NIS [241].

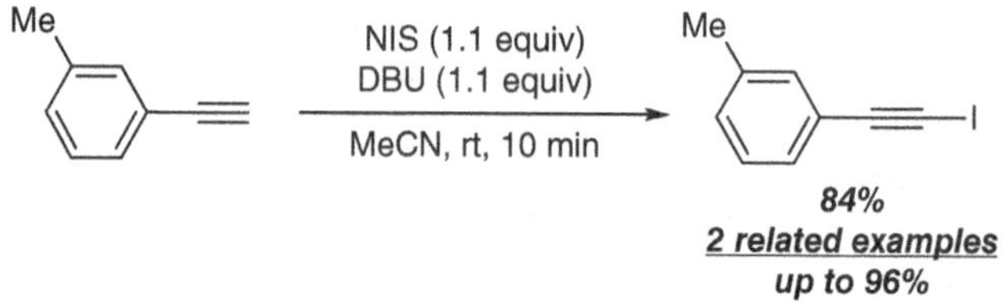

SCHEME 7.164   Synthesis of haloalkynes using DBU/NXS combinations [240].

increases the operational complexity; however, they are all injections into the same flask. The authors did note that best results were obtained with organohalides that were either recently obtained from commercial sources or freshly prepared.

The synthesis of iodoalkynes has also been achieved using silver nitrate and NIS (Scheme 7.163) [241]. This chemistry has broad scope and is operationally trivial to setup and carryout. A wide range of functional groups including alcohols have been tolerated by this approach. The reactions are quite fast and take advantage of all commercially available reagents. If this chemistry is an option, it is an excellent first attempt at the synthesis of an iodoalkyne. The conversion of alkynyl silanes into alkynyl halides has also been accomplished [259].

The metal-free conversion of terminal alkynes into iodoalkynes has been achieved using DBU as a base along with NIS (Scheme 7.164) [240]. The chemistry was quite fast, and outstanding yields of the iodoalkynes were obtained in 1 min at room temperature. Given the trivial nature of the setup as well as the metal-free aspect of the chemistry, this is an attractive approach to the synthesis of these compounds. In a variation on this approach, DMAP was successfully used as the base along with elemental iodine as the iodinating agent [264].

Alkynyl iodides have also been prepared through a two-step process beginning with aldehydes [214]. The reaction proceeded through the initial generation of a *gem*-diiodoalkene through the addition of iodoform, triphenylphosphine, and potassium *tert*-butoxide to a solution of the aldehyde. Once the diiodo complex was formed, addition of an excess of KOtBu at −78 °C afforded the iodoalkynes in moderate to excellent yields (up to 97%). Due the use of such strong bases, the functional group tolerance is low, but the operational simplicity is high since it is a one-pot synthesis.

7.4.1 Troubleshooting the Synthesis of Alkynyl Halides and Related Compounds

While it will not be possible to anticipate all of the needs of the synthetic chemist over the next few decades, this list will provide some insight into issues commonly encountered when trying to prepare alkynyl halides. This chemistry can be particularly confusing as several protocols will work for a couple of the halogens, but not for all of them. Care must be taken to make sure that the chemistry will be successful for the substrate under investigation.

Problem/Issue/Goal	Possible Solution/Suggestion
Chloroalkyne: highly functionalized substrates that are sensitive to bases/nucleophiles such as Grignard reagents	The silver nitrate/NCS/TBAF system would be a reasonable approach for a terminal alkyne or a TMS-protected alkyne [242]
Chloroalkyne: need a metal-free approach	If starting from a terminal alkyne, the DBU/NCP system would be a good place to start [240]
Chloroalkyne: substrate has an alcohol that would be challenging or time consuming to protect/deprotect	Several of these approaches are quite tolerant to the presence of alcohols and other electrophiles [242, 247]
Chloroalkyne: need to use mostly water as the solvent	If starting from a terminal alkyne, the combination of sodium hypochlorite and sodium hydroxide has been successful [236]
Chloroalkyne: need to start from a silane-protected alkyne	Consider AgNO$_3$/NCS combination with the addition of TBAF [242]
Bromoalkyne: highly functionalized substrates containing groups that are sensitive to bases/nucleophiles	The silver nitrate/NBS combination has shown great tolerance to a wide variety of functional groups [242]. The typical starting point is a terminal alkyne
Bromoalkyne: substrate will not dissolve without water present	Several of the methods are tolerant to water. Consider an Et$_3$N-promoted Hunsdiecker decarboxylation [172]
Bromoalkyne: need to use mostly water as the solvent	The Strauss approach would be attractive for this chemistry [244]
Bromoalkyne: would be convenient to start from a carboxylic acid	Treatment of the carboxylic acid with TBAFTFA/NBS in DCE [246] or simply with NEt$_3$/NBS have been reported [172]

(Continued)

(Continued)

Problem/Issue/Goal	Possible Solution/Suggestion
Bromoalkyne: substrate has an alcohol that would be challenging or time consuming to protect/deprotect	A number of the protocols are tolerant to the presence of alcohols in the substrate. NBS/AgNO$_3$ has been shown to generate the bromoalkynes in excellent yields [250]
Bromoalkyne: needs a metal-free approach	If starting from a terminal alkyne, the DBU/NBS system would be a good place to start [240]
Bromoalkyne: need to start from a silane-protected alkyne	The AgF/NBS system has been shown to be effective for these substrates [253, 254]
Iodoalkyne: highly functionalized substrates that are sensitive to bases/nucleophiles such as Grignard reagents	The silver nitrate/NCS/TBAF system would be a reasonable approach for a terminal alkyne or a TMS-protected alkyne [242]
Iodoalkyne: need an easy and fast protocol for workup and purification since iodoalkynes can degrade in the light	While several of the approaches are fast, the copper-catalyzed iodination of terminal alkynes using *N*-iodomorpholine-hydrogen iodide is fast and requires only minimal purifications to isolate pure material [261]
Iodoalkyne: substrate will not dissolve without a small amount of water present	Several of the methods are tolerant to water. Consider an Et$_3$N-promoted Hunsdiecker decarboxylation [172]
Iodoalkyne: need to use mostly water as the solvent	The Strauss approach would be attractive for this chemistry [236, 244]
Iodoalkyne: starting from a terminal alkyne	The copper-catalyzed iodination using *N*-iodomorpholine-hydrogen iodide is a good place to start. No strong bases or harsh conditions are required [261]
Iodoalkyne: it would be convenient to start from a carboxylic acid	Treatment of the carboxylic acid with TBAFTFA/NIS in DCE [246] or simply with NEt$_3$/NIS have been reported [172]
Iodooalkyne: need a metal-free approach	If starting from a terminal alkyne, the DBU/NIS system would be a good place to start [240]

REFERENCES

[1] Harsanyi, A.; Sandford, G. *Org. Process Res. Dev.* **2014**, *18*, 981–992.

[2] Campbell, M. G.; Ritter, T. *Org. Process Res. Dev.* **2014**, *18*, 474–480.

[3] Denmark, S. E.; Kuester, W. E.; Burk, M. T. *Angew. Chem. Int. Ed.* **2012**, *51*, 10938–10953.

[4] da Silva, F. M.; Jones, J. J.; de Mattos, M. C. S. *Curr. Org. Synth.* **2005**, *2*, 393–414.

[5] Britton, R.; Kang, B. *Nat. Prod. Rep.* **2013**, *30*, 227–236.

[6] Liang, T.; Neumann, C. N.; Ritter, T. *Angew. Chem. Int. Ed.* **2013**, *52*, 8214–8264.

[7] Appel, R. *Angew. Chem. Int. Ed.* **1975**, *14*, 801–811.

[8] Dang, H.; Mailig, M.; Lalic, G. *Angew. Chem. Int. Ed.* **2014**, *53*, 6473–6476.

[9] Reddy, D. S.; Shibata, N.; Nagai, J.; Nakamura, S.; Toru, T.; Kanemasa, S. *Angew. Chem. Int. Ed.* **2008**, *47*, 164–168.

[10] Kitamura, T.; Muta, K. *J. Org. Chem.* **2014**, *79*, 5842–5846.

[11] Zhang, Z.; Wang, F.; Mu, X.; Chen, P.; Liu, G. *Angew. Chem. Int. Ed.* **2013**, *52*, 7549–7553.

[12] Hull, K. L.; Anani, W. Q.; Sanford, M. S. *J. Am. Chem. Soc.* **2006**, *128*, 7134–7135.

[13] Bloom, S.; McCann, M.; Lectka, T. *Org. Lett.* **2014**, *16*, 6338–6341.

[14] Cantillo, D.; de Frutos, O.; Rincon, J. A.; Mateos, C.; Kappe, C. O. *J. Org. Chem.* **2014**, *79*, 8486–8490.

[15] Pérez-Perarnau, A.; Preciado, S.; Palmeri, C. M.; Moncunill-Massaguer, C.; Iglesias-Serret, D.; González-Gironès, D. M.; Miguel, M.; Karasawa, S.; Sakamoto, S.; Cosialls, A. M.; Rubio-Patiño, C.; Saura-Esteller, J.; Ramón, R.; Caja, L.; Fabregat, I.; Pons, G.; Handa, H.; Albericio, F.; Gil, J.; Lavilla, R. *Angew. Chem. Int. Ed.* **2014**, *53*, 10150–10154.

[16] Pitts, C. R.; Ling, B.; Woltornist, R.; Liu, R.; Lectka, T. *J. Org. Chem.* **2014**, *79*, 8895–8899.

[17] Alvarado, J.; Herrmann, A. T.; Zakarian, A. *J. Org. Chem.* **2014**, *79*, 6206–6220.

[18] Yang, X.; Phipps, R. J.; Toste, F. D. *J. Am. Chem. Soc.* **2014**, *136*, 5225–5228.

[19] McMurtrey, K. B.; Racowski, J. M.; Sanford, M. S. *Org. Lett.* **2012**, *14*, 4094–4097.

[20] Zi, W.; Wang, Y.-M.; Toste, F. D. *J. Am. Chem. Soc.* **2014**, *136*, 12864–12867.

[21] Lu, D.-F.; Liu, G.-S.; Zhu, C.-L.; Yuan, B.; Xu, H. *Org. Lett.* **2014**, *16*, 2912–2915.

[22] Cui, J.; Jia, Q.; Feng, R. Z.; Liu, S.-S.; He, T.; Zhang, C. *Org. Lett.* **2014**, *16*, 1442–1445.

[23] Bloom, S.; Pitts, C. R.; Miller, D. C.; Haselton, N.; Holl, M. G.; Urheim, E.; Lectka, T. *Angew. Chem. Int. Ed.* **2012**, *51*, 10580–10583.

[24] Pitts, C. R.; Bloom, S.; Woltornist, R.; Auvenshine, D. J.; Ryzhkov, L. R.; Siegler, M. A.; Lectka, T. *J. Am. Chem. Soc.* **2014**, *136*, 9780–9791.

[25] Halperin, S. D.; Fan, H.; Chang, S.; Martin, R. E.; Britton, R. *Angew. Chem. Int. Ed.* **2014**, *53*, 4690–4693.

[26] Talbot, E. P. A.; de A. Fernandes, T.; McKenna, J. M.; Toste, F. D. *J. Am. Chem. Soc.* **2014**, *136*, 4101–4104.

[27] Li, Z.; Wang, Z.; Zhu, L.; Tan, X.; Li, C. *J. Am. Chem. Soc.* **2014**, *136*, 16439–16443.

[28] Shigehisa, H.; Nishi, E.; Fujisawa, M.; Hiroya, K. *Org. Lett.* **2013**, *15*, 5158–5161.

[29] Yang, Q.; Mao, L.-L.; Yang, B.; Yang, S.-D. *Org. Lett.* **2014**, *16*, 3460–3463.

[30] St. Denis, J. D.; Zajdlik, A.; Tan, J.; Trinchera, P.; Lee, C. F.; He, Z.; Adachi, S.; Yudin, A. K. *J. Am. Chem. Soc.* **2014**, *136*, 17669–17673.

[31] Swatschek, J.; Grothues, L.; Bauer, J. O.; Strohmann, C.; Christmann, M. *J. Org. Chem.* **2014**, *79*, 976–983.

[32] Romanov-Michailidis, F.; Pupier, M.; Guenee, L.; Alexakis, A. *Chem. Commun.* **2014**, *50*, 13461–13464.

[33] Li, H.; Zhang, F.-M.; Tu, Y.-Q.; Zhang, Q.-W.; Chen, Z.-M.; Chen, Z.-H.; Li, J. *Chem. Sci.* **2011**, *2*, 1839–1841.

[34] Romanov-Michailidis, F.; Guenee, L.; Alexakis, A. *Org. Lett.* **2013**, *15*, 5890–5893.

[35] Zhang, H.; Song, Y.; Zhao, J.; Zhang, J.; Zhang, Q. *Angew. Chem. Int. Ed.* **2014**, *53*, 11079–11083.

[36] Yin, F.; Wang, Z.; Li, Z.; Li, C. *J. Am. Chem. Soc.* **2012**, *134*, 10401–10404.

[37] Mizuta, S.; Stenhagen, I. S. R.; O'Duill, M.; Wolstenhulme, J.; Kirjavainen, A. K.; Forsback, S. J.; Tredwell, M.; Sandford, G.; Moore, P. R.; Huiban, M.; Luthra, S. K.; Passchier, J.; Solin, O.; Gouverneur, V. *Org. Lett.* **2013**, *15*, 2648–2651.

[38] Danilkina, N. A.; Kulyashova, A. E.; Khlebnikov, A. F.; Brase, S.; Balova, I. A. *J. Org. Chem.* **2014**, *79*, 9018–9045.

[39] Lu, H.; Zhang, F.-M.; Pan, J.-L.; Chen, T.; Li, Y.-F. *J. Org. Chem.* **2014**, *79*, 546–558.

[40] Dai, C.; Narayanam, J. M. R.; Stephenson, C. R. J. *Nat. Chem.* **2011**, *3*, 140–145.

[41] Sandler, S. R. *J. Org. Chem.* **1970**, *35*, 3967–3968.

[42] De Luca, L.; Giacomelli, G.; Porcheddu, A. *Org. Lett.* **2002**, *4*, 553–555.

[43] Pouliot, M.-F.; Mahe, O.; Hamel, J.-D.; Desroches, J.; Paquin, J.-F. *Org. Lett.* **2012**, *14*, 5428–5431.

[44] Denton, R. M.; An, J.; Adeniran, B.; Blake, A. J.; Lewis, W.; Poulton, A. M. *J. Org. Chem.* **2011**, *76*, 6749–6767.

[45] Hunsdiecker, H.; Hunsdiecker, C. *Chem. Ber.* **1942**, *75*, 291–297.

[46] Kochi, J. K. *J. Am. Chem. Soc.* **1965**, *87*, 2500–2502.

[47] Barton, D. H. R.; Crich, D.; Motherwell, W. B. *Tetrahedron Lett.* **1983**, *24*, 4979–4982.

[48] Concepcion, J. I.; Francisco, C. G.; Freire, R.; Hernandez, R.; Salazar, J. A.; Suarez, E. *J. Org. Chem.* **1986**, *51*, 402–404.

[49] Wang, Z.; Zhu, L.; Yin, F.; Su, Z.; Li, Z.; Li, C. *J. Am. Chem. Soc.* **2012**, *134*, 4258–4263.

[50] Moriya, T.; Yoneda, S.; Kawana, K.; Ikeda, R.; Konakahara, T.; Sakai, N. *Org. Lett.* **2012**, *14*, 4842–4845.

[51] Wells, A. S. *Org. Process Res. Dev.* **2010**, *14*, 484–484.

[52] Sakai, N.; Nakajima, T.; Yoneda, S.; Konakahara, T.; Ogiwara, Y. *J. Org. Chem.* **2014**, *79*, 10619–10623.

[53] Finkelstein, H. *Ber. Dtsch. Ges.* **1910**, *43*, 1528–1532.

[54] Kim, D. W.; Song, C. E.; Chi, D. Y. *J. Org. Chem.* **2003**, *68*, 4281–4285.

[55] Shibatomi, K.; Soga, Y.; Narayama, A.; Fujisawa, I.; Iwasa, S. *J. Am. Chem. Soc.* **2012**, *134*, 9836–9839.

[56] Tripathi, C. B.; Mukherjee, S. *Org. Lett.* **2014**, *16*, 3368–3371.

[57] Liu, H.; Jiang, G.; Pan, X.; Wan, X.; Lai, Y.; Ma, D.; Xie, W. *Org. Lett.* **2014**, *16*, 1908–1911.

[58] Chen, J. C.; Zhou, L.; Tan, C. K.; Yeung, Y.-Y. *J. Org. Chem.* **2012**, *77*, 999–1009.

[59] Zhang, M.-Z.; Sheng, W.-B.; Jiang, Q.; Tian, M.; Yin, Y.; Guo, C.-C. *J. Org. Chem.* **2014**, *79*, 10829–10836.

[60] Toda, Y.; Pink, M.; Johnston, J. N. *J. Am. Chem. Soc.* **2014**, *136*, 14734–14737.

[61] Jaganathan, A.; Staples, R. J.; Borhan, B. *J. Am. Chem. Soc.* **2013**, *135*, 14806–14813.

[62] Schmidt, V. A.; Quinn, R. K.; Brusoe, A. T.; Alexanian, E. J. *J. Am. Chem. Soc.* **2014**, *136*, 14389–14392.

[63] Okoromoba, O. E.; Han, J.; Hammond, G. B.; Xu, B. *J. Am. Chem. Soc.* **2014**, *136*, 14381–14384.

[64] D'Oyley, J. M.; Aliev, A. E.; Sheppard, T. D. *Angew. Chem. Int. Ed.* **2014**, *53*, 10747–10750.

[65] Zhao, Y.; Kang, J.; Park, C.-M.; Bagdon, P. E.; Peng, B.; Xian, M. *Org. Lett.* **2014**, *16*, 4536–4539.

[66] Giri, R.; Chen, X.; Yu, J.-Q. *Angew. Chem. Int. Ed.* **2005**, *44*, 2112–2115.

[67] Liu, L.; Zhang-Negrerie, D.; Du, Y.; Zhao, K. *Org. Lett.* **2014**, *16*, 436–439.

[68] Jing, Y.; Daniliuc, C. G.; Studer, A. *Org. Lett.* **2014**, *16*, 4932–4935.

[69] Billing, P.; Brinker, U. H. *J. Org. Chem.* **2012**, *77*, 11227–11231.

[70] Adams, D. J.; Clark, J. H. *Chem. Soc. Rev.*. **1999**, *28*, 225–231.

[71] Vints, I.; Gatenyo, J.; Rozen, S. *J. Org. Chem.* **2013**, *78*, 11794–11797.

[72] Olah, G. A.; Welch, J. T.; Vankar, Y. D.; Nojima, M.; Kerekes, I.; Olah, J. A. *J. Org. Chem.* **1979**, *44*, 3872–3881.

[73] Laali, K. K.; Gettwert, V. J. *J. Fluor. Chem.* **2001**, *107*, 31–34.

[74] Lal, G. S.; Pez, G. P.; Syvret, R. G. *Chem. Rev.* **1996**, *96*, 1737–1756.

[75] Balz, G.; Schiemann, G. *Chem. Ber.* **1927**, *60*, 1186–1190.

[76] Campbell, M. G.; Ritter, T. *Chem. Rev.* **2015**, *115*, 612–633.

[77] Furuya, T.; Klein, J. E. M. N.; Ritter, T. *Synthesis* **2010**, 1804–1821.

[78] Mu, X.; Liu, G. *Org. Chem. Front.* **2014**, *1*, 430–433.

[79] Anbarasan, P.; Neumann, H.; Beller, M. *Angew. Chem. Int. Ed.* **2010**, *49*, 2219–2222.

[80] Yamada, S.; Gavryushin, A.; Knochel, P. *Angew. Chem. Int. Ed.* **2010**, *49*, 2215–2218.

[81] Yamada, S.; Knochel, P. *Synthesis* **2010**, 2490–2494.

[82] Furuya, T.; Strom, A. E.; Ritter, T. *J. Am. Chem. Soc.* **2009**, *131*, 1662–1663.

[83] Tang, P.; Furuya, T.; Ritter, T. *J. Am. Chem. Soc.* **2010**, *132*, 12150–12154.

[84] Ye, Y.; Sanford, M. S. *J. Am. Chem. Soc.* **2013**, *135*, 4648–4651.

[85] Watson, D. A.; Su, M.; Teverovskiy, G.; Zhang, Y.; Garcia-Fortanet, J.; Kinzel, T.; Buchwald, S. L. *Science* **2009**, *325*, 1661–1664.

[86] Maimone, T. J.; Milner, P. J.; Kinzel, T.; Zhang, Y.; Takase, M. K.; Buchwald, S. L. *J. Am. Chem. Soc.* **2011**, *133*, 18106–18109.

[87] Lee, H. G.; Milner, P. J.; Buchwald, S. L. *Org. Lett.* **2013**, *15*, 5602–5605.

[88] Noël, T.; Maimone, T. J.; Buchwald, S. L. *Angew. Chem. Int. Ed.* **2011**, *50*, 8900–8903.

[89] Lee, H. G.; Milner, P. J.; Buchwald, S. L. *J. Am. Chem. Soc.* **2014**, *136*, 3792–3795.

[90] Mu, X.; Zhang, H.; Chen, P.; Liu, G. *Chem. Sci.* **2014**, *5*, 275–280.

[91] Tang, P.; Ritter, T. *Tetrahedron* **2011**, *67*, 4449–4454.

[92] Furuya, T.; Ritter, T. *Org. Lett.* **2009**, *11*, 2860–2863.

[93] Fier, P. S.; Luo, J.; Hartwig, J. F. *J. Am. Chem. Soc.* **2013**, *135*, 2552–2559.

[94] Ye, Y.; Schimler, S. D.; Hanley, P. S.; Sanford, M. S. *J. Am. Chem. Soc.* **2013**, *135*, 16292–16295.

[95] Mazzotti, A. R.; Campbell, M. G.; Tang, P.; Murphy, J. M.; Ritter, T. *J. Am. Chem. Soc.* **2013**, *135*, 14012–14015.

[96] Ichiishi, N.; Canty, A. J.; Yates, B. F.; Sanford, M. S. *Org. Lett.* **2013**, *15*, 5134–5137.

[97] Ichiishi, N.; Brooks, A. F.; Topczewski, J. J.; Rodnick, M. E.; Sanford, M. S.; Scott, P. J. H. *Org. Lett.* **2014**, *16*, 3224–3227.

[98] Tang, P.; Wang, W.; Ritter, T. *J. Am. Chem. Soc.* **2011**, *133*, 11482–11484.

[99] Fujimoto, T.; Becker, F.; Ritter, T. *Org. Process Res. Dev.* **2014**, *18*, 1041–1044.

[100] Wang, X.; Mei, T.-S.; Yu, J.-Q. *J. Am. Chem. Soc.* **2009**, *131*, 7520–7521.

[101] Lou, S.-J.; Xu, D.-Q.; Xia, A.-B.; Wang, Y.-F.; Liu, Y.; Du, X.-H.; Xu, Z.-Y. *Chem. Commun.* **2013**, *49*, 6218–6220.

[102] Lou, S.-J.; Xu, D.-Q.; Xu, Z. *Angew. Chem. Int. Ed.* **2014**, *53*, 10330–10335.

[103] Racowski, J. M.; Gary, J. B.; Sanford, M. S. *Angew. Chem. Int. Ed.* **2012**, *51*, 3414–3417.

[104] Furuya, T.; Benitez, D.; Tkatchouk, E.; Strom, A. E.; Tang, P.; Goddard, W. A. III.; Ritter, T. *J. Am. Chem. Soc.* **2010**, *132*, 3793–3807.

[105] Lee, E.; Kamlet, A. S.; Powers, D. C.; Neumann, C. N.; Boursalian, G. B.; Furuya, T.; Choi, D. C.; Hooker, J. M.; Ritter, T. *Science* **2011**, *334*, 639–642.

[106] Furuya, T.; Kaiser, H. M.; Ritter, T. *Angew. Chem. Int. Ed.* **2008**, *47*, 5993–5996.

[107] Chan, K. S. L.; Wasa, M.; Wang, X.; Yu, J.-Q. *Angew. Chem. Int. Ed.* **2011**, *50*, 9081–9084.

[108] Chen, C.; Wang, C.; Zhang, J.; Zhao, Y. *J. Org. Chem.* **2015**, *80*, 942.

[109] Hodgson, H. H. *Chem. Rev.* **1947**, *40*, 251–277.

[110] Hubbard, A.; Okazaki, T.; Laali, K. K. *J. Org. Chem.* **2008**, *73*, 316–319.

[111] Barbero, M.; Degani, I.; Dughera, S.; Fochi, R. *J. Org. Chem.* **1999**, *64*, 3448–3453.

[112] Mitchell, R. H.; Lai, Y.-H.; Williams, R. V. *J. Org. Chem.* **1979**, *44*, 4733–4735.

[113] Jana, N.; Verkade, J. *Org. Lett.* **2003**, *5*, 3787–3790.

[114] Zhang, Y.; Lu, B. Z.; Li, G.; Rodriguez, S.; Tan, J.; Wei, H.-X.; Liu, J.; Roschangar, F.; Ding, F.; Zhao, W.; Qu, B.; Reeves, D.; Grinberg, N.; Lee, H.; Heckmann, G.; Niemeier, O.; Brenner, M.; Tsantrizos, Y.; Beaulieu, P. L.; Hossain, A.; Yee, N.; Farina, V.; Senanayake, C. H. *Org. Lett.* **2014**, *16*, 4558–4561.

[115] Castanet, A.-S.; Colobert, F.; Broutin, P. E. *Tetrahedron Lett.* **2002**, *43*, 5047–5048.

[116] Gilow, H. M.; Burton, D. E. *J. Org. Chem.* **1981**, *46*, 2221–2225.

[117] Ku, H.; Barrio, J. R. *J. Org. Chem.* **1981**, *46*, 5239–5241.

[118] Arshad, N.; Hashim, J.; Kappe, C. O. *J. Org. Chem.* **2008**, *73*, 4755–4758.

[119] Rajender Reddy, K.; Venkateshwar, M.; Uma Maheswari, C.; Santhosh Kumar, P. *Tetrahedron Lett.* **2010**, *51*, 2170–2173.

[120] Bonnafoux, L.; Gramage-Doria, R.; Colobert, F.; Leroux, F. R. *Chem. Eur. J.* **2011**, *17*, 11008–11016.

[121] Valasek, M.; Edelmann, K.; Gerhard, L.; Fuhr, O.; Lukas, M.; Mayor, M. *J. Org. Chem.* **2014**, *79*, 7342–7357.

[122] Menzel, K.; Fisher, E.-L.; DiMichele, L.; Frantz, D. E.; Nelson, T. D.; Kress, M. H. *J. Org. Chem.* **2006**, *71*, 2188–2191.

[123] de Almeida, L. S.; de Mattos, M. C. S.; Esteves, P. M. *Synlett* **2013**, 603–606.

[124] Rajesh, K.; Somasundaram, M.; Saiganesh, R.; Balasubramanian, K. K. *J. Org. Chem.* **2007**, *72*, 5867–5869.

[125] Bliman, D.; Pettersson, M.; Bood, M.; Grøtli, M. *Tetrahedron Lett.* **2014**, *55*, 2929–2931.

[126] Prakash, G. K. S.; Mathew, T.; Hoole, D.; Esteves, P. M.; Wang, Q.; Rasul, G.; Olah, G. A. *J. Am. Chem. Soc.* **2004**, *126*, 15770–15776.

[127] Wu, H.; Hynes, J. Jr. *Org. Lett.* **2010**, *12*, 1192–1195.

[128] Molander, G. A.; Cavalcanti, L. N. *J. Org. Chem.* **2011**, *76*, 7195–7203.

[129] Kabalka, G. W.; Mereddy, A. R. *Organometallics* **2004**, *23*, 4519–4521.

[130] Thiebes, C.; Prakash, G. K. S.; Petasis, N. A.; Olah, G. A. *Synlett* **1998**, 141–142.

[131] Murphy, J. M.; Liao, X.; Hartwig, J. F. *J. Am. Chem. Soc.* **2007**, *129*, 15434–15435.

[132] Partridge, B. M.; Hartwig, J. F. *Org. Lett.* **2013**, *15*, 140–143.

[133] Qiu, D.; Mo, F.; Zheng, Z.; Zhang, Y.; Wang, J. *Org. Lett.* **2010**, *12*, 5474–5477.

[134] Bedford, R. B.; Haddow, M. F.; Mitchell, C. J.; Webster, R. L. *Angew. Chem. Int. Ed.* **2011**, *50*, 5524–5527.

[135] Kakiuchi, F.; Kochi, T.; Mutsutani, H.; Kobayashi, N.; Urano, S.; Sato, M.; Nishiyama, S.; Tanabe, T. *J. Am. Chem. Soc.* **2009**, *131*, 11310–11311.

[136] MacNeil, S. L.; Familoni, O. B.; Snieckus, V. *J. Org. Chem.* **2001**, *66*, 3662–3670.

[137] Nack, W. A.; He, G.; Zhang, S.-Y.; Lu, C.; Chen, G. *Org. Lett.* **2013**, *15*, 3440–3443.

[138] Li, J.-J.; Giri, R.; Yu, J.-Q. *Tetrahedron* **2008**, *64*, 6979–6987.

[139] Du, B.; Jiang, X.; Sun, P. *J. Org. Chem.* **2013**, *78*, 2786–2791.

[140] Sadhu, P.; Alla, S. K.; Punniyamurthy, T. *J. Org. Chem.* **2013**, *78*, 6104–6111.

[141] Sun, X.; Shan, G.; Sun, Y.; Rao, Y. *Angew. Chem. Int. Ed.* **2013**, *52*, 4440–4444.

[142] Kalyani, D.; Dick, A. R.; Anani, W. Q.; Sanford, M. S. *Org. Lett.* **2006**, *8*, 2523–2526.

[143] Schroder, N.; Wencel-Delord, J.; Glorius, F. *J. Am. Chem. Soc.* **2012**, *134*, 8298–8301.

[144] Yu, D.-G.; Gensch, T.; de Azambuja, F.; Vasquez-Cespedes, S.; Glorius, F. *J. Am. Chem. Soc.* **2014**, *136*, 17722–17725.

[145] Sedelmeier, J.; Bolm, C. *J. Org. Chem.* **2005**, *70*, 6904–6906.

[146] Klapars, A.; Buchwald, S. L. *J. Am. Chem. Soc.* **2002**, *124*, 14844–14845.

[147] Chen, M.; Ichikawa, S.; Buchwald, S. L. *Angew. Chem. Int. Ed.* **2015**, *54*, 263–266.

[148] Cant, A. A.; Bhalla, R.; Pimlott, S. L.; Sutherland, A. *Chem. Commun.* **2012**, *48*, 3993–3995.

[149] Mei, T.-S.; Giri, R.; Maugel, N.; Yu, J.-Q. *Angew. Chem. Int. Ed.* **2008**, *47*, 5215–5219.

[150] Pan, J.; Wang, X.; Zhang, Y.; Buchwald, S. L. *Org. Lett.* **2011**, *13*, 4974–4976.

[151] Imazaki, Y.; Shirakawa, E.; Ueno, R.; Hayashi, T. *J. Am. Chem. Soc.* **2012**, *134*, 14760–14763.

[152] Wang, X.-C.; Hu, Y.; Bonacorsi, S.; Hong, Y.; Burrell, R.; Yu, J.-Q. *J. Am. Chem. Soc.* **2013**, *135*, 10326–10329.

[153] Rodriguez-Lojo, D.; Cobas, A.; Pena, D.; Perez, D.; Guitian, E. *Org. Lett.* **2012**, *14*, 1363–1365.

[154] Gao, D.-W.; Gu, Q.; You, S.-L. *ACS Catal.* **2014**, *4*, 2741–2745.

[155] Yin, X.; Chen, J.; Lalancette, R. A.; Marder, T. B.; Jäkle, F. *Angew. Chem. Int. Ed.* **2014**, *53*, 9761–9765.

[156] Dudnik, A. S.; Chernyak, N.; Huang, C.; Gevorgyan, V. *Angew. Chem. Int. Ed.* **2010**, *49*, 8729–8732.

[157] Merkul, E.; Boersch, C.; Frank, W.; Mueller, T. J. J. *Org. Lett.* **2009**, *11*, 2269–2272.

[158] Huang, Z.; Li, F.; Chen, B.; Lu, T.; Yuan, Y.; Yuan, G. *ChemSusChem* **2013**, *6*, 1337–1340.

[159] Zhdankin, V. V.; Stang, P. J. *Tetrahedron* **1998**, *54*, 10927–10966.

[160] Bielawski, M.; Zhu, M.; Olofsson, B. *Adv. Synth. Catal.* **2007**, *349*, 2610–2618.

[161] Bielawski, M.; Olofsson, B. *Chem. Commun.* **2007**, *2*, 2521–2523.

[162] Bielawski, M.; Aili, D.; Olofsson, B. *J. Org. Chem.* **2008**, *73*, 4602–4607.

[163] Champagne, P. A.; Desroches, J.; Hamel, J.-D.; Vandamme, M.; Paquin, J.-F. *Chem. Rev.* **2015**, *115*, 9073–9174.

[164] Arvela, R.; Leadbeater, N. *Synlett* **2003**, 1145–1148.

[165] Yoneda, N. *Tetrahedron*, **1991**, *47*, 5329–5365.

[166] Yang, M.-H.; Matikonda, S. S.; Altman, R. A. *Org. Lett.* **2013**, *15*, 3894–3897.

[167] Zhao, Y.; Huang, W.; Zhu, L.; Hu, J. *Org. Lett.* **2010**, *12*, 1444–1447.

[168] Hu, M.; He, Z.; Gao, B.; Li, L.; Ni, C.; Hu, J. *J. Am. Chem. Soc.* **2013**, *135*, 17302–17305.

[169] Sun, H.; DiMagno, S. G. *J. Am. Chem. Soc.* **2005**, *127*, 2050–2051.

[170] L'Heureux, A.; Beaulieu, F.; Bennett, C.; Bill, D. R.; Clayton, S.; LaFlamme, F.; Mirmehrabi, M.; Tadayon, S.; Tovell, D.; Couturier, M. *J. Org. Chem.* **2010**, *75*, 3401–3411.

[171] Ueda, T.; Konishi, H.; Manabe, K. *Org. Lett.* **2013**, *15*, 5370–5373.

[172] Das, J. P.; Roy, S. *J. Org. Chem.* **2002**, *67*, 7861–7864.

[173] Pandey, A. K.; Sharma, R.; Shivahare, R.; Arora, A.; Rastogi, N.; Gupta, S.; Chauhan, P. M. S. *J. Org. Chem.* **2013**, *78*, 1534–1546.

[174] Ye, C.; Shreeve, J. M. *J. Org. Chem.* **2004**, *69*, 8561–8563.

[175] Maity, P.; Klos, M. R.; Kazmaier, U. *Org. Lett.* **2013**, *15*, 6246–6249.

[176] Kabalka, G. W.; Reddy Mereddy, A. *Tetrahedron Lett.* **2004**, *45*, 1417–1419.

[177] Wang, C.; Tobrman, T.; Xu, Z.; Negishi, E. *Org. Lett.* **2009**, *11*, 4092–4095.

[178] Kuhl, N.; Schroder, N.; Glorius, F. *Org. Lett.* **2013**, *15*, 3860–3863.

[179] Yin, J.; Gallis, C. E.; Chisholm, J. D. *J. Org. Chem.* **2007**, *72*, 7054–7057.

[180] Salih, M. Q.; Beaudry, C. M. *Org. Lett.* **2014**, *16*, 4964–4966.

[181] Kawaguchi, S.; Ogawa, A. *Org. Lett.* **2010**, *12*, 1893–1895.

[182] Lemay, A. B.; Vulic, K. S.; Ogilvie, W. W. *J. Org. Chem.* **2006**, *71*, 3615–3618.

[183] Ye, L.; Zhang, L. *Org. Lett.* **2009**, *11*, 3646–3649.

[184] Gao, F.; Hoveyda, A. H. *J. Am. Chem. Soc.* **2010**, *132*, 10961–10963.

[185] Chen, Z.; Li, J.; Jiang, H.; Zhu, S.; Li, Y.; Qi, C. *Org. Lett.* **2010**, *12*, 3262–3265.

[186] Li, X.; Shi, X.; Fang, M.; Xu, X. *J. Org. Chem.* **2013**, *78*, 9499–9504.

[187] Yao, M.-L.; Quick, T. R.; Wu, Z.; Quinn, M. P.; Kabalka, G. W. *Org. Lett.* **2009**, *11*, 2647–2649.

[188] Pawluc, P.; Hreczycho, G.; Szudkowska, J.; Kubicki, M.; Marciniec, B. *Org. Lett.* **2009**, *11*, 3390–3393.

[189] Snelders, D. J. M.; Dyson, P. J. *Org. Lett.* **2011**, *13*, 4048–4051.

[190] Okitsu, T.; Nakata, K.; Nishigaki, K.; Michioka, N.; Karatani, M.; Wada, A. *J. Org. Chem.* **2014**, *79*, 5914–5920.

[191] Hu, Y.; Yi, R.; Wang, C.; Xin, X.; Wu, F.; Wan, B. *J. Org. Chem.* **2014**, *79*, 3052–3059.

[192] Reddy, M. S.; Thirupathi, N.; Hari Babu, M.; Puri, S. *J. Org. Chem.* **2013**, *78*, 5878–5888.

[193] Chen, Z.; Huang, G.; Jiang, H.; Huang, H.; Pan, X. *J. Org. Chem.* **2011**, *76*, 1134–1139.

[194] Gabriele, B.; Mancuso, R.; Salerno, G.; Larock, R. C. *J. Org. Chem.* **2012**, *77*, 7640–7645.

[195] Zhang, Z.; Ouyang, L.; Wu, W.; Li, J.; Zhang, Z.; Jiang, H. *J. Org. Chem.* **2014**, *79*, 10734–10742.

[196] Wilking, M.; Muck-Lichtenfeld, C.; Daniliuc, C. G.; Hennecke, U. *J. Am. Chem. Soc.* **2013**, *135*, 8133–8136.

[197] Murai, K.; Shimizu, N.; Fujioka, H. *Chem. Commun.* **2014**, *50*, 12530–12533.

[198] Huang, F.; Wu, P.; Wang, L.; Chen, J.; Sun, C.; Yu, Z. *Chem. Commun.* **2014**, *50*, 12479–12481.

[199] Jeong, Y.; Kim, B.-I.; Lee, J. K.; Ryu, J.-S. *J. Org. Chem.* **2014**, *79*, 6444–6455.

[200] Xu, S.; Li, C.; Jia, X.; Li, J. *J. Org. Chem.* **2014**, *79*, 11161–11169.

[201] Egi, M.; Ota, Y.; Nishimura, Y.; Shimizu, K.; Azechi, K.; Akai, S. *Org. Lett.* **2013**, *15*, 4150–4153.

[202] Barton, D. H. R.; Obrien, R. E.; Sternhell, S. *J. Chem. Soc.* **1962**, 470.

[203] Furrow, M. E.; Myers, A. G. *J. Am. Chem. Soc.* **2004**, *126*, 5436–5445.

[204] Ojha, D. P.; Prabhu, K. R. *Org. Lett.* **2015**, *17*, 18–21.

[205] Mofford, D. M.; Reddy, G. R.; Miller, S. C. *J. Am. Chem. Soc.* **2014**, *136*, 13277–13282.

[206] Dharuman, S.; Vankar, Y. D. *Org. Lett.* **2014**, *16*, 1172–1175.

[207] Bull, J. A.; Mousseau, J. J.; Charette, A. B. *Org. Lett.* **2008**, *10*, 5485–5488.

[208] Ji, X.; Huang, H.; Xiong, W.; Huang, K.; Wu, W.; Jiang, H. *J. Org. Chem.* **2014**, *79*, 7005–7011.

[209] Chelucci, G. *Chem. Rev.* **2012**, *112*, 1344–1462.

[210] Wang, Z.; Campagna, S.; Xu, G.; Pierce, M. E.; Fortunak, J. M.; Confalone, P. N. *Tetrahedron Lett.* **2000**, *41*, 4007–4009.

[211] Madabhushi, S.; Jillella, R.; Mallu, K. K. R.; Godala, K. R.; Vangipuram, V. S. *Tetrahedron Lett.* **2013**, *54*, 3993–3996.

[212] Desai, N. B.; McKelvie, N.; Ramirez, F. *J. Am. Chem. Soc.* **1962**, *84*, 1745–1747.

[213] Gung, B. W.; Kumi, G. *J. Org. Chem.* **2003**, *68*, 5956–5960.

[214] Michel, P.; Rassat, A. *Tetrahedron Lett.* **1999**, *40*, 8579–8581.

[215] Cloarec, J.-M.; Charette, A. B. *Org. Lett.* **2004**, *6*, 4731–4734.

[216] Dumele, O.; Wu, D.; Trapp, N.; Goroff, N.; Diederich, F. *Org. Lett.* **2014**, *16*, 4722–4725.

[217] Xiang, J.; Yuan, R.; Wang, R.; Yi, N.; Lu, L.; Zou, H.; He, W. *J. Org. Chem.* **2014**, *79*, 11378–11382.

[218] Liu, J.; Li, W.; Wang, C.; Li, Y.; Li, Z. *Tetrahedron Lett.* **2011**, *52*, 4320–4323.

[219] Pavlinac, J.; Zupan, M.; Stavber, S. *Org. Biomol. Chem.* **2007**, *5*, 699–707.

[220] Chen, Z.; Jiang, H.; Li, Y.; Qi, C. *Chem. Commun.* **2010**, *46*, 8049–8051.

[221] Zhu, C.; Zuo, Y.; Wang, R.; Liang, B.; Yue, X.; Wen, G.; Shang, N.; Huang, L.; Chen, Y.; Du, J.; Bu, X. *J. Med. Chem.* **2014**, *57*, 6364–6382.

[222] Schmidt, A. H.; Russ, M.; Grosse, D. *Synthesis* **1981**, 216–218.

[223] Hoffmann, H. M. R.; Haase, K. *Synthesis* **1981**, 715–719.

[224] Froyen, P. *Phosphorus Sulfur Silicon Relat. Elem.* **1995**, *102*, 253–259.

[225] Kang, D. H.; Joo, T. Y.; Lee, E. H.; Chaysripongkul, S.; Chavasiri, W.; Jang, D. O. *Tetrahedron Lett.* **2006**, *47*, 5693–5696.

[226] Aizpurua, J. M.; Cossio, F. P.; Palomo, C. *J. Org. Chem.* **1986**, *51*, 4941–4943.

[227] Tanaka, R.; Miller, S. I. *J. Org. Chem.* **1971**, *36*, 3856–3861.

[228] Vaughn, T. H.; Nieuwland, J. A. *J. Am. Chem. Soc.* **1932**, *54*, 787–791.

[229] Ku, Y.-Y.; Grieme, T.; Sharma, P.; Pu, Y.-M.; Raje, P.; Morton, H.; King, S. *Org. Lett.* **2001**, *3*, 4185–4187.

[230] Perkins, C.; Libri, S.; Adams, H.; Brammer, L. *CrystEngComm* **2012**, *14*, 3033–3038.

[231] Wu, W.; Jiang, H. *Acc. Chem. Res.* **2014**, *47*, 2483–2504.

[232] Hanamoto, T.; Koga, Y.; Kawanami, T.; Furuno, H.; Inanaga, J. *Angew. Chem. Int. Ed.* **2004**, *43*, 3582–3584.

[233] Besset, T.; Poisson, T.; Pannecoucke, X. *Chem. Eur. J.* **2014**, *20*, 16830–16845.

[234] Liu, C.; Ma, H.; Nie, J.; Ma, J. *Chin. J. Chem.* **2012**, *30*, 47–52.

[235] Fujii, A.; Miller, S. I. *J. Am. Chem. Soc.* **1971**, *93*, 3694–3700.

[236] Saalfrank, R. W.; Welch, A.; Haubner, M.; Bauer, U. *Liebigs Ann.* **1996**, 171–181.

[237] Martins, M. A. P.; Emmerich, D. J.; Pereira, C. M. P.; Cunico, W.; Rossato, M.; Zanatta, N.; Bonacorso, H. G. *Tetrahedron Lett.* **2004**, *45*, 4935–4938.

[238] Sud, D.; Wigglesworth, T. J.; Branda, N. R. *Angew. Chem. Int. Ed.* **2007**, *46*, 8017–8019.

[239] Wada, T.; Iwasaki, M.; Kondoh, A.; Yorimitsu, H.; Oshima, K. *Chem. Eur. J.* **2010**, *16*, 10671–10674.

[240] Li, M.; Li, Y.; Zhao, B.; Liang, F.; Jin, L.-Y. *RSC Adv.* **2014**, *4*, 30046–30049.

[241] Hofmeister, H.; Annen, K.; Laurent, H.; Wiechert, R. *Angew. Chem. Int. Ed.* **1984**, *23*, 727–729.

[242] Gulia, N.; Pigulski, B.; Charewicz, M.; Szafert, S. *Chem. Eur. J.* **2014**, *20*, 2746–2749.

[243] Straus, F.; Kollek, L.; Heyn, W. *Chem. Ber.* **1930**, *63*, 1868–1885.

[244] Marino, J. P.; Nguyen, H. N. *J. Org. Chem.* **2002**, *67*, 6841–6844.

[245] Zhang, Y.; Hsung, R. P.; Tracey, M. R.; Kurtz, K. C. M.; Vera, E. L. *Org. Lett.* **2004**, *6*, 1151–1154.

[246] Naskar, D.; Roy, S. *J. Org. Chem.* **1999**, *64*, 6896–6897.

[247] Zheng, G.; Lu, W.; Cai, J. *J. Nat. Prod.* **1999**, *64*, 626–628.

[248] Kohnen, A. L.; Dunetz, J. R.; Danheiser, R. L. *Organic Synth.* **2007**, *84*, 88–101.

[249] Gung, B. W.; Dickson, H. *Org. Lett.* **2002**, *4*, 2517–2519.

[250] Maleczka, R. E.; Terrell, L. R.; Clark, D. H.; Whitehead, S. L.; Gallagher, W. P.; Terstiege, I. *J. Org. Chem.* **1999**, *64*, 5958–5965.

[251] Webb, J. A.; Klijn, J. E.; Hill, P. A.; Bennett, J. L.; Goroff, N. S. *J. Org. Chem.* **2004**, *69*, 660–664.

[252] Leyva-Peìrez, A.; Rubio-Marqueìs, P.; Al-Deyab, S.; Al-Resayes, S. I.; Corma, A. *ACS Catal.* **2011**, *1*, 601–606.

[253] Kim, S.; Kim, S.; Lee, T.; Ko, H.; Kim, D. *Org. Lett.* **2004**, *6*, 3601–3604.

[254] Kim, S.; Lee, Y. M.; Kang, H. R.; Cho, J.; Lee, T.; Kim, D. *Org. Lett.* **2007**, *9*, 2127–2130.

[255] Rao, M. L. N.; Periasamy, M. *Synth. Commun.* **1995**, *25*, 2295–2299.

[256] Luithle, J. E. A.; Pietruszka, J. *Eur. J. Org. Chem.* **2000**, *2000*, 2557–2562.

[257] Corey, E. J.; Fuchs, P. L. *Tetrahedron Lett.* **1972**, *13*, 3769–3772.

[258] Hollingworth, G. J.; Sweeney, J. B. *Synlett* **1993**, 463–465.

[259] Nishikawa, T.; Shibuya, S.; Hosokawa, S.; Isobe, M. *Synlett* **1994**, 485–486.

[260] Wang, Y.-P.; Danheiser, R. L. *Tetrahedron Lett.* **2011**, *52*, 2111–2114.

[261] Hein, J. E.; Tripp, J. C.; Krasnova, L. B.; Sharpless, K. B.; Fokin, V. V. *Angew. Chem. Int. Ed.* **2009**, *48*, 8018–8021.

[262] Southwick, P. L.; Kirchner, J. R. *J. Org. Chem.* **1962**, *27*, 3305–3308.

[263] Pelletier, G.; Lie, S.; Mousseau, J. J.; Charette, A. B. *Org. Lett.* **2012**, *14*, 5464–5467.

[264] Meng, L.-C.; Cai, P.-J.; Guo, Q.-X.; Xue, S. *Synth. Commun.* **2008**, *38*, 225–231.

INDEX

Practical Functional Group Synthesis, First Edition. Robert A. Stockland, Jr.
© 2016 John Wiley & Sons, Inc. Published 2016 by John Wiley & Sons, Inc.